Lecture Notes in Artificial Intelligence 2083

Subseries of Lecture Notes in Computer Science
Edited by J. G. Carbonell and J. Siekmann

Lecture Notes in Computer Science

Edited by G. Goos, J. Hartmanis and J. van Leeuwen

T0189572

Springer
Berlin
Heidelberg
New York
Barcelona
Hong Kong
London
Milan
Paris
Singapore
Tokyo

Rajeev Goré   Alexander Leitsch
Tobias Nipkow (Eds.)

Automated
Reasoning

First International Joint Conference, IJCAR 2001
Siena, Italy, June 18-22, 2001
Proceedings

 Springer

Series Editors

Jaime G. Carbonell,Carnegie Mellon University, Pittsburgh, PA, USA
Jörg Siekmann, University of Saarland, Saarbrücken, Germany

Volume Editors

Rajeev Goré
Australian National University
Automated Reasoning Project and Department of Computer Science
Canberra, ACT, 0200, Australia
E-mail: Rajeev.Gore@arp.anu.edu.au

Alexander Leitsch
Technische Universität Wien
AG Theoretische Informatik und Logik, Institut für Computersprachen
Favoritenstr. 9, E185-2, 1040 Wien, Austria
E-mail: leitsch@logic.at

Tobias Nipkow
Technische Universität München, Institut für Informatik
80290 München, Germany
E-mail: nipkow@in.tum.de

Cataloging-in-Publication Data applied for

Die Deutsche Bibliothek - CIP-Einheitsaufnahme

Automated reasoning : first international joint conference ; proceedings /
IJCAR 2001, Siena, Italy, June 18 - 23, 2001. Rajeev Goré ... (ed.). -
Berlin ; Heidelberg ; New York ; Barcelona ; Hong Kong ; London ; Milan ;
Paris ; Singapore ; Tokyo : Springer, 2001
      (Lecture notes in computer science ; Vol. 2083 : Lecture notes in
      artificial intelligence)
      ISBN 3-540-42254-4

CR Subject Classification (1998): I.2.3, F.4.1, F.3, F.4, D.2.4

ISBN 3-540-42254-4 Springer-Verlag Berlin Heidelberg New York

Springer-Verlag Berlin Heidelberg New York
a member of BertelsmannSpringer Science+Business Media GmbH

http://www.springer.de

© Springer-Verlag Berlin Heidelberg 2001
Printed in Germany

Typesetting: Camera-ready by author, data conversion by PTP Berlin, Stefan Sossna
Printed on acid-free paper   SPIN 10839304   06/3142   5 4 3 2 1 0

Foreword

The last ten years have seen a gradual fragmentation of the Automated Reasoning community into various disparate groups, each with its own conference: the Conference on Automated Reasoning (CADE), the International Workshop on First-Order Theorem Proving (FTP), and the International Conference on Automated Reasoning with Analytic Tableau and Related Methods (TABLEAUX) to name three. During 1999, various members of these three communities discussed the idea of holding a joint conference in 2001 to bring our communities together again. The plan was to hold a one-off conference for 2001, to be repeated if it proved a success. This volume contains the papers presented at the resulting event: the first International Joint Conference on Automated Reasoning (IJCAR 2001), held in Siena, Italy, from June 18–23, 2001.

We received 88 research papers and 24 systems descriptions as submissions. Each submission was fully refereed by at least three peers who were asked to write a report on the quality of the submissions. These reports were accessible to members of the programme committee via a web-based system specially designed for electronic discussions. As a result we accepted 37 research papers and 19 system descriptions, which make up these proceedings. In addition, this volume contains full papers or extended abstracts from the five invited speakers.

Ten one-day workshops and four tutorials were held during IJCAR 2001. The automated theorem proving system competition (CASC) was organized by Geoff Sutcliffe to evaluate the performance of sound, fully automatic, classical, first-order automated theorem proving systems. The third Workshop on Inference in Computational Semantics (ICoS-3) and the 9th Symposium on the Integration of Symbolic Computation and Mechanized Reasoning (CALCULEMUS-2001) were co-located with IJCAR 2001, and held their own associated workshops and produced their own separate proceedings.

We would like to acknowledge the enormous amount of work put in by the members of the program committee, the various steering committees, the IJCAR officials, and additional referees named on the following pages. In particular, we would like to thank Fabio Massacci and Marco Baioletti for organizing the conference itself, Gernot Salzer for installing and maintaining the software for our web-based reviewing procedure, and Gertrud Bauer for assembling these proceedings. Finally, we thank the sponsors named on the following pages for their financial support.

Rajeev Goré, Alexander Leitsch and Tobias Nipkow
April 2001

Conference Chair and Local Organisation

Fabio Massacci (Univ. di Siena, Italy)

Programme Committee Chairs

Rajeev Goré   (Australian National Univ., Canberra)
Alexander Leitsch (TU Wien)
Tobias Nipkow   (TU München)

Programme Committee

Rajeev Alur   (Philadelphia)
Franz Baader   (Aachen)
Matthias Baaz   (Wien)
Bernhard Beckert   (Karlsruhe)
Ricardo Caferra   (Grenoble)
Roy Dyckhoff   (St Andrews)
Ulrich Furbach   (Koblenz)
Rajeev Goré   (Canberra)
Didier Galmiche   (Nancy)
Harald Ganzinger   (MPI Saarbrücken)
Jean Goubault-Larrecq (INRIA Rocq.)
Reiner Hähnle   (Chalmers)
John Harrison   (Intel, Hillsboro)
Deepak Kapur   (New Mexico)
Henry Kautz   (ATT, Florham Park)
Michael Kohlhase   (Saarbrücken)
Alexander Leitsch   (TU Wien)
Zohar Manna   (Stanford)
Tobias Nipkow   (TU München)
Peter Patel-Schneider   (Bell Labs)
Frank Pfenning   (Pittsburgh)
Andreas Podelski   (MPI Saarbrücken)
Wolfgang Reif   (Augsburg)
Gernot Salzer   (Wien)
Moshe Vardi   (Houston)

Invited Speakers

Neil D. Jones   (DIKU, Denmark)
Lawrence C. Paulson   (Univ. of Cambridge, UK)
Helmut Schwichtenberg (Univ. München, Germany)
Andrei Voronkov   (Univ. of Manchester, UK)
Doron Zeilberger   (Temple Univ., USA)

IJCAR Officials

Conference Chair: Fabio Massacci (Univ. di Siena, Italy)
Programme Committee Chairs:
  Rajeev Goré   (Australian National Univ., Canberra)
  Alexander Leitsch (TU Wien)
  Tobias Nipkow   (TU München)
Workshop Chair: Dieter Hutter (DFKI Saarbrücken, Germany)
Tutorial Chair: Toby Walsh (Univ. of York, U.K.)
Publicity Chair: Peter Baumgartner (Univ. Koblenz, Germany)
Treasurer: Enrico Giunchiglia (Univ. di Genova, Italy)
IJCAR Steering Committee:
  Fabio Massacci   (IJCAR 2001 Conference Chair)
  Ulrich Furbach   (President of CADE Inc.)
  Frank Pfenning   (Vice-President of CADE Inc.)
  Peter Schmitt   (Vice-President of TABLEAUX)
  Maria Paola Bonacina (President of FTP)
  Ricardo Caferra   (Representative of FTP)
Local Organisation: Marco Baioletti, Fabio Massacci (Univ. di Siena, Italy)

IJCAR Sponsors

IJCAR gratefully acknowledge the sponsorship of:
  Association for Automated Reasoning
  Consiglio Nazionale delle Ricerche (CNR)
  European Research Consortium on Informatics and Mathematics (ERCIM)
  Monte dei Paschi di Siena SpA - Banca dal 1947
  Univ. degli Studi di Siena
  Associazione Italiana per l'Intelligenza Artificiale (AI*IA)
  CADE Inc.
  European Association for Theoretical Computer Science (EATCS)
  European Coordinating Committee for Artificial Intelligence (ECCAI)
  German Informatics Society (GI)

Additional Reviewers

Andreas Abel	Erich Grädel	Norbert Preining
Wolfgang Ahrendt	Bernhard Gramlich	Aarne Ranta
Farid Ajili	Elmar Habermalz	Greg Restall
Thomas Baar	Miki Hermann	Mark Reynolds
Peter Baumgartner	Joshua Hodas	Muriel Roger
Andrew Bernard	Kahlil Hodgson	Ulrike Sattler
Steven Bird	Jacob M. Howe	Renate Schmidt
Alexander Bockmayr	Tomi Janhunen	Klaus Schneider
Thierry Boy de la Tour	Fairouz Kamareddine	Tobias Schröder
Stefan Brass	Manfred Kerber	Carsten Schürmann
Krysia Broda	Alexander Koller	John Slaney
Chad E Brown	Miyuki Koshimura	Andrew Slater
Ernst Buchberger	Dominique Larchey-Wendling	Gernot Stenz
Diego Calvanese	Reinhold Letz	Frieder Stolzenburg
Domenico Cantone	Carsten Lutz	Lutz Strassburger
Serena Cerrito	Heiko Mantel	Armando Tacchella
Kaustuv Chaudhuri	William McCune	Tanel Tammet
Agata Ciabattoni	Andreas Meier	Cesare Tinelli
Koen Classen	Daniel Mery	Stephan Tobies
Evelyne Contejean	Aart Middeldorp	Christian Urban
Veronica Dahl	Georg Moser	Stéphane Vaillant
Hans de Nivelle	Paliath Narendran	Hans P. van Ditmarsch
Stéphane Demri	Robert Nieuwenhuis	Vincent van Oostrom
Jörg Denzinger	Jean-Marc Notin	Femke van Raamsdonk
Michael Dierkes	Hans-Jürgen Ohlbach	Jaco van de Pol
Devdatt Dubhashi	Enno Ohlebusch	Helmut Veith
Niklas Een	Jens Otten	Andrei Voronkov
Uwe Egly	Lawrence C. Paulson	Uwe Waldmann
Wolfgang Faber	Nicolas Peltier	Joe Warren
Christian Fermüller	Brigitte Pientka	Kevin Watkins
Wan Fokkink	Jeff Polakow	Adnan Yahya
Martin Giese	Alberto Policriti	Jürgen Zimmer
Jürgen Giesl		

Table of Contents

Invited Talks

Program Termination Analysis by Size-Change Graphs 1
  Neil D. Jones

SET Cardholder Registration: the Secrecy Proofs 5
  Lawrence C. Paulson

Algorithms, datastructures and other issues in efficient automated deduction 13
  Andrei Voronkov

Description, Modal and Temporal Logics

The Description Logic $\mathcal{ALCNH}_{R^+}$ Extended with Concrete Domains: A
Practically Motivated Approach 29
  Volker Haarslev, Ralf Möller, Michael Wessel

NExpTime-complete Description Logics with Concrete Domains......... 44
  Carsten Lutz

Exploiting Pseudo Models for TBox and ABox Reasoning in Expressive
Description Logics 59
  Volker Haarslev, Ralf Möller, Anni-Yasmin Turhan

The Hybrid μ-Calculus 74
  Ulrike Sattler, Moshe Y. Vardi

The Inverse Method Implements the Automata Approach for Modal Satisfiability 89
  Franz Baader, Stephan Tobies

Deduction-based Decision Procedure for a Clausal Miniscoped Fragment
of FTL 104
  Regimantas Pliuškevičius

Tableaux for temporal description logic with constant domains.......... 119
  Carsten Lutz, Holger Sturm, Frank Wolter, Michael Zakharyaschev

Free-Variable Tableaux for Constant-Domain Quantified Modal Logics with
Rigid and Non-Rigid Designation 134
  Serenella Cerrito, Marta Cialdea Mayer

Saturation Based Theorem Proving, Applications and Data Structures

Instructing equational set-reasoning with Otter 149
  Andrea Formisano, Eugenio G. Omodeo, Marco Temperini

NP-Completeness of Refutability by Literal-Once Resolution 164
  Stefan Szeider

Ordered Resolution vs. Connection Graph Resolution 178
  Reiner Hähnle, Neil V. Murray, Erik Rosenthal

A Model-based Completeness Proof of Extended Narrowing And Resolution 191
  Jürgen Stuber

A Resolution-Based Decision Procedure for the Two-Variable Fragment
with Equality 206
  Hans de Nivelle, Ian Pratt-Hartmann

Superposition and Chaining for Totally Ordered Divisible Abelian Groups . 221
  Uwe Waldmann

Context Trees 236
  Harald Ganzinger, Robert Nieuwenhuis, Pilar Nivela

On the Evaluation of Indexing Techniques for Theorem Proving 251
  Robert Nieuwenhuis, Thomas Hillenbrand, Alexandre Riazanov, Andrei Voronkov

Logic Programming and Nonmonotonic Reasoning

Preferred Extensions of Argumentation Frameworks: Query Answering and
Computation 266
  Sylvie Doutre, Jerome Mengin

Bunched Logic Programming 281
  Pablo A. Armelín, David J. Pym

A Top-down Procedure for Disjunctive Well-founded Semantics 296
  Kewen Wang

System Description: A Second-order Theorem Prover applied to Circumscription 309
  Michael Beeson

System Description: NoMoRe: A System for Non-Monotonic Reasoning
with Logic Programs under Answer Set Semantics 316
  Christian Anger, Kathrin Konczak, Thomas Linke

Propositional Satisfiability and Quantified Boolean Logic

Conditional Pure Literal Graphs 321
  Marco Benedetti

Evaluating search heuristics and optimization techniques in propositional
satisfiability 336
    Enrico Giunchiglia, Massimo Maratea, Armando Tacchella, Davide Zambonin

System Description: QuBE: A system for deciding Quantified Boolean
Formulas Satisfiability 351
    Enrico Giunchiglia, Massimo Narizzano, Armando Tacchella

System Description: E 0.61 356
    Stephan Schulz

System Description: Vampire 1.1 361
    Alexandre Riazanov, Andrei Voronkov

System Description: DCTP: A Disconnection Calculus Theorem Prover 366
    Reinhold Letz, Gernot Stenz

Logical Frameworks, Higher-Order Logic, Interactive Theorem Proving

More On Implicit Syntax 371
    Marko Luther

Termination and Reduction Checking for Higher-Order Logic Programs . . . 386
    Brigitte Pientka

System Description: P.rex: An Interactive Proof Explainer 401
    Armin Fiedler

System Description: JProver: Integrating Connection-based Theorem
Proving into Interactive Proof Assistants 406
    Stephan Schmitt, Lori Lorigo, Christoph Kreitz, Aleksey Nogin

Semantic Guidance

The eXtended Least Number Heuristic 411
    Gilles Audemard, Laurent Henocque

System Description: SCOTT-5 426
    Kahlil Hodgson, John Slaney

System Description: Combination of distributed search and multi-search
in Peers-mcd.d 431
    Maria Paola Bonacina

System Description: Lotrec: The Generic Tableau Prover for Modal and
Description Logics 436
    Luis Farinas del Cerro, David Fauthoux, Olivier Gasquet, Andreas Herzig, Dominique Longin, Fabio Massacci

System Description: The ModProf Theorem Prover 441
  Jens Happe

System Description: A New System and Methodology for Generating
Random Modal Formulae 446
  Peter F. Patel-Schneider, Roberto Sebastiani

Equational Theorem Proving and Term Rewriting

Decidable Classes of Inductive Theorems 451
  Jürgen Giesl, Deepak Kapur

Automated Incremental Termination Proofs for Hierarchically Defined Term
Rewriting Systems 466
  Xavier Urbain

Decidability and Complexity of Finitely Closable Linear Equational Theories 481
  Christopher Lynch, Barbara Morawska

A New Meta-Complexity Theorem for Bottom-up Logic Programs 496
  Harald Ganzinger, David McAllester

Tableau, Sequent, Natural Deduction Calculi and Proof Theory

Canonical Propositional Gentzen-Type Systems 511
  Arnon Avron, Iddo Lev

Incremental Closure of Free Variable Tableaux 526
  Martin Giese

Deriving Modular Programs from Short Proofs 541
  Uwe Egly, Stephan Schmitt

A General Method for Using Schematizations in Automated Deduction ... 556
  Nicolas Peltier

Automata, Specification, Verification and Logics of Programs

Approximating Dependency Graphs using Tree Automata Techniques 571
  Aart Middeldorp

On the Use of Weak Automata for Deciding Linear Arithmetic with Integer
and Real Variables 588
  Bernard Boigelot, Sébastien Jodogne, Pierre Wolper

A Sequent Calculus for First-order Dynamic Logic with Trace Modalities . 603
  Bernhard Beckert, Steffen Schlager

Flaw Detection in Formal Specifications............................... 618
   W. Reif, G. Schellhorn, A. Thums

System Description: CCE: Testing Ground Joinability 633
   Jürgen Avenhaus, Bernd Löchner

System Description: RDL: Rewrite and Decision procedure Laboratory . 638
   Alessandro Armando, Luca Compagna, Silvio Ranise

Nonclassical Logics

lolliCoP – A Linear Logic Implementation of a Lean Connection-Method
Theorem Prover for First-Order Classical Logic 645
   Joshua S. Hodas, Naoyuki Tamura

System Description: MUSCADET 2.3: A Knowledge-based Theorem Prover
Based on Natural Deduction 660
   Dominique Pastre

System Description: Hilberticus - a Tool Deciding an Elementary Sublanguage of Set Theory 665
   Jörg Lücke

System Description: STRIP: Structural sharing for efficient proof-search 670
   D. Larchey-Wendling, D. Méry, D. Galmiche

System Description: RACER 675
   Volker Haarslev, Ralf Möller

Author Index.. 681

Program Termination Analysis by Size-Change Graphs (Abstract)

Neil D. Jones

DIKU, University of Copenhagen, e-mail: neil@diku.dk

Size-change analysis is based on *size-change graphs* giving local approximations to parameter size changes derivable from program syntax. The "size-change termination" principle for a first-order functional language with well-founded data is: a program terminates on all inputs if *every infinite call sequence* (following program control flow) would cause an infinite descent in some data values. Two termination detection algorithms are given in [9]: one involving Büchi automata that directly realizes the definition; and a more useful one involving a closure algorithm on a set of size-change graphs.

Termination analysis based on this principle seems simpler, more general and more automatic than other work in the literature: lexicographic orders, mutually recursive function calls and permuted arguments are all handled *automatically and without special treatment*, with no need for human-supplied argument orders, or theorem-proving search methods not certain to terminate at analysis time.

Finally, the problem's *intrinsic complexity* is surprisingly high: complete for PSPACE. An interesting consequence: many other analyses found in the termination and quasi-termination literature are also PSPACE hard.

Some examples of terminating programs.

1. Program with permuted parameters:

```
p(m,n,r) = if r>0 then 1:p(m, r-1, n) else
                 if n>0 then 2:p(r, n-1 ,m)
                 else m
```

2. Program with permuted and possibly discarded parameters:

```
f(x,y) = if y=[] then x else
               if x=[] then 1:f(y, tl y)
               else 2:f(y, tl x)
```

3. Function with lexically ordered parameters:

```
a(m,n) = if m=0 then n+1 else
             if n=0 then 1:a(m-1, 1)
             else 2:a(m-1, 3:a(m,n-1))
```

Claim: These programs all terminate for a common reason: any infinite call sequence (regardless of test outcomes) causes infinite descent in one or more values. Examples 1, 2 seem to possess no natural lexical descent. In fact, the reasoning is necessarily tricky, since the problem is PSPACE-hard. Two algorithms are given in [9] to perform the test automatically.

Theorem 1. *Size-change termination is decidable in polynomial space.*

R. Goré, A. Leitsch, and T. Nipkow (Eds.): IJCAR 2001, LNAI 2083, pp. 1–4, 2001.

Theorem 2. *Size-change termination is* PSPACE-*hard.*

A known PSPACE-complete problem: *Given* a Boolean program b, *To decide* whether b terminates. Proof idea: Given Boolean program b, construct a program p of size polynomial in the size of b such that b terminates if and only if p is *not* size-change terminating.

Corollary 1. *The termination and quasi-termination criteria of [2,5,7,10,13] all are* PSPACE-*hard.*

Proof. Point: These analyses all give correct results when applied to programs whose data flow is similar to that of p above. The proof is essentially the same, with the construction modified as necessary to make the program fail the condition tested by the respective method, just when the Boolean program terminates.

Related Work. The PSPACE lower bound is the first such result of which we are aware. The termination algorithm, though, has counterparts in other areas.

Typed functional programs. Abel and Altenkirch [1] developed a system called **foetus** that accepts as input mutual recursive function definitions over strict positive datatypes. It returns a lexical ordering on the arguments of the program's functions, if one exists. The method of [9] handles programs with or without such a lexical ordering.

Term rewriting systems. TRS termination analyses often perform expensive searches for a suitable ordering to solve a set of inequalities; e.g., in [15], a heuristic is given for automatically generating a general class of *transformation orderings*, which includes the lexical order. In the present work, it has not been the aim to *look for* orderings. Size-change termination naturally subsumes an interesting class of orderings, including the lexical ordering, and the ordering for the example with permuted and discarded parameters, which is not obvious.

One TRS application is to model semantics of functional programs. A functional program is easily translated into a TRS whose termination implies that of the subject program. Unfortunately, the result is often *non-simply-terminating*, which means the usual approach (find an order so the LHS of each rewrite rule is strictly greater than the RHS), does not work. To treat such TRS, Arts and Giesl [3,4,6] applied programming intuition to develop stronger methods. Unfortunately, this required extending existing techniques for TRS termination; and expensive searches for suitable orderings. For a term-rewriting perspective, these methods are able to a handle a larger class of TRS than dealt with before. For analyzing functional programs, our dataflow approach seems less circuitous.

Finally, for TRS corresponding to programs, the polynomial interpretation method for discovering orderings [6] can sometimes provide an alternative to size analysis by appropriately interpreting function symbols in the subject program (when it succeeds). The approach in [9] is instead to factor out size analysis as an orthogonal concern, and focus on the size-change termination principle and its application. This appears to give a natural separation of concerns when analyzing termination of programs.

Logic programs. There has been extensive research on automatic termination analysis for logic programs. As explained in [13], it is not always obvious that a predicate will terminate when executed with unusual instantiation patterns, or that a predicate always terminates on backtracking. For interpreters that have a choice of evaluation orders, termination analysis is especially important.

Some analyses that have been described for logic programs (e.g., in [11,14]) use a simple criterion: for every recursive invocation of a predicate, determine that the sum over a subset of input fields (fixed for each predicate) is strictly decreased. This does not allow handling of lexical descent. The strength of these methods derives from aggressive size analysis, which enables, in particular, sorting routines (quicksort and insertion sort) to be handled automatically. It is also possible to incorporate size analysis into the present approach, but our aim has been to investigate the size-change termination principle by itself.

Some logic program termination analyzers use a termination criterion compatible with size-change termination [10,5]. The analysis in [13] has been extended to a termination analyzer for Prolog programs called *Termilog* [10]. It turns out that Termilog can solve size-change termination problems precisely via a suitable encoding. In fact, our graph-based algorithm, although devised independently, is in essence a functional programming counterpart of the Termilog algorithm. Thus the PSPACE hardness result applies to Termilog's Analysis.

All the works on Prolog termination that we are aware of devote much attention to orthogonal issues such as uninstantiated variables and size analysis. While no doubt important in practice, an impression is created that the complexity of Prolog termination stems from these concerns; but our complexity result sayss that the core *size-change termination principle* is intrinsically hard.

Quasi-termination. The PSPACE hardness construction can be modified to show that the in-situ descent criterion for quasi-termination in [2] is PSPACE hard. Glenstrup [7] shows one way that quasitermination analysis techniques can be used for termination.

References

1. Andreas Abel and Thorsten Altenkirch. A semantical analysis of structural recursion. In *Abstracts of the Fourth International Workshop on Termination WST'99*, pages 24–25. unpublished, May 1999.
2. Peter Holst Andersen and Carsten Kehler Holst. Termination analysis for offline partial evaluation of a higher order functional language. In *Static Analysis, Proceedings of the Third International Symposium, SAS '96, Aachen, Germany, Sep 24–26, 1996*, volume 1145 of *Lecture Notes in Computer Science*, pages 67–82. Springer, 1996.
3. Thomas Arts. *Automatically Proving Termination and Innermost Normalisation of Term Rewriting Systems*. PhD thesis, Universiteit Utrecht, 1997.
4. Thomas Arts and Jürgen Giesl. Proving innermost termination automatically. In *Proceedings Rewriting Techniques and Applications RTA'97*, volume 1232 of *Lecture Notes in Computer Science*, pages 157–171. Springer, 1997.

5. Michael Codish and Cohavit Taboch. A semantic basis for termination analysis of logic programs and its realization using symbolic norm constraints. In Michael Hanus, Jan Heering, and Karl Meinke, editors, *Algebraic and Logic Programming, 6th International Joint Conference, ALP '97–HOA '97, Southampton, U.K., September 3–5, 1997*, volume 1298 of *Lecture Notes in Computer Science*, pages 31–45. Springer, 1997.
6. Jürgen Giesl. Termination analysis for functional programs using term orderings. In Alan Mycroft, editor, *Proc. 2nd Int'l Static Analysis Symposium (SAS), Glasgow, Scotland*, volume 983 of *Lecture Notes in Computer Science*, pages 154–171. Springer-Verlag, September 1995.
7. Arne J. Glenstrup. *Terminator II: Stopping partial evaluation of fully recursive programs*. Master's thesis, DIKU, University of Copenhagen, Denmark, 1999.
8. Chin Soon Lee. *Program termination analysis and the termination of off-line partial evaluation*. Ph.D. thesis, University of Western Australia.
9. Chin Soon Lee and Neil D. Jones and Amir Ben-Amram. The Size-Change Principle for Program Termination. In Hanne Riis Nielson, editor, *ACM SIGPLAN Symposium on Principles of Programming Langages, London, England*, pages 81–92. ACM Press, 2000.
10. Naomi Lindenstrauss and Yehoshua Sagiv. Automatic termination analysis of Prolog programs. In Lee Naish, editor, *Proceedings of the Fourteenth International Conference on Logic Programming*, pages 64–77, Leuven, Belgium, Jul 1997. MIT Press.
11. Lutz Plümer. *Termination Proofs for Logic Programs*, volume 446 of *Lecture Notes in Artificial Intelligence*. Springer-Verlag, 1990.
12. S. Safra. On the complexity of omega-automata. In *Proceedings of the 29th IEEE Symposium on Foundations of Computer Science*, pages 319–327, IEEE, 1988.
13. Yehoshua Sagiv. A termination test for logic programs. In Vijay Saraswat and Kazunori Ueda, editors, *Logic Programming, Proceedings of the 1991 International Symposium, San Diego, California, USA, Oct 28–Nov 1, 1991*, pages 518–532. MIT Press, 1991.
14. Chris Speirs, Zoltan Somogyi, and Harald Søndergaard. Termination analysis for Mercury. In Pascal Van Hentenryck, editor, *Static Analysis, Proceedings of the 4th International Symposium, SAS '97, Paris, France, Sep 8–19, 1997*, volume 1302 of *Lecture Notes in Computer Science*, pages 160–171. Springer, 1997.
15. Joachim Steinbach. Automatic termination proofs with transformation orderings. In Jieh Hsiang, editor, *Rewriting Techniques and Applications, Proceedings of the 6th International Conference, RTA-95, Kaiserslautern, Germany, April 5–7, 1995*, volume 914 of *Lecture Notes in Computer Science*, pages 11–25. Springer, 1995.

SET Cardholder Registration: The Secrecy Proofs
(Extended Abstract)

Lawrence C. Paulson

Computer Laboratory, Univ. of Cambridge, Cambridge, England
lcp@cl.cam.ac.uk

1 Introduction

Security protocols aim to protect the honest users of a network from the dishonest ones. *Asymmetric* (public key) cryptography is valuable, though it is normally used in conjunction with *symmetric* cryptography, where two users share a secret key, Asymmetric cryptography is typically used to securely exchange symmetric keys, which carry the bulk of the traffic. This mode of operation is faster than using expensive public-key encryption exclusively. It is also more secure, since the symmetric keys can be changed frequently. However, the protocol used to set up of this communication must be designed with care. For example, each message typically includes a *nonce*: a freshly-generated number that the other party must include in his response; the first party then knows that the response was not an old message replayed by an intruder. Many flaws have been discovered in security protocols [5].

Security protocol verification technologies have progressed in recent years. A variety of tools are available for analyzing protocols. Model checking is excellent for debugging a protocol, finding attacks in seconds [6,7]. Theorem proving is valuable too: it can analyze protocols in more detail and handles the protocols that are too big for model checking. Subgoals presented to the user suggest possible failure modes and give insights into how the protocol operates.

Past work on protocol verification has focused on protocols arising from the academic community. Only seldom have deployed protocols been investigated, such as Kerberos [3], SSL [8] and SSL's successor, TLS [12]. Past work has largely focused on *key exchange* protocols. Such protocols allow two participants (invariably called Alice and Bob) to agree on a *session key*: a short-term symmetric key. In this paper, I would like to describe a project, joint with Bella, Massacci and Tramontano, to verify a very large commercial protocol: SET, or Secure Electronic Transactions [15].

2 The SET Protocol

People normally pay for goods purchased over the Internet using a credit card. They give their card number to the merchant, who claims the cost of the goods

R. Goré, A. Leitsch, and T. Nipkow (Eds.): IJCAR 2001, LNAI 2083, pp. 5–12, 2001.

against it. To prevent eavesdroppers from stealing the card number, the transaction is encrypted using the SSL protocol. This arrangement requires the customer and merchant to trust each other: an undesirable requirement even in face-to-face transactions, and across the Internet it admits unacceptable risks.

- The cardholder is protected from eavesdroppers but not from the merchant himself. Some merchants are dishonest: pornographers have charged more than the advertised price, expecting their customers to be too embarrassed to complain. Some merchants are incompetent: a million credit card numbers have recently been stolen from Internet sites whose managers had not applied patches (available free from Microsoft) to fix security holes [9].
- The merchant has no protection against dishonest customers who supply an invalid credit card number or who claim a refund from their bank without cause. Contrary to popular belief, it is not the cardholder but the merchant who has the most to lose from fraud. Legislation in most countries protects the consumer.

The SET protocol aims to reduce fraud by introducing a preliminary registration phase. Both cardholders and merchants must register with a *certificate authority* (CA) before they can engage in transactions. The cardholder thereby obtains electronic credentials to prove that he is trustworthy. The merchant similarly registers and obtains credentials. These credentials do not contain sensitive details such as credit card numbers. Later, when the customer wants to make purchases, he and the merchant exchange their credentials. If both parties are satisfied then they can proceed with the transaction. Credentials must be renewed every few years, and presumably are not issued to known fraudsters.

SET comprises 15 subprotocols, or *transactions*, in all. Some observers, noting its extreme complexity, predict that it will never be deployed. However, the recent large rise in credit card fraud [1] suggests that current arrangements are unsustainable. SET or a derivative protocol may well be deployed in the next several years. To a researcher, SET has a further attraction: it makes heavy use of primitives such as digital envelopes that protocol verifiers have not examined before now.

3   Cardholder Registration

As described above, each cardholder must register before he is allowed to make purchases. He proves his identity by supplying personal information previously shared with his issuing bank. He chooses a private key, which he will use later to sign orders for goods, and registers the corresponding public key, which merchants can use to verify his signature. In keeping with normal practice, SET requires each participant to have separate key pairs for signature and encryption.

Cardholder registration comprises six messages:

1. The cardholder contacts the CA to request registration.

2. The CA replies, returning its public key certificates. These contain the CA's public keys (which the cardholder needs for the next phase) and are signed by the Root Certificate Authority (so that the cardholder knows they are genuine).

3. The cardholder requests a registration form. In this message, he submits his credit card number to the CA.

4. The CA uses the credit card number to determine the cardholder's issuing bank and returns an appropriate registration form.

5. The cardholder chooses an asymmetric public/private key pair. He submits the public key along with the completed registration form to the CA, who forwards it to the bank.

6. The bank checks the various details, and if satisfied, authorises the CA to issue credentials. The CA signs a certificate that includes the cardholder's public signature key and the cryptographic hash of a number — the **PAN-Secret** — known only to the CA and cardholder. Finally the cardholder receives the credentials and is ready to go shopping.

Does verifying cardholder registration serve any purpose? The payment phase performs the actual E-commerce, and protocol verifiers often assume that participants already possess all needed credentials. However, cardholder registration is a challenging protocol, particularly when it comes to proving that the PANSecret is actually secret.

The most interesting feature of cardholder registration, from the viewpoint of verification, is its use of *digital envelopes.* To send a long message to the CA, the cardholder generates a fresh symmetric key and encrypts the message, using public key encryption only to deliver the session key to the CA. As mentioned at the start of this paper, this combination of symmetric and asymmetric encryption is more efficient and secure than using asymmetric encryption alone. However, the two-stage process makes a protocol harder to analyze. The most complicated case is with the last message exchange, where the cardholder sends the CA *two* session keys. One of these keys encrypts the cardholder's message and the other encrypts the CA's reply.

We could simplify the protocol by eliminating digital envelopes and removing unnecessary encryption. However, the resulting protocol would be trivial. Experience shows that simplifying out implementation details can hide major errors [14]. Cardholder registration is valuable preparation for the eventual verification of the purchase phase.

4   The Secrecy Proofs

We use the inductive method of protocol verification, which has been described elsewhere [11,13]. This operational semantics assumes a population of honest agents obeying the protocol and a dishonest agent (the Spy) who can steal messages intended for other agents, decrypt them using any keys at his disposal and send new messages as he pleases. Some of the honest agents are compromised,

8   L.C. Paulson

meaning the Spy has full access to their secrets. A protocol is modelled by the
set of all possible traces of events that it can generate. Events are of three forms:

- Says $A\,B\,X$ means A sends message X to B.
- Gets $A\,X$ means A receives message X.
- Notes $A\,X$ means A stores X in its internal state.

The model of Cardholder Registration is largely the work of Bella, Massacci
and Tramontano, who devoted many hours to decrypting 1000 pages of SET
documentation [2]. We have flattened the hierarchy of certificate authorities.
The Root Certificate Authority is responsible for certifying all the other CAs.
Our model includes compromised CAs — as naturally it should — though we
assume that the root is uncompromised. The compromised CAs complicate the
proofs considerably, since large numbers of session keys and other secrets fall
into the hands of the Spy. Here is a brief summary of the notation:

- *set_cr* is the set of traces allowed by Cardholder Registration
- *used* is the set of items appearing in the trace, to express freshness
- *symkeys* is the set of symmetric keys[1]
- *Nonce*, *Key*, *Agent*, *Crypt* and *Hash* are message constructors
- $\{X_1, \ldots, X_n\}$ is an n-component message

Here is part of the specification, the inductive rule for message 5. Variable
evs5 refers to the current event trace:

```
[evs5 ∈ set_cr;   C = Cardholder k;
Nonce NC3 ∉ used evs5;   Nonce CardSecret ∉ used evs5; NC3≠CardSecret;
Key KC2 ∉ used evs5; KC2 ∈ symKeys;
Key KC3 ∉ used evs5; KC3 ∈ symKeys; KC2≠KC3;
cardSK ∉ symKeys;  ...
Gets C ... ∈ set evs5;
Says C (CA i) ... ∈ set evs5]
⟹ Says C (CA i)
      {Crypt KC3 {Agent C, Nonce NC3, Key KC2, Key cardSK,
                  Crypt (invKey cardSK)
                      (Hash{Agent C, Nonce NC3, Key KC2,
                            Key cardSK, Pan(pan C), Nonce
CardSecret})},
        Crypt EKi {Key KC3, Pan (pan C), Nonce CardSecret}}
# evs5 ∈ set_cr
```

Much has been elided from this rule, but we can see several things:

- the generation of two fresh nonces, *NC3* and *CardSecret*
- the generation of two fresh symmetric keys, *KC2* and *KC3*, to be used as session
  keys
- a message encrypted using *EKi* (the CA's public key) and containing the
  credit card number (*pan C*) and the key *KC3*

---

[1] In an implementation, a symmetric key occupies 8 bytes while an asymmetric one
occupies typically 128 bytes, so the two types are easily distinguishable.

- a message encrypted using *KC3* and containing the symmetric key *KC2* and the cardholder's public signature key, *cardSK*

The two encrypted messages constitute a digital envelope.

The PANSecret mentioned in §3 above is computed as the exclusive-OR of other secret numbers generated by the cardholder and the CA. Do these numbers really remain secret? Since they are encrypted using symmetric keys, the proof requires a lemma that symmetric keys remain secret. Two complications are that some symmetric keys do *not* remain secret, namely those involving a compromised CA, and that some symmetric keys are used to encrypt others. The latter point means that the loss of one key can compromise a second key, leading possibly to unlimited losses.

The problem of one secret depending on another has occurred previously, with the Yahalom [10] and Kerberos [3] protocols. Both of these are simple: the dependency relation links only two items. Cardholder registration has many dependency relationships. It also has a dependency chain of length three: in the last message, a secret number is encrypted using a key (*KC2*) that was itself encrypted using another key (*KC3*).

Fortunately, the method described in earlier work generalizes naturally to this case and to chains of any length. While the definitions become more complicated than before, they follow a uniform pattern. The idea is to define a relation, for a given trace, between pairs of secret items: (K, X) are related if the loss of the key K leads to the loss of the key or nonce X. Two new observations can be made about the dependency relation:

- It should ignore messages sent by the Spy, since nothing belonging to him counts as secret. This greatly simplifies some proofs.
- It must be transitive, since a dependency chain leading to a compromise could have any length. Past protocols were too simple to reveal this point.

Secrecy of session keys is proved as it was for Kerberos IV [3], by defining the relation *KeyCryptKey DK K evs*. This relation captures instances of message 5 in which somebody other than the Spy uses *KC3* to encrypt *KC2* in the event trace *evs*. The *session key compromise theorem* states that a given key can be lost only by the keys related to it by *KeyCryptKey*. The form of this lemma has been discussed elsewhere [10]; it handles cases of the induction in which some session keys are compromised. Using this lemma, we can prove that no symmetric keys are lost in a communication between honest participants:

⟦*CA i ∉ bad; K ∈ symKeys; evs ∈ set_cr;*
  Says (Cardholder k) (CA i) X ∈ set evs; Key K ∈ parts {X}⟧
⟹ *Key K ∉ analz (knows Spy evs)*

Any symmetric key that is part of a message *X* sent by a cardholder (that is, *Key K ∈ parts {X}*) is not derivable from material visible to the Spy (that is, *Key K ∉ analz (knows Spy evs)*).

Given that the session keys are secure, we might hope to find a simple proof that nonces encrypted using those keys remain secret. However, secrecy proofs for nonces require the same treatment as secrecy proofs for keys. We must define the dependency relation between keys and nonces and prove a lemma analogous to the one shown above.

Secrecy of nonces is proved as it was for Yahalom [10], except that there are many key-nonce relationships rather than one. Note also the occurrences of *KeyCryptKey*, which allow for longer dependency chains.

```
KeyCryptNonce DK N (ev # evs) =
 (KeyCryptNonce DK N evs ∨
  (case ev of
    Says A B Z ⟹
     A ≠ Spy ∧
     ((∃X Y. Z = {|Crypt DK {|Agent A, Nonce N, X|}, Y|}) ∨
      (∃K i X Y.
        Z = Crypt K {|sign (priSK i) {|Agent B, Nonce N, X|}, Y|} ∧
        (DK=K ∨ KeyCryptKey DK K evs)) ∨
      (∃K i NC3 Y.
        Z = Crypt K
              {|sign (priSK i) {|Agent B, Nonce NC3, Agent(CA i), Nonce N|},
                Y|} ∧
        (DK=K ∨ KeyCryptKey DK K evs)) ∨
      (∃i. DK = priEK i))
  | Gets A' X ⟹ False
  | Notes A' X ⟹ False))
```

Finally, we can show that the secrets exchanged by the parties in the final handshake remain secure.

```
[|CA i ∉ bad;
 Says (Cardholder k) (CA i)
   {|X, Crypt EKi {|Key KC3, Pan p, Nonce CardSecret|}|} ∈ set evs;
 ...; evs ∈ set_cr|]
⟹ Nonce CardSecret ∉ analz (knows Spy evs)
```

This theorem concerns the cardholder's secret. There is an analogous one for the CA's secret.

G. Bella has proved that the credit card number also remains secret. It looks straightforward: the number is encrypted using the CA's public key, which is secure provided the CA is uncompromised. As usual, however, the proof is harder than it looks. It requires a lemma stating that no symmetric keys are of any use to the spy for stealing a credit card number. This lemma looks obvious too, but both it and the main theorem are non-trivial inductions. Their proofs together require about one CPU minute.

Why are proofs so difficult and slow? The digital envelopes and digital signature conventions are to blame. Compared with other protocols analyzed using the inductive method, cardholder registration has nested encryption, resulting in huge case splits. The verifier is sometimes presented with a giant subgoal spanning many pages of text. One should not attempt to prove such a monstrosity but instead to improve the simplification so that it does not occur again.

5   Observations about Cardholder Registration

The proofs suggest that cardholder registration is secure. However, some anomalous features come to light. These do not derive from the formal analysis but merely by a close inspection of the protocol.

There is unnecessary encryption. The cardholder's signature verification key is encrypted, when it is a public key! The cardholder certificate is also encrypted, when it is of no use to anyone but the cardholder. Public-key certificates are nearly always sent in clear; this encryption is presumably intended to strengthen confidence in SET and to reassure cardholders. Nonces whose purpose is to ensure freshness do not have to be encrypted, but in SET they usually are. This forces KeyCryptNonce to take them into account, increasing the expression blow-up in secrecy proofs. We have a paradox: protocol designers who are concerned about security will include additional encryption, but that encryption actually makes the protocol more difficult to verify.

I observed two insecurities. The cardholder is not required to generate a fresh signature key pair, but may register an old one. There is a risk that this old one could be compromised. SET accordingly includes a further security measure: a secret number known to the cardholder, which he later uses as a password. This PANSecret is the exclusive-OR of numbers chosen by the two parties (see §4), and the cardholder chooses his number before the CA does. Since exclusive-OR is invertible, a criminal working for a CA can give every cardholder the same PANSecret.

This combination of insecurities introduces some risk that a criminal could impersonate the cardholder. The cardholder's implementation of SET can repair the first defect by always generating a fresh signature key pair. The second defect is, in principle, easy to fix: simply change the computation of the PANSecret, replacing the exclusive-OR by cryptographic hashing. But that unfortunately is a change to the protocol itself.

6   Conclusions

Our joint work has been fruitful. We had been able to specify and verify cardholder registration. Our model is abstract but retains much detail. We can prove secrecy in the presence of digital envelopes. We have strengthened our previous work on the relationships between secrets. There must be a connection with Cohen's secrecy invariant [4], though I am not sure of the details. We look forward to analyzing the remainder of SET.

Acknowledgements. Thanks above all to my colleagues G. Bella, F. Massacci and P. Tramontano for their many months devoted to understanding the SET specifications. (By contrast, the secrecy proofs reported above took only days.) This work was funded by the EPSRC grant GR/R01156/01 *Verifying Electronic Commerce Protocols*.

12   L.C. Paulson

References

1. Credit card fraud rises by 50%. On the Internet at
   http://news.bbc.co.uk/hi/english/business/newsid_1179000/1179590.stm,
   February 2001. BBC News (Business).
2. Giampaolo Bella, Fabio Massacci, Lawrence C. Paulson, and Piero Tramontano.
   Formal verification of cardholder registration in SET. In F. Cuppens, Y. Deswarte,
   D. Gollman, and M. Waidner, editors, *Computer Security — ESORICS 2000*,
   LNCS 1895, pages 159–174. Springer, 2000.
3. Giampaolo Bella and Lawrence C. Paulson. Kerberos version IV: Inductive analysis
   of the secrecy goals. In J.-J. Quisquater, Y. Deswarte, C. Meadows, and D. Goll-
   mann, editors, *Computer Security — ESORICS 98*, LNCS 1485, pages 361–375.
   Springer, 1998.
4. Ernie Cohen. TAPS: A first-order verifier for cryptographic protocols. In *13th
   Computer Security Foundations Workshop*, pages 144–158. IEEE Computer Society Press, 2000.
5. Gavin Lowe. Breaking and fixing the Needham-Schroeder public-key protocol using
   CSP and FDR. In T. Margaria and B. Steffen, editors, *Tools and Algorithms for
   the Construction and Analysis of Systems: second international workshop, TACAS
   '96*, LNCS 1055, pages 147–166. Springer, 1996.
6. Gavin Lowe. Casper: A compiler for the analysis of security protocols. *Journal of
   Computer Security*, 6:53–84, 1998.
7. Catherine Meadows. Analysis of the Internet Key Exchange protocol using the
   NRL Protocol Analyzer. In *Symposium on Security and Privacy*, pages 216–231.
   IEEE Computer Society, 1999.
8. John C. Mitchell, Vitaly Shmatikov, and Ulrich Stern. Finite-state analysis of
   SSL 3.0 and related protocols. In Hilarie Orman and Catherine Meadows, editors, *Workshop on Design and Formal Verification of Security Protocols*. DIMACS,
   September 1997.
9. Alan Paller. Alert: Large criminal hacker attack on Windows NTE-banking and
   E-commerce sites. On the Internet at
   http://www.sans.org/newlook/alerts/NTE-bank.htm, March 2001.
10. Lawrence C. Paulson. Relations between secrets: Two formal analyses of the Ya-
    halom protocol. *Journal of Computer Security*. in press.
11. Lawrence C. Paulson. The inductive approach to verifying cryptographic protocols.
    Journal of Computer Security, 6:85–128, 1998.
12. Lawrence C. Paulson. Inductive analysis of the Internet protocol TLS. *ACM
    Transactions on Information and System Security*, 2(3):332–351, August 1999.
13. Lawrence C. Paulson. Proving security protocols correct. In *14th Annual Sympo-
    sium on Logic in Computer Science*, pages 370–381. IEEE Computer Society Press,
    1999.
14. Peter Y. A. Ryan and Steve A. Schneider. An attack on a recursive authentication
    protocol: A cautionary tale. *Information Processing Letters*, 65(1):7–10, January
    1998.
15. SETCo. *SET Secure Electronic Transaction Specification: Business Description*,
    May 1997. On the Internet at http://www.setco.org/set_specifications.html.

Algorithms, Datastructures, and Other Issues in Efficient Automated Deduction

Andrei Voronkov

University of Manchester
voronkov@cs.man.ac.uk

Abstract. Algorithms and datastructures form the kernel of any efficient theorem prover. In this abstract we discuss research on algorithms and datastructures for efficient theorem proving based on our experience with the theorem prover Vampire. We also briefly overview other works related to algorithms and datastructures, and to efficient theorem proving in general.

1 Introduction

To implement an efficient automatic theorem prover, one has to put together at least three ingredients: good theory, efficient algorithms and datastructures, and clever heuristics.[1] In the recent years, a considerable progress has been made in the theory of resolution-based systems. This theory is build upon completeness theorems for resolution calculi with notions of redundant inferences and redundant derivations [6,63]. The theory is well-understood, but a good theory alone is not enough to implement an efficient prover.

The progress in theory can be characterized by the following observation: a prover based on the theory known in 1970 would now be hopelessly inefficient (unrestricted resolution and paramodulation, use of function reflexivity axioms). A prover based on the theory known in 1980 would outperform a 1970 prover by several orders of magnitude on difficult problems (mainly because of the use of simplification orderings). Compared to a 1980 prover, a prover based on the theory of 1990 would be several orders of magnitude faster on many difficult problems and moreover more flexible due to some new theoretical results (the general theory of redundancy, selection functions).

Since the first paper on resolution theorem proving [77], many theorem provers were developed by various researchers and groups. The most consistent implementation efforts were undertaken at Argonne National Laboratory; they resulted in the development of a series of systems, including Logic Machine Architecture [44,43] and Otter [51]. The nature of research at Argonne was formulated in [42]: *controlling redundancy in large search spaces*. Recently, several new efficient first-order theorem provers emerged, including the resolution-based provers Spass [96,95], E [78,79], Gandalf [89], Vampire [70,75], Bliksem [16], and

---

[1] See [56] for a similar observation.

R. Goré, A. Leitsch, and T. Nipkow (Eds.): IJCAR 2001, LNAI 2083, pp. 13–28, 2001.

SCOTT [82]; the model-elimination based prover Setheo [40], and the equational provers Waldmeister [34] and Fiesta (see [59]).

These provers are extremely efficient. For example the Steamroller problem discussed some time ago in the literature [87] is now a trivial problem for all of these systems. Several open problems solved by Otter in 1993 [50] can now be routinely solved by some of these provers. However, there are many first-order problems coming from various applications, which are still beyond the capabilities of the state-of-the-art provers. Since first-order logic is undecidable, it is unreasonable to expect new systems to efficiently solve all of them in the near future. However, if we can increase performance of the modern provers by several orders of magnitude for a large number of such problems, many of these problems will be routinely solved, thus saving time for application developers.

In our opinion, such a drastic increase in efficiency in the near future will be mainly based on the development of new algorithms and datastructures, and understanding how the theory developed so far can be efficiently implemented on top of the existing architectures of theorem provers.

2   How Do Efficient Datastructures Influence Performance

Before implementing Vampire, I implemented several more or less functioning theorem provers in REFAL [91], LISP, and Prolog. Efficiency was not an aim. I was interested in comparing their behavior when different methods were implemented (for example, [92] implemented the inverse method).

In 1993 Dominique Bolignano invited me to visit his research group at Bull, near Paris, for about two months. As part of my visit to Bull I gave a talk on theorem proving by the inverse method and seemed to convince the audience that the inverse method is worth trying. However, my old implementation of the inverse method was not maintained, while the new implementation did not exist. Then I decided to implement an *efficient* prover. The development of the new prover Vampire has changed considerably my perception of automated theorem proving. The first surprise came when I compared the behaviour of my newly implemented prover with that of Otter. When both provers used hyperresolution for solving the same problems, in the first few seconds of proof-search their performance was comparable. But after a few seconds Otter kept up making inferences at the same pace, while Vampire seemed to enter a deadlock. A simple profiling has shown that nearly all of the running time was spent on forward subsumption. So my first exercise in efficient theorem proving was to implement efficiently forward subsumption.[2] Search for literature on efficient subsumption did not help very much: although I could find a paper on efficient subsumption [29], the algorithm of this paper was not very useful when every newly generated clause had to be checked for subsumption with over 10^5 clauses in the current search space.

Search for an efficient subsumption algorithm resulted in discovery of the code tree indexing technique described in [93] and improved in [72]. After implemen-

---

[2] The value of subsumption was also observed by the Argonne group [97].

tation of code trees for forward subsumption, the performance of Vampire when running hyperresolution was comparable to that of Otter. From 1998 Vampire is implemented and maintained by Alexandre Riazanov. The rest of this paper describes some aspects of what Vampire has taught me about efficient theorem proving.

3 Saturation-Based Resolution Theorem Proving

All of the modern resolution-based first-order theorem provers implement some variant of the *given clause algorithm* [51,42]. Several variants of this algorithm are overviewed in [71,95]. One of the versions, roughly corresponding to those used in Otter and Fiesta, is shown in Figure 1 (taken from [71]). It is parametrized by several procedures explained below:

- *select* is the clause selection function. It decides which clause should be selected for activation.
- *infer* is the function that performs inferences between the current clause *current* and the set of active clauses *active*. This function returns the set of clauses obtained by all such possible inferences. This function varies from system to system. Usually, *infer* applies inferences in some complete inference system of resolution with paramodulation.
- *simplify(set, by)* is a procedure that performs simplification. It deletes redundant clauses from *set* and simplifies some clauses in *set* using the clauses in *by*. To preserve completeness, the simplified clauses are always moved to *passive*. Typically, deleted clauses include tautologies and those clauses subsumed by clauses in *by*. A typical example of simplification is rewriting by unit equalities in *by*.
- Likewise, *inner_simplify* simplifies clauses in *new* using other clauses in *new*.

When we simplify *new* using the clauses in *active* ∪ *passive*, we speak of *forward simplification*; when we simply *active* and *passive* using the clauses in *new*, we speak of *backward simplification*. The name *given clause algorithm* is due to the fact that the clause *current* is called the given clause in Otter's terminology.

It is instructive to explain several problems facing implementors of the given clause algorithm, to show the gap between the theory and practice of resolution-based theorem proving. In theory, it is not hard to prove that this algorithm is complete, provided that the underlying logical calculus is complete. In practice, at least two problems arise. Both problems are due to the fast growth of the search space. The first problem is how to select the right current clause in a huge set of passive clauses. This problem can be solved only by a large number of experiments over a large collection of problems. The most common approach is to maintain several priority queues on passive clauses, and pick up clauses from this priority queues. A very popular technique is to maintain two queues in which the clauses are prioritized by their weight and age, respectively. The clauses are

```
input: init : set of clauses ;
var active, passive, new : sets of clauses ;
var current : clause ;
active := ∅ ;
passive := init ;
while passive ≠ ∅ do
  current := select(passive) ;
  passive := passive − {current} ;
  active := active ∪ {current} ;
  new := infer(current, active) ;
  if goal_found(new) then return provable ;
  inner_simplify(new) ;
  simplify(new, active ∪ passive) ;
  if goal_found(new) then return provable ;
  simplify(active, new) ;
  simplify(passive, new) ;
  if goal_found(active ∪ passive) then return provable ;
  passive := passive ∪ new
od ;
return unprovable
```

Fig. 1. A Given Clause Algorithm

picked from the queues using the so-called *age-weight ratio* (also called the pick-given ratio); for example if the ratio is 1:5, than out of each 6 picked clauses, 1 will be selected as the oldest clause, and 5 as the lightest clauses. The prover E maintains more than two queues. Even this simple clause selection scheme creates many problems, when clauses must be deleted from the search space (for example, when the available memory is exhausted), since every clause may be kept in several priority queues. The best clause selection strategies are not yet well-understood.

Another serious problem is caused by the proliferation of passive clauses. A large number of passive clauses results in deterioration of the proof-search speed and huge memory consumption. To illustrate this, we provide statistics on an unsuccessful run of Vampire with the time limit of 1 minute on the TPTP problem ANA003-1. During this run, 261,573 clauses were generated. The overall number of active clauses was 1,967, the overall number of passive clauses 236,389. To cope with the problems of passive clauses, several solutions have been proposed. For example, to impose a weight limit on clauses and increase the weight limit, if the prover unsuccessfully terminates with the current weight limit.

The most radical solution to the passive clauses problem was originally implemented in DISCOUNT [2] and is now used in Waldmeister and E (and can be used as an option in Spass and Vampire). These provers use a different main loop, in which passive clauses do not participate in backward simplifications until they are selected as the current clause. In fact, these provers do not store the passive clauses at all, but only store information about the inference rule used to obtain a passive clause. This implementation scheme requires double forward simplification, but is believed to be space- and time-efficient. The provers implementing the DISCOUNT loop are essentially based on the principle: *process active clauses efficiently*. The price to pay is that some very useful simplifying inferences involving a passive clause can be considerably delayed until the clause has been selected as current. The main principle of Vampire with respect to the main loop is: *do simplifications eagerly and non-simplifying inferences lazily*. Therefore, Vampire's default option is the Otter main loop. However, to work with the large among of passive clauses efficiently, the so-called *limited resource strategy* was invented [71]. This strategy is applicable when the time limit on solving a problem is specified in advanced. Vampire tries to estimate which clauses cannot be processed by the end of the time limit at all, and discards such clauses as useless. Experiments reported in [71] show high efficiency of the limited resource strategy as compared to other approaches.

4   Term Indexing

To be able to process hundreds of thousands clauses in less than a minute, all most important operations should ideally be implemented on the set-at-a-time basis. For example, subsumption is NP-complete, and checking every newly generated clause for subsumption against several hundred thousand of kept clauses sequentially is hopelessly slow. There is a growing number of papers on *term indexing* (see [32,81] for an overview): an approach that allows one to implement efficiently expensive massive operations on terms and clauses, for example, subsumption of one clause by a large database of clauses.

The problem of term indexing can be formulated abstractly as follows (see [81]). Given a set L of *indexed terms* (or clauses), a binary relation R over terms (called the *retrieval condition*) and a term t (called the *query term*), identify the subset M of L that consists of the terms l such that $R(l, t)$ holds. Terms in M will be called the *candidate terms*. Typical retrieval conditions used in first-order theorem proving are matching, generalization, unifiability, subsumption, syntactic equality, variance etc. Such a retrieval of candidate terms in theorem proving is interleaved with insertion of terms to L, and deletion of them from L.

In order to support rapid retrieval of candidate terms, we need to process the indexed set into a data structure called the *index*. Indexing data structures are well-known to be crucial for the efficiency of the current state-of-the-art theorem provers. Term indexing is also used in logic and functional programming languages implementation, but indexing in theorem provers has several distinctive features:

1. Indexes in theorem provers frequently store 10^5–10^6 complex terms, unlike a typically small number of shallow terms in functional and logic programs.
2. In logic or functional language implementation the index is usually constructed during compilation. On the contrary, indexes in theorem proving are highly dynamic, since terms are frequently inserted in and deleted from

indexes. *Index maintenance* operations start with an index for an initial set
of terms L, and incrementally construct an index for another set L' that is
obtained by insertion or deletion of terms to or from L.

3. In many applications it is desirable for several retrieval operations to work
   on the same index structure in order to share maintenance overhead and
   memory consumption.

Therefore, along the last two decades a significant number of results on new
indexing techniques for theorem proving have been published and successfully
applied in different provers [81,32,33,69,84,64,49,13,30,31,93,76,72,26,57].

For every retrieval condition used in theorem provers, it is not necessary to
retrieve exactly all candidates satisfying this retrieval condition. If the retrieval
condition is used for inferences (as, e.g., unification is used for inferences by
resolution and paramodulation), then it is enough to retrieve a superset of candidates, and then check the retrieval condition for every member of this superset.
If the retrieval condition is used for simplifications (as, e.g., retrieval of generalizations is used for simplification by unit equalities or forward subsumption
by unit clauses), then it is enough to retrieve a subset of candidates only, since
simplifications only reduce the search space, but do not influence completeness.
If a particular term indexing technique retrieves exactly all candidates, then
this technique is said to perform *perfect filtering*. It is always an issue of debate
in automated deduction, whether perfect filtering should be implemented for a
particular retrieval condition. For example, [90] advocates the use of imperfect
filtering for subsumption.

Term indexing is one of the main research directions for Vampire. We believe
that perfect filtering is desirable for all operations for which indexing is required
at all. As a consequence, Vampire stores a large number of indexes for various
operations: partially adaptive code trees [93,72] for forward subsumption, subsumption resolution, and variance check; code trees with precompiled ordering
constraint for forward rewriting (also called demodulation) by unit equalities;
path indexing with compiled database joins implemented using skip lists for
backward subsumption and backward demodulation by unit equalities [73]; tries
for unification used to implement resolution and paramodulation; and another
kind of tries for storing perfectly shared terms. We learned that it pays off to
spend time for implementing new indexing techniques: for nearly every retrieval
condition there are problems for which this retrieval condition contributes to the
running time considerably. We believe that term indexing will be in the heart of
theorem proving research in the future.

It is due to term indexing that the modern provers can quickly solve problems which require search in a space of several million complex clauses. But it is
also due to term indexing that implementation of new features on top of existing
architectures requires non-trivial efforts and invention of completely new techniques. Every new feature brings in a tradeoff between the time/space gained by
the use of the feature on one hand, and the time spent on checking applicability
of this feature and space needed to implement the feature efficiently. In addition,
some promising features require a non-trivial implementation.

As a result, some enhancements of theorem provers well-known in theory have never been implemented. A typical example is the basic strategy [17,7,60]. Although in theory basic superposition saves from performing some redundant inferences, in practice implementation of basic superposition requires considerable changes in *all* algorithms and datastructures. Therefore, all indexing techniques and all algorithms for term retrieval must be adapted to the basic strategy. As a consequence, the basic strategy has never been fully implemented.

5   Building-In Equational Theories

We are in search of techniques that can speed up the provers by several orders of magnitude. Built-in equational theories is one of such techniques. Many problems coming from applications are theorems about structures axiomatized by a set of equations. The most common equational theory is AC: the theory containing associativity and commutativity axioms for a function symbol.

Although the idea of built-in equational theories have been around at least from 1972 [68], the first implementation of AC was undertaken more than 20 years after in the EQP equational theorem prover [48]. This implementation resulted in probably the most celebrated event in the automated deduction community: the automatic solution of the Robbins problem [47] by EQP.

Although EQP ran for 8 days to obtain the proof, the proof-search for the Robbins problem was remarkably small. The total number of equations processed during the proof was less than 50,000. Modern provers often process such a number of clauses in a matter of a few seconds. This shows that the Robbins problem may in fact be not very difficult for a prover in which AC is implemented efficiently, including term indexing modulo AC.

So far research on built-in equational theories was mainly built around equational unification (see e.g., [4] for a recent overview). The special case of AC-unification was discussed in a number of papers, dating back to [85,86], but only recently efficient algorithms have been described [1]. The experience with the non-AC theorem proving shows that efficient matching is more important than efficient unification,[3] but there is essentially no literature on efficient AC-matching. Moreover, there are essentially no publications on term indexing in presence of built-in theories. The only paper we know about discusses a special case of indexing for AC-matching of linear terms [5]. Vampire has commutativity built-in in some term indexes and retrieval algorithms (see e.g., [73]).

We conclude that efficient algorithms and datastructures for built-in equational theories should become a major topic for research in automated deduction.

6   Cheap Substitutes

When the price to pay for implementing a particular algorithm is too high, cheap substitutes can be used. This is especially true for the cases when com-

---

[3] Indeed, only 0.1% of the EQP running time on the Robbins problem was spent for AC-unification.

pleteness is not an issue, for example when implementing simplification rules. There are several examples of successful implementation of "cheap substitutes" for expensive operations. For example, Waldmeister implements specialized algorithms for AC-completion [3]. In resolution theorem proving it is desirable to quickly check whether one can apply a resolvent of two clauses which subsumes one of these clauses. Such an application simplifies the search space, since the resolvent will then replace one of its parents. However, no efficient algorithm is known for finding simplified resolvents. There is an incomplete but relatively cheap operation called *subsumption resolution* (see e.g. [6]) which can be implemented essentially without extra overhead, provided that efficient subsumption is implemented. Vampire implements subsumption resolution using the indexes for subsumption, but slightly modified algorithms. Another example of a cheap substitute is the implementation of splitting in Vampire [74] in the form of *splitting without backtracking*. Though it does not have full power of the splitting rule as implemented, e.g., in Spass [95], splitting without backtracking can be implemented without radical changes in the architecture of a theorem prover.

7   Constraints

The use of symbolic constraints in automated deduction gives stronger notions of redundancy than those formulated without constraints [62]. In addition, it is known that symbolic constraints may give a compact representation of large search spaces, for example when constraints modulo AC are used [61] to encode a doubly exponential number of AC-unifiers [36].

It is conjectured in [56] that deduction with symbolic constraints will be a major research topic in efficient automated deduction and can result in a major breakthrough in efficiency. However, ten years after the first publication on deduction with symbolic constraint [37] there is still no implementation. The problem is that algorithms and datastructures for solving symbolic constraints are not yet developed, except for some particular operations. The most advanced results on symbolic constraints are in the area of solving constraints over simplification orderings. The first algorithm for solving RPO ordering constraints were described in [14,35], followed by a number of results on solving RPO ordering constraints [55,15,54], but only recently an efficient algorithm was designed [58]. In the case of Knuth-Bendix ordering constraints, the decidability of constraint solving was proved in 1999 [38,39], but no simple algorithms have yet been described.

As shown in [94], good algorithms for solving ordering constraints are not enough for implementing constraint-based deduction. Efficient algorithms and datastructures should be developed also for constraint simplification and approximation. Moreover, first-order theories of ordering constraints are to be better understood.

8   Comparison of Algorithms and Datastructures

One of the main problems in research on algorithms and datastructures for automated deduction is that the algorithms are performed on terms and clauses, i.e., tree-like structures. It was observed that the worst-case complexity results for such structures can be inadequate in practice. For example, most provers use Robinson's unification algorithm [77] having worst-case exponential complexity instead of efficient linear algorithms [65,46], since overhead on maintaining datastructures for these algorithms does not pay off in practice, where the exponential behavior of Robinson's algorithm does not show up. Subsumption is NP-complete [9,27], but the modern provers often make subsumption-checks of 10^5 clauses against a database of 10^4 clauses in a few seconds.

A practical approach to comparing implementation of algorithms used in first-order automated deduction was recently undertaken in [57]. The essence of this approach is that the implementation techniques are compared on benchmarks taken from runs of theorem provers on real problems. It is likely that the methodology of [57] will be used for comparison of other important algorithms used in automated deduction.

9   Other Aspects of Efficient Theorem Proving

In this section we briefly overview aspects of efficient theorem proving other than those directly related to algorithms and datastructures.

9.1   Non-resolution Theorem Proving

Resolution is not the only automated reasoning method used in first-order automated deduction. Model elimination [41] implemented in SETHEO [40] often performs very well on problems difficult for resolution-based provers. A recent adaptation of propositional splitting to the full first-order case [8] seems to be promising. However, non-resolution based procedures often have difficulties with equality and other built-in theories, as witnessed by the results overviewed in [21]. They have been several proposals on combination of paramodulation-based reasoning and tableau-based reasoning [18,19,52], but none of them was implemented.

9.2   Parallel and Agent-Based Reasoning

Parallel computing is becoming cheaper. Networks of computers are now readily available. This makes parallel and agent-based theorem proving attractive. There were early projects aiming at parallelizing theorem provers, both in the context of model elimination [80] and resolution [45]. Distributed theorem proving was considered e.g., in [53]. However, parallelization of theorem proving, especially resolution-based, requires further investigation and experiments.

Theorem proving in which provers are considered as communicating agents was considered in [22]. It was shown that one can obtain a considerable speedup by running several provers in parallel and making them communicate by sending each other heuristically selected derived clauses.

There are several projects which put together several provers in different forms, for example, proving a common interface for running them over the Web. Examples are MathWeb [24,23], MBASE [25], SystemOnTPTP [88]. However, the emphasis in such systems was so far on providing a graphical user interface or interactive theorem proving but not so much on efficient automated deduction.

9.3 Other Research Directions

There are many other aspects of efficient theorem proving not considered in this paper. We will mention some of them very briefly.

In the future modern first-order theorem prover will be tightly integrated with other systems for automated reasoning, for example inductive theorem provers (see [12] for an overview), and proof assistants such as Isabelle [66] and HOL [28], and maybe model checkers. It is likely that first-order provers integrated in such systems will also partially implement proof-search specific to these systems, for example restricted forms of induction, proofs about inductively defined types, or even restricted forms of higher-order theorem proving. This will require new lines of research, for example

- saturation-based higher-order theorem proving [10,11];
- intelligent work with definitions [67,20];
- built-in data types;
- recognition of irrelevant axioms;
- propositional reasoning;
- built-in theories (not only equational) [83];
- reasoning with non-standard quantifiers, for example "there exists at most n";
- finite domain reasoning.

References

1. F. Ajili and E. Contejean. Avoiding slack variables in the solving of linear diophantine equations and inequations. *Theoretical Computer Science*, 173(1):183–208, 1997.
2. J. Avenhaus, J. Denzinger, and M. Fuchs. DISCOUNT: a system for distributed equational deduction. In J. Hsiang, editor, *Proceedings of the 6th International Conference on Rewriting Techniques and Applications (RTA-95)*, volume 914 of *Lecture Notes in Computer Science*, pages 397–402, Kaiserslautern, 1995.
3. J. Avenhaus and B. Löchner. Cce: Testing ground joinability. In R. Gore, A. Leitsch, and T. Nipkow, editors, *IJCAR 2001*, 2001. this volume.
4. F. Baader and W. Snyder. Unification theory. In A. Robinson and A. Voronkov, editors, *Handbook of Automated Reasoning*, volume I, chapter 8, pages 445–532. Elsevier Science, 2001.

5. L. Bachmair, T. Chen, and I.V. Ramakrishnan. Associative-commutative discrimination nets. In M.-C. Gaudel and J.-P. Jouannaud, editors, *Proceedings of the 4th International Joint Conference on Theory and Practice of Software Development (TAPSOFT)*, volume 668 of *Lecture Notes in Computer Science*, pages 61–74, Orsay, France, April 1993. Springer Verlag.

6. L. Bachmair and H. Ganzinger. Resolution theorem proving. In A. Robinson and A. Voronkov, editors, *Handbook of Automated Reasoning*, volume I, chapter 2, pages 19–99. Elsevier Science, 2001.

7. L. Bachmair, H. Ganzinger, C. Lynch, and W. Snyder. Basic paramodulation and superposition. In D. Kapur, editor, *11th International Conference on Automated Deduction*, volume 607 of *Lecture Notes in Artificial Intelligence*, pages 462–476, Saratoga Springs, NY, USA, June 1992. Springer Verlag.

8. P. Baumgartner. FDPLL — a first-order Davis-Putnam-Logemann-Loveland procedure. In D. McAllester, editor, *17th International Conference on Automated Deduction (CADE-17)*, volume 1831 of *Lecture Notes in Artificial Intelligence*, pages 200–219, Pittsburgh, 2000. Springer Verlag.

9. L.D. Baxter. The NP-completeness of subsumption. Unpublished manuscript, 1977.

10. C. Benzmüller and M. Kohlhase. Extensional higher-order resolution. In C.Kirchner and H. Kirchner, editors, *Automated Deduction — CADE-15. 15th International Conference on Automated Deduction*, volume 1421 of *Lecture Notes in Artificial Intelligence*, pages 56–71, Lindau, Germany, 1998. Springer Verlag.

11. C. Benzmüller and M. Kohlhase. LEO—a higher-order theorem prover. In C.Kirchner and H. Kirchner, editors, *Automated Deduction — CADE-15. 15th International Conference on Automated Deduction*, volume 1421 of *Lecture Notes in Artificial Intelligence*, pages 139–144, Lindau, Germany, 1998. Springer Verlag.

12. A. Bundy. The automation of proof by mathematical induction. In A. Robinson and A. Voronkov, editors, *Handbook of Automated Reasoning*, volume I, chapter 13, pages 845–911. Elsevier Science, 2001.

13. J. Christian. Flatterms, discrimination nets, and fast term rewriting. *Journal of Automated Reasoning*, 10(1):95–113, February 1993.

14. H. Comon. Solving symbolic ordering constraints. *International Journal of Foundations of Computer Science*, 1(4):387–411, 1990.

15. H. Comon and R. Treinen. Ordering constraints on trees. In S. Tison, editor, *Trees in Algebra and Programming: CAAP'94*, volume 787 of *Lecture Notes in Computer Science*, pages 1–14. Springer Verlag, 1994.

16. H. de Nivelle. *Bliksem 1.10 User's Manual*. MPI für Informatik, Saarbrücken, 2000.

17. A. Degtyarev. On the forms of inference in calculi with equality and paramodulation. In Yu.V. Kapitonova, editor, *Automation of Research in Mathematics*, pages 14–26. Institute of Cybernetics, Kiev, Kiev, 1982.

18. A. Degtyarev and A. Voronkov. Equality elimination for semantic tableaux. UP-MAIL Technical Report 90, Uppsala University, Computing Science Department, December 1994.

19. A. Degtyarev and A. Voronkov. Equality elimination for the tableau method. In J. Calmet and C. Limongelli, editors, *Design and Implementation of Symbolic Computation Systems. International Symposium, DISCO'96*, volume 1128 of *Lecture Notes in Computer Science*, pages 46–60, Karlsruhe, Germany, September 1996.

20. A. Degtyarev and A. Voronkov. Stratified resolution. In D. McAllester, editor, *17th International Conference on Automated Deduction (CADE-17)*, volume 1831 of *Lecture Notes in Artificial Intelligence*, pages 365–384, Pittsburgh, 2000. Springer Verlag.

21. A. Degtyarev and A. Voronkov. Equality reasoning in sequent-based calculi. In A. Robinson and A. Voronkov, editors, *Handbook of Automated Reasoning*, volume I, chapter 10, pages 609–704. Elsevier Science, 2001.

22. J. Denzinger and D. Fuchs. Cooperation of heterogeneous provers. In T. Dean, editor, *Proc. of the Sixteenth International Joint Conference on Artificial Intelligence (IJCAI-99)*, volume 1, pages 10–15, Stockholm, 1999.

23. A. Franke, S.M. Hess, C.G. Jung, M. Kohlhase, and V. Sorge. Agent-oriented integration of distributed mathematical services. *Journal of Universal Computer Science*, 5(3):156–187, 1999.

24. A. Franke and M. Kohlhase. Mathweb, an agent-based communication layer for distributed automated theorem proving. In H. Ganzinger, editor, *Automated Deduction—CADE-16. 16th International Conference on Automated Deduction*, volume 1632 of *Lecture Notes in Artificial Intelligence*, pages 217–221, Trento, Italy, July 1999.

25. A. Franke and M. Kohlhase. Mbase, an open mathematical knowledge base. In D. McAllester, editor, *17th International Conference on Automated Deduction (CADE-17)*, volume 1831 of *Lecture Notes in Artificial Intelligence*, pages 455–459, Pittsburgh, 2000. Springer Verlag.

26. H. Ganzinger, R. Nieuwenhuis, and P. Nivela. Context trees. In R. Gore, A. Leitsch, and T. Nipkow, editors, *IJCAR 2001*, 2001. this volume.

27. M.R. Garey and D.S. Johnson. *Computers and Intractability*. Freeman, San Francisco, 1979.

28. M. Gordon and T. Melham. *Introduction to HOL*. Cambridge University Press, 1993.

29. G. Gottlob and A. Leitsch. On the efficiency of subsumption algorithms. *Journal of the Association for Computing Machinery*, 32(2):280–295, April 1987.

30. P. Graf. Extended path-indexing. In A. Bundy, editor, *Automated Deduction — CADE-12. 12th International Conference on Automated Deduction*, volume 814 of *Lecture Notes in Artificial Intelligence*, pages 514–528, Nancy, France, June/July 1994.

31. P. Graf. Substitution tree indexing. In J. Hsiang, editor, *Proceedings of the 6th International Conference on Rewriting Techniques and Applications (RTA-95)*, volume 914 of *Lecture Notes in Computer Science*, pages 117–131, Kaiserslautern, 1995.

32. P. Graf. *Term Indexing*, volume 1053 of *Lecture Notes in Computer Science*. Springer Verlag, 1996.

33. C. Hewitt. *Description and theoretical analysis of Planner: a language for proving theorems and manipulating models in a robot*. PhD thesis, Department of Mathematics, MIT, Cambridge, Mass., January 1971.

34. T. Hillenbrand, A. Buch, R. Vogt, and B. Löchner. Waldmeister: High-performance equational deduction. *Journal of Automated Reasoning*, 18(2):265–270, 1997.

35. J.-P. Jouannaud and M. Okada. Satisfiability of systems of ordinal notations with the subterm property is decidable. In J.L. Albert, B. Monien, and M. Rodríguez-Artalejo, editors, *Automata, Languages and Programming, 18th International Colloquium, ICALP'91*, volume 510 of *Lecture Notes in Computer Science*, pages 455–468, Madrid, Spain, 1991. Springer Verlag.

36. D. Kapur and P. Narendran. Double-exponential complexity of computing a complete set of *AC*-unifiers. In *Proc. IEEE Conference on Logic in Computer Science (LICS)*. IEEE Computer Society Press, 1992.

37. C. Kirchner, H. Kirchner, and M. Rusinowitch. Deduction with symbolic constraints. *Revue Francaise d'Intelligence Artificielle*, 4(3):9–52, 1990. Special issue on automated deduction.

38. K. Korovin and A. Voronkov. A decision procedure for the existential theory of term algebras with the Knuth-Bendix ordering. In *Proc. 15th Annual IEEE Symp. on Logic in Computer Science*, pages 291–302, Santa Barbara, California, June 2000.

39. K. Korovin and A. Voronkov. Knuth-Bendix constraint solving is NP-complete. Preprint CSPP-8, Department of Computer Science, University of Manchester, November 2000. to appear in ICALP 2001.

40. R. Letz, J. Schumann, S. Bayerl, and W. Bibel. SETHEO: A high-performance theorem prover. *Journal of Automated Reasoning*, 8(2):183–212, 1992.

41. D.W. Loveland. Mechanical theorem proving by model elimination. *Journal of the Association for Computing Machinery*, 15:236–251, 1968.

42. E.L. Lusk. Controlling redundancy in large search spaces: Argonne-style theorem proving through the years. In A. Voronkov, editor, *Logic Programming and Automated Reasoning. International Conference LPAR'92.*, volume 624 of *Lecture Notes in Artificial Intelligence*, pages 96–106, St.Petersburg, Russia, July 1992.

43. E.L. Lusk, W. McCune, and R.A. Overbeek. Logic machine architecture: Inference mechanisms. In D.W. Loveland, editor, *Proceedings 6th Conference on Automated Deduction (CADE '82)*, number 138 in Lecture Notes in Computer Science, pages 85–108, New York, 1982. Springer Verlag.

44. E.L. Lusk, W. McCune, and R.A. Overbeek. Logic machine architecture: Kernel functions. In D.W. Loveland, editor, *Proceedings 6th Conference on Automated Deduction (CADE '82)*, number 138 in Lecture Notes in Computer Science, pages 70–84, New York, 1982. Springer Verlag.

45. E.W. Lusk, W.W. McCune, and J. Slaney. ROO: a parallel theorem prover. In D. Kapur, editor, *11th International Conference on Automated Deduction (CADE)*, volume 607 of *Lecture Notes in Artificial Intelligence*, pages 731–734, Saratoga Springs, NY, USA, June 1992. Springer Verlag.

46. A. Martelli and U. Montanari. An efficient unification algorithm. *ACM Transactions on Programming Languages and Systems*, 4(2):258–282, 1982.

47. W. McCune. Solution of the robbins problem. *Journal of Automated Reasoning*, 19(3):263–276, 1997.

48. W. McCune. Automatic proofs and counterexamples for some ortholattice identities. *Information Processing Letters*, 65(6):285–291, 1998.

49. W.W. McCune. Experiments with discrimination-tree indexing and path indexing for term retrieval. *Journal of Automated Reasoning*, 9(2):147–167, 1992.

50. W.W. McCune. Single axioms for the left group and the right group calculi. *Notre Dame J. of Formal Logic*, 34(1):132–139, 1993.

51. W.W. McCune. OTTER 3.0 reference manual and guide. Technical Report ANL-94/6, Argonne National Laboratory, January 1994.

52. M. Moser, C. Lynch, and J. Steinbach. Model elimination with basic ordered paramodulation. Technical Report AR-95-11, Fakultät für Informatik, Technische Universität München, München, 1995.

53. W.W. McCune M.P. Bonacina. Distributed theorem proving by peers. In A. Bundy, editor, *Automated Deduction — CADE-12. 12th International Conference on Automated Deduction*, volume 814 of *Lecture Notes in Artificial Intelligence*, pages 841–845, Nancy, France, June/July 1994.

54. P. Narendran, M. Rusinowitch, and R. Verma. RPO constraint solving is in NP. In G. Gottlob, E. Grandjean, and K. Seyr, editors, *Computer Science Logic, 12th International Workshop, CSL'98*, volume 1584 of *Lecture Notes in Computer Science*, pages 385–398. Springer Verlag, 1999.

55. R. Nieuwenhuis. Simple LPO constraint solving methods. *Information Processing Letters*, 47:65–69, 1993.

56. R. Nieuwenhuis. Rewrite-based deduction and symbolic constraints. In H. Ganzinger, editor, *Automated Deduction—CADE-16. 16th International Conference on Automated Deduction*, Lecture Notes in Artificial Intelligence, pages 302–313, Trento, Italy, July 1999.

57. R. Nieuwenhuis, T. Hillenbrand, A. Riazanov, and A. Voronkov. On the evaluation of indexing techniques for theorem proving. In R. Gore, A. Leitsch, and T. Nipkow, editors, *IJCAR 2001*, 2001. this volume.

58. R. Nieuwenhuis and J.M. Rivero. Solved forms for path ordering constraints. In *In Proc. 10th International Conference on Rewriting Techniques and Applications (RTA)*, volume 1631 of *Lecture Notes in Computer Science*, pages 1–15, Trento, Italy, 1999.

59. R. Nieuwenhuis, J.M. Rivero, and M.Á. Vallejo. The Barcelona prover. *Journal of Automated Reasoning*, 18(2):171–176, 1997.

60. R. Nieuwenhuis and A. Rubio. Basic superposition is complete. In *ESOP'92*, volume 582 of *Lecture Notes in Computer Science*, pages 371–389. Springer Verlag, 1992.

61. R. Nieuwenhuis and A. Rubio. AC-superposition with constraints: no AC-unifiers needed. In A. Bundy, editor, *Automated Deduction—CADE-12. 12th International Conference on Automated Deduction*, volume 814 of *Lecture Notes in Artificial Intelligence*, pages 545–559, Nancy, France, June/July 1994.

62. R. Nieuwenhuis and A. Rubio. Theorem proving with ordering and equality constrained clauses. *Journal of Symbolic Computations*, 19:321–351, 1995.

63. R. Nieuwenhuis and A. Rubio. Paramodulation-based theorem proving. In A. Robinson and A. Voronkov, editors, *Handbook of Automated Reasoning*, volume I, chapter 7, pages 371–443. Elsevier Science, 2001.

64. H.J. Ohlbach. Abstraction tree indexing for terms. In H.-J. Bürkert and W. Nutt, editors, *Extended Abstracts of the Third International Workshop on Unification*, pages 131–135. Fachbereich Informatik, Universität Kaiserslautern, 1989. SEKI-Report SR 89-17.

65. M. Paterson and M. Wegman. Linear unification. *Journal of Computer and System Sciences*, 16:158–167, 1978.

66. L.C. Paulson. *Isabelle: A Generic Theorem Prover*. Springer-Verlag LNCS 828, 1994.

67. D. Plaisted and Y. Zhu. Replacement rules with definition detection. In R. Caferra and G. Salzer, editors, *Automated Deduction in Classical and Non-Classical Logics*, volume 1761 of *Lecture Notes in Artificial Intelligence*, pages 80–94. Springer Verlag, 1999.

68. G. Plotkin. Building-in equational theories. In B. Meltzer and D. Michie, editors, *Machine Intelligence*, volume 7, pages 73–90. Edinburgh University Press, 1972.

69. P.W. Purdom and C.A. Brown. Fast many-to-one matching algorithms. In J.-P. Jouannaud, editor, *Rewriting Techniques and Applications, First International Conference, RTA-85*, volume 202 of *Lecture Notes in Computer Science*, pages 407–416, Dijon, France, 1985. Springer Verlag.

70. A. Riazanov and A. Voronkov. Vampire. In H. Ganzinger, editor, *Automated Deduction—CADE-16. 16th International Conference on Automated Deduction*, volume 1632 of *Lecture Notes in Artificial Intelligence*, pages 292–296, Trento, Italy, July 1999.

71. A. Riazanov and A. Voronkov. Limited resource strategy in resolution theorem proving. Preprint CSPP-7, Department of Computer Science, University of Manchester, October 2000.

72. A. Riazanov and A. Voronkov. Partially adaptive code trees. In M. Ojeda-Aciego, I.P. de Guzmán, G. Brewka, and L.M. Pereira, editors, *Logics in Artificial Intelligence. European Workshop, JELIA 2000*, volume 1919 of *Lecture Notes in Artificial Intelligence*, pages 209–223, Málaga, Spain, 2000. Springer Verlag.

73. A. Riazanov and A. Voronkov. An efficient algorithm for backward subsumption using path indexing and database joins. Preprint, Department of Computer Science, University of Manchester, 2001. To appear.

74. A. Riazanov and A. Voronkov. Splitting without backtracking. Preprint CSPP-10, Department of Computer Science, University of Manchester, January 2001. to appear at IJCAI'2001.

75. A. Riazanov and A. Voronkov. Vampire 1.1. (system description). In R. Gore, A. Leitsch, and T. Nipkow, editors, *IJCAR 2001*, 2001. this volume.

76. J.M.A. Rivero. *Data Structures and Algorithms for Automated Deduction with Equality*. Phd thesis, Universitat Politècnica de Catalunya, Barcelona, May 2000.

77. J.A. Robinson. A machine-oriented logic based on the resolution principle. *Journal of the Association for Computing Machinery*, 12(1):23–41, 1965.

78. S. Schulz. System abstract: E 0.3. In H. Ganzinger, editor, *Automated Deduction—CADE-16. 16th International Conference on Automated Deduction*, Lecture Notes in Artificial Intelligence, pages 297–301, Trento, Italy, July 1999.

79. S. Schulz. *Learning Search Control Knowledge for Equational Deduction*, volume 230 of *Dissertationen zur künstliche Intelligenz*. Akademische Verlagsgesellschaft Aka GmmH, 2000.

80. J. Schumann and R. Letz. PARTHEO: a high performance parallel theorem prover. In M.E. Stickel, editor, *Proc. 10th Int. Conf. on Automated Deduction*, volume 449 of *Lecture Notes in Artificial Intelligence*, pages 40–56, 1990.

81. R. Sekar, I.V. Ramakrishnan, and A. Voronkov. Term indexing. In A. Robinson and A. Voronkov, editors, *Handbook of Automated Reasoning*, volume II, chapter 26, pages 1853–1964. Elsevier Science, 2001.

82. J.K. Slaney, E.L. Lusk, and W. McCune. SCOTT: Semantically Constrained Otter (system description). In A. Bundy, editor, *Automated Deduction—CADE-12. 12th International Conference on Automated Deduction*, volume 814 of *Lecture Notes in Artificial Intelligence*, pages 764–768, Nancy, France, 1994.

83. Stickel. Automated deduction by theory resolution. *Journal of Automated Reasoning*, 1:333–355, 1985.

84. M. Stickel. The path indexing method for indexing terms. Technical Report 473, Artificial Intelligence Center, SRI International, Menlo Park, CA, October 1989.

85. M.E. Stickel. A complete unification algorithm for associative-commutative functions. In *Advance Papers of the Fourth International Joint Conference on Artificial Intelligence*, pages 71–76, Tbilisi, USSR, 1975.

86. M.E. Stickel. A complete unification algorithm for associative-commutative functions. *Journal of the Association for Computing Machinery*, 28(3):423–434, 1981.
87. M.E. Stickel. Schubert's Steamroller problem: Formulation and solutions. *Journal of Automated Reasoning*, 2(1):89–101, 1986.
88. G. Sutcliffe. Systemon tptp. In D. McAllester, editor, *17th International Conference on Automated Deduction (CADE-17)*, volume 1831 of *Lecture Notes in Artificial Intelligence*, pages 406–410, Pittsburgh, 2000. Springer Verlag.
89. T. Tammet. Gandalf. *Journal of Automated Reasoning*, 18(2):199–204, 1997.
90. T. Tammet. Towards efficient subsumption. In C.Kirchner and H. Kirchner, editors, *Automated Deduction — CADE-15. 15th International Conference on Automated Deduction*, volume 1421 of *Lecture Notes in Artificial Intelligence*, pages 427–441, Lindau, Germany, 1998. Springer Verlag.
91. V. Turchin. *Refal-5, Programming Guide and Reference Manual*. New England Publishing Co., 1989.
92. A. Voronkov. LISS — the logic inference search system. In M. Stickel, editor, *Proc. 10th Int. Conf. on Automated Deduction*, volume 449 of *Lecture Notes in Computer Science*, pages 677–678, Kaiserslautern, Germany, 1990. Springer Verlag.
93. A. Voronkov. The anatomy of Vampire: Implementing bottom-up procedures with code trees. *Journal of Automated Reasoning*, 15(2):237–265, 1995.
94. A. Voronkov. Formulae over reduction orderings: some solved and yet unsolved problems. Invited talk at RTA'2000, unpublished, 2000.
95. C. Weidenbach. Combining superposition, sorts and splitting. In A. Robinson and A. Voronkov, editors, *Handbook of Automated Reasoning*, volume II, chapter 27, pages 1965–2013. Elsevier Science, 2001.
96. C. Weidenbach, B. Afshordel, U. Brahm, C. Cohrs, T. Engel, E. Keen, C. Theobalt, and D. Topic. System description: SPASS version 1.0.0. In H. Ganzinger, editor, *Automated Deduction—CADE-16. 16th International Conference on Automated Deduction*, volume 1632 of *Lecture Notes in Artificial Intelligence*, pages 378–382, Trento, Italy, July 1999.
97. L. Wos, R. Overbeek, and E. Lusk. Subsumption, a sometimes undervalued procedure. In J.-L. Lassez and G. Plotkin, editors, *Computational Logic. Essays in Honor of Alan Robinson*, pages 3–40. The MIT Press, Cambridge, MA, 1991.

The Description Logic $\mathcal{ALCNH}_{R+}$ Extended with Concrete Domains: A Practically Motivated Approach

Volker Haarslev, Ralf Möller, and Michael Wessel

University of Hamburg, Computer Science Department
Vogt-Kölln-Str. 30, 22527 Hamburg, Germany

Abstract. In this paper the description logic $\mathcal{ALCNH}_{R+}(\mathcal{D})^-$ is introduced. Prominent language features beyond conjunction, full negation, and quantifiers are number restrictions, role hierarchies, transitively closed roles, generalized concept inclusions, and concrete domains. As in other languages based on concrete domains (e.g. $\mathcal{ALC}(\mathcal{D})$) a so-called existential predicate restriction is provided. However, compared to $\mathcal{ALC}(\mathcal{D})$ only features and no feature chains are allowed in this operator. This results in a limited expressivity w.r.t. concrete domains but is required to ensure the decidability of the language. We show that the results can be exploited for building practical description logic systems for solving e.g. configuration problems.

1 Introduction

In the field of knowledge representation, description logics (DLs) have been proven to be a sound basis for solving application problems. An application domain where DLs have been successfully applied is *configuration* (see [9] for an early publication). The main notions for domain modeling are concepts (unary predicates) and roles (binary predicates). Furthermore, a set of axioms (also called TBox) is used for modeling the terminology of an application. Knowledge about specific individuals and their interrelationships is modeled with a set of additional axioms (so-called ABox).

Experiences with description logics in applications indicate that negation, existential and universal restrictions, transitive roles, role hierarchies, and number restrictions are required to solve practical modeling problems without resorting to ad hoc extensions. A description logic which provides these language constructs is, for instance, $\mathcal{ALCNH}_{R+}$ [5]. The optimized DL knowledge representation system RACE [4] provides an optimized implementation for ABox reasoning in $\mathcal{ALCNH}_{R+}$. With the optimized implementation of RACE, practical systems based on description logics can be built. However, it is well-known that, in addition to the language constructs mentioned above, reasoning about objects from other domains (so-called concrete domains, e.g. for the reals) is very important for practical applications as well. In [1] the description logic $\mathcal{ALC}(\mathcal{D})$ is investigated and it is shown that, provided a decision procedure for the concrete

R. Goré, A. Leitsch, and T. Nipkow (Eds.): IJCAR 2001, LNAI 2083, pp. 29–44, 2001.

domain $\mathcal{D}$ exists, the logic $\mathcal{ALC}(\mathcal{D})$ is decidable. In this paper, an extension of the $\mathcal{ALCNH}_{R^+}$ knowledge representation system RACE with concrete domains is investigated.

Unfortunately, adding concrete domains (as proposed in the original approach) to expressive description logics might lead to undecidable inference problems. For instance, in [2] it is proven that the logic $\mathcal{ALC}(\mathcal{D})$ plus an operator for the transitive closure of roles can be undecidable if expressive concrete domains are considered. $\mathcal{ALCNH}_{R^+}$ offers transitive roles but no operator for the transitive closure of roles. In [8] it is shown that $\mathcal{ALC}(\mathcal{D})$ with generalized inclusion axioms (GCIs) can be undecidable. Even if GCIs were not allowed in $\mathcal{ALCNH}_{R^+}$, $\mathcal{ALCNH}_{R^+}$ with concrete domains would be undecidable (in general) because $\mathcal{ALCNH}_{R^+}$ offers role hierarchies and transitive roles, which provide the same expressivity as GCIs. With role hierarchies it is possible to (implicitly) declare a universal role, which can be used in combination with a value restriction to achieve the same effect as with GCIs. Decidability results can only be obtained for "trivial" concrete domains, which are hardly useful in practical applications. Thus, if termination and soundness of, for instance, a concept consistency algorithm are to be retained, there is no way extending an $\mathcal{ALCNH}_{R^+}$ DL system such as RACE with concrete domains as in $\mathcal{ALC}(\mathcal{D})$ without losing completeness.

Thus, $\mathcal{ALCNH}_{R^+}$ can only be extended with concrete domain operators with limited expressivity. In order to support practical modeling requirements at least to some extent, we pursue a pragmatic approach by supporting only features (and no feature chains as in $\mathcal{ALC}(\mathcal{D})$, for details see [1] and below). The resulting language is called $\mathcal{ALCNH}_{R^+}(\mathcal{D})^-$. By proving soundness and completeness (and termination) of a tableaux calculus, the decidability of inference problems w.r.t. the language $\mathcal{ALCNH}_{R^+}(\mathcal{D})^-$ is proved. As shown in this paper, $\mathcal{ALCNH}_{R^+}(\mathcal{D})^-$ can be used, for instance, as a basis for building practical application systems for solving configuration problems.

2   The Description Logic $\mathcal{ALCNH}_{R^+}(\mathcal{D})^-$

The description logic $\mathcal{ALCNH}_{R^+}(\mathcal{D})^-$ provides conjunction, full negation, quantifiers, number restrictions, role hierarchies, transitively closed roles and concrete domains. In addition to the operators known from $\mathcal{ALCNH}_{R^+}$, a restricted existential predicate restriction operator for concrete domains is supported. Furthermore, we assume that the unique name assumption holds for the individuals explicitly mentioned in an ABox.

We briefly introduce the syntax and semantics of the DL $\mathcal{ALCNH}_{R^+}(\mathcal{D})^-$. We assume five disjoint sets: a set of concept names C, a set of role names R, a set of feature names F, a set of individual names O and a set of names for (concrete) objects O_C. The mutually disjoint subsets P and T of R denote non-transitive and transitive roles, respectively ($R = P \cup T$). The language $\mathcal{ALCNH}_{R^+}$ is introduced in Figure 1 using a standard Tarski-style semantics with an interpretation $\mathcal{I}_{\mathcal{D}} = (\Delta_{\mathcal{I}}, \Delta_{\mathcal{D}}, \cdot^{\mathcal{I}})$ where $\Delta_{\mathcal{I}} \cap \Delta_{\mathcal{D}} = \emptyset$ holds. A variable assignment α maps concrete objects to values in $\Delta_{\mathcal{D}}$.

Syntax	Semantics
Concepts ($R \in R$, $S \in S$, $f \in F$)	
A	$A^{\mathcal{I}} \subseteq \Delta_{\mathcal{I}}$
$\neg C$	$\Delta_{\mathcal{I}} \setminus C^{\mathcal{I}}$
$C \sqcap D$	$C^{\mathcal{I}} \cap D^{\mathcal{I}}$
$C \sqcup D$	$C^{\mathcal{I}} \cup D^{\mathcal{I}}$
$\exists R . C$	$\{a \in \Delta_{\mathcal{I}} \mid \exists b \in \Delta_{\mathcal{I}} : (a,b) \in R^{\mathcal{I}}, b \in C^{\mathcal{I}}\}$
$\forall R . C$	$\{a \in \Delta_{\mathcal{I}} \mid \forall b \in \Delta_{\mathcal{I}} : (a,b) \in R^{\mathcal{I}} \Rightarrow b \in C^{\mathcal{I}}\}$
$\exists_{\geq n} S$	$\{a \in \Delta_{\mathcal{I}} \mid \|\{b \in \Delta_{\mathcal{I}} \mid (a,b) \in S^{\mathcal{I}}\}\| \geq n\}$
$\exists_{\leq m} S$	$\{a \in \Delta_{\mathcal{I}} \mid \|\{b \in \Delta_{\mathcal{I}} \mid (a,b) \in S^{\mathcal{I}}\}\| \leq m\}$
$\exists f_1, \ldots, f_n . P$	$\{a \in \Delta_{\mathcal{I}} \mid \exists x_1, \ldots, x_n \in \Delta_{\mathcal{D}} : (a,x_1) \in f_1^{\mathcal{I}}, \ldots, (a,x_n) \in f_n^{\mathcal{I}},$ $(x_1, \ldots, x_n) \in P^{\mathcal{I}}\}$
$\forall f . \perp_{\mathcal{D}}$	$\{a \in \Delta_{\mathcal{I}} \mid \neg \exists x_1 \in \Delta_{\mathcal{D}} : (a,x_1) \in f^{\mathcal{I}}\}$
Roles and Features	
R	$R^{\mathcal{I}} \subseteq \Delta_{\mathcal{I}} \times \Delta_{\mathcal{I}}$
f	$f^{\mathcal{I}} : \Delta_{\mathcal{I}} \to \Delta_{\mathcal{D}}$ (features are partial functions)

A is a concept name and $\|\cdot\|$ denotes the cardinality of a set $(n, m \in \mathbb{N}, n > 0)$.

Axioms		Assertions ($a, b \in O_O, x, x_i \in O_C$)	
Syntax	Satisfied if	Syntax	Satisfied if
$R \in T$	$R^{\mathcal{I}} = (R^{\mathcal{I}})^+$	$a : C$	$a^{\mathcal{I}} \in C^{\mathcal{I}}$
$R \sqsubseteq S$	$R^{\mathcal{I}} \subseteq S^{\mathcal{I}}$	$(a, b) : R$	$(a^{\mathcal{I}}, b^{\mathcal{I}}) \in R^{\mathcal{I}}$
$C \sqsubseteq D$	$C^{\mathcal{I}} \subseteq D^{\mathcal{I}}$	$(a, x) : f$	$(a^{\mathcal{I}}, \alpha(x)) \in f^{\mathcal{I}}$
		$(x_1, \ldots, x_n) : P$	$(\alpha(x_1), \ldots, \alpha(x_n)) \in P^{\mathcal{I}}$

Fig. 1. Syntax and Semantics of $\mathcal{ALCNH}_{R+}(\mathcal{D})^-$.

If $R, S \in R$ are role names, then $R \sqsubseteq S$ is called a *role inclusion* axiom. A *role hierarchy* $\mathcal{R}$ is a finite set of role inclusion axioms. Then, we define $\sqsubseteq^*$ as the reflexive transitive closure of $\sqsubseteq$ over such a role hierarchy $\mathcal{R}$. Given $\sqsubseteq^*$, the set of roles $R^{\downarrow} = \{S \in R \mid S \sqsubseteq^* R\}$ defines the *sub-roles* of a role R and $R^{\uparrow} = \{S \in R \mid R \sqsubseteq^* S\}$ defines the *super-roles* of a role. We also define the set $S := \{R \in P \mid R^{\downarrow} \cap T = \emptyset\}$ of *simple* roles that are neither transitive nor have a transitive role as a sub-role.

The concept language of $\mathcal{ALCNH}_{R+}$ is syntactically restricted with respect to the combination of number restrictions and transitive roles. Number restrictions are only allowed for simple roles. This restriction is motivated by a known undecidability result in case of an unrestricted syntax [7]. The set of individuals is divided into two subsets, the set of so-called "old" individuals O_O and set the of "new" individuals O_N. Every individual name from O is mapped to a single element of $\Delta_{\mathcal{I}}$ in a way such that for $a, b \in O_O$, $a^{\mathcal{I}} \neq b^{\mathcal{I}}$ if $a \neq b$ *(unique name assumption)*. Only old individuals may be mentioned in an ABox (new individual are generated by the completion rules introduced below).

In accordance with [1] we also define the notion of a concrete domain. A *concrete domain* $\mathcal{D}$ is a pair $(\Delta_{\mathcal{D}}, \Phi_{\mathcal{D}})$, where $\Delta_{\mathcal{D}}$ is a set called the domain, and

$\Phi_{\mathcal{D}}$ is a set of predicate names. The interpretation function maps each predicate name P from $\Phi_{\mathcal{D}}$ with arity n to a subset $P^{\mathcal{I}}$ of $\Delta_{\mathcal{D}}^n$. Concrete objects from O_C are mapped to an element of $\Delta_{\mathcal{D}}$. We assume that $\bot_{\mathcal{D}}$ is the negation of the predicate $\top_{\mathcal{D}}$.

A concrete domain $\mathcal{D}$ is called *admissible* iff the set of predicate names $\Phi_{\mathcal{D}}$ is closed under negation and $\Phi_{\mathcal{D}}$ contains a name $\top_{\mathcal{D}}$ for $\Delta_{\mathcal{D}}$, and the satisfiability problem $P_1^{n_1}(x_{11}, \dots, x_{1n_1}) \wedge \dots \wedge P_m^{n_m}(x_{m1}, \dots, x_{mn_m})$ is decidable (m is finite, $P_i^{n_i} \in \Phi_{\mathcal{D}}$, n_i is the arity of P, and x_{jk} is a concrete object).

If C and D are concept terms, then $C \sqsubseteq D$ (*generalized concept inclusion* or *GCI*) is a terminological axiom. A finite set of terminological axioms $\mathcal{T}_{\mathcal{R}}$ is called a *terminology* or *TBox* w.r.t. a given role hierarchy $\mathcal{R}$. For brevity, the reference to $\mathcal{R}$ is omitted in the following. An *ABox* $\mathcal{A}$ is a finite set of assertional axioms as defined in Figure 1.

An interpretation $\mathcal{I}$ is a *model* of a concept C (or *satisfies* a concept C) iff $C^{\mathcal{I}} \neq \emptyset$. An interpretation is a model of a TBox $\mathcal{T}$ iff it satisfies all axioms in $\mathcal{T}$. See Figure 1 for the satisfiability conditions. An interpretation is a model of an ABox $\mathcal{A}$ w.r.t. a TBox iff it is a model of $\mathcal{T}$ and satisfies all assertions in $\mathcal{A}$. Different individuals are mapped to different domain objects (unique name assumption). Note that features are interpreted differently from features in [1].

A concept C is called *consistent* (w.r.t. a TBox $\mathcal{T}$) iff there exists a model of C (that is also a model of $\mathcal{T}$). An ABox $\mathcal{A}$ is consistent (w.r.t. a TBox $\mathcal{T}$) iff $\mathcal{A}$ has model $\mathcal{I}$ (which is also a model of $\mathcal{T}$). A *knowledge base* $(\mathcal{T}, \mathcal{A})$ is called consistent iff there exists a model.

3 Solving an Application Problem with $\mathcal{ALCNH}_{R^+}(\mathcal{D})^-$

According to [3] configuration problem solving processes can be formalized as synthesis inference tasks. Following this approach, a solution of a configuration task is defined to be a (logical) model of the given knowledge base consisting of both the conceptual domain model (TBox) as well as the task specification (ABox). The TBox and the role hierarchy describe the configuration space.

For instance, in a technical domain, the concept of a cylinder might be defined as follows. A Cylinder is required to be a Motorpart, to be part_of a Motor, to have a displacement of 1 to 1000ccm, and to have a set of 4 to 6 parts (role has_part) which are all instances of Cylinderpart and it consists of exactly 1 Piston, exactly 1 Piston_Rod, and 2 to 4 Valves. This expression can be transformed to a terminological inclusion axiom of a description logic providing concrete domains. Let the concrete domain $\Re$ be defined as in [1]: $\Re = (\mathbb{R}, \Phi_{\Re})$ where $\Phi_{\Re}$ is a set of predicates which are based on polynomial equations or inequations. The concrete domain $\Re$ is admissible (see also [1]). A TBox $\mathcal{T}$ is defined as follows:

$$\text{has\_cylinder\_part} \sqsubseteq \text{has\_part}, \qquad \text{has\_piston\_part} \sqsubseteq \text{has\_part}$$
$$\text{has\_piston\_rod\_part} \sqsubseteq \text{has\_part}, \qquad \text{has\_valve\_part} \sqsubseteq \text{has\_part}$$

$$\top \sqsubseteq \forall \text{ has\_cylinder\_part . Cylinder,} \qquad \top \sqsubseteq \forall \text{ has\_piston\_part . Piston}$$
$$\top \sqsubseteq \forall \text{ has\_piston\_rod\_part . Piston\_Rod,} \qquad \top \sqsubseteq \forall \text{ has\_valve\_part . Valve}$$

In the first block, relationships between roles are declared. Then, in the second block, range restrictions for certain roles are imposed. Below, in the third block for Cylinderpart a so-called cover axiom is given. Moreover, additional axioms ensure the disjointness of more specific subconcepts of Cylinderpart (D is a subconcept of C iff C subsumes D).

$$\textbf{Cylinderpart} \sqsubseteq \text{Piston} \sqcup \text{Piston\_Rod} \sqcup \text{Valve,} \qquad \textbf{Piston} \sqsubseteq \neg\text{Piston\_Rod} \sqcap \neg\text{Valve}$$
$$\textbf{Piston\_Rod} \sqsubseteq \neg\text{Piston} \sqcap \neg\text{Valve,} \qquad \textbf{Valve} \sqsubseteq \neg\text{Piston} \sqcap \neg\text{Piston\_Rod}$$

The cylinder example is translated as follows (the term $\lambda_{\text{Vol}}\, c.\,(\dots)$ is a unary predicate of a numeric concrete domain for the dimension *Volume* with unit m^3).

$$\begin{aligned}
\textbf{Cylinder} \sqsubseteq\ & \text{Motorpart} \sqcap \exists_{=1} \text{ part\_of} \sqcap \\
& \exists \text{ displacement} . \lambda_{\text{Vol}}\, c.\,(0.001 \le c \le 1) \sqcap \\
& \forall \text{ has\_part . Cylinderpart} \sqcap \\
& \exists_{\ge 4} \text{ has\_cylinder\_part} \sqcap \exists_{\le 6} \text{ has\_cylinder\_part} \sqcap \\
& \exists_{=1} \text{ has\_piston\_part} \sqcap \exists_{=1} \text{ has\_piston\_rod\_part} \sqcap \\
& \exists_{\ge 2} \text{ has\_valve\_part} \sqcap \exists_{\le 4} \text{ has\_valve\_part}
\end{aligned}$$

We assume that displacement is declared as a feature. Furthermore, let $\exists_{=1}\, R$ be an abbreviation for $\exists_{\ge 1}\, R \sqcap \exists_{\le 1}\, R$. In our example, the ABox being used is very simple: $\mathcal{A} = \{a : \text{Cylinder} \sqcap \exists \text{ displacement} . \lambda_{\text{Vol}}\, c.\,(c \ge 0.5)\}$.

In order to solve the problem to construct a Cylinder, the knowledge base $(\mathcal{T}, \mathcal{A})$ is tested for consistency. If the knowledge base is consistent, there exists a model which can be considered as a solution (see [3]). Note that $(\mathcal{T}, \mathcal{A})$ is only a very simplified example for a representation of a configuration problem. For instance, using an ABox with additional assertions it is possible to explicitly specify some required cylinder parts etc. In order to actually compute a solution to a configuration problem, a sound and complete calculus for the $\mathcal{ALCNH}_{R+}(\mathcal{D})^-$ knowledge base consistency problem is required that terminates on any input.

4   A Tableaux Calculus for $\mathcal{ALCNH}_{R+}(\mathcal{D})^-$

In the following a calculus to decide the consistency of an $\mathcal{ALCNH}_{R+}(\mathcal{D})^-$ knowledge base $(\mathcal{T}, \mathcal{A})$ is devised. As a first step, the original ABox $\mathcal{A}$ of the knowledge base is transformed w.r.t. the TBox $\mathcal{T}$. The idea is to derive an ABox $\mathcal{A}_{\mathcal{T}}$ that is consistent (w.r.t. an empty TBox) iff $(\mathcal{T}, \mathcal{A})$ is consistent. The calculus introduced below is applied to $\mathcal{A}_{\mathcal{T}}$.

In order to define the transformation steps for deriving $\mathcal{A}_{\mathcal{T}}$, we have to introduce a few technical terms. First, for any concept term we define its negation normal form. A concept is in *negation normal form* iff negation signs may occur only in front of concept names.

Every $\mathcal{ALCNH}_{R^+}(\mathcal{D})^-$ concept term C can be transformed into negation normal form $nnf(C)$ by recursively applying the following transformation rules to subconcepts from left to right:

$$\neg(C \sqcap D) \to \neg C \sqcup \neg D, \quad \neg(C \sqcup D) \to \neg C \sqcap \neg D, \quad \neg \forall R . C \to \exists R . \neg C,$$
$$\neg \exists R . C \to \forall R . \neg C, \quad \neg \neg C \to C, \quad \neg \exists_{\geq n} S \to \exists_{\leq n-1} S, \quad \neg \exists_{\leq m} S \to \exists_{\geq m+1} S,$$
$$\neg \forall f . \bot_\mathcal{D} \to \exists f . \top_\mathcal{D}, \neg \exists f_1, \ldots, f_n . P \to \exists f_1, \ldots, f_n . \overline{P} \sqcup \forall f_1 . \bot_\mathcal{D} \sqcup \ldots \sqcup \forall f_n . \bot_\mathcal{D}$$

where $\overline{P}$ is the negation of P.

If no rule is applicable, the resulting concept is in negation normal form and all models of C are also models of $nnf(C)$ and vice versa. The transformation is possible in linear time.

Definition 1 (Additional ABox Assertions). *Let C be a concept term, $a, b \in O$ be individual names, and $x \notin O \cup O_C$, then the following expressions are also assertional axioms: $\forall x . x : C$ (universal concept assertion),[1] $a \neq b$ (inequality assertion).*

An interpretation $\mathcal{I}_\mathcal{D}$ satisfies an assertional axiom $\forall x . x : C$ iff $C^\mathcal{I} = \Delta_\mathcal{I}$ and $a \neq b$ iff $a^\mathcal{I} \neq b^\mathcal{I}$.

Definition 2 (Fork, Fork Elimination). *If it holds that $\{(a, x_1) : f, (a, x_2) : f\} \subseteq \mathcal{A}$ then there exists a fork in $\mathcal{A}$. In case of a fork w.r.t. x_1, x_2, the replacement of every occurrence of x_2 in $\mathcal{A}$ by x_1 is called fork elimination.*

Definition 3 (Augmented ABox). *For an initial ABox $\mathcal{A}$ we define its augmented ABox $\mathcal{A}_\mathcal{T}$ w.r.t a TBox $\mathcal{T}$ by applying the following transformation rules to $\mathcal{A}$. First of all, all forks in $\mathcal{A}$ are eliminated (note that the unique name assumption is not imposed on concrete objects). Then, for every GCI $C \sqsubseteq D$ in $\mathcal{T}$ the assertion $\forall x . x : (\neg C \sqcup D)$ is added to $\mathcal{A}$. Every concept term occurring in $\mathcal{A}$ is transformed into its negation normal form. Let $O_\mathcal{A} = \{a_1, \ldots, a_n\}$ be the set of individuals mentioned in $\mathcal{A}$, then the set of inequality assertions $\{a_i \neq a_j \mid a_i, a_j \in O_\mathcal{A}, i, j \in 1..n, i \neq j\}$ is added to $\mathcal{A}$.*

In order to check the consistency of an $\mathcal{ALCNH}_{R^+}(\mathcal{D})^-$ knowledge base $(\mathcal{T}, \mathcal{A})$, the augmented ABox $\mathcal{A}_\mathcal{T}$ is computed. Then, a set of so-called completion rules (see below) is applied to the augmented ABox $\mathcal{A}_\mathcal{T}$. The rules are applied in accordance with a completion strategy.

Lemma 1. *A knowledge base $(\mathcal{T}, \mathcal{A})$ is consistent if and only if $\mathcal{A}_\mathcal{T}$ is consistent (w.r.t. an empty TBox).*

The proof is straightforward, for details see [6].

The tableaux rules require the notion of blocking their applicability. This is based on so-called concept sets, an ordering for new individuals and concrete objects, and the notion of a blocking individual.

---

[1] $\forall x . x : C$ is to be read as $\forall x . (x : C)$.

Definition 4 (Ordering). *We define an* individual ordering '$\prec$' *for new individuals (elements of O_N) occurring in an ABox $\mathcal{A}$. If $b \in O_N$ is introduced in $\mathcal{A}$, then $a \prec b$ for all new individuals a already present in $\mathcal{A}$. A* concrete object ordering '$\prec_C$' *for elements of O_C occurring in an ABox $\mathcal{A}$ is defined as follows. If $y \in O_C$ is introduced in $\mathcal{A}$, then $x \prec_C y$ for all concrete objects x already present in $\mathcal{A}$.*

Definition 5 (Concept Set, Blocking Individual, Blocked by). *Given an ABox $\mathcal{A}$ and an individual a occurring in $\mathcal{A}$, we define the* concept set *of a as $\sigma(\mathcal{A}, a) := \{C \mid a : C \in \mathcal{A}\}$. Let $\mathcal{A}$ be an ABox and $a, b \in O_N$ be individuals in $\mathcal{A}$. We call a the* blocking individual *of b if the following conditions hold: $\sigma(\mathcal{A}, a) \supseteq \sigma(\mathcal{A}, b)$ and $a \prec b$. If a is a blocking individual for b, then b is said to be* blocked by *a. An individual b mentioned in an ABox $\mathcal{A}$ is said to be* blocked *(in $\mathcal{A}$) iff there exists a blocking individual for b in $\mathcal{A}$.*

4.1 Completion Rules

We are now ready to define the *completion rules* that are intended to generate a so-called completion (see also below) of an ABox $\mathcal{A}_{\mathcal{T}}$. From this point on, if we refer to an ABox $\mathcal{A}$, we always consider ABoxes derived from $\mathcal{A}_{\mathcal{T}}$.

Definition 6 (Completion Rules).

    $R\sqcap$ *The conjunction rule.*
    if *1.* $a : C \sqcap D \in \mathcal{A}$*, and*
       2. $\{a : C, \ a : D\} \not\subseteq \mathcal{A}$
    then $\mathcal{A}' = \mathcal{A} \cup \{a : C, \ a : D\}$

    $R\sqcup$ *The disjunction rule (nondeterministic).*
    if *1.* $a : C \sqcup D \in \mathcal{A}$*, and*
       2. $\{a : C, \ a : D\} \cap \mathcal{A} = \emptyset$
    then $\mathcal{A}' = \mathcal{A} \cup \{a : C\}$ *or* $\mathcal{A}' = \mathcal{A} \cup \{a : D\}$

    $R\forall C$ *The role value restriction rule.*
    if *1.* $a : \forall R . C \in \mathcal{A}$*, and*
       2. $\exists b \in O, S \in R^{\downarrow} : (a, b) : S \in \mathcal{A}$*, and*
       3. $b : C \notin \mathcal{A}$
    then $\mathcal{A}' = \mathcal{A} \cup \{b : C\}$

    $R\forall_+ C$ *The transitive role value restriction rule.*
    if *1.* $a : \forall R . C \in \mathcal{A}$*, and*
       2. $\exists b \in O, T \in R^{\downarrow}, T \in T, S \in T^{\downarrow} : (a, b) : S \in \mathcal{A}$*, and*
       3. $b : \forall T . C \notin \mathcal{A}$
    then $\mathcal{A}' = \mathcal{A} \cup \{b : \forall T . C\}$

    $R\forall_x$ *The universal concept restriction rule.*
    if *1.* $\forall x . x : C \in \mathcal{A}$*, and*
       2. $\exists a \in O:$ *a mentioned in $\mathcal{A}$, and*
       3. $a : C \notin \mathcal{A}$
    then $\mathcal{A}' = \mathcal{A} \cup \{a : C\}$

R∃C *The role exists restriction rule (generating).*
if1. a:∃R.C ∈ $\mathcal{A}$, *and*
   2. a *is not blocked, and*
   3. ¬∃b ∈ O, S ∈ R↓ : {(a, b):S, b:C} ⊆ $\mathcal{A}$
then $\mathcal{A}' = \mathcal{A} \cup \{(a, b):R,\ b:C\}$ *where* b ∈ O_N *is not used in* $\mathcal{A}$

R∃$_{\geq n}$ *The number restriction exists rule (generating).*
if1. a:∃$_{\geq n}$R ∈ $\mathcal{A}$, *and*
   2. a *is not blocked, and*
   3. ¬∃b$_1$, ..., b$_n$ ∈ O_N, S$_1$, ..., S$_n$ ∈ R↓ :
     {(a, b$_k$):S$_k$ | k ∈ 1..n} ∪ {b$_i$ ≠ b$_j$ | i, j ∈ 1..n, i ≠ j} ⊆ $\mathcal{A}$
then $\mathcal{A}' = \mathcal{A} \cup \{(a, b_k):R \mid k \in 1..n\} \cup \{b_i \neq b_j \mid i, j \in 1..n, i \neq j\}$
     where b$_1$, ..., b$_n$ ∈ O_N *are not used in* $\mathcal{A}$

R∃$_{\leq n}$ *The number restriction merge rule (nondeterministic).*
if1. a:∃$_{\leq n}$R ∈ $\mathcal{A}$, *and*
   2. ∃b$_1$, ..., b$_m$ ∈ O, S$_1$, ..., S$_m$ ∈ R↓: {(a, b$_1$):S$_1$, ..., (a, b$_m$):S$_m$} ⊆ $\mathcal{A}$
     with m > n, *and*
   3. ∃b$_i$, b$_j$ ∈ {b$_1$, ..., b$_m$} : i ≠ j, b$_i$ ≠ b$_j$ ∉ $\mathcal{A}$
then $\mathcal{A}' = \mathcal{A}[b_i/b_j]$, *i.e. replace every occurrence of* b$_i$ *in* $\mathcal{A}$ *by* b$_j$

R∃P *The predicate exists rule (generating).*
if1. a:∃f$_1$, ..., f$_n$.P ∈ $\mathcal{A}$, *and*
   2. ¬∃x$_1$, ..., x$_n$ ∈ O_C : {(a, x$_1$):f$_1$, ... (a, x$_n$):f$_n$, (x$_1$, ..., x$_n$):P} ⊆ $\mathcal{A}$
then $\mathcal{A}' = \mathcal{A} \cup \{(a, x_1):f_1, ... (a, x_n):f_n, (x_1, ..., x_n):P\}$
     where x$_1$, ..., x$_n$ ∈ O_C *are not used in* $\mathcal{A}$,
     eliminate all forks {(a, x):f$_i$, (a, x$_i$):f$_i$} ⊆ $\mathcal{A}$
     such that (a, x):f$_i$ *remains in* $\mathcal{A}$ *if* x≺$_C$x$_i$, i ∈ 1..n

We call the rules R⊔ and R∃$_{\leq n}$ *nondeterministic* rules since they can be applied in different ways to the same ABox. The remaining rules are called *deterministic* rules. Moreover, we call the rules R∃C, R∃$_{\geq n}$ and R∃P *generating* rules since they can introduce new individuals or concrete objects.

Given an ABox $\mathcal{A}$, more than one rule might be applicable to $\mathcal{A}$. This is controlled by a completion strategy in accordance to the ordering for new individuals (see Definition 4).

Definition 7 (Completion Strategy). *We define a* completion strategy *that must observe the following restrictions:*

- *Meta rules:*
  - *Apply a rule to an individual* b ∈ O_N *only if no rule is applicable to an individual* a ∈ O_O.
  - *Apply a rule to an individual* b ∈ O_N *only if no rule is applicable to another individual* a ∈ O_N *such that* a ≺ b.
- *The completion rules are always applied in the following order. A step is skipped in case the corresponding set of applicable rules is empty.*
  1. *Apply all nongenerating rules (R⊓, R⊔, R∀C, R∀$_+$C, R∀$_x$, R∃$_{\leq n}$) as long as possible.*

  2. Apply a generating rule ($R\exists C$, $R\exists_{\geq n}$, $R\exists P$) and restart with step 1 as long as possible.

In the following we always assume that rules are applied in accordance to this strategy. It ensures that the rules are applied to new individuals w.r.t. the ordering '$\prec$' which guarantees a breadth-first order. No rules are applied if a so-called clash is discovered.

Definition 8 (Clash, Clash Triggers, Completion). *We assume the same naming conventions as used above. An ABox $\mathcal{A}$ contains a* clash *if one of the following* clash triggers *is applicable. If none of the clash triggers is applicable to $\mathcal{A}$, then $\mathcal{A}$ is called* clash-free.

- Primitive clash: $\{a:C, a:\neg C\} \subseteq \mathcal{A}$
- Number restriction merging clash:
  $\exists S_1, \ldots, S_m \in R^\downarrow : \{a:\exists_{\leq n} R\} \cup \{(a, b_i):S_i \mid i \in 1..m\} \cup$
  $\{b_i \neq b_j \mid i, j \in 1..m, i \neq j\} \subseteq \mathcal{A}$ *with* $m > n$
- No concrete domain feature clash: $\{(a, x):f, a:\forall f . \bot_{\mathcal{D}}\} \subseteq \mathcal{A}$.
- Concrete domain predicate clash: $(x_1^{(1)}, \ldots, x_{n_1}^{(1)}):P_1 \in \mathcal{A}, \ldots,$
  $(x_1^{(k)}, \ldots, x_{n_k}^{(k)}):P_k \in \mathcal{A}$ *and the conjunction* $\bigwedge_{i=1}^k P_i(x_1^{(i)}, \ldots, x_{n_i}^{(i)})$ *is not satisfiable in $\mathcal{D}$. Note that this can be decided since $\mathcal{D}$ is required to be admissible.*

A clash-free ABox $\mathcal{A}$ is called complete *if no completion rule is applicable to $\mathcal{A}$. A complete ABox $\mathcal{A}'$ derived from an ABox $\mathcal{A}$ is also called a* completion *of $\mathcal{A}$.*

Any ABox containing a clash is obviously unsatisfiable. The purpose of the calculus is to generate a completion for an initial ABox $\mathcal{A}_T$ that proves the consistency of $\mathcal{A}_T$ or its inconsistency if no completion can be found.

4.2   Decidability of the $\mathcal{ALCNH}_{R+}(\mathcal{D})^-$ ABox Consistency Problem

In order to show that the calculus introduced above is correct, first the local correctness of the rules is proven.

Proposition 1 (Invariance). *Let $\mathcal{A}$ and $\mathcal{A}'$ be ABoxes. Then:*

1. *If $\mathcal{A}'$ is derived from $\mathcal{A}$ by applying a deterministic rule, then $\mathcal{A}$ is consistent iff $\mathcal{A}'$ is consistent.*
2. *If $\mathcal{A}'$ is derived from $\mathcal{A}$ by applying a nondeterministic rule, then $\mathcal{A}$ is consistent if $\mathcal{A}'$ is consistent. Conversely, if $\mathcal{A}$ is consistent and a nondeterministic rule is applicable to $\mathcal{A}$, then it can be applied in such a way that it yields an ABox $\mathcal{A}'$ which is consistent.*

Proof. **1.** "$\Leftarrow$" Due to the structure of the deterministic rules one can immediately verify that $\mathcal{A}$ is a subset of $\mathcal{A}'$. Therefore, $\mathcal{A}$ is consistent if $\mathcal{A}'$ is consistent.

"$\Rightarrow$" In order to show that $\mathcal{A}'$ is consistent after applying a deterministic rule to the consistent ABox $\mathcal{A}$, we examine each applicable rule separately. We assume that $\mathcal{I}_{\mathcal{D}} = (\Delta_{\mathcal{I}}, \Delta_{\mathcal{D}}, \cdot^{\mathcal{I}})$ satisfies $\mathcal{A}$. Then, by definition of $\sqsubseteq^*$ it holds that $R^{\mathcal{I}} \subseteq S^{\mathcal{I}}$ if $(R, S) \in \sqsubseteq^*$.

If the conjunction rule is applied to $a:C \sqcap D \in \mathcal{A}$, then we get a new Abox $\mathcal{A}' = \mathcal{A} \cup \{a:C, a:D\}$. Since $\mathcal{I}_{\mathcal{D}}$ satisfies $a:C \sqcap D$, $\mathcal{I}_{\mathcal{D}}$ satisfies $a:C$ and $a:D$ and therefore $\mathcal{A}'$.

If the role value restriction rule is applied to $a:\forall R.C \in \mathcal{A}$, then there must be a role assertion $(a,b):S \in \mathcal{A}$ with $S \in R^{\downarrow}$ and $\mathcal{A}' = \mathcal{A} \cup \{b:C\}$. $\mathcal{I}_{\mathcal{D}}$ satisfies $\mathcal{A}$, hence it holds that $(a^{\mathcal{I}}, b^{\mathcal{I}}) \in S^{\mathcal{I}}, S^{\mathcal{I}} \subseteq R^{\mathcal{I}}$. Since $\mathcal{I}_{\mathcal{D}}$ satisfies $a:\forall R.C$, $b^{\mathcal{I}} \in C^{\mathcal{I}}$ must hold. Thus, $\mathcal{I}_{\mathcal{D}}$ satisfies $b:C$ and therefore $\mathcal{A}'$.

If the transitive role value restriction rule is applied to $a:\forall R.C \in \mathcal{A}$, there must be an assertion $(a,b):S \in \mathcal{A}$ with $S \in T^{\downarrow}$ for some $T \in \mathcal{T}$ and $T \in R^{\downarrow}$ such that we get $\mathcal{A}' = \mathcal{A} \cup \{b:\forall T.C\}$. Since $\mathcal{I}_{\mathcal{D}}$ satisfies $\mathcal{A}$, we have $a^{\mathcal{I}} \in (\forall R.C)^{\mathcal{I}}$ and $(a^{\mathcal{I}}, b^{\mathcal{I}}) \in S^{\mathcal{I}}, S^{\mathcal{I}} \subseteq T^{\mathcal{I}} \subseteq R^{\mathcal{I}}$. It holds that $b^{\mathcal{I}} \in (\forall T.C)^{\mathcal{I}}$ unless there is some $z \in \Delta_{\mathcal{I}}$ with $(b^{\mathcal{I}}, z) \in T^{\mathcal{I}}$ and $z \notin C^{\mathcal{I}}$. Since T is transitive, $(a^{\mathcal{I}}, z) \in T^{\mathcal{I}}$ and $a^{\mathcal{I}} \notin (\forall R.C)^{\mathcal{I}}$ in contradiction to the assumption that $\mathcal{I}$ satisfies $\mathcal{A}$. Hence, $\mathcal{I}$ must satisfy $b:\forall T.C$ and therefore $\mathcal{I}_{\mathcal{D}}$ is a model for $\mathcal{A}'$.

If the universal concept restriction rule is applied to an individual a in $\mathcal{A}$ because of $\forall x.x:C \in \mathcal{A}$, then $\mathcal{A}' = \mathcal{A} \cup \{a:C\}$. Since $\mathcal{I}_{\mathcal{D}}$ satisfies $\mathcal{A}$, it holds that $C^{\mathcal{I}} = \Delta_{\mathcal{I}}$. Thus, it holds that $a^{\mathcal{I}} \in C^{\mathcal{I}}$ and $\mathcal{I}_{\mathcal{D}}$ satisfies $\mathcal{A}'$.

If the role exists restriction rule is applied to $a:\exists R.C \in \mathcal{A}$, then we get the ABox $\mathcal{A}' = \mathcal{A} \cup \{(a,b):R, b:C\}$. Since $\mathcal{I}_{\mathcal{D}}$ satisfies $\mathcal{A}$, there exists a $y \in \Delta_{\mathcal{I}}$ such that $(a^{\mathcal{I}}, y) \in R^{\mathcal{I}}$ and $y \in C^{\mathcal{I}}$. We define the interpretation function $\cdot^{\mathcal{I}'}$ such that $b^{\mathcal{I}'} := y$ and $x^{\mathcal{I}'} := x^{\mathcal{I}}$ for $x \neq b$. Hence, $\mathcal{I}'_{\mathcal{D}} = (\Delta_{\mathcal{I}}, \Delta_{\mathcal{D}}, \cdot^{\mathcal{I}'})$ satisfies $\mathcal{A}'$.

If the number restriction exists rule is applied to $a:\exists_{\geq n} R \in \mathcal{A}$, then we get $\mathcal{A}' = \mathcal{A} \cup \{(a,b_k):R \mid k \in 1..n\} \cup \{b_i \neq b_j \mid i,j \in 1..n, i \neq j\}$. Since $\mathcal{I}_{\mathcal{D}}$ satisfies $\mathcal{A}$, there must exist n distinct individuals $y_i \in \Delta_{\mathcal{I}}, i \in 1..n$ such that $(a^{\mathcal{I}}, y_i) \in R^{\mathcal{I}}$. We define the interpretation function $\cdot^{\mathcal{I}'}$ such that $b_i^{\mathcal{I}'} := y_i$ and $x^{\mathcal{I}'} := x^{\mathcal{I}}$ for $x \notin \{b_1, \dots, b_n\}$. Hence, $\mathcal{I}'_{\mathcal{D}} = (\Delta_{\mathcal{I}}, \Delta_{\mathcal{D}}, \cdot^{\mathcal{I}'})$ satisfies $\mathcal{A}'$.

If the predicate exists rule is applied to $a:\exists f_1, \dots, f_n.P \in \mathcal{A}$, then we get the ABox $\mathcal{A}' = \mathcal{A} \cup \{(x_1, \dots, x_n):P, (a,x_1):f_1, \dots, (a,x_n):f_n\}$. After fork elimination, some x_i may be replaced by z_i with $z_i \prec_C x_i$. Since $\mathcal{I}_{\mathcal{D}}$ satisfies $\mathcal{A}$, there exist $y_1, \dots, y_n \in \Delta_{\mathcal{D}}$ such that $\forall i \in \{1, \dots, n\} : (a^{\mathcal{I}}, y_i) \in f_i^{\mathcal{I}}$ and $(y_1, \dots, y_n) \in P^{\mathcal{I}}$. We define the interpretation function $\cdot^{\mathcal{I}'}$ such that $x_i^{\mathcal{I}'} := y_i$ for all x_i not replaced by z_i and $(y_1, \dots, y_n) \in P^{\mathcal{I}'}$. The fork elimination strategy used in the R$\exists$P rule guarantees that concrete objects introduced in previous steps are not eliminated. Thus, it is ensured that the interpretation of x_i is not changed in $\mathcal{I}'_{\mathcal{D}}$. It is easy to see that $\mathcal{I}'_{\mathcal{D}} = (\Delta_{\mathcal{I}}, \Delta_{\mathcal{D}}, \cdot^{\mathcal{I}'})$ satisfies $\mathcal{A}'$.

2. "$\Leftarrow$" Assume that $\mathcal{A}'$ is satisfied by $\mathcal{I}'_{\mathcal{D}} = (\Delta_{\mathcal{I}}, \Delta_{\mathcal{D}}, \cdot^{\mathcal{I}'})$. By examining the nondeterministic rules we show that $\mathcal{A}$ is also consistent.

If $\mathcal{A}'$ is obtained from $\mathcal{A}$ by applying the disjunction rule, then $\mathcal{A}$ is a subset of $\mathcal{A}'$ and therefore satisfied by $\mathcal{I}'_{\mathcal{D}}$.

If $\mathcal{A}'$ is obtained from $\mathcal{A}$ by applying the number restriction merge rule to $a:\exists_{\leq n} R \in \mathcal{A}$, then there exist b_i, b_j in $\mathcal{A}$ such that $\mathcal{A}' = \mathcal{A}[b_i/b_j]$. We define the interpretation function $\cdot^{\mathcal{I}}$ such that $b_i^{\mathcal{I}} := b_j^{\mathcal{I}'}$ and $x^{\mathcal{I}} := x^{\mathcal{I}'}$ for every $x \neq b_i$. Obviously, $\mathcal{I}_{\mathcal{D}} = (\Delta_{\mathcal{I}}, \Delta_{\mathcal{D}}, \cdot^{\mathcal{I}})$ satisfies $\mathcal{A}$.

"$\Rightarrow$" We suppose that $\mathcal{I}_{\mathcal{D}} = (\Delta_{\mathcal{I}}, \Delta_{\mathcal{D}}, \cdot^{\mathcal{I}})$ satisfies $\mathcal{A}$ and a nondeterministic rule is applicable to an individual a in $\mathcal{A}$.

If the disjunction rule is applicable to $a:C \sqcup D \in \mathcal{A}$ and $\mathcal{A}$ is consistent, it holds $a^{\mathcal{I}} \in (C \sqcup D)^{\mathcal{I}}$. It follows that either $a^{\mathcal{I}} \in C^{\mathcal{I}}$ or $a^{\mathcal{I}} \in D^{\mathcal{I}}$ (or both). Hence, the disjunction rule can be applied in a way that $\mathcal{I}_{\mathcal{D}}$ also satisfies the ABox $\mathcal{A}'$.

If the number restriction merge rule is applicable to $a:\exists_{\leq n} R \in \mathcal{A}$ and $\mathcal{A}$ is consistent, it holds $a^{\mathcal{I}} \in (\exists_{\leq n} R)^{\mathcal{I}}$ and $\|\{b \mid (a^{\mathcal{I}}, b^{\mathcal{I}}) \in R^{\mathcal{I}}\}\| \leq n$. However, it also holds $\|\{b \mid (a^{\mathcal{I}}, b^{\mathcal{I}}) \in R^{\mathcal{I}}\}\| \geq m$ with $m > n$. Without loss of generality we only need to consider the case that $m = n + 1$. Thus, we can conclude by the Pigeonhole Principle that there exist at least two R-successors b_i, b_j of a such that $b_i^{\mathcal{I}} = b_j^{\mathcal{I}}$. Since $\mathcal{I}_{\mathcal{D}}$ satisfies $\mathcal{A}$, it must have been possible to map b_i and b_j to the same domain object, i.e. at least one of the two individuals must be a new individual. Let us assume $b_i \in O_N$, then $\mathcal{I}_{\mathcal{D}}$ obviously satisfies $\mathcal{A}[b_i/b_j]$.

In order to define a canonical interpretation from a completion $\mathcal{A}$, the notion of a specific blocking individual is introduced. We call a the *witness* of b iff b is blocked by a and $\neg \exists c$ in $\mathcal{A} : c \in O_N, c \prec a, \sigma(\mathcal{A}, c) \supseteq \sigma(\mathcal{A}, b)$. The witness for a blocked individual is unique (see [6]). Note that the canonical interpretation is constructed differently from the one describe in [7].

Definition 9. *Let $\mathcal{A}$ be a complete ABox that has been derived by the calculus from an augmented ABox $\mathcal{A}_T$. Since $\mathcal{A}$ is clash-free, there exists a variable assignment α that satisfies (the conjunction of) all occurring assertions $(x_1, \dots, x_n):P \in \mathcal{A}$. We define the canonical interpretation $\mathcal{I}_C = (\Delta_{\mathcal{I}_C}, \Delta_{\mathcal{D}}, \cdot^{\mathcal{I}_C})$ w.r.t. $\mathcal{A}$ as follows:*

1. *$\Delta_{\mathcal{I}_C} := \{a \mid a$ is mentioned in $\mathcal{A}\}$*
2. *$a^{\mathcal{I}_C} := a$ iff a is mentioned in $\mathcal{A}$*
3. *$x^{\mathcal{I}_C} := \alpha(x)$ iff x is mentioned in $\mathcal{A}$*
4. *$a \in A^{\mathcal{I}_C}$ iff $a:A \in \mathcal{A}$ and A is a concept name*
5. *$(a, \alpha(x)) \in f^{\mathcal{I}_C}$ iff $(a, x):f \in \mathcal{A}$*
6. *$(a, b) \in R^{\mathcal{I}_C}$ iff $\exists c_0, \dots, c_n, d_0, \dots, d_{n-1}$ mentioned in $\mathcal{A}$:[2],*

   a) $n \geq 1, c_0 = a, c_n = b$, and

   b) $(a, c_1):S_1, (d_1, c_2):S_2, \dots (d_{n-2}, c_{n-1}):S_{n-1}, (d_{n-1}, b):S_n \in \mathcal{A}$, and

   c) $\forall i \in 1..n-1$:
      $d_i = c_i$ or
      d_i is a witness for c_i, and $(d_i, c_{i+1}):S_{i+1} \in \mathcal{A}$, and

   d) if $n > 1$
      $\forall i \in 1..n : \exists R' \in T, R' \in R^{\downarrow}, S_i \in R'^{\downarrow}$
      else
      $S_1 \in R^{\downarrow}$.

The construction of the canonical interpretation for the case 6 is illustrated with an example in Figure 2. The following cases can be seen as special cases of case 6 introduced above ($n = 1, c_0 = a, c_1 = b$):

---

[2] Note that the variables $c_0, \dots, c_n, d_0, \dots, d_{n-1}$ not necessarily denote different individual names.

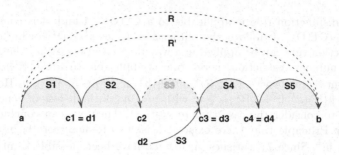

Fig. 2. Construction of the canonical interpretation. In the lower example we assume that the individual d2 is a witness for c2 (see text).

- $c_0 = d_0$: $(a, b) \in R^{\mathcal{I}c}$ iff $(c_0, c_1) : S_1 \in \mathcal{A}$ for a role $S_1 \in R^{\downarrow}$.
- $c_0 \neq d_0$: $(a, b) \in R^{\mathcal{I}c}$ iff d_0 is a witness for c_0, and
  $(d_0, c_1) : S_1 \in \mathcal{A}$, for a role $S_1 \in R^{\downarrow}$.

Since the witness of an individual is unique, the canonical interpretation is well-defined because there exists a unique blocking individual (witness) for each individual that is blocked.

Lemma 2 (Soundness). *Let $\mathcal{A}$ be a complete ABox that has been derived by the calculus from an augmented ABox $\mathcal{A}_{\mathcal{T}}$, then $\mathcal{A}_{\mathcal{T}}$ has a model.*

Proof. Let $\mathcal{I}_{\mathcal{C}} = (\Delta_{\mathcal{I}_c}, \Delta_{\mathcal{D}}, \cdot^{\mathcal{I}c})$ be the canonical interpretation for the ABox $\mathcal{A}$ constructed w.r.t. the TBox $\mathcal{T}$. $\mathcal{A}$ is clash-free.

Features are interpreted in the correct way: There can be no forks in $\mathcal{A}$ because (i) there are no forks in the augmented ABox $\mathcal{A}_{\mathcal{T}}$ and (ii) forks are immediately eliminated after an application of the R∃P rule. This rule is the only rule that introduces new assertions of the form $(a, x) : f \in \mathcal{A}$. Note that forks cannot be introduced by the R∃$_{\leq n}$ rule due to the completion strategy. Thus, $\mathcal{I}_{\mathcal{C}}$ maps features to (partial) functions because the variable assignment α is a function.

All role inclusions in the role hierarchy are satisfied: For every $S \sqsubseteq R$ it holds that $S^{\mathcal{I}c} \subseteq R^{\mathcal{I}c}$ This can be shown as follows. If $(a^{\mathcal{I}c}, b^{\mathcal{I}c}) \in S^{\mathcal{I}c}$, case 6 of Definition 9 must be applicable. Hence, there exists a chain of sub-roles possibly with gaps and witnesses (see Definition 9, case 6). Thus, the corresponding construction for $\mathcal{I}_{\mathcal{C}}$ adding $(a^{\mathcal{I}c}, b^{\mathcal{I}c})$ to $S^{\mathcal{I}c}$ is also applicable to R since $S \in R^{\downarrow}$ (see 6d). Therefore, there is also a tuple $(a^{\mathcal{I}c}, b^{\mathcal{I}c}) \in R^{\mathcal{I}c}$.

All (implicit) transitivity axioms are satisfied, i.e. transitive roles are interpreted in the correct way: $\forall R \in T : R^{\mathcal{I}c} = (R^{\mathcal{I}c})^{+}$. If there exist $(a^{\mathcal{I}c}, b^{\mathcal{I}c}) \in R^{\mathcal{I}c}$ and $(b^{\mathcal{I}c}, c^{\mathcal{I}c}) \in R^{\mathcal{I}c}$ then case 6 in Definition 9 must have been applied for each tuple. But then, a chain of roles from a to c exists as well (possibly with gaps and witnesses) such that $(a^{\mathcal{I}c}, c^{\mathcal{I}c})$ is added to $R^{\mathcal{I}c}$ as well.

In the following we prove that $\mathcal{I}_{\mathcal{C}}$ satisfies every assertion in $\mathcal{A}$.

For any $a \neq b \in \mathcal{A}$ or $(a, b) : R \in \mathcal{A}$, $\mathcal{I}_{\mathcal{C}}$ satisfies them by definition.

For any $(a,x){:}f \in \mathcal{A}$, $\mathcal{I}_{\mathcal{C}}$ satisfies them by definition.

For any $(x_1,\ldots,x_n){:}P \in \mathcal{A}$, $\mathcal{I}_{\mathcal{C}}$ satisfies them by definition. Since $\mathcal{A}$ is clash-free there exists a variable assignment such that the conjunction of all predicate assertions is satisfied. The variable assignment can be computed because the concrete domain is required to be admissible.

Next we consider assertions of the form $a{:}C$. We show by induction on the structure of concepts that $a{:}C \in \mathcal{A}$ implies $a^{\mathcal{I}c} \in C^{\mathcal{I}c}$.

If C is a concept name, then $a^{\mathcal{I}c} \in C^{\mathcal{I}c}$ by definition of $\mathcal{I}_{\mathcal{C}}$.

If $C = \neg D$, then D is a concept name since all concepts are in negation normal form (see Definition 3). $\mathcal{A}$ is clash-free and cannot contain $a{:}D$. Thus, $a^{\mathcal{I}c} \notin D^{\mathcal{I}c}$, i.e. $a^{\mathcal{I}c} \in \Delta_{\mathcal{I}_{\mathcal{C}}} \setminus D^{\mathcal{I}c}$. Hence $a^{\mathcal{I}c} \in (\neg D)^{\mathcal{I}c}$.

If $C = C_1 \sqcap C_2$ then (since $\mathcal{A}$ is complete) $a{:}C_1 \in \mathcal{A}$ and $a{:}C_2 \in \mathcal{A}$. By induction hypothesis, $a^{\mathcal{I}c} \in C_1{}^{\mathcal{I}c}$ and $a^{\mathcal{I}c} \in C_2{}^{\mathcal{I}c}$. Hence $a^{\mathcal{I}c} \in (C_1 \sqcap C_2)^{\mathcal{I}c}$.

If $C = C_1 \sqcup C_2$ then (since $\mathcal{A}$ is complete) either $a{:}C_1 \in \mathcal{A}$ or $a{:}C_2 \in \mathcal{A}$. By induction hypothesis, $a^{\mathcal{I}c} \in C_1{}^{\mathcal{I}c}$ or $a^{\mathcal{I}c} \in C_2{}^{\mathcal{I}c}$. Hence $a^{\mathcal{I}c} \in (C_1 \sqcup C_2)^{\mathcal{I}c}$.

If $C = \forall R.D$, then it must be shown that for all $b^{\mathcal{I}c}$ with $(a^{\mathcal{I}c}, b^{\mathcal{I}c}) \in R^{\mathcal{I}c}$ it holds that $b^{\mathcal{I}c} \in D^{\mathcal{I}c}$. If $(a^{\mathcal{I}c}, b^{\mathcal{I}c}) \in R^{\mathcal{I}c}$, then according to Definition 9, b is a successor of a via a chain of roles $S_i \in R^{\downarrow}$ or there exists corresponding witnesses as domain elements of $S_i \in R^{\downarrow}$, i.e. the chain might contain "gaps" with associated witnesses (see Figure 2). Since $(a^{\mathcal{I}c}, b^{\mathcal{I}c}) \in R^{\mathcal{I}c}$ and $S_i{}^{\mathcal{I}c} \subseteq R^{\mathcal{I}c}$ there exists tuples $(c_i{}^{\mathcal{I}c}, c_{i+1}{}^{\mathcal{I}c}) \in S_i{}^{\mathcal{I}c}$. Due to Definition 9 it holds that $\forall i \in 1..n : \exists R' \in T, R' \in R^{\downarrow}, S_i \in R'^{\downarrow}$. Therefore $c_k{:}\forall R'.D \in \mathcal{A}$, $(k \in 1..n-1)$ because $\mathcal{A}$ is complete. For the same reason $b{:}D \in \mathcal{A}$. By induction hypothesis it holds that $b^{\mathcal{I}c} \in D^{\mathcal{I}c}$. As mentioned before, the chain of roles can have one or more "gaps" (see Figure 2). However, due to Definition 9 in case of a "gap" there exists a witness such that a similar argument as in case 6 can be applied, i.e. in case of a gap between c_i and c_{i+1} with witness d_i for c_i, the blocking condition ensures that the concept set of the witness is a superset of the concept set of the blocked individual. Since it is assumed that $(d_i, c_{i+1}){:}S_{i+1} \in \mathcal{A}$ and $\mathcal{A}$ is complete it holds that $c_{i+1}{:}\forall R'.D \in \mathcal{A}$. Applying the same argument inductively, we can conclude that $c_{n-1}{:}\forall R'.D \in \mathcal{A}$ and again, we have $b^{\mathcal{I}c} \in D^{\mathcal{I}c}$ by induction hypothesis.

If $C = \exists R.D$, then it must be shown that there exists an individual $b^{\mathcal{I}c} \in \Delta_{\mathcal{I}_{\mathcal{C}}}$ with $(a^{\mathcal{I}c}, b^{\mathcal{I}c}) \in R^{\mathcal{I}c}$ and $b^{\mathcal{I}c} \in D^{\mathcal{I}c}$. Since ABox $\mathcal{A}$ is complete, we have either $(a,b){:}S \in \mathcal{A}$ with $S \in R^{\downarrow}$ and $b{:}D \in \mathcal{A}$ or a is blocked by an individual c and $(c,b){:}S \in \mathcal{A}$ (again $S \in R^{\downarrow}$). In the first case we have $(a^{\mathcal{I}c}, b^{\mathcal{I}c}) \in R^{\mathcal{I}c}$ by the definition of $\mathcal{I}_{\mathcal{C}}$ (case 6, $n = 1, c_i = d_i$) and $b^{\mathcal{I}c} \in D^{\mathcal{I}c}$ by induction hypothesis. In the second case there exists the witness c with $c{:}\exists R.D \in \mathcal{A}$. By definition c cannot be blocked, and by hypothesis $\mathcal{A}$ is complete. So we have an individual b with $(c,b){:}S \in \mathcal{A}$ ($S \in R^{\downarrow}$) and $b{:}D \in \mathcal{A}$. By induction hypothesis we have $b^{\mathcal{I}c} \in D^{\mathcal{I}c}$, and by the definition of $\mathcal{I}_{\mathcal{C}}$ (case 6, $n = 1, c_i \neq d_i$, d_i is a witness for c_i, and $a = c_i, c = d_i$) we have $(a^{\mathcal{I}c}, b^{\mathcal{I}c}) \in R^{\mathcal{I}c}$.

If $C = \exists_{\geq n} R$, we prove the hypothesis by contradiction. We assume that $a^{\mathcal{I}c} \notin (\exists_{\geq n} R)^{\mathcal{I}c}$. Then there exist at most m $(0 \leq m < n)$ distinct S-successors of a with $S \in R^{\downarrow}$. Two cases can occur: (1) the individual a is not blocked in $\mathcal{I}_{\mathcal{C}}$. Then we have less than n S-successors of a in $\mathcal{A}$, and the $R\exists_{\geq n}$-rule is applicable to a. This contradicts the assumption that $\mathcal{A}$ is complete. (2) a is blocked by an individual c but the same argument as in case (1) holds and leads to the same contradiction.

For $C = \exists_{\leq n} R$ we show the goal by contradiction. Suppose that $a^{\mathcal{I}c} \notin (\exists_{\leq n} R)^{\mathcal{I}c}$. Then there exist at least $n + 1$ distinct individuals $b_1{}^{\mathcal{I}c}, \ldots, b_{n+1}{}^{\mathcal{I}c}$ such that $(a^{\mathcal{I}c}, b_i{}^{\mathcal{I}c}) \in R^{\mathcal{I}c}$, $i \in 1..n + 1$. The following two cases can occur. (1) The individual a is not blocked: We have $n+1$ $(a, b_i) : S_i \in \mathcal{A}$ with $S_i \in R^{\downarrow}$ and $S_i \notin T$, $i \in 1..n+1$. The $R\exists_{\leq n}$ rule cannot be applicable since $\mathcal{A}$ is complete and the b_i are distinct, i.e. $b_i \neq b_j \in \mathcal{A}$, $i, j \in 1..n+1$, $i \neq j$. This contradicts the assumption that $\mathcal{A}$ is clash-free. (2) There exists a witness c for a with $(c, b_i) : S_i \in \mathcal{A}$, $S_i \in R^{\downarrow}$, and $S_i \notin T$, $i \in 1..n+1$. This leads to an analogous contradiction. Due to the construction of the canonical interpretation in case of a blocking condition (with c being the witness) and a non-transitive role R (R is required to be a simple role, see the syntactic restrictions for number restrictions and role boxes), there is no $(a^{\mathcal{I}c}, b_k{}^{\mathcal{I}c}) \in R^{\mathcal{I}c}$ if there is no $(c^{\mathcal{I}c}, b_k{}^{\mathcal{I}c}) \in R^{\mathcal{I}c}$ $(k \in 1..n + 1)$.

If $C = \exists f_1, \ldots, f_n . P$ we show that there exist concrete objects $y_1, \ldots, y_n \in \Delta_{\mathcal{D}}$ such that $(a^{\mathcal{I}c}, y_1) \in f_1{}^{\mathcal{I}c}, \ldots, (a^{\mathcal{I}c}, y_n) \in f_n{}^{\mathcal{I}c}$ and $(y_1, \ldots, y_n) \in P^{\mathcal{I}c}$. The $R\exists P$ rule generates assertions $(a, x_1) : f_1, \ldots, (a, x_n) : f_n, (x_1, \ldots, x_n) : P$. Since $\mathcal{A}$ is clash-free there is no concrete domain clash. Hence there exists a variable assignment α that maps $x_1, \ldots, x_n$ to elements of $\Delta_{\mathcal{D}}$. The conjunction of concrete domain predicates is satisfiable and $(x_1{}^{\mathcal{I}c}, \ldots, x_n{}^{\mathcal{I}c}) \in P^{\mathcal{I}c}$. By definition of $\mathcal{I}_C$ it holds that $(a^{\mathcal{I}c}, x_1{}^{\mathcal{I}c}) \in f_1{}^{\mathcal{I}c}, \ldots, (a^{\mathcal{I}c}, x_n{}^{\mathcal{I}c}) \in f_n{}^{\mathcal{I}c}$. Thus, there exist $y_1, \ldots, y_n$ such that the above-mentioned requirements are fulfilled and therefore $a^{\mathcal{I}c} \in (\exists f_1, \ldots, f_n . P)^{\mathcal{I}c}$.

If $C = \forall f . \perp_{\mathcal{D}}$ then we show that $a^{\mathcal{I}c} \in (\forall f . \perp_{\mathcal{D}})^{\mathcal{I}c}$. Because $\mathcal{A}$ is clash-free, there cannot be an assertion $(a, x) : f \in \mathcal{A}$ for some x in O_c and an $f \in F$. Thus, it does not hold that there exists $(a^{\mathcal{I}c}, y) \in f^{\mathcal{I}c}$ and hence $a^{\mathcal{I}c} \in (\forall f . \perp_{\mathcal{D}})^{\mathcal{I}c}$.

If $\forall x . x : D \in \mathcal{A}$, then –due to the completeness of $\mathcal{A}$– for each individual a in $\mathcal{A}$ we have $a : D \in \mathcal{A}$ and, by the previous cases, $a^{\mathcal{I}c} \in D^{\mathcal{I}c}$. Thus, $\mathcal{I}_C$ satisfies $\forall x . x : D$. Finally, since $\mathcal{I}_C$ satisfies all assertions in $\mathcal{A}$, $\mathcal{I}_C$ satisfies $\mathcal{A}$.

Lemma 3 (Completeness). *Let $\mathcal{A}_{\mathcal{T}}$ be an augmented ABox be a role box. If $\mathcal{A}_{\mathcal{T}}$ is consistent, then there exists at least one completion $\mathcal{A}'$ being computed by applying the completion rules.*

Proof. By contraposition: Obviously, an ABox containing a clash is inconsistent. If there does not exists a completion of $\mathcal{A}_{\mathcal{T}}$, then it follows from Proposition 1 that the ABox $\mathcal{A}_{\mathcal{T}}$ is inconsistent.

Lemma 4 (Termination). *The calculus described above terminates on every (augmented) input ABox.*

Proof. The termination of the calculus is shown by specifying an upper limit on the number of assertions that can result from an (augmented) input ABox of a certain length n. Compared to $\mathcal{ALCNH}_{R^+}$ in the termination proof for $\mathcal{ALCNH}_{R^+}(\mathcal{D})^-$ the additional constructs for concrete domains have to be considered. Basically, since features do not "interact" with value and number restrictions (see the completion rules), the same upper limit $O(2^{4n})$ for a completion can be derived. For details see [6].

Theorem 1 (Decidability). *Let $\mathcal{D}$ be an admissible concrete domain. Checking whether an $\mathcal{ALCNH}_{R^+}(\mathcal{D})^-$ knowledge base $(\mathcal{T}, \mathcal{A})$ is consistent is a decidable problem.*

Proof. Given a knowledge base $(\mathcal{T}, \mathcal{A})$, an augmented ABox $\mathcal{A}_\mathcal{T}$ can be constructed in linear time. The claim follows from Lemmas 1, 2, 3, and 4.

5   Conclusion

We presented a tableaux calculus deciding the knowledge base consistency problem for the description logic $\mathcal{ALCNH}_{R^+}(\mathcal{D})^-$. Applications of the logic in the context of configuration problems have been sketched. The Cylinder example demonstrates that some requirements of a model-based configuration system are fulfilled by $\mathcal{ALCNH}_{R^+}(\mathcal{D})^-$. The calculus presented in this paper can be used to solve "simple" configuration problems in which the configuration space can be described by an $\mathcal{ALCNH}_{R^+}(\mathcal{D})^-$ knowledge base (see [6] for an analysis of the models resulting from the canonical interpretation). We conjecture that concrete domains without features chains can also be included in description logics with inverse roles and qualified number restrictions.

A highly optimized variant of the calculus for the sublogic $\mathcal{ALCNH}_{R^+}$ is already implemented in the ABox description logic system RACE. RACE is available at http://kogs-www.informatik.uni-hamburg.de/~race/. RACE will be extended with support for reasoning with concrete domains in the near future. With this paper we provide a sound basis for practical extensions of expressive DL systems such that, for instance, construction problems can be effectively solved with description logic reasoning techniques.

References

1. F. Baader and P. Hanschke. A scheme for integrating concrete domains into concept languages. In *Twelfth International Conference on Artificial Intelligence, 1991*, pages 452–457, 1991. A longer version appeared as Tech. Report DFKI-RR-91-10.
2. F. Baader and P. Hanschke. Extensions of concept languages for a mechanical engineering application. In H.J. Ohlbach, editor, *Proceedings, GWAI-92: 16th German Conference on Artificial Intelligence*, pages 132–143. Springer, 1992.
3. M. Buchheit, R. Klein, and W. Nutt. Configuration as model construction: The constructive problem solving approach. In F. Sudweeks and J. Gero, editors, *Proc. 4th Int. Conf. on Artificial Intelligence in Design*. Kluwer, Dordrecht, 1994.
4. V. Haarslev and R. Möller. Consistency testing: The RACE experience. In *Proceedings International Conference Tableaux'2000*. Springer-Verlag, 2000.

5. V. Haarslev and R. Möller. Expressive ABox reasoning with number restrictions, role hierachies, and transitively closed roles. In A.G. Cohn, F. Giunchiglia, and B. Selman, editors, *Proceedings of the Seventh International Conference on Principles of Knowledge Representation and Reasoning (KR'2000)*, 2000.
6. V. Haarslev, R. Möller, and M. Wessel. The description logic $\mathcal{ALCNH}_{R+}$ extended with concrete domains. Technical Report FBI-HH-M-290/00, University of Hamburg, Computer Science Department, August 2000.
7. I. Horrocks, U. Sattler, and S. Tobies. Practical reasoning for expressive description logics. In H. Ganzinger, D. McAllester, and A. Voronkov, editors, *Proceedings of the 6th International Conference on Logic for Programming and Automated Reasoning (LPAR'99)*, number 1705 in LNAI, pages 161–180. Springer, 1999.
8. C. Lutz. The complexity of reasoning with concrete domains (revised version). LTCS-Report 99-01, LuFG Theoretical Computer Science, RWTH Aachen, 1999.
9. J.R. Wright, E.S. Weixelbaum, G.T. Vesonder, K. Brown, S.R. Palmer, J.I. Berman, and H.H. Moore. A knowledge-based configurator that supports sales, engineering, and manufacturing at AT&T network systems. *AI Magazine*, 14(3):69–80, 1993.

NExpTime-Complete Description Logics with Concrete Domains

Carsten Lutz

LuFG Theoretical Computer Science
RWTH Aachen, Germany
lutz@cs.rwth-aachen.de

Abstract. Concrete domains are an extension of Description Logics (DLs) allowing to integrate reasoning about conceptual knowledge with reasoning about "concrete properties" of objects such as sizes, weights, and durations. It is known that reasoning with $\mathcal{ALC}(\mathcal{D})$, the basic DL admitting concrete domains, is PSpace-complete. In this paper, it is shown that the upper bound is not robust: we give three examples for seemingly harmless extensions of $\mathcal{ALC}(\mathcal{D})$—namely acyclic TBoxes, inverse roles, and a role-forming concrete domain constructor—that make reasoning NExpTime-hard. As a corresponding upper bound, we show that reasoning with all three extensions *together* is in NExpTime.

1 Introduction

Description Logics (DLs) are a family of logical formalisms for the representation of and reasoning about conceptual knowledge. The knowledge is represented on an abstract logical level, i.e., by means of concepts (unary predicates), roles (binary predicates), and logical constructors. This makes it difficult to adequately represent knowledge concerning "concrete properties" of real-world entities such as their sizes, weights, and durations. Since, for many knowledge representation applications, it is essential to integrate reasoning about such concrete properties with reasoning about knowledge represented on an abstract logical level, Baader and Hanschke extended Description Logics by so-called *concrete domains* [1]. A concrete domain consists of a set called the domain and a set of predicates with a fixed interpretation over this domain. For example, one could use the real numbers as the domain and then define predicates such as the unary $=_{23}$, the binary "<" and "=", and the ternary "+" and "·" [1]. Or one could use the set of all intervals over, say, the rationals as the domain and then define "temporal" predicates such as *during*, *meets*, and *before* [15]. Baader and Hanschke propose to extend the basic Description Logic $\mathcal{ALC}$ with concrete domains which yields the logic $\mathcal{ALC}(\mathcal{D})$. The interface between $\mathcal{ALC}$ and the concrete domain is provided by a concrete domain concept constructor. To illustrate the use of concrete domains for knowledge representation, consider the example $\mathcal{ALC}(\mathcal{D})$-concept

$$\forall subprocess. Drilling \sqcap \exists workpiece.(\exists height.=_{5cm} \sqcap \exists height, length.>)$$

R. Goré, A. Leitsch, and T. Nipkow (Eds.): IJCAR 2001, LNAI 2083, pp. 45–60, 2001.

which describes a process all of whose subprocesses are drilling processes and which involves a workpiece with height 5cm and hight strictly greater than its length. Here, $=_{5cm}$ is a unary predicate from the concrete domain and $>$ is a binary predicate. The subconcept in brackets is a conjunction of two concrete domain concept constructors. Other DLs with concrete domains can be found in [3,8,12], while applications of such logics are described in [2,8].

In this paper, we are interested in the complexity of reasoning with Description Logics providing for concrete domains. The complexity of $\mathcal{ALC}(\mathcal{D})$ itself is determined in [14], where reasoning with $\mathcal{ALC}(\mathcal{D})$ is proved to be PSPACE-complete if reasoning with the concrete domain $\mathcal{D}$ is in PSPACE. However, for many applications, the expressivity of $\mathcal{ALC}(\mathcal{D})$ is not sufficient which makes it quite natural to consider extensions of this logic with additional means of expressivity. We consider three such extensions—all of them frequently used in the area of Description Logics—and show that, although all these extensions are seemingly "harmless", reasoning in the extended logics is considerably harder than in $\mathcal{ALC}(\mathcal{D})$ itself. Hence, the PSPACE upper bound of $\mathcal{ALC}(\mathcal{D})$ cannot be considered robust.

More precisely, we consider the extension of $\mathcal{ALC}(\mathcal{D})$ with (1) acyclic TBoxes, (2) inverse roles, and (3) a role-forming concrete domain constructor. TBoxes are used for representing terminological knowledge and background knowledge of application domains [5,13], inverse roles are present in most expressive Description Logics [5,10], and the role-forming constructor is a natural counterpart to the concept-forming concrete domain constructor [8]. By introducing a NEXPTIME-complete variant of the Post Correspondence Problem [17,9], we identify a large class of concrete domains $\mathcal{D}$ such that reasoning with each of the above three extensions of $\mathcal{ALC}(\mathcal{D})$ (separately) is NEXPTIME-hard. This dramatic increase in complexity is rather surprising since, from a computational point of view, all of the proposed extensions look harmless. For example, in [13], it is shown that the extension of many PSPACE Description Logics with acyclic TBoxes does not increase the complexity of reasoning. Moreover, it is well-known that the extension with inverse roles does usually not change the complexity class. For example, $\mathcal{ALC}$ extended with inverse roles is still in PSPACE [11]. As a corresponding upper bound, we show that, if reasoning with a concrete domain $\mathcal{D}$ is in NP, then reasoning with $\mathcal{ALC}(\mathcal{D})$ and all three above extensions (simultaneously) is in NEXPTIME. We argue that this upper bound captures a large class of interesting concrete domains. This paper is accompanied by a technical report containing full proofs [16].

2   Description Logics with Concrete Domains

We introduce the Description Logics we are concerned with in the remainder of this paper. First, $\mathcal{ALCI}(\mathcal{D})$ is defined which extends $\mathcal{ALC}(\mathcal{D})$ with inverse roles. In a second step, we add a role-forming concrete domain constructor and obtain the logic $\mathcal{ALCRPI}(\mathcal{D})$. This two-step approach is pursued since the definition of

$\mathcal{ALCRPI}(\mathcal{D})$ involves some rather unusual syntactic restrictions which we like to keep separated from the more straightforward syntax of $\mathcal{ALCI}(\mathcal{D})$.

Definition 1 (Concrete Domain). *A concrete domain $\mathcal{D}$ is a pair $(\Delta_\mathcal{D}, \Phi_\mathcal{D})$, where $\Delta_\mathcal{D}$ is a set called the domain, and $\Phi_\mathcal{D}$ is a set of predicate names. Each predicate name $P \in \Phi_\mathcal{D}$ is associated with an arity n and an n-ary predicate $P^\mathcal{D} \subseteq \Delta_\mathcal{D}^n$.*

With $\overline{P}$, we denote the negation of the predicate P, i.e. $\overline{P}^\mathcal{D} = \Delta_\mathcal{D} \setminus P^\mathcal{D}$. Based on concrete domains, we introduce the syntax of $\mathcal{ALCI}(\mathcal{D})$.

Definition 2 (Syntax). *Let N_C, N_R, and N_{cF} be mutually disjoint sets of concept names, role names, and concrete feature names, respectively, and let N_{aF} be a subset of N_R. Elements of N_{aF} are called abstract features. The set of $\mathcal{ALCI}(\mathcal{D})$ roles $\widehat{N_R}$ is $N_R \cup \{R^- \mid R \in N_R\}$. An expression $f_1 \cdots f_n g$, where $f_1, \ldots, f_n \in N_{aF}$ $(n \geq 0)$ and $g \in N_{cF}$, is called a path. The set of $\mathcal{ALCI}(\mathcal{D})$-concepts is the smallest set such that*

1. *every concept name is a concept*
2. *if C and D are concepts, R is a role, g is a concrete feature, $P \in \Phi$ is a predicate name with arity n, and $u_1, \ldots, u_n$ are paths, then the following expressions are also concepts: $\neg C$, $C \sqcap D$, $C \sqcup D$, $\exists R.C$, $\forall R.C$, $\exists u_1, \ldots, u_n.P$, and $g\uparrow$.*

An $\mathcal{ALCI}(\mathcal{D})$-concept which uses only roles from N_R is called an $\mathcal{ALC}(\mathcal{D})$-concept. With $sub(C)$, we denote the set of subconcepts of a concept C which is defined in the obvious way. Throughout this paper, we denote concept names with A and B, concepts with C and D, roles with R, abstract features with f, concrete features with g, paths with u, and predicates with P. As usual, we write $\top$ for $A \sqcup \neg A$, $\bot$ for $A \sqcap \neg A$ (where A is some concept name), and $\exists f_1 \cdots f_n.C$ (resp. $\forall f_1 \cdots f_n.C$) for $\exists f_1. \cdots \exists f_n.C$ (resp. $\forall f_1. \cdots \forall f_n.C$).

The syntactical part of a Description Logic is usually given by a concept language and a so-called TBox formalism. The TBox formalism is used to represent terminological knowledge of the application domain.

Definition 3 (TBoxes). *Let A be a concept name and C be a concept. Then $A \doteq C$ is a concept definition. Let $\mathcal{T}$ be a finite set of concept definitions. A concept name A directly uses a concept name B in $\mathcal{T}$ if there is a concept definition $A \doteq C$ in $\mathcal{T}$ such that B appears in C. Let uses be the transitive closure of "directly uses". $\mathcal{T}$ is called acyclic if there is no concept name A such that A uses itself in $\mathcal{T}$. If $\mathcal{T}$ is acyclic, and the left-hand sides of all concept definitions in $\mathcal{T}$ are unique, then $\mathcal{T}$ is called a TBox.*

TBoxes can be thought of as sets of macro definitions, i.e., the left-hand side of every concept definition is an abbreviation for the right-hand side of the concept definition. There also exist more general TBox formalisms allowing for arbitrary equations over concepts [5,10]. However, we will see that admitting these general TBoxes makes reasoning with $\mathcal{ALC}(\mathcal{D})$ (and hence also $\mathcal{ALCI}(\mathcal{D})$) undecidable.

Definition 4 (Semantics). *An* interpretation $\mathcal{I}$ *is a pair* $(\Delta_{\mathcal{I}}, \cdot^{\mathcal{I}})$, *where* $\Delta_{\mathcal{I}}$ *is a set called the* domain *and* $\cdot^{\mathcal{I}}$ *the* interpretation function. *The interpretation function maps each concept name* C *to a subset* $C^{\mathcal{I}}$ *of* $\Delta_{\mathcal{I}}$, *each role name* R *to a subset* $R^{\mathcal{I}}$ *of* $\Delta_{\mathcal{I}} \times \Delta_{\mathcal{I}}$, *each abstract feature* f *to a partial function* $f^{\mathcal{I}}$ *from* $\Delta_{\mathcal{I}}$ *to* $\Delta_{\mathcal{I}}$, *and each concrete feature* g *to a partial function* $g^{\mathcal{I}}$ *from* $\Delta_{\mathcal{I}}$ *to* $\Delta_{\mathcal{D}}$. *If* $u = f_1 \cdots f_n g$ *is a path, then* $u^{\mathcal{I}}(a)$ *is defined as* $g^{\mathcal{I}}(f_n^{\mathcal{I}} \cdots (f_1^{\mathcal{I}}(a)) \cdots)$. *The interpretation function is extended to arbitrary roles and concepts as follows:*

$$(R^-)^{\mathcal{I}} := \{(a, b) \mid (b, a) \in R^{\mathcal{I}}\}$$

$$(C \sqcap D)^{\mathcal{I}} := C^{\mathcal{I}} \cap D^{\mathcal{I}} \quad (C \sqcup D)^{\mathcal{I}} := C^{\mathcal{I}} \cup D^{\mathcal{I}} \quad (\neg C)^{\mathcal{I}} := \Delta_{\mathcal{I}} \setminus C^{\mathcal{I}}$$

$$(\exists R.C)^{\mathcal{I}} := \{a \in \Delta_{\mathcal{I}} \mid \{b \mid (a, b) \in R^{\mathcal{I}}\} \cap C^{\mathcal{I}} \neq \emptyset\}$$

$$(\forall R.C)^{\mathcal{I}} := \{a \in \Delta_{\mathcal{I}} \mid \{b \mid (a, b) \in R^{\mathcal{I}}\} \subseteq C^{\mathcal{I}}\}$$

$$(\exists u_1, \ldots, u_n.P)^{\mathcal{I}} := \{a \in \Delta_{\mathcal{I}} \mid (u_1^{\mathcal{I}}(a), \ldots, u_n^{\mathcal{I}}(a)) \in P^{\mathcal{D}}\}$$

$$(g{\uparrow})^{\mathcal{I}} := \{a \in \Delta_{\mathcal{I}} \mid g^{\mathcal{I}}(a) \text{ undefined}\}$$

An interpretation $\mathcal{I}$ *is called a* model *for a concept* C *iff* $C^{\mathcal{I}} \neq \emptyset$ *and a* model *for a TBox* $\mathcal{T}$ *iff* $A^{\mathcal{I}} = C^{\mathcal{I}}$ *for all* $A \doteq C \in \mathcal{T}$.

We call elements from $\Delta_{\mathcal{I}}$ *abstract objects* and elements from $\Delta_{\mathcal{D}}$ *concrete objects*. Our definition of $\mathcal{ALC}(\mathcal{D})$ differs slightly from the original version in [1]: Instead of separating concrete and abstract features, Baader and Hanschke define only one type of feature which is interpreted as a partial function from $\Delta_{\mathcal{I}}$ to $\Delta_{\mathcal{I}} \cup \Delta_{\mathcal{D}}$. We choose the separated approach since it allows clearer proofs. Moreover, it is not hard to see that the combined features can be "simulated" using pairs of concrete and abstract features.

Definition 5 (Inference Problems). *Let* C *and* D *be concepts.* C *subsumes* D *w.r.t. a TBox* $\mathcal{T}$ *(written* $D \sqsubseteq_{\mathcal{T}} C$*) iff* $D^{\mathcal{I}} \subseteq C^{\mathcal{I}}$ *for all models* $\mathcal{I}$ *of* $\mathcal{T}$. C *is* satisfiable *w.r.t. a TBox* $\mathcal{T}$ *iff there exists a model of both* $\mathcal{T}$ *and* C.

Both inferences are also considered without reference to TBoxes, i.e., with reference to the empty TBox. It is well-known that (un)satisfiability and subsumption can be mutually reduced to each other: $C \sqsubseteq_{\mathcal{T}} D$ iff $C \sqcap \neg D$ is unsatisfiable w.r.t. $\mathcal{T}$, and C is satisfiable w.r.t. $\mathcal{T}$ iff we do not have $C \sqsubseteq_{\mathcal{T}} \bot$. We call two concepts C and D *equivalent* iff C subsumes D and D subsumes C.

Let us now further extend $\mathcal{ALCI}(\mathcal{D})$ with a role-forming concrete domain constructor, i.e., with a constructor that allows the definition of complex roles with reference to the concrete domain. Such a constructor was first defined in [8], where it is motivated as an appropriate tool for spatial reasoning.

Definition 6 ($\mathcal{ALCRPI}(\mathcal{D})$ Syntax and Semantics). *A* predicate role *is an expression of the form* $\exists(u_1, \ldots, u_n), (v_1, \ldots, v_m).P$ *where* P *is an* $n + m$-*ary predicate. The semantics of predicate roles is*

$$(\exists(u_1, \ldots, u_n), (v_1, \ldots, v_m).P)^{\mathcal{I}} := \{(a, b) \in \Delta_{\mathcal{I}} \times \Delta_{\mathcal{I}} \mid$$
$$(u_1^{\mathcal{I}}(a), \ldots, u_n^{\mathcal{I}}(a), v_1^{\mathcal{I}}(b), \ldots, v_m^{\mathcal{I}}(b)) \in P^{\mathcal{D}}\}.$$

With $\mathcal{R}$, we denote the set of predicate roles. The set of $\mathcal{ALCRPI}(\mathcal{D})$ roles $\widehat{\mathcal{R}}$ is defined as $\widehat{N_R} \cup \mathcal{R} \cup \{R^- \mid R \in \mathcal{R}\}$. A role which is either a predicate role or the inverse of a predicate role is called complex role. An $\mathcal{ALCI}(\mathcal{D})$-concept with roles from $N_R \setminus N_{aF}$ replaced with roles from $\widehat{\mathcal{R}}$ is called $\mathcal{ALCRPI}(\mathcal{D})$-concept.

An $\mathcal{ALCRPI}(\mathcal{D})$-concept not using the inverse role constructor is called an $\mathcal{ALCRP}(\mathcal{D})$-concept. For example, the following is an $\mathcal{ALCRPI}(\mathcal{D})$-concept

$$Error \sqcap \exists time, next\ time.< \sqcap \forall next.\forall(\exists(time),(time).<)^-.\neg Error$$

where $Error$ is a concept, $time$ is a concrete feature and $next$ is an abstract feature. This concept is unsatisfiable since every domain object satisfying it would have to be both in $Error$ and $\neg Error$ which is impossible. In [7], it is proved that satisfiability of $\mathcal{ALCRP}(\mathcal{D})$-concepts is undecidable. However, as shown in [8], there exists a decidable fragment of $\mathcal{ALCRP}(\mathcal{D})$ that is still a useful extension of $\mathcal{ALC}(\mathcal{D})$. In the following, we introduce an analogous fragment of the logic $\mathcal{ALCRPI}(\mathcal{D})$.

Definition 7 (Restricted $\mathcal{ALCRPI}(\mathcal{D})$-concept). *Let C be an $\mathcal{ALCRPI}(\mathcal{D})$-concept, and $sub(C)$ the set of subconcepts of C. Then C is called restricted iff it fulfills the following conditions:*

1. *For any $\forall R.D \in sub(C)$, where R is a complex role, $sub(D)$ does not contain any concepts of the form $\exists u_1, \ldots, u_n.P$ or $\exists S.E$, where S is a complex role.*
2. *For any $\exists R.D \in sub(C)$, where R is a complex role, $sub(D)$ does not contain any concepts of the form $\exists u_1, \ldots, u_n.P$ or $\forall S.E$, where S is a complex role.*

Intuitively, these restrictions enforce the finite model property which leads to decidability, see [8,16] for details. In the remainder of this paper, we assume all $\mathcal{ALCRPI}(\mathcal{D})$ concepts to be restricted without further notice. Note that the set of restricted $\mathcal{ALCRPI}(\mathcal{D})$-concepts is closed under negation, and, hence, subsumption can be reduced to satisfiability.

3   A NExpTime-Complete Variant of the PCP

The Post Correspondence Problem (PCP), as introduced 1946 by Emil Post [17], is an undecidable problem frequently employed in undecidability proofs. In this section, we define a NExpTime-complete variant of the PCP together with a concrete domain $\mathcal{P}$ that is suitable for reducing PCPs to the satisfiability problem of Description Logics with concrete domains.

Definition 8 (PCP). *A Post Correspondence Problem (PCP) P is given by a finite, non-empty list $(\ell_1, r_1), \ldots, (\ell_k, r_k)$ of pairs of non-empty words over some alphabet Σ. A sequence of integers $i_1, \ldots, i_m$, with $m \geq 1$, is called a solution for P iff $\ell_{i_1} \cdots \ell_{i_m} = r_{i_1} \cdots r_{i_m}$. Let $f(n)$ be a mapping from $\mathbb{N}$ to $\mathbb{N}$ and let $|P|$ denote the sum of the lengths of all words in the PCP P. A solution $i_1, \ldots, i_m$ for P is called an $f(n)$-solution iff $m \leq f(|P|)$. With $f(n)$-PCP, we denote the version of the PCP that admits only $f(n)$-solutions.*

Analogous to the undecidability result for the general PCP given by Hopcroft and Ullman in [9], we may prove the following result.

Theorem 1. *It is* NExpTime-*complete to decide whether a* $2^n + 1$-*PCP has a solution.*

Hence, a reduction of the $2^n + 1$-PCP is a candidate for proving NExpTime lower bounds for Description Logics with concrete domains. As we will see now, the problem is in fact well-suited for this task since it is possible to define an appropriate concrete domain. It follows from the proof of the above theorem that it is sufficient to consider some fixed, finite alphabet Σ_U whose cardinality is the number of symbols needed to define a universal Turing machine.

Definition 9 (Concrete Domain $\mathcal{P}$). *The concrete domain $\mathcal{P}$ is defined by setting $\Delta_{\mathcal{P}} := \Sigma_U^*$ and defining $\Phi_{\mathcal{P}}$ as the smallest set containing the following predicates:*

- *unary predicates word and nword with $word^{\mathcal{P}} = \Delta_{\mathcal{P}}$ and $nword^{\mathcal{P}} = \emptyset$,*
- *unary predicates $=_\epsilon$ and $\neq_\epsilon$ with $=_\epsilon^{\mathcal{P}} = \{\epsilon\}$ and $\neq_\epsilon^{\mathcal{P}} = \Sigma_U^+$,*
- *a binary equality predicate $=$ and a binary inequality predicate $\neq$, and*
- *for each $w \in \Sigma_U^+$, two binary predicates $conc_w$ and $nconc_w$ with*
$$conc_w^{\mathcal{P}} = \{(u, v) \mid v = uw\} \text{ and } nconc_w^{\mathcal{P}} = \{(u, v) \mid v \neq uw\}.$$

The complexity of reasoning with a Description Logic providing a concrete domain $\mathcal{D}$ does obviously depend on the complexity of reasoning with $\mathcal{D}$. More precisely, most satisfiability algorithms involve checking the satisfiability of finite conjunctions of concrete domain predicates

$$\bigwedge_{1 \leq i \leq k} (x_0^{(i)}, \ldots, x_{n_i}^{(i)}) : P_i,$$

where each P_i is an n_i-ary predicate and the $x_j^{(i)}$ are variables from some fixed set [1]. This is also the case for the tableau algorithm that used to prove the upper bound in Section 7. Hence, we are interested in the complexity of this task which is called $\mathcal{D}$-*satisfiability* in what follows. By devising an algorithm that is based on repeated normalization combined with tests for obvious inconsistencies, the following result can be obtained.

Proposition 1. *$\mathcal{P}$-satisfiability is decidable in deterministic polynomial time.*

On first sight, the concrete domain $\mathcal{P}$ may look somewhat unnatural in the context of knowledge representation. However, it is straightforward to encode words as natural numbers and to define the operations on words as rather simple operations on the naturals [2]: Words over the alphabet Σ_U can be interpreted as numbers written at base $|\Sigma_U| + 1$ (assuming that the empty word represents 0); the concatenation of two words v and w can then be expressed as $vw = v * (|\Sigma_U| + 1)^{|w|} + w$, where $|w|$ denotes the length of the word w. Hence, each concrete domain (Δ, Φ), where Δ contains the natural numbers and Φ contains predicates for (in)equality, (in)equality to zero, addition, and multiplication may also be used for the reductions. A concrete domain with these properties is called *arithmetic*.

$$Ch[u_1, u_2, u_3, u_4] = (\exists(u_1, u_2). = \sqcap \exists(u_3, u_4). =)$$

$$\sqcup \bigsqcup_{(\ell_i, r_i) \text{ in } P} (\exists(u_1, u_2).conc_{\ell_i} \sqcap \exists(u_3, u_4).conc_{r_i})$$

$$C_0 \doteq \exists\ell.C_1 \sqcap \exists r.C_1$$

$$\sqcap Ch[\ell r^{n-1}g_\ell, r\ell^{n-1}g_\ell, \ell r^{n-1}g_r, r\ell^{n-1}g_r]$$

$$\vdots$$

$$C_{n-2} \doteq \exists\ell.C_{n-1} \sqcap \exists r.C_{n-1}$$

$$\sqcap Ch[\ell r g_\ell, r\ell g_\ell, \ell r g_r, r\ell g_r]$$

$$C_{n-1} \doteq Ch[\ell g_\ell, r g_\ell, \ell g_r, r g_r]$$

$$C_P \doteq C_0$$

$$\sqcap \exists\ell^n g_\ell. =_\epsilon \sqcap \exists\ell^n g_r. =_\epsilon$$

$$\sqcap \exists r^n y.\exists g_\ell, g_r. = \sqcap \exists r^n y g_\ell. \neq_\epsilon$$

$$\sqcap Ch[r^n g_\ell, r^n x g_\ell, r^n g_r, r^n x g_r]$$

$$\sqcap Ch[r^n x g_\ell, r^n y g_\ell, r^n x g_r, r^n y g_r]$$

Fig. 1. The $\mathcal{ALC}(\mathcal{P})$ reduction TBox $\mathcal{T}_P$ $(n = |P|)$.

4 Satisfiability of $\mathcal{ALC}(\mathcal{P})$-Concepts w.r.t. TBoxes

In this section, we show that the satisfiability of $\mathcal{ALC}(\mathcal{P})$-concepts w.r.t. TBoxes is NExpTime-hard. As already mentioned, this result is rather surprising since (1) satisfiability of $\mathcal{ALC}(\mathcal{D})$-concepts without reference to TBoxes is known to be PSpace-complete if reasoning with the concrete domain $\mathcal{D}$ is in PSpace [14], and (2) admitting acyclic TBoxes does "usually" not increase the complexity of reasoning [13].

The proof is by a reduction of the $2^n + 1$-PCP using the concrete domain $\mathcal{P}$ introduced in the previous section. Given a $2^n + 1$-PCP $P = (\ell_1, r_1), \ldots, (\ell_k, r_k)$, we define a TBox $\mathcal{T}_P$ of size polynomial in $|P|$ and a concept (name) C_P such that C_P is satisfiable w.r.t. $\mathcal{T}_P$ iff P has a solution. Figure 1 contains the reduction TBox and Figure 2 an example model for $|P| = 2$. In the figures, ℓ, r, x, and y denote abstract features and g_ℓ and g_r denote concrete features. The first equality in Figure 1 is not a concept definition but an abbreviation: Replace every occurrence of $Ch[u_1, u_2, u_3, u_4]$ in the lower three concept definitions by the right-hand side of the first identity substituting $u_1, \ldots, u_4$ appropriately.

The idea behind the reduction is to define $\mathcal{T}_P$ such that models of C_P and $\mathcal{T}_P$ have the form of a binary tree of depth $|P|$ whose leaves are connected by two "chains" of $conc_w$ predicates. Pairs of corresponding objects (x_i, y_i) on the chains represent partial solutions of the PCP P. More precisely, the first line of the definitions of the $C_0, \ldots, C_{n-1}$ concepts ensures that models have the form of a binary tree of depth n (with $n = |P|$) whose left edges are labeled with the abstract feature ℓ and whose right edges are labeled with the abstract feature r. Let the abstract objects $a_{n,0}, \ldots a_{n,2^n-1}$ be the leaves of this tree. By the second line of the definitions of the $C_0, \ldots, C_{n-1}$ concepts, every $a_{n,i}$ has

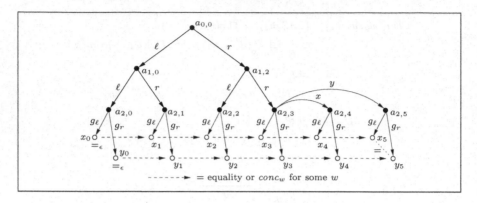

Fig. 2. An example model of C_P and T_P for $n = 2$.

a g_ℓ-successor x_i and a g_r-successor y_i. These second lines also ensure that the x_i and y_i objects are connected via two predicate chains, where the predicates on the chains are either equality or $conc_w$. More precisely, for $0 \leq i < 2^n - 1$, either $x_i = x_{i+1}$ and $y_i = y_{i+1}$, or there exists a $j \in \{1, \ldots, k\}$ such that $(x_i, x_{i+1}) \in conc_{\ell_j}^P$ and $(y_i, y_{i+1}) \in conc_{r_j}^P$. Furthermore, by the second line of the definition of C_P, we have $x_1 = y_1 = \epsilon$. Hence, pairs (x_i, y_i) are partial solutions for P. Since we must consider solutions of a length up to $2^n + 1$, the 2^n objects on the fringe of the tree with their $2^n - 1$ connecting predicate edges are not sufficient, and we need to "add" two more objects $a_{n,2^n}$ and $a_{n,2^n+1}$ which behave analogously to the objects $a_{n,0}, \ldots a_{n,2^n-1}$. This is done by the last two lines of the definition of C_P. Finally, the third line of the definition of C_P ensures that $x_{2^n+1} = y_{2^n+1} \neq \epsilon$ and hence that (x_{2^n+1}, y_{2^n+1}) is in fact a full solution.

Obviously, the size of T_P is polynomial in $|P|$ and T_P can be constructed in time polynomial in $|P|$ which, together with the fact that $\mathcal{P}$ may be replaced by any arithmetic concrete domain, yields the following theorem.

Theorem 2. *For every arithmetic concrete domain $\mathcal{D}$, satisfiability of $\mathcal{ALC}(\mathcal{D})$-concepts w.r.t TBoxes is* NExpTime-*hard.*

We also obtain a lower bound for subsumption since satisfiability can be reduced to subsumption. With some slight modifications, the reduction just presented can also be applied to the Description Logic $\mathcal{ALCR}(\mathcal{P})$, i.e., $\mathcal{ALC}(\mathcal{P})$ enriched with a role conjunction constructor [6]. Hence, reasoning with this logic is also NExpTime-hard. The corresponding reduction concept can be found in [16].

One may ask why we are interested in the relatively weak acyclic TBoxes instead of using a more general TBox formalism. The answer is that using general TBoxes leads to undecidability.

Definition 10 (General TBox). *A* general concept inclusion (GCI) *has the form $C \sqsubseteq D$, where both C and D are concepts. An interpretation $\mathcal{I}$ is a model for a GCI $C \sqsubseteq D$ iff $C^{\mathcal{I}} \subseteq D^{\mathcal{I}}$. Finite sets of GCIs are called* general TBoxes.

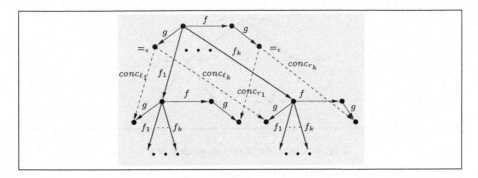

Fig. 3. An example model of C_P w.r.t. $\mathcal{T}_P$

An interpretation $\mathcal{I}$ is a model for a general TBox $\mathcal{T}$ iff $\mathcal{I}$ is a model for all GCIs in $\mathcal{T}$.

Using the concrete domain $\mathcal{P}$ and a reduction of the general PCP, the following theorem can be obtained.

Theorem 3. *For every arithmetic concrete domain $\mathcal{D}$, satisfiability of $\mathcal{ALC}(\mathcal{D})$-concepts w.r.t. general TBoxes is undecidable.*

Proof Let P be an instance of the PCP. Define a concept C_P and a general TBox $\mathcal{T}_P$ as follows:

$$C_P := \exists g. =_\epsilon \sqcap \exists fg. =_\epsilon$$
$$\mathcal{T}_P := \{\exists f.\top \sqsubseteq \bigsqcap_{(\ell_i, r_i) \in P} \exists g, f_i g.conc_{\ell_i} \sqcap \exists fg, f_i fg.conc_{r_i}$$
$$\top \sqsubseteq \exists g. =_\epsilon \sqcup \neg \exists g, fg.=\}$$

An example model of C_P w.r.t. $\mathcal{T}_P$ can be found in Figure 3. The first GCI ensures that models of C_P and $\mathcal{T}_P$ represent *all* possible solutions of the PCP P. Additionally, the last GCI ensures that no potential solution is a solution. It is hence straightforward to prove that C_P is satisfiable w.r.t. $\mathcal{T}_P$ iff P has no solution, i.e., we have reduced the general, undecidable PCP [17,9] to the satisfiability of $\mathcal{ALC}(\mathcal{D})$-concepts w.r.t. general TBoxes.   ❑

5   Satisfiability of $\mathcal{ALCI}(\mathcal{P})$-Concepts

We now show that satisfiability of $\mathcal{ALCI}(\mathcal{P})$-concepts—without reference to TBoxes—is NExpTime-hard. As in the previous section, it is surprising that a rather small change in the logic, i.e., adding inverse roles, causes a dramatic increase in complexity.

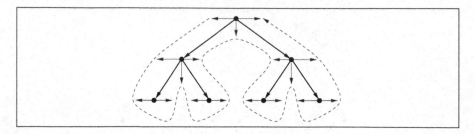

Fig. 4. Predicate chains in models of C_P.

The reduction is similar to the one used in the previous section: it is a reduction of the $2^n + 1$-PCP and uses the concrete domain $\mathcal{P}$. However, we need a slightly different strategy since, in the case of inverse roles, it is not possible to enforce chains of predicates connecting the leaves of the tree. Instead, the predicate chains emulate the structure of the tree following the scheme indicated in Figure 4. Given a PCP $P = (\ell_1, r_1), \ldots, (\ell_k, r_k)$, we define a concept C_P of size polynomial in $|P|$ which has a model iff P has a solution. The concept C_P can be found in Figure 5. In the figure, $h_\ell, h_r, x_\ell, x_r, y_\ell, y_r, z_\ell,$ and z_r are concrete features. Note that the equalities are not concept definitions but abbreviations. As in the previous section, replace every occurrence of $Ch[u_1, u_2, u_3, u_4]$ in the lower three concept definitions by the right-hand side of the first identity substituting $u_1, \ldots, u_4$ appropriately and similarly for every occurrence of X.

Let us discuss the structure of models of C_P. Due to the first line in the definition of C_P and the $\exists f^-$ quantifiers in the definition of X, models of C_P have the form of a tree of depth $|P|-1$ in which all edges are labeled with f^-. This edge labelling scheme is possible since the inverse of an abstract feature is not a feature. Additionally, we establish two chains of concrete domain predicates as indicated in Figure 4. Again, corresponding objects on the two chains represent partial solutions of the PCP P. A more detailed clipping from a model of C_P can be found in Figure 6. The existence of the chains is ensured by the definition of X and the second line in the definition of C_P: The concept X establishes the edges of the predicate chains as depicted in Figure 6 (in fact, Figure 6 is a model of the concept X) while the second line of C_P establishes the edges "leading around" the leaves. Edges of the latter type and the dotted edges in Figure 6 are labeled with the equality predicate. To see why this is the case, let us investigate the length of the chains.

The length of the two predicate chains is twice the number of edges in the tree plus the number of leaves, i.e., $2 * (2^{|P|} - 2) + 2^{|P|-1}$. To eliminate the factor 2 and the summand $2^{|P|-1}$, C_P is defined such that every edge in the predicate chains leading "up" in the tree and every edge "leading around" a leaf is labeled with the equality predicate. To extend the chains to length $2^{|P|} + 1$, we need to add three additional edges (definition of C_P, lines three, four, and five). Finally, the last two lines in the definition of C_P ensure that the first concrete object on

$$Ch[u_1, u_2, u_3, u_4] = (\exists(u_1, u_2). = \sqcap \exists(u_3, u_4). =)$$
$$\sqcup \bigsqcup_{(\ell_i, r_i) \text{ in } P} \exists(u_1, u_2).conc_{\ell_i} \sqcap \exists(u_3, u_4).conc_{r_i}$$

$$X = \exists f^-.(Ch[fg_\ell, g_\ell, fg_r, g_r] \sqcap \exists(h_\ell, fp_\ell). = \sqcap \exists(h_r, fp_r). =)$$
$$\sqcap \exists f^-.(Ch[fp_\ell, g_\ell, fp_r, g_r] \sqcap \exists(h_\ell, fh_\ell). = \sqcap \exists(h_r, fh_r). =)$$

$$C_P = X \sqcap \forall f^-.X \sqcap \cdots \sqcap \forall(f^-)^{n-1}.X$$
$$\sqcap \forall(f^-)^n.(\exists(g_\ell, h_\ell). = \sqcap \exists(g_r, h_r). =)$$
$$\sqcap Ch[h_\ell, x_\ell, h_r, x_r]$$
$$\sqcap Ch[x_\ell, y_\ell, x_r, y_r]$$
$$\sqcap Ch[y_\ell, z_\ell, y_r, z_r]$$
$$\sqcap \exists g_\ell, =_\epsilon \sqcap \exists g_r, =_\epsilon$$
$$\sqcap \exists z_\ell, z_r. = \sqcap \exists z_\ell. \neq_\epsilon$$

Fig. 5. The $\mathcal{ALCI}(\mathcal{P})$ reduction concept C_P ($n = |P| - 1$).

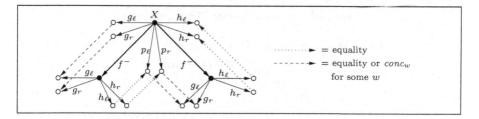

Fig. 6. A clipping from a model of C_P.

both chains represents the empty word and that the last objects on the chains represent a (non-empty) solution for P.

Theorem 4. *For every arithmetic concrete domain $\mathcal{D}$, satisfiability of $\mathcal{ALCI}(\mathcal{D})$-concepts is NExpTime-hard.*

6 Satisfiability of $\mathcal{ALCRP}(\mathcal{P})$-Concepts

In this section, we prove that satisfiability of $\mathcal{ALCRP}(\mathcal{P})$-concepts without reference to TBoxes is NExpTime-hard. Hence, adding the role-forming concrete domain constructor yields another extension of $\mathcal{ALC}(\mathcal{D})$ in which reasoning is much harder than in $\mathcal{ALC}(\mathcal{D})$ itself.

Given a PCP $P = (\ell_1, r_1), \ldots, (\ell_k, r_k)$, we define a concept C_P of size polynomial in $|P|$ which has a model iff P has a solution. The concept C_P can be found in Figure 7, where x and y denote abstract features and p denotes a predicate (written in lowercase to avoid confusion with the PCP P). Again, the equalities

$$DistB[k] = \bigcap_{i=0}^{k}((B_i \rightarrow \forall R.B_i) \sqcap \neg B_i \rightarrow \forall R.\neg B_i)$$

$$Tree = \exists R.B_0 \sqcap \exists R.\neg B_0$$
$$\sqcap \forall R.(DistB[0] \sqcap \exists R.B_1 \sqcap \exists R.\neg B_1)$$
$$\vdots$$
$$\sqcap \forall R^{n-1}.(DistB[n-1] \sqcap \exists R.B_{n-1} \sqcap \exists R.\neg B_{n-1})$$

$$S[g,p] = \exists(g),(g).\overline{p}$$

$$Edge[g,p] = \left(\bigsqcup_{k=0}^{n-1}\left(\bigsqcup_{j=0}^{k-1}\neg B_j\right) \sqcap (B_k \rightarrow \forall S[g,p].\neg B_k) \sqcap (\neg B_k \rightarrow \forall S[g,p].B_k)\right.$$
$$\left.\sqcup \bigsqcup_{k=0}^{n-1}\left(\bigcap_{j=0}^{k-1}B_j\right) \sqcap (B_k \rightarrow \forall S[g,p].B_k) \sqcap (\neg B_k \rightarrow \forall S[g,p].\neg B_k)\right)$$

$$DEdge = (Edge[g_\ell, =] \sqcap Edge[g_r, =]) \sqcup$$
$$\bigsqcup_{(\ell_i, r_i)\ \text{in}\ P}(Edge[g_\ell, conc_{\ell_i}] \sqcap Edge[g_r, conc_{r_i}])$$

$$Ch[u_1, u_2, u_3, u_4] = (\exists(u_1, u_2). = \sqcap \exists(u_3, u_4). =)$$
$$\sqcup \bigsqcup_{(\ell_i, r_i)\ \text{in}\ P}(\exists(u_1, u_2).conc_{\ell_i} \sqcap \exists(u_3, u_4).conc_{r_i})$$

$$C_P = Tree \sqcap \forall R^n.\exists g_\ell.word \sqcap \forall R^n.\exists g_r.word$$
$$\sqcap \forall R^n.[(\neg B_0 \sqcap \cdots \sqcap \neg B_{n-1}) \rightarrow (\exists g_\ell. =_\epsilon \sqcap \exists g_r. =_\epsilon)$$
$$\sqcap \neg(B_0 \sqcap \cdots \sqcap B_{n-1}) \rightarrow DEdge$$
$$\sqcap (B_0 \sqcap \cdots \sqcap B_{n-1}) \rightarrow$$
$$\left(Ch(g_\ell, xg_\ell, g_r, xg_r) \sqcap Ch(xg_\ell, yg_\ell, xg_r, yg_r)\right)]$$

Fig. 7. The $\mathcal{ALCRP}(\mathcal{P})$ reduction concept C_P ($n = |P|$).

in the figure serve as abbreviations. Moreover, we use $C \rightarrow D$ as an abbreviation for $\neg C \sqcup D$. Note that $S[g,p]$ denotes a predicate role and not a concept, i.e., $S[g,p]$ is an abbreviation for the role-forming concrete domain constructor $\exists(g),(g).\overline{p}$.

Figure 8 contains an example model of C_P with $|P| = n = 2$. Obviously, the models of C_P are rather similar to the ones from the $\mathcal{ALC}(\mathcal{D})$ reduction in Section 4: models have the form of a binary tree of depth n whose edges are labelled with the role R and whose leaves (together with two "extra" nodes) are connected by two predicate chains of length $2^n + 1$. The $Tree$ concept enforces the existence of the binary tree. The concept names $B_0, \ldots, B_{n-1}$ are used for a binary numbering (from 0 to $2^n - 1$) of the leaves of the tree. More precisely, for a domain object $a \in \Delta^{\mathcal{I}}$, set

$$pos(a) = \Sigma_{i=0}^{n-1}\beta_i(a) * 2^i \quad \text{where} \quad \beta_i(a) = \begin{cases} 1 \text{ if } a \in B_i^{\mathcal{I}} \\ 0 \text{ otherwise.} \end{cases}$$

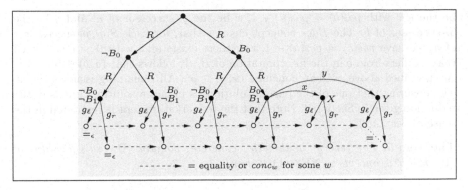

Fig. 8. An example model of C_P with $|P| = 2$.

The *Tree* and *DistB* concepts ensure that, if two leaves a and a' are reachable via different paths from the root node, then we have $pos(a) \neq pos(a')$. Due to the first line of the C_P concept, every leaf has (concrete) g_ℓ- and g_r-successors. The last two lines of C_P guarantee the existence of the two extra nodes which are connected by predicate edges due to the use of the *Ch* concepts. Hence, it remains to describe how the edges between the leaf nodes are established.

There are two main ideas underlying the establishment of these edges: (i) use the role-forming predicate constructor to establish single edges and (ii) use the position $pos()$ of leaf nodes together with the fact that counting modulo 2^n can be expressed by $\mathcal{ALC}$-concepts to do this with a concept of size polynomial in $|P|$. We first illustrate Point (i) in an abstract way. Assume that we have two abstract objects a and b, a has g_ℓ-successor x and b has g_ℓ-successor y. Moreover, let $b \in X^\mathcal{I}$ for some concept X. We may then establish a p-edge (for some binary predicate $p \in \Phi_P$) between x and y as follows: we enforce that $a \in (\forall S[g_\ell, p].\neg X)^\mathcal{I}$; since $b \in X^\mathcal{I}$, it follows that $(a, b) \notin S[g_\ell, p]^\mathcal{I}$, i.e., $(a, b) \notin (\exists(g_\ell), (g_\ell).\overline{p})^\mathcal{I}$ and thus $(x, y) \notin \overline{p}^\mathcal{P}$, which obviously implies that $(x, y) \in p^\mathcal{P}$.

In the third line of the C_P-concept, the *DEdge* concept is used to establish edges between the leaf nodes. The *DEdge* concept itself is just a disjunction over the various edge types while the *Edge* concept actually establishes the edges. In principle, the *Edge* concept establishes the edges as described above. However, it does this not only for two fixed nodes as in the description above but for *all* neighboring leaf nodes. To see how this is achieved, note that *Edge* is essentially the negation of the well-known propositional formula

$$\bigwedge_{k=0}^{n-1} (\bigwedge_{j=0}^{k-1} x_j = 1) \to (x_k = 1 \leftrightarrow x_k' = 0) \; \wedge \; \bigwedge_{k=0}^{n-1} (\bigvee_{j=0}^{k-1} x_j = 0) \to (x_k = x_k')$$

which encodes incrementation modulo 2^n, i.e., if t is the number (binarly) encoded by the propositional variables $x_0, \ldots, x_{n-1}$ and t' is the number encoded by the propositional variables $x_0', \ldots, x_{n-1}'$, then we have $t' = t + 1$ modulo 2^n, c.f. [4]. Assume $a \in (Edge[g_\ell, p])^\mathcal{I}$ (where p is either "$=$" or $conc_{\ell_i}$) and let b

be the leaf with $pos(b) = pos(a) + 1$, x be the g_ℓ-successor of a, and y be the g_ℓ-successor of b. The *Edge* concept ensures that, for each $S[g_\ell, p]$-successor c of a, we have $pos(c) \neq pos(a) + 1$, i.e., there exists an i with $0 \leq i \leq n$ such that c differs from b in the interpretation of B_i. It follows that $(a, b) \notin S[g_\ell, p]^\mathcal{I}$. As described above, we can conclude $(x, y) \in p^\mathcal{I}$. All remaining issues such as, e.g., ensuring that one of the partial solutions is in fact a solution, are as in the reduction given in Section 4. Note that the reduction concept is restricted in the sense of Section 2.

Theorem 5. *For every arithmetic concrete domain $\mathcal{D}$, satisfiability of $\mathcal{ALCRP}(\mathcal{D})$-concepts is NExpTime-hard.*

7   Upper Bounds

Due to space limitations, we can only give a short sketch of the proof of the upper bound and refer to [16] for details. First, a tableau algorithm for deciding the satisfiability of $\mathcal{ALCRPI}(\mathcal{D})$-concepts without reference to TBoxes is devised. This algorithm combines techniques from [8] for reasoning with $\mathcal{ALCRP}(\mathcal{D})$ with techniques from [10] for reasoning with inverse role. Second, the tableau algorithm is modified to take into account TBoxes by performing "on the fly unfolding" of the TBox as described in [13]. A complexity analysis yields the following theorem.

Theorem 6. *If $\mathcal{D}$-satisfiability is in NP, satisfiability of $\mathcal{ALCRPI}(\mathcal{D})$-concepts w.r.t. TBoxes can be decided in nondeterministic exponential time.*

This also gives an upper bound for subsumption since, as mentioned in Section 2, subsumption can be reduced to satisfiability. It should be noted that the above theorem only applies to so-called *admissible* concrete domains, where a concrete domain $\mathcal{D}$ is admissible if the set $\Phi_\mathcal{D}$ if closed under negation and contains a predicate name $\top_\mathcal{D}$ for $\Delta_\mathcal{D}$ [16]. Nevertheless, the given theorem captures a large class of interesting concrete domains such as $\mathcal{P}$ itself and concrete domains for temporal and spatial reasoning [8,15]. In contrast to the upper bound for $\mathcal{ALC}(\mathcal{D})$ established in [14], the above theorem if concerned with concrete domains for which $\mathcal{D}$-satisfiability is in NP instead of in PSPACE. For concrete domains of this latter type, the tableau algorithm in [16] yields an ExpSpace upper bound. A matching lower bound, however, is yet to be proved.

8   Related and Future Work

We demonstrated that the PSPACE upper bound for $\mathcal{ALC}(\mathcal{D})$-concept satisfiability is not robust: complexity shifts to NExpTime if seemingly harmless constructors are added and is even undecidable if we admit general TBoxes. However, the situation is not hopeless in all cases. Although the class of arithmetic concrete domains is quite large and captures many interesting concrete domains, there still exist non-trivial concrete domains for which reasoning with general TBoxes is decidable and the NExpTime lower bound obtained in this paper do presumably not hold. An example is presented in [15], where a temporal Description Logic based on concrete domains is defined.

As future work, it would be interesting to extend the obtained logics by additional means of expressivity such as transitive roles and qualifying number restrictions [11]. There are at least two ways to go: In [14] it is proved that reasoning with $\mathcal{ALCF}(\mathcal{D})$, i.e., the extension of $\mathcal{ALC}(\mathcal{D})$ with feature agreements and disagreements, is PSpace-complete (if reasoning with $\mathcal{D}$ is in PSpace). Hence, one could define extensions of $\mathcal{ALCF}(\mathcal{D})$ trying to obtain an expressive logic for which reasoning is still in PSpace. The second approach is to define extensions of $\mathcal{ALCI}(\mathcal{D})$ which means that the obtained logics are at least NExpTime-hard. Moreover, feature (dis)agreements—which are very closely related to concrete domains—cannot be considered since, in [16], we prove that the combination of inverse roles and feature (dis)agreements leads to undecidability.

Acknowledgements. My thanks go to Franz Baader, Ulrike Sattler, and Stephan Tobies for inspiring discussions. The work in this paper was supported by the DFG Project BA1122/3-1 "Combinations of Modal and Description Logics".

References

1. F. Baader and P. Hanschke. A scheme for integrating concrete domains into concept languages. In *Proc. of IJCAI-91*, pp. 452–457, Sydney, Australia, 1991. Morgan Kaufmann Publ., 1991.

2. F. Baader and P. Hanschke. Extensions of concept languages for a mechanical engineering application. In *Proc. of GWAI-92*, volume 671 of LNCS, pp. 132–143, Bonn (Germany), 1993. Springer–Verlag.

3. F. Baader and U. Sattler. Description logics with concrete domains and aggregation. In *Proc. of ECAI-98*, Brighton, August 23–28, 1998. John Wiley & Sons, New York, 1998.

4. E. Börger, E. Grädel, and Y. Gurevich. *The Classical Decision Problem*. Perspectives in Mathematical Logic. Springer-Verlag, Berlin, 1997.

5. D. Calvanese, G. De Giacomo, M. Lenzerini, and D. Nardi. Reasoning in expressive description logics. In *Handbook of Automated Reasoning*. Elsevier Science Publishers (North-Holland), Amsterdam, 1998.

6. F. Donini, M. Lenzerini, D. Nardi, and W. Nutt. The complexity of concept languages. DFKI Research Report RR-95-07, German Research Center for Artificial Intelligence, Kaiserslautern, 1995.

7. V. Haarslev, C. Lutz, and R. Möller. Defined topological relations in description logics. In *Proc. of KR'98*, pp. 112–124, Trento, Italy, 1998. Morgan Kaufmann Publ., 1998.

8. V. Haarslev, C. Lutz, and R. Möller. A description logic with concrete domains and role-forming predicates. *Journal of Logic and Computation*, 9(3), 1999.

9. J. E. Hopcroft and J. D. Ullman. *Introduction to Automata Theory, Languages and Computation*. Addison-Wesley, Reading, Mass., 1979.

10. I. Horrocks and U. Sattler. A description logic with transitive and inverse roles and role hierarchies. *Journal of Logic and Computation*, 9(3):385–410, 1999.

60   C. Lutz

11. I. Horrocks, U. Sattler, and S. Tobies. Practical reasoning for expressive description logics. In *Proc. of LPAR'99*, number 1705 in LNAI, pp. 161–180. Springer-Verlag, 1999.
12. G. Kamp and H. Wache. CTL - a description logic with expressive concrete domains. Technical Report LKI-M-96/01, Labor für Künstliche Intelligenz, Universität Hamburg, Germany, 1996.
13. C. Lutz. Complexity of terminological reasoning revisited. In *Proc. of LPAR'99*, number 1705 in LNAI, pp. 181–200. Springer-Verlag, 1999.
14. C. Lutz. Reasoning with concrete domains. In *Proc. of IJCAI-99*, pp. 90–95, Stockholm, Sweden, July 31 – August 6, 1999. Morgan-Kaufmann Publishers.
15. C. Lutz. Interval-based temporal reasoning with general TBoxes. In *Proc. of IJCAI-01*, Seattle, 2001.
16. C. Lutz. NExpTime-complete description logics with concrete domains. LTCS-Report 00-01, LuFG Theoretical Computer Science, RWTH Aachen, Germany, 2000. See http://www-lti.informatik.rwth-aachen.de/Forschung/Reports.html.
17. E. M. Post. A variant of a recursively unsolvable problem. *Bull. Am. Math. Soc.*, 52:264–268, 1946.

Exploiting Pseudo Models for TBox and ABox Reasoning in Expressive Description Logics

Volker Haarslev[1], Ralf Möller[1], and Anni-Yasmin Turhan[2]

[1] University of Hamburg
Computer Science Department
Vogt-Kölln-Str. 30
22527 Hamburg, Germany
[2] RWTH Aachen
LuFG Theoretical Computer Science
Ahornstr. 55
52074 Aachen, Germany

Abstract. This paper investigates optimization techniques and data structures exploiting the use of so-called *pseudo models*. These techniques are applied to speed up TBox and ABox reasoning for the description logics $\mathcal{ALCNH}_{R^+}$ and $\mathcal{ALC}(\mathcal{D})$. The advances are demonstrated by an empirical analysis using the description logic system RACE that implements TBox and ABox reasoning for $\mathcal{ALCNH}_{R^+}$.

1 Introduction

We introduce and analyze optimization techniques for reasoning in expressive description logics exploiting so-called pseudo models. The new techniques being investigated are called *deep model merging* and *individual model merging*. The presented algorithms are empirically evaluated using TBoxes and ABoxes derived from actual applications. The model merging technique is also developed for the logic $\mathcal{ALC}(\mathcal{D})$ [1] which supports so-called concrete domains. This is motivated by a proposal which extends $\mathcal{ALCNH}_{R^+}$ with a restricted form of concrete domains [4].

1.1 The Language $\mathcal{ALCNH}_{R^+}$

We briefly introduce the description logic (DL) $\mathcal{ALCNH}_{R^+}$ [3] (see the tables in Figure 1) using a standard Tarski-style semantics based on an interpretation $\mathcal{I} = (\Delta^{\mathcal{I}}, \cdot^{\mathcal{I}})$ $\mathcal{ALCNH}_{R^+}$ extends the basic description logic $\mathcal{ALC}$ by role hierarchies, transitively closed roles, and number restrictions. Note that the combination of transitive roles and role hierarchies implies the expressiveness of so-called general inclusion axioms (GCIs). The language definition is slightly extended compared to the one given in [3] since we additionally support the declaration of "native" features. This allows additional optimizations, e.g. an efficient treatment of features by the model merging technique (see below). The concept name $\top$ ($\bot$) is used as an abbreviation for $C \sqcup \neg C$ ($C \sqcap \neg C$). We assume

R. Goré, A. Leitsch, and T. Nipkow (Eds.): IJCAR 2001, LNAI 2083, pp. 61–75, 2001.

(a)

Syntax	Semantics
Concepts	
A	$A^\mathcal{I} \subseteq \Delta^\mathcal{I}$
$\neg C$	$\Delta^\mathcal{I} \setminus C^\mathcal{I}$
$C \sqcap D$	$C^\mathcal{I} \cap D^\mathcal{I}$
$C \sqcup D$	$C^\mathcal{I} \cup D^\mathcal{I}$
$\exists R.C$	$\{a \in \Delta^\mathcal{I} \mid \exists b \in \Delta^\mathcal{I} : (a,b) \in R^\mathcal{I}, b \in C^\mathcal{I}\}$
$\forall R.C$	$\{a \in \Delta^\mathcal{I} \mid \forall b : (a,b) \in R^\mathcal{I} \Rightarrow b \in C^\mathcal{I}\}$
$\exists_{\geq n} S$	$\{a \in \Delta^\mathcal{I} \mid \|\{b \in \Delta^\mathcal{I} \mid (a,b) \in S^\mathcal{I}\}\| \geq n\}$
$\exists_{\leq m} S$	$\{a \in \Delta^\mathcal{I} \mid \|\{b \in \Delta^\mathcal{I} \mid (a,b) \in S^\mathcal{I}\}\| \leq m\}$
Roles	
R	$R^\mathcal{I} \subseteq \Delta^\mathcal{I} \times \Delta^\mathcal{I}$

(b)

Terminol. Axioms	
Syntax	Satisfied if
$R \in T$	$R^\mathcal{I} = (R^\mathcal{I})^+$
$F \in F$	$\Delta^\mathcal{I} \subseteq (\exists_{\leq 1} F)^\mathcal{I}$
$R \sqsubseteq S$	$R^\mathcal{I} \subseteq S^\mathcal{I}$
$C \sqsubseteq D$	$C^\mathcal{I} \subseteq D^\mathcal{I}$

(c)

Assertions	
Syntax	Satisfied if
$a : C$	$a^\mathcal{I} \in C^\mathcal{I}$
$(a,b) : R$	$(a^\mathcal{I}, b^\mathcal{I}) \in R^\mathcal{I}$

Fig. 1. Syntax and Semantics of $\mathcal{ALCNH}_{R^+}$ ($n, m \in \mathbb{N}$, $n > 0$, $\| \cdot \|$ denotes set cardinality, and $S \in S$).

a set of concept names C, a set of role names R, and a set of individual names O. The mutually disjoint subsets F, P, T of R denote features, non-transitive, and transitive roles, respectively ($R = F \cup P \cup T$).

If $R, S \in R$ are role names, then the terminological axiom $R \sqsubseteq S$ is called a *role inclusion axiom*. A *role hierarchy* $\mathcal{R}$ is a finite set of role inclusion axioms. Then, we define $\sqsubseteq^*$ as the reflexive transitive closure of $\sqsubseteq$ over such a role hierarchy $\mathcal{R}$. Given $\sqsubseteq^*$, the set of roles $R^\downarrow = \{S \in R \mid S \sqsubseteq^* R\}$ defines the *descendants* of a role R. $R^\uparrow = \{S \in R \mid R \sqsubseteq^* S\}$ is the set of *ancestors* of a role R. We also define the set $S = \{R \in P \mid R^\downarrow \cap T = \emptyset\}$ of *simple* roles that are neither transitive nor have a transitive role as descendant. Every descendant G of a feature F must be a feature as well ($G \in F$).

A syntactic restriction holds for the combinability of number restrictions and transitive roles in $\mathcal{ALCNH}_{R^+}$. Number restrictions are only allowed for *simple* roles. This restriction is motivated by an undecidability result in case of an unrestricted combinability [8].

If C and D are concept terms, then $C \sqsubseteq D$ (*generalized concept inclusion* or *GCI*) is a terminological axiom. A finite set of terminological axioms $\mathcal{T_R}$ is called a *terminology* or *TBox* w.r.t. to a given role hierarchy $\mathcal{R}$.[1]

An *ABox* $\mathcal{A}$ is a finite set of assertional axioms as defined in Figure 1c. The set O of object names is divided into two disjoint subsets, O_O and O_N.[2] An initial ABox $\mathcal{A}$ may contain only assertions mentioning old individuals (from O_O). Every individual name from O is mapped to a single element of $\Delta^\mathcal{I}$ in a way such that for $a, b \in O_O$, $a^\mathcal{I} \neq b^\mathcal{I}$ if $a \neq b$ (*unique name assumption* or *UNA*). This ensures that different individuals in O_O are interpreted as different elements. The UNA does not hold for elements of O_N, i.e. for $a, b \in O_N$, $a^\mathcal{I} = b^\mathcal{I}$ may hold even if $a \neq b$, or if we assume without loss of generality that $a \in O_N, b \in O_O$.

---

[1] The reference to $\mathcal{R}$ is omitted in the following.

[2] The set of "old" individuals names characterizes all individuals for which the unique name assumption holds while the set of "new" names denotes individuals which are constructed during a proof.

The *ABox consistency problem* is to decide whether a given ABox $\mathcal{A}$ is consistent w.r.t. a TBox $\mathcal{T}$. Satisfiability of concept terms can be reduced to ABox consistency as follows: A concept term C is *satisfiable* iff the ABox $\{a\!:\!C\}$ is consistent. The *instance problem* is to determine whether an individual a is an instance of a concept term C w.r.t. an ABox $\mathcal{A}$ and a TBox $\mathcal{T}$, i.e. whether $\mathcal{A}$ entails $a\!:\!C$ w.r.t. $\mathcal{T}$. This problem can be reduced to the problem of deciding if the ABox $\mathcal{A} \cup \{a\!:\!\neg C\}$ is inconsistent w.r.t. $\mathcal{T}$.

1.2   A Tableaux Calculus for $\mathcal{ALCNH}_{R^+}$

In the following we present a *tableaux algorithm* to decide the consistency of $\mathcal{ALCNH}_{R^+}$ ABoxes. The algorithm is characterized by a set of tableaux or *completion rules* and by a particular *completion strategy* ensuring a specific order for applying the completion rules to assertional axioms of an ABox. The strategy is essential to guarantee the completeness of the ABox consistency algorithm. The purpose of the calculus is to generate a so-called completion for an initial ABox $\mathcal{A}$ in order to prove the consistency of $\mathcal{A}$ or its inconsistency if no completion can be found.

First, we introduce new assertional axioms needed to define the augmentation of an initial ABox. Let C be a concept term, $a, b \in O$ be individual names, and $x \notin O$, then the following expressions are also assertional axioms: (1) $\forall x \,.\, (x\!:\!C)$ *(universal concept assertion)*, (2) $a \neq b$ *(inequality assertion)*. An interpretation $\mathcal{I}$ satisfies an assertional axiom $\forall x \,.\, (x\!:\!C)$ iff $C^{\mathcal{I}} = \Delta^{\mathcal{I}}$ and $a \neq b$ iff $a^{\mathcal{I}} \neq b^{\mathcal{I}}$.

We are now ready to define an augmented ABox as input to the tableaux rules. For an initial ABox $\mathcal{A}$ w.r.t a TBox $\mathcal{T}$ and a role hierarchy $\mathcal{R}$ we define its *augmented* ABox or its *augmentation* $\mathcal{A}'$ by applying the following rules to $\mathcal{A}$. For every feature name F mentioned in $\mathcal{A}$ the assertion $\forall x \,.\, (x\!:\!(\exists_{\leq 1} F))$ is added to $\mathcal{A}'$. For every GCI $C \sqsubseteq D$ in $\mathcal{T}$ the assertion $\forall x \,.\, (x\!:\!(\neg C \sqcup D))$ is added to $\mathcal{A}'$. Every concept term occurring in $\mathcal{A}$ is transformed into its usual negation normal form. Let $O'_O = \{a_1, \ldots, a_n\} \subseteq O_O$ be the set of individuals mentioned in $\mathcal{A}$, then the following set of inequality assertions is added to $\mathcal{A}'$: $\{a_i \neq a_j \,|\, a_i, a_j \in O'_O, \ i, j \in 1..n, \ i \neq j\}$. Obviously, if $\mathcal{A}'$ is an augmentation of $\mathcal{A}$ then $\mathcal{A}'$ is consistent iff $\mathcal{A}$ is consistent.

$\mathcal{ALCNH}_{R^+}$ supports transitive roles and GCIs. Thus, in order to guarantee the termination of the tableaux calculus, the notion of *blocking* an individual for the applicability of tableaux rules is introduced as follows. Given an ABox $\mathcal{A}$ and an individual a occurring in $\mathcal{A}$, we define the *concept set* of a as $\sigma(\mathcal{A}, a) := \{\top\} \cup \{C \,|\, a\!:\!C \in \mathcal{A}\}$. We define an *individual ordering* '$\prec$' for new individuals (elements of O_N) occurring in an ABox $\mathcal{A}$. If $b \in O_N$ is introduced into $\mathcal{A}$, then $a \prec b$ for all new individuals a already present in $\mathcal{A}$. Let $\mathcal{A}$ be an ABox and $a, b \in O$ be individuals in $\mathcal{A}$. We call a the blocking individual of b if all of the following conditions hold: (1) $a, b \in O_N$, (2) $\sigma(\mathcal{A}, a) \supseteq \sigma(\mathcal{A}, b)$, (3) $a \prec b$. If there exists a blocking individual a for b, then b is said to be *blocked* (by a).

We are now ready to define the *completion rules* that are intended to generate a so-called completion (see also below) of an initial ABox $\mathcal{A}$ w.r.t. a TBox $\mathcal{T}$.

R⊓ The conjunction rule.

   if $a:C \sqcap D \in \mathcal{A}$, and $\{a:C, \ a:D\} \not\subseteq \mathcal{A}$

   then $\mathcal{A}' = \mathcal{A} \cup \{a:C, \ a:D\}$

R⊔ The disjunction rule.

   if $a:C \sqcup D \in \mathcal{A}$, and $\{a:C, \ a:D\} \cap \mathcal{A} = \emptyset$

   then $\mathcal{A}' = \mathcal{A} \cup \{a:C\}$ **or** $\mathcal{A}' = \mathcal{A} \cup \{a:D\}$

R∀C The role value restriction rule.

   if $a:\forall R . C \in \mathcal{A}$, and $\exists b \in O, S \in R^{\downarrow} : (a,b):S \in \mathcal{A}$, and $b:C \notin \mathcal{A}$

   then $\mathcal{A}' = \mathcal{A} \cup \{b:C\}$

R∀₊C The transitive role value restriction rule.

   if 1. $a:\forall R . C \in \mathcal{A}$, and $\exists b \in O, \ T \in R^{\downarrow}, \ T \in T, \ S \in T^{\downarrow} : (a,b):S \in \mathcal{A}$, and

       2. $b:\forall T . C \notin \mathcal{A}$

   then $\mathcal{A}' = \mathcal{A} \cup \{b:\forall T . C\}$

R∀ₓ The universal concept restriction rule.

   if $\forall x . (x:C) \in \mathcal{A}$, and $\exists a \in O$: a mentioned in $\mathcal{A}$, and $a:C \notin \mathcal{A}$

   then $\mathcal{A}' = \mathcal{A} \cup \{a:C\}$

R∃C The role exists restriction rule.

   if 1. $a:\exists R . C \in \mathcal{A}$, and a is not blocked, and

       2. $\neg \exists b \in O, \ S \in R^{\downarrow} : \{(a,b):S, \ b:C\} \subseteq \mathcal{A}$

   then $\mathcal{A}' = \mathcal{A} \cup \{(a,b):R, \ b:C\}$ where $b \in O_N$ is not used in $\mathcal{A}$

R∃≥n The number restriction exists rule.

   if 1. $a:\exists_{\geq n} R \in \mathcal{A}$, and a is not blocked, and

       2. $\neg \exists b_1, \ldots, b_n \in O, \ S_1, \ldots, S_n \in R^{\downarrow} :$

          $\{(a,b_k):S_k \,|\, k \in 1..n\} \cup \{b_i \neq b_j \,|\, i,j \in 1..n, i \neq j\} \subseteq \mathcal{A}$

   then $\mathcal{A}' = \mathcal{A} \cup \{(a,b_k):R \,|\, k \in 1..n\} \cup \{b_i \neq b_j \,|\, i,j \in 1..n, i \neq j\}$

       where $b_1, \ldots, b_n \in O_N$ are not used in $\mathcal{A}$

R∃≤n The number restriction merge rule.

   if 1. $a:\exists_{\leq n} R \in \mathcal{A}$, and

       2. $\exists b_1, \ldots, b_m \in O, \ S_1, \ldots, S_m \in R^{\downarrow} : \{(a,b_1):S_1, \ldots, (a,b_m):S_m\} \subseteq \mathcal{A}$

       with $m > n$, and

       3. $\exists b_i, b_j \in \{b_1, \ldots, b_m\} : i \neq j, \ b_i \neq b_j \notin \mathcal{A}$

   then $\mathcal{A}' = \mathcal{A}[b_i/b_j]$, i.e. replace every occurrence of b_i in $\mathcal{A}$ by b_j

Given an ABox $\mathcal{A}$, more than one rule might be applicable to $\mathcal{A}$. The order is determined by the *completion strategy* which is defined as follows.

A *meta rule* controls the priority between individuals: Apply a tableaux rule to an individual $b \in O_N$ only if no rule is applicable to an individual $a \in O_O$ and if no rule is applicable to another individual $c \in O_N$ such that $c \prec b$.

The completion rules are always applied in the following order. (1) Apply all non-generating rules (R⊓, R⊔, R∀C, R∀₊C, R∀ₓ, R∃≤n) as long as possible. (2) Apply a generating rule (R∃C, R∃≥n) once and continue with step 1.

In the following we always assume that the completion strategy is observed. This ensures that rules are applied to new individuals w.r.t. the ordering '$\prec$'.

We assume the same naming conventions as used above. An ABox $\mathcal{A}$ is called *contradictory* if one of the following *clash triggers* is applicable. If none of the clash triggers is applicable to $\mathcal{A}$, then $\mathcal{A}$ is called *clash-free*. The clash triggers have to deal with so-called primitive clashes and with clashes caused by number restrictions:

- *Primitive clash*: $a : \bot \in \mathcal{A}$ or $\{a : A, a : \neg A\} \subseteq \mathcal{A}$, where A is a concept name.
- *Number restriction merging clash*: $\exists S_1, \ldots, S_m \in R^{\downarrow} : \{(a, b_i) : S_i \mid i \in 1..m\} \cup \{a : \exists_{\leq n} R\} \cup \{b_i \neq b_j \mid i, j \in 1..m, i \neq j\} \subseteq \mathcal{A}$ with $m > n$.

Any ABox containing a clash is obviously unsatisfiable. A clash-free ABox $\mathcal{A}$ is called *complete* if no completion rule is applicable to $\mathcal{A}$. A complete ABox $\mathcal{A}'$ derived from an initial ABox $\mathcal{A}$ is called a *completion* of $\mathcal{A}$. The purpose of the calculus is to generate a completion for an initial ABox $\mathcal{A}$ to prove the consistency of $\mathcal{A}$. An augmented ABox $\mathcal{A}$ is said to be inconsistent if no completion can be derived. For a given initial ABox $\mathcal{A}$, the calculus applies the completion rules. It stops the application of rules, if a clash occurs. The calculus answers *"yes"* if a completion can be derived, and *"no"* otherwise. Based on these notions we introduce and evaluate the new optimization techniques in the next sections.

2   Deep Models for TBox Reasoning in $\mathcal{ALCNH}_{R^+}$

Given a set of concepts representing a conjunction whose satisfiability is to be checked, the *model merging strategy* tries to avoid a satisfiability test which relies on the "expensive" tableaux technique due to non-deterministic rules.[3] This idea was first introduced in [5] for the logic $\mathcal{ALCHf}_{R^+}$. A model merging test is designed to be a "cheap" test comparing cached "concept models." It is a *sound* but incomplete satisfiability tester for a set of concepts. The achievement of minimal computational overhead and the avoidance of any indeterminism are important characteristics of such a test. If the test returns false, a tableaux calculus based on the rules as defined in Section 1.2 is applied. In order to be more precise, we use the term *pseudo model* instead of "concept model."

For testing whether the conjunction of a set of concepts $\{C_1, \ldots, C_n\}$ is satisfiable, we present and analyze a technique called *deep model merging* that generalizes the original model merging approach [5] in two ways: (1) we extend the model merging technique to the logic $\mathcal{ALCNH}_{R^+}$, i.e. this technique also deals with number restrictions; (2) we introduce *deep* pseudo models for concepts that are *recursively* traversed and checked for possible clashes.

Let A be a concept name, R a role name, and C a concept. The consistency of the initial ABox $\mathcal{A} = \{a : C\}$ is tested. If $\mathcal{A}$ is inconsistent, the *pseudo model*[4] of C is defined as $\bot$. If $\mathcal{A}$ is consistent, then there exists a non-empty set of completions $\mathcal{C}$. A completion $\mathcal{A}' \in \mathcal{C}$ is selected and a pmodel M for a concept C

---

[3] In our case the rules R$\sqcup$ and R$\exists_{\leq n}$.
[4] For brevity a pseudo model is also called a *pmodel*.

is defined as the tuple $\langle M^{\mathsf{A}}, M^{\neg\mathsf{A}}, M^{\exists}, M^{\forall}\rangle$ of concept sets using the following definitions.

$$M^{\mathsf{A}} = \{\mathsf{A}\,|\,\mathsf{a}{:}\mathsf{A} \in \mathcal{A}', \mathsf{A} \in C\}, \qquad M^{\neg\mathsf{A}} = \{\mathsf{A}\,|\,\mathsf{a}{:}\neg\mathsf{A} \in \mathcal{A}', \mathsf{A} \in C\}$$

$$M^{\exists} = \{\exists\mathsf{R}\,.\,\mathsf{C}\,|\,\mathsf{a}{:}\exists\mathsf{R}\,.\,\mathsf{C} \in \mathcal{A}'\} \cup \{\exists_{\geq n}\,\mathsf{R}\,|\,\mathsf{a}{:}\exists_{\geq n}\,\mathsf{R} \in \mathcal{A}'\}$$

$$M^{\forall} = \{\forall\mathsf{R}\,.\,\mathsf{C}\,|\,\mathsf{a}{:}\forall\mathsf{R}\,.\,\mathsf{C} \in \mathcal{A}'\} \cup \{\exists_{\leq n}\,\mathsf{R}\,|\,\mathsf{a}{:}\exists_{\leq n}\,\mathsf{R} \in \mathcal{A}'\} \cup$$

$$\{\exists\mathsf{R}\,.\,\mathsf{C}\,|\,\mathsf{a}{:}\exists\mathsf{R}\,.\,\mathsf{C} \in \mathcal{A}', \mathsf{R} \in F\}$$

Note that pmodels are based on complete ABoxes. In contrast to the theoretical calculus presented above, model merging deals directly with features instead of representing them with at most restrictions. Therefore concept exists restrictions mentioning features are also included in the sets $M^{\forall}$ of pmodels. This guarantees that a possible "feature interaction" between pmodels is detected.

Procedure 1 mergable($MS, VM, D?$)

1: **if** $MS = \emptyset \vee MS \in VM$ **then**
2:     **return** *true*
3: **else if** $\bot \in MS \vee \neg$**atoms_mergable**(MS) **then**
4:     **return** *false*
5: **else**
6:     **for all** $M \in MS$ **do**
7:         **for all** $\mathsf{C} \in M^{\exists}$ **do**
8:             **if** **critical_at_most**(C, M, MS) **then**
9:                 **return** *false*
10:             **else**
11:                 $MS' \leftarrow$ **collect_successor_pmodels**(C, MS)
12:                 **if** $(\neg D? \wedge MS' \neq \emptyset) \vee \neg$**mergable**($MS', VM \cup \{\mathsf{MS}\}, D?$) **then**
13:                     **return** *false*
14: **return** *true*

The procedure **mergable** shown in Procedure 1 implements the flat and deep model merging test. The test has to discover potential clashes which might occur if all pmodels in MS are merged, i.e. their corresponding concepts are conjunctively combined. The test starts with a set of pmodels MS, an empty set of visited pmodel sets VM, and a parameter $D?$ controlling whether the deep or flat mode (see below) of **mergable** will be used. The test recursively traverses the pmodel structures. In case of a potential clash, **mergable** terminates and returns *false*, otherwise it continues its traversal and returns *true* if no potential clash can be discovered. Testing whether the actual pmodel set MS is already a member of the set VM (line 1) is necessary to ensure termination (in analogy to blocking an individual) for the deep mode. A potential primitive clash is checked in line 3. If no primitive clash is possible for the "root individual" of the pmodel, it is tested whether a clash might be caused by interacting concept exists, concept value, at least, and at most restrictions for the same successor individuals. Two nested loops (lines 6-13) check for every pmodel $M \in MS$ and every concept in the set $M^\exists$ whether an at most restriction might be violated by the other pmodels (line 8). If this is not the case,[5] the set MS' of "R-successor pmodels" is computed (line 11). If the flat mode is enabled and $MS' \neq \emptyset$, this indicates a potential interaction via the role R and **mergable** returns *false* (lines 12-13). If the deep mode is enabled, **mergable** continues and traverses the pmodels in MS'. Observe that the procedure **mergable** is sound but not complete, i.e. even if **mergable** returns *false* for a pmodel set the corresponding concept conjunction can be satisfiable.

The procedure **atoms_mergable** tests for a possible primitive clash between pairs of pmodels. It is applied to a set of pmodels MS and returns *false* if there exists a pair $\{M_1, M_2\} \subseteq MS$ with $(M_1^A \cap M_2^{\neg A}) \neq \emptyset$ or $(M_1^{\neg A} \cap M_2^A) \neq \emptyset$. Otherwise it returns *true*.

The procedure **critical_at_most** checks for a potential number restriction clash in a set of pmodels and tries to avoid positive answers which are too conservative. It is applied to a concept C of the form $\exists S.D$ or $\exists_{\geq n} S$, the current pmodel M and a set of pmodels $MS = \{M_1, \ldots, M_k\}$. Loosely speaking, it computes the maximal number of potential S-successors and returns *true* if this number exceeds the applicable at most bound m. More precisely, **critical_at_most** returns *true* if there exists a pmodel $M' \in (MS \setminus M)$ and a role $R \in S^\uparrow$ with $\exists_{\leq m} R \in M'^\forall$ such that $\sum_{E \in N} num(E, RS) > m$, $N = \cup_{i \in 1..k} M_i^\exists$, $RS = S^\uparrow \cap R^\downarrow$. In all other cases **critical_at_most** returns *false*. The function $num(E, RS)$ returns 1 for concepts of the form $E = \exists R'.D$ and n for $E = \exists_{\geq n} R'$, if $R' \in RS$, and 0 otherwise.

The procedure **collect_successor_pmodels** is applied to a concept C of the form $\exists S.D$ or $\exists_{\geq n} S$ and a set of pmodels MS. It computes the set Q containing all S-successor pmodels (by considering (transitive) superroles of S). We define $Q_{aux} = \{D\}$ if $C = \exists S.D$ and $Q_{aux} = \emptyset$ otherwise. Observe that $\exists R.E \in M^\forall$ implies that R is a feature. The procedure **collect_successor_pmodels** returns the pmodel set $\{M_C \mid C \in Q\}$.

$$Q = Q_{aux} \cup \{E \mid \exists M \in MS, R \in S^\uparrow : (\forall R.E \in M^\forall \vee \exists R.E \in M^\forall)\} \cup$$
$$\{\forall T.E \mid \exists M \in MS, R \in S^\uparrow, T \in T \cap S^\uparrow \cap R^\downarrow : \forall R.E \in M^\forall\}$$

Note that **mergable** depends on the clash triggers of the particular tableaux calculus chosen since it has to detect potential clashes in a set of pmodels. The structure and composition of the completion rules might vary as long as the clash triggers do not change and the calculus remains sound and complete.

Proposition 1 (Soundness of mergable). *Let $D?$ have either the value* true *or* false, $CS = \{C_1, \ldots, C_n\}$, $M_{C_i} = $ get_pmodel(C_i), *and* $PM = \{M_{C_i} \mid i \in 1..n\}$. *If the procedure call* **mergable**$(PM, \emptyset, D?)$ *returns* true, *the concept* $C_1 \sqcap \ldots \sqcap C_n$ *is satisfiable.*

Proof. This is proven by contradiction and induction. Let us assume that the call **mergable**$(PM, \emptyset, D?)$ returns *true* but the initial ABox $\mathcal{A} = \{a : (C_1 \sqcap \ldots \sqcap C_n)\}$ is inconsistent, i.e. there exists no completion of $\mathcal{A}$. Every concept C_i must be

---

[5] In the following let us assume that the concept C mentions a role name R.

satisfiable, otherwise we would have $\perp \in PM$ and **mergable** would return *false* due to line 3 in Procedure 1. Let us assume a finite set $\mathcal{C}$ containing all contradictory ABoxes encountered during the consistency test of $\mathcal{A}$. Without loss of generality we can select an arbitrary $\mathcal{A}' \in \mathcal{C}$ and make a case analysis of its possible clash culprits.

1. We have a primitive clash for the "root" individual a, i.e. $\{\mathsf{a}:\mathsf{D}, \mathsf{a}:\neg\mathsf{D}\} \subseteq \mathcal{A}'$. Thus, $\mathsf{a}:\mathsf{D}$ and $\mathsf{a}:\neg\mathsf{D}$ have not been propagated to a via role assertions and there have to exist $\mathsf{C}_i, \mathsf{C}_j \in CS$, $i \neq j$ such that $\mathsf{a}:\mathsf{D}$ ($\mathsf{a}:\neg\mathsf{D}$) is derived from $\mathsf{a}:\mathsf{C}_i$ ($\mathsf{a}:\mathsf{C}_j$) due to the satisfiability of the concepts C_i, $i \in 1..n$. It holds for the associated pmodels $M_{C_i}, M_{C_j} \in PM$ that $\mathsf{D} \in M_{C_i}^{\mathsf{A}} \cap M_{C_j}^{\neg\mathsf{A}}$. However, due to our assumption the call of **mergable**$(PM, \emptyset, \mathsf{D}?)$ returned *true*. This is a contradiction since **mergable** called **atoms_mergable** with PM (line 3 in Procedure 1) which returned *false* since $\mathsf{D} \in M_{C_i}^{\mathsf{A}} \cap M_{C_j}^{\neg\mathsf{A}}$.

2. A number restriction clash in $\mathcal{A}'$ is detected for a, i.e. $\mathsf{a}:\exists_{\leq m} \mathsf{R} \in \mathcal{A}'$ and there exist $l > m$ distinct R-successors of a.[6] These successors can only be derived from assertions of the form $\mathsf{a}:\exists \mathsf{S}_j . \mathsf{E}_j$ or $\mathsf{a}:\exists_{\geq n_j} \mathsf{S}_j$ with $\mathsf{S}_j \in \mathsf{R}^{\downarrow}$, $j \in 1..p$. The concepts $\mathsf{C}_i \in CS$, $i \in 1..n$ are satisfiable and there has to exist a subset $CS' \subseteq CS$ such that $\exists_{\leq m} \mathsf{R} \in \cup_{C \in CS'} M_C^{\forall}$ and $\sum_{E' \in N} \text{num}(\mathsf{E}', RS) \geq l$, $N = \cup_{C \in CS'} M_C^{\exists}$, $RS = (\cup_{j \in 1..p} \mathsf{S}_j^{\uparrow}) \cap \mathsf{R}^{\downarrow}$. However, due to our assumption the call of **mergable**$(PM, \emptyset, \mathsf{D}?)$ returned *true*. This is a contradiction since there exists a pmodel M_C', $\mathsf{C} \in CS'$ and a concept $\mathsf{E}' \in M_C'^{\exists}$ such that **mergable** called **critical_at_most**(E', M_C', PM) (lines 6-8 in Procedure 1) which returned *true* since $\sum_{E' \in N} \text{num}(\mathsf{E}', RS) \geq l > m$.

3. Let the individual a_n be a successor of a_0 via a chain of role assertions $(\mathsf{a}_0, \mathsf{a}_1):\mathsf{R}_1, \ldots, (\mathsf{a}_{n-1}, \mathsf{a}_n):\mathsf{R}_n$, $n > 0$ and we now assume that a clash for a_n is discovered.

   a) In case of a primitive clash we have $\{\mathsf{a}_n:\mathsf{D}, \mathsf{a}_n:\neg\mathsf{D}\} \subseteq \mathcal{A}'$. Without loss of generality we may assume that the clash culprits can only be derived from assertions of the form $\mathsf{a}_{n-1}:\exists_{\geq m} \mathsf{R}_n$ or $\mathsf{a}_{n-1}:\exists \mathsf{R}_n . \mathsf{E}_1$ in combination with $\mathsf{a}_{n-1}:\exists \mathsf{S}' . \mathsf{E}_2$ (if R_n and $\mathsf{S}' \in \mathsf{R}_n^{\uparrow}$ are features), and/or $\mathsf{a}_{n-1}:\forall \mathsf{S}'' . \mathsf{E}_3$ with $\mathsf{S}'' \in \mathsf{R}_n^{\uparrow}$. Due to the clash there exists a pair $\mathsf{E}', \mathsf{E}'' \subseteq \{\mathsf{E}_1, \mathsf{E}_2, \mathsf{E}_3\}$ with $\mathsf{D} \in M_{E'}^{\mathsf{A}} \cap M_{E''}^{\neg\mathsf{A}}$. Each role assertion in the chain between a_0 and a_{n-1} can only be derived from assertions of the form $\mathsf{a}_{k-1}:\exists \mathsf{R}_k . \mathsf{E}_k$ or $\mathsf{a}_{k-1}:\exists_{\geq m_k} \mathsf{R}_k$ with $k \in 1..n - 1$. The call graph of **mergable**$(PM, \emptyset, \mathsf{D}?)$ contains a chain of calls resembling the chain of role assertions. By induction on the call graph we know that the node resembling a_{n-1} of this call graph chain contains the call **mergable**$(PM', VM', true)$ such that $\{M_{E'}, M_{E''}\} \subseteq PM'$ and **atoms_mergable** has been called with a set MS' and $\{M_{E'}, M_{E''}\} \subseteq MS'$. The call of **atoms_mergable** has returned *false* since $\mathsf{D} \in M_{E'}^{\mathsf{A}} \cap M_{E''}^{\neg\mathsf{A}}$. This contradicts our assumption that **mergable**$(PM, \emptyset, \mathsf{D}?)$ returned *true*.

   b) In case of a number restriction clash we can argue in an analogous way to case 2 and 3a. Again, we have a chain of role assertions where a number restriction clash is detected for the last individual of the chain.

---

[6] Due to our syntax restriction, the elements of $\mathsf{R}^{\downarrow}$ are not transitive.

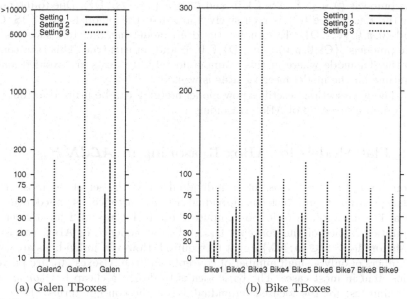

(a) Galen TBoxes   (b) Bike TBoxes

Fig. 2. Evaluation of model merging techniques (runtime in seconds, 3 runs for each TBox, left-right order corresponds to top-bottom order in the legend).

It exists a corresponding call graph chain where by induction the last call of **mergable** called **critical_at_most** with a set of pmodels for which **critical_at_most** returned *true*. This contradicts the assumption that **mergable**$(PM, \emptyset, D?)$ returned *true*.   □

It is easy to see that this proof also holds if the value of $D?$ is *false* since the "flat mode" is more conservative than the "deep" one, i.e. it will always return *false* instead of possibly *true* if the set of collected pmodels M' is not empty (line 12 in Procedure 1).

The advantage of the deep vs. the flat mode of the model merging technique is demonstrated by empirical tests using a set of "quasi-standard" application TBoxes [7,6,2]. Figure 2 shows the runtimes for computing the subsumption lattice of these TBoxes. Each TBox is iteratively classified using three different parameter settings. The first setting has the deep mode of model merging enabled, the second one has the deep mode of model merging disabled but the flat mode still enabled, and the third one has model merging completely disabled. The comparison between setting one and two indicates a speed up in runtimes of a factor $1.5 - 2$ if the deep mode is enabled. The result for setting three clearly demonstrate the principal advantage of model merging.

The principal advantage of the deep vs. the flat model merging mode is due to the following characteristics. If the flat model merging test is (recursively) applied during tableaux expansion and repeatedly returns *false* because of interacting value and exists restrictions, this test might be too conservative. This effect is illustrated by an example: The deep model merging test starts with the pmodels $\langle \emptyset, \emptyset, \{\exists R . \exists S . C\}, \emptyset \rangle$ and $\langle \emptyset, \emptyset, \emptyset, \{\forall R . \forall S . D\} \rangle$. Due to interaction on the role R, the test is recursively applied to the pmodels $\langle \emptyset, \emptyset, \{\exists S . C\}, \emptyset \rangle$ and $\langle \emptyset, \emptyset, \emptyset, \{\forall S . D\} \rangle$. Eventually, the deep model merging test succeeds with the pmodels $\langle \{C\}, \emptyset, \emptyset, \emptyset \rangle$ and $\langle \{D\}, \emptyset, \emptyset, \emptyset \rangle$ and returns true. This is in contrast to the flat mode where in this example no tableaux tests are avoided and the runtime for the model merging tests is wasted.

The next section describes how model merging can be utilized for obtaining a dramatic speed up of ABox reasoning.

3   Flat Models for ABox Reasoning in $\mathcal{ALCNH}_{R+}$

Computing the *direct types* of an individual a (i.e. the set of the most specific concepts from C of which an individual a is an instance) is called *realization* of a. For instance, in order to compute the direct types of a for a given subsumption lattice of the concepts $D_1, \ldots, D_n$, a sequence of ABox consistency tests for $\mathcal{A}_{D_i} = \mathcal{A} \cup \{a : \neg D_i\}$ might be required. However, individuals are usually members of only a small number of concepts and the ABoxes $\mathcal{A}_{D_i}$ are proven as consistent in most cases. The basic idea is to design a cheap but *sound* model merging test for the focused individual a and the concept terms $\neg D_i$ without explicitly considering role assertions and concept assertions for all the other individuals mentioned in $\mathcal{A}$. These "interactions" are reflected in the "individual pseudo model" of a. This is the motivation for devising the novel *individual model merging* technique.

A pseudo model for an individual a mentioned in a consistent initial ABox $\mathcal{A}$ w.r.t. a TBox $\mathcal{T}$ is defined as follows. Since $\mathcal{A}$ is consistent, there exists a set of completions $\mathcal{C}$ of $\mathcal{A}$. Let $\mathcal{A}' \in \mathcal{C}$. An *individual pseudo model* M for an individual a in $\mathcal{A}$ is defined as the tuple $\langle M^A, M^{\neg A}, M^\exists, M^\forall \rangle$ w.r.t. $\mathcal{A}'$ and $\mathcal{A}$ using the same definitions from the previous section for the components $M^A, M^{\neg A}, M^\forall$ and the following definition.

$$M^\exists = \{\exists R . C \mid a : \exists R . C \in \mathcal{A}'\} \cup \{\exists_{\geq n} R \mid a : \exists_{\geq n} R \in \mathcal{A}'\} \cup \{\exists_{\geq 1} R \mid (a, b) : R \in \mathcal{A}\}$$

Note the distinction between the initial ABox $\mathcal{A}$ and its completion $\mathcal{A}'$. Whenever a role assertion exists, which specifies a role successor for the individual a in the initial ABox, a corresponding at least restriction is added to the set $M^\exists$. This is based on the rationale that the cached pmodel of a cannot refer to individual names. However, it is sufficient to reflect a role assertion $(a, b) : R \in \mathcal{A}$ by adding a corresponding at least restriction to $M^\exists$. This guarantees that possible interactions via the role R are detected. Note that individual model merging is only defined for the *flat* mode of model merging.

Proposition 2 (Soundness of individual_model_merging). *Let M_a be the pmodel of an individual* a *mentioned in a consistent initial ABox $\mathcal{A}$, $M_{\neg C}$ be the pmodel of a satisfiable concept* $\neg C$, *and $PM = \{M_a, M_{\neg C}\}$. If the procedure call* **mergable**$(PM, \emptyset, false)$ *returns* true, *the ABox $\mathcal{A} \cup \{a : \neg C\}$ is consistent, i.e.* a *is not an instance of* C.

Proof. This is proven by contradiction. Let us assume that the procedure call **mergable**($\{M_a, M_{\neg C}\}, \emptyset, false$) returns *true* but the ABox $\mathcal{A}' = \mathcal{A} \cup \{a : \neg C\}$ is inconsistent, i.e. there exists no completion of $\mathcal{A}'$. Let us assume a finite set $\mathcal{C}$ containing all contradictory ABoxes encountered during the consistency test of $\mathcal{A}'$. Without loss of generality we can select an arbitrary $\mathcal{A}'' \in \mathcal{C}$ and make a case analysis of its possible clash culprits.

1. In case of a primitive clash for a we have $\{a : D, a : \neg D\} \subseteq \mathcal{A}''$. Since $\mathcal{A}$ is consistent and the concept $\neg C$ cannot indirectly refer to the old individual a via a role chain, we know that either $a : D$ or $a : \neg D$ must be derived from $a : \neg C$ and we have $D \in (M_a^A \cap M_{\neg C}^{\neg A}) \cup (M_a^{\neg A} \cap M_{\neg C}^A)$. This contradicts the assumption that the call **mergable**($\{M_a, M_{\neg C}\}, \emptyset, false$) returned *true* since **mergable** called **atoms_mergable**($\{M_a, M_{\neg C}\}$) which returned *false* (line 3 in Procedure 1) since $D \in (M_a^A \cap M_{\neg C}^{\neg A}) \cup (M_a^{\neg A} \cap M_{\neg C}^A)$.

2. A number restriction clash in $\mathcal{A}''$ is detected for a, i.e. $a : \exists_{\leq m} R \in \mathcal{A}''$ and there exist $l > m$ distinct R-successors of a in $\mathcal{A}''$. This implies that the set $N = M_a^{\exists} \cup M_{\neg C}^{\exists}$ contains concepts of the form $\exists S_j . E_j$ or $\exists_{\geq n_j} S_j$,[7] $S_j \in R^{\downarrow}$, $j \in 1..k$, such that $\sum_{E' \in N} \text{num}(E', RS) \geq l$, $RS = (\cup_{j \in 1..k} S_j^{\uparrow}) \cap R^{\downarrow}$. This contradicts the assumption that **mergable**($\{M_a, M_{\neg C}\}, \emptyset, false$) returned *true* since mergable called **critical_at_most** (lines 6-8 in Procedure 1) which returned *true* since $\sum_{E' \in N} \text{num}(E', RS) \geq l > m$.

3. A clash is detected for an individual b in $\mathcal{A}''$ that is distinct to a. Since $\mathcal{A}$ is consistent the individual b must be a successor of a via a chain of role assertions $(a, b_1) : R_1, \ldots, (b_n, b) : R_{n+1}, n \geq 0$, and one of the clash culprits must be derived from the newly added assertion $a : \neg C$ and propagated to b via the role assertion chain originating from a with $(a, b_1) : R_1$. Since $\neg C$ is satisfiable and $\mathcal{A}$ is consistent we have an "interaction" via the role or feature R_1. This implies for the associated pmodels $M_a, M_{\neg C}$ that $(M_a^{\exists} \cap M_{\neg C}^{\forall}) \cup (M_a^{\forall} \cap M_{\neg C}^{\exists}) \neq \emptyset$. This contradicts the assumption that **mergable**($\{M_a, M_{\neg C}\}, \emptyset, false$) returned *true* since **mergable** eventually called **collect_successor_pmodels** for $M_a, M_{\neg C}$ which returned a non-empty set (line 11 in Procedure 1). □

The performance gain by the individual model merging technique is empirically evaluated using a set of five ABoxes containing between 15 and 25 individuals. Each of these ABoxes is realized w.r.t. the application TBoxes Bike7-9 derived from a bike configuration task. The TBoxes especially vary on the degree of explicit disjointness declarations between atomic concepts. Figure 3 shows the runtimes for the realization of the ABoxes 1-5. Each ABox is realized with two different parameter settings. The first setting has the individual model merging technique enabled, the second one has it disabled. The comparison between both settings reveals a speed gain of at least one order of magnitude if the individual model merging technique is used. Note the use of a logarithmic scale.

---

[7] Any role assertion of the form $(a, b) : R \in \mathcal{A}$ implies that $\exists_{\geq 1} R \in M_a^{\exists}$. This takes care of implied at least restrictions due to the UNA for old individuals.

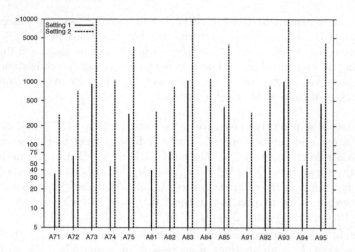

Fig. 3. Bike ABoxes: Evaluation of model merging techniques (runtime in seconds, 2 runs for each ABox, left-right order corresponds to top-bottom order in the legend).

4   Pseudo Models for Reasoning with Concrete Domains

The requirements derived from practical applications of DLs ask for more expressiveness w.r.t. reasoning about objects from other domains (so-called concrete domains, e.g. for the real numbers). Thus, in [4] the logic $\mathcal{ALCNH}_{R^+}$ is extended with a restricted form of reasoning about concrete domains. However, the classification of non-trivial TBoxes is only feasible, if the model merging technique can be applied. Therefore, we extend the model merging technique to the basic DL with concrete domains, the language $\mathcal{ALC}(\mathcal{D})$ [1]. We conjecture that the results from this approach can be directly transferred to the logic presented in [4]. First, we have to briefly introduce $\mathcal{ALC}(\mathcal{D})$.

4.1   The Language $\mathcal{ALC}(\mathcal{D})$

A *concrete domain* $\mathcal{D}$ is a pair $(\Delta_{\mathcal{D}}, \Phi_{\mathcal{D}})$, where $\Delta_{\mathcal{D}}$ is a set called the domain, and $\Phi_{\mathcal{D}}$ is a set of predicate names. Each predicate name $\mathsf{P}_{\mathcal{D}}$ from $\Phi_{\mathcal{D}}$ is associated with an arity n and an n-ary predicate $\mathsf{P}_{\mathcal{D}} \subseteq \Delta_{\mathcal{D}}^n$. A concrete domain $\mathcal{D}$ is called *admissible* iff (1) the set of predicate names $\Phi_{\mathcal{D}}$ is closed under negation and $\Phi_{\mathcal{D}}$ contains a name $\top_{\mathcal{D}}$ for $\Delta_{\mathcal{D}}$; (2) the satisfiability problem $\mathsf{P}_1^{n_1}(x_{11}, \dots, x_{1n_1}) \wedge \dots \wedge \mathsf{P}_m^{n_m}(x_{m1}, \dots, x_{mn_m})$ is decidable (m is finite, $\mathsf{P}_i^{n_i} \in \Phi_{\mathcal{D}}$, and x_{jk} is a name for an object from $\Delta_{\mathcal{D}}$).

Let S and F $(R = S \cup F)$ be disjoint sets of role and feature names, respectively. A composition of features (written $\mathsf{F}_1 \cdots \mathsf{F}_n$) is called a *feature chain*. A simple feature is a feature chain of length 1. Let C be a set of concept names which is disjoint from R. Any element of C is a *concept term*. If C and D are concept terms, $\mathsf{R} \in R$ is a role or feature name, $\mathsf{P} \in \Phi_{\mathcal{D}}$ is a predicate name from an admissible concrete domain, u_i's are feature chains, then the following expressions are also concept terms: $C \sqcap D$, $C \sqcup D$, $\neg C$, $\forall R . C$, $\exists R . C$, $\exists u_1, \ldots, u_n . P$. A concept term of the last kind is called *predicate exists restriction*.

An *interpretation* $\mathcal{I}_\mathcal{D} = (\Delta^\mathcal{I}, \Delta^\mathcal{D}, \cdot^\mathcal{I})$ consists of a set $\Delta^\mathcal{I}$ (the abstract domain), a set $\Delta^\mathcal{D}$ (the domain of an admissible 'concrete domain' $\mathcal{D}$) and an interpretation function $\cdot^\mathcal{I}$. Besides for feature and predicate names the interpretation function is defined as in Figure 1a. The function maps each feature name F from F to a partial function $F^\mathcal{I}$ from $\Delta^\mathcal{I}$ to $\Delta^\mathcal{I} \cup \Delta^\mathcal{D}$, and each predicate name P from $\Phi_\mathcal{D}$ with arity n to a subset $P^\mathcal{I}$ of $\Delta_\mathcal{D}^n$. For a feature chain $u = F_1 \cdots F_n$, $u^\mathcal{I}$ denotes the composition $F_1^\mathcal{I} \circ \cdots \circ F_n^\mathcal{I}$ of partial functions $F_1^\mathcal{I}, \ldots, F_n^\mathcal{I}$. Let $u_1, \ldots, u_n$ be feature chains and let P be a predicate name. Then, the interpretation function can be extended to concept terms as in Figure 1a. The semantics for the predicate exists restrictions is given by:

$$(\exists u_1, \ldots, u_n . P)^\mathcal{I} := \{ a \in \Delta^\mathcal{I} \mid \exists x_1, \ldots, x_n \in \Delta^\mathcal{D} : (a, x_1) \in u_1^\mathcal{I}, \ldots, (a, x_n) \in u_n^\mathcal{I},$$
$$(x_1, \ldots, x_n) \in P^\mathcal{I} \}$$

Note that in a concept term elements of $\Delta^\mathcal{D}$ can be used only as feature fillers.

A TBox $\mathcal{T}$ is a finite set of non-cyclic axioms of the form $A \sqsubseteq D$ or $A \doteq D$ where A must be a concept name. An interpretation $\mathcal{I}$ is a *model* of a TBox $\mathcal{T}$ iff it satisfies $A^\mathcal{I} \subseteq D^\mathcal{I}$ ($A^\mathcal{I} = D^\mathcal{I}$) for all $A \sqsubseteq D$ ($A \doteq D$) in $\mathcal{T}$.

An *ABox* $\mathcal{A}$ is a finite set of assertional axioms which are defined as follows: Let O be a set of individual names and let X be a set of names for concrete objects ($X \cap O = \emptyset$). If C is a concept term, $R \in R$, $F \in F$, $a, b \in O$ and $x, x_1, \ldots, x_n \in X$, then the following expressions are *assertional axioms*: $a : C$, $(a, b) : R$, $(a, x) : F$ and $(x_1, \ldots, x_n) : P$.

The interpretation function additionally maps every individual name from O to a single element of $\Delta^\mathcal{I}$ and names for concrete objects from X are mapped to elements of $\Delta^\mathcal{D}$. (The UNA does not necessarily hold in $\mathcal{ALC}(\mathcal{D})$.) An interpretation satisfies an assertional axiom $a : C$ iff $a^\mathcal{I} \in C^\mathcal{I}$, $(a, b) : R$ iff $(a^\mathcal{I}, b^\mathcal{I}) \in R^\mathcal{I}$, $(a, x) : F$ iff $(a^\mathcal{I}, x^\mathcal{I}) \in F^\mathcal{I}$, and $(x_1, \ldots, x_n) : P$ iff $(x_1^\mathcal{I}, \ldots, x_n^\mathcal{I}) \in P^\mathcal{I}$. An interpretation $\mathcal{I}$ is a *model* of an ABox $\mathcal{A}$ w.r.t. a TBox $\mathcal{T}$ iff it is a model of $\mathcal{T}$ and furthermore satisfies all assertional axioms in $\mathcal{A}$.

4.2   Pseudo Models for TBox Reasoning in $\mathcal{ALC}(\mathcal{D})$

By analogy to the previous sections, we assume a tableaux calculus which decides the ABox consistency problem for $\mathcal{ALC}(\mathcal{D})$ (see [1]). The clash triggers in this calculus are the primitive clash, two triggers for feature fillers with membership to both domains, and one clash trigger indicating inconsistencies between concrete domain objects.

In the following we assume the same naming conventions as used above. In order to obtain a *flat pseudo model* for a concept C the consistency of $\mathcal{A} = \{a : C\}$ is tested. If $\mathcal{A}$ is inconsistent, the *pseudo model* of C is defined as $\perp$. If $\mathcal{A}$ is consistent, then there exists a set of completions $\mathcal{C}$. A completion $\mathcal{A}' \in \mathcal{C}$ is selected and a pmodel M for a concept C is defined as the tuple

$\langle M^{\mathsf{A}}, M^{\neg\mathsf{A}}, M^{\exists}, M^{\forall}, M^{\exists\mathsf{F}}, M^{\forall\mathsf{F}}\rangle$ using the following definitions (let $\mathsf{u} = \mathsf{F}_1 \cdots \mathsf{F}_n$ be a feature chain, then $\mathit{first}(\mathsf{u}) = \mathsf{F}_1$).

$$M^{\mathsf{A}} = \{\mathsf{A} \mid \mathsf{a}{:}\mathsf{A} \in \mathcal{A}'\}, \qquad M^{\neg\mathsf{A}} = \{\mathsf{A} \mid \mathsf{a}{:}\neg\mathsf{A} \in \mathcal{A}'\},$$
$$M^{\exists} = \{\mathsf{R} \mid \mathsf{a}{:}\exists\mathsf{R}.\mathsf{C} \in \mathcal{A}'\}, \qquad M^{\forall} = \{\mathsf{R} \mid \mathsf{a}{:}\forall\mathsf{R}.\mathsf{C} \in \mathcal{A}'\},$$
$$M^{\exists\mathsf{F}} = \{\mathsf{F} \mid \mathsf{a}{:}\exists\mathsf{F}.\mathsf{C} \in \mathcal{A}'\} \cup$$
$$\{\mathsf{F} \mid \mathsf{F} = \mathit{first}(\mathsf{u}_\mathsf{j}), \mathsf{u}_\mathsf{j} \text{ used in } \exists\mathsf{u}_1,\dots,\mathsf{u}_n.\mathsf{P}, \mathsf{a}{:}\exists\mathsf{u}_1,\dots,\mathsf{u}_n.\mathsf{P} \in \mathcal{A}'\},$$
$$M^{\forall\mathsf{F}} = \{\mathsf{F} \mid \mathsf{a}{:}\forall\mathsf{F}.\mathsf{C} \in \mathcal{A}'\}$$

Note that sets from a flat pseudo model for an $\mathcal{ALC}(\mathcal{D})$ concept contain only concept, role, and/or feature names. In order to correctly deal with the semantics of features, the pmodel also contains separate sets $M^{\forall\mathsf{F}}$ and $M^{\exists\mathsf{F}}$. The set $M^{\exists\mathsf{F}}$ contains all feature names mentioned in exists restrictions and all feature names being first element of a feature chain in predicate exists restrictions, and the set $M^{\forall\mathsf{F}}$ contains all feature names mentioned in value restrictions.

The following procedure $\mathcal{ALC}(\mathcal{D})$-**mergable** implements the flat model merging test for $\mathcal{ALC}(\mathcal{D})$ for a given non-empty set of pmodels MS.

Procedure 2 $\mathcal{ALC}(\mathcal{D})$-**mergable**($MS$)
---
if $\bot \in MS \vee \neg$**atoms_mergable**(MS) then
  return *false*
else
  for all pairs $\{M_1, M_2\} \subseteq MS$ do
    if $(M_1^{\exists} \cap M_2^{\forall}) \neq \emptyset \vee (M_1^{\forall} \cap M_2^{\exists}) \neq \emptyset$ then
      return *false*
    else if $(M_1^{\exists\mathsf{F}} \cap M_2^{\forall\mathsf{F}}) \neq \emptyset \vee (M_1^{\forall\mathsf{F}} \cap M_2^{\exists\mathsf{F}}) \neq \emptyset \vee (M_1^{\exists\mathsf{F}} \cap M_2^{\exists\mathsf{F}}) \neq \emptyset$ then
      return *false*
return *true*

The idea of this test is to check for possible primitive clashes at the "root individual" of the pmodels in MS using **atoms_mergable**. Then the procedure $\mathcal{ALC}(\mathcal{D})$-**mergable** checks for possible references to the same direct role or feature filler by more than one pmodel in MS.

This easy, but conservative test handles, besides primitive clashes, the three $\mathcal{ALC}(\mathcal{D})$-specific clash triggers, because they can only appear at feature fillers. A proof for the soundness of $\mathcal{ALC}(\mathcal{D})$-**mergable** can therefore be easily adapted from the one given in Section 2. Due to lack of space, we cannot present the model merging technique for *deep* pseudo models which is described in [9] where this technique is also extended for other DLs with concrete domains. Full proofs for flat and deep model merging for $\mathcal{ALC}(\mathcal{D})$ can be found in [9].

5   Conclusion and Future Work

In this paper we have analyzed optimization techniques for TBox and ABox reasoning in the expressive description logic $\mathcal{ALCNH}_{R^+}$. These techniques exploit the traversal of flat and/or deep pmodels extracted from ABox consistency tests. A moderate speed gain using deep models for classification of concepts and a dramatic gain for realization of ABoxes is empirically demonstrated. The model merging technique has also been investigated for the logic $\mathcal{ALC}(\mathcal{D})$ with concrete domains. We conjecture that individual model merging for $\mathcal{ALC}(\mathcal{D})$ can be developed in analogy to Section 3. The model merging technique for $\mathcal{ALC}(\mathcal{D})$ is a prerequisite in order to apply model merging to $\mathcal{ALCNH}_{R^+}$ extended by concrete domains.

It is easy to see that an enhanced version of the individual model merging technique for $\mathcal{ALCNH}_{R^+}$ can be developed, which additionally exploits the use of deep models. This is immediately possible if only ABoxes containing no joins for role assertions are encountered. In case an ABox $\mathcal{A}$ contains a join (e.g. $\{(a, c): R, (b, c): R\} \subseteq \mathcal{A}$), one has to consider a graph-like instead of a tree-like traversal of pseudo models reflecting the dependencies caused by joins.

References

1. F. Baader and P. Hanschke. A scheme for integrating concrete domains into concept languages. In *Twelfth International Joint Conference on Artificial Intelligence, Darling Harbour, Sydney, Australia, Aug. 24-30, 1991*, pages 452–457, August 1991.
2. V. Haarslev and R. Möller. An empirical evaluation of optimization strategies for ABox reasoning in expressive description logics. In P. Lambrix et al., editor, *Proceedings of the International Workshop on Description Logics (DL'99), July 30 - August 1, 1999, Linköping, Sweden*, pages 115–119, June 1999.
3. V. Haarslev and R. Möller. Expressive ABox reasoning with number restrictions, role hierarchies, and transitively closed roles. In A.G. Cohn, F. Giunchiglia, and B. Selman, editors, *Proceedings of Seventh International Conference on Principles of Knowledge Representation and Reasoning (KR'2000), Breckenridge, Colorado, USA, April 11-15, 2000*, pages 273–284, April 2000.
4. V. Haarslev, R. Möller, and M. Wessel. The description logic $\mathcal{ALCNH}_{R^+}$ extended with concrete domains: A practically motivated approach. In *Proceedings of the International Joint Conference on Automated Reasoning, IJCAR'2001, June 18-23, 2001, Siena, Italy*, LNCS. Springer-Verlag, Berlin, June 2001.
5. I. Horrocks. *Optimising Tableaux Decision Procedures for Description Logics*. PhD thesis, University of Manchester, 1997.
6. I. Horrocks and P. Patel-Schneider. Optimising description logic subsumption. *Journal of Logic and Computation*, 9(3):267–293, June 1999.
7. I. Horrocks and P.F. Patel-Schneider. DL systems comparison. In *Proceedings of DL'98, International Workshop on Description Logics*, pages 55–57, Trento(Italy), 1998.
8. I. Horrocks, U. Sattler, and S. Tobies. Practical reasoning for expressive description logics. In H. Ganzinger, D. McAllester, and A. Voronkov, editors, *Proceedings of the 6th International Conference on Logic for Programming and Automated Reasoning (LPAR'99)*, number 1705 in Lecture Notes in Artificial Intelligence, pages 161–180. Springer-Verlag, September 1999.
9. A.-Y. Turhan. Optimierungsmethoden für den Erfüllbarkeitstest bei Beschreibungslogiken mit konkreter Domäne Optimization methods for the satisfiability test for description logics with concrete domains (in German). Master's thesis, Universität Hamburg, Fachbereich Informatik University of Hamburg, Computer Science Department, April 2000.

The Hybrid μ-Calculus

Ulrike Sattler[1][*] and Moshe Y. Vardi[2][**]

[1] LuFG Theor. Informatik, RWTH Aachen, Germany, sattler@cs.rwth-aachen.de
[2] Department of Computer Science, Rice University, Houston, TX, vardi@rice.edu

Abstract. We present an ExpTime decision procedure for the full μ-Calculus (including converse programs) extended with nominals and a universal program, thus devising a new, highly expressive ExpTime logic. The decision procedure is based on tree automata, and makes explicit the problems caused by nominals and how to overcome them. Roughly speaking, we show how to reason in a logic lacking the tree model property using techniques for logics with the tree model property. The contribution of the paper is two-fold: we extend the family of ExpTime logics, and we present a technique to reason in the presence of nominals.

1 Introduction

Description Logics (DLs) are a family of knowledge representation formalisms designed for the representation of and reasoning about *terminological knowledge* [34,28,2]. Over the last years, they turned out to be also well-suited for the representation of and reasoning about, e.g., ontologies [31,16] and database schemata, where they can support schema design, evolution, and query optimisation [7], source integration in heterogeneous databases/data warehouses [6], and conceptual modeling of multidimensional aggregation [18].

The basic notions of DLs are concepts (classes, unary predicates) and roles (binary predicates). A specific DL is mainly characterised by a set of constructors that allow to form complex concepts and roles from atomic ones. A standard DL knowledge base consists of two parts: in the *TBox*, the vocabulary of a given application domain is fixed. Some TBox formalisms only allow to introduce names for complex concepts, whereas others allow, additionally, to state general axioms such as $C \doteq D$ or $C \sqsubseteq D$ for two (possibly complex) concepts [11,22]. The second part of a DL knowledge base, the *ABox*, states facts concerning concrete individuals. Using the vocabulary fixed in the TBox, we can state in an ABox that the individual a is an instance of, e.g., the concept CMReactor, and that it is related via the role has-part to an individual b. Given such a "hybrid" knowledge base, interesting reasoning problems include the computation of the taxonomy (i.e., the hierarchy w.r.t. the subsumption relation) of those concepts defined in the TBox, finding inconsistent concepts defined in the TBox, and

---

[*] Part of this work was carried out while the second author was visiting Rice University on a DAAD Grant.
[**] Work partially supported by NSF grants CCR-9700061 and CCR-9988322

R. Goré, A. Leitsch, and T. Nipkow (Eds.): IJCAR 2001, LNAI 2083, pp. 76–91, 2001.

finding, for an individual a in the ABox, the most specific concepts defined in the TBox that a is an instance of.

To be of use in a specific application, a DL must provide the means to describe properties of objects that are relevant for this application. Unsurprisingly, the more expressive power a DL provides, the more complex the reasoning algorithms for this DL are. As a consequence, a variety of DLs were introduced together with investigations of the complexity of the corresponding reasoning algorithms/problems (see, e.g., [26,34,13]).

In 1991, Schild described the close relationship between DLs and modal logics or dynamic logics [32]. For example, it turned out that $\mathcal{ALC}$ is a notational variant of multi modal **K**. Following that, numerous new DLs with corresponding complexity results emerged by (extensions of) translations into modal and dynamic logics [9,33,10]. Due to its high expressive power, the full μ-calculus (i.e., propositional μ-calculus extended with converse programs) can be viewed as (one of) the "queens" of ExpTime modal/dynamic/temporal logics [23,35, 40]. It is able to capture, for example, converse-PDL, CTL*, and other highly expressive modal/dynamic/temporal logics, and thus also highly expressive DLs [5]. Unfortunately, the μ-calculus lacks two features that are of great importance for it being also a "queen" for DLs: it does not provide an analogue for concept definition/general axioms that are provided by TBoxes, and it has no equivalent to ABox individuals. The first point is not a serious one since we could "internalise" general axioms using a greatest fixpoint formula even though the μ-calculus does not provide (constructors to build) a universal program [32]. The second one is more serious since, for example, the extension of the μ-calculus with individuals no longer has the tree model property. Moreover, in the presence of individuals, internalisation becomes more subtle.

In this paper, we extend the μ-calculus with a *universal role/program* to enable direct internalisation of TBoxes [32], and with a generalised form of ABox individuals, namely *nominals*, thus devising a logic where all standard inference problems concerning TBoxes and ABoxes can be reduced to satisfiability. In contrast to ABox individuals, nominals can be used inside complex formulae in the same place as atomic propositions. We are able to show that the complexity of the full μ-calculus, when extended with a universal program and nominals, does not increase, but remains in ExpTime. To prove this upper bound, we reduce satisfiability to the emptiness of alternating automata on infinite trees—a family of automata that can be viewed as abstractions of tableau algorithms. This technique is rather elegant in that it separates the logic from the algorithmics [39]. For example, a tableau-based algorithm might require sophisticated blocking techniques to guarantee termination [22]. Using the automata-theoretic technique, termination is not an issue since we can work on infinite trees. Moreover, this technique makes explicit which problems arise when reasoning in the presence of nominals and universal roles, and how to deal with them. We have chosen to deal with nominals by explicitly guessing most of the relevant information concerning nominals—a choice that will be explained in the sequel.

Besides being of interest by itself and, once again, showing the power of the automata-theoretic approach, the complexity result presented here broadens the range description/modal/dynamic logics that have ExpTime decision procedures. Over the last few years, it was shown that tableau-based algorithms for certain ExpTime-complete reasoning problems are amenable to optimisation and behave quite well in practise [21,29,19,22]. Thus, establishing an ExpTime upper bound is a first step in developing a practical decision procedure for the hybrid μ-calculus or, at least, for fragments of this logic. We return to the practicality issue at the end of the paper.

Unfortunately, this new "queen" logic is still not "the queen" since it is missing a prominent feature, namely number restrictions/graded modalities [17, 12,38]. This is due to the fact that, in the presence of converse roles and universal programs/roles (or any other means to internalise axioms), nominals and number restrictions/graded modalities lead to NExpTime-hardness [37].

From the tense logic perspective [4], the hybrid μ-calculus can also be viewed as one of the "queen" hybrid logics with ExpTime-complete reasoning problems: our result extends ExpTime-completeness results for, e.g., Priorean tense logic over transitive frames (which can be viewed as a notational variant of multi-modal **K**4 with converse modalities) or converse-PDL with nominals in [1].

2   Preliminaries

In this section, we introduce syntax and semantics of the hybrid μ-calculus as well as two-way automata. It is the extension of the propositional μ-calculus with converse programs [40], a universal role, and nominals [30,1], i.e., atomic formulae to refer to single points.

Definition 1. *Let* AP *be a set of atomic propositions,* Var *a set of propositional variables,* Nom *a set of nominals, and* Prog *a set of atomic programs with the universal program* $o \in$ Prog. *A* program *is either an atomic program or the* converse a^- *of an atomic program* $a \in$ Prog. *The set of formulae of the* hybrid μ-calculus *is the smallest set such that*

- **true, false,** *p and $\neg p$ are formulae for $p \in$ AP $\cup$ Nom,*
- *$x \in$ Var is a formula,*
- *if φ_1 and φ_2 are formulae, α is a program, and x is a propositional variable, then $\varphi_1 \wedge \varphi_2$, $\varphi_1 \vee \varphi_2$, $\langle \alpha \rangle \varphi_1$, $[\alpha] \varphi_1$, $\mu x.\varphi_1(x)$ and $\nu x.\varphi_1(x)$ are formulae.*

A propositional variable $x \in$ Var is said to occur free *in a formula if it occurs outside the scope of a fixpoint operator. A* sentence *is formula that contains no free propositional variable, i.e., each occurrence of a variable x is in the scope of a fixpoint operator μ or ν. We use λ to denote a fixpoint operator μ or ν. For a λ-formula $\lambda x.\varphi(x)$, we write $\varphi(\lambda x.\varphi(x))$ to denote the formula that is obtained by replacing each free occurrence of x in φ with $\lambda x.\varphi(x)$.*

Semantics is defined by means of a Kripke structure and, in the presence of variables and fixpoints, a valuation that associates a set of points with each variable. Readers not familiar with fixpoints might want to look at [23,35] for instructive examples and explanations of the semantics of the μ-calculus.

Definition 2. *Semantics of the hybrid μ-calculus is given by means of a Kripke structure $K = (W, R, L)$, where*

- *W is a set of points,*
- *$R : \mathsf{Prog} \longrightarrow 2^{W \times W}$ assigns to an atomic program a binary relation on W,*
- *$R(o) = W \times W$, and*
- *$L : \mathsf{AP} \cup \mathsf{Nom} \longrightarrow 2^W$ assigns to each atomic proposition or nominal the set of points in which it holds, such that $L(n)$ is a singleton for each nominal n.*

R is extended to converse programs as follows: $R(a^-) = \{(v, u) \mid (u, v) \in R(a)\}$.

Given a Kripke structure $K = (W, R, L)$ and variables $x_1, \ldots, x_m$, a valuation $\mathcal{V} : \{x_1, \ldots, x_m\} \longrightarrow 2^W$ maps each variable to a subset of W. For a valuation $\mathcal{V}$, a variable x, and a set of points $W' \subseteq W$, $\mathcal{V}[x/W']$ is the valuation that is obtained from $\mathcal{V}$ by assigning W' to x.

A formula φ with free variables among $x_1, \ldots, x_m$ is interpreted over a Kripke structure $K = (W, R, L)$ as a mapping φ^K that associates, with each valuation $\mathcal{V}$, a subset $\varphi^K(\mathcal{V})$ of W. This mapping is defined inductively as follows:

- $\mathbf{true}^K(\mathcal{V}) = W$, $\mathbf{false}^K(\mathcal{V}) = \emptyset$,
- *for $p \in \mathsf{AP} \cup \mathsf{Nom}$, we have $p^K(\mathcal{V}) = L(p)$ and $(\neg p)^K(\mathcal{V}) = W \setminus L(p)$*
- $(\varphi_1 \wedge \varphi_2)^K(\mathcal{V}) = (\varphi_1)^K(\mathcal{V}) \cap (\varphi_2)^K(\mathcal{V})$,
  $(\varphi_1 \vee \varphi_2)^K(\mathcal{V}) = (\varphi_1)^K(\mathcal{V}) \cup (\varphi_2)^K(\mathcal{V})$,
    $(\langle \alpha \rangle \varphi)^K(\mathcal{V}) = \{u \in W \mid \text{ there is a } v \text{ with } (u, v) \in R(\alpha) \text{ and } v \in \varphi^K(\mathcal{V})\}$,
    $([\alpha] \varphi)^K(\mathcal{V}) = \{u \in W \mid \text{ for all } v, (u, v) \in R(\alpha) \text{ implies } v \in \varphi^K(\mathcal{V})\}$,
- $(\mu x.\varphi(x))^K(\mathcal{V}) = \bigcap \{W' \subseteq W \mid \varphi^K(\mathcal{V}[x/W']) \subseteq W'\}$
  $(\nu x.\varphi(x))^K(\mathcal{V}) = \bigcup \{W' \subseteq W \mid \varphi^K(\mathcal{V}[x/W']) \supseteq W'\}$

For a sentence ψ, a Kripke structure $K = (W, R, L)$, and $w \in W$, we write $K, w \models \psi$ iff $w \in \psi^K$, and call K a model *of ψ.[1] A sentence that has a model is called* satisfiable.

Remark 1. All formulae are by definition in negation normal form, i.e., negation occurs only in front of atomic propositions or nominals.

In the following, we will sometimes write $\psi(n_1, \ldots, n_\ell)$ to emphasize that $n_1, \ldots, n_\ell$ are exactly the nominals occurring in ψ.

Since we will treat atomic programs and their converse symmetrically, we will use $\overline{\alpha}$ to denote the converse of a program, i.e., a^- if $\alpha = a$ for some atomic program a, and b if $\alpha = b^-$ for some atomic program b. We use Prog_ψ to denote all (possibly negated) programs occurring in ψ.

In many decidable hybrid logics, we find formulae of the form $\varphi@n$ (to be read as " the formula φ holds at the nominal n") with the semantics

$$(\varphi@n)^K(\mathcal{V}) = \begin{cases} W & \text{if } n \in \varphi^K(\mathcal{V}) \\ \emptyset & \text{otherwise .} \end{cases}$$

---

[1] The interpretation of a sentence is independent of valuations.

We did not provide this operator since, in the presence of the universal role o, we can make use of the equivalence $\varphi@n \equiv [o](\neg n \vee \varphi)$.

We note that the formula $[o]n$ is satisfied only by a structure with a single state. This formula cannot be expressed without the use of both nominals and the universal program.

Finally, we introduce two-way alternating automata on infinite trees. This family of automata generalises non-deterministic tree automata in two ways: firstly, they allow for the rather elegant and succinct alternation [27], which allows for transitions such as "being in state q and seeing letter σ, the automaton either has an accepting run with q_1 from the left successor *and* an accepting run with q_2 from the right successor, *or* it has an accepting run with q' from the left successor." To express this kind of transitions, the transition functions involves positive boolean formulae instead of, e.g., sets of tuples of states as for non-deterministic automata. Secondly, being two-way allows runs to go up and down the input tree, similar to converse programs, which allow following programs in both directions. When running on a k-ary tree, a two-way automaton can have transitions going to the ith child and switching to state q' (denoted (i, q') with $1 \leq i \leq k$), staying at the same node switching to state q' (denoted $(0, q')$), or going to its (unique) predecessor and switching to state q' (denoted $(-1, q')$). For an introduction to two-way alternating automata and their application to the full μ-calculus, see [40].

Definition 3. *For $k \geq 1$ an integer, $(\{1, \dots, k\}^*, V)$ is a k-ary Σ-labelled tree if V is a mapping that associates, with each node $x \in \{1, \dots, k\}^*$, its label $V(x) \in \Sigma$. Intuitively, for $1 \leq i \leq k$, $x \cdot i$ is the ith child of x.*

*Let $\mathcal{B}^+(X)$ be the set of positive Boolean formulae (i.e., formulae built using $\wedge$ and $\vee$ only) over the set X. For $X' \subseteq X$, we say that X' satisfies a formula $\Theta \in \mathcal{B}^+(X)$ iff assigning **true** to all elements in X' and **false** to all elements in $X \setminus X'$ makes Θ true.*

Let $[k] = \{-1, 0, 1, \dots, k\}$. A two-way alternating automaton on k-ary Σ-labelled trees is a tuple $\mathbf{A} = (\Sigma, Q, \delta, q_0, F)$, where Q is a finite set of states, $q_0 \in Q$ is the initial state, $\delta : Q \times \Sigma \to \mathcal{B}^+([k] \times Q)$ is the transition relation, and F is the acceptance condition.

A run of $\mathbf{A}$ on a Σ-labelled k-ary tree (T, V) is a $(T \times Q)$-labelled tree (T_r, r) that satisfies the following conditions:

- $\epsilon \in T_r$ *and* $r(\epsilon) = (\epsilon, q_0)$,
- *If $y \in T_r$ with $r(y) = (x, q)$ and $\delta(q, V(x)) = \Theta$, then there is a (possibly empty) set $S \subseteq [k] \times Q$ that satisfies Θ such that, for each $(c, q') \in S$, there is a node $y \cdot i \in T_r$ satisfying the following conditions:*
  - *If $c = \epsilon$, then $r(y \cdot i) = (x, q')$.*
  - *If $c \geq 1$, then $r(y \cdot i) = (x \cdot c, q')$.*
  - *If $c = -1$, then $x = x' \cdot i$ for some $1 \leq i \leq k$, and $r(y \cdot i) = (x', q')$.*

A run (T_r, r) is accepting iff all its infinite paths satisfy the acceptance condition. Since we use tree automata for the μ-calculus, we consider the parity *condition*

[36]. *A parity condition is given by an ascending chain of states of sets $F = (F_0, \ldots, F_k)$ with $F_i \subseteq F_{i+1}$. Given a path P in (T_r, r), let $\inf(P)$ denote the states that are infinitely often visited by P. Then P is accepted iff the minimal i with $\inf(P) \cap F_i \neq \emptyset$ is even.*

For two-way alternating automata, the *emptiness problem* is the following question: given a two-way alternating automaton $\mathbf{A}$, is there a tree (T, V) such that $\mathbf{A}$ has an accepting run on (T, V)? It was shown in [40] that this problem is solvable in time that is exponential in the number of $\mathbf{A}$'s states, where the exponent is a polynomial in the length of the parity condition.

3   Hybrid μ-Calculus Has a Tree Model Property

As usual, when proving a tree model property for the hybrid μ-calculus, we want to "unravel" a given model to a tree model. In the presence of nominals, this is clearly not possible since, for example, the formula $n \wedge \langle \alpha \rangle (m \wedge \langle \beta \rangle n)$ with $n, m \in \mathsf{Nom}$ has no model in the form of a tree. However, we will show that we can unravel each model to a forest, i.e., a collection of trees. When unravelling, we must choose "good" points that witness diamond formulae (i.e., a point y with $y \in \varphi^K$ and $(x, y) \in R(\alpha)$ for $x \in (\langle \alpha \rangle \varphi)^K$)—where being "good" is rather tricky in the presence of fixpoints. To this purpose, we define a *choice function* that chooses the "good" witnesses. Essentially, this choice function is a memoryless strategy whose existence is guaranteed for parity games [14]. Definition 4 is the extension of the standard ones to nominals, see, e.g., [35,40].

Definition 4. *The closure $\mathsf{cl}(\psi)$ of a sentence ψ is the smallest set of sentences that satisfies the following:*

- $\psi \in \mathsf{cl}(\psi)$,
- *if $\varphi_1 \wedge \varphi_2 \in \mathsf{cl}(\psi)$ or $\varphi_1 \vee \varphi_2 \in \mathsf{cl}(\psi)$, then $\{\varphi_1, \varphi_2\} \subseteq \mathsf{cl}(\psi)$,*
- *if $\langle \alpha \rangle \varphi \in \mathsf{cl}(\psi)$ or $[\alpha] \varphi \in \mathsf{cl}(\psi)$, then $\varphi \in \mathsf{cl}(\psi)$, and*
- *if $\lambda x. \varphi(x) \in \mathsf{cl}(\psi)$, then $\varphi(\lambda x. \varphi(x)) \in \mathsf{cl}(\psi)$.*

An atom $\mathbf{A} \subseteq \mathsf{cl}(\psi)$ of ψ is a set of formulae that satisfies the following:

- *if $p \in \mathsf{AP} \cup \mathsf{Nom}$ occurs in ψ, then, exclusively, either $p \in \mathbf{A}$ or $\neg p \in \mathbf{A}$,*
- *if $\varphi_1 \wedge \varphi_2 \in \mathsf{cl}(\psi)$, then $\varphi_1 \wedge \varphi_2 \in \mathbf{A}$ iff $\{\varphi_1, \varphi_2\} \subseteq \mathbf{A}$,*
- *if $\varphi_1 \vee \varphi_2 \in \mathsf{cl}(\psi)$, then $\varphi_1 \vee \varphi_2 \in \mathbf{A}$ iff $\{\varphi_1, \varphi_2\} \cap \mathbf{A} \neq \emptyset$, and*
- *if $\lambda x. \varphi(x) \in \mathsf{cl}(\psi)$, then $\lambda x. \varphi(x) \in \mathbf{A}$ iff $\varphi(\lambda x. \varphi(x)) \in \mathbf{A}$.*

The set of atoms of ψ is denoted $\mathsf{at}(\psi)$.

A pre-model (K, π) for a sentence ψ consists of a Kripke structure $K = (W, R, L)$ and a mapping $\pi : W \longrightarrow \mathsf{at}(\psi)$ that satisfies the following properties:

- *there is a $u_0 \in W$ with $\psi \in \pi(u_0)$,*
- *for $p \in \mathsf{AP} \cup \mathsf{Nom}$, if $p \in \pi(u)$, then $u \in L(p)$, and if $\neg p \in \pi(u)$, then $u \notin L(p)$,[2]*

---

[2] Hence if a nominal n is in $\pi(u)$, then $L(n) = \{u\}$.

- if $\langle \alpha \rangle \varphi \in \pi(u)$, then there is a $v \in W$ with $(u,v) \in R(\alpha)$ and $\varphi \in \pi(v)$, and
- if $[\alpha] \varphi \in \pi(u)$, then $\varphi \in \pi(v)$ for each $v \in W$ with $(u,v) \in R(\alpha)$.

A choice function $\mathsf{ch} : W \times \mathsf{cl}(\psi) \longrightarrow \mathsf{cl}(\psi) \cup W$ for a pre-model (K, π) of ψ is a partial function that, for each $u \in W$,

(i) if $\varphi_1 \vee \varphi_2 \in \pi(u)$, then $\mathsf{ch}(u, \varphi_1 \vee \varphi_2) \in \{\varphi_1, \varphi_2\} \cap \pi(u)$ and
(ii) if $\langle \alpha \rangle \varphi \in \pi(u)$, then $\mathsf{ch}(u, \langle \alpha \rangle \varphi) = v$ for some v with $(u,v) \in R(\alpha)$ and $\varphi \in \pi(v)$.

An adorned pre-model (K, π, ch) consists of a pre-model (K, π) and a choice function ch.

For an adorned pre-model $(W, R, L, \pi, \mathsf{ch})$ of ψ, the derivation relation $\rightsquigarrow \subseteq (\mathsf{cl}(\psi), W)^2$ is defined as follows:

- if $\varphi_1 \vee \varphi_2 \in \pi(u)$, then $(\varphi_1 \vee \varphi_2, u) \rightsquigarrow (\mathsf{ch}(\varphi_1 \vee \varphi_2), u)$
- if $\varphi_1 \wedge \varphi_2 \in \pi(u)$, then $(\varphi_1 \wedge \varphi_2, u) \rightsquigarrow (\varphi_i, u)$ for each $i \in \{1, 2\}$,
- if $\langle \alpha \rangle \varphi \in \pi(u)$, then $(\langle \alpha \rangle \varphi, u) \rightsquigarrow (\varphi, \mathsf{ch}(\langle \alpha \rangle \varphi, u))$
- if $[\alpha] \varphi \in \pi(u)$, then $([\alpha] \varphi, u) \rightsquigarrow (\varphi, v)$ for each v with $(u,v) \in R(\alpha)$
  (for $\alpha = o$, that means that $([o] \varphi, u) \rightsquigarrow (\varphi, v)$ for each $v \in W$)
- if $\lambda x.\varphi(x) \in \pi(u)$, then $(\lambda x.\varphi(x), u) \rightsquigarrow (\varphi(\lambda x.\varphi(x)), u)$

A least-fixpoint sentence $\mu x.\varphi(x)$ is said to be regenerated from point u to point v in an adorned pre-model (K, π, ch) if there is a sequence $(\rho_1, u_1), \ldots, (\rho_k, u_k)$ with $k \geq 1$ such that $\rho_1 = \rho_k = \mu x.\varphi(x)$, $u = u_1$ and $v = u_k$, for each $1 \leq i < k$, we have $(\rho_i, u_i) \rightsquigarrow (\rho_{i+1}, u_{i+1})$, and $\mu x.\varphi(x)$ is a sub-sentence of each ρ_i. We say that (K, π, ch) is well-founded if there is no least fixpoint sentence $\mu x.\varphi(x) \in \mathsf{cl}(\psi)$ and an infinite sequence $u_0, u_1, \ldots$ such that, for each $i \geq 0$, $\mu x.\varphi(x)$ is regenerated from u_i to u_{i+1}.

Lemma 1. A sentence ψ has a model K iff ψ has a well-founded adorned pre-model (K, π, ch).

Proof. The construction of a model from a well-founded adorned pre-model and, vice versa, of a well-founded adorned pre-model from a model, are analogous to the constructions that can be found in [35]. These constructions are, as mentioned in [40], insensitive to converse programs, and—due to the according modifications of the technical details—also insensitive to nominals. Indeed, nominals behave simply like atomic propositions provided that $L(n)$ is guaranteed to be interpreted as a singleton. □

Definition 5. The relaxation of a pre-model (W, R, L, π) of a sentence $\psi(n_1, \ldots, n_\ell)$ consists of mappings R^r and π^r, where

$$R^r : \mathsf{Prog} \rightarrow W \times W \text{ and}$$
$$R^r : \alpha \mapsto R(\alpha) \setminus \{(u,v) \mid \text{for some } 1 \leq i \leq \ell, L(n_i) = \{v\}\}$$
$$\pi^r : W \rightarrow \{G \mid G = G_1 \cup G_2, \quad G_1 \in \mathsf{at}(\psi), \text{ and}$$
$$G_2 \subseteq \{\xrightarrow{\alpha} n_i \mid \alpha \text{ occurs in } \psi, \alpha \neq o, \text{ and } 1 \leq i \leq \ell\}\}$$
$$\pi^r : u \mapsto \pi(u) \cup \{\xrightarrow{\alpha} n \mid (u,v) \in R(\alpha), \ \alpha \neq o, \text{ and } L(n) = \{v\}\}$$

A relaxation is a forest if R^r forms a forest.

Lemma 2. *If a sentence ψ is satisfiable, then it has a well-founded adorned pre-model whose relaxation is a forest and has ψ in the label of one of its roots.*

Proof. Let ψ be satisfiable. Hence there is a well-founded adorned pre-model (K, π, ch) with $K = (W, R, L)$ for ψ due to Lemma 1. Using a technique similar to the one in [40], we construct from (K, π, ch) a well-founded adorned pre-model (K', π', ch') whose relaxation is a forest. Please note that, due to the presence of converse programs, we cannot simply unravel K. However, we can use the choice function to do something similar that yields the desired result also in the presence of converse programs.

Let $\psi = \psi(n_1, \dots, n_\ell)$ and $w_0 \in W$ such that $w_0 \in \psi^K$. Let $|\psi| = n$, let $\langle \alpha_1 \rangle \varphi_1, \dots, \langle \alpha_k \rangle \varphi_{k'}$ be all diamond formulae in $\mathsf{cl}(\psi)$, and let k be the maximum of k' and $\ell + 1$. Hence we have $k \leq n$. We define a mapping $\tau : \{1, \dots, k\}^+ \longrightarrow W \cup \{\perp\}$ inductively, together with an adorned pre-model (K', π', ch') where $K' = (W', R', L')$, $W' = \mathsf{dom}(\tau) \setminus \{x \mid \tau(x) = \perp\}$, and

- for $p \in \mathsf{AP} \cup \mathsf{Nom}$, $x \in L'(p)$ iff $\tau(x) \in L(p)$,
- $\pi'(x) = \pi(\tau(x))$,
- $\mathsf{ch}'(x, \varphi_1 \vee \varphi_2) = \mathsf{ch}(\tau(x), \varphi_1 \vee \varphi_2)$, and
- R' and $\mathsf{ch}'(x, \varphi)$ for diamond formulae φ are defined inductively together with τ.

(Fix the first level) For j with $1 \leq j \leq \ell$, let $v_{f(1)}, \dots, v_{f(\ell)} \in W$ be such that $L(n_j) = \{v_{f(j)}\}$ and $f(1) \leq \cdots \leq f(\ell) \leq \ell$—since it is possible that $L(n) = L(n')$ for nominals $n \neq n'$, f need not be injective. For $1 \leq j \leq \ell$, set $\tau(f(j)) = v_{f(j)}$.
For $w_0 \in W$ with $w_0 \in \psi^K$, if $w_0 \notin \{v_{f(1)}, \dots, v_{f(\ell)}\}$, then set $\tau(f(\ell) + 1) = w_0$. Set $\tau(j) = \perp$ for each $1 \leq j \leq k$ not yet defined.

(Fix the rest) For the induction, let i be such that $\tau(x)$ is already defined for each $x \in \{1, \dots, k\}^i$, and j with $1 \leq j \leq k$ such that $\tau(x1), \dots, \tau(x(j-1))$ is already defined for each $x \in \{1, \dots, k\}^i$. Then, for each $x \in \{1, \dots, k\}^i$, do the following:
(1) if $\langle \alpha_j \rangle \varphi_j \notin \pi'(x)$ or $\tau(x) = \perp$, then define $\tau(xj) = \perp$.
(2) if $\langle \alpha_j \rangle \varphi_j \in \pi'(x)$, then (since (K, π, ch) is a pre-model and $\pi'(x) = \pi(\tau(x))$), there is some $v \in W$ with $\mathsf{ch}(\tau(x), \langle \alpha_j \rangle \varphi_j) = v$ and $(\tau(x), v) \in R(\alpha_j)$.
  - If $\{v\} = L(n_{\ell'})$ for some $1 \leq \ell' \leq \ell$, then (since we have already fixed the first level) there is some r with $1 \leq r \leq \ell$ with $\tau(r) = v$. Add (x, r) to $R'(\alpha_j)$, and set $\mathsf{ch}'(x, \langle \alpha_j \rangle \varphi_j) = r$ and $\tau(xj) = \perp$.
  - Otherwise, add (x, xj) to $R'(\alpha_j)$, set $\tau(xj) = v$ and $\mathsf{ch}'(x, \langle \alpha_j \rangle \varphi_j) = xj$.

Since we started from an adorned pre-model, (K', π', ch') is obviously an adorned pre-model. Moreover, if a sentence $\mu x.\varphi(x)$ is regenerated from x to y in (K', π', ch'), then $\mu x.\varphi(x)$ is also regenerated from $\tau(x)$ to $\tau(y)$ in (K, π, ch). Since the latter is well-founded, we thus have that (K', π', ch') is well-founded. Next, its relaxation R'^r is a forest (consisting of trees starting at the first level) since the only edges in R' that "go back", i.e., that are not of the form (x, xi), are exactly those that are eliminated in R'^r. Finally, ψ is satisfied in one of the root nodes since, by definition of (K', π', ch'), we have $j \in \psi^{K'}$ for some $1 \leq j \leq f(\ell) + 1$.

$\square$

Remark 2. Please note that in this construction, if x satisfies a diamond formula $\langle \alpha \rangle \, \varphi$, then either a successor xj of x or one of the first level nodes representing nominals satisfies φ.

4   Deciding Existence of Forest Models

It remains to devise a procedure that decides, for a sentence ψ, whether it has a well-founded adorned pre-model whose relaxation is a forest. To this purpose, we define a two-way alternating tree automaton that accepts exactly the forest-relaxations of ψ's pre-models—provided that we added a new dummy node whose successors are the root nodes of the forest relaxation.

The automaton depends on a *guess* which contains relevant information concerning the interpretation of nominals. The guess makes sure that the following kind of situation is handled correctly: suppose a nominal n must satisfy a formula of the form $[\alpha] \, \varphi$, and we have a point x with $(x, n) \in R(\overline{\alpha})$, but this relationship is only implicit since we work on relaxations of pre-models, i.e., $(x, n) \notin R^r(\overline{\alpha})$ and $\xrightarrow{\overline{\alpha}} n \in \pi^r(x)$. In that case, the guess makes sure that x satisfies φ since it determines which box formulae are satisfied by nominals. Moreover, the guess determines which nominals are interpreted as the same objects, and how nominals are related to each other by programs.

It is possible to refer all this "guessing" directly to the automaton—hence we had only one automaton instead of one per guess. We have chosen, however, to work with explicit guesses since, on the one hand, it makes explicit the additional non-determinism one has to cope with in the presence of nominals and how it can be dealt with. On the other hand and more importantly, referring the guessing into the automaton would yield a quadratic blow-up of the state space. Let n be the number of states and m be the length of the acceptance condition of a two-way alternating tree automaton. When deciding emptiness of a two-way alternating tree automaton [40], it is transformed into a non-deterministic (one-way) parity tree automaton whose state space is of size $(nm^2)^{nm^2}$, and whose acceptance condition is of length nm^2. Emptiness of the latter automaton can be decided in time $2^{O((n^2m^4)(\log n + 2\log m))}$ [25]. Hence a (quadratic) blow-up of the state space of our initial two-way alternating tree automaton would further increase the degree of the polynomial in the exponent of the runtime, and thus be rather expensive.

Formally, a guess consists of three components, the first one consisting, for each nominal n, of a set γ of formulae satisfied by a point u with $L(n) = \{u\}$. Since one point may represent several nominals, we use a second component f to relate a nominal n_i to "its" set of formulae $\gamma_{f(i)}$. The third component describes how two points representing nominals are interrelated via (interpretations of) programs, making sure that, if one is an α-successor of the other, then the other is an $\overline{\alpha}$-successor of the first one.

Definition 6. *A guess* $\mathcal{G} = (G, f, C)$ *for a hybrid μ-calculus sentence* $\psi(n_1,$ $\dots, n_\ell)$ *consists of a guess list* $G = (\gamma_1, \dots, \gamma_\ell)$ *together with connections* $C \subseteq$ $\mathsf{Nom} \times \mathsf{Prog}_\psi \times \mathsf{Nom}$ *and a guess mapping* $f : \{1, \dots, \ell\} \longrightarrow \{1, \dots, \ell\}$, *where, for each* $1 \le i, j \le \ell$, *we have* $\emptyset \subsetneq \gamma_i \subseteq \mathsf{cl}(\psi)$ *or* $\gamma_i = \bot$, $n_i \in \gamma_{f(i)}$, $n_i \notin \gamma_j$ *for all* $j \ne f(i)$, $\mathsf{Nom} \cap \gamma_i = \emptyset$ *implies* $\gamma_i = \bot$, *and* $(n_i, \alpha, n_j) \in C$ *iff* $(n_j, \overline{\alpha}, n_i) \in C$.

Theorem 1. *Let* ψ *be a hybrid μ-calculus sentence. For each guess* $\mathcal{G}$ *for* ψ, *we define a two-way alternating tree automaton* $\mathcal{B}(\psi, \mathcal{G})$, *such that*

1. *if* ψ *is satisfiable, then there exists a guess* $\mathcal{G}'$ *for* ψ *such that the language accepted by* $\mathcal{B}(\psi, \mathcal{G}')$ *is non-empty,*
2. *if a tree is accepted by* $\mathcal{B}(\psi, \mathcal{G})$, *then eliminating its root node yields a forest relaxation of a well-founded adorned pre-model of* ψ, *and*
3. *the number of* $\mathcal{B}(\psi, \mathcal{G})$ *'s states is linear in* $|\psi|$.

Proof. For ease of presentation, we assume that all input trees are full trees, i.e., all non-leaf nodes have the same number of children. As we have seen in the proof of Lemma 2, we can simply "fill" a tree with additional nodes labelled $\bot$ to make it a full tree. Moreover, we assume a "dummy" root node whose direct successors are exactly the root nodes of trees in the forest relaxation.

For a sentence $\psi(n_1, \dots, n_\ell)$ with k' diamond subformulae in $\mathsf{cl}(\psi)$ as specified in the proof of Lemma 2 and a guess $\mathcal{G}$, we define two alternating automata, $\mathcal{A}(\psi, \mathcal{G})$ and $\tilde{\mathcal{A}}(\psi, \mathcal{G})$, and then define $\mathcal{B}(\psi, \mathcal{G})$ as the intersection of $\mathcal{A}(\psi, \mathcal{G})$ and $\tilde{\mathcal{A}}(\psi, \mathcal{G})$. For alternating automata, intersection is trivial (basically, we introduce a new initial state $\tilde{q}$ with $\delta(\tilde{q}, \sigma) = (0, q_0) \wedge (0, q_0')$ for the former initial states q_0, q_0'), and the size of $\mathcal{B}(\psi, \mathcal{G})$ is the sum of the sizes of $\mathcal{A}(\psi, \mathcal{G})$ and $\tilde{\mathcal{A}}(\psi, \mathcal{G})$.

The automaton $\tilde{\mathcal{A}}(\psi, \mathcal{G})$ is rather simple and guarantees that the structure of the input tree is as required, whereas $\mathcal{A}(\psi, \mathcal{G})$ really makes sure that the input tree (more precisely, the sub-forest of the input tree obtained by eliminating the root and all nodes labelled with $\bot$) is a relaxation of a well-founded adorned pre-model.

Both automata work on the same alphabet Σ, which is defined as follows: For $\mathsf{Prog}^+ = \{p_\alpha, p_{\overline{\alpha}}, \overline{p}_\alpha, \overline{p}_{\overline{\alpha}} \mid \alpha$ is a program in ψ different from $o\}$,

$$\Sigma = \{\bot, \mathsf{root}\} \cup \{\sigma \mid \sigma \subseteq \mathsf{AP} \cup \mathsf{Nom} \cup \mathsf{Prog}^+ \cup \{\xrightarrow{\alpha_j} n_i \mid 1 \le j \le m \text{ and } 1 \le i \le \ell\},$$
$$\sigma \text{ contains, for each } \alpha, \text{ exclusively, either } p_\alpha \text{ or } \overline{p}_\alpha, \text{ and,}$$
$$\text{exclusively, either } p_{\overline{\alpha}} \text{ or } \overline{p}_{\overline{\alpha}}\}$$

The intuition of the additional symbols are as follows: Nodes not representing points in a Kripke structure are labelled root and $\bot$, where root labels the root node. Nodes having n_i (i.e., the node labelled with the corresponding guess $\gamma_{f(i)}$) as an α-successor are marked $\xrightarrow{\alpha} n_i$, just like in relaxations. A node label contains p_α ($p_{\overline{\alpha}}$) if this node is an α-successor ($\overline{\alpha}$-successor) of its (unique) predecessor. We do allow that a node is both an α- and a β-successor, or that no program can be associated to the edge between two nodes. Analogously, $\overline{p}_\alpha$ ($\overline{p}_{\overline{\alpha}}$) are used to mark those nodes that are *not* α-successors ($\overline{\alpha}$-successors).

The "simple" automaton $\tilde{\mathcal{A}}(\psi, \mathcal{G})$ guarantees that root is only found at the root label, the nominals in γ_i are only found at the ith successors of the root, the first level nodes contain no p_α or $p_{\overline{\alpha}}$ and that, if a nominal n_i has another nominal n_j as its α-successor (i.e., if $\overset{\alpha}{\to} n_j$ is in the label of the node representing n_i), then n_j has n_i as its $\overline{\alpha}$-successor (i.e., $\overset{\overline{\alpha}}{\to} n_i$ is in the label of the node representing n_j). More precisely, $\tilde{\mathcal{A}}(\psi, \mathcal{G}) = (\Sigma, \{q_0, q_1, \ldots, q_\ell, q', q\}, \delta', q_0)$ is a safety one-way alternating automaton (i.e., each state is accepting and thus every run is an accepting run), and δ' is defined as follows for $\sigma \in \Sigma$:

$$\delta'(q_0, \sigma) = \begin{cases} \bigwedge_{i=1}^{\ell}(i, q_i) \wedge \bigwedge_{i=\ell+1}^{k}(i, q) \wedge \bigwedge_{i=1}^{k}(i, q') & \text{if root} = \sigma \\ \textbf{false} & \text{otherwise} \end{cases}$$

$$\delta'(q', \sigma) = \begin{cases} \textbf{true} & \text{if } p_\alpha \notin \sigma \text{ and } p_{\overline{\alpha}} \notin \sigma \text{ for each } \alpha \neq o \text{ in } \psi \\ \textbf{false} & \text{otherwise} \end{cases}$$

for $1 \leq i \leq \ell$:

$$\delta'(q_i, \sigma) = \begin{cases} \bigwedge_{i=1}^{k}(i, q) & \text{if } \gamma_i \cap (\text{Nom} \cup \text{AP}) = \sigma \cap (\text{Nom} \cup \text{AP}), \text{ root} \neq \sigma, \text{ and,} \\ & \text{for each } n \in \text{Nom} \cap \sigma \text{ and } (n, \alpha, n') \in C, \overset{\alpha}{\to} n' \in \sigma \\ \textbf{false} & \text{otherwise} \end{cases}$$

$$\delta'(q, \sigma) = \begin{cases} \bigwedge_{i=1}^{k}(i, q) & \text{if } \sigma \cap \text{Nom} = \emptyset \text{ and root} \neq \sigma \\ \textbf{false} & \text{otherwise} \end{cases}$$

Due to the symmetry in the definition of the connection component in a guess and the way $\delta'(q_i, \sigma)$ is defined, if $\tilde{\mathcal{A}}(\psi, \mathcal{G})$ accepts a tree, $\{n_i, \overset{\alpha}{\to} n_j\} \subseteq \sigma$, and $n_j \in \sigma'$, then $\overset{\overline{\alpha}}{\to} n_i \in \sigma'$, and σ, σ' label direct successors of the root node.

The two-way alternating tree automaton $\mathcal{A}(\psi, \mathcal{G})$ verifies that the input tree is indeed a relaxation of a well-founded adorned pre-model. To this purpose, (most of) its states correspond to formulae in $\text{cl}(\psi)$, and the transition relation basically follows the semantics.

The first conjunct in the definition of $\delta(q_0', \sigma)$ guarantees that the ith successor of the root node indeed satisfies all formulae in γ_i, and that one of the root node successors satisfies ψ.

An additional state q' that "travels" once through the whole input tree makes sure that, whenever a node has a nominal n_i as its implicit $\overline{\alpha}$-successor (i.e., its label contains $\overset{\overline{\alpha}}{\to} n_i$), then this node satisfies indeed all formulae φ with $[\alpha]\varphi \in \gamma_{f(i)}$.

Finally, the diamond and box formulae on the universal role are treated separately since they apply to all but the root node, regardless of marks p_α or $\overline{p}_\alpha$. Please note that, since the root node does not represent any point of a Kripke structure, $\delta([o]\varphi, \text{root})$ is defined such that only all root successors satisfy $[o]\varphi$, but not the root node itself. More precisely, we have

$$\mathcal{A}(\psi, \mathcal{G}) = (\Sigma, Q, \delta, q_0', F), \text{ with}$$
$$Q = \{\bot, q_0', q'\} \cup \text{cl}(\psi) \cup \text{Prog}^+.$$

The transition relation δ is defined as follows: firstly, for $q \in Q$ and $\sigma \in \Sigma$ let

$$\delta(q, \bot) = \begin{cases} \textbf{true} & \text{if } q = \bot \\ \textbf{false} & \text{otherwise} \end{cases} \qquad \delta(\bot, \sigma) = \begin{cases} \textbf{true} & \text{if } \sigma = \bot \\ \textbf{false} & \text{otherwise} \end{cases}$$

Secondly, for $1 \leq i \leq \ell$ and $\sigma \in \Sigma$, let

$$\Gamma(i) = \begin{cases} (i, \bot) & \text{if } \gamma_i = \bot \\ \bigwedge_{\varphi \in \gamma_i}(i, \varphi) & \text{if } \gamma_i \subseteq \text{cl}(\psi) \end{cases} \qquad N(\sigma) = \bigwedge_{\substack{\xrightarrow{\alpha} n_i \in \sigma \text{ and} \\ [\alpha]\,\varphi \in \gamma_{f(i)}}} (0, \varphi)$$

Thirdly, for $\sigma \in \Sigma$, $\sigma \neq \bot$, and α a program, we define δ as follows:

$$\delta(q_0', \sigma) = \bigwedge_{i=1}^{\ell} \Gamma(i) \wedge \bigvee_{i=1}^{k}(i, \psi) \wedge \bigwedge_{i=1}^{k}((i, q') \vee (i, \bot))$$
$$\delta(q', \sigma) = N(\sigma) \wedge \bigwedge_{i=1}^{k}((i, q') \vee (i, \bot))$$

for $p \in \text{AP} \cup \text{Nom} \cup \text{Prog}^+$:
$$\delta(p, \sigma) = \begin{cases} \textbf{true} & \text{if } p \in \sigma \\ \textbf{false} & \text{otherwise} \end{cases}$$

for $p \in \text{AP} \cup \text{Nom}$:
$$\delta(\neg p, \sigma) = \begin{cases} \textbf{true} & \text{if } p \notin \sigma \text{ and } \sigma \neq \text{root} \\ \textbf{false} & \text{otherwise} \end{cases}$$

$$\delta(\varphi_1 \wedge \varphi_2, \sigma) = (0, \varphi_1) \wedge (0, \varphi_2)$$
$$\delta(\varphi_1 \vee \varphi_2, \sigma) = (0, \varphi_1) \vee (0, \varphi_2)$$
$$\delta(\lambda x.\varphi(x), \sigma) = (0, \varphi(\lambda x.\varphi(x)))$$

for $\alpha \notin \{o, o^-\}$:
$$\delta(\langle \alpha \rangle\,\varphi, \sigma) = \begin{cases} \textbf{true} & \text{if } \xrightarrow{\alpha} n_i \in \sigma \text{ and } \varphi \in \gamma_{f(i)} \\ \bigvee_{j=1}^{k}((j, \varphi) \wedge (j, p_\alpha)) & \text{otherwise} \end{cases}$$

for $\alpha \notin \{o, o^-\}$:
$$\delta([\alpha]\,\varphi, \sigma) = \begin{cases} \textbf{false} & \text{if } \xrightarrow{\alpha} n_i \in \sigma \text{ and } \varphi \notin \gamma_{f(i)} \\ ((-1, \varphi) \vee (0, \overline{p_\alpha})) \wedge & \text{otherwise} \\ \bigwedge_{j=1}^{k}((j, \varphi) \vee (j, \overline{p}_\alpha) \vee (j, \bot)) \end{cases}$$

for $\alpha \in \{o, o^-\}$:
$$\delta(\langle \alpha \rangle\,\varphi, \sigma) = \begin{cases} \textbf{true} & \text{if } \varphi \in \gamma_{f(i)} \\ \bigvee_{j=1}^{k}(j, \varphi) & \text{otherwise} \end{cases}$$

for $\alpha \in \{o, o^-\}$:
$$\delta([\alpha]\,\varphi, \sigma) = \begin{cases} (0, \varphi) \wedge (-1, [\alpha]\,\varphi) \wedge & \text{if root} \neq \sigma \\ \bigwedge_{j=1}^{k}((j, [\alpha]\,\varphi) \vee (j, \bot)) \\ \bigwedge_{j=1}^{k}((j, [\alpha]\,\varphi) \vee (j, \bot)) & \text{otherwise} \end{cases}$$

Please note that, following the construction in the proof of Lemma 2, satisfaction of diamond formulae (including those on the universal program) needs to be tested for only in direct successors and in the nodes representing nominals.

Moreover, since $\psi = \psi(n_1, \ldots, n_\ell)$ and due to the definition of $\delta(q_0', \sigma)$ and $\Gamma(i)$, δ checks whether the node representing n_i satisfies indeed all formulae in $\gamma_{f(i)}$.

The acceptance condition F is defined analogously to the one in [15,24], and given here for the sake of completeness. Firstly, for a fixpoint formula $\varphi \in \mathrm{cl}(\psi)$, define the alternation level of φ to be the number of alternating fixpoint formulae one has to "wrap φ with" to reach a sub-*sentence* of ψ. More precisely, the *alternation level* $\mathrm{al}_\psi(\varphi)$ of $\varphi = \lambda x.\varphi'(x) \in \mathrm{cl}(\psi)$ is defined as follows [3]: if φ is a sentence, then $\mathrm{al}_\psi(\varphi) = 1$. Otherwise, let $\rho = \lambda'y.\rho'(y)$ be the innermost fixpoint formula in $\mathrm{cl}(\psi)$ that contains φ as a proper sub-formula. If $\lambda = \lambda'$, then $\mathrm{al}_\psi(\varphi) = \mathrm{al}_\psi(\rho)$, otherwise $\mathrm{al}_\psi(\varphi) = \mathrm{al}_\psi(\rho) + 1$. Let d be the maximal alternation level of (fixpoint) subformulae of ψ, and define

$$G_i = \{\nu x.\varphi(x) \in \mathrm{cl}(\psi) \mid \mathrm{al}_\psi(\nu x.\varphi(x)) = i\}$$
$$L_i = \{\mu x.\varphi(x) \in \mathrm{cl}(\psi) \mid \mathrm{al}_\psi(\mu x.\varphi(x)) \leq i\}$$

Now we are ready to define the acceptance condition $F = \{F_1, \ldots, F_{2d}\}$ with $F_i = \emptyset$ for $i = 0$, $F_i = F_{i-1} \cup L_i$ for odd $i \geq 1$, and $F_i = F_{i-1} \cup G_i$ for even $i \geq 1$. Obviously, $F_i \subseteq F_{i+1}$ for each $1 \leq i \leq 2d$. As mentioned in Definition 3, a path r_p of a run r is accepting if the minimal i with $\inf(r_p) \cap F_i \neq \emptyset$ is even—this i corresponds to the outermost fixpoint formula that was infinitely often visited/postponed. A run r is accepting if each of its paths are accepting. Intuitively, the acceptance condition makes sure that, if a fixpoint formula was visited infinitely often, then this was a greatest fixpoint formulae, and that all of its least fixpoint super-formulae were visited only finitely many times.

It remains to verify the three claims in Theorem 1. The proof of the first one uses Lemma 1 and a straightforward construction of a guess $\mathcal{G}$ from a forest relaxation of a well-founded adorned pre-model, and then shows how an input forest similar to the one constructed in the proof of Lemma 1 is accepted by $\mathcal{B}(\psi, \mathcal{G})$. The second claim can be proved by taking an accepting run of $\mathcal{B}(\psi, \mathcal{G})$ on some input tree, and verifying that the input tree indeed satisfies all properties of relaxations of well-founded adorned pre-models. Finally, the third claim is by definition of $\mathcal{B}(\psi, \mathcal{G})$. $\square$

Theorem 2. *Satisfiability of hybrid μ-calculus is decidable in exponential time.*

Proof. As we have mentioned in the beginning of Section 4, emptiness of $\mathcal{B}(\psi, \mathcal{G})$ can be decided in time $2^{\mathcal{O}(n^6 \log n)}$ for $n = |\psi|$. Let ℓ be the number of nominals and m the number of programs different from o in ψ. Since, for a guess $\mathcal{G} = (G, f, C)$, the mapping f is determined by G, the number of guesses is bound by the number of connections and guess lists, i.e., by $2^{\ell^2 m} \cdot 2^{\ell n}$. Hence we have to test at most an exponential number of automata $\mathcal{B}(\psi, \mathcal{G})$ for emptiness. Combining these results with Lemma 1, Lemma 2, and Theorem 1 concludes the proof. $\square$

5   Conclusion

We have shown that satisfiability of the hybrid μ-calculus can be decided in exponential time, thus partially answering an open question in [5]. Deciding satisfiability of a logic that lacks the tree model property using tree automata was possible using a certain abstraction of models, *relaxations*, and involved an additional non-determinism, *guesses*. Then, we were able to use the emptiness algorithm in [40] as a sub-routine. For an input sentence, the algorithm presented constructs a family of tree automata, each of which depends on a guess that determines relevant information concerning the interpretation of nominals. We have chosen this explicit guess since, on the one hand, it directly shows how nominals can be dealt with. On the other hand, when referring the guessing into the automaton, we would blow up its state space quadratically. Since deciding emptiness of this family of automata is exponential in the size of its state space, it is clearly preferable to avoid even such a polynomial blow-up. The complexity of the hybrid μ-calculus with deterministic programs[3] remains an interesting open problem. As a consequence of NExpTime-hardness results in [37], this extension leads to NExpTime-hardness. Another interesting research problem is the development of practical decision procedures for (fragments of) the hybrid μ-calculus. To the best of our knowledge, automata-theoretic methods are the only known methods for the μ-calculus, and, so far, such methods have been implemented successfully only for linear temporal logic, see, e.g., [8,20].

References

1. C. Areces, P. Blackburn, and M. Marx. The computational complexity of hybrid temporal logics. *Logic Journal of the IGPL*, 8(5), 2000.
2. F. Baader and B. Hollunder. A terminological knowledge representation system with complete inference algorithm. In *Proc. of PDK-91*, vol. 567 of *LNAI*. Springer-Verlag, 1991.
3. G. Bhat and R. Cleaveland. Efficient local model-checking for fragments of the modal μ-calculus. In *Proc. of TACAS*, vol. 1055 of *LNCS*. Springer-Verlag, 1996.
4. P. Blackburn. Nominal tense logic. *Notre Dame Journal of Formal Logic*, 34, 1993.
5. D. Calvanese, G. De Giacomo, and M. Lenzerini. Reasoning in expressive description logics with fixpoints based on automata on infinite trees. In *Proc. of IJCAI'99*, 1999.
6. D. Calvanese, G. De Giacomo, M. Lenzerini, D. Nardi, and R. Rosati. Description logic framework for information integration. In *Proc. of KR-98*, 1998.
7. D. Calvanese, M. Lenzerini, and D. Nardi. Description logics for conceptual data modeling. In *Logics for Databases and Information Systems*. Kluwer Academic Publisher, 1998.
8. E.M. Clarke, O. Grumberg, and K. Hamaguchi. Another look at LTL model checking. In *Proc. of CAV'94*, vol. 818 of *LNCS*, pages 415–427. Springer-Verlag, 1994.
9. G. De Giacomo and M. Lenzerini. Boosting the correspondence between description logics and propositional dynamic logics. In *Proc. of AAAI-94*, 1994.
10. G. De Giacomo and M. Lenzerini. Concept language with number restrictions and fixpoints, and its relationship with μ-calculus. In *Proc. of ECAI-94*, 1994.
11. G. De Giacomo and M. Lenzerini. Tbox and Abox reasoning in expressive description logics. In *Proc. of KR-96*. Morgan Kaufmann, 1996.

---

[3] Or Description Logic's number restrictions or Modal Logic's graded modalities.

12. F. Donini, M. Lenzerini, D. Nardi, and W. Nutt. The complexity of concept languages. In *Proc. of KR-91*. Morgan Kaufmann, 1991.
13. F. M. Donini, M. Lenzerini, D. Nardi, and W. Nutt. The complexity of concept languages. *Information and Computation*, 134, 1997.
14. E. A. Emerson and C. S. Jutla. Tree automata, μ-calculus, and determinacy. In *Proc. of FOCS-91*. IEEE, 1991.
15. E. A. Emerson, C. S. Jutla, and A. P. Sistla. On model checking for fragments of the μ-calculus. In *Proc. of CAV'93*, vol. 697 of *LNCS*. Springer-Verlag, 1993.
16. D. Fensel, I. Horrocks, F. van Harmelen, S. Decker, M. Erdmann, and M. Klein. OIL in a nutshell. In *Proc. EKAW-2000*, vol. 1937 of LNAI, 2000. Springer-Verlag.
17. K. Fine. In so many possible worlds. *Notre Dame J. of Formal Logics*, 13, 1972.
18. E. Franconi and U. Sattler. A data warehouse conceptual data model for multidimensional aggregation: a preliminary report. *AI*IA Notizie*, 1, 1999.
19. V. Haarslev and R. Möller. Expressive abox reasoning with number restrictions, role hierarchies, and transitively closed roles. In *Proc. of KR-00*, 2000.
20. Gerard J. Holzmann. The spin model checker. *IEEE Trans. on Software Engineering*, 23(5), 1997.
21. I. Horrocks. Using an Expressive Description Logic: FaCT or Fiction? In *Proc. of KR-98*, 1998.
22. I. Horrocks, U. Sattler, and S. Tobies. Practical reasoning for very expressive description logics. *Logic Journal of the IGPL*, 8(3), May 2000.
23. D. Kozen. Results on the propositional μ-calculus. In *Proc. of ICALP'82*, vol. 140 of *LNCS*. Springer-Verlag, 1982.
24. O. Kupferman and M. Y. Vardi. μ-calculus synthesis. In *Proc. MFCS'00*, LNCS. Springer-Verlag, 2000.
25. O. Kupferman and M.Y. Vardi. Weak alternating automata and tree automata emptiness. In *Proc. of STOC-98*, 1998.
26. H. Levesque and R. J. Brachman. Expressiveness and tractability in knowledge representation and reasoning. *Computational Intelligence*, 3, 1987.
27. D. E. Muller and P. E. Schupp. Alternating automata on infinite trees. *Theoretical Computer Science*, 54(1-2), 1987.
28. B. Nebel. *Reasoning and Revision in Hybrid Representation Systems*. LNAI. Springer-Verlag, 1990.
29. P. F. Patel-Schneider and I. Horrocks. DLP and FaCT. In *Proc. TABLEAUX-99*, vol. 1397 of *LNAI*. Springer-Verlag, 1999.
30. A. Prior. *Past, Present and Future*. Oxford University Press, 1967.
31. A. Rector and I. Horrocks. Experience building a large, re-usable medical ontology using a description logic with transitivity and concept inclusions. In *Proc. of the AAAI Spring Symposium on Ontological Engineering*. AAAI Press, 1997.
32. K. Schild. A correspondence theory for terminological logics: Preliminary report. In *Proc. of IJCAI-91*, 1991.
33. K. Schild. Terminological cycles and the propositional μ-calculus. In *Proc. of KR-94*, 1994. Morgan Kaufmann.
34. M. Schmidt-Schauß and G. Smolka. Attributive concept descriptions with complements. *Artificial Intelligence*, 48(1), 1991.
35. R. S. Streett and E. A. Emerson. An automata theoretic decision procedure for the propositional μ-calculus. *Information and Computation*, 81(3), 1989.
36. W. Thomas. Languages, automata, and logic. In *Handbook of Formal Language Theory*, vol 1. Springer-Verlag, 1997.
37. S. Tobies. The complexity of reasoning with cardinality restrictions and nominals in expressive description logics. *J. of Artificial Intelligence Research*, 12, 2000.

38. S. Tobies. PSPACE reasoning for graded modal logics. *J. of Logic and Computation*, 2001. To appear.
39. M. Y. Vardi. What makes modal logic so robustly decidable? In *Descriptive Complexity and Finite Models*, American Mathematical Society, 1997.
40. M. Y. Vardi. Reasoning about the past with two-way automata. In *Proc. of ICALP'98*, vol. 1443 of *LNCS*, 1998. Springer-Verlag.

The Inverse Method Implements the Automata Approach for Modal Satisfiability

Franz Baader and Stephan Tobies

LuFG Theoretical Computer Science, RWTH Aachen, Germany
{baader,tobies}@cs.rwth-aachen.de

Abstract. This paper ties together two distinct strands in automated reasoning: the tableau- and the automata-based approach. It shows that the inverse tableau method can be viewed as an implementation of the automata approach. This is of interest to automated deduction because Voronkov recently showed that the inverse method yields a viable decision procedure for the modal logic K.

1 Introduction

Decision procedures for (propositional) modal logics and description logics play an important rôle in knowledge representation and verification. When developing such procedures, one is both interested in their worst-case complexity and in their behavior in practical applications. From the theoretical point of view, it is desirable to obtain an algorithm whose worst-case complexity matches the complexity of the problem. From the practical point of view it is more important to have an algorithm that is easy to implement and amenable to optimizations, so that it behaves well on practical instances of the decision problem. The most popular approaches for constructing decision procedures for modal logics are i) semantic tableaux and related methods [10,2]; ii) translations into classical first-order logics [15,1]; and iii) reductions to the emptiness problem for certain (tree) automata [17,14].

Whereas highly optimized tableaux and translation approaches behave quite well in practice [11,12], it is sometimes hard to obtain exact worst-case complexity results using these approaches. For example, satisfiability in the basic modal logic K w.r.t. global axioms is known to be ExpTime-complete [16]. However, the "natural" tableaux algorithm for this problem is a NExpTime-algorithm [2], and it is rather hard to construct a tableaux algorithm that runs in deterministic exponential time [6]. In contrast, it is folklore that the automata approach yields a very simple proof that satisfiability in K w.r.t. global axioms is in ExpTime. However, the algorithm obtained this way is not only worst-case, but also best-case exponential: it first constructs an automaton that is always exponential in the size of the input formulae (its set of states is the powerset of the set of subformulae of the input formulae), and then applies the (polynomial) emptiness test to this large automaton. To overcome this problem, one must try to construct the automaton "on-the-fly" while performing the emptiness test. Whereas this

R. Goré, A. Leitsch, and T. Nipkow (Eds.): IJCAR 2001, LNAI 2083, pp. 92–106, 2001.
© Springer-Verlag Berlin Heidelberg 2001

idea has successfully been used for automata that perform model checking [9,5], to the best of our knowledge it has not yet been applied to satisfiability checking.

The original motivation of this work was to compare the automata and the tableaux approaches, with the ultimate goal of obtaining an approach that combines the advantages of both, without possessing any of the disadvantages. As a starting point, we wanted to see whether the tableaux approach could be viewed as an on-the-fly realization of the emptiness test done by the automata approach. At first sight, this idea was persuasive since a run of the automaton constructed by the automata approach (which is a so-called looping automaton working on infinite trees) looks very much like a run of the tableaux procedure, and the tableaux procedure does generate sets of formulae on-the-fly. However, the polynomial emptiness test for looping automata does *not* try to construct a run starting with the root of the tree, as done by the tableaux approach. Instead, it computes inactive states, i.e., states that can never occur on a successful run of the automaton, and tests whether all initial states are inactive. This computation starts "from the bottom" by locating obviously inactive states (i.e., states without successor states), and then "propagates" inactiveness along the transition relation. Thus, the emptiness test works in the opposite direction of the tableaux procedure. This observation suggested to consider an approach that inverts the tableaux approach: this is just the so-called inverse method. Recently, Voronkov [19] has applied this method to obtain a bottom-up decision procedure for satisfiability in K, and has optimized and implemented this procedure.

In this paper we will show that the inverse method for K can indeed be seen as an on-the-fly realization of the emptiness test done by the automata approach for K. The benefits of this result are two-fold. First, it shows that Voronkov's implementation, which behaves quite well in practice, is an optimized on-the-fly implementation of the automata-based satisfiability procedure for K. Second, it can be used to give a simpler proof of the fact that Voronkov's optimizations do not destroy completeness of the procedure. We will also show how the inverse method can be extended to handle global axioms, and that the correspondence to the automata approach still holds in this setting. In particular, the inverse method yields an ExpTime-algorithm for satisfiability in K w.r.t. global axioms.

2   Preliminaries

First, we briefly introduce the modal logic K and some technical definitions related to K-formulae, which are used later on to formulate the inverse calculus and the automata approach for K. Then, we define the type of automata used to decide satisfiability (w.r.t. global axioms) in K. These so-called looping automata [18] are a specialization of Büchi tree automata.

Modal Formulae. We assume the reader to be familiar with the basic notions of modal logic. For a thorough introduction to modal logics, refer to, e.g., [4].

K-formulae are built inductively from a countably infinite set $\mathcal{P} = \{p_1, p_2, \dots\}$ of propositional atoms using the Boolean connectives $\wedge$, $\vee$, and $\neg$

and the unary modal operators $\Box$ and $\Diamond$. The semantics of K-formulae is define as usual, based on Kripke models $\mathcal{M} = (W, R, V)$ where W is a non-empty set, $R \subseteq W \times W$ is an accessibility relation, and $V : \mathcal{P} \to 2^W$ is a valuation mapping propositional atoms to the set of worlds they hold in. The relation $\models$ between models, worlds, and formulae is defined in the usual way. Let G, H be K-formulae. Then G is *satisfiable* iff there exists a Kripke model $\mathcal{M} = (W, R, V)$ and a world $w \in W$ with $\mathcal{M}, w \models G$. The formula G is *satisfiable w.r.t. the global axiom H* iff there exists a Kripke model $\mathcal{M} = (W, R, V)$ and a world $w \in W$ such $\mathcal{M}, w \models G$ and $\mathcal{M}, w' \models H$ for all $w' \in W$. K-satisfiability is PSPACE-complete [13], and K-satisfiability w.r.t. global axioms is ExpTime-complete [16].

A K-formula is in *negation normal form* (NNF) if $\neg$ occurs only in front of propositional atoms. Every K-formula can be transformed (in linear time) into an equivalent formula in NNF using de Morgan's laws and the duality of the modal operators.

For the automata and calculi considered here, sub-formulae of G play an important role and we will often need operations going from a formula to its superior sub-formulae. As observed in [19], this becomes easier when dealing with "addresses" of sub-formulae in G rather than with the sub-formulae themselves.

Definition 1 (G-Paths). *For a K-formula G in NNF, the set of G-paths Π_G is a set of words over the alphabet $\{\vee_l, \vee_r, \wedge_l, \wedge_r, \Box, \Diamond\}$. The set Π_G and the sub-formula $G|_\pi$ of G addressed by $\pi \in \Pi_G$ are defined inductively as follows:*

- $\epsilon \in \Pi_G$ and $G|_\epsilon = G$
- *if $\pi \in \Pi_G$ and*
  - $G|_\pi = F_1 \wedge F_2$ *then* $\pi\wedge_l, \pi\wedge_r \in \Pi_G$, $G|_{\pi\wedge_l} = F_1$, $G|_{\pi\wedge_r} = F_2$, *and π is called $\wedge$-path*
  - $G|_\pi = F_1 \vee F_2$ *then* $\pi\vee_l, \pi\vee_r \in \Pi_G$, $G|_{\pi\vee_l} = F_1$, $G|_{\pi\vee_r} = F_2$, *and π is called $\vee$-path*
  - $G|_\pi = \Box F$ *then* $\pi\Box \in \Pi_G$, $G|_{\pi\Box} = F$ *and π is called $\Box$-path*
  - $G|_\pi = \Diamond F$ *then* $\pi\Diamond \in \Pi_G$, $G|_{\pi\Diamond} = F$ *and π is called $\Diamond$-path*
- *Π_G is the smallest set that satisfies the previous conditions.*

We use of $\wedge_*$ and $\vee_*$ as placeholders for $\wedge_l, \wedge_r$ and $\vee_l, \vee_r$, resp. Also, we use λ and $\bar{\lozenge}$ as placeholders for $\wedge, \vee$ and $\Box, \Diamond$, resp. If π is an $\wedge$- or and $\vee$-path then π is called λ-path. If π is a $\Box$- or a $\Diamond$-path then π is called $\bar{\lozenge}$-path. Fig. 1 shows an example of a K-formula G and the corresponding set Π_G, which can be read off the edge labels. For example, $\wedge_r\wedge_r$ is a G-path and $G|_{\wedge_r\wedge_r} = \Box(\neg p_2 \vee p_1)$.

Looping Automata. For a natural number n, let $[n]$ denote the set $\{1, \ldots, n\}$. An *n-ary infinite tree* over the alphabet Σ is a mapping $t : [n]^* \to \Sigma$. An *n-ary looping tree automaton* is a tuple $\mathfrak{A} = (Q, \Sigma, I, \Delta)$, where Q is a finite set of states, Σ is a finite alphabet, $I \subseteq Q$ is the set of initial states, and $\Delta \subseteq Q \times \Sigma \times Q^n$ is the transition relation. Sometimes, we will view Δ as a function from $Q \times \Sigma$ to 2^{Q^n} and write $\Delta(q, \sigma)$ for the set $\{\mathbf{q} \mid (q, \sigma, \mathbf{q}) \in \Delta\}$. A *run* of $\mathfrak{A}$ on a tree t is a n-ary infinite tree r over Q such that $(r(p), t(p), (r(p1), \ldots, r(pn))) \in \Delta$ for

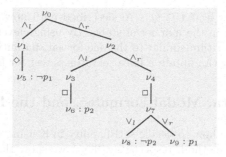

Fig. 1. The set Π_G for $G = \Diamond \neg p_1 \wedge (\Box p_2 \wedge \Box(\neg p_2 \vee p_1))$

every $p \in [n]^*$. The automaton $\mathfrak{A}$ *accepts* t iff there is a run r of $\mathfrak{A}$ on t such that $r(\epsilon) \in I$. The set $L(\mathfrak{A}) := \{t \mid \mathfrak{A} \text{ accepts } t\}$ is the *language* accepted by $\mathfrak{A}$.

Since looping tree automata are special Büchi tree automata, emptiness of their accepted language can effectively be tested using the well-known (quadratic) emptiness test for Büchi automata [17]. However, for looping tree automata this algorithm can be specialized into a simpler (linear) one. Though this is well-known, there appears to be no reference for the result.

Intuitively, the algorithm works by computing inactive states. A state $q \in Q$ is *active* iff there exists a tree t and a run of $\mathfrak{A}$ on t in which q occurs; otherwise, q is *inactive*. It is easy to see that a looping tree automaton accepts at least one tree iff it has an active initial state. How can the set of inactive states be computed? Obviously, a state from which no successor states are reachable is inactive. Moreover, a state is inactive if every transition possible from that state involves an inactive state. Thus, one can start with the set

$$Q_0 := \{q \in Q \mid \forall \sigma \in \Sigma.\Delta(q, \sigma) = \emptyset\}$$

of obviously inactive states, and then propagate inactiveness through the transition relation. We formalize this propagation process in a way that allows for an easy formulation of our main results.

A *derivation* of the emptiness test is a sequence $Q_0 \triangleright Q_1 \triangleright \ldots \triangleright Q_k$ such that $Q_i \subseteq Q$ and $Q_i \triangleright Q_{i+1}$ iff $Q_{i+1} = Q_i \cup \{q\}$ with

$$q \in \{q' \in Q \mid \forall \sigma \in \Sigma.\forall(q_1, \ldots, q_n) \in \Delta(q, \sigma).\exists j.q_j \in Q_i\}.$$

We write $Q_0 \triangleright^* P$ iff there is a $k \in \mathbb{N}$ and a derivation $Q_0 \triangleright \ldots \triangleright Q_k$ with $P = Q_k$. The emptiness test answers "$L(\mathfrak{A}) = \emptyset$" iff there exists a set of states P such that $Q_0 \triangleright^* P$ and $I \subseteq P$.

Note that $Q \triangleright P$ implies $Q \subseteq P$ and that $Q \subseteq Q'$ and $Q \triangleright P$ imply $Q' \triangleright^* P$. Consequently, the *closure* $Q_0^\triangleright$ of Q_0 under $\triangleright$, defined by $Q_0^\triangleright :=: \bigcup \{P \mid Q_0 \triangleright^* P\}$, can be calculated starting with Q_0, and successively adding states q to the current set Q_i such that $Q_i \triangleright Q_i \cup \{q\}$ and $q \notin Q_i$, until no more states can be added. It is easy to see that this closure consists of the set of inactive states, and thus $L(\mathfrak{A}) = \emptyset$ iff $I \subseteq Q_0^\triangleright$. As described until now, this algorithm runs in time polynomial in the number of states. By using clever data structures and a propagation algorithm similar to the one for satisfiability of propositional Horn formulae [7], one can obtain a linear emptiness test for looping tree automata.

3   Automata, Modal Formulae, and the Inverse Calculus

We first describe how to decide satisfiability in K using the automata approach and the inverse method, respectively. Then we show that both approaches are closely connected.

3.1   Automata and Modal Formulae

Given a K-formula G, we define an automaton $\mathfrak{A}_G$ such that $L(\mathfrak{A}_G) = \emptyset$ iff G is not satisfiable. In contrast to the "standard" automata approach, the states of our automaton $\mathfrak{A}_G$ will be subsets of Π_G rather than sets of subformulae of G. Using paths instead of subformulae is mostly a matter of notation. We also require the states to satisfy additional properties (i.e., we do not allow for arbitrary subsets of Π_G). This makes the proof of correctness of the automata approach only slightly more complicated, and it allows us to treat some important optimizations of the inverse calculus within our framework. The next definition introduces these properties.

Definition 2 (Propositionally expanded, clash). *Let G be a K-formula in NNF, Π_G the set of G-paths, and $\Phi \subseteq \Pi_G$. An $\wedge$-path $\pi \in \Phi$ is propositionally expanded in Φ iff $\{\pi\wedge_l, \pi\wedge_r\} \subseteq \Phi$. An $\vee$-path $\pi \in \Phi$ is propositionally expanded in Φ iff $\{\pi\vee_l, \pi\vee_r\} \cap \Phi \neq \emptyset$. The set Φ is propositionally expanded iff every $\mathbb{X}$-path $\pi \in \Phi$ is propositionally expanded in Φ. We use "p.e." as an abbreviation for "propositionally expanded".*

The set Φ' is an expansion of the set Φ if $\Phi \subseteq \Phi'$, Φ' is p.e. and Φ' is minimal w.r.t. set inclusion with these properties. For a set Φ, we define the set of its expansions as $\langle\!\langle \Phi \rangle\!\rangle := \{\Phi' \mid \Phi' \text{ is an expansion of } \Phi\}$.

Φ contains a clash iff there are two paths $\pi_1, \pi_2 \in \Phi$ such that $G|_{\pi_1} = p$ and $G|_{\pi_2} = \neg p$ for a propositional variable p. Otherwise, Φ is called clash-free.

For a set of paths Ψ, the set $\langle\!\langle \Psi \rangle\!\rangle$ can effectively be constructed by successively adding paths required by the definition of p.e. A formal construction of the closure can be found in the proof of Lemma 4. Note that $\emptyset$ is p.e., clash-free, and $\langle\!\langle \emptyset \rangle\!\rangle = \{\emptyset\}$.

Definition 3 (Formula Automaton). *For a K-formula G in NNF, we fix an arbitrary enumeration $\{\pi_1, \ldots, \pi_n\}$ of the $\Diamond$-paths in Π_G. The n-ary looping automaton $\mathfrak{A}_G$ is defined by $\mathfrak{A}_G := (Q_G, \Sigma_G, \langle\!\langle\{\epsilon\}\rangle\!\rangle, \Delta_G)$, where $Q_G := \Sigma_G := \{\Phi \subseteq \Pi_G \mid \Phi \text{ is p.e.}\}$ and the transition relation Δ_G is defined as follows:*

– Δ_G contains only tuples of the form $(\Phi, \Phi, \ldots)$.

- If Φ is clash-free, then we define $\Delta_G(\Phi, \Phi) := \langle\!\langle \Psi_1 \rangle\!\rangle \times \cdots \times \langle\!\langle \Psi_n \rangle\!\rangle$, where

$$
\Psi_i = \begin{cases} \{\pi_i \Diamond\} \cup \{\pi\Box \mid \pi \in \Phi \text{ is a } \Box\text{-path }\} & \text{if } \pi_i \in \Phi \text{ for the } \Diamond\text{-path } \pi_i \\ \emptyset & \text{else} \end{cases}
$$

- If Φ contains a clash, then $\Delta_G(\Phi, \Phi) = \emptyset$, i.e., there is no transition from Φ.

Note, that this definition implies $\Delta_G(\emptyset, \emptyset) = \{(\emptyset, \ldots, \emptyset)\}$ and only states with a clash have no successor states.

Theorem 1. For a K-formula G, G is satisfiable iff $L(\mathfrak{A}_G) \neq \emptyset$.

This theorem can be proved by showing that i) every tree accepted by $\mathfrak{A}_G$ induces a model of G; and ii) every model $\mathcal{M}$ of G can be turned into a tree accepted by $\mathfrak{A}_G$ by a) unraveling $\mathcal{M}$ into a tree model $\mathcal{T}$ for G; b) labeling every world of $\mathcal{T}$ with a suitable p.e. set depending on the formulae that hold in this world; and c) padding "holes" in $\mathcal{T}$ with $\emptyset$.

Together with the emptiness test for looping tree automata, Theorem 1 yields a decision procedure for K-satisfiability. To test a K-formula G for unsatisfiability, construct $\mathfrak{A}_G$ and test whether $L(\mathfrak{A}_G) = \emptyset$ holds using the emptiness test for looping tree automata: $L(\mathfrak{A}_G) = \emptyset$ iff $\langle\!\langle \{\epsilon\} \rangle\!\rangle \subseteq Q_0^{\triangleright}$, where $Q_0 \subseteq Q_G$ is the set of states containing a clash. The following is a derivation of a superset of $\langle\!\langle \{\epsilon\} \rangle\!\rangle$ from Q_0 for the example formula from Fig. 1:

$$
Q_0 = \underbrace{\{\{\nu_5, \nu_6, \nu_7, \nu_8\}, \ \{\nu_5, \nu_6, \nu_7, \nu_9\}, \ldots\}}_{= \ \langle\!\langle \nu_5, \nu_6, \nu_7 \rangle\!\rangle} \triangleright Q_0 \cup \underbrace{\{\{\nu_0, \nu_1, \nu_2, \nu_3, \nu_4\}\}}_{= \ \langle\!\langle \{\epsilon\} \rangle\!\rangle}
$$

3.2 The Inverse Calculus

In the following, we introduce the inverse calculus for K. We stay close to the notation and terminology used in [19].

A *sequent* is a subset of Π_G. Sequents will be denoted by capital greek letters. The union of two sequents Γ and Λ is denote by Γ, Λ. If Γ is a sequent and $\pi \in \Pi_G$ then we denote $\Gamma \cup \{\pi\}$ by Γ, π. If Γ is a sequent that contains only $\Box$-paths then we write $\Gamma\Box$ to denote the sequent $\{\pi\Box \mid \pi \in \Gamma\}$. Since states of $\mathfrak{A}_G$ are also subsets of Π_G and hence sequents, we will later on use the same notational conventions for states as for sequents.

Definition 4 (The inverse path calculus). Let G be a formula in NNF and Π_G the set of paths of G. Axioms of the inverse calculus are all sequents $\{\pi_1, \pi_2\}$ such that $G|_{\pi_1} = p$ and $G|_{\pi_2} = \neg p$ for some propositional variable p. The rules of the inverse calculus are given in Fig. 2, where all paths occurring in a sequent are G-paths and, for every $\Diamond^+$ inference, π is a $\Diamond$-path. We refer to this calculus by IC_G.[1]

---

[1] G appears in the subscript because the calculus is highly dependent of the input formula G: only G-paths can be generated by IC_G.

$$(\vee)\frac{\Gamma_l, \pi\vee_l \quad \Gamma_r, \pi\vee_r}{\Gamma_l, \Gamma_r, \pi} \qquad (\wedge_l)\frac{\Gamma, \pi\wedge_l}{\Gamma, \pi} \qquad (\wedge_r)\frac{\Gamma, \pi\wedge_r}{\Gamma, \pi}$$

$$(\Diamond)\frac{\Gamma\square, \pi\Diamond}{\Gamma, \pi} \qquad (\Diamond^+)\frac{\Gamma\square}{\Gamma, \pi}$$

Fig. 2. Inference rules of IC_G

We define $\mathcal{S}_0 := \{\Gamma \mid \Gamma$ is an axiom $\}$. A derivation of IC_G is a sequence of sets of sequents $\mathcal{S}_0 \vdash \cdots \vdash \mathcal{S}_m$ where $\mathcal{S}_i \vdash \mathcal{S}_{i+1}$ iff $\mathcal{S}_{i+1} = \mathcal{S}_i \cup \{\Gamma\}$ such that there exist sequents $\Gamma_1, \ldots \Gamma_k \in \mathcal{S}_i$ and $\dfrac{\Gamma_1 \quad \cdots \quad \Gamma_k}{\Gamma}$ is an inference.

We write $\mathcal{S}_0 \vdash^* \mathcal{S}$ iff there is a derivation $\mathcal{S}_0 \vdash \cdots \vdash \mathcal{S}_m$ with $\mathcal{S} = \mathcal{S}_m$. The *closure* $\mathcal{S}_0^\vdash$ of $\mathcal{S}_0$ under $\vdash$ is defined by $\mathcal{S}_0^\vdash = \bigcup\{\mathcal{S} \mid \mathcal{S}_0 \vdash^* \mathcal{S}\}$. Again, the closure can effectively be computed by starting with $\mathcal{S}_0$ and then adding sequents that can be obtained by an inference until no more new sequents can be added.

As shown in [19], the computation of the closure yields a decision procedure for K-satisfiability:

Fact 1. *G is unsatisfiable iff $\{\epsilon\} \in \mathcal{S}_0^\vdash$.*

Fig. 3 shows the inferences of IC_G that lead to $\nu_0 = \epsilon$ for the example formula from Fig. 1.

3.3 Connecting the Two Approaches

The results shown in this subsection imply that IC_G can be viewed as an on-the-fly implementation of the emptiness test for $\mathfrak{A}_G$. In addition to generating states on-the-fly, states are also represented in a compact manner: one sequent generated by IC_G represents several states of $\mathfrak{A}_G$.

Definition 5. *For the formula automaton $\mathfrak{A}_G$ with states Q_G and a sequent $\Gamma \subseteq \Pi_G$ we define $[\![\Gamma]\!] := \{\Phi \in Q_G \mid \Gamma \subseteq \Phi\}$, and for a set $\mathcal{S}$ of sequents we define $[\![\mathcal{S}]\!] := \bigcup_{\Gamma \in \mathcal{S}}[\![\Gamma]\!]$.*

The following theorem, which is one of the main contributions of this paper, establishes the correspondence between the emptiness test and IC_G. Its proof will be sketched in the remainder of this section (see [3] for details).

Theorem 2 (IC_G and the emptiness test mutually simulate each other). *Let Q_0, $\mathcal{S}_0$, $\rhd$, and $\vdash$ be defined as above.*

1. *Let Q be a set of states such that $Q_0 \rhd^* Q$. Then there exists a set of sequents $\mathcal{S}$ with $\mathcal{S}_0 \vdash^* \mathcal{S}$ and $Q \subseteq [\![\mathcal{S}]\!]$.*
2. *Let $\mathcal{S}$ be a set of sequents such that $\mathcal{S}_0 \vdash^* \mathcal{S}$. Then there exists a set of states $Q \subseteq Q_G$ with $Q_0 \rhd^* Q$ and $[\![\mathcal{S}]\!] \subseteq Q$.*

$$(\vee)\dfrac{\wedge_l\Diamond,\ \wedge_r\wedge_r\Box\vee_r\quad|\quad \wedge_r\wedge_l\Box,\ \wedge_r\wedge_r\Box\vee_l}{(\Diamond)\dfrac{\wedge_l\Diamond,\ \wedge_r\wedge_l\Box,\ ,\wedge_r\wedge_r\Box}{(\wedge_r)\dfrac{\wedge_l,\ \wedge_r\wedge_l,\ \wedge_r\wedge_r}{(\wedge_l)\dfrac{\wedge_l,\ \wedge_r,\ \wedge_r\wedge_l}{(\wedge_r)\dfrac{\wedge_l,\ \wedge_r}{(\wedge_l)\dfrac{\epsilon,\ \wedge_l}{\epsilon}}}}}}$$

Fig. 3. An example of inferences in IC_G

The first part of the theorem shows that IC_G can simulate each computation of the emptiness test for $\mathfrak{A}_G$. The set of states represented by the set of sequents computed by IC_G may be larger than the one computed by a particular derivation of the emptiness test. However, the second part of the theorem implies that all these states are in fact inactive since a possibly larger set of states can also be computed by a derivation of the emptiness test. In particular, the theorem implies that IC_G can be used to calculate a compact representation of $Q_0^\triangleright$. This is an on-the-fly computation since $\mathfrak{A}_G$ is never constructed explicitly.

Corollary 1. $Q_0^\triangleright = [\![S_0^\vdash]\!]$.

The proof of the second part of Theorem 2 is the easier one. It is a consequence of the next three lemmata. First, observe that the two calculi have the same starting points.

Lemma 1. *If S_0 is the set of axioms of IC_G, and Q_0 is the set of states of $\mathfrak{A}_G$ that have no successor states, then $[\![S_0]\!] = Q_0$.*

Second, since states are assumed to be p.e., propositional inferences of IC_G do not change the set of states represented by the sequents.

Lemma 2. *Let $S \vdash T$ be a derivation of IC_G that employs a $\wedge_l$-, $\wedge_r$-, or a $\vee$-inference. Then $[\![S]\!] = [\![T]\!]$.*

Third, modal inferences of IC_G can be simulated by derivations of the emptiness test.

Lemma 3. *Let $S \vdash T$ be derivation of IC_G that employs a $\Diamond$- or $\Diamond^+$-inference. If Q is a set of states with $[\![S]\!] \cup Q_0 \subseteq Q$ then there exists a set of states P with $Q \triangleright^* P$ and $[\![T]\!] \subseteq P$.*

Given these lemmata, proving Theorem 2.2 is quite simple.

Proof of Theorem 2.2. The proof is by induction on the length m of the derivation $S_0 \vdash S_1 \cdots \vdash S_m = S$ of IC_G. The base case $m = 0$ is Lemma 1. For the induction step, S_{i+1} is either inferred from S_i using a propositional inference, which is dealt with by Lemma 2, or by a modal inference, which is dealt with by Lemma 3. Lemma 3 is applicable since, for every set of states Q with $Q_0 \triangleright^* Q$, $Q_0 \subseteq Q$. $\square$

Proving the first part of Theorem 2 is more involed because of the calculation of the propositional expansions implicit in the definition of $\mathfrak{A}_G$.

Lemma 4. *Let $\Phi \subseteq \Pi_G$ be a set of paths and $\mathcal{S}$ a set of sequents such that $\langle\!\langle \Phi \rangle\!\rangle \subseteq [\![\mathcal{S}]\!]$. Then there exists a set of sequents $\mathcal{T}$ with $\mathcal{S} \vdash^* \mathcal{T}$ such that there exists a sequent $\Lambda \in \mathcal{T}$ with $\Lambda \subseteq \Phi$.*

Proof. If Φ is p.e., then this is immediate, as in this case $\langle\!\langle \Phi \rangle\!\rangle = \{\Phi\} \subseteq [\![\mathcal{S}]\!]$.

If Φ is not p.e., then let select be an arbitrary *selection function*, i.e., a function that maps every set Ψ that is not p.e. to a X-path $\pi \in \Psi$ that is not p.e. in Ψ. Let T_Φ be the following, inductively defined tree:

- The root of T_Φ is Φ.
- If a node Ψ of T_Φ is not p.e., then
    - if select$(\Psi) = \pi$ is an $\wedge$-path, then Ψ has the successor node $\Psi, \pi\wedge_l, \pi\wedge_r$ and Ψ is called an $\wedge$-node.
    - if select$(\Psi) = \pi$ is an $\vee$-path, then Ψ has the successor nodes $\Psi, \pi\vee_l$ and $\Psi, \pi\vee_l$ and Ψ is called an $\vee$-node.
- If a node Ψ of T_Φ is p.e., then it is a leaf of the tree.

Obviously, the construction is such that the set of leaves of T_Φ is $\langle\!\langle \Phi \rangle\!\rangle$.

Let $\Upsilon_1, \ldots \Upsilon_\ell$ be a post-order traversal of this tree, so the sons of a node occur before the node itself and $\Upsilon_\ell = \Phi$. Along this traversal we will construct a derivation $\mathcal{S} = \mathcal{T}_0 \vdash^* \cdots \vdash^* \mathcal{T}_\ell = \mathcal{T}$ such that, for every $1 \leq i \leq j \leq \ell$, $\mathcal{T}_j$ contains a sequent Λ_i with $\Lambda_i \subseteq \Upsilon_i$. Since the sets $\mathcal{T}_j$ grow monotonically, it suffices to show that, for every $1 \leq i \leq \ell$, $\mathcal{T}_i$ contains a sequent Λ_i with $\Lambda_i \subseteq \Upsilon_i$.

Whenever Υ_i is a leaf of T_Φ, then $\Upsilon_i \in \langle\!\langle \Phi \rangle\!\rangle \subseteq [\![\mathcal{S}]\!]$. Hence there is already a sequent $\Lambda_i \in \mathcal{T}_0$ with $\Lambda_i \subseteq \Upsilon_i$ and no derivation step is necessary. Particularly, in a post-order traversal, Υ_1 is a leaf.

We now assume that the derivation has been constructed up to $\mathcal{T}_i$. We restrict our attention to the case where Υ_{i+1} is an $\vee$-node since the case where Υ_{i+1} is an $\wedge$-node can be treated similarly and the case where Υ_{i+1} is a leaf as above.

Thus, assume that Υ_{i+1} is an $\vee$-node with selected $\vee$-path $\pi \in \Upsilon_{i+1}$. Then, the successors of Υ_{i+1} in T_Φ are $\Upsilon_{i+1}, \pi\vee_l$ and $\Upsilon_{i+1}, \pi\vee_r$, and by construction there exist sequences $\Lambda_l, \Lambda_r \in \mathcal{T}_i$ with $\Lambda_* \subseteq \Upsilon_{i+1}, \pi\vee_*$. If $\pi\vee_l \notin \Lambda_l$ or $\pi\vee_r \notin \Lambda_r$, then $\Lambda_l \subseteq \Upsilon_{i+1}$ or $\Lambda_r \subseteq \Upsilon_{i+1}$ holds and hence already $\mathcal{T}_i$ contains a sequent Λ with $\Lambda \subseteq \Upsilon_{i+1}$.

If $\Lambda_l = \Gamma_l, \pi\vee_l$ and $\Lambda_r = \Gamma_r, \pi\vee_r$ with $\pi\vee_* \notin \Gamma_*$ then IC_G can use the inference

$$(\vee)\frac{\Gamma_l, \pi\vee_l \qquad \Gamma_r, \pi\vee_r}{\Gamma_l, \Gamma_r, \pi}$$

to derive $\mathcal{T}_i \vdash \mathcal{T}_i \cup \{\Gamma_l, \Gamma_r, \pi\} = \mathcal{T}_{i+1}$, and $\Gamma_l, \Gamma_r, \pi \subseteq \Upsilon_{i+1}$ easily follows.   □

Proof of Theorem 2.1. We show this by induction on the number k of steps in the derivation $Q_0 \rhd \ldots \rhd Q_k = Q$. Again, Lemma 1 yields the base case.

For the induction step, let $Q_0 \rhd \ldots \rhd Q_i \rhd Q_{i+1} = Q_i \cup \{\Phi\}$ be a derivation of the emptiness test and $\mathcal{S}_i$ a set of sequents such that $\mathcal{S}_0 \vdash^* \mathcal{S}_i$ and $Q_i \subseteq [\![\mathcal{S}_i]\!]$. If already $\Phi \in Q_i$ then $Q_{i+1} \subseteq [\![\mathcal{S}_i]\!]$ and we are done.

If $\Phi \notin Q_i$, then $Q_0 \subseteq Q_i$ implies that $\Delta_G(\Phi, \Phi) \neq \emptyset$. Since $\emptyset$ is an active state, we know that $\emptyset \notin Q_i$, and for $Q_i \rhd Q_{i+1}$ to be a possible derivation of the emptiness test, $\Delta_G(\Phi, \Phi) = \langle\!\langle \Psi_1 \rangle\!\rangle \times \cdots \times \langle\!\langle \Psi_n \rangle\!\rangle \neq \{(\emptyset, \ldots, \emptyset)\}$ must hold, i.e., there must be a $\Psi_j \neq \emptyset$ such that $\langle\!\langle \Psi_j \rangle\!\rangle \subseteq Q_i \subseteq [\![S_i]\!]$. Hence $\pi_j \in \Phi$ and $\Psi_j = \{\pi_j \Diamond\} \cup \{\pi \Box \mid \pi \in \Phi \text{ is a } \Box\text{-path}\}$.

Lemma 4 yields the existence of a set of sequents T_i with $S_i \vdash^* T_i$ containing a sequent Λ with $\Lambda \subseteq \Psi_j$. This sequent is either of the form $\Lambda = \Gamma\Box, \pi_j \Diamond$ or $\Lambda = \Gamma\Box$ for some $\Gamma \subseteq \Phi$. In the former case, IC_G can use a $\Diamond$-inference and in the latter case a $\Diamond^+$-inference to derive $S_0 \vdash^* S_i \vdash^* T_i \vdash T_i \cup \{\Gamma, \pi_j\} = S$ and $\Phi \subseteq [\![\Gamma, \pi_j]\!]$ holds.   □

4   Optimizations

Since the inverse calculus can be seen as an on-the-fly implementation of the emptiness test, optimizations of the inverse calculus also yield optimizations of the emptiness test. We use the connection between the two approaches to provide an easier proof of the fact that the optimizations of IC_G introduced by Voronkov [19] do not destroy completeness of the calculus.

4.1   Unreachable States / Redundant Sequents

States that cannot occur on any run starting with an initial state have no effect on the language accepted by the automaton. We call such states *unreachable*. In the following, we will determine certain types of unreachable states.

Definition 6. *Let* $\pi, \pi_1, \pi_2 \in \Pi_G$.

- *The* modal length *of* π *is the number of occurrences of* $\Box$ *and* $\Diamond$ *in* π.
- $\pi_1, \pi_2 \in \Pi_G$ *form a* $\vee$-fork *if* $\pi_1 = \pi \vee_l \pi_1'$ *and* $\pi_2 = \pi \vee_r \pi_2'$ *for some* π, π_1', π_2'.
- π_1, π_2 *are* $\Diamond$-separated *if* $\pi_1 = \pi_1' \Diamond \pi_1''$ *and* $\pi_2 = \pi_2' \Diamond \pi_2''$ *such that* π_1', π_2' *have the same modal length and* $\pi_1' \neq \pi_2'$.

Lemma 5. *Let* $\mathfrak{A}_G$ *be the formula automaton for a* K-*formula* G *in NNF and* $\Phi \in Q$. *If* Φ *contains a* $\vee$-fork, *two* $\Diamond$-separated paths, *or two paths of different modal length, then* Φ *is unreachable.*

The lemma shows that we can remove such states from $\mathfrak{A}_G$ without changing the accepted language. Sequents containing a $\vee$-fork, two $\Diamond$-separated paths, or two paths of different modal length represent only unreachable states, and are thus redunant, i.e., inferences involving such sequents need not be considered.

Definition 7 (Reduced automaton). *Let* $\bar{Q}$ *be the set of states of* $\mathfrak{A}_G$ *that contain a* $\vee$-fork, *two* $\Diamond$-separated paths, *or two paths of different modal length.* *The* reduced automaton $\mathfrak{A}_G' = (Q_G', \Sigma_G, \langle\!\langle \{\epsilon\} \rangle\!\rangle, \Delta_G')$ *is defined by*

$$Q_G' := Q_G \setminus \bar{Q} \quad \text{and} \quad \Delta_G' := \Delta_G \cap (Q_G' \times \Sigma_G \times Q_G' \times \cdots \times Q_G').$$

Since the states in $\bar{Q}$ are unreachable, $L(\mathfrak{A}_G) = L(\mathfrak{A}_G')$. From now on, we consider $\mathfrak{A}_G'$ and define $[\![\cdot]\!]$ relative to the states on $\mathfrak{A}_G'$: $[\![\Gamma]\!] = \{\Phi \in Q_G' \mid \Gamma \subseteq \Phi\}$.

4.2 G-Orderings / Redundant Inferences

In the following, the applicability of the propositional inferences of the inverse calculus will be restricted to those where the affected paths are maximal w.r.t. a total ordering of Π_G. In order to maintain completeness, one cannot consider arbitrary orderings in this context.

Two paths π_1, π_3 are *brothers* iff there exists a $\mathbb{X}$-path π such that $\pi_1 = \pi \mathbb{X}_l$ and $\pi_3 = \pi \mathbb{X}_r$ or $\pi_1 = \pi \mathbb{X}_r$ and $\pi_3 = \pi \mathbb{X}_l$.

Definition 8 (G-ordering). *Let G be a K-formula in NNF. A total ordering $\succ$ of Π_G is called a G-ordering iff*

1. *$\pi_1 \succ \pi_2$ whenever*
    a) *the modal length of π_1 is strictly greater than the modal length of π_2; or*
    b) *π_1, π_2 have the same modal length, the last symbol of π_1 is $\mathbb{X}_*$, and the last symbol of π_2 is $\lozenge$; or*
    c) *π_1, π_2 have the same modal length and π_2 is a prefix of π_1*
2. *There is no path between brothers, i.e., there exist no G-paths π_1, π_2, π_3 such that $\pi_1 \succ \pi_2 \succ \pi_3$ and π_1, π_3 are brothers.*

For the example formula G of Fig. 1, a G-ordering $\succ$ can be defined by setting $\nu_9 \succ \nu_8 \succ \cdots \succ \nu_1 \succ \nu_0$. Voronkov [19] shows that G-orderings exist for every K-formula G in NNF. Using an arbitrary, but fixed G-ordering $\succ$, the applicability of the propositional inferences is restricted as follows.

Definition 9 (Optimized Inverse Calculus). *For a sequent Γ and a path π we write $\pi \succ \Gamma$ iff $\pi \succ \pi'$ for every $\pi' \in \Gamma$.*

– *An inference* $(\wedge_*)\dfrac{\Gamma, \pi\wedge_*}{\Gamma, \pi}$ *respects $\succ$ iff $\pi\wedge_* \succ \Gamma$.*

– *An inference* $(\vee)\dfrac{\Gamma_l, \pi\vee_l \quad \Gamma_r, \pi\vee_r}{\Gamma_l, \Gamma_r, \pi}$ *respects $\succ$ iff $\pi\vee_l \succ \Gamma_l$ and $\pi\vee_r \succ \Gamma_r$.*

– *The $\lozenge$- and $\lozenge^+$-inferences always respect $\succ$.*

The optimized inverse calculus $\mathsf{IC}_G^{\succ}$ works as IC_G, but for each derivation $\mathcal{S}_0 \vdash \cdots \vdash \mathcal{S}_k$ the following restrictions must hold:

– *For every step $\mathcal{S}_i \vdash \mathcal{S}_{i+1}$, the employed inference respects $\succ$, and*
– *$\mathcal{S}_i$ must not contain $\vee$-forks, $\lozenge$-separated paths, or paths of different modal length.*

To distinguish derivations of IC_G and $\mathsf{IC}_G^{\succ}$, we will use the symbol $\vdash_{\succ}$ in derivations of $\mathsf{IC}_G^{\succ}$. In [19], correctness of $\mathsf{IC}_G^{\succ}$ is shown.

Fact 2 ([19]). *Let G be a K-formula in NNF and $\succ$ a G-ordering. Then G is unsatisfiable iff $\{\epsilon\} \in \mathcal{S}_0^{\vdash_{\succ}}$.*

Using the correspondence between the inverse method and the emptiness test of $\mathfrak{A}'_G$, we will now give an alternative, and in our opinion simpler, proof of this fact. Since $\mathsf{IC}_G^{\succ}$ is merely a restriction of IC_G, soundness (i.e., the if-direction of the fact) is immediate.

Completeness requires more work. In particular, the proof of Lemma 4 needs to be reconsidered since the propositional inferences are now restricted: we must show that the $\mathbb{X}$-inferences employed in that proof respect (or can be made to respect) $\succ$. To this purpose, we will follow [19] and introduce the notion of $\succ$-compactness. For $\succ$-compact sets, we can be sure that all applicable $\mathbb{X}$-inferences respect $\succ$. To ensure that all the sets $\varUpsilon_i$ constructed in the proof of Lemma 4 are $\succ$-compact, we again follow Voronkov and employ a special selection strategy.

Definition 10 ($\succ$-compact, select$_\succ$). *Let G be a K-formula in NNF and $\succ$ a G-ordering. An arbitrary set $\varPhi \subseteq \varPi_G$ is $\succ$-compact iff, for every $\mathbb{X}$-path $\pi \in \varPhi$ that is not p.e. in $\varPhi$, $\pi\mathbb{X}_* \succ \varPhi$.*

The selection function select$_\succ$ is defined as follows: if $\varPhi$ is not p.e., then let $\{\pi_1, \ldots, \pi_m\}$ be the set of $\mathbb{X}$-paths that are not p.e. in $\varPhi$. From this set, select$_\succ$ selects the path π_i such that the paths $\pi_i\mathbb{X}_*$ are the two smallest elements in $\{\pi_j\mathbb{X}_* \mid 1 \leq j \leq m\}$.

The function select$_\succ$ is well-defined because of Condition (2) of G-orderings. The definition of compact ensures that $\mathbb{X}$-inferences applicable to not propositionally expanded sequents respect $\succ$.

Lemma 6. *Let G be a K-formula in NNF, $\succ$ a G-ordering, and select$_\succ$ the selection function as defined above. Let $\varPhi = \{\epsilon\}$ or $\varPhi = \varGamma\square, \pi_i\Diamond$ with $\square$-paths $\varGamma$ and a $\Diamond$-path π, all of equal modal length. If $\mathsf{T}_\varPhi$, as defined in the proof of Lemma 4, is generated using select$_\succ$ as selection function, then every node $\varPsi$ of $\mathsf{T}_\varPhi$ is $\succ$-compact.*

The proof of this lemma can be found in [3]. It is similar to the proof of Lemma 5.8.3 in [19]. Given this lemma, it is easy to show that the construction employed in the proof of Lemma 4 also works for $\mathsf{IC}_G^\succ$, provided that we restrict the set $\varPhi$ as in Lemma 6:

Lemma 7. *Let $\varPhi = \{\epsilon\}$ or $\varPhi = \varGamma\square, \pi_i\Diamond$ with $\square$-paths $\varGamma$ and a $\Diamond$-path π all of equal modal length and $\mathcal{S}$ a set of sequents such that $\langle\!\langle\varPhi\rangle\!\rangle \subseteq [\![\mathcal{S}]\!]$. Then there exists a set of sequents $\mathcal{T}$ with $\mathcal{S} \vdash_\succ^* \mathcal{T}$ such that there exists $\varLambda \in \mathcal{T}$ with $\varLambda \subseteq \varPhi$.*

Alternative Proof of Fact 2. As mentioned before, soundness (the if-direction) is immediate. For the only-if-direction, if G is not satisfiable, then $L(\mathfrak{A}_G') = \emptyset$ and there is a set of states Q with $Q_0 \rhd^* Q$ and $\langle\!\langle\{\epsilon\}\rangle\!\rangle \subseteq Q$. Using Lemma 7 we show that there is a derivation of $\mathsf{IC}_G^\succ$ that simulates this derivation, i.e., there is a set of sequents $\mathcal{S}$ with $\mathcal{S}_0 \vdash_\succ^* \mathcal{S}$ and $Q \subseteq [\![\mathcal{S}]\!]$.

The proof is by induction on the length m of the derivation $Q_0 \rhd \ldots \rhd Q_m = Q$ and is totally analogous to the proof of Theorem 2. The base case is Lemma 1, which also holds for $\mathsf{IC}_G^\succ$ and the reduced automaton. The induction step uses Lemma 7 instead of Lemma 4, but this is the only difference.

Hence, $Q_0 \rhd^* Q$ and $\langle\!\langle\{\epsilon\}\rangle\!\rangle \subseteq Q$ implies that there exist a derivation $\mathcal{S}_0 \vdash_\succ^* \mathcal{S}$ such that $\langle\!\langle\{\epsilon\}\rangle\!\rangle \subseteq [\![\mathcal{S}]\!]$. Lemma 7 yields a derivation $\mathcal{S} \vdash_\succ^* \mathcal{T}$ with $\{\epsilon\} \in \mathcal{T} \subseteq \mathcal{S}_0^\models$.

$\square$

5 Global Axioms

When considering satisfiability of G w.r.t. the global axiom H, we must take subformulae of G and H into account. We address subformulae using paths in G and H.

Definition 11 ((G, H)-Paths). *For K-formulae G, H in NNF, the set of (G, H)-paths $\Pi_{G,H}$ is a subset of $\{\epsilon_G, \epsilon_H\} \cdot \{\vee_l, \vee_r, \wedge_l, \wedge_r, \Box, \Diamond\}^*$. The set $\Pi_{G,H}$ and the subformula $(G, H)|_\pi$ of G, H addressed by a path $\pi \in \Pi_{G,H}$ are defined inductively as follows:*

- *$\epsilon_G \in \Pi_{G,H}$ and $(G, H)|_{\epsilon_G} = G$, and $\epsilon_H \in \Pi_{G,H}$ and $(G, H)|_{\epsilon_H} = H$*
- *if $\pi \in \Pi_{G,H}$ and $(G, H)|_\pi = F_1 \wedge F_2$ then $\pi\wedge_l, \pi\wedge_r \in \Pi_{G,H}$, $(G, H)|_{\pi\wedge_l} = F_1$, $(G, H)|_{\pi\wedge_r} = F_2$, and π is called $\wedge$-path.*
- *The other cases are defined analogously (see also Definition 1).*
- *$\Pi_{G,H}$ is the smallest set that satisfies the previous conditions.*

The definitions of *p.e.* and *clash* are extended to subsets of $\Pi_{G,H}$ in the obvious way, with the *additional requirement* that, for $\Phi \neq \emptyset$ to be p.e., $\epsilon_H \in \Phi$ must hold. This additional requirement enforces the global axiom.

Definition 12 (Formula Automaton w. Global Axiom). *For K-formulae G, H in NNF, let $\{\pi_1, \ldots, \pi_n\}$ be an enumeration of the $\Diamond$-paths in $\Pi_{G,H}$. The n-ary looping automaton $\mathfrak{A}_{G,H}$ is defined by $\mathfrak{A}_G := (Q_{G,H}, \Sigma_{G,H}, \langle\!\langle \{\epsilon_G\} \rangle\!\rangle, \Delta_{G,H})$, where $Q_{G,H} := \Sigma_{G,H} := \{\Phi \in \Pi_{G,H} \mid \Phi$ is p.e.$\}$ and the transition relation $\Delta_{G,H}$ is defined as for the automaton $\mathfrak{A}_G$ in Definition 3.*

Theorem 3. *G is satisfiable w.r.t. the global axiom H iff $L(\mathfrak{A}_{G,H}) \neq \emptyset$.*

Definition 13 (The Inverse Calculus w. Global Axiom). *Let G, H be K-formula in NNF and $\Pi_{G,H}$ the set of paths of G, H. Sequents are subsets of $\Pi_{G,H}$, and operations on sequents are defined as before.*

In addition to the inferences from Fig. 2, the inverse calculus for G w.r.t. the global axiom H, $IC^{ax_{G,H}}$, employs the inference

$$(ax)\frac{\Gamma, \epsilon_H}{\Gamma}.$$

From now on, $[\![\cdot]\!]$ is defined w.r.t. the states of $\mathfrak{A}_{G,H}$, i.e., $[\![\Gamma]\!] := \{\Phi \in Q_{G,H} \mid \Gamma \subseteq \Phi\}$.

Theorem 4 ($IC^{ax_{G,H}}$ and the emptiness test for $\mathfrak{A}_{G,H}$ simulate each other). *Let $\vdash_{ax}$ denote derivation steps of $IC^{ax_{G,H}}$, and $\triangleright$ derivation steps of the emptiness test for $\mathfrak{A}_{G,H}$.*

1. *Let $Q \subseteq Q_{G,H}$ be a set of states such that $Q_0 \triangleright^* Q$. Then there exists a set of sequents S with $S_0 \vdash_{ax}^* S$ and $Q \subseteq [\![S]\!]$.*

2. *Let $\mathcal{S}$ be a set of sequents such that $\mathcal{S}_0 \vdash_{ax}^* \mathcal{S}$. Then there exists a set of states $Q \subseteq Q_G$ with $Q_0 \rhd^* Q$ and $[\![\mathcal{S}]\!] \subseteq Q$.*

Lemma 1, 2, and 3, restated for $\mathfrak{A}_{G,H}$ and $\mathsf{IC}^{ax_{G,H}}$, can be shown as before. The following lemma deals with the ax-inference of $\mathsf{IC}^{ax_{G,H}}$.

Lemma 8. *Let $\mathcal{S} \rhd \mathcal{T}$ be a derivation of $\mathsf{IC}^{ax_{G,H}}$ that employs an ax-inference. Then $[\![\mathcal{S}]\!] = [\![\mathcal{T}]\!]$.*

The proof of Theorem 4.2 is now analogous to the proof of Theorem 2.2. For the proof of Theorem 4.1, Lemma 4 needs to be re-proved because the change in the definition of p.e. now also implies that $\epsilon_H \in \varPhi$ holds for every set $\varPhi \in \langle\!\langle \varPsi \rangle\!\rangle$ for any $\varPsi \neq \emptyset$. This is where the new inference ax comes into play. In all other respects, the proof of Theorem 4.1 is analogous to the proof of Theorem 2.1.

Corollary 2. *$\mathsf{IC}^{ax_{G,H}}$ yields an EXPTIME decision procedure for satisfiability w.r.t. global axioms in K.*

The following algorithm yields the desired procedure:

Algorithm 1. *Let G, H be K-formulae in NNF. To test satisfiability of G w.r.t. H, calculate $\mathcal{S}_0^{\overleftarrow{ax}}$. If $\{\emptyset, \{\epsilon_G\}\} \cap \mathcal{S}_0^{\overleftarrow{ax}} \neq \emptyset$, then answer "not satisfiable," and "satisfiable" otherwise.*

Correctness of this algorithm follows from Theorem 3 and 4. If G is not satisfiable w.r.t. H, then $L(\mathfrak{A}_{G,H}) = \emptyset$, and there exists a set of states Q with $Q_0 \rhd^* Q$ and $\langle\!\langle \{\epsilon_G\} \rangle\!\rangle \subseteq Q$, there exists a set of sequents $\mathcal{S}$ with $\mathcal{S}_0 \vdash_{ax}^* \mathcal{S}$ such that $Q \subseteq [\![\mathcal{S}]\!]$. With (the appropriately reformulated) Lemma 4 there exists a set of sequents $\mathcal{T}$ with $\mathcal{S} \vdash_{ax}^* \mathcal{T}$ such that there is a sequent $\varLambda \in \mathcal{T}$ with $\varLambda \subseteq \{\epsilon_G\}$. Consequently, $\varLambda = \emptyset$ or $\varLambda = \{\epsilon_G\}$.

Conversely, since $\mathcal{S}_0 \vdash_{ax}^* \mathcal{S}_0^{\overleftarrow{ax}}$, there exists a set of (inactive) states Q such that $Q_0 \rhd^* Q$ and $[\![\mathcal{S}_0^{\overleftarrow{ax}}]\!] \subseteq Q$. Since $\langle\!\langle \{\epsilon_G\} \rangle\!\rangle \subseteq [\![\{\epsilon_G\}]\!] \subseteq [\![\emptyset]\!]$, we know that $\{\emptyset, \{\epsilon_G\}\} \cap \mathcal{S}_0^{\overleftarrow{ax}} \neq \emptyset$ implies $\langle\!\langle \{\epsilon_G\} \rangle\!\rangle \subseteq Q$. Consequently, $L(\mathfrak{A}_{G,H}) = \emptyset$ and thus G is not satisfiable w.r.t. H.

For the complexity, note that there are only exponentially many sequents. Consequently, it is easy to see that the saturation process that leads to $\mathcal{S}_0^{\overleftarrow{ax}}$ can be realized in time exponential in the size of the input formulae.

6   Future Work

There are several interesting directions in which to continue this work. First, satisfiability in K (without global axioms) is PSPACE-complete whereas the inverse method yields only an EXPTIME-algorithm. Can suitable optimizations turn this into a PSPACE-procedure? Second, can the optimizations considered in Section 4 be extended to the inverse calculus with global axioms? Third, Voronkov considers additional optimizations. Can they also be handled within our framework? Finally, can the correspondence between the automata approach and the inverse method be used to obtain inverse calculi and correctness proofs for other modal or description logics?

References

1. C. Areces, R. Gennari, J. Heguiabehere, and M. de Rijke. Tree-based heuristics in modal theorem proving. In W. Horn, editor, *Proc. of ECAI2000*, Berlin, Germany, 2000. IOS Press Amsterdam.
2. F. Baader and U. Sattler. An overview of tableau algorithms for description logics. *Studia Logica*, 2001. To appear.
3. F. Baader and S. Tobies. The inverse method implements the automata approach for modal satisfiability. LTCS-Report 01-03, LuFG Theoretical Computer Science, RWTH Aachen, Germany, 2001.
   See http://www-lti.informatik.rwth-aachen.de/Forschung/Reports.html.
4. P. Blackburn, M. de Rijke, and Y. Venema. *Modal Logic*. Cambridge University Press, 2001. Publishing date May 2001, preliminary version available online from http://www.mlbook.org/.
5. C. Courcoubetis, M. Y. Vardi, P. Wolper, and M. Yannakakis. Memory efficient algorithms for the verification of temporal properties. In E. M. Clarke and R. P. Kurshan, editors, *Proc. of Computer-Aided Verification (CAV '90)*, volume 531 of *LNCS*, pages 233–242. Springer Verlag, 1991.
6. F. M. Donini and F. Massacci. EXPTIME tableaux for ALC. *Artificial Intelligence*, 124(1):87–138, 2000.
7. W. F. Dowling and J. H. Gallier. Linear-time algorithms for testing the satisfiability of propositional horn formulae. *Journal of Logic Programming*, 1(3):267–28, 1984.
8. R. Dyckhoff, editor. *Proc. of TABLEAUX 2000*, number 1847 in LNAI, St Andrews, Scotland, UK, 2000. Springer Verlag.
9. R. Gerth, D. Peled, M. Y. Vardi, and P. Wolper. Simple on-the-fly automatic verification of linear temporal logic. In *Proc. of the 15th International Symposium on Protocol Specification, Testing, and Verification*, pages 3–18, Warsaw, Poland, 1995. Chapman & Hall.
10. R. Goré. Tableau methods for modal and temporal logics. In M. D'Agostino, D. M. Gabbay, R. Hähnle, and J. Posegga, editors, *Handbook of Tableau Methods*. Kluwer, Dordrecht, 1998.
11. I. Horrocks. Benchmark analysis with FaCT. In Dyckhoff [8], pages 62–66.
12. U. Hustadt and R. A. Schmidt. MSPASS: Modal reasoning by translation and first-order resolution. In Dyckhoff [8], pages 67–71.
13. R. E. Ladner. The computational complexity of provability in systems of modal propositional logic. *SIAM Journal on Computing*, 6(3):467–480, 1977.
14. C. Lutz and U. Sattler. The complexity of reasoning with boolean modal logic. In Wolter F., H. Wansing, M. de Rijke, and M. Zakharyaschev, editors, *Preliminary Proc. of AiML2000*, Leipzig, Germany, 2000.
15. R. A. Schmidt. Resolution is a decision procedure for many propositional modal logics. In M. Kracht, M. de Rijke, H. Wansing, and M. Zakharyaschev, editors, *Advances in Modal Logic, Volume 1*, volume 87 of *Lecture Notes*, pages 189–208. CSLI Publications, Stanford, 1998.
16. E. Spaan. *Complexity of Modal Logics*. PhD thesis, Univ. van Amsterdam, 1993.
17. M. Y. Vardi and P. Wolper. Automata-theoretic techniques for modal logics of programs. *Journal of Computer and System Sciences*, 32:183–221, 1986.
18. M. Y. Vardi and P. Wolper. Reasoning about infinite computations. *Information and Computation*, 115:1–37, 1994.
19. A. Voronkov. How to optimize proof-search in modal logics: new methods of proving redundancy criteria for sequent calculi. *ACM Transactions on Computational Logic*, 1(4):35pp, 2001.

Deduction-Based Decision Procedure for a Clausal Miniscoped Fragment of FTL

Regimantas Pliuškevičius

Institute of Mathematics and Informatics,
Akademijos 4, Vilnius 2600, LITHUANIA,
regis@ktl.mii.lt

Abstract. A simple decision deductive-based procedure for the so-called clausal miniscoped fragment of a first-order linear temporal logic with temporal operators Next and Always is presented. The soundness and completeness of the proposed decision procedure is proved.

1 Introduction

A temporal logic has been found valuable for specifications of various computer and multi-agent systems. To use such specifications, however, it is necessary to have techniques for reasoning on temporal logic formulas. Model-checking methods are effective and automatic for temporal formulas that are propositional. For more complex systems, however, it is necessary or convenient to employ a first-order temporal logic (FTL, in short). FTL is a very expressive language. Unfortunately, FTL is incomplete, in general [11]. But it becomes complete [5, 12] after adding an ω-type rule.

In some particular cases, the FTL (and, of course, in the propositional case) is finitary complete and/or decidable. Recently in [4] the decidability of a so-called monodic fragment of FTL has been proved. In this fragment all formulas of FTL (without function symbols!) beginning with a temporal operator $\bigcirc, \square$ (Next or Always)have at most one free variable (monodic condition).

In this paper, we consider a so-called miniscoped fragment of FTL. A formula A of FTL is in miniscoped form if all negative (positive) occurrences of $\forall$ ($\exists$, correspondingly) in A occur only in the formula of the shape $Q\bar{x}E(\bar{x})$ (where $\bar{x} = x_1, \ldots, x_n$) and $E(\bar{x})$ is an elementary formula. The objects of consideration of the proposed decision procedure $CMSat$ are so-called CM-sequents with an n-place ($n \geq 1$) predicate and function symbols. In this sense the presented fragment of FTL is non-monodic. Other decidable non-monodic fragments of FTL are considered in [8, 9, 10]. CM-sequents are a "miniscoped" version of Fisher's normal form [2]. Since we consider the miniscoped fragment of FTL based on clauses from Fisher's normal form, the considered here subclass of FTL we call a clausal miniscoped fragment of FTL.

The proposed decision procedure $CMSat$ is based on the saturation method [6–10]. In the saturation method the notions of calculus and deduction-based

R. Goré, A. Leitsch, and T. Nipkow (Eds.): IJCAR 2001, LNAI 2083, pp. 107–120, 2001.
© Springer-Verlag Berlin Heidelberg 2001

decision/semi-decision procedure are identical. A derivation in the proposed procedure $CMSat$ is constructed in a finite tree form. The tree is constructed automatically and satisfies the so-called loop property (see Lemma 11), which is the main characteristic peculiarity of the saturation method. Namely, each leaf S_i of the tree is either a traditional logical axiom or there exists a vertex S_j^* of the tree such that S_i and S_j^* satisfy some similarity relation. We can see a similar situation in Fisher's resolution method (see, e.g., [1, 3]) for a propositional temporal logic.

2   Description of Infinitary Sequent Calculi $G_{L\omega}$ and $G_{L\omega}^*$

The proposed decision procedure $CMSat$ is justified by infinitary calculi $G_{L\omega}$, $G_{L\omega}^*$ containing the ω-type rule.

Definition 1 (term, elementary formula, formula). *We assume that all predicate symbols are flexible (i.e., change their value in time), and all constants and function symbols are rigid (i.e., with time-independent meanings). A term is defined as usual. An elementary formula is either the truth constant T, or an expression of the form $P(t_1, \ldots, t_m)$ where P is a predicate symbol, t_i $(1 \leq i \leq n)$ is a term. Formulas are defined as usual.*

In the first order linear temporal logic over infinite sequences we have that $\bigcirc(A \odot B) \equiv \bigcirc A \odot \bigcirc B$ ($\odot \in \{\supset, \wedge, \vee\}$) and $\bigcirc \sigma A \equiv \sigma \bigcirc A$ ($\sigma \in \{\neg, \square, \forall x, \exists x\}$). Relying on these equivalences we can consider occurrences of the "next" operator $\bigcirc$ only entering the formula $\bigcirc^k E$ (k-time "next" elementary formula E). For the sake of simplicity, we "eliminate" the "next" operator and the formula $\bigcirc^k E$ is abbreviated as E^k (i.e., as an elementary formula with the index k). We also use the notation A^k for an arbitrary formula A in the following meaning.

Definition 2 (index, atomic formula). *1) If E is an elementary formula, $i, k \in \omega$, $k \neq 0$, then $(E^i)^k := E^{i+k}(E^0 := E)$; $E^l(l \geq 0)$ is called an atomic formula, and E^l becomes elementary if $l = 0$; 2) $(A \odot B)^k := A^k \odot B^k$ if $\odot \in \{\supset, \vee\}$; $(\sigma A)^k := \sigma A^k$, if $\sigma \in \{\square, \forall x, \exists x\}$. In case we want to indicate the dependence of an atomic formula on the terms $t_1, \ldots, t_n$ we write $E^k(t_1, \ldots t_n)$ instead of E^k. For example the expression $\forall x(P^1(x) \supset Q^3(x))^1$ means the formula $\forall x(\bigcirc \bigcirc P(x) \supset \bigcirc \bigcirc \bigcirc \bigcirc Q(x))$, or $\forall x(P^2(x) \supset Q^4(x))$.*

Definition 3 (sequent, miniscoped sequent, quasi-atomic formula and quasi-elementary formula). *A sequent is an expression of the form $\Gamma \to \Delta$, where we assume that Γ, Δ are arbitrary finite multisets (i.e., not sequences or sets) of formulas. A sequent S is a miniscoped sequent if all negative (positive) occurrences of $\forall$ ($\exists$, correspondingly) in S occur only in formulas of the shape $Q\bar{x}E(\bar{x})$ (where $\bar{x} = x_1, \ldots, x_n$, $E(\bar{x})$ $(n \geq 0)$ is an atomic (elementary) formula. This formula is called a quasi-atomic (quasi-elementary, respectively) formula; if $Q\bar{x} = \varnothing$, then a quasi-atomic formula becomes an atomic one; if $Q\bar{x} = \varnothing$ and E is an elementary formula, then a quasi-atomic formula becomes an elementary one.*

Now we shall consider some special form of a miniscoped version of Fisher's normal form [2]. First we define so-called kernel formulas.

Definition 4 (regular and non-regular kernel formula, indexed regular and non-regular kernel formula). *A formula A is a regular kernel formula, if $A = \Box(\bigwedge_{i=1}^{m} \forall \bar{x} \sigma_i E_i(\bar{x}) \supset \bigvee_{j=1}^{n} \exists \bar{z} \sigma_j P_j^1(\bar{z}))$, where $\sigma_i, \sigma_j \in \{\varnothing, \neg\}$, $E_i(\bar{x})$, $P_j(\bar{z})$ are some quasi-elementary formulas, $\bar{x} = x_1, \ldots, x_k$; $\bar{z} = z_1, \ldots, z_k$ $(k \geq 0)$; A formula B is a non-regular kernel formula, if $B = \Box(\bigwedge_{i=1}^{m} \forall \bar{x} \sigma_i E_i(\bar{x}) \supset \neg \Box \sigma P)$, where $\sigma_i, \sigma \in \{\varnothing, \neg\}$; $E_i(\bar{x}), P$ are some quasi-elementary formulas; $\bar{x} = x_1, \ldots, x_k$ $(k \geq 0)$. Let A be a regular (non-regular) kernel formula. Then A^1 is an indexed regular (non-regular, respectively) kernel formula.*

Definition 5 (CM-sequent, indexed CM-sequent, parametrical formulas of CM-sequent, induction-free CM-sequent and indexed CM-sequent). *A miniscoped sequent S is a CM-sequent, if $S = \Sigma_1, \Box \Omega \to \Sigma_2, \Box^0 \Delta$, where $\Sigma_i = \varnothing$ $(i \in \{1, 2\})$ or consist of quasi-elementary formulas (which are called parametrical formulas of CM-sequents); $\Box \Omega$ consists of regular/non-regular kernel formulas; $\Box^0 \in \{\varnothing, \Box\}$; $\Box^0 \Delta = \varnothing$ or consists of formulas of the shape $\Box \sigma E$, where $\sigma \in \{\varnothing, \neg\}$, E is a quasi-elementary formula; $\Delta = \varnothing$ or consists of formulas of the shape σE, where $\sigma = \{\varnothing, \neg\}$, E is a quasi-elementary formula. A miniscoped sequent S is an indexed CM-sequent, if $S = \Sigma_{11}, \Sigma_{12}^1, \Box \Omega^1 \to \Sigma_{21}, \Sigma_{22}^1, \Box^0 \Delta^1$, where $\Sigma_{i1}, \Sigma_{i2} = \varnothing$ $(i \in \{1, 2\})$ or consist of quasi-elementary formulas; $\Box \Omega$, $\Box^0 \Delta$ mean the same as in the case of CM-sequents. If $\Box^0 = \varnothing$ and S does not contain non-regular kernels, then the CM-sequent S (indexed CM-sequent) is an induction-free CM-sequent (indexed CM-sequent, respectively).*

Definition 6 (calculus $G_{L\omega}$). *The calculus $G_{L\omega}$ is defined by the following postulates.*
Axioms: $\Gamma, A \to A, \Delta$; $\Gamma \to \Delta, T$.
Rules:
1) temporal rules

$$\frac{A, \Box A^1, \Gamma \to \Delta}{\Box A, \Gamma \to \Delta}(\Box \to) \qquad \frac{\Gamma \to \Delta, A; \ldots; \Gamma \to \Delta, A^k, \ldots}{\Gamma \to \Delta, \Box A}(\to \Box_\omega),$$

where $k \in \omega$; here and below A^k means A^{l+k}, if $A = A^l$ $(l \geq 1)$.
2) logical rules consist of the traditional invertible rules for $\supset, \wedge, \vee, \neg, \forall, \exists$.

Theorem 1. *(a) (soundness and completeness of G_ω). Let S be a sequent, then the sequent is universally valid iff $G_{L\omega} \vdash S$*
(b) (admissibility of (cut)). $G_{L\omega} + (cut) \vdash S \Rightarrow G_{L\omega} \vdash S$.

Proof. Point (a) is proved by the Shutte method. Point (b) follows from point (a).

Definition 7 (calculi $G^*_{L\omega}$, G^*). *A calculus $G^*_{L\omega}$ is obtained from the calculus $G_{L\omega}$ by following transformations: (1) by dropping the rules $(\to\supset)$, $(\wedge\to)$, $(\to\vee)$, $(\forall\to)$, $(\to\exists)$ and (2) replacing the axiom $\Gamma, A \to A, \Delta$ by the following axioms:*

$$\Gamma, \Box A \to \Delta, \Box A;$$
$$\Gamma, E^k(t_1,\dots,t_n) \to \Delta, \exists x_1 \dots x_n E^k(x_1,\dots,x_n);$$
$$\Gamma, \forall x_1 \dots x_n E^k(x_1,\dots,x_n) \to \Delta, E^k(t_1,\dots,t_n);$$
$$\Gamma, \forall x_1 \dots x_n E^k(t_1(x_1),\dots,t_n(x_n)) \to \Delta, \exists y_1 \dots y_n E^k(p(y_1),\dots,p_n(y_n)),$$

where $k \geq 0$, E is a predicate symbol; $\forall i$ $(1 \leq i \leq n)$ terms $t_i(x_i)$ and $p_i(y_i)$ are unifiable.

A calculus G^* is obtained from $G^*_{L\omega}$ by dropping the rule $(\to \Box_\omega)$.

Theorem 2. *Let S be a CM-sequent. Then $G_{L\omega} \vdash S \iff G^*_{L\omega} \vdash S$.*

Proof. Follows from the definition of the CM-sequent using traditional proof-theoretical transformations.

3 Some Auxiliary Tools of Decidable Procedure $CMSat$

First we present the following notions.

Definition 8 (reduction of a sequent S to sequents $S_1,\dots,S_n$). *Let $\{i\}$ denote a set of rules of a calculus. Then the $\{i\}$-reduction (or briefly, reduction) of S to a set of sequents $S_1,\dots,S_n$ (denoted by $R(S)\{i\} \Rightarrow \{S_1,\dots,S_n\}$ or briefly by $R(S)$, is defined to be a tree of sequents with the root S and leaves $S_1,\dots,S_n$, and, possibly, axioms of the calculus $G^*_{L\omega}$, such that each sequent in $R(S)$, different from S, is the "upper sequent" of the rule from $\{i\}$ whose "lower sequent" also belongs to $R(S)$.*

Now we define rules by which the reduction of a CM-sequent S to a set of indexed CM-sequents is carried out.

Definition 9 (reduction rules, parametrically identical formulas). *The following rules will be called reduction ones (all these rules will be applied in the bottom-up manner):*

1) all logical rules of the calculus $G^*_{L\omega}$, i.e., the rules $(\supset\to),(\vee \to)$, $(\to \neg),(\neg \to),(\exists \to),(\to \forall)$;

2) the following temporal rules:

$$\frac{A,\Box A^1, \Gamma \to \Delta}{\Box A, \Gamma \to \Delta}(\Box \to) \qquad \frac{\Gamma \to \Delta, A; \; \Gamma \to \Delta, A^1}{\Gamma \to \Delta, \Box A}(\to \Box^1)$$

where $A \neq A^1$;

3) the following contraction rules:

$$\frac{E,\ \Gamma \to \Delta}{E,\ E^*,\ \Gamma \to \Delta}\ (C_E \to) \qquad \frac{\Gamma \to \Delta, E}{\Gamma \to \Delta, E, E^*}\ (\to C_E),$$

where E, E^* are quasi-atomic formulas such that either $E = E^*$ or E, E^* are congruent, or E, E^* differ only by corresponding occurrences of eigen-variables of the rules $(\to \forall)$, $(\exists \to)$.

$$\frac{\Gamma \to \Delta,\ \Box A}{\Gamma \to \Delta,\ \Box A,\ \Box A^*}\ (\to C_\Box),$$

where either $A = A^*$ or A, A^* are congruent, or A, A^* differ only by corresponding occurrences of eigen-variables of the rules $(\to \forall)$, $(\exists \to)$, such formulas A and A^* are parametrically identical.

Lemma 1. *The rule $(\to \Box^1)$ and contraction rules $(C_E \to)$, $(\to C_E)$, $(\to C_\Box)$ are admissible and invertible in $G_{L\omega}^*$.*

Proof. The admissibility and invertibility of $(\to \Box^1)$ follow from the fact that $G_{L\omega}^* \vdash \Box A \equiv A \land \Box A^1$ and Theorems 1(b), 2. The admissibility and invertibility of contraction rules are obvious.

Lemma 2 (reduction to a set of indexed CM-sequents). *Let S be a CM-sequent. Then one can construct $R(S)\{i\} \Rightarrow \{S_1, \ldots, S_n\}$, where $\forall j (1 \le j \le n)$ is an indexed CM-sequent; $\{i\}$ is the set of reduction rules; moreover, $G_{L\omega}^* \vdash S \Rightarrow G_{L\omega}^* \vdash S_j$, $j \in \{1, \ldots, n\}$.*

Proof. Using Lemma 1 and bottom-up applying reduction rules from $\{i\}$.

Definition 10 (proper reduction and proper reduction-tree to indexed CM-sequents: $R^*(S)$ and $R^*T(S)$, successful construction of $R^*T(S)$). *Let $R(S) = \{S_1, \ldots, S_n\}$ be the reduction of a CM-sequent S to the set of indexed CM-sequents $\{S_1, \ldots, S_n\}$. Then the set $R^*(S)$ obtained from $R(S)$ by dropping axioms of $G_{L\omega}^*$ is a proper reduction of the CM-sequent S to indexed CM-sequents $S_1, \ldots, S_n$. A derivation consisting of bottom-up applications of reduction rules and having a CM-sequent as the root and the set $R^*(S)$ as the leaves is a proper reduction-tree of S to indexed CM-sequents, and it is denoted by $R^*T(S)$. If $R^*(S)$ does not contain an induction-free indexed CM-sequent, then we say that the construction of $R^*T(S)$ is successful (in notation $R^*T(S) \ne \bot$). In the opposite case, we say that the construction of $R^*T(S)$ is not successful (in symbols $R^*T(S) = \bot$).*

Lemma 3 (decidability of $R^*T(S)$). *Let S be a CM-sequent, then the problem of construction of $R^*T(S)$ is decidable. Moreover, if $G_{L\omega}^* \vdash S$, then $\forall i (G_{L\omega}^* \vdash S_i)$, where $S_i \in R^*(S)$.*

Proof. Let S be a CM-sequent, then using Lemma 2 one can automatically reduce S to a set $R(S)$ of indexed CM-sequents. Afterwards, using the decidability of the axioms of $G^*_{L\omega}$ from $R(S)$, one can automatically get $R^*T(S)$. Then, using the syntactical notion of an induction-free indexed CM-sequent, one can automatically verify whether $R^*T(S) \neq \bot$ or $R^*T(S) = \bot$. We get from Lemma 2 that, if $G^*_{L\omega} \vdash S$, then $G^*_{L\omega} \vdash S_i$ where $S_i \in R^*(S)$.

Definition 11 (separation rule: (S)). *Let us introduce the following rule (which will be applied in the bottom-up manner):*

$$\frac{S^+ = \Pi_1, \Box\Omega \to \Pi_2, \Box\Delta}{S^* = \Sigma_1, \Pi_1^1, \Box\Omega^1 \to \Sigma_2, \Pi_2^1, \Box\Delta^1} \ (S),$$

where S^* is an indexed CM-sequent and $\Sigma_1, \Pi_1^1 \to \Sigma_2, \Pi_2^1$ is not an axiom of $G^*_{L\omega}$; S^+ is the CM-sequent.

Lemma 4 (invertibility of (S)). *Let $S^* = \Sigma_1, \Pi_1^1, \Box\Omega^1 \to \Sigma_2, \Pi_2^1, \Box\Delta^1$ be an indexed CM-sequent and $\Sigma_1, \Pi_1^1 \to \Sigma_2, \Pi_2^1$ is not an axiom of $G^*_{L\omega}$. Then $G^*_{L\omega} \vdash S^* \Rightarrow G^*_{L\omega} \vdash \Pi_1, \Box\Omega \to \Pi_2, \Box\Delta$.*

Proof. By proving (using induction on the height) the invertibility of (S) in $G_{L\omega}$ and using Theorem 2.

4   Description of a Decidable Calculus $CMSat$

In this section, the decidable calculus $CMSat$ for CM-sequents will be described. First let us introduce the following rule.

Definition 12 (structural rule (W^*), parametrically identical CM-sequents). *CM-sequents S and S' are parametrically identical (in symbols $S \approx S'$) if S, S' differ only by parametrically identical formulas. Let us introduce the following structural rule:*

$$\frac{\Gamma \to \Delta}{\Pi, \Gamma' \to \Delta', \theta} \ (W^*),$$

where $\Gamma \to \Delta \approx \Gamma' \to \Delta'$.

Definition 13 (subsumed CM-sequent). *We say that a CM-sequent S_1 subsumes a CM-sequent S_2 or S_2 is subsumed by S_1 (in symbols $S_1 \succcurlyeq S_2$) if $\frac{S_1}{S_2}(W^*)$ (in a special case, $S_1 = S_2$).*

Definition 14 (subsumption rule (Sub^+), subsumption-tree (ST^+), active and passive parts of (ST^+)). *The subsumption rule is the following rule (it is applied in the bottom-up manner):*

$$\frac{S_1^0, \ldots, S_{j-1}^0, S_{j+1}^0, \ldots, S_n^0}{S_1, \ldots, S_j, \ldots, S_n} (Sub^+),$$

where $S_1, \ldots, S_n$ are CM-sequents and $S_i \succcurlyeq S_j$ ($i \in \{1, \ldots, j-1, j+1, \ldots, n\}$); $+ \in \{\varnothing, *\}$, $S_i^0 = \varnothing$ if $+ = *$, otherwise $S_i^0 = S_i$.

The subsumption-tree of CM-sequents $S_1, \ldots, S_n$ is defined by the following bottom-up deduction (and denoted by (ST^+));

$$\frac{\dfrac{S_1^*, \ldots, S_k^*}{} (Sub^+)}{\dfrac{\cdots \qquad \cdots}{S_1, \ldots, S_n}} (Sub^+),$$

where the set of CM-sequents $M = \{S_1^*, \ldots, S_k^*\}$ is such that it is impossible to bottom-up apply (Sub^+) to the set M. The sequents from (ST^+) which subsumes some sequents from (ST^+) will be called an active part of (ST^+), and the sequents of (ST^+) which are subsumed will be called a passive part of (ST^+).

Definition 15 (resolvent-tree $ReT(S)$ and resolvent $Re(S)$ of CM-sequent S). *The resolvent-tree of a CM-sequent S is defined by the following bottom-up deduction (denoted by $ReT(S)$):*

$$\frac{\dfrac{S_1, \ldots, S_k}{} (ST)}{\dfrac{\dfrac{S_1^+}{S_1^*}(S) \ldots \dfrac{S_m^+}{S_m^*}(S)}{S}} R^*T(S),$$

where $\{S_1^*, \ldots, S_m^*\}$ is the set of indexed CM-sequents; $\{S_1^+, \ldots, S_m^+\}$ and $\{S_1, \ldots, S_k\}$ are the set of CM-sequents. The latter set is a resolvent of a CM-sequent and is denoted by $Re(S)$. If $R^*T(S) = \bot$ (i.e., if the set $\{S_1^*, \ldots, S_m^*\}$ contains an induction-free CM-sequent), then we say that the construction of $ReT(S)$ is not successful (in symbols $ReT(S) = \bot$). If $\forall i (1 \leq i \leq m)$ S_i^* is an axiom, then $Re(S) = \varnothing$.

Lemma 5 (decidability of $ReT(S)$). *Let S be a CM-sequent, then the problem of construction of $ReT(S)$ is decidable. If $G_{L\omega}^* \vdash S$, then either $Re(S) = \varnothing$ or $\forall i \ (G_{L\omega}^* \vdash S_i)$, where $S_i \in Re(S)$. Moreover, if S is an induction-free CM-sequent, then instead of the calculus $G_{L\omega}^*$ we have the calculus G^*.*

Proof. Analogously as in Lemma 3, using Lemmas 3, 4.

By one of the saturation-based paradigms, i.e., "calculus = deductive decision/semi-decision procedure", we can consider the procedure $ReT(S)$ as a calculus.

Definition 16 (induction-free CM-sequent S derivable in $ReT(S)$). *An induction-free CM-sequent S is derivable in $ReT(S)$ (in symbols: $ReT(S) \vdash S$) if $Re(S) = \varnothing$.*

Lemma 6. *Let S be an induction-free CM-sequent. Then $ReT(S) \vdash S \iff G^* \vdash S$.*

Proof. Let $ReT(S) \vdash S$, i.e., $Re(S) = \varnothing$. It means that all indexed CM-sequents S_i^* $(1 \le i \le n)$ are axioms, i.e., $G^* \vdash S$. Let $G^* \vdash S$, then, using the invertibility of reduction rules, we get that all indexed CM-sequents S_i^* $(1 \le i \le n)$ are axioms, i.e., $Re(S) = \varnothing$.

Lemma 7 (decision procedure for induction-free CM-sequent). *Let S be an induction-free CM-sequent. Then the problem derivability in $ReT(S)$ is decidable.*

Proof. Follows from Lemma 5.

To define the main deductive procedure of the proposed decidable saturation-based procedure $CMSat$ let us define some auxiliary notions and prove some lemmas that are of own interest.

Definition 17 (resolvent subformulas of kernel formulas, of a set of kernel formulas and of a CM-sequent; parametrically finite set of formulas). *Let $A = \Box(\bigwedge_{i=1}^{m} \forall \bar{x} \sigma_i E_i(\bar{x}) \supset \bigvee_{j=1}^{n} \exists \bar{z} \sigma_j P_j^1(\bar{z}))$ (where $\sigma_i, \sigma_j \in \{\varnothing, \neg\}$, $\bar{x} = x_1, \ldots, x_l$; $\bar{z} = z_1, \ldots, z_l (l \ge 0)$, $E_j(\bar{x})$, $P_j(\bar{z})$ are some quasi-elementary formulas) be a regular kernel formula. Then the resolvent subformulas of A (denoted by $RSub(A)$) are a set of $\{P_1(\bar{b}_1), \ldots, P_n(\bar{b}_n)\}$, where $\bar{b}_j = b_{j1}, \ldots, b_{jl}$ $(1 \le j \le n; l \ge 0)$ and any b_{ji} $(1 \le i \le l)$ is a new variable. Let $A = \Box(\bigwedge_{i=1}^{m} \forall \bar{x} \sigma_i E_i(\bar{x}) \supset \neg \Box \sigma E)$ (where $\sigma_i \in \{\varnothing, \neg\}$; $\sigma \in \{\varnothing, \neg\}$; $E_i(\bar{x})$ $(1 \le i \le n)$, E are some quasi-elementary formulas) be a non-regular kernel formula. Then the resolvent subformula of A is the formula of the shape $\Box \sigma E$. Let $\Box \Omega$ be a set of kernel formulas $A_1, \ldots, A_n$. Then resolvent subformulas of $\Box \Omega$ (denoted by $RSub(\Box \Omega)$) are the set $\bigcup_{i=1}^{n} RSub(A_i)$. Let $S = \Sigma_1, \Box \Omega \to \Sigma_2, \Box A_1, \ldots, \Box A_n$ be a CM-sequent. Then resolvent subformulas of S (denoted by $RSub(S)$) are the set $RSub(\Box \Omega) \cup \bigcup_{i=1}^{n} \Box A_i$. $R^*Sub(S)$ means the set obtained from $RSub(S)$ by dropping the truth constant T and merging the formulas that are parametrically identical (see Definition 9). Let M be a set of formulas and M^* be the set obtained from M by merging the formulas that are parametrically identical. Then the set M is parametrically finite if M^* is finite.*

Lemma 8. *Let* $S = \Sigma_1, \Box\Omega \rightarrow \Sigma, \Box\Delta$ *be a CM-sequent. Then the sets* $R^*Sub(\Box\Omega)$ *and* $R^*Sub(S)$ *are parametrically finite.*

Proof. Follows from Definition 17.

Definition 18 (saturated CM-sequent). *Let S be a non-induction-free CM-sequent not containing non-regular kernels. Then S is saturated if $S = \Sigma_1, \Box\Omega \rightarrow \Sigma_2, \Box\Delta$ and $\Sigma_1, \Sigma_2 \subseteq R^*Sub(\Box\Omega)$. Let S be a CM-sequent containing non-regular kernels and let $Re(S) = \{S_1, \ldots, S_n\}$. Then $\forall i$ $(1 \leq i \leq n)$ S_i is a saturated CM-sequent.*

Lemma 9 (reducing to saturated CM-sequents). *Let $S = \Sigma_1, \Box\Omega \rightarrow \Sigma_2, \Box^\circ\Delta$ (where $\Box^\circ\Delta \in \{\varnothing, \Box^\circ\Delta\}$) be a non-induction-free CM-sequent. Then either S does not contain a non-regular kernel and $S = \Sigma_1, \Box\Omega \rightarrow \Sigma_2, \Box\Delta$ (where $\Sigma_1, \Sigma_2 \subseteq R^*Sub(\Box\Omega)$) or $ReT(S) = \bot$, or $Re(S) = \varnothing$ or S contains non-regular kernels and $Re(S) = \{S_1, \ldots, S_n\}$, where $\forall i(1 \leq i \leq n)$ S_i is a saturated CM-sequent and $S_i = \Sigma_{1i}, \Box\Omega \rightarrow \Sigma_{2i}, \Box\theta, \Box^0\Delta$, where $\Sigma_{i1}, \Sigma_{2i}, \Box\theta \subseteq R^*Sub(\Box\Omega)$.*

Proof. Follows from the definitions of $Re(S)$ and $R^*Sub(\Box\Omega)$.

Lemma 10. *Let S be a non-induction-free CM-sequent. Then the problem of reduction of the CM-sequent S to a set of saturated CM-sequents is decidable.*

Proof. Follows from Lemma 5.

We define now the main deductive tool of the proposed decision saturation-based procedure $CMSat$, which will be applied to a saturated CM-sequent.

Definition 19 (k-th resolvent-tree $Re^kT(S)$ and k-th resolvent $Re^k(S)$ of a saturated CM-sequent S). *Let $S = \Sigma_1, \Box\Omega \rightarrow \Sigma_2, \Box^\circ\theta, \Box^\circ\Delta$ (where $\Sigma_1, \Sigma_2, \Box^\circ\theta \in \Box\Omega$ and $\Box^\circ\theta \in \{\varnothing, \Box\theta\}$, $\Box^\circ\Delta \in \{\varnothing, \Box^\circ\Delta\}$, $\Box^\circ\theta, \Box^\circ\Delta \neq \varnothing$) be a saturated CM-sequent. Then $Re^\circ(S) = Re^\circ T(S) = S$. Let $Re^k(S) = \{S_1, \ldots, S_n\}$, then $Re^{k+1}T(S)$ and $Re^{k+1}(S)$ are defined by the following bottom-up deduction:*

$$
\cfrac{Re^{k+1}(S)}{\cfrac{R^+e^{k+1}(S)}{\underbrace{\cfrac{Re(S_1)}{S_1} ReT(S_1) \ldots \cfrac{Re(S_n)}{S_n} ReT(S_n)}_{Re^k(S)}} (ST) \qquad \overset{k}{\underset{i=0}{\cup}} Re^i(S)} (ST^*)
$$

The bottom-up application of (ST) reduces the set $\overset{n}{\underset{i=1}{\cup}} Re(S_i)$ to the set $R^+e^{k+1}(S)$ of saturated CM-sequents. The bottom-up application of (ST^*) reduces the set $R^+e^{k+1}(S) \cup \overset{k}{\underset{i=0}{\cup}} Re^i(S)$ to the set $Re^{k+1}(S)$ which will be called

a $k + 1$-th resolvent of a saturated CM-sequent S. Moreover, the members
of $\bigcup\limits_{i=0}^{k} Re^i(S)$ are the active part of the application of (ST^*) and members of
$R^+e^{k+1}(S)$ are the passive ones. If $\exists i$ such that $ReT(S_i) = \bot$, then $Re^{k+1}(S) =$
$\bot$ and $Re^{k+1}T(S) = \bot$, i.e., the construction of $Re^{k+1}(S)T(S)$ is unsuccessful.
The set $Re^{k+1}(S)$ is empty in two following cases: (1) $\forall i$ $(1 \le i \le n)$ $Re(S_i) = \varnothing$
or (2) the bottom-up application of (ST^*) in $Re^{k+1}T(S)$ yields an empty set.

The notation $Re^k(S) \ne \bot$ $(k \in \omega)$ means that the construction of $Re^k(S)$ is
successful for all $k \in \omega$.

Now we establish the main property of the procedure $Re^k(S)$.

Lemma 11 (loop property). *Let $Re^k(S) \ne \bot$ $(k \in \omega)$. Then there exists a*
finite natural number p such that $p < |R^*Sub(S)|$ and $Re^p(S) = \varnothing$.

Proof. Let $S = \Sigma_1, \Box\Omega \to \Sigma_2, \Box^\circ\theta, \Box^\circ\Delta$ and $Re^kT(S) \ne \bot$ $(k \in \omega)$. Then
from the definition of $Re^k(S)$ it follows that either $\forall i(i \ge 1)$ $Re^i(S) = \varnothing$, or
$Re^i(S) = \{S_1, \ldots, S_n\}$, where S_l $(1 \le l \le n) = \Sigma_{l1}, \Box\Omega \to \Sigma_{l2}, \Box^\circ\theta_l, \Box^\circ\Delta_l$,
where $\Sigma_{l1}, \Sigma_{l2}, \Box^\circ\theta_l \in RSub(\Box\Omega)$ and $\Box^\circ\Delta_l \subseteq \Box\Delta$. Lemma 8 implies that the set
$RSub(S)$ is parametrically finite. Therefore, there must be a finite number l such
that for all i, if $S_i \in R^*e^l(S)$, then $\exists j$ $(0 \le j \le l-1)$ such that if $S_j \in Re^j(S)$,
then $S_j \succcurlyeq S_i$, i.e., $Re^l(S) = \varnothing$. It is easy to verify that $l < |R^*Sub(S)|$.

Lemma 12 (decidability of relation $Re^p(S) = \varnothing$). *The relation*
$Re^p(S) = \varnothing$ is decidable.

Proof. Follows from Lemmas 5, 11.

Definition 20 (calculus $CMSat$, CM-sequent derivable in $CMSat$). *Let*
S be an induction-free CM-sequent. Then the calculus $CMSat$ consists of proce-
dure of construction $ReT(S)$ (see Definition 15). Let S be a non-induction-free
CM-sequent S, then the calculus $CMSat$ consists of two procedures: (1) a pre-
liminary procedure of reduction S to a set of saturated CM-sequents and (2)
the procedure of construction $Re^k(S)$. An induction-free CM-sequent S is deriv-
able in $CMSat$ (in symbols $CMSat \vdash S$) if $Re(S) = \varnothing$, in the opposite case,
$CMSat \nvdash S$. A non-induction-free CM-sequent S is derivable in $CMSat$, if ei-
ther $Re(S) = \varnothing$ or 1) $R(S) = \{S_1, \ldots, S_n\}$ (i.e., if S can be reduced to saturated
CM-sequents $S_1, \ldots, S_n$) and 2) $\forall i$ $(1 \le i \le n)$ $Re^l(S_i) = \varnothing$, in the opposite
case, $CMSat \nvdash S$.

Theorem 3 (decidability of $CMSat$). *The calculus $CMSat$ is decidable for*
any CM-sequent.

Proof. Follows from Lemmas 5, 12.

Let us introduce some notions that will be used in a so-called invariant calculus $CMIN$ (see below).

Definition 21 (saturation set, decomposition of saturation set, nuclear of decomposition of saturation set). *Let S be a saturated CM-sequent and $Re^n(S) = \varnothing$. Let us construct a set $\overset{n-1}{\underset{i=0}{\cup}} Re^i(S)$. Then a set, obtained from the set $\overset{n-1}{\underset{i=0}{\cup}} Re^i(S)$ applying the rule (Sub), is a saturated set of the saturated sequent S and is denoted by $Sat\{S\}$. A set $\{Sat\{S\}\}$ is a decomposition of the set $Sat\{S\}$, if: (1) $Sat\{S\} = \underset{i}{\cup} Sat^i\{S\}$; (2) $\forall ij(Sat^i\{S\} \cap Sat^j\{S\}) = \varnothing$; (3) if S_1, $S_2 \in Sat^i\{S\}$, then $S_1 = \Sigma_1, \square\Omega \to \Sigma_2, \square A, \square^\circ \Delta_1$; $S_2 = \Sigma_1^*, \square\Omega \to \Sigma_2^*, \square A, \square^\circ \Delta_2$; i.e., S_1, S_2 have a common succedent member $\square A$, which is a nuclear of $Sat^i\{S\}$.*

Now let us give some examples which demonstrate the main features of the presented decision procedure $CMSat$.

Example 1. (a) Let $S = \forall z P(z), \square\Omega \to$, where P is a one-place predicate symbol, $\square\Omega = \square(\forall x P(x) \supset \exists x P^1(x))$, $\square(\forall y P(y) \supset \neg\square\neg E)$ and E is a quasi-elementary formula of the shape $\forall y P(y)$. So, S is the CM-sequent containing one regular kernel and one non-regular kernel. First, let us construct a proper reduction of the CM-sequent S to the set of indexed CM-sequents, i.e., let us construct $R^*(S)$ and $R^*T(S)$ (see Definition 10). Bottom-up applying reduction rules, from S we get $R^*(S) = \{S_1, S_2\}$, where $S_1 = \forall z P(z), P^1(b_1), \square\Omega^1 \to$; $S_2 = \forall z P(z), P^1(b_2), \square\Omega^1 \to \square\neg\forall y P^1(y)$. Since the sequent S_1 is not an axiom, $R^*T(S) = \perp$ and $CMSat \nvdash S$.

(b) Let $S = \forall xy P(g(y), f(x, h(x), y)), \square\Omega \to \square E_5$, where $\square\Omega = \square(E_1 \supset \exists u E_2^1(u))$, $\square(E_3 \supset \neg\square E_4)$, where E_1 is a quasi-elementary formula of the shape $\exists x w z P(x, f(g(z), w, z))$; $E_2(u)$ is a quasi-elementary formula of the shape $\forall v P(u, v)$; E_3 is a quasi-elementary formula of the shape $\exists xy P(x, y)$; E_4 is a quasi-elementary formula of the shape $\exists uv P(u, v)$; E_5 is a quasi-elementary formula of the shape $\exists u_1 v_1 P(u_1, v_1)$. So, S is the CM-sequent containing one regular kernel and one non-regular kernel. Let us perform a preliminary procedure $CMSat$, i.e., reduce the given CM-sequent to the set of saturated ones, and construct $Re(S)$. Since the system of terms $g(y) = x$; $f(x, h(x), y) = f(g(z), w, z)$ is unifiable, bottom-up applying the reduction rules and rules (S), (Sub), we get $Re(S) = S_1 = \forall v P(b, v), \square\Omega \to \square \exists uv P(u, v)$. Next, construct sets $R^*Sub(\square\Omega)$ and $R^*Sub(S)$. By definition, $R^*Sub(\square\Omega) = R^*Sub(S) = \{\forall u P(b, v), \square \exists uv P(u, v)\}$ (and $|R^*Sub(S)| = 2$). Thus, S_1 is a saturated CM-sequent and we can calculate $Re^k(S_1)$. By definition, $Re^0(S_1) = S_1$. Bottom-up applying the reduction rules and rules (S), (Sub), we get $R^*e^1(S) = S_2 = \forall v P(b_1, v), \square\Omega \to \square \exists uv P(u, v)$. Since $S_1 \succcurlyeq S_2$, we have $Re^1(S_1) = \varnothing$, $Sat\{S_1\} = S_1$ and $CMSat \vdash S$.

(c) Let $S = P(b), \square\Omega \to \square \exists u P(u), \square \exists v Q(v)$, where $\square\Omega = \square(\neg E_1 \supset \neg T^1), \square(\neg E_2 \supset \neg T^1), \square(E_1 \supset (\exists u P^1(u) \vee \exists v Q^1(v))), \square(E_2 \supset (\exists u_1 P^1(u_1) \vee \exists v_1 Q(v_1)))$; $E_1 = \exists x P(x)$; $E_2 = \exists y Q(y)$ and T is the truth constant. So, S is a CM-sequent containing four regular kernels. Let us construct $R^*Sub(\square\Omega)$ and $R^*Sub(S)$. By definition, $R^*Sub(\square\Omega) = \{P(b_1), Q(b_2)\}$ and $R^*Sub(S) = R^*Sub(\square\Omega) \cup \{\square \exists u P(u), \square \exists v Q(v)\}$ (and $|R^*Sub(S)| = 4$). Since $P(b) \in R^*Sub(\square\Omega)$, we get S is a saturated CM-sequent. Therefore we can start

calculating $Re^k(S)$. By definition, $Re^0(S)$. It is easy to verify that $R^*e^1(S) =$
$\{S_1 = P(b_1), \Box\Omega \to \Box\exists uP(u), \Box\exists vQ(v); S_2 = Q(b_2), \Box\Omega \to \Box\exists uP(u), \Box\exists vQ(v)\}$.
Since $S \succcurlyeq S_1$, $Re^1(S) = S_2$. Analogously we get $R^*e^2(S) = \{S_3 = P(b_3), \Box\Omega \to$
$\Box\exists uP(u), \Box\exists vQ(v); S_4 = Q(b_4), \Box\Omega \to \exists uP(u), \Box\exists vQ(v)\}$. Since $S \succcurlyeq S_3$,
$S_2 \succcurlyeq S_4$, we have $Re^2(S) = \varnothing$, $CMSat \vdash S$ and $Sat\{S\} = \{S, S_2\}$.

In order to justify the decidable calculus $CMSat$, we introduce a so-called
invariant calculus $CMIN$. First we introduce some simple calculi.

Definition 22 (calculi Log and G^+). *A calculus Log is obtained from the
calculus G^* (see Definition 7) by dropping the rule ($\Box \to$). A calculus G^+ is
obtained from the calculus G^* by adding:*
   1) the following rule:

$$\frac{\Gamma \to \Delta, A; \Gamma \to \Delta, \Box A^1}{\Gamma \to \Delta, \Box A} \ (\to \Box^1);$$

   2) traditional invertible logical rules ($\wedge \to$), ($\vee \to$).

Definition 23 (invariant calculus $CMIN$). *The invariant calculus $CMIN$
is obtained from the calculus G^+ by adding the following rule:*

$$\frac{\Gamma \to \Delta, I; I \to I^1; I \to A}{\Gamma \to \Delta, \Box A} \ (\to \Box)$$

The rule ($\to \Box$) satisfies the following conditions:
   *1) the conclusion of ($\to \Box$), i.e., the sequent $S = \Gamma \to \Delta, \Box A$ is such
that (a) S is a saturated CM-sequent; (b) $S \in Sat^i\{S\} = \{\Sigma_{i1}, \Box\Omega \to
\Pi_{i1}, \Box^0\Delta_{i1}, \Box A; \ldots; \Sigma_{in}, \Box\Omega \to \Pi_{in}, \Box^0\Delta_{in}, \Box A\}$, where $\Box A$ is the nuclear of
$Sat^i\{S\}$ (see Definition 21); $\Box^0\Delta_{ij} \in \{\varnothing, \Box\Delta_{ij}\}$ $(1 \le j \le n)$;*
   *2) $I = \overset{n}{\underset{j=1}{\vee}}((\exists\Sigma_{ij})^\wedge \wedge \neg(\forall\Pi_{ij})^\vee \wedge \neg(\Box^0\Delta_{ij})^\vee \wedge (\Box\Omega)^\wedge)$, where $Q\Gamma =
QE_1, \ldots, QE_n$, if $\Gamma = E_1, \ldots, E_n$; $Q \in \{\forall\bar{x}, \exists\bar{x}\}$ (i.e., all the eigen-variables
are correspondingly bounded in Σ_{ij}, Π_{ij}); $\Gamma^\wedge(\Gamma^\vee)$ means the conjunction (disjunction, respectively) of formulas from Γ.*
   *(3) the left premise of ($\to \Box$), the sequent $S_1 = \Gamma \to \Delta, I$ is such that
$Log \vdash S_1$;*
   *(4) the middle premise of ($\to \Box$), the sequent $S_2 = I \to I^1$ is such that
$G^+ \vdash S_2$;*
   *(5) the right premise of ($\to \Box$), the sequent $S_3 = I \to A$ is such that $CMIN \vdash
S_3$.*

Lemma 13. *The problem of finding the invariant formula I in the rule ($\to \Box$)
is decidable.*

Proof. Follows from Lemma 12.

Example 2. Let S be the saturated CM-sequent from Example 1(c), i.e.,
$S = P(b), \Box\Omega \;\to\; \Box\exists u P(u), \Box\exists v Q(v)$, where $\Box\Omega = \Box(\neg E_1 \supset \neg T^1)$,
$\Box(\neg E_2 \supset \neg T^1), \Box(E_1 \supset (\exists u P^1(u) \lor \exists v Q^1(x))), \Box(E_2 \supset (\exists u_1 P^1(u_1) \lor \exists v_1 Q^1(v_1)))$.
From Example 1(c) we get that $Sat\{S\} = \{S, S_2\}$, where $S = Q(b_2), \Box\Omega \to$
$\Box\exists u P(u), \Box\exists v Q(v)$. Therefore, by definition, $Sat^1\{S\} = \{S, S_2\}$ with two alternative nuclei: (1) $\Box\exists P(u)$ or (2) $\Box\exists v Q(v)$. Let us take case (1). Then, by definition, the invariant formula has the following shape $I = (\exists x P(x) \lor \exists y Q(y)) \land$
$\neg\Box\exists v Q(v) \land (\Box\Omega)^\wedge$. It is easy to verify that
$Log \vdash P(b), \Box\Omega \to I, \Box\exists v Q(v)$ (1);
$G^+ \vdash I \to I^1$ (2);
$G^+ \vdash I \to \exists u P(u)$ (3).
Applying $(\to \Box)$ to (1), (2), (3), we get $CMIN \vdash S$. We get the same result
when we take the formula $\Box\exists v Q(v)$ as a nuclear.

In the same way as in [7] we get

Theorem 4. *Let S be a saturated CM-sequent. Then $CMSat \vdash S \iff$*
$CMIN \vdash S \iff G^*_{L\omega} \vdash S$.

Lemma 14. *The calculus $CMIN$ is decidable.*

Proof. Follows from Theorems 3, 4.

Theorem 5. *(a) The calculi $CMSat$ and $CMIN$ are sound and complete for*
the class of saturated CM-sequents.
  (b) The calculus $CMSat$ is sound and complete for the class of CM-sequents.

Proof. Point (a) follows from Theorems 2, 4. Point (b) follows from Theorems 2,
4 and Lemma 9.

Acknowledgements. I am greatly indebted to Professor M.Fisher and Dr. A.
Degtiarev for their useful scientific discussions and remarks during my fellowship
(May–July, 2000) at Manchester Metropolitan University. I would like to thank
also the anonymous reviewers for helpful comments.

References

1. M. Fisher, A resolution method for temporal logic, Proc. of the IJCAI, Sydney,
   99–104 (1991).
2. M. Fisher, A normal form for temporal logics and its applications in theorem
   proving and execution, *Journal of Logic and Computation*, 7(4), 429–456 (1997).
3. M. Fisher, C. Dixon, and M. Peim, Clausal temporal resolution (To appear in:
   ACM Transactions on Computational Logic).
4. I. Hodkinson, F. Wolter, Zakharyaschev M.: Decidable fragments of first-order
   temporal logics. (To appear in: *Annals of Pure and Applied Logic*).

5. H. Kawai, Sequential calculus for a first-order infinitary temporal logic, Zeitchr. fur Math. Logic und Grundlagen der Math. 33 (1987) 423–452.
6. R. Pliuškevičius, On saturated calculi for a linear temporal logic. *LNCS 711*, (1993) 640–650.
7. R. Pliuškevičius, The saturated tableaux for linear miniscoped Horn-like temporal logic, Journal of Automated Reasoning 13 (1994) 51–67.
8. R. Pliuškevičius, On ω-decidable and decidable deductive procedures for a restricted FTL with Unless, Proc. of FTP (2000) 194–205, St. Andrews, UK.
9. R. Pliuškevičius, A deductive decision procedure for a restricted *FTL*. Abstracts of Seventh Workshop on Automated Reasoning (2000), London.
10. R. Pliuškevičius, On an ω-decidable deductive procedure for non-Horn sequents of a restricted *FTL*, LNAI 1861, 523–537 (2000).
11. A. Szalas, Concerning the semantic consequence relation in first-order temporal logic, Theoretical Comput. Sci. 47 (1986) 329–334.
12. A. Szalas, A complete axiomatic characterization of first-order temporal logic of linear time, Theoretical Comput. Sci. 54 (1987) 199–214.

Tableaux for Temporal Description Logic with Constant Domains

Carsten Lutz[1], Holger Sturm[2], Frank Wolter[3], and Michael Zakharyaschev[4]

[1] LuFG Theoretical Computer Science, RWTH Aachen,
Ahornstraße 55, 52074 Aachen, Germany
lutz@cs.rwth-aachen.de

[2] Fachbereich Philosophie, Universität Konstanz,
78457 Konstanz, Germany
holger.sturm@uni-konstanz.de

[3] Institut für Informatik, Universität Leipzig,
Augustus-Platz 10-11, 04109 Leipzig, Germany
wolter@informatik.uni-leipzig.de

[4] Department of Computer Science, King's College,
Strand, London WC2R 2LS, U.K.
mz@dcs.kcl.ac.uk

Abstract. We show how to combine the standard tableau system for the basic description logic $\mathcal{ALC}$ and Wolper's tableau calculus for propositional temporal logic PTL (with the temporal operators 'next-time' and 'until') in order to design a terminating sound and complete tableau-based satisfiability-checking algorithm for the temporal description logic PTL$_{\mathcal{ALC}}$ of [20] interpreted in models with constant domains. We use the method of quasimodels [18,16] to represent models with infinite domains, and the technique of minimal types [11] to maintain these domains constant. The combination is flexible and can be extended to more expressive description logics or even to decidable fragments of first-order temporal logics.

1   Introduction

Temporal description logics (TDLs) are knowledge representation formalisms intended for dealing with *temporal conceptual knowledge*. In other words, TDLs combine the ability of description logics (DLs) to represent and reason about conceptual knowledge with the ability of temporal logics (TLs) to reason about time. A dozen TDLs designed in the last decade (see e.g. [15,14,2,20,3,10] and survey [1]) showed that the equation TDL = DL + TL may have different, often very complex solutions, partly because of the rich choice of DLs and TLs, but primarily because of principle difficulties in combining systems; see [7]. With rare exceptions, the work so far has been concentrated on theoretical foundations of TDLs (decidability and undecidability, computational complexity, expressive power). The investigation of 'implementable' algorithms is still at the embryo stage, especially for the TDLs with non-trivial interactions between their DL and

R. Goré, A. Leitsch, and T. Nipkow (Eds.): IJCAR 2001, LNAI 2083, pp. 121–136, 2001.
© Springer-Verlag Berlin Heidelberg 2001

TL components. The problem we are facing is as follows: is it possible to combine the existing implementable reasoning procedures for the interacting DL and TL components into a reasonably efficient (on 'real world problems') algorithm for their TDL hybrid? As the majority of the existing reasoning mechanisms for DLs are based on the tableau approach, a first challenging step would be to combine a tableau system for a DL with Wolper's tableaux [17] for the propositional temporal logic PTL.

The first TDL tableau system was constructed by Schild [14], who merged the basic description logic $\mathcal{ALC}$ with PTL by allowing applications of the temporal operator $\mathcal{U}$ (until) and its derivatives only to concepts. For example, he defines a concept Mortal by taking

$$\text{Mortal} = \text{Living\_being} \sqcap (\text{Living\_being} \ \mathcal{U} \ \square \neg \text{Living\_being}),$$

where $\square$ means 'always in the future.' The resulting language is interpreted in models based on the flow of time $\langle \mathbb{N}, < \rangle$ and, for each $n \in \mathbb{N}$, specifying an $\mathcal{ALC}$-model that describes the state of the knowledge base at moment n. Schild obtains his sound, complete and terminating tableau system (for checking concept satisfiability) simply by putting together the tableau rules of $\mathcal{ALC}$ and PTL. The reason behind this 'trivial' solution is that, in Schild's logic, there is no actual interaction between the temporal operators of PTL and the constructors of $\mathcal{ALC}$; the logic is the *fusion* or *independent join* of its components.

A more sophisticated combination PTL$_{\mathcal{ALC}}$ of $\mathcal{ALC}$ and PTL allowing applications of temporal and Boolean operators to both concepts and TBox axioms was constructed in [20]. Using PTL$_{\mathcal{ALC}}$, one can express statements like 'in all times all living beings are mortal' or 'living beings will never die out completely:'

$$\square(\text{Living\_being} \sqsubseteq \text{Mortal}), \quad \square \lozenge \neg (\text{Living\_being} = \bot),$$

where $\lozenge$ means 'some time in the future.' The degree of interaction between the DL and TL components in PTL$_{\mathcal{ALC}}$ depends on the 'domain assumption' the intended models comply with. A tableau system for PTL$_{\mathcal{ALC}}$ interpreted in models with *expanding* $\mathcal{ALC}$ domains (which means that when moving from earlier moments of time to later ones, the domains of $\mathcal{ALC}$-models can get larger and larger, but never shrink) was designed in [16]. The interaction between the components becomes even stronger if we consider models with constant domains, where an introduction of a domain element at moment n forces us to introduce the same element at all previous moments as well. This makes the problem of constructing tableaux for PTL$_{\mathcal{ALC}}$ with constant domains considerably more difficult.

The choice of the domain assumption—*expanding, varying, decreasing,* or *constant*—depends on the knowledge to be represented. One can argue, for instance, whether the domain element representing a living being A in a model exists before A's birth or after A's death. However, in many applications such as reasoning about temporal entity relationship (ER) diagrams [2,3], expanding domains do not suffice and must be replaced by constant ones. Apart from being appropriate for applications, the constant domain assumption is the most general case in the sense that reasoning with expanding (or varying) domains can be reduced to reasoning with constant domains (see e.g. [20]).

The main aim of this paper is to design a terminating, sound, and complete tableau system for checking satisfiability of PTL$_{\mathcal{ALC}}$*-formulas in models with constant domains.*

This is achieved by

- combining (in a modular way) the standard tableaux for $\mathcal{ALC}$ with Wolper's [17] tableaux for PTL,
- using so-called *quasimodel* representations of constraint systems, and
- using so-called *minimal type representations* of domain elements introduced in subsequent states.

Quasimodels [18,19,20] are abstractions of models representing elements by their types and the evolution of elements in time by certain functions called runs. As was shown in [16], quasimodels make it possible to cope with PTL$_{\mathcal{ALC}}$ models having infinite $\mathcal{ALC}$ domains (an example showing that PTL$_{\mathcal{ALC}}$ does not have the finite domain property can be found in Section 2). The concept of 'minimal partial types' is the main new idea of this paper which is used to maintain the $\mathcal{ALC}$ domains constant.

Although the formula-satisfiability problem for PTL$_{\mathcal{ALC}}$ is rather complex—as is shown in [3], it is ExpSpace-complete—we hope that the tableau system constructed in this paper will lead to a 'reasonably efficient' implementation of the PTL$_{\mathcal{ALC}}$ reasoning services. However, in order to achieve an acceptable runtime behavior, it is still necessary to devise suitable optimization strategies for the algorithm. We believe that such strategies can be found, since, as shown in e.g. [9], related tableau algorithms are amenable to optimization.

It is to be noted that the developed approach can be used to design tableau algorithms for other combinations of description and modal logics (in particular, temporal epistemic logics of [6]). For instance, [11] gives a solution to the open problem of Baader and Laux [4] by constructing tableaux for their combination of the modal logic K with $\mathcal{ALC}$ interpreted in models with constant domains.

The paper is accompanied by a technical report [12] containing full proofs of all theorems.

2   Basic Definitions

We begin by introducing the temporal description logic PTL$_{\mathcal{ALC}}$ of [20].

Let $N_C = \{C_0, C_1, \dots\}$, $N_R = \{R_0, R_1, \dots\}$, and $N_O = \{a_0, a_1, \dots\}$ be countably infinite sets of *concept names*, *role names*, and *object names*, respectively. PTL$_{\mathcal{ALC}}$-*concepts* are defined inductively: all the C_i as well as $\top$ are concepts, and if C, D are concepts and $R \subset N_R$, then $C \sqcap D$, $\neg C$, $\exists R.C$, $\bigcirc C$, and $C \mathcal{U} D$ are concepts.

$\mathsf{PTL}_{\mathcal{ALC}}$-*formulas* are defined as follows: if C, D are concepts and $a, b \in N_O$, then $C = D$, $a : C$, and aRb are atomic formulas; and if φ and ψ are formulas, then so are $\neg\varphi$, $\varphi \wedge \psi$, $\bigcirc\varphi$, and $\varphi\mathcal{U}\psi$.

The intended models of $\mathsf{PTL}_{\mathcal{ALC}}$ are natural *two-dimensional* hybrids of standard models of $\mathcal{ALC}$ and PTL. More precisely, a $\mathsf{PTL}_{\mathcal{ALC}}$-*model* is a triple $\mathfrak{M} = \langle \mathbb{N}, <, I \rangle$, where $<$ is the standard ordering of $\mathbb{N}$ and I a function associating with each $n \in \mathbb{N}$ an $\mathcal{ALC}$-model $I(n) = \left\langle \Delta, R_0^{I(n)}, \ldots, C_0^{I(n)}, \ldots, a_0^{I(n)}, \ldots \right\rangle$, in which Δ, the (constant) *domain* of $\mathfrak{M}$, is a non-empty set, the $R_i^{I(n)}$ are binary relations on Δ, the $C_i^{I(n)}$ subsets of Δ, and the $a_i^{I(n)}$ are elements of Δ such that $a_i^{I(n)} = a_i^{I(m)}$, for every $n, m \in \mathbb{N}$.

(Note that in the given definition, the object names are assumed to be *global*, while the concept names are interpreted *locally*. Neither of these assumptions is essential; in particular, global concepts can be defined via local ones and $\mathcal{U}$.)

The *extension* $C^{I(n)}$ of a concept C in $\mathfrak{M}$ at a moment n is defined in the following way:

$$\top^{I(n)} = \Delta;$$
$$(C \sqcap D)^{I(n)} = C^{I(n)} \cap D^{I(n)};$$
$$(\neg C)^{I(n)} = \Delta \setminus C^{I(n)};$$
$$(\exists R.C)^{I(n)} = \{d \in \Delta \mid \exists d' \in C^{I(n)} \; dR^{I(n)}d'\};$$
$$(C\mathcal{U}D)^{I(n)} = \{d \in \Delta \mid \exists m \geq n \; (d \in D^{I(m)} \;\&\; \forall k \; (n \leq k < m \to d \in C^{I(k)}))\};$$
$$(\bigcirc C)^{I(n)} = C^{I(n+1)}.$$

The *truth-relation* $\mathfrak{M}, n \models \varphi$ for the Boolean operators is standard and

$\mathfrak{M}, n \models C = D$ iff $C^{I(n)} = D^{I(n)}$;

$\mathfrak{M}, n \models a : C$ iff $a^{I(n)} \in C^{I(n)}$;

$\mathfrak{M}, n \models aRb$ iff $a^{I(n)} R^{I(n)} b^{I(n)}$;

$\mathfrak{M}, n \models \varphi\mathcal{U}\psi$ iff $\exists m \geq n \; (\mathfrak{M}, m \models \psi \;\&\; \forall k \; (n \leq k < m \to \mathfrak{M}, k \models \varphi))$;

$\mathfrak{M}, n \models \bigcirc\varphi$ iff $\mathfrak{M}, n + 1 \models \varphi$.

The only reasoning task we consider in this paper is satisfiability of $\mathsf{PTL}_{\mathcal{ALC}}$-formulas, a formula φ being *satisfiable* if there are a model $\mathfrak{M}$ and a moment $n \in \mathbb{N}$ such that $\mathfrak{M}, n \models \varphi$. Other standard inference problems for $\mathsf{PTL}_{\mathcal{ALC}}$—concept satisfiability, subsumption, ABox consistency, etc.—can be easily reduced to satisfiability of formulas.

There are two main difficulties in designing a tableau system for $\mathsf{PTL}_{\mathcal{ALC}}$. First, as was mentioned in the introduction, there exist formulas satisfiable only in models with infinite domains. For example, such is the conjunction of the formulas

$$\Box\neg((C \sqcap \bigcirc\neg C) = \bot), \quad \Box(\neg C \sqsubseteq \Box\neg C),$$

where $\Box C = \neg(\top\mathcal{U}\neg C)$ and $\bot = \neg\top$. To tackle this difficulty, we employ the standard tableaux for $\mathcal{ALC}$ for constructing *finite representations* of infinite models and keep track of the development of their elements in time by using quasimodels as introduced in [18,20,16].

The second difficulty is that at moment $n + 1$ the $\mathcal{ALC}$ tableau algorithm can introduce an element which does not exists at moment n. To ensure that all elements always have their immediate predecessors, at each time point we create certain 'marked' elements satisfying as few conditions as possible, and use them as those predecessors if necessary.

3 Constraint Systems

In this section, we introduce constraint systems which serve a two-fold purpose. First, they form a basis for defining quasimodels, which, in contrast to [20], are defined purely syntactically. Second, constraint systems are the underlying data structure of the tableau algorithm to be devised. Intuitively, a constraint system describes an $\mathcal{ALC}$-model.

In what follows, without loss of generality we assume that all equalities are of the form $C = \top$. ($C = D$ is clearly equivalent to $\big(\neg(C \sqcap \neg D) \sqcap \neg(D \sqcap \neg C)\big) = \top$.) Often we shall write $C \neq \top$ instead of $\neg(C = \top)$.

Constraint systems are formulated in the following language L_C. Let V be a fixed countably infinite set of (individual) *variables*. We assume V to be disjoint from the set N_O of object names. Elements of $V \cup N_O$ are called L_C-*terms*. If φ is a $\mathsf{PTL}_{\mathcal{ALC}}$-formula, C a concept, R a role, and x, y are L_C-terms, then φ, $x : C$, and xRy are called L_C-*formulas*.

We assume that V comes equipped with a well-order $<_V$. Let X be a non-empty subset of V. Then $\min(X)$ denotes the first variable in X with respect to $<_V$. Variables may occur in constraint systems either *marked* or *unmarked*; certain formulas may occur $\mathcal{U}$-*marked* or $\mathcal{U}$-*unmarked*. As we said above, marked variables are used to deal with constant domains. $\mathcal{U}$-markedness will be explained after the saturation rules have been introduced.

Definition 1. A *constraint system* S is a finite (non-empty) set of L_C-formulas such that

- each variable in S is either *marked* or *unmarked*,
- each formula in S of the form $\varphi \mathcal{U} \psi$ or $x : (C \mathcal{U} D)$ is either $\mathcal{U}$-*marked* or $\mathcal{U}$-*unmarked*,
- S contains $\min(V) : \top$.

We will say that a constraint system S is saturated if it satisfies a number of closure conditions. With a few exceptions, these conditions require that if S contains a formula φ of a certain form, then S contains some other formulas composed from subformulas and subconcepts of φ (possibly using additional negation and $\bigcirc$). For example, S is closed under conjunction if whenever S contains $\psi_1 \wedge \psi_2$, then it contains both conjuncts ψ_1 and ψ_2 as well. We formulate the closure conditions as the *saturation rules* in Fig. 1–3. Later these rules will also be used as rules of our tableau algorithm. A constraint system S is called *saturated* if none of the saturation rules can be applied to it.

$\mathcal{ALC}$-rules for formulas

$S \longrightarrow_{\neg\neg} \{\varphi\} \cup S$ if

$\neg\neg\varphi \in S$ and $\varphi \notin S$

$S \longrightarrow_{\wedge} \{\varphi, \psi\} \cup S$ if

$\varphi \wedge \psi \in S$ and $\{\varphi, \psi\} \not\subseteq S$

$\qquad S \longrightarrow_{\neg\wedge} \{\neg\theta\} \cup S$ if

$\qquad \neg(\varphi \wedge \psi) \in S$, $\neg\varphi \notin S$, and $\neg\psi \notin S$

$\qquad \theta \doteq \varphi$ or $\theta \doteq \psi$

Temporal rules for formulas

$S \longrightarrow_{\neg\bigcirc} \{\bigcirc\neg\varphi\} \cup S$ if

$\neg\bigcirc\varphi \in S$ and $\bigcirc\neg\varphi \notin S$

$S \longrightarrow_{\mathcal{U}} X \cup S$ if

$\varphi\mathcal{U}\psi$ appears $\mathcal{U}$-unmarked in S

$X = \{\psi\}$ or $X = \{\varphi, \bigcirc(\varphi\mathcal{U}\psi)\}$

$\varphi\mathcal{U}\psi$ is $\mathcal{U}$-marked in $X \cup S$

$\qquad S \longrightarrow_{\neg\mathcal{U}} X \cup S$ if

$\qquad \neg(\varphi\mathcal{U}\psi) \in S$, $\{\neg\psi, \neg\varphi\} \not\subseteq S$, and $\{\neg\psi, \bigcirc\neg(\varphi\mathcal{U}\psi)\} \not\subseteq S$

$\qquad X = \{\neg\psi, \neg\varphi\}$ or $X = \{\neg\psi, \bigcirc\neg(\varphi\mathcal{U}\psi)\}$

Fig. 1. Saturation rules for formulas.

A few remarks below will help the reader to understand the rules. As the temporal part of our tableaux is based on Wolper's [17] algorithm for PTL, the temporal saturation rules resemble those of Wolper's. Note also that the saturation rules $\longrightarrow_{\neg\wedge}$, $\longrightarrow_{\mathcal{U}}$, $\longrightarrow_{\neg\mathcal{U}}$, $\longrightarrow_{\neg\sqcap}$, $\longrightarrow_{\mathcal{U}c}$, and $\longrightarrow_{\neg\mathcal{U}c}$ are *disjunctive*: they have more than one possible outcome. In this section, it is convenient to view these rules as nondeterministic. Later, when the saturation rules are regarded as tableau rules, we will apply them deterministically, i.e., consider *all* of their possible outcomes. Unless otherwise stated, we assume rules to introduce $\mathcal{U}$-unmarked formulas. Intuitively, $\mathcal{U}$-markedness is needed to ensure that the $\longrightarrow_{\mathcal{U}}$ and $\longrightarrow_{\mathcal{U}c}$ rules are applied exactly once to each formula $\varphi\mathcal{U}\psi$ and $x : C\mathcal{U}D$, respectively. For example, we must ensure that the $\longrightarrow_{\mathcal{U}}$ rule is applied (once) to $\varphi\mathcal{U}\psi$ even if the constraint system under consideration already contains φ and $\bigcirc(\varphi\mathcal{U}\psi)$. This is required to make the tableau algorithm complete (see [17, 16] for an example and a more detailed discussion).

As was already noted, marked variables are needed to cope with constant domains. For now, we just observe that the disjunctive rules treat marked and unmarked variables differently. Intuitively, in case of marked variables it is not sufficient to consider only one of the possible outcomes of the disjunctive rule application per constraint system, but we must additionally consider both possible outcomes together. For example, if we have $S = \{v : E\mathcal{U}F, v : \neg(C \sqcap D)\}$ and v is marked in S, then we should consider not only the obvious saturations $S_1 = S \cup \{v : \neg C\}$ and $S_2 = S \cup \{v : \neg D\}$, but also

$$S_3 = \{v : E\mathcal{U}F, v : \neg(C \sqcap D), v : \neg C, v' : E\mathcal{U}F, v' : \neg(C \sqcap D), v' : \neg D\},$$

$\mathcal{ALC}$-rules for concepts

$S \longrightarrow_{\neg\neg c} \{x : C\} \cup S$ if

$x : \neg\neg C \in S$ and $x : C \notin S$

$S \longrightarrow_{\sqcap} \{x : C, x : D\} \cup S$ if

$x : C \sqcap D \in S$ and $\{x : C, x : D\} \not\subseteq S$

$S \longrightarrow_{\neg\sqcap} X \cup S$ if

$x : \neg(C \sqcap D) \in S$, $x : \neg C \notin S$ and $x : \neg D \notin S$
$X = \{x : \neg C\}$ or $X = \{x : \neg D\}$ or
$\quad x$ marked in S and $X = (\text{copy}(S, x, v) \cup \{x : \neg C, v : \neg D\})$
$\quad$ where v is marked in $X \cup S$ and the first new variable from V

$S \longrightarrow_{=} \{x : C\} \cup S$ if

$C = \top \in S$, x occurs in S, and $x : C \notin S$

$S \longrightarrow_{\neg\exists} \{y : \neg C\} \cup S$ if

$x : \neg\exists R.C \in S$, $xRy \in S$, and $y : \neg C \notin S$

Temporal rules for concepts

$S \longrightarrow_{\neg\bigcirc c} \{x : \bigcirc\neg C\} \cup S$ if

$x : \neg\bigcirc C \in S$ and $x : \bigcirc\neg C \notin S$

$S \longrightarrow_{\mathcal{U} c} X \cup S$ if

$x : C\mathcal{U}D$ appears $\mathcal{U}$-unmarked in S
$X = \{x : D\}$ or $X = \{x : C, x : \bigcirc(C\mathcal{U}D)\}$ or
$\quad x$ marked in S and $X = (\text{copy}(S, x, v) \cup \{x : D, v : C, v : \bigcirc(C\mathcal{U}D)\})$
$\quad$ where v is marked in $X \cup S$ and the first new variable from V
$x : C\mathcal{U}D$ and $v : C\mathcal{U}D$ (if introduced) are $\mathcal{U}$-marked in $X \cup S$

$S \longrightarrow_{\neg\mathcal{U} c} X \cup S$ if

$x : \neg(C\mathcal{U}D) \in S$, $\{x : \neg D, x : \neg C\} \not\subseteq S$, and $\{x : \neg D, x : \bigcirc\neg(C\mathcal{U}D)\} \not\subseteq S$
$X = \{x : \neg D, x : \neg C\}$ or $X = \{x : \neg D, x : \bigcirc\neg(C\mathcal{U}D)\}$ or
$\quad x$ marked in S and $X = (\text{copy}(S, x, v) \cup \{x : \neg D, x : \neg C, v : \neg D, v : \bigcirc\neg(C\mathcal{U}D)\})$
$\quad$ where v is marked in $X \cup S$ and the first new variable from V

Fig. 2. Non-generating saturation rules for concepts.

$S \longrightarrow_{\neq} \{v : \neg C\} \cup S$ if

$C \neq \top \in S$ and there exists no y with $y : \neg C \in S$
v is the first new variable from V

$S \longrightarrow_{\exists} \{v : C, xRv\} \cup S$ if

$x : \exists R.C \in S$, there is no y such that $\{xRy, y : C\} \subseteq S$ and x is not blocked in S
by an unmarked variable; v is unmarked and the first new variable from V

Fig. 3. Generating saturation rules.

where, v is marked in S_1, S_2, S_3 and v' is marked in S_3. In S_3, we created a 'marked copy' v' of v and saturated v in one possible way and v' in the other. In the formulation of the rules, copies are made by using $\mathrm{copy}(S, v, v')$ which denotes the set $\{v' : C \mid v : C \in S\}$, where v is marked and v' is a fresh variable (not used in S). Note that by definition of L_C-formulas, marked variables do not occur in complex formulas such as $x : C \wedge x : D$ and thus such formulas need not be considered for copy. We generally assume that copies preserve $\mathcal{U}$-markedness: in the example above, $v' : E\mathcal{U}F$ is $\mathcal{U}$-marked in S_3 iff $v : E\mathcal{U}F$ is $\mathcal{U}$-marked in S.

To ensure termination of repeated applications of the saturation rules, we use a 'blocking' technique, c.f. [5]. Blocked variables are defined as follows. For now, assume that each constraint system is equipped with a strict partial order $\ll$ on the set of terms. Say that a variable v in a constraint system S is *blocked by a variable* v' in S if $v' \ll v$ and $\{C \mid v : C \in S\} \subseteq \{C \mid v' : C \in S\}$. Later, when we consider sequences of constraint systems obtained by repeated rule applications, $\ll$ will denote the order of introduction of terms. Note that only variables, rather than object names, may block terms. Also, only variables can be blocked.

A constraint system S is said to be *clash-free* if it contains no formulas $\neg\top$ and $x : \neg\top$ and neither a pair of the form $x : C$, $x : \neg C$, nor a pair of the form $\varphi, \neg\varphi$. We write $S \longrightarrow_\bullet S'$ to say that the constraint system S' can be obtained from S by an application of the saturation rule $\longrightarrow_\bullet$.

Let $S_0, \ldots, S_n$ be a sequence of constraint systems such that, for every $i < n$, there is a saturation rule $\longrightarrow_\bullet$ for which $S_i \longrightarrow_\bullet S_{i+1}$ and in case $\longrightarrow_\bullet$ is a generating rule, no non-generating rule is applicable to S_i (where non-generating rules are from Fig. 1 and 2 while generating rules are from Fig. 3). Then we say that $S_0, \ldots, S_n$ is built *according to the saturation strategy*. If this is the case and no saturation rule is applicable to S_n, then we call S_n a *saturation* of S_0.

4 Quasimodels

As was already said, $\mathsf{PTL}_{\mathcal{ALC}}$ does not have the finite domain property, and so our tableau algorithm constructs abstractions of models, called quasimodels, rather than models themselves.

Quasimodels are based on the idea of *concept types*. A concept type is simply a set of concepts that are 'relevant' to the tested formula and satisfied by an element of the domain. The 'fragment' of relevant concepts and formulas is defined as follows. Let Φ be a set of formulas. Denote by $Sb(\Phi)$ the set of all subformulas of formulas in Φ, by $ob(\Phi)$ the set of all object names that occur in Φ, by $rol(\Phi)$ the set of all roles in Φ, and by $con(\Phi)$ the set of all concepts in Φ. If $\#$ is a unary operator, say, $\neg$ or $\bigcirc$, then $\#(\Phi)$ is the union of Φ and $\{\#\varphi \mid \varphi \in \Phi\}$. The *fragment* $Fg(\Phi)$ *generated by* Φ is defined as the union of the following four sets: $ob(\Phi)$, $rol(\Phi)$, $\bigcirc(\neg con(\Phi \cup \{\top\}))$ and $\bigcirc(\neg Sb(\Phi \cup \{\top\}))$.

Roughly, a quasimodel is a sequence $(S_n \mid n \in \mathbb{N})$ of saturated constraint systems that satisfies certain conditions which control interactions between the S_n and ensure that quasimodels can be reconstructed into real models. Unlike standard tableaux, where a variable usually represents an element of a model, a variable in a quasimodel represents a concept type. More precisely, if a constraint system contains a variable v, then the corresponding $\mathcal{ALC}$-models contain at least one—but potentially (infinitely) many—elements of the type represented by v. As our $\mathsf{PTL}_{\mathcal{ALC}}$-models have constant domains, we need some means to keep track of the types representing the same element at different moments of time. This can be done using a function r, called a *run*, which associates with each $n \in \mathbb{N}$ a term $r(n)$ from S_n. Thus $r(0), r(1), \ldots$ are type representations of one and the same element at moments $0, 1, \ldots$.

We are in a position now to give precise definitions. Fix a $\mathsf{PTL}_{\mathcal{ALC}}$-formula ϑ.

Definition 2. A *quasiworld* for ϑ is a saturated clash-free constraint system S satisfying the following conditions:

 - $\{a \mid \exists C \ (a : C) \in S\} = ob(\vartheta)$,
 - $con(S) \subseteq Fg(\vartheta)$ and $rol(S) \subseteq Fg(\vartheta)$,
 - for every formula $\varphi \in S$, if φ is a $\mathsf{PTL}_{\mathcal{ALC}}$-formula then $\varphi \in Fg(\vartheta)$,
 - all variables in S are unmarked.

One should not be confused by that all variables in quasiworlds are unmarked. Marked variables are—as we shall see later on—important for the *construction* of a quasimodel. After the construction, marked variables can simply be 'unmarked' (note that this operation preserves saturatedness of constraint systems).

Definition 3. A sequence $Q = (S_n \mid n \in \mathbb{N})$ of quasiworlds for ϑ is called a ϑ-*sequence*. A *run* in Q is a function r associating with each $n \in \mathbb{N}$ a term $r(n)$ from S_n such that

 - for every $m \in \mathbb{N}$ and every concept C, if $(r(m) : \bigcirc C) \in S_m$ then we have $(r(m+1) : C) \in S_{m+1}$,
 - for all $m \in \mathbb{N}$, if $(r(m) : C\mathcal{U}D) \in S_m$ then there is $k \geq m$ such that $(r(k) : D) \in S_k$ and $(r(i) : C) \in S_i$ whenever $m \leq i < k$.

Definition 4. A ϑ-sequence Q is called a *quasimodel* for ϑ if the following hold:

 - for every object name a in Q, the function r_a defined by $r_a(n) = a$, for all $n \in \mathbb{N}$, is a run in Q,
 - for every $n \in \mathbb{N}$ and every variable v in S_n, there is a run r in Q such that $r(n) = v$,
 - for every $n \in \mathbb{N}$ and every $\bigcirc \varphi \in S_n$, we have $\varphi \in S_{n+1}$,
 - for every $n \in \mathbb{N}$ and every $(\varphi \mathcal{U} \psi) \in S_n$, there is $m \geq n$ such that $\psi \in S_m$ and $\varphi \in S_k$ whenever $n \leq k < m$.

We say that ϑ is *quasi-satisfiable* if there are a quasimodel $Q = (S_n \mid n \in \mathbb{N})$ for ϑ and $n \in \mathbb{N}$ such that $\vartheta \in S_n$.

Theorem 1. *A $\mathsf{PTL}_{\mathcal{ALC}}$-formula ϑ is satisfiable iff ϑ is quasi-satisfiable.*

5   The Tableau Algorithm

In this section, we present a tableau algorithm for checking satisfiability of $\mathsf{PTL}_{\mathcal{ALC}}$-formulas in models with constant domains. Before going into technical details, we explain informally how quasimodels for an input formula ϑ are constructed and, in particular, how marked variables help to maintain constant domains.

Intuitively, marked variables represent so-called 'minimal types.' If a constraint system S contains marked variables $v_1, \ldots, v_k$ then *every* element of an $\mathcal{ALC}$-model corresponding to S is described by one of the v_i. It should now be clear why the disjunctive saturation rules must be applied in a special way to marked variables. Consider, for example, the $\longrightarrow_{\neg\sqcap}$ rule and assume that there is a single marked variable v_m in S and that $v_m : \neg(C \sqcap D) \in S$. In the context of minimal types, this means that every element in corresponding $\mathcal{ALC}$-models satisfies $\neg(C \sqcap D)$. From this, however, it does not follow that every element satisfies $\neg C$ or that every element satisfies $\neg D$. Hence, the $\longrightarrow_{\neg\sqcap}$ rule cannot be applied in the same way as for unmarked variables.

Here is a simple example illustrating the construction of quasimodels with minimal types. Consider the formula

$$\vartheta = \left((\neg(\bigcirc C \sqcap \bigcirc \neg C)) = \top\right) \wedge a : \bigcirc \exists R.C.$$

With this formula we associate the initial constraint system $S_\vartheta = \{\vartheta, v_m : \top\}$ containing ϑ and a single marked variable v_m. By applying saturation rules, we obtain then the constraint system $S_0 = \{a : \bigcirc \exists R.C, v_m : \bigcirc C, v_m' : \bigcirc \neg C\}$ (slightly simplified for brevity) that describes the $\mathcal{ALC}$-model for time moment 0. The constraint system for moment 1 is $\{a : \exists R.C, v_1 : C, v_2 : \neg C, v_m : \top\}$ (where v_m is the only marked variable) which can then be extended to the system $S_1 = \{a : \exists R.C, v_m : \top, v_1 : C, v_2 : \neg C, aRv, v : C\}$ by the saturation rules. Note that we introduced a new (unmarked) variable v. Every element d which is of type v at moment 1 must—according to the constant domain assumption—also exist at moment 0. But what is the type of d at that moment (i.e., the 'predecessor type' of d at 1)? By the definition of minimal types, we must only choose among marked variables. So either d is of type v_m at 0, which means that we must add $v : C$ to S_1, or d is of type v_m' at 0, and so we must add $v : \neg C$ to S_1. The former choice yields an (initial fragment of a) quasimodel, while the latter leads to a clash. For a more detailed discussion we refer the reader to [11].

We can now define the tableau algorithm. In general, tableau algorithms try to construct a (quasi)model for the input formula by repeatedly applying *tableau rules* to an appropriate data structure. Let us first introduce this data structure.

Definition 5. A *tableau* for a $\mathsf{PTL}_{\mathcal{ALC}}$-formula ϑ is a triple $\mathcal{G} = (G, \prec, l)$, where $(G, \prec)$ is a finite tree and l a labelling function associating with each $g \in G$ a constraint system $l(g)$ for ϑ such that $S_\vartheta = \{\vartheta\} \cup \{\min(V) : \top\} \cup \{a : \top \mid a \in ob(\vartheta)\}$ is associated with the root of $\mathcal{G}$, where $\min(V)$ is marked and ϑ is $\mathcal{U}$-unmarked if it is of the form $\varphi \mathcal{U} \psi$ or $x : (C\mathcal{U}D)$.

To decide whether ϑ is satisfiable, the tableau algorithm for $\mathsf{PTL}_{\mathcal{ALC}}$ goes through two phases. In the first phase, the algorithm starts with an initial tableau $\mathcal{G}_\vartheta$ and exhaustively applies the tableau rules to be defined below. Eventually we obtain a tableau $\mathcal{G}$ to which no more rule is applicable; it is called a *completion* of $\mathcal{G}_\vartheta$. In the second phase, we eliminate those parts of $\mathcal{G}$ that contain obvious contradictions or eventualities which are not realized. After that we are in a position to deliver a verdict: ϑ is satisfiable iff the resulting tableau $\mathcal{G}'$ is not empty, i.e., iff the root of $\mathcal{G}$ has not been eliminated.

Let us first concentrate on phase 1. The initial tableau $\mathcal{G}_\vartheta$ associated with ϑ is defined as $(\{g^r\}, \prec^r, l)$, where $\prec^r = \emptyset$ and $l(g^r) = S_\vartheta$. To define the tableau rules, we require a number of auxiliary notions. Let S be a constraint system and x a term occurring in S. Denote by $A_x(S)$ the set $\{C \mid (x : \bigcirc C) \in S\}$ and define an equivalence relation $\sim_S$ on the set of variables (not terms) in S by taking $v \sim_S u$ iff $A_v(S) = A_u(S)$. The equivalence class generated by v is denoted by $[v]_S$. Finally, let $[S]_\sim$ denote the set of all equivalence classes $[v]_S$.

Similar to the local blocking strategy on variables of constraint systems, we need a global blocking strategy on the nodes of tableaux. To define this kind of blocking, it is convenient to abstract from variable names.

Let S and S' be constraint systems. S' is called a *variant* of S if there is a bijective function π from the variables occurring in S onto the variables occurring in S' which respects markedness (i.e., unmarked variables are mapped to unmarked variables and marked variables to marked variables) and S' is obtained from S by replacing each variable v from S with $\pi(v)$. In this case π is called a *renaming*.

Like constraint systems, tableaux are equipped with a strict partial order $\ll$ on the set of nodes which indicates the order in which the nodes of the tableau have been introduced. The tableau rules are shown in Fig. 4. Intuitively, the $\Longrightarrow_\bigcirc$ rule generates a new time point, while the other rules infer additional knowledge about an already existing time point. For every saturation rule $\longrightarrow_s$ we have a corresponding tableau rule $\Longrightarrow_s$. The $\Longrightarrow_\downarrow$ and $\Longrightarrow_{\downarrow'}$ rules deal with constant domains and use the notion of ancestor which is defined as follows.

Let $\mathcal{G} = (G, \prec, l)$ be a tableau for ϑ. A node $g \in G$ is called a *state* if only the $\Longrightarrow_\bigcirc$ rule is applicable to g. The node g is an *ancestor* of a node $g' \in G$ if there is a sequence of nodes $g_0, \ldots, g_n$ such that $g_0 = g$, $g_n = g'$, $g_i \prec g_{i+1}$ for $i < n$, and g_0 is the only state in the sequence.

As to the $\Longrightarrow_\bigcirc$ rule, recall that variables represent types rather than elements. In view of this, when constructing the next time point, we 'merge' variables satisfying the same concepts (by using the equivalence classes). Actually, this idea is crucial for devising a terminating tableau algorithm despite the lack of the finite domain property. The $\Longrightarrow_\downarrow$ rule formalizes the choice of a predecessor type as was sketched in the example above. Since we have to *choose* a predecessor type, the rule behaves similar to a disjunctive saturation rule, which means that we must apply the rule in a different way for marked variables. That is why we need the $\Longrightarrow_{\downarrow'}$ rule: for marked variables, it considers arbitrary combinations of choices of predecessor types.

$(G, \prec, l) \Longrightarrow_s (G', \prec', l')$

if g is a leaf in G, the saturation rule $\longrightarrow_s$ is applicable to $l(g)$,
$S_1, \ldots, S_n$ are the possible outcomes of the application of $\longrightarrow_s$ to $l(g)$,
$G' = G \uplus \{g_1, \ldots, g_n\}$ and, for $1 \le i \le n$, $\prec' = \prec \cup \{(g, g_i)\}$ and $l'(g_i) = S_i$

$(G, \prec, l) \Longrightarrow_\bigcirc (G', \prec', l')$

if $G' = G \uplus \{g'\}$, $\prec' = \prec \cup \{(g, g')\}$ for some leaf $g \in G$,
$l'(g')$ is the union of the following sets:

$$\{a : \top\} \cup \{a : C \mid (a : \bigcirc C) \in l(g)\}, \text{ for } a \in ob(l(g)),$$
$$\{\psi \mid \bigcirc \psi \in l(g)\},$$
$$\{\min([v]_{l(g)}) : \top\} \cup \{\min([v]_{l(g)}) : C \mid (\min([v]_{l(g)}) : \bigcirc C) \in l(g)\},$$
$$\text{for} \quad [v]_{l(g)} \quad \in$$
$[l(g)]_\sim,$
$$\{v' : \top\},$$

where v' is the only marked variable in $l(g')$,
and there is no $g'' \in G$ with $g'' \ll g$ such that $l(g'')$ is a variant of $l(g)$
(i.e., the rule is not blocked)

$(G, \prec, l) \Longrightarrow_\downarrow (G', \prec', l')$

if g is a leaf in G, v is an unmarked variable in $l(g)$, g' is the ancestor of g,
for no term x in $l(g')$ do we have

$$\{C \mid (x : \bigcirc C) \in l(g')\} \subseteq \{C \mid (v : C) \in l(g)\},$$

$v_1, \ldots, v_n$ are the marked variables in $l(g')$, $G' = G \uplus \{g_1, \ldots, g_n\}$, and,
for $1 \le i \le n$, we have $\prec' = \prec \cup \{(g, g_i)\}$ and

$$l'(g_i) := l(g) \cup \{v : C \mid (v_i : \bigcirc C) \in l(g')\}.$$

$(G, \prec, l) \Longrightarrow_{\downarrow'} (G', \prec', l')$

if g is a leaf in G, v is a marked variable in $l(g)$, g' is the ancestor of g,
for no term x in $l(g')$ do we have

$$\{C \mid (x : \bigcirc C) \in l(g')\} \subseteq \{C \mid (v : C) \in l(g)\},$$

$X = \{\min([v']_{l(g')}) \mid v' \text{ is a marked variable in } l(g')\}$,
Y_i is the ith subset of X (for some ordering),
$G' = G \uplus \{g_1, \ldots, g_{2^{|X|}}\}$, and, for $1 \le i \le 2^{|X|}$, we have $\prec' = \prec \cup \{(g, g_i)\}$ and
$l'(g_i)$ is the union of $l(g)$ and the following sets, where we assume $Y_i = \{v_1, \ldots, v_n\}$:

$$\{v : C \mid (v_1 : \bigcirc C) \in l(g)\}$$
$$\text{copy}(l(g), v, v'_j) \text{ for } 1 < j \le n$$
$$\{v'_j : C \mid (v_j : \bigcirc C) \in l(g')\} \text{ for } 1 < j \le n$$

Here, all newly introduced variables v'_j are marked in $l'(g_i)$.

Note: For all rules, we assume that $l'(g) = l(g)$ for all $g \in G$. $A \uplus B$ denotes the disjoint union of A and B.

Fig. 4. Tableau rules.

The tableau rules are applied until no further rule application is possible. To ensure termination, we must follow a certain strategy of rule applications.

Definition 6. A tableau is *complete* if no tableau rule is applicable to it. Let $\mathcal{G}_0, \ldots, \mathcal{G}_n$ be a sequence of tableaux such that the associated orders $\ll_0, \ldots, \ll_n$ describe the order of node introduction and, for every $i < n$, there is a tableau rule $\Longrightarrow_\bullet$ such that $\mathcal{G}_i \Longrightarrow_\bullet \mathcal{G}_{i+1}$ and

- if the rule is one of the generating rules $\Longrightarrow_{\neq}$ or $\Longrightarrow_\exists$, then no tableau rule different from $\Longrightarrow_{\neq}$, $\Longrightarrow_\exists$, and $\Longrightarrow_\bigcirc$ is applicable to $\mathcal{G}_i$,
- if the rule is $\Longrightarrow_\bigcirc$, then no other tableau rule is applicable to $\mathcal{G}_i$.

Then $\mathcal{G}_0, \ldots, \mathcal{G}_n$ is said to be built *according to the tableau strategy*. If this is the case, $\mathcal{G}_0 = \mathcal{G}_\vartheta$, and $\mathcal{G}_n$ is complete, then $\mathcal{G}_n$ is called a *completion* of ϑ.

The following lemma claims that the tableau strategy ensures termination.

Theorem 2. *If the tableau rules are applied according to the tableau strategy, then a completion is reached after finitely many steps.*

Let us now turn to the second phase of the algorithm, i.e., to the elimination phase. We begin by defining which nodes are blocked.

Definition 7. Let $\mathcal{G} = (G, \prec, l)$ be a tableau for ϑ. A state $g \in G$ is *blocked* by a state $g' \in G$ if $g' \ll g$ and $l(g')$ is a variant of $l(g)$. We define a new relation $\preceq$ by taking $g \preceq g'$ if either $g \prec g'$, or g has a successor g'' that is blocked by g'.

An important part of the elimination process deals with so-called eventualities. An L_C-formula $\alpha \in S$ is called an *eventuality* for a constraint system S if α is either of the form $x : C\mathcal{U}D$ or of the form $\varphi\mathcal{U}\psi$. An eventuality is said to be *unmarked* if it is not of the form $v : C\mathcal{U}D$ for any marked variable v. All eventualities occurring in the tableau have to be 'realized' in the following sense.

Definition 8. Let $\mathcal{G} = (G, \prec, l)$ be a tableau for ϑ, $g \in G$, and let α be an eventuality for $l(g)$. Then α is *realized* for g in $\mathcal{G}$ if there is a sequence of unblocked nodes $g_0 \preceq g_1 \ldots \preceq g_n$ in G with $g = g_0$, $n \geq 0$, such that the following holds:

(1) if α is $\varphi\mathcal{U}\psi$ then $\psi \in l(g_n)$;

(2) if α is $v : C\mathcal{U}D$, with v unmarked or marked variable, then there are variables v_i from $l(g_i)$, $i \leq n$, with $v_0 = v$, $v_1, \ldots, v_n$ unmarked, $(v_n : D) \in l(g_n)$, and, for all i, $0 < i \leq n$, we have

- if g_{i-1} is a state, then $\{C \mid (v_{i-1} : \bigcirc C) \in l(g_{i-1})\} \subseteq \{C \mid (v_i : C) \in l(g_i)\}$,
- if g_{i-1} is not a state, then $\{C \mid (v_{i-1} : C) \in l(g_{i-1})\} \subseteq \{C \mid (v_i : C) \in l(g_i)\}$;

(3) if α is $a : C\mathcal{U}D$, for some object name a, then $(a : D) \in l(g_n)$.

Intuitively, the variables $v_0, \ldots, v_n$ in (2) describe the same element at different moments of time. It should be clear that in a tableau representing a quasimodel, all eventualities have to be realized. Apart from removing nodes that contain clashes, to remove nodes with non-realized eventualities is the main aim of the elimination phase.

Definition 9. Let $\mathcal{G} = (G, \prec, l)$ be a tableau for ϑ. We use the following rules to eliminate points in $\mathcal{G}$:

(e_1) if $l(g)$ contains a clash, eliminate g and all its $\prec^*$-successors
(where '$\prec^*$-successor' is the transitive closure of '$\prec$-successor');

(e_2) if all $\preceq$-successors of g have been eliminated, eliminate g as well;

(e_3) if $l(g)$ contains an unmarked eventuality not realized for g, eliminate g and all its $\preceq^*$-successors.[1]

The elimination procedure is as follows. Say that a tableau $\mathcal{G}_1 = (G_1, \prec_1, l_1)$ is a *subtableau* of $\mathcal{G}_2 = (G_2, \prec_2, l_2)$ if $G_2 \supseteq G_1$ and $\mathcal{G}_1$ is the restriction of $\mathcal{G}_2$ to G_1. Obviously, if $\mathcal{G}_2$ is a tableau for ϑ and G_1 contains the root of G_2, then $\mathcal{G}_1$ is a tableau for ϑ. Suppose now that $\mathcal{G} = (G, \prec, l)$ is a completion of ϑ. We construct a decreasing sequence of subtableaux $\mathcal{G} = \mathcal{G}_0, \mathcal{G}_1, \ldots$ by iteratively eliminating nodes from G according to rules (e_1)–(e_3), with (e_1) being used only at the first step. (The two other rules are used in turns.) Since we start with a finite tableau, this process stops after finitely many steps, i.e., we reach a subtableau $\mathcal{G}' = (G', \prec', l')$ of $\mathcal{G}$ to which none of the elimination rules can be applied. We say that the root of $\mathcal{G}$ is *not eliminated* iff $G' \neq \emptyset$.

Theorem 3. *A* PTL$_{\mathcal{ALC}}$*-formula ϑ is satisfiable iff there is a completion of ϑ of which the root is not eliminated.*

As a consequence of Theorems 2 and 3 we obtain

Theorem 4. *There is an effective tableau procedure which, given a* PTL$_{\mathcal{ALC}}$*-formula ϑ, decides whether ϑ is satisfiable.*

6   Conclusion

This paper—a continuation of the series [14,4,16,11]—develops a tableau reasoning procedure for the temporal description logic PTL$_{\mathcal{ALC}}$ interpreted in two-dimensional models with constant $\mathcal{ALC}$ domains. As shown in [12], the algorithm runs in double exponential time—thus paralleling the complexity of Wolper's original PTL-algorithm [17] which solves a PSPACE-complete problem using exponential time. Despite the high complexity, we believe that the devised tableau algorithm is an important first step towards the use of TDLs as KR&R tools. A prototype implementation of the described algorithm is currently underway. Based on the experiences with this implementation, possible optimization startegies will be investigated using the work in [9] as a starting point.

An important feature of the developed algorithm is that the DL component can be made considerably more expressive, provided that the extension is also supported by a reasonable tableau procedure. One idea we are working on now

---

[1] Of course, eventualities which are marked also have to be realized. However, the fact that all unmarked eventualities in a tableau are realized implies that all other eventualities are also realized (see proofs).

is to extend this component to expressive fragments of first-order logic, thereby obtaining tableau procedures for fragments of first-order temporal logic (cf. [8]) having potential applications in a growing number of fields such as specification and verification of reactive systems, model-checking, query languages for temporal databases, etc.

Another interesting aspect of this paper is that, with minor modifications, the constructed tableaux can be used as a satisfiability checking procedure for the Cartesian product of S5 and PTL (cf. [13]), thus contributing to a new exciting field in modal logic studying the behavior of multi-dimensonal modal systems [7].

References

1. A. Artale and E. Franconi. Temporal description logics. In L. Vila *et al.* eds., *Handbook of Time and Temporal Reasoning in AI.* MIT Press, 2001. (To appear.)
2. A. Artale and E. Franconi. Temporal ER modeling with description logics. In *Proceedings of ER'99*, 1999. Springer–Verlag.
3. A. Artale, E. Franconi, M. Mosurovic, F. Wolter, and M. Zakharyaschev. Temporal description logics for conceptual modelling: expressivity and complexity. Submitted, 2001.
4. F. Baader and A. Laux. Terminological logics with modal operators. In *Proceedings of IJCAI'95*, pp. 808–814, 1995. Morgan Kaufmann.
5. F. Baader and U. Sattler. Tableau algorithms for description logics. In R. Dyckhoff, ed., *Proceedings of Tableaux 2000*, vol. 1847 of LNAI, pp. 1–18, Springer, 2000.
6. R. Fagin, J. Halpern, Y. Moses, and M. Vardi. *Reasoning about Knowledge.* MIT Press, 1995.
7. D. Gabbay, A. Kurucz, F. Wolter, and M. Zakharyaschev. *Many-Dimensional Modal Logics: Theory and Applications.* Elsevier, 2001. (To appear.)
8. I. Hodkinson, F. Wolter, and M. Zakharyaschev. Decidable fragments of first-order temporal logics. *Annals of Pure and Applied Logic*, 106:85–134, 2000.
9. I. Horrocks and P. F. Patel-Schneider. Optimising description logic subsumption. *Journal of Logic and Computation*, 9(3):267–293, 1999.
10. C. Lutz. Interval-based temporal reasoning with general TBoxes. In *Proceedings of IJCAI'01*, Morgan-Kaufmann, 2001.
11. C. Lutz, H. Sturm, F. Wolter, and M. Zakharyaschev. A tableau decision algorithm for modalized $\mathcal{ALC}$ with constant domains. Submitted, 2000.
12. C. Lutz, H. Sturm, F. Wolter, and M. Zakharyaschev. A tableau calculus for temporal description logic: The constant domain case. See http://www-lti.informatik.rwth-aachen.de/Forschung/Reports.html.
13. M. Marx, Sz. Mikulas, and S. Schlobach. Tableau calculus for local cubic modal logic and its implementation. *Journal of the IGPL*, 7:755–778, 1999.
14. K. Schild. Combining terminological logics with tense logic. In *Proceedings of the 6th Portuguese Conference on AI*, pp. 105–120, Porto, 1993.
15. A. Schmiedel. A temporal terminological logic. In *Proceedings AAAI'90*, pp. 640–645, 1990.
16. H. Sturm and F. Wolter. A tableau calculus for temporal description logic: The expanding domain case. Journal of Logic and Computation, 2001. (In print.)
17. P. Wolper. The tableau method for temporal logic: An overview. *Logique et Analyse*, 28:119–152, 1985.

18. F. Wolter and M. Zakharyaschev. Satisfiability problem in description logics with modal operators. In A. Cohn, *et al.* eds., *KR'98*, pp. 512–523, 1998.
19. F. Wolter and M. Zakharyaschev.   Multi-dimensional description logics.   In D. Thomas, ed., *Proceedings of IJCAI'99*, pp. 104–109, 1999.
20. F. Wolter and M. Zakharyaschev. Temporalizing description logic. In D. Gabbay and M. de Rijke, eds., *Frontiers of Combining Systems 2*, pp. 379–402. Studies Press/Wiley, 2000.

Free-Variable Tableaux for Constant-Domain Quantified Modal Logics with Rigid and Non-rigid Designation

Serenella Cerrito[1] and Marta Cialdea Mayer[2]

[1] Université de Paris-Sud, L.R.I.
[2] Università di Roma Tre, Dipartimento di Informatica e Automazione

Abstract. This paper presents a sound and complete free-variable tableau calculus for constant-domain quantified modal logics, with a propositional *analytical* basis, i.e. one of the systems **K, D, T, K4, S4**. The calculus is obtained by addition of the classical free-variable γ-rule and the "liberalized" δ^+-rule [14] to a standard set of propositional rules. Thus, the proposed system characterizes proof-theoretically the constant-domain semantics, which cannot be captured by "standard" (non-prefixed, non-annotated) ground tableau calculi. The calculi are extended so as to deal also with non-rigid designation, by means of a simple numerical annotation on functional symbols, conveying some semantical information about the worlds where they are meant to be interpreted.

1 Introduction

Quantified modal logic (QML) can be given a model-theoretical characterization by extending the propositional Kripke semantics: a first-order modal structure is a set of first-order classical interpretations (the "possible worlds"), connected by a binary relation (the accessibility relation). However, things are not just so simple, and several issues have to be addressed (see for example [12] for an overview). Among them, possible restrictions on the designation of terms and the object domains associated to the possible worlds distinguish different "variants" of QML. When the interpretation of a ground term is required to be the same in every world, then it is said to be a *rigid* term, otherwise it is *non-rigid*. Rigid and non-rigid designation can in principle coexists within the same logic, whenever some symbols are given a rigid interpretation and others are not. On the contrary, requirements about possible relations between the universes of different worlds necessarily characterize different logics, the same way as restrictions on the accessibility relation do, on the propositional side. The most commonly considered variants of QML, in this respect, are the *constant-domain* variant, where the object domain is the same for all worlds, and the *cumulative-domain* (or *increasing-domain*) variant, where the object domains can vary, but monotonically, i.e. if w' is accessible from w, then the object domain of w is included in the domain of w'; when the domains of different worlds are independent one of the other, then domains are said to be *varying*.

R. Goré, A. Leitsch, and T. Nipkow (Eds.): IJCAR 2001, LNAI 2083, pp. 137–151, 2001.

Cumulative-domain QML, with rigid designation only, can easily be given proof theoretic characterizations, obtained by addition of the principles of classical first-order logic to a propositional modal system. In fact, in such systems the converse of the Barcan Formula, that characterizes cumulative domains, is provable: $\Box \forall x A \rightarrow \forall x \Box A$. Rigid designation, on the other hand, is a consequence of the instantiation rule of classical logic. Quantified modal logics with cumulative domains and rigid designation have been given sequent and tableau calculi [9,10,15], natural deduction proof systems [3], matrix proof procedures [21], resolution style calculi [1,6], and have been treated by means of translation methods [2,19].

Constant-domain logics with rigid designation can be treated axiomatically, by addition of the Barcan formula, $\forall x \Box A \rightarrow \Box \forall x A$, to the axioms and rules of classical first-order logic and modal propositional logics. Beyond the axiomatic approach, translation methods are general enough to treat constant-domain logics, and rigid as well as non-rigid designation (see, for instance, [2,19]). Constant-domain logics with rigid designation have been formalized in the tableau style, but with the addition of prefixes labelling tableau nodes [9], as well as by means of matrix proof methods [21]: in both kinds of calculi in fact it is possible to analyse more than one possible world at a time, and this allows the proof to "go back and forth" (the same mechanism solves a similar problem raised by symmetric logics). In [9] all the variants of QML concerning the object domains of possible worlds are treated by prefixed tableau methods. A different approach is represented by modal display calculi, where the addition of the classical sequent rules for quantifiers captures constant-domain logics [4], and constant, increasing and decreasing-domain modal logics can all be presented as cut-free display sequent calculi, by use of structure-dependent rules for quantifiers [22]. A further direct approach dealing with both varying domains and non-rigid symbols has a representative in [16], which defines a resolution method for epistemic logics, where terms can be annotated by a "bullet" constructor distinguishing rigid terms from non-rigid ones. A rather different direction is followed in [11], where the language of modal logic is enriched by means of a predicate abstraction operator, in order to capture differences on the denotation of terms, and a tableau proof procedure is presented for such a logic, with no restriction on the domains of possible worlds.

The constant-domain requirement bears some relationship with the symmetry of the accessibility relation: if a model is symmetric and its domains are cumulative, then it is a constant-domain model. Constant-domain and symmetric logics also share a proof-theoretical difficulty: it is provable that constant-domain logics cannot be captured by "standard" tableau methods, or cut-free sequent calculi. In fact, the Craig Interpolation Theorem does not hold for such logics [8], while it is a consequence, in some cases, of the existence of sound and complete cut-free Gentzen systems [9]:

It follows that there can be no "reasonable" cut-free tableau system for such logics for if there were we could use it, or the related symmetric Gentzen system, to prove an Interpolation result ([9], p.383).

This paper shows that, however, constant-domain logics – with a propositional *analytical* basis, i.e. one of the systems **K**, **D**, **T**, **K4**, **S4**, whose Gentzen calculi enjoy the cut-elimination property – can be given a simple proof-theoretical characterization by means of free-variable tableau systems (obviously, the proof of the Interpolation Theorem given in [9] for cumulative-domain QML does not extend to such systems. See also Section 5). Moreover, non-rigid designation can also be treated, by means of a simple numerical annotation on functional symbols, conveying some semantical information about the worlds where they are meant to be interpreted. A similar annotation mechanism is used in the tableau calculi for cumulative and varying domain QML, with either rigid or non-rigid designation, presented in [7].

The paper is organized as follows: Section 2 gives an informal presentation of the free-variable calculus for constant-domain logics with rigid designation only, and the main intuitions justifying its appropriateness. The syntax and formal semantics of constant-domain logics with both rigid and non-rigid designation are presented in Section 3, and the tableau systems for such logics in Section 4, where the main lines of the soundness and completeness proofs are also given. Section 5 concludes this work.

2   The Role of the δ-Rule in Modal Calculi

Ground tableaux systems for QML with increasing domains and rigid designation are easily obtained by addition of the classical quantifier rules to the modal systems [9]:

$$(\gamma_G) \ \frac{\forall x A, S}{A[t/x], \forall x A, S} \qquad (\delta_G) \ \frac{\exists x A, S}{A[c/x], S}$$

In the γ_G-rule, t is any ground term occurring in the branch. In the δ_G-rule, c is a new constant, that does not occur in $\{\exists x A\} \cup S$.

The ground approach cannot be adapted to the constant-domain case. A naive and intuitive account of the difficulties that are encountered can be given as follows. In order to be complete, the expansions of a universal formula $\forall x A$ should include the instances $A[c/x]$ where c is a constant denoting an object known to exist in a further accessible world. For instance, if we attempt to prove the Barcan formula (with the π-rule for system **K**):

$$\frac{\frac{\frac{\frac{\neg(\forall x \Box A \to \Box \forall x A)}{\forall x \Box A, \ \neg \Box \forall x A} \ (\alpha)}{\Box A[a/x], \ \forall x \Box A, \ \neg \Box \forall x A} \ (\gamma_G)}{A[a/x], \ \neg \forall x A} \ (\pi_K)}{A[a/x], \ \neg A[c/x]} \ (\delta_G)$$

we observe that, in order to obtain a closed tableau, we need to get a leaf containing also $A[c/x]$. But, after the application of the π-rule, the universal formula is not available any more. So, in principle, the formula $\forall x \Box A$ should be expanded to $\Box A[c/x]$, before the application of the π-rule. This is reasonable, since the object denoted by c, that will further on be known to exist, belongs to the domain of the presently considered world, too. But, if we change the tableau, replacing every occurrence of the constant a with c, then the constant c cannot be used any more in the application of the δ_G-rule, since it is not "new" to the node.

Free-variable tableau calculi for QML with increasing domains and rigid designation can be obtained in the same straightforward manner as the ground calculi are, by addition of the classical free-variable rules, and the substitution rule, to a modal propositional system [10]. The quantifier rules are:

$$(\gamma) \ \frac{\forall x A, S}{A[v/x], \forall x A, S} \qquad (\delta) \ \frac{\exists x A, S}{A[f(v_1, ..., v_k)/x], S}$$

In the γ-rule, v is a new free variable (also called *parameter*). In the δ-rule, f is a new functional symbol, and $v_1, ..., v_k$ are all the parameters occurring in $\{\exists x A\} \cup S$.

It has been proved [14] that the classical calculus stays sound if, in the δ-rule, only the parameters actually occurring in $\exists x A$ are required to be the arguments of the Skolem function. Such a "liberalized" rule is called the δ^+-rule. Although in the classical case the δ^+-rule is a more efficient - but equivalent - reformulation of the δ-rule, it is not sound with respect to the cumulative-domain variant of QML. In fact, the following tableau shows that the δ^+-rule makes the Barcan formula provable in a modal calculus (with the π-rule for system **K**):

$$\frac{\neg(\forall x \Box A \rightarrow \Box \forall x A)}{\cfrac{\forall x \Box A, \ \neg \Box \forall x A}{\cfrac{\Box A[v/x], \ \forall x \Box A, \ \neg \Box \forall x A}{\cfrac{A[v/x], \ \neg \forall x A}{A[v/x], \ \neg A[c/x]} \ (\delta^+)} \ (\pi_K)} \ (\gamma)} \ (\alpha)$$

The tableau above is closed by application of the substitution $\{c/v\}$.

In this paper we show that the free-variable modal tableau calculi obtained by replacing the δ-rule with the δ^+-rule are sound and complete with respect to the constant-domain variants of QML, on a propositional basis that is one of the analytical systems **K**, **D**, **T**, **K4** or **S4**. Note, however, that this fact does not imply that Skolemization preserves satisfiability in constant-domain QML, because quantifiers cannot always cross the modal operators. In fact, $\forall x \Diamond A \rightarrow \Diamond \forall x A$ is not valid, even in constant-domain QML. As a consequence, although run-time Skolemization is allowed, formulae cannot be initially skolemized.

Intuitively, the shift from cumulative to constant domains caused by the "liberalization" on the arguments of Skolem functions corresponding to the δ^+-rule is due to the following reasons. The key remark is that the role of the parameters $v_1, ..., v_k$ in the new Skolem term $f(v_1, ..., v_k)$, introduced by the δ-rule in free-variable calculi, is to prevent the unification of such a term with any of $v_1, ..., v_k$. Thus, the effect of reducing the set of parameters in Skolem terms is to make more unifications possible, in particular also the unification of a parameter v with a Skolem term that still has to be introduced when v is generated.

Clearly, in order to preserve soundness, the unification of a parameter v with a term $f(v_1, ..., v_k)$ must be forbidden when the existential quantifier "generating" $f(v_1, ..., v_k)$ is in the scope of the universal quantifier generating v, otherwise the dependence $\forall$-$\exists$ is lost. As a consequence, in that case v must be one of $v_1, ..., v_k$, so that v and $f(v_1, ..., v_k)$ are not unifiable. With this exception, v and $f(v_1, ..., v_k)$ can be unified: if the two terms are introduced "in the same world", the reason is the same as in the classical case. Also, if v ranges on the domain of a previous world w, but the corresponding quantifier does not dominate the existential quantifier corresponding to the Skolem term, then it can be unified with such a term, because, in the constant-domain variant of QML, the object existing in any world accessible from w belongs to the domain of w too.

On the contrary, in the free-variable calculus for cumulative-domain logics the parameters occurring in a Skolem term $f(v_1, ..., v_k)$ must include all the parameters introduced in "previous" worlds, because a universal quantifier varying on the domain of a world w must not be instantiated with a term denoting an object possibly belonging only to the domain of another world (accessible from w). Cumulative-domain QML can be given a free-variable tableau calculus with the liberalized δ^+-rule, but only with the addition of numerical annotations on functional symbols, in the style of [7]. In fact, in that case, the unification of a parameter related to a world w with a term that is not guaranteed to belong to the domain of w is prevented by a suitable restriction on the notion of substitution, that takes into account symbol annotations.

The tableau system for constant-domain QML considered above is a special case of the calculus presented in detail in Section 4, where both rigid and non-rigid designation are allowed.

3   Constant-Domain QML with Rigid and Non-rigid Terms

A first order modal language L is constituted by logical symbols (propositional connectives, quantifiers, modal operators and a countable set X of individual variables), a non empty set L_P of predicate symbols, and a set L_F of functional symbols with an associated arity. The set L_F is partitioned into a set L_{F_R} of rigid functional symbols, and a disjoint set $L_{F_{NR}}$ of non-rigid functional symbols. The set L_F is the union of L_{F_R} and $L_{F_{NR}}$. Constants are considered as functional symbols with null arity. Terms are built by use of symbols from L_F and X in the usual way. We consider modal formulae in negation normal form, i.e. built out of literals (atoms and negated atoms) by use of $\wedge$, $\vee$, $\square$, $\diamond$ and the quantifiers $\forall$ and $\exists$. Negation over non-atomic formulae and implication are considered as defined symbols.

In the case of constant domain QML, a first-order modal interpretation $\mathcal{M}$ of a language L is a tuple $\langle W, w_0, R, D, \phi, \pi \rangle$ such that:

- W is a non empty set (the set of "possible worlds");
- w_0 is a distinguished element of W (the "initial world");[1]
- R is a binary relation on W (the *accessibility relation*); wRw' abbreviates $\langle w, w' \rangle \in R$;
- D is a non empty set (the object domain);
- ϕ represents the interpretation of constants and functional symbols in the language: for every world $w \in W$ and k-ary functional symbol $f \in L_F$ (with $k \geq 0$),
$$\phi(w, f) \in D^k \to D$$
  Moreover, if $f \in L_{F_R}$, then for all $w, w' \in W$, $\phi(w, f) = \phi(w', f)$.
- π is the interpretation of predicate symbols: if p is a k-ary predicate symbol and $w \in W$, then $\pi(w, p) \subseteq D^k$ is a set of k-tuples of elements in D.

The interpretation function ϕ is extended to terms in the usual way, and, by an abuse of notation, $\phi(w, t)$ denotes the interpretation of t in w.

If $\mathcal{M} = \langle W, w_0, R, D, \phi, \pi \rangle$ is an interpretation of the language L, the *language of the model $\mathcal{M}$*, $L(D)$, is obtained from L by addition of a "name" for each $d \in D$, i.e. a new constant $\overline{d}$. It is assumed that for every $d \in D$ and $w \in W$, $\phi(w, \overline{d}) = d$. Note that the interpretation of names $\overline{d}$ is always rigid, i.e. $\overline{d} \in L(D)_{F_R}$

The relation $\models$ between an interpretation $\mathcal{M} = \langle W, w_0, R, D, \phi, \pi \rangle$, a world $w \in W$ and a closed formula in $L(D)$ is defined inductively as follows:

1. $\mathcal{M}, w \models p(t_1, ..., t_n)$ iff $\langle \phi(w, t_1), ..., \phi(w, t_n) \rangle \in \pi(w, p)$.
2. $\mathcal{M}, w \models \neg A$ iff $\mathcal{M}, w \not\models A$.
3. $\mathcal{M}, w \models A \wedge B$ iff $\mathcal{M}, w \models A$ and $\mathcal{M}, w \models B$.
4. $\mathcal{M}, w \models A \vee B$ iff $\mathcal{M}, w \models A$ or $\mathcal{M}, w \models B$.
5. $\mathcal{M}, w \models \forall x A$ iff for all $d \in D$, $\mathcal{M}, w \models A[\overline{d}/x]$
6. $\mathcal{M}, w \models \exists x A$ iff there exists $d \in D$ such that $\mathcal{M}, w \models A[\overline{d}/x]$
7. $\mathcal{M}, w \models \Box A$ iff for all $w' \in W$ such that wRw', $\mathcal{M}, w' \models A$
8. $\mathcal{M}, w \models \Diamond A$ iff there is a $w' \in W$ such that wRw' and $\mathcal{M}, w' \models A$

A closed formula A is true in $\mathcal{M}$ iff $\mathcal{M}, w_0 \models A$, and it is valid iff it is true in all interpretations.

The accessibility relation R of a modal structure can be required to satisfy additional properties, characterizing different logics: we consider seriality (**D**), reflexivity (**T**), transitivity (**K4**), both reflexivity and transitivity (**S4**). When no additional assumption on R is made, the logic is **K**.

4   The Free-Variable Tableau System

The language of any tableau for a set of formulae in the language L extends L with a denumerable set of parameters (or free variables) and a denumerable set

---

[1] The semantics with a distinguished initial world dates back to [17]. Obviously, the notion of validity (truth in every model) coincides with the semantics where no initial world is singled out.

of Skolem functional symbols. Moreover it contains annotated symbols f^n, for every $f \in L_{F_{NR}}$ and $n \in \mathbb{N}$.

Definition 1. *Let L be a modal language, $V = \{v_1, v_2, ...\}$ an infinite denumerable set of new symbols, called free variables or parameters, and F^{Sk} a denumerable set of new function symbols such that, for each $k \in \mathbb{N}$, F^{Sk} contains infinitely many function symbols of arity k. Then the labelled free-variable extension of L is the language L^* such that:*

$$L_P^* = L_P$$
$$L_{F_R}^* = L_{F_R} \cup F^{Sk} \cup \{f^n \mid f \in L_{F_{NR}}, n \in \mathbb{N}\}$$
$$L_{F_{NR}}^* = L_{F_{NR}}$$

Terms in L^* are built up by use of L_F^*, X and V in the usual way.

Note that, in the definition above, Skolem functions and annotated functional symbols are considered as rigid symbols.

Definition 2.

1. *A symbol occurrence in a formula or set of formulae is called a non-modal occurrence if it is in the scope of no modal operators.*
2. *An n-annotated modal formula is a modal formula where non-modal occurrences of non-rigid functional symbols are all annotated with n. A formula is annotated if it is n-annotated for some n.*
3. *A term t is completely annotated iff every non-rigid functional symbol in t is annotated.*
4. *A modal substitution σ is a function from the set of free-variables V to completely annotated terms in L^*, that is the identity almost everywhere. Substitutions are denoted as usual by expressions of the form $\{t_1/v_1, ..., t_m/v_m\}$. Unifiers, i.e. solutions of unification problems, and most general unifiers (m.g.u.) are defined as in the classical case.*
5. *If A is a modal formula and $n \in \mathbb{N}$, then A^n is obtained from A by annotating each non-modal occurrence of a non-rigid functional symbol with n. If S is a set of modal formulae, then $S^n = \{A^n \mid A \in S\}$.*
   For instance, if $f \in L_{F_R}$, $c, g \in L_{F_{NR}}$, and $A = \forall x(p(f(x, c), g(x)) \land \Box q(g(c)))$, then $A^n = \forall x(p(f(x, c^n), g^n(x)) \land \Box q(g(c)))$.
6. *If S is a set of modal formulae and $n \in \mathbb{N}$, then the single node S^n is an initial tableau for S.*

The main intuition behind the annotation of non-rigid symbols is that, since a symbol in $L_{F_{NR}}$ can occur with different annotations in a tableau branch, its annotations distinguish the designations of such a symbol in the different worlds corresponding to the tableau nodes. Non-modal occurrences of non-rigid symbols are initialized with a given annotation n (for instance 0), that identifies the initial world of the searched model. Symbols of $L_{F_{NR}}$ occurring in the scope of a modal operator receive their annotations only when, by application of a modal rule, they come to the surface (by means of operation 5 in Definition 2).

Classical propositional rules	
$(\alpha)\quad \dfrac{A \wedge B, S}{A, B, S}$	$(\beta)\quad \dfrac{A \vee B, S}{A, S \qquad B, S}$
ν-rules	
$(\nu_D)\quad \dfrac{\Box S, S'}{S^m}$ where $m \in \mathbb{N}$ is new in the whole tableau	$(\nu_T)\quad \dfrac{\Box A, S}{A^n, \Box A, S}$ where $n \in \mathbb{N}$ either occurs as an annotation of some symbol in the premisse or it is new in the whole tableau
π-rules	
$(\pi_K)\quad \dfrac{\Diamond A, \Box S, S'}{A^m, S^m}$ where $m \in \mathbb{N}$ is new in the whole tableau	$(\pi_4)\quad \dfrac{\Diamond A, \Box S, S'}{A^m, S^m, \Box S}$ where $m \in \mathbb{N}$ is new in the whole tableau

Fig. 1. Propositional expansion rules

We consider a simple set of propositional tableau rules (others may be chosen as well), shown in Figure 1, where S, S' are sets of modal formulae (in L^*), $\Box S$ stands for $\{\Box A \mid A \in S\}$, S' is a set of non-boxed modal formulae, comma is set union. The propositional part of the tableau systems we consider consists of the classical rules, and: the rule π_K in the systems **K**, **D** and **T**, the rule π_4 in **K4** and **S4**, the rule ν_T in **T** and **S4**, the rule ν_D in **D**. The *quantifier expansion rules* are:

Quantifier rules	
$(\gamma)\quad \dfrac{\forall x A, S}{A[v/x], \forall x A, S}$ where v is a new parameter	$(\delta^+)\quad \dfrac{\exists x A, S}{A[f(v_1, ..., v_k)/x], S}$ where f is a new Skolem function, i.e. a symbol in F^{Sk} that does not occur elsewhere in the tableau, and $v_1, ..., v_k$ are all the parameters occurring in $\exists x A$

Adopting a terminology from [13], the rules ν_D, π_K and π_4 are called *dynamic*, the others are called *static*. So, a non-annotated non-rigid symbol can get an annotation only with the application of dynamic rules. Note, moreover, that only non-modal occurrences of non-rigid functional symbols in a tableau node can be annotated. As a consequence, every tableau node is a set of n-annotated formulae, for some $n \in \mathbb{N}$. In fact, the expansions obtained by application of a dynamic rule never contain symbols with different annotations (i.e. if both f^n and g^k occur in the expansion, then $n = k$). Such a property is preserved by applications of the ν_T-rule. Therefore, no tableau node contains symbols with different annotations.

The substitution rule we adopt here requires the following preliminary definitions:

- If P and Q are two atomic formulae in L^*, then $P = Q$ if P and Q are identical, including their annotations.
- Let $S_1, ..., S_k$ be all the leaves of a free-variable tableau $\mathcal{T}$, and, for each $i = 1, ..., k$, let P_i and $\neg Q_i$ be literals in S_i. If the modal substitution σ is a solution of the unification problem $P_1 = Q_1, ..., P_k = Q_k$, then σ is an *atomic closure substitution for* $\mathcal{T}$. If it is a most general solution for such a unification problem, then it is a *most general atomic closure substitution for* $\mathcal{T}$.

The substitution rule is then the following:

Most General Atomic Closure Substitution Rule
If $\mathcal{T}$ is a tableau for a set S of sentences in L and the modal substitution σ is a most general atomic closure substitution for $\mathcal{T}$, then the tree $\mathcal{T}\sigma$, obtained by applying σ to $\mathcal{T}$, is a tableau for S.

A tableau is closed iff each of its leaves contains a pair of complementary literals, i.e. literals P and $\neg P$. A closed tableau for a formula $\neg A$ is a tableau proof of A, and a closed tableau for a set of formulae S is a refutation of S.

As a first example, consider the language L with the unary predicates p and q, the rigid constant c and the non-rigid unary functional symbol f. The following tableau shows that $\{\Box p(f(c)), \forall x \Box (p(x) \rightarrow q(x)), \neg \Box q(f(c))\}$ is refutable in **K**, since it can be closed by the substitution $\{f^0(c)/v\}$:

$$\frac{\cfrac{\cfrac{\Box p(f(c)), \forall x \Box (p(x) \rightarrow q(x)), \neg \Box q(f(c))}{\Box p(f(c)), \Box (p(v) \rightarrow q(v)), \forall x \Box (p(x) \rightarrow q(x)), \neg \Box q(f(c))} \; (\gamma)}{p(f^0(c)), p(v) \rightarrow q(v), \neg q(f^0(c))} \; (\pi_K)}{p(f^0(c)), \neg p(v), \neg q(f^0(c)) \qquad p(f^0(c)), q(v), \neg q(f^0(c))} \; (\beta)$$

And now a tableau that should not and indeed does not close, thanks to the different annotations of f in different worlds:

$$\frac{\cfrac{p(f^0(c)), \; \forall x (p(x) \rightarrow \Box q(x)), \; \neg \Box q(f(c))}{p(f^0(c)), \; p(v) \rightarrow \Box q(v), \; \forall x (p(x) \rightarrow \Box q(x)), \; \neg \Box q(f(c))} \; (\gamma)}{\begin{array}{c} p(f^0(c)), \neg p(v), \\ \forall x (p(x) \rightarrow \Box q(x)), \\ \neg \Box q(f(c)) \end{array} \qquad \cfrac{\begin{array}{c} p(f^0(c)), \; \Box q(v), \\ \forall x (p(x) \rightarrow \Box q(x)), \; \neg \Box q(f(c)) \end{array}}{q(v), \; \neg q(f^1(c))} \; (\pi_k)} \; (\beta)$$

The tableau above is in fact a failed attempt to give a refutation of the satisfiable set $S = \{p(f(c)), \forall x(p(x) \rightarrow \Box q(x)), \neg\Box q(f(c))\}$, where c is rigid and f is non rigid, in any of the systems **K**, **D** or **T**. The tableau cannot be closed, since the unification problem $p(f^0(c)) = p(v)$, $q(v) = q(f^1(c))$ has no solution (note that S is satisfiable because f is non-rigid). And no matter how many applications of the γ-rule are added to the tableau above, the set S cannot be refuted.

Note that, in the absence of non-rigid symbols, i.e. when $L_{F_{NR}} = \emptyset$, the calculus presented in this sections results exactly from the addition of the classical free-variable γ and δ^+-rules to a standard tableau calculus for any of the considered propositional modal logics.

4.1   Soundness and Completeness of the Free-Variable Calculus

In this section we give a sketchy account of the soundness and completeness proofs. The proofs make use of a modal Substitution Theorem, stating that, if $\mathcal{M} = \langle W, w_0, R, D, \phi, \pi, \rangle$ is an interpretation of a modal language L, $w \in W$, A is a formula with only the variable x free, and t, t' are ground terms of L, then:

1. if the interpretations of t and t' are rigid and equal, i.e. for all $w' \in W$, $\phi(w', t) = \phi(w', t')$, then $\mathcal{M}, w \models A[t/x]$ iff $\mathcal{M}, w \models A[t'/x]$.
2. If x does not occur in A in the scope of any modal operator and $\phi(w, t) = \phi(w, t')$, then $\mathcal{M}, w \models A[t/x]$ iff $\mathcal{M}, w \models A[t'/x]$.

Theorem 1 (Soundness). *If there is a closed tableau for S, then S is unsatisfiable.*

The soundness proof runs along standard lines, except for the fact that the following notion of tableau satisfiability must be considered:

Definition 3. *Let $\mathcal{T}$ be a tableau, with parameters among $v_1, ..., v_k$, and let $\mathcal{M} = \langle W, w_0, R, D, \phi, \pi \rangle$ be a modal interpretation. Then $\mathcal{M} \models \mathcal{T}$ iff for all $d_1, ..., d_k \in D$ there is a leaf S of $\mathcal{T}$ and a world $w \in W$ such that*

$$\mathcal{M}, w \models S\{\overline{d_1}/v_1, ..., \overline{d_k}/v_k\}$$

$\mathcal{T}$ is satisfiable iff $\mathcal{M} \models \mathcal{T}$ for some interpretation $\mathcal{M}$.

Soundness follows from the fact that, if S is a satisfiable set of formulae, then any tableau for S is satisfiable. The proof of this fact is an induction on tableaux. We just note here that it is the case of the substitution rule that makes essential use of the hypothesis that the object domain is the same for each world.

Theorem 2 (Completeness). *If S is an unsatisfiable set of modal formulae, then there exists a closed tableau for S.*

The completeness proof also follows the standard approach consisting of the construction of a canonical model of a set of formulae having no closed tableau. In general, we deal with possibly infinite sets of formulae, and make use of the following notion of *tab-consistency*: S is tab-consistent iff for every finite subset $S' \subseteq S$ there is no closed tableau for S'.

The reformulation of the basic notion of downward saturated sets of formulae is the following: a (finite or infinite) set S of n-annotated modal formulae, possibly containing parameters, is n-downward saturated iff

1. S does not contain any pair of literals P_1 and $\neg P_2$ such that P_1 and P_2 are unifiable;
2. if $A \wedge B \in S$ then $A \in S$ and $B \in S$
3. if $A \vee B \in S$ then either $A \in S$ or $B \in S$;
4. if $\forall x A \in S$ then $A[v/x] \in S$ for every parameter $v \in V$;
5. if $\exists x A \in S$, and $v_1, ..., v_k$ are all the parameters occurring in A, then for some Skolem functional symbol f, $A[f(v_1, ..., v_k)/x] \in S$;
6. if the logic is either **T** or **S4** and $\Box A \in S$ then $A^n \in S$.

It can easily be proved that if a set S of n-annotated modal formulae is tab-consistent, then:

1. If $A \wedge B \in S$, then $S \cup \{A, B\}$ is tab-consistent.
2. If $A \vee B \in S$, then either $S \cup \{A\}$ or $S \cup \{B\}$ is tab-consistent.
3. If $\forall x A \in S$ then

$$S \cup \bigcup_{v \in V} \{A[v/x]\}$$

   is tab-consistent.
4. If $\exists x A \in S$, f is a functional symbol that does not occur in S and $v_1, ..., v_n$ are all the parameters occurring in A, then $S \cup \{A[f(v_1, ..., v_n)/x]\}$ is tab-consistent.
5. If the logic is either **T** or **S4** and $\Box A \in S$, then $S \cup \{A^n\}$ is tab-consistent.

The result stated above is used to prove the following:

Lemma 1. *Let L be a modal language, $n \in \mathbb{N}$ and S a (finite or infinite) set of n-annotated modal formulae in L^*. If H is a set containing an infinite number of functional symbols (for each arity) that do not occur in S and S is tab-consistent, then there exists a tab-consistent set $S^\infty \supseteq S$ of n-annotated formulae (possibly containing also symbols from H), such that S^∞ is n-downward saturated.*

The main idea behind the proof of the lemma above is standard and consists of the construction of a (pseudo-)tableau $\mathcal{T}^\infty$, rooted at S, in such a way that $\mathcal{T}^\infty$ is closed with respect to the application of static expansion rules. Such a construction uses Skolem functions in H in the applications of the δ^+-rule. The tree $\mathcal{T}^\infty$ is not properly a tableau, since its nodes may be infinite sets of formulae. In presence of such infinite objects, some care has to be taken in defining a "fair" rule application strategy.

The canonical model of a tab-consistent set of formulae S in L^* is built along standard lines, using Lemma 1. The initial world w_0 is a tab-consistent and 0-downward saturated superset of S^0, and each world w in the model is a tab-consistent and n-downward saturated set of n-annotated formulae, for some

$n \in \mathbb{N}$ that is uniquely associated to w (and is called its "name"): $name(w) = n$. To each world w is also associated a set of Skolem functions $H(w)$ to be used in the application of Lemma 1, in such a way that if $w \neq w'$ then $H(w)$ and $H(w')$ are disjoint. The interpretation of the language is "syntactical": its domain D is the set of completely annotated ground terms in L^*, and the interpretation of terms is "almost" Herbrand-like: for all $w \in W$

- if $f \in L_{F_R} \cup F^{Sk}$ is either a rigid symbol in L or a Skolem function, then for all $t_1, ..., t_k \in D$: $\phi(w, f)(t_1, ..., t_k) = f(t_1, ..., t_k)$
- If $f \in L_{F_{NR}}$ is a non-rigid symbol in L, then for all $t_1, ..., t_k \in D$:
  for all $m \in \mathbb{N}$: $\phi(w, f^m)(t_1, ..., t_k) = f^m(t_1, ..., t_k)$
  and, if $name(w) = n$: $\phi(w, f) = \phi(w, f^n)$

So, the interpretation of annotated symbols is always rigid and, for all $w \in W$ and completely annotated ground term t in L^*, $\phi(w, t) = t$; i.e. if $t \in D$ then $\phi(w, t) = t$. In particular, the interpretation of completely annotated terms is rigid.

Finally, let σ be any surjective function from the set of parameters V to the set of completely annotated ground terms of the extended language: for each completely annotated ground term t there exists $v \in V$ such that $\sigma(v) = t$. If A is a formula, then $\sigma(A)$ denotes the formula obtained from A by replacement of each parameter v in A by $\sigma(v)$. If S is a set of formulae, $\sigma(S) = \{\sigma(A) \mid A \in S\}$. The interpretation function π is then defined as follows: for all $w \in W$ and k-ary predicate symbols p in L^*:

$$\pi(w, p) = \{\langle t_1, ..., t_k \rangle \mid p(t_1, ..., t_k) \in \sigma(w)\}$$

It can be proved that, for all $w \in W$ and for every formula A:

$$\text{if } A \in w \text{ then } \mathcal{M}, w \models \sigma(A)$$

Hence, in particular, $\mathcal{M}, w_0 \models \sigma(w_0)$ and, since S^0 contains no parameters and $S^0 \subseteq w_0$, $\mathcal{M}, w_0 \models S^0$. This easily implies that $\mathcal{M}, w_0 \models S$.

5   Concluding Remarks

This work shows that the addition of the free-variable γ-rule and δ^+-rule, rather than the more constraining δ-rule, to standard propositional modal tableau calculi captures proof-theoretically the constant domain semantics for QML in the case of rigid designation.[2] Moreover, non-rigid designation can also be answered for, by means of simple annotations on non-rigid functional symbols. Thus, it makes it apparent the sensitivity of modal tableau free-variable calculi to alternative formulations of the δ-rule, contrarily to the classical case, and similarly

---

[2] Although complete proofs have been carried out only for logics whose propositional basis is among **K, D, T, K4, S4**, the extension to other cut-free propositional calculi should be straightforward.

to what happens for other non classical logics. For instance, in [5], where a free-variable sequent system which is sound and complete for linear logic is proposed, it is shown that an analogous liberalization of the $\forall_R$ and the $\exists_L$ rules is unsound: in that case, too, restricting the set of arguments of Skolem functions results in a shift to a different "logic". [20] and [18] make a fine analysis of the relationship between the arguments of Skolem functions in run-time Skolemization and rule permutability in cut-free sequent calculi for intuitionist and linear logic, respectively.

As already observed, even in the case of rigid designation, "standard" (i.e. non-annotated, non-prefixed) *ground* tableau systems for constant-domain QML cannot be given, as a consequence of the failure of Craig's Interpolation Theorem [8,9]. However, the use of unification in the quantifier rules, which is the essential feature of free-variable tableaux, makes it possible to give a proof-theoretical characterization of constant domains in the Fitting-Gentzen style.

In Figure 2, some of the inference rules of a sequent calculus for QML which is sound and complete with respect to constant domains and rigid designation are provided. This calculus is formulated in a completely standard Gentzen style, but for the fact that unification and run-time Skolemization are embedded in the quantifier rules. A *proof* of a sequent S in such a formal system is given by a deduction tree $\mathcal{T}$, whose root is S, and a *substitution* σ such that all the leaves of $\mathcal{T}\sigma$ are axioms, i.e. sequents of the form $S, A \vdash A, S'$. For the sake of concision, we present here only the calculus for system **T**.

Such sequent systems can obviously be reformulated as *symmetric* calculi (in the sense of [9], i.e. calculi where formulae never cross the sequent arrow), taking negation as a defined connective and eliminating the rules for negation and implication. In this case, axioms include also sequents of the form $S, A, \neg A \vdash S'$ and $S \vdash A, \neg A, S'$. The equivalence of such symmetric calculi and the tableau calculi presented in Section 4 is immediate: if S and S' are sets of formulae in negation normal form, a sequent $S \vdash S'$ is provable in the symmetric sequent calculus if and only if the set of formulae $S \cup \{\neg A \mid A \in S'\}$ has a refutation in the (corresponding) tableau system of Section 4.

We conclude this section showing where an attempt to adapt Fitting's proof of the Interpolation Theorem for "ordinary" cut-free sequent calculi ([9]) to the calculus presented in Figure 2 fails. In fact, the rules $\exists_L$ and $\forall_R$ (corresponding to the δ^+-rule of the free variable tableau system) do not preserve the existence of interpolants. This is not surprising: such rules are not locally sound, since "liberalized Herbrandization" (the δ^+-rule) does not correspond to the *eigenvariable* condition in sequent systems, like standard Herbrandization (the δ-rule) does. As a simple example, let us consider the following deduction tree in the (symmetric) sequent calculus:

$$\cfrac{\cfrac{\cfrac{p(v) \vdash p(c)}{p(v) \vdash \forall x\, p(x)}\ (\forall_R)}{\forall x \Box p(x), \Box p(v) \vdash \Box \forall x\, p(x)}\ (\Box_R)}{\forall x \Box p(x) \vdash \Box \forall x\, p(x)}\ (\forall_L)$$

$\dfrac{S \vdash S', A}{S, \neg A \vdash S'}$ $(\neg L)$	$\dfrac{S, A \vdash S'}{S \vdash S', \neg A}$ $(\neg R)$
$\dfrac{S, A \vdash S' \quad S, B \vdash S'}{S, A \vee B \vdash S'}$ $(\vee L)$	$\dfrac{S \vdash S', A, B}{S \vdash S', A \vee B}$ $(\vee R)$
$\dfrac{S, \forall x A, A[v/x] \vdash S'}{S, \forall x A \vdash S'}$ $(\forall L)$ where v is a new parameter	$\dfrac{S \vdash S', A[f(v_1, ..., v_n)/x]}{S \vdash \forall x A}$ $(\forall R)$ where $v_1, ..., v_n$ are all the parameters in A
$\dfrac{S, A[f(v_1, ..., v_n)/x] \vdash S'}{S, \exists x A \vdash S'}$ $(\exists L)$ where $v_1, ..., v_n$ are all the parameters in A	$\dfrac{S \vdash S', \exists x A, A[v/x]}{S \vdash S', \exists x A}$ $(\exists R)$ where v is a new parameter
$\dfrac{S, A, \Box A \vdash S'}{S, \Box A \vdash S'}$ $(\Box L)$	$\dfrac{S \vdash S', A}{S_0, \Box S \vdash \Diamond S', \Box A, S_1}$ $(\Box R)$

Fig. 2. Rules of a Free-Variable Sequent Calculus for **T-QML** with Constant Domains and Rigid Designation

An application of the substitution $\{c/v\}$ produces a proof of the valid sequent (in **T-QML** with constant domains) $\forall x \Box p(x) \vdash \Box \forall x p(x)$ (corresponding to the Barcan formula), starting with the inference:

$$\dfrac{p(c) \vdash p(c)}{p(c) \vdash \forall x\, p(x)} \ (\forall R)$$

Now, following the lines of Fitting's proof, we immediately find the trivial interpolant $p(c)$ for the axiom. Such a formula should be an interpolant also for the sequent $p(c) \vdash \forall x p(x)$, derived via an application of the $\forall R$-rule (corresponding to a δ^+-rule in the tableaux systems). But this is clearly false.

References

1. M. Abadi and Z. Manna. Modal theorem proving. In *Proc. of the 8th Int. Conf. on Automated Deduction*, number 230 in LNCS, pages 172–189, Berlin, 1986. Springer.
2. Y. Auffray and P. Enjalbert. Modal theorem proving: an equational viewpoint. *Journal of Logic and Computation*, 2:247–297, 1992.

3. D. Basin, M. Matthews, and L. Viganò. Labelled modal logics: Quantifiers. *Journal of Logic Language and Information*, 7(3):237–263, 1998.
4. N. Belnap. Display logic. *J. of Philosophical Logic*, 11:375–417, 1982.
5. S. Cerrito. Herbrand methods for sequent calculi; unification in LL. In K. Apt, editor, *Proc. of the Joint International Conference and Symposium on Logic Programming*, pages 607–622, 1992.
6. M. Cialdea. Resolution for some first order modal systems. *Theoretical Computer Science*, 85:213–229, 1991.
7. M. Cialdea Mayer and S. Cerrito. Variants of first-order modal logics. In R. Dyckhoff, editor, *Proc. of Tableaux 2000*, LNCS. Springer Verlag, 2000.
8. K. Fine. Failures of the interpolation lemma in quantified modal logic. *Journal of Symbolic Logic*, 44(2), 1979.
9. M. Fitting. *Proof Methods for Modal and Intuitionistic Logics*. Reidel Publishing Company, 1983.
10. M. Fitting. First-order modal tableaux. *Journal of Automated Reasoning*, 4:191–213, 1988.
11. M. Fitting. On quantified modal logic. *Fundamenta Informaticae*, 39:105–121, 1999.
12. J. W. Garson. Quantification in modal logic. In D. Gabbay and F. Guenthner, editors, *Handbook of Philosophical Logic*, volume II, pages 249–307. D. Reidel Publ. Co., 1984.
13. R. Goré. Tableau methods for modal and temporal logics. In M. D'Agostino, G. Gabbay, R. Hähnle, and J. Posegga, editors, *Handbook of tableau methods*. Kluwer, 1999.
14. R. Hähnle and P. H. Schmitt. The liberalized δ-rule in free variable semantic tableaux. *Journal of Automated Reasoning*, 13:211–222, 1994.
15. P. Jackson and H. Reichgelt. A general proof method for first-order modal logic. In *Proc. of the 10th Joint Conf. on Artificial Intelligence (IJCAI '87)*, pages 942–944. Morgan Kaufmann, 1987.
16. K. Konolige. Resolution and quantified epistemic logics. In J. H. Siekmann, editor, *Proc. of the 8th Int. Conf. on Automated Deduction (CADE 86)*, number 230 in LNCS, pages 199–208. Springer, 1986.
17. S. A. Kripke. Semantical analysis of modal logic I: Normal propositional calculi. *Zeitschrift für Mathematische Logik und Grundlagen der Mathematik*, 9:67–96, 1963.
18. P. D. Lincoln and N. Shankar. Proof search in first order linear logic and other cut-free sequent calculi. In S. Abramsky, editor, *Proc. of the Ninth Annual IEEE Symposium on Logic In Computer Science (LICS)*, pages 282–291. IEE Computer Societ Press, 1994.
19. H. J. Ohlbach. Semantics based translation methods for modal logics. *Journal of Logic and Computation*, 1(5):691–746, 1991.
20. N. Shankar. Proof search in the intuitionist sequent calculus. In D. Kapur, editor, *Proc. of the Eleventh International Conference on Automated Deduction (CADE)*, number 607 in LNCS, pages 522–536. Springer Verlag, 1992.
21. L. A. Wallen. *Automated Deduction in Nonclassical Logics: Efficient Matrix Proof Methods for Modal and Intuitionistic Logics*. MIT Press, 1990.
22. H. Wansing. Predicate logics on display. *Studia Logica*, 62(1):49–75, 1962.

Instructing Equational Set-Reasoning with Otter[*]

Andrea Formisano[1], Eugenio G. Omodeo[2], and Marco Temperini[3]

[1] Univ. di Perugia, Dip. di Matematica e Informatica formis@dipmat.unipg.it
[2] Univ. di L'Aquila, Dip. di Matematica Pura ed Applicata omodeo@univaq.it
[3] Univ. 'La Sapienza' di Roma, DIS marte@dis.uniroma1.it

Abstract. An experimentation activity in automated set-reasoning is reported. The methodology adopted is based on an equational re-engineering of ZF set theory within the ground formalism $\mathcal{L}^\times$ developed by Tarski and Givant. On top of a kernel axiomatization of map algebra we develop a layered formalization of basic set-theoretical concepts. A first-order theorem prover is exploited to obtain automated certification and validation of this layered architecture.

Keywords. Set reasoning, map algebra, first-order theorem proving.

Introduction

Any basic mathematical concept can be suitably formulated within axiomatic Set Theory, which can hence be regarded as the most promising, as well as challenging, arena for automated theorem-provers.

Sustained efforts have been devoted to experimentation with state-of-the-art theorem provers, to get automated proofs of common set-theoretical theorems. Different axiomatic systems of set theory have been tried, to determine which one offers the best support to resolution-based theorem provers (such as Otter, cf. [12,16]).

In much of the experimentation activities carried out in the past (cf., e.g., [3, 11]), the von Neumann-Gödel-Bernays axiomatization of set theory has been preferred, because NGB offers a *finite* first-order axiomatization. On the other hand, to mention just one alternative approach, [9] and [10] resort to higher order features of Isabelle to deal with the Zermelo-Fraenkel set theory. The recourse to higher order logic turns out to be mandatory, because ZF cannot be finitely axiomatized in first-order logic.

Deepening the same approach proposed in [6], this paper will focus on an equational rendering of ZF. Our formulation of the axioms is based on the formalism $\mathcal{L}^\times$ of [15], which is equational and devoid of variables. A theory stated in $\mathcal{L}^\times$ can easily be emulated through a first-order system, simply by treating the meta-variables that occur in the schematic formulation of its axioms (both

---

[*] This research was partially funded by the Italian IASI-CNR (coordinated project log(SETA)) and by MURST (PGR-2000—Automazione del ragionamento in teorie insiemistiche).

R. Goré, A. Leitsch, and T. Nipkow (Eds.): IJCAR 2001, LNAI 2083, pp. 152–167, 2001.
© Springer-Verlag Berlin Heidelberg 2001

the logical axioms and the ones endowed with a genuinely set-theoretic content) as if they were first-order variables. In practice, this means treating ZF as if it were an extension of the theory of relation algebras [14]. As an immediate consequence, we obtain a finite axiomatization (stronger but retaining the traits) of ZF; this can be achieved since variables are not supposed to range over sets but over the dyadic relations on the universe of sets.

By taking the results presented in [6] as a starting point, we report on the beginnings of an experimentation activity exploratorily based on the first-order theorem-prover Otter. This prover will be exploited to provide an inferential apparatus for $\mathcal{L}^\times$. In turn, this apparatus will serve as the basis on which an inferential machinery for set-reasoning will grow.

1   A Layered Development of Map-Reasoning

$\mathcal{L}^\times$ is a ground equational language where one can state properties of dyadic relations —MAPS— over a domain $\mathcal{U}$ of discourse. The basic ingredients of $\mathcal{L}^\times$ are three *constants* $\emptyset$, $\mathbb{1}$, ι; three dyadic constructs $\cap$, $\triangle$, $\circ$ of map *intersection*, map *symmetric difference*, and map *composition*, respectively; and the monadic construct $^{-1}$ of map *conversion*. Then, a *map expression* is any term of the following signature:

symbol :	$\emptyset$	$\mathbb{1}$	ι	$\in$	$\cap$	$\triangle$	$\circ$	$^{-1}$	$^-$	$\backslash$	$\cup$	$\dagger$
degree :	0	0	0	0	2	2	2	1	1	2	2	2
priority :					5	3	6	7		2	2	4

The map whose properties we intend to specify is the membership relation $\in$ over the class $\mathcal{U}$ of all sets. Hence, $\in$ is the only primitive map letter of $\mathcal{L}^\times$.

The language $\mathcal{L}^\times$ consists of *map equalities* $Q = R$, where Q and R are *map expressions*. A number of derived constructs and shorthands for map equalities can be easily introduced (cf. Fig. 1).

For an *interpretation* of $\mathcal{L}^\times$, one fixes a nonempty subset $\in^\Im$ of $\mathcal{U}^2 =_{\mathrm{Def}} \mathcal{U} \times \mathcal{U}$. Then each map expression P designates a map $P^\Im$ on the basis of the usual evaluation rules, e. g.:

$(Q \circ R)^\Im =_{\mathrm{Def}} \{\, [a, b] \in \mathcal{U}^2 : \text{there are } cs \text{ in } \mathcal{U} \text{ for which } [a, c] \in Q^\Im \text{ and } [c, b] \in R^\Im \}$.

Accordingly, an equality $Q = R$ turns out to be either true or false in $\Im$.

The logical axioms characterizing the derivability notion $\vdash$ for $\mathcal{L}^\times$ (cf. Fig. 1) will be supplemented with proper axioms reflecting one's conception of $\mathcal{U}$ as being a hierarchy of nested sets over which $\in$ behaves as membership.

It must be said that there is no representation theorem that plays for map algebras a role analogous to the Stone theorem for Boolean algebras (cf. [2]). In other words, there exist equalities that are true in all algebras of dyadic relations over a fixed $\mathcal{U}$ but which are false in some structure which, though fulfilling the axioms of map algebra, does not consist of relations. This defect will presumably propagate to any set theory formulated as an extension of the map algebra; but anyway, even in first-order logic, a set theory never reflects the intended semantics univocally, and hence the map-algebraic formulation and the

$$P \cup Q \equiv_{\text{Def}} P \triangle Q \triangle P \cap Q \qquad\qquad P \setminus Q \equiv_{\text{Def}} P \triangle P \cap Q$$

$$\overline{P} \equiv_{\text{Def}} P \triangle \mathbb{1} \qquad P \dagger Q \equiv_{\text{Def}} \overline{\overline{P} \circ \overline{Q}} \qquad P \subseteq Q \equiv_{\text{Def}} P \cap Q = P$$

$$\text{funcPart}(P) \equiv_{\text{Def}} P \setminus P \circ \bar{\iota} \qquad\qquad \text{Func}(P) \equiv_{\text{Def}} P^{-1} \circ P \subseteq \iota$$

$$\text{IAbs}(P) \equiv_{\text{Def}} P = \mathbb{1} \circ P \qquad\qquad \text{Total}(P) \equiv_{\text{Def}} P \circ \mathbb{1} = \mathbb{1}$$

$$P \cap Q = Q \cap P$$
$$P \cap (Q \triangle R) \triangle P \cap Q = P \cap R$$
$$(P \star_1 Q) \star_1 R = P \star_1 (Q \star_1 R)$$
$$\iota \circ P = P$$
$$P^{-1^{-1}} = P$$
$$(P \star_2 Q)^{-1} = Q^{-1} \star_2 P^{-1}$$
$$((P \triangle Q) \triangle P \cap Q) \circ R = (Q \circ R \triangle P \circ R) \triangle Q \circ R \cap P \circ R$$
$$P^{-1} \circ (R \cap (P \circ Q \triangle \mathbb{1})) \cap Q = \emptyset$$
$$\mathbb{1} \cap P = P$$

$\star_1 \in \{\triangle, \cap, \circ\}$ and $\star_2 \in \{\cap, \circ\}$

Fig. 1. Derived constructs and axioms for map algebra.

logical one can, with their limitations, be on a par. The results reported in [4], which we will briefly review in Sec. 2, constitute a verification of this fact.

Otter is a resolution-style theorem prover developed at the Argonne National Laboratory (cf. [8]). It can manipulate statements written in full first-order logic with equality. The inference rules available in Otter are: binary resolution, (ordered) hyperresolution, UR-resolution, and binary paramodulation. Otter's main features include: forward and backward demodulation, forward and backward subsumption, (a variant of) Knuth-Bendix completion method, weight functions and lexical ordering, etc.. Moreover, Otter offers a large number of parameters and options to help the user in guiding the inference process. In what follows we briefly illustrate those we found more useful in our experimentation. This will be done by giving the reader a description of the basic strategy we adopted in proving theorems with Otter. As we will see, in most cases this strategy worked well, whereas we needed some kind of tuning in order to successfully cope with a few theorems.

Since we are dealing with equality, we selected the Knuth-Bendix completion procedure; whenever non-unit clauses or non-equational predicates entered into play, we enabled hyperresolution and binary resolution. Paramodulation was employed. We usually exploited the default strategies for ordering, demodulation, and weighting. Nevertheless, we made systematic use of the parameters devoted to limit the search space. In particular, all theorems were proved by imposing bounds on the maximum number of literals and distinct variables occurring in any derived clause. Moreover, we often imposed a threshold on the weight of derived clauses. We also adopted Otter's default weighting strategy (cf. [8]); in some cases we found it useful to give extra weight to certain terms or literals in order reduce the time spent for finding a proof. Here are the Otter settings we used in almost all experiments we report on (for the parameters and flags not mentioned here, we kept the values adopted by Otter's autonomous mode):

% Strategy:		% Limits on the search space:
set(knuth_bendix). set(para_from). set(para_into). set(dynamic_demod_all).	set(back_demod). set(hyper_res). set(binary_res).	assign(max_distinct_vars,3). assign(max_literals,1). assign(max_weight,18).

Notice that the value assigned to `max_weight` was usually 'guessed' by taking into account the syntactical structural complexity of the theorem to be proved.

Initial experimentation in map reasoning with Otter has been described in [1]; in [6] an equational re-engineering of set theories is presented. Automated set reasoning based on this equational formulation of ZF set theory was explored in [4]. In particular, in [4] the authors obtained a (semi-)automated proof of a fundamental result: by assuming the axioms of a weak set theory (namely, extensionality, null-set, single-element addition and removal) it was possible to derive the existence of a pair of projections satisfying the pairing axiom (cf. Sec. 2, to be seen). This result guarantees the equipollence in means of proof of the equational formulation of ZF w.r.t. its first-order version (cf. [15]). We will briefly survey this result in Sec. 2.

The experimentation reported in [4] was essentially carried out by exploiting the autonomous mode supplied by Otter and by always adopting the default settings. The explicit tuning of parameters and flags was avoided in order to obtain a higher independence of the approach from the specific theorem prover. Since the syntactic complexity of the theorems tackled in [4] was quite low, this approach represented a viable choice.

The experimentation activity we are going to describe here, is aimed at proving theorems that involve set-theoretical concepts whose syntactical and semantical complexity grow as the experimentation proceeds. This fact can easily be grasped by considering the higher level of abstraction of notions such as totality or functionality w.r.t. the basic map constructs. To reflect this growth in complexity, we will develop a layered hierarchy of lemmas. Starting with a 'kernel' consisting of the constructs and axioms of Fig. 1, we will proceed systematically by defining new set-theoretical concepts and by proving groups of laws that characterize the new set-constructs. Each one of these extension steps will be a (potential) part of the basis for the next extension. Moreover, in proving a generic theorem, it will be possible to select a subset of the available constructs, together with their laws. This, actually, will help the search for the proof in two orthogonal ways: firstly, Otter will deal only with the part of the global environment that the user judges to be relevant and related to the theorem to be proved; and secondly, the inference activity will be better focused at the most suitable level of abstraction. For instance, in proving a law that infers the totality of the composition of maps from the totality of the components (cf. Fig. 8), a deep treatment of 'low level' concepts such as the intrinsic properties of symmetric difference should not be needed.

The first step in the development of our layers consists in proving a series of auxiliary laws for the kernel constructs (namely, $\triangle, \cap, \circ, ^{-1}$). From the theoretical point of view, these laws are not necessary to prove any (provable) theorem of map calculus. Nevertheless, experimentation revealed that Otter was unable to

law	premises	length	timing	generated	kept
I_1 $P \cap \emptyset = \emptyset$	Ax	20	7	1120	185
$P \cap P = P$	Ax	20	13	2304	382
$P \cap (P \cap Q) = P \cap Q$	Ax	27	13	2157	318
I_2 $P \cap Q = P \wedge Q \cap P = Q \to Q = P$	Ax, I_1	1	< 1	2	24
$P \cap Q = Q \wedge Q \cap R = Q \to P \cap R = P$	Ax, I_1	2	3	162	62
S_1 $P \triangle Q = Q \triangle P$	Ax	7	2	195	52
$P \triangle (Q \triangle R) = Q \triangle (P \triangle R)$	Ax	8	4	258	54
$\emptyset \triangle P = P$	Ax	20	8	1124	190
$P \triangle P = \emptyset$	Ax	16	5	1110	180
$P \triangle (P \triangle Q) = Q$	Ax	5	2	234	52
$\iota \cap (P \triangle P^{-1}) = \emptyset$	Ax, S_1	199	5m 30s	6360755	13842
$P \cap (Q \triangle R) = (P \cap Q) \triangle (P \cap R)$	Ax, I_1, S_1	2	2	120	45
G_1 $\emptyset^{-1} = \emptyset$	Ax	22	8	1434	226
$1^{-1} = 1$	Ax	4	< 1	85	40
$\iota^{-1} = \iota$	Ax	3	< 1	38	22
$(P \triangle 1)^{-1} = P^{-1} \triangle 1$	Ax, S_1	43	1.33s	24972	2033
$(P \triangle Q)^{-1} = P^{-1} \triangle Q^{-1}$	Ax, S_1, G_1	89	1.12s	17147	1554
C_1 $\emptyset \circ P = \emptyset$	Ax	26	9	1447	231
$P \circ \emptyset = \emptyset$	Ax	17	8	1378	219
$P \circ \iota = P$	Ax	4	2	38	23
$1 \circ 1 = 1$	Ax	29	20	3215	526
$((P \circ P^{-1}) \cap \iota) \circ P = P$	Ax, G_1, C_1	66	18.53s	221080	8774
$P \circ ((P \circ P^{-1}) \cap \iota) = P$	Ax, G_1, C_1	71	19.02s	227467	8844
$P \cap (P \circ 1) = P$	Ax	62	6.36s	68558	6734
$P \cap (1 \circ P) = P$	Ax	61	6.08s	67926	6646

Fig. 2. Laws on the primitive map constructs: $\cap$, $\triangle$, $^{-1}$, and $\circ$

prove several simple theorems in a reasonable amount of time, unless by employing these auxiliary laws. A conspicuous part of the laws regarding the primitive
constructs are shown in Fig. 2.

The laws are divided into groups because each group usually corresponds to
an input file that could be loaded into Otter; moreover, the laws in the same
group were usually proved by adopting similar settings for parameters and search
controls, and often by using the same groups of premises as hypotheses.

For each law in the tables, we indicated: a) the groups of formulas given to
Otter as input; b) the length of the proof found by Otter; c) the time spent (if
not differently specified, it is expressed in hundredth of seconds); d) the number
of clauses generated during the inference process: e) the number of clauses being
kept (i.e., the generated clauses that fulfill all restrictions on weight, number of
variables, number of literals, etc.). In our experimentation we used Otter 3.0.6
running under Linux on a PC (Pentium III-450, with 128Mbyte of RAM).

Notice that sometimes there are more kept clauses than generated clauses.
This is because the former include all clauses obtained by processing the input set
of formulas. The writing '**Ax**' reported for most of the laws, does not necessarily
mean that all of the axioms of Fig. 1 have been fed into Otter; usually this is
the case only when no other group of laws is employed in the proof; otherwise,
just (part of) the axioms regarding the constructs occurring in the theorem have
been given in input. For instance, to prove the law

$$((P \circ P^{-1}) \cap \iota) \circ P = P \tag{1}$$

law	premises	length	timing	generated	kept
N_1 $\quad \overline{P} = P$	Ax	5	2	195	53
$\overline{\emptyset} = 1$	Ax	21	9	1229	318
$\overline{1} = \emptyset$	Ax	17	9	1215	308
$\overline{P \cap Q} = Q \triangle (P \cap Q)$	Ax	11	4	361	77
$\overline{P} \triangle Q = \overline{P \triangle Q}$	Ax	9	2	257	57
$\overline{P} \triangle P = 1$	Ax	2	<1	40	24
$\overline{P} \cap P = \emptyset$	Ax	18	15	2210	496
N_2 $\quad \overline{P^{-1}} = \overline{P}^{-1}$	Ax, N_1, S_1, I_1, G_1	1	2	0	40
$P \triangle \overline{P} = 1$	"	1	2	0	40
$P \cap Q = P \to P \cap \overline{Q} = \emptyset$	"	4	3	164	68
$P \cap \overline{Q} = \emptyset \to P \cap Q = P$	"	8	4	181	71
$\iota \cap \overline{P^{-1} \circ \overline{P}} = \iota$	"	20	17	2336	467
$P \triangle Q = \overline{\overline{P \cap Q} \cap \overline{\overline{P} \cap \overline{Q}}}$	"	18	37	5012	1435
$\overline{P \triangle Q} = \overline{P \cap Q} \cap \overline{\overline{P} \cap \overline{Q}}$	"	42	$10m\,36s$	11780356	13860
$P \triangle Q = \overline{\overline{P \cap Q} \cap \overline{\overline{P} \cap \overline{Q}}}$	Ax, $N_1, S_1, I_1, G_1, N_2.6$	7	10	1645	385
$\overline{P^{-1} \cap \overline{Q^{-1}}} = (\overline{Q} \cap \overline{P})^{-1}$	"	5	4	560	182
$(P \triangle Q)^{-1} = \overline{\overline{P \cap Q} \cap \overline{\overline{P} \cap \overline{Q}}}^{-1}$	"	3	2	0	43

Fig. 3. Laws on map complementation

of group C_1, we exploited the laws of G_1 and those of C_1 (meaning with this that Otter was allowed to use the laws listed before (1) in C_1); moreover, we loaded the portion of **Ax** relative to $\circ$ and to $^{-1}$.

Figures 3 and 4 list the laws on map complementation and map union, respectively. The definitions of these constructs in term of the primitive ones are listed in Fig. 1, together with the map formalization of other notions that will come into play in the sequel.

Other laws on map composition and expressing properties of ι are listed in Fig. 6. In order to prove these laws, Otter needed to employ the defined map constructs of complementation and union, together with their laws. It should be noticed that Otter was not able to prove, in a reasonable amount of time, several of the laws of Fig. 6 without using the laws in $I_1, C_1, G_1, U_{1,2,3,4}$.

Next come the laws on map inclusion and left-absoluteness. This extension of the signature can be considered as preparatory for the study on totality and functionality of maps. In turn, the laws on totality and functionality will play a crucial role in proving the set-theoretical theses we will report on in later sections.

A few remarks on the behavior of Otter confronted with map calculus are due. Firstly, experimentation revealed that, in general, proving a theorem/law seems to be more challenging (with our inference machinery) when the map ι or some of its properties are involved. Consider, for instance, the penultimate law in Fig. 2, and the laws involving ι in C_1 or C_2. The same can be said for those laws that correspond to deep intrinsic characteristics of ι, such as the property:

$$\text{for each } P \subseteq \iota \text{ it holds that } P^{-1} = P \tag{2}$$

This phenomenon could be intuitively explained by observing that statements such as (2) assert properties that do not concern the map as a single object, but predicate on a relationship holding between the components of each pair

law	premises	length	timing	generated	kept
U_1 $P \cup Q = Q \cup P$	Ax	8	< 1	107	46
$\emptyset \cup P = P$	Ax	19	3	675	122
$1 \cup P = 1$	Ax	6	3	210	65
$P \cup P = P$	Ax	24	13	1746	478
$P \cap (Q \cap (P \cup R)) = P \cap Q$	Ax	37	17	1951	604
$P \cup (P \cap Q) = P$	Ax	33	16	1916	559
$(\overline{P} \cap Q) \cup (P \cap Q) = Q$	Ax	35	18	1996	624
$\overline{P} \cup P = 1$	Ax, N_1	9	2	0	28
$\overline{P \cup Q} = \overline{P} \cap \overline{Q}$	Ax	19	11	1298	448
U_2 $P \cup (P \cup Q) = P \cup Q$	Ax, U_1	6	2	101	68
$(P \cup Q) \cup R = P \cup (Q \cup R)$	Ax, I_1, C_1, U_1	6	2.74s	69861	1047
$P \cup (Q \cup R) = Q \cup (P \cup R)$	//	4	2.62s	68421	1035
$(P \cup Q) \cap (P \cup R) = P \cup (Q \cap R)$	//	13	1.41s	39504	709
$\overline{P} \cup (Q \cup (P \cap R)) = \overline{P} \cup (Q \cup R)$	//	14	81	23781	582
$(P \cup Q) \cup (\overline{P} \cap R) = P \cup (Q \cup R)$	//	11	11	2232	300
U_3 $P \cup Q = \emptyset \rightarrow P = \emptyset$	Ax, U_2	2	4	233	68
$P \bigtriangleup Q = (P \cap \overline{Q}) \cup (\overline{P} \cap Q)$	//	82	1.84s	26090	2116
$(P \cup Q) \cap (\overline{P \cap Q}) = (P \cap \overline{Q}) \cup (\overline{P} \cap Q)$	//	53	37	7033	792
$P \bigtriangleup Q = (P \cup Q) \cap (\overline{P \cap Q})$	//	43	1.44s	25517	1802
$\iota \cap ((P \cap \overline{P^{-1}}) \cup (\overline{P} \cap P^{-1})) = \emptyset$	//	35	9.60s	101784	9462
$\iota \cap ((P \cap \overline{P^{-1}}) \cup (\overline{P} \cap P^{-1})) = \emptyset$	Ax, U_2, U_3	6	5	0	94
U_4 $(P \cup Q) \circ R = (P \circ R) \cup (Q \circ R)$	Ax	9	2	288	144
$(P \circ (Q \cup R))^{-1} = ((P \circ Q) \cup (P \circ R))^{-1}$	Ax, G_1	42	42	5959	1508
$P \circ (Q \cup R) = (P \circ Q) \cup (P \circ R)$	Ax, U_4	2	4	377	141

Fig. 4. Some of the laws on map union proved by Otter

law	premises	length	timing	generated	kept
Y_1 $P \circ Q \cap R = \emptyset \rightarrow P^{-1} \circ R \cap Q = \emptyset$	Ax	56	13	2104	328
T_1 $P = \emptyset \vee 1 \circ P \circ 1 = 1$	Simpl, Ax	13	22	6252	362
$P \circ 1 = 1 \vee 1 \circ \overline{P} = 1$	Simpl, Ax	2	2	240	62

Fig. 5. Cycle law and some consequences of simplicity

belonging to the map. In a sense, this kind of statements can be thought of as having a 'deeper character', or, in other words, to model a sort of deep knowledge on the domain(s) of discourse.

Secondly, simple syntactical changes (preserving the semantics) in the thesis to be proved sometimes badly affect Otter's performances.

For instance, the proof of (3) (see also Fig. 3) was relatively easy if compared with the one of (4), which is obtainable from (3) by just applying the rule $P = Q \not\vDash \overline{P} = \overline{Q}$ and by exploiting the double-negation law $\overline{\overline{P}} = P$.

$$P \bigtriangleup Q = \overline{\overline{P} \cap Q} \cap \overline{P \cap \overline{Q}} \tag{3}$$

$$\overline{P \bigtriangleup Q} = \overline{\overline{P} \cap Q} \cap \overline{P \cap \overline{Q}} \tag{4}$$

To find a possible justification of this 'unstable' behavior, we have to consider that Otter adopts a default lexicographic ordering of terms (whenever the user does not supply his own criterion), in order to orient the rewriting rules (recall that Knuth-Bendix completion is employed), and to handle demodulation and weighting. In the above-mentioned case, the default ordering is the same for both theses, but it works better with the former of them. Changing the crite-

law		premises	length	timing	generated	kept
C_2	$P \cap (P \circ (Q \cap \iota)) = P \circ (Q \cap \iota)$	$\mathbf{Ax}, \mathbf{I}_1, \mathbf{C}_1, \mathbf{G}_1, \mathbf{U}_i, \mathbf{Y}_1$	21	20.61s	236370	13644
	$P \cap ((Q \cap \iota) \circ P) = (Q \cap \iota) \circ P$	"	21	40.52s	584457	15052
	$P \cap \iota = P^{-1} \cap \iota$	"	76	40.34s	568993	14885
	$(P \cap \iota)^{-1} = P^{-1} \cap \iota$	"	3	7	946	160
	$(P \cap \iota)^{-1} = P \cap \iota$	"	74	43.47s	616878	15167
$\overline{C_3}$	$P^{-1} \circ \overline{P} \cap \iota = \iota$	$\mathbf{Ax}, \mathbf{I}_1, \mathbf{C}_{1,2}, \mathbf{G}_1, \mathbf{U}_i, \mathbf{Y}_1$	13	4.78s	59433	6707
	$(P^{-1} \circ ((P \circ Q) \bigtriangleup \mathbb{1})) \cap Q = \emptyset$	$\mathbf{Ax}$	5	9	1217	241
$\overline{C_3'}$	$(P^{-1} \circ (R \cap (\mathbb{1} \bigtriangleup (P \circ Q)))) \cap Q = \emptyset$	$\mathbf{Ax}$	34	15	2472	442
	$(P^{-1} \circ \overline{P \circ Q}) \cap Q = \emptyset$	$\mathbf{Ax}, \mathbf{I}_1, \mathbf{C}_{1,3}, \mathbf{G}_1, \mathbf{Y}_1$	2	2	204	46
	$(P^{-1} \circ (R \cap \overline{P \circ Q})) \cap Q = \emptyset$	$\mathbf{Ax}, \mathbf{I}_1, \mathbf{C}_{1,3}, \mathbf{G}_1, \mathbf{Y}_1$	4	9	2335	192

Fig. 6. More laws on map composition and ι

rion for lexicographic ordering (in proving (4)) would have determined a better performance.

As a last remark on this phenomenon, notice that, as one expects, the proof of (4) turns out to be extremely easy (cf. Fig. 3) when (3) is included among hypotheses.

There are also cases of laws whose proofs become easier if some additional lemmas are given in input (cf., for instance, $\mathbf{U}_3$ or $\mathbf{lAbs}_1$). This is a motivation for our choice of splitting in several groups the laws regarding a particular map construct.

Otter exhibited different behaviors even in proving the same thesis when formulated at different levels of our 'layered architecture'. For example, consider the couple of laws $\overline{\mathbb{1} \circ P} \cap P = \emptyset$ and $P \subseteq \mathbb{1} \circ P$, or the following two
$\mathbb{1} \circ P = P \to (R \circ Q) \cap P = R \circ (Q \cap P)$ and $\mathsf{lAbs}(P) \to (R \circ Q) \cap P = R \circ (Q \cap P)$
(cf. Fig. 7). Experimentation revealed that, in general, the proof turns out to be easier when the thesis is expressed by employing the constructs of the higher layer (e.g. $\subseteq$ instead of $^-$ and $\cap$, or $\mathsf{lAbs}(\cdot)$ instead of '$\mathbb{1} \circ \cdot$'). Clearly, this is because the higher the layer, the greater is the expressiveness of the constructs/operators involved and, obviously, the larger is the set of previously proved laws that can be usefully used by Otter. This fact strongly supports our choice of developing experimentations in a 'layered' fashion.

It is sometimes customary to add to the axiomatization of Fig. 1 the axiom:
(Simpl) $\qquad\qquad\qquad R \neq \emptyset \to \mathbb{1} \circ R \circ \mathbb{1} = \mathbb{1}$
It can be shown that any theorem that is proved under this 'simplicity' assumption is also provable without it. Fig. 5 lists some of the consequences of simplicity, proved by Otter.

2 Set-Reasoning in Map Calculus

Often, a particular class of interpretations can be characterized by imposing a collection of map equalities that will serve as proper axioms of the specific application of interest. A task of this nature has been undertaken in [6], where an equational re-engineering of ZF is developed.

	law	premises	len.	timing	gen.	kept
$\mathbf{Inc_1}$	$P \subseteq P$	$\mathbf{Ax,I_1,C_1,G_1,N_1,U_1}$	1	3	46	58
	$P \subseteq Q \to (Q \subseteq R \to P \subseteq R)$	"	8	4	362	107
	$P \subseteq Q \to P^{-1} \subseteq Q^{-1}$	"	7	7	1582	229
	$P \subseteq Q \to (R \subseteq S \to (P \cap R \subseteq Q \cap S))$	"	16	74	19377	1638
	$P \subseteq Q \to P \cap Q = P$	"	1	2	0	50
	$\emptyset \subseteq P$	"	1	3	32	50
$\mathbf{Inc_2}$	$P \subseteq Q \to (R \subseteq S \to (P \circ R \subseteq Q \circ S))$	$\mathbf{Ax,I_1,C_1,G_1,N_1,}$ $\mathbf{U_{1,4},Y_1,Inc_1}$	16	$1m\,16s$	$2.1 \cdot 10^6$	3425
$\mathbf{Inc_3}$	$P \cap Q = P \to P \subseteq Q$	$\mathbf{Ax,I_1,C_1,G_1,N_1,}$ $\mathbf{U_{1,4},Y_1,Inc_{1,2}}$	1	3	1	54
	$P \subseteq Q \to (Q \subseteq P \to P = Q)$	"	3	3	205	65
	$\iota \cap P \subseteq P$	"	4	5	413	90
	$\iota \cap P \subseteq P^{-1}$	"	24	$2m\,30s$	$3.3 \cdot 10^6$	25386
	$P \subseteq \overline{Q} \to Q \subseteq \overline{P}$	"	9	8	1641	268
	$P \subseteq Q \to (R \subseteq S \to (P \cap \overline{S} \subseteq Q \cap \overline{R}))$	"	10	$11.46s$	76721	28971
	$P \subseteq \mathbf{1}$	"	2	3	210	65
	$P \subseteq Q \to (P \subseteq R \to (P \subseteq Q \cap R))$	"	2	15	3381	730
	$P \subseteq Q \to P \circ P^{-1} \subseteq Q \circ Q^{-1}$	"	2	25	6067	1586
$\mathbf{Inc_4}$	$P \subseteq \mathbf{1} \circ P$	$\mathbf{Ax,I_{1,2},C_{1,2,3'},G_1,}$ $\mathbf{N_{1,2},U_i,Y_1}$	1	4	201	104
	$P \subseteq P \circ \mathbf{1}$	"	1	6	164	99
	$P \cap Q \subseteq (\mathbf{1} \circ P) \cap Q$	$\mathbf{Ax,Inc_{1,2,3}}$	3	5	818	235
	$P \circ ((\mathbf{1} \circ Q) \cap R) = (\mathbf{1} \circ Q) \cap (P \circ R)$	$\mathbf{Ax,1Abs_1.10}$	77	$1.57s$	17442	2695
$\overline{\mathbf{Inc_5}}$	$(P \cap Q) \circ R \subseteq P \circ R \cap Q \circ R$	$\mathbf{Ax,Inc_{1,2,3}}$	6	18	5199	281
	$P \circ (Q \cap R) \subseteq P \circ Q \cap P \circ R$	$\mathbf{Ax,Inc_{1,2,3}}$	6	18	5199	281
$\mathsf{1Abs_1}$	$\mathsf{1Abs}(\mathbf{1})$	$\mathbf{Ax,I_1,C_1,G_1,}$ $\mathbf{N_1,U_{1,4},Y_1}$	1	1	48	48
	$\mathsf{1Abs}(\emptyset)$	"	1	2	11	47
	$\mathsf{1Abs}(\mathbf{1} \circ P)$	"	3	6	958	188
	$\mathsf{1Abs}(P) \to \mathsf{1Abs}(\overline{P})$	"	16	$24.38s$	257235	10844
	$\mathsf{1Abs}(P) \to \mathsf{1Abs}(\overline{P})$	$\mathbf{Ax,I_1,C_1,G_1,}$ $\mathbf{N_1,U_{1,4},Y_1,1Abs_1}$	21	76	18640	1525
	$\mathsf{1Abs}(P) \to \mathsf{1Abs}(P \circ Q)$	"	6	12	2831	314
	$\mathsf{1Abs}(P) \wedge \mathsf{1Abs}(Q) \to \mathsf{1Abs}(P \cup Q)$	$\mathbf{Ax,U_4,1Abs_1}$	5	99	8229	5234
	$\mathsf{1Abs}(P) \wedge \mathsf{1Abs}(Q) \to \mathsf{1Abs}(P \cap Q)$	$\mathbf{Ax,N_4,U_4,1Abs_1}$	4	21	3114	2159
	$\mathbf{1} \circ P = P \to (R \circ Q) \cap P = R \circ (Q \cap P)$	$\mathbf{Ax,C_1,G_1,}$ $\mathbf{N_1,U_{1,4},Y_1,1Abs_1}$	139	$18.75s$	172397	13368
	$\mathsf{1Abs}(P) \to (R \circ Q) \cap P = R \circ (Q \cap P)$	"	6	65	7659	4056
	$\mathsf{1Abs}(P) \wedge \mathsf{1Abs}(Q) \to \mathbf{1} \circ (P \cap Q) = P \cap Q$	$\mathbf{Ax,1Abs_1}$	2	32	4942	4733

Fig. 7. Laws on map inclusion and left absoluteness of maps

Two derived constructs are of great help in stating the properties of membership:[1] $\partial(P) \equiv_{\mathrm{Def}} \overline{P \circ \not\ni}$,   and $\mathcal{F}(P) \equiv_{\mathrm{Def}} \partial(P) \setminus \overline{P} \circ \in$.

By means of these constructs it is possible to express within the map calculus a number of axioms of ZF set theory as briefly summarized in Fig. 9, where we listed the map formulations of *extensionality* (**E**), *power-set* ($\mathcal{P}ow$), *union-set* ($\mathcal{U}n$), *transitive embedding* (**T**), *separation* (**S**), *pairing* (**Pair**), *finiteness* (**F**), *foundation* (**R**), *infinity* (**I**), and *replacement* (**Repl**).

A detailed treatment of this map formulation of ZF can be found in [6]. As an example, let us consider the EXTENSIONALITY axiom. It states that *sets whose elements are the same are identical* (see also Fig. 10). This can be rendered by the following map equality: $\mathcal{F}(\ni) = \iota$.

---

[1] Plainly, $a\partial(Q)^{\Im}b$ and $a\mathcal{F}(R)^{\Im}b$ hold in an interpretation if and only if, respectively,
- all cs in $\mathcal{U}$ for which $aQ^{\Im}c$ holds are 'elements' of b (in the sense that $c \in^{\Im} b$);
- the elements of b are precisely those c in $\mathcal{U}$ for which $aR^{\Im}c$ holds.

	law	premises	len.	timing	gen.	kept
$\mathbf{Tot_1}$	$\mathrm{Total}(\mathbf{1})$	$\mathbf{Ax},\mathbf{I_1},\mathbf{C_1},\mathbf{G_1},\mathbf{Y_1}$	1	<1	99	34
	$\mathrm{Total}(\iota)$	$\mathbf{Ax},\mathbf{I_1},\mathbf{C_1},\mathbf{G_1},\mathbf{Y_1},\mathbf{Tot_1}$	1	<1	98	33
	$\mathrm{Total}(\bar{\iota})$	$\mathbf{Ax},\mathbf{N_4},\mathbf{U_{1,4}},\mathbf{lAbs_1}$	5	$1.37s$	21280	2311
	$\mathrm{Total}(P\cap Q)\to\mathrm{Total}(Q)$	$\mathbf{Ax},\mathbf{I_1},\mathbf{C_1},\mathbf{G_1},$ $\mathbf{Y_1},\mathbf{N_1},\mathbf{U_1},\mathbf{Tot_1}$	7	12	3530	133
	$\mathrm{Total}(P\circ Q)\to\mathrm{Total}(P)$	$''$	8	11	3530	128
	$\mathrm{Total}(P\cup\overline{P\circ\mathbf{1}})$	$''$	22	$1.07s$	25650	1791
	$\mathrm{Total}(P\bigtriangleup\overline{P\circ\mathbf{1}})$	$''$	53	85	9111	1277
	$\mathrm{Total}(P^{-1})\vee\mathrm{Total}(\overline{P})$	$\mathbf{Ax},\mathbf{C_1},\mathbf{G_1},$ $\mathbf{N_1},\mathbf{Tot_1},\mathbf{Simpl}$	4	2	275	92
	$\mathrm{Total}(P)\vee\mathrm{Total}(\overline{P}^{-1})$	$''$	4	2	349	107
	$\mathrm{Total}(P)\vee\mathrm{Total}(\mathbf{1}\circ P^{-1})$	$''$	6	5	531	132
	$P\cap P^{-1}=\emptyset\to\mathrm{Total}(\overline{P})$	$\mathbf{Ax},\mathbf{I_1},\mathbf{C_1},\mathbf{G_1},$ $\mathbf{Y_1},\mathbf{N_1},\mathbf{U_1},\mathbf{Tot_1}$	7	6	1148	225
	$\mathrm{Total}(P)\wedge\mathrm{Total}(Q)\to\mathrm{Total}(P\circ Q)$	$''$	7	11	1584	419
	$\mathrm{Total}(P)\wedge\mathrm{Total}(Q)\to\mathrm{Total}((P\circ\mathbf{1})\cap(Q\circ\mathbf{1}))$	$''$	3	2	8	40
	$\mathrm{Total}(P)\wedge\mathrm{Total}(Q)\wedge\mathrm{Total}(R)$ $\to\mathrm{Total}((P\circ Q)\cap(R\circ\mathbf{1}))$	$''$	5	13	1705	651
	$P\circ Q=\mathbf{1}\to\mathrm{Total}(P)\vee\mathrm{Total}(Q)$	$''$	2	<1	80	50
	$P\circ Q^{-1}=\mathbf{1}\to\mathrm{Total}(P)\wedge\mathrm{Total}(Q)$	$''$	5	56	3130	1718
	$P\circ Q=\mathbf{1}\to\mathrm{Total}(P)\wedge\mathrm{Total}(Q^{-1})$	$''$	5	5	334	114
	$P\cap Q=P\wedge\mathrm{Total}(P)\to\mathrm{Total}(Q)$	$''$	2	4	89	76
	$P\circ Q^{-1}=\mathbf{1}\wedge\mathrm{Total}(R)\to Po(Q^{-1}\circ R^{-1})=\mathbf{1}$	$''$	5	3	189	83
	$P\circ Q^{-1}=\mathbf{1}\to\mathrm{Total}(P\cap Q)$	$''$	2	1	11	8
	$P\circ Q^{-1}=\mathbf{1}\wedge\mathrm{Total}(R)\to\mathrm{Total}(P\cap(R\circ Q))$	$''$	7	31	10191	568
	$\mathrm{Total}(P)\wedge Q\circ(R\circ S)=\mathbf{1}$ $\to\mathrm{Total}(P\circ(Q\circ(R\circ S)))$	$''$	2	1	8	23
	$P\circ Q^{-1}=\mathbf{1}\wedge\mathrm{Total}(R)\wedge\mathrm{Total}(S)$ $\to\mathrm{Total}((S\circ P)\cap(R\circ Q))$	$''$	45	$9m\ 12s$	$6.6\cdot10^6$	30429
	$\mathrm{lAbs}(P)\to(P=\emptyset)\vee\mathrm{Total}(P)$	$\mathbf{Ax},\mathbf{C_1},\mathbf{lAbs_1},\mathbf{Simpl}$	7	1.91	44040	1809
$\mathbf{Func_1}$	$\mathrm{Func}(\emptyset)$	$\mathbf{Ax},\mathbf{I_1},\mathbf{C_1},\mathbf{G_1}$	2	2	92	39
	$\mathrm{Func}(\iota)$	$\mathbf{Ax},\mathbf{I_1},\mathbf{C_1},\mathbf{G_1}$	2	2	110	44
	$\mathrm{Func}(P)\to\mathrm{Func}(P\cap Q)$	$\mathbf{Ax},\mathbf{I_1},\mathbf{Inc_{1,2,3}}$	9	74	20065	913
	$\mathrm{Func}(P)\wedge\mathrm{Func}(Q)\wedge P\subseteq Q\wedge Q\subseteq P\circ\mathbf{1}$ $\to P=Q$	$\mathbf{Ax},\mathbf{I_{1,2}},\mathbf{C_{1,2}},$ $\mathbf{S_1},\mathbf{N_{1,2}},\mathbf{Y_1}$	288	$51m\ 36s$	$3.4\cdot10^7$	24052

Fig. 8. Totality and functionality of maps

An alternative formulation of extensionality. A useful variant of the extensionality axiom is the scheme $\mathrm{Func}(\mathcal{F}(P))$, where P ranges over all map expressions. Our first task in automated set-reasoning consists in proving the equivalence of the two formulations of $(\mathbf{E})$, i.e., that: $\mathcal{F}(\ni)=\iota \; \dashv\vDash \; \mathrm{Func}(\mathcal{F}(P))$.

Otter was unable to prove this theorem in a single shot. Hence we had to split the theorem into two. First, we got a proof of $\mathrm{Func}(\mathcal{F}(P)) \vDash \mathcal{F}(\ni)=\iota$, via the following sequence of intermediate results:

law	length	timing	note
$\iota\subseteq\mathcal{F}(\ni)$	3	4	by using $\mathbf{I_1},\mathbf{C_{1,3}},\mathbf{G_1},\mathbf{N_1},\mathbf{U_{1,2,3,4}},\mathbf{Y_1}$
$\mathrm{Func}(\mathcal{F}(\ni))$	--	--	immediately from the hypotheses
$\mathcal{F}(\ni)\subseteq\iota\circ\mathbf{1}$	3	2	by $\mathbf{Ax},\mathbf{Inc_1},\mathbf{Func_1}$
$\mathcal{F}(\ni)=\iota$	0	<1	immediately from $\mathbf{Func_1}$

(E)	$\mathcal{F}(\ni) = \iota$
($\mathcal{P}ow$)	$\mathsf{Total}\big(\partial(\overline{\not\ni\circ\in})\big)$
($\mathcal{U}n$)	$\mathsf{Total}\big(\partial(\ni\circ\ni)\big)$
(T)	$\mathsf{Total}\big(\in\circ\iota \cap \partial(\ni\circ\ni)\big)$
(S)	$\mathsf{Total}\big(\mathcal{F}\big(\mathsf{funcPart}(Q)\circ\ni\cap P\big)\big)$
(Pair)$_{1,2,3,4}$	$\pi_0^{-1}\circ\pi_1 = \mathbb{1},\quad \mathsf{Func}(\pi_0),\quad \mathsf{Func}(\pi_1),\quad \in\circ\ni = \mathbb{1}$
(Pair)$_5$	$\pi_0\circ\pi_0^{-1} \cap \pi_1\circ\pi_1^{-1} \setminus \iota = \emptyset$
(F)	$\iota \subseteq \mathbb{1}\circ\Big(\in\cap((\iota\cup\not\ni\circ\in)\dagger\not\in)\Big)\dagger\not\ni$
(R)	$\mathsf{Total}(\overline{\ni\circ\mathbb{1}}\ni \setminus \ni\circ\in)$
(I)	$\mathsf{Total}\Big(\mathbb{1}\circ\big(\partial(\ni\circ\ni)\cap\partial(\ni\circ\ni)^{-1}\setminus\in\setminus\ni\setminus\iota\setminus\ni\circ\overline{\in}\triangle\ni\circ\in\big)\Big)$
(Repl)	$\mathsf{Total}\Big(\partial\big((\pi_0\circ\ni\circ\pi_0^{-1}\cap\pi_1\circ\pi_1^{-1})\circ\mathsf{funcPart}(Q)\big)\Big)$

Fig. 9. Axioms of set theory within map calculus.

(E) $\forall v\,(v\in X \leftrightarrow v\in Y) \to X = Y,$	**(E)** $\overline{\ni\circ\not\in} \cap \overline{\not\ni\circ\in} \subseteq \iota$
(N) $\exists z\,\forall v\, v\not\in z,$	**(N)** $\overline{\mathbb{1}\circ\in}\circ\mathbb{1} = \mathbb{1}$
(W) $\exists w\,\forall v\,(v\in w \leftrightarrow v\in X \vee v = Y),$	**(WL)** $(\not\subseteq\circ\in \cap \mathsf{valve}(\in\circ\in,\overline{\not\ni\circ\in}))\circ\ni = \mathbb{1}$
(L) $\exists\ell\,\forall v\,(v\in\ell \leftrightarrow v\in X \wedge v\neq Y)$	with $\mathsf{valve}(P,Q) \equiv_{\mathrm{Def}} P\setminus\overline{\iota}\circ(P\setminus Q)$

Fig. 10. Specification of a weak set theory in first-order logic and in map algebra.

The converse, i.e. $\mathcal{F}(\ni) = \iota \;\not\vDash\; \mathsf{Func}(\mathcal{F}(P))$, was proved as follows:

law	length	timing	note
$\mathcal{F}(P)^{-1}\circ\mathcal{F}(P) \subseteq \overline{\not\ni\circ P^{-1}}\circ\overline{P}\circ\in$	10	$12.29s$	by $\mathbf{Ax}, \mathbf{G}_1, \mathbf{N}_1$
$\mathcal{F}(P)^{-1}\circ\mathcal{F}(P) \subseteq \overline{\not\ni\circ P^{-1}}\circ\overline{P}\circ\not\in$	9	$12.38s$	by $\mathbf{Ax}, \mathbf{G}_1, \mathbf{N}_1$
$\mathcal{F}(P)^{-1}\circ\mathcal{F}(P) \subseteq \mathcal{F}(\ni)$	3	2	by $\mathbf{Inc}_i$
$\mathcal{F}(P)^{-1}\circ\mathcal{F}(P) \subseteq \iota$	1	<1	by $\mathbf{Inc}_i$

Designing pairs of conjugated projections. In [4] a possible choice was proposed for two maps π_0, π_1 which fulfill the pairing axiom, provided that the axioms of a weak theory of sets —i.e., *extensionality*, *null set*, single-element *addition* and *removal*— are assumed (cf. Fig. 10, where uppercase letters stand for variables ruled by universal quantifiers). The authors presented an Otter-based proof that those two specific maps satisfy **(Pair)**$_{1,2,3}$. In that context, the approach to experimentation was aimed at 'miniaturizing' the obtained proofs, i.e., at developing the proofs by starting with the raw axiomatization of Fig. 1, without the explicit introduction of defined constructs, and by strictly interacting with and guiding Otter, to make it perform only the essential inference steps.

It is claimed in [7] that **(W)** is not, when taken alone, expressible in map algebra.[2] Nevertheless, in [4] it is shown that taken together with **(N)** these axioms enable one to build the pair $\big\{\,\{Y\}\setminus\{X\},\ \{Y\}\cup\{X\}\,\big\}$ out of any given sets X and Y. This fact, thanks to **(E)**, can be stated as

$$\exists d\,\big(Y{\in}d \wedge \forall u\,\big(u{=}X \leftrightarrow \exists v\,\exists w\,(u{\in}v{\in}d \wedge u{\notin}w{\in}d)\big)\big),$$

and in turn (again with the contribution of **(E)**) it yields **(N)**, **(W)**, and **(L)**. Consequently, the set axioms of Fig. 10.a can be translated as shown in Fig. 10.b. The key point consists in observing that the map $\nu \equiv_{\mathrm{Def}} \not\in{\in} \cap \mathsf{valve}(\in\circ\in, \overline{\not\in}\circ\in)$ occurring in **(WL)**, enables a quick implementation of the projections π_i:

$$\pi_0 \equiv_{\mathrm{Def}} \nu^{-1}, \qquad\qquad \pi_1 \equiv_{\mathrm{Def}} \mathsf{valve}^{-1}(\in\circ\in, \nu).$$

The main result of [4] consisted in proving *within map algebra* (under minimal assumptions on membership), that π_0 and π_1 designate *conjugated projections*. As mentioned, the important consequence is that the equational specification of our assumptions on membership has the same deductive power as its counterpart formulated in quantified first-order logic; this follows from results in [15].

The experimentation reported in [4] proceeded by proving a number of intermediate results ultimately yielding the desired proof. The most relevant of them is the following lemma:

Lemma. (Functionality) $Q\circ Q^{-1} \subseteq \iota$ entails $\mathsf{valve}(P,Q)\circ\mathsf{valve}^{-1}(P,Q) \subseteq \iota$.

This lemma mainly relies on various elementary Boolean identities, and on some obvious consequences of the Peircean axioms (i.e., the logical axioms regarding $\circ$, $^{-1}$, and ι). The only non-obvious laws on maps needed are the so-called *cycle law* (cf. Fig. 5) and *Dedekind law* (cf. [13]):

$$P\circ Q\cap R \subseteq (P\cap R\circ Q^{-1})\circ(Q\cap P^{-1}\circ R).$$

A 'miniaturized' derivation of the Dedekind law was obtained from the bare axioms in Fig. 1. It consists in 25 verifications of the average CPU-time cost of 6 to 8 seconds (depending on the machine).[3] It is worth stressing that these 25 steps included the proofs of basic facts such as some of the laws on symmetric difference, intersection, and composition already seen in Figures 2 and 2.

While the functionality lemma easily allowed Otter to prove **(Pair)**$_{2,3}$, in order to proof that **(Pair)**$_1$ it was necessary to proceed as follows. First, a temporary assumption was made that a singleton set $\{a\}$ can be formed out of any given a. This assumption can be stated more precisely as follows:

$$\textbf{(Sng)} \qquad \mathsf{sng}\circ \mathbb{1} = \mathbb{1}, \qquad \text{where } \mathsf{sng} \equiv_{\mathrm{Def}} \in\setminus\overline{\iota}\circ\in,$$

holds along with **(WL)**. Hence we have the following lemma:

Lemma.   Assume **(Sng)** and **(WL)**. It follows that $\nu\circ\pi_1 = \mathbb{1}$.

To prove this result it turned out that Otter had to extensively use map-inclusion laws drawn from the list shown in Sec. 1.

The subsequent step consists in proving that it is actually possible to do without a postulate of singleton formation. Verifying this claim amounted to getting an automated proof of the derivability of **(Sng)** from **(WL)** and **(N)**.

---

[2] A proof of this fact, which unfortunately remains quite obscure to the authors, is supplied by [7].

[3] These verifications were run on a G3 Macintosh and under Linux.

In this case, an analysis of Otter's proof showed that the most useful intermediate results (implicitly proved in the main proof) were the laws on totality.

Totality of some elementary relations on sets. By using the laws of Sec. 1, Otter was able to prove the totality of a number of relations on sets. We exhibit below an excerpt of the results we obtained. The laws of Fig. 8 intervene crucially in these proofs.

- $\mathsf{Total}(\in \circ \ni)$. Thanks to **(Pair)**, it reduces to prove that $\mathsf{Total}(\mathbb{1})$ holds. It was immediately proved from the laws on totality.
- $\mathsf{Total}(\in \circ \mathbb{1})$. It follows from the previous result and from the laws in Fig. 8. It was proved in 0.02 seconds, the proof-length is 3.
- $\mathsf{Total}(\in)$. It follows from the previous results and from the laws in Fig. 8. It was proved in 0.02 seconds, the proof-length is 1.

A general technique for proving totality of set constructors. The next task consists in obtaining the proof of a general law for deriving the totality of expressions of the form $\mathsf{Total}(\mathcal{F}(R))$. This law will give us the capability of defining a number of set-constructs (cf. [5]). Let us start with two useful lemmas.
Lemma. For any P, Q such that

$$P^{-1} \circ Q \subseteq \ni \quad \text{and} \quad \mathsf{Func}(\pi_1) \tag{5}$$

it holds that:

$$(P \circ \pi_0^{-1} \cap \pi_1^{-1}) \circ \mathcal{F}(\pi_0 \circ \ni \cap \pi_1 \circ Q) \subseteq \mathcal{F}(Q). \tag{6}$$

In the following we describe Otter's proof. The thesis (6) can be rewritten as

$$(P \circ \pi_0^{-1} \cap \pi_1^{-1}) \circ \mathcal{F}(\pi_0 \circ \ni \cap \pi_1 \circ Q) \subseteq \overline{\overline{Q} \circ \notin} \cap \overline{\overline{Q} \circ \in} \tag{7}$$

By assuming the hypothesis (5).1, Otter was able to prove the following intermediate result: $(\pi_0 \circ P^{-1} \cap \pi_1) \circ Q) \subseteq \pi_0 \circ \ni \cap \pi_1 \circ Q$. Otter proved this result in 0.31 seconds, it generated 4162 clauses (the number of kept clauses was 915). The proof-length was 4. The proof was easily obtained by extensive use of the map-inclusion laws (cf. Fig. 7). The main settings used to drive Otter imposed any generated clause consisting of more than two literals, or having more than two distinct variables, to be discarded. From (7), by exploiting the cycle law and the laws on inclusion, Otter easily proved that:

$$(P \circ \pi_0^{-1} \cap \pi_1^{-1}) \circ \mathcal{F}(\pi_0 \circ \ni \cap \pi_1 \circ Q) \subseteq \overline{Q \circ \notin} \tag{8}$$

The proof was found in 1.30 seconds (its length was 9), by generating 13729 unit clauses (`max_literals=1` and `max_distinct_vars=3`) and keeping 2652 clauses.

On the other hand, the following map inclusion was proved by assuming the functionality of π_1 (cf. hypothesis (5).2), in 0.81 seconds. The proof-length was 13 (the generated and the kept clauses were 9848 and 2097, respectively):

$$(P \circ \pi_0^{-1} \cap \pi_1^{-1}) \circ \mathcal{F}(\pi_0 \circ \ni \cap \pi_1 \circ Q) \subseteq \overline{\overline{Q} \circ \in} \tag{9}$$

Putting together the two results (8) and (9), in order to obtain the thesis (6), took 0.08 seconds (two inferences, by hyper-resolution).

Lemma. Assume $(\mathbf{Pair})_{1,2}$ and $(\mathbf{S})$. Then for any P, Q

$$\mathsf{Total}(P) \overset{\times}{\vdash} \mathsf{Total}((P \circ \pi_0^{-1} \cap \pi_1^{-1}) \circ \mathcal{F}(\pi_0 \circ \ni \cap \pi_1 \circ Q)).$$

Otter proved this lemma (by proving two intermediate results) in a total time of 0.24 seconds. On this ground, the following proposition was proved.

Proposition. Assume $(\mathbf{Pair})_{1,2,3}$ and $(\mathbf{S})$. Then for any P, Q

$$\mathsf{Total}(P),\ P^{-1} \circ Q \subseteq\, \ni\ \overset{\times}{\vdash} \mathsf{Total}(\mathcal{F}(Q)). \tag{10}$$

This proposition was proved in two stages. We first drew from the hypotheses a series of intermediate lemmas yielding $(P \circ \pi_0^{-1} \cap \pi_1^{-1}) \circ \mathcal{F}(\pi_0 \circ \ni \cap \pi_1 \circ Q) \subseteq \mathcal{F}(Q)$. The thesis then readily followed, with the help of the laws on totality. The overall time of this proof was 3.57 seconds.

By using this general tactic, Otter proved the totality of several map expressions, certifying in this way that these expressions characterize legal operations on sets:

- $\mathsf{Total}(\mathcal{F}(\iota))$. It defines the singleton operation $a \mapsto \{a\}$. Its totality was proved in 0.05 seconds (length:7, generated:768, kept:108), by using the result previously obtained: $\mathsf{Total}(\in)$ (Otter instantiated $P \equiv\ \in$ and $Q \equiv \iota$ in proposition (10)).

- $\mathsf{Total}(\mathcal{F}(\emptyset))$. It characterizes the nullset constructor: $a \mapsto \{\ \}$. As in the previous case, its totality was proved in 0.04 seconds (length:3, generated:335, kept:52). Notice that this thesis was proved also without resorting to the above proposition, but in this case Otter's task was more difficult: the proof was produced in much more time: 1.15 seconds. Otter used the laws in $\mathbf{C}_1, \mathbf{I}_{1,2}, \mathbf{G}_1, \mathbf{N}_{1,2}$ and in particular those in $\mathbf{Tot}_1$; it generated 21521 clauses, keeping 343 of them.

- Consider the two axioms $\mathsf{Total}(\partial(\overline{\not\ni} \circ \in))$ and $\mathsf{Total}(\partial(\ni \circ \ni))$ (cf. Fig. 9). Otter was able to prove their stronger version: $\mathsf{Total}(\mathcal{F}(\overline{\not\ni} \circ \in))$ and $\mathsf{Total}(\mathcal{F}(\ni \circ \ni))$ by using, in particular, the law (10) and the cycle law. The first proof was generated in 0.11 seconds (length:4, generated:2616, kept:265). The strong version of the second axiom was proved in 17.88 seconds (length:6, generated:386130, kept:5070).

- A more general result was also proved. Namely, under the axioms $(\mathbf{Pair})$ and $(\mathbf{S})$, Otter proved this property of totality: $\mathsf{Total}(\partial(P)) \overset{\times}{\vdash} \mathsf{Total}(\mathcal{F}(P))$. The proof was found in 0.12 seconds (length:4, generated:2616, kept:265) by using the above proposition, the cycle law, and the laws of Fig. 8.

Conclusions

We reported on an initial experimentation activity in automated set-reasoning, based on a formalization of axiomatic ZF theory within a ground equational framework. This approach made it possible to profitably exploit traditional first-order theorem provers for experimenting in set reasoning. The main efforts were devoted to develop a structured methodology for experimentation. This approach allowed Otter to prove several theorems of map algebra as well as of set theory that it was not possible to prove (with Otter) in absence of a layered methodology. Clearly, this approach is not specific to Otter and can be adapted to other (first-order) automated theorem provers as well.

We moved the first steps in equational set-reasoning by setting the ground for further studies and experimentation. Analogous proposals for the automation of set-reasoning have been developed by other researchers (cf. [3,11,12], among others). A comparison with these approaches certainly deserves further investigation and is matter of ongoing work.

The ultimate purpose of the research we presented here should consist in assessing what can be achieved nowadays by applying unspecialized proof methods of automated deduction in the set-theoretical context. The knowledge we gain can be used both to refine our approach based on first-order theorem provers, and as an aid to single out which would be the most promising specialized methods to be employed in the realization of an *ad hoc* basic inference machinery for $\mathcal{L}^{\times}$ (and hence, for set theory).

Moreover, the assessment of the exact kinship of our own formulation of ZF with ZF proper on the one hand, and with NGB on the other, is a crucial theoretical issue and constitutes a challenging and promising starting point for future research.

References

[1] F. Aureli, A. Formisano, E. G. Omodeo, and M. Temperini. Map calculus: Initial application scenarios and experiments based on Otter. Rep. 466, IASI-CNR, 1998.

[2] J. L. Bell and A. B. Slomson. *Models and ultraproducts: An introduction.* North-Holland, *Studies in Logic and the Foundations of Mathematics*, 3rd printing, 1974.

[3] R. Boyer, E. Lusk, W. McCune, R. Overbeek, M. Stickel, and L. Wos. Set theory in first-order logic: Clauses for Gödel's axioms. *JAR*, 2(3), 1986.

[4] A. Chiacchiaretta, A. Formisano, and E. G. Omodeo. Benchmark #1 for equational set theory. *Giornata "Analisi Sperimentale di Algoritmi per l'Intelligenza Artificiale"* (Rome, 16, December 1999).

[5] A. Chiacchiaretta, A. Formisano, and E. G. Omodeo. *Map reasoning through existential multigraphs.* Rep. 05/00, Dipartimento di Matematica Pura ed Applicata, Università di L'Aquila, April 2000.

[6] A. Formisano and E. G. Omodeo. An equational re-engineering of set theories. In Caferra and Salzer eds., *Automated Deduction in Classical and Non-Classical Logics*, LNCS 1761 (LNAI), pages 175–190. Springer, 2000.

[7] M. K. Kwatinetz. *Problems of expressibility in finite languages.* PhD thesis, University of California, Berkeley, 1981.

[8] W. W. McCune. *OTTER 3.0 Reference Manual and Guide.* Argonne National Laboratory/IL, USA, 1994.

[9] P. A. J. Noël. Experimenting with Isabelle in ZF set theory. *JAR*, 10(1):15–58, 1993.

[10] L. C. Paulson. Set Theory for verification. II: Induction and recursion. *JAR*, 15(2):167–215, 1995.

[11] A. Quaife. Automated deduction in von Neumann-Bernays-Gödel Set Theory. *JAR*, 8(1):91–147, 1992.

[12] A. Quaife. *Automated development of fundamental mathematical theories*. Kluwer Academic Publishers, 1992.

[13] G. Schmidt and T. Ströhlein. Relation algebras: Concept of points and representability. *Discrete Mathematics*, 54:83–92, 1985.

[14] E. Schröder. *Vorlesungen über die Algebra der Logik (exakte Logik)*. B. Teubner, Leipzig, 1891-1895. [Reprinted by Chelsea Publishing Co., New York, 1966.]

[15] A. Tarski and S. Givant. *A formalization of set theory without variables*. Vol. 41 of *Colloquium Publications*. AMS, 1987.

[16] L. Wos. *Automated reasoning. 33 basic research problems*. Prentice Hall, 1988.

NP-Completeness of Refutability by Literal-Once Resolution*

Stefan Szeider

Institute of Discrete Mathematics
Austrian Academy of Sciences
Sonnenfelsgasse 19, 1010 Vienna, Austria
stefan.szeider@oeaw.ac.at

Abstract. A boolean formula in conjunctive normal form (CNF) F is refuted by *literal–once resolution* if the empty clause is inferred from F by resolving on each literal of F at most once. Literal–once resolution refutations can be found nondeterministically in polynomial time, though this restricted system is not complete. We show that despite of the weakness of literal–once resolution, the recognition of CNF-formulas which are refutable by literal–once resolution is NP-complete. We study the relationship between literal–once resolution and *read-once resolution* (introduced by Iwama and Miyano). Further we answer a question posed by Kullmann related to minimal unsatisfiability.

1 Introduction

Resolution is a method for establishing the unsatisfiability of formulas in conjunctive normal form (CNF), based on the *resolution rule*: if $C_1 \cup \{\ell\}$ and $C_2 \cup \{\overline{\ell}\}$ are clauses, then the clause $C_1 \cup C_2$ may be inferred, *resolving on* the literal ℓ. A *resolution refutation* of a CNF-formula F is a derivation of the empty clause $\square$ from F, using the resolution rule. It is well-known that resolution is *sound* and *complete*, i.e., a CNF-formula is unsatisfiable if and only if there is a resolution refutation of it ([14]). Resolution refutations can be represented as binary trees, where the leaves are labeled by clauses of F (see Figure 1 for an example). Unfortunately, the size of a shortest resolution refutation of a CNF-formula F

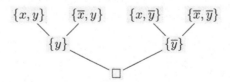

Fig. 1. A resolution refutation of $F = \{\{x, y\}, \{x, \overline{y}\}, \{\overline{x}, y\}, \{\overline{x}, \overline{y}\}\}$.

---

* This work has been supported by the Austrian Science Fund, P13417-MAT.

R. Goré, A. Leitsch, and T. Nipkow (Eds.): IJCAR 2001, LNAI 2083, pp. 168–181, 2001.

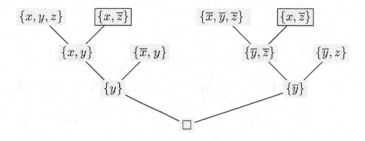

Fig. 2. A resolution refutation which is not read–once.

can be exponential in the number of clauses of F ([6,7]). Therefore, considerable effort has been made to identify restricted (and incomplete) classes of resolution refutations where the size of refutations is polynomially bounded by the size of input formulas (see [10] for a survey). One of the best known examples is *unit resolution*, where the resolution rule is only applied to pairs of clauses C_1, C_2 if C_1 or C_2 is a unit clause (i.e., a singleton). Unit resolution is not complete any more, but the class of formulas which can be refuted by unit resolution can be recognized in linear time (see, eg., [10]).

Iwama and Miyano ([8]) considered *read–once resolution*, where each clause of the input formula must be used at most once in a refutation; i.e., two leaves of the resolution tree may not be labeled by the same clause. (In [8] also resolution refutations are considered, where clauses of the input formula may used more than once, but the number of repetitions is restricted.) For example, the refutation exhibited in Figure 2 is not read-once, since the clause $\{x, \overline{z}\}$ occurs at two leaves (in fact, it can be shown that for $F = \{\{x, y, z\}, \{x, \overline{z}\}, \{\overline{x}, y\}, \{\overline{x}, \overline{y}, \overline{z}\}, \{\overline{y}, z\}\}$ no read–once resolution exists, despite F being unsatisfiable; see [8] or Proposition 1 below). It is easy to see that the size of a read–once resolution refutation is polynomially bounded by the size of the input formula. However, in [8] it is shown that—in spite of the shortness of read–once resolution refutations—it is NP-complete to recognize formulas which can be refuted by read–once resolution.

If we modify the above example by adding two clauses $\{w, x, \overline{z}\}$ and $\{\overline{w}, x, \overline{z}\}$ to F, then we get a read–once resolution refutation (exhibited in Figure 3). There are still two occurrence of $\{x, \overline{z}\}$, but one occurrence became an interior vertex of the tree, and so the refutation became read-once. Thus, it is natural to consider resolution trees where no clause appears more than once *at any position* in the resolution tree. We call such refutations *strict read-once*. It can be shown that there are CNF-formulas which are refutable by read–once resolution, but not by strict read-once resolution (see Proposition 1 below). Since strict read–once resolution is therefore weaker than read–once–resolution, it is conceivable that refutability by strict read-once resolution can be decided in polynomial time. We will show, however, that recognition of formulas refutable by strict read-once resolution is NP-complete.

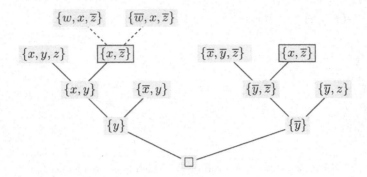

Fig. 3. A resolution refutation obtained from Figure 3; it is read–once, but not strict read–once.

Going one step further, we also consider a type of resolution which is even weaker than strict read–once resolution: a resolution tree is *literal–once* if it does not contain two or more vertices whose clauses are inferred by resolving on the same literal. For example, the resolution refutation depicted in Figure 1 is strict read–once, but it is not literal–once, since clauses at two positions are inferred by resolving on the same literal x. However, it is easy to see that every literal–once resolution refutation is a (strict) read-once resolution. The main result of this paper is the intractability of literal–once resolution; i.e., it is NP-complete to recognize CNF-formulas which are refutable by literal–once resolution.

Furthermore, we show that intractability of read–once resolution can be obtained as corollary of our main result. This fact may be of interest, since Iwama and Miyano obtain the quoted result solely by presenting a single example without giving an accurate proof.

In [11] Kullmann asked for the computational complexity of finding a subset F' of a given formula F such that

(i) F' is minimal unsatisfiable (F' is unsatisfiable, but every proper subset of F' is satisfiable), and

(ii) F' has exactly one more clause than variables.

We denote by MU(1) the class of formulas F' satisfying (i) and (ii). This class is of special interest; for example, every minimal unsatisfiable Horn formula belongs to MU(1) ([4]). We show that F has a subset $F' \in$ MU(1) if and only if F is refutable by literal–once resolution. Whence the intractability of Kullmann's problem follows from the NP–completeness of refutability by literal–once resolution.

2 Notation

2.1 Digraphs

We denote a digraph D by an ordered pair (V, A) consisting on a finite nonempty set V of *vertices* and a set A of *arcs*; an arc is an ordered pair (u, v) of distinct vertices $u, v \in V$. Let $D = (V, A)$ be a digraph and $v \in V$. We denote the sets of *incoming* and *outgoing* arcs of v by $\text{out}(v) = \{\, (u, w) \in A \mid u = v \,\}$ and $\text{in}(v) = \{\, (u, w) \in A \mid w = v \,\}$, respectively. For $(u, v) \in A$ we say that u is a *predecessor of v* and that v *is a successor of u*.

A digraph $T = (V, A)$ is an *in–tree* if there is exactly one vertex v without successors (the *root* of T), and for every vertex $w \in V$ there is exactly one (directed) path P_w from w to v. Consequently, every vertex which is different from the root has exactly one successor. A vertex without predecessors is a *leaf*. An in–tree T is *binary* if every non-leaf has exactly two predecessors. Note that a binary in–tree with k leaves has $2k - 1$ vertices. For graph theoretic terminology not defined here, the reader is referred to [2].

2.2 CNF-Formulas

Let var be a set of boolean variables. A *literal* ℓ is an object of the form x or $\overline{x}$ for $x \in$ var; in the first case we call ℓ *positive*, in the second case *negative*; for a negative literal $\ell = \overline{x}$, $x \in$ var, we put $\overline{\ell} = x$. Literals ℓ and $\overline{\ell}$ are *complements* of each other. If x is a variable and $\ell \in \{x, \overline{x}\}$, then we call x the *variable of ℓ* and write $\text{var}(\ell) = x$. A *clause* is a finite set of literals without complements. The empty clause is denoted by $\square$. For a clause C we put $\text{var}(C) = \{\, \text{var}(\ell) \mid \ell \in C \,\}$. A *CNF-formula* (or *formula*, for short) is a finite set of clauses. For a formula F we put $\text{var}(F) = \bigcup_{C \in F} \text{var}(C)$. A literal ℓ is a *pure literal* of F if $\ell \in \bigcup_{C \in F} C \not\ni \overline{\ell}$. A formula F is *Horn* if every clause in F contains at most one positive literal.

A *truth assignment* t to a formula F is a map $t : \text{var}(F) \to \{0, 1\}$. Let t be a truth assignment to F; we put $t(\overline{x}) = 1 - t(x)$ for $x \in \text{var}(F)$, and we say that t *satisfies* a clause $C \in F$ if $t(\ell) = 1$ for at least one literal $\ell \in C$. Furthermore, we say that t *satisfies* F if t satisfies all clauses of F. A formula F is *satisfiable* if there is a truth assignment which satisfies F; otherwise F is called *unsatisfiable*. We denote the set of all unsatisfiable formulas by UNSAT.

2.3 Resolution

Let C_1, C_2 be two clauses. If there is *exactly one* literal ℓ such that $\ell \in C_1$ and $\overline{\ell} \in C_2$ then we call the clause $C = (C_1 \setminus \{\ell\}) \cup (C_2 \setminus \{\overline{\ell}\})$ the *resolvent of C_1 and C_2*; in this case we also say that C is obtained from C_1, C_2 by *resolving on ℓ*.

Let $T_0 = (V, A)$ be an in–tree and λ a labeling of its vertices such that $\lambda(v)$ is a clause for every $v \in V$. We call $T = (V, A, \lambda)$ a *resolution tree* if for every vertex $v \in V$ with predecessors v_1, v_2 it holds that $\lambda(v)$ is the resolvent of $\lambda(v_1)$ and $\lambda(v_2)$. Let $T = (V, A, \lambda)$ be a resolution tree and $v \in V$. If v is a leaf, then we put $\text{rlit}(v) = \emptyset$; otherwise v has two predecessors, say v_1 and v_2; we put

$\mathsf{rlit}(v) = (\lambda(v_1) \cup \lambda(v_2)) \setminus \lambda(v)$. We call the elements of $\mathsf{rlit}(v)$ *resolution literals of v*. A clause C is a *premise* of a resolution tree T if $\lambda(v) = C$ for some leaf v of T. We write $\mathsf{pre}(T)$ for the set of all premises of T. A clause C is the *conclusion* of T if $\lambda(v) = C$ for the root v of T; in this case we write $\mathsf{con}(T) = C$. A resolution tree T is a *resolution refutation* if $\mathsf{con}(T) = \square$. Let F be a formula and T a resolution refutation. If $\mathsf{pre}(T) \subseteq F$ then we say that F is *refuted* by T, or that T is a resolution refutation *of F*. A resolution tree $T = (V, A, \lambda)$ is *trivial* if $|V| = 1$. Clearly, a formula F is refuted by the trivial resolution tree $T = (\{v\}, \emptyset, \lambda)$ if and only if $\lambda(v) = \square \in F$.

For a resolution tree $T = (V, A, \lambda)$ and $v \in V$ we define T_v to be the resolution tree (V', A', λ') where (V', A') is the maximal subtree of (V, A) with root v and λ' is the restriction of λ to V'.

It is well–known that a formula F is unsatisfiable if and only if it can be refuted by some resolution refutation T.

3   Restricted Types of Resolution

Read–Once Resolution. A resolution tree $T = (V, A, \lambda)$ is *read–once* if $\lambda(v) \neq \lambda(w)$ for any two distinct leaves v, w of T. We denote by ROR the class of all formulas refutable by read–once resolution refutations. (ROR corresponds to the class which is denoted by $R(0)$ in [8].)

Strict Read–Once Resolution. A resolution tree $T = (V, A, \lambda)$ is *strict read–once* if $\lambda(v) \neq \lambda(w)$ for any two distinct vertices v, w of T. We denote by SROR the class of all formulas refutable by strict read–once resolution refutations.

Literal–Once Resolution. A resolution tree $T = (V, A, \lambda)$ is *literal–once* if $\mathsf{rlit}(v) \neq \mathsf{rlit}(w)$ for any two distinct non–leaves v, w of T. We denote by LOR the class of all formulas refutable by literal–once resolution refutations.

Proposition 1 LOR $\subsetneq$ SROR $\subsetneq$ ROR $\subsetneq$ UNSAT.

Proof. If a resolution refutation is literal–once, then it is obviously strict read–once; thus LOR $\subseteq$ SROR. Consider the formula $F = \{\{x, y\}, \{x, \overline{y}\}, \{\overline{x}, y\}, \{\overline{x}, \overline{y}\}\}$. Figure 1 shows a strict read-once resolution refutation T of F, hence $F \in$ SROR. (We note in passing that F belongs to a subclass of minimal unsatisfiable formulas characterized in [9].) However, T is not literal-once. It is easy to see that there is no literal–once resolution refutation of F at all. Whence LOR $\subsetneq$ SROR follows.

We have SROR $\subseteq$ ROR by definition. Consider the formula $F = \{C_1, \ldots, C_5\}$ with

$$\begin{aligned} C_1 &= \{x, \overline{z}\}, & C_4 &= \{x, y, z\}, \\ C_2 &= \{\overline{x}, y\}, & C_5 &= \{\overline{x}, \overline{y}, \overline{z}\}, \\ C_3 &= \{\overline{y}, z\}. \end{aligned}$$

Figure 2 exhibits a resolution refutation of F, hence $F \in$ UNSAT. We show that $F \notin$ ROR. Consider a resolution refutation T of F with root v, and let v_1, v_2 the predecessors of v. Clearly $|\mathsf{con}(T_{v_1})| = |\mathsf{con}(T_{v_2})| = 1$. However, no pair of clauses $C', C'' \in F$ have a resolvent C with $|C| = 1$. Thus $|\mathsf{pre}(T_{v_1})|, |\mathsf{pre}(T_{v_2})| \geq 3$. Since $|F| = 5$ it follows that $\mathsf{pre}(T_{v_1}) \cap \mathsf{pre}(T_{v_2}) \neq \emptyset$. Consequently, T is not read–once. Hence $F \notin \mathrm{ROR}$ and so $\mathrm{ROR} \neq \mathrm{UNSAT}$.

Let $W_1 = \{w, x, \overline{z}\}$, $W_2 = \{\overline{w}, x, \overline{z}\}$, and consider $F^* = F \cup \{W_1, W_2\}$. Observe that C_1 is the resolvent of W_1 and W_2. The resolution tree exhibited in Figure 3 shows that $F^* \in \mathrm{ROR}$. Consider a read–once resolution refutation T of F^*. We show that T is not strict read–once. Again, let v_1, v_2 be the predecessors of the root of T. W.l.o.g., we assume $|\mathsf{pre}(T_{v_1})| \leq |\mathsf{pre}(T_{v_2})|$. Similarly as above, $|\mathsf{pre}(T_{v_1})|, |\mathsf{pre}(T_{v_2})| \geq 3$ follows. Since T is assumed to be read–once, $|\mathsf{pre}(T_{v_1})| + |\mathsf{pre}(T_{v_2})| \leq |F^*|$; thus $|\mathsf{pre}(T_{v_1})| = 3$. It can be verified that there is no resolution tree T' with $\mathsf{pre}(T') \subseteq F^*$, $|\mathsf{pre}(T')| = 3$ and $|\mathsf{con}(T')| = 1$, such that $W_1 \in \mathsf{pre}(T')$ or $W_2 \in \mathsf{pre}(T')$. However, $W_1, W_2 \in \mathsf{pre}(T)$ since $F \notin \mathrm{ROR}$. It follows that $W_1, W_2 \in \mathsf{pre}(T_{v_2})$ and $|\mathsf{pre}(T_{v_2})| = 4$. Hence we have $\mathsf{pre}(T_{v_2}) = \{W_1, W_2, D_1, D_2\}$ for some $D_1, D_2 \in \{C_2, \ldots, C_5\}$. Checking all possibilities for D_1, D_2 shows that either $\{D_1, D_2\} = \{C_2, C_4\}$ or $\{D_1, D_2\} = \{C_3, C_5\}$. In both cases, the two vertices u_1, u_2 of T_{v_2} which are labeled by W_1 and W_2, respectively, have a common successor u. Evidently u is labeled by C_1. Since $C_1 \in \mathsf{pre}(T)$, it follows that T is not strict read–once. Whence $\mathrm{SROR} \neq \mathrm{ROR}$. □

4 NP-Completeness Results

Let F be a formula with m clauses and $T = (V, A, \lambda)$ a read–once (strict read–once, literal–once, respectively) resolution refutation of F. Clearly T has at most m leaves, and so $|V| \leq 2m - 1$. Thus one can guess such resolution refutation T of F and verify in deterministic polynomial time whether T is indeed read–once (strict read–once, literal–once, respectively). Hence the following holds.

Lemma 1 *The recognition problems for LOR, SROR, and ROR are in NP.*

Next we state our main result whose proof we present in Section 6.

Theorem 1 *Recognition of LOR is NP-complete.*

We are going to show that recognition of ROR and recognition of SROR are both NP-complete problems as well. We proceed by reducing recognition of LOR to recognition of SROR and ROR, respectively. For these reductions, the following construction is crucial.

Let F be a formula. For each $x \in \mathsf{var}(F)$ we take two new variables $x[1]$, $x[2]$, and for every clause $C \in F$ we define

$$C^{\circ} = \{ \ \overline{x[1]} \mid \overline{x} \in C \ \} \cup \{ \ \overline{x[2]} \mid x \in C \ \}.$$

We put

$$F^{\circ} = \{ \ C^{\circ} \mid C \in F \ \} \cup \{ \ \{x[1], x[2]\} \mid x \in \mathsf{var}(F) \ \}.$$

Observe that $F°$ is satisfiable if and only if F is satisfiable; furthermore, for every $x[i] \in \mathsf{var}(F°)$ there is exactly one clause $C \in F°$ with $x[i] \in C$.

The following result is a direct consequence of Lemmas 4, 5, and 7, which are more technical and will be presented in the Appendix.

Proposition 2 *For every formula F the following statements are equivalent.*

$$F \in \mathrm{LOR}; \quad F° \in \mathrm{ROR}; \quad F° \in \mathrm{SROR}.$$

The next two results follow from Theorem 1 and Proposition 2.

Theorem 2 *Recognition of* SROR *is* NP-*complete.*

Theorem 3 (Iwama and Miyano [8]) *Recognition of* ROR *is* NP-*complete.*

5   Literal–Once Resolution and Minimal Unsatisfiable Formulas

In this section we apply Theorem 1 to answer a question posed by Kullmann ([11]). A formula F is *minimal unsatisfiable* if F is unsatisfiable but $F \setminus \{C\}$ is satisfiable for every $C \in F$. The *deficiency* $\delta(F)$ of a formula F is defined by

$$\delta(F) = |F| - |\mathsf{var}(F)|\,.$$

Let k be an integer; we write $\mathrm{MU}(k)$ for the class of minimal unsatisfiable formulas F with $\delta(F) = k$. By a result due to Tarsi ([1]), $\mathrm{MU}(k) = \emptyset$ for $k \leq 0$. Recognition of minimal unsatisfiable formulas is D^P-complete ([12]); however, for every fixed k, the class $\mathrm{MU}(k)$ can be recognized in polynomial time ([11,5]). In [11], Kullmann asked whether recognizing

$$\mathcal{C} = \{\, F \mid \text{there is some } F' \subseteq F \text{ with } F' \in \mathrm{MU}(1) \,\}$$

is NP-complete. We answer this question positively: in the next lemma we show $\mathcal{C} = \mathrm{LOR}$; hence NP-completeness of $\mathcal{C}$ follows from Theorem 1.

Proposition 3 *Let F be a formula. Then $F \in \mathrm{MU}(1)$ if and only if there is a literal–once resolution refutation T with $\mathsf{pre}(T) = F$. Consequently $\mathrm{LOR} = \mathcal{C}$.*

Proof. We apply the following results from [4].

(i) If $F \in \mathrm{MU}(1)$ and $F \neq \square$ then there is a literal ℓ and clauses $C_1, C_2 \in F$ such that C_1 is the only clause of F containing ℓ; C_2 is the only clause of F containing $\bar{\ell}$.

(ii) Let F be a formula and ℓ a literal such that there are unique clauses $C_1, C_2 \in F$ with $\ell \in C_1$ and $\bar{\ell} \in C_2$; let $C_{1,2}$ be the resolvent of C_1 and C_2. Then $F \in \mathrm{MU}(1)$ if and only if $(F \setminus \{C_1, C_2\}) \cup \{C_{1,2}\} \in \mathrm{MU}(1)$.

We proceed by induction on $|F|$. The proposition evidently holds if $|F| = 1$; hence consider $|F| > 1$. Assume $F \in \text{MU}(1)$ and choose ℓ, C_1, and C_2 according to (i). It follows now from (ii) that $F^* = (F \setminus \{C_1, C_2\}) \cup \{C_{1,2}\} \in \text{MU}(1)$. By induction hypothesis, there is a literal–once resolution refutation T^* with $C_{1,2} \in \text{pre}(T^*) = F^*$. We extend T^* to a a literal–once resolution refutation T with $\text{pre}(T) = F$ by adding leaves v_1, v_2 (labeled by C_1 and C_2, respectively) to T^*.

Conversely, assume that there is a literal–once resolution refutation $T = (V, A, \lambda)$ with $\text{pre}(T) = F$. Choose two leaves v_1, v_2 of T which have a common successor v. Put $C_i = \lambda(v_i)$, $i = 1, 2$ and $C_{1,2} = \lambda(v)$. Consequently, there is a literal ℓ such that $\ell \in C_1$ and $\overline{\ell} \in C_2$. Hence removing v_1 and v_2 from T yields a literal–once resolution refutation T^* with $\text{pre}(T^*) = (F \setminus \{C_1, C_2\}) \cup \{C_{1,2}\}$; $\text{pre}(T^*) \in \text{MU}(1)$ by induction hypothesis. It follows now from (ii) that $F \in \text{MU}(1)$. $\qquad\square$

In [4] it is shown that every minimal unsatisfiable Horn formula belongs to $\text{MU}(1)$. Since every unsatisfiable Horn formula contains a minimal unsatisfiable Horn formula, Proposition 3 implies the following.

Proposition 4 *Every unsatisfiable Horn formula is refutable by literal–once resolution.*

6   Proof of Theorem 1

This section is devoted to a proof of Theorem 1. We reduce 3-SAT to recognition of LOR (in fact we could reduce SAT as well, but we choose 3-SAT to keep notation simpler). In a first step we reduce 3-SAT to the problem of finding a "satisfying path" in a digraph D, i.e., a path which does not run through prescribed pairs of vertices. In a second step we mimic this path problem by constructing a formula F such that literal–once resolution refutations of F and satisfying paths of D correspond to each other.

First we prove two short lemmas which we will need below.

Lemma 2 *Let T be a literal–once resolution tree and $C_1, C_2 \in \text{pre}(T)$ with $\ell \in C_1$ and $\overline{\ell} \in C_2$ such that $\text{rlit}(v) = \{\ell, \overline{\ell}\}$ for the root of T. Then $C_1 \cap C_2 \subseteq \text{con}(T)$.*

Proof. Let v be the root of T and v_1, v_2 the predecessors of v. Consider $\ell' \in C_1 \cap C_2$. Since T is literal–once, it follows that ℓ' cannot be an element of both $\text{rlit}(T_{v_1})$ and $\text{rlit}(T_{v_2})$. Hence $\ell' \in \lambda(v) = \text{con}(T)$. $\qquad\square$

Lemma 3 *Let $T = (V, A, \lambda)$ be a literal–once resolution refutation and $C_1, C_2 \in \text{pre}(T)$. Then there cannot be distinct literals $\ell, \ell' \in C_1$ such that $\overline{\ell}, \overline{\ell'} \in C_2$.*

Proof. We observe that there are vertices $v, v_1, v_2 \in V$ such that v_1, v_2 are predecessors of v and $\ell \in \text{rlit}(v)$. W.l.o.g., assume $\ell \in \lambda(v_1)$ and $\overline{\ell} \in \lambda(v_2)$. It follows that $C_1 \in \text{pre}(T_{v_1})$ and $C_2 \in \text{pre}(T_{v_2})$. Since $\text{rlit}(T_{v_1}) \cap \text{rlit}(T_{v_2}) = \emptyset$, ℓ is the only literal with $\ell \in C_1$ and $\overline{\ell} \in C_2$. $\qquad\square$

Construction I. Let $F_3 = \{C_1, \ldots, C_n\}$ be a formula with $C_i = \{\ell_{i,1}, \ell_{i,2}, \ell_{i,3}\}$ for $1 \le i \le n$. We write L for the set of literals ℓ such that $\mathrm{var}(\ell) \in \mathrm{var}(F_3)$. Further, for $\ell \in L$ we put

$$q(\ell) = \{\, i \mid \ell \in C_i,\ 1 \le i \le n \,\}.$$

Observe that $i \notin q(\bar\ell)$ for every $\ell \in C_i$, $1 \le i \le n$, since clauses do not contain complementary pairs of literals. We assume w.l.o.g. that F_3 has no pure literals; i.e., $|q(\ell)| \ge 1$ for every $\ell \in L$.

We construct a digraph $D = (V, A)$ as follows. We take a set of $n + 1$ vertices $\{u_0, \ldots, u_n\}$, and for $i = 1, \ldots, n$ we join u_{i-1} and u_i by three (directed) paths $P_{i,1}$, $P_{i,2}$, $P_{i,3}$ of length $|q(\bar\ell_{i,1})| + 1$, $|q(\bar\ell_{i,2})| + 1$, $|q(\bar\ell_{i,3})| + 1$, respectively. We denote the set of inner vertices of $P_{i,j}$ by $V_{i,j}$ ($1 \le i \le n$, $1 \le j \le 3$). Hence we have $|V_{i,j}| = |q(\bar\ell_{i,j})|$ for $0 \le i \le n$, $1 \le j \le 3$. Now we form a set S of pairs (v, v') of vertices $v, v' \in V \setminus \{u_0, \ldots, u_n\}$ such that

- there is a pair $(v, v') \in S$ with $v \in V_{i,j}$ and $v' \in V_{i',j'}$ ($1 \le i < i' \le n$, $1 \le j, j' \le 3$) if and only if $\ell_{i,j} = \bar\ell_{i',j'}$, and
- every vertex in $V \setminus \{u_0, \ldots, u_n\}$ is contained in exactly one pair of S.

Note that such set S exists and can be obtained efficiently. We call a directed path in D *satisfying* if it runs from u_0 to u_n and contains at most one vertex of each pair in S. Observe that each satisfying path has to pass through all of the vertices $u_0, \ldots, u_n$ in increasing order.

Claim 1 F_3 *is satisfiable if and only if D has a satisfying path.*

Proof. If F_3 is satisfied by some truth assignment t, then we can choose $\sigma(i) \in \{1, 2, 3\}$ for $0 \le i \le n$ such that $t(\ell_{i,\sigma(i)}) = 1$. We observe that

$$P = P_{0,\sigma(0)} \cdots P_{n,\sigma(n)} \tag{1}$$

is a satisfying path. Conversely, by definition, every satisfying path P is of the form (1) for some $\sigma : \{0, \ldots, n\} \to \{1, 2, 3\}$. Thus, if P is a satisfying path, then putting $t(\ell_{i,\sigma(i)}) = 1$ for $0 \le i \le n$ induces a truth assignment t which satisfies F_3. $\square$

Note that the above construction is closely related to the *connection method* (see, e.g., [13,3,10]).

Construction II. Let $D = (V, A)$ be the digraph obtained from a given 3-CNF formula F_3 according to Construction I. We consider a portion of distinct boolean variables: for $0 \le i \le n$ we take a new variable ν_i; for each arc $a \in A$ we take a new variable α_a; for each pair $p \in S$ we take three distinct new variables $\beta_p, \gamma_p, \delta_p$. We define a formula F with

$$\mathrm{var}(F) = \{\nu_0, \ldots, \nu_n\} \cup \{\, \alpha_a \mid a \in A \,\} \cup \{\, \beta_p, \gamma_p, \delta_p \mid p \in S \,\}$$

by

$$F = \{\{\nu_0\}, \{\overline{\nu_n}\}\} \cup \bigcup_{v \in V} F(v)$$

and the following definitions (recall that in(v) and out(v) denote the sets of arcs incoming to and outgoing from v, respectively). For $0 \le i \le n$ let

$$F(u_i) = \{\ \{\overline{\alpha_a}, \nu_i\} \mid a \in \text{in}(u_i)\ \} \cup \\ \{\ \{\alpha_b, \overline{\nu_i}\} \mid b \in \text{out}(u_i)\ \}. \tag{2}$$

For $p = (v, v') \in S$ with

$$\text{in}(v) = \{a\}, \quad \text{out}(v) = \{b\}, \quad \text{in}(v') = \{a'\}, \quad \text{out}(v') = \{b'\} \tag{3}$$

we put

$$\begin{array}{llll} F(v) = \{\{\overline{\alpha_a}, \beta_p, \gamma_p\}, & \text{and} & F(v') = \{\{\overline{\alpha_{a'}}, \beta_p, \overline{\gamma_p}\}, \\ \quad \{\overline{\beta_p}, \gamma_p\}, & & \quad \{\overline{\beta_p}, \overline{\gamma_p}\}, \\ \quad \{\overline{\gamma_p}, \delta_p\}, & & \quad \{\gamma_p, \delta_p\}, \\ \quad \{\alpha_b, \overline{\gamma_p}, \delta_p\}\} & & \quad \{\alpha_{b'}, \gamma_p, \delta_p\}\}, \end{array}$$

and write $F(p) = F(v) \cup F(v')$.

Claim 2 *Let $T = (V, A, \lambda)$ be a literal–once resolution refutation of F and $p = (v, v') \in S$. If $F(p) \cap \text{pre}(T) \ne \emptyset$ then either $F(p) \cap \text{pre}(T) = F(v)$ or $F(p) \cap \text{pre}(T) = F(v')$.*

Proof. Let $a, a', b, b' \in A$ according to (3). We use the shorthands

$$\begin{array}{ll} C_1 = \{\overline{\alpha_a}, \beta_p, \gamma_p\}, & C_1' = \{\overline{\alpha_{a'}}, \beta_b, \overline{\gamma_p}\}, \\ C_2 = \{\overline{\beta_p}, \gamma_p\}, & C_2' = \{\overline{\beta_p}, \overline{\gamma_p}\}, \\ C_3 = \{\delta_p, \overline{\gamma_p}\}, & C_3' = \{\delta_p, \gamma_p\}, \\ C_4 = \{\alpha_b, \overline{\gamma_p}, \delta_p\}, & C_4' = \{\alpha_{b'}, \gamma_p, \delta_p\} \end{array}$$

so that $F(v) = \{C_1, \ldots, C_4\}$ and $F(v') = \{C_1', \ldots, C_4'\}$. First we show

$$\{C_1, C_1'\} \not\subseteq \text{pre}(T). \tag{4}$$

Suppose to the contrary that $\{C_1, C_1'\} \subseteq \text{pre}(T)$. Consequently, there is some $v \in V$ such that $\gamma_p \in \text{rlit}(v)$. Thus $C_1, C_1' \in \text{pre}(T_v)$. By Lemma 2 it follows that $\beta_p \in \lambda(v)$. Hence there must be a clause $C \in \text{pre}(T) \setminus \text{pre}(T_v)$ with $\overline{\beta_p} \in C$. By construction of F, C_2 and C_2' are the only clauses of F which contain $\overline{\beta_p}$. Observe that $\gamma_p \in C_2$ and $\overline{\gamma_p} \in C_2'$. Thus $\gamma_p \in \text{rlit}(T)$, since $\text{con}(T) = \square$. It follows that $\gamma_p \in \text{rlit}(T) \setminus \text{rlit}(T_v)$. However, $\gamma_p \in \text{rlit}(T_v)$, and therefore we have a contradiction to the assumption T being literal–once. Whence (4) holds. By analogous arguments one can show

$$\begin{array}{l} \{C_4, C_4'\} \not\subseteq \text{pre}(T), \\ \{C_2, C_2'\} \not\subseteq \text{pre}(T), \\ \{C_3, C_3'\} \not\subseteq \text{pre}(T). \end{array} \tag{5}$$

We show that

$$C_1 \in \mathsf{pre}(T) \quad \Leftrightarrow \quad C_2 \in \mathsf{pre}(T). \tag{6}$$

Assume $C_1 \in \mathsf{pre}(T)$. Since $\beta_p \in C_1$, there must be a clause $C \in \mathsf{pre}(T)$ with $\overline{\beta_p} \in C$; C_2 and C_2' are the only clauses of F which contain $\overline{\beta_p}$. By Lemma 3 we conclude that $C_2' \notin \mathsf{pre}(T)$; thus $C_2 \in \mathsf{pre}(T)$. Whence we have shown one direction of (6). The converse can be shown similarly applying Lemma 3. Moreover, one can show by analogous arguments that

$$\begin{aligned}
C_1' \in \mathsf{pre}(T) \quad &\Leftrightarrow \quad C_2' \in \mathsf{pre}(T), \\
C_3 \in \mathsf{pre}(T) \quad &\Leftrightarrow \quad C_4 \in \mathsf{pre}(T), \\
C_3' \in \mathsf{pre}(T) \quad &\Leftrightarrow \quad C_4' \in \mathsf{pre}(T).
\end{aligned} \tag{7}$$

Finally we observe that

$$\mathsf{pre}(T) \cap \{C_1, C_2, C_3', C_4'\} \neq \emptyset \quad \Leftrightarrow \quad \mathsf{pre}(T) \cap \{C_1', C_2', C_3, C_4\} \neq \emptyset. \tag{8}$$

Claim 2 now follows from (4)–(8). □

Claim 3 *D has a satisfying path if and only if $F \in \mathrm{LOR}$.*

Proof. Assume that D has a satisfying path P. We denote by $V(P)$ and $A(P)$ the vertices and arcs of P, respectively. For $0 \le i \le n$ we put

$$\begin{aligned}
F_P(u_i) = \{\, &\{\overline{\alpha_a}, \nu_i\} \mid a \in \mathsf{in}(u_i) \cap A(P) \,\} \cup \\
\{\, &\{\alpha_b, \overline{\nu_i}\} \mid b \in \mathsf{out}(u_i) \cap A(P) \,\}
\end{aligned}$$

and for $v \in V(P) \setminus \{u_0, \ldots, u_n\}$ we put $F_P(v) = F(v)$. We show that

$$F(P) = \{\{\nu_0\}, \{\overline{\nu_n}\}\} \cup \bigcup_{v \in V(P)} F_P(v)$$

can be refuted by literal–once resolution (observe that $F(P) \subseteq F$). Consider a vertex $v \in V(P)$ with $p = (v, v') \in S$. Using the same notation as in the proof of Claim 2, we have $F(v) = \{C_1, C_2, C_3, C_4\} \subseteq F(P)$. Now $C_{1,2} = \{\overline{\alpha_a}, \gamma_p\}$ is a resolvent of C_1 and C_2; $C_{3,4} = \{\alpha_b, \overline{\gamma_p}\}$ is a resolvent of C_3 and C_4. Further, $C_v = \{\overline{\alpha_a}, \alpha_b\}$ is a resolvent of $C_{1,2}$ and $C_{3,4}$. Hence finding a literal–once resolution refutation of $F(P)$ reduces to finding a literal–once resolution refutation of $(F(P) \setminus F(v)) \cup \{\{\overline{\alpha_a}, \alpha_b\}\}$. Similarly, if $v' \in V(P)$ with $p = (v, v') \in S$, then it suffices to find a literal–once resolution refutation of $(F(P) \setminus F(v')) \cup \{\{\overline{\alpha_{a'}}, \alpha_{b'}\}\}$. By multiple applications of this argument, $F(P)$ can be reduced to a formula of the form

$$F_{\mathrm{lin}} = \{\{\ell_1\}, \{\overline{\ell_1}, \ell_2\}, \{\overline{\ell_2}, \ell_3\}, \ldots, \{\overline{\ell_{r-1}}, \ell_r\}, \{\overline{\ell_r}\}\}.$$

It is easy to construct a literal–once resolution refutation T_{lin} for F_{lin}. Now T_{lin} can be extended by the above considerations to a literal–once resolution refutation of $F(P)$. Whence $F \in \mathrm{LOR}$ follows.

Conversely, assume that $F \in \mathrm{LOR}$. We show that D has a satisfying path. Let T be a literal–once resolution refutation of F and put $F' = \mathsf{pre}(T)$. Let W

be the set of vertices $w \in W$ such that there is at least one arc $a \in \mathsf{in}(w) \cup \mathsf{out}(w)$ with $\alpha_a \in \mathsf{var}(F')$. Clearly $W \neq \emptyset$. Since F' has no pure literals, it follows that for every $w \in W \setminus \{u_0, u_n\}$ there are arcs $a \in \mathsf{in}(w)$, $b \in \mathsf{out}(w)$ such that $\alpha_a, \alpha_b \in \mathsf{var}(F')$ (if $w = u_i$ for some $1 \le i \le n - 1$ this is obvious; on the other hand, if w belongs to some pair in S, then it follows by Claim 2). Thus, for every $w \in W$, at least one predecessor and at least one successor belongs to W. Consider the subdigraph D_W of D induced by W. Clearly D_W is acyclic, since D is acyclic by construction. Every nonempty acyclic digraph has at least one vertex s without incoming arcs and at least one vertex t without outgoing arcs. For D_W the only possibility is $s = u_0$ and $t = u_n$. We conclude that D_W contains a path from u_0 to u_n. By Claim 2 it follows that for every $(v, v') \in S$ at most one of v, v' belongs to W. Thus P must be a satisfying path necessarily. This completes the proof of the claim. $\square$

In view of Lemma 1, Theorem 1 now follows from Claims 1, 3, and the NP-completeness of the 3-SAT problem.

Appendix: Technical Lemmas

Lemma 4 $F \in \mathrm{LOR}$ *implies* $F^\circ \in \mathrm{LOR}$ *for every formula F.*

Proof. We show by induction on $|V|$ that for every literal–once resolution tree $T = (V, A, \lambda)$ there is a literal–once resolution tree T' with $\mathsf{pre}(T') = \mathsf{pre}(T)^\circ$, $\mathsf{con}(T') = \mathsf{con}(T)^\circ$, and $\mathsf{rlit}(T') = \{\ x[i], \overline{x[i]} \mid x \in \mathsf{var}(\mathsf{rlit}(T)),\ i = 1, 2\ \}$. If $|V| = 1$, then there is nothing to show. Assume $|V| > 1$ and let v be the root of T and x the variable in $\mathsf{rlit}(v)$. Moreover, let v_1, v_2 the predecessors of v such that $\overline{x} \in \lambda(v_1)$ and $x \in \lambda(v_2)$. For $i = 1, 2$ let T_i' be a literal–once resolution tree obtained from T_{v_i} as supplied by the induction hypothesis. Since $\mathsf{rlit}(T_{v_1}) \cap \mathsf{rlit}(T_{v_2}) = \emptyset$, it follows that $\mathsf{rlit}(T_1') \cap \mathsf{rlit}(T_2') = \emptyset$. Now $\overline{x[i]} \in \mathsf{con}(T_i') = \mathsf{con}(T_{v_i})^\circ$. It is obvious how T_1' and T_2' can be assembled to literal–once resolution tree T' with the desired properties by adding two non–leaves and a leaf w with $\lambda(w) = \{x[1], x[2]\}$. $\square$

The following Lemma is due to an observation by Kullmann.

Lemma 5 $F^\circ \in \mathrm{ROR}$ *implies* $F^\circ \in \mathrm{LOR}$ *for every formula F.*

Proof. Observe that for every resolution tree $T = (V, A, \lambda)$ and two distinct vertices $v, v' \in V$ with $\mathsf{rlit}(v) = \mathsf{rlit}(v') = \{x, \overline{x}\}$, there must be at least four distinct leaves $u_1, u_2, u_1', u_2' \in V$ such that $x \in \lambda(u_1) \cap \lambda(u_1')$ and $\overline{x} \in \lambda(u_2) \cap \lambda(u_2')$. (Every vertex $v \in V$ with $\mathsf{rlit}(v) = \{x, \overline{x}\}$ "consumes" at least one leaf u_1 with $x \in \lambda(u_1)$ and one leaf u_2 with $x \in \lambda(u_2)$.) However, for every variable $x[i]$ of F° there is exactly one clause $C \in F^\circ$ such that $x[i] \in C$. Hence every read–once resolution refutation of F° is literal–once. $\square$

Lemma 6 *Let F be a formula and T a resolution refutation with $\mathsf{pre}(T) \subseteq F^\circ$. Then $\mathsf{pre}(T) = F_1^\circ$ for some $F_1 \subseteq F$.*

Proof. Follows from the fact that $\mathsf{pre}(T)$ has no pure literals.   □

Lemma 7 *$F^\circ \in \mathrm{LOR}$ implies $F \in \mathrm{LOR}$ for every formula F.*

Proof. We show by induction on $|F|$ that if T is a literal–once resolution refutation with $\mathsf{pre}(T) = F^\circ$, then there is a literal–once resolution refutation T' with $\mathsf{pre}(T') = F$; the lemma will follow by Lemma 6. If $|F| = 1$, then $F = F^\circ = \{\square\}$, and the result follows by taking $T' = T$. Now assume $|F| > 1$ and let T be a literal–once resolution refutation with $\mathsf{pre}(T) = F^\circ$. We call a vertex v' of T *mistimed* if there is a predecessor v_1 of v' with $\lambda(v_1) = \{x[1], x[2]\}$, $x \in \mathsf{var}(F)$, and a successor v of v' such that $\mathsf{rlit}(v) \cap \{x[1], x[2]\} = \emptyset$. Mistimed vertices can be successively eliminated as follows (roughly speaking, we shift leaves labeled by clauses of the form $\{x[1], x[2]\}$ towards the root). Consider a mistimed vertex v' of T with predecessors v_1 and v_2 such that $\lambda(v_2) = \{x[1], x[2]\}$, $x \in \mathsf{var}(F)$. Let v be the successor of v' such that v' and v'' are the predecessors of v. We remove the arcs (v_1, v') and (v'', v) from T and add instead the arcs (v_1, v) and (v'', v'). Clearly $\lambda(v')$ and $\lambda(v)$ can be modified appropriately such that the result is still a read–once resolution refutation with same set of premises. Hence we can assume, w.l.o.g., that T has no mistimed vertices.

We write L_1 for the set of leaves v of T with $\lambda(v) = C^\circ$ for some $C \in F$, and we write L_2 for the set of leaves of T not in L_1 (i.e., if $v \in L_2$, then $\lambda(v) = \{x[1], x[2]\}$ for some $x \in \mathsf{var}(F)$). Observe that for any two leaves v_1, v_2 of T which have the same successor, either $v_1 \in L_1$ and $v_2 \in L_2$, or vice versa. Therefore, if T is nontrivial, then the height of T (i.e., the length of a longest path in T) is at least 2.

We choose a vertex v of T such that T_v has height 2. Since T has no mistimed vertices by assumption, we conclude that exactly one leaf of T_v is in L_1. Hence v has two predecessors v' and v'' such that v' has two predecessors $v_1 \in L_1$ and $v_2 \in L_2$, and $v'' \in L_1$. Let $Q, R \in F$ and $x \in \mathsf{var}(F)$ such that $\lambda(v_2) = \{x[1], x[2]\}$, $\lambda(v_1) = Q^\circ$, and $\lambda(v'') = R^\circ$. It follows for $\{i, j\} = \{1, 2\}$ that $x[i] \in \mathsf{rlit}(v')$ and $x[j] \in \mathsf{rlit}(v)$. Observe that $x[i] \notin \mathsf{var}(R^\circ)$; otherwise there would be a leaf $v_2^* \neq v_2$ with $\lambda(v_2^*) = \lambda(v_2)$. We conclude that $x[1], x[2] \notin \mathsf{var}(\lambda(v))$. Thus Q and R have a resolvent C with $\lambda(v) = C^\circ$. Let T_0 be the resolution tree obtained from T by removing v_1, v_2, v', v''. We have

$$\mathsf{pre}(T_0) = (\mathsf{pre}(T) \setminus \{\{x[1], x[2]\}, Q^\circ, R^\circ\}) \cup \{C^\circ\}.$$

Clearly T_0 is literal–once, hence the induction hypothesis applies. Thus, there is a literal–once resolution refutation T_0' with $\mathsf{pre}(T_0')^\circ = \mathsf{pre}(T_0)$; in particular, $C \in \mathsf{pre}(T_0')$. Let w be the leaf of T_0' labeled by C. It is now obvious how a literal–once resolution refutation $T' = (V', A', \lambda')$ can be obtained from T_0': we add two vertices w_1, w_2, the arcs (w_1, w), (w_2, w) to T_0', and we put $\lambda'(w_1) = Q$ and $\lambda'(w_2) = R$. Hence the lemma follows.   □

References

1. R. Aharoni and N. Linial. Minimal non-two-colorable hypergraphs and minimal unsatisfiable formulas. *J. Combin. Theory Ser. A*, 43:196–204, 1986.
2. J. Bang-Jensen and G. Gutin. *Digraphs: theory, algorithms, applications.* Springer Monographs in Mathematics. Springer Verlag, 2000.
3. W. Bibel. On matrices with connections. *Journal of the ACM*, 28(4):633–645, 1981.
4. G. Davidov, I. Davydova, and H. Kleine Büning. An efficient algorithm for the minimal unsatisfiability problem for a subclass of CNF. *Annals of Mathematics and Artificial Intelligence*, 23:229–245, 1998.
5. H. Fleischner, O. Kullmann, and S. Szeider. Polynomial-time recognition of minimal unsatisfiable formulas with fixed clause-variable difference. Submitted, 2000.
6. Z. Galil. On the complexity of regular resolution and the Davis and Putnam procedure. *Theoretical Computer Science*, 4:23–46, 1977.
7. A. Haken. The intractability of resolution. *Theoretical Computer Science*, 39:297–308, 1985.
8. K. Iwama and E. Miyano. Intractability of read-once resolution. In *Proceedings of the Tenth Annual Structure in Complexity Theory Conference*, pages 29–36, Minneapolis, Minnesota, 1995. IEEE Computer Society Press.
9. H. Kleine Büning. On subclasses of minimal unsatisfiable formulas. To appear in Discr. Appl. Math., 2001.
10. H. Kleine Büning and T. Lettman. *Propositional logic: deduction and algorithms.* Cambridge University Press, Cambridge, 1999.
11. O. Kullmann. An application of matroid theory to the SAT problem. In *Fifteenth Annual IEEE Conference of Computational Complexity*, pages 116–124, 2000.
12. C. H. Papadimitriou and D. Wolfe. The complexity of facets resolved. *J. Comput. Syst. Sci.*, 37(1):2–13, 1988.
13. D. Prawitz. Advances and problems in mechanical proof procedures. In *Machine Intelligence*, volume 4, pages 59–71. American Elsevier, New York, 1969.
14. J. A. Robinson. A machine-oriented logic based on the resolution principle. *Journal of the ACM*, 12(1):23–41, 1965.

Ordered Resolution vs. Connection Graph Resolution

Reiner Hähnle[1], Neil V. Murray[2], and Erik Rosenthal[3]

[1] Chalmers University of Technology, Dept. of Computing Science, S-41296
Gothenburg, Sweden, reiner@cs.chalmers.se
[2] Department of Computer Science, State University of New York, Albany,
NY 12222, USA, nvm@cs.albany.edu
[3] Department of Mathematics, University of New Haven, West Haven, CT 06516,
USA, brodsky@charger.newhaven.edu

Abstract. Connection graph resolution (cg-resolution) was introduced
by Kowalski as a means of restricting the search space of resolution. Several researchers expected unrestricted connection graph (cg) resolution
to be strongly complete until Eisinger proved that it was not. In this paper, ordered resolution is shown to be a special case of cg-resolution, and
that relationship is used to prove that ordered cg-resolution is strongly
complete. On the other hand, ordered resolution provides little insight
about completeness of first order cg-resolution and little about the establishment of strong completeness from completeness. A first order version
of Eisinger's cyclic example is presented, illustrating the difficulties with
first order cg resolution. But resolution with selection functions does yield
a simple proof of strong cg-completeness for the unit-refutable class.

1 Introduction

Connection graph resolution (cg-resolution) was introduced by Kowalski
in 1975 [10] as a means of restricting the search space of resolution. Proving completeness turned out to be non-trivial, the first proof appearing in Bibel's sem-
inal paper [3] in 1981. Several researchers expected cg-resolution to be *strongly
complete* — any sequence of cg resolution steps applied to a propositional formula would eventually terminate — because it is a *destructive* calculus: Links
are deleted upon activation. More than one author[1] tried to prove strong completeness of cg-resolution until Eisinger's famous example [7] dispelled that idea.
He showed that even a quite restrictive notion of fairness cannot prevent cyclic
derivations.

An example of a strongly complete calculus is *path dissolution* [12], which
operates by deleting all paths (and thus the link itself) through a given link,
decreasing the number of paths in the formula. Since there are finitely many
paths, the process terminates in a linkless formula. That formula is empty if
the original formula was unsatisfiable; otherwise the remaining paths are models of the original formula. What is interesting for the discussion here is that

---

[1] The intersection of that club with the authors of this paper is non-empty.

R. Goré, A. Leitsch, and T. Nipkow (Eds.): IJCAR 2001, LNAI 2083, pp. 182–194, 2001.

dissolution has a metric — the number of paths — that is reduced each time a link is activated. Such behavior guarantees termination regardless of the order in which links are selected. Any destructive calculus might exhibit this behavior, i.e., might reduce some metric, guaranteeing termination. This is what many researchers hoped would be true for cg-resolution. There is no obvious metric with that property for cg-resolution: While links are deleted, the number of links does not in general decrease because additional links are inherited. Nonetheless, there might have been some metric that did decrease — perhaps the number of links would eventually decrease monotonically. Of course, Eisinger's work dispelled that notion as well.

It should be noted that *any* complete calculus together with breadth first search is strongly complete. To be interesting, strong completeness should arise from some intrinsic property of the calculus itself. Moreover, it should not require a search that is so broad as to virtually guarantee long derivations. For example, Bibel's original proof of cg-resolution completeness activated every link associated with one atom, and then, when every occurrence of that atom was unlinked (due to link deletions), links to the next atom are activated. Adding that link selection strategy to cg-resolution does make it strongly complete, but not in a very interesting way because it forces a large search space.

In [3] Bibel ordered literals (and, in effect, links) and used the ordering to prove that cg-resolution is complete. The key to his proof is that selecting and activating links based on the ordering (along with judicial use of tautology deletion) preserves spanning if the initial connection graph has the full set of links. It is the initial full set of links, and in turn tautology deletion, that enables not only the completeness result but the observation that the technique is in fact strongly complete, albeit not in a very interesting way, as noted above.

Since the desired result — that unrestricted cg-resolution is strongly complete — simply is not true, one might ask the question, is there a restriction on cg-resolution that is strongly complete? One approach is to look for the weakest possible restriction with the goal of obtaining the most general result. A second approach is to build up from Bibel's ordering, which is the approach adopted in this paper. That is, an ordered strategy for link selection that is more general (and more interesting from a strong completeness point of view) than the one Bibel employed is shown to be strongly complete. It turns out that ordered resolution is a special case of cg-resolution, so the strong completeness result presented in Section 2 is not really a new result. But the proofs seem to be substantially simplified, and some of the theorems have apparently never appeared in print.

On the other hand, ordered resolution provides little insight about completeness of first order cg-resolution and little about the establishment of strong completeness from completeness. A first order version of Eisinger's cyclic example is presented, illustrating the difficulties with first order cg-resolution. The problem in each case is that liftable literal orderings are not total on the first order level.

However, cg-resolution restricted to unit-refutable clause sets is strongly complete. There is, a simple proof that employs resolution with selection functions; see section 4.2.

2 Preliminaries

A *multiset* over a set L is a mapping M from L to the non-negative integers; M is finite if $M(l) = 0$ for all but finitely many $l \in L$. Conceptually, a multiset is a collection of any number (including zero) of occurrences of the elements of L. The set L may be treated as a multiset by letting $M(l) = 1$ for all $l \in L$. If $\prec$ is a *strict partial order* (i.e., an irreflexive, transitive relation) on L, then $\prec$ can be extended to multisets over L. If M and M' are distinct multisets over L, then $M' \prec M$ if whenever there is an l in L with $M(l) \prec M'(l)$, there is an l' in L with $l \prec l'$ and $M'(l') \prec M(l')$. In other words, to diminish a multiset, remove one occurrence of an element and replace it with any finite number of occurrences of smaller elements. Well-foundedness and totality of $\prec$ are inherited by a multiset extension; a good source for this material is [6].

Let Σ be a countable propositional signature (i.e., atom set). A *literal* is either an atom or a symbol of the form $-p$, where p is an atom, and a *clause* is a finite set of literals. Hence, since a clause may be treated as a multiset of literals, any order on literals defines an order on clauses. If l is a literal then $-l$ denotes the literal $-l$ if $l \in \Sigma$, and it denotes p if $l = -p$ for some $p \in \Sigma$. As usual, $\square$ denotes the empty clause.

A *link* in a finite set of clauses S is a set $\{l, -l\}$, where the complementary literals l and $-l$ occur in distinct clauses of S. Links are ordered as sets if literals are ordered, and this ordering can be extended to multisets of links. An occurrence of the literal l in the clause C may be written l_C, and $\{l_C, -l_D\}$ denotes a link, where $l \in C$, $-l \in D$. A link $\{l_C, -l_D\}$ is said to be *ordered* if l_C is the maximum literal in C (with respect to $\prec$), and l_D is the maximum literal in D. All orderings are well-founded because the number of distinct clauses and links is finite.

To define connection graph resolution, which is the focus of this paper, we first define ordinary resolution. Let S be a set of clauses, and let $\{l_C, -l_D\}$ be a link in S. Then the clause $E = (C - \{l_C\}) \cup (D - \{-l_D\})$ is the *resolvent* of S with respect to the link $\{l_C, -l_D\}$. The clause E is said to be an *ordered resolvent* if $\{l_C, -l_D\}$ is an ordered link, and the resolution procedure that requires every resolvent be ordered is called *ordered resolution*. Typically, one additional assumption is made for ordered resolution: For each atom p, $-p$ is the immediate successor of p.

The diagram below illustrates resolution.

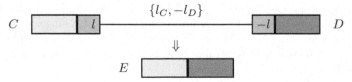

A *resolution derivation* of a clause E from a clause set S is a finite sequence of clauses $D_1, D_2, \ldots, D_n = E$ such that D_i is a resolvent of $S \cup \{D_1, \ldots, D_{i-1}\}$ for all $1 \le i \le n$. Such a derivation may also be viewed as the sequence of clause sets $S, S \cup \{D_1\}, S \cup \{D_1, D_2\}, \ldots$. An *ordered resolution derivation* is one in which each resolution step is ordered.

Connection graph resolution is essentially ordinary resolution with link deletion. As a result, it is necessary to keep track of the links that are present. Thus, a *connection graph* (*c-graph*) is defined to be a pair $(S, \mathcal{L})$, where S is a finite set of clauses, and $\mathcal{L}$ is a set of links in S. As usual, $\square$ denotes any c-graph whose clause set contains the empty clause.[2] The connection graph containing all possible links is called the *full graph* of the clause set S.

The notions of path and spanning are key to cg-resolution. A (*conjunctive*) *path* through a set of clauses S is a set containing exactly one literal from each clause. A c-graph $G = (S, \mathcal{L})$ is said to be *spanned* by $\mathcal{L}$ if each path through S contains a link from $\mathcal{L}$. Observe that a spanned formula is unsatisfiable, and a formula is unsatisfiable if and only if it is spanned by the full set of links.

To define connection graph resolution, let $G = (S, \mathcal{L})$ be a connection graph, let $L = \{l_C, -l_D\}$ be a link in $\mathcal{L}$, and let E be the clause obtained by resolving on L. Then *connection graph resolution* on the link L produces the connection graph $G' = (S', \mathcal{L}')$, where $S' = S \cup \{E\}$, and

$$\mathcal{L}' = (\mathcal{L} - \{L\}) \cup \{\{l'_E, -l'_F\} \mid l' \ne l \text{ and } (\{l'_C, -l'_F\} \in \mathcal{L} \text{ or } \{l'_D, -l'_F\} \in \mathcal{L})\} .$$

The clause E is called a *cg-resolvent*. If $G = G_0, G_1, \ldots, G_n = H$ is a sequence of c-graphs such that each G_i is produced from G_{i-1} by cg-resolution, then H is said to be obtained from G by a *cg-resolution derivation*. The diagram below illustrates cg-resolution.

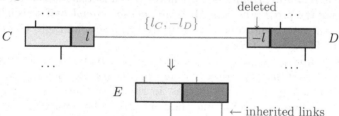

Let us emphasize that what distinguishes cg-resolution from ordinary resolution is, given a c-graph $G = (S, \mathcal{L})$, only the links in $\mathcal{L}$ may be cg-resolved upon. A key property of cg-resolution is that it preserves spanning — see Lemma 1 below.

A literal occurrence l in a c-graph $G = (S, \mathcal{L})$ is called *pure* if l is unlinked; i.e., if no link in $\mathcal{L}$ contains the occurrence l. The clause C is *pure* it it contains a pure literal. A clause that contains complementary literals is called a *tautology*. Pure clauses and tautologies are of interest because of the *pure rule* and the *TAUT lemma*, which delete clauses under some circumstances. The next three lemmas are from Bibel's 1981 paper [3].

---

[2] In the presence of a subsumption rule, this graph can be assumed to be $(\{\square\}, \emptyset)$.

Lemma 1 (*Spanning Lemma*). If a link present in a spanned connection graph is cg-resolved upon, then the resulting connection graph is spanned.   □

Lemma 2 (*Pure Rule*). If G is a spanned connection graph, and if a pure clause together with the associated links are deleted from G, then the resulting connection graph is spanned.   □

Lemma 3 (*TAUT Lemma*). If $G = (S, \mathcal{L})$ is a spanned connection graph, and if $C \in S$ is a tautology containing the literals l and $-l$, and if C together with the associated links are deleted from G, then the resulting connection graph is spanned provided that the following condition is met.

$$\{\{l_D, -l_E\} \mid \{\{l_D, -l_C\}, \{l_C, -l_E\}\} \subseteq \mathcal{L}\} \subseteq \mathcal{L}.$$

     □

Propositional logic is decidable, so one expects a proof procedure to be able to determine both satisfiability and unsatisfiability for a ground clause set. In the following, *affirmation* properties deal with satisfiable clause sets, *refutation* properties with unsatisfiable clause sets. Some additional terminology will be useful: A clause set or c-graph is *saturated* with respect to (standard or ordered or cg-) resolution iff no new resolvent can be generated from it.[3] Formally, a proof procedure is *refutation complete* if whenever S is unsatisfiable, □ can be derived from S using the proof procedure. It is *affirmation complete* if satisfiability of S implies that a saturated clause set (saturated connection graph) can be derived from S (the full graph of S).

Strong completeness adds to completeness the requirement that some *concrete and deterministic* procedure actually refutes (or affirms) a given clause set after a finite number of resolution steps. In resolution theorem proving the task of such a procedure is to determine the next link to be resolved away.

There are several possibilities for what is meant by a "concrete and deterministic procedure." An obvious but trivial way to obtain strong completeness is simply to make a breadth first search for all possible derivations. In that case there is essentially no distinction between strong and standard completeness. Similarly, one can enumerate all possible derivations via backtracking. In practice, one is interested in depth first, backtracking-free proof procedures. Moreover, selecting the link for the next resolution step should be reasonably cheap, typically polynomial in the size of the given clause set or c-graph.

A link selection rule is said to be *local* if it selects links only from the last clause set or c-graph in a derivation. This property seems to be implicitly assumed in much of the resolution literature, and in standard resolution systems

---

[3] With the pure rule, a saturated c-graph must then be $(\emptyset, \emptyset)$.

it often comes cheap — i.e., almost automatically. If, for example, a new clause set is obtained always by adding a resolvent to the previous clause set, then any selection rule is essentially local. Sometimes, however, a selection rule is desired in which the behavior is dependent upon the the derivation history and the point in that history when clauses were introduced. A classic example is the *level saturation strategy* of Wos et. al. [20]. It can easily be implemented if one stamps each clause with its *level*: Clauses in the input set S have level 0; if E is a resolvent of C and D, then $\text{level}(E) = \max\{\text{level}(C), \text{level}(D)\} + 1$. Then a strongly complete, local selection rule is obtained by choosing a link whose resolvent has minimal level and is not already present.

Even when destructive operations such as subsumption deletion are introduced, selection rules have typically been assumed to be local. The consequences for completeness, strong or otherwise, can be subtle. Removal of clause B when it is subsumed by C from a set containing both is obviously unsatisfiability preserving. This was often assumed to be essentially completeness preserving. Sibert [15] was the first to point out that this semantic observation, combined with completeness of resolution, guaranteed only that a refutation was available at any point, *provided that only resolution was employed from that point onward*. These issues were subsequently investigated in great detail — see [11]. Of course, an expensive global selection rule that made use of previous clause sets containing subsumed clauses was simply never contemplated (as far as we know).

A local selection rule is called a *filter* in the connection graph literature (an excellent reference is [7][4]). As with ordinary resolution combined with subsumption, cg-resolution is *destructive*. This means that a deduction step not only derives new information but alters the present state *by removing information about the previous state*. In particular, any clause and link may vanish during a cg-derivation. Not surprisingly, reconciling completeness issues with the destructive nature of the calculus has proved to be considerably more elusive for cg-resolution than for the more standard resolution systems. Other examples of destructive calculi are free variable tableaux [2] and dissolution [12].

Level saturation is an instance of what is often called a *fair* selection rule. In cg-resolution a fair rule guarantees that each link in each derivation vanishes within a finite amount of time (called a *coveringthree filter* by Siekmann and Stephan [16]). It can easily be implemented locally by an implicit or explicit queue and, in non-destructive calculi (for example, standard resolution), implies strong completeness. In destructive calculi, however, it is not obvious that loops cannot occur, and, as Eisinger showed — see Theorem 1 below — this can in fact happen with cg-resolution. Concern for such issues led to consideration of more restrictive fairness conditions. Siekmann and Stephan color each link differently in the initial graph and colors are inherited. Then, in addition to the fairness condition described above, in each derivation state each color must

---

[4] There, locality is not explicitly enforced, but seems to be intended, because a loop check on the derivation history for isomorphic c-graphs would deal with the cyclic counter example.

188   R. Hähnle, N.V. Murray, and E. Rosenthal

be resolved upon within a finite number of steps (this was called a *coveringtwo filter*). Eisinger's example [7] shows that even this version of fairness is not enough.

Theorem 1. For ground cg-resolution there are fair selection rules with the coloring restriction that are not strongly refutation complete. □

Ordered resolution is one refinement of resolution that is strongly complete. Let $\prec$ be a total, well-founded ordering of literals with the property that the complement of a literal is the literal's immediate predecessor or successor. Then the extension of $\prec$ to clauses is well-founded and total, and the resolvent of any ordered resolution step is smaller than each of its parents. By well-foundedness, there are only finitely many smaller clauses than the greatest clause in S. This observation ensures that ordered resolution terminates. In Section 3), termination is combined with completeness to obtain strong completeness. Kowalski and Hayes [9] used a very different approach to show that ordered resolution is complete: a semantic tree argument.

3   Strong Completeness

We begin this section by showing that ordered resolution is a special case of cg-resolution.[5] The key is Lemma 5, which says, in effect, that any total ordering of literals can be used as a guide for link selection with cg-resolution. It is very easy to see that ordered links are present in any spanned connection graph since the path containing the maximal element from each clause must contain a link. The lemma says that *every* ordered link not yet activated must be present.

First, we point out that ground clauses become pure only because of links deleted after activation.

Lemma 4. A ground clause may become pure only from a cg-resolution step activating the sole link to one of its literals.

Proof: Given clauses $C = \{p_C\} \cup A$ and $D = \{\bar{p}_D\} \cup B$, and link $L = \{p_C, \bar{p}_D\}$, consider a cg-resolution step activating L, producing the cg-resolvent $E = A \cup B$. Clauses C and D were not pure prior to activating L. If L is the only link to p_C or to $\bar{p}_D$, then one or both parents may become pure and be deleted. All links to the deleted clause(s) will also be deleted.

To see that not other clause has become pure, consider links to literals in A or in B. All such links are inherited in the cg-resolvent. As a result, any clause linked to a deleted parent has gained exactly one link for each link removed due to deletion of the parent. Now consider links (other than L) to p_C. If there are any such, then C did not become pure, and those links remain. If there are none, then none (other than L) were removed. The same argument applies to links to $\bar{p}_D$. □

---
[5] This observation was made independently by Harald Ganzinger.

Lemma 4 essentially states that "cascading purity" is not possible at the ground level. Deletion of a pure clause and its links happens only when that clause is the parent of a resolvent and cannot cause another clause to become pure.

Lemma 5. Let G be a connection graph with the full set of links, and suppose that the literals of G have a total ordering. Suppose further that $(S, \mathcal{L})$ is obtained by a sequence of ordered cg-resolution steps. Then every ordered link in S must either be in $\mathcal{L}$ or have been directly activated.

Proof. Suppose to the contrary that the ordered link $\{l_E, -l_F\}$ in S is not present in $\mathcal{L}$ and has not been activated. Since the initial graph G was a full graph, this link must have been deleted in an earlier cg-resolution step. Hence at least one of the two clauses E, F is a resolvent, say $E = \{l\} \cup E'$ is the resolvent of C and D. Then l cannot be maximal in its parent, nor can any occurrence of it be maximal in any of its ancestors. Thus, the link $\{l_E, -l_F\}$ can never have been activated and hence could never have been deleted. □

One way to interpret the lemma is that for ordered resolution, link deletions are no more than a convenient means of preventing links from being activated more than once. Specifically, we have

Theorem 2. Let G be any set of clauses with the full set of links. If the literals of G are totally ordered, then, with respect to that ordering, ordered resolution and ordered cg-resolution are identical. □

We will demonstrate the (affirmation and refutation) strong completeness of ordered resolution by proving two theorems. First, as long as no link is activated more than once and tautologies are not used as parent clauses, any sequence of ordered resolutions must eventually terminate, producing a saturated clause set. Secondly, if a formula is unsatisfiable, the saturated clause set contains the empty clause.

We assume the atom set is totally ordered, and that the literal set has been ordered by making $-p$ the immediate successor of p. That ordering (and its extension to links and clauses) will be referred to as the *atom ordering*. Observe that the clause ordering is total and well founded. Observe also that under this ordering, an ordered resolvent precedes both of its parents (as long as neither parent is a tautology). We use *head of a clause* to denote the largest literal in the clause.

Theorem 3. If G is a clause set, and if the literals of G have the atom ordering, then ordered resolution must terminate; that is, any sequence of ordered resolution steps in which no link is activated more than once and tautologies are not used as parent clauses will eventually produce a clause set with the property that every ordered link has been activated once.

Proof. Let G be a set of clauses, and let $\mathcal{A} = \{p_1, p_2, ..., p_n\}$ be the atom set of G, where $p_i \prec p_j$ if $i < j$. No ordered resolvent that does not use a tautology can ever produce a new p_n link, so the number of p_n links is fixed, say there are $N_n p_n$ links. New (ordered) p_{n-1} links can be produced by activations of (some of) those $N_n p_n$ links. Thus, the number of (ordered) p_{n-1} links that can ever be produced is fixed, regardless of the order in which links are activated. Similarly, new p_{n-2} links can be produced only if p_n or p_{n-1} links are activated. Since only finitely many such activations are possible, only finitely many new p_{n-2} links can be produced. Continuing inductively, only finitely many ordered links can ever exist, so only finitely many ordered cg-resolvents are possible. We emphasize that the order in which links are activated has no impact on this analysis, and the proof is complete.   □

The proof method for the next theorem[6] is due to Bachmair and Ganzinger [1]; the proof presented below is a refinement of an elegant exposition by Paliath Narendran. Interpretations in the proof will be described as a set of atoms, indicating that atoms in the set are assigned *true* and atoms not in the set are assigned *false*.

Theorem 4. Let S be an unsatisfiable set of clauses that is saturated with respect to ordered resolution. Then S contains the empty clause.

Proof. Suppose to the contrary that S does not contain the empty clause. Let $S = \{C_1, C_2, ..., C_k\}$ where $C_i \prec C_j$ if $i < j$. For each C_i, iteratively define an interpretation I_i, where $I_0 = \emptyset$, and the interpretation I_i is constructed from I_{i-1} as follows: $I_i = I_{i-1}$ if the maximal literal in C_i is negative or if I_{i-1} satisfies C_i. Otherwise, $I_i = I_{i-1} \cup \{q\}$ where q is the maximal literal in C_i. In that case, C_i is said to *produce* q. Let $I = I_k$.

Now, since S is unsatisfiable, there is a first clause C_i that is falsified by $\tilde{I}$. Then the head of C_i cannot be positive, since if it were and since $\tilde{I}$ falsifies the other literals in C_i, $\tilde{I}$ would assign true to the head. Let $-p$ be the head of C_i. Then p must occur in some earlier clause C_j that produced p. All other literals in C_j are *false* since otherwise C_j would not have produced p. This implies that neither C_j nor C_i is a tautology; in particular, every non-head literal in both clauses precedes p and is different from p and is falsified by $\tilde{I}$. Hence $\{p_{C_j}, -p_{C_i}\}$ is an ordered link. Resolving produces a clause C that is falsified by $\tilde{I}$. Moreover, C precedes both C_j and C_i, contradicting the fact that C_i is the first clause falsified by $\tilde{I}$.   □

Theorem 4 is known but is based on the assumption of saturation. In combination with Theorem 3, it can be seen that for ordered cg-resolution, saturation criteria need never be tested: The supply of ordered links is simply guaranteed to run out, regardless of the order in which they are selected.

---

[6] Ordered resolution is attributed to J. Reynolds [13]. Variations have been proven complete by Kowalski and Hayes [9] and by Joyner [8].

4   First Order Issues

In the first order case, each literal in a resolvent must of course be instantiated with the most general unifier (mgu) of the activated link. One side effect could be that the literals of "inherited links" are no longer unifiable, in which case there is no inherited link. Even when these literals are unifiable, the mgu may be incompatible with other links in the resolvent — see [10]. Such incompatibility enables deletion of the link. The point is that the situation is quite different from that described by Lemma 4: Literals other than those resolved on may become pure, resulting in cascaded pure clauses (see, for example, [7, p. 4]).

Consider the example in Figure 1. The variables are $x, y, z, v,$ and w.

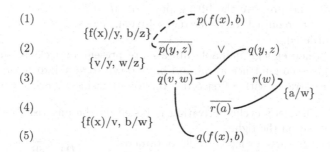

Fig. 1. Cascading purity at the first order level.

Activating the dashed link produces (5), and (2) becomes pure. Removing (2) does not immediately purify (3) because of the inherited link to the resolvent. But the mgu for this link binds w to b, whereas the mgu for the other link to (3) binds w to a; due to this incompatibility, either link may be deleted, rendering (3) pure. Note also that if the negated literal in (3) were $\overline{q(v,a)}$, the link to (2) would not have been inherited at all.

This makes lifting with purity difficult, but as will be seen, lifting is a big problem in cg-resolution even without purity.

First we observe that completeness for standard resolution without self-resolution (both parent clauses are variants of the same clause) is not open as indicated in [7, p. 123]. To see this, it is sufficient to consider P1-resolution[14]. Since one parent is positive and one is negative (later called "mixed") in each resolution step, they cannot be lifted to the same clause. This renders unnecessary some but not all applications of the "copy rule" (in which a new variant of a clause together with suitable link occurrences is generated).

Unfortunately, ordered resolution does not substantially improve Eisinger's copying strategy. The reason is that the total ordering on literals required in Theorems 2 and 3 above, does not lift to the first order level. These results hinge on the fact that there is exactly one maximal literal per clause, which is not the case on the first order level (see the following section). To see that literal orders do not work, the example[7] of [7, p. 125] suffices with the ground atom order, where $Pab > Pba > Paa > Pbb$. Hence, the lifting problems identified by Eisinger apply here as well; the only known remedy is to copy enough clauses and links to ensure that lifting works.

4.1 Strong Completeness

Judicious use of the copy rule can guarantee completeness. Even so, ordered resolution does not help prove first order strong completeness. An ordered resolution proof on the first order level is in general *not* a cg-resolution proof. The culprit here is the same as in the previous section: One cannot totally order first order literals and preserve the lifting property ($l < l'$ implies $l\sigma < l'\sigma$ for all substitutions σ) required for ordered resolution; for example, unifiable literals cannot be ordered.

In fact, when enough unifiable literals are present in a clause set, ordered resolution gives no guidance at all and boils down to standard resolution. This can be demonstrated with a first order version of Eisinger's cyclic example.

Example 1. Eisinger's cyclic derivation is based on the unsatisfiable ground c-graph displayed on the right.

Consider the first order c-graph obtained by replacing each positive literal P with $m(x_P, c_P^+)$, each negative literal $-P$ with $m(x_P, c_P^-)$ and adding the clauses $-m(x_P, c_P^+) \vee$ $- m(x_P, c_P^-)$ for all atoms P. The x_p are variables and the c_p^+, c_p^- are constants. In the resulting c-graph:

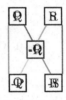

- There is an unsatisfiable clause set,
- all literals in both the graph and all of its resolvents are maximal wrt any literal order,[8]
- and each ground cg-resolution step from Eisinger's example can be simulated by two admissible ordered resolution steps on the first order level involving one of the additional clauses.

Hence no literal order can exclude an "Eisinger-cycle". This destroys all hope for achieving strong completeness through orders.

Just as in the propositional case, the root of the problem is in the destructive nature of the calculus: In standard first order resolution, a clause is unaffected

---

[7] The example is too lengthy to be reproduced here.

[8] Depending on the order, some polarities of literals may have to be adjusted or arguments of predicates exchanged; the present transformation works for orders in which negative literals are larger than positive ones and the first argument has precedence over the second one.

when it becomes the parent of a resolvent, whereas in cg-resolution one of its links is destroyed. This lifts only from ground proofs in which *every* instance of the activated link is deleted. On the other hand, the copy rule can be restricted to clauses that contain a literal which might resolve with more than one other clause.

An alternative to lifting would be to find a termination order that works directly on first order connection graphs. However, this seems more difficult than for the ground case (see Section 3), in part because the factoring rule copies links excessively.

4.2   Resolution with Selection Functions

Because total ground orderings become partial when extended to the first order case, ordered resolution is too weak a restriction for approximating first order cg-resolution. A resolution refinement which does not have this drawback is *resolution with selection functions*. A selection function f selects exactly one literal per clause and must be liftable. This works fine, because the proof of Theorem 2 can be lifted and literal selection is not ambiguous even for first order clauses. There is a drawback, of course: Resolution with arbitrary selection functions is incomplete in general.

On the other hand, resolution with selection functions *is* complete when no factoring is required[9], for example, for Horn or unit refutable clause sets. Strong completeness (without copying) for these classes was already known by [7], but at least the proof via selection functions is much simpler.

Acknowledgement. Bernhard Beckert noticed an error in an earlier version and gave several useful and simplifying suggestions. A discussion with Harald Ganzinger helped clarify several issues.

References

[1] Bachmair, L. and Ganzinger, H. Rewrite-based equational theorem proving with selection and simplification. *Journal of Logic and Computation*, 4(3):217–247, 1994.

[2] Beckert, B. and Hähnle, R. Analytic tableaux. In W. Bibel and P. Schmitt, editors, *Automated Deduction: A Basis for Applications*, volume I, chapter 1, pp. 11–41. Kluwer, 1998.

[3] Bibel, W. On matrices with connections. *JACM*, 28:633–645, 1981.

[4] Bibel, W. and Eder, E. Decomposition of tautologies into regular formulas and strong completeness of connection graph resolution. *Journal of the ACM*, 44(2): 320–344, 1997.

[5] de Nivelle, Hans. Ordering Refinements of Resolution. Ph.D. Thesis, Technische Universiteit Delft.

---

[9] This observation has been made by Hans de Nivelle in his thesis [5] and by Harald Ganzinger.

[6] Dershowitz, N. Termination of rewriting. In J.-P. Jouannaud, editor, *Rewriting Techniques and Applications*, pages 69–115. Academic Press, 1987. Reprinted from *Journal of Symbolic Computation*.

[7] Eisinger, N. *Completeness, confluence, and related properties of clause graph resolution*. Research notes in artificial intelligence. Pitman Publishing, 1991.

[8] Joyner, W.H. Resolution strategies as decision procedures. *J. ACM* **23**,*1* (July 1976), 398-417.

[9] Kowalski, R. and Hayes, P.J. Semantic trees in automatic theorem proving. In *Machine Intelligence*, volume 4, pages 87–101. Edinburgh University Press, 1969. Reprinted in [17].

[10] Kowalski, R. A proof procedure using connection graphs. *Journal of the ACM*, 22(4):572–595, 1975.

[11] Loveland, D.W., *Automated Theorem Proving: A Logical Basis*, North-Holland, New York (1978).

[12] Murray, N.V. and Rosenthal, E. Dissolution: Making paths vanish. *Journal of the ACM*, 40(3):504–535, 1993.

[13] Reynolds, J. Unpublished seminar notes, Stanford University, Palo Alto, CA, 1966.

[14] Robinson, J.A., Automatic deduction with hyper-resolution, *International Journal of Computer Mathematics*, **1** (1965), 227-234.

[15] Sibert, E.E., A machine-oriented logic incorporating the equality relation, *Machine Intelligence 4* (Meltzer and Michie, eds.), Edinburgh University Press, Edinburgh, 1969, 103-133.

[16] Siekmann, J. and Stephan, W. Completeness and soundness of the connection graph procedure. Interner Bericht 7/76, Institut I, Fakultät für Informatik, Universität Karlsruhe, 1976.

[17] Siekmann, J. and Wrightson, G., editors. *Automation of Reasoning: Classical Papers in Computational Logic 1967–1970*, volume 2. Springer-Verlag, 1983.

[18] Siekmann, J. and Wrightson, G., editors. *Automation of Reasoning: Classical Papers in Computational Logic 1957–1966*, volume 1. Springer-Verlag, 1983.

[19] Siekmann, J. and Wrightson, G. Erratum: a counterexample to W. Bibel's and E. Eder's strong completeness results for connection graph resolution. Journal of the ACM, to appear.

[20] Wos, L., Carson, D., and Robinson, G.A. The unit preference strategy in theorem proving. In *Fall Joint Computer Conference, AFIPS, Washington D.C.*, pages 615–621. Spartan Books, 1964. Reprinted in [18].

A Model-Based Completeness Proof of Extended Narrowing and Resolution

Jürgen Stuber

INRIA LORIA, 615 Rue Jardin Botanique, 54600 Villers-les-Nancy, France.
stuber@loria.fr, http://www.loria.fr/~stuber

Abstract. We give a proof of refutational completeness for Extended Narrowing And Resolution (ENAR), a calculus introduced by Dowek, Hardin and Kirchner in the context of Theorem Proving Modulo. ENAR integrates narrowing with respect to a set of rewrite rules on propositions into automated first-order theorem proving by resolution. Our proof allows to impose ordering restrictions on ENAR and provides general redundancy criteria, which are crucial for finding nontrivial proofs. On the other hand, it requires confluence and termination of the rewrite system, and in addition the existence of a well-founded ordering on propositions that is compatible with rewriting, compatible with ground inferences, total on ground clauses, and has some additional technical properties. Such orderings exist for hierarchical definitions of predicates. As an example we provide such an ordering for a fragment of set theory.

1 Introduction

Dowek, Hardin and Kirchner [6] introduce Theorem Proving Modulo and in that context the calculus Extended Narrowing And Resolution (ENAR). They show completeness of ENAR by transforming proofs in a sequent calculus modulo a congruence on formulas into ENAR proofs with respect to the same congruence represented by a term rewriting system, using cut elimination for the sequent calculus in the process. Dowek and Werner [7] show the cut elimination property for the cases of HOL-$\lambda\sigma$, quantifier-free theories and positive theories.

Here we give an alternate completeness proof based on the reduction-of-counterexamples method developed over recent years [3]. This allows to impose ordering restrictions on the calculus and provides a strong notion of redundancy, which is crucial for solving larger problems. The proof requires a well-founded ordering on propositions with certain properties such as compatibility with the rewrite relation. Such orderings exist for hierarchical definitions of predicates. As an example we define such an ordering for a small fragment of set theory.

From the viewpoint of automated theorem proving it is interesting to study how the technique for proving refutational completeness can be extended to handle skolemization or even quantifiers in formulas. Since logical equivalence is lost by skolemization, we have to adapt the notion of soundness of the calculus accordingly. By capturing the effect of skolemization in the addition of *skolemization axioms*, we can keep logical equivalence for most of the proof.

R. Goré, A. Leitsch, and T. Nipkow (Eds.): IJCAR 2001, LNAI 2083, pp. 195–210, 2001.

Finally, it is interesting to study calculi with built-in theories in order to improve the efficiency of automated theorem provers. It is generally recognized that automated theorem provers have problems proving theorems in theories with permutative axioms like associativity, commutativity, distributivity and the inverse law which are common in algebra, and there have been various approaches to the integration of these axioms into provers [17,11,16,5,2,8,13,14,15]. A similar argument holds for the use of equivalences on the level of logical formulas. State-of-the-art resolution theorem provers such as SPASS do a clause normal form transformation once at the beginning, which destroys in particular the equivalences. With some effort it is possible to reconstruct the equivalences [12] and take advantage of them, but to us it seems more fruitful to work towards using them directly. There has been some work on nonclausal resolution by Bachmair and Ganzinger [1], but this does not cover formulas with quantifiers.

2 Preliminaries

We consider first-order logic without equality with respect to fixed sets $\mathcal{P}$ of predicate symbols and $\mathcal{F}$ of function symbols. We assume that $\mathcal{F}$ contains countably many function symbols of each arity, in order to provide sufficiently many fresh function symbols for skolemization. An *atom* is a formula $p(t_1, \ldots, t_n)$ where $p \in \mathcal{P}$ and $t_1, \ldots, t_n$ are terms. Propositions are built from atoms, $\top$ (truth), $\bot$ (falsity), by the junctors $\wedge$, $\vee$, $\neg$, $\rightarrow$ (implication), $\leftrightarrow$ (equivalence), and the quantifiers $\forall$ and $\exists$. We use the double arrow for rewriting: $\Rightarrow$ for a rule or a single step and $\overset{*}{\Rightarrow}$ for the reflexive-transitive closure of $\Rightarrow$. $\overset{!}{\Rightarrow}$ rewrites to normal form, i.e. $s \overset{!}{\Rightarrow} t$ if $s \overset{*}{\Rightarrow} t$ and t is irreducible. We write $P|_\pi$ for the subproposition or subterm of a proposition P at the position π, and $P[Q]_\pi$ or $P[t]_\pi$ for the proposition P where we have replaced the subformula or subterm at position π by Q or t, respectively.

A *literal* is either an atom or the negation of an atom, and a *clause* is a disjunction of literals. We will use *constrained clauses* of the form $C\,[\mathcal{C}]$ where C is a clause and $\mathcal{C}$ is a *constraint*. A *syntactic equality constraint* $s \approx t$ is *satisfied* for those ground substitutions σ that unify s and t, i.e. where $s\sigma = t\sigma$. Analogously, for a fixed given ordering $\succ$ on ground terms, σ satisfies an ordering constraint $s \succ t$ if $s\sigma \succ t\sigma$. We will use constraints that are conjunctions of these atomic constraints. The meaning of a constrained clause is the set of ground instances obtained by substitutions that satisfy the constraint.

A (finite) *multiset* M over a set S is a function from S into the natural numbers such that $M(x) > 0$ only for finitely many x in S. For each x in S, $M(x)$ denotes the number of occurrences of x in M. The *multiset extension* $\succ_{mul}$ of a strict partial ordering $\succ$ is the strict partial ordering on multisets over S that is defined by $M \succ_{mul} N$ if and only if $M \neq N$ and for all x in S such that $N(x) > M(x)$ there exists an y in S such that $y \succ x$ and $M(y) > N(y)$. We will use that the multiset extension of a total ordering is total, that the multiset extension preserves well-foundedness, and that the ordering on multisets is dominated by the ordering on their maximal elements.

We consider propositions to be modulo associativity and commutativity (AC) for $\vee$ and $\wedge$. In particular, clauses that differ only in the order of their literals are identical. We write $\{t_1/x_1, \ldots, t_n/x_n\}$ for the substitution that replaces x_i by t_i for $i \in \{1, \ldots, n\}$.

We use the following rules for the transformation to clause normal form:

$$\neg\bot \Rightarrow \top \tag{1}$$
$$\neg\top \Rightarrow \bot \tag{2}$$
$$\bot \wedge P \Rightarrow \bot \tag{3}$$
$$\top \wedge P \Rightarrow P \tag{4}$$
$$\bot \vee P \Rightarrow P \tag{5}$$
$$\top \vee P \Rightarrow \top \tag{6}$$
$$P \leftrightarrow Q \Rightarrow (P \to Q) \wedge (Q \to P) \tag{7}$$
$$P \to Q \Rightarrow \neg P \vee Q \tag{8}$$
$$\neg\neg P \Rightarrow P \tag{9}$$
$$\neg(P \vee Q) \Rightarrow \neg P \wedge \neg Q \tag{10}$$
$$\neg(P \wedge Q) \Rightarrow \neg P \vee \neg Q \tag{11}$$
$$\neg(\forall x\, P) \Rightarrow \exists x \neg P \tag{12}$$
$$\neg(\exists x\, P) \Rightarrow \forall x \neg P \tag{13}$$
$$P \vee (Q_1 \wedge Q_2) \Rightarrow (P \vee Q_1) \wedge (P \vee Q_2) \tag{14}$$
$$\forall x\, P \Rightarrow P\{z/x\} \tag{15}$$
$$\exists x\, P \Rightarrow P\{f(y_1, \ldots, y_n)/x\} \tag{16}$$

where in (15) z is a new variable, and in (16) f is a fresh function symbol and $x, y_1, \ldots, y_n$ are the free variables of P. A *clause normal form* of a proposition P is obtained by exhaustively applying these rules, with the restriction that (1)–(13) must be applied before (15) and (16), in order to apply the quantifier rules only below positive contexts. The clause normal form transformation is nondeterministic by the choice of new variables and fresh function symbols. This is not a problem, as in any context where a clause normal form is needed any one will do, i.e. this is don't-care-nondeterminism. Note that by this definition clauses are not sets, but are formed from the same symbols as logical propositions. In particular, the empty clause is $\bot$. Also, by equivalence modulo AC these clauses behave as multisets, where the same element may occur several times.

If we consider free variables to be universally quantified then (1)–(15) are logical equivalences, while (16) is only an implication from right to left. It becomes an equivalence if we add the implication in the other direction:

$$\forall y_1, \ldots, y_n.(\exists x.P) \to P\{f(y_1, \ldots, y_n)/x\} \tag{17}$$

We call (17) the *skolem axiom* for the *skolem function symbol* f with respect to $\exists x.P$. We call a set of skolem axioms S *fresh* with respect to a set of propositions N if no skolem function symbol of S occurs in N, and there is only one skolem axiom for every skolem function symbol in S. A fresh set of skolem axioms is always obtained when fresh function symbols are used for skolemization.

Lemma 1 *Let N be a set of propositions and S a set of skolem axioms that is fresh with respect to N, and let I be a model of N. Then there exists a model I' of $N \cup S$.*

Proof: Sketch: define the interpretation of skolem functions in I' so that they provide witnesses for the true instances of their corresponding existential formula. Since S is fresh this is possible without changing the truth value of N.

Lemma 1 isolates the argument that the clause normal form transformation preserves satisfiability. By adding skolem axioms to our theory at the beginning, we get logical equivalence for all later steps. This simplifies our arguments below.

3   Rewriting on Propositions

Let R_p be a set of rewrite rules on first-order propositions such that left-hand sides are atomic, let R_t be a set of rewrite rules on terms, and let $R = R_p \cup R_t$. The right-hand side of a rule in R may contain only free variables that also occur in the left-hand side. We write T_{R_p} for the logical meaning of R_p, which is the set of logical equivalences $\{l \leftrightarrow r \mid l \Rightarrow r \in R\}$. The intended meaning of the rules in R_t is equality. However, equality is not directly available to us, since we use first-order logic without built-in equality. As an alternative, we may apply Leibniz' equality to atomic propositions to obtain a set of equivalences that capture the logical meaning of R_t. That is, we let

$$T_{R_t} = \{A[l]_\pi \leftrightarrow A[r]_\pi \mid l \Rightarrow r \in R,\ A \text{ an atom, and } \pi \text{ a position in } A\}.$$

This is adequate, since any model that satisfies T_{R_t} can be factored through the congruence induced by R_t to obtain a model of R_t with respect to first-order logic with equality, while preserving the truth value of propositions. Finally, we let $T_R = T_{R_p} \cup T_{R_t}$. The theory T_R is compatible with the rewrite rules in R in the sense of Dowek, Hardin and Kirchner [6]. It is somewhat smaller than the one given there, as it includes equivalences only for single rewrite steps and relies on the properties of logical equivalence for reflexive-transitive closure and for closure under contexts of logical operators and substitutions.

We assume that R is confluent and terminating modulo AC for $\wedge$ and $\vee$, and write $R(P)$ for the normal form of a proposition P with respect to $\Rightarrow_R$.

4   The Inference System ENAR

An *inference system* is a set of inferences on constrained clauses. Each *inference* has a *main premise* C, zero or more *side premises* $C_1, \ldots, C_n$, and a *conclusion* D, and is written

$$\frac{C \quad C_1 \quad \ldots \quad C_n}{D}.$$

The main premise and the side premises have different roles in the completeness proof and in the resulting notion of redundancy for inferences.

The calculus of Extended Narrowing And Resolution (**ENAR**) consists of the following two rules operating on constrained clauses:

$$\textit{Extended Resolution} \quad \frac{\neg A_1 \vee \ldots \vee \neg A_n \vee C\,[\mathcal{C}_1] \qquad B_1 \vee \ldots \vee B_m \vee D\,[\mathcal{C}_2]}{C \vee D\,[\mathcal{C}_1 \wedge \mathcal{C}_2 \wedge A_1 \approx \ldots \approx A_n \approx B_1 \approx \ldots \approx B_m]}$$

where $A_1 \approx \ldots \approx A_n \approx B_1 \approx \ldots \approx B_m$ is an abbreviation for $A_1 \approx A_2 \wedge \ldots \wedge A_1 \approx A_n \wedge A_1 \approx B_1 \wedge \ldots \wedge A_1 \approx B_m$. The main premise of Extended Resolution is $\neg A_1 \vee \ldots \vee \neg A_n \vee C\,[\mathcal{C}_1]$.

$$\textit{Extended Narrowing} \quad \frac{U\,[\mathcal{C}]}{\mathrm{cl}(U[r]_\pi)\,[\mathcal{C} \wedge (U|_\pi \approx l)]}$$

where $l \Rightarrow r$ is a rule in R and $U|_\pi$ is not a variable.

Here $\mathrm{cl}(P)$ denotes one of the clauses in a clause normal form of P.

This calculus is slightly different from the original one [6], as it doesn't use equality modulo a congruence. On the other hand, we allow the use of rewrite rules on terms. This is somewhat closer to an implementation, as in general the solving of the constraint in the original calculus would also be done by some form of narrowing. Note that by the redundancy criteria which we introduce below and by the use of equality constraints we cover also refined variants of narrowing such as normalized and basic narrowing. To keep the exposition simple we do not cover the case of narrowing modulo associativity and commutativity (AC). The extension of our calculus to AC can be done using well-known techniques [11].

5   The Ordering on Propositions

We say that an ordering has the *multiset property* for $\vee$ if $L \succ L'$ for any literal L' in a clause C implies $L \succ C$. We assume an ordering $\succ$ on propositions that is well-founded, total on ground clauses, that has the multiset property for $\vee$, that satisfies $\neg A \succ A$ for any atom A, that is compatible with rewriting, i.e., $(\Rightarrow_R) \subseteq (\succ)$, and $A \Rightarrow_R P$ implies $A \succ B$ for every ground instance B of an atom in P. The latter implies compatibility with Extended Narrowing, i.e. $C \succ D$ for every ground inference with main premise C and conclusion D. Compatibility with Extended Resolution follows by the multiset property. Note that we have replaced compatibility with rewrite rules and with contexts by the somewhat weaker compatibility with rewriting, i.e. the application of rewrite rules under contexts. We have done this because it is difficult to obtain compatibility with contexts for quantifiers under negative contexts.

The property that A is greater than any ground instance of some B with free variables is not satisfied by typical term orderings, in particular not by simplification orderings. The separation of propositions and terms can be used to avoid this problem by giving predicate symbols precedence over terms. This technique is applicable in particular for hierarchical definitions of predicates by equivalences, e.g. in set theory. We present an example below.

6   Constructing Herbrand Models That Satisfy the Equivalences

We now define a function, called closure$_R$, that maps a Herbrand interpretation H_i for ground atoms that are irreducible by R to a Herbrand interpretation for all ground atoms. The mapping is defined in such a way that the interpretation of irreducible atoms is not changed and T_R becomes true in closure$_R(H_i)$.

Let H_i be a set of ground atoms irreducible by R. We construct a tree from each closed proposition, whose inner nodes are labeled by $\wedge$, $\vee$ and $\neg$ and whose leaves are irreducible ground atoms. $\leftrightarrow$ and $\rightarrow$ are always expanded using rules (7) and (8).

1. The tree for a reducible proposition P (w.r.t. R) is the tree for the normal form of P with respect to R.
2. The tree for an irreducible ground atom A is a leaf labeled A.
3. The tree for an irreducible proposition $P \wedge Q$ is labeled $\wedge$ at the root and has as children the trees for P and Q.
4. The tree for an irreducible proposition $P \vee Q$ is labeled $\vee$ at the root and has as children the trees for P and Q.
5. The tree for an irreducible proposition $\neg P$ is labeled $\neg$ at the root and has the tree for P as the only child.
6. The tree for an irreducible proposition $\forall x.P$ is labeled $\wedge$ at the root and has as children all trees for $P\{t/x\}$ where t is a ground term.
7. The tree for an irreducible proposition $\exists x.P$ is labeled $\vee$ at the root and has as children all trees for $P\{t/x\}$ where t is a ground term.

Lemma 2 *All the branches of the tree are finite.*

Proof: The only possible source of nontermination is the interaction of rewriting and instantiation of quantifiers. One of the properties of $\succ$ is that a rewrite step followed by instantiation decreases all propositions in the ordering. Since $\succ$ is a well-founded ordering this implies termination.   □

Now we label the nodes of the tree with truth values true or false from the bottom up. A leaf A is labeled true if $A \in H_i$, and false otherwise. A node labeled with $\wedge$ is labeled true if all its children are labeled true, and false otherwise. A node labeled with $\vee$ is labeled true if some of its children is labeled true, and false otherwise. A node labeled $\neg$ is labeled true if its child is labeled false and vice-versa. Since all branches are finite all the nodes are labeled. We let A belong to closure$_R(H_i)$ if the root of the tree for A is labeled true.

Lemma 3 *Let T be a tree for a closed proposition P. Then the root of T is labeled true if and only if P is true in* closure$_R(H_i)$.

Proof: By structural induction over the proposition.   □

Lemma 4 $closure_R(H_i) \models T_R$.

Proof: Consider some equivalence $A \leftrightarrow P$ in T_R. The tree for A is identical to the tree for P, because both are the tree for the normal form of A with respect to R, which is unique by confluence and termination of R. Hence their truth value is equal and the equivalence holds.   □

7 Refutational Completeness

To be able to lift narrowing steps on terms to constrained narrowing we use the standard technique that considers only reduced ground instances on the ground level. This technique was originally used to show completeness of basic narrowing [9], and later for constrained or basic first-order calculi [4,10]. Formally, a substitution σ is *reduced* if $x\sigma$ is irreducible with respect to R_t for all variables x. An instance is called *reduced* if it is obtained by a reduced substitution. We write $\mathrm{gnd}(N)$ for the set of ground instances of clauses in N, and $\mathrm{rgnd}_R(N)$ for the subset of reduced ground instances of clauses in N. Note that ground instances have to satisfy the constraint. We also consider reduced ground inferences. Since premises can always be made ground, the restriction to reduced instances of inferences restricts only the instantiation of newly introduced variables in conclusions. We will show that **ENAR** is refutationally complete by showing that **ENAR** has the reduction property for counterexamples, using the approach of Bachmair and Ganzinger [3].

We will define mutually recursive functions I and P that map a set of ground clauses to a set of ground atoms. Here I stands for "interpretation" and P for "produced". Let M be a set of ground clauses. $P(M)$ defines the interpretation of ground atoms that are irreducible with respect to R, and $I(M)$ extends $P(M)$ to all ground atoms, using the function $closure_R$ defined above.

The definition is with respect to the well-ordering $\succ$ on ground clauses, considering the ground clauses in M in turn. For some ground clause C we write $M^{\prec C}$ ($M^{\preceq C}$) for the ground clauses in M that are smaller than C (less or equal to C).

$$P(M) = \bigcup_{C \in M} \Delta(M, C)$$

$$\Delta(M, C) = \begin{cases} \{A\} & \text{if } C \text{ is false in } I(M^{\prec C}), \\ & \quad C = C' \vee A \vee \ldots \vee A \text{ where } A \succ C', \text{ and} \\ & \quad A \text{ is irreducible by } R; \text{ or} \\ \emptyset & \text{otherwise.} \end{cases}$$

$$I(M) = closure_R(P(M))$$

Finally we define the interpretation of a set of (non-ground) clauses N by

$$I(N) = I(\mathrm{rgnd}_R(N)).$$

We write N_C for $\mathrm{rgnd}_R(N)^{\prec C}$. Note that $\Delta(M,C) = \Delta(M^{\prec C}, C) = \Delta(N_C, C)$, since this is the part of M that is used recursively. Since we start the definition of $I(N)$ with the set of reduced ground instances of N, the set $\Delta(\mathrm{rgnd}_R(N), C)$ is the increment that takes us from $P(\mathrm{rgnd}_R(N)^{\prec C})$ to $P(\mathrm{rgnd}_R(N)^{\preceq C})$.

We say that a ground clause C *produces* A if $A \in \Delta(\mathrm{rgnd}_R(N), C)$. We say that a ground clause C is a *counterexample* for $I(N)$ if it is in $\mathrm{rgnd}_R(N)$ and false in $I(N)$. Let C be the least counterexample for $I(N)$ and let $\mathcal{I}$ be a ground inference with main premise C and conclusion D such that $C \succ D$. We say that $\mathcal{I}$ *reduces the counterexample* C (with respect to $I(N)$) if $I(N) \models \neg D$. An inference system Calc has the *reduction property for counterexamples* (with respect to I) if there is a reduced ground instance of an inference in Calc that reduces C with respect to $I(N)$ for any set N of ground clauses such that $I(N)$ has the least counterexample $C \neq \bot$.

Lemma 5 ENAR *has the reduction property for counterexamples.*

Proof: Let N be a set of clauses, let C be the least counterexample in $I(N)$ and suppose $C \neq \bot$. Then C is the reduced ground instance of a clause $\hat{C}$ in N and has the form $C' \vee L \vee \ldots \vee L$ where $L \succ C'$ for some literal L.

(1) Suppose L is reducible by R. We have $L = A$ or $L = \neg A$ and A is reducible by some rule $B \Rightarrow \hat{P}$ in R, where $A = B\sigma$ and $P = \hat{P}\sigma$. We consider the single R-step $A \Rightarrow P$ applied to $C[A]_\pi$, resulting in the formula $C[P]_\pi$ that is false in $I(N)$, since the equivalence $A \leftrightarrow P$ holds in $I(N)$. Any clause normal form of $C[P]_\pi$ implies $C[P]_\pi$, as skolemization is an implication in the reverse direction. Since $C[P]_\pi$ is false in $I(N)$, the clause normal form is also false in $I(N)$, and there is a ground instance of a clause in the CNF that is false in $I(N)$. As C is a reduced ground instance of some clause $\hat{C}$ in N, the position π can not be a variable position of $\hat{C}$. Hence there exists an Extended Narrowing inference

$$\frac{\hat{C}[\hat{A}]_\pi \, [\mathcal{C}]}{\hat{D}_i \, [\mathcal{C} \wedge \hat{A} \approx B]}$$

such that $\hat{D}_1 \wedge \ldots \wedge \hat{D}_n$ is a clause normal form of $\hat{C}[\hat{P}]$ and $\hat{D}_i \, [\mathcal{C} \wedge \hat{A} \approx B]$ has a ground instance D that is false in $I(N)$ for some $i \in \{1, \ldots, n\}$. Since newly introduced variables are unconstrained we may even choose D to be a reduced ground instance of $\hat{D}_i$. Hence the inference above reduces the counterexample C.

(2) Otherwise L is irreducible by R.

(2.1) Suppose L is positive, i.e. $L = A$ for some ground atom A. Since C is false in $I(N)$, A is false in $I(N)$ and in $P(N)$. All other literals are smaller than A, hence their truth value depends only on the truth values of irreducible atoms smaller than A, which are the same in $I(N)$ and in $I(N_C)$. Therefore C is false in $I(N_C)$. Hence C produces A and is true in I_N, a contradiction.

(2.2) Otherwise L is negative, that is, $L = \neg A$ for some atom A. Since L is false, A is true in $I(N)$, and by irreducibility A must be in $P(N)$. This A is produced by some reduced ground instance D of a clause $\hat{D}$ in N with $C \succ D$.

Then $D = D' \vee A \vee \ldots \vee A$, $A \succ D'$, and D' is false in $I(N_D)$. No clause greater than or equal to D can make a literal in D' become true, hence D' is false in $I(N)$. We can resolve C and D:

$$\frac{C' \vee \neg A \vee \ldots \vee \neg A \qquad D' \vee A \vee \ldots \vee A}{C' \vee D'}$$

This is a reduced ground instance of Extended Resolution that reduces C.   □

Lemma 6 *Let N be a set of clauses that is closed under* ENAR. *Then either* $\perp \in N$ *or* $I(N) \models \mathrm{rgnd}_R(N)$.

Proof: Suppose $C \neq \perp$ is the least counterexample in N for $I(N)$. Then by the reduction property there exists a reduced ground inference

$$\frac{C \qquad C_1 \quad \ldots \quad C_n}{D}$$

that is an instance of an inference in ENAR that reduces C. That implies that D is false in $I(N)$, and since N is closed under ENAR the ground clause D is a reduced ground instance of N and hence a smaller counterexample, a contradiction to the minimality of C.   □

A reduced ground instance $C\sigma$ of a constrained clause C is called *redundant* in N (with respect to R) if there exist reduced ground instances $C_1\sigma_1, \ldots, C_k\sigma_k$ of clauses $C_1, \ldots, C_k$ in N such that $C\sigma \succ C_i\sigma_i$ for $i = 1, \ldots, k$, and $T_R \cup \{C_1\sigma_1, \ldots, C_k\sigma_k\} \models C\sigma$.
    A reduced ground instance

$$\frac{C_1\sigma \quad \ldots \quad C_n\sigma}{C\sigma} \qquad \text{of an inference} \qquad \frac{C_1 \quad \ldots \quad C_n}{C}$$

where $C_n\sigma$ is the main premise is called *redundant* in N (with respect to R) if either one of the premises $C_1\sigma, \ldots, C_n\sigma$ is redundant, or if there exist reduced ground instances $D_1\sigma_1, \ldots, D_k\sigma_k$ of N such that $C_n\sigma \succ D_i\sigma_i$ for $i = 1, \ldots, k$ and $T_R \cup \{D_1\sigma_1, \ldots, D_k\sigma_k\} \models C\sigma$. A non-ground clause or inference is redundant if all its reduced ground instances are redundant. The following well-known lemma allows to delete clauses without losing redundancy.

Lemma 7 *Let N be a set of constrained clauses and M the set of redundant clauses in N. If a constrained clause C is redundant in N then it is redundant in $N \setminus M$.*

Proof: Consider the least reduced ground instance of C that violates this property and derive a contradiction.   □

A set N of clauses is called *saturated up to redundancy* (with respect to ENAR) if all inferences in ENAR from premises in N are redundant.

Lemma 8 *Let N be a set of clauses that is saturated up to redundancy with respect to* ENAR. *Then either* $\perp \in N$ *or* $I(N) \models \mathrm{rgnd}_R(N)$.

Proof: Suppose N is saturated up to redundancy, $\perp \notin N$, and $I(N) \not\models \mathrm{rgnd}_R(N)$. Let C be the least counterexample for $I(N)$ among the reduced ground instances of N. Since ENAR has the reduction property for counterexamples, there exists a reduced ground instance of an inference in ENAR that reduces C to some clause D that is also false in $I(N)$. Since N is saturated w.r.t. ENAR, this inference is redundant, hence D follows from T_R and reduced ground instances smaller than C. By minimality of C these are true in $I(N)$, and D must be true in $I(N)$ as well, a contradiction.   □

A *theorem proving derivation* is a sequence of sets of clauses $N_0 \vdash N_1 \vdash \ldots$ such that for all steps $N_i \vdash N_{i+1}$, $i \geq 0$, either (1) $N_{i+1} = N_i \cup \{C\}$ for some constrained clause C such that $T_R \cup N_i \cup S_i \models \{C\}$ where S_i is a set of skolem axioms, or (2) $N_{i+1} = N_i \setminus \{C\}$ for some constrained clause C which is redundant in N_i, and $\bigcup_i S_i$ is fresh with respect to $T_R \cup N_0$. In case (1) we call the step a *deduction step* and in case (2) a *deletion step*. For such a derivation the set of *persistent clauses* N_∞ is defined as $N_\infty = \bigcup_{i \geq 0} \bigcap_{j \geq i} N_j$.

A simplification may be viewed as two derivation steps $\{C\} \cup N \vdash \{C, D\} \cup N \vdash \{D\} \cup N$. That is, a clause $C \in N$ may be simplified to a clause D if $T_R \cup \{C\} \cup N \models D$ and C is redundant in $\{D\} \cup N$.

For example, tautologies are always redundant, and reduction with R is a simplification. A subsumed clause that has at least one literal more than the clause that subsumes it is also redundant by this definition. However, the case where a clause subsumes one of its instances is not covered, as the individual ground instances do not decrease with respect to $\succ$. At the price of some technical complications it is possible to extend the definition to also cover that case.

We will now show that certain properties are preserved when going from N_0 to the limit N_∞ or vice-versa.

Lemma 9 *Let N be a set of constrained clauses, let I be a model of T_R, and let $N_0 \vdash N_1 \vdash \ldots$ be a theorem proving derivation. If all reduced ground instances of N_∞ are true in I then all reduced ground instances of $\bigcup_i N_i$ are true in I.*

Proof: Consider some reduced ground instance C of some clause $\hat{C}$ in $\bigcup_i N_i$. If $\hat{C}$ is in N_∞ then C is true by assumption. Otherwise $\hat{C}$ has been removed by some deletion step $N_i \vdash N_{i+1}$, hence it is redundant in N_i and thus in $\bigcup_i N_i$. Let M be the set of all redundant clause in $\bigcup_i N_i$, then by the above argument $(\bigcup_i N_i) \setminus M \subseteq N_\infty$. Hence C is redundant in N_∞ by Lemma 7, and there exist reduced ground instances of N_∞ that together with T_R imply C and are true in I. We conclude that C is true in I.   □

This implies in particular that all reduced ground instances of N_0 are true in I. We say that a clause is *unconstrained* if its constraint is $\top$. A set of clauses is *unconstrained* if all its clauses are unconstrained.

Lemma 10 *Let N be a set of unconstrained clauses. Then $\mathrm{rgnd}_R(N) \cup T_{R_t} \models \mathrm{gnd}(N)$.*

Proof: Consider some ground instance C of a clause $\hat{C}$ in $\mathrm{gnd}(N)$. Then $\hat{C} = U[\top]$ and $C = U\sigma$. By normalizing σ with respect to R_t we obtain the ground substitution $\tau = \{x\sigma \Downarrow_{R_t} /x \mid x \neq x\sigma\}$. The instance $C' = U\tau$ is reduced and is also a ground instance of $\hat{C}$ that solves the trivial constraint $\top$, hence it is in $\mathrm{rgnd}_R(N)$. All the changes of atoms by the reduction from C to C' are covered by equivalences in T_{R_t}, hence C is a consequence of $\mathrm{rgnd}_R(N) \cup T_{R_t}$. $\square$

Corollary 11 *Let N be a set of unconstrained clauses, and let I be a model of T_R. If I is a model of all reduced ground instances of N. Then I is a model of all ground instances of N.*

Lemma 12 *Let $N_0 \vdash N_1 \vdash \ldots$ be a theorem proving derivation. If $T_R \cup N_0$ is satisfiable then $T_R \cup N_\infty$ is satisfiable.*

Proof: Suppose we are given a model I_0 of $T_R \cup N_0$. Let $S = \bigcup_i S_i$ be the set of skolem axioms used in the derivation. Then by Lemma 1 there exists a model I of $T_R \cup N_0 \cup S$. Furthermore, $T_R \cup N_0 \cup S$ is logically equivalent to $T_R \cup N_i \cup S$ for $i \geq 0$, hence $I \models N_i$ for all $i \geq 0$. We conclude $I \models N_\infty$. $\square$

A theorem proving derivation is called *fair* (with respect to **ENAR**) if all inferences in **ENAR** from clauses in N_∞ are redundant in N_i for some $i \geq 0$. Since the conclusions of ground inferences are always smaller than the main premise, and since they imply themselves, an inference can be made redundant by a deduction step that adds its conclusion. Thus a fair derivation is obtained by considering inferences in a fair way, i.e. not delaying an inference ad infinitum, and adding their conclusion if they are not already known to be redundant by some suitable sufficient criterion.

Lemma 13 *Let $N_0 \vdash N_1 \vdash \ldots$ be a fair theorem proving derivation. Then N_∞ is saturated up to redundancy.*

Proof: By fairness we get that every inference from premises in N_∞ is redundant in $\bigcup_i N_i$, and by Lemma 7 in N_∞. $\square$

Theorem 14 *Let $N_0 \vdash N_1 \vdash \ldots$ be a fair theorem proving derivation such that N_0 is unconstrained. Then N_0 is inconsistent if and only if N_∞ contains the empty clause.*

Proof: Since the derivation is fair, N_∞ is saturated up to redundancy. Thus either $\bot \in N_\infty$ or $I(N_\infty) \models T_R \cup \mathrm{rgnd}_R(N_\infty)$ by Lemma 8. Let S be the set of skolem axioms used in the derivation. If $\bot \in N_\infty$ then $T_R \cup N_0 \cup S$ is inconsistent, and in turn $T_R \cup N_0$ is inconsistent by Lemma 1, since S is fresh. Otherwise $I(N_\infty) \models \mathrm{rgnd}_R(N_\infty)$, $I(N_\infty) \models \mathrm{rgnd}_R(N_0)$ by Lemma 9, and $I(N_\infty) \models \mathrm{gnd}(N_0)$ by Lemma 10, and $T_R \cup N_0$ is consistent. $\square$

That is, **ENAR** is refutationally complete with respect to T_R.

Looking back at the proof, in particular Lemma 5, we see that we have indeed proved refutational completeness of a more restricted inference system that constrains inferences to maximal atoms:

Extended Resolution
$$\frac{\neg A_1 \vee \ldots \vee \neg A_n \vee C\,[\mathcal{C}_1] \qquad B_1 \vee \ldots \vee B_m \vee D\,[\mathcal{C}_2]}{C \vee D\,[\mathcal{C}_1 \wedge \mathcal{C}_2 \wedge A_1 \approx \ldots \approx B_m \wedge A_1 \succ C \wedge A_1 \succ D]}$$

Extended Narrowing
$$\frac{A \vee C\,[\mathcal{C}]}{\mathrm{cl}(A[r]_\pi \vee C)\,[\mathcal{C} \wedge A|_\pi \approx l \wedge A \succ C]}$$

where $l \Rightarrow r$ is a rule in R and $A|_\pi$ is not a variable.

The ordering constraints are interpreted by the given well-ordering $\succ$. To make this useful in practice it is of course necessary to provide a constraint solver. Note however, that it is sound to discard the ordering constraints whenever they are too hard to solve.

8 Example

We consider an example from set theory given by Plaisted and Zhu [12]. Suppose we have the rewrite rules

$$x \approx y \Rightarrow x \subseteq y \wedge y \subseteq x \tag{18}$$
$$x \subseteq y \Rightarrow \forall z(z \in x \to z \in y) \tag{19}$$
$$x \in y \cap z \Rightarrow x \in y \wedge x \in z \tag{20}$$

that describe a fragment of set theory.

We define the ordering on formulas as follows. We start by an ordering on atoms by letting $(s_1 \approx t_1) \succ (s_2 \subseteq t_2) \succ (s_3 \in t_3)$ for all terms $s_1, s_2, s_3, t_1, t_2, t_3$. On terms we assume some simplification ordering that is total on ground terms, for example a lexicographic path ordering. We extend it lexicographically to atoms with the same predicate symbol. Literals are ordered first with respect to their atom and then to their polarity. That is, $A \succ B$ implies $[\neg]A \succ [\neg]B$ and $\neg A \succ A$ for any atoms A and B. This is extended to clauses by the multiset extension of the literal ordering and to propositions in clause normal form by the multiset extension of the clause ordering.

This ordering is total on ground clauses, since the term ordering is total, and it is extended to atoms, literals and clauses so that this property is preserved. By the same argument it is well-founded.

We have to show that the ordering is compatible with the rewrite relation followed by instantiation. We first consider the effect of applying a rewrite rule to a single literal in a clause, and compare the instantiated clause normal forms.

There are six cases, and we easily see that in each case the ordering holds:

$$x \approx y \vee C \succ (x \subseteq y \vee C) \wedge (y \subseteq x \vee C)$$
$$\neg x \approx y \vee C \succ \neg x \subseteq y \vee \neg y \subseteq x \vee C$$
$$x \subseteq y \vee C \succ \neg t \in x \vee t \in y \vee C \quad \text{for any ground term } t$$
$$\neg x \subseteq y \vee C \succ (f(x,y) \in x \vee C) \wedge (\neg f(x,y) \in y \vee C)$$
$$x \in y \cap z \vee C \succ (x \in y \vee C) \wedge (x \in z \vee C)$$
$$\neg x \in y \cap z \vee C \succ \neg x \in y \vee \neg x \in z \vee C$$

The first four are covered by the precedence of the predicate symbols, and the last two by the subterm property of the term ordering. This ordering extends to a context containing additional clauses. An atom in a proposition may lead to several occurrences of the atom in the clause normal form. Rewriting such an atom thus leads to the replacement of several atoms. We may obtain its effect on the clause normal form by chaining together several of the simple replacements. Then by transitivity the ordering is compatible with any rewrite step. Since our ordering includes the transformation to clause normal form, compatibility also holds for Extended Narrowing inferences.

Now suppose that in this theory we want to prove idempotency of intersection, i.e., $\forall x. x \cap x \approx x$. To simplify the presentation, we prove only the direction $\forall x. x \cap x \subseteq x$. We negate and skolemize and obtain $N_0 = \{\neg a \cap a \subseteq a\}$. To illustrate the model construction we also give the candidate models corresponding to the clause sets in the derivation. Since the only clause in N_0 has no positive literal, we get $P_{N_0} = \emptyset$. However, $I(N)$ is not empty, as it contains atoms such as $a \subseteq a$. We check that $C_0 = \neg(a \cap a \subseteq a)$ is false in $I(N)$ by rewriting $A_0 = a \cap a \subseteq a$:

$$a \cap a \subseteq a \Rightarrow (\forall x. x \in a \cap a \rightarrow x \in a)$$
$$\overset{*}{\Rightarrow} \forall x. (x \in a \wedge x \in a) \rightarrow x \in a$$

We easily see that the normal form is a tautology, hence it is true in particular in $I(N_0)$. Then by definition A_0 is true and C_0 is false in $I(N_0)$. Thus C_0 is the least counterexample, as it is the only ground instance of a clause in N_0. This counterexample can be reduced by Extended Narrowing: By rewriting C_0 we get

$$\neg(\forall x. x \in a \cap a \rightarrow x \in a)$$

which we have to transform to clause normal form:

$$\neg(\forall x. x \in a \cap a \rightarrow x \in a) \overset{*}{\Rightarrow} \exists x. (x \in a \cap a \wedge \neg(x \in a))$$
$$\Rightarrow b \in a \cap a \wedge \neg b \in a$$

where b is the skolem constant introduced for x. For each clause in the CNF we have an Extended Narrowing inference that has it as its conclusion:

$$\frac{\neg a \cap a \subseteq a}{b \in a \cap a} \tag{21}$$

and

$$\frac{\neg a \cap a \subseteq a}{\neg b \in a} \tag{22}$$

We pick the first inference as it is false in $I(N_0)$, and let $N_1 = N_0 \cup \{b \in a \cap a\}$. We notice that $C_1 = b \in a \cap a$ is not productive, since it is reducible to $b \in a \wedge b \in a$, hence we get another Extended Narrowing inference

$$\frac{b \in a \cap a}{b \in a}$$

which originates from both clauses of the reduct. We let $N_2 = N_1 \cup \{b \in a\}$. Now $b \in a$ is productive in $I(N_2)$, i.e., $P_{N_2} = \{b \in a\}$, and the least counterexample for $I(N_1)$ is once again C_0. To reduce it we now need the second inference above (22), and let $N_3 = N_2 \cup \{\neg b \in a\}$. The new clause becomes the least counterexample. It is irreducible and can be resolved with $b \in a$:

$$\frac{b \in a \qquad \neg b \in a}{\bot}$$

In this example we have used the model construction as our guide, choosing always the inference that reduces the least counterexample. We have done this to illustrate the model construction. In practice, however, it is usually not feasible to construct the model explicitly.

9 Conclusions and Further Work

We have proven the refutational completeness for Extended Narrowing and Resolution with ordering restrictions, under the proviso that a suitable ordering for the given theory exists. We have avoided the problem that skolemization does not preserve logical equivalence by adding axioms implying that equivalence. It remains to investigate how useful ENAR is in practice, with or without the ordering restrictions. To this end we are currently working on a prototype implementation of ENAR in ELAN[1].

This work may be viewed as a first step towards calculi that integrate clause normal form computation with inferences. In particular for logical equivalences it seems preferable to use them for rewriting, instead of destroying their structure by an initial transformation to clause normal form. ENAR is limited by the fact that the rewrite system is fixed. It will be interesting to see whether it is possible to find good inference systems where equivalences are added dynamically, to perform a kind of paramodulation on the level of propositions.

In the case where the rewrite system is positive in the sense of Dowek and Werner [7] the closure operation in our model construction resembles their construction of a premodel as a fixpoint of a functional derived from the rewrite rules. It will be interesting to further investigate this connection for the other cases in order to better understand which congruences lead to sequent calculi with the cut elimination property.

---

[1] http://www.loria.fr/equipes/protheo/SOFTWARES/ELAN/

Acknowledgments. I thank Claude Kirchner for the many discussions on this paper.

References

1. Leo Bachmair and Harald Ganzinger. Non-clausal resolution and superposition with selection and redundancy criteria. In *Intl. Conf. on Logic Programming and Automated Reasoning*, LNCS 624, pages 273–284. Springer, 1992.
2. Leo Bachmair and Harald Ganzinger. Associative-commutative superposition. In *Proc. 4th Int. Workshop on Conditional and Typed Rewriting*, LNCS 968, pages 1–14, Jerusalem, 1994. Springer.
3. Leo Bachmair and Harald Ganzinger. Equational reasoning in saturation-based theorem proving. In *Automated Deduction - A Basis for Applications. Volume I*, chapter 11, pages 353–397. Kluwer, Dordrecht, The Netherlands, 1998.
4. Leo Bachmair, Harald Ganzinger, Christopher Lynch, and Wayne Snyder. Basic paramodulation. *Information and Computation*, 121(2):172–192, 1995.
5. Leo Bachmair, Harald Ganzinger, and Jürgen Stuber. Combining algebra and universal algebra in first-order theorem proving: The case of commutative rings. In *Proc. 10th Workshop on Specification of Abstract Data Types*, LNCS 906, pages 1–29, Santa Margherita, Italy, 1995. Springer.
6. Gilles Dowek, Thérèse Hardin, and Claude Kirchner. Theorem proving modulo. Rapport de Recherche 3400, INRIA, 1998.
7. Gilles Dowek and Benjamin Werner. Proof normalization modulo. Rapport de Recherche 3542, INRIA, 1998. Also in Types for proofs and programs 98, T. Altenkirch, W. Naraschewski, B. Rues (Eds.), LNCS 1657, Springer 1999, pp. 62-77.
8. Harald Ganzinger and Uwe Waldmann. Theorem proving in cancellative abelian monoids (extended abstract). In *13th Int. Conf. on Automated Deduction*, LNAI 1104, pages 388–402, New Brunswick, NJ, USA, 1996. Springer.
9. Jean-Marie Hullot. Canonical forms and unification. In *Proc. 5th Conf. on Automated Deduction*, LNCS 87, pages 318–334, Les Arcs, France, 1980. Springer.
10. Robert Nieuwenhuis and Albert Rubio. Theorem proving with ordering constrained clauses. In *Proc. 11th Int. Conf. on Automated Deduction*, LNCS 607, pages 477–491, Saratoga Springs, NY, 1992. Springer.
11. Robert Nieuwenhuis and Albert Rubio. Paramodulation with built-in AC-theories and symbolic constraints. *Journal of Symbolic Computation*, 23:1–21, 1997.
12. David A. Plaisted and Yunshan Zhu. Replacement rules with definition detection. In *Automated Deduction in Classical and Non-Classical Logics*, LNAI 1761. Springer, 1999.
13. Jürgen Stuber. Superposition theorem proving for abelian groups represented as integer modules. *Theoretical Computer Science*, 208(1–2):149–177, 1998.
14. Jürgen Stuber. Superposition theorem proving for commutative rings. In Wolfgang Bibel and Peter H. Schmitt, editors, *Automated Deduction - A Basis for Applications. Volume III. Applications*, chapter 2, pages 31–55. Kluwer, Dordrecht, The Netherlands, 1998.
15. Jürgen Stuber. *Superposition Theorem Proving for Commutative Algebraic Theories*. Dissertation, Naturwissenschaftlich-Technische Fakultät I, Universität des Saarlandes, Saarbrücken, 2000.

16. Laurent Vigneron. Associative-commutative deduction with constraints. In *Proc. 12th Int. Conf. on Automated Deduction*, LNCS 814, pages 530–544, Nancy, France, 1994. Springer.
17. Ulrich Wertz. First-order theorem proving modulo equations. Technical Report MPI-I-92-216, Max-Planck-Institut für Informatik, Saarbrücken, April 1992.

A Resolution-Based Decision Procedure for the Two-Variable Fragment with Equality

Hans de Nivelle[1] and Ian Pratt-Hartmann[2]

[1] Max Planck Institut für Informatik
Stuhlsatzenhausweg 85
66123 Saarbrücken, Germany
nivelle@mpi.mpg-sb.de
[2] Department of Computer Science,
University of Manchester,
Manchester M13 9PL, UK
ipratt@cs.man.ac.uk

Abstract. The two-variable-fragment $\mathcal{L}^2_\approx$ of first order logic is the set of formulas that do not contain function symbols, that possibly contain equality, and that contain at most two variables. This paper shows how resolution theorem-proving techniques can be used to provide an algorithm for deciding whether any given formula in $\mathcal{L}^2_\approx$ is satisfiable. Previous resolution-based techniques could deal only with the equality-free subset $\mathcal{L}^2$ of the two-variable fragment.

1 Introduction

The two-variable-fragment $\mathcal{L}^2_\approx$ is the set of formulas that do not contain function symbols, that possibly contain equality ($\approx$), and that use only two variables. The *two-variable fragment without equality* $\mathcal{L}^2$ is the subset of $\mathcal{L}^2_\approx$ not involving the predicate $\approx$. For example, the formula $\forall x \exists y[r(x,y) \wedge \forall x(r(y,x) \to x \approx y)]$, stating that every element is r-related to some element whose only r-successor is itself, is in $\mathcal{L}^2_\approx$ (but not in $\mathcal{L}^2$). Note in particular the 're-use' of the variable x by nested quantifiers in this example. In the same way, it is possible to translate modal formulas into $\mathcal{L}^2$, (without equality) by reusing variables. For example, the modal formula $\square\Diamond\square a$ can be translated into $\forall y(r(x,y) \to \exists x(r(y,x) \wedge \forall y(r(x,y) \to a(y))))$. No equality is needed for translating modal formulas.

Both two-variable fragments are known to be decidable. That is: an algorithm exists which, given any formula $\phi \in \mathcal{L}^2_\approx$, will determine whether ϕ is satisfiable. In [GKV97], decidability of $\mathcal{L}^2_\approx$ is proven by analyzing the structure of possible models, and showing that if a formula ϕ has any models at all, then it has a model of size $O(2^{|\phi|})$. This gives in principle a decision procedure in nondeterministic exponential time. However this procedure is probably inefficient in practice. This is caused by inherent problems of backtracking. Any backtracking procedure will be spending time retrying details, that are irrelevant for the truth of the formula. Intelligent backtracking can decrease this problem, but cannot

R. Goré, A. Leitsch, and T. Nipkow (Eds.): IJCAR 2001, LNAI 2083, pp. 211–225, 2001.
© Springer-Verlag Berlin Heidelberg 2001

completely remove it. Moreover, a backtracking procedure cannot be prevented from redoing the same work in different branches. Improved implementations might decrease this problem, but it cannot be removed completely.

Opposed to this, a resolution procedure works bottom up, starting with the formula in which one is interested. This means that every clause that is derived is related to the original formula, and hence more likely to be relevant. Additionally, a clause can be seen as a lemma, which can be used many times, and which can be seen as representing the set of its instances. Because of this, the risk of repeated work is decreased.

Another advantage, of a more practical nature, is that resolution decision procedures are close to the standard methods of full first-oder automated theorem-proving, so that existing implementations and optimizations can be used. Indeed only a small modification in a standard prover is needed in order to obtain a resolution decision procedure.

To date, resolution-based decision procedures have been developed for various fragments of first-order logic, including the guarded fragment with equality, the Gödel class, and the two-variable fragment without equality $\mathcal{L}^2$. (See [dN95],[dN00a],[GdN99]).

A resolution theorem prover (for unrestricted first order logic) works as follows: The first order formula is translated into a set of clauses through a clausal normal form-transformation. After that, the resolution prover derives new clauses, using derivation rules. Examples of derivation rules are *resolution, factoring* and *paramodulation*. The prover can also delete *redundant* clauses, using certain deletion rules. Common deletion rules are *subsumption, demodulation* and *tautology elimination*. This process terminates when either the empty clause is derived, or a stable set of clauses is reached. This is a set for which every clause that can be derived, can be immediately deleted by one of the deletion rules. For full first-order logic, there are formulas for which the process will not terminate.

Resolution decision procedures are obtained by first identifying an appropriate clause fragment. Then it is shown that certain restrictions of resolution are complete for the given clause fragment, and that all newly derived clauses are within the clause fragment. After that it is shown that the first order formulas of the fragment under consideration can be translated into the given clause fragment. Finally it is shown that there exists only a finite number of non-redundant clauses in the given fragment.

The strategy that we give in this paper is different from usual decision procedures in the fact that it consists of two stages. First, the clauses are partially saturated under a restricted form of resolution. After that, it is shown that clauses containing equality can be replaced by clauses without equality, without affecting satisfiability. The result is a clause set that corresponds to $\mathcal{L}^2$, the two-variable fragment without equality.

At this point, one could continue the work by using any decision procedure for $\mathcal{L}^2$; we prefer to stay within the resolution framework. we present a novel resolution decision procedure based on a *liftable* order.

The plan of the paper is as follows. Section 2 motivates the search for an efficient decision procedure for $\mathcal{L}_{\approx}^2$. Section 3 shows how equality can be removed from a formula ϕ of $\mathcal{L}_{\approx}^2$ without affecting its satisfiability. Section 4 then presents the new resolution-based algorithm for determining satisfiability of formulas of $\mathcal{L}^2$. We assume familiarity with the standard terminology and the basic techniques of resolution theorem-proving. The reader is referred to [FLTZ93] Ch. 2 for the relevant definitions.

2   Motivation

A logic is said to have the *finite-model property* if any satisfiable formula in that logic is satisfiable in a finite structure. It is easy to see that any fragment of first-order logic having the finite model property is decidable; and indeed, most of the known decidable fragments of first-order logic have the finite model property. (For a comprehensive survey, see Börger, Grädel and Gurevitch [BGG97].) One such fragment of particular interest here is the so-called Gödel class: the set of first-order formulas *without equality* which, when put in prenex form, have quantifier prefixes matching the pattern $\exists^*\forall\forall\exists^*$. Gödel [G33] showed that the Gödel class has the finite model property, and is thus decidable. In the same paper, Gödel claimed that allowing $\approx$ in formulas of the Gödel class would not affect the finite model property, a claim which was later shown to be false by Goldfarb [G84]. Between these two discoveries, Scott [S62] showed that any formula of the two-variable fragment can be transformed into a formula in the Gödel fragment which is equisatisfiable. Relying on Gödel's incorrect claim, Scott concluded decidability for $\mathcal{L}_{\approx}^2$. Of course, what Scott actually showed was the decidability for $\mathcal{L}^2$ only. That the full two-variable fragment does indeed have the finite model property was eventually established by Mortimer [M75].

Most proofs of the finite model property actually yield a bound on the size of a smallest model of a satisfiable formula ϕ in terms of the size of (number of symbols in) ϕ, and the result for the two-variable fragment is a case in point. A satisfiable formula $\phi \in \mathcal{L}_{\approx}^2$ of size n has a model with at most 2^{cn} elements, for some constant c (see Bgg97, pp. 377–381). Therefore, we can determine the satisfiability of ϕ by enumerating all such models, a process which can evidently be completed in nondeterministic exponential time. Moreover, it can be shown using standard techniques that satisfiability in $\mathcal{L}_{\approx}^2$ is in fact NEXPTIME-complete (as, indeed, is satisfiability in $\mathcal{L}^2$). Hence, the complexity of model enumeration agrees with the known worst-case complexity of determining satisfiability in $\mathcal{L}_{\approx}^2$.

Nevertheless, enumeration of models up to a certain size is in practice an inefficient method for determining satisfiability in $\phi \in \mathcal{L}_{\approx}^2$, especially if no model exists. A much more promising method is to adapt a resolution-based theorem prover so that termination on formulas of $\mathcal{L}_{\approx}^2$ is guaranteed. Such an approach has already been employed for other decidable fragments. In particular, the Gödel class can be decided using *ordered resolution*; moreover, as was pointed out in [FLTZ93], by applying Scott's reduction from $\mathcal{L}^2$ to the Gödel class, the same technique can be used to decide $\mathcal{L}^2$ as well. Unfortunately however, this method does not apply to the whole fragment $\mathcal{L}_\approx^2$, since adding equality to the Gödel class leads to undecidability. The main contribution of the present paper is to show that, with the aid of some technical manœuvres, this approach can nevertheless be extended to the whole fragment $\mathcal{L}_\approx^2$. To the best of the authors' knowledge, this is the first really practical decision procedure that has been proposed for the full two-variable fragment.

The fragment $\mathcal{L}_\approx^2$ is of particular interest when dealing with natural language input, because many simple natural language sentences translate into $\mathcal{L}_\approx^2$. To give a somewhat fanciful example, the sentence

Every meta-barber shaves every man who shaves no man who shaves himself

translates to the two-variable formula

$\forall x(\text{meta-barber}(x) \rightarrow$
$\quad \forall y((\text{man}(y) \wedge \forall x((\text{man}(x) \wedge \text{shave}(x,x)) \rightarrow \neg\text{shave}(y,x))) \rightarrow \text{shave}(x,y))).$

Although by no means all of English translates to the two-variable fragment (just think, for example, of ditransitive verbs), a useful subset of English can nevertheless be treated in this way. In particular, Pratt-Hartmann [PH00] gives a naturally circumscribed fragment of English which is shown to have exactly the expressive power of $\mathcal{L}_\approx^2$. Certainly, the two-variable fragment is more useful when dealing with natural language than other well-known decidable fragments, such as the guarded fragment [AvBN98] or any of the quantifier prefix fragments, whose formulas do not fit translations of natural language constructs easily.

3   Making Equality Disappear

In this section, we give give a method for removing equality from a formula in $\mathcal{L}_\approx^2$, based on resolution. Let ϕ to be some formula of $\mathcal{L}_\approx^2$. We can assume that ϕ contains only unary and binary predicates, since predicates of higher arity—as long as they feature only two variables—can be removed by a transformation (see [GKV97] for details). Throughout this section, if $\psi(x)$ is an $\mathcal{L}_\approx^2$-formula whose only free variable is x, we use the abbreviation $\exists! x\, \psi(x)$ for the $\mathcal{L}_\approx^2$-formula

$$\exists x[\psi(x) \wedge \forall y(\, \psi(y) \rightarrow x \approx y)],$$

asserting that ψ is satisfied by exactly one object.

Occurrences of the $\approx$-symbol fall into two groups. Negative occurrences can be 'simulated' without recourse to equality. Positive occcurrences can be restricted to those belonging to a $\exists!$ quantifier. This is done in the key step of our procedure, which is described in Lemma 4. The remaining occurrences of $\approx$ can axiomatized within $\mathcal{L}^2$. This enables us to remove all occurrences of $\approx$.

Definition 1. *An atom is defined as usual. A literal is an atom, or its negation. A formula α is in* conjunctive normal form *if it has form $c_1 \wedge \cdots \wedge c_n$, where each $c_i, (1 \leq i \leq n)$ is a disjunction $c_i = \beta_{i,1} \vee \cdots \vee \beta_{i,l_i}$ of literals.*

A formula in CNF is not the same as a set of clauses, because the clauses can be universally quantified. We begin by removing individual constants from our formula ϕ.

Lemma 1. *Let $\phi \in \mathcal{L}_{\approx}^2$, and let the sets of individual constants, unary predicates and binary predicates occurring in ϕ be D, P and R, respectively. Let ϕ' be the result of replacing any atoms in ϕ involving individual constants with formulas according to the following table:*

Atom	Replacement formula
$p(d)$	$\exists x (p(x) \wedge p_d(x))$
$r(d, x)$	$\exists y (r(y, x) \wedge p_d(y))$
$r(d, y)$	$\exists x (r(x, y) \wedge p_d(x))$
$r(x, d)$	$\exists y (r(x, y) \wedge p_d(y))$
$r(y, d)$	$\exists x (r(y, x) \wedge p_d(x))$
$r(d, d')$	$\exists x \exists y (r(x, y) \wedge p_d(x) \wedge p_{d'}(y))$

where $p \in P$, $r \in R$, $d \in D$, and where the unary predicates $p_d(x)$ (for $d \in D$) are all new. Then ϕ is equisatisfiable with

$$\phi_0 := \phi' \wedge \bigwedge_{d \in D} \exists! x \; p_d(x).$$

In fact, individual constants will be reintroduced later; however removing them at this point makes the key step described in Lemma 4 much easier to follow. Next, we convert to Scott normal form. The following result is standard (see, e.g. [BGG97] lemma 8.1.2).

Lemma 2. *Let ϕ be a formula in $\mathcal{L}_{\approx}^2$. There is an equisatisfiable formula ϕ_1 with form $\phi_1 = \beta_1 \wedge \cdots \wedge \beta_n$, where each β_i has one of the following three forms:*

$$\exists x \; \alpha_i, \qquad \forall x \exists y \; \alpha_i, \qquad or \; \forall x \forall y \; \alpha_i.$$

Each α_i is a formula in conjunctive normal form. We call the types of the β_i from left to right Type 1, Type 2, and Type 3. If β_i is a Type 1 formula, then α_i contains at most the variable x. If β_i is a Type 2 or a Type 3 formula, then α_i contains at most the variables x and y. There are no constants in the α_i.

The transformation in question introduces some new predicate letters, but no new individual constants or function symbols. Next, we move all occurrences of $\approx$ out of the α_i into the quantifiers.

Lemma 3. *Let ϕ_1 be as defined in Lemma 2. Formula ϕ_1 can be transformed into a formula $\phi_2 = \beta_1 \wedge \cdots \wedge \beta_n$, where each β_i has one of the following forms:*

$$\exists x \; \alpha_i,$$

$$\forall x \exists y \; (x \not\approx y \wedge \alpha_i),$$

or

$$\forall x \forall y \; (x \approx y \vee \alpha_i).$$

Now each of the α_i is a formula without equality, in conjunctive normal form.

If ϕ_2 contains more than one Type 3 formula, then they can be merged. In the sequel we will assume that this is done, and that there is only one Type 3 formula in ϕ_2.

We come to the key idea of the present transformation—the elimination of the positive occurrence of $\approx$ in a disjunction $\forall x \forall y (\alpha(x, y) \lor \beta(x) \lor \gamma(y) \lor x \approx y)$ of ϕ_2. Our solution is to saturate ϕ_2 partially under resolution, and than to eliminate the disjunctions that have a non-empty $\alpha(x, y)$. When this is done, we have only disjunctions of the form $\forall x \forall y (x \approx y \lor \beta(x) \lor \gamma(y))$. These can be read as: If $\beta(x)$ does not hold everywhere, and $\gamma(y)$ does not hold everywhere, then there is one point c, which is the only point on which $\beta(c)$ does not hold, and also the only point on which $\gamma(c)$ does not hold. This provides the transformation of $\approx$ into $\exists!$.

Definition 2. *We need the following, restricted version of the resolution rule. It is restricted because we allow resolution only between two-variable literals.*

Let $\beta(x, y) \lor r_1$ and $\neg\beta(x, y) \lor r_2$ be disjunctions, occurring in an α_i of ϕ_2. Then $r_1 \lor r_2$ is a resolvent. The r_1 and r_2 are disjunctions of literals. We implicitly assume that $r_1 \lor r_2$ is normalized after the resolution step. That means that multiple occurrences of the same literal are removed.

We allow swapping of variables, but we do not allow proper instantiation. So $\beta(y, x) \lor r_1$ and $\neg\beta(x, y) \lor r_2$ can resolve into $r_1[x \leftrightarrow y] \lor r_2$.

Inside a formula ϕ_2, defined as in Lemma 3, we allow resolution as follows: Resolution is allowed between disjunctions inside the Type 3 formula. The results are added to the Type 3 formula. We also allow resolution between a disjunction inside a Type 2 formula and a disjunction inside a Type 3 formula. The result is added to the Type 2 disjunction that was used.

It is easily checked that the resolution rules are sound. If ϕ' is obtained from ϕ by a resolution step, then $\phi' \leftrightarrow \phi$. Termination follows from the fact that only finitely many normalized disjunctions exist over a given signature.

Lemma 4. *Let ϕ_2 be defined as in Lemma 3. Let ϕ_3 be its closure under resolution, as defined in Definition 2. Let ϕ_4 be obtained from ϕ_3 by removing all disjunctions containing a two-variable literal (other than $\approx$) from the Type 3 formula. Then ϕ_3 has a model iff ϕ_4 has a model*

Proof. It is clear that if ϕ_3 has a model, then ϕ_4 has a model, since resolution is a sound inference rule.

For the other direction, let $\mathfrak{B}$ be a structure in which ϕ_4 is true. We will modify $\mathfrak{B}$ into a new structure $\mathfrak{B}^*$, in which ϕ_3 holds. We use B for the domain of $\mathfrak{B}$. The new structure $\mathfrak{B}^*$ will have the same domain B. It will be obtained from $\mathfrak{B}$ by changing the truth-values of the two-variable predicates. For the rest, $\mathfrak{B}^*$ will be identical to $\mathfrak{B}$.

We assume that both ϕ_3 and ϕ_4 are decomposed as in Lemma 3. We write $\beta_1 \land \cdots \land \beta_n$ for the decomposition of ϕ_3, and $\beta'_1 \land \cdots \land \beta'_n$ for the decomposition of ϕ_4. We define α_i and α'_i accordingly. We use t for the index of both Type 3 subformulas. Obviously the β'_i can be arranged in such a way that $(i \neq t) \Rightarrow (\beta_i = \beta'_i)$.

As said before, we intend to reinterpret the two-variable predicates in such a way that β_t becomes true. When doing so, we run the risk of making a β_i of Type 2 false. In order to avoid this we need the following:

For each Type 2 subformula β_i of ϕ_4, we assume a *choice function* f_i of type $B \to B$. It is defined as follows: Because $\beta_i = \forall x \exists y (x \not\approx y \wedge \alpha_i)$ is true in $\mathfrak{B}$, for each $b_1 \in B$, there exists a $b_2 \neq b_1$ in B, s.t. $\mathfrak{B}^* \models \alpha_i(b_1, b_2)$. The choice function f_i is defined by choosing one such b_2 for each b_1.

Let b_1 and b_2 be two distinct elements of B. We define *the pattern* of $\mathfrak{B}$ on $\{b_1, b_2\}$ as the vector of truth values for the binary predicates involving both of b_1 and b_2. So, if $p_1, \ldots, p_r$ are all the binary predicates symbols of ϕ_3, then for each $\{b_1, b_2\}$, the pattern determines whether or not $\mathfrak{B} \models p_j(x, y)(b_1, b_2)$ and whether or not $\mathfrak{B} \models p_j(y, x)(b_1, b_2)$. It does not say anything about the unary predicates, nor about $\mathfrak{B} \models p_j(x, y)(b_1, b_1)$ or $\mathfrak{B} \models p_j(x, y)(b_2, b_2)$.

The intuition of the construction is as follows: If ϕ_3 is not true in $\mathfrak{B}$, this is caused by the fact that there are $(b_1, b_2) \in B$, for which $\mathfrak{B} \not\models \alpha_t(b_1, b_2)$. This must be caused by the fact there is a disjunction $\gamma(x, y) \vee \delta(x) \vee \eta(y) \in \alpha_t$, for which $\mathfrak{B} \not\models \delta(x)(b_1)$, $\mathfrak{B} \not\models \eta(y)(b_2)$, and $\mathfrak{B} \not\models \alpha(x, y)$. We will change the pattern on $\{b_1, b_2\}$ in such a way that $\alpha(x, y)$ will become true.

Before we can proceed, we need to define the subformulas that are involved: Let $\lambda_1, \ldots, \lambda_i, \ldots, \lambda_p$, $p \geq 0$ be the indices of the Type 2 subformulas of ϕ_3, for which $f_{\lambda_i}(b_1) = b_2$, Similarly, let $\mu_1, \ldots, \mu_j, \ldots, \mu_q$, $q \geq 0$ be the indices of the Type 2 subformulas of ϕ_3, for which $f_{\mu_j}(b_2) = b_1$.

The α_t and the α_{λ_i} and α_{μ_j} are formulas in conjunctive normal form, i.e. conjunctions of disjunctions. We are going to select the disjunctions, whose truth depends on the binary predicates on $\{b_1, b_2\}$. For each $i, (1 \leq i \leq p)$, let $\alpha^1_{\lambda_i}$ be obtained from α_{λ_i} by selecting those disjunctions of which the literals involving one variable are false in $\mathfrak{B}$ on (b_1, b_2). Let $\alpha^2_{\lambda_i}$ be obtained by removing the one-variable predicates from $\alpha^1_{\lambda_i}$. For $j, (1 \leq j \leq q)$, we define $\alpha^1_{\mu_j}$ and $\alpha^2_{\mu_j}$ analogeously.

For α_t we need two copies, because of the two directions involved. Let $\alpha^1_{t_1}$ be obtained from α_t be selecting those disjunctions of which the literals involving one variable are false in $\mathfrak{B}$ on (b_1, b_2). Let $\alpha^2_{t_1}$ be obtained by deleting the one-variable predicates from $\alpha^1_{t_1}$. Similarly, let $\alpha^1_{t_2}$ be obtained by selecting the disjunctions, of which the one-variable predicates are false on (b_2, b_1). Then $\alpha^2_{t_2}$ is obtained by deleting the one-variable predicates from $\alpha^1_{t_2}$.

It is clear that in order to obtain $\mathfrak{B}^*$, it is sufficient to replace in $\mathfrak{B}$ the patterns on each $\{b_1, b_2\}, b_1 \neq b_2$ by a pattern making the $\alpha^2_{\lambda_i}$, $\alpha^2_{\mu_j}$, $\alpha^2_{t_1}$, $\alpha^2_{t_2}$ true on (b_1, b_2). The patterns can be replaced in $\mathfrak{B}$ in parallel.

It remains to show that the patterns exist. This is guaranteed by the fact that α_t and the α_i of Type 2 are sufficiently closed under resolution. Since we are considering a fixed (b_1, b_2), we are dealing with a propositional problem. We apply Lemma 11, using $r = \alpha^2_{t_1} \cup \alpha^2_{t_2}$, and using $c_1 \wedge \cdots \wedge c_m = \alpha^2_{\lambda_i}$, $\alpha^2_{\mu_j}$.

It is easily checked that all necessary resolvents are allowed by Definition 2. The $\alpha^2_{\lambda_i}$, $\alpha^2_{\mu_j}$ are consistent, because they are true in $\mathfrak{B}$ on (b_1, b_2). It is also clear $\alpha^2_{t_1} \cup \alpha^2_{t_2}$ cannot contain the empty clause, since it would originate from a disjunction in α_t for which the one-variable literals are false in $\mathfrak{B}$ on (b_1, b_2). But this disjunction would be in α'_t as well, as it does not contain a two-variable literal. This contradicts the assumption that ϕ_4 is true in $\mathfrak{B}$.

Thus, Lemma 4 tells us that, if we saturate ϕ_3 under resolution, then we can delete all the disjunctions $\alpha(x, y) \vee \beta(x) \vee \gamma(y) \vee x \approx y$, for which $\alpha(x, y)$ is non-empty. Although exhaustive application of resolution is computationally expensive, in the context of determining satisfiability in the two-variable fragment, the transformation step from ϕ_3 to ϕ_4 in fact comes for free. This is because existing resolution-based approaches to determining satisfiability in $\mathcal{L}^2$ begin with conversion to Scott normal form, followed by clausification and exhaustive application of ordered resolution *anyway*. All the present procedure requires is that we perform a resolution version of resolution first, and then pause to delete the inseparable clauses, before resuming (pretty well) what we would have done even if no equality were present. It hardly needs mentioning that the elimination of a whole set of clauses requires no computation whatever: in particular, the model-theoretic manipulations used to prove Lemma 4 form no part of the decision procedure for $\mathcal{L}^2_\approx$.

The negative occurrences can be easily deleted by introducing a new, non-reflexive predicate neq. We add a formula $\forall x (\neg \mathrm{neq}(x, x))$, and replace each negative equality $x \not\approx y$ by $\mathrm{neq}(x, y)$. We will do this later, we now first concentrate on the positive equalities in the Type 3 formulas.

The remaining steps in our equality-deletion procedure are all straightforward. There is (at most) one positive occurrence of $\approx$ left. In occurs in the Type 3 subformula α_t and the other literals in α_t are unary. The following logical equivalence is simple to verify:

Lemma 5. *Let $\gamma(x)$ be a formula not involving the variable y and let $\delta(y)$ a formula not involving the variable x. Then the formulas*

$$\forall x \forall y (\gamma(x) \vee \delta(y) \vee x \approx y)$$

and

$$\forall x \; \gamma(x) \vee \forall x \; \delta(x) \vee (\exists! x \; \neg \; \gamma(x) \wedge \forall x (\gamma(x) \leftrightarrow \delta(x)))$$

are logically equivalent.

Using this, we can use the splitting rule to decompose the disjunctions of the Type 3 formula. The result is a formula, in all positive occurrences of $\approx$ belong to a $\exists!$ quantifier. These can be eliminated by introducing new individual constants.

Lemma 6. *Let η be a conjunction of constant-free $\mathcal{L}^2$-formulas in prenex form with quantifier prefixes $\forall x$ and $\forall x \exists y$, and for each i $(1 \leq i \leq m)$ let ζ_i be a quantifier- and constant-free formula of $\mathcal{L}^2$ not involving the variable y. Define*

$$\theta_i := \zeta_i(e_i) \wedge \bigwedge_p \forall x(\zeta_i(x) \rightarrow (p(x) \leftrightarrow p(e_i))) \wedge$$
$$\bigwedge_q \forall x \forall y(\zeta_i(x) \rightarrow (q(x,y) \leftrightarrow q(e_i, y))) \wedge$$
$$\bigwedge_q \forall x \forall y(\zeta_i(x) \rightarrow (q(y,x) \leftrightarrow q(y, e_i)))$$

where $e_1, \ldots e_m$ are (new) individual constants, p ranges over all unary predicates mentioned in η or any of the ζ_i, and q ranges over all binary predicates mentioned in η or any of the ζ_i. Then the formulas

$$\psi := \eta \wedge \bigwedge_{1 \leq i \leq m} \exists! x \zeta_i(x)$$

and

$$\psi' := \eta \wedge \bigwedge_{1 \leq i \leq m} \theta_i$$

are equisatisfiable.

Proof. If $\mathfrak{B} \models \psi$, then it is easy to expand $\mathfrak{B}$ to a structure $\mathfrak{B}'$ such that $\mathfrak{B}' \models \psi'$. Conversely, suppose that $\mathfrak{B}' \models \psi'$, where the domain of $\mathfrak{B}'$ is B. Let $b_1, \ldots b_m$ be the denotations of the constants $e_1, \ldots e_m$ in $\mathfrak{B}'$, respectively. Now define a function f on B as follows:

$$f(b) = \begin{cases} b_i \text{ if } \mathfrak{B}' \models \zeta_i[b] \\ b \text{ otherwise.} \end{cases}$$

And define the structure $\mathfrak{B}$ with domain $f(B)$ as follows:

$$\mathfrak{B} \models p[f(b)] \text{ iff } \mathfrak{B}' \models p[b]$$
$$\mathfrak{B} \models q[f(b), f(b')] \text{ iff } \mathfrak{B}' \models q[b, b']$$

where b ranges over B, p over the unary predicates mentioned in ψ' and q over the binary predicates mentioned in ψ'. Since $\mathfrak{B}' \models \psi'$, $\mathfrak{B}$ is well-defined. It is then easy to see that $\mathfrak{B} \models \psi$.

Theorem 1. *Let ϕ be any formula of $\mathcal{L}^2_{\approx}$. Then the steps described in this section allow us to compute an equisatisfiable formula ϕ^* in the class $\mathcal{L}^2$. Indeed, if ϕ^* is satisfiable over some domain, then ϕ is satisfiable over a subset of that domain. Moreover, ϕ^* can be written as a conjunction of prenex formulas with quantifier prefixes $\exists x$, $\forall x \forall y$ and $\forall x \exists y$.*

Thus, the satisfiability problem for $\mathcal{L}^2_{\approx}$ has been reduced to that for $\mathcal{L}^2$. Moreover, as we have observed, no significant extra computational cost is incurred in this reduction.

Finally, we note that theorem 1 allows us to infer that $\mathcal{L}^2_{\approx}$ has the finite model property from the corresponding fact about the Gödel fragment. This constitutes an alternative to the proofs cited in Section 2.

4   The 2-Variable Fragment without Equality

In this section, we provide a new decision procedure for the two-variable fragment without equality. Practically, the method does not differ from the method given in [dN00a], but the theoretical foundation is different. The method that we give here is based on a liftable order, i.e. an order that is preserved by substitution. The advantage of liftable orders, is that they are better understood theoretically. It is the (far) hope of the authors that this will eventually leed to an understanding of what makes the resolution decision procedures work. At this moment, the termination/completeness proofs are a collection of tricks. One would hope for a real understanding of the relation between model based decision procedures, and resolution based decision procedures. Decision procedures based on liftable orders appear to be a step in this direction.

As said before, procedure makes use of indexed resolution [B71], also called *lock resolution.* It works as follows: one starts with some clause set upon which we want to apply resolution. First, integers are attached to the literals in the initial clause set. This can be done in any arbitrary way; and distinct occurrences of the same literal can be indexed with different integers. After that, the integers can be used by an order restriction for determining which literals can be resolved away. When a resolvent is formed, the literals in the resolvent simply inherit their indices from the parent clauses. In standard versions of lock-resolution, only the literals indexed by a maximal integer can be resolved away. Our procedure is slightly more general: we allow the selection function to look at both the index and the the literal itself. The key property possessed by this selection function is that is is obtained by *lifting* some order $\prec$ on the ground indexed literals, as we will explain below.

4.1   Indexed Resolution

Definition 3. *An* indexed literal *is a pair* (L, a), *where* L *is a literal, and* a *is an element of some index set. We use the index set* $\{0, 1\}$. *We write* $L{:}\,a$ *instead of* (L, a). *An* indexed clause *is a finite set of indexed literals.*

The effect of a substitution Θ on an indexed literal $A{:}\,a$ is defined by $(A{:}\,a)\Theta = A\Theta{:}\,a$. The effect of a substitution on a clause is defined memberwise.

In the sequel we use the term *clause* to mean either a clause (in the usual sense of a finite set of literals) or an indexed clause. It will be clear from the context what type of clause we mean. When present, the indices play no role in the semantics of the clause; however, they do play a role in determining which literals are selected. Extending the notion of a selection function to indexed clauses in the obvious way, we define:

Definition 4. *A* selection function Σ *is a function mapping indexed clauses to indexed clauses for which always* $\Sigma(c) \subseteq c$. *For a clause* c, *we call the indexed literals in* $\Sigma(c)$ selected.

Resolution. *Let* $c_1 = \{A_1:a_1\} \cup R_1$ *and* $c_2 = \{\neg A_2:a_2\} \cup R_2$ *be clauses, s.t.* A_1 *and* A_2 *are unifiable and selected. Let* Θ *be the most general unifier of* A_1 *and* A_2. *Then the clause* $R_1\Theta \cup R_2\Theta$ *is a resolvent of* c_1 *and* c_2.

Factoring. *Let* $c = \{A_1:a_1, A_2:a_2\} \cup R$ *be a clause, s.t.* $A_1:a_1$ *is selected, and* A_1, A_2 *have most general unifier* Θ. *Then* $\{A_2:a_2\Theta\} \cup R\Theta$ *is a factor of* c.

In addition to resolution and factoring, our decision procedure for $\mathcal{L}^2$ uses the following rules.

Subsumption. If for clauses c_1, c_2 there is a substitution Θ, such that $c_1\Theta \subseteq c_2$, then c_1 *subsumes* c_2. In that case c_2 can be deleted from the database.

Splitting. The splitting rule can be applied when a clause c can be partitioned into two non-empty parts that do not have overlapping variables. If c can be partitioned as $R_1 \vee R_2$, then the prover tries to refute R_1 and R_2 independently.

Observe that the subsumption rule has to take the indices into account. The clause $\{p(x,y):1\}$ does not subsume the clause $\{p(x,0):0\}$.

Definition 5. *Let* $\prec$ *be an order on ground indexed literals. Let* c *be a ground clause containing an indexed literal* $A:a$. *An indexed literal* $A:a$ *is* maximal *in* c, *if there is no literal* $B:b \in c$ *for which* $A:a \prec B:b$.

We say that a selection function Σ is obtained by lifting $\prec$ if it meets the condition that, for every clause c and indexed literal $A:a \in c$, if c has an instance $c\Theta$ in which $(A:a)\Theta$ is maximal, then $A:a \in \Sigma(c)$.

The significance of Definition 5 lies in the following completeness result, which has a simple proof. It can be obtained by modifying the completeness proof for lock resolution of [CL73], or the one in [FLTZ93].

Lemma 7. *Let* C *be a set of initial clauses. Let* Σ *be a selection function obtained by lifting some order* $\prec$. *Let* $\overline{C}$ *be a set of indexed clauses satisfying the following:*

- *For every clause* $c = \{A_1, \ldots, A_p\} \in C$, *there exists an indexed clause* $d \in \overline{C}$ *that subsumes some indexing* $\{A_1:a_1, \ldots, A_p:a_p\}$ *of* c.
- *For every clause* c *that can be derived from clauses in* $\overline{C}$, *either by resolution or by factoring, there is a clause* $d \in \overline{C}$ *that subsumes* c.

Then, if C *is unsatisfiable,* $\overline{C}$ *contains the empty clause.*

4.2   The S^2-Class

The S^2-class characterizes the format of clauses obtained from clausifications of formulas in the Gödel class. The equality-free formulas, produced by the transformation of Section 3, can be transformed directly into S^2 by Skolemization. Formula in full $\mathcal{L}^2$ can be transformed through the Gödel class, or directly through a structural clause transformation, as is done in [FLTZ93].

Definition 6. *A clause c is in S^2 if it meets the following conditions:*

1. *c contains at most 2 variables, and c contains no nested function symbols.*
2. *If c contains ground literals, then it is a ground clause.*
3. *Each functional term in c contains all variables occurring in c.*
4. *There is a literal in c that contains all variables of c.*

Note that conditions 2 and 4 can be ensured by application of the splitting rule.

The results of section 3 thus suffice to reduce satisfiability in $\mathcal{L}^2_\approx$ to satisfiability in S^2. However, closer examination of the transformation allows us to be slightly more specific about the class of clauses we need to consider. Starting with an arbitrary $\mathcal{L}^2_\approx$-formula ϕ, theorem 1 guarantees the existence of an equisatisfiable formula ϕ^* which is a conjunction of $\mathcal{L}^2$-formulas in prenex form with quantifier prefixes $\forall x$, $\forall x \forall y$ and $\forall x \exists y$. Skolemizing and clausifying ϕ^* thus yields a collection of clauses in which all function symbols have arity 1. Furthermore, given that the clauses in question are to be indexed in some way, there is nothing to stop us indexing two-variable clauses with index 1 and clauses with fewer than two variables with index 0. This leads us to the definition:

Definition 7. *The class S^{2+i} is defined as the class S^2, but with additional conditions:*

1. *All functions are 1-place.*
2. *If c contains 2 variables, then the indexed literals with 2 variables are exactly the literals with index 1.*

Notice, incidentally, that Conditions 3 and 1 immediately imply that all two-variable clauses are function-free.

At first sight, it might appear that we are making no use of the possibility of giving different indices to distinct occurrences of the same literal. However, resolving $\{p(x,y)\!:\!1, q(x,y)\!:\!1, q(x,x)\!:\!0\}$ with $\{\neg p(x,x)\!:\!1\}$, yields the clause $\{q(x,x)\!:\!1, q(x,x)\!:\!0\}$, in which different indices are used for distinct occurrences of the same literal. To obtain a decision procedure for S^{2+i}, it suffices to define a selection function which is obtained by lifting, and which ensures that all derived clauses are within S^{2+i}. Correctness follows from the soundness of resolution and lemma 7; termination follows from the fact that for a given finite signature, there exist only finitely many non-equivalent clauses in S^{2+i}.

4.3   A Decision Procedure for the S^{2+i}-Class

We begin by establishing a suitable order on ground literals.

Definition 8. *Let the order $\prec_2$ on ground indexed literals be defined as follows:*

– *If literal A is strictly less deep than B, then $A\!:\!a \prec_2 B\!:\!b$.*
– *If literal A and B have the same depth, and $a < b$, then $A\!:\!a \prec_2 B\!:\!b$.*

Next, we give the selection function Σ_2 used in the second phase of resolution.

Definition 9. *Every indexed literal A: a in a clause c is selected, unless one of the following two conditions holds:*

- *A has no functional terms, $a = 0$, and there is a literal B: b in c with $b = 1$.*
- *A: a has no functional terms, and there are literals with functional terms in c.*

Lemma 8. *Σ_2 is obtained by lifting $\prec_2$.*

Proof. Let c be an arbitrary clause in S^{2+i}. Let A: a be a literal in c that is not selected. We need to show that in every instance $c\Theta$ of c, the literal $A\Theta$: a is non-maximal. First observe that A cannot have any functional terms.

If there is a literal B: b with functional terms in c, then this functional term contains all variables of A. So, for every substitution Θ it is the case that $B\Theta$ is deeper than $A\Theta$. As a consequence, $A\Theta$: a is non-maximal in $c\Theta$.

If there is no literal B: b with functional terms, then $a = 0$ and there must be a B: $b \in c$ with $b = 1$. If c is a one-variable clause, then $A\Theta$ and $B\Theta$ always have the same depth, for every substitution Θ. Because of this $A\Theta$: $0 \prec_2 B\Theta$: 1. Otherwise Condition 2 applies to c. Literal A contains 1 variable and literal B contains 2 variables. Write $A[X]$: 0 and $B[X,Y]$: 1. Let Θ be a substitution. If $Y\Theta$ is deeper than $X\Theta$, then $B[X,Y]\Theta$ is deeper than $A[X]\Theta$ and $A[X]\Theta$: $0 \prec_2 B[X,Y]\Theta$: 1. Otherwise $A[X]\Theta$ and $B[X,Y]\Theta$ have equal depth and also $A[X]\Theta$: $0 \prec_2 B[X,Y]\Theta$: 1. This completes the proof.

Note how the indices play an essential role in the definition of Σ_2: no selection function obatined by lifting an ordering on unindexed literals could ensure that $p(x,y)$ is always prefered over $p(x,x)$, since the literals have a common instance.

The next step is to show that resolution with Σ_2 never leads outside S^{2+i}.

Lemma 9. *Each literal selected by Σ_2 contains all variables of its clause, and contains a deepest occurrence of each variable in the clause.*

Proof. If the clause is ground, then the lemma is trivial. If the clause has one variable, then all literals have one variable, by condition 2. Moreover, if there exist any functional literals in the clause at all, any selected clause is functional, and so contains a deepest occurrence of its literal by condition 1. If the clause is two-variable, it must be non-functional by conditions 3 and 1. Thus, any selected literal must have index 1, and hence contains both variables by condition 2.

Lemma 10. *The strategy keeps clauses inside S^{2+i}.*

Proof. It is quite standard to show that every selection function that satisfies the conditions of Lemma 9 preserves Conditions 1-4 of Definition 6. See for example the remark on p. 115 of [FLTZ93]. Condition 1 is obviously preserved by resolution. The only difficulty lies in showing that condition 2 applies to the resolvent of any two clauses in S^{2+i}. Let c by such a resolvent, then, and assume that c is a two-variable clause.

Since all functions are unary, it is obvious that any two-variable literal in c must have come from a two-variable literal in one of the parent clauses, and hence must have index 1.

Conversely, we must show that any 1-indexed literal in c is two-variable. Let the parents of c be c_1 and c_2, let $B_1 \in c_1$ and $B_2 \in c_2$ be the literals resolved upon, and let Θ be the substitution used in this resolution. Without loss of generality, any 1-indexed literal in c may be written $A\Theta{:}1$, where $A{:}1 \in c_1$. By lemma 9, $\mathrm{Vars}(c_1) = \mathrm{Vars}(B_1)$ and $\mathrm{Vars}(c_2) = \mathrm{Vars}(B_2)$. Hence $\mathrm{Vars}(c_1\Theta) = \mathrm{Vars}(B_1\Theta)$ and $\mathrm{Vars}(c_2\Theta) = \mathrm{Vars}(B_2\Theta)$. But we also have $\mathrm{Vars}(B_1\Theta) = \mathrm{Vars}(B_2\Theta)$, whence $\mathrm{Vars}(c_1\Theta) = \mathrm{Vars}(c_2\Theta)$. Thus, $\mathrm{Vars}(c) \subseteq \mathrm{Vars}(c_1\Theta) \cup \mathrm{Vars}(c_2\Theta) = \mathrm{Vars}(c_1\Theta)$. Since c_1 is in S^{2+i}, condition 2 implies $\mathrm{Vars}(A) = \mathrm{Vars}(c_1)$, whence $\mathrm{Vars}(c) \subseteq \mathrm{Vars}(A)$ as required.

Gathering together the lemmas in this section, we have

Theorem 2. *The rules of resolution, factoring and subsumption give us a decision procedure for sets of clauses in the class S^{2+i}.*

This completes the description of the resolution procedure for $\mathcal{L}^2$.
We end the section with a technical lemma that was needed for the proof of Lemma 4.

Lemma 11. *Let $c_1, \ldots, c_m$ and r be sets of propositional clauses. Let r be closed under resolution. Furthermore assume that each possible resolvent between a clause of a c_k and a clause of r is in c_k. Then, if $c_1 \wedge \cdots \wedge c_m$ is consistent and r does not contain the empty clause, then $c_1 \wedge \cdots \wedge c_m \wedge r$ is consistent.*

Proof. The result can be easily obtained from the completeness of semantic resolution. Semantic resolution is obtained by fixing an interpretation I, and by forbidding resolution steps between two clauses, that are both true in I. Semantic resolution is proven complete in [CL73].

Since $c_1 \wedge \cdots \wedge c_m$ is consistent, there is an interpretation I that makes $c_1 \wedge \cdots \wedge c_m$ true. If $c_1 \wedge \cdots \wedge c_m \wedge r$ were inconsistent, then r would contain the empty clause. All clauses that are false in I, must be in r. Hence a resolution step involving a false clause is always allowed.

5   Conclusions

In this paper we have given a practical procedure for deciding satisfiablity in the two-variable fragment with equality. This procedure involves two new contributions. The first is the use of resolution to transform formulas in the two-variable fragment with equality to the two-variable fragment without equality. The second is a new resolution-based procedure for deciding satisfiability in the two-variable fragment without equality based on a selection function obtained by lifting an order on ground indexed literals.

At this moment, handling of constants is unsatisfactory. We hope to be able to adapt Definition 2 and Lemma 4 in such a way that they can handle constants.

References

[AvBN98]   Andréka, H., van Benthem, J. Németi, I.: Modal languages and bounded fragments of predicate logic. *Journal of Philosophical Logic*, 27(3):217–274, (1998).

[BGG97]   Börger, E., Grädel, E., Gurevich, Y.: The Classical Decision Problem. Springer Verlag, Berlin Heidelberg New York, (1997).

[B71]   Boyer, R.S.: Locking: A Restriction of Resolution (Ph. D. Thesis). University of Texas at Austin, (1971).

[CL73]   C-L. Chang, R. C-T. Lee, Symbolic Logic and Mechanical Theorem Proving, Academic Press, New York, 1973.

[FLTZ93]   Fermüller, C., Leitsch, A., Tammet, T., Zamov, N.: Resolution Methods for the Decision Problem. LNCS 679, Springer Verlag Berlin Heidelberg New York, (1993).

[GdN99]   Ganzinger, H., de Nivelle, H.: A superposition procedure for the guarded fragment with equality. LICS **14**, IEEE Computer Society Press, (1999), 295–303.

[G84]   Goldfarb, W.: the unsolvability of the Gödel Class with Identity. Journal of Symbolic Logic, **49**, (1984), 1237–1252.

[G33]   Gödel, K.: Collected Works, Volume 1: Publications 1929-1936. Oxford University Press. Edited by Feferman, S., Dawson, J. jr., Kleene, S., Moore, G., Solovay, R., van Heijenoort, J., (1986).

[GKV97]   Grädel, E., Kolaitis, Ph. G., Vardi, M.Y.: On the Decision Problem for Two-Variable First-Order Logic. The Bulletin of Symbolic Logic **3-1**, (1997) 53–68.

[HdNS00]   Hustadt, U., de Nivelle, H., Schmidt, R.: Resolution-Based Methods for Modal Logics, Journal of the IGPL **8-3**, (2000), 265-292.

[M75]   Mortimer, M.: On languages with two variables, Zeitschrift für mathematische Logik und Grundlagen der Mathematik **21**, (1975), 135–140.

[dN95]   de Nivelle, H. Ordering Refinements of Resolution, PhD Thesis. Delft University of Technology, (1995).

[dN00]   de Nivelle, H.: Deciding the E^+-Class by an A Posteriori, Liftable Order. Annals of Pure and Applied Logic **104-(1-3)**, (2000), 219-232.

[dN00a]   de Nivelle, H.: An Overview of Resolution Decision Procedures, Formalizing the Dynamics of Information, CSLI **91**, (2000), 115–130.

[PH00]   Pratt-Hartmann, I: On the Semantic Complexity of some Fragments of English. University of Manchester, Department of Computer Science Technical Report UMCS–00–5–1, (2000).

[S62]   Scott, D.: A Decision Method for Validity of Sentences in two Variables. Journal of Symbolic Logic **27**, (1962), 477.

Superposition and Chaining for Totally Ordered Divisible Abelian Groups (Extended Abstract)

Uwe Waldmann

Max-Planck-Institut für Informatik, Stuhlsatzenhausweg 85,
66123 Saarbrücken, Germany, uwe@mpi-sb.mpg.de

Abstract. We present a calculus for first-order theorem proving in the presence of the axioms of totally ordered divisible abelian groups. The calculus extends previous superposition or chaining calculi for divisible torsion-free abelian groups and dense total orderings without endpoints. As its predecessors, it is refutationally complete and requires neither explicit inferences with the theory axioms nor variable overlaps. It offers thus an efficient way of treating equalities and inequalities between additive terms over, e. g., the rational numbers within a first-order theorem prover.

1 Introduction

Most real life problems for an automated theorem prover contain both uninterpreted function and predicate symbols, that are specific for a particular domain, and standard algebraic structures, such as numbers or orderings. General theorem proving techniques like resolution or superposition are notoriously bad at handling algebraical theories involving axioms like associativity, commutativity, or transitivity, since explicit inferences with these axioms lead to an explosion of the search space. To deal efficiently with such structures, it is therefore necessary that specialized techniques are built tightly into the prover.

AC-superposition (Bachmair and Ganzinger [1], Wertz [12]) is a well-known example of such a technique. It incorporates associativity and commutativity into the standard superposition calculus using AC-unification and extended clauses. In this way, inferences with the theory axioms and certain inferences involving variables are rendered unnecessary. Still, reasoning with the associativity and commutativity axioms remains difficult for an automated theorem prover, even if explicit inferences with the AC axioms can be avoided. This is not only due to the NP-completeness of the AC-unifiability problem, but it stems also from the fact that AC-superposition requires an inference between literals $u_1 + \cdots + u_k \approx s$ and $v_1 + \cdots + v_l \approx t$ (via extended clauses) whenever *some* u_i is unifiable with *some* v_j. Consequently, a variable in a sum can be unified with any part of any other sum – in this situation unification is completely unable to limit the search space.

The inefficiency inherent in the theory of associativity and commutativity can be mitigated by integrating further axioms into the calculus. In abelian

R. Goré, A. Leitsch, and T. Nipkow (Eds.): IJCAR 2001, LNAI 2083, pp. 226–241, 2001.

groups (or even in cancellative abelian monoids) the ordering conditions of the inference rules can be refined in such a way that summands u_i and v_j have to be overlapped only if they are *maximal* with respect to some simplification ordering $\succ$ (Ganzinger and Waldmann [4,8], Marché [5], Stuber [7]). In this way, the number of variable overlaps can be greatly reduced; however, inferences with unshielded, i. e., potentially maximal, variables remain necessary.

In non-trivial divisible torsion-free abelian groups (e. g., the rational numbers and rational vector spaces), the abelian group axioms are extended by the torsion-freeness axiom $\forall k \in \mathbf{N}^{>0} \; \forall x, y \colon kx \approx ky \Rightarrow x \approx y$, the divisibility axiom $\forall k \in \mathbf{N}^{>0} \; \forall x \; \exists y \colon ky \approx x$, and the non-triviality axiom $\exists y \colon y \not\approx 0$. In such structures every clause can be transformed into an equivalent clause without unshielded variables. Integrating this variable elimination algorithm into cancellative superposition results in a calculus that requires neither extended clauses, nor variable overlaps, nor explicit inferences with the theory axioms. Furthermore, using full abstraction even AC unification can be avoided (Waldmann [10]).

When we want to work with a transitive relation $>$ in a theorem prover, we encounter a situation that is surprisingly similar to the one depicted above. Just as associativity and commutativity, the transitivity axiom is fairly prolific. It allows to derive a new clause whenever the left hand side of a literal $r > s$ overlaps with the right hand side of another literal $s' > t$. As such an overlap is always possible if s or s' is a variable, unification is not an effective filter to control the generation of new clauses. The use of the chaining inference rule makes explicit inferences with the transitivity axiom superfluous (Slagle [6]). Since this inference rule can be equipped with the restriction that the overlapped term s must be maximal with respect to a simplification ordering $\succ$, overlaps with shielded variables become again unnecessary. Only inferences with unshielded, i. e., potentially maximal, variables have to be computed.

Once more, the number of unshielded variables in a clause can be reduced if further axioms are available. In particular, in dense total orderings without endpoints, unshielded variables can be eliminated completely (Bachmair and Ganzinger [3]).

There are two facts that suggest to investigate the combination of the theory of divisible torsion-free abelian groups and the theory of dense total orderings without endpoints. On the one hand, the vast majority of applications of divisible torsion-free abelian groups (and in particular of the rationals or reals) requires also an ordering; so the combined calculus is likely to be much more useful in practice than the DTAG-superposition calculus on which it is based. On the other hand, these two theories are closely related: An abelian group $(G, +, 0)$ can be equipped with a total ordering that is compatible with $+$ if and only if it is torsion-free; furthermore divisibility and compatibility of the ordering imply that the ordering is dense and has no endpoints. One can thus assume that the two calculi fit together rather smoothly. We show in this paper that this is in fact true. The resulting calculus splits again into two parts: The first one is a base calculus, that works on clauses without unshielded variables, but whose rules may produce clauses with unshielded variables. This calculus has the property that saturated sets of clauses are unsatisfiable if and only if they contain the empty clause, but it can not be used to effectively saturate a given set of clauses. The second part of the calculus is a variable elimination algorithm that makes it possible to get rid of unshielded variables, and thus renders the base calculus effective. The integration of these two components happens in essentially the same way as in the equational case (Waldmann [10]).

In this extended abstract, we can only sketch the main ideas of our work. The reader is referred to the full version [11] for the proofs.

2   The Base Calculus

2.1   Preliminaries

We work in a many-sorted framework and assume that the function symbol $+$ is declared on a sort G. If t is a term of sort G and $n \in \mathbf{N}$, then nt is an abbreviation for the n-fold sum $t + \cdots + t$; in particular, $0t = 0$ and $1t = t$. Analogously, $\sum_{i \in \{1,\ldots,n\}} t_i$ is an abbreviation for the sum $t_1 + \cdots + t_n$.

Without loss of generality we assume that the equality relation $\approx$ and the semantic ordering $>$ are the only predicates of our language. Hence a literal is either an equation $t \approx t'$, or a negated equation $t \not\approx t'$, where t and t' have the same sort, or an inequation $t > t'$, or a negated inequation $t \not> t'$, where t and t' have sort G. Occasionally we write $t' < t$ instead of $t > t'$. The symbol $\gtrless$ denotes either $>$ or $<$, the symbol $\sim$ stands for $\gtrless$ or $\approx$, and $\dot\sim$ denotes $\gtrless$ or $\approx$ or $\not\approx$. The equality symbol is supposed to be symmetric. Multiple occurrences of one of the symbols $\gtrless$, $\sim$, or $\dot\sim$ within a single inference rule denote consistently the same relation. A clause is a finite multiset of literals, usually written as a disjunction.

The clauses

$$(x + y) + z \approx x + (y + z) \qquad \text{(Associativity (A))}$$
$$x + y \approx y + x \qquad \text{(Commutativity (C))}$$
$$x + 0 \approx x \qquad \text{(Identity (U))}$$
$$(-x) + x \approx 0 \qquad \text{(Inverse (Inv))}$$
$$n \; divided\text{-}by_n(x) \approx x \qquad \text{(Divisibility (Div))}$$
$$a_0 \not\approx 0 \qquad \text{(Non-Triviality (Nt))}$$
$$x \not> x \qquad \text{(Irreflexivity (Ir))}$$
$$x \not> y \vee y \not> z \vee x > z \qquad \text{(Transitivity (Tr))}$$
$$x \not> y \vee x + z > y + z \qquad \text{(Monotonicity (Mon))}$$
$$x > y \vee y > x \vee x \approx y \qquad \text{(Totality (Tot))}$$

plus the equality axioms[1] are the axioms ODAG of totally ordered divisible abelian groups.

---

[1] including the congruence axiom $x \not\approx y \vee y \not\gtrless z \vee x \gtrless z$ for the predicate $>$.

The following clauses are consequences of these axioms (for every $\psi \in \mathbf{N}^{>0}$):

$$x + z \not\approx y + z \lor x \approx y \qquad \text{(Cancellation (K))}$$
$$\psi x \not\approx \psi y \lor x \approx y \qquad \text{(Torsion-Freeness (T))}$$
$$x + z \not> y + z \lor x > y \qquad \text{(>-Cancellation (K}^>\text{))}$$
$$\psi x \not> \psi y \lor x > y \qquad \text{(>-Torsion-Freeness (T}^>\text{))}$$

We write OTfCAM for the union of the clauses A, C, U, Ir, Tr, Mon, K, T, $\text{K}^>$, $\text{T}^>$ and the equality axioms.

We denote the entailment relation modulo ODAG by $\models_{\text{ODAG}}$, and the entailment relation modulo OTfCAM by $\models_{\text{OTfCAM}}$. That is, $\{C_1, \ldots, C_n\} \models_{\text{ODAG}} C_0$ if and only if $\{C_1, \ldots, C_n\} \cup \text{ODAG} \models C_0$, and $\{C_1, \ldots, C_n\} \models_{\text{OTfCAM}} C_0$ if and only if $\{C_1, \ldots, C_n\} \cup \text{OTfCAM} \models C_0$.

A function symbol is called free, if it is different from 0 and +. A term is called atomic, if it is not a variable and its top symbol is different from +. We say that a term t occurs at the top of s, if there is a position $o \in \text{pos}(s)$ such that $s|_o = t$ and for every proper prefix o' of o, $s(o')$ equals +; the term t occurs in s below a free function symbol, if there is an $o \in \text{pos}(s)$ such that $s|_o = t$ and $s(o')$ is a free function symbol for some proper prefix o' of o. A variable x is called shielded in a clause C, if it occurs at least once below a free function symbol in C, or if it does not have sort G. Otherwise, x is called unshielded.

A clause C is called fully abstracted, if no non-variable term of sort G occurs below a free function symbol in C. Every clause C can be transformed into an equivalent fully abstracted clause $\text{abs}(C)$ by iterated rewriting

$$C[f(\ldots, t, \ldots)] \rightarrow x \not\approx t \lor C[f(\ldots, x, \ldots)],$$

where x is a new variable and t is a non-variable term of sort G occurring immediately below the free function symbol f in C.

We say that an ACU-compatible ordering $\succ$ has the multiset property, if whenever a ground atomic term u is greater than v_i for every i in a finite nonempty index set I, then $u \succ \sum_{i \in I} v_i$. Every reduction ordering over terms not containing + that is total on ground terms and for which 0 is minimal can be extended to an ordering that is ACU-compatible and has the multiset property (Waldmann [9]).[2]

From now on we will work only with ACU-congruence classes, rather than with terms. So all terms, equations, substitutions, inference rules, etc., are to be taken modulo ACU, i.e., as representatives of their congruence classes. The symbol $\succ$ will always denote an ACU-compatible ordering that has the multiset property, is total on ground ACU-congruence classes, and satisfies $t \not\succ s[t]_o$ for every term $s[t]_o$.

Let A be a ground literal. Then the largest atomic term occurring on either side of A is denoted by $\text{mt}(A)$.

---

[2] In fact, we use the extended ordering only as a theoretical device; as we work with fully abstracted clauses, the original reduction ordering is sufficient for actual computations.

The ordering $\succ_L$ on literals compares lexicographically first the maximal atomic terms of the literals, then the polarities (negative $\succ$ positive), then the kinds of the literals (inequation $\succ$ equation), then the number of the sides of the literals on which the maximal atomic term occurs, then the multisets of all non-zero terms occurring at the top of the literals, and finally the multisets $\{\{s\}, \{t\}\}$ (for equations $[\neg]\, s \approx t$) or $\{\{s, s\}, \{t\}\}$ (for inequations $[\neg]\, s > t$). The ordering $\succ_C$ on clauses is the multiset extension of the literal ordering $\succ_L$. Both $\succ_L$ and $\succ_C$ are noetherian and total on ground literals/clauses.

2.2 Superposition and Chaining

We present the ground versions of the inference rules of the base calculus $OCInf$. The non-ground versions can be obtained by lifting in a rather straightforward way (see below).

Let us start the presentation of the inference rules with a few general conventions: Every term occurring in a sum is assumed to have sort G. The letters u and v, possibly with indices, denote atomic terms, unless explicitly said otherwise. In an expression like $mu + s$, m is a natural number, s may be zero.

If an inference involves a literal, then it must be maximal in the respective clause (except for the last but one literal in *factoring* inferences). A positive literal that is involved in a *superposition* or *chaining* inference must be strictly maximal in the respective clause. In all *superposition* or *chaining* inferences, the left premise is smaller than the right premise.

Cancellation
$$\frac{C' \vee mu + s \mathrel{\dot\sim} m'u + s'}{C' \vee (m - m')u + s \mathrel{\dot\sim} s'}$$
if $m \geq m' \geq 1$, $u \succ s$, $u \succ s'$.

Equality Resolution
$$\frac{C' \vee u \not\approx u}{C'}$$
if u either equals 0 or does not have sort G.

Inequality Resolution
$$\frac{C' \vee 0 > 0}{C'}$$

Canc. Superposition
$$\frac{D' \vee nu + t \approx t' \qquad C' \vee mu + s \mathrel{\dot\sim} s'}{D' \vee C' \vee ns + mt' \mathrel{\dot\sim} ns' + mt}$$
if $n \geq 1$, $m \geq 1$, $u \succ s$, $u \succ s'$, $u \succ t$, $u \succ t'$.[3]

Canc. Chaining
$$\frac{D' \vee t' \gtrsim nu + t \qquad C' \vee mu + s \gtrsim s'}{D' \vee C' \vee ns + mt' \gtrsim ns' + mt}$$
if $n \geq 1$, $m \geq 1$, $u \succ s$, $u \succ s'$, $u \succ t$, $u \succ t'$.

---

[3] If $\gcd(m, n) > 1$, then the conclusion of this inference can be simplified to $D' \vee C' \vee \psi s + \chi t' \mathrel{\dot\sim} \psi s' + \chi t$, where $\psi = n/\gcd(m, n)$ and $\chi = m/\gcd(m, n)$ (and similarly for the following inference rules). To enhance readability, we leave out this optimization in the sequel.

Std. Superposition

$$\frac{D' \vee u \approx u' \qquad C' \vee s[u] \stackrel{\sim}{\sim} s'}{D' \vee C' \vee s[u'] \stackrel{\sim}{\sim} s'}$$

if u occurs in a maximal atomic subterm of s and does not have sort G, $u \succ u'$, $s[u] \succ s'$.

Canc. Eq. Factoring

$$\frac{C' \vee nu + t \approx t' \vee mu + s \approx s'}{C' \vee mt + ns' \not\approx mt' + ns \vee nu + t \approx t'}$$

if $n \geq 1$, $m \geq 1$, $u \succ s$, $u \succ s'$, $u \succ t$, $u \succ t'$.

Canc. Ineq. Factoring (I)

$$\frac{C' \vee nu + t \gtrsim t' \vee mu + s \gtrsim s'}{C' \vee mt + ns' \gtrsim mt' + ns \vee mu + s \gtrsim s'}$$

if $n \geq 1$, $m \geq 1$, $u \succ s$, $u \succ s'$, $u \succ t$, $u \succ t'$.

Canc. Ineq. Factoring (II)

$$\frac{C' \vee nu + t \gtrsim t' \vee mu + s \gtrsim s'}{C' \vee mt' + ns \gtrsim mt + ns' \vee nu + t \gtrsim t'}$$

if $n \geq 1$, $m \geq 1$, $u \succ s$, $u \succ s'$, $u \succ t$, $u \succ t'$.

Std. Eq. Factoring

$$\frac{C' \vee u \approx v' \vee u \approx u'}{C' \vee u' \not\approx v' \vee u \approx v'}$$

if u, u' and v' do not have sort G, $u \succ u'$, $u \succ v'$.

The inference rules of the calculus *OCInf* do not handle negative inequality literals. We assume that in the beginning of the saturation process every literal $s \not> t$ in an input clause is replaced by the two literals $t > s \vee t \approx s$, which are equivalent to $s \not> t$ by the totality, transitivity and irreflexivity axioms. Note that the inference rules of *OCInf* do not produce any new negative inequality literals.

Example 1. Let the ordering on constant symbols be given by $b \succ c \succ d$. We will use the inference rules of *OCInf* to show that the following three clauses are contradictory with respect to ODAG. (The maximal parts of every clause are underlined.)

$$\underline{3b} > 2d \tag{1}$$

$$\underline{b} > 2c \tag{2}$$

$$\underline{2b} \approx c + d \tag{3}$$

Cancellative superposition of (3) and (1) yields

$$\underline{3c} + 3d > 4d \tag{4}$$

Cancellative superposition of (3) and (2) yields

$$\underline{c} + d > \underline{4c} \tag{5}$$

By *cancellation* of (5) we obtain

$$d > \underline{3c} \tag{6}$$

Cancellative chaining of (6) and (4) produces

$$\underline{4d} > \underline{4d} \tag{7}$$

which yields the empty clause by *cancellation* and *inequality resolution*.

In the standard superposition calculus, lifting means replacing equality in the ground inference by unifiability. As long as all variables in our clauses are shielded, the situation is similar here: For instance, in the second premise $C' \vee A_1$ of a *cancellative superposition* inference the maximal literal A_1 need no longer have the form $mu + s \mathrel{\dot\sim} s'$ with a unique maximal atomic term u. Rather, it may contain several (distinct but ACU-unifiable) maximal atomic terms u_k with multiplicities m_k, where k ranges over some finite non-empty index set K. We obtain thus $A_1 = \sum_{k \in K} m_k u_k + s \mathrel{\dot\sim} s'$. In the inference rule, the substitution σ that unifies all u_k (and the corresponding terms v_l from the other premise) is applied to the conclusion. Consequently, the *cancellative superposition* rule has now the following form:

$$\frac{D' \vee \sum_{l \in L} n_l v_l + t \approx t' \qquad C' \vee \sum_{k \in K} m_k u_k + s \mathrel{\dot\sim} s'}{(D' \vee C' \vee ns + mt' \mathrel{\dot\sim} ns' + mt)\sigma}$$

where

(i) $m = \sum_{k \in K} m_k \geq 1$, $n = \sum_{l \in L} n_l \geq 1$.
(ii) σ is a most general ACU-unifier of all u_k and v_l ($k \in K, l \in L$).
(iii) u is one of the u_k ($k \in K$).
(iv) $u\sigma \not\preceq s\sigma$, $u\sigma \not\preceq s'\sigma$, $u\sigma \not\preceq t\sigma$, $u\sigma \not\preceq t'\sigma$.

The other inference rules can be lifted in a similar way, again under the condition that all variables in the clauses are shielded. As usual, the *standard superposition* rule is equipped with the additional restriction that the subterm of s that is replaced during the inference is not a variable. For clauses with unshielded variables, lifting would be significantly more complicated; however, as we will combine the base calculus with an algorithm that eliminates unshielded variables, we need not consider this case.

Theorem 2. *The inference rules of the calculus OCInf are sound with respect to* $\models_{\text{ODAG}}$.

Definition 3. Let N be a set of clauses, let $\overline{N}$ be the set of ground instances of clauses in N. An inference is called *OCRed*-redundant with respect to N if for each of its ground instances with conclusion $C_0\theta$ and maximal premise $C\theta$ we have $\{ D \in \overline{N} \mid D \prec_{\text{c}} C\theta \} \models_{\text{OTfCAM}} C_0\theta$. A clause C is called *OCRed*-redundant with respect to N, if for every ground instance $C\theta$, $\{ D \in \overline{N} \mid D \prec_{\text{c}} C\theta \} \models_{\text{OTfCAM}} C\theta$.

2.3   Rewriting on Equations

To prove that the inference system described so far is refutationally complete we have to show that every saturated clause set that does not contain the empty clause has a model. The traditional approach to construct such a model is rewrite-based: First an ordering is imposed on the set of all ground instances of clauses in the set. Starting with an empty interpretation all such instances are inspected in ascending order. If a reductive clause is false and irreducible in the partial interpretation constructed so far, its maximal positive literal is turned into a rewrite rule and added to the interpretation. If the original clause set is saturated and does not contain the empty clause, then the final interpretation is a model of all ground instances, and thus of the original clause set (Bachmair and Ganzinger [2]).

In order to be able to treat cancellative superposition we have modified this scheme in [4] in such a way that the rewrite relation operates on equations rather than on terms. But if we also have to deal with inequations, a further extension is necessary: We need to be able to rewrite inequations *with inequations*; and unlike rewriting with equations, this does of course not produce logically equivalent formulae.

Definition 4. A ground equation or inequation e is called a cancellative rewrite rule with respect to $\succ$, if $mt(e)$ does not occur on both sides of e.

We will usually drop the attributes "cancellative" and "with respect to $\succ$", speaking simply of "rewrite rules".

Every rewrite rule has either the form $mu + s \sim s'$, where u is an atomic term, $m \in \mathbf{N}^{>0}$, $u \succ s$, and $u \succ s'$, or the form $u \approx s'$, where $u \succ s'$ and u (and thus s') does not have sort G. This is an easy consequence of the multiset property of $\succ$.

Definition 5. Given a set R of rewrite rules, the four binary relations $\to_{\gamma,R}$, $\to_{\delta,R}$, $\to_{o,R}$, and $\to_\kappa$ on ground equations and inequations are defined (modulo ACU) as follows:[4]

(i) $mu + t \sim t' \to_{\gamma,R} s' + t \sim t' + s$,
   if $mu + s \approx s'$ is a rule in R.

(ii) $t[s] \sim t' \to_{\delta,R} t[s'] \sim t'$,
   if (i) $s \approx s'$ is a rule in R and (ii) s does not have sort G or s occurs in t below some free function symbol.

(iii) $mu + t \gtrless t' \to_{o,R} s' + t \gtrless t' + s$,
   if $mu + s \gtrless s'$ is a rule in R.

(iv) $u + t \sim u + t' \to_\kappa t \sim t'$,
   $u \approx u \to_\kappa 0 \approx 0$,
   if u is atomic and different from 0.

---

[4] While we have the restriction $u \succ s$, $u \succ s'$ for the rewrite rules, there is no such restriction for the (in-)equations to which rules are applied.

The union of $\to_{\gamma,R}$, $\to_{\delta,R}$, $\to_{o,R}$, and $\to_\kappa$ is denoted by $\to_R$.[5]

If $e \to_R e'$ using a γ-, δ- or κ-step, then e and e' are equivalent modulo OTfCAM and the applied rewrite rule. If $s \gtrless s' \to_{o,R} t \gtrless t'$, then both $t \gtrless t'$ and $t \approx t'$ imply $s \gtrless s'$ modulo OTfCAM and the applied rewrite rule.

We say that an (in-)equation e is γ-reducible, if $e \to_\gamma e'$ (analogously for δ, o, and κ). It is called reducible, if it is γ-, δ-, o-, or κ-reducible.

Unlike o- and κ-reducibility, γ- and δ-reducibility can be extended to terms: A term t is called γ-reducible, if $t \approx t' \to_\gamma e'$, where the rewrite step takes place at the left-hand side (analogously for δ). It is called reducible, if it is γ- or δ-reducible.

Lemma 1. *The relation $\to_R$ is contained in $\succ_L$ and thus noetherian.*

Definition 6. Given a set R of rewrite rules, the relation $\to_R^\circ$ is defined by
$$\to_R^\circ = (\to_R^* \circ \to_{o,R} \circ \to_R^*).$$

Definition 7. Given a set R of rewrite rules, the truth set $\mathrm{tr}(R)$ of R is the set of all equations $s \approx s'$ for which there exists a derivation $s \approx s' \to_R^* 0 \approx 0$, and the set of all inequations $s \gtrless s'$ for which there exists a derivation $s \gtrless s' \to_R^\circ 0 \gtrless 0$. The Ψ-truth set $\mathrm{tr}_\Psi(R)$ of R is the set of all equations or inequations $e = s \sim s'$, such that either $e \in \mathrm{tr}(R)$ and s does not have sort G, or $\psi s \sim \psi s' \in \mathrm{tr}(R)$ for some $\psi \in \mathbf{N}^{>0}$.

All (in-)equations in $\mathrm{tr}_\Psi(R)$ are logical consequences of the rewrite rules in R and the theory axioms OTfCAM.

2.4 Model Construction

Definition 8. A ground clause $C' \vee e$ is called reductive for e, if e is a cancellative rewrite rule and strictly maximal in $C' \vee e$.

Definition 9. Let N be a set of (possibly non-ground) clauses that does not contain the empty clause, and let $\overline{N}$ the set of all ground instances of clauses in N. Using induction on the clause ordering we define sets of rules R_C, R_C^Ψ, E_C, and E_C^Ψ, for all clauses $C \in \overline{N}$. Let C be such a clause and assume that R_D, R_D^Ψ, E_D, and E_D^Ψ have already been defined for all $D \in \overline{N}$ such that $C \succ_c D$. Then the set R_C of primary rules and the set R_C^Ψ of secondary rules are given by

$$R_C = \bigcup_{D \prec_c C} E_D \quad \text{and} \quad R_C^\Psi = \bigcup_{D \prec_c C} E_D^\Psi.$$

---

[5] As we deal only with ground terms and as there are no non-trivial contexts around (in-)equations, this operation does indeed satisfy the definition of a rewrite relation, albeit in an unorthodox way.

E_C is the singleton set $\{e\}$, if C is a clause $C' \vee e$ such that (i) C is reductive for e, (ii) C is false in $\mathrm{tr}(R_C^\Psi)$, (iii) C' is false in $\mathrm{tr}_\Psi(R_C^\Psi \cup \{e\})$, and (iv) $\chi\, \mathrm{mt}(e)$ is $\gamma\delta$-irreducible with respect to R_C^Ψ for every $\chi \in \mathbf{N}^{>0}$. Otherwise, E_C is empty.

If $E_C = \{e\}$, then E_C^Ψ is the set of all rewrite rules $e' \in \mathrm{tr}_\Psi(R_C^\Psi \cup E_C)$ such that $\mathrm{mt}(e') = \mathrm{mt}(e)$ and e' is $\delta\kappa$-irreducible with respect to R_C^Ψ. Otherwise, E_C^Ψ is empty.

Finally, the sets R_∞ and R_∞^Ψ are defined by

$$R_\infty = \bigcup_{D \in \overline{N}} E_D \quad \text{and} \quad R_\infty^\Psi = \bigcup_{D \in \overline{N}} E_D^\Psi .$$

Our goal is to show that, if N is saturated with respect to $OCInf$, then $\mathrm{tr}(R_\infty^\Psi)$ is a model of the axioms of totally ordered divisible abelian groups and of the clauses in N.

2.5   Refutational Completeness of $OCInf$

The relations $\to_{R_C^\Psi}$ and $\to_{R_\infty^\Psi}$ are in general not confluent, not even in the purely equational case. One can merely show that that $\to_{R_C^\Psi}$ is confluent on equations in $\mathrm{tr}(R_C^\Psi)$, that is, that any two derivations starting from an equation e can be joined, provided that there is a derivation $e \to^+ 0 \approx 0$. But even this kind of restricted confluence does not hold for inequations, and in particular, not for o-rewriting. We can only prove that two derivations starting from the same inequation can be joined, if one of them leads to $0 > 0$ and if the other one does not use o-steps. This property will be sufficient for our purposes, however.

Definition 10. Let E be a set of equations and/or inequations. We say that the relation $\to_R$ is partially confluent on E, if for all equations $e_0 \in E$ and e_1, e_2 with $e_1 \leftarrow_R^* e_0 \to_R^* e_2$ there exists an equation e_3 such that $e_1 \to_R^* e_3 \leftarrow_R^* e_2$, and if for all inequations $e_0' \in E$ and e_1' with $e_1' \leftarrow_{\gamma\delta\kappa,R}^* e_0' \to_R^* 0 > 0$ there is a derivation $e_1' \to_R^* 0 > 0$.

There is one important technical difference between the equational case developed in (Waldmann [8]) and the inequational case that we consider here: In the equational case, one can show that $\to_{R_\infty^\Psi}$ is confluent on $\mathrm{tr}(R_\infty^\Psi)$, and hence that $\mathrm{tr}(R_\infty^\Psi)$ is a model of the theory axioms, without requiring that the set N of clauses is saturated. Saturation is only necessary to prove that $\mathrm{tr}(R_\infty^\Psi)$ is also a model of $\overline{N}$. In the inequational case, such a separation does not work: Proving partial confluence of $\to_{R_\infty^\Psi}$ is only possible if we require that *cancellative chaining* inferences are redundant. For this reason, the proof that $\mathrm{tr}(R_\infty^\Psi)$ is partially confluent and the proof that $\mathrm{tr}(R_\infty^\Psi)$ is a model of $\overline{N}$ must be combined within a single induction.

The following two lemmas are copied almost verbatim from (Waldmann [8]).

Lemma 2. *The relation $\to_{R_C^\Psi}$ is confluent on the equations in $\mathrm{tr}(R_C^\Psi)$ for every $C \in \overline{N}$. The relation $\to_{R_\infty^\Psi}$ is confluent on the equations in $\mathrm{tr}(R_\infty^\Psi)$.*

Corollary 11. *For every $C \in \overline{N}$, $\mathrm{tr}(R_C^\Psi)$ and $\mathrm{tr}(R_\infty^\Psi)$ satisfy ACUKT and the equality axioms (except the congruence axiom for the predicate $>$).*

In a similar way as Lemma 2, we obtain by a rather tedious case analysis over various kinds of critical pairs:

Lemma 3. *If for every pair of rules $mu + s > s'$ and $nu + t < t'$ from $E_C \cup R_C$ the inequation $ns + mt' > ns' + mt$ is contained in $\mathrm{tr}(R_C^\Psi)$, then $\rightarrow_{R_C^\Psi \cup E_C^\Psi}$ is partially confluent on $\mathrm{tr}(R_C^\Psi \cup E_C^\Psi)$.*

Using the same techniques as in (Waldmann [8]) and (Bachmair and Ganzinger [3]) we can now prove the following theorem. Note in particular that in the presence of the totality axiom *cancellative inequality factoring (I)/(II)* inferences are simplifications, hence clauses where the maximal atomic term occurs on the same side of two ordering literals do not produce primary rules.

Theorem 12. *Let N be a set of clauses without negative inequality literals and without unshielded variables; suppose that N is saturated up to redundancy and contains the theory axiom Div, Inv, Nt, and all ground instances of Tot. If all clauses of N, except the ground instances of Tot, are fully abstracted, and if N does not contain the empty clause, then we have for every ground clause $C\theta \in \overline{N}$:*

(i) $E_{C\theta} = \emptyset$ if and only if $C\theta$ is true in $\mathrm{tr}(R_{C\theta}^\Psi)$.

(ii) $C\theta$ is true in $\mathrm{tr}(R_\infty^\Psi)$ and in $\mathrm{tr}(R_D^\Psi)$ for every $D \succ_c C\theta$.

(iii) The relation $\rightarrow_{R_{C\theta}^\Psi}$ is partially confluent on $\mathrm{tr}(R_{C\theta}^\Psi)$ and the relation $\rightarrow_{R_{C\theta}^\Psi \cup E_{C\theta}^\Psi}$ is partially confluent on $\mathrm{tr}(R_{C\theta}^\Psi \cup E_{C\theta}^\Psi)$.

(iv) $\mathrm{tr}(R_{C\theta}^\Psi)$ and $\mathrm{tr}(R_{C\theta}^\Psi \cup E_{C\theta}^\Psi)$ satisfy the axioms Ir, Tr, Mon, $\mathrm{K}^>$, $\mathrm{T}^>$, and the congruence axiom for the predicate $>$.

(v) The relation $\rightarrow_{R_\infty^\Psi}$ is partially confluent on $\mathrm{tr}(R_\infty^\Psi)$ and $\mathrm{tr}(R_\infty^\Psi)$ satisfies the axioms Ir, Tr, Mon, $\tilde{\mathrm{K}}^>$, $\mathrm{T}^>$, and the congruence axiom for the predicate $>$.

Theorem 13. *Let N be a set of clauses without negative inequality literals and without unshielded variables; suppose that N is saturated up to redundancy and contains the theory axiom Div, Inv, Nt, and all ground instances of Tot. Suppose that all clauses of N, except the ground instances of Tot, are fully abstracted. Then $N \cup \mathrm{ODAG}$ is unsatisfiable if and only if N contains the empty clause.*

We may assume without loss of generality that the constant a_0 does not occur in non-theory input clauses and that the function symbols $-$ and *divided-by$_n$* are eliminated eagerly from all non-theory input clauses. In this case, no inferences are possible with the axioms Div, Inv, and Nt. Furthermore, one can show that inferences with the totality axiom Tot are always redundant (analogously to Bachmair and Ganzinger [3]).

3 The Extended Calculus

3.1 Variable Elimination

As we have mentioned in the introduction, the calculus *OCInf* works on clauses without unshielded variables, but its inference rules may produce clauses with unshielded variables. To make it effectively saturate a given set of clauses, it has to be supplemented by a variable elimination algorithm.

In the equational case, every clause with unshielded variables can be transformed into an equivalent clause without unshielded variables. However, in the presence of ordering literals, this does no longer hold.

Example 14. Consider the clause $C = x > a \lor x \approx b \lor x < c$. This clause is true for every value of x, if either $c > a$ or both $a \approx b$ and $c \approx b$. So C can be replaced by the clause normal form of $c > a \lor (a \approx b \land c \approx b)$, that is, by the two clauses $c > a \lor a \approx b$ and $c > a \lor c \approx b$, but C is not equivalent to a single clause without unshielded variables.

For any disjunction of conjunctions of literals F let $\mathrm{CNF}(F)$ be the clause normal form of F (represented as a multiset of clauses).

Let x be a variable of sort G. We define a binary relation $\to_x$ over multisets of clauses by

CancelVar $M \cup \{C' \lor mx + s \mathbin{\dot\sim} m'x + s'\} \;\to_x$
$M \cup \{C' \lor (m-m')x + s \mathbin{\dot\sim} s'\}$
if $m \geq m' \geq 1$.

ElimNeg $M \cup \{C' \lor mx + s \not\approx s'\} \;\to_x$
$M \cup \{C'\}$
if $m \geq 1$ and x does not occur in C', s, s'.

ElimPos $M \cup \{C' \lor \bigvee_{i \in I} l_i x + r_i \approx r'_i \lor \bigvee_{j \in J} m_j x + s_j > s'_j$
$\lor \bigvee_{k \in K} n_k x + t_k < t'_k\} \;\to_x$
$M \cup \mathrm{CNF}(C' \lor \bigvee_{j \in J} \bigvee_{k \in K} (n_k s_j + m_j t'_k > n_k s'_j + m_j t_k$
$\lor \bigvee_{i \in I}(l_i s_j + m_j r'_i \approx l_i s'_j + m_j r_i \land l_i t_k + n_k r'_i \approx l_i t'_k + n_k r_i)))$
if $I \cup J \cup K \neq \emptyset$, $l_i \geq 1$, $m_j \geq 1$, $n_k \geq 1$ and x does not occur in
$C', r_i, r'_i, s_j, s'_j, t_k, t'_k$, for $i \in I$, $j \in J$, $k \in K$.

Coalesce $M \cup \{C' \lor mx + s \not\approx s' \lor nx + t \mathbin{\dot\sim} t'\} \;\to_x$
$M \cup \{C' \lor mx + s \not\approx s' \lor mt + ns' \mathbin{\dot\sim} mt' + ns\}$
if $m \geq 1$, $n \geq 1$, and x does not occur in s, s', t, t'.

It is easy to show that $\to_x$ is noetherian. We define the relation $\to_{\mathrm{elim}}$ over multisets of clauses in such a way that $M \cup \{C\} \to_{\mathrm{elim}} M \cup M'$ if and only if C contains an unshielded variable x and M' is a normal form of $\{C\}$ with respect to $\to_x$.

The relation $\to_{\mathrm{elim}}$ is again noetherian. For a clause C, $\mathrm{elim}(C)$ denotes some (arbitrary but fixed) normal form of $\{C\}$ with respect to the relation $\to_{\mathrm{elim}}$.

Corollary 15. *For any C, the clauses in* $\mathrm{elim}(C)$ *contain no unshielded variables.*

Lemma 4. *For every C,* $\{C\} \models_{\mathrm{ODAG}} \mathrm{elim}(C)$ *and* $\mathrm{elim}(C) \cup \mathrm{Tot} \models_{\mathrm{OTfCAM}} C$. *For every ground instance $C\theta$,* $\mathrm{elim}(C)\theta \cup \mathrm{Tot} \models_{\mathrm{OTfCAM}} C\theta$.

3.2   Integration of the Elimination Algorithm

Using the technique sketched so far, every clause C_0 can be transformed into a set of clauses $\mathrm{elim}(C_0)$ that do not contain unshielded variables, follow from C_0 and the axioms of totally ordered divisible abelian groups, and imply C_0 modulo OTfCAM $\cup$ Tot. Obviously, we can perform this transformation for all initially given clauses *before* we start the saturation process. However, when clauses with unshielded variables are produced during the saturation process, then logical equivalence is not sufficient to eliminate them. We have to require that the transformed set of clauses $\mathrm{elim}(C_0)$ makes the inference ι producing C_0 redundant. Unfortunately, it may happen that the clauses in $\mathrm{elim}(C_0)$ or the instances of the totality axiom needed in Lemma 4 are too large, at least for some instances of ι. To integrate the variable elimination algorithm into the base calculus, it has to be supplemented by a case analysis technique.

Let $k \in \{1, 2\}$, let $C_1, \ldots, C_k$ be clauses without unshielded variables and let ι be an *OCInf*-inference

$$\frac{C_k \ \ldots \ C_1}{C_0\sigma}$$

We call the unifying substitution σ that is computed during ι and applied to the conclusion the pivotal substitution of ι. (For ground inferences, the pivotal substitution is the identity mapping.) If the last premise C_1 has the form $C_1' \vee A$ where A is maximal (and the replacement or cancellation takes place at A) then we call $A\sigma$ the pivotal literal of ι. Finally, if u_0 is the atomic term that is cancelled out in ι, or in which some subterm is replaced,[6] then we call $u_0\sigma$ the pivotal term of ι.

Two properties of pivotal terms are important for us: First, whenever an inference ι from clauses *without* unshielded variables produces a conclusion *with* unshielded variables, then all these unshielded variables occur in the pivotal term of ι. Second, no atomic term in the conclusion of ι can be larger than the pivotal term of ι.

One can now show that, if the clauses in $\mathrm{elim}(C_0)$ or the instances of the totality axiom needed in Lemma 4 are too large to make the *OCInf*-inference ι redundant, then there must be an atomic term in some clause in $\mathrm{elim}(C_0)$ that is unifiable with the pivotal term. If we apply the unifier to the conclusion of the *OCInf*-inference, then the result does no longer contain unshielded variables, and

---

[6] More precisely, u_0 is the maximal atomic subterm of s containing u in *standard superposition* inferences, and the term u in all other inferences.

moreover it subsumes the critical instances of ι. Using this result, we can now transform the inference system $OCInf$ into a new inference system that operates on clauses without unshielded variables and produces again such clauses. The new system $ODInf$ is given by two meta-inference rules:

Eliminating Inference

$$\frac{C_n \quad \ldots \quad C_1}{C'}$$

if the following conditions are satisfied:

(i) $\dfrac{C_n \quad \ldots \quad C_1}{C_0}$ is a $OCInf$-inference.

(ii) $C' \in \mathrm{elim}(C_0)$.

Instantiating Inference

$$\frac{C_n \quad \ldots \quad C_1}{C_0\tau}$$

if the following conditions are satisfied:

(i) $\dfrac{C_n \quad \ldots \quad C_1}{C_0}$ is a $OCInf$-inference with pivotal literal A and pivotal term u.

(ii) $\mathrm{elim}(C_0) \neq \{C_0\}$.

(iii) A literal A_1 with the same polarity as A occurs in some clause in $\mathrm{elim}(C_0)$.

(iv) An atomic term u_1 occurs at the top of A_1.

(v) τ is contained in a minimal complete set of ACU-unifiers of u and u_1.

We define the redundancy criterion for the new inference system in such a way, that an $ODInf$-inference is redundant, if the appropriate instances of its parent $OCInf$-inference are redundant. Then a set of clauses without unshielded variables that is saturated with respect to $ODInf$ up to redundancy is also saturated with respect to $OCInf$ up to redundancy. $ODInf$ can thus be used for effective saturation of a given set of input clauses:

Theorem 16. *Let N_0 be a set of clauses without negative inequality literals and without unshielded variables; let N_0 contain the theory axiom Div, Inv, Nt, and all ground instances of Tot. Suppose that all clauses of N_0, except the ground instances of Tot, are fully abstracted. Let $N_0 \vdash N_1 \vdash N_2 \vdash \ldots$ be a fair $ODInf$-derivation. Let N_∞ be the limit of the derivation. Then $N_0 \cup \mathrm{ODAG}$ is unsatisfiable if and only if N_∞ contains the empty clause.*

4   Conclusions

We have presented a superposition-based calculus for first-order theorem proving in the presence of the axioms of totally ordered divisible abelian groups. It is based on the DTAG-superposition calculus from (Waldmann [10]) and the ordered chaining calculus for dense total orderings without endpoints (Bachmair and Ganzinger [3]), and it shares the essential features of these two calculi: It is refutationally complete, it does not require explicit inferences with the theory clauses, and due to the integrated variable elimination algorithm it does not require variable overlaps. It offers thus an efficient way of treating equalities and inequalities between additive terms over, e.g., the rational numbers within a first-order theorem prover.

Acknowledgments. I would like to thank the anonymous IJCAR referees for helpful comments on this paper.

References

1. Leo Bachmair and Harald Ganzinger.   Associative-commutative superposition. In Nachum Dershowitz and Naomi Lindenstrauss, eds., *Conditional and Typed Rewriting Systems, 4th International Workshop, CTRS-94*, Jerusalem, Israel, July 13–15, 1994, LNCS 968, pp. 1–14. Springer-Verlag.
2. Leo Bachmair and Harald Ganzinger. Rewrite-based equational theorem proving with selection and simplification. *Journal of Logic and Computation*, 4(3):217–247, 1994.
3. Leo Bachmair and Harald Ganzinger. Ordered chaining calculi for first-order theories of transitive relations. *Journal of the ACM*, 45(6):1007–1049, November 1998.
4. Harald Ganzinger and Uwe Waldmann. Theorem proving in cancellative abelian monoids (extended abstract). In Michael A. McRobbie and John K. Slaney, eds., *Automated Deduction – CADE-13, 13th International Conference on Automated Deduction*, New Brunswick, NJ, USA, July 30–August 3, 1996, LNAI 1104, pp. 388–402. Springer-Verlag.
5. Claude Marché. Normalized rewriting: an alternative to rewriting modulo a set of equations. *Journal of Symbolic Computation*, 21(3):253–288, March 1996.
6. James R. Slagle. Automatic theorem proving with built-in theories including equality, partial ordering, and sets. *Journal of the ACM*, 19(1):120–135, January 1972.
7. Jürgen Stuber. Superposition theorem proving for abelian groups represented as integer modules. In Harald Ganzinger, ed., *Rewriting Techniques and Applications, 7th International Conference, RTA-96*, New Brunswick, NJ, USA, July 27–30, 1996, LNCS 1103, pp. 33–47. Springer-Verlag.
8. Uwe Waldmann. *Cancellative Abelian Monoids in Refutational Theorem Proving.* Dissertation, Universität des Saarlandes, Saarbrücken, Germany, 1997. http://www.mpi-sb.mpg.de/~uwe/paper/PhD.ps.gz.
9. Uwe Waldmann.   Extending reduction orderings to ACU-compatible reduction orderings. *Information Processing Letters*, 67(1):43–49, July 16, 1998.

10. Uwe Waldmann. Superposition for divisible torsion-free abelian groups. In Claude Kirchner and Hélène Kirchner, eds., *Automated Deduction – CADE-15, 15th International Conference on Automated Deduction*, Lindau, Germany, July 5–10, 1998, LNAI 1421, pp. 144–159. Springer-Verlag.
11. Uwe Waldmann. Superposition and chaining for totally ordered divisible abelian groups. Technical Report MPI-I-2001-2-001, Max-Planck-Institut für Informatik, Saarbrücken, Germany, 2001.
12. Ulrich Wertz. First-order theorem proving modulo equations. Technical Report MPI-I-92-216, Max-Planck-Institut für Informatik, Saarbrücken, Germany, April 1992.

Context Trees[*]

Harald Ganzinger[1], Robert Nieuwenhuis[2], and Pilar Nivela[2]

[1] Max-Planck-Institut für Informatik, 66123 Saarbrücken, Germany.
hg@mpi-sb.mpg.de
[2] Technical University of Catalonia, Jordi Girona 1, 08034 Barcelona, Spain.
roberto@lsi.upc.es   nivela@lsi.upc.es

Abstract. Indexing data structures have a crucial impact on the performance of automated theorem provers. Examples are discrimination trees, which are like tries where terms are seen as strings and common prefixes are shared, and substitution trees, where terms keep their tree structure and all common contexts can be shared. Here we describe a new indexing data structure, called context trees, where, by means of a limited kind of context variables, also common subterms can be shared, even if they occur below different function symbols. Apart from introducing the concept, we also provide evidence for its practical value. We describe an implementation of context trees based on Curry terms and on an extension of substitution trees with equality constraints, where one also does not distinguish between internal and external variables. Experiments with matching benchmarks show that our preliminary implementation is already competitive with tightly coded current state-of-the-art implementations of the other main techniques. In particular space consumption of context trees is significantly less than for other index structures.

1 Introduction

Indexing data structures have a crucial impact on the performance of theorem provers. The indexes have to store a large number of terms and to support the fast retrieval, for any given *query* term t, of all terms in the index satisfying a certain relation with t, such as matching, unifiability, or syntactic equality. Indexing for matching, where, to check for forward redundancy, one searches in the index for a generalization of the query term, is well-known to be the most limiting bottleneck in practice. Another aspect which is becoming more and more crucial is memory consumption. During the last years processor speed has been growing much faster than memory capacity and one may assume that this gap will become even wider in the coming years. At the same time memory access bandwidth is also becoming an important bottleneck. Excessive memory consumption leads to more cache faults, which become the dominant factor for

---

[*] The second and third author are partially supported by the Spanish CICYT project HEMOSS ref. TIC98-0949-C02-01. All test programs, implementations and benchmarks mentioned in this paper are available at www.lsi.upc.es/~roberto.

R. Goré, A. Leitsch, and T. Nipkow (Eds.): IJCAR 2001, LNAI 2083, pp. 242–256, 2001.

time, instead of processor speed. Therefore, in what follows we will mainly focus on matching retrieval operations and on memory consumption.

One important aspect makes indexing techniques in theorem proving essentially different from indexing in other contexts like functional or logic programming: the index is subject to insertions and deletions. Therefore, during the last two decades a significant number of results on new specific indexing techniques for theorem proving have been published and applied in different provers. The currently best-known and most frequently used indexing techniques for matching are *discrimination trees* [1,4], the compiled variant of discrimination trees, called *code trees* [9], and *substitution trees* [2].

Discrimination trees are like tries where terms are viewed as strings and where common prefixes are shared. A substitution tree has, in each node, a substitution, a list of pairs $x_i \mapsto t$ where each x_i is an *internal* variable and t is a term that may contain other internal variables as well as *external* variables which are the variables in the terms to be stored.

Example 1. The two terms $f(a, g(x), h(y))$ and $f(h(b), g(y), h(y))$ will be stored in a substitution tree and discrimination tree, respectively, as shown:

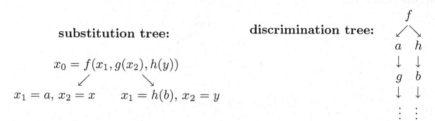

In a substitution tree all terms $x_0\sigma$ are stored such that σ is the composition of the substitutions on some path from the root to a leaf of the tree. In the example, after inserting the first term in an empty substitution tree we obtain the single node $x_0 = f(a, g(x), h(y))$. When inserting the second term, internal variables are placed at the points of disagreement, and children are created with the "remaining" substitutions of both. Therefore all common contexts can be shared.   □

Example 2. It clear that the additional sharing in substitution trees avoids repeated work (which is the main goal of all indexing techniques). Assume one has two terms $f(c, x, t)$ and $f(x, c, t)$ in the index, and a query $f(c, c, s)$, where s and t are terms such that s is not an instance of t. Then two attempts to match s against t will be made in a discrimination tree, and only one in a substitution tree. But, on the other hand, in substitution trees the basic traversal algorithms are significantly more costly.   □

Here we describe a new indexing data structure, called *context trees*, where, by means of a limited kind of context variables, certain common subterms can be shared, even if they occur below different function symbols. Roughly, the idea is that $f(s)$ and $g(s,t)$ can be represented as $F(s,t)$, with children $F = f$ and $F = g$. Function variables such as F stand for single function symbols only (although extensions to allow for more complex forms of second-order terms are possible).

Example 3. Assume one has three terms $h(x, f(t))$, $h(x, g(t))$, and $h(b, f(t))$ in the index. Then, in a discrimination tree, t will occur three times. In a substitution tree, we will have:

$$x0 = h(x1, x2)$$

$x1 = x$ $x1 = b,\ x2 = f(t)$

$x2 = f(t)$ $x2 = g(t)$

and with a query $h(b, f(s))$, the terms s and t will be matched twice against each other (at the leftmost and rightmost leaves). In a context tree, the term t occurs only once:

$$x0 = h(x1, F(t))$$

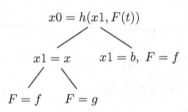

$x1 = x$ $x1 = b,\ F = f$

$F = f$ $F = g$

and if s does not match t, the failure with the query $h(b, f(s))$ will be found at the root.   □

In addition to proposing the concept of context trees, in this paper we will also provide some evidence for its practical value. First, we show how they can be adequately implemented. In order to be able to reuse some of the main ideas for efficient implementation of substitution trees, we will consider terms built from a single pairing constructor and constants. These terms will also be called *Curry terms*. We describe an implementation based on these Curry terms and an extension of substitution trees by equality constraints and by not distinguishing *internal* and *external* variables. The second evidence for its practical value is empirical. Experiments with matching show that our preliminary implementation (which does not yet include several important enhancements) is already competitive with tightly coded state-of-the-art implementations, namely the implementation of discrimination trees of the Waldmeister prover [3] and the code trees of the Vampire prover [9].

For the experiments, we adopted the methods for evaluation of indexing techniques described in [5]: (i) we use 30 very large benchmarks containing the exact sequence of (update and retrieval) operations on the matching index that take place when running three well-known state-of-the-art provers on a selected set of

10 problems; (ii) comparisons are made with the discrimination tree implementation of the Waldmeister prover [3], and the code trees of the Vampire prover [9], as provided by their own implementors using the test driver of [5].

This paper is structured as follows. Section 2 introduces some basic concepts of indexing, discrimination trees and substitution trees. In Section 3 we outline some problems with direct implementations of context trees and explain how one can use Curry terms to solve theses problems. We also show that the use of Curry terms has several additional advantages. In Section 4 we describe our implementation in a certain detail. Finally, Sections 5 and 6 describe the experimental results and some promising directions for future work.

2   Discrimination Trees and Substitution Trees

Discrimination trees can be made very efficient if query terms are linear (as the terms in the trees are). Usually, queries are the so-called *flatterms* of [1], which are linked lists with additional pointers to jump over subterms t when a variable of the index gets instantiated by t.

In *standard discrimination trees*, all variables are represented by a single variable symbol $*$, so that different terms such as $f(x,y)$ and $f(x,x)$ are both represented by $f(*,*)$, and the corresponding path in the tree is common to both. This increases the amount of sharing, and also the retrieval speed, because the low-level operations (basically symbol comparison and variable instantiation) are very simple. But it is only a *prefilter*: once a possible match has been found, additional equality tests have to be performed between the query subterms by which the variables of terms like $f(x,x)$ have been instantiated. Nodes are usually arrays of pointers indexed by the function symbols, plus one additional pointer for $*$. If, during matching, the query symbol currently treated is f, then one can directly jump to the child for f, if it exists, or to the one of $*$. Especially for larger signatures, this kind of nodes lead to high memory consumption. Note that the case where children for both f and $*$ exist is the only situation where backtracking points are created.

In *perfect discrimination trees*, variables are not collapsed into a single symbol. Instead, nodes of different sizes exist: apart form the function symbols, each node can have a child for any of the variables that already occurred along the path in the tree, plus an additional child for a possible new variable. Hence even more memory is needed in this approach. Also there is less sharing in the index. On the other hand, the equality tests are not delayed (which is good according to the first-fail principle; see also below), all matches found are correct and no later equality tests are needed. The Waldmeister prover [3] uses these perfect discrimination trees for matching.

2.1   Implementation Techniques for Substitution Trees

Let us now consider substitution trees in more detail. They were introduced by Peter Graf [2], who also developed an implementation that is still used in the

Spass prover [10]. (A more efficient implementation was given in the context of the Dedam (Deduction abstract machine) kernel of data structures [6], and has served as a basis for our implementation of context trees as well.)

As for discrimination trees, it is important to deal with an adequate representation of query terms. In Dedam, Prolog-like terms are used: each term $f(t_1, \ldots, t_n)$ is represented by $n + 1$ contiguous *heap cells* with a *tag* and an *address* field:

$$
\begin{array}{r|c|c|}
a & f & \\
\cline{2-3}
a+1 & \texttt{ref} & a_1 \\
\cline{2-3}
& \vdots & \vdots \\
\cline{2-3}
a+n & \texttt{ref} & a_n \\
\cline{2-3}
\end{array}
$$

where each address field a_i points to the subterm t_i, and (uninstantiated) variables are `ref`'s pointing to themselves. In this setting, contiguous heap cell blocks of different sizes co-exist, and traversal of terms requires controlling arities. Term-to-term operations like matching or unification only instantiate self-referencing `ref` positions. If these instantiated positions are pushed on a stack, called the *refstack*, then undoing the operation amounts to restoring the positions in the refstack to self-references again.

Substitutions in substitution trees are always pairs of heap addresses; each right hand side points to a term; each left hand side points to an internal variable (i.e., a self-ref position) occurring exactly once in some term at the right hand side of a substitution along the path to the root.

The basic idea for all retrieval operations (finding a term, matching, unification) in substitution trees is the same: one instantiates the internal variable x_0 at the root with the query term, and traverses the tree, where at each visited node with a substitution $x_1 = t_1, \ldots, x_n = t_n$, one performs the basic term-to-term operation (syntactic equality, matching, unification) between each (already instantiated) x_i and its corresponding t_i. The term-to-term operations only differ in which variables are allowed to be instantiated, and which variables are considered as constants: for finding terms (syntactic equality), only the internal `ref`'s (called `intref`) can be instantiated; for matching, also the external `ref`'s of the index (but not of the query); for unification, all `ref`'s can be instantiated.

Upon failure, backtracking occurs. After successfully visiting a node, before continuing with its first child, its next sibling is stored in the *backtracking stack*, together with the current height of the refstack. Therefore, for backtracking, one pops the next node to visit from the backtracking stack, together with its corresponding refstack height, and restores all `ref` positions above this height. A failure occurs when trying to backtrack on an empty backtracking stack.

Due to space limitations, we cannot go into details here about the update operations for substitution trees. Let us only mention that several insertion strategies are possible (first-fit, best-fit), and that the basic operation for insertion is the computation of the *common* part and *remainders* of two substitutions. For deletions, one sometimes needs to apply the reverse operation, namely to *merge* two substitutions into a single one.

Example 4. Let σ_1 be the substitution $\{x_1 = g(a, h(b)), x_2 = a\}$ and let σ_2 be $\{x_1 = g(b, h(c)), x_2 = b\}$. Their common part is $\{x_1 = g(x_3, h(x_4))\}$. The remainders of both substitutions are $\{x_2 = a, x_3 = a, x_4 = b\}$ and $\{x_2 = b, x_3 = b, x_4 = c\}$ respectively.   □

2.2   Substitution Trees for Matching

In Dedam a special version of substitution trees for matching has been developed, which is about three times faster than the general-purpose implementation in Spass and Dedam.

Example 5. Suppose the query is of the form $f(s, t)$ and consider a substitution tree with the two terms $f(x, x)$ and $f(a, x)$: the root is $x_0 = f(x_1, x)$, with children $x_1 = x$ and $x_1 = a$. When matching, at the root x_1 gets instantiated with s, and x with t; then, at the leftmost child, the terms s and t are matched against each other. Note that one has to keep track of whether or not x has already been instantiated, i.e., one has to keep a refstack.   □

The idea for improving this procedure is similar to the one of the *standard* variant of discrimination trees: external variables are all considered to be different. But in substitution trees the advantages are more effective: the refstack becomes unnecessary (and hence also the information about its height in the backtracking stack), because one can always override the values of the internal and external variables and restauration becomes unnecessary. Matching operations between query subterms, like s and t in the previous example, are replaced by a cheaper syntactic equality test of the *equality constraints* at the leaves.

3   Context Trees

We start by illustrating the increased amount of sharing in context trees as intuitively described in Section 1 compared with substitution trees.

Example 6. Assume in a context tree we have a subtree T below a node $x_i = f(x_j, t)$ (depicted below at the left) where we have to insert $x_i = g(s, t, u)$. Then the common part is $x_i = F(x_j, t, u)$, the remaining parts are $\{F = f\}$ and $\{F = g, x_j = s\}$, respectively, and we obtain:

During retrieval on the new tree, the term-to-term operations have to be guided by the arities of the query: if x_i is instantiated with a query term headed with f, when arriving at the node $x_i = F(x_j, t, u)$, then, since the arity of f is 2, one can simply ignore the term u of this common part.   □

It is not difficult to see that, with the restricted kind of function variables F that stand for single function symbols, the common part of two terms s and t, as in substitution trees, will contain the entire common context, and additionally also those subterms u that occur at the same position p in both terms, that is, for which $u = s|_p = t|_p$.

Example 7. The common part of the two terms $f(g(b,b), a, c)$ and $h(h(b,c), d)$ is $F(G(b, x_1), x_2, x_3)$. Indeed, the subterm b at position 1.1 is the only term occurring at the same position in both terms.   □

To implement context trees for matching by an extension of the specialized substitution trees for matching requires to deal with the specific properties of the context variables.

Example 8. Consider again Example 6. The term $f(x_j, t)$ consists of three contiguous heap cells. The first contains f, the second is an `intref` corresponding to x_j, and the third is a `ref` pointing to the subterm t. Initially, in the subtree T below that node, along each path to a leaf x_j appears once as a left hand side in a substitution. After inserting $x_i = g(s, t, u)$, the common part is $x_i = F(x_j, t, u)$, and the new term $F(x_j, t, u)$ needs four contiguous heap cells instead of three.

A serious implementation problem now is that, if we allocate a new block of size four, all left hand sides pointing to x_j in the subtree T have to be changed to point to the new address of x_j.   □

3.1   Context Trees through Curry Terms

A simple solution to the previous problem would be to always use blocks corresponding to the maximal arity of symbols, but this is too expensive in memory consumption. Here we propose a different solution, which is conceptually appealing and at the same time turns out to be very efficient since it completely avoids the need for checking arities. We suggest to represent all terms in Curry form. Curry terms are formed with a single binary *apply* symbol @, and all other function symbols are considered (second-order) constants to be treated much like their first-order counterparts. This idea is standard in the context of functional programming, but, surprisingly, does not seem to have been considered for term indexing data structures for automated deduction before.

Example 9. Consider again the terms of Example 6, where we saw that one can share the term t in $f(x_j, t)$ and $g(s, t, u)$ by having $F(x_j, t, u)$. In Curry form, these terms become @(@(f, x_j), t) and @(@(@(g, s), t), u) and the t cannot be shared. But in the Curry form the same amount of sharing exists: still all arguments that are in the same position are shared, assuming that *positions are counted from right to left.* Consider the arguments of the same terms in reverse order. The we have $f(t, x_j)$ and $g(u, t, s)$, which in Curry form become @(@(f, t), x_j) and @(@(@(g, u), t), s). The common part, which was $F(u, t, x_j)$, can be computed on the Curry terms exactly as it was done for common contexts of first-order terms in substitution trees. In the example we get $@(@(x_k, t), x_j)$, where the remaining parts are $\{x_k = f\}$ and $\{x_k = @(g, u), x_j = s\}$.

It is not difficult to see that in this way one obtains exactly the same amount of sharing as with context variables: all common contexts and all subterms u that occur at the same position in both terms (but remember: if positions are computed from right to left; for instance, the shared t in $f(t, x_j)$ and $g(u, t, s)$ is at position 2 in both terms).   □

An important additional advantage is that the basic algorithms do not depend on the arities of the symbols anymore. Moreover, since it is obviously not necessary to store any apply symbols, all memory blocks contain exactly two heap cells.

Example 10. The term $f(b, g(x))$ becomes $@(@(f, b), @(g, x))$, which can be written in pair notation simply as $((f, b), (g, x))$. Compare the Prolog format with how the Curry term can be stored:

```
    Prolog term:                    Curry Term:

    10 f                      100 ref -->  120 ref -----------> 150 f
    11 ref --> 40 b                        121 ref --> 140 g    151 b
    12 ref -------> 60 g                               141 var
                61 ref --> 80 ref
```

Note that in the Prolog term, constants are a block of heap cells on their own such as the b at address 40. Alternatively, constants can also be placed directly at pointer position for minimizing space. But with Prolog terms this makes the algorithms slower as a uniform treatment for all function symbols (constants or not) becomes impossible. But in Curry terms, since *all* function symbols are constants, this space optimization can be used without any cost.   □

In Curry terms each cell is either a constant, a variable var or a ref to a subterm. Curry terms are always headed by a single heap cell, and all other blocks consist of two contiguous heap cells. This makes the basic algorithms very efficient, as exemplified by the following recursive algorithm for testing the equality of two (ground) Curry terms:

```
int TermEqual(_HeapAddr s, _HeapAddr t){
    if (HeapTag(s)!=HeapTag(t)) return(0);
    if (HeapIsRef(s)){
        if (!TermEqual(HeapAddr(s),  HeapAddr(t)  )) return(0);
        if (!TermEqual(HeapAddr(s)+1,HeapAddr(t)+1)) return(0);}
    return(1);}
```

4   Implementation

4.1   Equality Constraints

In order to exploit the idea of equality constraints in its full power, it is important to perform the equality tests not only at the leaves, but as high up as possible in the tree without decreasing the amount of sharing (see [6]). For example, if we have $f(a, x, x, x, a)$, $f(b, x, x, x, a)$, and $f(c, y, y, x, a)$, then the tree can be:

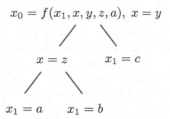

$$x_0 = f(x_1, x, y, z, a), \quad x = y$$

Note that placing the equality tests $x = y$ and $x = z$ in the leaves would frequently lead to repeated work during retrieval time. Also, according to the first-fail principle (which is strongly recommended in indexing techniques), it is important to impose strong restrictions like the equality of two whole subterms as soon as possible. Below we outline some details about our implementation of equality constraints, their evaluation during retrieval time and their creation during insertions by means of MF-sets (merge-find sets).

4.2   Internal vs. External Variables

We have seen that in our Curry terms we only consider heap cells that are a constant, a **ref**, or a variable **var**. Indeed, it turns out that the usual distinction between internal and external variables can also be dropped. (A variable that is not instantiated represents an external variable.) This leads to even more sharing in the index and increases matching retrieval speed, however, at the price of significantly more complex update operations (see below).

Example 11. If the tree contains the two terms $f(a, a)$ and $f(x, b)$, we have:

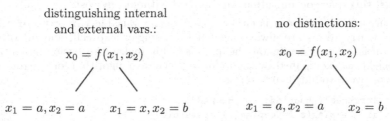

distinguishing internal
and external vars.:

no distinctions:

Note that in the second tree the variable x_1 plays the role of an internal variable in the leftmost branch and of an external variable in the other one.   □

In this setting, the term-to-term matching operation can be implemented as follows:

```
int TermMatch(_HeapAddr query, _HeapAddr set){
    if (HeapIsVar(set)) { HeapSetAddr(set,query); return(1); }
    if (HeapTag(query)!=HeapTag(set)) return(0);
    if (HeapIsRef(set)){
        if (!TermMatch(HeapAddr(query),  HeapAddr(set)  )) return(0);
        if (!TermMatch(HeapAddr(query)+1,HeapAddr(set)+1)) return(0);}
    return(1);}
```

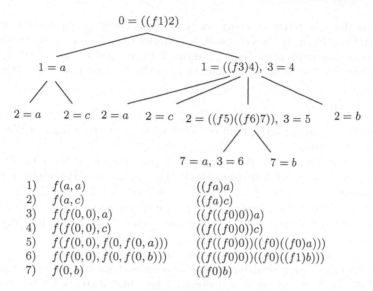

1)	$f(a,a)$	$((fa)a)$
2)	$f(a,c)$	$((fa)c)$
3)	$f(f(0,0),a)$	$((f((f0)0))a)$
4)	$f(f(0,0),c)$	$((f((f0)0))c)$
5)	$f(f(0,0),f(0,f(0,a)))$	$((f((f0)0))((f0)((f0)a)))$
6)	$f(f(0,0),f(0,f(0,b)))$	$((f((f0)0))((f0)((f1)b)))$
7)	$f(0,b)$	$((f0)b)$

Fig. 1. Context tree for the terms 1)–7)

4.3 Matching Retrieval

In Fig. 1 we show a tree as it would have been generated in our implementation
after inserting the seven terms given both in standard representation and Curry
form, respectively. Variables are written as numbers. Note that the equality
constraints $3 = 4$ and $3 = 5$ are shared among several branches. Given the
term-to-term operations for equality and matching from above, the remaining
code needed for matching retrieval on a context tree is very simple. One needs
a function for checking the substitution of a context tree node during matching:

```
int SubstMatch(_Subst subst){
  while (subst){
    if (subst->IsEqualityConstraint)
      { if (!TermEqual(subst->lhs,subst->rhs)) return(0); }
    else
      { if (!TermMatch(subst->lhs,subst->rhs)) return(0); }
    subst = subst->next;}
  return(1);}
```

Finally, the general traversal algorithm of the tree is the one presented below (assuming that the root variable x_0 has already been instantiated with the query):

```
int CTreeMatch(_CTree tree){
  if (!SubstMatch(tree->subst))
    if (tree->nextSibling) return(CTreeMatch(tree->nextSibling));
    else return(0);
  if (!tree->firstChild ) return(1);
  if (!tree->nextSibling) return(CTreeMatch(tree->firstChild));
  return(CTreeMatch(tree->nextSibling)||CTreeMatch(tree->firstChild));}
```

This is the only retrieval algorithm that is not coded in our implementation as shown here. In the implementation, it is iterative and uses a backtracking stack. The other recursive algorithms for term-to-term equality and matching are also recursive in our current implementation, in the form we have given them above.

4.4 Updates

Updates are significantly more complex in context trees than in the standard substitution trees.

For insertion, one starts with a linearized term, together with several MF-sets for keeping the information about the equivalence classes of the variables. For instance, the term $f(x, y, x, y, y)$ is inserted as $f(x_1, x_2, x_3, x_4, x_5)$ with the associated information that x_1 and x_3 are in the same class, and that x_2, x_4, and x_5 are in the same class. Hence if this term is inserted in an empty tree, we obtain a tree with one node containing the substitution:

$$x_0 = f(x_1, x_2, x_3, x_4, x_5), \quad x_1 = x_3, \quad x_2 = x_4, \quad x_2 = x_5$$

(or with other, equivalent but always non-redundant, equality constraints).

The insertion process in a non-empty tree first searches for the node where insertion will take place. This search process is like matching, except for two aspects. Firstly, the external variables of the index are only allowed to be instantiated with variables of the inserted term. But since a variable x in the tree sometimes plays the role of an internal and external variable at the same time (see Example 11), one cannot know in advance which situation applies until a leaf is reached: if x has no occurrence as the left hand side of a substitution (not an equality constraint) along the path to the leaf, then it plays the role of an external variable for this leaf. Secondly, during insertion the equality constraints are checked on the associated information about the variables classes, instead of checking the syntactic equality of subterms.

If a siblings list is reached where no sibling has a total agreement with the inserted term then two different situations can occur. If there is a sibling with a partial agreement with the inserted term, then one takes the first such sibling (first-fit, this is what we do) or the sibling with the maximal (in some sense) agreement (best-fit). The substitution of this node is replaced with the common part (including the common equality constraints), and two new nodes are created with the remaining substitutions. If a point is reached where all sibling nodes have an empty common part with the inserted substitution, then the inserted substitution is added to the siblings list. In both situations, the remaining substitution of the inserted term is built including the equality relations that have not been covered by the equality constraints encountered along the path from the root.

Deletion is also tricky, mainly because finding the term to be deleted requires again to control the equality constraints and the instantiation of external variables only with variables of the term to be found. Moreover, unlike what happens in insertion, backtracking is needed for finding.

It is important to be aware of the fact that updates cost little time in practice, because updates are relatively infrequent compared with retrieval operations.

Experiments seem to confirm that there are only one or two updates per thousand retrievals. On the benchmarks of [5] (see below) in none of the benchmarks updates took more than 5 percent of the time.

5 Experiments

In our experiments we have adopted the methodology described in [5]. (In that paper one can find a detailed discussion of how to design experiments for the evaluation of indexing techniques so that they can be repeated and validated by others without difficulty.) For the purposes of the present paper, we ran 30 very large benchmarks, each containing the exact sequence of (thousands of update and millions of retrieval) operations on the matching index as they are executed when running one of three well-known state-of-the-art provers on certain problems drawn from various subsets of the TPTP problem date base [8]. Comparisons are made between our preliminary implementation (column "Cont." in Figure 5 below) with the discrimination tree implementation (column "Disc.") of the Waldmeister prover [3], and the code trees (column "Code") of the Vampire prover [9], as provided and run by their own implementors.

The figure 5 shows that, in spite of the fact that our implementation can be much further improved (see Section 6), it is already quite competitive in time. Moreover, context trees are, as expected, best in space, except for the very small indexes (mostly coming from the Waldmeister prover). (A substantial further space improvement can be expected from a compiled implementation as sketched in section 6.2.) Code trees are, conceptually, a refined form of standard discrimination trees. In their latest version [7], code trees apply a similar treatment of the equality tests as the one of [6] we use here. The faster speed of code trees is, in our opinion, by and large due to the compilation of the index (see Section 6). We do not include here the results of the aforementioned Spass and Dedam implementations of substitution trees, because, with a similar degree of refinement of coding as our current implementation of context trees, they are at least a factor three slower and need much more space.

6 Conclusions and Future Work

The concept of context trees has been introduced and we have shown (and experimentally verified) that large space savings can be possible compared with substitution trees and discrimination trees. We have described in detail how these trees can be efficiently implemented. By representing terms in Curry form, an implementation can be based on a simplified variant of substitution trees. Already from the performance of our first (unfinished) implementation it can be seen that context trees have a great potential for applications in automated theorem proving. Due to the high degree of sharing, they allow for efficient matching, they require much less memory, and yet the time needed for the somewhat more complex updates remains negligible.

TPTP problem	benchmark from prover	time in seconds			space in KB		
		Code	Disc.	Cont.	Code	Disc.	Cont.
COL002-5	Fiesta	1.30	1.55	2.61	925	6090	922
COL004-3	Fiesta	0.96	1.22	2.39	80	727	86
LAT023-1	Fiesta	1.10	1.49	1.92	198	1646	210
LAT026-1	Fiesta	1.11	1.49	1.79	373	2813	371
LCL109-2	Fiesta	0.47	0.65	0.80	508	2285	466
RNG020-6	Fiesta	2.26	3.19	5.33	544	2435	517
ROB022-1	Fiesta	0.92	1.20	2.23	119	1086	101
GRP164-1	Fiesta	17.60	24.25	32.40	5823	28682	5352
GRP179-2	Fiesta	18.34	24.25	32.40	5597	29181	5207
GRP196-1	Fiesta	6.96	11.92	15.45	1	543	1
BOO015-4	Waldmeister	0.25	0.31	0.46	11	575	11
GRP024-5	Waldmeister	3.54	4.82	7.44	19	591	22
GRP187-1	Waldmeister	10.44	11.68	17.64	96	903	97
LAT009-1	Waldmeister	3.78	4.97	5.97	19	591	20
LAT020-1	Waldmeister	17.74	24.97	29.87	30	631	31
LCL109-2	Waldmeister	0.49	0.66	0.82	16	591	15
RNG028-5	Waldmeister	4.19	6.66	9.08	28	607	29
RNG035-7	Waldmeister	8.20	12.10	18.55	36	647	37
ROB006-2	Waldmeister	9.88	14.31	21.60	128	1142	116
ROB026-1	Waldmeister	8.52	13.55	17.34	69	807	68
COL079-2	Vampire	5.46	8.41	7.24	2769	9158	2138
LAT002-1	Vampire	5.83	7.72	9.48	3164	14603	2554
LCL109-4	Vampire	5.62	7.65	13.02	6703	24403	4986
CIV003-1	Vampire	7.57	7.13	15.57	3754	22664	3081
RNG034-1	Vampire	3.27	4.95	6.86	2545	8330	2125
SET015-4	Vampire	2.54	2.69	4.53	314	1373	258
HEN011-2	Vampire	3.36	3.39	5.18	221	2069	211
CAT001-4	Vampire	3.28	5.74	7.11	3859	13786	3109
CAT002-3	Vampire	2.90	5.51	6.67	2483	9281	2021
CAT003-4	Vampire	3.21	5.82	6.90	3826	13595	3086

Fig. 2. Experimental results

With respect to our implementation, more work remains to be done regarding a tighter coding of the four (two of them recursive) algorithms used for retrieval — those we have seen in this paper. Experience with other implementations of term indexes has shown that this can give substantial factors of speedup.

Apart from these low-level aspects we also believe that there are at least two other directions for further work from which further substantial improvements will be obtained. We are going to describe them briefly now.

6.1 Exact Computation of Backtracking Nodes

Far more information than we have discussed so far can be precomputed at update time on the index. We describe one of the more promising ideas that should help to considerably reduce the amount of nodes visited at retrieval time and to eliminate the need of a backtracking stack.

Consider an occurrence p of a substitution pair $x_i = t$ in a substitution of the tree. Denote by $accum(p)$ the term that is, roughly speaking, the accumulated substitution from the root to the pair p, including p itself. If, during matching, a failure occurs just after the pair p, then the query term is an instance of $accum(p)$ (and this is the most general statement one can make at that point for all possible queries). This knowledge can be exploited to exactly determine the node to which one should backtrack. Let p' be first pair after p in preorder traversal of the tree whose associated term $accum(p')$ is unifiable with $accum(p)$. Then $accum(p')$ is the "next" term in the tree that can have a common instance with $accum(p)$. Therefore, $accum(p')$ is precisely the next term in the tree of which the query can be an instance as well! Hence the backtracking node to which one should jump in this situation is the one just after p'.

It seems possible to recompute locally, upon each update of the tree, the backtracking pointers associated to each substitution pair, and store these pointers at the pair itself, thus actually minimizing (in the strictest sense of the word) the search during matching. We are currently working out this idea in more detail.

6.2   Compiled Context Trees

One of the conclusions that can be drawn from the experiments of [5] is that it does in fact pay off to compile an index into a form of interpreted abstract instructions as suggested by the code trees method of [9] (similar findings have also been obtained in the field of logic programming). For context trees, consider for example again the SubstMatch loop we saw before. Instead of such a loop, one can simply use a linked list of abstract code instructions like TermEqual(adress1,addres2,FailureAddress) where FailureAddress is the address to jump to in case of failure. The main advantage of this approach is that no control (like the outermost if statement of SubstMatch) has to be looked up, and, since the correct address arguments are already part of the abstract code, no instructions like subst = subst->next are needed and many indirect accesses like subst->lhs can be avoided.

In addition, one can use instructions decomposing operations like TermMatch into sequences of instructions for the concrete second argument, which is known at compile (i.e., index update) time. Assume we specialized the TermMatch function for matching with the index term $(f(gx))$, that is,

```
10 ref -->   20 f
             21 ref    --> 30 g
                           31 var
```

This would give code such as the following sequence of 7 one-argument instructions:

```
if (!HeapIsRef(query))     goto fail;
query = HeapAddr(query);
if (HeapTag(query)!='f')   goto fail;
```

```
if (!HeapIsRef(query+1))    goto fail;
query = HeapAddr(query+1);
if (HeapTag(query)!='g')    goto fail;
HeapSetAddr(31,query+1)
```

By simple instruction counting, this code is easily shown to be far more efficient on an average query term than the general-purpose two-argument `TermMatch` function with $(f(gx))$ as second argument. All these advantages give more speedup than what has to be paid for in overhead arising from the need for interpreting the operation code of the abstract instructions. The latter is just a `switch` statement that, in all modern compilers, produces constant time code.

References

[1] Jim Christian. Flatterms, Discrimination Nets, and Fast Term rewriting. *Journal of Automated Reasoning*, 10:95–113, 1993.

[2] Peter Graf. Substitution Tree Indexing. In J. Hsiang, editor, *6th RTA*, LNCS 914, pages 117–131, Kaiserslautern, Germany, April 4–7, 1995. Springer-Verlag.

[3] Thomas Hillenbrand, Arnim Buch, Roland Vogt, and Bernd Löchner. WALDMEISTER—high-performance equational deduction. *Journal of Automated Reasoning*, 18(2):265–270, April 1997.

[4] William McCune. Experiments with discrimination tree indexing and path indexing for term retrieval. *Journal of Automated Reasoning*, 9(2):147–167, October 1992.

[5] Robert Nieuwenhuis, Thomas Hillenbrand, Alexandre Riazanov, and Andrei Voronkov. On the evaluation of indexing data structures. 2001. This proceedings.

[6] Robert Nieuwenhuis, José Miguel Rivero, and Miguel Ángel Vallejo. A kernel of data structures and algorithms for automated deduction with equality clauses (system description). In William McCune, editor, *14th International Conference on Automated Deduction (CADE)*, LNAI 1249, pages 49–53, Jamestown, Australia, 1997. Springer-Verlag. Long version at www.lsi.upc.es/~roberto.

[7] Alexandre Riazanov and Andrei Voronkov. Partially adaptive code trees. In *Proceedings of JELIA 2000*, Malaga, Spain, 2000.

[8] Geoff Sutcliffe, Christian Suttner, and Theodor Yemenis. The TPTP problem library. In Alan Bundy, editor, *Proceedings of the 12th International Conference on Automated Deduction*, volume 814 of *LNAI*, pages 252–266, Berlin, June/July 1994. Springer.

[9] Andrei Voronkov. The anatomy of vampire implementing bottom-up procedures with code trees. *Journal of Automated Reasoning*, 15(2):237–265, October 1995.

[10] Christoph Weidenbach. SPASS—version 0.49. *Journal of Automated Reasoning*, 18(2):247–252, April 1997.

On the Evaluation of Indexing Techniques for Theorem Proving

Robert Nieuwenhuis[1]*, Thomas Hillenbrand[2], Alexandre Riazanov[3], and
Andrei Voronkov[3]

[1] Technical University of Catalonia
roberto@lsi.upc.es
[2] Universität Kaiserslautern
hillen@informatik.uni-kl.de
[3] University of Manchester
{riazanov,voronkov}@cs.man.ac.uk

1 Introduction

The problem of *term indexing* can be formulated abstractly as follows (see [19]).
Given a set L of *indexed terms*, a binary relation R over terms (called the *retrieval
condition*) and a term t (called the *query term*), identify the subset M of L that
consists of the terms l such that $R(l, t)$ holds. Terms in M will be called the
candidate terms. Typical retrieval conditions used in first-order theorem proving
are matching, generalization, unifiability, and syntactic equality. Such a retrieval
of candidate terms in theorem proving is interleaved with insertion of terms to
L, and deletion of them from L.

In order to support rapid retrieval of candidate terms, we need to process
the indexed set into a data structure called the *index*. Indexing data structures
are well-known to be crucial for the efficiency of the current state-of-the-art
theorem provers. Term indexing is also used in logic and functional programming languages implementation, but indexing in theorem provers has several
distinctive features:

1. Indexes in theorem provers frequently store 10^5–10^6 complex terms, unlike
   a typically small number of shallow terms in functional and logic programs.
2. In logic or functional language implementation the index is usually constructed during compilation. On the contrary, indexes in theorem proving
   are highly dynamic, since terms are frequently inserted in and deleted from
   indexes. *Index maintenance* operations start with an index for an initial set
   of terms L, and incrementally construct an index for another set L' that is
   obtained by insertion or deletion of terms to or from L.
3. In many applications it is desirable for several retrieval operations to work
   on the same index structure in order to share maintenance overhead and
   memory consumption.

---

* Partially supported by the Spanish CICYT project HEMOSS ref. TIC98-0949-C02-
01.

R. Goré, A. Leitsch, and T. Nipkow (Eds.): IJCAR 2001, LNAI 2083, pp. 257–271, 2001.

Therefore, along the last two decades a significant number of results on new indexing techniques for theorem proving have been published and successfully applied in different provers [19,6,8,18,24,16,11,2,4,5,26,22,21]. In spite of this, important improvements of the existing indexing techniques are still possible and needed, and other techniques for previously not considered retrieval operations need to be developed [14]. But implementors of provers need to know which indexing technique is likely to behave best for his/her applications, and developers of new indexing techniques need to be able to compare techniques in order to get intuition about where to search for improvements, and in order to provide scientific evidence of the superiority of new techniques over other previous ones[1][2].

Unfortunately, practice has revealed that an asymptotic worst-case or average-case complexity analysis of indexing techniques is not a very realistic enterprise. Even if such analysis was done, it would be hardly useful in practice. For example, very efficient linear- or almost linear-time algorithms exist for unification [17,10] but in practice these algorithms proved to be inefficient *for typical applications*, and quadratic- or even exponential-time unification algorithms are used instead in the modern provers. Thus, theoretically worse (w.r.t. asymptotic complexity analysis) algorithms in this area frequently behave better in practice than other optimal ones. For many techniques one can design bad examples whose (worst-case) computational complexity is as bad as for completely naive methods. An average-case analysis is also very difficult to realize, among other reasons because in most applications no realistic predictions can be made about the distribution of the input data.

Similar phenomena take place when randomly generated data are used for benchmarking. For example, in propositional satisfiability attempts to find hard problems resulted in discovering random distributions of clauses which guarantee the existence of a *phase transition* [1,13]. Experimentally it has been discovered that problems resulting from the random clause generation in the phase transition region are hard for all provers, but the provers best for these problems proved to be not very efficient for more structured problems coming from practical applications [7].

---

[1] In fact, in the recent past two of the authors recommended rejection for CADE of each other's papers on improvements of code tree and substitution tree indexing due to the lack of evidence for a better performance.

[2] Due to the lack of evidence of superiority of some indexing techniques over other ones, system implementors have to take decisions about implementing a particular indexing technique based on criteria not directly relevant to the efficiency of the technique. As an example, we cite [23]:

> Following the extremely impressive results of Waldmeister, we have chosen a *perfect discrimination tree* ... as the core data structure for our indexing algorithms.

Here the overall results of Waldmeister were considered enough to conclude that it implements the best indexing techniques.

Hence the only reasonable evaluation method is to apply a statistical analysis of the *empirical* behaviour of the different techniques on benchmarks corresponding to real runs of real systems on real problems. But one has to be careful because one should not draw too many conclusions about the efficiency of different *techniques* based on a comparison between different concrete *implementations* of them; many times these implementations are rather incomparable due to different degrees of optimization and refinement[3].

Our main contribution here is the *design of the first method for comparing different implementations* based on a virtually unlimited supply of large real-world benchmarks for indexing. The basic requirements to such benchmarks are as follows. First, since different provers may impose different requirements on the indexing data structures, a general means should be given for obtaining realistic benchmarks for any given prover. Second, it should be possible to easily create benchmarks by running this prover on any problem, and do this for a significant number of different problems from different areas. Third, these benchmarks have to reproduce real-life sequences of operations on the index, where updates (deletions and insertions of terms) are interleaved with (in general far more frequent) retrieval operations.

The method we use for creating such benchmarks for a given prover is to add instructions making the prover write to a log file a trace each time an operation on the index takes place, and then run it on the given problem. For example, each time a term t is inserted (deleted, unified with), a trace like $+t$ (resp. $-t$, ut) is written to the file. Moreover, we require to store the traces along with information about the result of the operation (e.g., success/failure), which allows one to detect cases of incorrect behaviour of the indexing methods being tested. Ideally, there should be enough disk space to store all traces (possibly in a compressed form) of the whole run of the prover on the given problem (if the prover terminates on the problem; otherwise it should run at least for enough time to make the benchmark representative for a usual application of the prover).

The main part of the evaluation process is to test a given implementation of indexing on such a benchmark file. This given implementation is assumed to provide operations for querying and updating the indexing data structure, as well as a translation function for creating terms in its required format from the benchmark format. In order to avoid overheads and inexact time measurements due to translations and reading terms from the disk, the evaluation process first reads a large block of traces, storing them in main memory. After that, all terms read are translated into the required format. Then time measuring is switched on, and a loop is started which calls the corresponding sequence of operations, and time is turned off before reading the next block of traces from disk, and so on.

---

[3] Unfortunately, in the literature one frequently encounters papers where a very tightly coded implementation of the author's new method is compared with other relatively naive implementations of the previously existing methods, sometimes even run on machines with different characteristics that are difficult to compare.

This article is structured as follows. Section 2 discusses some design decisions taken after numerous discussions of the authors. Section 3 gives benchmarks for term retrieval and index maintenance for the problem of retrieval of generalizations (matching a query term by index terms), generated by running our provers Vampire [20], Fiesta [15] and Waldmeister [9] (three rather different, we believe quite representative, state-of-the-art provers) on a selection of carefully chosen problems from different domains of the TPTP library [25].

In Section 4 we describe the evaluation on these benchmarks of code trees, context trees, and discrimination trees as they are provided and integrated in the test package by their own implementors, and run under identical circumstances. As far as we know, this is the first time that different indexing data structures for deduction are compared under circumstances which, we believe, guarantee that experiments are not biased in any direction. Moreover, the implementations of code trees and discrimination trees we consider are the ones of the Vampire and Waldmeister provers, which we believe to be among the fastest (if not the fastest) current implementations for each one of these techniques. Hence for these two implementations it is unlikely that there are any differences in quality of coding. Although context trees are a new concept (see [3]) and the implementation used in the experiments (the only one that exists) was finished only one week before, the implementation is not naive since it is based on earlier, quite refined implementations of substitution trees.

All test programs, implementations and benchmarks mentioned in this paper are publicly available at `http://www.lsi.upc.es/~roberto`.

2   Some Design Decisions

In this section we discuss some decisions we had to take, since they may be helpful for the design of similar experiments for other term indexing techniques in the future.

We decided to concentrate in this paper on the measurement of three main aspects of indexing techniques (but the same methodology is applicable to other algorithms and data structures, see Section 6 for examples).

The first aspect we focus on is the time needed for queries where the retrieval condition is generalization: given a query term t, is there any indexed term l such that for some substitution σ we have $l\sigma = t$? This retrieval condition has at least two uses in first-order theorem provers: forward subsumption for unit clauses and forward demodulation (simplification by rewriting with unit equalities). It is also closely related (as an ingredient or prefilter) to general forward subsumption. It is well-known to be the main bottleneck in many provers (especially, but not only, in provers with built-in equality). Unlike some other retrieval conditions, search for generalizations is used in both general theorem provers and theorem provers for unit equalities. Some provers do not even index the other operations.

Though time is considered to be the main factor for system comparison, memory consumption is also crucial, especially for long-running tests. During the last years processor speed has grown much faster than memory capacity and it is foreseen that this trend will continue in the coming years. Moreover, memory access speed is also becoming an important bottleneck. Excessive memory consumption leads to more cache faults, which become the dominant factor for time, instead of processor speed. In experiments described in [27], run on a relatively slow computer with 1Gbyte of RAM memory, it was observed that some provers consume 800Mbytes of memory in the first 30 minutes. The best modern computers are already about 10 times faster than the computer used in these experiments, which means that memory problems arise very quickly, a problem that will become more serious in the coming years. For all these reasons, memory consumption is the second object of measurement in this paper.

The third and last aspect we focus on here is frequency of updates (insertions and deletions) and the time needed for them. In this field there exists an amount of *folk* knowledge (but which, unfortunately, differs among researchers), about the following questions. How frequent are updates in real applications? Is it really true that the time needed for updates is negligible? Is it worth or feasible to restructure larger parts of the index at update time (i.e., more than what is currently done)?

Perfect filtering or not? Our definition of term indexing corresponds to *perfect filtering*. In some cases implementation of *imperfect filtering*, when a subset or a superset of candidate terms is retrieved, does not affect soundness and completeness of a prover. In particular, neither soundness nor completeness are affected if only a subset of candidate terms is retrieved, when the retrieved generalizations are only used for subsumption or forward demodulation.

It was decided that only perfect filtering techniques should be compared. Of course, this includes implementations based on imperfect filters retrieving a superset of candidate terms, combined with a final term-to-term correctness test. Indeed, comparing imperfect indexing techniques alone does not make much sense in our context, since the clear winner in terms of time would be the system reporting "substitution not found" without even considering the query term.

Should the computed substitution be constructed in an explicit form? When search for generalizations is used for forward subsumption, computing the substitution is unnecessary. When it is used for forward demodulation, the computed substitution is later used for rewriting the query term. We decided that explicit representation of the computed substitution is unnecessary since all indexing techniques build such a substitution in an implicit form, and this implicit representation of the substitution is enough to perform rewriting.

One or all generalizations? In term indexing, one can search for one, some, or all indexed candidate terms. When indexing is used for forward subsumption or forward demodulation, computing all candidates is unnecessary. In the case of subsumption, if any candidate term is found, the query term is subsumed and can be discarded. In the case of forward demodulation, the first substitution found will be used for rewriting the query term, and there is no need either for finding more of them. Hence we decided to only search for one candidate. Note that for other retrieval conditions computing all candidate terms can be more appropriate. Examples are unification, used for inference computation, and (backward) *matching*, used for backward subsumption and backward demodulation.

How should problems be selected? It was decided for this particular experiment that every participant contributes with an equal number of benchmarks, and that the problems should be selected by each participant individually. There was a common understanding that the problems should be as diverse as possible, and that benchmarks should be large enough. Among the three participants, Vampire is the only prover that can run on nonunit problems, so to achieve diversity benchmarks generated by Vampire were taken from nonunit problems only. Nonunit problems tend to have larger signatures than the unit ones. Quite unexpectedly, benchmarks generated by Fiesta and Waldmeister happened to be quite diverse too, since Fiesta generates large terms much more often than Waldmeister.

But even if one collects such a diverse set of benchmarks, one cannot expect that all practical situations are covered by (statistically speaking) sufficiently many of them. For example, if one wants to check how a particular indexing technique behaves on large signatures, our benchmarks suite would be inappropriate since it contains only two benchmarks in which signatures are relatively large. Our selection of 30 problems is also inadequate if one wants to check how a particular technique behaves when the index contains over 10^5 terms, since the greatest number of terms in our indices is considerably less.

But it is one of the advantages of the proposed methodology that a potentially unlimited number of new benchmarks with any given properties can easily be generated, provided that these properties correspond to those occurring in real problems.

Input file format. Input file format was important for a reason not having direct relation to the experiments. The benchmark files, written in any format, are huge. Two input formats have been proposed: one uses structure sharing and refers to previously generated terms by their numbers; the other one uses a stringterm representation of query terms. In the beginning it was believed that the first format would give more compact files, but in practice this happened to be not the case since a great majority of query terms proved to be small. In addition, files with structure sharing did not compress well, so a stringterm representation was chosen, see Figure 1. Finally, stringterms are easy to read for humans which helped us a lot when one indexing data structure produced strange results because an input file was mistakenly truncated.

But even compressed files with benchmarks occupy hundreds of megabytes, which means that difficulties can arise to only store them in a publicly accessible domain and transfer them over the Web. Fortunately, this problem was easy to solve, since these files have a relatively low Kolmogorov complexity, an observation that, this time, can also be used in practice: they have been produced by three much smaller generators (provers) from very short inputs, so for reproducing the files it is enough to store the generators and their input files.

We also reconsidered our initial ideas about how to store the terms internally before calling the indexing operations. We experimented with complicated shared internal representations, in order to keep the memory consumption of the test program small (compared with the memory used by the indexing data structures), thus avoiding noise in the experiments due to cache faults caused by our test driver, since the driver itself might heavily occupy the cache. But finally we found that it is better for minimising cache faults to simply read relatively small blocks (of 2MB) of input from disk at a time, and store terms without any sharing, but *contiguously* in the same order as they will be considered in the calls to the indexing data structure. Of course the number of terms (i.e., of operations on the index) read from disk at a time should not be too small, because otherwise timing is turned on and off too often, which also produces noise.

This is an extract from a benchmark file generated by Waldmeister from the TPTP problem LCL109-2. Comments have been added by the authors.

```
a/2                 #    each benchmark file starts with the
b/0                 #    signature symbols with respective arities
c/1                 #
...
?ab0                #    query term a(b,x0), "?" signals failure
?b                  #    query term b
+ab0                #    insert term a(b,x0) to the index
...
!ab5                #    query term a(b,x5), "!" signals success
...
-accbb              #    delete term a(c(c(b)),b) from index
...
```

Fig. 1. An example benchmark file

Is the query term creation time included in the results? Despite the simplicity of this question, it was not easy to answer. Initially, it was thought that time for doing this should be negligible and essentially equal for the different indexing implementations. However, in actual provers the query terms are already available in an appropriate format as a result of some previous operation, so there is no need to measure the time spent on copying the query term. In addition, some, but not all, participants use the flatterm representation for query terms and it was believed that creating flatterm query terms from similarly structured stringterms could be several times faster than creating terms in the tree form. Hence it was decided that the time for creating query terms should not be included in the results.

However, this decision immediately created another problem. An expensive single operation for matching the query term against a particular indexed term is the comparison of two subterms of the query term. For example, to check if the term $f(x,x)$ is a generalization of a query term $f(s,t)$, one has to check whether s and t are the same term. Such a check can be done in time linear in the size of s,t. All other operations used in term-to-term matching can be performed in constant time (for example, comparison of two function symbols or checking if the query term is a variable). Now suppose that everyone can create any representation of the query term at the expense of system time. Then one can create a perfectly shared representation in which equal subterms will be represented by the same pointer, and checking for subterm equality becomes a simple constant time pointer comparison. Clearly, in practice one has to pay for creating a perfectly shared representation, so doing this in system time would be hardly appropriate[4]. The solution we have agreed upon is this: the representation of the query term should not allow for a constant-time subterm comparison, and for each participating system the part of the code which transforms the stringterm into the query term should be clear and easy to localize (e.g., included in the test driver itself).

3   Generation of Benchmarks

We took 30 problems from TPTP (10 by each participant), which created 30 benchmarks. The table of problems and simple quantitative characteristics of the resulting benchmarks is given in Table 1. The first two columns contain the name of the problem and the system that generated the problem. The third column contains the number of symbols in the signature of the problem, for example 3+4 means that the signature contains 3 nonconstant function symbols plus 4 constants. In the following four columns we indicate the total number of operations (*Total* in the table), insertions in and deletions from the index (*Ins* and *Del* in the table), and the maximal size of the index in number of terms (*Max* in the table) during the experiment. In the last four columns we show the average size and depth of the indexed and query terms, respectively. Here by size we mean the number of symbols in the term, and the depth is measured so that the depth of a constant is 0.

4   Evaluation

We ran each indexing data structure on each benchmark, i.e., we did 90 experiments. The results are given in Table 2. We measured time spent and memory used by each system. In parentheses we put the time measured for index maintenance only (i.e. insertions and deletions). This time was measured in a second run of the 90 experiments, on the same benchmarks, but with all retrieval requests removed.

---

[4] Note that the (forward) matching operation is applied to formulae just after they have been generated, with the purpose of eliminating or simplifying them. Hence it is unlikely that they are in perfectly shared form, even in systems in which retained formulae are kept in a perfectly shared representation.

Table 1. Benchmark characteristics

problem and generator		sig	operations				indexed terms		query terms	
			Total	Ins	Del	Max	Size	Depth	Size	Depth
BOO015-4	wal	3+4	275038	228	228	179	13.8	5.8	5.1	2.2
CAT001-4	vam	3+4	2203880	18271	3023	16938	26.0	9.9	7.1	2.9
CAT002-3	vam	4+4	2209777	12934	4181	11191	29.2	10.4	7.1	3.1
CAT003-4	vam	3+4	2151408	18159	4476	16606	28.3	10.6	7.1	2.9
CIV003-1	vam	6+14	3095080	70324	22757	47567	14.1	5.1	4.3	1.6
COL002-5	fie	2+7	940127	13399	5329	8353	25.4	9.3	7.5	2.8
COL004-3	fie	2+5	1176507	765	28	737	18.1	6.7	9.3	3.2
COL079-2	vam	3+2	2143156	14236	4619	9633	38.8	11.8	7.3	2.7
GRP024-5	wal	3+4	2686810	506	506	296	16.3	6.6	7.7	2.9
GRP164-1	fie	5+4	11069073	53934	3871	50063	16.4	6.1	5.9	2.4
GRP179-2	fie	5+2	10770018	52825	2955	49870	16.8	6.1	6.1	2.4
GRP187-1	wal	4+3	9999990	2714	1327	1387	12.0	5.3	5.0	2.0
GRP196-1	fie	2+2	18977144	3	0	3	26.3	7.0	10.8	4.8
HEN011-2	vam	1+10	4313282	4408	439	3969	10.7	4.1	2.7	0.8
LAT002-1	vam	2+6	2646466	26095	1789	24306	17.5	6.2	5.6	2.0
LAT009-1	wal	2+3	2514005	596	596	291	19.2	7.4	7.8	2.8
LAT020-1	wal	2+3	9999992	910	493	417	14.8	5.6	9.3	3.2
LAT023-1	fie	3+4	822539	4088	2065	2499	18.4	6.6	6.0	2.3
LAT026-1	fie	3+4	772413	6162	3509	4770	20.8	7.1	5.6	2.0
LCL109-2	fie	3+4	312992	4465	519	3947	16.7	6.1	6.0	2.6
LCL109-2	wal	2+1	463493	196	196	165	19.2	7.9	5.7	2.3
LCL109-4	vam	4+3	1944335	40949	3135	37817	20.0	5.7	8.1	2.7
RNG020-6	fie	6+6	2107343	4872	960	3912	17.4	5.4	6.3	2.4
RNG028-5	wal	5+4	3221510	304	304	218	31.4	9.0	14.0	3.8
RNG034-1	vam	4+4	2465088	15068	4589	11685	26.4	7.0	6.1	2.4
RNG035-7	wal	3+5	5108975	482	482	360	21.7	9.0	13.9	4.7
ROB006-2	wal	3+4	9999990	1182	34	1148	19.1	7.7	12.4	4.9
ROB022-1	fie	3+4	922806	2166	826	1341	21.4	9.3	6.5	2.8
ROB026-1	wal	2+4	9999991	648	15	633	16.9	7.9	12.6	5.0
SET015-4	vam	4+6	3664777	3256	995	2261	16.3	5.8	3.6	1.4

Time. Vampire's implementation of code trees is on the average 1.39 times than Waldmeister's implementation of discrimination trees and 1.91 times faster than the current implementation of context trees[5]. These factors roughly hold for almost all problems, with a few exceptions, for example on problem CIV003-1 Waldmeister is slightly faster than Vampire. The time spent for index mainte-

---

[5] Note that in these average times problems with large absolute run-times predominate. Furthermore, context trees are new (see [3]) and several important optimizations for them have not yet been implemented. According to the first author, they are at least 3 times faster than the substitution trees previously used in Fiesta. We welcome anyone who has an efficient implementation of substitution trees (or any other indexing data structure) to compare her or his technique with ours on the same benchmarks.

nance is in all cases negligible compared to the retrieval time, and there are essentially no fluctuations from the average figures: Waldmeister spends on the index maintenance 1.18 times less than Vampire and 1.49 times less than the context trees implementation.

Memory. In memory consumption, differences are more important. On average, the implementation of context trees used 1.18 times less memory than Vampire's implementation of code trees and 5.39 times less memory than Waldmeister's implementation of discrimination trees.

5   A Short Interpretation of the Results

Although the main aim of this work was to design a general-purpose technique for measuring and comparing the efficiency of indexing techniques in time and space requirements, in this section we very briefly and globally describe why we believe these concrete experiments have produced these concrete results.

5.1   Waldmeister's Discrimination Trees

Waldmeister uses *discrimination trees*, which are like tries where terms are seen as strings and common prefixes are shared, in its so-called *perfect* variant. Nodes are arrays of pointers and can have different sizes: each node can have a child for each one of the function symbols and for each one of the variables that already occurred along the path in the tree, plus an additional child for a possible new variable. During matching one can index the array by the currently treated query symbol f, and directly jump to the child for f, if it exists, or to one of the variable children. Note that the case where children for both f and some variable exist (or for more than one variable), is the only situation where backtracking points are created. Usually, queries are represented as the so-called *flatterms* of [2], which are linked lists with additional pointers to jump over subterms t when a variable of the index gets instantiated by t.

The results of our experiments for Waldmeister's discrimination trees are not unexpected. The implementation is very tightly coded, and in spite of the lower amount of sharing than in other techniques, the retrieval speed is very high because the low-level operations (essentially, symbol comparison and variable instantiation) are very simple. It is clear that, especially for larger signatures and deep terms, the kind of nodes used leads to high memory consumption. Memory consumption is even higher because the backtracking stack needed for retrieval is inscribed into the tree nodes, enlarging them even more. This speeds up retrieval at the cost of space.

5.2   Context Trees

As we have mentioned, in discrimination trees common prefixes are shared. In *substitution trees* [6], terms keep their tree structure and all common *contexts*

Table 2. Results. In this table, we list the benchmarks, named after the TPTP problem they come from, along with the prover run on the given problem. Vam stands for Vampire's code tree implementation, Wal for Waldmeister's discrimination trees, and Con for the context trees implementation. Between parentheses, times without retrievals, i.e., only insertions and deletions.

problem	from	time (in seconds)			memory (in Kbytes)		
		Vam	Wal	Con	Vam	Wal	Con
BOO015-4	wal	0.25 (0.00)	0.31 (0.01)	0.46 (0.01)	11	575	11
CAT001-4	vam	3.28 (0.34)	5.74 (0.31)	7.11 (0.41)	3859	13786	3109
CAT002-3	vam	2.90 (0.22)	5.51 (0.31)	6.67 (0.33)	2483	9281	2021
CAT003-4	vam	3.21 (0.34)	5.82 (0.34)	6.90 (0.46)	3826	13595	3086
CIV003-1	vam	7.57 (0.93)	7.13 (0.65)	15.57 (1.16)	3754	22664	3081
COL002-5	fie	1.30 (0.17)	1.55 (0.21)	2.61 (0.28)	925	6090	922
COL004-3	fie	0.96 (0.00)	1.22 (0.00)	2.39 (0.02)	80	727	86
COL079-2	vam	5.46 (0.29)	8.41 (0.26)	7.24 (0.43)	2769	9158	2138
GRP024-5	wal	3.54 (0.01)	4.82 (0.00)	7.44 (0.01)	19	591	22
GRP164-1	fie	17.60 (0.72)	24.60 (0.61)	32.06 (0.89)	5823	28682	5352
GRP179-2	fie	18.34 (0.71)	24.25 (0.60)	32.40 (0.87)	5597	29181	5207
GRP187-1	wal	10.44 (0.02)	11.68 (0.03)	17.64 (0.02)	96	903	97
GRP196-1	fie	6.96 (0.00)	11.92 (0.00)	15.45 (0.00)	1	543	1
HEN011-2	vam	3.36 (0.03)	3.39 (0.02)	5.18 (0.04)	221	2069	211
LAT002-1	vam	5.83 (0.32)	7.72 (0.29)	9.48 (0.44)	3164	14603	2554
LAT009-1	wal	3.78 (0.01)	4.97 (0.01)	5.97 (0.01)	19	591	20
LAT020-1	wal	17.73 (0.01)	24.97 (0.01)	29.87 (0.01)	30	631	31
LAT023-1	fie	1.10 (0.04)	1.49 (0.03)	1.92 (0.07)	198	1646	210
LAT026-1	fie	1.11 (0.09)	1.49 (0.07)	1.79 (0.12)	373	2813	371
LCL109-2	fie	0.47 (0.04)	0.65 (0.06)	0.80 (0.05)	508	2285	466
LCL109-2	wal	0.49 (0.00)	0.66 (0.00)	0.82 (0.00)	16	591	15
LCL109-4	vam	5.62 (0.70)	7.65 (0.46)	13.02 (0.72)	6703	24403	4986
RNG020-6	fie	2.25 (0.07)	3.19 (0.05)	5.33 (0.08)	544	2435	517
RNG028-5	wal	4.19 (0.01)	6.66 (0.01)	9.08 (0.01)	28	607	29
RNG034-1	vam	3.27 (0.33)	4.95 (0.21)	6.86 (0.34)	2545	8330	2125
RNG035-7	wal	8.19 (0.01)	12.10 (0.01)	18.55 (0.01)	36	647	37
ROB006-2	wal	9.88 (0.01)	14.31 (0.02)	21.60 (0.02)	128	1142	116
ROB022-1	fie	0.92 (0.03)	1.20 (0.03)	2.23 (0.03)	119	1086	101
ROB026-1	wal	8.52 (0.01)	13.35 (0.01)	17.34 (0.01)	69	807	68
SET015-4	vam	2.54 (0.02)	2.69 (0.02)	4.53 (0.05)	314	1373	258
		total			total		
		161.09 (5.48)	224.39 (4.64)	308.31 (6.90)	44258	201835	37248

can be shared (note that this includes the comon prefixes of the terms seen as strings). *Context trees* are a new indexing data structure, where, by means of a limited kind of context variables, also common subterms can be shared, even if they occur below different function symbols (see [3] for all details). More sharing allows one to avoid repeated work (this is of course the key to all indexing techniques).

The basic idea is the following. Assume one has three terms $h(x, f(t))$, $h(x, g(t))$, and $h(b, f(t))$ in the index, and let the query be $h(b, f(s))$ where the terms s and t are large and s is not an instance of t. In a discrimination tree (and in a substitution tree) t will occur three times, and two repeated attempts will be made for matching s against t. In a context tree, the root contains $h(x_1, F(t))$ (where F can be instantiated by a single function symbol of any arity) and an immediate failure will occur without exploring any further nodes.

These experiments were run on a first implementation (which does not yet include several important enhancements, see [3]), based on curried terms and on an extension of substitution trees with equality constraints and where one does not distinguish between *internal* and *external* variables. Due to the high amount of sharing, context trees need little space, while still being suitable for compiling (see below).

5.3 Vampire's Code Trees

On one hand, Vampire's code trees can be seen as a form of compiled discrimination trees. One of the important conclusions that can be drawn from the experiments is that it pays off to compile an index into a form of interpreted abstract instructions (similar findings have also been obtained in the field of logic programming). Consider for instance a typical algorithm for term-to-term matching `TermMatch(query,set)`. It is clear that the concrete term `set` is known at compile (i.e. index update) time, and that a specialized algorithm `TermMatchWithSet(query)` can be much more efficient. If the whole index tree is compiled, then also all control instructions become unnecessary (asking whether a child exists, or whether a sibling exists, or whether the node is leaf, etc.). These advantages give far more speedup than the overhead for interpreting the operation code of the abstract instructions (which can be a constant time computed `switch` statement).

But code trees are more than compiled discrimination trees. Let us first consider *standard* discrimination trees, where all variables are represented as a general variable symbol $*$, e.g., different terms like $f(x, y)$ and $f(x, x)$ are both seen as $f(*, *)$, and the corresponding path in the tree is common to both. This increases the amount of sharing, and also the retrieval speed, because the low-level operations are simpler. But it is only a *prefilter*: once a possible match has been found, additional equality tests have to be performed between the query subterms by which the variables of terms like $f(x, x)$ have been instantiated. One important optimization is to perform the equality tests not always in the leaves, but as high up as possible in the tree *without decreasing the amount of sharing*. This is how it is done in partially adaptive code trees [21]. Doing the equality tests in the leaves would frequently lead to repeated work during retrieval time. Also, according to the first-fail principle (which is often beneficial in indexing techniques), it is important to impose strong restrictions like the equality of two whole subterms as soon as possible. Due to the fact that it is conceptually easier to move instructions than to restructure a discrimination tree, code trees are very adequate for this purpose. Finally, code is also compact and hence space-efficient, because one complex instruction can encode (matching with) a relatively specific term structure.

6   Conclusions and Related Work

In Graf's book on term indexing [6], the benchmarks are a small number of sets of terms (each set having between 500 and 10000 terms), coming from runs with Otter [12]. In other experiments also randomly generated terms are used. In all of Graf's experiments first the index is built from one set and then retrieval operations are done using as queries all terms of another set over the same signature. As said, the drawback of such an analysis is that it is unclear how representative the sets of Otter terms are and how frequent updates are in relation with retrieval operations. In addition, in real provers updates are interleaved with queries, which makes it difficult to construct optimal indexes, especially in the case of highly adaptive structures such as substitution trees or context trees. Furthermore, in such experiments it is usually unclear whether the quality of coding of the different techniques is comparable. Although the quality of code in our provers can also be questioned, at least the authors of the discrimination tree and code tree implementations believe that their code is close to optimal. Note that, unlike all previously published papers, the systems participating in the experiments are competitors, and benchmarks were generated by each of them independently, so it is unlikely that the results are biased toward a particular technique.

We expect more researchers to use our benchmarks and programs, which are available at http://www.lsi.upc.es/~roberto, and report on the behaviour of new indexing techniques. A table of results will be maintained as well at that web site.

Other algorithms and data structures for indexing could be compared using our framework; let us mention but a few:

1. retrieval of instances (used in backward subsumption by unit clause);
2. retrieval of instances on the level of subterms (used in backward demodulation);
3. retrieval of unifiable terms;
4. forward subsumption on multiliteral clauses;
5. backward subsumption on multiliteral clauses.

Other interesting algorithms are related to the use of orderings:

1. comparison of terms or literals in the lexicographic path order or the Knuth-Bendix order.

S

2. retrieval of generalizations together with checking that some ordering conditions are satisfied.

We suggest anyone interested to contact the authors on the design of benchmark suites for these problems within the framework of this paper.

Acknowledgments. We thank Jürgen Avenhaus for valuable remarks to yesterday's version of this paper which was as a result substantially rewritten today.

References

P. Cheeseman, B. Kanefsky, and W. M. Taylor. Where the really hard problems are. In R. Reiter J. Mylopoulos, editor, *IJCAI 1991. Proceedings of the 12th International Joint Conference on Artificial Intelligence*, pages 331–340, Sydney, Australia, 1991. Morgan Kaufmann.
2. J. Christian. Flatterms, discrimination nets, and fast term rewriting. *Journal of Automated Reasoning*, 10(1):95–113, February 1993.
3. H. Ganzinger, R. Nieuwenhuis, and P. Nivela. Context trees. In *IJCAR 2001, Proceedings of the International Joint Conference on Automated Reasoning*, Lecture Notes in Artificial Intelligence, Siena, Italy, June 2001. Springer Verlag.
4. P. Graf. Extended path-indexing. In A. Bundy, editor, *CADE-12. 12th International Conference on Automated Deduction*, volume 814 of *Lecture Notes in Artificial Intelligence*, pages 514–528, Nancy, France, June/July 1994.
5. P. Graf. Substitution tree indexing. In J. Hsiang, editor, *Procs. 6th International Conference on Rewriting Techniques and Applications (RTA-95)*, volume 914 of *Lecture Notes in Computer Science*, pages 117–131, Kaiserslautern, 1995.
6. P. Graf. *Term Indexing*, volume 1053 of *Lecture Notes in Computer Science*. Springer Verlag, 1996.
7. J. Gu, P.W. Purdom, J.Franco, and B.W. Wah. *Algorithms for the Satisfiability Problems*. Cambridge University Press, 2001.
8. C. Hewitt. *Description and theoretical analysis of Planner: a language for proving theorems and manipulating models in a robot*. PhD thesis, Department of Mathematics, MIT, Cambridge, Mass., January 1971.
9. T. Hillenbrand, A. Buch, R. Vogt, and B. Löchner. Waldmeister: High-performance equational deduction. *Journal of Automated Reasoning*, 18(2):265–270, 1997.
10. A. Martelli and U. Montanari. An efficient unification algorithm. *ACM Transactions on Programming Languages and Systems*, 4(2):258–282, 1982.
11. W.W. McCune. Experiments with discrimination-tree indexing and path indexing for term retrieval. *Journal of Automated Reasoning*, 9(2):147–167, 1992.
12. W.W. McCune. OTTER 3.0 reference manual and guide. Technical Report ANL-94/6, Argonne National Laboratory, January 1994.
13. D.G. Mitchell, B. Selman, and H.J. Levesque. Hard and easy distributions of SAT problems. In W.R. Swartout, editor, *Procs. 10th National Conference on Artificial Intelligence*, pages 459–465, San Jose, CA, January 1992. AAAI Press/MIT Press.
14. R. Nieuwenhuis. Rewrite-based deduction and symbolic constraints. In H. Ganzinger, editor, *CADE-16. 16th Int. Conf. on Automated Deduction*, Lecture Notes in Artificial Intelligence, pages 302–313, Trento, Italy, July 1999.
15. R. Nieuwenhuis, J.M. Rivero, and M.Á. Vallejo. The Barcelona prover. *Journal of Automated Reasoning*, 18(2):171–176, 1997.

16. H.J. Ohlbach. Abstraction tree indexing for terms. In H.-J. Bürkert and W. Nutt, editors, *Extended Abstracts of the Third International Workshop on Unification*, pages 131–135. Universität Kaiserslautern, 1989. SEKI-Report SR 89-17.
17. M. Paterson and M. Wegman. Linear unification. *Journal of Computer and System Sciences*, 16:158–167, 1978.
18. P.W. Purdom and C.A. Brown. Fast many-to-one matching algorithms. In J.-P. Jouannaud, editor, *Rewriting Techniques and Applications, First International Conference, RTA-85*, volume 202 of *Lecture Notes in Computer Science*, pages 407–416, Dijon, France, 1985. Springer Verlag.
19. I.V. Ramakrishnan, R. Sekar, and A. Voronkov. Term indexing. In A. Robinson and A. Voronkov, editors, *Handbook of Automated Reasoning*, pages 1–97. Elsevier Science and MIT Press, 2001. To appear.
20. A. Riazanov and A. Voronkov. Vampire. In H. Ganzinger, editor, *CADE-16. 16th International Conference on Automated Deduction*, volume 1632 of *Lecture Notes in Artificial Intelligence*, pages 292–296, Trento, Italy, July 1999.
21. A. Riazanov and A. Voronkov. Partially adaptive code trees. In M. Ojeda-Aciego, I.P. de Guzmán, G. Brewka, and L.M. Pereira, editors, *Logics in Artificial Intelligence. European Workshop, JELIA 2000*, volume 1919 of *Lecture Notes in Artificial Intelligence*, pages 209–223, Málaga, Spain, 2000. Springer Verlag.
22. J.M.A. Rivero. *Data Structures and Algorithms for Automated Deduction with Equality*. Phd thesis, Universitat Politècnica de Catalunya, Barcelona, May 2000.
23. S. Schulz. *Learning Search Control Knowledge for Equational Deduction*, volume 230 of *Dissertationen zur künstliche Intelligenz*. Akademische Verlagsgesellschaft Aka GmmH, 2000.
24. M. Stickel. The path indexing method for indexing terms. Technical Report 473, Artificial Intelligence Center, SRI International, Menlo Park, CA, October 1989.
25. G. Sutcliffe and C. Suttner. The TPTP problem library — CNF release v. 1.2.1. *Journal of Automated Reasoning*, 21(2), 1998.
26. A. Voronkov. The anatomy of Vampire: Implementing bottom-up procedures with code trees. *Journal of Automated Reasoning*, 15(2):237–265, 1995.
27. A. Voronkov. CASC $16\frac{1}{2}$. Preprint CSPP-4, Department of Computer Science, University of Manchester, February 2000.

Preferred Extensions of Argumentation Frameworks: Query, Answering, and Computation

Sylvie Doutre and Jérôme Mengin

Institut de Recherche en Informatique de Toulouse
Université Paul Sabatier
118 route de Narbonne – F-31062 Toulouse cedex 4
{doutre,mengin}@irit.fr

Abstract. The preferred semantics for argumentation frameworks seems to capture well the intuition behind the stable semantics while avoiding several of its drawbacks. Although the stable semantics has been thoroughly studied, and several algorithms have been proposed for solving problems related to it, it seems that the algorithmic side of the preferred semantics has received less attention. In this paper, we propose algorithms, based on the enumeration of some subsets of a given set of arguments, for the following tasks: 1) deciding if a given argument is in a preferred extension of a given argumentation framework; 2) deciding if the argument is in all the preferred extensions of the framework; 3) generating the preferred extensions of the framework.

1 Introduction

Argumentation frameworks of [Dun95,BDKT97] abstract many logical systems that have been used to formalize common-sense reasoning or to give a meaning to logic programs. Argumentation frameworks provide a unifying tool for the study of several aspects of these systems, notably their semantics.

Underlying these systems is some notion of deduction, often nonmonotonic. One important consequence of nonmonotonicity is that the set of consequences that can be deduced from such a system is usually inconsistent in the sense of classical logic. The semantics is then given in terms of *extensions*, where each extension corresponds to a possible interpretation of the world. All extensions of a theory are maximally consistent subsets of the set of nonmonotonic consequences of the theory, but various semantics accept more or less of these subsets as extensions. Probably the most widespread semantics is the stable semantics, which is also the most restrictive one. Its definition is very intuitive, but it gives problematic results for many theories; in particular, there are theories which have no stable extension. [Dun95] has proposed another semantics, the *preferred semantics* for argumentation frameworks, which seems to capture the intuition behind the stable semantics while avoiding several of its drawbacks.

The stable semantics has been thoroughly studied, and several algorithms have been proposed for the computation of stable extensions of nonmonotonic

R. Goré, A. Leitsch, and T. Nipkow (Eds.): IJCAR 2001, LNAI 2083, pp. 272–288, 2001.

theories. Algorithms by e.g. [Rei87,Lev91,DMP97,Nie95] rely on some underlying monotonic deduction system that provides information about the possible conflicts between arguments: they then compute extensions so as to avoid these conflicts within each single extension. It has been shown by e.g. [Ino92] that there is a strong connection between the computation of the possible conflicts and consequence finding. These conflicts can be captured in a directed graph, and [CL91,DT93,DM94] have proved that stable extensions of an argumentation framework correspond to particular subsets of this graph called *kernels* in graph theory [Ber73]. This connection between graphs and nonmonotonic theories is the basis for an algorithm in [DMP97] that computes stable extensions of a default theory.

It seems that the algorithmic side of the preferred semantics has received less attention. The worst-case analysis of [DNT99,DNT00] show that the preferred semantics is at least as hard to compute as the stable semantics, and often harder (depending on what are accepted as consequences of argumentation frameworks: the union of the extensions or their intersection). One question related to the preferred semantics has been studied by a number of authors: given an argumentation framework and an argument, is there a preferred extension of the framework that contains the argument? Vreeswijk and Prakken [VP00] propose a proof theory to answer that question, but stop short of actually giving an algorithm that would implement their proof theory. [DKT96] propose a proof procedure to answer the same question but in the restricted case where the argumentation framework corresponds to a logic program. [DM00] propose an algorithm that computes all the preferred extensions of an argumentation framework. This algorithm is based on a technique called set-enumeration in [Rym92], which can be used to generate all the subsets of a given set. This technique has been applied to many problems studied in Artificial Intelligence, like propositional deduction or the computation of stable extensions of some nonmonotonic logics. In this paper, we propose algorithms, also based on the enumeration of the subsets of a given set of arguments, for the following tasks: 1) deciding if a given argument is in an extension of a given argumentation framework; 2) deciding if the argument is in all the extensions of the framework; 3) generating the extensions of the framework. In particular, the third algorithm improves on that presented in [DM00].

The paper is built as follows: the next section presents the preferred semantics of argumentation frameworks, and a characterization in terms of graph-theoretic concepts. Set-enumeration is described in section 3, where we also exhibit properties of preferred extensions that can be used to reduce the number of generated sets of arguments. The algorithms are described in section 4. Section 5 concludes the paper with a comparison of our algorithms with related works and a discussion of possibilities for further enhancements. Proofs of the results can be found in [DM01].

2 The Preferred Semantics for Argumentation Frameworks

In this section, we present Dung's argumentation framework [Dun95] and the preferred semantics.

Definition 1. [Dun95] An *argumentation framework* is a pair (A,R) where A is a set of arguments and R is a binary relation over arguments, i.e. $R \subseteq A \times A$. Given two arguments a and b, $(a,b) \in R$ or $a\,R\,b$ means a *attacks* b (a is said to be an *attacker* of b). Moreover, we say that a set $S \subseteq A$ of arguments attacks an argument a if some argument b in S attacks a.

Thus an argumentation framework can be simply represented as a directed graph, where vertices are the arguments and edges correspond to the elements of R. Given an argument $a \in A$, we denote by $R^+(a) = \{b \in A \mid (a,b) \in R\}$ the set of the *successors* of a, by $R^-(a) = \{b \in A \mid (b,a) \in R\}$ the set of its *predecessors*, and by $R^{\pm}(a)$ the set $R^+(a) \cup R^-(a)$. Moreover, given a set $S \subseteq A$ of arguments and $\varepsilon \in \{+,-,\pm\}$, $R^\varepsilon(S) = \cup_{a \in S} R^\varepsilon(a)$.

Example 1. Let $AF_1 = (A,R)$ with $A = \{a,b,c,d,e,f,g,h,j,k\}$ and R as indicated on the graph below. We use this example throughout this section to illustrate our definitions and propositions, and then as a running example for our algorithms.

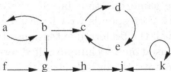

Definition 2. Let (A,R) be an argumentation framework. An argument $a \in A$ *is defended by* a set $S \subseteq A$ of arguments (or S defends a) if and only if $\forall b \in A$, if $b\,R\,a$ then S attacks b, i.e. $\exists c \in S$ such that $c\,R\,b$. A set $S \subseteq A$ is *conflict-free* if and only if there are no arguments a and b in S such that a attacks b. A set $S \subseteq A$ is *admissible* if and only if S is conflict-free and $\forall x \in S$, S defends x.

Dung [Dun95] defines the preferred semantics of an argumentation framework by the set of preferred extensions. We recall below Dung's definition, and give a characterization of preferred extensions in terms of graph-theoretic concepts:

Definition 3. Given an argumentation framework (A,R), a set $S \subseteq A$ is a *preferred extension* if and only if: 1) S is conflict-free; 2) S defends every argument it contains; 3) S is $\subseteq$-maximal such that 1 and 2. The set of the preferred extensions of (A,R) is denoted by $\text{Pref}(A,R)$.

Proposition 1. *Given an argumentation framework (A,R), a subset S of A is a preferred extension if and only if the following conditions hold:* 1) $R^+(S) \cap S = \emptyset$; 2) $R^-(S) \subseteq R^+(S)$; 3) *for every non-empty* $X \subseteq A - S$, $X \cap R^+(S \cup X) \neq \emptyset$ *or* $R^-(X) \not\subseteq R^+(S \cup X)$.

Example 2. $S_1 = \{f, g\}$, $S_2 = \{f, j\}$ and $S_3 = \{f, h\}$ defend h against the attack of g. S_2 and S_3 are conflict-free, not S_1. Then S_1 cannot be an admissible set, nor is S_2 which cannot defend its argument j against the attack of h. S_3 is able to defend every argument it contains, thus S_3 is admissible. AF_1 possesses two preferred extensions: $\{b, d, f, h\}$ and $\{a, f, h\}$.

Dung [Dun95] exhibits interesting properties of the preferred semantics: in particular, every admissible set is contained in a preferred extension, every argumentation framework possesses at least one preferred extension. Moreover, a finite argumentation framework without cycle has exactly one preferred extension. Working with frameworks containing cycles is difficult, since they lead generally to multiple extensions. These difficulties will be illustrated with our example AF_1.

We want to answer in this paper two important questions on preferred extensions: given an argument and an argumentation framework (A, R), is the argument in at least one preferred extension of (A, R)? Is it in every preferred extension? Or equivalently, is the argument a credulous or a sceptical consequence of (A, R)? We define formally these notions:

Definition 4. Given an argumentation framework (A, R) and an argument $a \in A$, a is a *credulous consequence* of (A, R) if and only if a is contained in the union of the preferred extensions of (A, R); a is a *sceptical consequence* of (A, R) if and only if a is contained in the intersection of all the preferred extensions of (A, R).

Example 3. b, a, d, f and h are credulous consequences of AF_1, any other argument of AF_1 is not. The only sceptical consequences of AF_1 are f and h.

3   Extension Enumeration

Before we describe our algorithms in the next section, we present here the general technique on which they are based, and formal properties that will be used to speed-up the computations.

The enumeration of the subsets of a given set A can be performed by exploring a binary tree, the nodes of which are labeled by a partition of A into three sets I, O, and U. If $U = \emptyset$, the node is a leaf, corresponding to the subset I of A. More generally, at any given node n, I is a set of elements of A that will be *In* every subset of A found in the subtree the root of which is n, and O is a set of elements of A that will be *Out* of every subset of A found in the same subtree, while $U = A - (I \cup O)$; thus U is a set of elements that are *Undecided* at that stage, they can end up in some of the sets of the subtree rooted at the current node and out of some other sets in that subtree. If n is such that $U \neq \emptyset$, then U has at most two children: one is labelled with the partition $(I \cup \{x\}, O, U - \{x\})$, the other is labelled with the partition $(I, O \cup \{x\}, U - \{x\})$ for some $x \in U$.

Since we will be interested in the preferred extensions of some argumentation framework (A, R), only some subsets of A are of interest. Even in this case, it would be possible to enumerate all subsets of A, and test for each of them if it verifies a given property. However, it can happen that it is sufficient to generate only one child, or no child at all, for some nodes.

Before we investigate more closely the enumeration of preferred extensions, let us introduce notations that will lighten the presentation of our results. Preferred extensions are conflict-free, so it is legitimate to generate only nodes labelled by triples (I, O, U) such that $R^{\pm}(I) \subseteq O$: since every preferred extension S found in the subtree rooted at that node must be conflict-free (cf. Def. 3) and verifies $I \subseteq S$, it cannot contain any element of $R^{\pm}(I)$; since $S \subseteq I \cup U$, one way to ensure that S will be conflict-free is to explore only nodes such that $R^{\pm}(I) \subseteq O$. We call R-*candidate* a triple of disjoint sets (I, O, U) such that $R^{\pm}(I) \subseteq O$. Given such an R-candidate $C = (I, U, O)$, and an element x of U, we denote by $C + x$ the triple $(I \cup \{x\}, O \cup R^{\pm}(x), U - (\{x\} \cup R^{\pm}(x)))$, and by $C - x$ the triple $(I, O \cup \{x\}, U - \{x\})$. Given a binary relation R and an R-candidate $C = (I, O, U)$, we denote by $\mathrm{Pref}^*(C, R)$ the set of preferred extensions that are in the subtree rooted at a node labelled by (I, O, U):

$$\mathrm{Pref}^*((I, O, U), R) = \{S \in \mathrm{Pref}(I \cup O \cup U, R) \mid I \subseteq S \subseteq I \cup U\}.$$

Our first result shows that we can have a complete enumeration while exploring only nodes that are R-candidates:

Proposition 2. *Let R be a binary relation, let $C = (I, O, U)$ be an R-candidate, and let $x \in U$. If $x \notin R^+(x)$, then $C + x$ and $C - x$ are both R-candidates, and $\mathrm{Pref}^*(C, R) = \mathrm{Pref}^*(C + x, R) \cup \mathrm{Pref}^*(C - x, R)$. If $x \in R^+(x)$, then $C - x$ is an R-candidate, and $\mathrm{Pref}^*(C, R) = \mathrm{Pref}^*(C - x, R)$.*

Essentially, our algorithms will therefore select, at each stage, an undecided argument x and generate two new R-candidates, by putting x in I or in O (and adding predecessors and successors of x to O in the first case), thereby emptying U. Note that arguments that attack themselves are particular, since they can be in no preferred extension. In the sequel, we denote by $\mathrm{Refl}(A, R)$ the set of these arguments:

$$\mathrm{Refl}(A, R) = \{x \in A \mid x \in R^+(x)\}.$$

When U is empty, we need to check if the leaf corresponds to a preferred extension. Let $(I, O, \emptyset)$ be the R-candidate that we find at the leaf, then I is conflict-free because $R^{\pm}(I) \subseteq O$. We need to check that I defends itself, so that I is admissible, and some maximality condition: it must be the case that for every subset X of O, $I \cup X$ is not admissible. We define the following property:

$$\mathrm{Max}(I, O, U, R) \equiv \forall X \subseteq O, X = \emptyset \vee X \cap R^+(X) \neq \emptyset \vee R^-(X) \not\subseteq R^+(I \cup U \cup X).$$

Proposition 3. *Let R be a binary relation, and let $C = (I, O, U)$ be an R-candidate such that $U = \emptyset$. If $R^-(I) \subseteq R^+(I)$, and if $\mathrm{Max}(I, O - R^{\pm}(I), \emptyset, R)$ holds, then $\mathrm{Pref}^*(C, R) = \{I\}$, otherwise $\mathrm{Pref}^*(C, R) = \emptyset$.*

In the remainder of the section, we study properties that can be applied to prune the binary search. First we have strong properties that can be used to stop the search at a given stage, because they guarantee the existence or the absence of preferred extensions that verify certain conditions. These properties will be used in the query answering algorithms, that do not need to exhibit extensions, but only test if there is some or no extension that contains a particular argument.

Our first condition guarantees that there can be no preferred extension in the subtree rooted at some node: since a preferred extension must defend itself, every element of I at a given stage must be defensible; so every predecessor of I must have predecessors that have not already been put out of the extensions currently explored; if this is not the case, then there is no hope of finding a preferred extension in the subtree whose root is the current node. Conversely, there is a condition that guarantees that there exists a preferred extension that contains a set I: if every element of I is defended by I, that is, if every predecessor of I is also a successor of I, then I is admissible, therefore contained in at least one preferred extension (which may not always be found in the subtree rooted at the current node, since too many elements may have already been put in O, thereby preventing I from growing until it becomes maximally admissible). Formally, the following holds:

Proposition 4. Let R be a binary relation, let $C = (I, O, U)$ be an R-candidate. If there exists $x \in R^-(I) - R^+(I)$ such that $R^-(x) \subseteq O$, then $\mathrm{Pref}^*(C, R) = \emptyset$. Otherwise, if $R^-(I) - R^+(I) = \emptyset$, then $\mathrm{Pref}(I \cup O \cup U, R) \neq \emptyset$ and I is admissible.

In particular, these conditions can be used to check if a given argument $a \in I$ is contained in at least one preferred extension. However, if we want to check if there is some extension that does not contain a given argument $a \in O$, then we need a stronger property:

Proposition 5. Let R be a binary relation, (I, O, U) be an R-candidate such that $R^-(I) - R^+(I) = \emptyset$. If $a \in R^+(I)$, or if $a \in O - R^\pm(I)$ and $\mathrm{Max}(I, O - R^\pm(I), U, R)$ holds, then there exists $S \in \mathrm{Pref}(I \cup O \cup U, R)$ such that $a \notin S$.

Let us now turn to properties that will enable us to prune one half of the subtrees rooted at some nodes. That is, we exhibit conditions that guarantee that a given $x \in U$ is such that every extension S such that $I \subseteq S \subseteq I \cup U$ contains x, or such that none of these extensions contains x. We define the following sets:

$$\mathrm{App}(I, U, R) = \{x \in U \mid x \notin R^\pm(I) \cup R^+(U) \text{ and } R^-(x) \subseteq R^+(I \cup \{x\})\}$$
$$\mathrm{Undef}(I, U, R) = \{x \in U \mid R^-(x) \not\subseteq R^+(I \cup U)\}$$
$$\mathrm{Heroes}(I, U, R) = \{x \in U \mid \exists z \in I, \exists y \in R^-(z), \{x\} = R^-(y) \cap (U \cup I)\}$$
$$\mathrm{Traitors}(I, O, U, R) = \{x \in U \mid \exists y \in R^+(x), \exists z \in O \cap R^+(y),$$
$$z \notin R^+(I \cup U \cup \{z\}) \text{ and } R^-(z) - y \subseteq R^+(I)\}$$

$\mathrm{App}(I, U, R)$ is the set of undecided elements that cannot be in conflict with any extension S found in the subtree rooted at the current node (since $x \notin$

$R^{\pm}(I \cup U) \supseteq R^{\pm}(S))$, and who are already defended by I or defend themselves $(R^-(x) \subseteq R^+(I \cup \{x\}))$: these elements will be in every preferred extension in that subtree. Similarly, the elements of $\mathrm{Undef}(I, U, R)$ will be in no extension S in that subtree, since they cannot be defended anymore: not all their predecessors can be attacked $(R^-(x) \not\subseteq R^+(I \cup U) \supseteq R^+(S))$.

Elements of $\mathrm{App}(I, U, R)$ and of $\mathrm{Undef}(I, U, R)$ will be added to I and O respectively, and their fate will be sealed (at least in the current subtree). However, when no such rule can be applied, we may have to create two branches, in which case we will add an element z to I in one branch, to O in the other, without any certainty that it truly belongs to where it is put; in fact, the presence of that element in one set or the other may only become fully justified when the fate of other elements is decided. In particular, if $z \in I$ only has one potential defender x left, then x must be added to I. $\mathrm{Heroes}(I, U, R)$ is the set of such last defenders of some elements of I. Conversely, if $z \in O$ has no attacker in I, then it must not be defensible by I; so if z is already defended against all its attackers except one, whose attackers are not already in O, then these last potential defender of z must not be in I. $\mathrm{Traitors}(I, O, U, R)$ is the set of those undecided elements who may have gone to I, but go to O in order to support another element of O.

Proposition 6. *Let R be a binary relation, let $C = (I, O, U)$ be an R-candidate, and let $x \in U$ such that $x \notin R^+(x)$. Then:*

1. *if $x \in \mathrm{App}(I, U, R) \cup \mathrm{Heroes}(I, U, R)$, then $\mathrm{Pref}^*(C, R) = \mathrm{Pref}^*(C + x, R)$;*
2. *if $x \in \mathrm{Undef}(I, U, R) \cup \mathrm{Traitors}(I, O, U, R)$, then $\mathrm{Pref}^*(C, R) = \mathrm{Pref}^*(C - x, R)$.*

Prop. 6 gives conditions under which only one child of a given node can be generated: this interesting feature relies on the possibility to select an undecided element that has some particular property. If no such undecided element can be found, then we may have to generate two branches. The question that arises is if this can be further avoided, since one can expect the enumeration in a subtree whose root has two children to be on average twice as hard as in a subtree whose root only has one child. Recall that an argumentation framework that has no cycle only has one preferred extension. So it is natural to check if the set of undecided elements at a given node is cycle-free: we can hope that in this case there can only be one extension in the corresponding subtree. It is indeed the case:

Proposition 7. *Let R be a finite binary relation, and let $C = (I, O, U)$ be an R-candidate such that $\mathrm{App}(I, U, R) = \mathrm{Undef}(I, U, R) = \emptyset$ and U is cycle-free. Then $\mathrm{Pref}^*(C, R) = \mathrm{Pref}^*((I \cup U, O, \emptyset), R)$.*

4   The Algorithms

The general extension enumeration algorithm is quite straightforward. We describe it by means of a recursive function PrefEnum, which, given a binary relation R and an R-candidate $C = (I, O, U)$, returns true if $\mathrm{Pref}^*(C, R) \neq \emptyset$, and returns false otherwise. At each step, the function tests if the answer can be found at the current node: StopCondTrue(C, R) tests if the answer is true, StopCondFalse(C, R) tests if it is false. If there is no easy answer, then the function generally calls itself on $(C + x, R)$ and on $(C - x, R)$, for a selected element x of U. This selection of the branching x is crucial, since it may be the case that for a good choice of x, only one branch needs to be explored (cf. Prop 6): for this reason, the selection function Select(C, R) returns a triple (x, b_I, b_O), where b_I and b_O are two booleans, indicating respectively if the branch in which x is added to I, and the branch in which x is added to O, have to be explored. When U is empty or contains no more cycle (Prop 7), the selection function returns b_I and b_O false and it only remains to test if $I \cup U$ is a preferred extension of $(I \cup U \cup O, R)$; this is the role of the condition FinalTest$(I \cup U, O, R)$.

Function PrefEnum(R, C)
Param. *a binary relation R, an R-candidate $C = (I, O, U)$*
Result $\top$ *if an extension is found, $\bot$ otherwise*

1. if StopCondTrue(C, R) then $\top$;
2. else if StopCondFalse(C, R) then $\bot$;
3. else
    a) $(x, b_I, b_O) \leftarrow$ Select(C, R);
    b) if $\neg b_I \wedge \neg b_O$ then FinalTest$(I \cup U, O, R)$
    c) else $(b_I \wedge \mathrm{PrefEnum}(C + x, R)) \vee (b_O \wedge \mathrm{PrefEnum}(C - x, R))$;

Step 3c of the algorithm leaves unspecified the order in which the branches are explored when two branches have to be explored. The choice of the order may have a big influence on the efficiency of the algorithm, since if a positive answer is found on the first branch that is explored, then the other one does not have to be explored.

This skeleton of algorithm is used to answer the credulous and the sceptical query answering problems (that is, checking if an argument is a credulous or a sceptical consequence of an argumentation framework), and also to compute preferred extensions. Note that a self-attacking argument does not belong to any preferred extension: we directly answer false to a credulous or a sceptical query on such an argument. For the computation, and for all the other credulous and sceptical queries, we describe precisely the call of PrefEnum, the conditions StopCondTrue, StopCondFalse and FinalTest and the function Select. For this last function, we use a function Choose which, applied to a non-empty set (of arguments or of sets of arguments), returns an unspecified element of the set. We also use a function Cycles which, given a set X of arguments, returns the set of the sets of arguments which compose cycles in X.

4.1 Credulous Query Answering

Given an argumentation framework (A, R) and an element a of A, we want to check if a is in a preferred extension of (A, R). Since we want to find an extension that contains a, we call the extension enumeration function PrefEnum on R and on the R-candidate $(\emptyset, \emptyset, A) + a$. The function returns $\top$ if an extension is found, $\bot$ otherwise. According to Prop. 4, the stop conditions are:

StopCondTrue$(C, R) \equiv R^-(I) - R^+(I) = \emptyset$ (I is an admissible set)
StopCondFalse$(C, R) \equiv \exists x \in R^-(I) - R^+(I)$ such that $R^-(x) \subseteq O$
    (I cannot defend all its elements)

Our strategy to answer this problem is to try to empty the set $R^-(I) - R^+(I)$, that we denote in the sequel by O_p (O_p is a subset of O). This set contains predecessors of I that are not (yet) successors of I. To empty it, we look at the predecessors of its elements and we try to make them go in I. This is possible only if the predecessors are not self-attacking (since I must be conflict-free): self-attacking predecessors are put in O. Otherwise, we check if undecided predecessors of O_p belong to Heroes(I, U, R) or App(I, U, R), in which case they go in I (Prop 6). If none of these pruning properties is applicable, we select an undecided predecessor of O_p. The Select function is the following:

Function SelectCred(C, R)
Param. *a binary relation R, an R-candidate $C = (I, O, U)$*
Result *an argument and two booleans*

1. if Refl$(U \cap R^-(O_p)) \neq \emptyset$ then (Choose(Refl$(U \cap R^-(O_p))), \bot, \top)$
2. else if Heroes$(I, U, R) \neq \emptyset$ then (Choose(Heroes$(I, U, R)), \top, \bot)$
3. else if App$(I, U, R) \cap R^-(O_p) \neq \emptyset$ then (Choose(App$(I, U, R) \cap R^-(O_p)), \top, \bot)$
4. else (Choose$(R^-(O_p)), \top, \top)$;

As the Select function never returns b_I and b_O false, the function FinalTest will never be used. Let us now run the algorithm on two examples. We describe the call of PrefEnum and we show on the graph representation of the argumentation framework which arguments are in I (circled), in O (crossed) or in U at the end of the run.

Example 4. Given the argumentation framework AF_1, is there a preferred extension containing d? We call PrefEnum on R and on $(\emptyset, \emptyset, A) + d$. Since d is in I, c and e go in O, because they are respectively predecessor and successor of d. Argument c belongs to O_p and it has only one undecided predecessor: b. Argument b is in fact the only potential defender of d: it belongs to Heroes(I, U, R). So b goes in I and consequently its successor-predecessor a goes in O. Now there is no more argument in O without predecessor in I. StopCondTrue$((I, O, U), R)$ is true, we have found an admissible set : $\{d, b\}$, so d belongs to a preferred extension.

Given AF_1, is there a preferred extension containing c? We call PrefEnum on R and on $(\emptyset, \emptyset, A) + c$. Since c is in I, arguments b, d and e go in O. Argument e is in O_p because it is predecessor but not successor of I. In fact, its only predecessor is in O: we are in the case where I cannot defend all its elements. StopCondFalse$(((I, O, U), R)$ is true, so there is no preferred extension containing c.

4.2 Sceptical Query Answering

Given an argumentation framework (A, R) and an element a of A, we want to check if a is in every preferred extension of (A, R). To solve this problem, we look at its complementary: we look for an extension not containing a. To this end, we call the extension enumeration function PrefEnum on R and on the R-candidate $(\emptyset, \emptyset, A) - a$, and we try to build an extension. If PrefEnum$(R, (\emptyset, \emptyset, A) - a) = \bot$, then a is in every preferred extension, otherwise it is not.

We keep the notation $O_p = R^-(I) - R^+(I)$ and we introduce a new one: $O_u = O - R^\pm(I)$. The stop conditions are the following, according to Prop. 4:

StopCondTrue$(C, R) \equiv O_p = \emptyset \wedge (a \in R^+(I) \vee (a \in O_u \wedge \mathrm{Max}(I, O_u, U, R)))$ (a is attacked by an admissible set or a cannot be defended by any preferred extension containing I);

StopCondFalse$(C, R) \equiv (\exists x \in O_p, \ R^-(x) \subseteq O) \vee \neg\mathrm{Max}(I, O_u, \emptyset, R)$ (I cannot defend all its elements or I will never be maximal).

Our strategy to answer a credulous query is the following: we try to show that a is attacked by an admissible set, and if it is not possible, we try to show that it is not defensible. We give the Select function and then we comment it:

Function SelectScept$_a(C, R)$
Param. *a binary relation R, an R-candidate $C = (I, O, U)$*
Result *an argument and two booleans*

1. if $a \in R^+(I)$ then SelectCred(C, R)
2. else if Traitors$(I, O, U, R) \neq \emptyset$ then (Choose(Traitors(I, U, R)), $\bot, \top$)
3. else if Refl$(U \cap R^-(O_u), R) \neq \emptyset$ then (Choose(Refl$(U \cap R^-(O_u), R)$), $\bot, \top$)
4. else if Undef$(I, U, R) \cap R^-(O_u) \neq \emptyset$ then (Choose(Undef$(I, U, R) \cap R^-(O_u)$), $\bot, \top$)
5. else if App$(I, U, R) \cap R^-(O_u) \neq \emptyset$ then (Choose(App$(I, U, R) \cap R^-(O_u)$), $\top, \bot$)
6. else if Cycles$(U) = \emptyset$ then (Choose$(U), \bot, \bot$)
7. else if $\exists e \in$ Cycles(U) such that $e \cap R^-(O_u) \neq \emptyset$ then (Choose$(e \cap R^-(O_u)$), $\top, \top$)

8. else if $U \cap R^-(O_u) \neq \emptyset$ then $(\text{Choose}(U \cap R^-(O_u)), \top, \top)$
9. else $(\text{Choose}(\text{Choose}(\text{Cycles}(U))), \top, \top)$;

When we call the function PrefEnum, a is in O_u: it is rejected, but it is neither successor nor predecessor of I. First of all, we try to show that a is not defensible. To this end, we use the pruning conditions Traitors, Refl and Undef (instructions 2, 3 and 4). Traitors directly acts on O_u predecessors, whereas we must choose in Refl and Undef arguments which are predecessors of O_u. If we cannot show that a is not defensible, then we try to make a a successor of I, thanks to the pruning condition App (instruction 5). If we succeed, then we try to build an admissible set just like we did for the credulous query answering (instruction 1). Finally, if no pruning condition can be applied, we check if U is cycle-free (instruction 6). Then the function PrefEnum will do the following final test on the set $I \cup U$:

$$\text{FinalTest}(I \cup U, O, R) \equiv \text{Max}(I \cup U, O_u, \emptyset, R)$$

If U is not cycle-free, then we choose an undecided predecessor of O_u, possibly cutting a cycle (instructions 7 and 8). If O_u is empty, we choose an undecided argument cutting a cycle.

Example 5. Given the argumentation framework AF_1, is argument d contained in every preferred extension? We call PrefEnum on R and on $(\emptyset, \emptyset, A) - d$. Argument d is in O_u. Its only predecessor, c, cannot be defended against its attacker e, because its only defender was d: c belongs to $\text{Undef}(I, U, R) \cap R^-(O_u)$, so it goes in O, precisely in O_u. Argument e is a predecessor of c, and its only defender against d was c: e belongs to $\text{Undef}(I, U, R) \cap R^-(O_u)$ and goes in O_u. Now argument d has still a potential defender against c, the undecided argument b. This argument must go in O, since we want to build an extension not containing d: b belongs to $\text{Traitors}(I, U, R)$. O_p is empty and no argument of the set O_u is defensible. $\text{StopCondTrue}((I, O, U), R)$ is true, there exists a preferred extension not containing d.

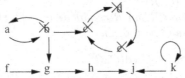

Given AF_1, is argument h contained in every preferred extension? We call PrefEnum on R and on $(\emptyset, \emptyset, A) - h$. Argument h has only one predecessor, g. We cannot apply any pruning property to make g go in I or O. So we build two branches, one where g goes in I (branch 1), the other one where it goes in O (branch 2). In branch 1, f and b go in O because they are predecessor of g. But now f belongs to O_p, and it does not have any predecessor: g will never be defended against f. It means there is no extension in this branch. The graph representation at the end of the run on branch 1 is the following:

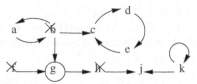

In branch 2, f and b belong to Traitors(I, U, R) because they are potential defenders of h against g. If we select f, then f goes in O. But $\{f\} \subseteq O_u$ is conflict-free and $R^-(\{f\}) \subseteq R^+(I)$ since f has no predecessor: StopCondFalse$(((I, O, U), R)$ is true. This branch cannot lead to a preferred extension because argument f should be in I for I to be a maximal admissible set at the end of the run. The graph representation at the end of the run on branch 2 is the following:

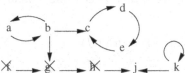

Consequently, since PrefEnum returns false in each branch, h is in every preferred extension of AF_1.

4.3   Computation of the Preferred Extensions

We call PrefEnum on R and on the R-candidate $(\emptyset, \emptyset, A)$. The stop conditions are the following:

StopCondTrue$(C, R) \equiv \bot$;
StopCondFalse$(C, R) \equiv \exists x \in O_p,\ R^-(x) \subseteq O$ (I cannot defend all its elements).

StopCondTrue(C, R) is always false, because we can say we have found a preferred extension only when the status of every argument is decided; this test is done by FinalTest$(I \cup U, O, R)$ at the end of the run. Our strategy is to apply at first the pruning conditions that make arguments go in I or O on a justified way (looking at the arguments of App(I, U, R) and Undef(I, U, R)). According to Prop 7, if these two conditions are not applicable, then we check if U is cycle-free. If U is not cycle-free, we look for Heroes and Traitors to justify the presence of arguments put in I or O at a branching point. Finally, if none of these pruning conditions are applicable, we choose an undecided argument that cuts a cycle.

Function Select(C, R)
Param. *a binary relation R, an R-candidate $C = (I, O, U)$*
Result *an argument and two booleans*

1. if Refl$(U, R) \neq \emptyset$ then (Choose(Refl$(U, R)), \bot, \top$)
2. else if App$(I, U, R) \neq \emptyset$ then (Choose(App$(I, U, R)), \top, \bot$)
3. else if Undef$(I, U, R) \neq \emptyset$ then (Choose(Undef$(I, U, R)), \bot, \top$)

4. else if Cycles(U) = $\emptyset$ then (Choose(U), $\bot$, $\bot$)
5. else if Heroes(I, U, R) $\neq \emptyset$ then (Choose(Heroes(I, U, R)), $\top$, $\bot$)
6. else if Traitors(I, O, U, R) $\neq \emptyset$ then (Choose(Traitors(I, U, R)), $\bot$, $\top$)
7. else (Choose(Choose(Cycles(U))), $\top$, $\top$);

Since we want to enumerate all the extensions, FinalTest must always be $\bot$, otherwise the function PrefEnum could stop before having finished to enumerate all the extensions. FinalTest must be written as a function that, as a side effect, tests if I is a preferred extension, and if true, outputs I. To this end, we introduce the function Print.

Function FinalTest
Param. *a triple of disjoint sets* (I, U, O), *and a binary relation* R
Result $\bot$

1. if Max($I \cup U, O_u, \emptyset, R$) then Print($I$);
2. $\bot$.

Example 6. Let us compute all the preferred extensions of AF_1. We call PrefEnum on R and on the R-candidate $(\emptyset, \emptyset, A)$. Argument k is self-attacking, so it goes in O. Argument f is not attacked: it belongs to App(I, U, R), so it goes in I. Since it is in I, its successor g goes in O. Argument h is attacked by g but it is defended by f; moreover, it is conflict-free with I, so it belongs to App(I, U, R) and then goes in I. Consequently, h's successor (argument j) goes in O. Now, no more pruning property is applicable and U is not cycle-free. The graph representation of AF_1 is the following:

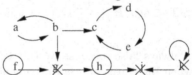

Then the Select function returns an argument that cuts a cycle, for example argument d. We build two branches, one where d goes in I, the other one where d goes in O. When d goes in I, we have the same reasoning on arguments a, b, c and e as in example 4. The status of every argument being decided, a call to FinalTest shows that $I = \{b, d, f, h\}$ is a preferred extension. The graph representation of AF_1 at the end of the run on the first branch is:

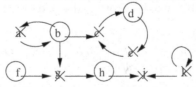

When d goes in O, we have the same reasoning on arguments b, c and e as on example 5. Argument a is not in conflict with $I \cup U$ and defends itself against b: a belongs to App(I, U, R), so it goes in I. The status of every argument being decided, a call to FinalTest shows that $I = \{a, f, h\}$ is a preferred extension of AF_1. The graph representation of AF_1 at the end of the run on the second branch is:

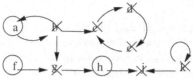

5   Discussion

Since preferred extensions have been put forward by [Dun95] as a remedy to features of stable extensions that are sometimes undesired, it is interesting to compare the algorithms presented in the previous sections to similar ones that have been designed for the computation of stable extensions.

Recall that, by definition, a subset S of the set of vertices A of some argumentation framework (A, R) is a stable extension of (A, R) if and only if it is conflict-free and dominant, that is if every element of A which is not in S has a predecessor in S. In terms of graph-theoretic concepts, S is a stable extension of (A, R) if and only if $R^+(S) \cap S = \emptyset$ and $A - S \subseteq R^+(S)$. A well-known result of [Dun95], is that every stable extension is also a preferred one. So all the properties that have been used to prune the enumeration of subsets of A when looking for preferred extensions can also be used when computing the stable extensions. In particular, the algorithms of [Nie95,DMP97] also explore the equivalent of our R-candidates. Although other properties that we have put forward in section 3 could be used for the computation of stable extensions, these algorithms use other properties that are too strong to be used for the computation of preferred extensions. In particular, in the case of stable extensions, we can add an element x to I as soon as all its predecessors are in O, but in the case of preferred extension we need to ensure that the predecessors of x have predecessors in I.

Our algorithm can be used to compute the stable extensions of an argumentation framework, since these are preferred extensions as well: we only need to check for every leaf if $O \subseteq R^+(I)$: every argument that is not in a stable extension must have a predecessor in this extension. A detailed comparison of our algorithm with that of [DMP97] can be found in [DM01].

The idea of looking for an element of a cycle to branch at a stage when no pruning property can be applied (cf Prop. 7) was first put forward by [DMP97]. However, [DMP97] propose to compute a feedback vertex set first, that is, a set F of vertices of a graph (A, R) such that the graph restricted to $A - F$ contains no cycle. They then run an enumeration-like algorithm, choosing an undecided element of F every time a branching point is reached. Notice that finding a minimal feedback vertex set of a given graph (A, R) is known to be an NP-complete problem, although heuristics can be used to find a small one. However, there are cases where elements of a feedback vertex set, even a minimal one, are of no use. This is why it may be more efficient to look for a cycle only when necessary. Moreover, in order not to find several times the same cycle in different branches of the algorithm, we can maintain a set of all cycles that have been found so far during the execution of the algorithm.

Example 7. Let $A = \{a, b, c, d\}$ and let $R = \{(a, b), (b, c), (c, d), (d, b)\}$. Since a has no predecessor, $a \in I$; then, the successor of a, that is b, has to be in O; then d cannot be defended, so it has to be in O, whereas c has to be in I since it is defended by a and it is not adjacent to it. Although there is a cycle here $(bRcRdRb)$, the tree explored by the algorithm has a single branch.

It is not difficult to see that, in the case of the preferred semantics, strongly connected components of an argumentation framework can be treated separately (see e.g. [DM01]). This is not new, and is particularly interesting when generating the preferred extensions: the first call to PrefEnum should then be split into as many calls as there are strongly connected components. When answering a sceptical query, only the component to which the query belongs should be explored. In the case of the credulous query answering algorithm only the strongly connected component to which the query belongs will be explored, since in this case the Select function returns predecessors of the set O_p, that is, predecessors of predecessors of the current set I.

The worst-case analysis of [DNT99] shows that, under the preferred semantics, the sceptical reasoning problem is in a complexity class above that of the credulous reasoning problem in the polynomial hierarchy. The extra cost of sceptical reasoning appears in our algorithm in the stop conditions and in the final test, which involve maximality tests. Such a test is expensive since it implies some enumeration of the parts of the set O of the arguments which are Out of the extension being built: the algorithm needs to check that none of these parts could be added to the admissible set I found so far while preserving the admissibility property. Notice that we can use our enumeration algorithm in order to perform this check: we would call it on the R-candidate $(I, R^{\pm}(I), O - R^{\pm}(I))$ and look for an admissible set different from I.

[VP00] and [CDM01] propose dialectical proof theories to answer credulous queries on preferred semantics. These proof theories have the form of a dialogue between a proponent and an opponent. An important aspect of such proofs is that they are quite natural, in the sense that they give an easy way to understand the implications of the underlying notions of acceptability. The two dialectical proofs theories of [CDM01] are directly inspired by the credulous query answering algorithm presented here. They improve on the one of [VP00], since the proofs that they produce are generally shorter than the proofs of [VP00].

[VP00] also propose a dialectical proof theory to answer sceptical queries, but it works only in a particular case: when preferred extensions coincide with stable extensions. This is a very restrictive condition, because stable extensions do not always exist. This implies that the sceptical proof theory of [VP00] is able to answer no sceptical query on the argumentation framework of example 1 (this framework has no stable extension).

The good results of [CDM01] suggest to design a dialectical proof theory to answer sceptical queries, inspired by the sceptical query answering algorithm presented here. This is the purpose of some future work.

Acknowledgements. We would like to thank the referees for helpful comments on an earlier version of the paper. In particular, one of them suggested the call to PrefEnum in order to perform the maximality checks.

References

[BDKT97]   A. Bondarenko, P. Dung, R. Kowalski, and F. Toni.   An abstract, argumentation-theoretic approach to default reasoning. *Art. Int.*, 93:63–101, 1997.

[Ber73]   C. Berge. *Graphs and Hypergraphs*, volume 6 of *North-Holland Mathematical Library*. North-Holland, 1973.

[CDM01]   C. Cayrol, S. Doutre and J. Mengin. Dialectical proof theories for the credulous preferred semantics of argumentation frameworks. Submitted, 2001.

[CL91]   G. Chaty and F. Levy. Default logic and kernel in digraph. Tech. Rep. 91-9, LIPN, 1991.

[DKT96]   P. Dung, R. Kowalski, and F. Toni. Synthesis of Proof Procedures for Default Reasoning. In *Proceedings of LOPSTR'96*, vol. 1207 of LNCS. Springer, 1996.

[DM94]   Y. Dimopoulos and V. Magirou. A graph-theoretic approach to default logic. *Information and computation*, 112:239–256, 1994.

[DM00]   S. Doutre and J. Mengin. An Algorithm that Computes the Preferred Extensions of Argumentation Frameworks. In *ECAI'2000, Third International Workshop on Computational Dialectics (CD'2000)*, pages 55–62, August 2000.

[DM01]   S. Doutre and J. Mengin. Preferred extensions of argumentation frameworks: computation and query answering. Tech. Rep. IRIT/2001-03-R, March 2001.

[DMP97]   Y. Dimopoulos, V. Magirou, and C. Papadimitriou. On kernels, Defaults and Even Graphs. *Annals of Mathematics and AI*, 1997.

[DNT99]   Y. Dimopoulos, B. Nebel, and F. Toni. Preferred Arguments are Harder to Compute than Stable Extensions. In *Proceedings of IJCAI'99*, pages 36–41, 1999.

[DNT00]   Y. Dimopoulos, B. Nebel, and F. Toni. Finding Admissible and Preferred Arguments Can be Very Hard. In *Proceedings of KR'2000*, pages 53–61, 2000.

[DT93]   Y. Dimopoulos and A. Torres. Graph-theoretic structures in logic programs and default theories. Tech. Rep. 93-264, Max-Planck-Institut für Informatik, November 1993.

[Dun95]   P. Dung. On the acceptability of arguments and its fundamental role in non-monotonic reasoning, logic programming and n-person games. *Art. Int.*, 77:321–357, 1995.

[Ino92]   K. Inoue. Linear resolution for consequence finding. *Art. Int.*, 56:301–353, 1992.

[Lev91]   F. Levy. Computing Extensions of Default Theories. In R. Kruse and
            P. Siegel, editors, *Proceedings of ECSQAU'91*, vol. 548 of LNCS, pages
            219–226. Springer, 1991.
[Nie95]   I. Niemelä. Towards Efficient Default Reasoning. In *IJCAI'95*, pages 312–
            318, 1995.
[Rei87]   R. Reiter. A Theory of Diagnosis from First Principles. *Art. Int.*, 32:57–95,
            1987.
[Rym92]   R. Rymon. Search through Systematic Set Enumeration. In B. Nebel,
            C. Rich, and W. Swartout, editors, In *Proc. of KR'92*, pages 539–550.
            Morgan Kaufmann, 1992.
[VP00]   G. Vreeswijk and H. Prakken. Credulous and Sceptical Argument Games
            for Preferred Semantics. In *Proceedings of JELIA'2000*, vol. 1919 of LNAI.
            Springer, 2000.

Bunched Logic Programming
(Extended Abstract)

Pablo A. Armelín and David J. Pym

Queen Mary
University of London
England, UK
{pablo,pym}@dcs.qmw.ac.uk

Abstract. We give an operational semantics for the logic programming language BLP, based on the hereditary Harrop fragment of the logic of bunched implications, **BI**. We introduce **BI**, explaining the account of the sharing of resources built into its semantics, and indicate how it may be used to give a logic programming language. We explain that the basic input/output model of operational semantics, used in linear logic programming, will not work for bunched logic. We show how to obtain a complete, goal-directed proof theory for hereditary Harrop **BI** and how to reformulate the operational model to account for the interaction between multiplicative and additive structure. We give a prototypical example of how the resulting programming language handles, in contrast with Prolog, sharing and non-sharing use of resources purely logically.

1 Bunched Logic and Logic Programming

The logic of bunched implications, **BI**, freely combines an additive (intuitionistic) and a multiplicative (linear) implication as connectives of equal status [12,14, 15]. Thus it stands in stark contrast with linear logic [4], in which intuitionistic implication is available via an exponential [12,14,15].

The semantics of **BI** may be motivated directly by modelling the notion of *resource*. Consider the following, very simple, axiomatization of the notion of resource (clearly, refinements are possible):

- An underlying *set* of resources, M;   – A way of *combining* resources, $\cdot$;
- A representative for *zero* resources, e;   – A way of *comparing* resources, $\sqsubseteq$.

Mathematically, we recognize that we have naturally identified a (for now, commutative) preordered monoid $\mathcal{M} = (M, e, \cdot, \sqsubseteq)$ of resources.

First, we may exploit the presence of the monoidal combining operation to define the following multiplicative conjunction [17,12,14,15], in the possible-worlds style [9]:

$$m \models \phi * \psi \quad \text{iff} \quad \text{there are } n \text{ and } n' \text{ such that } m \sqsubseteq n \cdot n', \, n \models \phi \text{ and } n' \models \psi$$

R. Goré, A. Leitsch, and T. Nipkow (Eds.): IJCAR 2001, LNAI 2083, pp. 289–304, 2001.

This conjunction is interpreted as follows: the resource required to establish $\phi * \psi$ is obtained by combining the resources required to establish each of ϕ and ψ. Similarly, we can define the corresponding implication

$$m \models \phi \mathbin{-\!\!*} \psi \quad \text{iff} \quad \text{for all } n \text{ such that } n \models \phi,\, m \cdot n \models \psi.$$

This implication is interpreted as follows: if the resource required to establish the "function", $\phi \mathbin{-\!\!*} \psi$, is m and the resource required to establish the "argument" ϕ, is n, then the resource required to establish the result is $m \cdot n$. Thus the function and the argument do not share resources. Rather, their respective resources are taken from distinct worlds.

Second, the presence of the preorder suggests the possibility of a satisfaction relation for the intuitionistic connectives, using worlds $m, n \in M$ as usual:

$$
\begin{aligned}
m &\models \phi \wedge \psi & \text{iff} & \quad m \models \phi \text{ and } m \models \psi \\
m &\models \phi \vee \psi & \text{iff} & \quad m \models \phi \text{ or } m \models \psi \\
m &\models \phi \to \psi & \text{iff} & \quad \text{for all } n \sqsubseteq m,\, n \models \phi \text{ implies } n \models \psi.
\end{aligned}
$$

The conjunction (and the disjunction) are interpreted as follows: each of the conjuncts (disjuncts) may share resources with the other. The implication is interpreted as a "function" which may share resources with its argument. We refer to the meaning of the semantics described here as the *sharing interpretation* [12, 15]. Similarly, the additives may be seen as describing *local* properties whereas the multiplicatives are *global*. We return to this point in § 4, in which we give a concrete, and implemented, programming example.

Proof-theoretically, the presence of the two implications is, at first sight, problematic. To see this consider that whilst we may easily distinguish between multiplicative and additive elimination (or left) rules,

$$\frac{\Gamma \vdash \phi \mathbin{-\!\!*} \psi \quad \Delta \vdash \phi}{\Gamma, \Delta \vdash \psi} \;\; {-\!\!*}E \qquad \frac{\Gamma \vdash \phi \to \psi \quad \Gamma \vdash \phi}{\Gamma \vdash \psi} \;\; {\to}E,$$

how are we to distinguish the corresponding introduction rules ? We may write

$$\frac{\Gamma, \phi \vdash \psi}{\Gamma \vdash \phi \mathbin{-\!\!*} \psi} \;\; {-\!\!*}I \quad \text{but how then to write a rule for } {\to}I, \quad \frac{\Gamma \,?\vdash \psi}{\Gamma \vdash \phi \to \psi} \;?$$

A semantically clean solution is provided by moving to sequents built not out of finite sequences of hypotheses but rather out of *bunches of hypotheses*, *i.e.*, finite trees, with the leaf nodes being formulæ, and the internal nodes denoted by either "," or ";", and referred to as *bunches*. The grammar of bunches is given as follows:

$$
\begin{array}{llll}
\Gamma ::= & \phi & \text{propositional assumption} & \quad \mid \Gamma, \Gamma \quad \text{multiplicative combination} \\
& \mid \emptyset_m & \text{multiplicative unit} & \quad \mid \emptyset_a \qquad\quad\;\; \text{additive unit} \\
& & & \quad \mid \Gamma; \Gamma \qquad\; \text{additive combination}
\end{array}
$$

We write $\Gamma(\Delta)$ to denote that Δ is a sub-bunch of Γ in the evident sense. Equality of bunches, $\equiv$, is given by the commutative monoid laws for "," and

";", together with substitution congruence: if $\Delta \equiv \Delta'$, then $\Gamma(\Delta) \equiv \Gamma(\Delta')$. A bunch is said to be *multiplicative* if its top-level combinator is "," and *additive* if its top-level combinator is ";". Contraction and Weakening are permitted for ";" but not for ",":

$$\frac{\Gamma(\Delta \,; \Delta) \vdash \phi}{\Gamma(\Delta) \vdash \phi} \quad Contraction \qquad \frac{\Gamma(\Delta) \vdash \phi}{\Gamma(\Delta \,; \Delta') \vdash \phi} \quad Weakening.$$

In much of what follows, we regard "," and ";" as multi-ary operators, and refer to their operands as *components*.

The introduction and elimination rules for the multiplicative and additive implications now go as follows:

$$\frac{\Gamma, \phi \vdash \psi}{\Gamma \vdash \phi \mathbin{-\!\!*} \psi} \quad \mathbin{-\!\!*} I \qquad \frac{\Gamma; \phi \vdash \psi}{\Gamma \vdash \phi \to \psi} \quad \to I$$

$$\frac{\Gamma \vdash \phi \mathbin{-\!\!*} \psi \quad \Delta \vdash \phi}{\Gamma, \Delta \vdash \psi} \quad \mathbin{-\!\!*} E \qquad \frac{\Gamma \vdash \phi \to \psi \quad \Delta \vdash \phi}{\Gamma; \Delta \vdash \psi} \quad \to E.$$

Turning to predication, consider that we can express a first-order sequent over a collection X of first-order variables as $(X)\Gamma \vdash \phi$. Given this point of view, we can see that it is possible to allow not only Γ to be bunched but also X. So for each propositional rule, we have two possible forms of variable maintenance, *i.e.*, additive and multiplicative. For example, the two choices for the predicate $*R$ rule are

$$\frac{(X)\Gamma \vdash \phi \quad (Y)\Delta \vdash \psi}{(X;Y)\Gamma, \Delta \vdash \phi * \psi} \quad * R_a \quad \text{and} \quad \frac{(X)\Gamma \vdash \phi \quad (Y)\Delta \vdash \psi}{(X,Y)\Gamma, \Delta \vdash \phi * \psi} \quad * R_m.$$

The former choice is the one taken in linear logic and in this paper. It may be simplified, via Weakening and Contraction, to

$$\frac{(X)\Gamma \vdash \phi \quad (X)\Delta \vdash \psi}{(X)\Gamma, \Delta \vdash \phi * \psi} \quad * R_a$$

The latter is the one taken in the basic version of **BI** [12,14,15].

The presence of bunched variables has one very significant consequence: It permits the definition of both additive, or extensional, *and* multiplicative, or intensional, quantifiers. For example,

$$\frac{(X;x)\Gamma \vdash \phi}{(X)\Gamma \vdash \forall x.\phi} \quad \forall I \qquad \frac{(X,x)\Gamma \vdash \phi}{(X)\Gamma \vdash \forall_{\text{new}} x.\phi} \quad \forall_{\text{new}} I,$$

where x is not free in Γ, and

$$\frac{(X)\Gamma \vdash \forall x.\phi \quad Y \vdash t : \text{Term}}{(X;Y)\Gamma \vdash \phi[t/x]} \quad \forall E \qquad \frac{(X)\Gamma \vdash \forall_{\text{new}} x.\phi \quad Y \vdash t : \text{Term}}{(X,Y)\Gamma \vdash \phi[t/x]} \quad \forall_{\text{new}} E$$

Here, we assume a simple bunched calculus of term formation [12,14,15] and, as before, the additive case may be simplified to use just one bunch of variables. The corresponding existentials are similar.

Semantically, the additive quantifier is handled intuitionistically,

$$m \models \forall x.\phi \text{ iff for all } n \sqsubseteq m \text{ and all } t \text{ defined at } n, n \models \phi[t/x],$$

i.e., the resource required to establish each instance of the quantified proposition must be available at the starting world. The semantics of the multiplicative $\forall_{new}$ explains our use of the term "new":

$$m \models \forall_{new} x.\phi \text{ iff for all } n \text{ and all } t \text{ defined at } n, m \cdot n \models \phi[t/x],$$

i.e., the resource required to instantiate the proposition is taken from a *new* world or location. Again, the existentials are similar.

Our notion of *logic programming* is that introduced in [11,10], based on the sequent calculus. We start with the fragment of the logic for which *uniform proofs* are complete for logical consequence. Reading proofs from the root upwards, uniform proofs require that right rules be applied whenever possible, so that left rules are applied only when the right-hand side is atomic. Uniform proofs are said to be *simple* just in case the implicational left rules are restricted to be essentially unary. For example, in first-order intuitionistic logic, we get

$$\frac{\Gamma \vdash \phi[t/x] \quad \alpha[t/x] \vdash \beta[t/x]}{\Gamma, \phi \supset \alpha \vdash \beta} \supset L,$$

with α, β atomic and $\alpha[t/x] = \beta[t/x]$ (often, $\phi \supset \alpha$ is retained in the left-hand premiss).

In intuitionistic logic, simple uniform proofs, which are *goal-directed* and in which the non-determinism is confined to the choice of implicational formula, are complete for hereditary Harrop sequents [11,10]. Simple uniform proofs amount to the analytic notion of *resolution*. Taking all this together, we interpret hereditary Harrop sequents $\Gamma \vdash_o \exists x.\phi$ as a logic program, Γ, together with a query, or goal, ϕ, in which there is a logical variable x [8]. We use $\vdash_o$ to denote the simple, uniform, *i.e.*, resolution, proof, read from root to leaves.

In **BI**, the corresponding class of sequents may be defined. *Bunched hereditary Harrop formulæ* are given by the following grammar, in which A denotes atoms (we simplify a bit, for brevity, omitting, for example, universal goals):

Definite formulæ $D ::= A \mid D \wedge D \mid G \rightarrow A \mid D * D \mid G \mathbin{-\!\!*} A \mid \forall x.D \mid \forall_{new} x.D$

Goal formulæ $\quad G ::= \top \mid A \mid G \wedge G \mid D \rightarrow G \mid G * G \mid D \mathbin{-\!\!*} G \mid G \vee G$
$\qquad\qquad\qquad \mid \exists x.G \mid \exists_{new} x.G$

A *bunched hereditary Harrop sequent* is a sequent $\mathcal{P} \vdash G$, where $\mathcal{P}$ is a bunch of definite formulæ. Such sequents, for now without $\forall_{new}$ and $\exists_{new}$, are the basis of the bunched logic programming language, **BLP**, to which we give an operational semantics in § 3.

In intuitionistic logic, each of the reduction operators used in the execution of goal-directed search is additive. In **BI**, however, as in linear logic [16,10,5], we have multiplicatives which introduce a computationally significant difficulty. The typical case is $*R$:

$$\frac{\Gamma_1 \vdash_o \phi_1 \quad \Gamma_2 \vdash_o \phi_2}{\Gamma \vdash_o \phi_1 * \phi_2} \quad \Gamma = \Gamma_1, \Gamma_2.$$

Faced with $\Gamma \vdash_o \phi_1 * \phi_2$, the division of Γ into Γ_1 and Γ_2 must be calculated.

The basic solution, described for linear logic in [7,6,16,5], is the so-called *input/output* model. First pass all of Γ to the left-hand branch of the proof, leaving the right-hand branch undetermined. Proceed with the left-hand branch until it is completed. Then calculate which of the formulæ in Γ have been used to complete the left-hand branch and collect them into a finite set, Γ_{left}; The remaining, unused formulæ may now be passed to the right-hand branch:

$$\frac{\vdots \\ \overline{\Gamma_{\text{left}} \vdash_o \phi_1} \quad \Gamma \backslash \Gamma_{\text{left}} \vdash_o \phi_2}{\Gamma \vdash_o \phi_1 \otimes \phi_2} \quad \Gamma = \Gamma_{\text{left}}, \Gamma \backslash \Gamma_{\text{left}}.$$

We refer to $\backslash$ as a "remainder operator" because it removes from Γ the consumed formulæ and passes the remainder to the next branch.

In **BI**, the problem is made *much* more complex by the mixing of additive and multiplicative structure enforced by the presence of bunches and the basic input/output idea will not work. To see this, consider the following search for a proof of the (provable) sequent $\phi, \psi \vdash (\chi \to \psi \wedge \chi) * \phi$ (note that it is convenient to put the remainder operator, read as "without", in the "current" computation):

$$\frac{\dfrac{\overline{} \; (\text{not provable})}{(\chi; (\phi, \psi)) \backslash \phi \vdash_o \psi \wedge \chi}}{\dfrac{(\phi, \psi) \backslash \phi \vdash_o \chi \to (\psi \wedge \chi)}{(\phi, \psi) \backslash \emptyset_m \vdash_o (\chi \to (\psi \wedge \chi)) * \phi}} \to R \qquad \dfrac{\overline{}}{\phi \backslash \emptyset_m \vdash_o \phi}} *R$$

Consider the left-hand branch of the candidate $*R$ rule (the actual operational rules are defined formally in § 3). In order to get an axiom of the form $\psi \vdash \psi$, we must first remove the χ from the program by performing a Weakening and then perform a subtraction of ϕ, which is required on the right-hand branch of the $*R$, so that the result of $\chi; (\phi, \psi) \backslash \phi$ is ψ, *i.e.*, the remainder operator first throws away, via Weakening, the additive bunch surrounding the multiplicative bunch within which ψ, the formula which must be removed, is contained. Now, the remaining bunch is sufficient to form, after an $\wedge R$ reduction, the axiom $\psi \vdash \psi$ but insufficient to form the necessary axiom for χ.

At first sight, thinking of axioms of the form $\Gamma; \alpha \vdash \alpha$, it might seem like we need a subtraction operation which does not perform Weakening. This would solve the problem in the particular case above but would worsen it in other cases. From being incomplete the system would become unsound. To see this, consider a search for a proof of the unprovable sequent $\tau; (\phi, \psi) \vdash (\chi \to (\psi \wedge \tau)) * \phi$. Once the $*R$ rule is applied, after doing $\to R$, the propositions τ and χ are at the same level in $(\chi; \tau; (\phi, \psi)) \backslash \phi$ and, if this is taken to be equal to $\chi; \tau; \psi$, both χ and τ are provable.

Again it looks like this could be fixed, by requiring, for example, that the *R rule be applied only on multiplicative bunches. Aside from the unpleasant non-determinism introduced by this solution, it doesn't quite solve the problem either. To see this, it is enough to consider a slight modification of the last search. Consider the unprovable sequent $v, (\tau; (\phi, \psi)) \vdash (\tau \to (\psi \wedge \tau)) * (\phi * v)$. Here, after the application of *R and $\to R$ it would be possible to prove both ψ and τ, which is unsound. It seems clear that the root of the difficulty lies with the interaction between the additives, in particular $\to R$, and the multiplicatives.

The basic idea for the solution, though far from simple in detail, is to introduce stacks which keep a record of which resources have been added to the program as a result of $\to R$ and which manage their interaction with the formation of axioms, and with subtraction, and with passing by continuations. The detailed formulation of the continuation-passing style (CPS) operational semantics is rather complex and before giving it, in § 3, we must look, in § 2, at uniform, and simple, proofs in **BI**.

2   Uniform Proofs in BI

So far we have discussed **BI** semantically, and informally considered its use as a logic programming language. Formally, as we have indicated, the basis of logic programming in **BI** relies on the availability of *goal-directed* proofs.

Proofs in **BI** may be presented as a sequent calculus, here given in Definition 1. The Cut-elimination theorem holds [15] and semantic completeness theorems are available [12,14,15,13]. In order to explain uniform proofs, and so also the subsequent operational semantics, we restrict our attention to propositional **BI**: although our logic programming language uses predicate **BI**, it exploits, in its core language, only additive predication and quantification, just as in Prolog [2,1], so that its use of logical variables and unification is completely standard. It follows that our presentation involves no significant loss of generality.

The use of multiplicative predication and quantification is possible but more complex. We conjecture that its main use will be in BLP's module system, adapted to **BI** from the basic ideas presented in [10], in which we conjecture the importing of functions from one module to another may exploit the additive–multiplicative distinction to valuable effect for the programmer. This topic is beyond our present scope.

Definition 1. *The sequent calculus* **LBI** *is defined as follows:*

IDENTITIES

$$\frac{}{\phi \vdash \phi} \; Axiom \qquad \frac{\Gamma \vdash \phi \quad \Delta(\phi) \vdash \psi}{\Delta(\Gamma) \vdash \psi} \; Cut$$

STRUCTURALS

$$\frac{\Gamma(\Delta) \vdash \phi}{\Gamma(\Delta; \Delta') \vdash \phi} \; W \qquad \frac{\Gamma(\Delta; \Delta) \vdash \phi}{\Gamma(\Delta) \vdash \phi} \; C \qquad \frac{\Gamma \vdash \phi}{\Delta \vdash \phi} \; (\Delta \equiv \Gamma) \; E$$

UNITS

$$\dfrac{\Gamma(\emptyset_m) \vdash \phi}{\Gamma(I) \vdash \phi} \; IL \qquad \dfrac{}{\emptyset_m \vdash I} \; IR \qquad \dfrac{}{\bot \vdash \phi} \; \bot L \qquad \dfrac{\Gamma(\emptyset_a) \vdash \phi}{\Gamma(\top) \vdash \phi} \; TL \qquad \dfrac{}{\emptyset_a \vdash \top} \; TR$$

MULTIPLICATIVES

$$\dfrac{\Gamma \vdash \phi \quad \Delta(\Delta',\psi) \vdash \chi}{\Delta(\Delta',\Gamma,\phi \mathbin{-\!*} \psi) \vdash \chi} \; \mathbin{-\!*}L \qquad \dfrac{\Gamma,\phi \vdash \psi}{\Gamma \vdash \phi \mathbin{-\!*} \psi} \; \mathbin{-\!*}R$$

$$\dfrac{\Gamma(\phi,\psi) \vdash \chi}{\Gamma(\phi * \psi) \vdash \chi} \; *L \qquad \dfrac{\Gamma \vdash \phi \quad \Delta \vdash \psi}{\Gamma,\Delta \vdash \phi * \psi} \; *R$$

ADDITIVES

$$\dfrac{\Gamma \vdash \phi \quad \Delta(\Delta';\psi) \vdash \chi}{\Delta(\Delta';\Gamma;\phi \to \psi) \vdash \chi} \; \to L \qquad \dfrac{\Gamma;\phi \vdash \psi}{\Gamma \vdash \phi \to \psi} \; \to R$$

$$\dfrac{\Gamma(\phi_1;\phi_2) \vdash \psi}{\Gamma(\phi_1 \wedge \phi_2) \vdash \psi} \; \wedge L \qquad \dfrac{\Gamma \vdash \phi \quad \Delta \vdash \psi}{\Gamma;\Delta \vdash \phi \wedge \psi} \; \wedge R$$

$$\dfrac{\Gamma(\phi) \vdash \chi \quad \Delta(\psi) \vdash \chi}{\Gamma(\phi \vee \psi);\Delta(\phi \vee \psi) \vdash \chi} \; \vee L \qquad \dfrac{\Gamma \vdash \phi}{\Gamma \vdash \phi \vee \psi} \; \vee R \qquad \dfrac{\Gamma \vdash \psi}{\Gamma \vdash \phi \vee \psi} \; \vee R'$$

The predicate substitution and quantifier rules may be given sequentially.   □

Most of the permutations, $\rightsquigarrow$, necessary to show that uniform proofs are complete for bunched hereditary Harrop formulæ are very straightforward. For example,

$$\dfrac{\Gamma \vdash \phi \quad \dfrac{\Delta(\psi) \vdash \chi \quad \Delta(\psi) \vdash \chi'}{\Delta(\psi) \vdash \chi \wedge \chi'} \wedge R}{\Delta(\Gamma,\phi \mathbin{-\!*} \psi) \vdash \chi \wedge \chi'} \mathbin{-\!*}L \;\; \rightsquigarrow \;\; \dfrac{\dfrac{\Gamma \vdash \phi \quad \Delta(\psi) \vdash \chi}{\Delta(\Gamma,\phi \mathbin{-\!*} \psi) \vdash \chi} \mathbin{-\!*}L \quad \dfrac{\Gamma \vdash \phi \quad \Delta(\psi) \vdash \chi'}{\Delta(\Gamma,\phi \mathbin{-\!*} \psi) \vdash \chi'} \mathbin{-\!*}L}{\Delta(\Gamma,\phi \mathbin{-\!*} \psi) \vdash \chi \wedge \chi'} \wedge R$$

5
or

$$\dfrac{\Gamma \vdash \phi \quad \dfrac{\Delta(\psi) \vdash \chi \quad \Delta' \vdash \chi'}{\Delta(\psi),\Delta' \vdash \chi * \chi'} *R}{\Delta(\Gamma,\phi \mathbin{-\!*} \psi),\Delta' \vdash \chi * \chi'} \mathbin{-\!*}L \;\; \rightsquigarrow \;\; \dfrac{\dfrac{\Gamma \vdash \phi \quad \Delta(\psi) \vdash \chi}{\Delta(\Gamma,\phi \mathbin{-\!*} \psi) \vdash \chi} \mathbin{-\!*}L \quad \Delta' \vdash \chi'}{\Delta(\Gamma,\phi \mathbin{-\!*} \psi),\Delta' \vdash \chi * \chi'} *R.$$

Note, however, that uniform proofs in **BI** must always perform any possible (and trivial) $*Ls$ (and $\wedge Ls$) before performing any right rule: to see this, consider that we should like to have a uniform proof of $\phi * \psi \vdash \phi * \psi$.

Weakening is, at first sight, a source of difficulty. Weakenings may be permuted above all rules *except* $*R$. To see this, consider that $*R$ must divide a multiplicative bunch between its two premisses. So if a $*R$ has a Weakening immediately below it, then there is no way in which $*R$ may be applied directly to the resulting sequent. However, it turns out that this difficulty may be handled within the operational semantics by examining in turn each of the multiplicative bunches below the ";" introduced by the Weakening. To make this work, we must consider a canonical form for bunches: a bunch is in *canonical form* iff

1. its left-hand branch is either a proposition, a unit or a canonical bunch of the opposite (additive or multiplicative) type, and
2. its right-hand branch is canonical.

For example, if Γ and Γ' are additive and canonical and if Δ is canonical, then $(\Gamma, \Gamma', \Delta)$ is canonical. Our operational semantics, in § 3, assumes that bunches are in canonical form.

Lemma 1. *Every bunch may be written in canonical form.*   □

Proposition 1. *Uniform proofs in* **BI** *are complete for bunched hereditary Harrop sequents.*   □

As we have seen, uniform proofs do not, however, characterize resolution. For that, we must ensure that the choice of clause in each implicational left rule is also goal-directed. For that we require proofs which are not merely uniform but simple, *i.e.*, in which all instances of implicational left rules are of the following, essentially unary, forms (I is the unit of $*$):

$$\frac{\Gamma \vdash \phi \quad \Delta \vdash I}{\Delta, (\Gamma; \phi \rightarrow \alpha) \vdash \alpha} \rightarrow L \qquad \frac{\Gamma \vdash \phi}{\Gamma, \phi \twoheadrightarrow \alpha \vdash \alpha} \twoheadrightarrow L$$

These rules are clearly admissible in **LBI** but to understand why the $\twoheadrightarrow L$ rule is complete we must consider not only the canonical form for bunches but also that we may replace the basic axiom sequent with the following CutAxiom rule:

$$\frac{\Gamma \vdash I}{\Gamma, \alpha \vdash \alpha} \quad \text{CutAxiom (CA).}$$

The effect of this rule, which may be seen as a form of garbage collection, is to absorb, trivially, unused multiplicative bunches. Then we can make the following transformation of proof figures ($\emptyset_m$ is the unit of ",", *i.e.*, I on the left):

$$\cfrac{\Gamma_1 \vdash \phi \quad \cfrac{\Gamma_2 \vdash I}{\Gamma_2, \alpha \vdash \alpha}\,\text{CA}}{\Gamma_1, \Gamma_2, \phi \twoheadrightarrow \alpha \vdash \alpha}\twoheadrightarrow L \quad \rightsquigarrow \quad \cfrac{\cfrac{\Gamma_2 \vdash I \quad \cfrac{\cfrac{\Gamma_1 \vdash \phi}{\Gamma_1, \emptyset_m \vdash \phi}\,\text{Unit of ","}}{\Gamma_1, \Gamma_2 \vdash \phi}\text{Cut}}{}\quad \cfrac{}{\alpha \vdash \alpha}\text{Ax}}{\Gamma_1, \Gamma_2, \phi \twoheadrightarrow \alpha \vdash \alpha}\twoheadrightarrow L$$

in which the right-hand premiss of the $\twoheadrightarrow L$ is rendered trivial.

This step, together with an analysis of the permutations of $\twoheadrightarrow L$ and $\rightarrow L$ rules with respect to one another, is sufficient to give us the following:

Proposition 2. *Simple uniform proofs in* **BI** *are complete for bunched hereditary Harrop formulæ.*

3   An Operational Semantics for BLP

As we have suggested, the operational semantics of BLP may be seen as a development of the *input/output model*, described in [7], to account for the bunched structure of sequents. However, this development represents a substantial generalization of the method and involves a degree of technical complexity.

Starting from uniform proofs, presented in § 2, we see that the remaining source of non-determinism is, essentially, the splitting of bunches in the $*R$ operator. Thus a remainder can be valid only if it is calculated on the left-hand branch of a search above a $*R$ operator. However, additional operational complexity arises from the bunched structure itself, *i.e.*, from the interaction between the multiplicative and additive implications. (Notation: for clarity, we use just Γs, ϕs, *etc.*, rather than the more cumbersome Ds and Gs notation. We distinguish atomic formulæ as αs. Nevertheless, we are working with hereditary Harrop sequents.)

It follows that operational sequents have the form $\langle \Gamma | s \rangle_n \backslash_n^{n'} \langle \Gamma' | s' \rangle \vdash_o \phi$, which should be interpreted as follows:

- Γ is the bunch which is passed *up* the current branch of the search tree;
- Γ' is the bunch which is passed *down* the current branch of the search tree;
- n and n' are counters which keep track of the number of $\top$ goals found. This is necessary since a $\top$ goal makes the logic locally affine. We use a counter instead of a simple flag so when applying the $-*R$ rule to a goal $\phi-*\psi$, if a $\top$ goal is found while proving ψ (*i.e.*, if $n' > n$) then any left-over of ϕ can be removed instead of failing;
- s and s' denote stacks of bunches and are used to manage the interaction between the reduction operators used and the formulæ available:
  - s manages upward propagation via *open boxes* and *locked boxes*;
  - s' manages downward propagation via *full boxes*, *open boxes* and *locked boxes*.

Informally, full, open and locked boxes, which are used in the definitions of the $*R$, $\wedge R$ and $\rightarrow R$ operators, arise as follows:

Full box:	$\{\Delta, \xi\}$	introduced downwards, arising from assignment to an unknown remainder — creating a full box corresponds to an assignment in the CPS execution
Open box:	$\boxed{\xi}^{\varepsilon}$	introduced upwards, arising from a search-figure in which a $*R$ occurs below an $\rightarrow R$. Here ε is the "theorem flag", explained below
Locked box:	$\boxtimes$	introduced upwards, arising from a $*R$

Their meanings will be made clearer in our description (below) of the operational semantics;

- Finally, $\backslash$ is a remainder operator: informally, it is read as "without" or "leaves unused". Formally, its meaning is defined by the operators for $\vdash_o$.

Overall, an operational sequent, $\langle \Gamma | s \rangle_n \backslash^{n'} \langle \Gamma' | s' \rangle \vdash_o \phi$, is read as "given a program Γ with a stack s as input, after proving ϕ we are left with a program Γ' and stack s' for the subsequent computation". An $n > 0$ indicates that a theorem or $\top$ was found as a goal, so that we are (locally) in affine **BI**.

At each occurrence of a $*R$ rule, a box is created and a flag, represented by ε (for εmpty flag), is placed at the left-hand end of the bunch which is the remainder, is set to indicate whether remainders from the left-hand branch of the search are permitted (if ε is present, then remainders are not permitted, if it is absent they are; it is rather like I). Initially, the box is locked, and is put at the top of the upward stack.

When an $\rightarrow R$ operator is used, the box is unlocked and the antecedent of the implication $\phi \rightarrow \psi$ is put into it: so we get a stack of the form $\boxed{\phi}^\varepsilon_{::}s$. If the box is unlocked already, with previous implicational antecedents in it, then the new antecedent is additively combined with the existing ones: so we get a stack of the form $\boxed{\Delta; \phi}^\varepsilon_{::}s$. Also, the empty flag is reset, thereby allowing no remainder to pass except through the box. At this stage the open box carries the "theorem flag" ε which indicates that $\phi \rightarrow \psi$ is regarded as possibly an intuitionistic theorem, failure of which will be detected at a CutAxiom rule. This procedure is necessary because theorems behave like $\top$ with respect to multiplicative resources.

A box that contains a remainder is denoted by $\{\Delta, \phi\}::s$. Such boxes occur in downward stacks and indicate that the computation was performed under the wrapping of ϕ, arising from additives, and leaves a remainder Δ.

Finally, in order to give the operators of the operational semantics, we need to formally define a *subtraction operator*. We are now using $\backslash$ to calculate remainders but we still need a basic way of removing a bunch from a super-bunch within which it lives. The subtraction of Γ from Δ is definable just in case Γ is a sub-bunch of Δ, written $\Gamma \subseteq \Delta$ and is defined, exploiting the canonical form for bunches, as follows (here $\overset{.}{\Gamma}$ means that Γ is additive and $\overset{.}{\Gamma}$ means that Γ is multiplicative; we label the components of a bunch Δ as Δ_is):

- If $\Gamma = \emptyset_m$, then $\Gamma \subseteq \Delta$, any Δ;

- $\overset{.}{\Gamma} \subseteq \overset{.}{\Delta}$ iff $\overset{.}{\Gamma} \equiv \overset{.}{\Delta}$ or $\overset{.}{\Gamma} \subseteq \overset{.}{\Delta_i}$, some i;

- $\overset{.}{\Gamma} \subseteq \overset{.}{\Delta}$ iff $\overset{.}{\Gamma} \subseteq \overset{.}{\Delta_i}$, some i;

- $\overset{.}{\Gamma} \subseteq \overset{.}{\Delta}$ iff $\overset{.}{\Gamma} \subseteq \overset{.}{\Delta_i}$, some i;

- $\overset{.}{\Gamma} \subseteq \overset{.}{\Delta}$ iff $\overset{.}{\Gamma} \equiv \overset{.}{\Delta}$ or $\overset{.}{\Gamma} \subseteq \overset{.}{\Delta_i}$, some i.

We then write $\Delta - \Gamma$ to denote the following subtraction operation: find the additive/multiplicative sub-bunch of Δ within which Γ lives and delete all of that sub-bunch. For example, $(\phi, ((\psi, \psi'); \chi)) - (\psi, \psi') = \phi$. (Formally, $\Delta - \Gamma$ is defined by recursion over additive/multiplicative cases.)

The operators, read from conclusion to premisses, which describe the operational semantics of **BLP**, are summarized in Table 1. The presentation relies on the execution of the left-hand branches before the right-hand. We describe the continuation-passing style execution of the operational semantics by considering the key cases of the operational reductions, given in Table 1, in turn.

CutAxiom: We are given a bunch Γ which contains the atom α, *i.e.*, the principal formula of the axiom, in any position, and a stack s. There is exactly one way of performing the minimum number of Weakening reductions, reducing $\Gamma(\alpha) \vdash \alpha$ to Γ', $\alpha \vdash \alpha$, on Γ so as to bring α to top-level. This is done and we give to the success continuation the resulting bunch *without* α, *i.e.*, the remainder Γ'. In fact, since the only possible reduction above a CutAxiom is one of the Unit reductions, we can be more specfic about what the form of Δ and s' must be. For example, if Δ is ε, Γ' is not equal to $\emptyset_m$ and s has an open box on top, then s' must have a full box on top. Otherwise, Δ must be equal to Γ' and s' must be equal to s. If the top of the stack is an open box containing ξ, it is necessary to check whether α is in ξ. If not, then the theorem flag of the open box must be removed because it means that the $\phi \rightarrow \psi$ in a previous $\rightarrow R$ which created the open box is not a theorem.

$\rightarrow\!\!*L$: We start with a bunch in which the clause $\phi \rightarrow\!\!* \alpha$ occurs in an arbitrary position. The unary, or "resolution", version of $\rightarrow\!\!*L$ is invoked but the bunch taken in the premiss is that which is obtained, as in the CutAxiom case, by performing the minimum number of Weakening reductions on Γ so as to bring α to top-level.

$*R$: We start with Γ and s given and try to prove ϕ from Γ and $\boxtimes::s$. Upon success, we get a remainder, Γ'. At this point, we can try to prove ψ from Γ' and s. Upon success, we get as a result a remainder Δ together with an arbitrary modification of the stack. Because we managed to prove $\phi * \psi$ from Γ, leaving Δ, this result is given to the final success continuation. The special case in which Δ is ε works in the same way but for the fact that Γ' and s, which must be $[]$, will prove ψ only when Γ' proves ψ without leaving any remainder. So, Γ and s are given, Γ' and Δ are calculated, and s is modified as necessary, depending on the reductions encountered above. An example which includes this case follows after the description of the operators. If a $\top$ goal or an intuitionistic theorem was found, the counter will be greater than zero and the computation will succeed even if ε is set and there is a non-empty remainder.

$\wedge Rs$: There are five subcases. In the first two, a remainder is allowed, which may occur only if we are on some multiplicative branch of the search. In the first, handled by $\wedge R$, the bunch Γ and the stack s are given. Notice that the stack s is the same on both sides of the left-hand branch: modifications to the stack before passing to the next branch occur only in operators in which remainders are not permitted. Moreover, notice that the right-hand branch gets an empty stack: it has already all the information needed to search for a proof. For example, this case of the rule is used for the proof of $(\phi; \psi), \chi \vdash_o (\phi \wedge \psi) * \chi$ and, indeed, the same program, with this rules, proves $(\psi \wedge \phi) * \chi$, $(\phi \wedge \phi) * \chi$, $(\psi \wedge \psi) * \chi$, *etc.*. In each case, the context given to the left-hand branch will be equivalent to $(\phi; \psi), \chi$ and the subtraction, "$\Gamma - \Delta$", will leave χ for the right-hand branch. The second case, $\wedge R^i$ takes care of the case in which $\top$ is found as a goal in both conjuncts. Of course, the $\wedge R$ rules do not propagate the presence of $\top$ from the

Table 1. Summary of the Operational Semantics of BLP

$$\frac{}{\langle\Gamma|s\rangle_n\backslash^{n+1}\langle\Gamma|s\rangle \vdash_o \top}\ \text{TUnit} \qquad \frac{}{\langle\Gamma|s\rangle_n\partial^0\langle\Gamma|s\rangle \vdash_o I}\ \text{Unit}$$

$$\frac{}{\langle\Gamma|s\rangle_n\backslash^n\langle\varepsilon|s\rangle \vdash_o I}\ \text{Unit}^i(n>0) \qquad \frac{}{\langle\phi;\Gamma|\boxed{\phi}^\varepsilon::s\rangle_n\backslash^n\langle\varepsilon|\{\Gamma,\phi\}::s\rangle \vdash_o I}\ \text{Unit}^{ii}$$

$$\frac{\langle\Gamma'|s\rangle_n\backslash^{n'}\langle\Delta|s'\rangle \vdash_o I}{\langle\Gamma(\alpha)|s\rangle_n\backslash^{n'}\langle\Delta|s'\rangle \vdash_o \alpha}\ \text{CutAxiom}^\dagger \qquad \frac{\langle\Gamma'|\boxed{\xi}::s\rangle_n\backslash^{n'}\langle\Delta|s'\rangle \vdash_o I}{\langle\xi;\Gamma(\alpha)|\boxed{\xi}^\varepsilon::s\rangle_n\backslash^{n'}\langle\Delta|s'\rangle \vdash_o \alpha}\ \text{CutAxiom}^{ii\dagger}$$

$$\frac{\langle\Gamma'|s\rangle_n\backslash^{n'}\langle\Delta|s'\rangle \vdash_o \phi}{\langle\Gamma(\phi\!-\!\!*\alpha)|s\rangle_n\backslash^{n'}\langle\Delta|s'\rangle \vdash_o \alpha}\ -\!\!*L^\dagger \qquad \frac{\langle\Theta;\phi\!\rightarrow\!\alpha|[]\rangle\partial^m\langle\varepsilon|[]\rangle \vdash_o \phi \quad \langle\Gamma'|s\rangle_n\backslash^{n'}\langle\Delta|s'\rangle \vdash_o I}{\langle\Gamma(\Theta;\phi\!\rightarrow\!\alpha)|s\rangle_n\backslash^{n'}\langle\Delta|s'\rangle \vdash_o \alpha}\ \rightarrow\!L^\dagger$$

$$\frac{\langle\Gamma|\boxtimes::s\rangle_n\backslash^{n'}\langle\Gamma'|\boxtimes::s\rangle \vdash_o \phi \quad \langle\Gamma'|s\rangle_n\backslash^{n''}\langle\Delta|s'\rangle \vdash_o \psi}{\langle\Gamma|s\rangle_n\backslash^{n''}\langle\Delta|s'\rangle \vdash_o \phi*\psi}\ *R \qquad \frac{\langle\phi,\Gamma|s\rangle_n\backslash^{n'}\langle\Delta|s'\rangle \vdash_o \psi}{\langle\Gamma|s\rangle_n\backslash^{n'}\langle\Delta'|s'\rangle \vdash_o \phi\!-\!\!*\psi}\ -\!\!*R^\ddagger$$

$$\frac{\langle\Gamma|s\rangle_n\backslash^n\langle\Delta|s\rangle \vdash_o \phi \quad \langle\Gamma-\Delta|[]\rangle\partial^m\langle\varepsilon|[]\rangle \vdash_o \psi}{\langle\Gamma|s\rangle_n\backslash^n\langle\Delta|s\rangle \vdash_o \phi\wedge\psi}\ \wedge R$$

$$\frac{\langle\Gamma|s\rangle_n\backslash^{n'}\langle\Delta|s\rangle \vdash_o \phi \quad \langle\Gamma|s\rangle_n\backslash^{n''}\langle\Delta'|s\rangle \vdash_o \psi}{\langle\Gamma|s\rangle_n\backslash^{n'}\langle\Delta\cap\Delta'|s\rangle \vdash_o \phi\wedge\psi}\ \wedge R^i(n',n''>n)$$

$$\frac{\langle\Gamma|s\rangle_n\backslash^{n'}\langle\varepsilon|s\rangle \vdash_o \phi \quad \langle\Gamma|s\rangle_n\backslash^{n''}\langle\varepsilon|s\rangle \vdash_o \psi}{\langle\Gamma|s\rangle_n\backslash^{n'''}\langle\varepsilon|s\rangle \vdash_o \phi\wedge\psi}\ \wedge R^{ii\sharp}$$

$$\frac{\langle\xi;\Gamma|\boxed{\xi}^\varepsilon::s\rangle_n\backslash^{n'}\langle\varepsilon|\{\Delta,\xi\}::s\rangle \vdash_o \phi \quad \langle\xi;(\Gamma-\Delta)|[]\rangle\partial^{n''}\langle\varepsilon|[]\rangle \vdash_o \psi}{\langle\xi;\Gamma|\boxed{\xi}^\varepsilon::s\rangle_n\backslash^{n'''}\langle\varepsilon|\{\Delta,\xi\}::s\rangle \vdash_o \phi\wedge\psi}\ \wedge R^{iii\sharp}$$

$$\frac{\langle\xi;\Gamma|\boxed{\xi}^\varepsilon::s\rangle_n\backslash^{n'}\langle\varepsilon|\boxed{\xi}^\varepsilon::s\rangle \vdash_o \phi \quad \langle\xi;\Gamma|\boxed{\xi}^\varepsilon::s\rangle_n\backslash^{n''}\langle\varepsilon|\{\Delta,\xi\}::s\rangle \vdash_o \psi}{\langle\xi;\Gamma|\boxed{\xi}^\varepsilon::s\rangle_n\backslash^{n'''}\langle\varepsilon|\{\Delta,\xi\}::s\rangle \vdash_o \phi\wedge\psi}\ \wedge R^{iv\sharp}$$

$$\frac{\langle\phi;\Gamma|[]\rangle_n\backslash^{n'}\langle\varepsilon|[]\rangle \vdash_o \psi}{\langle\Gamma|[]\rangle_n\backslash^{n'}\langle\varepsilon|[]\rangle \vdash_o \phi\rightarrow\psi}\ \rightarrow\!R \qquad \frac{\langle\phi;\Gamma|\boxed{\phi}^\varepsilon::s\rangle_n\backslash^{n'}\langle\varepsilon|\{\Delta,\phi\}::s\rangle \vdash_o \psi}{\langle\Gamma|\boxtimes::s\rangle_n\backslash^{n'}\langle\Delta|\boxtimes::s\rangle \vdash_o \phi\rightarrow\psi}\ \rightarrow\!R^i$$

$$\frac{\langle\phi;\xi;\Gamma|\boxed{\xi;\phi}^\neg::s\rangle_n\backslash^{n'}\langle\varepsilon|\{\Delta,(\xi;\phi)\}::s\rangle \vdash_o \psi}{\langle\xi;\Gamma|\boxed{\xi}^\neg::s\rangle_n\backslash^{n'}\langle\varepsilon|\{\Delta,\xi\}::s\rangle \vdash_o \phi\rightarrow\psi}\ \rightarrow\!R^{ii\natural} \qquad \frac{\langle\phi;\Gamma|\boxed{\phi}^\varepsilon::s\rangle_n\backslash^{n'}\langle\varepsilon|\boxed{\phi}^\varepsilon::s\rangle \vdash_o \psi}{\langle\Gamma|\boxtimes::s\rangle_n\backslash^{n+1}\langle\Gamma|\boxtimes::s\rangle \vdash_o \phi\rightarrow\psi}\ \rightarrow\!R^{iii}$$

$$\frac{\langle\Gamma|s\rangle_n\backslash^{n'}\langle\Delta|s'\rangle \vdash_o \phi}{\langle\Gamma|s\rangle_n\backslash^{n'}\langle\Delta|s'\rangle \vdash_o \phi\vee\psi}\ \vee R \qquad \frac{\langle\Gamma|s\rangle_n\backslash^{n'}\langle\Delta|s'\rangle \vdash_o \psi}{\langle\Gamma|s\rangle_n\backslash^{n'}\langle\Delta|s'\rangle \vdash_o \phi\vee\psi}\ \vee R$$

† Γ' is obtained uniquely by performing the minimal number of Weakenings required to bring α, $\phi\!-\!\!*\alpha$ or $(\Theta;\phi\!\rightarrow\!\alpha)$ to top-level.

‡ If $n'>n$ then $\Delta'=\Delta-\phi$, else $\Delta'=\Delta$.

♯ If $n',n''>n$, then $n'''=n+1$, else $n'''=n$.

♮ Here $\boxed{\ }::s$ indicates that here the theorem flag may be present or not.

left branch to the right branch. An increased counter should be passed on to the rest of the computation since any left over found later could have been given to the $\phi \wedge \psi$ subproof and weakened away at this point. In the other three, handled by $\wedge R^{ii}$, $\wedge R^{iii}$ and $\wedge R^{iv}$, no remainder is permitted. Here we have three subcases, depending on whether the stack is full on return of the left, right or neither branch. An increased counter will be passed to the rest of the computation only if $\top$ is found in *both* left and right subproofs. An example which includes one of these cases follows after the description of the operators.

$\rightarrow R$s: There are four subcases. All four share the property that the antecedent *of the implication* leaves no remainder. Note that we assume that any occurrences of $\wedge$ or $*$ in the antecendents ϕ in $\phi \rightarrow \psi$ which are (inductively) principal connectives are immediately removed using operators corresponding to the operationally trivial $\wedge L$ or $*L$ rules. An example which illustrates $\rightarrow R$ follows after the description of the operators. The last rule is used when the implication is an intuitionistic theorem, when it should behave in similar ways to $\top$. The remaining cases are similar. In all cases, failure invokes backtracking.

A worked example, as mentioned above, will clarify these complex constructions (here, "CA" denotes CutAxiom and the counters are 0 throughout):

$$\cfrac{\cfrac{\overline{\langle\phi|\boxed{\chi}\rangle\backslash\langle\varepsilon|\{\phi,\chi\}\rangle \vdash I}\;\text{Unit}^{ii}}{\langle\chi;(\phi,\psi)|\boxed{\chi}^\varepsilon\rangle\backslash\langle\varepsilon|\{\phi,\chi\}\rangle \vdash \psi}\;CA^i \quad \cfrac{\overline{\langle\varnothing_m|[]\rangle\backslash\langle\varepsilon|[]\rangle \vdash I}\;\text{Unit}}{\langle\chi;\psi|[]\rangle\backslash\langle\varepsilon|[]\rangle \vdash \chi}\;CA}{\cfrac{\langle\chi;(\phi,\psi)|\boxed{\chi}^\varepsilon\rangle\backslash\langle\varepsilon|\{\phi,\chi\}\rangle \vdash \psi\wedge\chi}{\langle(\phi,\psi)|\boxtimes\rangle\backslash\langle\phi|\boxtimes\rangle \vdash \chi\rightarrow(\psi\wedge\chi)}\;\rightarrow R^i}\;\wedge R^{iii} \quad \cfrac{\overline{\langle\varnothing_m|[]\rangle\backslash\langle\varepsilon|[]\rangle \vdash I}\;\text{Unit}}{\langle\phi|[]\rangle\backslash\langle\varepsilon|[]\rangle \vdash \phi}\;CA}{\langle(\phi,\psi)|[]\rangle\backslash\langle\varepsilon|[]\rangle \vdash (\chi\rightarrow(\psi\wedge\chi))*\phi}\;*R$$

Thus we revisit our earlier, problematic example, showing how the use of boxes manages the interaction between the multiplicatives and additives, specially $\rightarrow R$.

In order to establish the soundness and completeness, with respect to logical consequence, of the operational semantics, we need to relate intermediate states of a computation, *i.e.*, operational sequents, to logical consequences, *i.e.*, **BI** sequents. To this end we introduce a mapping $\backslash\!\backslash$ from pairs of bunch-stack pairs, together with the goal, *i.e.*, $\langle\Gamma,s\rangle_n^{n'}\langle\Gamma',s'\rangle \vdash_o \phi$, to bunches. The idea is to extend the basic subtraction operation, defined for bunches, to bunch-stack pairs. There are three cases.

1. Remainders are not allowed and a full box is on the top of the downward stack: $\langle\xi;\Gamma|\boxed{\xi}^\varepsilon::s\rangle\backslash\!\backslash\langle\varepsilon|\{\Delta,\xi\}::s\rangle = f(\xi);(\Gamma - \Delta)$, where $f(\xi)$ is ξ if ϕ is an additive conjunction or implication, and is $\top$ otherwise.
2. Remainders are not allowed and there is something other than a full box on top of the downward stack: $\langle\Gamma|s\rangle\backslash\!\backslash\langle\varepsilon|s\rangle = \Gamma \quad (= \Gamma - \varnothing_m)$.
3. Finally, in case neither (1) nor (2): $\langle\Gamma|s\rangle\backslash\!\backslash\langle\Delta|s\rangle = \Gamma - \Delta$.

Then we get, by induction on the structure of proofs, the following:

Lemma 2. *The stacks used in the operational semantics do not corrupt the logical consequence relation:*

$$\langle \Gamma, s \rangle_n \backslash^{n'} \langle \Gamma', s' \rangle \vdash_o \phi \quad \text{if and only if} \quad \langle \overline{\Gamma}, s \rangle \backslash \langle \overline{\Gamma'}, s' \rangle \vdash \phi,$$

where if $n > 0$, then $\overline{\Gamma}$ and $\overline{\Gamma'}$ may be sub-bunches of Γ and Γ', and if $n = 0$, then $\overline{\Gamma} = \Gamma$ and $\overline{\Gamma'} = \Gamma'$. □

Soundness and completeness follow as a corollary of Lemma 2.

Theorem 1. *The operational semantics is sound and complete with respect to the bunched sequent calculus,* **LBI***:*

$$\langle \Gamma, [] \rangle \, \lambda \text{-} \langle \varepsilon, [] \rangle \vdash_o \phi \quad \text{if and only if} \quad \Gamma \vdash \phi.$$ □

We conjecture that our operational techniques will be applicable to a wide range of substructural logics.

4   Programming in BLP: An Example

Recall that our semantics for **BI**'s connectives, as set out in § 1, is couched in terms of *sharing*. Here, we give a quite generic, yet small, example of the type of problems for which bunched logic programming is well-suited. Consider the bunch $(p(a1); p(a2)), (p(b1); p(b2))$. Here, $p(x)$ means "x is a person". The bunch structure shows that $a1$ and $a2$ belong to the same group and that $a1$ and $b1$ belong to beligerent groups. To say that two individuals are possibly in a fighting relation we say simply $\forall x, y.p(x) * p(y) \twoheadrightarrow fight(x, y)$, which is to say that x and y may fight if they belong to different groups. A complete BLP program (we write on one line to save space) would be (here, T is ⊤, the unit of ∧)

```
(p(a1);p(a2)), (p(b1);p(b2)), [x,y]fight(x,y)*- p(x)*p(y)*T
```

Notice that the definition of fight has been slightly modified to take into account that there might be more than two groups; but they may be disregarded.

An alternative solution would be to decorate each group with a multiplicative unit to signal that it can be ignored. So we might have, for example,

```
(p(a1);p(a2);I), (p(b1);p(b2);I), (p(a5);p(a6);I)
```

However, the first approach is to be recommended since it doesn't produce redundant solutions. Adding a unit to each group allows the unit operation to be performed in different places, but without changing the solution.

The following is an equivalent Prolog program for this problem. It uses tags to distinguish the groups:

```
p(a1,t1). p(a2,t1). p(b1,t2). p(b2,t2).
fight(X,Y):- p(X,T),p(Y,U),T\=U.
```

Thinking of political parties, sometimes they split into rival factions but each faction in turn might want to keep its former allies. This situation might be represented by the bunch $(p(a1); p(a2)), (p(b1); (p(b21); p(b22)), (p(b23); p(b24)))$. Notice that $b21$ fights with $a1$ and $a2$ but also with $b23$ and $b24$. If we call x and y allies if they don't fight, then despite $b1$'s being an ally of $b21$, and also of $b23$, $b21$ and $b23$ are not allies. The modification of the BLP program to reflect this state of affairs is straightforward:

```
(p(a1);p(a2)), (p(b1); (p(b21);p(b22)), (p(b23);p(b24))),
[x,y]fight(x,y)*- p(x)*p(y)*T
```

Notice that *the defining clause doesn't need any modification.*

To modify the Prolog program we could start by adding an extra tag to reflect the structure of the problem like this

```
p(a1,t1,_). p(a2,t1,_).    fight(X,Y):- p(X,T,_),p(Y,U,_),T\=U.
p(b1,t2,_).                fight(X,Y):- p(X,T,V),p(Y,U,W),T=U,V\=W.
p(b21,t2,t1). p(b22,t2,t1).
p(b23,t2,t2).p(b24,t2,t2).
```

and we should be aware that the whole program has had to be modified to account for the extra tag. Or a new, more flexible implementation may be dreamed up, like using lists of tags as a second argument:

```
p(a1,[t1]).     fight(X,Y):- p(X,U),p(Y,S),mismatch(U,S).
p(a2,[t1]).     mismatch([H1|_],[H2|_]):- H1\=H2.
p(b1,[t2]).     mismatch([H1|T1],[H2|T2]):- H1=H2,mismatch(T1,T2).
p(b21,[t2,t1]).
p(b22,[t2,t1]). p(b23,[t2,t2]). p(b24,[t2,t2]).
```

Please compare the heavy machinery used in this example with the simplicity of the BLP version.

The bunched structure also helps to give fine control over the scope of predicates, specially implications. In the example above, we can think of a variety of ways in which constants can be predicated. For example $a2$ might be a special kind of person. It would be possible to modify the program in the following way:

```
(p(a1);q(a2);[x]p(x) <- q(x)),
(p(b1);(p(b21);q(b22)),(p(b23);p(b24))),
[x,y]fight(x,y)*- p(x)*p(y)*T
```

Now this program says that $a2$ is a q but also that all qs are ps. However, this relation between ps and qs holds only for the group formed by $a1$ and $a2$, *i.e.*, is *local* to that world. Other qs appearing in other places in the program, for example $b22$, will not be picked up by the *local* implication, or $\rightarrow$, (which matches the ";" combining $p(a1)$ and $q(a2)$).

The language BLP has been implemented, in the continuation-passing style, by Armelín using the OCaml system [3].

Acknowledgements. Armelín is supported by an EPSRC Research Studentship. Pym acknowledges the support of the EPSRC via an Advanced Fellowship.

References

1. W. Clocksin. *Clause and effect.* Springer-Verlag, 1997.
2. W. Clocksin and C. Mellish. *Programming in Prolog.* Springer-Verlag, 1994.
3. G. Cousineau and M. Mauny. *The Functional Approach to Programming.* Cambridge University Press, 1998.
4. J.-Y. Girard. Linear logic. *Theoretical Computer Science*, pages 1–102, 1987.
5. J.A. Harland, D.J. Pym, and M. Winikoff. Programming in Lygon: an overview. In M. Wirsing and M. Nivat, editors, LNCS 1101: 391–405, 1996.
6. I. Cervesato J. Hodas and F. Pfenning. Efficient resource management for linear logic proof search. *Theoretical Computer Science*, 232:133–163, 2000.
7. J.S. Hodas and D. Miller. Logic programming in a fragment of intuitionistic linear logic. *Information and Computation*, 110(2):327–365, 1 May 1994.
8. R. Kowalski. *Logic for Problem-solving.* North-Holland, Elsevier, 1979.
9. S. A. Kripke. Semantical analysis of intuitionistic logic I. In J. N. Crossley and M. A. E. Dummett, editors, *Formal Systems and Recursive Functions*, pages 92–130. North-Holland, Amsterdam, 1965.
10. D. Miller. A logical analysis of modules in logic programming. *J. Logic. Programming*, 6(1& 2):431–483, 1981.
11. D. Miller, G. Nadathur, F. Pfenning, and A. Ščedrov. Uniform proofs as a foundation for logic programming. *Annals of Pure and Applied Logic*, 51:125–157, 1991.
12. P.W. O'Hearn and D.J. Pym. The logic of bunched implications. *Bull. Symb. Logic*, 5(2):215–244, June 1999.
13. P.W. O'Hearn, D.J. Pym, and H. Yang. Possible worlds and resources: The semantics of **BI**. Submitted. Manuscript at http://www.dcs.qmw.ac.uk/∼pym, 2000.
14. D.J. Pym. On bunched predicate logic. In *Proc. LICS'99*, pages 183–192. IEEE Computer Society Press, 1999.
15. D.J. Pym. The Semantics and Proof Theory of the Logic of the Logic of Bunched Implications. Draft of research monograph, manuscript at http://www.dcs.qmw.ac.uk/~pym, 2000.
16. D.J. Pym and J.A. Harland. A uniform proof-theoretic investigation of linear logic programming. *J. Logic. Computat.*, 4:175–207, 1994.
17. A. Urquhart. Semantics for relevant logics. *J. Symb. Logic*, 1059–1073, 1972.

A Top-Down Procedure for Disjunctive Well-Founded Semantics

Kewen Wang*

Institut für Informatik, Universität Potsdam
Postfach 60 15 53, D–14415 Potsdam, Germany
kewen@cs.uni-potsdam.de

Abstract. Skepticism is one of the most important semantic intuitions in artificial intelligence. The semantics formalizing skeptical reasoning in (disjunctive) logic programming is usually named *well-founded semantics*. However, the issue of defining and computing the well-founded semantics for disjunctive programs and databases has proved to be far more complex and difficult than for normal logic programs. The argumentation-based semantics WFDS is among the most promising proposals that attempts to define a natural well-founded semantics for disjunctive programs. In this paper, we propose a top-down procedure for WFDS called D-SLS Resolution, which naturally extends the Global SLS-resolution and SLI-resolution. We prove that D-SLS Resolution is sound and complete with respect to WFDS. This result in turn provides a further yet more powerful argument in favor of the WFDS.

1 Introduction

Disjunctive logic programming (DLP) has gained wide acceptance as an important tool for knowledge representation. One critical reason is that DLP is more expressive and natural to use than normal (i.e. non-disjunctive) logic programming. The additional expressive power allows direct encodings of a great number of application domains into logic programs. However, the issue of defining and computing semantics for disjunctive programs and databases has proved to be far more complex and difficult than for normal logic programs. The skepticism and credulism are two major semantic intuitions for knowledge representation. A skeptical reasoner does not infer any conclusion in uncertainty conditions while a credulous reasoner tries to give conclusions as much as possible. Therefore, a skeptical reasoner usually get more feasible conclusions. In normal logic programming, these two opposite semantic intuitions are suitably captured by the well-founded semantics [11] and the stable semantics [6], respectively. There has already been a widely accepted stable semantics for disjunctive programs [9]. To date, there is no widely accepted well-founded semantics for DLP and no consensus has been reached about what constitutes an intended semantics for skeptical reasoning in DLP. Based on a comparative study of some recent approaches to defining well-founded semantics for disjunctive programs in [2,9,5,

---

* On leave from Tsinghua University, Beijing.

R. Goré, A. Leitsch, and T. Nipkow (Eds.): IJCAR 2001, LNAI 2083, pp. 305–317, 2001.

7,12], it has been proved in [13] that these approaches become equivalent when some "minor" modifications are made on them. Specifically, there exists a semantics (i. e. WFDS*) for well-founded reasoning in DLP which can be equivalently characterized by argumentation, program transformations and unfounded sets.

In the same style as the D-WFS defined in [1,2], a bottom-up computation procure has also been provided in [13]. In this paper, we investigate the problem of top-down computation for disjunctive well-founded semantics. Specifically, we propose a top-down procedure for disjunctive well-founded semantics called D-SLS Resolution, which naturally extends the Global SLS-resolution and SLI-resolution. We prove that D-SLS Resolution is sound and complete with respect to WFDS*.

Since logic programming is essentially goal-oriented, the existence of an elegant top-down procedure is surely a significant feature for query answering under any semantics. Our results in turn provide further yet more powerful arguments in favor of the semantics WFDS*.

The paper is organized as follows. In the next section we briefly recall related definitions in logic programming and specify our notations. In Section 3 we give the argumentative definition of the semantics WFDS*. Then in Section 4 we present the D-SLS Resolution procedure. Our procedure is not only a combination of Ross's Global SLS-resolution and SLI-resolution, it also elegantly incorporates the intuition of resolving default negation with disjunctive information. To illustrate our resolution procedure and its relation to some other semantic intuitions, two examples are given in Section 5. In Section 6 we state the soundness and completeness of D-SLS Resolution with respect to WFDS*. Finally, in Section 7 we conclude the paper.

2   Preliminaries

We assume the existence of an arbitrary, but fixed propositional language, generated from a selected set of propositional symbols (atoms). An expression (disjunction, formula, rule, or set of rules, etc) with variables is understood as an abbreviation for the set of all its grounded instances. If S is an expression, $atoms(S)$ denotes the set of all atoms appearing in S. A *general disjunctive logic program* (simply, *disjunctive program*) P is defined as a finite set of rules of the form:

$$p_1 \vee \cdots \vee p_t \leftarrow p_{t+1}, \ldots, p_s, not\ p_{s+1}, \ldots, not\ p_n. \tag{1}$$

Here, $n \geq s \geq t > 0$ and p_i's are atoms for $i = 1, \ldots, n$. The symbols '$\vee$' and '*not*' denote (non-classical) disjunction and default negation, respectively.

A literal is either an atom p or its default negation $not\ p$ while $not\ p$ is called a *negative* literal.

The informal meaning of rule (1) is that "if $p_{t+1}, \ldots, p_s$ are true and $p_{s+1}, \ldots, p_n$ are all not provable, then one of $\{p_1, \ldots, p_t\}$ is true". For example, $male(greg) \vee female(greg) \leftarrow animal(greg), not\ ab(greg)$ means, informally, that if *greg* is an animal and it is not provable that *greg* is abnormal, then *greg* is either male or female.

If $t = 1$, rule (1) is said to be *normal*. P is a *normal program* if each rule of P is normal.

If $n = s$, rule (1) is said to be *positive*. P is a *positive disjunctive program* if each rule of P is positive.

If $t = s$, rule (1) is said to be *negative*. P is a *negative disjunctive program* if each rule of P is negative.

As usual, B_P is the Herbrand base of disjunctive program P (i. e. the set of all ground atoms in P). A *positive (negative) disjunction* is a disjunction of atoms (negative literals) in P. A *pure disjunction* is either a positive one or a negative one. The *disjunctive base* of P is $DB_P = DB_P^+ \cup DB_P^-$ where DB_P^+ is the set of all positive disjunctions in P and DB_P^- is the set of all negative disjunctions in P. If α and $\beta = \alpha \vee \alpha'$ are two disjunctions, then we say α is a *sub-disjunction* of β.

A *model state* of disjunctive program P is a subset of DB_P. Usually, a well-founded semantics for disjunctive logic programs is defined as a mapping such that each disjunctive program P is assigned a model state.

For simplicity, we also express a rule of form (1) as $\Sigma \leftarrow \Pi_1, not.\Pi_2$, where Σ is a disjunction of atoms in B_P, Π_1 a finite subset of B_P denoting a conjunction of atoms, and $not.\Pi_2 = \{not\ q \mid q \in \Pi_2\}$ for $\Pi_2 \subseteq B_P$ denoting a conjunction of negative literals.

3   Skeptical Argumentation

As illustrated in [12], argumentation can be used to define a unifying semantic framework for DLP. In this section, we first briefly recall the well-founded extension semantics WFDS in [12] and give a minor modification WFDS* of WFDS. The basic idea of the argumentation-based approach for DLP is to translate each disjunctive logic program into an argument framework $\mathbf{F}_P = \langle P, DB_P^-, \leadsto_P \rangle$. Here, an *assumption* of P is a negative disjunction of P, and a *hypothesis* is a set of assumptions; $\leadsto_P$ is an attack relation among the hypotheses. An *admissible hypothesis* Δ is one that can attack every hypothesis which attacks it.

The intuitive meaning of an assumption $not\ a_1 \vee \cdots \vee not\ a_m$ is that $a_1 \wedge \cdots \wedge a_m$ can not be proved from the disjunctive program.

Given a hypothesis Δ of disjunctive program P, similar to the GL-transformation [6], we can easily reduce P into another disjunctive program without default negation.

Definition 1. *Let Δ be a hypothesis of disjunctive program P, then the reduct of P with respect to Δ is the disjunctive program*

$$P_\Delta^+ = \{A \leftarrow B \mid \text{there is a rule of form } A \leftarrow B, not.C \text{ in } P \text{ s.t. } not.C \subseteq \Delta\}.$$

Based on Definition 1, we will first introduce a special resolution $\vdash_P$ which resolves default-negation literals with a disjunction and can be intuitively illustrated by the following principle:

K. Wang

If there is an agent who holds the assumptions not $b_1, \ldots, not\ b_m$ and can infer the disjunctive information $b_1 \vee \cdots \vee b_m \vee b_{m+1} \vee \cdots \vee b_n$, then the agent should be able to infer $b_{m+1} \vee \cdots \vee b_n$.

The following definition precisely formulates this principle in the setting of DLP.

Definition 2. *Let Δ be a hypothesis of disjunctive program P and $\alpha \in DB_P^+$. If there exists $\beta \in DB_P^+$ and not $b_1, \ldots, not\ b_m \in \Delta$ such that $\beta = \alpha \vee b_1 \vee \cdots \vee b_m$ and $P_\Delta^+ \vdash \beta$. Then Δ is said to be a supporting hypothesis for α, denoted $\Delta \vdash_P \alpha$. Here $\vdash$ is the classical inference; P_Δ^+ is considered as a classical logic theory while β is considered as a formula in classical logic.*

The consequence set of Δ consists of all positive disjunctions that are supported by Δ:
$$cons_P(\Delta) = \{\alpha \mid \alpha \in DB_P^+, \Delta \vdash_P \alpha\}.$$

For example, if $P = \{a \vee b \leftarrow c, not\ d;\ c \leftarrow\}$ and $\Delta = \{not\ b, not\ d\}$, then $\Delta \vdash_P a$.

The task of defining a semantics for a disjunctive logic program P is to determine the state that can represent the intended meaning of P. Here we first specify the negative information in the semantics and then derive the positive part. To derive suitable hypotheses for a given disjunctive program, some constraints will be required to filter out unintuitive hypotheses.

Definition 3. *Let Δ and Δ' be two hypotheses of disjunctive program P. If at least one of the following two conditions holds:*

1. there exists $\beta = not\ b_1 \vee \cdots \vee not\ b_m \in \Delta'$, $m > 0$, such that $\Delta \vdash_P b_i$, for all $i = 1, \ldots, m$; or

2. there exist $not\ b_1, \ldots, not\ b_m \in \Delta', m > 0$, such that $\Delta \vdash_P b_1 \vee \cdots \vee b_m$,

then we say Δ attacks Δ', and denoted $\Delta \rightsquigarrow_P \Delta'$.

Intuitively, $\Delta \rightsquigarrow_P \Delta'$ means that Δ causes a direct contradiction with Δ' and the contradiction may come from one of the above two cases.

Example 1.

$$a \vee b \leftarrow d$$
$$c \leftarrow d, not\ a, not\ b$$
$$d \leftarrow$$
$$e \leftarrow not\ e$$

Let $\Delta' = \{not\ c\}$ and $\Delta = \{not\ a, not\ b\}$, then $\Delta \rightsquigarrow_P \Delta'$.

The next definition specifies what is an acceptable hypothesis.

Definition 4. *Let Δ be a hypothesis of disjunctive program P. An assumption $\beta \in DB_P^-$ is admissible with respect to Δ if $\Delta \rightsquigarrow_P \Delta'$ holds for any hypothesis Δ' of P such that $\Delta' \rightsquigarrow_P \{\beta\}$.*

Denote $\mathcal{A}_P(\Delta) = \{\alpha \in DB_P^- \mid \alpha \text{ is admissible with respect to } \Delta\}.$

For any disjunctive program P, $\mathcal{A}_P$ is a monotonic operator: $\Delta \subseteq \Delta'$ implies $\mathcal{A}_P(\Delta) \subseteq \mathcal{A}_P(\Delta')$ for any two hypotheses Δ and Δ' of P. Thus, $\mathcal{A}_P$ has the least fixpoint $lfp(\mathcal{A}_P)$. Since B_P is finite in this paper, the fixpoint can be obtained in finite steps by iterating $\mathcal{A}_P$ from the emptyset. That is, $lfp(\mathcal{A}_P) = \mathcal{A}_P^k(\emptyset)$ where $\mathcal{A}_P^{k+1}(\emptyset) = \mathcal{A}_P^k(\emptyset)$.

Definition 5. *The well-founded disjunctive hypothesis WFDH(P) of disjunctive program P is defined as the least fixpoint of the operator $\mathcal{A}_P$. That is,*
$$WFDH(P) = \mathcal{A}_P \uparrow \omega.$$

The well-founded extension semantics *WFDS for P is defined as the model state $WFDS(P) = WFDH(P) \cup cons_P(WFDH(P))$.*

By the above definition, WFDS(P) is uniquely determined by WFDH(P).

For the program P in Example 1, $WFDS(P) = \{a \vee b, d, not\ c, not\ a \vee not\ b\}$. To compare with different semantics, $\mathcal{A}_P$ can be modified by defining

$$\mathcal{A}_P(\Delta) = \{\beta \in DB_P^- \mid not\ q \text{ is admissible w.r.t. } \Delta \text{ for some literal } not\ q \text{ in } \beta\}.$$

Parallel to the definition of WFDS, we can get a new well-founded semantics denoted WFDS* for disjunctive programs, which is a modification of WFDS. For instance, let P be the program given in Example 1, then $WFDS^*(P) = \{a \vee b, d, not\ c\}$. Now $not\ a \vee not\ b$ is no longer in $WFDS^*(P)$.

We can prove that WFDS is no less strong than WFDS* in the following sense.

Proposition 1. *For any disjunctive program P and $\alpha \in DB_P$, if $\alpha \in WFDS^*(P)$, then $\alpha \in WFDS(P)$.*

As we have seen above, the converse of this proposition is not true in general. Specifically, WFDS allows more negative disjunctions to be inferred. However, this is not a big difference as the following results show. In fact, except for this difference, these two semantics coincide.

Proposition 2. *Let Δ be an admissible hypothesis of disjunctive program P. If $\beta \in DB_P^-$ but is not a literal, then*

1. *a hypothesis α is admissible w.r.t. Δ iff it is admissible w.r.t. $\Delta - \{\beta\}$.*
2. *for any positive disjunction D, $D \in cons_P(\Delta)$ iff $D \in cons_P(\Delta - \{\beta\})$.*

An interesting result is the equivalence of WFDS and WFDS*.

Theorem 1. *Let P be a disjunctive program. Then*

1. *$not\ p \in WFDS^*(P)$ iff $not\ p \in WFDS(P)$ for any atom p.*
2. *$\alpha \in WFDS^*(P)$ iff $\alpha \in WFDS(P)$ for any positive disjunction α.*

This theorem convinces that the difference of WFDS* from WFDS is only in that they derive different sets of true negative disjunctions.

4   D-SLS Resolution

In this section, we will define a top-down procedure, called D-SLS Resolution, for disjunctive well-founded semantics. This procedure combines the idea of Global SLS-resolution in [10] with a linear resolution procedure. The linear resolution is a generalization of the SLI-resolution presented in [8]. One key part in our procedure is the incorporation of resolving default negation with disjunctive information into SLS-resolution. D-SLS Resolution will be based on the notion of D-SLS *tree*, which in turn depends on the notion of *positive trees*. In the next section, we prove that D-SLS Resolution is sound and complete with respect to the disjunctive well-founded semantics WFDS and WFDS*. To achieve completeness of D-SLS Resolution, we adopt the so-called *positivistic* computation rule, that is, we always select positive literals ahead of negative ones.

A goal G is of the form $\leftarrow D_1, \ldots, D_r, \neg b_1, \ldots, \neg b_m, not\ c_1, \ldots, not\ c_n$, where each D_i is a positive disjunction; all b_i and c_i are atoms. To distinguish from default literals, we shall say that l is a classic literal if $l = p$ or $l = \neg p$.

In our resolution-like procedure, given a rule $C : \Sigma \leftarrow \Pi_1, not.\Pi_2$, we transform C into a goal $gt(C) : \leftarrow \neg\Sigma, \Pi_1, not.\Pi_2$ and call it the *goal transformation* of C, where $\neg\Sigma = \{\neg p \mid p \in \Sigma\}$.

Since our resolution is to resolve literals in both heads and bodies of rules, this transformation allows a unifying and simple approach to defining resolution-like procedure for disjunctive logic programs as we shall see.

The special goal $\leftarrow$ is called an *empty goal*. The empty goal $\leftarrow$ is also written as the familiar symbol $\square$. A non-empty goal of form $\leftarrow \neg\Sigma, not.\Pi$ is said to be a negative goal.

Given a disjunctive program P, set $gt(P) = \{gt(C) : C \in P\}$. The traditional goal resolution can be generalized as follows.

Disjunctive Goal Resolution (DGR) If $G : \leftarrow b_1 \vee \cdots \vee b_r, G_1$ and $G' : \leftarrow \neg b_1, \ldots, \neg b_s, G_2$ are two goals with $s \leq r$, then the DGR-*resolvent* of G with G' on selected disjunction $b_1 \vee \cdots \vee b_r$ is the goal $\leftarrow G_1, G_2$.

It should be noted that resolution rule DGR incorporates several resolution rules including *Goal resolution*, *Ancestor resolution* and *Body literal resolution* [8,15]. Since we allow positive disjunctions in goals and the resolution rule, DGR is more powerful than the above mentioned three resolution rules as the following example shows.

Example 2. Let P be the following disjunctive program:

$$a \vee b \leftarrow$$
$$c \leftarrow not\ a, not\ b$$

Then P can be transformed into the following set $gt(P)$ of goals:

$$\leftarrow \neg a, \neg b$$
$$\leftarrow \neg c, not\ a, not\ b$$

Then the DGR-resolvent of $G : \leftarrow c$ with the second goal is $\leftarrow not\ a, not\ b$, which can be obtained by the ordinary goal resolution; however, the DGR-resolvent of goal $\leftarrow a \vee b$ with the first goal is $\Box$, which can not be obtained by any of those three resolution rules.

Definition 6. *Let P be a disjunctive program and G a goal. A positive tree T_G^+ for G is defined as follows:*

1. *The root of T_G^+ is G.*
2. *For each node $G' : \leftarrow \neg \Sigma', p, \Pi_1', not.\Pi_2'$, and each goal G_i in $gt(P)$, if G_i' is the DGR-resolvent of G' with G_i on p and G_i' is different from all nodes in the branch of G', then G' has a child G_i'.*

A node labeled $\leftarrow \neg p_1, \ldots, \neg p_n, not\ p_{n+1}, \ldots, not\ p_m (m \geq n \geq 0)$ is called an active node.

Thus, an active node is either the empty goal or a negative goal. For a negative goal, its success/failure has to be decided in subsequent stages. Now we define the D-SLS tree for a goal in terms of positive trees.

Definition 7. *(D-SLS Tree)   Let P be a disjunctive logic program and G a goal. The D-SLS tree Γ_G for G is a tree whose nodes are of two types: negation nodes and tree nodes. Tree nodes are actually positive trees for intermediate goals. The nodes of Γ_G is defined inductively as follows:*

1. *The root of Γ_G is the positive tree T_G^+ for the goal G.*
2. *For any tree node T_H^+ of Γ_G, The children of T_H^+ are negation nodes, one corresponding to each active leaf of T_H^+ (there will be a negation node corresponding to an empty active leaf).*
3. *Let J be a negation node corresponding to the active leaf $\leftarrow Q$ where $Q = \{not\ q_1, \ldots, not\ q_n\}$ and $n \geq 0$. J is denoted $N(\leftarrow Q)$. Then, if $n > 0$, J has one child which is the positive tree $T_{q_1 \vee \cdots \vee q_n}^+$.*

We distinguish three types of leaves in a D-SLS tree (successful nodes, failed nodes and intermediate nodes) according to the following rules. Successful and failed nodes also have an associated level.

1 *For negation node J,*
   (a) *if the child of a negation node J is a successful tree node, then we say J is failed. The level is the level of its successful child.*
   (b) *if the child of a negation node J is a failed tree node, or if J has no children, then we say J is successful. The level of J is the level of the child of J (if J has no children, the level of J is 0).*
2 *For tree node T,*
   (a) *if every child of a tree node T is a failed negation node, or if T is a leaf of Γ_G (i. e. T has no active leaves) then we say T is failed. T has the level 1 if T is a leaf; the level of T is $k+1$ if the maximum level of levels of the children of T is k.*

(b) *if some child of a tree node T is a successful negation node, then we say T is successful. A non-root tree node T has level $k + 1$ if the minimum level of all its successful children is k. The root tree node may have several associated levels, one for each successful child; the level of the root tree node with respect to such a successful child is one more than the level of the child.*

3 We say a node is well determined *if it is either successful or failed. Otherwise, we say the node is* indeterminate.

Let L be an active leaf of a tree node in Γ_G. We may say that L is successful (resp. failed or indeterminate) if the corresponding negation node is successful (resp. failed or indeterminate). We may also say that the goal G is successful (resp. failed or indeterminate) if T_G^+ is successful (resp. failed or indeterminate). Compared to Global SLS-resolution for normal logic programs, D-SLS Resolution has the following major features:

1. the underlying reasoning mechanism for D-SLS Resolution is a generalization of SLI-resolution while the underlying reasoning mechanism for SLS-resolution is SLD-resolution;

2. each negation node has just one child in D-SLS Resolution while a negation node may have several children in Global SLS-resolution because disjunctions are allowed now. This makes a simpler form of D-SLS Resolution.

To guarantee the termination of D-SLS Resolution, we also assume that every node is not repeated. That is, whenever a repeated node in D-SLS tree is found, the extending of the tree will be stopped.

5   Examples

Let us look at some illustrating examples.

Example 3. Consider the following disjunctive program P:

$$a \vee b \leftarrow c, not\ d$$
$$e \leftarrow not\ a, not\ b$$
$$e \leftarrow not\ a, g$$
$$c \leftarrow$$
$$a \leftarrow not\ c$$

It can be verified that $not\ e \in \text{WFDS}^*(P)$ and thus $not\ e \in \text{WFDS}(P)$.

Now let us see how $not\ e$ is inferred by D-SLS Resolution.

First, P is transformed into $gt(P)$ which consists of the following goals:

$$G_1 : \leftarrow \neg a, \neg b, c, not\ d$$
$$G_2 : \leftarrow \neg e, not\ a, not\ b$$
$$G_3 : \leftarrow \neg e, not\ a, g$$
$$G_4 : \leftarrow \neg c$$
$$G_5 : \leftarrow \neg a, not\ c$$

In fact, we have the following D-SLS tree $\Gamma_{\leftarrow e}$ for $\leftarrow e$:

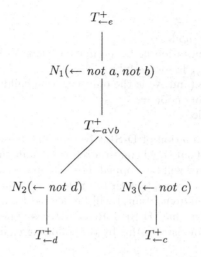

The positive tree $T^+_{\leftarrow e}$ for $\leftarrow e$ is as follows:

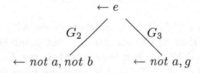

The positive tree $T^+_{\leftarrow a \vee b}$ is:

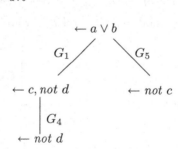

The positive tree $T^+_{\leftarrow d}$ consists of only the root node $\leftarrow d$.
The positive tree $T^+_{\leftarrow c}$ is

$$\leftarrow c$$
$$\Big|\ G_4$$
$$\leftarrow$$

314   K. Wang

By Definition 7,
$T^+_{\leftarrow d}$ is a leaf of $\Gamma_{\leftarrow e}$ ⇒
$T^+_{\leftarrow d}$ is a failed tree node ⇒
N_2 is a successful negation node (no matter what N_3 is) ⇒
the tree node $T^+_{\leftarrow a \vee b}$ is successful ⇒
N_1 is a failed node (and N_1 is the only negation child of $T^+_{\leftarrow e}$) ⇒
$T^+_{\leftarrow e}$ is a failed (root) node ⇒
the goal $\leftarrow e$ is failed.

To guarantee the termination of D-SLS Resolution, we also assume that every node is not repeated. That is, whenever a repeated node in D-SLS tree is found, the extending of the tree will be stopped. For example, if we replace the last rule $a \leftarrow not\ c$ in the above example with the rule $a \leftarrow not\ a, not\ b$, then $N_3 = N_1$ and thus N_3 and its children (if any) will be deleted from $\Gamma_{\leftarrow e}$.

It should be noted that D-SLS Resolution is different from the SLIN-resolution [14]. We demonstrate this by the following example.

Example 4. Let P consist of two rules:

$$a \vee b \leftarrow$$
$$c \leftarrow not\ a, not\ b$$

Although $not\ c \in \text{WFDS}^*(P)$, c is indeterminate with respect to the SLIN-resolution. This means that SLIN-resolution is not complete for the disjunctive well-founded semantics WFDS*. However, D-SLS tree $\Gamma_{\leftarrow c}$ for the goal $\leftarrow c$ is failed (to save space, the tree is figured in one line because each of its internal nodes has the unique child):

$$T^+_{\leftarrow c} \text{ --- } N_1(\leftarrow not\ a, not\ b) \text{ --- } T^+_{\leftarrow a \vee b} \text{ --- } N_2(\leftarrow) \text{ --- } T^+_{\leftarrow}$$

Here, $T^+_{\leftarrow}$ has the unique node $\leftarrow$; $T^+_{\leftarrow c}$ and $T^+_{\leftarrow a \vee b}$ are as follows, respectively:

It is easy to see that $T^+_{\leftarrow a}$ is an indeterminate node of the D-SLS tree for the goal $\leftarrow a$.

It should be noted that, in D-SLS tree, an active node containing classic negative literals can not be ignored[1]. That is, there may be negation node having classic negative literals as child in D-SLS tree. Consider the following program:

$$b \vee l \leftarrow not\ p$$
$$l \vee p \leftarrow$$

---
[1]   This question is proposed by one referee.

The D-SLS tree $\Gamma_{\leftarrow b}$ for the goal $\leftarrow b$ is as follows:

$$T^+_{\leftarrow b} \; \text{---} \; N(\leftarrow \neg l, not \; p) \; \text{---} \; T^+_{\leftarrow l \vee p}.$$

It can be verified that the goal $\leftarrow b$ is failed.

6   Soundness and Completeness of D-SLS Resolution

In this section, we address the soundness and completeness of D-SLS Resolution. We first show that D-SLS Resolution is sound and complete w.r.t. the argumentative semantics WFDS. Then, by Theorem 1, we get the soundness and completeness of D-SLS Resolution w.r.t. WFDS*. Although we allow a goal to have a very general form in our D-SLS Resolution, each goal G considered in this section actually has one of the two forms: either $\leftarrow a_1 \vee \ldots \vee a_r$ or $\leftarrow a_1, \ldots, a_r, \neg b_1, \ldots, \neg b_m, not \; c_1, \ldots, not \; c_n$, where all a_i, b_i and c_i are atoms. Thus, from now on we will always mean either of the above form when a goal is mentioned. The detailed proofs of results in this section are not difficult but tedious, thus we omit them here.

Theorem 2. *(Soundness of* D-SLS *Resolution w.r.t. WFDS)*
Let P be a disjunctive logic program. Then

1. *If goal $G : \leftarrow q_1, \ldots, q_n$ is failed, then not $q_1 \vee \cdots \vee not \; q_n \in WFDS(P)$.*
2. *If goal $G : \leftarrow q_1 \vee \cdots \vee q_n$ is successful, then $q_1 \vee \cdots \vee q_n \in WFDS(P)$.*

To prove Theorem 2, we need only to show the following lemma.

Lemma 1. *Let P be a disjunctive logic program. Then*

1. *If goal $G : \leftarrow q_1, \ldots, q_n$ is failed, then not $q_1 \vee \cdots \vee not \; q_n \in WFDH(P)$.*
2. *If goal $G : \leftarrow q_1 \vee \cdots \vee q_n$ is successful, then $WFDH(P) \vdash_P q_1 \vee \cdots \vee q_n$.*

Sketch of Proof   It suffices to prove the following two propositions hold by using simultaneous induction on the level $l(\Gamma_G)$ of D-SLS tree Γ_G:

S1 *not $q_1 \vee \cdots \vee not \; q_n \in \mathbf{A}^k_P(\emptyset)$ if the goal $G : \leftarrow q_1, \ldots, q_n$ is failed and $l(\Gamma_G) = k \geq 1$;*
S2 $\mathbf{A}^k_P(\emptyset) \vdash_P q_1 \vee \cdots \vee q_n$ *if the goal $G : \leftarrow q_1 \vee \cdots \vee q_n$ is successful and $l(\Gamma_G) = k \geq 0$.*

Theorem 3. *(Completeness of* D-SLS *Resolution w.r.t. WFDS)*
Let P be a disjunctive logic program. Then

1. *If $q_1 \vee \cdots \vee q_n \in WFDS(P)$, then the goal $\leftarrow q_1 \vee \cdots \vee q_n$ is successful.*
2. *If not $q_1 \vee \cdots \vee not \; q_n \in WFDS(P)$, then the goal $G :\leftarrow q_1, \ldots, q_n$ is failed.*

This theorem follows directly from the next lemma.

Lemma 2. *Let P be a disjunctive logic program and G a goal of P. Then*

1. If $WFDH(P) \vdash_P q_1 \vee \cdots \vee q_n$, then the goal $G : \leftarrow q_1 \vee \cdots \vee q_n$ is successful.
2. If not $q_1 \vee \cdots \vee$ not $q_n \in WFDH(P)$, then the goal $G : \leftarrow q_1, \ldots, q_n$ is failed.

Sketch of Proof   It is enough to show that both of the following C1 and C2 hold by using simultaneous induction on the level $l(\Gamma_G) = k$:

C1 For $k \geq 1$, if $\mathbf{A}^{k-1}(\emptyset) \vdash_P q_1 \vee \cdots \vee q_n$, then the goal $G : \leftarrow q_1 \vee \cdots \vee q_n$ is successful and its level is k.
C2 For $k \geq 1$, if not $q_1 \vee \cdots \vee$ not $q_n \in \mathbf{A}_P^k(\emptyset)$, then the goal $G : \leftarrow q_1, \ldots, q_n$ is failed and its level is no more than $k + 1$.

By the definition of WFDS*, for any atoms $q_1, \ldots, q_n$, not $q_1 \vee \cdots \vee$ not $q_n \in$ WFDS*(P) if and only if $q_i \in$ WFDS*(P) for some q_i. Thus, the following two theorems follows directly from Theorem 1 and the two theorems above.

Theorem 4. *(Soundness of* D-SLS *Resolution w.r.t.* WFDS*)*
Let P be a disjunctive logic program. Then

1. *If goal* $G : \leftarrow q_i$ *is failed for some* q_i, *then not* $q_1 \vee \cdots \vee$ *not* $q_n \in$ WFDS*(P).
2. *If goal* $G : \leftarrow q_1 \vee \cdots \vee q_n$ *is successful, then* $q_1 \vee \cdots \vee q_n \in$ WFDS*(P).

Theorem 5. *(Completeness of* D-SLS *Resolution w.r.t.* WFDS*)*
Let P be a disjunctive logic program. Then

1. *If* $q_1 \vee \cdots \vee q_n \in$ WFDS*(P), *then the goal* $\leftarrow q_1 \vee \cdots \vee q_n$ *is successful.*
2. *If not* $q_1 \vee \cdots \vee$ *not* $q_n \in$ WFDS*(P), *then the goal* $G : \leftarrow q_i$ *is failed for some* q_i.

7   Conclusion

The main contribution of this paper is that we have proposed a top-down procedure D-SLS Resolution for disjunctive well-founded semantics. This resolution-like procedure extends both the Global SLS-resolution [10] and SLI-resolution [8]. We prove that D-SLS Resolution is sound and complete with respect to the disjunctive well-founded semantics WFDS and WFDS*. We know that the Global SLS-resolution is a classic procedure for the well-founded semantics of normal logic programs while SLI-resolution is the most important procedure for positive disjunctive programs. D-SLS Resolution is actually a novel characterization for WFDS* and thus provides a further yet powerful argument in favor of the semantics WFDS*. On the other hand, the results in this paper pave a promising way to implement the WFDS* by employing some existing theorem provers. Although this point has been made clear for Brass and Dix's D-WFS in [3], no top-down procedure is provided for their semantics. It is worth noting that D-SLS Resolution in the current form is not efficient yet. We are currently working on more efficient algorithm for D-SLS Resolution by employing some techniques such as the tabling method [4].

Acknowledgments. The author would like to thank Peter Baumgartner, Ph. Besnard, J. Dix, Th. Linke and T. Schaub for useful comments and discussions on (the topic of) this work. The author would also like to thank the four anonymous referees for their detailed comments. This work was supported in part by the German Science Foundation (DFG) within Project "Nichtmonotone Inferenzsysteme zur Verarbeitung konfligierender Regeln" under grant FOR 375/1-1, TP C, the Natural Science Foundation of China under the projects 69883008, the National Foundation Research Programme of China under G1999032704.

References

1. S. Brass, J. Dix. Characterizations of the Disjunctive Well-founded Semantics: Confluent Calculi and Iterated GCWA. *Journal of Automated Reasoning*, 20(1):143–165, 1998.
2. S. Brass, J. Dix. Semantics of disjunctive logic programs based on partial evaluation. *Journal of Logic programming*, 38(3):167-312, 1999.
3. S. Brass, J. Dix, I. Niemelä, T. Przymusinski. On the equivalence of the Static and Disjunctive Well-founded Semantics and its computation. *Theoretical Computer Science*, 251 (to appear), 2001.
4. W. Chen, D. Warren. Efficient top-down computation of queries under the well-founded semantics. *J. ACM*, 43(1): 20-74, 1996.
5. T. Eiter, N. Leone and D. Sacca. On the partial semantics for disjunctive deductive databases. *Annals of Math. and AI.*, 19(1-2): 59-96, 1997.
6. M. Gelfond, V. Lifschitz. The stable model semantics for logic programming. In: *Proceedings of the 5th Symposium on Logic Programming*, MIT Press, pages 1070-1080, 1988.
7. N. Leone, P. Rullo and F. Scarcello. Disjunctive stable models: unfounded sets, fixpoint semantics, and computation. *Information and Computation*, 135(2), 1997.
8. J. Lobo, J. Minker and A. Rajasekar. *Foundations of Disjunctive Logic Programming*. MIT Press, 1992.
9. T. Przymusinski. Static semantics of logic programs. *Annals of Math. and AI.*, 14: 323-357, 1995.
10. K. Ross. A procedural semantics for well-founded negation in logic programs. *Journal of Logic programming*, 13(1): 1-22, 1992.
11. A. Van Gelder, K. A. Ross and J. Schlipf. The well-founded semantics for general logic programs. *J. ACM*, 38(3): 620-650, 1991.
12. K. Wang. Argumentation-based abduction in disjunctive logic programming. *Journal of Logic programming*, 45(1-3):105-141, 2000.
13. K. Wang. Disjunctive well-founded semantics revisited. In: *Proceedings of the 5th Dutch-German Workshop on Nonmonotonic Reasoning Techniques and their Applications (DGNMR'01)*, Potsdam, April 4-6, pages 193-203, 2001.
14. K. Wang, F. Lin. Closed world reasoning and query evaluation in disjunctive deductive databases. In: *Proceedings of the International Conference on Applications of Prolog /DDLP'99*, 1999.
15. J. You, L. Yuan and R. Goebel. An abductive approach to disjunctive logic programming. *Journal of Logic programming*, 44(1-3):101-127, 2000.

A Second-Order Theorem Prover Applied to Circumscription

Michael Beeson

Department of Mathematics and Computer Science
San Jose State University

Abstract. Circumscription is naturally expressed in second-order logic, but previous implementations all work by handling cases that can be reduced to first-order logic. Making use of a new second-order unification algorithm introduced in [3], we show how a theorem prover can be made to find proofs in second-order logic, in particular proofs by circumscription. We work out a blocks-world example in complete detail and give the output of an implementation, demonstrating that it works as claimed.

1 Introduction

Circumscription was introduced by John McCarthy [11] as a means of formalizing "common-sense reasoning" for artificial intelligence. It served as the foundation of his theory of non-monotonic reasoning. The essential idea is to introduce, when axiomatizing a situation, a predicate ab for "abnormality", and to axiomatize the ab predicate by saying it is the least predicate such that the other axioms are valid. Some other predicates may be allowed to "vary" in the minimization as well. There are several technical difficulties with McCarthy's idea: First, the circumscription principle is most naturally expressed in second-order logic, where we have variables over predicates of objects. Second, unless the rest of the axioms contain ab only positively, the circumscription principle is not an ordinary inductive definition, and there may not even be a (unique) least solution for the ab predicate, so the circumscription principle can be inconsistent. McCarthy's ultimate goal was implementation of software using the circumscription principle to construct artificial intelligence. Believing that implementation of second-order logic was not a practical approach, many researchers have tried various methods of reducing special cases of the circumscription principle to first-order logic; see [7] for a summary of these efforts. Some of these reductions were in turn implemented.

In this paper we take the other path, and exhibit a direct implementation of second-order logic which is capable of handling some circumscription problems. The key to making this work is a new notion of second-order unification. This notion of unification was introduced in [3], where some theorems about it are proved. In that paper, I pointed out the possibility of converting your favorite first-order theorem prover to a second-order theorem prover by adding second-order unification. This paper shows explicitly how this can be done, and that the

R. Goré, A. Leitsch, and T. Nipkow (Eds.): IJCAR 2001, LNAI 2083, pp. 318–324, 2001.

resulting second-order prover can indeed find circumscription proofs. Note that it would already be interesting if the resulting proof-checker could accept and verify circumscription proofs, but the essential point of this paper is that the use of the new unification algorithm of [3] enables a simple theorem-prover to find circumscription proofs by itself. The hard part of this, of course, is finding the correct values of the second-order predicates involved. These are generally give by λ terms involving an operator for definition by cases. It is therefore essential to use a formalization of second-order logic which has terms for definition by cases.

A longer version of this paper is available on the Web [1]. It includes additional background and details, and the complete computer-produced proof of the example treated here. In particular the exact syntax of our version of second-order logic, including application terms and lambda terms, is given there. Second-order unification, and its application to circumscription, both depend on the use of conditional terms, or case-terms. These are terms of the form

$$\left\{ \begin{array}{lll} P(x) & \text{if} & x = y \\ Q(x) & \text{ow} & \end{array} \right.$$

There are several different notations for such terms in use, including the form used in the C and Java programming languages:

$$x = y \ ? \ P(x) \ : Q(x)$$

and the form used in [3] and in the theories of Feferman [2]:

$$\mathbf{d}(x, y, P(x), Q(x)).$$

The form with $\mathbf{d}$ is the one in the official syntax, but the other two forms are both more readable. The syntax used by our computer implementation allows a more general kind of case term in which there can be several cases, instead of just one, before the "otherwise" term. For representing such terms the notation with a brace is more readable, so the output of the prover is presented in that notation. For writing papers, the notation with a question mark is more compact and equally readable, so we will use it in the paper.

We use a Gentzen-sequent formulation of second-order logic. We simply take the usual Gentzen rules (e.g. G3 as in [9]) for both predicate and object quantifiers. The G3 rules need to be supplemented with rules corresponding to the formation of λ-terms and $\mathbf{ap}$-terms, as well as with rules corresponding to the introduction of case terms in both antecedent and succedent. We do not repeat the G3 rules here, but here are the other rules:

$$\frac{t = s, A \Rightarrow C, \Gamma \qquad t \neq s, B, \Gamma \Rightarrow C}{\mathbf{d}(t, s, A, B)), \Gamma \Rightarrow C}$$

$$\frac{t = s \Rightarrow A}{\Gamma \Rightarrow \mathbf{d}(t, s, A, B))}$$

$$\frac{t \neq s \Rightarrow B}{\Gamma \Rightarrow \mathbf{d}(t, s, A, B))}$$

$$\frac{\Gamma \Rightarrow A[t/x]}{\Gamma \Rightarrow (\lambda x.A)t}$$

$$\frac{\Gamma, A[t/x] \Rightarrow \phi}{\Gamma, (\lambda x.A)t \Rightarrow \phi}$$

We will be using the notion of unification introduced in [3]. This definition is also reviewed in [1], where it is also compared to Huet's notion.

2   Circumscription

If U and V are predicate expressions of the same arity, then $U \leq V$ stands for $\forall x(U(x) \rightarrow V(x))$. If $U = U_1, \ldots, U_n$ and $V = V_1, \ldots, V_n$ are similar tuples of predicate expressions, i.e. U_i and V_i are of the same arity, $1 \leq i \leq n$, then $U \leq V$ is an abbreviation for $\wedge_{i=0}^{n} U_i \leq V_i$. We write $U = V$ for $U \leq V \wedge V \leq U$, and $U < V$ for $U \leq V \wedge \neg V \leq U$.

Definition 1 (Second-Order Circumscription). *Let P be a tuple of distinct predicate constants, S be a tuple of distinct function and/or predicate constants disjoint from P, and let $T(P; S)$ be a sentence. The second-order circumscription of P in $T(P; S)$ with variable S, written $Circ(T; P; S)$, is given in [7] as*

$$T(P; S) \wedge \forall \Phi \Psi \neg [T(\Phi, \Psi) \wedge \Psi < P]$$

where Φ and Ψ are tuples of variables similar to P and S, respectively. This can equivalently be stated in the form

$$T(P; S) \wedge \forall \Phi \Psi [T(\Phi, \Psi) \wedge \Psi \leq P \rightarrow P \leq \Psi],$$

which is the form our prover uses.

3   Blocks World Example

We treat the first example from [7] as a typical circumscription problem.

Let $\Gamma(Ab, On)$ be the theory

$$c \neq b \wedge \neg On(c) \wedge \forall x(\neg ab(x) \rightarrow On(x))$$

where the variables range over "blocks" and $On(x)$ means "x is on the table". Circumscription enables us to conclude that a is the only block not on the table. For simplicity, we first consider the problem without the predicate B, i.e. we assume all variables range only over blocks. The idea is that normal blocks are on the table, and since c is the only abnormal block, b is a normal block and hence is on the table. Circumscription should enable us to prove $On(b)$.

Circumscription in this example is taken to minimize ab with variable On, so in the general schema above, we take P to be ab and S to be On.

$$c \neq b \tag{1}$$

$$\forall x(\neg ab(x) \rightarrow On(x)) \tag{2}$$

$$\neg On(c) \tag{3}$$

$$\forall \Phi \Psi [\forall x(\neg \Psi(x) \rightarrow \Phi(x)) \wedge \neg \Phi(c) \wedge \Psi \leq ab \rightarrow ab \leq \Psi] \tag{4}$$

We first present a human-produced proof, for later comparison to the proof found by our program. We take as the goal to prove $On(b)$. Backchaining from (2) produces the new goal $\neg ab(b)$. The human then suggests the values

$$\Psi = \lambda x.(x = c \ ? \ \textbf{true} : \textbf{false}) \tag{5}$$

$$\Phi = \lambda x.(x = c \ ? \ \textbf{false} : \textbf{true}) \tag{6}$$

With these values of Φ and Ψ, we want to prove $ab(b) \rightarrow \textbf{false}$, so we need to verify $\psi(b) = \textbf{false}$. But $\psi(b) = (b = c \ ? \ \textbf{true} : \textbf{false})$, and $b = c$ evaluates to **false** since $b \neq c$ is in the antecedent, so $\psi(b)$ evaluates to **false**. It therefore suffices to verify the hypothesis of (4), namely

$$\forall x(\neg \Psi(x) \rightarrow \Phi(x)) \wedge \neg \Phi(c) \wedge \Psi \leq ab.$$

Fix an x, and suppose $\neg \Psi(x)$. Then $x \neq c$, from which $\Phi(x)$ follows, which proves the first conjunct. The second conjunct, $\neg \Phi(c)$, follows immediately by reduction to **true**. The third conjunct, $\Psi \leq ab$, is proved as follows: suppose $\Psi(x)$. Then $x = c$ and so we must prove $ab(c)$. But by (3) we have $\neg On(c)$, and so by (2) we have $ab(c)$. That completes the proof.

We now explain how the prover attacks this problem. We want to prove $\neg ab(b)$. (Officially that goal is the succedent of a sequent whose antecedent is the list of axioms.) So the prover assumes $ab(b)$, and the new goal is $ab(b) \Rightarrow \textbf{false}$. (Of course officially the axioms should appear in the antecedent of the goal sequent, too, but we do not write them.) This causes (3) to be "opened up", introducing metavariables $\textbf{P}$ and $\textbf{Q}$. The formula $ab \leq \textbf{Q}$ is really $\forall w(ab(w) \rightarrow \textbf{Q}(w))$, so a metavariable W is introduced for w as well, but soon it is instantiated to b to unify $ab(b)$ with $ab(W)$, in the hopes of proving $ab(W) \Rightarrow \textbf{Q}(W)$ from $ab(b) \Rightarrow \textbf{false}$. Thus the prover tries to unify $\textbf{Q}(b)$ with **false**. This gives

$$\textbf{Q} = \lambda y.(y = b \ ? \ \textbf{false} : Y(y))$$

where Y is a new variable. The next goal is the conjunction of the three formulae on the left of the implication in 4. These are taken in order; the first one is $\forall v(\neg \textbf{Q}(v) \rightarrow \textbf{P}(v))$. Fixing v the goal is $\neg \textbf{Q}(v) \rightarrow \textbf{P}(v)$; writing out the current value of $\textbf{Q}$ and β-reducing, the goal is

$$\neg(v = b \ ? \ \textbf{false} : Y(v)) \rightarrow \textbf{P}(v)$$

There is a simplification rule for pushing a negation into a cases term, namely

$$\neg(v = p \ ? \ q \ : r) = (v = p \ ? \ \neg q \ : \neg r).$$

So the goal becomes

$$(v = b \ ? \ \textbf{true} : \neg Y(v)) \rightarrow \mathbf{P}(v).$$

This is solved by second-order unification, taking

$$\mathbf{P} = \lambda v.((v = b? \ \textbf{true} : \neg Y(v)) \vee Zv).$$

The next goal is $\neg\mathbf{P}(c)$. That is, after a beta reduction,

$$\neg((c = b \ ? \ \textbf{true} : \neg Y(c)) \vee Z(c))).$$

Now we can apply a simplification rule using the axiom $c \neq b$, reducing the cases term to $\neg Y(z)$ and hence the whole goal to $\neg(\neg Y(c) \vee Z(c))$. Using rewrite rules appropriate to classical logic we simplify this to $Y(z) \wedge \neg Z(c)$. Splitting the conjunction into two subgoals, the first one to be proved is $Y(c)$. This is solved by second-order unification, taking

$$Y = \lambda u.(u = c? \ ab(b) : A(u))$$

where A is a new metavariable. You might think we should get **true** in place of $ab(b)$ in the value of Y, but when the prover has to prove a goal of the form $Y(c)$, it does not try to unify $Y(c)$ with **true**, but rather with one of the assumptions (formulas in the antecedent). It tries the most recently-added ones first, and it finds $ab(b)$ there, which explains the value given for Y.

The second goal is $\neg Z(c)$. Then $Z(c)$ is assumed, leading to a goal $Z(c) \Rightarrow \textbf{false}$. Unifying $Z(c)$ with **false** gives Z the value

$$Z = \lambda r.(r = c \ ? \ On(c) : B(r)),$$

where B is a new metavariable. Again, you might expect **false** to occur in place of $On(c)$ in the value of Z, but the prover finds the value given, which is equivalent since $\neg On(c)$ is an axiom.

At this point, the values of $\mathbf{P}$ has become

$$\mathbf{P} = \lambda v.(v = b \ ? \ \textbf{true} : \neg(v = c? \ \textbf{true} : A(z)$$

which simplifies to

$$\mathbf{P} = \lambda v.(v = b \ ? \ \textbf{true} : v = c \ ? \ On(c) : \neg A(z))$$

The value of $\mathbf{Q}$ is now given by

$$\mathbf{Q} = \lambda y.(y = b \ ? \ \textbf{false} : (\lambda u.(u = c \ ? \ ab(b) : A(u)))y)$$

which reduces to

$$\mathbf{Q} = \lambda y.(y = b \ ? \ \textbf{false} : y = c \ ? \ ab(b) : A(y))$$

The next goal is $\mathbf{Q} \leq ab$, that is $\forall z(\mathbf{Q}(z) \to ab(z))$. Fixing z, the goal is $\mathbf{Q}(z) \to ab(z)$. Using the Gentzen rule for introducing $\to$ on the right, and writing out the current value of $\mathbf{Q}$, our goal is the sequent

$$z = b \ ? \ \mathbf{false} : z = c \ ? \ a(b) : W(x) \Rightarrow ab(x).$$

This is proved by cases, specifically by the cases-left rule.

Case 1, $z = b$. The goal reduces to $\mathbf{false} \to ab(z)$ which is immediate.

Case 2, $z \neq b$ and $z = c$. Then $\mathbf{Q}(a)$ reduces to

$$z = c \wedge z \neq b \wedge ab(b)$$

so the goal becomes

$$z = c, \quad z \neq b, \quad ab(b) \Rightarrow ab(z).$$

The human can note that $ab(c)$ follows from $\forall x(\neg ab(x) \to On(x))$ and $\neg On(c)$, and from $ab(c)$ the goal follows quickly. This is a relatively simple problem in first-order logic with equality, the difficulties of which are irrelevant to circumscription and second-order logic. Weierstrass is able to prove the goal.

Case 3, $z \neq b$ and $z \neq c$. Then $\mathbf{Q}(z)$ reduces to $W(z)$, so the goal becomes

$$W(z) \Rightarrow ab(z).$$

This goal is proved by instantiating the metavariable W:

$$W = \lambda z.(ab(z) \vee T(z))$$

where T is a new metavariable. The final values of $\mathbf{P}$ and $\mathbf{Q}$ are thus

$$\mathbf{P} = \lambda v.(v = b \ ? \ \mathbf{true} : v = c \ ? \ On(c) : (\neg ab(v) \wedge \neg T(v)))$$

$$\mathbf{Q} = \lambda y.(y = b \ ? \ \mathbf{false} : y = c \ ? \ a(b) : (ab(y) \vee T(y)))$$

To achieve the stated goal $On(b)$, the prover has only needed to deduce that b is not abnormal. Unlike the human, it has not gone ahead to deduce anything about other objects than a and b– the uninstantiated metavariable T remains as "undetermined". Of course, the constant b might as well have been a variable; the prover can prove $\forall x(x \neq a \to On(x))$ just as well as it can prove $On(b)$. But that proof, like the one above, will still use instantiations of $\mathbf{P}$ and $\mathbf{Q}$ involving free metavariables.

The proof as produced (and typeset) by our prover can be found in [1].

References

1. Beeson, M., Implementing circumscription using second-order unification, http://www.mathcs.sjsu.edu/faculty/beeson/Papers/pubs.html
2. Beeson, M., *Foundations of Constructive Mathematics*, Springer-Verlag, Berlin/ Heidelberg/ New York (1985).
3. Beeson, M., Unification in Lambda Calculus with if-then-else, in: Kirchner, C., and Kirchner, H. (eds.), *Automated Deduction-CADE-15. 15th International Conference on Automated Deduction, Lindau, Germany, July 1998 Proceedings*, pp. 96-111, Lecture Notes in Artificial Intelligence **1421**, Springer-Verlag (1998).
4. Beeson, M., Automatic generation of epsilon-delta proofs of continuity, in: Calmet, Jacques, and Plaza, Jan (eds.) *Artificial Intelligence and Symbolic Computation: International Conference AISC-98, Plattsburgh, New York, USA, September 1998 Proceedings*, pp. 67-83. Springer-Verlag (1998).
5. Beeson, M., Automatic generation of a proof of the irrationality of e, in Armando, A., and Jebelean, T. (eds.): *Proceedings of the Calculumus Workshop, 1999, Electronic Notes in Theoretical Computer Science* **23** 3, 2000. Elsevier. Available at http://www.elsevier.nl/locate/entcs. This paper has also been accepted for publication in a special issue of *Journal of Symbolic Computation* which should appear in the very near future.
6. Beeson, M., Some applications of Gentzen's proof theory to automated deduction, in P. Schroeder-Heister (ed.), *Extensions of Logic Programming*, Lecture Notes in Computer Science **475** 101-156, Springer-Verlag (1991).
7. Doherty, P., Lukaszewicz, W., And Szalas, A., Computing circumscription revisited: a reduction algorithm, *J. Automated Reasoning* **18**, 297-334 (1997).
8. Ginsberg, M. L., A circumscriptive theorem prover, *Artificial Intelligence* **39** pp. 209-230, 1989.
9. *Introduction to Metamathetics*, van Nostrand, Princeton, N.J. (1950).
10. Lifschitz, V., Computing circumscription, in: *Proceedings of the 9th International Joint Conference on Artificial Intelligence*, volume 1, pages 121-127, 1985.
11. McCarthy, J., Circumscription, a form of non-monotonic reasoning, *Artificial Intelligence*, **13** (1-2), pp. 27-39, 1980.
12. Przymusinski, T., An algorithm to compute circumscription, *Artificial Intelligence*, **38**, pp. 49-73, 1991.

NoMoRe: A System for Non-monotonic Reasoning with Logic Programs under Answer Set Semantics

Christian Anger, Kathrin Konczak, and Thomas Linke

Universität Potsdam, Institut für Informatik, Am Neuen Palais 10,
D-14469 Potsdam, {canger,konczak,linke}@cs.uni-potsdam.de

1 Introduction

The noMoRe system (first prototype) implements answer set semantics for propositional normal logic programs. It uses an alternative implementation paradigm to compute answer sets by computing non-standard graph colorings of labeled directed graphs associated with logic programs. Therefore noMoRe is an interesting experimental tool for scientists working with logic programs on a theoretical or practical basis. Furthermore, we have included a tool for visualization of those graphs corresponding to programs.

2 General Information

The noMoRe-system is implemented in the programming language Prolog; it has been developed under the ECLiPSe Constraint Logic Programming System [1] and it was also successfully tested with SWI-Prolog [11]. The source code, test cases and documentation are available at http://www.cs.uni-potsdam.de/~linke/nomore. In order to use the system, ECLiPSe- or SWI-Prolog is needed [1,11]. Both Prolog systems are freely available for scientific use. Clearly, noMoRe works under each platform under which one of the above Prolog systems is available. The total number of lines of code is only about 2700, i.e. noMoRe is very transparent and it nicely reflects the underlying theory.

3 Description of the System

The experimental prototype of the noMoRe system implements nonmonotonic reasoning with propositional normal logic programs under answer set semantics [5]. Originally, answer set semantics was defined for extended logic programs[1] [5] as a generalization of the stable model semantics [4] of normal logic programs. We consider rules r of the form

$$p \leftarrow q_1, \ldots, q_n, not\ s_1, \ldots, not\ s_k \qquad (1)$$

---

[1] Extended logic programs are logic programs with classical negation.

R. Goré, A. Leitsch, and T. Nipkow (Eds.): IJCAR 2001, LNAI 2083, pp. 325–330, 2001.
© Springer-Verlag Berlin Heidelberg 2001

where p, q_i ($0{\leq}i{\leq}n$) and s_j ($0{\leq}j{\leq}k$) are ground atoms, $head(r) = p$, $body^+(r) = \{q_1,\ldots,q_n\}$, $body^-(r) = \{s_1,\ldots,s_k\}$ and $body(r) = body^+(r) \cup body^-(r)$. Intuitively, the head p of a rule $p \leftarrow q_1,\ldots,q_n, not\ s_1,\ldots,not\ s_k$ is in some answer set A if $q_1,\ldots,q_n$ are in A and none of $s_1,\ldots,s_k$ is in A. Look at the following normal logic program

$$P = \{a \leftarrow b, not\ e. \quad b \leftarrow d. \quad c \leftarrow b. \quad d \leftarrow . \quad e \leftarrow d, not\ f. \quad f \leftarrow a.\} \quad (2)$$

Let us call the rules of program (2) r_a, r_b, r_c, r_d, r_e, and r_f, respectively. Then P has two different answer sets $A_1 = \{d,b,c,a,f\}$ and $A_2 = \{d,b,c,e\}$. It is easy to see that the application of r_f blocks the application of r_e wrt A_1, because if r_f contributes to A_1, then $f \in A_1$ and thus r_e cannot be applied. Analogously, r_e blocks r_a wrt answer set A_2.

3.1 Syntax

The syntax accepted by noMoRe is Prolog-like (without variables). For example, program (2) is represented through the following rules:

```
a :- b, not e.    b :- d.         c :- b.
d.                e :- d, not f.   f :- a.
```

NoMoRe also accepts ground formulas which are treated as propositional atoms.

3.2 Block Graphs and A-Colorings

NoMoRe implements a novel paradigm to compute answer sets by computing non-standard graph colorings of the so-called *block graph* [6] associated with a given program P. A set of rules S of the form (1) is *grounded* iff there exists an enumeration $\langle r_i \rangle_{i \in I}$ of S such that for all $i \in I$ we have that $body^+(r_i) \subseteq head(\{r_1,\cdots,r_{i-1}\})^2$. With this terminology, the block graph of P is defined as follows:

Definition 1. *([6]) Let P be a logic program and let $P' \subseteq P$ be maximal grounded.*[3] *The block graph $\Gamma_P = (V_P, A_P^0 \cup A_P^1)$ of P is a directed graph with vertices $V_P = P$ and two different kinds of arcs defined as follows*

$$A_P^0 = \{(r',r) \mid r',r \in P' \text{ and } head(r') \in body^+(r)\}$$
$$A_P^1 = \{(r',r) \mid r',r \in P' \text{ and } head(r') \in body^-(r)\}.$$

Figure 1 shows the block graph of program (2). Observe, that the rules of P are the nodes of Γ_P. Since groundedness (by definition) ignores negative bodies, there exists a unique maximal grounded set $P' \subseteq P$ for each program P, that is, Γ_P is well-defined. Definition 1 captures the conditions under which a rule r' blocks another rule r (e.g. $(r',r) \in A^1$). We also gather all groundedness information

---

[2] The definition of the head of a rule is generalized to sets of rules in the usual way.
[3] A maximal grounded set P' is a grounded set that is maximal wrt set inclusion.

in Γ_P, due to the restriction to rules in the maximal grounded part of P. This is essential because a block relation between two rules r' and r becomes effective only if r' is groundable through other rules. In all, Γ_P captures all information necessary for computing the answer sets of program P.

Answer sets then are characterized as special non-standard graph colorings of block graphs. We denote 0-predecessors, 0-successors, 1-predecessors and 1-successors of Γ_P by $\gamma_0^-(v)$, $\gamma_0^+(v)$, $\gamma_1^-(v)$ and $\gamma_1^+(v)$ for $v \in V$, respectively.

Definition 2. *([6]) Let P be a logic program, s.t. $|body^+(r)| \leq 1$ for each $r \in P$, let $\Gamma_P = (P, A_P^0 \cup A_P^1)$ be the corresponding block graph and let $c : P \mapsto \{\ominus, \oplus\}$ be a mapping. Then c is an a-coloring (application-coloring) of Γ_P iff the following conditions hold for each $r \in P$*

A1 $c(r) = \ominus$ *iff one of the following conditions holds*
   a. *$\gamma_0^-(r) \neq \emptyset$ and for each $r' \in \gamma_0^-(r)$ we have $c(r') = \ominus$*
   b. *there is some $r'' \in \gamma_1^-(r)$ s.t. $c(r'') = \oplus$.*
A2 $c(r) = \oplus$ *iff both of the following conditions hold*
   a. *$\gamma_0^-(r) = \emptyset$ or it exists grounded 0-path[4] G_r s.t. $c(G_r) = \oplus$[5]*
   b. *for each $r'' \in \gamma_1^-(r)$ we have $c(r'') = \ominus$.*

For the generalization of condition $|body^+(r)| \leq 1$ (for $r \in P$) see [6]. There you can also find further details on a-colorings and the algorithm to compute them.

Observe, that there are programs like $P = \{p \leftarrow not\ p\}$ s.t. no a-coloring exists for Γ_P. Intuitively, each node of the block graph (corresponding to some rule) is colored with one of two colors, representing application ($\oplus$) or non-application ($\ominus$) of the corresponding rule. The coloring presented in Figure 1 corresponds to answer set A_1 of P. Node (rule) r_e has to be colored $\ominus$ (not applied), because there is some 1-predecessor of r_e colored $\oplus$ (applied). In other words, r_f blocks r_e.

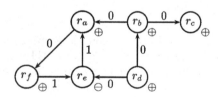

Fig. 1. Block graph of program (2 with a-coloring corresponding to answer set A_1).

---

[4] A subset of rules $G_r \subseteq P$ is a *grounded 0-path* for $r \in P$ if G_r is a 0-path from some fact to r in Γ_P.

[5] For a set of rules $S \subseteq P$ we write $c(S) = \oplus$ or $c(S) = \ominus$ if for each $r \in S$ we have $c(r) = \oplus$ or $c(r) = \ominus$, respectively.

3.3 Architecture

NoMoRe computes the answer sets of a logic program P in three steps (see Figure 2). First, the block graph Γ_P is computed. Second, Γ_P is compiled into Prolog code in order to obtain an efficient coloring procedure. The compilation borrows ideas from techniques utilized in e.g. [10,9] for efficient theorem proving. To read logic programs we use a parser and there is a separate part for interpretation of a-colorings into answer sets. For information purpose there is yet another part for visualizing block graphs using the graph drawing tool DaVinci [7].

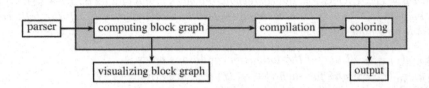

Fig. 2. The architecture of noMoRe.

4   Applying the System

The noMoRe system is used for purposes of research on the underlying paradigm. One has to keep in mind, that in the current state noMoRe is just a prototype with limited application. However, considering the short amount of time it took to develop the system and the progress made concurrently with further developments it is save to assume that a more useful version will be available shortly. But even in this early state, usability for anybody familiar with the logic programming paradigm is given.

5   Evaluating the System

As a first benchmark, we used two NP-complete problems proposed in [2]: the problem of finding a Hamiltonian path in a graph (**Ham**) and the independent set problem (**Ind**). In terms of time used for computing answer sets, our first prolog implementation (development time 8 months) is not comparable with state of the art C/C++ implementations, e.g. smodels [8] and dlv [3]. Therefore we compare the number of used choice points, because it reflects how an algorithm deals with the exponential part of a problem. Unfortunately, only smodels gives information about its choice points. For this reason, we have concentrated on comparing our approach with smodels. Results are given for finding all solutions of different instances of **Ham** and **Ind**. Table 1 shows results for some **Ham**-encodings of complete graphs K_n where n is the number of nodes[6]. Surprisingly,

---

[6] In a complete graph each node is connected to each other node.

Table 1. Number of choice points for **HAM**-problems.

	all solutions				one solution							
	K_7	K_8	K_9	K_{10}	K_5	K_6	K_7	K_8	K_9	K_{10}	K_{11}	K_{12}
smodels	4800	86364	1864470	45168575	3	4	30	8	48	1107	18118	398306
noMoRe	15500	123406	1226934	12539358	17	21	32	34	59	70	80	109

it turns out that noMoRe performs very well on this problem class. That is, with growing problem size we need less choice points than smodels. This can also be seen in Table 2 which shows the corresponding time measurements. For finding all Hamiltonian cycles of a K_{10} we need less time than the current smodels version. To be fair, for **Ind**-problems of graphs Cir_n[7] we need twice the choice points (and much more time) smodels needs, because we have not yet implemented backward-propagation. However, even with the same number of choice points smodels is faster than noMoRe, because noMoRe uses general backtracking of prolog, whereas smodels backtracking is highly specialized for computing answer sets. The same applies to dlv. Measurements with smodels and dlv are made with all optimizations (e.g. lookahead, heuristics to select next choice point) activated, whereas noMoRe currently has no such optimizations.

Table 2. Time measurements in seconds for **HAM**- and **IND**-problems on a SUN Ultra2 with two 300MHz Sparc processors.

	Ham for K_n											**Ind for** Cir_n		
	all solutions			one solution								all solutions		
$n =$	8	9	10	5	6	7	8	9	10	11	12	40	50	60
smodels	54	1334	38550	0.01	0.02	0.04	0.04	0.11	1.61	24	526	8	219	4052
dlv	4	50	493	0.02	0.03	0.03	0.05	0.06	0.07	0.09	0.15	13	259	4594
noMoRe	198	2577	34775	0.07	0.16	0.57	2.10	5.17	11.28	18.05	52.30	38	640	11586

References

1. A. Aggoun, D. Chan, P. Dufresne, and other. Eclipse user manual release 5.0. ECLiPSe is available at http://www.icparc.ic.ac.uk/eclipse, 2000.
2. P. Cholewiński, V. Marek, A. Mikitiuk, and M. Truszczyński. Experimenting with nonmonotonic reasoning. In *Proceedings of the International Conference on Logic Programming*, pages 267–281. MIT Press, 1995.

---

[7] A so-called circle graph Cir_n has n nodes $\{v_1, \cdots, v_n\}$ and arcs $A = \{(v_i, v_{i+1}) \mid 1 \leq i \leq n\} \cup \{(v_n, v1)\}$.

3. T. Eiter, N. Leone, C. Mateis, G. Pfeifer, and F. Scarcello. A deductive system for nonmonotonic reasoning. In J. Dix, U. Furbach, and A. Nerode, editors, *Proceedings of the Fourth International Conference on Logic Programming and Non-Monotonic Reasoning*, volume 1265 of *Lecture Notes in Artificial Intelligence*, pages 363–374. Springer Verlag, 1997.

4. M. Gelfond and V. Lifschitz. The stable model semantics for logic programming. In *Proceedings of the International Conference on Logic Programming*, 1988.

5. M. Gelfond and V. Lifschitz. Classical negation in logic programs and deductive databases. *New Generation Computing*, 9:365–385, 1991.

6. Th. Linke. Graph theoretical characterization and computation of answer sets. In *Proceedings of the International Joint Conference on Artificial Intelligence*, 2001. to appear.

7. M.Werner. davinci v2.1.x online documentation. daVinci is available at http://www.tzi.de/ davinci/doc_V2.1/, University of Bremen, 1998.

8. I. Niemelä and P. Simons. Smodels: An implementation of the stable model and well-founded semantics for normal logic programs. In J. Dix, U. Furbach, and A. Nerode, editors, *Proc. of the Fourth International Conference on Logic Programming and Nonmonotonic Reasoning*, pages 420–429. Springer, 1997.

9. T. Schaub and S. Brüning. Prolog technology for default reasoning: Proof theory and compilation techniques. *Artificial Intelligence*, 109(1):1–75, 1998.

10. M. Stickel. A Prolog technology theorem prover. *New Generation Computing*, 2:371–383, 1984.

11. Jan Wielemaker. Swi-prolog 3.4.3 reference manual. SWI-Prolog is available at http://www.swi.psy.uva.nl/projects/SWI-Prolog/Manual/, 1990–2000.

Conditional Pure Literal Graphs

Marco Benedetti

DIS, Dipartimento di Informatica e Sistemistica
Facoltà di Ingegneria
Università di Roma "La Sapienza"
mabe@dis.uniroma1.it

Abstract. *Conditional Pure Literal Graphs* (CPLG) characterize the
set of models of a propositional formula and are introduced to help understand connections among formulas, models and autarkies. They have
been applied to the SAT problem within the framework of refutation-
based algorithms. Experimental results and comparisons show that the
use of CPLGs is a promising direction towards efficient propositional
SAT solvers based upon model elimination. In addition, they open a
new perspective on hybrid search/resolution schemes.

1 Introduction

Propositional satisfiability is a many-sided problem, that captures theoretical
and practical interests. We address both of them, by (1) introducing a tool
called *Conditional Pure Literal Graph* and (2) showing its promising practical
application in speeding up some refutation procedures. With respect to the former point, CPLGs are introduced to help us gain insights into the connections
among formulas, models and autarkies. In particular, the concept of autarky [24]
is investigated and analyzed in detail. As to the latter point, we present an algorithm based on resolution which heavily exploits CPLGs. It is shown that redundancy in the search can be greatly reduced using CPLGs within a scheme that
merges concepts from direct model search and refutation procedures. Refutation
procedures constitute one of the two main categories of *complete* algorithms
for deciding propositional satisfiability (we focus on *complete algorithms* as opposed to *local search* techniques). The other category of complete approaches is
the one of *direct model search* algorithms [16,19,8,4,27,9], which are based upon
the Davis-Putnam (DP) procedure [22]. Almost all of the best performing and
widest used complete algorithms for satisfiability [15,6] fall in the *model search*
category.

Unlike DP-like algorithms, refutation strategies cope with the satisfiability
problem trying to derive a contradiction from a given formula by means of propositional resolution. If a sound and complete procedure derives a contradiction
from a formula $\mathcal{F}$, then $\mathcal{F}$ is guaranteed to be unsatisfiable, and vice-versa.
Many such strategies have been proposed, and we here focus on *model elimination* (ME) [20,21,26,23]. Model elimination is a subgoal-reduction strategy that

R. Goré, A. Leitsch, and T. Nipkow (Eds.): IJCAR 2001, LNAI 2083, pp. 331–346, 2001.

works as a backward-reasoning by reducing a goal to a set of subgoals and recursively working on the new subgoals. It shares redundancy-related problems
with other subgoal-reduction approaches used in automated deduction, as a pure
subgoal-reduction mechanism may run into the same sub-problem many times,
with no memory of the previously found solution. It so happens that reduction
of redundancy is an important issue in the field of automated deduction [7].

A well known way to prevent redundancy in ME is by caching intermediate
results occuring while doing refutation (succeed or fail in some sub-refutation
attempt). As far as succeeded sub-refutations are concerned, redundancy is prevented by lemmaizing [2,14,1,3]. Conversely, failures in sub-refutation attempts
in the propositional framework have been reduced [13] by using autarkies [24] as a
caching mechanism [12]. A more general framework that comprises both caching
mechanisms in the form of meta-level rules for inference has been proposed [7],
and theoretical analyses on the effectiveness of these methods have been performed [25,11]. Together with a reduction of redundancy, a failure-related caching
mechanism can help in extracting a model for satisfiable formulas [10] from a
ME procedure. The key idea is to *extract a model relying on the explanation
of why a refutation attempt failed*. CPLGs get into this framework by allowing
us (1) to thoroughly understand the concept of autarky and its relation with
formulas, models and caching devices, (2) to heavily reduce redundancy in the
search and (3) to extract a model for a formula using a refutation procedure. In
this sense, our work extends that of A. Van Gelder and F. Okushi [13].

CPLGs are graphs that represent properties of models for a given formula
and extend the concept of autarky to a finer degree of granularity. They are
built and exploited while the refutation procedure goes on to avoid redundancy
by memorizing sets of partial assignments which surely belong to a model under
some conditions (together with explanations of why things go that way).

From a different and intriguing perspective, we can consider CPLG-based algorithms as an attempt to overcome the dichotomy between direct model search
procedures and refutation based approaches. Previous approaches in this direction have shown their strength [17], and our algorithm substantially increases the
degree of coupling between search-related and resolution-related machinery. In
this sense, the CPLG-algorithm as a whole can be seen either as a model search
procedure, in which branching on unassigned variables is made exploiting clause
structure via resolution, or as a resolution-based approach that builds a candidate model as a consequence of failed refutation caching. This kind of bridge
seems likely to bring better efficiency results than similar previous approaches.
Moreover, such results pay no duty to clearness.

The outline of the paper is as follows: Section 2 gives preliminaries on both
propositional formulas and graphs and introduces notations. Section 3 formally
introduces CPLGs and their properties and presents a working example. The
practical use of CPLG within a ME-procedure is discussed in Section 4. Experimental results are reported in Section 5 to show how CPLGs are more effective
than previous approaches in reducing redundancy. Finally, in Section 6 we draw

our conclusions and present directions for future work. Further discussions about CPLGs and formal proofs of all the results presented below can be found in [5].

2   Preliminaries

We consider formulas in *conjunctive normal form* (CNF). To our extent, a propositional formula is a set of clauses, every clause being a non-trivial set of literals (a set of literals is *trivial* if it contains a literal and its negation). Double negation is absent. Formulas are indicated by calligraphic letters ($\mathcal{E}, \mathcal{F}, \dots$), clauses (and set of literals) by uppercase greek letters ($\Delta, \Phi, \dots$), literals by lowercase greek letters ($\varphi, \gamma, \dots$) and variables by lowercase roman letters ($a, b, \dots$).

An *assignment* for $\mathcal{F}$ is a non-trivial set of literals Δ containing only variables in $\mathcal{F}$; it is *total* if every variable in $\mathcal{F}$ appears in Δ, *partial* otherwise. The variable in a literal δ is referred to as $var(\delta)$, and the set of variables appearing in a set of literals Δ is denoted as $VAR(\Delta) = \{var(\delta)|\delta \in \Delta\}$. The notation $VAR(\mathcal{F})$ is similarly used to refer to the set $\bigcup_i VAR(\Delta_i)$ of variables appearing in $\mathcal{F} = \{\Delta_1, \Delta_2, ..., \Delta_n\}$. The set of every possible positive and negative literal on the variables $VAR(\mathcal{F})$ is written as $LIT(\mathcal{F})$. The complement set of Δ is defined as $\overline{\Delta} \stackrel{\text{def}}{=} \{\neg\delta \mid \delta \in \Delta\}$. The set $PAS(\mathcal{F})$ of partial assignments on the variables of $\mathcal{F}$ is defined as $PAS(\mathcal{F}) \stackrel{\text{def}}{=} \{\Delta \in 2^{LIT(\mathcal{F})} \mid \Delta \cap \overline{\Delta} = \emptyset\}$.

Definition 1 (star). *Given a formula $\mathcal{F}$ and an assignment $\Delta \in PAS(\mathcal{F})$, we define $\mathcal{F} * \Delta \stackrel{\text{def}}{=} (\mathcal{F} *^s \Delta) *^r \Delta$, where $\mathcal{F} *^s \Delta \stackrel{\text{def}}{=} \{\Gamma \in \mathcal{F} \mid \Gamma \cap \Delta = \emptyset\}$ and $\mathcal{F} *^r \Delta \stackrel{\text{def}}{=} \{\Gamma \setminus \overline{\Delta} \mid \Gamma \in \mathcal{F}\}$.*

$\mathcal{F} * \Delta$ is the formula resulting from $\mathcal{F}$ after the assignment Δ is made. The $*^s$ operator represents *unit subsumption* with all the literals in Δ considered as unit clauses, while $*^r$ represents *unit resolution* with the same unit clauses. When $\Delta = \{\delta\}$ we write $\mathcal{F} * \delta$ instead of $\mathcal{F} * \{\delta\}$. When the star operator is applied to two sets of literals, its meaning is assumed to be the same as the union operator, hence $(\mathcal{F} * \Delta) * \gamma = \mathcal{F} * (\Delta * \gamma) = \mathcal{F} * \Delta * \gamma$. We denote with $\mathcal{F}|_\Delta$ the set of clauses in $\mathcal{F}$ which contain at least one of the literals in Δ. The formula $\mathcal{F}|_\Delta$ is called *projection* of $\mathcal{F}$ onto Δ. We write $\mathcal{F}|_\delta$ instead of $\mathcal{F}|_{\{\delta\}}$ when $\Delta = \{\delta\}$.

Two literal symbols $\top$ and $\bot$ are introduced, in so as $\neg\top = \bot$, $\mathcal{F} * \top = \mathcal{F}$ for every $\mathcal{F}$ and $\mathcal{F}|_\bot = \mathcal{F}$ for every $\mathcal{F}$ ($\bot$ and $\top$ are the unit elements for the projection and star). By reversing the order in which star and projection are applied, we get the same result, provided arguments are consistent, i.e.: $(\mathcal{F} * \Delta)|_\Gamma = (\mathcal{F}|_\Gamma) * \Delta$ for any $\Delta, \Gamma \in PAS(\mathcal{F})$ such that $\Delta \cap \overline{\Gamma} = \emptyset$.

Any formula $\mathcal{F}' = \mathcal{F} * \Delta$ for some Δ is called sub-formula of $\mathcal{F}$, and a refutation of some sub-formula of $\mathcal{F}$ is called sub-refutation of $\mathcal{F}$.

We work on direct graphs and indicate them by uppercase roman letters and their nodes by lowercase roman letters. We call *source* node a node with no incoming arcs (wrt a graph W). Similarly, a subgraph G' of a graph G is a *source-subgraph* if no direct arc from $G - G'$ to G' exists. The set $\mathbf{par}_G(w)$ of nodes in G from which a direct arc to $w \in G$ exists is called the *parent set* of w

M. Benedetti

in G. Given two nodes $w, w' \in W$, w' is said to be in the *scope* of w if a path from w to w' exists. We call *transitive-removal* of the node w in W the operation of removing from W the node w together with every node w' in its scope (and together with every arc that looses its root or destination node).

2.1 Autarkies

Definition 2 (autarky). *A (partial) truth assignment $\Delta \in PAS(\mathcal{F})$ that satisfies a subset $\mathcal{S} \subseteq \mathcal{F}$ of a propositional formula $\mathcal{F}$, and contains no variable in $\mathcal{F} - \mathcal{S}$, is said to be an* autarky *for $\mathcal{F}$.*

This definition ensures that the set of models of $\mathcal{F} - \mathcal{S}$ is not shrunk by an autarky Δ, every clause in $\mathcal{F} - \mathcal{S}$ being untouched.

Lemma 1. $\Delta \in PAS(\mathcal{F})$ *is an autarky for $\mathcal{F}$ iff $\mathcal{F}|_{\overline{\Delta}} \subseteq \mathcal{F}|_{\Delta}$.*

The above lemma concisely characterizes an autarky as a (partial) assignment in which clauses involved in resolution are a subset of those involved in subsumption. As an example, consider the following formula.

$$\mathcal{F} = \{\{a, \neg b\}, \{\neg a, \neg b, f\}, \{\neg a, c, d\}, \{e, a, \neg c\}, \{\neg e, d, \neg f\}, \{\neg d, \neg f\}\}$$

According to Lemma 1, it is $\mathcal{F}|_{\overline{\Delta}} \subseteq \mathcal{F}|_{\Delta}$ for the autarky $\Delta = \{a, \neg b, c\}$:

$$\{\{\neg a, \neg b, f\}, \{\neg a, c, d\}, \{e, a, \neg c\}\} \subseteq \{\{a, \neg b\}, \{\neg a, \neg b, f\}, \{\neg a, c, d\}, \{e, a, \neg c\}\}$$

A *model* is a special case of autarky: if the subset $\mathcal{S} \subseteq \mathcal{F}$ of the formula $\mathcal{F}$ satisfied by the autarky is equal to the formula itself, the autarky is a model. This is guaranteed to happen when the autarky is a total assigment. However, it is not necessary for an assignment (and hence for an autarky) to be total in order to be a model. We say that a literal φ in a model Δ of $\mathcal{F}$ is *essential* if $\Delta - \{\varphi\}$ is not a model for $\mathcal{F}$.

If we find an autarky Δ for a given formula $\mathcal{F}$, we are not guaranteed that $\mathcal{F}$ is satisfiable, unless Δ is a model itself. However, we have already pointed out that the set of models of the sub-formula $\mathcal{F} - \mathcal{F}|_{\Delta}$ is not shrunk by Δ. These results can be summarized as follows.

Theorem 1. *Given any autarky Δ on $\mathcal{F}$, it is $\mathcal{F} \longleftrightarrow \mathcal{F} * \Delta$.*

3 Understanding CPLG

We now introduce the key concept of *Conditional Pure Literal Graph*, starting from the well known definition of pure literal.

Definition 3 (PL). *A literal φ is said to be* pure *in a formula $\mathcal{F}$ when it occurs in $\mathcal{F}$ and its negation $\neg\varphi$ does not.*

Definition 4 (CPL). *A literal φ is said to be a* conditional pure literal *(CPL) in $\mathcal{F}$ with respect to a partial assignment Δ if it is a pure literal in $\mathcal{F} * \Delta$, i.e. $(\mathcal{F} * \Delta)|_{\neg\varphi} = \emptyset$. The assignment Δ is called the* condition *or the* premise *for φ to be pure.*

We use the shorthand $\Delta \gg_{\mathcal{F}} \varphi$ (or simply $\Delta \gg \varphi$ when no confusion arises) in place of $\mathcal{F} * \Delta |_{\neg \varphi} = \emptyset$ to denote a literal φ that is pure under the condition Δ with respect to the formula $\mathcal{F}$. If a conditional pure literal has an empty premise, then it is a pure literal.

Definition 5 (CPLG). *A* Conditional Pure Literal Graph W *on a propositional formula* $\mathcal{F}$ *is a direct graph with the following properties:*

- *every node is labeled with a literal in* $LIT(\mathcal{F})$, *and every variable in* $VAR(\mathcal{F})$ *is represented at most once;*
- *the nodes are partitioned into* hypothesis-nodes *and* CPL-nodes, *every hypothesis-node being a source for the graph;*
- *every CPL-node* $w \in W$ *is labeled by a literal* φ *such that* $\Delta_w \gg_{\mathcal{F}} \varphi$, *where* Δ_w *is the set of literals labeling nodes in* $\mathbf{par}_W(w)$.

The set of literals labeling CPL-nodes in a CPLG W is denoted by $CPL(W)$, while $HYP(W)$ is the set of literals labeling hypothesis-nodes. We pose $NODES(W) = CPL(W) \cup HYP(W)$. Notice that every hypothesis-node has to be a source for the graph according to Definition 5, but CPL-nodes may be sources as well.

Definition 6 (proper and minimal CPLs). *A CPL* $\Delta \gg \varphi$ *is said to be:*

1. proper *iff* $\Delta' \gg \varphi$ *holds for no premise* $\Delta' \subset \Delta$;
2. of minimal size *iff* $\Delta' \gg \varphi$ *holds for no condition* Δ' *with* $|\Delta'| < |\Delta|$.

Definition 7 (proper, self-contained and complete CPLG). *A CPLG* W *on a formula* $\mathcal{F}$ *is:*

1. proper *iff every CPL-node in* W *is proper;*
2. self-contained *iff* $HYP(W) = \emptyset$;
3. complete *iff* $VAR(\mathcal{F}) = VAR(NODES(W))$.

As CPLs are bricks to build a CPLG-wall, it is straight to introduce two operators to explain how walls are built and dismantled.

Definition 8 (extension step). *Let* $\mathcal{G}^{\Gamma}$ *be the set of CPLGs on the set of variables* Γ *and* $\mathcal{C} = PAS(\Gamma) \times LIT(\Gamma)$ *the set of CPLs on* Γ. *We define a partial binary function*

$$ext : \mathcal{G}^{\Gamma} \times \mathcal{C} \longrightarrow \mathcal{G}^{\Gamma}$$

such that $W' = ext(W, \Delta \gg \varphi)$ *is defined only when* $var(\varphi) \notin VAR(CPL(W))$ *and* $NODES(W) \cap \overline{\Delta} = \emptyset$, *and is obtained from* W *as follows:*

- *a new hypotesis-node is added to* W *for every literal in* $\Delta - NODES(W)$;
- *if* $\varphi \in HYP(W)$, *the node labeled by* φ *is turned into a CPL-node; if* $\varphi \notin HYP(W)$ *a CPL-node labeled by* φ *is added to* W;
- *for every literal* $\delta_i \in \Delta$ *an arc from the node labeled by* δ_i *to the node labeled by* φ *is added to* W.

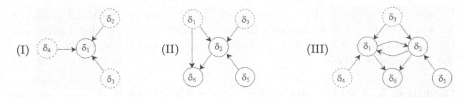

Fig. 1. Graphical representation and composition of CPLs

G' is said to extend G by $\Delta \gg \varphi$.

The reflexive and transitive closure of the just defined relation can be considered. This way, a CPLG W' is said to *extend* a CPLG W if W' can be obtained from W by zero or more extension steps.

As opposite to the extension operation, a *prune* operation is defined to capture the way a CPLG changes by removing some hypothesis-node. The intuition is that when a hypothesis-node is removed, all the CPL-nodes that (directly or not) rely on the removed hypothesis must be removed as well.

Definition 9 (prune step). *Let $\mathcal{G}^\Gamma$ be the set of CPLGs on the set of variables Γ. We define a partial binary function prune : $\mathcal{G}^\Gamma \times LIT(\Gamma) \longrightarrow \mathcal{G}^\Gamma$ such that $W' = prune(W, \varphi)$ is defined only when $\varphi \in HYP(W)$ and is obtained by a transitive-removal of φ from W.*

A simple graphical representation for CPLs and CPLGs is adopted. Figure 1(I) illustrates how a CPL $\{\delta_2, \delta_3, \delta_4\} \gg \delta_1$ is represented. The conditions under which δ_1 is pure are drawn in dotted circles, while the CPL itself is represented within a continuous border. This kind of representation naturally evolves into a graph structure when CPLG are considered instead of single CPL. Figure 1(II) represents a sample CPLG. Figure 1(III) illustrates an extension step: the CPLG (II) is extended by the CPL (I) to obtain the CPLG (III).

3.1   An Example

Let us consider the following formula:

$$\mathcal{E} = \{ \ \{a, b\}, \{a, \neg b, d\}, \{\neg a, \neg f\}, \{\neg a, \neg b, \neg d\}, \{\neg a, d, e\},$$
$$\{\neg b, \neg c\}, \{c, e\}, \{c, \neg e\}, \{\neg c, \neg f\}, \{\neg c, \neg e\}, \{d, \neg f\} \ \}$$

and the CPLG $W_\mathcal{E}$ on $\mathcal{E}$ represented in Figure 2(I). A number of observations can be made even in this simple case.

- The graph contains four nodes (equivalently, four variables and four literals). Three of them are CPL-nodes, and one is a hypothesis-node.
- The hypothesis-node d is present in $W_\mathcal{E}$ as a premise of a. It is a source node.
- The CPL-node $\neg f$ has no premise, so it actually is a pure literal. This can be quickly verified by looking at $\mathcal{E}$.
- The node $\neg b$ is a pure literal conditioned to $\Delta_{\neg b} = \{a\}$. This means that all the clauses in $\mathcal{E}$ that contain b are satisfied by the assignment $\Delta_{\neg b}$.

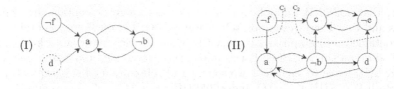

Fig. 2. Two CPLGs on $\mathcal{E}$

- The node a is a pure literal conditioned to $\Delta_a = \{\neg f, d, \neg b\}$. This means that all the clauses in $\mathcal{E}$ that contain $\neg a$ are satisfied by the assignment Δ_a.
- Every CPLG may be cyclic, and $W_{\mathcal{E}}$ is cyclic. There is no contradiction between the definitions of CPLG and the fact that a CPL φ' is in the premise of a CPL φ'' while - at the same time - φ'' belongs to the premise of φ'. The nodes a and $\neg b$ are involved in one such "one-length" cycle in Figure 2(I).
- $W_{\mathcal{E}}$ is a proper, not self-contained, not complete CPLG.

The CPLG $W_{\mathcal{E}}$ in Figure 2(I) can be *extended* to the complete and self-contained graph $W'_{\mathcal{E}}$ in Figure 2(II). It is easy to check that $W'_{\mathcal{E}}$ is still a CPLG on $\mathcal{E}$.

3.2   Properties

The condition Δ_φ under which a literal φ not pure in $\mathcal{F}$ becomes pure in $\mathcal{F} * \Delta$ has an obvious meaning in terms of models for subsets of $\mathcal{F}$.

Lemma 2. *If φ is a CPL in $\mathcal{F}$ conditioned to Δ_φ, Δ_φ is a model for $\mathcal{F}|_{\neg \varphi}$.*

Lemma 2 can be exploited to obtain the following results about CPLGs.

Lemma 3. *For every CPLG W on a given formula $\mathcal{F}$it holds that:*

1. *$HYP(W)$ is a model for $(\mathcal{F} * CPL(W))|_{\overline{CPL(W)}}$;*
2. *$CPL(W)$ is a model for $(\mathcal{F} * HYP(W))|_{\overline{CPL(W)}}$;*
3. *$(\mathcal{F} * NODES(W))|_{\overline{CPL(W)}} = \emptyset$.*

As an example, Lemma 3 applied to the CPLG in Figure 2(I) says that (1) d is a model for $(\mathcal{F} * \{a, \neg b, \neg f\})|_{\{\neg a, b, f\}}$, (2) $\{a, \neg b, \neg f\}$ is a model for $(\mathcal{F} * d)|_{\{\neg a, b, f\}}$ and (3) $(\mathcal{F} * \{a, \neg b, \neg f, d\})|_{\{\neg a, b, f\}} = \emptyset$.

In the previous section, we recalled what an autarky is. Now we can explain autarkies in terms of CPLGs, according to the following lemma.

Lemma 4. *For every CPLG W on a given formula $\mathcal{F}$it holds that:*

1. *$CPL(W)$ is an autarky for $\mathcal{F} * HYP(W)$;*
2. *$NODES(W)$ is an autarky for $\mathcal{F}$ if W is self-contained;*
3. *$NODES(W)$ is a model for $\mathcal{F}$ if W is self-contained and complete.*

Notice that $NODES(W)$ may be an autarky even though W contains hypothesis-nodes and that every (even non self-contained) CPLG on $\mathcal{F}$ may contain many self-contained sub-graphs that are CPLGs on $\mathcal{F}$. So, many autarkies can be encoded into a single CPLG. For example, the two cuts c_1 and c_2 represented in Figure 2(II) leave to their left two source and self-contained subgraphs W_1 and W_2. Both $NODES(W_1) = \{\neg f\}$ and $NODES(W_2) = \{\neg f, a, \neg b, d\}$ are autarkies.

An essential result for integrating CPLG into SAT algorithms is given in the following theorem.

Theorem 2. *For any CPLG W on $\mathcal{F}$ it holds that $\mathcal{F} * HYP(W) \longleftrightarrow \mathcal{F} * NODES(W)$ equivalently, if $\mathcal{E} = \mathcal{F} * HYP(W)$, then $\mathcal{E} \longleftrightarrow \mathcal{E} * CPL(W)$*

Notice that Theorem 2 does not guarantee that $\mathcal{F}*HYP(G)$ and $\mathcal{F}*NODES(G)$ are logically equivalent. Nevertheless, it ensures that $\mathcal{F} * HYP(G)$ and $\mathcal{F} * NODES(G)$ are equivalent as to satisfiability. So, the idea for integration within SAT algorithms is that we can consider $\mathcal{F} * NODES(W)$ instead of $\mathcal{F} * HYP(W)$ as long as we are interested in deciding satisfiability for $\mathcal{F}$.

4   CPLG within Model Elimination

We can construct complete and self-contained CPLGs on $\mathcal{F}$ by inspecting the structure of $\mathcal{F}$, provided we know a model for such formula. Conversely, a SAT algorithm does not know any model in advance, its aim being to discover whether such a model exists. So, which is the utility of CPLGs in a SAT algorithm?

Reversing the perspective, we will use a CPLG as a tool to help building a model for $\mathcal{F}$ which may be constructed and used *before* the entire model (or even just an autarky) is known. The concept of CPLG advantageously meets the SAT problem within the framework of propositional model elimination, where it plays the role of *a caching device to store information about failed sub-refutation attempts, for later re-use*. Figure 3 shows a doubly-recursive algorithm written in a C-like pseudo code which attempts to refute a formula by propositional model elimination. It is the starting point for our work. The global variable $\mathcal{F}$ represents the formula to refute. The algorithm is activated by "`refuteGoal(∅,⊤)`". The first argument of both procedures is a set of literals called *ancestor literals* (or simply *ancestors*). They determine the current sub-refutation to be the refutation of $\mathcal{F} * \Delta$. The second argument of `refuteGoal` is a literal to be refuted called *sub-goal*, while the second argument of `refuteClause` is a clause to be refuted. Figure 4 shows a version of `refuteGoal` modified to introduce the use of a CPLG. The global variable W represents the (initially empty) CPLG on which the algorithm is working. The procedure `refuteClause` is unchanged.

Extension, prune and *use* are three different operations performed by the procedure in Figure 4 on the CPLG W. Let us consider them in turn.

ME attempts to show that no model exists, by systematically proving the refutability of sub-formulas $\mathcal{F}*\Delta$ for some suitably generated partial assignments Δ. Many sub-refutations may fail, even though the formula $\mathcal{F}$ is unsatisfiable (i.e. its top-level refutation succeeds). Figure 4 shows that failing a sub-refutation is

```
boolean refuteGoal(Δ, φ)
{
   Δ' ← Δ ∪ {φ};
   E ← F |₋φ *Δ';

   refutationSucceed ← false;
   while (not refutationSucceed and |E| > 0 )

      { Γ ← getAClause(E);
        refutationSucceed ← refuteClause(Δ',Γ);
        E ← E − Γ; }

   return refutationSucceed;
}
```

```
boolean refuteClause(Δ, Φ)
{
   refutationSucceed ← true;
   while (refutationSucceed and |Φ| > 0)

      { φ ← getALiteral(Φ);
        refutationSucceed ← refuteGoal(φ,Δ);
        Φ ← Φ − φ; }

   return refutationSucceed;
}
```

Fig. 3. Structure of the basic model elimination algorithm

```
boolean refuteGoal(Δ, φ)
{
   Δ' ← Δ ∪ {φ};
   E' ← F |₋φ *Δ';
use→   E ← E' * CPL(W);
   refutationSucceed ← false;

   while (not refutationSucceed and |E| > 0 )
      { Γ ← getAClause(E);
        refutationSucceed ← refuteClause(Δ',Γ);
use→    E ← E' * CPL(W); }

   if (refutationSucceed)
prune→    W ← prune(W,φ);
   else
         { Φ ← a sub-set of Δ ∪ CPL(W) which is a model for F |₋φ;
extension→ W ← ext(W,Φ≻φ); }

   return refutationSucceed;
}
```

Fig. 4. Structure of the `refuteGoal` procedure modified to introduce a CPLG

the precondition for succeeding in adding a CPL to the CPLG (the *extension* step). When a sub-refutation succeeds, a *prune* step takes place. The refuted sub-goal φ is for sure a hypothesis-node in the CPLG. The nodes of the graph in the scope of φ must be removed, to guarantee that only sound (wrt the use step) information is held in W. In other words, a transitive removal on φ is performed.

As a consequence of the way the extension and prune steps are performed, the CPLG W maintains two interesting properties (see Section 4.3). (1) The literals in $CPL(W)$ cannot be refuted given $HYP(W)$. (2) $HYP(W)$ is always contained in the ancestor set. Therefore, any literal in $CPL(W)$ is not refutable given the current set of ancestors: the *use* step exploits the CPLG as a caching device by avoiding - in any refutation attempt - the use of every clause containing literals known to be not refutable (it uses $\mathcal{F} * \Delta' * CPL(W)$ instead of $\mathcal{F} * \Delta'$).

4.1 Computing the Premise of a CPL

A sub-set of $\Delta \cup CPL(W)$ satisfying $\mathcal{F}|_{\neg\varphi}$ is chosen in Figure 4 to be the premise of the CPL on φ. After the relevant subset $\Delta'' \subseteq \Delta \cup CPL(W)$ of literals on variables involved in $\mathcal{F}|_{\neg\varphi}$ is isolated, the graph can be extended by $\Delta'' \gg \varphi$. However, this CPL is guaranteed to be neither minimal nor proper, whereas we would like to have such CPLs in our CPLG because of the way prune steps are performed. Minimal and proper CPLs are less prone to be removed than general CPLs are, because their survival relies on less hypotheses.

In general, given any CPL on φ with premise Γ, there exists at least one *proper* CPL on φ (and a CPL of minimal size) with premise $\Gamma' \subseteq \Gamma$. So, there is a tradeoff between the time spent looking for better CPLs (premises of little size) and the time spent rebuilding a CPLG pruned just because proper CPLs were not used. A *greedy* strategy can be used to incrementally build a premise for φ to be pure in $\mathcal{F}|_{\neg\varphi}$, starting from the empty set $\Gamma_0 = \emptyset$. It is sufficient to consider all the clauses $\Lambda_i \in \mathcal{F}|_{\neg\varphi}$, for $i = 1, ..., n$, in any order and put $\Gamma_i = \Gamma_{i-1}$ if Λ_i is satisfied by Γ_i and $\Gamma_i = \Gamma_{i-1} \cup \{\delta\}$ for some $\delta \in \Delta'' \cap \Lambda_i$ otherwise. A *heuristic* strategy can be introduced, using a search algorithm that uses Δ'' as a trusted guide towards a model, and makes a heuristic choice on literals aimed at finding a small subset of Δ'' that is still a model for $\mathcal{F}|_{\neg\varphi}$. Finally, an *exhaustive* strategy can resort to backtracking. As soon as a satisfying subset of Δ'' is found, the algorithm chronologically backtracks along the history of its choices and restarts the search towards an even smaller model.

Even if the last strategy is the only one that guarantees a proper CPL as a result, an empirical evaluation of the considered tradeoff led us to prefer the greedy strategy for the experiments presented in Section 5.

4.2   An Example of Building and Exploiting CPLG

We now describe how the modified refutation procedure may construct and use the CPLG in Figure 5(I) on the formula $\mathcal{E}$ of Section 3.1, through the intermediate steps in Figure 5(II). Figure 5(I) shows a portion of the generated refutation tree. Closed branches are marked by an hourglass-like symbol. Sub-goals have a left and a right numeric label. Numbers on the left side give the order in which nodes are expanded, while the right-ordering is the one in which they are exited (i.e.: it is the order in which CPLs are generated and represented in 5(II)).

The first CPL to be identified is the one on $\neg f$. When the recursion stops at the expansion step on the literal $\neg f$ with ancestors $A_{\neg f} = \{c\}$, $\mathcal{F} * A_{\neg f}$ does not contain clauses in which f appears (otherwise, a further reduction step would have been taken). Now it suffices to extract a condition $\Delta_{\neg f}$ for the CPL on $\neg f$, i.e. a model for $\mathcal{F}|_f$, which surely exists and is a subset of $A_{\neg f}$. In this particular case, the empty set is a well suited assignment, as f does not appear at all in $\mathcal{F}$, so that $\neg f$ is a pure literal and $\emptyset \gg \neg f$ is added to W (see Figure 5(II).1).

The second CPL to be extracted is the one on d. This time, the subset Δ_d of the set of ancestors $A_d = \{c, \neg b, a\}$ that is a model for $\mathcal{F}|_{\neg d}$ results to be $\{\neg b\}$. Hence, $\{\neg b\} \gg d$ is added to W (see Figure 5(II).2).

The third CPL found from the procedure is $\Delta_a \gg a$ with $\Delta_a = \{d, \neg f, \neg b\}$. This is an interesting case, because the condition Δ_a consists of an ancestor of a, a descendant of a and a CPL already in W (Figure 5(II).3).

When the fourth CPL ($\{a\} \gg \neg b$) is discovered (Figure 5(II).4) and added to W, the whole CPLG W becomes self-contained. At this point, $CPL(W) = NODES(W) = \{a, \neg b, d, \neg f\}$ is known to be an autarky for $\mathcal{F}$. The procedure goes on leaving the leaf-node $\neg e$ and adding the relative CPL (step 5 in Figure

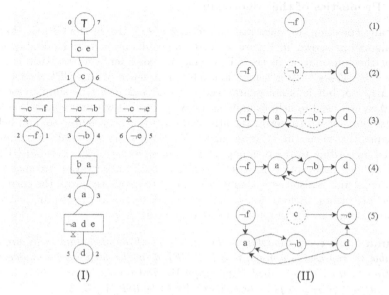

(I)   (II)

Fig. 5. Failed refutation tree and CPLG construction for $\mathcal{E}$

5(II)) and finally introducing the CPL $\{\neg f, \neg b, \neg e\} \succ c$ in the sixth step to obtain the complete graph in Figure 2(II).

Even in the short refutation trace just seen, the procedure takes a ready advantage from the presence of a CPLG. Not only ends the procedure up with a model codified in the CPLG, but it also exploits the unfinished graph structure to avoid doing some useless work. We now show how by means of some examples.

The tree in Figure 5 is meant to represent only the failure-related portion of the global tree generated by the ME procedure. Nevertheless, according to Figure 3, it seems to be incomplete. Below the sub-goal a the clause $\{\neg a, \neg f\}$ seems missing. The leaf-node $\neg e$ should not be a leaf-node because $\{\neg a, d, e\} \in \mathcal{F} * c$. We also expect that other clauses should be proven to be non-refutable below the root. These lacks are not an oversight. Some refutations did not take place as their outcomes are known to be failures, thank to the presence of the CPLG. The reduction of a by $\{\neg a, \neg f\}$ is not attempted because the CPLG number 3 proves that $\neg f$ is not refutable. In the same way, $\{\neg a, d, e\}$ is not chosen to reduce $\neg e$ because the CPLG (number 5) ensures that the current set of ancestors $\{c\}$ suffices to make $\neg e$ non-refutable. Finally, when $\{c, e\}$ is proven to be non-refutable, the CPLG (Figure 2(II)) by then encodes a complete model for $\mathcal{F}$ which makes it meaningless to proceed in further refutation attempts. All these refutations are skipped because the sub-formula $\mathcal{F} * \Delta' * CPL(W)$ considered in the modified algorithm does not contain the "redundant" clauses, due to the $CPL(W)$ factor, whereas $\mathcal{F} * \Delta'$ alone would have still contained them.

In this simple example, prune steps are never performed.

4.3   Properties of the Algorithm

Logically speaking, we have just considered a *safe* use of the CPLG. However, the algorithm shown in Figure 4 makes also an implicit and possibly *unsafe* use of the information in the CPLG. Let us consider just the basic case of a single CPL $\Delta \mathord{>}_{\mathcal{F}} \delta$. As a semantic-level consequence of this CPL we can write $\mathcal{F} * \Delta \not\models \neg\delta$, but it is not guaranteed to be $\mathcal{F} * \Delta \models \delta$ (the simple case when $\Delta = \emptyset$ can be used to help us understand why: models with pure literals assigned to false *may* exist). However, the algorithm in Figure 4 assumes so, and behaves consequently: when a literal φ belongs to $\mathit{CPL}(W)$, every sub-goal $\neg\varphi$ is implicitly refuted, according to the meaning of the star operator in $\mathcal{F} * \Delta' * \mathit{CPL}(W)$.

This lazy assumption is strictly related to the observation we made after Theorem 2 and is also the reason why we need to explicitly prove the correctness of the algorithm. Before doing this, we need to make precise an underlying assumption about the graph W the algorithm deals with.

Lemma 5. *At every time during the execution of the algorithm in Figure 4, the variable W represents a graph that is a CPLG on the formula $\mathcal{F}$ considered. In addition to this, if we consider the graph W and the set of ancestors Δ at any time during the search procedure, it results to be $\mathit{HYP}(W) \subseteq \Delta$.*

Theorem 3. *The algorithm in Figure 4 is sound and complete.*

The CPLG goes through many changes as extension and prune steps alternate. However, not everything is made to be broken. For example, when an autarky is found, it is never lost, no matter how many prune steps occur. The reason is that only pieces of the graph that are in the scope of some hypothesis-node can be removed, while autarkies are encoded by self-contained source sub-graphs. So, prune steps are not always destructive. In favorable cases they could even remove hypotheses upon which no CPL relies. An ultimate property of CPLGs comes out of this game of cut and extension steps: the procedure eventually finds a model if the formula is satisfiable.

Theorem 4. *If W is the CPLG remaining after the algorithm in Figure 4 terminates its execution on a satisfiable formula $\mathcal{F}$, $\mathit{NODES}(W)$ is a model for $\mathcal{F}$.*

As a final remark, we point out that CPLGs are built as separate data structures (not totally coupled with the search procedure) in so as autarkies (and even models) can be recognized while the search procedure is going on at any depth.

5   Experimental Results

Even though the algorithm in Figure 3 can suitably be combined with CPLGs, it is known to suffer of a high degree of redundancy, as Table 1(I) shows. The average number of subgoals expanded on average on very small uniform random formulas is reported. The number of subgoals expanded is chosen to measure the efficiency of the procedure as it is independent of the machine speed and it allows for easy comparisons between subgoal-reduction based algorithms. The *CV* ratio is the ratio between the number of clauses and variables in the formula. The middle value (4.27) is chosen according to an empirical complexity result [18,8] indicating this ratio as the hardest one for uniform random 3CNF formulas. The other two values are arbitrarily chosen to represent under-constrained formulas (likely-to-be-satisfiable) and hyper-constrained ones (likely-to-be-unsatisfiable).

Table 1. Subgoals expanded by (I) the basic and (II) the modified ME procedure

CV	\multicolumn{4}{c}{*Number of variables*}		*CV*	\multicolumn{4}{c}{*Number of variables*}
	6	*8*	*10*	*12*			*6*	*8*	*10*	*12*
3.27	2,251	12,991	57,045	272,982		*3.27*	45	71	120	261
4.27	5,059	43,019	298,011	1,912,228		*4.27*	27	43	90	141
5.27	10,082	93,741	921,880	>5,000,000		*5.27*	10	18	26	33

(I)   (II)

The use of CPLG helps avoiding such redundancy. Table 1(II) shows the effect of introducing CPLGs. We here concentrate on the relative advantage the algorithm achieves with respect to failed refutation caching. The promising results achieved are indicative of impressive potential performance. In this perspective, a significant measure can be obtained comparing the CPLG approach with MODOC (proposed by Van Gelder and Okushi [12,13]), which is another SAT solver based upon the procedure in Figure 3. It uses boosting mechanisms to perform both failed-subrefutation and succeeded-subrefutation caching. To directly compare their approach with our, we extracted from MODOC the failure caching machinery, and re-implemented it upon the usual basic algorithm as unique boosting mechanism. We call mini-MODOC the resulting procedure.

Table 2(I) shows the result of comparisons between mini-MODOC and our algorithm on (I) uniform, (II) chain and (III) tree random formulas. Each cell in the table contains a "$x \div y \to p\%$" value, where x and y are the number of subgoals expanded on average by mini-MODOC and our algorithm respectively, and $p = 100 * (1 - y/x)$ is the percentage of expansions saved using CPLGs.

Structured random generators are used to capture many properties of the instances coming from realistic domains, retaining the typical advantages of parametric and controllable random generators. We here focus on *chain* and *tree* generators. The key idea behind both of them is to generate several independent *sub-instances* (uniform random formulas), and then to obtain a single global instance by connecting these sub-instances. The chain generator obtains connections by introducing a supplementary amount of clauses containing variables from two sub-formulas (n independent sub-formulas $\mathcal{F}_i, i = 1, ..., n$ are connected in a linear chain by means of $n-1$ additional clauses containing only two literals, the first on a variable in $VAR(\mathcal{F}_i)$ and the second on a variable

Table 2. Confrontation on different random domains

(I) Uniform random formulas

CV	30 variables	40 variables	50 variables
3.27	2,081÷1,774 →15%	12,915÷9,437 →27%	127,321÷80,962 →36%
4.27	46,593÷42,119→10%	556,440÷485,151→13%	111,111÷9,983,028→15%
5.27	48,603÷46,647→4%	696,375÷654,848→6%	8,457,481÷8,015,975→5%

(II) Chain random formulas

#F	10 variables	20 variables	30 variables
10	479÷389 →19%	1,790÷1,048 →41%	165,600÷6,549 →96%
20	898÷799 →11%	42,056÷5,876 →86%	343,510÷12,019 →97%
30	1,227÷1,030 →16%	40,607÷11,785 →71%	815,020÷16,300 →98%

(III) Tree random formulas

#F	(3,6)-trees	(4,8)-trees	(5,10)-trees
40	3,604÷2,332 →35%	120,764÷63,626 →47%	433,398÷115,826 →73%
60	4,848÷7,877 →38%	290,685÷60,663 →86%	3,796,063÷1,848,278→51%
80	19,936÷8,267 →59%	453,676÷178,471→61%	9,510,065÷3,804,026→60%

in $VAR(\mathcal{F}_{i+1})$ for $i = 1, ..., n - 1$). The tree generator connects sub-formulas in a random tree structure. For every couple of sub-formulas that are neighbor in the tree, connection is guaranteed by imposing that the two formulas in the couple share a fixed percentage of their variables. Once fixed the number of literals in each clause (here always 3), these random generators need, as additional parameters, the number of sub-instances to be connected and the number of variables and clauses in each sub-instance. A *(k,m)-tree* is a formula where each sub-instance has $k + m$ variables, k of which shared with its neighbors. In our experiments with the chain generator, we consider chains of 10, 20 and 30 subformulas, each one containing 10, 20 and 30 variables and twice as many clauses as variables. Table 2(II) reports experimental results on chains, and shows an impressive advantage of the CPLG based algorithm. As to the tree generator, we considered trees with 40, 60 and 80 nodes, each one containing 9, 12 and 15 variables and twice as many clauses as variables. Results shown in Table 2(III) confirm the great advantage exhibited when dealing with structured domains.

6   Conclusions and Future Work

We presented a simple and powerful tool - called CPLG - to help understand connections among propositional formulas, models, autarkies and caching devices. Properties of this tool have been investigated and several examples have been presented to illustrate how it works. CPLGs proved to be useful when exploited within ME based algorithms. We showed in details how the bridge between CPLGs and ME is built, and presented a resulting algorithm for the satisfiability problem. Soundness, completeness and model extraction capability have been investigated for this algorithm.

Even if we heavily restricted ourselves in the use of complementary boosting mechanisms, results are quite promising. The use of CPLG in isolation from other devices, allowed us to highlight their contribution in comparison with similar techniques. In particular, we tested our approach against a recent failure caching mechanism built upon the same basic ME algorithm (namely, the one used in MODOC). Experimental results show our greater efficiency on both uniform and structured random instances. Encouraged by these results, we are extending our technique to an algorithm based upon a more general pruning device that generalizes CPLGs. Conditional Pure Literal Graphs are the first step towards this promising kind of integration of direct model search and refutation.

Acknowledgements. This work is partly supported by ASI funds and by MURST-COFIN funds (MOSES project). We thank Prof. Fabio Massacci for helpful discussions, and Prof. Luigia Carlucci Aiello for many hours of invaluable help, for advices and comments on preliminary versions of this paper and for believing in me.

References

1. O. L. Astrachan. Meteor: Exploring model elimination theorem proving. *JAR*, 13:283–296, 1994.
2. O. L. Astrachan and D. W. Loveland. The use of lemmas in the model elimination procedure. *JAR*, 19:117–141, 1997.
3. O. L. Astrachan and M. E. Stickel. Caching and lemmaizing in model elimination theorem provers. In *Proceedings of CADE-92*, pages 224–238, 1992.
4. R. Bayardo and R. Schrag. Using CSP look-back techniques to solve real-world SAT instances. *Proceedings of AAAI97*, 1997.
5. M. Benedetti. Conditional Pure Literal Graphs. Technical Report 18-00, Dipartimento Informatica e Sistemistica, Università di Roma "La Sapienza", ftp://ftp.dis.uniroma1.it/PUB/benedett/sat/techrep-18-00.ps.gz, 2000.
6. D. Le Berre. Sat live! page: a dynamic collection of links on sat-related research. http://www.satlive.org, 2000.
7. M. P. Bonacina and J. Hsiang. On semantic resolution with lemmaizing and contraction and a formal treatment of caching. *NGC*, 16(2):163–200, 1998.
8. S. Cook and D. Mitchell. Finding hard instances of the satisfiability problem: A survey. In *Proceedings of the DIMACS Workshop on Satisfiability Problems*, 1997.
9. J.W. Freeman. *Improvements to Propositional Satisfiability Search Algorithms*. PhD thesis, The University of Pennsylvania, 1995.
10. A. Van Gelder. Simultaneous construction of refutations and models for propositional formulas. Technical Report UCSC-CRL-95-61, UCSC, 1995.
11. A. Van Gelder. Complexity analysis of propositional resolution with autarky pruning. Technical Report UCSC-CRL-96, UC Santa Cruz, 1996.

12. A. Van Gelder. Autarky pruning in propositional model elimination reduces failure redundancy. *JAR*, 23:137–193, 1999.
13. A. Van Gelder and F. Okushi. A propositional theorem prover to solve planning and other problems. *AMAI*, 26:87–112, 1999.
14. A. Van Gelder and Y. K. Tsuji. Satisfiability testing with more reasoning and less guessing. Cliques, Coloring and Satisfiability: Second DIMACS Implementation Challenge, *DIMACS Series in Discrete Mathematics and Theoretical Computer Science*, 1996.
15. T. Stützle H. Hoos. Satlib - the satisfiability library. http://www.informatik.tu-darmstadt.de/AI/SATLIB, 1998.
16. M. Stickel H. Zhang. Implementing Davis-Putnam's method by tries. Technical report, The University of Iowa, 1994.
17. R. Dechter I. Rish. Resolution versus search: Two strategies for sat. In I. Gent et al., editor, *SAT2000*, pages 215–259. IOS Press, 2000.
18. J.Crawford and L. Auton. Experimental results on the cross-over point in satisfiability problems. In *AAAI-93*, pages 21–27, 1993.
19. C. Li and Anbulagan. Heuristics based on unit propagation for satisfiability problems. In *Proceedings of IJCAI-97*, 1997.
20. D. W. Loveland. Mechanical theorem-proving by model elimination. *Journal of the ACM*, 15:236–251, 1968.
21. D. W. Loveland. A simplified format for the model elimination theorem-proving procedure. *Journal of the ACM*, 16:349–363, 1969.
22. G. Logemann M. Davis and D. Loveland. A machine program for theorem proving. *Journal of the ACM*, 5:394–397, 1962.
23. J. Minker and G.Zanon. An extension to linear resolution with selection function. *Information Processing Letters*, 14:191–194, 1982.
24. B. Monien and E. Speckenmeyer. Solving satisfiability in less than 2^n steps. *Discrete Applied Mathematics*, 10:287–295, 1985.
25. D. A. Plaisted. The search efficiency of theorem proving strategies. In *Twelfth CADE*, volume 814 of *LNAI*, pages 57–71. Springer Verlag, 1994.
26. A. K. Smiley S. Fleisig, D. W. Loveland and D. L. Yarmush. An implementation of the model elimination proof procedure. *Journal of the ACM*, 21:124–139, 1974.
27. João P. Marques Silva. An overview of backtrack search satisfiability algorithms. In *Fifth International Symposium on Artificial Intelligence and Mathematics*, 1998.

Evaluating Search Heuristics and Optimization Techniques in Propositional Satisfiability

Enrico Giunchiglia, Massimo Maratea, Armando Tacchella, and
Davide Zambonin

DIST, Università di Genova, Viale Causa, 13 – 16145 Genova, Italy

Abstract. This paper is devoted to the experimental evaluation of several state-of-the-art search heuristics and optimization techniques in propositional satisfiability (SAT). The test set consists of random 3CNF formulas as well as real world instances from planning, scheduling, circuit analysis, bounded model checking, and security protocols. All the heuristics and techniques have been implemented in a new library for SAT, called SIM. The comparison is fair because in SIM the selected heuristics and techniques are realized on a common platform. The comparison is significative because SIM as a solver performs very well when compared to other state-of-the-art solvers.

1 Introduction

The problem of propositional satisfiability (SAT) is fundamental in many areas of computer science such as formal verification, planning and theorem proving. Despite the exponential worst-case complexity of all known algorithms for solving SAT [1], recent implementations of the Davis-Logemann-Loveland (DLL) algorithm are able to solve problems having thousands of propositions in a few seconds. Much of the success of these solvers is due to clever search heuristics and various optimization techniques that are implemented on top of the basic DLL algorithm. In general, it is difficult to assess the effectiveness of such heuristics and techniques because (*i*) the quality of the implementations varies greatly and, (*ii*) each solver is hard-coded with specific search strategies.

In this paper we present an experimental comparison of several state-of-the-art search heuristics and optimization techniques. The test set consists of random 3CNF formulas as well as real world instances from planning, scheduling, circuit analysis, bounded model checking, and security protocols. All the heuristics and techniques have been implemented in a new library for SAT, called SIM. The comparison is fair because in SIM the selected heuristics and techniques are realized on a common platform. Thus, our experimental evaluation is not biased by the differences due to the quality of the implementation, and provides a clear picture about the relative effectiveness of different search strategies. The comparison is significative because SIM itself performs very well when compared to other state-of-the-art solvers.

The paper is structured as follows. In Section 2 we characterize the test set that we have been using. We also experiment with various state-of-the-art SAT

R. Goré, A. Leitsch, and T. Nipkow (Eds.): IJCAR 2001, LNAI 2083, pp. 347–363, 2001.
© Springer-Verlag Berlin Heidelberg 2001

solvers on our test set with the purpose of showing that the chosen problems are indeed significative, and of getting an indication of the relative strength of the various SAT solvers. Section 3 is devoted to "basic SIM", i.e., to the presentation of SIM corresponding to the basic DLL algorithm. In Sections 4 and 5, we present the procedures implementing respectively backjumping and learning. The architecture, the branching heuristics, the backjumping and learning procedures are described in as many details as possible (given the space requirements). These detailed presentations are necessary in order to fully understand the procedures, but also the computational cost associated to each of them. As another added value of the paper, we describe all the procedures (except for the ones corresponding to the branching heuristics) by presenting the corresponding pseudo-code. We believe that our presentation is detailed enough to provide a good starting point for implementation. The presentation of the procedures in each Section is complemented with the experimental analysis, showing the relative efficiency of the different heuristics if implemented in basic SIM (Section 3), or combined with backjumping (Section 4) and learning (Section 5). In the final remarks (Section 6) we summarize the results obtained, and point out that SIM, used as a stand-alone solver, behaves very well.

Formal Preliminaries: We use the term *atom* as a shorthand for propositional letter. A *literal* is an atom or its negation. If l is a literal, (i) $|l|$ is the atom in l, and (ii) $\bar{l}$ is $\neg l$ if l is an atom, and is $|l|$ otherwise. A *clause* C is an n-ary disjunction of literals such that, for each pair of literals l, l' in C it is not the case that $|l|=|l'|$. A clause is *Horn* if it contains at most one atom not preceded by the negation symbol. A *formula* is an m-ary conjunction of clauses. As customary, we think of clauses as sets of literals, and formulas as sets of clauses. An *assignment* is a function mapping each atom into $\{T,F\}$. An assignment can be extended to formulas according to the standard truth tables of propositional logic. An assignment μ *satisfies* a formula φ if $\mu(\varphi) = T$. A formula φ is *satisfiable* if there exists an assignment which satisfies φ. The problem we are interested in is: "Given a formula, is it satisfiable?".

2   Experimenting with DLL Solvers

2.1   Designing the Test Set

Traditionally, SAT solvers are compared on instances corresponding to real world problems and/or randomly generated samples. It is well known that problems belonging to the two categories have different characteristics. Roughly speaking, real world instances present some "structure", corresponding somehow to the "structure" of the original problem coded in SAT. For example, in the case of a a SAT-instance corresponding to the verification problem of a circuit C, typically the SAT-instance consists of different parts, each part corresponding to a sub-circuit of C. Randomly generated tests became popular after [2], in which it is showed that, using the Fixed Clause Length model (FCL) [3], it is possible to generate very hard instances. Since then, they have been widely used to test SAT solvers' performances. However, they lack the structure of real world instances. Given their different characteristics, we decided to use both a test set consisting of instances corresponding to real world problems and a test set of randomly generated samples.

Our test set of real world problems consists of 200 instances, 97 satisfiable and 103 unsatisfiable, yielding an approximate 50% chance of a satisfiable (resp. unsatisfiable) instance. In selecting these instances, we wanted to have benchmarks (i) which are well known in the literature, and (ii) –if possible– that have already been used for comparing SAT-solvers. In particular, we chose:

- the "famous" 30 parity problems, see, e.g., [4];
- the 16 instances of the Bejing Competition held in 1996: 7 circuit equivalences, 6 scheduling problems, 3 planning problems;
- the 32 Data Encryption Standard (DES) problems, see [5,6];
- 34 formal equivalence verification problems, see [7];
- 31 instances of bounded model checking (BMC), see [8,6];
- 7 instances of formal verification properties of pipelined circuits from Velev's web page http://www.ece.cmu.edu/~mvelev;
- 11 quasigroup problems, see [9];
- 39 planning problems, including 2 hanoi* instances from SATLIB (www.satlib.org), and blocks world, logistics and rockets problems: most of these problems have been already used to test SAT solvers (see, e.g., [10]) and are available from SATLIB as well.

Notice that some of these instances are known to be very hard to solve for currently available SAT solvers. This is, e.g., the case for the 8 DES instances cnf-r4-*. We decided to have them on purpose, given that we plan to use this test set also for future experimental evaluations. All the 200 instances are available on SIM's web page.

For the random problems, we use the FCL model. Formulas are thus generated according to the number of propositions N, the number of literals per clause K, and the number of clauses L. Usually, K is fixed to 3 (see, e.g., [11]), N is fixed to a certain value, while L is varied in order to cover the "100% satisfiable – 100% unsatisfiable" transition. In our experiments, we considered $N = 300$. For $N = 300$, none of the solvers that we consider consistently exceeds the time limit, and the samples in the phase transition (i.e., for the value of L where the probability of a formula being satisfiable is 50%) are not trivial.

2.2 Experimental Results

There are many SAT solvers available. In our analysis, we restricted our attention to the DLL-based publicly available procedures. Among these, we considered some of the most effective. In particular: SATZ213, a new version of SATZ [11], POSIT ver 1.0 [12], EQSATZ [6], RELSAT ver. 2.00 [10], SATO ver 3.2 [13]. We do not describe these systems, for which we refer to the corresponding citation. Here, we only point out that these systems rely on very different data-structures and/or search heuristics and/or techniques. Indeed, some of them have been designed to solve specific classes of problems. This is the case, e.g., of EQSATZ, which has been designed to solve formulas with a lot of equivalences, like the parity problems.

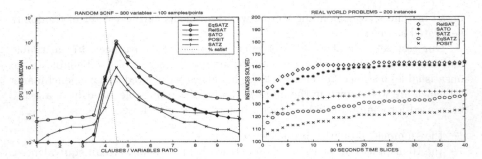

Fig. 1. POSIT, SATZ213, EQSATZ, RELSAT, and SATO. Left: random samples. Right: real world instances.

Of course, we decided to test these solvers (and not others) in a completely arbitrary way. However, the above solvers are state-of-the art, and present interesting features. SATZ and SATZ213 are well known for their dramatic performances on randomly generated tests. POSIT is interesting because of its differences from SATZ (and SATZ213). As reported in [11], the main difference is in the branching heuristic the two solvers use: both examine k propositions before choosing the one to assign, but while POSIT specifies an upper bound on k, SATZ does not. Almost quoting [11], SATZ examines many more propositions at each node than POSIT. Given the fundamental role played by the branching heuristic, we wanted to have a direct comparison between these two solvers on our tests. RELSAT and SATO are well known for their effectiveness on real world problems. EQSATZ is very recent, but has already showed impressive performances on problems with lots of equivalences [6].

Before discussing the experimental results, some more information. All the tests have been run on several identical Pentium III, 600MHz, 128MBRAM. The execution of a system on an instance (be it random or real world) is stopped after 1200s of CPU time. All the state-of-the-art solvers have been run in their default configuration on all the instances, with the only exception of SATO on randomly generated problems. As a matter of fact, SATO implements a form of "size learning" in which all the "reasons" of length ≤ 20 are indefinitely added to the set of input clauses. Given that learning does not help on randomly generated tests, and that RELSAT implements a form of "relevance learning" in which all the "reasons" of length ≤ 3 are indefinitely retained, we decided to change SATO's default when tested on random samples. In particular, we let SATO keep only the clauses with length ≤ 3, like RELSAT. Learning clauses of ≤ 3 on random problems does not slow down SATO in a significant way as it happens with its default value. In the case of SIM, options are changed in order to see the corresponding effects. For the plots showing systems' performances on random tests, the x-axis is the ratio L/N; the y-axis –in logarithmic scale– is the median CPU time on 100 samples per point for each solver. For the real-word tests, the y-axis is the number of instances solved by each solver within the CPU time specified on the x-axis.

Figure 1 shows the performances of POSIT, SATZ213, EQSATZ, RELSAT, and SATO on our test sets. As it can be seen, POSIT and SATZ213 are very effective on random tests, while RELSAT (with 164 instances solved) and SATO (with 163) are very effective on the real world ones. The bad performances of POSIT with respect to SATZ213 on the real world instances, point out that SATZ213's philosophy, i.e., exploring many propositions before choosing one, seems to pay off. Remarkably, EQSATZ is worse than SATZ213 on both random and real world tests. These data (with the exception of EQSATZ, about which not much is known yet) could have been expected. Indeed, POSIT and SATZ213 (resp. RELSAT and SATO) have been designed to be effective on random (resp. real world) problems. As we will see, the good performances of RELSAT and SATO on the real world instances are because of their learning mechanisms. About EQSATZ, its performances are somehow surprising: evidently, the data-structures and procedures that are at the basis of the big wins for problems with lots of equivalences (EQSATZ is the only system able to solve some of the `par32-*` parity instances in 1200s) have some overhead causing big losses for problems with (almost) no equivalences.

A more detailed data analysis shows that some instances which are solved by some solvers are not solved by the others, and the aother way around. We take this fact as an indication that our test set of real world problems is a good one. Indeed, the `cnf-r4-*` instances mentioned in section 2.1, are not solved by any solver.

3   Basic DLL in SIM

In this section we give a rather detailed description of the data structure, the search control, and the heuristics of SIM. These parts together, define a basic implementation of the DLL algorithm [14]. The conventions we use to present the data types and algorithms are those of [15], described at pages. 4,5. In particular, array elements are accessed by specifying the array name followed by the index in square brackets: for example, $A[i]$ indexes the i-th element of the array A. Compound data are organized into *objects* which are comprised of several *attributes* (or *fields*). A particular field is accessed using the field name followed by an instance of its object in square brackets: for example, $f[x]$ accesses the field f in the object x. Instances, i.e., variables representing arrays or objects, are treated as pointers to the data representing the arrays or objects. If a pointer does not refer to any object, we give it the special value NIL . Finally, stacks are considered a primitive data type and are accessed with the usual primitives PUSH, POP, TOP, etc.. We also assume to have a primitive FLUSH which flushes the stack.

3.1   Data Structures and Basic Primitives of SIM

We now introduce *propositions*, *literals*, *clauses*, and *states* by defining the corresponding data types. Assuming that these data types have been already defined (as we do below), a *proposition* for us is an instance of the proposition data type, and analogously for the others. In the following, we assume that, T, F, U, UN, LS, and RS are six pairwise distinct constants, each one being distinct from NIL . A *proposition data type* is comprised of the following attributes (in the following, p is a proposition):

– *value* is either T, F, or U: intuitively, $value[p]$ is the value assigned to p,

- *mode* is either UN, LS, or RS: intuitively, *mode*[*p*] is UN (resp. LS, resp. RS) if *p* is assigned by unit-propagation (resp. left split, resp. right split),
- *Pos* and *Neg* are arrays of clauses: intuitively, they are the clauses in which *p* occurs positively and negatively.

For each proposition *p*, we say that *p* is *open* if *value*[*p*] = U and *valued* otherwise. A *literal data type* has the attributes:

- *prop* is the proposition *associated to* (or *corresponding to*) the literal;
- *v* is the *sign* of the literal, either T or F.

For each literal *l*, we say that *l* is *open* or *valued* if the corresponding proposition *prop*[*l*] is open or valued, respectively; *l* is *positive* (resp. *negative*) if *v*[*l*] = T (resp. *v*[*l*] = F). A *clause data type* is comprised of the attributes:

- *open* is a non negative integer, representing the number of open literals in the clause;
- *sub* is the clause *subsumer* represented by a proposition;
- *Lits* is an array of literals.

For each clause *cl*, we say that *cl* is (*i*) empty if *open*[*cl*] = 0; (*ii*) unary (resp. binary) if *open*[*cl*] = 1 (resp. *open*[*cl*] = 2); (*iii*) open if *sub*[*cl*] = NIL , and valued (or *subsumed*) otherwise. A proposition *p* occurs in a clause *cl* if *p* = *prop*[*l*] for some literal *l* in *Lits*[*cl*]. The proposition occurs *positively* if *l* is positive and *negatively* otherwise. Finally, a *state data type* is comprised of the attributes:

- *Props*, an array of propositions;
- *Clauses*, an array of clauses;
- *Stack* and *Unit* are the *search* stack of propositions and the *unit propagation* stack of clauses, respectively;
- *open* is a non negative integer, representing the number of open clauses.

Each state *s*, in order to represent a valid state of the computation, must satisfy the following properties:

1. *Props*[*s*] stores precisely the propositions that occur in the clauses of *Clauses*[*s*];
2. for each proposition *p* in *Props*[*s*], *Pos*[*p*] stores precisely the clauses contained in *Clauses*[*s*] where *p* occurs positively, and *Neg*[*p*] stores precisely the clauses contained in *Clauses*[*s*] where *p* occurs negatively;
3. *open*[*s*] amounts to the number of open clauses in *Clauses*[*s*];
4. *Stack*[*s*] stores precisely the valued propositions in *Props*[*s*];
5. *Unit*[*s*] contains all the open unary clauses in *Clauses*[*s*].

If *s* represents an initial state of the computation, then for *s* it also true that:

6. *Clauses*[*s*] is not empty; each clause *cl* in *Clauses*[*s*] is open; its number of open literals is > 0 and is precisely the length of *Lits*[*cl*];
7. each proposition in *Props*[*s*] is open.

Intuitively, it is easy to see that propositions, literals and clauses data types are faithful representations of the corresponding propositional logic objects as defined in Section 1. The arrays *Clauses*[*s*] and *Props*[*s*], together with their contents of propositions and clauses, are a faithful representation of formulas. Notice that we explicitly disallow empty input formulas, i.e. {}, and input formulas containing empty clauses, e.g. {{}}, since the satisfiability problem for these formulas can be simply solved by inspection.

About the primitives of our propositional data structures, let *s* be a state, *p* a proposition in *Props*[*s*], *v* a *boolean* value (i.e., T or F) and *m* a mode (i.e., one of UN, LS, RS). EXTEND-PROP(*s*, *p*, *v*, *m*) is the function that extends the valuation *v* with mode *m* to the whole state *s* for the proposition *p*: it returns F if an empty clause is found in *Clauses*[*s*], and U otherwise. More precisely, EXTEND-PROP(*s*, *p*, *v*, *m*) does the following:

- set *value*[*p*] to *v* and *mode*[*p*] to *m*; push *p* in *Stack*[*s*];
- for each clause *cl* in *Pos*[*p*] (resp. *Neg*[*p*]), if *sub*[*cl*] = NIL and *v*=T (resp. *v*=F) then set *sub*[*cl*] = *p* and decrement *open*[*s*] (*unit subsumption*);
- for each clause *cl* in *Neg*[*p*] (resp. *Pos*[*p*]), if *sub*[*cl*] = NIL and *v*=T (resp. *v*=F) then decrement *open*[*cl*] (*unit resolution*). If the clause becomes unary, push it in *Unit*[*s*];
- return F if any clause became empty in the above step, and U otherwise.

Notice the use of *Unit*[*s*] to detect unary clauses, and of *open*[*s*] to keep track of subsumed clauses: both operations come for free as a result of performing EXTEND-PROP, while their corresponding simple-minded implementations are linear in the size of *Clauses*[*s*]. Since EXTEND-PROP destructively updates the state, we have a primitive, RETRACT-PROP, to revert the effects of EXTEND-PROP. For lack of space, we do not describe RETRACT-PROP here, but SIM features both primitives, thus avoiding the burden of copying and saving the state each time EXTEND-PROP is performed.

3.2   DLL-Solve, Look-Ahead, and Look-Back in SIM

In Figure 2 we present our implementation of DLL algorithm. In the figure, CHECK-SAT(*s*) returns T if *open*[*s*] = 0, and F otherwise; *s* in DLL-SOLVE has to be the initial state corresponding to the input formula. DLL-SOLVE returns T if the input formula is satisfiable and F otherwise.

3.3   Heuristic in SIM

A heuristic is a function that given an open state *s* and an open proposition *p* in *s*, computes a score for *p* and a branching order determining whether *p* should be satisfied or falsified first. The optimal heuristic is the one which chooses an open proposition and a branching order that will cause the exploration of less nodes. Unfortunately, deciding whether a proposition *p* is optimal is harder than deciding the satisfiability of the formula itself [16], so we resort to approximations: a good approximate heuristic does not require a lot of computation time and gives higher scores to the propositions that, once assigned, lead to simpler problems.

354   E. Giunchiglia et al.

```
LOOK-AHEAD(s)                              LOOK-BACK(s, var v, var m)
 1 r ← U                                    1 FLUSH(Unit[s])
 2 while r = U and                          2 repeat
        length[Unit[s]] > 0 do              3    p ← TOP(Stack[s])
 3    cl ← POP(Unit[s])                      4    if mode[p] ≠ LS then
 4    if sub[cl] = NIL then                  5       RETRACTPROP(s, p)
 5       i ← 1                               6 until length[Stack[s]] = 0 or
 6       repeat                                      mode[p] = LS
 7          l = Lits[cl][i]                  7 if length[Stack[s]] > 0 then
 8          i ← i + 1                        8    if value[p] = T then
 9       until value[prop[l]] = U            9       v ← F
10       r ← EXTEND-PROP(s, prop[l], v[l], UN)  10   else
11 if r = U and open[s] = 0 then            11       v ← T
12    return T                              12    m ← RS
13 else                                     13    RETRACT-PROP(s, p)
14    return r                              14 else
                                            15    p ← NIL
DLL-SOLVE(s)                                16 return p
 1 repeat
 2    case LOOK-AHEAD(s) of
 3       T : return T
 4       F : p ← LOOK-BACK(s, v, m)
 5       U : p ← HEURISTIC(s, v, m)
 6    if p ≠ NIL then
 7       EXTEND-PROP(s, p, v, m)
 8 until p = NIL
 9 return CHECK-SAT(s)
```

Fig. 2. Implementation of the DLL algorithm in SIM.

We now describe in details the branching heuristics that we have tested. Our interest lies in evaluating some of the search heuristics that have been proposed. We restrict our attention to RELSAT's, SATO's, and SATZ's default branching heuristics, and we compare them with a new heuristic (called UNITIE2) that pushes "SATZ's philosophy" to the limit: it always examines every open proposition at each branching node. The heuristics of RELSAT, SATO and SATZ have been implemented by looking at the available literature, but in particular by carefully inspecting the solvers' code. SATZ's heuristics is described in detail in [11].[1] An high-level description of SATO heuristic is given in [13]. Here we give a much more detailed presentation of it, as we understood it from the code and as we implemented it in SIM. We are not aware of any detailed description of RELSAT's heuristic, so we present it according to its implementation in RELSAT and SIM.

We start describing SATO's heuristic. Then, before discussing RELSAT's heuristic, we present the simpler UNITIE2, and reuse parts of its description also for RELSAT.

SATO's heuristic: before the search begins, one of the following options is selected:

---

[1] Notice however that the heuristic in SATZ213 has changed.

1. if the percentage of non horn clauses in the input clauses is bigger than 28.54 of the total, then the heuristic scores all the open propositions,
2. otherwise, the heuristic scores only the first 7 propositions occurring in the first 7 shortest non horn clauses.

The score of a proposition p is given by $(pos\text{-}p+1) \times (neg\text{-}p+1)$, where $pos\text{-}p$ is the number of open binary clauses in which p occurs, and analogously for $neg\text{-}p$. The branching is on the proposition with the highest score, its value determined in the following way: if the percentage of non-horn input clauses is less (resp. greater) than 2.36 of the total, then the value of p is T (resp. F) if $pos\text{-}w > neg\text{-}w$ and F (resp. T) otherwise.

UNITIE2 **heuristic**: For each open proposition p, this heuristic scores p on the basis of the effective simplifications that would be caused by assigning p on one side, and then $\neg p$ on the other. To compute the simplifications caused by assigning p, a "lean" version of EXTEND-PROP and LOOK-AHEAD is used. In particular the lean versions take care of unit resolutions only. UNITIE2 also incorporates a failed literal detection mechanism: if assigning p would cause a contradiction (in the LOOK-AHEAD), it safely assigns $\neg p$ as if it was a unit (analogously for $\neg p$). Assuming that there are no failed literals, and assuming p (resp. $\neg p$) leads to a set of clauses C^+ (resp. C^-), the score of p is given by $2 \times binposnew \times binnegnew + unitpos + unitneg$ where $binposnew$ (resp. $binnegnew$) is the number of binary clauses in C^+ (resp. C^-) and not in C, and $unitpos$ (resp. $unitneg$) is the number of unit propagations in the LOOK-AHEAD after p (resp. $\neg p$) is assigned. Notice that UNITIE2 always scores all the open propositions. SATZ [11], under certain conditions, scores only a subset of the open propositions. As we said, UNITIE2 pushes SATZ's philosophy to the limit. However, UNITIE2 uses a slightly different formula for computing the score of propositions. In fact, SATZ's formula (in our notation) is $1024 \times binposnew \times binnegnew + binposnew + binnegnew + 1$. The idea behind UNITIE2 is to use $unitpos + unitneg$ to break ties between propositions having the same $2 \times binposnew \times binnegnew$.

RELSAT's **heuristic**: Also RELSAT scores an open proposition p on the basis of the effective simplifications that would be caused by assigning p on one side, and then $\neg p$ on the other. However, it does this differently from UNITIE2. For each open proposition p, RELSAT first checks whether (*i*) the number of binary clauses in which p occurs positively is augmented from the previous branching node, and (*ii*) p has not already been assigned in the heuristic (i.e., if p has not been assigned by unit propagation while scoring another proposition). If this is the case, then RELSAT assigns $\neg p$ and does the consequent (lean) LOOK-AHEAD. If $\neg p$ fails, then $\neg p$ is immediately assigned causing the contradiction and the immediate backtrack.[2] If $\neg p$ does not fail, the number of unit propagations performed in the LOOK-AHEAD is counted as the score of p. If one of the conditions (*i*), (*ii*) is false, and if p occurs positively in some binary clauses, then the score $pos\text{-}p$ of p is the number of binary clauses in which p occurs positively. All the above, has to be analogously done for computing $neg\text{-}p$ corresponding to $\neg p$. The above rules produce some non-null $pos\text{-}p$ or $neg\text{-}p$ if there are binary clauses. If there are not,

---

[2] The contradiction arising by assigning $\neg p$ is explicitly generated in order to have the necessary interactions with backjumping and learning, see next Section.

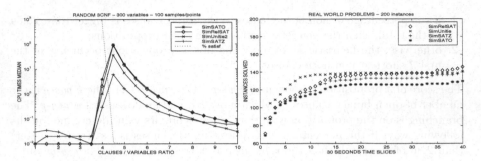

Fig. 3. Basic **SIM**. Left: random samples. Right: real world instances.

then RELSAT applies another strategy: for each proposition p, $pos\text{-}p$ is given by $256 \times cl_3 + 16 \times cl_4 + 4 \times cl_5 + cl_6$ where cl_i ($3 \leq i \leq 5$) is the number of clauses in which the number of open literals is i, and cl_6 is the number of clauses in which the number of open literals is greater or equal to 6. Analogously for $neg\text{-}p$. At this point, for some proposition p we have two scores $pos\text{-}p$, $neg\text{-}p$ one of which is distinct from 0. For all these propositions p, we compute a single score given by $2 \times pos\text{-}p \times neg\text{-}p + pos\text{-}p + neg\text{-}p + 1$. Finally, RELSAT chooses randomly among the propositions whose last computed score is ≥ 0.9 of the best score, and fixes its value to T or F randomly.

3.4   Experimental Comparison with SIM

We now look at the experimental analysis. Figure 3 shows SIM's performances when using the different heuristics. In the figure, SIMSATO means SIM using SATO's heuristic, and similarly for the others. Consider Figure 3. Our plots on the random problems are similar to the ones in Section 2: SIMSATZ performs better than SIMRELSAT and SIMSATO, these last two systems performing roughly in the same way. The good performances of SIMSATZ on random problems is highlighted by Table 1, showing the minimum, median, IQ-Range and maximum of the CPU time of the solvers at the phase transition.[3] Besides having the lowest median CPU time, SIMSATZ is also the system whose performances have less variations.

From the plots on the real world problems, we see that SIMSATO (with 130 problems solved) is much worse than SIMSATZ (140), SIMUNITIE2 (141) and, above all, SIMRELSAT (146). On the other hand, SIMSATO is the fastest on the 11 quasi-group problems: this is not surprising given that this heuristic has been tuned on these problems [13]. SIMUNITIE2 gives reasonable timings on the two test sets.

---

[3] Out of 100 numbers listed in ascending order, the Q-percentile is the Q-th in the list. The minimum is thus the 1-percentile, the median the 50-percentile, the maximum the 100-percentile, and the IQ-range is the 75-percentile minus the 25-percentile.

Table 1. Basic SIM. Random tests. Performances at the phase transition.

Data	SIMSATO	SIMRELSAT	SIMUNITIE2	SIMSATZ
Min	31.56	29.24	7.09	1.53
Median	97.18	89.95	36.27	6.00
IQ-range	70.33	66.38	25.34	3.60
Max	256.45	348.16	121.43	14.96

4   Conflict-Directed Backjumping in SIM

Since the basic DLL algorithm relies on simple chronological back-tracking, it is not infrequent for DLL implementations to keep exploring a possibly large subtree whose leaves are all dead-ends. This phenomenon occurs also when the formula is satisfiable, but some choice performed way up in the search tree is responsible for the constraints to be violated. The solution, borrowed from constraint network solving [17], is to jump back over the choices that were not at the root of the conflict, whenever one is found. The corresponding technique is widely known as (conflict-directed) backjumping (CBJ).

4.1   CBJ-Look-Back in SIM

The function CBJ-LOOK-BACK in Figure 4.left, implements conflict-directed backjumping in SIM. In the figure,

- *wr* is meant to store the subset of the propositions in $Stack[s]$ whose assignment is a reason for the discovered inconsistencies. Technically, *wr* is a clause initialized by INIT-WR(s) which returns a clause cl in $Clauses[s]$ having $open[cl] = 0$. At least one such clause exists, given that when CBJ-LOOK-BACK is invoked, at least one empty clause belongs to $Clauses[s]$.
- *reason* is a new attribute of the proposition data type and it is a clause.
- UPDATE-WR(p, wr) returns a clause cl such that:
  - $open[cl] = 0$;
  - $sub[cl] =$ NIL ;
  - $Lits[cl]$ is the array of literals obtained by merging (without duplications) $reason[p]$ and $Lits[wr]$. Then the literals $l \in Lits[cl]$ having $prop[l] = p$ are eliminated from $Lits[cl]$.

Indeed, the procedure LOOK-AHEAD in Figure 2 needs to be extended in order to set the reason of p when p is assigned by unit. The following line of code has to be written below line 9 of LOOK-AHEAD, at the same level of indentation of line 9:

$$reason[prop[l]] \leftarrow cl$$

For an high-level description of CBJ-LOOK-BACK, see [10].

CBJ-LOOK-BACK(s, **var** v, **var** m)	CBJ-LEARN-LOOK-BACK(s, **var** v, **var** m)
1 FLUSH($Unit[s]$)	1 FLUSH($Unit[s]$)
2 $wr \leftarrow$ INIT-WR(s)	2 $wr \leftarrow$ INIT-WR(s)
3 **repeat**	3 **repeat**
4 $p =$ TOP($Stack[s]$)	4 $p =$ TOP($Stack[s]$)
5 **if** $p \in wr$ **then**	5 **if** $p \in wr$ **then**
6 **if** $mode[p] \in \{$UN, RS$\}$ **then**	6 **if** $mode[p] \in \{$UN, RS$\}$ **then**
7 $wr \leftarrow$ UPDATE-WR(p, wr)	7 $wr \leftarrow$ UPDATE-WR(p, wr)
8 RETRACT-PROP(s, p)	8 LEARN(s, wr)
9 **else**	9 FORGET(s, p)
10 **if** $value[p] =$ T **then**	10 RETRACT-PROP(s, p)
11 $v \leftarrow$ F	11 **else**
12 **else**	12 **if** $value[p] =$ T **then**
13 $v \leftarrow$ T	13 $v \leftarrow$ F
14 $m \leftarrow$ RS	14 **else**
15 RETRACT-PROP(s, p)	15 $v \leftarrow$ T
16 $reason[p] \leftarrow wr$	16 $m \leftarrow$ RS
17 **return** p	17 RETRACT-PROP(s, p)
18 **else**	18 **case** LOOK-AHEAD(s) **of**
19 RETRACT-PROP(s, p)	19 T: **return** NIL
20 **until** $length[Stack[s]] = 0$	20 F: FLUSH($Unit[s]$)
21 **return** NIL	21 $wr \leftarrow$ INIT-WR(s)
	22 U: **if** $p =$ U **then**
	23 $reason[p] \leftarrow wr$
	24 **return** p
	25 **else**
	26 **return** HEURISTIC(s, v, m)
	27 **else**
	28 RETRACT-PROP(s, p)
	29 **until** $length[Stack[s]] = 0$
	30 **return** NIL

Fig. 4. Conflict-directed backjumping and learning in **SIM**.

4.2   Experimental Comparison with SIM

Figure 5 shows SIM's results when incorporating backjumping. As it can be observed from the figure on the left, backjumping does not produce benefits on the random tests. Indeed, on the random problems the reason of an inconsistency is almost never localized to a small subset of the input clauses. It is nevertheless interesting to have a look at Table 2, corresponding to Table 1 in Section 3. By comparing the Min of the two tables, we see that for SIMSATO and SIMRELSAT, i.e., the two systems whose initial branching literals are selected using a heuristic not based on LOOK-AHEAD, backjumping may produce some benefits. The fact that backjumping never helps for SIMUNITIE2 and SIMSATZ, are an indication that for random instances, LOOK-AHEAD-based heuristics are a good choice. Notice however that the overhead due to backjumping is limited: comparing the Max, we see that it is roughly in the order of the 20% or less. We take this as an indication that our implementation of backjumping is good.

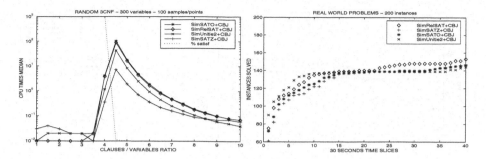

Fig. 5. SIM with backjumping. Left: random samples. Right: real world instances.

Table 2. SIM with backjumping. Random tests. Performances at the phase transition.

Data	SIMSATO	SIMRELSAT	SIMUNITIE2	SIMSATZ
Min	32.95	8.44	8.35	1.86
Median	102.95	88.61	42.59	7.23
IQ-range	73.32	75.42	30.61	4.42
Max	278.29	395.23	143.35	18.15

Looking at the results on the real world tests (Figure 5), we see that REL-SAT heuristic (153 instances solved) produces still some advantages, and that the differences with SATO heuristic (146 solved) is not as big as before. SIMSATZ and SIMUNITIE2 solve respectively 147 and 144 instances. SIMUNITIE2 has some advantage at the beginning, but at the end it looses it. Comparing the data with the corresponding ones in Section 3, we see that for each heuristic that we used, backjumping allows to solve more instances.

5 Learning in SIM

For backjumping can be very effective in "shaking" the solver from regions where no solutions can be found, but since the cause of the conflict is discarded as soon as it gets mended, the solver may get repeatedly stuck in such regions. To escape this pattern, some sort of global knowledge is needed: the negation of the causes of the conflicts may be added as an additional constraint (i.e., as a clause) that has to be satisfied. Adding all the clauses corresponding to the reasons of the discovered conflicts has the advantage that the same mistake is never repeated. However, this may cause an exponential blow up of the size of the formula. It is therefore necessary to introduce some limit to the number of stored clauses, i.e., a mechanism that enables the solver to "forget" clauses.

In *size learning*, a clause corresponding to a conflict is stored if it has no more than *Order* literals. Size learning is implemented in SATO, and is used by default with *Order*=20. In *relevance learning*, a clause corresponding to a conflict is always stored. A "forget" mechanism eliminates the learned clauses in which more than

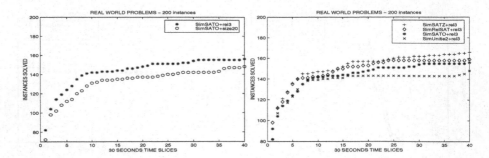

Fig. 6. SIM with learning. Real world instances. Left: size vs relevance learning. Right: different heuristics.

Order literals are open or have been assigned differently since the time they have been added to the set of input clauses. Relevance learning is implemented by RELSAT, and is used by default with *Order*=3.

5.1   CBJ-Learn-Look-Back in SIM

The procedure CBJ-LEARN-LOOK-BACK in Figure 4.right, implements learning in SIM. In the figure, *Order* is yet another attribute of a state (set initially by an input parameter), while the calls LEARN(s, wr) and FORGET(s, p) have different meaning depending on whether they do "relevance learning" as RELSAT, or "size learning" as SATO.

size learning: If wr is a clause with less than *Order*[s] literals, LEARN(s, wr) adds the clause wr to *Clauses*[s], and for each literal l in *Lits*[wr], adds wr to *Pos*[*prop*[l]] (resp. *Neg*[*prop*[l]]) if l is positive (resp. negative). FORGET(s, p) is the null instruction.

relevance learning: LEARN(s, wr) performs the above operations in any case, i.e., also if wr is a clause with more than *Order*[s] literals. FORGET(s, p) deletes the learned clauses cl such that $|\{l : l \in Lits[cl], value[prop[l]] \in \{v[l], \text{U}\}\}| > Order[s]$.

Notice the call to LOOK-AHEAD (line 18) and the following piece of code. Indeed, this call is necessary because of the possible presence of unary clauses among the ones that we learned.

5.2   Experimental Comparison with SIM

For lack of space we do not show the plots on the random instances. They have the same qualitative behavior of the ones we already showed, and are available at SIM's web site. Here we only remark that learning does not produce benefits on random instances, but its overhead is always quite small.

On the real world instances, [10] reports that relevance learning is more effective than size learning. Here we show only one plot (Figure 6.left) in which we compare SIMSATO with relevance learning *Order*=3, and SIMSATO with size

Table 3. Solvers performances on random and real world tests.

Random	POSIT	EQSATZ	SATZ213	SATO	RELSAT	SIM SATZ	SIM RELSAT	SIM SATO	SIM UNITIE2
Min	3.36	40.90	1.77	10.68	30.59	1.53	29.24	31.56	7.09
Median	9.88	112.52	4.26	81.31	88.15	6.00	89.95	97.18	36.25
IQ-range	7.39	73.76	2.46	60.74	63.99	3.60	66.38	70.33	25.34
Max	39.76	323.75	10.91	284.99	344.77	14.96	348.16	256.45	121.43
Real World #Solved	126	137	141	163	164	166	159	156	148

learning *Order*=20 as in SATO. As it can be seen, the plot with relevance learning dominates the one with size learning.

On the basis of the above, we tested the various heuristics only with relevance learning *Order*=3. The results are shown in Figure 6.right. We see that the heuristic which allows SIM to solve more instances is the one featured by SATZ: SIMSATZ solves 166 instances, SIMRELSAT 159, SIMSATO 156, SIMUNITIE2 148. Interestingly, we have seen that UNITIE2 heuristic –on the real world tests– was second only to RELSAT's heuristic when used in basic SIM. However, if we include backjumping and/or learning, we see that its performances gets worse in comparison with the other heuristics. Analogously, SIMSATZ's performances becomes better as soon as we introduce backjumping and/or learning. These are clear indications of the obvious fact that the benefits of a heuristics depend on its interactions with the other implemented techniques (see also [18]).

6   Final Remarks

We conclude the paper with Table 3, comparing SIM with the various settings, and the solvers that we considered in Section 2. From this table, we see that SIMSATZ solves more real world instances than all the other solvers, including SIM with other heuristics. This last fact shows that the good performances of RELSAT and SATO on real world instances (see Section 2) are due to their learning mechanisms. As we said, SIMSATZ is not always the best on all the classes of problems: different heuristics behave better than others on different problems. Tuning the heuristic on specific classes pays off. SIMSATZ is also the best, among the SIM-systems, for solving random problems, and its performances are not far away from SATZ213's performances. Having one good settings for both random and real world instances is indeed very positive.

The good performances of SIMSATZ also indicate that the quality of our implementation is comparable to the corresponding state-of-the-art solvers. Indeed, if we compare the performances of SIMSATZ on real world instances as shown in Figure 6.right, with RELSAT's and SATO's corresponding performances (see Figure 1.right), we see that both RELSAT and SATO are able to solve more instances than SIMSATZ in the first 700 seconds. Our explanation is that RELSAT and SATO

feature rather sophisticated forms of pre-processing of the input clauses, while
SIM has no pre-processing at all. Indeed, pre-processing allows for smaller solution
times, and we are working on its implementation.

Finally, we showed only a small part of the data that we have collected. For
example, we have a figure comparing SIMRELSAT and RELSAT on the random tests:
the plots overlaps and are distinguishable with difficulty. Analogously for SIMSATO
and SATO.[4] These plots show that our reconstruction of their branching heuristics
is accurate.

All the plots, SIM and more information about SIM are available at SIM web
page:

www.mrg.dist.unige.it/star/sim

Acknowledgements. We would like to thank Holger H. Hoos and Thomas
Stützle for the excellent work they do with SATLIB. This work is partially supported by MURST and Intel Corp.

References

1. J. Gu, P. W. Purdom, J. Franco, and B. W. Wah. Algorithms for the satisfiability (sat) problem: A survey. In *Satisfiability Problem: Theory and Applications*,
   DIMACS Series in Discrete Mathematics and Theoretical Computer Science, pages
   19–153. AMS, 1997.
2. D. G. Mitchell, B. Selman, and H. J. Levesque. Hard and Easy Distributions for
   SAT Problems. In *Proc. of AAAI*, pages 459–465. AAAI Press, 1992.
3. J. Franco and M. Paull. Probabilistic analysis of the Davis-Putnam procedure for
   solving the satisfiability problem. *Discrete Applied Mathematics*, 5:77–87, 1983.
4. Bart Selman, Henry Kautz, and David McAllester. Ten Challenges in Propositional
   Reasoning and Search. In *Proc. of IJCAI*, pages 50–54. Morgan-Kauffmann, 1997.
5. Massacci and Marraro. Logical Cryptanalysis as a SAT Problem. *JAR: Journal of
   Automated Reasoning*, 24, 2000.
6. Chu Min Li. Integrating Equivalency Reasoning into Davis-Putnam Procedure. In
   Proc. of AAAI. AAAI Press, 2000.
7. T. E. Uribe and M. E. Stickel. Ordered Binary Decision Diagrams and the Davis-
   Putnam Procedure. In *Proc. of the 1st International Conference on Constraints in
   Computational Logics*, 1994.
8. A. Biere, A. Cimatti, E. Clarke, and Y. Zhu. Symbolic Model Checking without
   BDDs. In *Proceedings of TACAS*, volume 1579 of *LNCS*, pages 193–207. Springer
   Verlag, 1999.
9. H. Zhang and M. E. Stickel. Implementing the Davis-Putnam Method. In *Highlights
   of Satisfiability Research in the Year 2000*. IOS Press, 2000.

---

[4] This holds for the median time. Indeed, by looking at Table 3, we see that SIMSATO's
and SATO's performances are not as similar as SIMRELSAT's and RELSAT's performances
are. This is due to the particular heuristic of SATO, which in some cases chooses the
first 7 propositions in the first 7 non Horn clauses. Indeed, the selected proposition
(and thus the behavior) depends also on the order in which clauses are stored, and
–in this– SIMSATO is different from SATO.

10. R. J. Bayardo, Jr. and R. C. Schrag. Using CSP Look-Back Techniques to Solve Real-World SAT instances. In *Proc. of AAAI*, pages 203–208. AAAI Press, 1997.
11. Chu Min Li and Anbulagan. Heuristics Based on Unit Propagation for Satisfiability Problems. In *Proc. of IJCAI*, pages 366–371. Morgan-Kauffmann, 1997.
12. Jon W. Freeman. *Improvements to propositional satisfiability search algorithms.* PhD thesis, University of Pennsylvania, 1995.
13. H. Zhang. SATO: An efficient propositional prover. In *Proc. of CADE*, volume 1249 of *LNAI*, pages 272–275. Springer Verlag, 1997.
14. M. Davis, G. Logemann, and D. Loveland. A machine program for theorem proving. *Journal of the ACM*, 5(7):394–397, 1962.
15. T. H. Cormen, C. E. Leiserson, and R. R. Rivest. *Introduction to Algorithms.* MIT Press, 1998.
16. P. Liberatore. On the complexity of choosing the branching literal in DPLL. *Artificial Intelligence*, 116(1–2):315–326, 2000.
17. R. Dechter, I. Meiri, and J. Pearl. Temporal Constraint Networks. *Artificial Intelligence*, 49:61–95, 1991.
18. F. Copty, L. Fix, E. Giunchiglia, G. Kamhi, A. Tacchella, and M. Vardi. Benefits of Bounded Model Checking at an Industrial Setting. In *Proc. of CAV*, LNCS. Springer Verlag, 2001. To appear.

QuBE: A System for Deciding Quantified Boolean Formulas Satisfiability*

Enrico Giunchiglia, Massimo Narizzano, and Armando Tacchella

DIST, Università di Genova, Viale Causa 13, 16145 Genova – Italy

1   Introduction

Deciding the satisfiability of a Quantified Boolean Formula (QBF) is an important research issue in Artificial Intelligence. Many reasoning tasks involving planning [1], abduction, reasoning about knowledge, non monotonic reasoning [2], can be directly mapped into the problem of deciding the satisfiability of a QBF.

In this paper we present QuBE, a system for deciding QBFs satisfiability. We start our presentation in § 2 with some terminology and definitions necessary for the rest of the paper. In § 3 we present a high level description of QuBE's basic algorithm. QuBE's available options are described in § 4. We end our presentation in § 5 with some experimental results showing QuBE effectiveness in comparison with other systems. QuBE, and more information about QuBE, are available at `www.mrg.dist.unige.it/star/qube`.

2   Formal Preliminaries

Consider a set P of propositional letters. An *atom* is an element of P. A *literal* is an atom or the negation of an atom. For each literal l, (i) $\bar{l}$ is x if $l = \neg x$, and is $\neg x$ if $l = x$; (ii) $|l|$ is the atom occurring in l. A *clause* C is an n-ary $(n \geq 0)$ disjunction of literals such that no atom occurs twice in C. A *propositional formula* is a k-ary $(k \geq 0)$ conjunction of clauses. As customary, we represent a clause as a set of literals, and a propositional formula as a set of clauses.

A *QBF* is an expression of the form

$$Q_1 x_1 \ldots Q_n x_n \Phi, \qquad (n \geq 0) \tag{1}$$

where every Q_i $(1 \leq i \leq n)$ is a quantifier, either existential $\exists$ or universal $\forall$; $x_1, \ldots, x_n$ are pairwise distinct atoms in P; and Φ is a propositional formula in the atoms $x_1, \ldots, x_n$. $Q_1 x_1 \ldots Q_n x_n$ is the *prefix* and Φ is the (quantifier-free) *matrix* of (1).

Consider a QBF of the form (1). A literal l occurring in Φ is:

---

* We wish to thank Marco Cadoli, Rainer Feldmann, Theodor Lettman, Jussi Rintanen, Marco Schaerf and Stefan Schamberger for providing us with their systems and helping us to figure them out during our experimental analisys. This work has been partially supported by ASI and MURST.

R. Goré, A. Leitsch, and T. Nipkow (Eds.): IJCAR 2001, LNAI 2083, pp. 364–369, 2001.
© Springer-Verlag Berlin Heidelberg 2001

```
1  φ = ⟨the input QBF⟩;
2  Stack= ⟨the empty stack⟩;

3  function Simplify() {
4  do {
5    φ' = φ;
6    if (⟨a contradictory clause is in φ⟩)
7      return FALSE;
8    if (⟨the matrix of φ is empty⟩)
9      return TRUE;
10   if (⟨l is unit in φ⟩)
11     { |l|.mode = UNIT; Extend(l); }
12   if (⟨l is monotone in φ⟩)
13     { |l|.mode = PURE; Extend(l); }
14 } while (φ' != φ);
15 return UNDEF; }

16 function Backtrack(res) {
17 while (⟨Stack is not empty⟩) {
18   l = Retract();
19   if (((|l|.mode == L-SPLIT) &&
20     ((res==FALSE && |l|.type ==∃) ||
21     (res==TRUE && |l|.type ==∀)))
22     {|l|.mode = R-SPLIT; return l; }}
23 return NULL; }

24 function QubeSolver() {
25 do {
26   res = Simplify();
27   if (res == UNDEF) l = ChooseLiteral();
28   else   l = Backtrack(res);
29   if (l != NULL)  Extend(l);
30 } while (l != NULL);
31 return res; }
```

Fig. 1. The algorithm of QuBE.

- *existential* if ∃|l| belongs to the prefix of (1), and is *universal* otherwise.
- *unit* in (1) if l is existential, and, for some $k \geq 0$,
  - a clause $\{l, l_1, \ldots, l_k\}$ belongs to Φ, and
  - each expression $\forall|l_i|$ $(1 \leq i \leq k)$ is at the right of $\exists|l|$ in the prefix of (1).
- *monotone* if either l is existential, l occurs in Φ, and $\bar{l}$ does not occur in Φ; or l is universal, l does not occur in Φ, and $\bar{l}$ occurs in Φ.

A clause C is *contradictory* if no existential literal belongs to C.

The semantics of a QBF φ can be defined recursively as follows. If φ contains a contradictory clause then φ is not satisfiable. If the matrix of φ is empty then φ is satisfiable. If φ is $\exists x \psi$ (resp. $\forall x \psi$), φ is satisfiable if and only if φ_x or (resp. and) $\varphi_{\neg x}$ are satisfiable. If $\varphi = Qx\psi$ is a QBF and l is a literal, φ_l is the QBF obtained from ψ by deleting the clauses in which l occurs, and removing $\bar{l}$ from the others. It is easy to see that if φ is a QBF without universal quantifiers, the problem of deciding φ satisfiability reduces to propositional satisfiability (SAT).

Notice that we allow only for propositional formulas in conjunctive normal form (CNF) as matrices of QBFs. Indeed, by applying standard CNF transformations (see, e.g., [3]) it is always possible to rewrite a QBF into an equisatisfiable one satisfying our restrictions.

3   QuBE Algorithm

QuBE is implemented in C on top of SIM, an efficient SAT decider developed by our group. A C-like high-level description of QuBE is shown in Figure 1. In this Figure,

- φ is a global variable initially set to the input QBF.

- *Stack* is a global variable storing the search stack, and is initially empty.
- ∃, ∀, FALSE, TRUE, UNDEF, NULL, UNIT, PURE, L-SPLIT, R-SPLIT are pairwise distinct constants.
- for each atom x in Φ, (i) $x.mode$ is a variable whose possible values are UNIT, PURE, L-SPLIT, R-SPLIT, and (ii) $x.type$ is ∃ if x is existential, and ∀ otherwise.
- *Extend*(l) first pushes l and φ in the stack; then deletes the clauses of φ in which l occurs, and removes $\bar{l}$ from the others.
- *Retract*() pops the literal and corresponding QBF that are on top of the stack: the literal is returned, while the QBF is assigned to φ.
- *Simplify*() simplifies φ till a contradictory clause is generated (line 6), or the matrix of φ is empty (line 8), or no simplification is possible (lines 5, 14).
- *ChooseLiteral*() returns a literal l occurring in φ such that for each atom x occurring to the left of $|l|$ in the prefix of φ, x does not occur in φ, or x is existential iff l is existential. *ChooseLiteral*() also sets $|l|.mode$ to L-SPLIT.
- *Backtrack*(res): pops all the literals and corresponding QBFs (line 18) from the stack, till a literal l is reached such that $|l|.mode$ is L-SPLIT (line 19), and either (i) l is existential and $res =$ FALSE (line 20); or (ii) l is universal and $res =$ TRUE (line 21). If such a literal l exists, $|l|.mode$ is set to R-SPLIT, and $\bar{l}$ is returned (line 22). If no such literal exists, NULL is returned (line 23).

QuBE returns TRUE if the input QBF is satisfiable, and FALSE otherwise. It is easy to see that QuBE, like other QBF procedures (see, e.g., [4,5,6]), is a generalization of the Davis, Logemann, Loveland procedure (DLL) for SAT: QuBE and DLL have the same behavior on QBFs without universal quantifiers.

4   QuBE Options

Consider Figure 1. QuBE ver. 1.0 features backjumping, trivial truth, six different branching heuristics, i.e., implementations of *ChooseLiteral*, and other control options.

The backjumping procedure implemented in QuBE [7] is a generalization of the conflict-direct backjumping procedure as implemented in SAT solvers. As far as we know, QuBE is the only QBF solver with backjumping. Because of the potential overhead, backjumping has to be enabled when compiling the system, while all the other heuristics and optimizations can be enabled/disabled using QuBE's command line.

QuBE's command line is:

```
qube [-tt] [-heuristics unit|bohm|jw2] [-length exists|all]
     [-verbose] [-timeout <n1>] [-memout <n2>] <file-name>.
```

By default, after the simplifications following the branch on an universal variable have been performed, QuBE checks whether the formula obtained from φ by deleting universal literals is satisfiable. If it is, then φ is satisfiable [4]. This optimization can produce dramatic speed-ups, particularly on randomly generated QBFs (see, e.g. [4]). The option -tt disables this check. Notice that ours is an optimized version of "trivial truth" as described in [4], where the check is performed at each branching node.

QuBE branching heuristics have been inspired by the SAT literature. Our current version includes böhm,jw2 and unit heuristics. The behavior of these heuristics depends on the notion of "length" of a clause. QuBE features two definitions of length of a clause C: the number of literals in C (-length all) as in [4,5], and the number of existential literals in C (-length exists) as in [6]. By combining these options, six different branching heuristics are possible.

The böhm and jw2 heuristics are, respectively, a generalization of Böhm's heuristic [8] and "two-sided Jeroslow-Wang" heuristic [9] for SAT. The idea behind böhm and jw2 is to choose literals that occur as often as possible in the shortest clauses of φ. The hope is that by assigning such literals, we will have the largest amount of simplification. The unit heuristic is based on the one implemented in SATZ [10]. As opposed to böhm and jw2, the unit heuristic tentatively assigns truth values to atoms in order to get the exact amount of simplification caused by such assignments.

Independently from the particular branching heuristic used, if the selected atom x is existential (resp. universal), QuBE tries first x if x has more (resp. less) positive than negative occurrences in the matrix of φ. The idea is to maximize the chances of showing φ satisfiability (resp. unsatisfiability) in the first branch.

The -verbose options enables printing search information during the execution, including the variable x being assigned, its mode (whether is a unit, a pure, ...), and –in case it is a L-SPLIT– whether x or $\neg x$ is tried first.

The -timeout <n1> and -memout <n2> options are used to limit the amount of resources used by QuBE. Whenever QuBE exceeds <n1> seconds of CPU time (resp. <n2> megs of RAM), its execution is halted.

5 QuBE Performances

To evaluate QuBE performances, we compare it with DECIDE [5], EVALUATE [4], QKN [11], and QSOLVE [6]. According to our preliminary experimental results, the options -heuristics bohm -length exists give good performances on all the problems, and are thus the default. The tests run on a Pentium III, 600MHz, 128MBRAM.

We consider sets of randomly generated QBFs. We generate QBFs according to model A as described in [12]. In this model, each QBF has the following 4 properties: (*i*) the prefix consists of k sequences, each sequence has n quantifiers, and any two quantifiers in a sequence, are of the same type, (*ii*) the rightmost quantifier is $\exists$, (*iii*) the matrix consists of l clauses, (*iv*) each clause consists of h literals of which at least 2 are existential. The Figure 2 shows the median, out of 100 samples, of the CPU times when $k = 2$, $n = 100$ (left) and $k = 3$, $n = 100$ (right). We fixed $h = 5$ because it yields harder QBFs than $h < 5$ (see [12]), and l (on the x-axis) is varied in such a way to cover the "100% satisfiable – 100% unsatisfiable" transition (shown in the background). Notice the logarithmic scale on the y-axis.

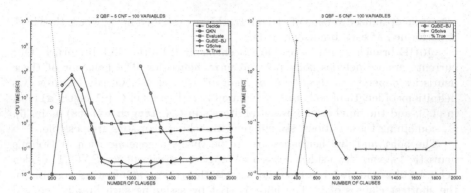

Fig. 2. CPU times, median, 100 samples/point. Background: satisfiability percentage. Left: $k = 2, n = 100$, Right: $k = 3, n = 100$,

Looking at Figure 2 (left) we immediatly see that QuBE and QSOLVE perform roughly the same (with QSOLVE being slightly better than QuBE) and better than all the other solvers that we tested. Still in Figure 2, for $k = 3$ we further observe that QuBE is always faster than QSOLVE, sometimes by orders of magnitude. Since QSOLVE runs trivial truth (and trivial falsity), but no backjumping, we take this as an evidence on the effectiveness of backjumping.

Our experimental analysis includes the 38 problems contributed by Rintanen in [5]. They are translations from planning problems into the language of QBFs and the best solver overall turns out to be DECIDE with 33 problems solved and 42.28s average running time (on solved samples), followed by QuBE, with 18 problems solved and 73.21s, and QSOLVE with 11 problems solved and 149.29s. QKN and EVALUATE with, respectively, 1 and 0 problems solved, trail the list. In this regard, we point out that DECIDE features "inversion of quantifiers" and "sampling" mechanisms which are particularly effective on these benchmarks .

References

1. Jussi Rintanen. Constructing conditional plans by a theorem prover. *Journal of Artificial Intelligence Research*, 10:323–352, 1999.
2. U. Egly, T. Eiter, H. Tompits, and S. Woltran. Solving advanced reasoning tasks using quantified boolean formulas. In *Proc. AAAI*, 2000.
3. D.A. Plaisted and S. Greenbaum. A Structure-preserving Clause Form Translation. *Journal of Symbolic Computation*, 2:293–304, 1986.
4. M. Cadoli, M. Schaerf, A. Giovanardi, and M. Giovanardi. An algorithm to evaluate quantified boolean formulae and its experimental evaluation. *Journal of Automated Reasoning*, 2000. To appear. Reprinted in [13].
5. Jussi T. Rintanen. Improvements to the evaluation of quantified boolean formulae. In Dean Thomas, editor, *Proceedings of the 16th International Joint Conference on Artificial Intelligence (IJCAI-99-Vol2)*, pages 1192–1197, S.F., July 31–August 6 1999. Morgan Kaufmann Publishers.
6. R. Feldmann, B. Monien, and S. Schamberger. A distributed algorithm to evaluate quantified boolean formulae. In *Proc. AAAI*, 2000.

7. Enrico Giunchiglia, Massimo Narizzano, and Armando Tacchella. Backjumping for quantified boolean logic satisfiability. In *Proc. of the International Joint Conference on Artificial Intelligence (IJCAI'2001)*, 2001.
8. M. Böhm and E Speckenmeyer. A fast parallel SAT-solver – efficient workload balancing. *Annals of Mathematics and Artificial Intelligence*, 17:381–400, 1996.
9. Robert G. Jeroslow and Jinchang Wang. Solving propositional satisfiability problems. *Annals of Mathematics and Artificial Intelligence*, 1:167–187, 1990.
10. Chu Min Li and Anbulagan. Heuristics based on unit propagation for satisfiability problems. In *Proceedings of the 15th International Joint Conference on Artificial Intelligence (IJCAI-97)*, pages 366–371, San Francisco, August 23–29 1997. Morgan Kaufmann Publishers.
11. Kleine-Büning, H. and Karpinski, M. and Flögel, A. Resolution for quantified boolean formulas. *Information and computation*, 117(1):12–18, 1995.
12. Ian Gent and Toby Walsh. Beyond NP: the QSAT phase transition. In *Proc. AAAI*, pages 648–653, 1999.
13. Ian P. Gent, Hans Van Maaren, and Toby Walsh, editors. *SAT2000. Highlights of Satisfiability Research in the Year 2000*. IOS Press, 2000.

System Abstract: E 0.61

Stephan Schulz

Institut für Informatik, Technische Universität München,
D-80290 München, Germany, schulz@informatik.tu-muenchen.de

Abstract. We describe the main characteristics of version 0.61 of the
E equational theorem prover. E is based on superposition (with literal
selection) and rewriting. A particular strength of E is the ability to control the proof search very well. This is reflected by a very powerful and
flexible interface for the specification of clause selection functions, and
by a wide variety of functions for the selection of inference literals. We
discuss some important aspects of the implementation and demonstrate
the performance of the prover by presenting experimental results on the
TPTP. Finally, we describe our future plans for the system.

1 Introduction

E is a fully automatic theorem prover for clausal logic with equality. It is based
on a variant of the superposition calculus [BG94] and the DISCOUNT loop
proof procedure. The prover can read (and write) proof problems in its native,
PROLOG-like E-LOP syntax or in TPTP syntax. E is completely implemented
in ANSI C and compiles cleanly on most common UNIX versions.

This paper describes E 0.61, the direct successor of the version that won
the MIX category of CASC-17 [Sut01]. The prover has significantly changed
since the previous published description [Sch99a]. Major changes include the
introduction of literal selection, the addition of new generic clause evaluation
functions, support for using the prover as a preprocessor or clause set normalizer,
proof output and checking, and a much improved automatic mode. Due to space
constraints, we will restrict this discussion to the core functionality of the prover.

The complete distribution of E 0.61 is available on the Internet at [Sch99b].

2 Calculus

We only introduce a few essential terms. Clauses are multi-sets of literals, usually written as disjunctions. Literals are either equations (positive literals) or
inequations (negative literals) over terms, a non-equational literal $P(t_1, \ldots, t_n)$
is encoded as $P(t_1, \ldots, t_n) \simeq \top$ for the special symbol $\top$.

The calculus **SP** [Sch00a,Sch00b] implemented by E is a variant of the standard superposition calculus with selection [BG94]. E implements the generating inference rules *Superposition* (restricted paramodulation), *equality factoring* (generalized factoring) and *equality resolution*. All generating inferences are

R. Goré, A. Leitsch, and T. Nipkow (Eds.): IJCAR 2001, LNAI 2083, pp. 370–375, 2001.

constrained by term ordering and literal selection as follows: If a clause has *no* selected literals, inferences are performed on maximal literals. If, on the other hand, a clause has at least one selected literal, inferences are restricted to selected literals. In this case, the clause is not used to paramodulate into other clauses (although selected literals are a target for being paramodulated into). In all cases, only the maximal sides of a literal are used in the inferences. Literal selection is arbitrary, with the restriction that at least one of the literals selected in a clause has to be negative. Surprisingly, many of the best selection schemes also use the strictly unnecessary selection of at least one positive literal.

In addition to generating inferences, simplification plays a major role in E. We use the obvious clause simplifications (deletion of duplicate and trivial literals), as well as unconditional rewriting, subsumption, tautology deletion, and the *simplify-reflect* inference that resolves a negative literal against a unit equation. The latest addition is *AC-redundancy elimination*, a special technique for dealing with associative and/or commutative function symbols. As shown in [AHL00] (and implemented in the Waldmeister prover), most consequences of the AC axioms are superfluous in Knuth-Bendix completion. This result carries over to full superposition by using the general redundancy notion described in [BG94]. In E, we recognize and delete the corresponding clauses. Additionally, we can delete all *negative* literals in which both terms are equal modulo the recognized AC theory from clauses.

3   Proof Procedure

E implements the saturation of the clause set using the variant of the *given-clause* algorithm that was introduced in DISCOUNT [DKS97]. The core idea is to split the set of all clauses into a subset P of processed clauses and a subset U of unprocessed clauses. P is maximally simplified (with respect to clauses in P), and all generating inferences between clauses in P have been performed. Clauses in U are not directly used for any operations.

Each traversal of the main loop of the algorithm pick the best clause from U and simplifies it with P. If the clause is not redundant, it is used to back-simplify clauses in P and to generate new clauses. It is then added to P. New clauses are simplified once (to improve heuristic evaluation) and are added to U.

Typically, U is several orders of magnitude bigger than P. Therefore, only cheap operations are performed on clauses in U. More expensive ones, like detection of semantic tautologies, or non-unit subsumption, are only applied if a clause is selected for processing. By concentrating on P, we achieve a high rate of inferences as well as a fairly good locality of references, i.e. only a small number of clauses are typically used for each loop traversal.

4   Search Control

E offers a very wide range of options for the control of the proof search. The three major choice points are the selection of the next clause to process, the potential

selection of literals within a clause, and the selection of the term ordering used
to constrain the proof search.

4.1 Clause Selection

The order in which clauses are selected for processing is the most important
choice point for any theorem prover based on the given-clause algorithm. In E,
this order is determined by a *clause selection heuristic.* Such a heuristic sets up
a variety of priority queues and a weighted round-robin scheme that determines
from which queue the next clause is to be selected. Precedence within each
queue is determined, in this order, by a *priority function* that can e.g. prefer all-
negative clauses, ground clauses, or initial axioms, and by an *evaluation function*
that is typically based on symbol counting. Evaluation functions are created by
instantiating one of about 15 different generic function templates. Completeness
of the prover can be guaranteed by careful selection of priority and evaluation
functions, or by simple addition of a fair queue (e.g. a FIFO-queue). In practice,
most successful heuristics in E combine two evaluation functions (one specializing
on potential goals, i.e. all negative clauses, and one for the remaining clauses)
based on refined symbol counting (which assigns a higher weight to maximal
terms and literals), and a FIFO queue.

4.2 Literal Selection

As described in the calculus section above, inferences can be restricted to certain
selected literals. E currently implements about 60 different strategies for the
selection of literals, many of which are (more or less successful) experiments.
Currently, these strategies are hard-coded in C. Since the necessary code changes
to add a new strategy are quite minimal, there is no strong pressure to find a
more abstract interface.

The most successful strategies select negative ground literals before non-
ground ones, and prefer literals with a large size difference between both terms.
Moreover, it often is useful to refrain from selecting literals in clauses with a
unique maximal literal or in range-restricted horn clauses.

4.3 Term Orderings

E uses two different ground-complete simplification orderings (which are lifted to
orderings on literals and clauses): The *Lexicographic Path Ordering* (LPO) and
the *Knuth-Bendix-Ordering* (KBO). Both are parameterized by a precedence
on the function symbols, the KBO additionally requires a weight function for
function symbols and variables. The precedence can either be specified explicitly
on the command line, or one of several predefined schemes can be used. Similar
schemes exist for selecting the symbol weights.

4.4 Automatic Mode

The large flexibility in specifying search strategies for E makes the selection of an adequate set of options for any given problem fairly hard. We have therefore implemented an automatic mode that determines a search strategy based on the proof problem. The code realizing this automatic mode is generated from test results of the prover on the TPTP problem library [SSY94] as follows: We manually determine a partition of the set of test problems, induced by variety of features (presence of Unit, Horn or general clauses, equality content, average term size,...) and order all heuristics by overall performance. A small program traverses the set of heuristics in descending order and assign to each class the first (i.e. the most general) heuristic that achieves optimal performance in this class. Output of this program is the C code implementing the automatic mode.

5 Implementation

As stated in the introduction, E is implemented in ANSI C. The most outstanding feature of the implementation is use of shared terms with shared rewriting. Except for short-lived temporary copies and terms with special restrictions imposed by the calculus, no subterm is ever represented twice in the prover. Rewriting is done directly on the shared structure.

Like almost all other high-performance saturating provers, E uses indexing techniques to speed up common operations. In particular, E uses perfect discrimination trees with age and size constraints to speed up most simplifying unit operations: Subsumption, forward-rewriting, and simplify-reflect. We use normal form dates and rewritability flags on the shared terms to avoid duplication of effort in both forward- and backward-rewriting. As current *hot spots* in the code do not involve unification or, for most proof problems, non-unit-subsumption, we have not yet implemented indexing for generating inferences and non-unit-subsumption.

Due to the shared term representation, the construction of a proof object is fairly hard, as each rewriting step may affect an arbitrary number of usually unknown clauses. We have added a post-processor that reconstructs the term/clause relationship during rewriting. The resulting proof objects are not yet very detailed, but can be checked for correctness using a proof checker. We can use either Otter, SPASS, or E itself to verify the validity of each deduction in a proof derivation.

6 Performance

The table below shows the number of proofs (and models) found by E 0.61 (prerelease version) on all clause normal form problems from TPTP 2.3.0. The prover was running in automatic mode on a SUN Ultra 60 Workstation at 300 MHz, with time limits of 100, 300 and 2000 seconds and a memory limit of 192 MB.

If we compare E with other state-of the art theorem provers, we find that E is particularly strong for Horn problems. It also is among the best general-purpose systems for unit-equality problems. For non-Horn problems without equality, however, the performance is below par. This may be due to the fact that currently neither analytic features nor special inference rules for non-equational literals are implemented. Finally, compared to e.g. SPASS, E is not very god at finding models.

Time limit Problem class	Size of class	100 s		300 s		2000 s		
		Proof	Models	Proof	Models	Proofs	Models	Total
Unit, no equality	11	8	3	8	3	8	3	11
Unit, equality	447	349	3	362	3	362	3	365
Horn, no equality	609	521	5	548	5	551	5	556
Horn, equality	507	357	3	389	3	397	3	400
General, no eq.	766	301	89	320	92	327	94	421
General, equality	1218	378	3	414	3	424	3	427
Overall	3558	1914	106	2041	109	2069	111	2180

7   Future Work

While the current prover already shows a quite satisfactory level of performance, our highest priority is still the improvement of the base prover. At the calculus level, we are planning to integrate clause splitting (as in Vampire) to achieve a better performance for non-Horn problems, and to integrate literal splitting for equational literals to simulate Waldmeister's multiple normal form strategy for unit problems. At the control level, we will try to move from the current feature-based classification of proof problems to the recognition of important algebraic substructures (e.g. groups, rings, set theory...) to achieve a better adaption to the problem at hand.

In the longer term, we are planning to integrate the saturating prover with an analytic top-down component and to work on issues of proof analysis and presentation.

References

[AHL00] J. Avenhaus, T. Hillenbrand, and B. Löchner. On Using Ground Joinable Equations in Equational Theorem Proving. *Proc. of the 3rd FTP, St. Andrews, Scottland,* Fachberichte Informatik. Universität Koblenz-Landau, 2000. (revised version submitted to the JSC).

[BG94] L. Bachmair and H. Ganzinger. Rewrite-Based Equational Theorem Proving with Selection and Simplification. *Journal of Logic and Computation,* 3(4):217–247, 1994.

[DKS97] J. Denzinger, M. Kronenburg, and S. Schulz. DISCOUNT: A Distributed and Learning Equational Prover. *Journal of Automated Reasoning,* 18(2):189–198, 1997.

[Sch99a] S. Schulz. System Abstract: E 0.3. *Proc. of the 16th CADE, Trento*, LNAI 1632 , 297–391. Springer, 1999.

[Sch99b] S. Schulz. The E Web Site. http://wwwjessen.informatik.tu-muenchen.de/~schulz/WORK/- eprover.html, 1999.

[Sch00a] S. Schulz. *The E Equational Theorem Prover – User Manual.* Automated Reasoning Group, Institut für Informatik, TU München, 2001. (to appear).

[Sch00b] S. Schulz. *Learning Search Control Knowledge for Equational Deduction.* DISKI 230. Akademische Verlagsgesellschaft Aka GmbH Berlin, 2000.

[SSY94] G. Sutcliffe, C.B. Suttner, and T. Yemenis. The TPTP Problem Library. *Proc. of the 12th CADE, Nancy*, LNAI 814, 252–266. Springer, 1994.

[Sut01] G. Sutcliffe. The CADE-17 ATP System Competition. *Journal of Automated Reasoning*, 2001 (to appear).

Vampire 1.1
(System Description)

Alexandre Riazanov and Andrei Voronkov

Computer Science Department, University of Manchester

Abstract. In this abstract we describe version 1.1 of the theorem prover Vampire. We give a general description and comment on Vampire's original features and differences with the previously described version 0.0.

From the very beginning, the main research principle of Vampire was efficiency. Vampire uses a large number of data structures for indexing terms and clauses. Efficiency is still the most distinctive feature of Vampire. Due to reimplementation of some algorithms and data structures, Vampire 1.1 is on the average considerably more efficient than Vampire 0.0.

However, the last year many efforts were invested in flexibility: several new inference and simplification rules were implemented, options for controlling the proof search process added, and new literal selection schemes designed.

For the remaining time before IJCAR 2001, we are going to concentrate on adding more flexibility to Vampire, both for experienced and inexperienced users.

1   General Description

Vampire is a completely automatic saturation-based theorem prover for first-order logic with or without equality.

Calculi. Two kinds of calculi are implemented:

1. binary resolution with superposition and negative selection;
2. positive and negative hyperresolution, but only for logic without equality.

Saturation algorithm and splitting. Most of the existing first-order theorem provers implement either the OTTER-style saturation algorithm [6] or the DIS-COUNT-style algorithm [1] (see [8] for details). Vampire implements both algorithms and in addition an original algorithm based on the so-called *limited resource strategy* [8]. The DISCOUNT algorithm has been implemented recently. The main feature of this algorithm is that unused clauses are absolutely passive until they are selected for performing inferences, i.e. they are not eligible for any simplifications.

Until recently, SPASS [13] was the only prover implementing a splitting rule on top of a saturation algorithm. To avoid high cost of implementing the standard

R. Goré, A. Leitsch, and T. Nipkow (Eds.): IJCAR 2001, LNAI 2083, pp. 376–380, 2001.

splitting, we have implemented *splitting without backtracking* [10]. If the search space contains a clause $C \vee D$, where the variables of C and D are disjoint, we replace this clause by two clauses $C \vee p$ and $D \vee \neg p$, where p is a new propositional symbol. There are several options controlling splitting in Vampire 1.1:

1. *Blocking* and *parallel* splitting are obtained by different modifications of the literal selection function to the clauses containing new predicate symbols. These versions allow us to simulate, to some extent, sequential and parallel case analysis. Suppose we split $C \vee D$ into $C \vee p$ and $D \vee \neg p$. The blocking extension of the selection function will select $\neg p$ in $D \vee \neg p$ thus "blocking" inferences with the literals from D until $\neg p$ is cut off by resolution. The ordering on literals is adjusted in such a way that p is less than any literal with a predicate from the original signature. Thus, unblocking D by resolving with $\neg p$ is postponed until the literals from C and literals introduced by resolving with them are all cut off. This roughly corresponds to the standard sequential analysis of cases. In the parallel extension of the selection function literal with input predicates are always selected before any literals with new predicates introduced upon splittings.
2. New literals can be used as *names* of clauses. For example, if we split $C \vee D$ into $C \vee p$ and $D \vee \neg p$, we then consider $\neg p$ as the name of C. The next time we encounter a clause $C \vee D'$ such that the variables of C and D' are disjoint, we simply replace this clause by $D' \vee \neg p$.
3. A simplification rule called *branch rewriting* can be used. This rule is essentially simplification by nonunit equalities of the form $s = t \vee P$, where P only consists of the new literals. Such clause can be used to rewrite a clause $C[s\theta]$ into $C[t\theta]$ under the condition $P \subset C[s\theta]$.

Literal selection. Several literal selection strategies are now available. For example, the user may choose selecting only the maximal literals, or first selecting the maximal literals and then trying to change it by negative selection if it gives (heuristically) better selection. We have also added an option to use *inherited negative selection* used in the prover E [11]: after paramodulating into a clause with a literal $\neg A$ that was selected by negative selection, in the resulting clause we will necessarily select the literal obtained from $\neg A$.

Orderings. Only one kind of term orderings is implemented in Vampire 1.1: a *nonrecursive* version of the Knuth-Bendix ordering. For two ground terms s, t we have $s \succ t$ if either the weight of s is greater than the weight of t, or the weights are equal but s is lexicographically greater than t. The lexicographic comparison is done on the words obtained by enumerating the symbols of the terms visited in the left-to-right depth-first traversal. For nonground terms, the ordering is defined in a similar way. Apart from some other pleasant properties, this ordering enables efficient approximation of algorithms for solving ordering constraints.

2   Other Original Features

Precompiled ordering constraints. In version 0.0 rewriting by a unit equality $s = t$ was allowed only when this equality was ordered, i.e. $s \succ t$. In the new version more simplifications by unit equalities can be done due to introduction of constrained rewrite rules. Now, if an equality $s = t$ is not preordered, we can still rewrite a clause $C[s\theta]$ into $C[t\theta]$ provided that $s\theta \succ t\theta$.

In general, the check $s\theta \succ t\theta$ can be expensive. To avoid this expensive check, we generate constraints which encode the sufficient and necessary conditions for $s\theta \succ t\theta$ to hold. For example, for the commutativity axiom $x \cdot y = y \cdot x$ and the Knuth-Bendix ordering $\succ$ instead of checking $x\theta \cdot y\theta \succ y\theta \cdot x\theta$ we can check a simpler but equivalent condition $x\theta \succ y\theta$. For the ordering used in Vampire, the corresponding condition will be a simple lexicographic comparison of x and y.

Commutativity optimization. Special treatment of commutative functors is implemented. Now, every subterm of a generated clause is normalized with respect to commutativity of certain functions. If f is commutative, the term $f(t_1, t_2)$ is replaced by $f(t_2, t_1)$ provided that t_2 is lexicographically smaller than t_1. Due to our choice of the term ordering, this normalization does not violate completeness since it can be interpreted as rewriting using the commutativity law with an ordering constraint. Doing the normalization explicitly spares us the necessity of using a general algorithm for solving the constraint.

Negative equality splitting. This rule may be applied on the preprocessing phase and allows us to simulate an algorithm used in Waldmeister [5]. If the term s is ground, the clause $s \neq t \vee C$ can be replaced by the following two clauses: $p(s)$ and $\neg p(t) \vee C$, where p is a new predicate symbol. This allows us to process s and the rest of the clause separately. Apart from other things, this may lead to the following addition-instead-of-multiplication effect. Suppose that s can be rewritten by paramodulation in m different ways resulting in the terms $s_1, \ldots, s_m$, and rewriting of t produces n versions $t_1, \ldots, t_n$. Without *negative equality splitting* we would have to keep $m \cdot n$ clauses $s_i \neq t_j \vee C, 1 \leq i \leq m, 1 \leq j \leq n$, while negative equality splitting allows us to keep only $m + n$ clauses, namely $p(s_i)$ and $\neg p(t_j) \vee C$.

Subsumption resolution. One of the simplification techniques that is heavily used in Vampire, and to which it owes a great part of its power, is *subsumption resolution* [3]. A clause $\neg A \vee C_1$ can be replaced by C_1 if among the kept clauses there is a clause C_2 subsuming $A \vee C_1$.

New algorithm for backward subsumption. One of the major improvements is a new indexing method for backward subsumption. Vampire 0.0 used an algorithm based on discrimination trees. This algorithm proved to be extremely slow on many problems. A new algorithm based on path indexing [12] and database joins was designed [9]. The first implementation of this algorithm was inefficient, but a couple of optimizations turned it into an extremely efficient one. The first key decision was to replace the set of (clause,literal,term) tuples in the leaves of the path index by a more suitable data structure called *skip lists* [7]. The second key decision was to change the order of evaluating joins depending on the sizes of the sets. In addition, we added an optimization for handling symmetric predicates.

3   Future Developments

In the future we will develop Vampire using the principles mentioned in the introduction: efficiency and flexibility, with the emphasis on the latter.

Efficiency. We will continue experiments with indexing techniques and new algorithms and datastructures for the most important problems. In particular, we will try to improve our indexing techniques for treating symbols with special properties, such as commutativity of functions and symmetry of predicates. Another expected change is a reimplementation of indexing for unification.

Flexibility. So far Vampire was very difficult to use. First, many options were inherited from rather old versions and have become obsolete. The user's manual was primitive. We are planning to make Vampire more user-friendly so it can be used for two kinds of applications: (i) interactive use by experienced users who can understand various proof-control options and use Vampire to prove hard theorems; (ii) as a built-in subsystem of automated proof-assistants, interactive provers, and verification systems. These will require enhancement of Vampire by various features, such as new selection strategies, new simplification orderings: ordinary Knuth-Bendix ordering and lexicographic path ordering (see e.g. [2]), built-in AC etc. We are going to provide interface to interactive provers and automatic proof-assistants, such as HOL and Isabelle.

Preprocessing. Currently, preprocessing of clauses is part of Vampire. We are going to implement preprocessing as a separate program that analyzes input, simplifies it, and calls Vampire with suitable options. A new clausifier will be implemented.

Other. We are going to reimplement memory management in Vampire since currently there are situations when Vampire will request memory from the system behind the specified limit. We are planning to implement *stratified resolution* [4].

4   Availability

Vampire is available free of charge. The authors will be glad to provide the newest versions of the system with necessary assistance to any interested party. The system can be run under Linux and Solaris. A temporarily unsupported port is also available for the Win32 platforms. The latest information about Vampire is available at http://www.cs.man.ac.uk/fmethods/vampire/.

References

1. J. Avenhaus, J. Denzinger, and M. Fuchs. DISCOUNT: a system for distributed equational deduction. In J. Hsiang, editor, *Proceedings of the 6th International Conference on Rewriting Techniques and Applications (RTA-95)*, volume 914 of *Lecture Notes in Computer Science*, pages 397–402, Kaiserslautern, 1995.

2. F. Baader and T. Nipkow. *Term Rewriting and and All That.* Cambridge University press, Cambridge, 1998.

3. L. Bachmair and H. Ganzinger. Resolution theorem proving. In A. Robinson and A. Voronkov, editors, *Handbook of Automated Reasoning.* Elsevier Science and MIT Press, 2001. To appear.

4. A. Degtyarev and A. Voronkov. Stratified resolution. In D. McAllester, editor, *17th International Conference on Automated Deduction (CADE-17)*, volume 1831 of *Lecture Notes in Artificial Intelligence*, pages 365–384, Pittsburgh, 2000. Springer Verlag.

5. Th. Hillenbrand, A. Buch, R. Vogt, and B. Löchner. Waldmeister: High-performance equational deduction. *Journal of Automated Reasoning*, 18(2):265–270, 1997.

6. W.W. McCune. OTTER 3.0 reference manual and guide. Technical Report ANL-94/6, Argonne National Laboratory, January 1994.

7. W. Pugh. Skip lists: A probabilistic alternative to balanced trees. *Communications of the ACM*, 33(6):668–676, 1990.

8. A. Riazanov and A. Voronkov. Limited resource strategy in resolution theorem proving. Preprint CSPP-7, Department of Computer Science, University of Manchester, October 2000.

9. A. Riazanov and A. Voronkov. Implementing backward subsumption and search for instances with path indexing and joins. 2001. To appear.

10. A. Riazanov and A. Voronkov. Splitting without backtracking. Preprint CSPP-10, Department of Computer Science, University of Manchester, January 2001. Accepted to IJCAI 2001.

11. S. Schulz. System abstract: E 0.3. In H. Ganzinger, editor, *Automated Deduction—CADE-16. 16th International Conference on Automated Deduction*, Lecture Notes in Artificial Intelligence, pages 297–301, Trento, Italy, July 1999.

12. M. Stickel. The path indexing method for indexing terms. Technical Report 473, Artificial Intelligence Center, SRI International, Menlo Park, CA, October 1989.

13. C. Weidenbach, B. Afshordel, U. Brahm, C. Cohrs, T. Engel, E. Keen, C. Theobalt, and D. Topic. System description: SPASS version 1.0.0. In H. Ganzinger, editor, *Automated Deduction—CADE-16. 16th International Conference on Automated Deduction*, volume 1632 of *Lecture Notes in Artificial Intelligence*, pages 378–382, Trento, Italy, July 1999.

DCTP – A Disconnection Calculus Theorem Prover – System Abstract

Reinhold Letz and Gernot Stenz

Institut für Informatik
Technische Universität München
D-80290 Munich, Germany
{letz,stenzg}@in.tum.de

Abstract. We describe the theorem prover DCTP, which is an implementation of the disconnection tableau calculus, a confluent tableau method, in which free variables are treated in a non-rigid manner. In contrast to most other free-variable tableau variants, the system can also be used for model generation. We sketch the underlying calculus and its refinements, and present the results of an experimental evaluation.

1 Introduction

In this paper we present the theorem prover DCTP. It is based on the disconnection tableau calculus [3], which is a clausal tableau calculus with some promising characteristics. The calculus is inherently cut-free, it provides a decision procedure for a larger class of formulae than most other first-order calculi and, most importantly, it is proof confluent, so that it can be used for model generation. The extraction of a model from a saturated branch is one of the main motivations and advantages of the traditional semantic tableau approach. It is important to emphasize that this advantage is lost in contemporary free-variable tableau calculi like connection tableaux or model elimination [6] or certain confluent variants of tableaux [1] or hypertableaux [2], in which free variables are treated in a rigid manner.

In the paper, we describe the underlying calculus and its relation to other methods like clause linking [4]. Furthermore, we expound how the basic proof system can significantly be improved by extending it with clause simplification and the generation of unit lemmas, both in a top-down and a bottom-up manner. This also permits the use of unit subsumption. Also, the main implementation decisions of the system are briefly sketched. We conclude with reporting on results of an experimental evaluation, which is quite encouraging for a first prototype.

2 The Disconnection Tableau Calculus

Essentially, the disconnection tableau calculus can be viewed as an integration of Plaisted's clause linking method [4] into a tableau control structure. The original clause linking method works by iteratively producing instances of the input

R. Goré, A. Leitsch, and T. Nipkow (Eds.): IJCAR 2001, LNAI 2083, pp. 381–385, 2001.
© Springer-Verlag Berlin Heidelberg 2001

clauses, which are occasionally tested for unsatisfiability by a separate propositional decision procedure. The use of a tableau as a control structure has two advantages. On the one hand, the tableau format restricts the number of clause linking steps that may be performed. On the other hand, the tableau method provides a propositional decision procedure for the produced clause instances, thus making a separate propositional decision procedure superfluous. For the description of the proof method, we use the standard terminology for clausal tableaux. The disconnection tableau calculus consists of a single complex inference rule, the so-called *linking rule*.

> **Linking rule.** Given a tableau branch B containing two literals K and L in tableau clauses c and d, respectively, if there exists a unifier for the complement of K and a variable-renamed variant $L\tau$ of L, then successively expand the branch with the two clauses $c\sigma$ and $d\tau\sigma$ as illustrated in Figure 1.

In other terms, we perform a clause linking step and attach the coupled instantiated clauses at the end of the current tableau branch. Afterwards, the respective connection cannot be used any more on the branches expanding B, which explains the naming "disconnection" tableau calculus for the proof method. Additionally, in order to be able to start the tableau construction, one must choose an arbitrary *initial active path* through all the input clauses, from which the initial connections can be selected. This initial active path has to be used as a common initial segment of all proper tableau branches considered later on.

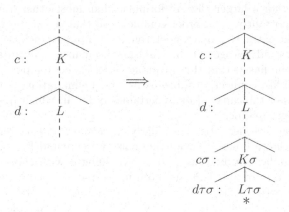

Fig. 1. Illustration of a linking step.

As branch closure condition we use the same notion as employed in the clause linking method. That is, a branch of a tableau is *ground closed* if it contains two literals K and L such that $K\sigma$ is the complement of $L\sigma$ where σ is a substitution mapping all variables in the tableau to a new constant. Applied to the tableau in Figure 1, this means that after the linking step at least the middle branch is ground closed, as indicated with an asterisk. The disconnection tableau calculus has the following properties. First of all, it is refutation complete for first-order clause logic. In order to find a proof, the linking rule must be applied in a fair manner. This is achieved by ensuring that any connection on a branch is eventually used on the branch. Second, the method provides a decision procedure for the Bernays-Schönfinkel class, i.e., for clause sets containing no function symbols of arity > 0, just like the clause linking method or hypertableau methods like Satchmo. If the set of connections on a branch is exhausted, the literals on the branch describe a model for the clause set. In this respect, the method is superior to most other free-variable tableau calculi like model elimination, in which the free variables are treated in a rigid manner. Furthermore, in contrast to hypertableau calculi, in the disconnection approach the instantiation of clauses is fully guided by connections and no hidden form of Smullyan's γ rule is needed.

3 Refinements

As usual in the development of theorem provers, implementing a simple calculus in its pure form will not result in a competitive system. In order to improve the performance of the system, we have integrated a number of refinements, which preserve completeness and increase the performance of the disconnection tableau calculus tremendously.

Pruning of clause variants. In a standard linking step two new clauses c and d are attached to the current tableau. In many cases, however, one of the attached clauses, say c, is a variant of a clause already on the branch. In this case, the branch can be expanded with the clause d only.

Pruning of redundant branches. Since connections on a branch have to be used in a fair manner, it may happen that a closed subtableau is produced in which the top literal L is irrelevant. In this case the brother literals of L need not be solved.

Unit simplification. The special treatment of unit clauses is one of the most successful methods to increase the performance of theorem provers. This also holds for the disconnection tableau calculus. When a clause has to be attached to a tableau branch, we can remove all literals whose complements are subsumed by unit clauses contained in the clause set. In the course of the tableau construction, this leads to shorter and shorter tableau clauses.

Unit lemma generation. In the tableau framework new unit clauses can be generated in two entirely different ways, in a top-down and a bottom-up manner. Top-down unit lemmas naturally result from unit simplifications, when the length of the tableau clause to be attached is 1. In such a case, instead of attaching the respective literal to the end of the current branch it is simply attached to the root of the tableau and may now be used by all other literals in the tableau. The bottom-up generation of units is more delicate. Assume a subtableau T with root literal L is ground closed without referring to literals on the branch above L. This amounts to the generation of a new unit lemma, namely, the complement of $L\sigma$ where σ identifies all variables in L—just taking the complement of L would be unsound because of the modified branch closure condition. This feature is a special case of the so-called *folding up* procedure [5].

Unit subsumption. Once a large number of unit clauses are available, we can strongly profit from unit subsumption, i.e., tableau clauses are not attached when they are subsumed by a unit clause. As a matter of fact, subsumption deletion between non-unit clauses cannot be performed, since then no linking step could be applied at all.

4   Implementation Issues

The disconnection calculus employed by DCTP is proof confluent. This means we do not have to backtrack on inferences, i.e. we do not enumerate possible proof trees. We must, however backtrack on the branches of our proof tree, as even though all inferences performed on a closed branch remain fixed, we still need to solve all remaining open subgoals in the tableau. Backtracking within the proof tree, as opposed to a simple tail recursive iteration over open subgoals, allows us to reconstruct the necessary environment, regarding for instance the existence and usability of the links on the branch, without explicitly storing this information for each subgoal. The main loop of the system has two choice points, the selection of an open branch and the selection of a link on the branch for a linking step extension. Both of these selections are heuristically guided. The branch selection uses a depth-first strategy that selects the new branch depending on the instantiatedness of the leaf literals. The link selection favours links connecting short clauses. This way, a unit preference strategy is implemented. Additionally, the link selection considers the term complexity of the linked literal and thus guarantees the fairness of the proof procedure. The evaluation of the main loop is continued until either no open subgoals are left or the available links on a branch are exhausted. In order to improve the efficiency when checking ground closedness, subsumption, or clause variants, we use indexing techniques for the storing of path literals, unit lemmas, and clause instances. All of the indexes use discrimination trees. In the current implementation, we use no unification index, the generation of new links is performed by checking the new open subgoal against the list of potentially linked path literals. The theorem prover DCTP is implemented in the Scheme dialect *bigloo* (http://kaolin.unice.fr/bigloo/bigloo.html). The system can be obtained from http://wwwjessen.informatik.tu-muenchen.de/ letz/dctp.html.

5   Future Extensions

We have not yet implemented special inference rules for equality literals. Consequently, the performance of the prover on problems containing equality is rather poor. While a satisfactory solution for equality handling using the model elimination calculus or classical tableaux has yet to be found, we have good reason to believe that the disconnection calculus is compatible with reasonable adaptations of orderings and demodulation, since the calculus does not use rigid variables and unification is strictly local. We intend to implement a version of ordered paramodulation in our system.

6   Evaluation

We have evaluated the system DCTP on the TPTP library. Since currently no reasonable equality handling is integrated in the system, we give results for all clause problems and problems without equality predicates. We compared DCTP with two other provers, the Scheme reimplementation of the Setheo prover and the latest version of the E prover [7]. The time limit for all tests was 300 seconds per problem on a Sun Ultra 60 with 300 MHz processors and 384 MB of main memory. Apart from highlighting the excellence of E, the results given in Table 1 confirm that model generation is one of the strengths of our new system. Considering the entire TPTP problem library, an entirely new prover of course cannot compete with refined state-of-the-art systems. Still, the reasonable performance on satisfiable or groundable formulae shows that a disconnection calculus prover has considerable potential.

Table 1. Solutions found by various provers within 300 seconds for all TPTP clause problems (left) and non-equality clause problems (right).

	DCTP	Scheme-Setheo	E 0.61
Total	3558	3558	3558
Proofs	906	1159	2041
Models	128	26	109

	DCTP	Scheme-Setheo	E 0.61
Total	1385	1385	1385
Proofs	618	701	873
Models	125	25	100

References

1. P. Baumgartner, N. Eisinger, and U. Furbach. A confluent connection calculus. In *Proceedings, CADE-16, Trento*, LNAI 1632, pp. 329–343. Springer, 1999.
2. P. Baumgartner, U. Furbach, and I. Niemelä. Hyper tableaux. In *Proceedings, JELIA-96*, LNAI 1126, pp. 1–17, Springer, 1996.
3. Jean-Paul Billon. The disconnection method: a confluent integration of unification in the analytic framework. In *Proceedings of the 5th TABLEAUX, Terrasini*, LNAI 1071, pp. 110–126, Springer, 1996.
4. S.-J. Lee and D. Plaisted. Eliminating duplication with the hyper-linking strategy. *Journal of Automated Reasoning*, pp. 25–42, 1992.
5. Reinhold Letz, Klaus Mayr, and C. Goller. Controlled integration of the cut rule into connection tableau calculi. *Journal of Automated Reasoning*, 13(3):297–338, December 1994.
6. Reinhold Letz, Johann Schumann, Stephan Bayerl, and Wolfgang Bibel. SETHEO: A high-performance theorem prover. *Journal of Automated Reasoning*, 8(2):183–212, 1992.
7. Stephan Schulz. System Abstract: E 0.61. In *Proc. of the 1st IJCAR, Siena*, LNAI. Springer, 2001.

More On Implicit Syntax*

Marko Luther

Universität Ulm
Fakultät für Informatik
D-89069 Ulm, Germany
luther@informatik.uni-ulm.de

Abstract. Proof assistants based on type theories, such as Coq and
Lego, allow users to omit subterms on input that can be inferred automatically. While those mechanisms are well known, ad-hoc algorithms
are used to suppress subterms on output. As a result, terms might be
printed identically although they differ in hidden parts. Such ambiguous
representations may confuse users. Additionally, terms might be rejected
by the type checker because the printer has erased too much type information. This paper addresses these problems by proposing effective erasure methods that guarantee successful term reconstruction, similar to
the ones developed for the compression of proof-terms in Proof-Carrying
Code environments. Experiences with the implementation in Typelab
proved them both efficient and practical.

1 Implicit Syntax

Type theories are powerful formal systems that capture both the notion of computation and deduction. Particularly the expressive theories, such as the Calculus of Constructions ($\mathcal{CC}$) [CH88] which is investigated in this paper, are used
for the development of mathematical and algorithmic theories since proofs and
specifications are representable in a very direct way using one uniform language.

There is a price to pay for this expressiveness: abstractions have to be deco-
rated with annotations, and type applications have to be written explicitly, because type abstraction and type application are just special cases of λ-abstraction
and application. For example, to form a list one has to provide the element type
as an additional argument to instantiate the polymorphic constructors *cons* and
nil as in (*cons* $\mathbb{N}$ 1 (*nil* $\mathbb{N}$))). Also, one has to annotate the abstraction of n in
$\lambda n{:}\mathbb{N} . n + 1$ with its type $\mathbb{N}$ although this type is determined by the abstraction
body. These excessive annotations and applications make terms inherently ver-
bose and thus hard to read and write. Without such explicit type information,
decidability of type-checking may be lost [Miq01,Wel99].

Proof assistants are programs that have primarily been built to support humans in constructing proofs represented as λ-terms. They combine a (mostly)
interactive proof-development system with a type-checker that can check those

---

* This research has partly been supported by the "Deutsche Forschungsgemeinschaft"
within the "Schwerpunktprogramm Deduktion".

R. Goré, A. Leitsch, and T. Nipkow (Eds.): IJCAR 2001, LNAI 2083, pp. 386–400, 2001.

proofs. Moreover, most proof assistants support a syntax that is more convenient to use than the language of the underlying type theory.

Besides some purely syntactic enhancements, such as the possibility to use operations in infix notation, most systems allow to leave out subterms that the system can infer as a remedy to the redundancy of explicit types. The *explicit language* (a variant of $\mathcal{CC}$ in our case) is complemented by an *implicit language* defined in terms of the underlying type theory through some inference mechanism. From the viewpoint of the user, the implicit language acts as an alternative grammar for the type theory. It improves the expressiveness of the original theory as more terms have types, but no additional types are inhabited. This pragmatic approach to simplify the user interface of proof assistants is referred to as *implicit syntax* [Pol90]. Motivated by the examples above, the inference mechanisms that we focus on in this paper are *argument synthesis*, to avoid explicit polymorphic instantiations, and *(partial) term reconstruction*, to suppress annotations on abstractions. For these, we use the term *elaboration*. The inverse process that removes redundant subterms is called *erasure*.

Ad-hoc argument synthesis, implemented in the proof assistant CoQ [Bar99], uses explicit placeholders to mark omitted subterms that should be inferred. The above example can be written in CoQ as (*cons* ? 1 (*cons* ? 2 (*nil* ?))) using the placeholder symbol "?". In addition, CoQ supports the automatic insertion of placeholders. This is done by analyzing the types of global constants. Parameters occurring free in the type of at least one of the succeeding parameters are assumed to be inferable (e. g., the first parameter of *cons*). We may write (*cons* 1 (*cons* 2 (*nil* ?)))[1]. To decide whether a term can be hidden on output or not, the same oversimplified[2] analysis is used. This might lead to representations of terms without unique elaborations. Especially if two terms are printed identically although they are not identical internally, resulting from different hidden arguments, users may get confused [Miq01]. Even in cases where the automatic detection and erasure work correctly, the system tends to hide arguments one wants to see explicitly although they would be inferable.

A finer control over implicit positions is possible through *uniform argument synthesis* as implemented in LEGO [LP92]. The user can mark parameter positions, using abstractions of a second 'color', at which arguments can be left implicit. Explicit arguments trigger the elaboration of arguments at preceding hidden positions[3]. To allow the specialization of polymorphic functions there is a syntactic facility to overwrite argument synthesis and to supply arguments 'forced' at implicit positions by preceding them with "!". At the internal representations of terms LEGO uses annotations that correspond to forced marks and colored parameters of the user language. These annotations are used to decide which arguments to hide on output. Unfortunately, there are cases in which also LEGO suppresses arguments that cannot be reconstructed. Furthermore, LEGO

---

[1] Note that we still have to apply a placeholder to the empty list.

[2] A free occurrence of a parameter in another parameter type does not generally guarantee a successful elaboration. One reason for this is that argument synthesis is based on unification, which is not decidable in the higher-order case [Gol81].

[3] This rules out the inference of the type argument of *nil*.

sometimes does not hide arguments at marked positions, even when they are inferable. Both defects result from difficulties to define reduction for a language with forced arguments properly since uniqueness of elaboration cannot be decided locally [HT95].

This paper solves the problems and limitations caused by the usage of implicit syntax in current proof assistants by proposing a stronger elaboration algorithm and by complementing it with an erasure algorithm that guarantees unique elaboration. The elaboration algorithm is stronger than the mentioned ones since it allows the inference of implicit arguments at marked positions triggered also by the outer term context, while doing universal argument synthesis (e. g., our algorithm accepts (*cons* 1 *nil*)). In addition, the algorithm avoids the inconvenience of having to attach type information to all abstracted variables by doing (partial) term reconstruction. This allows the omission of type annotations[4] that can be inferred by propagating type information (e. g., the annotation of n in $(\lambda n . n + 1)$). The erasure algorithm does a global analysis of terms and reconstructs forced marks (if necessary) instead of just propagating them.

The rest of this paper is organized as follows. After introducing a bicolored variant of $\mathcal{CC}$ in the next section, we present the elaboration algorithm (Sect. 3). Guided by the strategy of this elaboration algorithm, we develop in Sect. 4 an erasure algorithm that is both effective in removing most subterms and practical as shown by experimental results (Sect. 5). Finally, we report on the adoption to more realistic languages and comment on related work (Sect. 6).

2   Bicolored Calculus of Constructions

The bicolored Calculus of Constructions ($\mathcal{CC}^{\text{bi}}$) to be used as explicitly typed language, is a variant of the Calculus of Construction ($\mathcal{CC}$) [CH88]. Terms are built up from a set $\mathcal{V}$ of variables, sort constants $s \in \{Prop, Type\}$, dependent function types ($\Pi x{:}A . B$), λ-abstractions ($\lambda x{:}A . M$) and function applications ($M\ N$) as in $\mathcal{CC}$. To this we add abstractions of the form $\Pi x|A . B$ and $\lambda x|A . B$ using the vertical bar as a color to mark implicit argument positions. If the color is irrelevant we use the symbol "$\|$" to stand for either a colon or a bar and we abbreviate $\Pi x{:}A . B$ by $A \to B$ if x does not occur free in B.

We denote the set of free variables of a term M by $\mathcal{FV}(M)$, the term resulting from substituting N for each free occurrence of x in M by $M[x{:=}N]$, syntactic equality, the one-step β-reduction and the β-conversion relation by $\equiv$, $\to_\beta$ and $\simeq$, respectively. For a term M the weak head normal form (whnf) is denoted by $\overline{M}$, the normal form by $|M|$, and the application head by $head(M)$. As usual, we will consider terms up to α-conversions.

The colors of abstractions have no essential meaning in the explicit language, only the distinction is important. So, implicit λ-abstractions behave analogously to the corresponding explicit variant with respect to reduction.

$$((\lambda x|A . M)\ N) \to_\beta M[x{:=}N]$$

---

[4] Note that even if some annotations can be left implicit we demand all variables to be introduced explicitly, to avoid confusions caused by typos [GH98].

The typing rules of $\mathcal{CC}$ are augmented by the following rules which differ only in the coloring from the related uncolored rules of $\mathcal{CC}$.

$$\frac{\Gamma \vdash A{:}s_1 \quad \Gamma,x{:}A \vdash B{:}s_2}{\Gamma \vdash \Pi x|A.B : s_2} \qquad \frac{\Gamma,x{:}A \vdash M{:}B \quad \Gamma \vdash \Pi x|A.B : s}{\Gamma \vdash \lambda x|A.M : \Pi x|A.B} \qquad \frac{\Gamma \vdash N{:}A \quad \Gamma \vdash M : \Pi x|A.B}{\Gamma \vdash (M\ N) : B[x{:=}N]}$$

The consistency of $\mathcal{CC}^{\text{bi}}$ under β follows immediately from that of $\mathcal{CC}$. Note that while $\mathcal{CC}$ is still confluent with respect to $\beta\eta$-reduction [Geu92], this property is lost in the bicolored system. The term $(\lambda x{:}A.(\lambda y|A.y)\ x)$ yields the critical pair $\langle \lambda x{:}A.x , \lambda y|A.y \rangle$ under $\beta\eta$-reduction.

2.1 Unification Variables

The elaboration algorithm, to be described below, maps *partial terms*[5] (i. e., terms of the implicit language) to terms of $\mathcal{CC}^{\text{bi}}$. A partial term M' is translated to an *open term* M of the explicit language that contains unsolved *unification variables*. These are solved in turn by unification during the elaboration process.

The explicit language extended by unification variables, which are syntactically distinguished from other variables by a leading "?", is a variant of the one introduced by Strecker [SLvH98,Str99]. Unification variables are handled as constants with respect to $\mathcal{FV}$ and reduction. For a term M, $\mathcal{UV}(M)$ denotes the set of unification variables occurring in M. A unification variable depends on a context Γ and has a type A, as expressed by the suggestive notation $\Gamma \vdash ?n{:}A$. *Sort unification variables*, $\Gamma \vdash ?n{:}*$, are a special flavor of unification variables where instantiation is restricted to terms of type *Prop* or *Type*.

An *instantiation* is a function, mapping a finite set of unification variables to terms. When instantiating a unification variable, it is replaced by the solution term without renaming of bound variables. Instantiations are inductively extended to terms and contexts.

There can be complex dependencies among unification variables in calculi with dependent types. Therefore, a context and a type are not invariantly assigned to a unification variable $?n$, but they are determined by the *elaboration state* under consideration. An elaboration state $\mathcal{E}$ consists of a finite set $\Phi_\mathcal{E}$ of unification variables, a function $ctxt_\mathcal{E}$ assigning a context to each $?n \in \Phi_\mathcal{E}$, and a function $type_\mathcal{E}$ assigning a type to each $?n \in \Phi_\mathcal{E}$. Our elaboration algorithm will only produce *well-typed elaboration states* $\mathcal{E}$, where the dependencies among unification variables $?n \in \Phi_\mathcal{E}$ are not cyclic[6] and the typing constraints imposed by $ctxt_\mathcal{E}$ and $type_\mathcal{E}$ are 'internally consistent'.

For well-typed elaboration states reduction is confluent and strongly normalizing. Type inference and type checking are decidable and subject reduction holds [Str99].

---

[5] As a convention we prime metavariables that stand for terms of the implicit language, as M', to distinguish them syntactically from those standing for fully explicit terms.

[6] Note that unification variables do not have to be linearly ordered, as in the calculus of Muñoz [Muñ01], since this would restrict elaboration considerably.

2.2 Unification

Unification tries to find an instantiation that, when applied to two terms, makes them equal. Equality is taken modulo convertibility, thus, the unification problems we obtain will in general be higher-order. We use a 'colored' version of the unification algorithm defined by Strecker [Str99]. It essentially carries out first-order unification by structural comparison of weak head normal forms. Unification judgments are of the form $\langle \mathcal{E}_0; \Gamma \vdash t_1 \approx t_2 \rangle \Rightarrow \mathcal{E}_1; \iota$, which express that the unification problem $\langle \mathcal{E}_0; \Gamma \vdash t_1 \approx t_2 \rangle$ can be solved by instantiation ι, leaving open the unification variables of $\mathcal{E}_1$. The resulting elaboration state $\mathcal{E}_1$ is guaranteed to be well typed if the elaboration state $\mathcal{E}_0$ is well typed. For presentation purposes we use the simplified notation $\Gamma \vdash t_1 \approx t_2$ for unification judgments. We assume that all instantiations are immediately applied to all terms and we keep the elaboration states implicit.

For us, the key properties are that unification is decidable for unification problems that produce only disagreement pairs of the simple rigid-rigid kind, and that solvable unification problems of this kind have most general unifiers (MGUs). Stronger decidable unification algorithms computing MGUs for the pattern fragment of higher-order unification [Mil91] depend on the η-relation, which we have to rule out to keep $\mathcal{CC}^{\mathrm{bi}}$ confluent.

3 Elaboration

The implicit user language is an extension of the explicit language by Curry-style abstractions of both colors and applications with forced arguments.

$$\mathcal{T} \ ::= \ \ldots \ \mid \ \lambda \mathcal{V} \| . \mathcal{T} \ \mid \ \Pi \mathcal{V} \| . \mathcal{T} \ \mid \ (\mathcal{T} \ n! \mathcal{T}) \qquad n \in \mathbb{N}$$

The value of n indicates implicit arguments preceding the forced one. The notation $(M \ !N)$ abbreviates $(M \ 0!N)$ and we write $(M \ \|N)$ to subsume all variants of applications. We assume a function $@(M)$ that decomposes the term M in its application head and the list of its arguments in reverse order. For example, $@(f \ a \ 3!b \ c) = \langle f, \ c :: 3!b :: a :: \cdot \rangle$.

3.1 Bidirectional Elaboration Algorithm

We present the algorithm at an abstract level through five judgments making use of unification as introduced above. In- and output parameters are separated by "$\Rightarrow$" and the flow of typing information is indicated by up and down arrows.

Main Elaboration $\Gamma \vdash M' \Rightarrow M : N$
Synthesis Mode (SM) $\Gamma \vdash M' \Rightarrow M \uparrow N$
Argument Generation (AG) $\Gamma \vdash L'; l \Rightarrow M : N$
Checking Mode (CM) $\Gamma \vdash M' \downarrow N \Rightarrow M$
Coerce to Type (CT) $\Gamma \vdash M_1 : N \downarrow O \Rightarrow M_2$

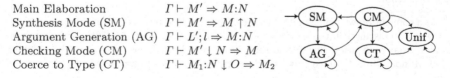

Elaboration works in two distinct modes: SM, where type information is propagated upward from subexpressions, and CM, where information is propagated downward from enclosing expressions[7]. Unification variables are generated for implicit arguments through the AG function and are solved later on by the CT function which essentially calls unification.

The main elaboration judgment is a partial function taking a context Γ and a partial term M', producing the elaboration M of M' and its type N. Elaboration always starts with an empty elaboration state $\mathcal{E}$, $\Phi_{\mathcal{E}} = \emptyset$, in synthesis mode. As for unification, we keep instantiations and elaboration states implicit in the presentation of rules and judgments.

In the rest of this section we define the four remaining judgments as a collection of syntax-directed inference rules. The judgments are mutually dependent according to the above control flow graph. All rules assume a valid context Γ and check that context extensions maintain validity.

Elaboration, Synthesis Mode: $\Gamma \vdash M' \Rightarrow M \uparrow N$

The synthesizing (partial) elaboration function (Fig. 1) takes a context Γ and a partial term M' and produces the corresponding explicit term M and its type N. For example, the following judgment is derivable.

$$\Gamma \vdash (cons\ 1\ nil) \Rightarrow (cons\ \mathbb{N}\ 1\ (nil\ \mathbb{N})) \uparrow (List\ \mathbb{N})$$

The function implements essentially a colored version of the most natural type inference algorithm for $\mathcal{CC}$ known as the Constructive Engine [Hue89] and is used if nothing is known about the expected type of an expression. It makes use of the abbreviating judgment $\Gamma \vdash M' \overset{wh}{\Rightarrow} M \uparrow N$, which is identical to the SM judgment but assumes N to be in whnf. Note that the syntactic test in the side condition $B \not\equiv Type$ is strong enough to ensure the Π-abstraction to be well typed even in our calculus extended with unification variables. The differences with respect to the Constructive Engine are the addition of the Curry-style abstraction rules, $(\lambda^*\uparrow)$ and $(\Pi^*\uparrow)$, which generate new unification variables for the missing abstraction types, and a modified application rule $(@\uparrow)$. This application rule embodies a simple heuristic: always synthesize the type of the function, and then use the resulting information to switch to checking mode for the argument expression by calling the argument generation function.

Argument Generation: $\Gamma \vdash L'; l \Rightarrow M{:}N$

Argument generation essentially calculates the type of the application head L' and elaborates the list of arguments l under the expected parameter types in checking mode (Fig. 2). If the type of the head is not functional (e. g., a unification variable) elaboration fails. Unification variables are introduced at hidden positions unless overwritten by a forced argument. The result is the elaborated application M of type N. The example derivation from the last paragraph contains a subderivation of the following argument generation judgment.

$$\Gamma \vdash cons; nil :: 1 :: \cdot \Rightarrow (cons\ \mathbb{N}\ 1\ (nil\ \mathbb{N})) : (List\ \mathbb{N})$$

---

[7] The basic idea of bidirectional checking is not new and is for example used by the ML type-inference algorithm known as Algorithm $\mathcal{M}$ [LY98].

$$(\textsc{Prop}\uparrow) \; \frac{}{\Gamma \vdash Prop \Rightarrow Prop \uparrow Type} \qquad (\textsc{Var}\uparrow) \; \frac{}{\Gamma_1, x{:}A, \Gamma_2 \vdash x \Rightarrow x \uparrow A}$$

$$(\lambda\uparrow) \; \frac{\Gamma \vdash A' \overset{wh}{\Rightarrow} A \uparrow s \qquad \Gamma, x{:}A \vdash M' \Rightarrow M \uparrow B}{\Gamma \vdash \lambda x\|A'.M' \Rightarrow \lambda x\|A.M \uparrow \Pi x\|A.B} \; [B \not\equiv Type]$$

$$(\lambda^*\uparrow) \; \frac{\Gamma \vdash ?a{:}* \qquad \Gamma, x{:}?a \vdash M' \Rightarrow M \uparrow B}{\Gamma \vdash \lambda x\|.M' \Rightarrow \lambda x\|?a.M \uparrow \Pi x\|?a.B} \; [B \not\equiv Type] \qquad (@\uparrow) \; \frac{@(M'\|N') = \langle L', l\rangle \qquad \Gamma \vdash L';l \Rightarrow U{:}V}{\Gamma \vdash (M'\|N') \Rightarrow U \uparrow V}$$

$$(\Pi\uparrow) \; \frac{\Gamma \vdash A' \overset{wh}{\Rightarrow} A \uparrow s_1 \qquad \Gamma, x{:}A \vdash B' \overset{wh}{\Rightarrow} B \uparrow s_2}{\Gamma \vdash \Pi x\|A'.B' \Rightarrow \Pi x\|A.B \uparrow s_2} \qquad (\Pi^*\uparrow) \; \frac{\Gamma \vdash ?a{:}* \qquad \Gamma, x{:}?a \vdash B' \overset{wh}{\Rightarrow} B \uparrow s}{\Gamma \vdash \Pi x\|.B' \Rightarrow \Pi x\|?a.B \uparrow s}$$

Fig. 1. Elaboration, Synthesis Mode

$$(\bullet) \; \frac{\Gamma \vdash M' \Rightarrow M \uparrow N}{\Gamma \vdash M'; \cdot \Rightarrow M{:}N} \qquad (@|) \; \frac{\Gamma \vdash M'; l \overset{wh}{\Rightarrow} L : \Pi x|A.B \qquad \Gamma \vdash ?n{:}A}{\Gamma \vdash M'; N' :: l \Rightarrow (L \; ?n) : B[x:=?n]}$$

$$(@{:}) \; \frac{\Gamma \vdash M'; l \overset{wh}{\Rightarrow} L : \Pi x{:}A.B \qquad \Gamma \vdash N' \downarrow \overline{A} \Rightarrow N}{\Gamma \vdash M'; N' :: l \Rightarrow (L \; N) : B[x := N]} \qquad (@!) \; \frac{\Gamma \vdash M'; l \overset{wh}{\Rightarrow} L : \Pi x|A.B \qquad \Gamma \vdash N' \downarrow \overline{A} \Rightarrow N}{\Gamma \vdash M'; 0!N' :: l \Rightarrow (L \; N) : B[x := N]}$$

$$(@n!) \; \frac{\Gamma \vdash M'; (n-1)!N' :: l \overset{wh}{\Rightarrow} L : \Pi x|A.B \qquad \Gamma \vdash ?n{:}A}{\Gamma \vdash M'; n!N' :: l \Rightarrow (L \; ?n) : B[x := ?n]} \; [n>0]$$

Fig. 2. Argument Generation

Elaboration, Checking Mode: $\Gamma \vdash M' \downarrow N \Rightarrow M$

The checking mode, described by the rules of Fig. 3, is used if the surrounding term context determines the type of the expression. The partial term M' is elaborated to M under the given expected type N, which is assumed to be in whnf. The resulting term M is guaranteed to be of type N. There is no side condition $B \not\equiv Type$ in the λ-abstraction rules, since the expected type is known to be valid. Note further, that the expected type in rule $(@\downarrow)$ cannot be propagated down to elaborate the function part of an application since the result type of a function in $\mathcal{CC}$ depends on the actual arguments and not only on their number. Thus, to ensure soundness a final unification, essentially done by a call to the coerce to type function, is necessary. The argument nil of the one-element list $(cons \; 1 \; nil)$ is elaborated in CM by a derivation of the following judgment.

$$\Gamma \vdash nil \downarrow (List \; \mathbb{N}) \Rightarrow (nil \; \mathbb{N})$$

Coerce to Type: $\Gamma \vdash M_1{:}N \downarrow O \Rightarrow M_2$

The coerce to type function tries to convert the given term M_1 of type N, with $\Gamma \vdash M_1{:}N$, to a related term M_2 of the expected type O (Fig. 4). The rule (Unif) just checks if the given and the expected type are unifiable and therefore,

$$(\text{Prop}\downarrow) \quad \frac{}{\Gamma \vdash Prop \downarrow Type \Rightarrow Prop}$$

$$(\text{Var}\downarrow) \quad \frac{\Gamma \vdash x \Rightarrow x \uparrow A \qquad \Gamma \vdash x{:}\overline{A} \downarrow B \Rightarrow M}{\Gamma \vdash x \downarrow B \Rightarrow M}$$

$$(\lambda\downarrow) \quad \frac{\Gamma \vdash A' \overset{wh}{\Rightarrow} A \uparrow s \qquad \Gamma \vdash A \approx A_e \qquad \Gamma, x{:}A \vdash M' \downarrow \overline{B} \Rightarrow M}{\Gamma \vdash \lambda x \| A' . M' \downarrow \Pi x \| A_e . B \Rightarrow \lambda x \| A . M}$$

$$(@\downarrow) \quad \frac{\Gamma \vdash (M' \, \| N') \Rightarrow U \uparrow V \qquad \Gamma \vdash U{:}\overline{V} \downarrow O \Rightarrow Q}{\Gamma \vdash (M' \, \| N') \downarrow O \Rightarrow Q}$$

$$(\lambda^*\downarrow) \quad \frac{\Gamma, x{:}A \vdash M' \downarrow \overline{B} \Rightarrow M}{\Gamma \vdash \lambda x \| . M' \downarrow \Pi x \| A . B \Rightarrow \lambda x \| A . M}$$

$$(\Pi^*\downarrow) \quad \frac{\Gamma \vdash ?a{:}* \qquad \Gamma, x{:}?a \vdash B' \downarrow s \Rightarrow B}{\Gamma \vdash \Pi x \| . B' \downarrow s \Rightarrow \Pi x \| ?a . B}$$

$$(\Pi\downarrow) \quad \frac{\Gamma \vdash A' \overset{wh}{\Rightarrow} A \uparrow s_1 \qquad \Gamma, x{:}A \vdash B' \downarrow s_2 \Rightarrow B}{\Gamma \vdash \Pi x \| A' . B' \downarrow s_2 \Rightarrow \Pi x \| A . B}$$

$$(\text{UV}\downarrow) \quad \frac{\Gamma \vdash M' \Rightarrow M \uparrow N \qquad \Gamma \vdash M; \overline{N} \downarrow ?n \Rightarrow O}{\Gamma \vdash M' \downarrow ?n \Rightarrow O}$$

Fig. 3. Elaboration, Checking Mode

$$(\text{Unif}) \quad \frac{\Gamma \vdash N \approx O}{\Gamma \vdash M{:}N \downarrow O \Rightarrow M}$$

$$(\text{NTI}) \quad \frac{\Gamma \vdash (\Pi x | A . B) \not\approx O \qquad \Gamma \vdash ?n{:}A \qquad \Gamma \vdash (M \; ?n) : \overline{B[x:=?n]} \downarrow O \Rightarrow P}{\Gamma \vdash M : \Pi x | A . B \downarrow O \Rightarrow P}$$

Fig. 4. Coerce to Type

the given term M_1 has not to be modified apart from resulting instantiations. If this fails and the given type is an implicit Π-abstraction, a newly created unification variable is applied to M_1 through *nil-type inference* using rule (NTI) and the result is recursively checked. In the other cases elaboration fails. While this strategy enables the inference of the type argument of *nil*, it rules out the possibility to collect unification constraints first and solve them later. Solving constraints immediately seems to be more efficient anyway [BN00].

The elaboration of the argument *nil* under the expected type (*List* $\mathbb{N}$) is done using the rule (NTI), deriving the following judgment.

$$\Gamma \vdash nil : \Pi T | Prop . (List \; T) \downarrow (List \; \mathbb{N}) \Rightarrow (nil \; \mathbb{N})$$

3.2   Properties

Proposition 1. *If* $\Gamma \vdash M_1{:}N \downarrow O_1 \Rightarrow M_2$ *then* $\Gamma \vdash M_2{:}O_2$ *with* $O_1 \simeq O_2$.

Proposition 2 (Soundness). *If* $\Gamma \vdash M' \Rightarrow M{:}N$ *then* $\Gamma \vdash M{:}N$.

Proof. By (mutual) induction on the derivation trees of the elaboration and argument generation judgments using Proposition 1, correctness of unification, subject reduction and correctness of types.

Proposition 3 (Partial Completeness). *If* $\Gamma \vdash M{:}N_1$ *and* M' *corresponds to the term* M *with forced applications at all implicit parameter positions then* $\Gamma \vdash M' \Rightarrow M{:}N_2$ *and* $N_1 \simeq N_2$.

Proof. Since the term M' has no missing subterms our algorithm never generates unification variables. Therefore, all derivations can be translated into derivations without CM judgments and unification reduces to conversion leading to derivations identically to those of the (bicolored) Constructive Engine, which is known to be complete [Hue89].

Partial completeness essentially enables the user to give just as much explicit type information as needed. This is necessary, because we cannot expect elaboration to be complete.

Generally, the elaboration algorithm calculates only one of several possible elaborations. For example, assuming a constant id of type $\Pi T|Prop\,.\,T \to T$ the partial term $(id\ id\ 1)$ has the two elaborations $(id\ (\mathbb{N} \to \mathbb{N})\ (id\ \mathbb{N})\ 1)$ and $(id\ (\Pi T|Prop\,.\,T \to T)\ id\ \mathbb{N}\ 1)$ which are not convertible. Our algorithm would generate the second elaboration.

4   Erasure

The erasure algorithm is supposed to remove as many type annotations and arguments at implicit positions from a given term as possible, without losing any information. We propose an algorithm (Fig. 5) that mimics elaboration in an abstract way to predict its behavior. The erasure judgments, of the form $\Gamma \vdash M'; \mathcal{C} \overset{\mathrm{M}}{\Leftarrow} M$, compute for a unification variable free term M, $\mathcal{UV}(M) = \emptyset$, the erasure M' and the set $\mathcal{C}$ of variables that are erased in CM on first occurrence. It works again in one of two modes, $\mathrm{M} \in \{\mathrm{SM}, \mathrm{CM}\}$, corresponding roughly to those of the bidirectional elaboration algorithm. In contrast to elaboration, erasure works also in synthesis mode if the expected type on elaboration would be an open term, conservatively assuming it not to contain any structural information. Only if the expected type on elaboration is known to be unification variable free, erasure works in checking mode.

Type annotations of λ-abstractions are always left implicit if erasure is in checking mode, since those can be read off the fully explicit expected type without even generating a unification variable using rule $(\lambda^* \downarrow)$ of Fig. 3. Other type annotations are only left implicit if the first reference to the corresponding abstraction variable x is erased in checking mode (i.e., $x \in \mathcal{C}$). Both cases are shown by the following two derivable judgments.

$$\Gamma, f{:}(\mathbb{N} \to \mathbb{N}) \to \mathbb{N} \vdash f\ (\lambda n\,.\,n); \emptyset \overset{\mathrm{SM}}{\Leftarrow} f\ (\lambda n{:}\mathbb{N}\,.\,n) \qquad \Gamma \vdash \lambda n\,.\,n{+}1; \{n\} \overset{\mathrm{SM}}{\Leftarrow} \lambda n{:}\mathbb{N}\,.\,n{+}1$$

Arguments at marked position are left implicit if they are determined by the first depending argument type in elaboration order or by the expected type, if in CM (Fig. 6). Otherwise, erasure represents these arguments explicit as forced arguments.

$$(\text{PROP}) \ \frac{}{\Gamma \vdash Prop;\emptyset \overset{\text{M}}{\Leftarrow} Prop} \qquad (\text{VAR}) \ \frac{}{\Gamma \vdash x;\emptyset \overset{\text{SM}}{\Leftarrow} x} \qquad (\text{VAR}^*) \ \frac{}{\Gamma \vdash x;\{x\} \overset{\text{CM}}{\Leftarrow} x}$$

$$(\lambda^*) \ \frac{\Gamma,x{:}A \vdash M';\mathcal{C} \overset{\text{CM}}{\Leftarrow} M}{\Gamma \vdash \lambda x\| \, . \, M';\mathcal{C} \overset{\text{CM}}{\Leftarrow} \lambda x\|A \, . \, M} \qquad\qquad (@) \ \frac{\Gamma \vdash L';P;i;\mathcal{S};\mathcal{C} \overset{\text{M}}{\Leftarrow} (M\ N);0}{\Gamma \vdash L';\mathcal{C} \overset{\text{M}}{\Leftarrow} (M\ N)}$$

$$\mathcal{Q} \in \{\lambda, \Pi\}$$

$$\begin{array}{c} x \in \mathcal{C} \\ (\mathcal{Q}^*) \ \dfrac{\Gamma,x{:}A \vdash M';\mathcal{C} \overset{\text{M}}{\Leftarrow} M}{\Gamma \vdash \mathcal{Q}x\| \, . \, M';\mathcal{C} \overset{\text{M}}{\Leftarrow} \mathcal{Q}x\|A \, . \, M} \end{array} \qquad (\mathcal{Q}) \ \dfrac{\begin{array}{c}\Gamma \vdash A';\mathcal{C}_1 \overset{\text{SM}}{\Leftarrow} A \\ \Gamma,x{:}A \vdash M';\mathcal{C}_2 \overset{\text{M}}{\Leftarrow} M \quad x \notin \mathcal{C}_2 \\ \mathcal{C} = \mathcal{C}_1 \cup (\mathcal{C}_2 \setminus \mathcal{FV}(A))\end{array}}{\Gamma \vdash \mathcal{Q}x\|A' \, . \, M';\mathcal{C} \overset{\text{M}}{\Leftarrow} \mathcal{Q}x\|A \, . \, M}$$

Fig. 5. Bidirectional Erasure

Argument erasure judgments are of the form $\Gamma \vdash M';N;i;\mathcal{S};\mathcal{C} \overset{\text{M}}{\Leftarrow} M;n$, where M is the application term to be erased and n is the number of additional arguments. It calculates the erasure M' and the type N of M, the number of preceding implicit argument positions i, a set $\mathcal{S}$ of positions which should be erased in SM since an implicit argument has to be inferred from the corresponding argument type and the set $\mathcal{C}$ as described above. Note, that arguments are identified here by the number of consecutive arguments. The erasure mode of an argument is computed from the set $\mathcal{S}$ as follows.

$$mode(n,\mathcal{S}) = \begin{cases} \text{SM} & \text{if } n \in \mathcal{S} \\ \text{CM} & \text{else} \end{cases}$$

On erasing the term $(cons\ \mathbb{N}\ 1\ (nil\ \mathbb{N}))$ a derivation of the following judgment is constructed. The resulting set $\mathcal{S} = \{1\}$ determines the second argument, 1, to be erased in SM while the last argument, $(nil\ \mathbb{N})$, can be erased in CM since nothing has to be inferred from its type.

$$\Gamma \vdash cons; \Pi T|Prop\,.\,T \rightarrow (List\ T) \rightarrow (List\ T); 1; \{1\}; \emptyset \overset{\text{SM}}{\Leftarrow} (cons\ \mathbb{N}); 2$$

The calculation of the determining information source, if any, for arguments at implicit parameter positions is done by the function $dpos_{\text{M}}$.

Definition 4 (Determining Position). *The function $dpos_{\text{M}}$ for a mode* M, *a context* Γ, *a term* $T \equiv \Pi x|A\,.\,B$ *with* $|B| \equiv \Pi x_1\|A_1, \ldots, x_l\|A_l\,.\,C$, $C \not\equiv \Pi y\|D\,.\,E$, *another term* N *with* $\Gamma \vdash N{:}A$ *and* $n \in \mathbb{N}$ *is specified as follows.*

$$dpos_{\text{M}}(\Gamma,T,N,n) = \begin{cases} n-r & \text{if } \exists r \in \mathbb{N}.\,0 < r \leq \min(l,n),\ x \notin \bigcup_{i=1}^{r-1} \mathcal{FV}(|A_i|) \\ & \text{and } x \in \mathcal{SF}(dom(\Gamma), A_r) \\ \blacksquare & \text{if } \text{M}=\text{CM},\ n \leq l,\ x \notin \bigcup_{i=1}^{n} \mathcal{FV}(|A_i|),\ T \not\equiv B[x{:=}N] \\ & \text{and } x \in \mathcal{SF}(dom(\Gamma), \Pi x_{n+1}\|A_{n+1}, \ldots, x_l\|A_l\,.\,C) \\ \bigstar & \text{else} \end{cases}$$

The result of $dpos_{\text{M}}$ is a number $d \in \mathbb{N}$ indicating the determining argument, or one of the symbols "$\blacksquare$", "$\bigstar$" if the argument can be inferred from the expected type or cannot be inferred at all, respectively. The condition $T \not\equiv B[x := N]$

$$\text{(NoArg)}\quad \frac{\Gamma \vdash M{:}N \qquad \Gamma \vdash M';\mathcal{C} \overset{\text{SM}}{\Leftarrow} M}{\Gamma \vdash M';\overline{N};0;\emptyset;\mathcal{C} \overset{M}{\Leftarrow} M;n}\quad {\scriptstyle [M \not\equiv (M_1\ M_2)]}$$

$$\text{(Visible)}\frac{\Gamma \vdash M';\Pi x{:}A\,.\,B;i;\mathcal{S};\mathcal{C}_1 \overset{M}{\Leftarrow} M;n+1 \qquad \Gamma \vdash N';\mathcal{C}_2 \overset{m}{\Leftarrow} N}{\Gamma \vdash (M'\ N');\overline{B[x:=N]};0;\mathcal{S};\mathcal{C}_1 \cup (\mathcal{C}_2 \setminus \mathcal{FV}(M)) \overset{M}{\Leftarrow} (M\ N);n}\quad {\scriptstyle [m=mode(n,\mathcal{S})]}$$

$$\text{(Forced)}\frac{\begin{array}{c}\Gamma \vdash M';\Pi x|A\,.\,B;i;\mathcal{S};\mathcal{C}_1 \overset{M}{\Leftarrow} M;n+1 \qquad \Gamma \vdash N';\mathcal{C}_2 \overset{m}{\Leftarrow} N \\ n \in \mathcal{S} \vee dpos_{\text{M}}(\Gamma, \Pi x|A\,.\,B, N, n) = \bigstar\end{array}}{\Gamma \vdash (M'\ i!N');\overline{B[x:=N]};0;\mathcal{S};\mathcal{C}_1 \cup (\mathcal{C}_2 \setminus \mathcal{FV}(M)) \overset{M}{\Leftarrow} (M\ N);n}\quad {\scriptstyle [m=mode(n,\mathcal{S})]}$$

$$\text{(Implicit)}\quad \frac{\begin{array}{c}\Gamma \vdash M';\Pi x|A\,.\,B;i;\mathcal{S};\mathcal{C} \overset{M}{\Leftarrow} M;n+1 \\ dpos_{\text{M}}(\Gamma, \Pi x|A\,.\,B, N, n) = d \qquad n \notin \mathcal{S}\end{array}}{\Gamma \vdash M';\overline{B[x:=N]};i+1;\mathcal{S} \cup \{d\};\mathcal{C} \overset{M}{\Leftarrow} (M\ N);n}\quad {\scriptstyle [d \neq \bigstar]}$$

Fig. 6. Argument Erasure

ensures the applicability of the rule (NTI) on elaboration by forcing the term M applied to the argument N to change its type. This is for example not the case for any term of type $\Pi T|Prop\,.\,T$ applied to its own type.

The definition of $dpos_{\text{M}}$ depends on the set $\mathcal{SF}$ of free variables of a term, solvable by unification with a fully explicit term. This set is defined as follows, assuming nothing is ever substituted to variables of the set $\mathcal{V}_0$ during unification.

Definition 5 (Solvable and Dangerous Free Variables). *The set of* solvable free variables, $\mathcal{SF}(\mathcal{V}_0, M)$, *of a term M relative to variables $\mathcal{V}_0$ is defined mutual dependent with the set of* dangerous free variables, $\mathcal{DF}(\mathcal{V}_0, M)$, *as follows.*

$$\mathcal{SF}(\mathcal{V}_0, M) = \mathcal{SF}^*(\mathcal{V}_0, \overline{M}, \emptyset) \qquad \mathcal{DF}(\mathcal{V}_0, M) = \mathcal{DF}^*(\mathcal{V}_0, \overline{M}, \emptyset)$$

with

$$\mathcal{SF}^*(\mathcal{V}_0, M, \mathcal{V}) = \begin{cases} \{x\} & \text{if } M \equiv x \text{ and } x \notin \mathcal{V}_0 \cup \mathcal{V} \\ \mathcal{SF}^*(\mathcal{V}_0, \overline{A}, \mathcal{V}) & \text{if } M \equiv \mathcal{Q}x\|A\,.\,N, \\ \quad \cup\, \mathcal{SF}^*(\mathcal{V}_0, \overline{N}, \mathcal{V} \cup \{x\}) \setminus \mathcal{DF}^*(\mathcal{V}_0, \overline{A}, \mathcal{V}) & \mathcal{Q} \in \{\lambda, \Pi\} \\ \mathcal{SF}^*(\mathcal{V}_0, M_1, \mathcal{V}) & \text{if } M \equiv (M_1\ M_2), \\ \quad \cup\, \mathcal{SF}^*(\mathcal{V}_0, \overline{M_2}, \mathcal{V}) \setminus \mathcal{DF}^*(\mathcal{V}_0, M_1, \mathcal{V}) & head(M_1) \in \mathcal{V}_0 \cup \mathcal{V} \\ \emptyset & \text{else} \end{cases}$$

and $\quad \mathcal{DF}^*(\mathcal{V}_0, M, \mathcal{V}) = \mathcal{FV}(|M|) \setminus \mathcal{SF}^*(\mathcal{V}_0, M, \mathcal{V})$

The definition of the set $\mathcal{SF}$ mimics unification in that terms are kept essentially in normal form through stepwise reduction to whnf. The condition $head(M_1) \in \mathcal{V}_0 \cup \mathcal{V}$ ensures that the heading of the term M remains unchanged under application of any substitution with the implication that all $x \in \mathcal{SF}$ have residuals in every reduction of every substitution instance of M.

To illustrate the last two definitions consider the erasure of $(nil\ \mathbb{N})$ in CM; the last argument of the above example. $dpos_{\text{CM}}(\Gamma, \Pi T|Prop\,.\,(List\ T), \mathbb{N}, 0)$ yields

■ since $\mathcal{SF}(dom(\Gamma), (List\ T)) = \{T\}$ assuming $List \in dom(\Gamma)$. This allows to derive the following argument erasure judgment using rule (IMPLICIT) of Fig. 6.

$$\Gamma \vdash nil; (List\ \mathbb{N}); 1; \{\blacksquare\}; \emptyset \overset{\text{CM}}{\Leftarrow} (nil\ \mathbb{N}); 0$$

4.1 Properties

Proposition 6. *If* $\Gamma \vdash M_1{:}N$, $\Gamma \vdash M_2{:}N$, $M_1 \simeq M_2$, $\mathcal{UV}(M_2) = \emptyset$, $\Gamma \vdash x{:}A$, $x \in \mathcal{SF}(dom(\Gamma), M_1), \Gamma \vdash ?n{:}A$ then $\Gamma \vdash M_1[x := ?n] \approx M_2$ yields the most general instantiation ι with $?n \in dom(\iota)$.

It is always possible to reconstruct the original term from the erasure using the elaboration algorithm of Section 3.

Proposition 7 (Invertibility). *If* $\Gamma \vdash M_1{:}N_1$ and $\Gamma \vdash M'; \mathcal{C} \overset{\text{SM}}{\Leftarrow} M_1$ then $\Gamma \vdash M' \Rightarrow M_2{:}N_2$ with $M_1 \simeq M_2$.

Invertibility essentially holds, since all unification variables generated by elaboration for erased subterms are solved by the first typing constraints they participate in. It can be verified that all generated unification problems are such that one of the terms to be unified does not contain any unification variable and the other term does only contain solvable ones, which are guarantied to be instantiated (Proposition 6).

Since type annotations on parameters can help to make expressions more readable serving as checked documentation, erasure prefers implicit arguments over implicit annotations. Consider an (explicit) polymorphic function f of type $\Pi T|Prop.(T \rightarrow Prop) \rightarrow Prop$. The algorithm above computes the erasure $(f\ (\lambda x{:}\mathbb{N}.\mathbb{N}))$ rather than $(f\ !\mathbb{N}\ (\lambda x.\mathbb{N}))$ for the explicit term $(f\ \mathbb{N}\ (\lambda x{:}\mathbb{N}.\mathbb{N}))$. Note further, that two explicit terms that are structurally equal, can still have different erasures and that different explicit terms can lead to the same erasure, but only if both terms occur in different term contexts.

5 Experimental Results

We have implemented and tested several variants of the elaboration and erasure algorithms discussed above as part of the proof assistant TYPELAB [vHLS97]. For evaluation purposes we analyzed terms of different sizes up to 15,000 abstract syntax tree nodes. The terms were arbitrarily selected from definitions and proofs of the standard TYPELAB library.

We found that compression factors are independent of the fully explicit term size. For that, we calculated the percentage reduction in the total size of all terms but separated the results for definitions from the results for proofs (Table 1). On average, our combined erasure algorithm almost reduces the representation in half the size while the compression is slightly more effective for proof terms.

It has to be asked whether one could find an erasure algorithm which yields much smaller representations. To answer this question we determined the arguments at implicit positions that our combined erasure algorithm presented explicitly. We found that on average, terms could only be reduced by another 1.1%

Table 1. Reduction in abstract syntax tree size

	Implicit Arguments	Implicit Annotations	Combination
Proofs	22.1%	32.3%	49,4%
Definitions	24.7%	26.3%	39.5%

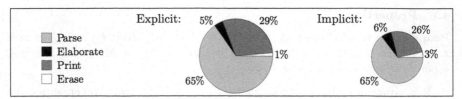

Fig. 7. Effect of implicit syntax on the full turnaround time

through blindly erasing all those arguments while only 20% of them had enough information to be reconstructed. Erasing in addition all remaining type annotations from abstractions reduced the representation by another 7,6%, but none of those implicit terms could be reconstructed. We conclude that our combined erasure algorithm removes the vast majority of redundant subterms, respecting the given color information, thus leaving little room for further improvement.

To analyze the performance benefits gained using implicit syntax we measured for all terms the times[8] needed for a full turnaround including parsing, elaboration, erasure and the final printing of the implicit representation. Since we found all timings to be linear in the size of the fully explicit term, we averaged the results again. Fig. 7 shows the results for the fully explicit representation compared with our combined implicit representation. The pies are sized by area, which corresponds to the absolute representation size. We can conclude that, in practice, the additional costs produced by the erasure algorithm are smaller than the savings gained from dealing with reduced representations.

6   Disscussion

We have presented algorithms that improve the usability of implicit syntax for proof assistants. Our inference algorithm is stronger than the one of CoQ or LEGO, since it allows to omit more subterms. Furthermore, our erasure algorithm generates only implicit representations that allow the reconstruction of the original terms, in contrast to the ad-hoc erasure algorithms implemented in CoQ and LEGO. The experimental results presented in the previous section provide evidence that our algorithms, while still being efficient, save considerable bandwidth between the user and the system.

To implement the algorithms of this paper for the assistant TYPELAB we had to consider additional aspects. Deciding when to expand *notational definitions* is subtle for unification algorithms. Stepwise expansion, as done by most proof assistants, may return a unifier which is not most general and hence renders

---

[8] We found similar results about the space requirements.

unification incomplete even for the first-order case [PS99] and thus would limit our elaboration algorithm. To adapt the erasure algorithm to the language of TYPELAB with *recursive definitions*, the set $\mathcal{SF}$ had to be restricted since applications with recursively defined heads are potentially 'instable' with respect to substitution at recursive parameter positions.

The local erasure algorithm in this paper hides only arguments that can be elaborated by local methods [PT98], while our elaboration algorithm allows the global distribution of unification variables. Consider the polymorphic operation *append* on lists of type $\Pi T | Prop . (List\ T) \rightarrow (List\ T) \rightarrow (List\ T)$ and a list l of type $(List\ \mathbb{N})$. The implicit term $(append\ nil\ l)$ is elaborated into the explicit term $(append\ \mathbb{N}\ (nil\ \mathbb{N})\ l)$, but erasure would produce the wordy representation $(append\ (nil\ !\mathbb{N})\ l)$. We have also implemented an erasure algorithm that does not force elaboration to solve implicit arguments on the first opportunity completely. It works with an additional erasure mode, CM*, where the expected type is allowed to be an open term. This considerably more complex algorithm[9] computes the optimal erasure $(append\ nil\ l)$ for the explicit term above.

One drawback of implicit syntax is the enlarged trusted code-base of the proof-checker. Fortunately, for sensible applications internal terms can always be rechecked by a small trusted or even verified checker for the base calculus.

6.1   Related Work

Berghofer and Nipkow describe a dynamic compression algorithm for proof terms in ISABELLE [BN00]. Their algorithm searches for optimal representations by essentially doing elaboration on erasure and seems not to be efficient enough for interactive usage.

The problem of redundancy has been addressed also by Necula and Lee [NL98] in the context of Proof-Carrying Code systems. They analyze the combination of ad-hoc argument synthesis with implicit type annotations for canonical first-order proof objects represented as fully applied LF terms in long $\beta\eta$-normal form, given a fully explicit expected type. This special setting enables the pre-computing of large parts of the erasure for constants.

While a language generated by implicit syntax is natural to humans, it seems rather difficult to give it a direct foundation. Hagiya and Toda [HT95] have implemented an implicit version of $\mathcal{CC}$ directly using a typed version of β-reduction defined on the implicit language. Several complicated syntactic restrictions have to be imposed to ensure decidability of type inference and to avoid dynamic type checking during reduction. On the theoretical side, Miquel defined a Curry-style version of $\mathcal{CC}$ [Miq01]. Although the metatheory of this calculus is not yet fully developed, it is strongly conjectured that type checking is undecidable. For that reason, this calculus seems to be a poor basis for proof assistants.

Acknowledgments I thank Martin Strecker for developing large parts of the TYPELAB system.

---

[9] Details are subject of current research.

References

[Bar99] B. Barras et al. The Coq proof assistant reference manual – Version 6.3.1. Technical report, INRIA, France, 1999.

[BN00] S. Berghofer, T. Nipkow. Proof terms for simply typed higher order logic. In *Proc. of TPHOLs'00*, volume 1869 of *LNCS*, pp. 38–53. Springer, 2000.

[CH88] Th. Coquand, G. Huet. The Calculus of Constructions. *Information and Computation*, 76(2/3):95–120, 1988.

[Geu92] H. Geuvers. The Church-Rosser property for $\beta\eta$-reduction in typed λ-calculi. In *Proc. of LICS'92*, pp. 453–460. IEEE Press, 1992.

[GH98] W. O. D. Griffioen, M. Huisman. A comparison of PVS and Isabelle/HOL. In *Proc. of TPHOLs'98*, volume 1479 of *LNCS*, pp. 123–142. Springer, 1998.

[Gol81] W. Goldfarb. The undecidability of the second-order unification problem. *Theoretical Computer Science*, 13:225–230, 1981.

[HT95] M. Hagiya, Y. Toda. On implicit arguments. Technical Report TR-95-1, Department of Information Science, University of Tokyo, 1995.

[Hue89] G. Huet. The Constructive Engine. In *A Perspective in Theoretical Computer Science*. World Scientific Publishing, Singapore, 1989.

[LP92] Z. Luo, R. Pollack. *LEGO Proof Development System*. University of Edingburgh, 1992. Technical Report ECS-LFCS-92-211.

[LY98] O. Lee, K. Yi. Proofs about a folklore let-polymorphic type inference algorithm. *ACM TOPLAS*, 20(4):707–723, 1998.

[Mil91] D. Miller. A logic programming language with lambda-abstraction, function variables, and simple unification. *J. Logic Comput.*, 1(4):497–536, 1991.

[Miq01] A. Miquel. The implicit calculus of constructions. In *Proc. of the Conf. TLCA'01, Krakow, Poland, May 2–5, 2001*, LNCS. Springer, Berlin, 2001.

[Muñ01] C. Muñoz. Proof-term synthesis on dependent-type systems via explicit substitutions. *Theoretical Computer Science*, 2001. To appear.

[NL98] G. Necula, P. Lee. Efficient representation and validation of proofs. In *Proc. of LICS'98*, pp. 93–104. IEEE Press, 1998.

[Pol90] R. Pollack. Implicit syntax. In G. Huet, G. Plotkin, eds., *Informal Proc. of the 1st Workshop on Logical Frameworks (LF'90), Antibes*. 1990.

[PS99] F. Pfenning, C. Schürmann. Algorithms for equality and unification in the presence of notational definitions. In T. Altenkirch, W. Naraschewski, B. Reus, eds., *Proc. of TYPES'98*, volume 1657 of *LNCS*, pp. 179–193. Springer, 1999.

[PT98] B. Pierce, D. Turner. Local type inference. In *Conf. Record of POPL'98*, pp. 252–265. ACM Press, 1998.

[SLvH98] M. Strecker, M. Luther, F. von Henke. Interactive and automated proof construction in type theory. In W. Bibel, P. Schmitt, eds., *Automated Deduction — A Basis for Applications*, volume I, chapter 3. Kluwer, 1998.

[Str99] M. Strecker. *Construction and Deduction in Type Theories*. Ph.D. thesis, Fakultät für Informatik, Universität Ulm, 1999.
<http://www.informatik.uni-ulm.de/ki/Strecker/phd.html>

[vHLS97] F. von Henke, M. Luther, M. Strecker. TYPELAB: An environment for modular program development. In M. Bidoit, M. Dauchet, eds., *Proc. of the 7th Intl. Conf. TAPSOFT'97*, volume 1214 of *LNCS*, pp. 851–854. Springer, 1997.

[Wel99] J. B. Wells. Typability and type checking in System F are equivalent and undecidable. *Ann. Pure & Appl. Logics*, 98(1–3):111–156, 1999.

Termination and Reduction Checking for Higher-Order Logic Programs

Brigitte Pientka

Department of Computer Science
Carnegie Mellon University
Pittsburgh, PA 15213, USA
bp@cs.cmu.edu

Abstract. In this paper, we present a syntax-directed termination and reduction checker for higher-order logic programs. The reduction checker verifies parametric higher-order subterm orderings describing relations between input and output of well-moded predicates. These reduction constraints are exploited during termination checking to infer that a specified termination order holds. To reason about parametric higher-order subterm orderings, we introduce a deductive system as a logical foundation for proving termination. This allows the study of proof-theoretical properties, such as consistency, local soundness and completeness and decidability. We concentrate here on proving consistency of the presented inference system. The termination and reduction checker are implemented as part of the Twelf system and enable us to verify proofs by complete induction.

1 Introduction

One of the central problems in verifying specifications and checking proofs about them is the need to prove termination. Several automated methods to prove termination have been developed for first-order functional and logic programs in the past years (for example [15,1]). One typical approach is to transform the program into a term rewriting system (TRS) such that the termination property is preserved. A set of inequalities is generated and the TRS is terminating if there exists no infinite chain of inequalities. This is usually done by synthesizing a suitable measure for terms. To show termination in higher-order simply-typed term rewriting systems (HTRS) mainly two methods have been developed (for a survey see [13]): the first approach relies on strict functionals by van de Pol [12], and the second one is a generalization of recursive path orderings to the higher order case by Jouannaud and Rubio [4].

In this paper, we present a syntax-directed method for proving termination of higher-order logic programs. First, the reduction checker verifies properties relating input and output of higher-order predicates. Using a deductive system to reason about reduction constraints, the termination checker then proves that the inputs of the recursive call are smaller than the inputs of the original call with respect to higher-order subterm orderings. Our method has been developed

R. Goré, A. Leitsch, and T. Nipkow (Eds.): IJCAR 2001, LNAI 2083, pp. 401–415, 2001.

for the higher-order logic programming language Twelf [8] which is based on the logical framework LF [3]. Although Twelf encompasses pure Prolog, it has been designed as a meta-language for the specification of deductive systems and proofs about them. In addition to Prolog it allows hypothetical and parametric subgoals. As structural properties play an important role in this setting, higher-order subterm orderings have been proven to be very powerful (see Section 5).

The principal contributions of this paper are two-fold: 1) We present a logical foundation for proving termination which is of interest in proving termination of first-order and higher-order programs. The logical perspective on reasoning about orders allows the study of proof-theoretical properties, such as consistency, local soundness and completeness and decidability. In this paper, we concentrate on proving consistency of the presented reasoning system by showing admissibility of cut. This implies soundness and completeness of the reasoning system. 2) We describe a practical syntax-directed system for proving termination of higher-order logic programs. Unlike most other approaches, we are interested in checking a given order for a program and not in synthesizing an order for a program. The advantage is that checking whether a given order holds is more efficient than synthesizing orders. In the case of failure, we can provide detailed error messages. These help the user to revise the program or to refine the specified order. The termination checker is implemented as part of the Twelf system and has been used successfully on examples from compiler verification (soundness and completeness proofs for stack semantics and continuation-based semantics), cut-elimination and normalization proofs for intuitionistic and classical logic, soundness and completeness proofs for the Kolmogorov translation of classical into intuitionistic logic (and vice versa).

The paper is organized as follows: In Section 2 we give a representative Twelf program taken from the domain of compiler verification. Using this example we illustrate the basic idea of the termination checker. We review the background (see Section 3) In Section 4 we outline the deductive system for reasoning about orders and prove consistency of the system. Finally, in Section 5 we discuss related work, summarize the results and outline future work.

2   Motivating Example

Our work on termination is motivated by induction theorem proving in the logical framework and its current limitations to handle proofs by complete induction. In this section, we consider a typical example from compiler verification [2] to illustrate our approach.

Compilation is the automatic transformation of a program written in a source language to a program in a target language. Typically, there are several stages of compilation. Starting with a high-level language, the computational description is refined in each step into low-level machine language. To prove correctness of a compiler, we need to show the correspondence between the source and target language. In this example, we consider Mini-ML as the source language, and the language of the abstract machine as the target language. We only consider a small subset of a programming language in this paper.

Mini-ML Syntax

expressions e ::= $e_1 e_2$|lam $x.e$

values   v ::= x|Lam $x.e$

Abstract Machine Syntax

instructions I ::= ret v|ev e|app$_1$ v e|app$_2$ $v_1 v_2$

stack   S ::= nil |$S; \lambda v.I$

The Mini-ML language consists of lambda-abstraction and application. To evaluate an application e_1 e_2, we need to evaluate e_1 to some value Lam $x.e'$, e_2 to some value v_2 and $[v_2/x]e'$ to the final value of the application. Note, the order of evaluation of these premises is left unspecified. The abstract machine has a more refined computation model which is reflected in the instruction set. We not only have instructions operating on expressions and values, but also intermediate mixed instructions such as app$_1$ v_1 e_2 and app$_2$ v_1 v_2. Computation in an abstract machine can be represented as a sequence of states. Each state is characterized by a stack S representing the continuation and an instruction I and written as $S\#I$. In contrast to the big-step semantics for Mini-ML, the small-step transition semantics precisely specifies that an application is evaluated from left to right.

Big step Mini-ML semantics:

$$\frac{}{\text{lam } x.e \hookrightarrow \text{Lam } x.e} \; ev\_lam \qquad \frac{e_1 \hookrightarrow \text{Lam } x.e_1' \quad e_2 \hookrightarrow v_2 \quad [v_2/x]e_1' \hookrightarrow v}{e_1 e_2 \hookrightarrow v} \; ev\_app$$

Small-step transition semantics (single step):

t_lam : $S\#(\text{ev lam } x.e) \longmapsto S\#(\text{ret Lam } x.e)$

t_app: $S\#(\text{ev } e_1 e_2) \longmapsto (S; \lambda v.\text{app}_1 \, v \, e_2)\#(\text{ev } e_1)$

t_app1: $S\#(\text{app}_1 \, v_1 e_2) \longmapsto (S; \lambda v_2.\text{app}_2 \, v_1 v_2)\#(\text{ev } e_2)$

t_app2: $S\#(\text{app}_2 \, (\text{Lam } x.e')v_2) \longmapsto S\#(\text{ev } [v_2/x]e')$

t_ret: $(S; \lambda v.I)\#(\text{ret } v_1) \longmapsto S\#[v_1/v]I$

A computation sequence

$$S\#(\text{ev } e_1 e_2) \overset{t\_app}{\longmapsto} (S; \lambda v.\text{app}_1 \, v \, e_2)\#(\text{ev } e_1) \overset{*}{\longmapsto} \text{nil} \#(\text{ret } w)$$

is represented in Twelf as (t_app @ D1) where t_app represents the first step of computation $S\#(\text{ev } e_1 e_2) \overset{t\_app}{\longmapsto} (S; \lambda v.\text{app}_1 \, v \, e_2)\#(\text{ev } e_1)$ while D1 describes the tail of the computation $(S; \lambda v.\text{app}_1 \, v \, e_2)\#(\text{ev } e_1) \overset{*}{\longmapsto} \text{nil} \#(\text{ret } w)$. We will sometimes mix multi-step transitions $\overset{*}{\longmapsto}$ with single step transitions $\longmapsto$ with the obvious meaning.

An evaluation tree in the big step semantics

$$\frac{\begin{array}{ccc} \mathcal{P}_1 & \mathcal{P}_2 & \mathcal{P}_3 \\ e_1 \hookrightarrow \text{Lam } x.e_1' & e_2 \hookrightarrow v_2 & [v_2/x]e_1' \hookrightarrow v \end{array}}{e_1 e_2 \hookrightarrow v} \; ev\_app$$

is implemented as (ev_app P1 P2 P3). The leaves of the evaluation tree are formed by applications of the *ev_lam* axiom which is implemented as a constant ev_lam in Twelf.

To show that the compiler works correctly, we need to show soundness and completeness of the two semantics. We will concentrate on the first property. To prove soundness we show the following: if we start in an arbitrary state $S\#(\text{ev } e)$ with a computation $S\#(\text{ev } e) \xmapsto{*} \text{nil } \#(\text{ret } w)$ then there exists an intermediate state $S\#(\text{ret } v)$ such that $e \hookrightarrow v$ in the Mini-ML semantics and $S\#(\text{ret } v) \xmapsto{*} \text{nil } \#(\text{ret } w)$.

Theorem 1 (Soundness).
For all computation sequences $\mathcal{D} : S\#(\text{ev } e) \xmapsto{*} \text{nil } \#(\text{ret } w)$ there exists an evaluation tree $\mathcal{P} : e \hookrightarrow v$ and a tail computation $\mathcal{D}' : S\#(\text{ret } v) \xmapsto{*} \text{nil } \#(\text{ret } w)$ such that $\mathcal{D}'$ is smaller than $\mathcal{D}$.

The proof follows by complete induction on $\mathcal{D}$. We consider each computation sequence $\mathcal{D}$ in the small step semantics and translate it into an evaluation tree $\mathcal{P}$ in the Mini-ML semantics and some tail computation $\mathcal{D}'$ which is a sub-derivation of the original computation $\mathcal{D}$. This translation can be described by a meta-predicate **sound** which takes a computation sequence as input and returns an evaluation tree and a tail computation sequence.

As a computation sequence can either start with t_lam or t_app transition, we need to consider two cases. If the computation sequence starts with a t_lam transition (t_lam @ D) then there exists an evaluation of lam $x.e$ to Lam $x.e$ by the ev_lam rule and a tail computation D. The interesting case is when the computation sequence starts with an t_app transition (t_app @ D1).

$$S\#(\text{ev } e_1 \ e_2) \xmapsto{t\_app} \underbrace{(S; \lambda v.\text{app}_1 \ v \ e_2)\#(\text{ev } e_1) \xmapsto{*} \text{nil } \#(\text{ret } w)}_{\text{D1}}$$

We recursively apply the translation to D1 and obtain P1 which represents an evaluation starting in $e_1 \hookrightarrow v_1$ and $(S; \lambda v.\text{app}_1 \ v \ e_2)\#(\text{ret } v_1) \xmapsto{*} \text{nil } \#(\text{ret } w)$ as the tail computation sequence $\mathcal{D}'$. By inversion using the *t_ret* and *t_app1* transition rules, we unfold $\mathcal{D}'$ and obtain the following tail computation sequence

$$(S; \lambda v.\text{app}_1 \ v \ e_2)\#(\text{ret } v_1) \xmapsto{t\_ret} S\#(\text{app}_1 \ v_1 e_2) \xmapsto{t\_app1}$$
$$\underbrace{(S; \lambda v.\text{app}_2 \ v_1 v)\#(\text{ev } e_2) \xmapsto{*} \text{nil } \#(\text{ret } w)}_{\text{D2}}$$

which is represented as (t_ret @ t_app1 @ D2). By applying the translation again to D2, we obtain an evaluation tree for $e_2 \hookrightarrow v_2$ described by P2 and some computation sequence $\mathcal{D}'' : (S; \lambda v.\text{app}_2 \ v_1 v)\#(\text{ret } v_2) \xmapsto{*} \text{nil } \#(\text{ret } w)$. By inversion using rules *t_ret* and *t_app2*, we know that the value v_1 represents some function Lam $x.e'$ and $\mathcal{D}''$ can be unfolded to obtain the tail computation

$$(S; \lambda v.\text{app}_2 \ (\text{Lam } x.e')v)\#(\text{ret } v_2) \xmapsto{t\_ret} S\#(\text{app}_2 \ (\text{Lam } x.e')v_2) \xmapsto{t\_app2}$$
$$\underbrace{S\#(\text{ev } [v_2/x]e') \xmapsto{*} \text{nil } \#(\text{ret } w)}_{\text{D3}}$$

which is represented as (t_ret @ t_app2 @ D3). Now we apply the translation for a final time to D3 and obtain an evaluation tree P3 starting in $[v_2/x]e' \hookrightarrow v$

and some tail computation $S\#(\text{ret } v) \overset{*}{\longmapsto} \text{nil } \#(\text{ret } w)$ which we refer to as D4. The final results of translating a computation sequence (t_app @ D1) are the following: The first result is an evaluation tree for $e_1 e_2 \hookrightarrow v$ which can be constructed by using the ev_app rule and the premises $e_1 \hookrightarrow (\mathsf{Lam}\ x.e')$, $e_2 \hookrightarrow v_2$ and $[v_2/x]e' \hookrightarrow v$. This step is represented in Twelf by (ev_app P1 P2 P3). As a second result, we return the tail computation sequence D4.

The following Twelf program implements the described translation. Throughout this example, we reverse the function arrows writing A_2 <- A_1, instead of A_1 -> A_2 following logic programming notation. Since -> is right associative, <- is left associative. A more detailed discussion of this example is given in [7].

```
sound : S # (ev E) =>* nil # (ret W) ->
          eval E V -> S # (ret V) =>* nil # (ret W) -> type.
%mode sound +D -P -D'.
s_lam : sound (t_lam @ D ) ev_lam D.
s_app : sound (t_app @ D1) (ev_app P3 P2 P1) D4
              <- sound D1 P1 (t_ret @ t_app1 @ D2)
              <- sound D2 P2 (t_ret @ t_app2 @ D3)
              <- sound D3 P3 D4.
```

First the type of the meta-predicate **sound** is defined. It has three arguments: the computation $S\#(\text{ev } E) \overset{*}{\longmapsto} \text{nil } \#(\text{ret } W)$ which is described as S # (ev E) =>* nil # (ret W), the evaluation $e \hookrightarrow v$ which is represented as eval E V and the tail computation sequence $S\#(\text{ret } E) \overset{*}{\longmapsto} \text{nil } \#(\text{ret } W)$ which is defined as S # (ret V) =>* nil # (ret W).

The mode declaration %mode sound +D -P -D' specifies inputs and outputs of the defined predicate. When executed this program translates computations on the abstract machine into Mini-ML evaluations. Dependent types underlying this implementation guarantee that only valid computation sequences and evaluations are generated. The mode checker [11] verifies that all inputs are known when the predicate is called and all output arguments are known after successful execution of the predicate. To check that this program actually constitutes a proof, meta-theoretic properties such as coverage and termination need to be established. Termination guarantees that the input of each recursive call (induction hypothesis) is smaller than the input of the original call (induction conclusion). For termination checking the program needs to be well-moded. In addition, the user specifies which input arguments to consider and in which order they diminish. In the given example, we specify that the predicate **sound** should terminate in the first argument by %terminates D (sound D P D'). For reduction checking we specify an explicit order relation between input and output elements. In the example we say %reduces D' < D (sound D E D'). In general, we allow atomic, lexicographic ($\{Arg_1, Arg_2\}$) or simultaneous ($[Arg_1, Arg_2]$) subterm orderings. To show that a given program satisfies a given reduction constraint pattern, we proceed for each clause in two stages: First we extract a set Δ of reduction constraints from the recursive calls which can be assumed and the reduction constraint P of the whole clause which needs to be satisfied. Second, we prove that the set Δ implies the reduction constraint P. For proving termination of a given program, we also proceed in two stages: For each clause, and for each recursive call we first extract a set Δ of reduction constraints which are valid and a termination constraint P which characterizes the relation between the recursive call and the original call. Second, we prove that the set Δ implies the termination constraint P. For example, to show that the predicate sound terminates, we show the following properties:

Reduction: %reduces D' < D (sound D P D')
if D4 $\prec$ D3, (t_ret @ t_app2 @ D3) $\prec$ D2 and (t_ret @ t_app1 @ D2) $\prec$ D1 then
D4 $\prec$ (t_app @ D1).
Termination: %terminates D (sound D P D')
1. D1 $\prec$ (app @ D1)
2. if (ret @ app1 @ D2) $\prec$ D1 then D2 $\prec$ (app @ D1)
3. if (ret @ app2 @ D3) $\prec$ D2 and (ret @ app1 @ D2) $\prec$ D1 then D3 $\prec$ (app @ D1).

We use $\prec$ to represent the subterm order relation. In general we might have nested clauses which need to be checked recursively. Moreover, we generate parametric reduction constraints for parametric sub-clauses. In Section 5 we give another example for checking termination and reduction. A more detailed explanation for extracting the termination and reduction properties can be found in [9]. In the remainder of the paper we will briefly explain the background theory and then discuss a deductive system for reasoning about structural orderings.

3   Background

The higher-order logic programming language we are working with is based on the logical framework LF [3]. The meta-language of LF is the $\lambda\Pi$-calculus. It is a three-level hierarchical calculus for *objects*, *families*, and *kinds*. Families are classified by kinds, and objects are classified by types, that is, families of kind type.

Kinds $K := \text{type} \mid \Pi x : A.K$ Signatures $\Sigma := \cdot \mid \Sigma, h : K \mid \Sigma, c : A$
Types $A := hM_1 \ldots M_n \mid \Pi x : A_1.A_2$ Context $\Gamma := \cdot \mid \Gamma, x : A$
Objects $M := c \mid x \mid \lambda x : A.M \mid M_1 M_2$

We will use h for type family constants, c for object constants, and x for variables. Constants are introduced through a signature. $\Pi x : A_1.A_2$ denotes the dependent function type or dependent product: the type A_2 may depend on an object x of type A_1. Whenever x does not occur free in A_2 we may abbreviate $\Pi x : A_1.A_2$ as $A_1 \rightarrow A_2$. Below we assume a fixed signature Σ. The types of free variables in a term M are provided by a context Γ. The equivalence $\equiv$ is equality modulo $\beta\eta$-conversion. We will rely on the fact that canonical (i.e. long $\beta\eta$-normal) forms of LF object are computable and that equivalent LF objects have the same canonical form up to α-conversion. We assume that constants and variables are declared at most once in a signature and context, respectively. As usual we apply tacit renaming of bound variables to maintain this assumption and to guarantee capture-avoiding substitutions.

To illustrate the use of basic notation, we consider the representation of the abstract machine which was introduced in the last section. The operations application and lambda abstraction can be represented as canonical LF objects of type `exp`. Values, continuations, instructions and states are defined in a similar fashion. The evaluation derivation $e \hookrightarrow v$ is represented by the judgement `eval` : `exp -> val -> type`. in Twelf. Similarly, we can encode the one-step transition relation and the multi-step transition relation as a judgements in Twelf.

```
exp:      type.                    eval:     exp -> val -> type.
lam:      (val -> exp) -> exp.     ev_lam :  eval (lam E) (lam* E).
app:      exp -> exp -> exp.       ev_app :  eval (app E1 E2) V
                                             <- eval E1 (lam* E1')
val:      type.                              <- eval E2 V2
lam*:     (val -> exp) -> val.               <- eval (E1' V2) V.
```

The capitalized identifiers that occur free in each declaration are implicitly Π-quantified. The appropriate type is deduced from the context during type reconstruction. The fully explicit form of the first declaration would be `ev_lam`: Π `E: val -> exp. eval (lam E) (lam* E)`.

4 A Logical Approach to Termination

4.1 Reasoning about Higher-Order Subterm Orderings

In Section 2 we sketched the analysis of higher-order logic programs for termination and reduction properties. Termination and reduction analysis is separated from reasoning about higher-order subterm relations. The analysis collects valid reduction properties as assumptions and states the ordering which needs to be satisfied under the assumptions. In this section we develop a formal inference system to check whether a set of valid reduction constraints implies an ordering constraint. For now, we consider only first-order subterm reasoning. An ordering constraint is either the $\prec$ subterm relation, the $\preceq$ subterm relation or structural equivalence relation $\equiv$. A context Δ is a set of ordering constraints.

$$
\begin{aligned}
&\text{Context} &\Delta &:= \cdot | \Delta, P \\
&\text{Ordering constraints } P &&:= Arg_1 \prec Arg_2 | Arg_1 \preceq Arg_2 | Arg_1 \equiv Arg_2 \\
&\text{Arg} &Arg &:= M | \{Arg_1, Arg_2\} | [Arg_1, Arg_2]
\end{aligned}
$$

The reasoning system should exhibit a minimal set of desired properties such as transitivity reasoning, congruence closure for structural equality reasoning, and reasoning about λ-terms. The system for first-order subterm reasoning is given in Figure 1. It is similar to the sequent calculus formulation with right and left rules for each ordering relation. $\preceq$ is defined in terms of $\prec$ and $\equiv$. If the rule $L\prec$ has no premises, i.e., N is a constant c with no arguments, the hypothesis is contradictory and the conclusion $\Delta, M \prec c \longrightarrow P$ is trivially true. Reasoning about structural orderings is inherently different from the usual reasoning with equality and inequality. Usually when reasoning about equalities/inequalities, we reason about the value of a term. For example, the value of $hM_1 \ldots M_n$ can be

$$\frac{}{\Delta, P \longrightarrow P} \; id$$

$$\frac{}{\Delta \longrightarrow M \equiv M} \; refl \qquad\qquad \frac{\Delta, M' \equiv M \longrightarrow P}{\Delta, M \equiv M' \longrightarrow P} \; sym$$

$$\frac{\Delta \longrightarrow M_1 \equiv N_1 \quad \ldots \quad \Delta \longrightarrow M_n \equiv N_n}{\Delta \longrightarrow hM_1 \ldots M_n \equiv hN_1 \ldots N_n} \; R{\equiv} \qquad \frac{\Delta, M_1 \equiv N_1, \ldots, M_n \equiv M_n \longrightarrow P}{\Delta, hM_1 \ldots M_n \equiv hN_1 \ldots N_n \longrightarrow P} \; L{\equiv}_1$$

$$\frac{\Delta[M], X \equiv M \longrightarrow P}{\Delta[X], X \equiv M \longrightarrow P} \; Lsubst \qquad\qquad \frac{h \neq g}{\Delta, hM_1 \ldots M_n \equiv gN_1 \ldots N_k \longrightarrow P} \; L{\equiv}_2$$

$$\frac{\Delta \longrightarrow M \prec M_1 \quad \Delta \longrightarrow M_1 \prec M'}{\Delta \longrightarrow M \prec M'} \; t{\prec} \qquad \frac{\Delta \longrightarrow M \prec M_1 \quad \Delta \longrightarrow M_1 \equiv M'}{\Delta \longrightarrow M \prec M'} \; t{\prec}{\equiv}$$

$$\frac{\Delta \longrightarrow M \equiv M_1 \quad \Delta \longrightarrow M_1 \prec M'}{\Delta \longrightarrow M \prec M'} \; t{\equiv}{\prec} \qquad \frac{\Delta \longrightarrow M \equiv M_1 \quad \Delta \longrightarrow M_1 \equiv M'}{\Delta \longrightarrow M \equiv M'} \; t{\equiv}{\equiv}$$

$$\frac{\Delta \longrightarrow M \preceq N_i}{\Delta \longrightarrow M \prec hN_1 \ldots N_n} \; R{\prec}_i \qquad \frac{\Delta, M \preceq N_1 \longrightarrow P \quad \ldots \quad \Delta, M \preceq N_n \longrightarrow P}{\Delta, M \prec hN_1 \ldots N_n \longrightarrow P} \; L{\prec}$$

$$\frac{\Delta \longrightarrow M \prec N}{\Delta \longrightarrow M \preceq N} \; R{\preceq}_1 \qquad\qquad \frac{\Delta \longrightarrow M \equiv N}{\Delta \longrightarrow M \preceq N} \; R{\preceq}_2$$

$$\frac{\Delta, M \prec N \longrightarrow P \quad \Delta, M \equiv N \longrightarrow P}{\Delta, M \preceq N \longrightarrow P} \; L{\preceq}$$

Fig. 1. First-order Subterm Relations ($\prec$, $\preceq$)

equal to the value of $gN_1 \ldots N_k$ where h and g denote different function symbols. When reasoning about subterms, we are only interested in the syntactic structure of a term. Therefore, a term $hM_1 \ldots M_n$ can never be structurally equivalent to $gN_1 \ldots N_k$, if $g \neq h$. If $hM_1 \ldots M_n \equiv gN_1 \ldots N_k$ occurs in our assumptions, we can infer anything ($L{\equiv}_2$).

This system is already expressive enough to prove termination of the translation of small-step semantics into big-step Mini-ML semantics which is implemented by the **sound** predicate (see p. 404). One of the claims we need to prove during termination checking is the following:

(ret @ app1 @ D2) $\prec$ D1 $\longrightarrow$ D2 $\prec$ (app @ D1)

The proof written in a bottom-up linear notation is as follows:

2. (ret @ app1 @ D2) $\prec$ D1 $\longrightarrow$ (ret @ app1 @ D2) $\prec$ D1 id
   (ret @ app1 @ D2) $\prec$ D1 $\longrightarrow$ D2 $\equiv$ D2 $refl$
   (ret @ app1 @ D2) $\prec$ D1 $\longrightarrow$ D2 $\preceq$ D2 $R{\preceq}_2$
   (ret @ app1 @ D2) $\prec$ D1 $\longrightarrow$ D2 $\prec$ (app1 @ D2) $R{\prec}_2$
   (ret @ app1 @ D2) $\prec$ D1 $\longrightarrow$ D2 $\preceq$ (app1 @ D2) $R{\preceq}_1$
1. (ret @ app1 @ D2) $\prec$ D1 $\longrightarrow$ D2 $\prec$ (ret @ app1 @ D2) $R{\prec}_2$
   (ret @ app1 @ D2) $\prec$ D1 $\longrightarrow$ D2 $\prec$ D1 $t{\prec}$ using 1,2
   (ret @ app1 @ D2) $\prec$ D1 $\longrightarrow$ D2 $\preceq$ D1 $R{\preceq}_1$
   (ret @ app1 @ D2) $\prec$ D1 $\longrightarrow$ D2 $\prec$ (app @ D1) $R{\prec}_2$

We can extend the system with rules for lexicographic orderings by defining left and right rules (see Figure 2). O_1 and O_2 are considered to be lexicographically smaller than O_1' and O_2' if either O_1 is smaller than O_1' or O_1 is structurally equivalent to O_1' and O_2 is smaller than O_2'. This disjunctive choice is reflected in the two rules $RLex\prec_1$ and $RLex\prec_2$. If we assume O_1 and O_2 to be lexicographically smaller than O_1' and O_2', then we need to be able to prove some ordering P under the assumption O_1 is smaller than O_1' and under the assumptions O_1 is structurally equivalent to O_1' and O_2 is smaller than O_2' (see $LLex\prec$). The rules for $\preceq$ and $\equiv$ are straightforward. Similarly, we can define extensions for simultaneous orderings. Although we do not pursue other more complex structural orderings for now, in general this approach can be also applied to define extensions for simplification orderings, multi-set orderings or recursive path orderings. In this paper, we focus on extending the system to higher-order subterm relations.

In the setting of a dependently typed calculus, we face two challenges: First, we need to reason about orders involving higher-order terms. Second, we might synthesize parametric order relations due to parametric subgoals. When considering higher-order terms, we need to find an appropriate interpretation for lambda-terms. This problem is illustrated by the following example. Assume the constructor `lam` is defined as `lam: (exp -> exp) -> exp`. We want to show that $E\,a$ is a subterm of `lam` $\lambda x. E\,x$ where a is a parameter. In the informal proof we might count the number of constructors and consider $E\,a$ an instance of $\lambda x. E\,x$. Therefore we consider a term M a subterm of $\lambda x. N\,x$ if there exists a parameter instantiation $\underline{a}$ for x s.t. M is smaller than $[\underline{a}/x]N$. We will use the convention that a will represent a new parameter, while $\underline{a}$ stands for an already defined parameter. To adopt a logical point of view, the λ-term on the left of a subterm relation can be interpreted as universally quantified and the λ-term on the right as existentially quantified.

$$\frac{\Delta \longrightarrow O_1 \prec O_1'}{\Delta \longrightarrow \{O_1, O_2\} \prec \{O_1', O_2'\}}\; RLex\prec_1 \qquad \frac{\Delta \longrightarrow O_1 \equiv O_1' \quad \Delta \longrightarrow O_2 \prec O_2'}{\Delta \longrightarrow \{O_1, O_2\} \prec \{O_1', O_2'\}}\; RLex\prec_2$$

$$\frac{\Delta \longrightarrow \{O_1, O_2\} \prec \{O_1', O_2'\}}{\Delta \longrightarrow \{O_1, O_2\} \preceq \{O_1', O_2'\}}\; RLex\preceq_1 \qquad \frac{\Delta \longrightarrow \{O_1, O_2\} \equiv \{O_1', O_2'\}}{\Delta \longrightarrow \{O_1, O_2\} \preceq \{O_1', O_2'\}}\; RLex\preceq_2$$

$$\frac{\Delta \longrightarrow O_1 \equiv O_1' \quad \Delta \longrightarrow O_2 \equiv O_2'}{\Delta \longrightarrow \{O_1, O_2\} \equiv \{O_1', O_2'\}}\; RLex\equiv$$

$$\frac{\Delta, O_1 \prec O_1' \longrightarrow P \quad \Delta, O_1 \equiv O_1', O_2 \prec O_2' \longrightarrow P}{\Delta, \{O_1, O_2\} \prec \{O_1', O_2'\} \longrightarrow P}\; LLex\prec$$

$$\frac{\Delta, \{O_1, O_2\} \prec \{O_1', O_2'\} \longrightarrow P \quad \Delta, \{O_1, O_2\} \equiv \{O_1', O_2'\} \longrightarrow P}{\Delta, \{O_1, O_2\} \preceq \{O_1', O_2'\} \longrightarrow P}\; LLex\preceq$$

$$\frac{\Delta, O_1 \equiv O_1', O_2 \equiv O_2' \longrightarrow P}{\Delta, \{O_1, O_2\} \equiv \{O_1', O_2'\} \longrightarrow P}\; LLex\equiv$$

Fig. 2. Lexicographic Extensions

Another example is taken from the representation of first-order logic [6]. We can represent formulas by the type family o. Individuals are described by the type family i. The constructor ∀ can be defined as `forall: (i -> o) -> o`. We might want to show that $A\,T$ (which represents $[t/x]A$) is smaller than `forall` $\lambda x.A\,x$ (which represents $\forall x.A$). Similarly, we might count the number of quantifiers and connectives in the informal proof, noting that a term t in first-order logic cannot contain any logical symbols. Thus we may consider $A\,T$ a subterm of `forall` $\lambda x.A\,x$ as long as there is no way to construct an object of type i from objects of type o. A term M is smaller than a λ-term $(\lambda x.N)$ if there exists an instantiation T for x s.t. M is smaller than $[T/x]N$ and the type of T is a subordinate to N. For a more detailed development of mutual recursion and subordination we refer the reader to R. Virga's PhD thesis [14].

$$\frac{}{\Delta, \lambda x.M \equiv hN_1 \ldots N_n \longrightarrow P}\ L{\equiv}_3 \qquad \frac{}{\Delta, a \equiv hN_1 \ldots N_n \longrightarrow P}\ L{\equiv}_4$$

$$\frac{\Delta \longrightarrow [a/x]M \equiv [a/x]N}{\Delta \longrightarrow \lambda x.M \equiv \lambda x.N}\ R{\equiv}\,\lambda \qquad \frac{\Delta, [\underline{a}/x]M \equiv [\underline{a}/x]N \longrightarrow P}{\Delta, \lambda x.M \equiv \lambda x.N \longrightarrow P}\ L{\equiv}\,\lambda$$

$$\frac{\Delta \longrightarrow [a/x]M \prec N}{\Delta \longrightarrow \lambda x : A.M \prec N}\ RL{\prec}\,\lambda^a \qquad \frac{\Delta, [\underline{a}/x]M \prec N \longrightarrow P}{\Delta, \lambda x : A.M \prec N \longrightarrow P}\ LL{\prec}\,\lambda$$

$$\frac{\Delta \longrightarrow M \prec [\underline{a}/x]N}{\Delta \longrightarrow M \prec \lambda x : A.N}\ RR{\prec}\,\lambda \qquad \frac{\Delta, M \prec [a/x]N \longrightarrow P}{\Delta, M \prec \lambda x : A.N \longrightarrow P}\ LR{\prec}\,\lambda^a$$

$$\frac{\Delta \longrightarrow [a/x]M \preceq N}{\Delta \longrightarrow \lambda x : A.M \preceq N}\ RL{\preceq}\,\lambda^a \qquad \frac{\Delta, [\underline{a}/x]M \preceq N \longrightarrow P}{\Delta, \lambda x : A.M \preceq N \longrightarrow P}\ LL{\preceq}\,\lambda$$

$$\frac{\Delta \longrightarrow M \preceq [\underline{a}/x]N}{\Delta \longrightarrow M \preceq \lambda x : A.N}\ RR{\preceq}\,\lambda \qquad \frac{\Delta, M \preceq [a/x]N \longrightarrow P}{\Delta, M \preceq \lambda x : A.N \longrightarrow P}\ LR{\preceq}\,\lambda^a$$

$$\frac{\Delta \longrightarrow [a/x]P}{\Delta \longrightarrow \Pi x.P}\ R\Pi^a \qquad \frac{\Delta, [\underline{a}/x]P \longrightarrow P'}{\Delta, \Pi x.P \longrightarrow P'}\ L\Pi$$

Fig. 3. Higher-order Extensions

Reasoning about λ-terms cannot be solely based $\prec$ and $\equiv$, as neither $[a/x]M \equiv \lambda x.M$ nor $[a/x]M \prec \lambda x.M$ is true. Therefore, we introduce a set of inference rules to reason about $\preceq$ which are similar to the $\prec$ rules. Extensions to higher-order subterm reasoning are presented in Figure 3. As we potentially need different instantiations of the relation $\lambda x.M \prec N$ when reading the inference rules bottom-up, we need to copy $\lambda x.M \prec N$ in Δ even after it has been instantiated. For simplicity, we assume all assumptions persist. Note that we only show the case for mutual recursive type families, but the case where type family a is a subordinate to the type family a' can be added in straightforward manner. For handling parametric order relations we add $R\Pi^a$ and $L\Pi$ which are similar to universal quantifier rules in the sequent calculus. Similar to instantiations of

$\lambda x.M \prec N$, we need to keep a copy of $\Pi x.P$ after it has been instantiated. The weakening and contraction property hold for the given calculus.

Reasoning about higher-order subterm relations is complex due to instantiating λ-terms and parametric orderings. Although soundness and decidability of the first-order reasoning system might still be obvious, this is non-trivial in the higher-order case. In this paper, we concentrate on proving consistency of the higher-order reasoning system. Consistency of the system implies soundness, i.e. any step in proving an order relation from a set of assumptions is sound. The proof also implies completeness i.e. anything which should be derivable from a set of assumptions is derivable.

4.2 Consistency of Higher-Order Subterm Reasoning

In general, the consistency of a logical system can be shown by proving cut admissible.

$$\frac{\Delta \longrightarrow P \qquad \Delta, P \longrightarrow P'}{\Delta \longrightarrow P'} \; cut$$

Δ usually consists of elements which are assumed to be true. Any P which can be derived from Δ is true and can therefore be added to Δ to prove P'. In our setting Δ consists of reduction orderings which have already been established. Hence, the reduction orderings are true independently from any other assumptions in Δ and they are assumed to be valid. The application of the cut-rule in the proof can therefore only introduce valid orderings as additional assumptions in Δ.

Theorem 2 (Admissibility of cut).

1. If $\mathcal{D} : . \longrightarrow M \equiv M'$ and $\mathcal{E} : \Delta, M \equiv M' \longrightarrow P'$ then $\mathcal{F} : \Delta \longrightarrow P'$.
2. If $\mathcal{D} : . \longrightarrow \sigma M \prec M'$ and $\mathcal{E} : \Delta, \lambda \overrightarrow{x}.M \prec M' \longrightarrow P'$ then $\mathcal{F} : \Delta \longrightarrow P'$.
3. If $\mathcal{D} : . \longrightarrow \sigma M \preceq M'$ and $\mathcal{E} : \Delta, \lambda \overrightarrow{x}.M \preceq M' \longrightarrow P'$ then $\mathcal{F} : \Delta \longrightarrow P'$.

The substitution σ maps free variables to new parameters. In general, we allow the cut between $\sigma M \prec N$ and $\lambda \overrightarrow{x}.M \prec N$ where σM is an instance of $\lambda \overrightarrow{x}.M$.

However, we will not be able to show admissibility of cut directly in the given calculus due to the non-deterministic choices introduced by λ-terms. Consider, for example, the cut between

$$\mathcal{D} = \frac{\mathcal{D}_1}{. \longrightarrow \sigma \circ [a/x]M \prec N} RL\prec\lambda. \qquad \overset{\mathcal{E}}{\Delta, \lambda x.M \prec N \longrightarrow P}$$

We would like to apply inversion on $\mathcal{E}$; therefore we need to consider all possible cases of previous inference steps which lead to $\mathcal{E}$. There are three possible cases we need to consider: $L\prec$, $LR\prec\lambda^a$ and $LL\prec\lambda$. Unfortunately, it is not possible to appeal to the induction hypothesis and finish the proof in the $L\prec$ and $LR\prec\lambda$ case. This situation does not arise in the first order case, because all the inversion steps were unique. In the higher-order case we have many choices and we are manipulating the terms by instantiating variables in λ-terms.

The simplest remedy seems to restrict the calculus in such a way, that we always first introduce all possible parameters, and then instantiate all Π quantified orders and $\lambda x.M$ which occur on the left side of a relation. This means, we push the instantiation with parameter variables as high as possible in the proof tree. This way, we can avoid the problematic case above, because we only instantiate a λ-term in $\lambda x.M \prec N$, if N is atomic.

Therefore, we proceed as follows: First, we define an inference system, in which we first introduce all new parameters. This means we restrict the application of the $R\preceq_1$, $R\preceq_2$, $R\prec_i$, $RR\prec\lambda$, $RR\preceq\lambda$ to only apply if the left hand side of the principal order relation $\prec$ or $\preceq$ is already of base type. Similarly, we restrict the application of $L\preceq$, $LL\preceq\lambda$, $LL\prec\lambda$, i.e. the rule only applies if the right hand side of the principal ordering relation is of base type. In addition, we show that the application of the identity rules can be restricted to atomic terms. Second, we show this restricted system is sound and complete with respect to the original inference system. Third, we show that cut is admissible in the restricted calculus. This implies that cut is also admissible in the original calculus. The proof proceeds by nested induction on the structure of P, the derivation $\mathcal{D}$ and $\mathcal{E}$. More precisely, we appeal to the induction hypothesis either with a strictly smaller order constraint P or P stays the same and one of the derivations is strictly smaller while the other one stays the same. For a more detailed development of the intermediate inference system and the proofs we refer to [9]

Using the cut-admissibility theorem, cut-elimination follows immediately. Therefore, our inference system is consistent. This implies that all derivation steps in the given reasoning system are sound. It also implies that the inference rules are strong enough to deduce as much as possible from the assumptions and hence the system is complete.

5 Related Work and Conclusion

Most work in automating termination proofs has focused on first-order languages. The most general method for synthesizing termination orders for a given term rewriting system (TRS) is by Arts and Giesl [1]. One approach to proving termination of logic programs is to translate it into a TRS and show termination of the TRS instead. However this approach has several drawbacks. In general, a lot of information is lost during the translation. In particular, if termination analysis fails for the TRS, it is hard to provide feedback and re-use this failure information to point to the error in the logic program. Moreover important structural information is lost during the translation and constructors and functions are indistinguishable. One of the consequences is that proving termination of the TRS often requires more complicated orders. This is illustrated using an example from arithmetic. Using logic programming we implement a straightforward version of minus and the quotient predicate quot.

```
minus : nat -> nat -> nat -> type.        quot : nat -> nat -> nat -> type.
%mode minus +X +Y -Z.                     %mode quot +X +Y -Z.
m_z : minus X z X.                        q_z : quot z (s Y) z.
m_s : minus (s X) (s Y) Z                 q_s : quot (s X) (s Y) (s Z)
      <- minus X Y Z.                            <- minus X Y X'
                                                 <- quot X' (s Y) Z.

%reduces Z <= X (minus X Y Z).
%terminates X (minus X Y Z).              %terminates X (quot X Y Z).
```

Proving termination of `quot` is straightforward with the presented method. We first prove termination of `minus`. In addition we show that `minus X Y Z` satisfies the reduction constraint `Z <= X`. When we prove termination of `quot`, we can assume the reduction constraint $X' \preceq X$. As the reduction constraint $X' \preceq X$ implies $X' \prec (s\,X)$, we proved termination of `quot`. Note that only subterm reasoning is required to prove termination of `quot` while other methods like Arts and Giesl's method for proving the corresponding term rewrite system needs a recursive path ordering. Another example is an algorithm to compute the negation normal form of a first-order logical formula and uses higher-order functions (see [9]). We implemented this algorithm using two mutual recursive predicates. Termination of this algorithm can be proven based on subterm ordering, while the corresponding term rewriting system given in [5] requires a more complicated ordering like recursive path ordering.

Although some of the underlying ideas in higher-order term rewriting system (HTRS) are shared with the logical framework, there are two principal differences: First, all arguments of a predicate are in canonical form and therefore are terminating. This additional restriction simplifies termination analysis in the logical framework. On the other hand, the dependently typed $\lambda\Pi$ calculus, on which the logical framework LF is based, allows the representation of hypothetical and parametric judgements which make termination and reduction analysis more challenging. Hypothetical and parametric judgements have in general no counterpart in HTRS and their translation to HTRS seems difficult.

One approach which analyzes logic programs directly has been developed by Plümer [10]. The idea is to construct a subgoal dependency graph and then show that this graph is acyclic according to some ordering. Although this approach works well for Prolog programs, it is not obvious how to extend this method in a higher-order setting with parametric and hypothetical subgoals. In this paper we propose a proof-theoretical foundation for termination checking of higher-order logic programs. To infer that a specified ordering holds under a set of assumptions, we introduced a deductive system to reason about structural orderings. We focused on consistency of the presented reasoning system. Consistency implies that anything we derive from the assumption is sound. Cut-elimination implies that the reasoning system is complete, i.e. everything which should be derivable from the assumptions is in fact derivable. A valuable advantage of this approach is its extensibility and its modularity. Similar to lexicographic extensions we can imagine extensions for simplification ordering, multi-set ordering and recursive path orderings. In addition our method allows us to combine different structural orderings for different predicates. This is unlike other termination methods which require one ordering for the whole dependency graph.

This paper builds on Rohwedder and Pfenning's work on mode and termination checking for higher-order logic programs [11]. Their termination checker requires a direct relationship between inputs of the recursive call and inputs of the original call without taking into account input and output relations. Reasoning about orderings allows us to check proofs by complete induction such as the soundness proof discussed in this paper. The emphasis of their work has been the correctness of the termination checker with respect to the operational semantics of Twelf programs. Although we have not proven the correctness of the extended termination checker, we are expecting the proof to be a straightforward extension of their proof.

One question not discussed in this paper is whether the system is decidable. This question is not trivial as we can potentially instantiate λ-terms and Π-quantified order relations which occur in the context multiple times. One approach for proving decidability would be to show that we can bound the number of instantiations needed.

Our system is implemented as part of Twelf, and efficiently checks programs and proofs. Currently multiplicity is restricted to one, i.e. we instantiate Π-quantified orderings and λ-terms occurring on the left hand side of a relation in the hypothesis just once. Although we can artificially construct examples which require multiplicity more than one, we have not encountered these cases in practice so far. If a higher multiplicity is needed, an appropriate warning is returned. As our algorithm analyzes program clauses directly, its behaviour is easy to understand. In the case of failure, our implementation will point to the clause and argument where the error occurred. This enables the user to either revise the program or strengthen the ordering. In practice we have used the termination and reduction checker on examples from compiler verification (soundness and completeness proofs for stack semantics and continuation-based semantics), cut-elimination and normalization proofs for intuitionistic and classical logic, soundness and completeness proofs for the Kolmogorov translation of classical into intuitionistic logic (and vice versa)[1]. Currently, Rohwedder and Pfenning's termination checker is used in the automatic induction theorem prover. In the future, we plan to incorporate the extended termination checker.

Acknowledgements. The author gratefully acknowledges numerous fruitful discussions with Frank Pfenning regarding the subject of this paper. His guidance and careful reading of this paper contributed greatly to its clarity and correctness.

---

[1] The code of all the examples mentioned in the paper can be found at
http://www.cs.cmu.edu/~bp/code.

References

1. Thomas Arts and Jürgen Giesl. Termination of term rewriting using dependency pairs. *Theoretical Computer Science*, 236:133–178, 2000.
2. John Hannan and Frank Pfenning. Compiler verification in LF. In Andre Scedrov, editor, *Seventh Annual IEEE Symposium on Logic in Computer Science*, pages 407–418, Santa Cruz, California, June 1992.
3. Robert Harper, Furio Honsell, and Gordon Plotkin. A framework for defining logics. *Journal of the Association for Computing Machinery*, 40(1):143–184, January 1993.
4. J.-P. Jouannaud and A. Rubio. The higher-order recursive path ordering. In G. Longo, editor, *Proceedings of the 14th Annual Symposium on Logic in Computer Science (LICS'99)*, pages 402–411, Trento, Italy, July 1999. IEEE Computer Society Press.
5. Olav Lysne and Javier Piris. A termination ordering for higher order rewrite systems. In Jieh Hsiang, editor, *Proceedings of the Sixth International Conference on Rewriting Techniques and Applications*, pages 26–40, Kaiserslautern, Germany, April 1995. Springer-Verlag LNCS 914.
6. Frank Pfenning. Structural cut elimination. In D. Kozen, editor, *Proceedings of the Tenth Annual Symposium on Logic in Computer Science*, pages 156–166, San Diego, California, June 1995. IEEE Computer Society Press.
7. Frank Pfenning. *Computation and Deduction*. Cambridge University Press, 2000. In preparation. Draft from April 1997 available electronically.
8. Frank Pfenning and Carsten Schürmann. System description: Twelf — a metalogical framework for deductive systems. In H. Ganzinger, editor, *Proceedings of the 16th International Conference on Automated Deduction (CADE-16)*, pages 202–206, Trento, Italy, July 1999. Springer-Verlag LNAI 1632.
9. Brigitte Pientka. Termination and reduction checking in the logical framework. Technical report cmu-cs-01-115, Carnegie Mellon University, 2001.
10. Lutz Plümer. *Termination Proofs for Logic Programs*. LNAI 446. Springer-Verlag, 1990.
11. Ekkehard Rohwedder and Frank Pfenning. Mode and termination checking for higher-order logic programs. In Hanne Riis Nielson, editor, *Proceedings of the European Symposium on Programming*, pages 296–310, Linköping, Sweden, April 1996. Springer-Verlag LNCS 1058.
12. J. van de Pol and H. Schwichtenberg. Strict functionals for termination proofs. In M. Dezani-Ciancaglini and G. Plotkin, editors, *Proceedings of the International Conference on Typed Lambda Calculi and Applications*, pages 350–364, Edinburgh, Scotland, April 1995. Springer-Verlag LNCS 902.
13. Femke van Raamsdonk. Higher-order rewriting. In *Proceedings of the 10th International Conference on Rewriting Techniques and Applications (RTA '99)*, pages 220–239, Trento, Italy, July 1999. Springer-Verlag LNCS 1631.
14. Roberto Virga. *Higher-Order Rewriting with Dependent Types*. PhD thesis, Department of Mathematical Sciences, Carnegie Mellon University, 2000.
15. Christoph Walther. On proving the termination of algorithms by machine. *Artificial Intelligence*, 71(1), 1994.

P.rex: An Interactive Proof Explainer

Armin Fiedler

Universität des Saarlandes, FR Informatik,
Postfach 15 11 50, D-66041 Saarbrücken, Germany
afiedler@cs.uni-sb.de

Abstract. This paper outlines the interactive proof explanation system
P.rex, which adapts its explanation to the user and allows him anytime
to utter questions or requests, to which it reacts flexibly. As a generic system, it can be connected to different theorem provers. The distribution is
available via the *P.rex* home page at http://www.ags.uni-sb.de/~prex.

1 The *P.rex* System

P.rex is an interactive proof explanation system that adapts its explanations
to the user and flexibly reacts to his questions or requests. An overview of its
architecture is provided in Figure 1.

As a generic system, *P.rex* can be connected to different theorem provers,
namely by means of the formal language TWEGA for specifying proofs and
mathematical theories (cf. Section 2). *Mathematical theories* are organized in
a hierarchical knowledge base. Each theory in it may contain, for example, axioms, definitions, and theorems along with proofs. A *proof* of a theorem can be
represented hierarchically in TWEGA such that the various levels of abstraction
are made explicit.

The central component of the system is the *dialog planner* (cf. Section 3).
It is implemented in ACT-R [1], a goal-directed production system that aims to
model human cognition. In ACT-R, declarative and procedural representations
of knowledge are explicitly separated into the declarative memory and the procedural production rule base. The plan operators of the dialog planner are defined
in terms of productions and the discourse plan is represented in the declarative
memory.

To explain a particular proof, the dialog planner first assumes the individual
user's supposed cognitive state by updating its declarative and procedural mem-
ories from the data base of user models. Then, the dialog planner sets the global
goal to show the proof. ACT-R tries to fulfill this goal by successively applying
productions that decompose or fulfill goals. Thereby, the dialog planner not only
produces a dialog plan, but also traces the user's cognitive states in the course
of the explanation. This allows the system both to always choose an explanation
adapted to the user, and to react to the user's interactions flexibly.

The dialog plan is passed on to the *presentation component*. Currently, we use
a derivate of *PROVERB*'s micro-planner [8] to plan the internal structure of the
sentences, which are then realized by the syntactic generator TAG-GEN [9]. The
uttered sentences are finally displayed on the *interface*. It also allows the user to

R. Goré, A. Leitsch, and T. Nipkow (Eds.): IJCAR 2001, LNAI 2083, pp. 416–420, 2001.
© Springer-Verlag Berlin Heidelberg 2001

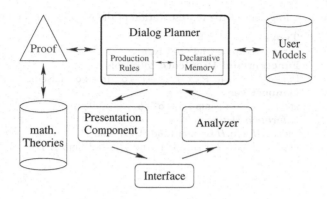

Fig. 1. The architecture of *P.rex*

enter remarks, requests and questions anytime. An *analyzer* receives the user's interactions and passes them on to the dialog planner. In the current stage, we use a simplistic analyzer that understands fifteen predefined quasi-natural language interactions.

2   The Representation of Mathematical Objects

The calculus of constructions (λC) [3] is a dependent typed lambda calculus that was devised as a formalism to represent mathematics. TWEGA is an implementation of λC with $\beta\eta$-conversion extended by two additional features: signatures and constant definitions.[1] Its abstract syntax is given as follows:

$$Terms\ \mathcal{T} ::= \mathcal{V} \mid \mathcal{C} \mid \mathcal{T}\mathcal{T} \mid \lambda\mathcal{V}\!:\!\mathcal{T}.\mathcal{T} \mid \Pi\mathcal{V}\!:\!\mathcal{T}.\mathcal{T}$$

where $\mathcal{V}$ and $\mathcal{C}$ are infinite collections of variables and constants, respectively. We write $A \to B$ for $\Pi x\!:\!A.B$ if x does not occur in B.

To describe the basic judgments, we consider signatures, which contain only constant declarations and definitions, and contexts, which contain only variable declarations. The type system stratifies terms into three levels: objects, types, and kinds. Let Σ be a signature, Γ a context, and A, B terms. Judgments are

$$\Gamma \vdash_\Sigma A : B \quad \text{A and B are valid terms and A is of type B}$$
$$\Gamma \vdash_\Sigma A \equiv B \quad \text{A is definitionally equal to B}$$

The notion of definitional equality we consider here is $\beta\eta$-conversion. [6] gives the complete definition of TWEGA.

In TWEGA, we employ a representation technique that is called *judgments-as-types* [7]. This technique is characterized by mapping judgments to types and their proofs to object terms, thus reducing the problem of proof checking to the problem of type checking. A special type family that is indexed by formulae is

---

[1] The implementation of TWEGA draws on Twelf [10], an implementation of the LF logical framework [7], which is contained in λC.

Table 1. A fragment of the representation of the ND calculus in TWEGA.

Types
i : type   o : type   nd : o→type
Function and Predicate Symbols
f : i→···→i→i   P : i→···→i→o   true : o   false : o
Connectives
and : o→o→o   imp : o→o→o
Inference Rules
ander : ΠA:o.ΠB:o.nd (and A B)→nd B
impi : ΠA:o.ΠB:o.(nd A→nd B)→nd (imp A B)

used to represent judgments as types. Inference rules are represented as functions from judgments to judgments. Constant definitions allow us to represent several levels of abstraction for a given proof.

Example 1. Table 1 gives the judgment-as-types representation of a fragment of the ND calculus in TWEGA. Note that the type family nd serves to represent ND judgments. The following constant definition represents a derived inference rule ∧*Comm* that expresses the commutativity of the conjunction:

andcomm ≡ λA:o.λB:o.λp:nd(and A B).andi B A (ander A B u) (andel A B u)
    : ΠA:o.ΠB:o.nd(and A B)→nd(and B A)

The λ-term is called the *expansion* of andcomm and represents the derivation of the inference rule, whereas its type (i.e., the Π-term) represents the inference rule itself. Now, let us consider the following ND proof of the theorem $P\wedge Q \supset Q\wedge P$:

$$\frac{\dfrac{[\vdash_{ND} P \wedge Q]^u}{\vdash_{ND} Q \wedge P} \wedge Comm}{\vdash_{ND} P \wedge Q \supset Q \wedge P} \supset I^u$$

This theorem and its proof are represented in TWEGA by the following judgment:

$\vdash_\Sigma \lambda P$:o.λQ:o.impi (and P Q) (and Q P)
        (λu:nd (and P Q).andcomm P Q u)
    : ΠP:o.ΠQ:o.nd (imp (and P Q) (and Q P))

Replacement of andcomm by its expansion renders a more detailed proof.

In the remainder of this paper, we mean $\vdash_\Sigma$ whenever we write $\vdash$. We often write $\Gamma \vdash \varphi$ when there is some $\mathcal{D}$ such that $\Gamma \vdash \mathcal{D} : \varphi$.

3   Discourse Planning

The dialog planner of *P.rex* plans the dialog by building a representation of the structure of the discourse that includes speech acts as well as relations among them. *Speech acts* are the primitive actions planned by the dialog planner. Each speech act can always be realized by a single sentence. The discourse structure is represented in the declarative memory.

The *plan operators* are defined as productions. Each production either fulfills the current goal directly or splits it into subgoals. Let us consider

$$\Gamma \vdash R c_1 \ldots c_m \mathcal{D}_1 \ldots \mathcal{D}_n : \psi$$

where R is an inference rule, $c_1, \ldots, c_m$ are parameters, and $\mathcal{D}_i$ with $\Gamma \vdash \mathcal{D}_i : \varphi_i$ is the derivation of φ_i for $1 \leq i \leq n$.

An example for a production is:

(P1)   IF   the current goal G is to show $\Gamma \vdash \psi$
              and R is the most abstract rule known to the user justifying G
              and $\Gamma \vdash \varphi_1, \ldots, \Gamma \vdash \varphi_n$ are known to the user
       THEN   produce the speech act
              (Derive :Reasons $(\varphi_1, \ldots, \varphi_n)$:Conclusion ψ :Method R)
              and pop G (thereby storing $\Gamma \vdash \psi$ in the declarative memory).

By producing the speech act (which may be verbalized as "Since $\varphi_1, \ldots, \varphi_n$, we obtain ψ by R.") the current goal is fulfilled and can be popped from the goal stack. An example for a production decomposing the current goal into several subgoals is:

(P2)   IF   the current goal G is to show $\Gamma \vdash \psi$
              and R is the most abstract rule known to the user justifying G
              and $\Phi = \{\varphi_i | \Gamma \vdash \varphi_i$ is unknown to the user for $1 \leq i \leq n\} \neq \emptyset$
       THEN   for each $\varphi_i \in \Phi$ push the goal to show $\Gamma \vdash \varphi_i$.

Note that the conditions of (P1) and (P2) only differ in the knowledge of the premises φ_i for rule R. (P2) introduces the subgoals to prove the unknown premises in Φ. As soon as those are derived, (P1) can apply and derive the conclusion. Moreover, note that these production are independent of the calculus that is represented in TWEGA. However, there are also calculus-specific productions. Cf. [4] for a more detailed discussion of plan operators in *P.rex*.

The ability for user interaction is an important feature of explanation systems. In *P.rex*, the user can interact with the system at any time. Each interaction is analyzed by the analyzer and passed on to the dialog planner, which then plans an appropriate reaction. [5] describes user interaction in *P.rex*.

Example 2. Consider again the proof from Example 1. The explanation of this proof by *P.rex* is displayed in Figure 2 with the user's utterances underlined.

Note that the user interrupted the system when he did not understand the step justified by $\wedge Comm$. The system entered a clarification dialog to pinpoint the reason why the explanation did not succeed and re-planned the step using the expansion of andcomm. After verifying that the user understood the expansion, the dialog planner resumed the explanation of the proof.

Similarly, the system is able to re-plan a derivation at a higher level of abstraction when the user complains that an explanation is too detailed.

4   Availability

P.rex has been successfully used for the explanation of proofs found by ΩMEGA [2] and Twelf [10]. The distribution is available via the *P.rex* home page at http://www.ags.uni-sb.de/~prex.

420   A. Fiedler

Theorem 1 $((P \wedge Q) \supset (Q \wedge P))$.

Proof:

Let $(P \wedge Q)$. That implies that $(Q \wedge P)$ by $\wedge Comm$.
This step is too difficult.
$(Q \wedge P)$ since $(P \wedge Q)$ by $\wedge Comm$.
Do you understand this step? No.
Do you understand the premises? Yes.
Then Q by $\wedge E_r$. P by $\wedge E_l$. That implies that $(Q \wedge P)$ by $\wedge I$.
Do you understand this step? Yes.
Then $((P \wedge Q) \supset (Q \wedge P))$ by $\supset I$.   ∎

Fig. 2. An example explanation.

Acknowledgments. The author wishes to thank Frank Pfenning and Carsten Schürmann for their help in the development and the implementation of TWEGA and Christian Lebiere for his help in the use of ACT-R.

References

[1] John R. Anderson and Christian Lebiere. *The Atomic Components of Thought.* Lawrence Erlbaum, 1998.
[2] C. Benzmüller, et al. ΩMEGA: Towards a mathematical assistant. In William McCune, ed., *Proc. of the 14th Conference on Automated Deduction*, number 1249 in LNAI, pages 252–255, Townsville, Australia, 1997. Springer Verlag.
[3] Thierry Coquand and Gérard Huet. The Calculus of Constructions. *Information and Computation*, 76(2/3):95–120, 1988.
[4] Armin Fiedler. Using a cognitive architecture to plan dialogs for the adaptive explanation of proofs. In Thomas Dean, ed., *Proc. of the 16th International Joint Conference on Artificial Intelligence (IJCAI)*, pages 358–363, Stockholm, Sweden, 1999. Morgan Kaufmann.
[5] Armin Fiedler. Dialog-driven adaptation of explanations of proofs. In *Proc. of the 17th International Joint Conference on Artificial Intelligence (IJCAI)*, Seattle, Washington, 2001. In press.
[6] Armin Fiedler. *User adaptive proof explanation.* PhD thesis, University of the Saarland, Saarbrücken, Germany, 2001. In preparation.
[7] Robert Harper, Furio Honsell, and Gordon Plotkin. A framework for defining logics. *Journal of the Association for Computing Machinery*, 40(1):143–184, 1993.
[8] Xiaorong Huang and Armin Fiedler. Proof verbalization as an application of NLG. In Martha E. Pollack, ed., *Proc. of the 15th International Joint Conference on Artificial Intelligence (IJCAI)*, pages 965–970, Nagoya, Japan, 1997. Morgan Kaufmann.
[9] Anne Kilger and Wolfgang Finkler. Incremental generation for real–time applications. Research Report RR-95-11, DFKI, Saarbrücken, Germany, July 1995.
[10] Frank Pfenning and Carsten Schürmann. System description: Twelf — a meta-logical framework for deductive systems. In Harald Ganzinger, ed., *Proc. of the 16th Conference on Automated Deduction*, number 1632 in LNAI, pages 202–206. Springer Verlag, 1999.

JProver: Integrating Connection-Based Theorem Proving into Interactive Proof Assistants

Stephan Schmitt[1], Lori Lorigo[2], Christoph Kreitz[2], and Aleksey Nogin[2]

[1] Department of Sciences & Engineering, Saint Louis University, Madrid, Spain
schmitts@spmail.slu.edu
[2] Department of Computer Science, Cornell-University, Ithaca, NY, U.S.A.
{lolorigo,kreitz,nogin}@cs.cornell.edu

Abstract. JProver is a first-order intuitionistic theorem prover that creates sequent-style proof objects and can serve as a proof engine in interactive proof assistants with expressive constructive logics. This paper gives a brief overview of JProver's proof technique, the generation of proof objects, and its integration into the Nuprl proof development system.

1 Introduction

In large scale applications of automated reasoning, interactive proof assistants such as Coq, HOL, Isabelle, Nuprl, and PVS are the tools of choice. Because of their expressive logics, they are more generally applicable than first-order tools, yet at a much lesser degree of automation.

JProver was developed in an effort to combine the expressive power of interactive proof assistants with the automatic capabilities of first-order theorem proving, both for reasoning about mathematics and for reasoning about programs. It provides a theorem prover for first-order intuitionistic and classical logic based on the connection method [3,10], a tool for generating proof objects in the style of sequent proofs [11], and is coupled with mechanisms for integrating the prover into the Nuprl proof/program development system [4,1] and the MetaPRL proof environment [8,9]. These components enable a user to invoke the automatic prover on proof goals that can be solved by first-order reasoning while using the expressive logic of the proof assistant for the more demanding proof parts. Furthermore, the proof information returned by JProver enables the proof assistant to build a valid proof in its own calculus.

As an example, Figure 1 describes the link between JProver and Nuprl, which is described in detail in Section 3. JProver is a stand-alone prover that communicates with a proof assistant through a logic module. Invoking JProver on a Nuprl subgoal sequent causes this sequent to be sent to JProver. The proof-search method in JProver will then generate a matrix proof from the corresponding formula tree (provided the sequent is valid), which then will be converted into a list of sequent rules that expresses a sequent proof for the formula. Upon receiving this list, Nuprl will build a sequent proof for the original goal sequent, thus confirming that the proof found is valid. Information about the relation between this sequent and the formula proven by JProver will be used during that step.

R. Goré, A. Leitsch, and T. Nipkow (Eds.): IJCAR 2001, LNAI 2083, pp. 421–426, 2001.

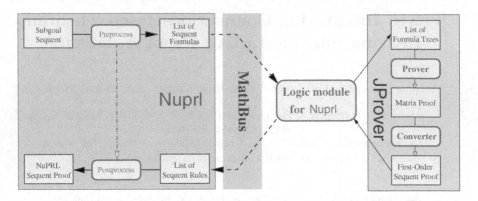

Fig. 1. Architecture of JProver in connection with Nuprl

Over the past years there have been various approaches to combining inter-active proof assistants with automatic proof tools [2,17]. Our application differs from these in that we provided a fully automatic theorem prover for classical *and* intuitionistic first-order logic with a very compact search space. A user may trust its results or expand them in order to inspect the proof. Furthermore, JProver supports constructive logic and is thus well suited for reasoning about programs.

Although this paper focuses on the integration of JProver into Nuprl and MetaPRL, the underlying mechanisms are quite general and might easily be adapted to integrate JProver into other proof assistants for constructive and classical logics. In the rest of this paper we shall briefly discuss JProver's proof search procedure, the tool for generating proof objects, and the mechanisms for integrating JProver into the Nuprl and MetaPRL proof development systems.

2   JProver: **Proof Search and Transformation**

JProver implements a full first-order theorem prover for classical and intuition-istic logic that realizes the connection-based proof procedure presented in [10]. It transforms a set of first-order sequent formulas into a set of formula trees, that will be annotated by tableau *types*, *polarities*, and so-called *prefixes*. During the proof process, JProver identifies *connections* between pairs of atoms and checks whether each *path* through the formulas contains such a connection. The formula is valid if each of these connections is *complementary*, that is if the connected atomic formulas can be unified by a global *term substitution* and – for intuitionistic validity – if their prefixes can be unified. To compute the *pre-fix substitution*, we use a specialized string unification algorithm based on [14]. The resulting *matrix proof* is a *reduction ordering* that consists of the original formula trees together with the connections and non-permutability constraints induced by the substitutions.

JProver's converter component uses the algorithms described in [11,15] to reconstruct a first-order sequent proof from the classical or intuitionistic matrix proof. It essentially transforms the reduction ordering into a linear order and constructs a sequent rule for each node, using the term substitution to instantiate quantified variables. Since additional proof knowledge from the matrix proof is exploited, proof reconstruction can be done without search [15,16].

The selection of the target sequent calculus for proof reconstruction depends on the calculus underlying the connected proof assistant. For the intuitionistic case, JProver first generates a *multiple-conclusioned* sequent proof [6] because of its proof-theoretical closeness to the matrix proof. If needed, this proof can further be transformed into a *single-conclusioned* sequent proof [7] using a second conversion step as described in [5]. Nuprl, for instance, requires a single-conclusioned proof whereas MetaPRL does not. The resulting sequent proofs can be used to generate proof objects in order to validate, check, or guide proof construction in the interactive proof assistants.

JProver is implemented in OCaml as a stand-alone theorem prover. However, it is embedded into the MetaPRL environment [9], which allows it to use MetaPRL's quantifier unification algorithm as well as its module system for communicating with interactive proof assistants.

3   Integration into Interactive Proof Assistants

JProver is implemented on top of the MetaPRL core, using MetaPRL as a *toolkit* that provides the basic functionality — term structure, substitution, unification, etc. JProver takes as its input a small JLogic module that represents the logic of the proof assistant with which JProver will cooperate. The JLogic module describes which terms implement logical connectives, how to access subterms from those connectives, and how to convert JProver's generic representation of a sequent proof into the internal data structures of the proof assistant.

In order to be able to call JProver from some proof assistant, one would need to write a *logic module* that consists of two components: a piece of OCaml code for communicating with that proof assistant (using whatever communication protocol developers would choose) and a JLogic module capable of decoding the sequent received from that communication code and of encoding JProver's response into a form the communication code expects.

Currently we have integrated JProver into the MetaPRL and Nuprl systems. The technical integration of JProver into MetaPRL is straightforward, as JProver is a module in MetaPRL's code base. MetaPRL can communicate with it simply by making a function call. The logical module of the MetaPRL type theory passes its formulas directly to JProver and the JLogic module for MetaPRL converts JProver's sequent proof into a MetaPRL tactic, which will generate a MetaPRL proof for the proof goal.

The integration into Nuprl (Figure 1) is not as straightforward. Calling JProver from a Nuprl sequent requires Nuprl to preprocess the goal and the list of hypotheses and to send them to a MetaPRL process running JProver. The preprocessing accounts for differences in the representation of variables and applications of terms, and also addresses differences in the type theory semantics.

For example, JProver, as a first-order intuitionistic prover, cannot understand type information contained in Nuprl's sequents. We can, however, encode the type information as a logical predicate which is understood, and then later reinterpret JProver's results to fit the original sequent. In most cases, however, the logical proof does not depend on type information. We simply discard it if the sequent mentions only a single type.

To communicate the processed sequent, the Nuprl /JProver link takes advantage of the Nuprl Logical Programming Environment's [1] open architecture, which supports communication with external proof tools by sending terms in MathBus format [13] over an INET socket. Since most of the terms in the sequent are left unchanged, the common MathBus format is valuable in communicating and understanding contrasting syntax of the linked systems. Once the sequent is sent, the JLogic module for Nuprl describes how JProver can access the semantical information of its terms and also how to convert JProver's resulting sequent proof into a list of sequent rules with parameters, that Nuprl can then interpret. From this list of rules, Nuprl then builds a proof tree for the original sequent in a depth-first, left-to-right fashion.

Neither MetaPRL nor Nuprl rely on the correctness of JProver or the processing. Instead, JProver's output provides these systems with a proof strategy, which is then executed on the original sequent in the respective environment.

4   Progress and Availability

The connection between JProver and Nuprl is an example in which *hybrid proofs*, i.e. proofs created by multiple provers with different formalisms, have been successfully and *verifiably* generated. It gives a user the full expressive power of the proof assistant when dealing with complex proofs and verifications, while at the same time taking advantage of well-understood and efficient proof techniques for subproblems that only depend on first-order reasoning.

A snapshot from a proof of the "Agatha Murder Puzzle" is depicted in Figure 2 and illustrates the cooperation of JProver with Nuprl. After the first step the user invokes JProver through a Nuprl tactic, which completely proves the goal (left window). To inspect proof details, the user may request the complete sequent proof with elementary rules to be displayed (right window). Experience has shown that this option has considerable educational value.

It should be noted that JProver is not restricted to the syntax of first-order logic: unknown terms are simply treated as uninterpreted function or predicate symbols. This allows us to apply JProver to proof problems that are usually outside the range of first-order provers and to combine it with other proof techniques that are available to proof assistants.

In the future we intend to extend JProver's capabilities by coupling it with Nuprl tactics and decision procedures. We also intend to strengthen the prover component by adding mechanisms for inductive theorem proving described in [12] and modules for handling modal logics and fragments of linear logic [10,11].

Fig. 2. The Nuprl /JProver link: proving the "Agatha Murder Puzzle"

These modules will make JProver valuable for a variety of other proof assistants. We plan to build the corresponding interfaces as well.

Although JProver's main emphasis is not high-performance but bringing the advantages of connection-based theorem proving such as complete and efficient search into tactic-based proof assistants, we plan to incorporate well-known techniques for speeding up automated theorem provers in order to improve JProver's performance as a stand-alone prover.

JProver is a part of the MetaPRL code base and can be downloaded from MetaPRL's home page [9]. An executable copy of Nuprl running under Linux is available at

`http://www.cs.cornell.edu/Info/Projects/NuPrl/nuprl5/index.html`

References

1. S. Allen, R. Constable, R. Eaton, C. Kreitz & L. Lorigo. The Nuprl open logical environment. In D. McAllester, ed., *CADE-17*, LNAI 1831, pages 170–176. Springer, 2000.
2. C. Benzmüller et. al. *Ωmega: Towards a mathematical assistant.* In W. McCune, ed., *CADE-14*, LNAI 1249, pages 252–256. Springer, 1997.
3. W. Bibel. *Automated Theorem Proving.* Vieweg, 1987.
4. R. L. Constable et. al. *Implementing Mathematics with the Nuprl proof development system.* Prentice Hall, 1986.
5. U. Egly & S. Schmitt. On intuitionistic proof transformations, their complexity, and application to constructive program synthesis. *Fundamenta Informaticae* 39(1–2):59–83, 1999.
6. M. C. Fitting. *Intuitionistic logic, model theory and forcing.* Studies in logic and the foundations of mathematics. North–Holland, 1969.

7. G. Gentzen. Untersuchungen über das logische Schließen. *Mathematische Zeitschrift*, 39:176–210, 405–431, 1935.
8. J. Hickey & A. Nogin. Fast tactic-based theorem proving. In J. Harrison & M. Aagaard, eds., *TPHOLs 2000*, LNCS 1869, pages 252–266. Springer, 2000.
9. Jason J. Hickey, Aleksey Nogin, et al. MetaPRL home page. http://metaprl.org/.
10. C. Kreitz & J. Otten. Connection-based theorem proving in classical and non-classical logics. *Journal for Universal Computer Science* 5(3):88–112, 1999.
11. C. Kreitz & S. Schmitt. A uniform procedure for converting matrix proofs into sequent-style systems. *Journal of Information and Computation* 162(1–2):226–254, 2000.
12. C. Kreitz & B. Pientka. Matrix-based inductive theorem proving. In R. Dyckhoff, ed., *TABLEAUX-2000*, LNAI 1847, pages 294–308. Springer, 2000.
13. The MathBus Term Structure. http://www.cs.cornell.edu/simlab/papers/mathbus/mathTerm.htm
14. J. Otten & C. Kreitz. T-string-unification: unifying prefixes in non-classical proof methods. In U. Moscato, ed., *TABLEAUX'96*, LNAI 1071, pages 244–260. Springer, 1996.
15. S. Schmitt. *Proof reconstruction in classcial and non-classical logics, Dissertationen zur Künstlichen Intelligenz 239*. Infix, 2000.
16. S. Schmitt. A tableau-like representation framework for efficient proof reconstruction. In R. Dyckhoff, ed., *TABLEAUX-2000*, LNAI 1847, pages 398–414. Springer, 2000.
17. K. Slind, M. Gordon, R. Boulton & A. Bundy. An interface between CLAM and HOL. In C. Kirchner & H. Kirchner, eds., *CADE-15*, LNAI 1421, pages 134–138. Springer, 1998.

The eXtended Least Number Heuristic

Gilles Audemard and Laurent Henocque

LSIS equipe INCA - LIM
Centre de Mathématiques et d'Informatique
39, Rue Joliot Curie - 13453 Marseille cedex 13 - France
Tel: 33 4 91 82 85 16 - Fax : 33 4 91 11 36 02
email: {audemard, henocque}@lim.univ-mrs.fr

Abstract. This paper presents an algorithm, XLNH, to generate finite models of first order equational theories. Unlike conventional methods, which focus on using as few individual constants as possible to preserve symmetries, XLNH heuristically selects then fully generates the functions that appear in the problem, using a weighted directed graph of functional dependency. One key issue here is to constructively generate isomorphic partial models then further exploit the resulting symmetries. This algorithm proves very efficient on problems involving a unary bijective function f (like the additive inverse in a group or ring theory). When such a bijection is fully instantiated, XLNH statically exploits remaining isomorphic subspaces. These ideas are implemented using the public domain SEM software framework, and give order of magnitude improvements on many problems. These results are interesting on their own but potentially generalize to many practical CSP applications.

1 Introduction

Equational theories provide a great number of difficult problems. Zhang in [9] defines a set of problems which can form a challenge of finite model search systems. Several open mathematic problems were solved with different approaches: FALCON [10], FINDER [6], MGTP-G [3], LDPP, SATO [8], FMC [5] and MACE [4].

An equational theory is a set of axioms: first order logic formulas involving equality (e.g. $\forall x, \forall y, \forall z : h(f(x,y)) = f(z,x)$). We consider here theories in which all the variables are universally quantified. Finding a finite model for such a theory amounts to finding an interpretation of functional symbols over a finite domain D_n which satisfies all axioms. The existence of a model demonstrates the consistency of the theory. The existence of a counter model may refute a conjecture.

Finding a model of an equational theory can be viewed as a special kind of constraint program, where the constraints are highly symmetrical. Symmetries arise because all constraints are universally quantified. Known approaches to finite model search have explored ways to tackle those symmetries. MGTP-G [3] uses ad hoc axioms to filter out some symmetries statically. FALCON [10],

R. Goré, A. Leitsch, and T. Nipkow (Eds.): IJCAR 2001, LNAI 2083, pp. 427–442, 2001.

and SEM [11] use a dynamic cut and heuristic procedure (LNH: Least Number Heuristic) to avoid exploring symmetrical subspaces during search. On the other hand, many constraint programs of practical or industrial interest exhibit a subproblem having functional semantics.

Our approach generalizes the LNH heuristic to avoid exploring isomorphic subspaces. The new heuristic can be used with many difficult problems, and gives impressive performance improvements in all cases.

The paper is organized as follows: section 2 defines equational theories. Section 3 describes the model equivalence proposition. The basic principles of the enumeration procedure are discussed in section 4. In section 5 we describe the function selection strategy. Experimental results are listed in section 6. Section 7 gives a conclusion.

2   Equational Theories

2.1   Syntax

We use a subset $\mathcal{L}$ of first order logic, without existential quantifiers, with equality as the only predicate $\{=\}$. In $\mathcal{L}$, all the variables are universally quantified. The disequality symbol $\{\neq\}$ denotes the negation of equality. The set of variable names is $\{x, y, z, x_1, x_2 \ldots\}$. Constants are either integers from the set $\{0, 1, 2 \ldots\}$ or identifiers (most often a letter from the set $\{a, b, c, k, k_1, k_2 \ldots\}$). A functional symbol can be any identifier not ambiguous with one of the previous categories, most often a letter from the set $\{f, g, h, \ldots r, s\}$. A term is recursively built upon functional symbols, variable names and constants.

$$h(x, 0) = x$$
$$h(0, x) = x$$
$$h(x, g(x)) = 0$$
$$h(g(x), x) = 0$$
$$h(h(x, y), z) = h(x, h(y, z))$$
$$h(x, y) = h(y, x)$$

Fig. 1. Abelian Group Axioms

Since all variables are universally quantified, universal quantifiers are usually omitted in the axioms for simplicity. Figure 1 illustrates the possibilities offered by the language. $\mathcal{L}$ is rich enough to formulate the axioms of mathematical objects like abelian groups or unit rings. Because $\mathcal{L}$ has only one predicate, equality, sets of $\mathcal{L}$ axioms are commonly called "equational theories". It is of considerable interest to mathematicians to prove or refute the existence of finite structures satisfying axioms in $\mathcal{L}$. Hence $\mathcal{L}$ is at the same time an excellent experimentation basis and a field of application. The concepts introduced in this paper can be extended to richer languages like the many sorted first order language used as input to the first order finite model generator SEM [11].

2.2   Semantics

We use traditional naming conventions in the field of CSP-based finite model generation (FALCON [10], SEM [11]). Without loss of generality, individuals are taken from the set $N = \{0, 1, 2, \dots\}$ of natural numbers. Since we are only interested in finite models, we interpret a theory T in $\mathcal{L}$ on a finite set $D_n = \{0, 1, 2, \dots n - 1\}$. Constants (integers) are interpreted as themselves. We call a *cell* the *ground term* $f(e_1, \dots e_k)$ where all e_i belong to D_n. Cells map to constraint variables in the associated constraint problem. D_n is called the *domain* of these variables. The members of D_n are called *individuals*. An *interpretation* I_n (or simply I) of a theory T maps each cell to a value from D_n. The resulting structure defines an operation table for every function that appears in T (for instance the set: $\{h(0,0) = 0, h(0,1) = 2, h(0,2) = 3 \dots\}$). A *model I of order n* of a theory T is an interpretation on D_n which satisfies all the theory axioms.

Let f be a function, and I an interpretation. We naturally define the interpretation I_f of f as the restriction of I to f cells. We often use $f(e_1, \dots e_k)$ to denote $I(f(e_1, \dots e_k))$ when not ambiguous. Let g be a unary function, we may also use $g^i(x)$ to denote $\underbrace{I(g(I(g(\dots(x)))))}_{i}$

2.3   A CSP Approach to Model Generation

An equational theory can be viewed as a special kind of constraint program, a triple (V, D, C) where V is the set of constraint variables, D is the domain (or set of possible values) for these variables, and C is a set of constraints, i.e. relations listing possible combinations of variables values.

Here, the set V of variables is the set of function cells and the domain D is the set D_n. Different approaches exist to implement the constraints (more efficiently than as extensive lists of compatible tuples). Enumerative model generators usually rely upon constraint propagation algorithms, with a tradeoff between propagation efficiency (i.e. the completeness of the decisions made by the propagation algorithm alone) and the cost of maintaining the associated data structures. FMSET [2] experimented using boolean propagation and a clause flattening technique (to achieve ultimate propagation efficiency), at the expense of additional memory costs. SEM (the public domain tool we based our experiments upon) uses the terminal instances of the axioms and propagates newly known values of function cells upwards in the structure. SEM compensates the loss of most downward propagations by more concise and efficient data structures and indexing. A potentially useful source of (non symmetry aware) improvement of the finite model generation may be achieved using lookahead strategies as shown in [1].

Before the search starts, SEM generates all the terminal instances corresponding to every axiom in the theory. For instance, the axiom $h(x, g(x)) = 0$ (group inverse) expands to $h(0, g(0)) = 0, h(1, g(1)) = 0, h(2, g(2)) = 0 \dots$. These terminal axioms are stored in memory using a pointer based representation that allows for fast upward propagation. Whenever a leaf cell value becomes known (e.g. $g(0)$), its actual value is substituted in all the constraints where it appears (which may generate a new cell value: here $h(0,0) = 0$), and the process repeats to fix point or failure as long as new cell values are introduced.

3   Model Isomorphism

When building an equational theory model, some isomorphic branches in the search tree can be cut by observing that all individuals i from D_n that were not used as a cell index or as a cell value in previous choice points are interchangeable. This intuition led to the implementation of the Least Number Heuristic in FALCON ([10]), a program that solved several open problems for the first time.

Because equational theories are highly symmetrical, the LNH alone does not cut all unwanted search branches. This research focuses on that issue, exploring ways to retrieve part of the original symmetries, even after all individuals have been used. Most equational theories involve a unary bijection (as in group or ring axioms), or can be adapted to involve one (like in quasi groups). This section proves a proposition that leads to an improved model generation procedure: the search starts by generating a model of a unary function if it exists. After this model was computed, the search can proceed from a state where remaining symmetries can be deterministically suppressed and thus require no dynamic tests.

Definition 1. *Let T be a theory, I an interpretation, E a subset of D_n, and f a (unary) functional symbol. We define $f(E)$ as the set $\{I(f(e))|e \in E\}$. As usual, we also define $f^{-1}(E)$ as the set $\{e \in D_n|I(f(e)) \in E\}$.*

Definition 2. *Let T be a theory, g a unary function and I_g an interpretation of g on $D_n = \{0, \ldots n-1\}$. An inclusion minimal subset c of D_n such that $g(c) = c$ is called a cycle. An element $i \in D_n$ appears in at most one cycle, called c_i. We define the size $size(c)$ of a cycle c as $|c| - 1$.*

Definition 3. *Let g be a unary function and I_g an interpretation of g on D_n. The inclusion maximal subset D_{I_g} of D_n such that $g(D_{I_g}) = D_{I_g} = g^{-1}(D_{I_g})$ is the bijective restriction of g.*

Note that such a bijective restriction is not the union of all the cycles, except when I_g interprets g as a bijection. In that case, we even have $D_{I_g} = D_n$.

Example 1. Let g be a unary function under the following interpretation I_g.

$$\frac{g\begin{vmatrix} 0\ 1\ 2\ 3\ 4\ 5\ 6\ 7\ 8\ 9\ 10\ 11 \end{vmatrix}}{1\ 2\ 2\ 3\ 4\ 6\ 5\ 8\ 7\ 9\ 9\ \ 1}$$

Under the interpretation I_g, the elements of D_{I_g} (here the set $\{3, \ldots 8\}$) belonging to cycles of equal sizes remain interchangeable (e.g. 5 and 7). Obviously however, the individual 2 is not interchangeable with 3.

These intuitions lead to the proposition 1 below.

Proposition 1. *Let T be an axiom system with a unary function g, and I_n a model of T. Let h be a function in T ($h \neq g$), $h(i_1, \ldots i_k)$ a cell, and $v \in D_{I_g}$ such that $I_n(h(i_1, \ldots i_k)) = v$ and $v \notin c_{i_j}$ for all i_j. Then for every $w \neq v$ s.t. $|c_w| = |c_v|$ and w belongs to none of all c_{i_j} there exists an isomorphism transforming I_n to a model I'_n of T in which $I'_n(h(i_1, \ldots i_k)) = w$.*

Proof. Let c_v and c_w be the (non empty) cycles of v and w. There are two cases:

- $c_v \neq c_w$: let $\sigma : D \mapsto D$ be the isomorphism equal to the identity everywhere but on c_v and c_w which maps $g^i(v)$ to $g^i(w)$ and $g^i(w)$ to $g^i(v)$ for all i in $[0..|c_v|)$.
- $c_v = c_w = c$: let k be the isomorphism equal to the identity everywhere but on c which maps $g^i(v)$ to $g^i(w)$ for all i in $[0..|c|[$.

By definition, σ is such that $\sigma(g(i)) = g(\sigma(i))$. σ naturally extends to a model isomorphism by mapping any ground assignment $h(i_1, \ldots i_n) = v$ (where $h \neq g$) to $h(\sigma(i_1), \ldots \sigma(i_n)) = \sigma(v)$ so that $\sigma(h(i_1, \ldots i_n)) = h(\sigma(i_1), \ldots \sigma(i_n))$ for all $i_1, \ldots i_n$ in D_n. These conditions, together with the fact that σ is bijective and that all universally quantified axioms are valid under I_n, ensure that every terminal instance of any axiom $t_1 = t_2$ of the theory T is valid under $\sigma(I_n)$. □

Example 2. Assume we want to generate abelian groups (cf. figure 1 axioms) of order 5. Let I_n be a model of AG that interprets g as I_g below:

g	0	1	2	3	4
	0	1	2	4	3

If I_n interprets $h(1, 2)$ as 3, we know that there exists an isomorphic model I'_n where $h(1, 2) = 4$. This property can be used in the enumeration procedure to avoid exploring symmetrical search spaces.

4   Enumeration Procedure

Proposition 1 shows that some model isomorphisms remain when a partial interpretation for a unary function g has been computed. The best situation occurs if the theory axioms ensure that g is bijective, since in that case $D_{I_g} = D_n$ and is maximal in size. This suggests to try starting the model generation by completely producing a model for a bijective function g, if it exists, to exploit the remaining symmetries further. In the rich domain of group and ring theories, the group inverse function g is not only bijective, but satisfies $g^2(i) = i$ for all $i \in D_n$. Cycles in that case are of size 0 or 1, which even further reduces the number of different g interpretations.

It is easy to generate only non isomorphic (canonic) interpretations of a bijective unary function. The idea simply is to generate the function so that its cycles are increasing, or decreasing in sizes, as suggests the following proposition:

Proposition 2. *Let g be a unary function and I_g an interpretation of g on D_n. There exists an integer l in D_n and a permutation σ on D_n mapping I_g to an interpretation $\sigma(I_g)$ such that in $\sigma(I_g)$:*

- *$D_{I_g} = \{0, \ldots l\}$*
- *for every cycle c, c elements are consecutive integers and only the highest element e in c is such that $g(e) \leq e$.*
- *for every two consecutive cycles c_1 and c_2, $|c_1| \leq |c_2|$*

This proposition is easily proved by iteratively building σ as the appropriate renaming on D. Because an equational theory involves constants (like the neutral element in group axioms), for completeness reasons, proposition 2 cannot be used as such in an enumeration procedure. Constants appearing in equalities must be interpreted first by the algorithm. The chosen values are not interchangeable with any other. In addition, these individuals may belong to a cycle of g, or not (in that case, the function g is not bijective). Technically, the individuals selected to interpret p constants k_p may without loss of generality satisfy $I(k_j) < j - 1$ for $j \in \{0..p - 1\}$. This leads to the following proposition:

Proposition 3. *Let $k_1, \ldots, k_p$ be p constants, I_{k_i} their interpretation, and g a unary function. Let I_g interpret g on D_n. There exist two integers m and l in D_n ($m \leq l$) and a permutation σ on D_n mapping I_g to an interpretation $\sigma(I_g)$ such that in $\sigma(I_g)$:*

- *$g^n(\{I_{k_i}\}) = \{0, \ldots m - 1\}$*
- *$D_{I_g} - g^n(\{I_{k_i}\}) = \{m, \ldots, l\}$*
- *for every cycle $c \subset \{m, \ldots, l\}$, c elements are consecutive integers and only the highest element e in c is such that $g(e) \leq e$.*
- *for every two consecutive cycles $c_1 \subset \{m, \ldots, l\}$ and $c_2 \subset D_{I_g}$, $|c_1| \leq |c_2|$*

Definition 4. *An interpretation I_g satisfying proposition 3 requirements is called canonic. Let c be a cycle in I_g. The smallest element s in c is called start(c), and the highest element e in c is called end(c).*

According to the above statements, given g a unary function and I_g a canonic interpretation, it is clear that $size(c) = end(c) - start(c)$ for every cycle c. In the example 1, if we have one constant interpreted as 0, we have $m = 3$, $l = 8$, $\{3, \ldots, 8\}$ contains 4 cycles of respective lengths 0, 0, 1, 1. Note that two cycles are not in $\{3, \ldots, 8\}$: $g(2) = 2$ and $g(9) = 9$.

Model generation uses the two propositions 1 and 3 to exclude isomorphic subspaces when building models of a theory T involving a unary function g. It operates in three steps.

4.1   Step 0

The algorithm interprets the p constants appearing in equalities so that $I(k_0) = 0$ and $I(k_j) \leq I(k_{j-1}) + 1$. This constructs a set $\{0, \ldots, q\}$ of integers ($q \leq p$).

4.2   Step 1

It then selects a function g from the problem statement and iteratively generates all its canonical models, as described in proposition 3. The idea simply is to construct I_g so that

- $g^n(\{0, \dots, q\}) = \{0, \dots, m-1\}$
- $D_{I_g} - g^n(\{0, \dots, q\}) = \{m, \dots, l\}$ for some m, l (note that there may exist cycles in $\{0, \dots, m-1\}$).
- $\{m, \dots, l\}$ cycles only contain consecutive elements (note that this property cannot be ensured for the cycles that appear over $\{0, \dots, m-1\}$)
- for all i in $\{m, \dots, l\}$, either $g(i) = i+1$ or i is the end of a cycle and thus $g(i) = i - size(c_i)$
- the cycles in $\{m, \dots, l\}$ are increasing in size (note again that this property cannot be ensured for the cycles that appear over $\{0, \dots, m-1\}$)

This procedure allows to generate only a limited number of interpretations having isomorphic bijective restriction. In the case of a bijection, only non isomorphic models are generated. For example, using this strategy, the generation of bijective functions of order 6 only produces the eleven non isomorphic models whereas using the LNH heuristic produces 32 models.

4.3   Step 2

For every such generated I_g, the algorithm enumerates possible models for the theory using an almost standard CSP enumeration procedure, described as algorithm 1. We say that an index i has been *hit* by the algorithm when either

- the value i has been assigned to a cell
- the value of a cell $h(i_1, \dots i_n)$ has been chosen, or is currently under consideration, and $i = i_j$ for some j in $\{1, \dots n\}$.

We say that a cycle has been *hit* when one of its members has been *hit*. The algorithm keeps track of a high water mark called mdn_s for every cycle size s. The value mdn_s represents the end index of the highest cycle of size s that has been hit. Remember that cycles with equal sizes are consecutive. Let $v \in D_{I_g}$, we note $mdn(v)$ the value $mdn_{|c_v|}$. By the proposition 1, when selecting a value v for a cell $h(i_1, \dots i_k)$, only values smaller than $mdn(v) + 1$ need to be tried.

After an interpretation of g has been computed, the index m identifying the start of the bijective restriction section is known, and thus all mdn values are set equal to $max(0, m-1)$.

Traditional implementations of the LNH heuristic (as in [10]) use only one mdn value. They attempt to favor the heuristic application by selecting new cells so that the mdn does not change, as long as possible. To achieve this, it is enough to select cells $h(i_1, \dots i_k)$ with indexes smaller than mdn. If impossible, a new cell is selected that yields the smallest possible change to mdn values. In practice, it suffices to choose $h(i_1, \dots i_k)$ so that $max(i_1, \dots i_k)$ is the smallest.

Algorithm 1 The enumeration procedure

---

function XLNH(S: set of assign., F: set of terminal axiom):boolean;
begin
       forall non propagated assignments a in S, Propagate(a, F, S)
       if S contains incompatible assignments **then return**(false)
       if F is empty **then return**(true)
       select unbound cell $b(i_1, \ldots i_n)$
            (so that $max(i_1, \ldots i_n)$ is the smallest)
       update $mdn(i_j)$ for all i_j
       forall $v \in D$ s.t. $v \leq mdn(v) + 1$
            if XLNH($S \cup b = v, F$) **then return**(true)
       return false
end

This strategy however picks candidate cells in all existing functions (including functions that depend upon the value of others, which thus should not be taken in consideration here).

Our approach focuses on the generation of an interpretation for all functions, one after another. The algorithm thus valuates all of a function cells before changing the function. This prevents spreading the model generation over all function symbols. During the generation of a function, we use an extended version of the LNH, that treats as equivalent the individuals belonging to the same cycle, or to cycles of equal sizes, as long as these individuals have not been "hit".

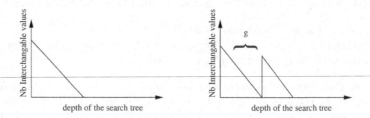

Fig. 2. LNH versus XLNH

The figure 2 illustrates the difference between the LNH and the XLNH heuristics. By using LNH, after a certain search tree depth is reached, all D_n values are used, and yet are no more interchangeable (as illustrated in the leftmost diagram in figure 2). By using XLNH, after a canonical interpretation of the unary function g was computed, existing cycles in the bijective restriction let some individuals become interchangeable again (as in the rightmost diagram in figure 2). Hence the XLNH heuristic allows to first compute in a deterministic way the canonical models of a unary function, then statically exploit remaining isomorphisms to prune the search. The implementation of the heuristic is thus entirely static, and requires no complex run time tests. The only tests performed are integer comparisons, at a null cost.

Example 3. Figure 3 illustrates the generation of all models of order 5 abelian groups (cf. figure 1). Only three canonic interpretations of g are generated. In figure 3, bracket surrounded values are the ones suppressed from the search tree by the XLNH heuristic (by proposition 1). Missing branches are suppressed by constraint propagation. The $\Box$ symbol represents inconsistency. The symbol M represents the obtention of a model.

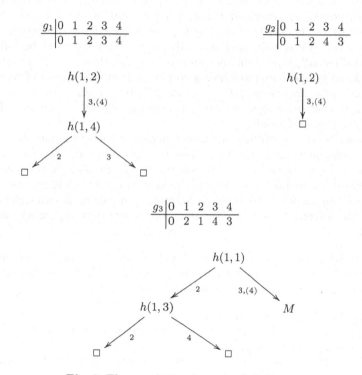

Fig. 3. The search tree for order 5 abelian groups

By looking at this diagram, we can observe that the first generated interpretation of g is the identity, all cycles having the same size zero. In that case, the rest of the search proceeds as with a LNH heuristic starting with $mdn = 0$. This results in suppressing many isomorphic interpretations.

5   Function Selection Strategy

Experimental results show that the sequence of selected functional symbols impacts on computation times. It can be clearly understood why when we observe that some functional symbols never appear as top-level term labels in the axioms (these functions are called "pure output") while others never appear as sub-terms (they are "pure input" functions). It is obvious that pure input functions are uniquely determined when all other functions are known, and thus should not be explicitly enumerated. On the other hand, pure output functions should be generated in priority, because SEM's propagation is essentially of a bottom up kind.

The other functions appearing both as top-level terms and as sub-terms in the problem axioms are generated in a statically computed intuitive order, based on the number of times a function has a previously generated function as its input. This heuristic is computed statically (before search starts) by building a weighted directed graph from the problem axioms where nodes are labelled with functional symbols, and weighted arrows represent the number of recursive function invocation in axioms ($f(g(...)...)$ eventually introduces the arrow from g to f or increments its weight by one). The existence of a static heuristic clearly improves program performance.

Constants deserve a distinct treatment depending whether they appear in equations or disequations. In the former case (like of the additive inverse "$zero$"), the constants are valuated before program starts, and result in introducing non interchangeable individuals (step zero of the algorithm). In the latter case, however, (like in an axiom introducing a counter example for associativity), the constants are better valuated once the functions where they appear are entirely generated.

Definition 5. *Let T be an equational theory, F its set of functional symbols (without constants) and C its set of axioms. Let $G = (X, W)$ be the weighted and oriented dependency graph defined as follows:*

- *the set of vertices is isomorphic to F.*
- *the edge $(f_i \rightarrow f_j)$ belongs to W if there exists an axiom in C where f_i appears as a sub-term of f_j.*
- *the weight of the edge $(f_i \rightarrow f_j)$ equals the number of times there exists an axiom in C where f_i appears as a sub-term of f_j.*

Example 4. The following theory defines the axioms of an unit ring: g, a is the group and m is the multiplicative law.

$$a(0, x) = x \qquad\qquad a(x, 0) = x$$
$$a(g(x), x) = 0 \qquad\qquad a(x, g(x)) = 0$$
$$m(x, 1) = x \qquad\qquad m(1, x) = x$$
$$a(x, a(y, z)) = a(a(x, y), z) \qquad m(x, m(y, z)) = m(m(x, y), z)$$
$$a(m(x, y), m(x, z)) = m(x, a(y, z)) \quad a(m(x, z), m(y, z)) = m(a(x, y), z)$$

This set of axioms yields the following dependency graph:

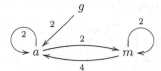

Note: the weighted graph construction is not semantical, and depends upon the theory syntax. Hence, two different formulations of the same problem could have different graphs. This issue is field of future research.

5.1 Function Selection Algorithm

The best experimental results are achieved by selecting the next function to instantiate according to the following preference order:

1. select a constant appearing in an equation (step 0)
2. select a pure output bijective unary function (step 1)
3. select a pure output unary function (step 1)
4. select a pure output n-ary function (step 2)
5. select a function that with maximal sum of weights of arrows coming from already generated functions (step 2)
6. select a constant appearing in already fully generated functions (step 2)

Note that all vertices having only input edges functionally depend upon the other functional symbols, and are uniquely determined as soon as the other functions have been entirely generated. Pure input functions are thus never generated explicitly.

Example 5. As an example, let us consider the theory *RNG041-1* from the TPTP problem collection [7]. We have the following dependency graph:

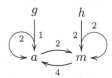

According to the previous selection strategy, we first select the constants which appear in equality equation, after g (pure input unary bijective), then h (pure input unary). Then, the choice of a before m is guided by the dependencies. Hence the instantiation order is: g, h, a, m.

The table 1 illustrates those concerns by comparing the results obtained with the TPTP problem *RNG041-1* at order 6 using different function orderings. This example suggests a few comments: the best node complexity is achieved by selecting function a (the additive law) just after its inverse g. However, the best execution times are obtained with our function selection strategy, because many nodes (cell value choices for h in that case) trigger very immediate fails.

Table 1. Comparing different orders for the *RNG041-1* problem

	g,a,h,m	$g,h,a,m,$	g,m,a,h
Time	0.13	0.06	0.15
Nodes	89	879	634
Models	0	0	0

Example 6. Let us consider the theory *RNG025-8* from the TPTP problem collection [7]. We have the following dependency graph:

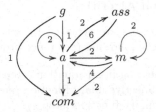

In this case, the statically computed ordering is the following one: g, a, ass, m, com. Three constants appear in the set of clauses as disequations literals, and are thus interpreted last.

Again, table 2 list the results obtained with different function orderings. The size of models is equal to 5.

Table 2. Comparing different orders for the *RNG025-8* problem

	g,a,ass,m,com	$g,a,m,ass,com,$	g,ass,a,m,com	g,m,a,ass,com
Time	0.14	0.33	$\geq$10 minutes	0.78
Nodes	3 149	1 280	-	6 398
Models	3 072	1 280	-	2 048

The choice of g first is obvious, because it is pure input, unary bijective. *com* needs not be generated because it is pure output. Choosing a after g is natural because it has g as input (and *com* is discarded). Then *ass* should be preferred over m because the axioms involve six occurrences of a as a sub-term of *ass*, instead of only four in m. Again, you may observe the existence of a node count per execution time trade off. As before, the best execution times are obtained at the expense of more choice nodes, which suggests that many choices lead to very quick failures in the best option.

6   Experimentation

We have implemented the heuristic XLNH on the SEM software, as a heuristic variant. We thus use exactly the same data structures and propagation algorithm. Our implementation currently only handles problems involving a unary bijective function, which are numerous. SEM's source code is available at the web address `www.cs.uiowa.edu/~hzhang/sem.html`. All results are obtained on a K6II 400Mhz with 128 Mb of RAM. We limit to two hours the maximum time to solve a problem.

We compare SEM + XLNH and classical SEM (+ LNH) on different mathematical problems: abelian groups first, then several ring problems from the TPTP collection [7]. We have tested XLNH on a large number of these problem instances and obtained very good results, and list them for three problems *RNG041-1* (rating 0.22 -very easy-), *RNG025-8* (rating 0.67 -medium-) and *RNG030-6* (rating 1 -difficult-).

Table 3. Abelian Groups

Order	SEM + XLNH			SEM + LNH		
	Models	Time	Nodes	Models	Time	Nodes
32	529	168	9 769	2 295	956	421 178
33	15	28	2 769	15	1 151	466 883
34	2	239	26 077	20	1 402	481 249
35	13	41	3 477	13	1 700	490 606
36	321	375	27 975	2 142	2 345	872 374
37	1	65	4 107	1	2 848	921 379
38	2	532	39 789	22	3 525	935 527
39	17	86	4 350	17	4 263	946 669
40	282	816	39 130	2 220	5 632	1 393 433
41	1	116	5 163	+	+	+
42	42	1 247	54 822	+	+	+
43	1	154	5 396	+	+	+
44	31	1 788	58 137	+	+	+
45	180	226	6 122	+	+	+
46	2	2 481	60 281	+	+	+
47	1	361	14 096	+	+	+
48	3 345	4 446	75 905	+	+	+
49	8	492	22 976	+	+	+
50	22	6 375	232 718	+	+	+
51	21	636	25 053	+	+	+
52	+	+	+	+	+	+

We generate all models of the abelian group and only solve the satisfiability problem on the TPTP's instances. The table 3 shows the comparison between LNH and XLNH for abelian groups. Our approach is 7 times faster for even orders and 70 times faster for odd orders. Our method shows that odd order abelian group generation is much easier than that of even order abelian groups.

440   G. Audemard and L. Henocque

This agrees with known results about the number of non isomorphic finite abelian group instances (it is known that every subgroup order divides the order of the group). This result shows the efficiency of XLNH. The results table 3 stops at order 52. However we can generate all odd order abelian groups up to order 63 (278 models obtained in 2470 seconds and 40 764 nodes). We can observe that XLNH always produces fewer models than LNH.

Table 4. Some TPTP problems

Problem	Size	XLNH			LNH		
		Model	Time	Nodes	Model	Time	Nodes
RNG041-1	8	0	0.05	1 236	0	24	175 672
	9	0	0.06	1 673	0	123	736 625
	10	0	0.09	2 049	0	632	3 061 678
	11	0	0.09	2 497	0	2 928	12 221 898
	12	0	0.16	2 880	-	-	-
	14	0	0.22	3 725	-	-	-
	16	0	0.4	4 584	-	-	-
RNG025-8	8	1	19	872 609	1	135	6 965 608
	9	1	2.8	86 629	1	12	92 396
	10	1	1.6	10 435	1	41	19 276
	11	1	2.5	14 889	1	221	67 835
	12	1	8.3	148 203	1	529	244 126
	13	1	5.4	28 901	1	6 462	985 767
	14	1	11	39 383	-	-	-
	15	1	9.5	51 683	-	-	-
	16	-	-	-	-	-	-
RNG030-6	11	0	0.76	1 465	0	216	576 977
	12	0	28.7	83 007	-	-	-
	13	0	1.5	2 379	-	-	-
	14	0	3.5	6 250			
	15	0	17	32 987			
	16	0	6 826	4 039 087			
	17	0	5	5 338			
	18	0	278	400 055			
	19	0	8	7 375			

Table 4 lists the results obtained for several TPTP problems, sorted by increasing difficulty, according to the TPTP collection difficulty ratings. The XLNH heuristic proves very efficient for the RNG class instance (rings with added axioms). The *RNG041-1* problem, an easy one, is solved in less than 1 second at order 16. The table stops here because of paper size limitations. However XLNH solves *RNG041-1* at order 30 in 3.5 seconds and 10 989 nodes. The program is limited by memory. The *RNG025-8* problem shows a difficulty peak at orders being a power of two (order 8). This is expected, as in the case of abelian groups, because there exist many finite groups of order a power of two. XLNH fails at order 16 but solves the problem at order 17 easily (25 seconds). The *RNG030-6*, a currently open instance of the TPTP collection (rating 1), exhibits comparable behavior at power of two orders. Here again, because the problem involves many functions, the limitation comes from processor memory rather than combinatorial complexity.

7   Conclusion

We demonstrate the efficiency of exploiting the underlying structure of equational theories to generate their finite models. The existence of a pure output unary function, specially if it is bijective, allows to avoid exploring many isomorphic subspaces. Our results generalize the least number heuristic in situations when a unary function exists in the theory. This generalization proves very efficient in reducing the number of search nodes explored by the enumerator, and thus produces a lot fewer isomorphic solutions.

Themes of future work include: the extension to non bijective unary functions, integration of dynamic symmetry tests, changes in the data structures to consume less memory and obtain solutions at higher orders, exhaustive generation of canonical solutions for finite abelian groups or other theories up to unprecedented orders, the adaptation of these results to practical CSP problems involving a functional subpart (and eventually a bijection like in the Travelling Sales Person problem).

References

1. Gilles Audemard, Belaid Benhamou, and Laurent Henocque. Two techniques to improve finite model search. In D. McAllester, editor, *Proceedings of the 17th International Conference on Automated Deduction (CADE-17)*, volume 1831 of *LNCS*, pages 302–308. Springer, June 2000.
2. Belaid Benhamou and Laurent Henocque. A hybrid method for finite model search in equational theories. *Fundamenta Informaticae*, 39(1-2):21–38, 1999.
3. Masayuki Fujita, John Slaney, and Franck Bennett. Automatic generation of some results in finite algebra. In *Proceedings of International Join Conference on Artificial Intelligence*, pages 52–57. Morgan Kaufmann, 1993.
4. William McCune. A Davis-Putnam program and its application to finite first-order model search: quasigroup existence problems. Technical Memorandum ANL/MCS-TM-194, Argonne National Laboratories, IL/USA, 1994.
5. Nicolas Peltier. A new method for automated finite model building exploiting failures and symmetries. *Journal of Logic and Computation*, 8(4):511–543, 1998.
6. John Slanley. Finder : Finite domain enumerator. version 3 notes and guides. Technical report, Austrian National University, 1993.
7. Christian B. Suttner and Geoff Sutcliffe. The TPTP problem library - v2.1.0. Technical Report JCU-CS-97/8, Department of Computer Science, James Cook University, 15 December 1997.
8. Hantao Zhang and Mark Stickel. Implemanting the Davis-Putnam method. *Journal of Automated Reasonning*, 24:277–296, 2000.

9. Jian Zhang. Problems on the Generation of Finite Models. In Alàn Bundy, editor, *12th International Conference on Automated Deduction*, LNAI 814, pages 753–757, Nancy, France, June 26–July 1, 1994. Springer-Verlag.
10. Jian Zhang. Constructing finite algebras with FALCON. *Journal of Automated Reasoning*, 17(1):1–22, August 1996.
11. Jian Zhang and Hantao Zhang. SEM: a system for enumerating models. In Chris S. Mellish, editor, *Proceedings of the Fourteenth International Joint Conference on Artificial Intelligence*, pages 298–303, 1995.

System Description: SCOTT-5

Kahlil Hodgson and John Slaney

Australian National University, Canberra 0200, Australia
{Kahlil.Hodgson, John.Slaney}@anu.edu.au

Abstract. This paper reports recent experimental work in the development and refinement of the first order theorem prover SCOTT-5. This is descended from the SCOTT (Semantically Constrained OTTER) prover (see *Proc. IJCAI* 1993, pp. 109–114) and uses the same combination of a saturation-based theorem prover and a finite domain constraint solver, but the architecture of SCOTT-5 is radically different from that of its ancestor. Here we briefly outline semantic guidance as it occurs in SCOTT-5, and give experimental evidence of an improvement in performance (in terms of efficiency) that we attribute to the guidance strategy.

1 Introduction

Question: What semantically oriented strategy can direct a reasoning program in its choice of clauses to which to apply the inference rule(s) being employed: what properties can be used other than the current criterion of simply finding a clause containing a literal with the appropriately signed predicate?

L. Wos [8] (problem #5)

In [8] Wos identifies "inadequate focus" as one of the primary obstacles to effective theorem proving. In this paper we report one line of attack on the focus problem for saturation methods of first order theorem proving, by injecting semantic information into heuristics for ordering the possible inferences. Preliminary work on this idea[1] resulted in the system SCOTT [1,5,7] which showed some modest performance gains relative to its parent OTTER. However, the main technique used in that prover was model resolution; the work on false preference (see below) remained unsystematic and lacked a theoretical basis. The new SCOTT rests on a new understanding of semantic guidance and shows much more stable behaviour over a wide range of problems. We present results on the TPTP problems and performance under fair conditions in CASC as compelling evidence that the effects exploited by our technique are real and useful.

Preliminary versions of the present work were presented in the workshops FTP-2000 and Reunion Workshop on Implementations on Logic (at LPAR-2000) but as these had no published proceedings, no account of the new SCOTT has yet been published.

---

[1] Ours is not, of course, the only approach represented in the literature. Semantically based refinements of resolution go back at least to Slagle's early work [4]. Mention should also be made of Plaisted's semantic hyper-linking [3] which was developed independently of SCOTT and also uses information from models in first order proof search.

R. Goré, A. Leitsch, and T. Nipkow (Eds.): IJCAR 2001, LNAI 2083, pp. 443–447, 2001.

2   Semantic Guidance

At any point during a saturation-based proof search, let S be a consistent subset of the clauses derived so far and let c be a particular clause inconsistent with S. Then there are derivations of the empty clause from $S \cup \{c\}$, and c occurs in *every* such derivation, including the shortest ones and those which contain no irrelevant excursions. Therefore if the next clause chosen as parent for inferences is c, at least one proof (and possibly many) will be extended by it. If we knew which were the maximal consistent subsets of the clauses so far derived, we could choose given clauses from the complements of these sets, thus guaranteeing at every step that some proof fragment is extended. Naturally, this would not guarantee that the proofs extended at a given point were globally the shortest ones, nor that one of them would be the first proof eventually found: screening out excess proofs is a different issue, which we do not claim to have addressed.

Unfortunately, we do *not* know which sets are maximal consistent, for consistency is famously undecidable. However, that does not mean we can never detect it: for instance, a formula may happen to have a model in a domain of just three elements, in which case it may be rather easy to find that model and so establish consistency. We hope to show that a search for (small) finite models can usefully *approximate* a satisfiability oracle. By relying on it, we give our prover access to *Near-Maximal Consistent Sets* (NMCS), which may be a little less than maximal, but which are still consistent and which can still focus the search for the most part on proofs rather than on useless sequences of inferences.

Evidently, there is a tradeoff between time spent in the semantic component, searching for models, and time spent in the syntactic component, searching for proofs. Investing time in modelling tends to improve the quality of semantic guidance and therefore to increase the efficiency of our proof search,[2] but on the other hand our goal is a proof not a model, so at some point we have to stop modelling and risk making some inferences. The tradeoff between the quality of guidance and its cost has three aspects. Firstly, the model generator must be forced to terminate by means of a bound on its search; the more generous this bound the closer the NMCS approximate MCS, but generosity costs time so the bound must be set judiciously. We find that sensible bounds can usually be set dynamically during the search using information from previous attempts at modelling clauses. Secondly, the probability that a given model can be improved within the search bound falls to near zero at some point, whereupon we can stop trying to improve that model and treat the set of clauses true in it as sufficiently near maximal to count as the NMCS for guidance purposes. Thirdly, it is important to bound the number of NMCS maintained. The more NMCS the system maintains the better its coverage of the search space, but each NMCS incurs severe overheads.

It may not be possible to resolve all these tradeoffs in a uniform way. The present program SCOTT-5 uses limits derived experimentally over three years of development work with successive versions of the algorithm.

---

[2] I.e. the proportion of given clause selections that actually occur in the final proof.

3   Implementation

We assume familiarity with the given clause algorithm used by SCOTT and by most contemporary high-performance theorem provers[3] As suggested above, SCOTT's technique is to select given clauses almost always from a collection of co-NMCS (complements of near-maximal consistent sets). To identify these NMCS and witness their consistency, it uses models generated at need by a finite domain constraint solver [6] which thus functions as a semi-oracle yielding no false positives but an unknown proportion of false negatives. The underlying theorem prover is OTTER.

Where S is any consistent set, clearly every derivation of the empty clause contains at least one complete branch of formulae, right up to an input formula as a leaf, which is disjoint from S. Hence, if the intersection of some co-NMCS with the passive set is small, it makes sense to take all the clauses in that intersection as the next few given clauses, so that we have activated as much as possible of at least one branch of every proof. We call this the "semantic queue" strategy. The meaning of "small" for this purpose is determined pragmatically; it seems from our observations that, roughly speaking, single-digit numbers are "small".

In order to enforce fairness, and because there is no easy way to know whether a given NMCS is a good one to use for guidance, SCOTT does not rely exclusively on the semantic queue strategy. Where all the co-NMCS are "large", it lets the choice of given clause cycle through them. Within each co-NMCS, the clauses are ordered by weight and age as normal, so that the choice among them is fair. Moreover, because the cost of maintaining *all* NMCS would be prohibitive, and again to ensure complete fairness, it occasionally chooses a given clause from among those which are in all known NMCS. Thus SCOTT's given clause selection strategy is fair if OTTER's is.

In principle, generating a NMCS is straightforward: simply scan the list of clauses, adding a clause whenever the resultant set can be shown consistent by finding a model. By listing the clauses in different orders, different NMCS may be obtained. In SCOTT this process is dynamic, since the NMCS change as more clauses are deduced, and so the witnessing models may also be changed during the search. It has always been part of the concept of SCOTT that semantic guidance should be fitted to the particular proof search, being driven by the actual clauses deduced rather than being set up in advance. The algorithm for generating models as a by-product of labelling the "kept" clauses is the same as that of the earlier SCOTT [1,5,7] and will not be repeated here for reasons of space. Nor will the algorithm and the refinements of it due to extensive experimentation be further elaborated.

---

[3] See, for instance, page 5 of [2]. For historical reasons, SCOTT uses the "OTTER loop", in which certain simplification inferences are performed eagerly within the passive set, rather than the "DISCOUNT (or WALDMEISTER) loop" in which all such inference is done lazily. We do not see the difference between these two versions of the algorithm as very important to SCOTT.

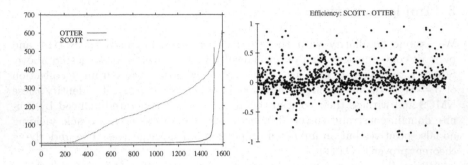

Fig. 1. Comparison of SCOTT and OTTER. Left plot shows time in seconds to solution. The problems are in each case those from TPTP that the prover solves within 600 seconds, and the times are plotted in monotone increasing order. Right plot shows difference in efficiency ratings (number of clauses in proof over number of given clauses) between the two provers. Above the line: SCOTT more efficient; below the line: OTTER more efficient. Problems are those solved in 600 seconds by *both* provers, omitting those solved by both with under 10 given clauses, in TPTP order.

4   Performance

Performance measures are inevitably rough, since SCOTT does not have a highly tuned autonomous mode because the algorithm is still under development and its parameters are so many and varied. Nonetheless we give comparisons against OTTER (actually against SCOTT with its semantic features turned off) to illustrate the effects of the guidance strategy, and a summary of the fair and independent comparison against the best contemporary provers in CASC-17.

The time plot shows SCOTT's comparatively gentle decline in performance as the problems get harder. Among high-performance provers to date, only SETHEO seems to show a similar performance profile. The scatter plot shows SCOTT emerging ahead of OTTER in terms of search efficiency, but also illustrates that the effect is far from uniform across problems. For reasons of space, we omit a more detailed breakdown of the results.

The above comparisons are exclusively with OTTER. However, it is also interesting to compare the performance of SCOTT with other, more recent, theorem provers. To that end, we instance some results from CASC, in which SCOTT has competed for the last four years. In the general MIX division, its performance has been little better that of OTTER, mainly because of the difficulty of devising an appropriate set of defaults for its many parameters. On the other hand, in the UEQ (unit equality) division, where problem sets are relatively homogeneous, it has performed rather well even in comparison with highly engineered provers such as E and WALDMEISTER. The full results of CASC-17 (July 2000) are available on the CASC web page[4] so here we reproduce only summaries of the MIX and UEQ results (Figures 2(a) and 2(b)).

---

[4] http://www.cs.jcu.edu.au/~tptp/CASC

System	Problems Solved	Average Time	System	Problems Solved	Average Time
E 0.6	57	79.31	Waldmeister 600	30	43.45
E-SETHEO 2000csp	57	160.53	SCOTT 5.0.0	12	186.27
Gandalf c-21	55	99.60	E 0.6	8	77.85
Vampire 1.0	45	48.06	E-SETHEO 2000csp	8	190.18
Vampire 0.0	37	54.72	Vampire 1.0	7	89.06
SCOTT 5.0.0	22	178.06	Otter 3.1b	6	33.42
Bliksem 1.10	18	65.33	Gandalf c-2.1	6	100.80
Otter 3.1b	8	55.86	Bliksem 1.10	4	28.05

  (a) MIX division   (b) UEQ division

Fig. 2. Summary of CASC-17 results

References

1. E. Lusk, J. Slaney, and W. McCune. SCOTT: Semantically constrained OTTER (system description). In A. Bundy, editor, *Proceedings of 12th International Conference on Automated Deduction*, number 814 in Lecture Notes in Artificial Intelligence, pages 764–768, Nancy, France, 1994. Springer-Verlag.
2. W.W. McCune. Otter 3.0 reference manual and guide. Technical Report ANL-94/6, Argonne National Laboratory, Argonne, USA, 1994.
3. D. Plaisted and Y. Zhu. Ordered semantic hyper linking. In *Proceedings of the 14th National Conference on Artificial Intelligence (AAAI-97)*, 1997.
4. J. R. Slagel. Automatic theorem proving with renamable and semantic resolution. *Journal of the ACM*, 14(4):687–697, 1967.
5. J. Slaney. SCOTT: A Model-Guided Theorem Prover. In R. Bajcsy, editor, *Proceedings of the 13th International Conference on Artificial Intelligence*, pages 109–114, Chambery, France, 1993. Morgan-Kaufman.
6. J. Slaney. Finder: Finite Domain Enumerator (System Description). In A. Bundy, editor, *Proceedings of 12th International Conference on Automated Deduction*, number 814 in Lecture Notes in Artificial Intelligence, pages 764–768, Nancy, France, 1994. Springer-Verlag.
7. J. Slaney and T. Surendonk. Combining finite model generation with theorem proving. In *Frontiers of Combining Systems*, pages 141–155, 1996.
8. L. Wos. *Automated Reasoning: 33 Basic Research Problems*. Pentice Hall, Englewood Cliffs, NewJersey, 1988.

Combination of Distributed Search and Multi-search in Peers-mcd.d
(System Description)

Maria Paola Bonacina*

Department of Computer Science – The University of Iowa
Iowa City, IA 52242-1419, USA
bonacina@cs.uiowa.edu

1   A Modified Clause-Diffusion Prover with Multi-search

Peers-mcd.d implements contraction-based strategies for equational logic, modulo associativity and commutativity, with paramodulation, simplification and functional subsumption. It is a new version of Peers-mcd [4], that parallelizes McCune's prover EQP (version 0.9d), according to the Modified Clause-Diffusion methodology (http://www.cs.uiowa.edu/~bonacina/cd.html).

In *parallel search* with peer processes (no master-slave hierarchy), multiple deductive processes search the space of the theorem-proving problem, each developing its own data base and derivation, and cooperate through communication of data, until one finds a proof and all halt. Within parallel search, we distinguish between *distributed search*, where the searches generated by the processes are differentiated by subdividing the inferences among them, and *multi-search*, where they are differentiated by assigning different search plans to the processes. Most approaches to parallel search in theorem proving adopted either one or the other: for instance, the systems based on Team-Work and combination of homogeneous provers emphasized multi-search, while the previous Clause-Diffusion provers emphasized distributed search (see [6] for a survey and references). A major difference between Peers-mcd.d and all its predecessors is that Peers-mcd.d implements both *distributed search* and *multi-search*, and their combination.

Peers-mcd.d can run in one of three modes:

- *Pure distributed-search mode:* the search space is subdivided among the processes; all processes execute the same search plan.
- *Pure multi-search mode:* the search space is not subdivided; every process executes a different search plan.
- *Hybrid mode:* the search space is subdivided, and the processes execute different search plans.

The basic structure of the search plan in a Peers-mcd.d process is to select premises for *expansion* (paramodulation), normalize the generated equations (*forward contraction*), and apply them to normalize pre-existing equations

---

* Supported in part by the National Science Foundation with grants CCR-97-01508 and EIA-97-29807.

R. Goré, A. Leitsch, and T. Nipkow (Eds.): IJCAR 2001, LNAI 2083, pp. 448–452, 2001.
© Springer-Verlag Berlin Heidelberg 2001

(*backward contraction*), in such a way to keep the data base inter-reduced. Peers-mcd.d offers three ways of diversifying the search plan:

- *Different premise selection mechanism,*
- *Different ratio of breadth-first search and best-first search,* and
- *Different heuristic function to sort equations for premise selection.*

Peers-mcd.d inherits from EQP two mechanisms to select the premises for paramodulation, the *given-clause* algorithm and the *pair* algorithm. The first one is a best-first search with the weight of equations as heuristic function: at every selection extract an equation of smallest weight and generate all its paramodulants with the already selected equations. The second one is a best-first search on pairs: at every selection extract a pair of equations of smallest weight and generate all their paramodulants. The most basic way of introducing multi-search is to have some processes execute the given-clause and some the pair algorithm: in Peers-mcd.d, when the flag `diverse-sel` is set, the even-numbered processes execute the pair algorithm, and the odd-numbered processes execute the given-clause algorithm.

If the parameter `pick-given-ratio` has value x, the given-clause/pair algorithm picks the oldest, rather than lightest, equation/pair once every $x + 1$ choices. The second way of diversifying search plans in Peers-mcd.d is to let each process use a different value of `pick-given-ratio`: when the flag `diverse-pick` is set, process p_k resets its `pick-given-ratio` to $x + k$.

The third ingredient to obtain different search plans is to let the processes do best-first search with different heuristic functions. The heuristic functions of [1,7] measure the syntactic similarity between an equation and the target theorem(s): the higher the similarity, the better the heuristic value, since an equation similar to the goal might reduce it. Peers-mcd.d implements the heuristic functions *occ-nest*, *CP-in-goal*[1] and *goal-in-CP* of [7], except that it uses the measure m_0 of [1] for the number of occurrences of a function symbol in a term, to take into account that AC operators are varyadic, since terms under AC operators are flattened. When the flag `heuristic-search` is set, process p_k executes the given-clause algorithm with heuristic function *occ-nest* if $k \bmod 3 = 0$, *CP-in-goal* if $k \bmod 3 = 1$, and *goal-in-CP* if $k \bmod 3 = 2$. The pair algorithm does not use these heuristic functions, because they are defined for equations, not pairs.

The search space is subdivided by subdividing the generated equations among the processes. This is achieved *without a top-level scheduler*: whenever a process generates and keeps an equation (i.e., the equation is not deleted by forward contraction), it gives it a process number, which becomes part of the equation's identifier (see [5] for details). This induces a subdivision of inferences, because each process skips the steps that it knows are done by others based on the identifiers of the premises. All inferences that generate new clauses, including backward-contraction, are thus subdivided, while deletions are not. Each process broadcasts the equations it has generated and kept after normalization. In Peers-mcd.d, the parameter `decide-owner-strat`, that controls the choice of

---

[1] CP stands for critical pair, hence equation.

subdivision criterion, may also have the value no-subdivide, meaning that no subdivision occurs, and a process broadcasts an equation only if its weight (its heuristic value if heuristic-search is set) is lower than a given parameter.

In summary, if decide-owner-strat = no-subdivide, and at least one of diverse-sel, diverse-pick and heuristic-search is set, Peers-mcd.d runs in pure multi-search mode; if decide-owner-strat ≠ no-subdivide, and none of diverse-sel, diverse-pick and heuristic-search is set, Peers-mcd.d runs in pure distributed-search mode; if decide-owner-strat ≠ no-subdivide, and at least one of diverse-sel, diverse-pick and heuristic-search is set, Peers-mcd.d runs in hybrid mode.

2   Proofs of the Moufang Identities without Cancellation

The first automated proofs of the Moufang identities in alternative (i.e., non-associative) rings by a general-purpose prover were presented in [2]. They used AC-UKB, the *inequality ordered-saturation* inference rule (i.e., superposition of an un-orientable equation into a goal to generate a new goal which is kept only if its normal form is not greater or equal than an already existing inequality), inference rules that *build the cancellation laws in* [8], and the heuristic measures of [1] to sort equations and delete those whose heuristic value is worse than a given threshold. These problems are still used as benchmarks (e.g., [3]) and in competitions (e.g., [9]). The TPTP library presents them in different formulations: some differ from [2] in choice of axioms and/or conjecture; those that follow [2] include the cancellation laws as implications, so that they are not equational. In the experiments reported here, the problems were formulated as in [2], but *without cancellation laws*, since EQP and Peers-mcd.d do not implement the rules of [8], and they are purely equational provers which cannot handle implications.

In the following tables, the first column tells the mode: D for pure distributed-search mode and H for hybrid mode. The second column tells the search plan, given-clause, or pair, or *diverse*, if diverse-sel was set. The h means that heuristic-search was used. The number at the front is the pick-given-ratio: x if it was x for all processes, xd, if diverse-pick was set and process p_k used $x + k$, nothing, if pick-given-ratio was not used. The number in parenthesis is the value of max-weight, if deletion by weight was used. The times (expressed in sec) are *average CPU times*. For each search plan, five subdivision criteria were tried, and the best result (among the averages) was retained. "T" means time-out after 3600 sec. The workstations were HP B2000 or C360, with 1G or 512M of memory, with EQP0.9d running on a B2000 with 1G, and *N-Peers* (Peers-mcd.d with N processes) on N workstations, one per process.

The first two problems, **moufang1** (*Middle Alternative Law*) and **moufang2** (*Skew-Symmetry Relation of the Associator*), are too easy for parallelization: EQP0.9d proved them in 4 and 1 sec, respectively, using the *pair* algorithm with pick-given-ratio = 4. However, with the default search plan, namely *given-clause* algorithm and no pick-given-ratio, EQP0.9d terminated abnormally[2],

---

[2] Some constant in the AC-matching or AC-unification code of EQP was exceeded.

whereas 1-Peer did both problems in 1 sec, due to the heuristic function used by the given-clause algorithm.

For the *Left Moufang Identity* (**moufang3**), EQP0.9d could not find a proof with the default search plan, while Peers-mcd.d did, thanks to distributed search:

Mode	Search plan	EQP0.9d	1-Peer	2-Peers	4-Peers	6-Peers	8-Peers
D	given(32)	T	T	598	91	187	40
H	given-h(32)	T	415	230	57	42	**9**
D	pair(32)	3,215	3,277	551	109	51	83
D	4-pair(32)	956	1,068	126	**38**	56	58
D	2-pair(32)	88	130	66	39	109	25
H	2d-diverse-h(32)	88	147	84	75	41	25

With `heuristic-search` on (second row), also 1-Peer found a proof, which shows the merit of the heuristic function, and all other times improved, up to a proof in only 9 sec with 8-Peers. EQP found a proof with the pair algorithm (third row), and the sequential time was reduced with `pick-given-ratio = 4` (fourth row), but Peers-mcd.d with more than one node sped-up with these search plans also, finding a proof in 38 sec with 4 processes. The best sequential time was obtained with `pick-given-ratio = 2` (last two rows): with this value, the parallel prover behaved more smoothly in hybrid mode.

For the *Right Moufang Identity* (**moufang4**), EQP found a proof only with the pair algorithm and `pick-given-ratio = 4`:

Mode	Search plan	EQP0.9d	1-Peer	2-Peers	4-Peers	6-Peers	8-Peers
H	given-h(32)	T	437	268	162	100	**28**
D	pair(32)	T	T	865	356	161	105
H	4d-diverse-h(32)	1,558	1,638	75	**32**	27	47

The problem proved to be elusive for the default search plan, but with `heuristic-search` on (first row), Peers-mcd.d solved it, with run-time decreasing down to 28 sec for 8-Peers. With the pair algorithm and `pick-given-ratio` not set (second row), 1-Peer did like EQP, since the pair algorithm does not use the heuristic function, but the parallel prover succeeded. With the hybrid search plan *4d-diverse-h*, Peers-mcd.d exhibited super-linear speed-up for all numbers of processes, with the best result for 4-Peers: the speed-up was $1,558/32 = 48.68$ and the efficiency $48.68/4 = 12.17$.

For the *Middle Moufang Identity* (**moufang5**), EQP could not find a proof within 3,600 sec with the default search plan, and took 572 sec with the pair algorithm, while Peers-mcd.d was much faster: using the default search plan, but with `heuristic-search` on, hence in hybrid mode, 1-Peer found a proof in 16 sec, 2-Peers took 9 sec and 4-Peers only 5 sec.

Problems *moufang3*, *moufang4* and *moufang5* were tried also in pure multi-search mode, with the same search plans tried in hybrid mode, but no subdivision and equation broadcasting limited by heuristic value. Almost no speed-up was observed. Thus, distributed search did much better than multi-search on these problems, and the combination of the two did even better. This suggests that a key factor in parallel search, possibly even more basic than limiting communication, is to differentiate the processes, so that they *do not overlap* and explore different parts of the search space. The statistics showed that a speed-up is typically accompanied by a strong reduction in number of equations generated (for Peers-mcd.d, the sum of the equations generated by all peers), hinting that the subdivision was effective, and led the processes to generate different searches and different from the sequential one.

Directions for future work include the development of a Modified Clause-Diffusion prover for first-order logic with equality, to allow application to a larger class of problems.

Acknowledgements. Thanks to Bill McCune for EQP0.9d, to my former student Javeed Chida, for implementing the heuristic functions in his master thesis, and to Gigina Carlucci Aiello of the Dipartimento di Informatica e Sistemistica, Università di Roma "La Sapienza," where part of this work was done.

References

1. Siva Anantharaman and Nirina Andrianarivelo. Heuristical criteria in refutational theorem proving. In Alfonso Miola, editor, *Proceedings of the 1st DISCO*, volume 429 of *LNCS*, pages 184–193. Springer Verlag, 1990.
2. Siva Anantharaman and Jieh Hsiang. Automated proofs of the Moufang identities in alternative rings. *Journal of Automated Reasoning*, 6(1):76–109, 1990.
3. Jürgen Avenhaus, Thomas Hillenbrand, and Bernd Löchner. On using ground joinable equations in equational theorem proving. In Peter Baumgartner and Hantao Zhang (ed.), Proceedings of FTP 2000, Technical Report 5/2000, Institut für Informatik, Universität Koblenz-Landau, 33–43, 2000.
4. Maria Paola Bonacina. The Clause-Diffusion theorem prover Peers-mcd. In William W. McCune, editor, *Proceedings of CADE-14*, volume 1249 of *LNAI*, pages 53–56. Springer, 1997.
5. Maria Paola Bonacina. Experiments with subdivision of search in distributed theorem proving. In Markus Hitz and Erich Kaltofen, editors, *Proceedings of PASCO97*, pages 88–100. ACM Press, 1997.
6. Maria Paola Bonacina. A taxonomy of parallel strategies for deduction. *Ann. of Math. and AI*, in press, 2000. Available as Tech. Rep., Dept. of Computer Science, Univ. of Iowa from http://www.cs.uiowa.edu/~bonacina/distributed.html.
7. Jörg Denzinger and Matthias Fuchs. Goal-oriented equational theorem proving using Team-Work. In *Proceedings of the 18th KI*, volume 861 of *LNAI*, pages 343–354. Springer, 1994.
8. Jieh Hsiang, Michaël Rusinowitch, and Ko Sakai. Complete inference rules for the cancellation laws. In *Proceedings of the 10th IJCAI*, pages 990–992, 1987.
9. Geoff Sutcliffe. The CADE-16 ATP system competition. *Journal of Automated Reasoning*, 24:371–396, 2000.

Lotrec: The Generic Tableau Prover for Modal and Description Logics

Luis Fariñas del Cerro[1], David Fauthoux[1], Olivier Gasquet[1], Andreas Herzig[1],
Dominique Longin[1], and Fabio Massacci[1,2]

[1] IRIT - Université Paul Sabatier — Toulouse (France)
{farinas|fauthoux|gasquet|herzig|longin}@irit.fr
[2] Dip. Ingegneria dell'Informazione - Univ. di Siena — Siena (Italy)
massacci@dii.unisi.it

1   A Manifesto for a *Generic Tableau Prover*

The last years have seen a renewed interest in modal and description logics
(MDLs). Better algorithms, coding, and technology have led to effective systems
based on tableau and constraint systems [6,7] to DPLL-based implementations
[5], first order provers [8] and the inverse method [13]. PSPACE problems such
as satisfiability are within reach for realistic instances [10] and potentially EX-
PTIME problems stemming from real applications can also be solved [3,7].

However, the comparisons now held at the Description Logic workshops and
at the TABLEAUX conferences have also shown a major problem: the emphasis
on performance is so strong that most implementors have restricted their prover
to few *fixed* logics, hacking logics and strategies in their systems.

Yet, there are infinitely many MDLs and the choice of one logic over another
is driven by modeling needs and computational constraints of one's applications.
A logic about actions and plans is likely to have different semantical and computational properties from a logic about database schemata. Even with the same
logic, different search strategies may be needed for different applications.

If a user wants to use logics or even search strategies slightly different from
those of the current systems, he must hack his own prover. "What if I use
this constructor", "What if I change the order of rules" experiments are almost
impossible for somebody who is not the implementor of the system.

To answer the needs of users wishing to *experiment and model with different
logics or strategies* there is a need of a *generic theorem prover for MDLs*. A
prover playing the same role as Isabelle [12] or PVS [11] for higher order logics,
while being less complex. If the user is not the same person as the programmer
of the prover, one needs (a) flexibility and portability of the implementation,
(b) high-level languages for tableau rules and strategy definition, and (c) user-
friendly interfaces.

Lotrec is such a generic tableau prover. It aims at covering all logics having
possible worlds semantics, in particular MDLs[1].

---

[1] Behind Lotrec is the work on modal tableaux with back/forth rules [9], graphs [1,2],
and its DL counterpart in [7]. Lotrec has been implemented by D. Fauthoux [4].

R. Goré, A. Leitsch, and T. Nipkow (Eds.): IJCAR 2001, LNAI 2083, pp. 453–458, 2001.
© Springer-Verlag Berlin Heidelberg 2001

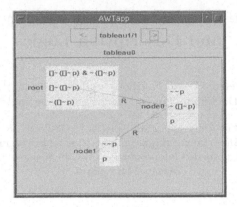

Fig. 1. Lotrec presentation of a tableau for logic K4

2 Architecture

The aims of flexibility, portability, and nice interfaces had motivated the choice of Java as the implementation language. Within such an object-based programming language, Lotrec raises Java's event-based architecture to a declarative approach.

Tableaux are usually presented in tree form. In **Lotrec**, they are generalized to *graphs* in order to enable complex MDLs such as the ML of confluence, or MLs with complex interactions between knowledge and action. Graphs also allow to visualize possible worlds models (e.g. after transitive or symmetric closure of accessibility relations). Graph nodes are labelled by formulae, and edges by any term (possibly containing variables). **Lotrec** graphically presents the tableaux it has generated (Fig. 1), and allows for "drag-and-drop" restructuring of its shape by the user. (It remains to implement "drag-and-drop" interfaces for defining rules, strategies... As in **Isabelle** and **PVS** this is currently done via textual files.)

3 Defining the Language of Your Own Pet Logic...

Before starting to define the rules, the user must define the *logical* connectives he wants to use. Let us take the definition for a logic of actions:

```
connector falsum      0   true   "FALSUM"   4
connector and         2   true   "_&_"      3
connector not         1   true   "~_"       5
connector feasible    2   true   "<_>_"     4
connector after       2   true   "[_]_"     4
```

Consider e.g. the last definition: `after` is the internal name of the connective, and 2 is the number of its arguments. The rest of the parameters defines the graphical presentation: `true` means that the connective is associative, `"[_]_"` stipulates that the internal (`after hit (feasible smash broken)`) will be

```
rule "K"
  descriptor links node0 node1    (variable R)
  descriptor hasElement node0     (nec (variable R) (variable A))
  action      add         node1   (variable A)
end

rule "diamond"
  descriptor hasElement  node0    (not (nec (variable R) (variable A)))
  descriptor isNotMarked node0    CONTAINED
  action newNode node0   node1
  action link node0      node1    (variable R)
  action add             node1    (not (variable A))
end

rule "InclusionTest"
  descriptor isAncestor node0 node1
  descriptor contains node0 node1
  descriptor isNotMarked node1   CONTAINED
  action mark node1              CONTAINED
end
```

Fig. 2. Some possible rules for the modal logic K4

written [hit]<smash>broken on the screen. 4 is the priority of the connective wrt the others.

4 Defining the Semantic-Tableau Rules of Your Pet Logic...

A rule consists of a *descriptor* and an *action* part. The former contains the applicability conditions, while the latter contains operations on tableaux.

A tableau rule is interpreted as mapping a pattern to a pattern, where patterns are connected fragments of a given tableau: if the descriptor part matches the pattern then that pattern is replaced by the result of the action part.

Consider the standard multi-modal logic $K4_n$, with modal operator **nec**. A handful of rules is in Fig. 2. The rule for handling formulae of the form **nec A** is the rule K. It says if some node **node0** of a Kripke structure is linked to a node **node1** via the relation R and contains a formula of the form **nec R A**, then A is added to **node1**. R and A must be variables in order to make the rule work as a schema. Constants are useful for specific formulae or relations.

Lotrec also allows for manipulating expressions on links. Thus one may easily have logics like dynamic logics where links are labelled by complex programs.

Sometimes the ordered application of rules via a strategy is not enough to ensure termination or completeness; or maybe a user just wants to test various strategies. Thus, nodes, links, and formulae in nodes can be *marked*. For instance, for logics with transitive accessibility relations such as $K4$, before creating a new node we may wish to check whether the current node is not included in some ancestor. The rules **diamond** and **InclusionTest** in Fig. 2 do that.

We may want to do more than just simple propositional reasoning and we may have what are called *concrete domains or quantitative domains*. Then we allow for *oracle calls* to programs exterior to **Lotrec**. These programs typically are rewriting procedures, constraint solvers, SAT provers, etc.

5 Defining Your Pet Search Strategies ...

After you have the rules defining the semantics of your logics you may want to say how to combine and apply them. *Search Strategies* do exactly that by mapping tableaux to tableaux (or sets thereof for disjunction-like rules) by repeatedly applying rules in some suitable ways.

If a user has a set of tableau rules {rule1, rule2, ..., ruleN} that has been proven to be complete for the logic under concern, then he can immediately implement a complete theorem prover for this logic via a *fair strategy*, which repeats applying all rules sequentially. Such a naive strategy is written:

```
repeat allRules rule1; rule2; ... ; ruleN end end
```

Here, to apply a rule means to *apply the rule simultaneously to every possible pattern in the tableau*. For our **rule** "K" this means simultaneous application to every formula of every node.

Your pet logic may require more sophisticated strategies for termination, or completeness, soundness etc. or, again, you may just want to experiment. Thus we allow for *search strategy programming* with the following constructs:

```
strategy ::= rule |
             repeat strategy end |
             allRules strategy1; strategy2; ... ; strategyN end |
             firstRule strategy1; strategy2; ... ; strategyN end
```

We use **firstRule** rule1; rule2; rule3 when we want to apply the first applicable rule, and we use **allRule** rule1; rule2; rule3 to apply all applicable rules among **rule1, rule2,** ... in that order. For instance, if **rule1** and **rule3** are the applicable rules then **firstRule** will only apply **rule1**, whereas **allRule** will apply first **rule1** and then **rule3** to the result of the first rule. There is a close similarity with **Isabelle** "tacticals" **FIRST** and **EVERY** for combining tactics.

In Fig. 3 we show an example of a correct but inefficient strategy for $K4$. With **Lotrec** it is easy to experiment and see what happens and what we save if we move the **not and** rule outside the innermost **repeat** (which is one ofthe improvements to makethe strategy more efficient). More efficient versions are available at the **Lotrec** webpage.

At present, strategies are applied globally, to all nodes and formulae. We plan future refinements where users may wish to define orderings among nodes or formulae and strategies applying rules only to the first element in the order.

```
repeat firstRule
          "stop";
          // propositional rules
          repeat allRules
                  "not not"; "and"; "not and"
               end
          end;
          // generate and check successors
          allRules
              "diamond"; "K"; "4"; "InclusionTest"
          end
      end
end
```

Fig. 3. A possible strategy for the logic $K4$

6   Great, Where Can I Find Lotrec ?

Lotrec is available at http://www.irit.fr/ACTIVITES/LILaC/Lotrec

One can also find there a library containing the standard modal logics such as K, KD, KT, $K4$, $S4$, KB, PDL, the modal logic of density, and several logics of knowledge and action, as well as intuitionistic logic.

Acknowledgements. Thanks to the LILaC members who tested Lotrec and helped to improve it by their comments, in particular L. Aszalos, J.-F. Condottta, Th. Polacsek, and S. Suwanmanee. F. Massacci acknowledges the CNR Fellowship CNR-203-07-27.

References

1. M. Castilho, L. Fariñas del Cerro, O. Gasquet, and A. Herzig. Modal tableaux with propagation rules and structural rules. *Fund. Inf.*, 32(3):281–297, 1997.
2. L. Fariñas del Cerro, O. Gasquet. Tableaux Based Decision Procedures for Modal Logics of Confluence and Density. *Fund. Inf.*, 40(4): 317-333 (1999).
3. *Proc. 1998 Int. Workshop on Description Logics*, 1998. Technical rep. ITC-IRST 9805-03.
4. D. Fauthoux. Lotrec, un outil javanais de traitement formel sur les graphes. Tech. rep., IRIT, June 2000. Master Thesis (rapport de D.E.A).
5. E. Giunchiglia, F. Giunchiglia, R. Sebastiani, and A. Tacchella. Sat vs. translation based decision procedures for modal logics: a comparative evaluation. *JANCL*, 10(2):145–173, 2000.
6. I. Horrocks and P. F. Patel-Schneider. Optimizing description logic subsumption. *JLC*, 9(3):267–293, 1999.
7. I. Horrocks, U. Sattler, and S. Tobies. Practical reasoning for expressive description logics. *Log. J. of the IGPL*, 8(3):239–263, 2000.

8. U. Hustadt and R. A. Schmidt. An empirical analysis of modal theorem provers. *JANCL*, 9(4):479–522, 1999.
9. F. Massacci. Single step tableaux for modal logics: methodology, computations, algorithms. *JAR*, 24(3):319–364, 2000.
10. F. Massacci and F. M. Donini. Design and results of TANCS-00. In *Proc. of TABLEAUX 2000*. LNAI, 2000.
11. S. Owre, J. M. Rushby, and N. Shankar. PVS: A prototype verification system. In *Proc. of CADE'92*, LNAI, pp. 748–752, 1992.
12. L. C. Paulson. *Isabelle: A Generic Theorem Prover*, LNCS. Springer-Verlag, 1994.
13. A. Voronkov. Deciding k using inverse-k. In *Proc. of KR 2000*, pp. 198–209. 2000.

The MODPROF Theorem Prover

Jens Happe

School of Computing Science
Simon Fraser University
Burnaby, B.C., V5A 1S6, CANADA
jhappe@cs.sfu.ca

Abstract. This paper introduces MODPROF, a new theorem prover and model finder for propositional and modal logic **K**. MODPROF is based on labelled modal tableaux. Its novel feature is a sophisticated simplification algorithm using structural subsumption to detect redundancies. Further distinctive features are the use of syntactic branching, and an enhanced loop-checking algorithm using a cache of satisfiable worlds created in the course of the proof. Experimental results on problems of the TANCS 2000 Theorem Prover comparison are presented.

1 The MODPROF Prover

For the last few years, fast theorem proving in modal logics has been dominated by tableaux-based systems which employ semantic branching. However, recent research [5] suggests that syntactic branching may well be competitive, particularly in structured problems which often occur in real applications, provided simplification strategies are employed. MODPROF[1] has been developed to support this claim. MODPROF is a tableaux-based theorem prover which uses syntactic branching on disjunctions. It makes use of well-known optimization techniques, notably some of those employed in FaCT [3] and DLP [7], but in a novel, more general framework. Its simplification strategy consists of eliminating subsumptions, a technique borrowed from resolution-based theorem proving and adapted to nested **K** formulas. Formulas are stored in a compact form aiding the discovery of subsumptions. A dedicated cache holds previously expanded satisfiable worlds; the cache lookup also employs subsumption testing instead of identical matching.

2 Architecture and Algorithm

2.1 Data Structures

Modal formulas are represented in MODPROF as tuples, called *templates*, of the form $w = (P, N, D, c)$, where P and N are (ordered) lists of propositional variables representing positive and negative literals, D is a list of *dual templates*,

---

[1] MODPROF is available online at http://www.cs.sfu.ca/~cl/projects/Modprof.

R. Goré, A. Leitsch, and T. Nipkow (Eds.): IJCAR 2001, LNAI 2083, pp. 459–463, 2001.
© Springer-Verlag Berlin Heidelberg 2001

and c is a *constraint template*. Any two elements of D are *sister templates* of each other, and a template or dual template occurring inside w is called a *subtemplate* of w. We also define the *inconsistent* and the *empty template*, $\emptyset$ and 0.

Templates are interpreted as formulas via the following dual functions f, f^-:

$$f(p_1, \ldots, p_k, n_1, \ldots, n_l, d_1, \ldots, d_r, c) =$$
$$p_1 \wedge \ldots \wedge p_k \wedge \neg n_1 \wedge \ldots \wedge \neg n_l \wedge f^-(d_1) \wedge \ldots \wedge f^-(d_n) \wedge \Box f(c)$$
$$f^-(p_1, \ldots, p_k, n_1, \ldots, n_l, d_1, \ldots, d_r, c) =$$
$$\neg p_1 \vee \ldots \vee \neg p_k \vee n_1 \vee \ldots \vee n_l \vee f(d_1) \vee \ldots \vee f(d_n) \vee \Diamond f^-(c)$$
$$f(\emptyset) = \bot \qquad f^-(\emptyset) = \top \qquad f(0) = \top \qquad f^-(0) = \bot$$

For example, $f\big(((p), \text{nil}, \text{nil}, (\text{nil}, (q), \text{nil}, 0))\big) = p \wedge \Box(\neg q)$, where nil denotes the empty list. Thanks to the **K**-equivalence $\Box(p \wedge q) \leftrightarrow \Box p \wedge \Box q$, all the necessities in a conjunction can be gathered into one single constraint template. Due to the axiom $\Box \top$, we can set $c = 0$ for any formula without a necessity conjunct. Disjunctions and possibilities are represented implicitly using dual templates. For example, $(\text{nil}, \text{nil}, (\text{nil}, \text{nil}, \text{nil}, (\text{nil}, (p), \text{nil}, 0)), 0)$ represents the singleton possibility $\Diamond p$; we call it a *world template*, as a world satisfying p will be created when this template is expanded. The apparent storage overhead for world templates is compensated by the elimination of negation, disjunction and possibility, which reduces the number of fields in the tuple.

Observe that a formula and its negation are represented by syntactically identical templates, distinguished only by their depth within the template structure. This corresponds to the normalized form of FaCT [3], giving us the same advantages as mentioned there. Additionally, we remark that transformation rules obeying the duality principle can be implemented efficiently. These include all simplification rules described below. Only in the model creation stage will the algorithm need to distinguish between conjunctions and disjunctions.

Templates can be nested to arbitrary depth. Note that MODPROF does not transform formulas into modal CNF or a similar form. Instead, modal formulas are first converted into negation normal form, from which the template form can be readily obtained; both steps take linear time.

The *knowledge base* is just a conjunction of formulas and thus gets represented by a template, which we call the *root template*. (We also consider this a world template). Henceforth, we consider formulas and subformulas synonymous with the templates representing them.

2.2 Subsumption and Simplification

Prior to expanding templates into models, all input formulas to MODPROF undergo a *simplification* process. At the heart of this process lies a subsumption detection algorithm [4]. (A formula F *subsumes* a formula G iff all models of G are models of F.) Subsumption checking is well-established in resolution-based theorem provers for formulas in CNF. Also, complete structural subsumption detection algorithms exist for weak description logics such as $\mathcal{FL}^-$. The novel algorithm of MODPROF is also structural; while obviously not complete, it does cover all instances of the subsumption relation in the above two classes and extends further to disjunctions and modalities of arbitrary depth.

Within the simplification process, any subformula F of the knowledge base is checked for subsumption against any formula G which holds in its scope. (In the propositional case, G holds in the scope of F exactly when F is a sister template, or a subtemplate of a sister template of G. The modal case is omitted here for brevity.) If any of these subsumption tests succeeds, F is redundant and gets erased. Empty templates arising from the simplification process exhibit local contradictions and lead to the parent template being erased. Components in unary templates, except for world templates, get "flattened out" into their parent template; since this operation changes their scope within the knowledge base, additional subsumption checks against the components of the parent template must be performed. This makes the simplification process recursive; termination is guaranteed, as the number of atoms in the knowledge base strictly decreases with every subsumption deletion.

MODPROF's simplification algorithm can be seen as an extension of the clash detection used in DLP; it is also strictly more powerful than the simplification described in [5], thanks to the subsumption test replacing syntactic equivalence, and a pervasive application of the duality principle. It alone suffices to prove certain interesting problems such as all of the problem class k_lin_p in the Tableaux 1998 Theorem Prover Comparison without any branching.

2.3 Model Creation, Heuristics, and Caching

Upon user request, MODPROF tests the simplified knowledge base for satisfiability by *expanding* the root template. The disjunctive nodes, i.e. all dual subtemplates of the root template with at least two components, are instantiated depth-first; each alternative instantiated from a dual template is checked for subsumption in all other components of the root template; this ensures that the knowledge base remains in simplified form. If an alternative fails, its negation is added to the knowledge base as a lemma. Currently, MODPROF employs chronological backtracking. However, a backjumping strategy compatible with the simplification algorithm can be devised.

Once all disjunctions are expanded, the only remaining dual subtemplates are world templates. MODPROF proceeds to expand them depth-first into accessible worlds (or shows them unsatisfiable). Thus, an explicit Kripke model for the knowledge base is constructed recursively. MODPROF reuses previous work and reduces the likelihood of exponential-space Kripke models by keeping a cache of all world templates that have been accessed on the current branch; with each successfully expanded template, a satisfying model is stored. Unlike most other theorem provers, a cache entry *subsumed by* a query template constitutes a match, in which case the query template is marked as satisfiable, and a link to the cached model is created. Note that this is done even if the cached model is still being constructed; this implements the conventional loop checking algorithm, necessary for termination in the presence of global axioms.

Clauses	Depth	V=4			V=8			V=16		
		Total	Satisf	TO	Total	Satisf	TO	Total	Satisf	TO
10	4	8 0.39	8 0.39	0	8 0.97	8 0.97	0	8 2.67	8 2.67	0
20	4	8 2.08	6 1.62	0	8 5.77	8 5.77	0	8 14.05	8 14.05	0
30	4	8 9.07	2 3.31	0	8 20.50	8 20.50	0	8 144.77	8 144.77	0
40	4	8 17.02	1 14.98	0	8 273.93	2 47.75	0	8 759.31	7 759.31	1
50	4	8 18.18	0 -.-	0	8 754.63	0 -.-	0	8 -.-	0 -.-	8
10	6	8 0.66	8 0.66	0	8 1.43	8 1.43	0	8 4.97	8 4.97	0
20	6	8 4.21	8 4.21	0	8 13.39	8 13.39	0	8 58.17	8 58.17	0
30	6	8 11.55	7 9.78	0	8 77.46	8 77.46	0	8 275.08	8 275.08	0
40	6	8 32.67	2 14.40	0	8 724.03	7 497.58	1	8 -.-	0 -.-	8
50	6	8 48.15	0 -.-	0	8 1407.88	6 914.87	1	8 -.-	0 -.-	8

Fig. 1. MODPROF performance on the problem class qbf-cnfSSS-K4 of TANCS 2000

3 Implementation

MODPROF has been written and compiled under Allegro Common Lisp Version 4.3. The source code consists of roughly 1,500 non-comment lines, distributed over several modules. The standard user interface provides standard commands to maintain a knowledge base of formulas, test its consistency and print models.

The experiments shown in Fig. 1 report running times on a class of problems of the TANCS 2000 Theorem Prover comparison [6]. They were obtained from a compiled version of MODPROF on a Sparc Ultra 10 with 384 MB of main memory. Each problem class contains 8 problems for which two running times, measured in seconds, are reported: the first one is the geometric mean over all 8 instances, the second only over those found satisfiable. The number of satisfiable problems and those timed out is also given.

4 Strengths and Weaknesses

The experimental results indicate that MODPROF performs much better on satisfiable problems than on unsatisfiable problems. This is due to the fact that satisfiable world templates are cached, whereas unsatisfiable ones are discarded. This, in conjunction with the inferior chronological backtracking, leads to a large amount of thrashing, which is also the reason why MODPROF still lags behind the fastest currently existing theorem provers, FaCT and DLP. However, MODPROF's normalized performance is already comparable or better than that of two other tableau-based theorem provers, RACE [2] and *SAT [1], both of which are reported to make extensive use of caching, and one of which employs dependency-directed backtracking. (The comparison is based on data taken from [6].) All four systems use semantic branching or convert formulas to modal CNF.

We conjecture that MODPROF will perform better on structured formulas with repetitions of similar, but not necessarily equal, subformulas which can be simplified by the subsumption checker. These are the formulas which occur more commonly in real-world knowledge bases. On flat, random formulas such as random k-SAT, we expect MODPROF to perform poorly; the asymptotic number $O(n^2)$ of initial subsumption checks (where n is the number of clauses) would be attained, but no simplification would arise if all clauses are guaranteed to be distinct and free of multiple atoms.

5 Summary and Remaining Work

At this stage, MODPROF is only a research prototype. Many of its features, and definitely the code, have not been optimized for speed. Standard optimization techniques (caching, backjumping) [8] have not fully been implemented. Also, more extensive series of experiments will be necessary to make a final comparative statement about the performance of MODPROF. However, the current results suggest that a sophisticated simplification technique can drastically speed up tableau-based theorem proving.

Our original goal was to address the question whether syntactic branching can compete in efficiency with semantic branching. The problems from the TANCS 2000 comparison do not help here, as they are already in modal CNF. However, experiments on the Tableaux 1998 problem suite which features nested, structured formulas, indicate no difference in relative performance to other systems in their 1998 implementations, which suggests that the answer is positive.

The subsumption detection algorithm was shown to yield even greater improvements on problems involving global axioms [4]; this is because the loop checking algorithm only requires an existing world *subsumed* by the current world in order to discover a periodicity and close the loop; other systems have to "wait" for an *equivalent* world template and thus incur a lot more branching.

References

1. E. Giunchiglia, A. Tacchella, *SAT: a System for the Development of Modal Decision Procedures*, in: *Automated Deduction—CADE-17*, Springer, LNAI 1831, 2000.
2. V. Haarslev, R. Möller, *RACE System Description*, in: *Proc. of the International Workshop on Description Logics (DL '99)*, Linköping, pp. 130-132, 1999
3. I. Horrocks, *Optimising Tableaux Decision Procedures for Description Logics*. Ph.D. thesis, Univ. of Manchester, U.K. 1997
4. J. Happe, *A Subsumption-Aided Caching Technique*, to appear.
5. F. Massacci, *Simplification with Renaming: A General Proof Technique for Tableau and Sequent Based Provers*, Tech. Report 424, Computer Laboratory, Univ. of Cambridge, UK, 1997.
6. F. Massacci, *Design and Results of TANCS-2000 Non-classical (Modal) Systems Comparison*, in: *Automated Reasoning with Analytic Tableaux and Related Methods (TABLEAUX 2000)*, pp. 52–56, Springer, LNAI 1847, 2000.
7. P. F. Patel-Schneider, *DLP System Description*. In E. Franconi, G. de Giacomo, R.M. MacGregor, W. Nutt, C.A. Welty and F. Sebastiani (eds.) *Collected Papers from the International Description Logics Workshop (DL '98)*, pp. 87-89, 1998.
8. I. Horrocks, P. F. Patel-Schneider, *Optimizing Description Logic Subsumption*. Journal of Logic and Computation, 9:3, pp. 267-293, 1999.

A New System and Methodology for Generating Random Modal Formulae

Peter F. Patel-Schneider[1] and Roberto Sebastiani[2]

[1] Bell Labs Research, 600 Mountain Ave., 2A-427 Murray Hill, NJ 07974, USA
pfps@research.bell-labs.com
[2] Dept. of Mathematics, University of Trento, via Sommarive 14, I-38050 Trento,
Italy
rseba@science.unitn.it

Abstract. Previous methods for generating random modal formulae
(for the multi-modal logic $\mathbf{K}_{(m)}$) result either in flawed test sets or formulae that are too hard for current modal decision procedures and, also,
unnatural. We present here a new system and generation methodology
which results in unflawed test sets and more-natural formulae that are
better suited for current decision procedures.

Most empirical testing of decision procedures for propositional modal logics,
usually for the multi-modal logic $\mathbf{K}_{(m)}$, employs randomly generated formulae.
This style of testing was initially proposed by Giunchiglia *et al* [3] and later improved by them and also by Hustadt and Schmidt [6,1]. Other kinds of randomly
generated formulae have been proposed by Massacci [7]. Randomly generated
formulae have been used with all the recent, highly optimised modal decision
procedures, including DLP [8], FaCT [4], KSATC [1], *SAT [2], and TA [6], and
have been used in several comparisons of these systems [7].

The basic idea behind the random generator in [3,6,1] is to generate a number
of clauses. Each clause has the same number of disjuncts. As most testing as been
on clauses with three disjuncts, the generated formulae are often called 3CNF$_{\Box_m}$
formulae. Each disjunct in each clause is independently either a possibly negated
propositional variable or a modal formula consisting of a possibly negated box
over a clause. The embedded clauses are generated in the same way as the top-
level clauses, except that at some maximum modal depth all the disjuncts are
possibly negated propositional variables. Care is taken to not repeat disjuncts
in a clause, nor to have complementary disjuncts in a clause.

Testing consists of setting various parameters, usually the maximum modal
depth, d, the probability that a disjunct is a propositional literal, p, and the
number of propositional variables, N. Most tests use standard values for the
other parameters, namely three disjuncts in a clause, C, and only one kind of
modal box, m. Then several values are picked for the number of top-level clauses
in a formula, L, and for each value for L many (usually about 100) formulae are
generated and tested for satisfiability.

This testing methodology as described above provides a decent test for current modal decision procedures, allowing good comparisons of their behaviour.

R. Goré, A. Leitsch, and T. Nipkow (Eds.): IJCAR 2001, LNAI 2083, pp. 464–468, 2001.

However, problems have remained with these methods, most notably that the generated formulae can contain pieces that make the entire formula easy to solve. Because even the fastest current modal decision procedures can only handle formulae with a few propositional variables, the presence of even a small number of top-level clauses with only propositional disjuncts can easily cover all the combinations of the propositional literals and make the entire formula unsatisfiable. (We call such formulae trivially unsatisfiable.)

Previous attempts to eliminate this trivial unsatisfiability have concentrated on eliminating top-level propositional disjuncts by setting $p = 0$ [6]. However, formulae with propositional literals showing up only at the deepest modal depth are extremely hard to solve for modal depths greater than 1. Moreover, the resulting formule are not very natural [5].

We have devised a new generation methodology for generating modal formulae for $\mathbf{K}_{(\mathbf{m})}$ that eliminates or reduces the problems with the previous generation methods. The first new idea of our approach, first suggested in [5], is to eliminate strictly-propositional clauses except at the maximum modal depth by requiring that the number of propositional literals in each clause be less than 1 away from the average value, while still maintaining the overall ratio. For three disjuncts per clause ($C = 3$) and propositional probability one-half ($p = 0.5$) this means that half the clauses have 2 propositional literals and half have 1.

To show the benefit of this change, we present two experimental runs for maximum modal depth $d = 2$ and propositional probability $p = 0.5$. In the old method the time curves are dominated by a "half-dome" shape, whose steep side shows up where the number of trivially unsatisfiable formulae becomes large before the formulae become otherwise easy to solve, as shown in Figure 1 (left). In fact, nearly all the unsatisfiable formulae here are propositionally unsatisfiable. With our new method, as shown in Figure 1 (right), the formulae are much more difficult to solve than the old method, because there is no abrupt drop-off from propositional unsatisfiability, but they are much easier to solve than those generated with $p = 0$. Further, trivially unsatisfiable formulae do not appear at all in the interesting portion of the test sets.

The second new idea of our approach is to allow the number of disjuncts in a clause C to vary in a manner similar to the number of propositional disjuncts. We then determine the number of propositional literals in each clause based on the number of disjuncts in that particular clause.

We tested this generation methodology for several values of C. The results for $C = 2.5$ are given in Figure 2 (left). These formulae are much easier than those generated with $C = 3$, although they are still quite hard and form a reasonable source of testing data. Trivially unsatisfiable formulae appear in large numbers well after all the formulae are all unsatisfiable and relatively easy. By reducing the number of disjuncts we are thus able to create interesting and unflawed test sets for higher modal depths.

The last new idea of our approach is that the generator allows (optionally) direct specification of the probability distribution of the number of propositional atoms in a clause, and allows the distribution to be different for each modal depth

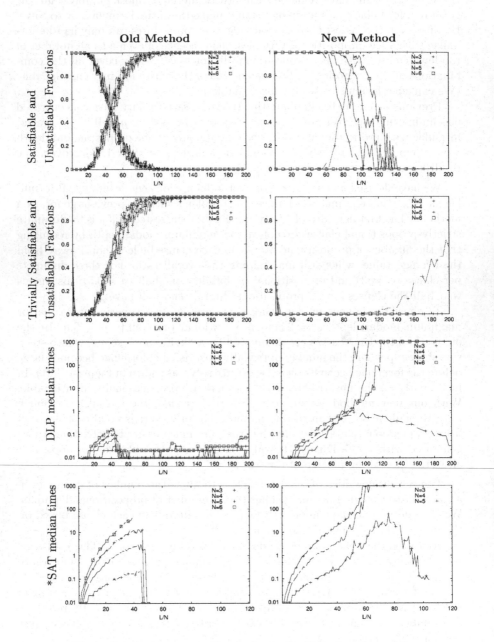

Fig. 1. Results for $m = 1$, $C = 3$, $d = 2$, $N = 3, ..., 6$, and $p = 0.5$.

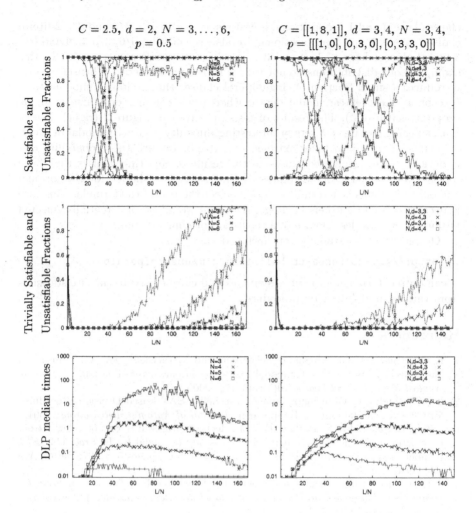

Fig. 2. New method. Left column: $m = 1$, $C = 2.5$, $d = 2$, $N = 3, ..., 6$, and $p = 0.5$. Right column: $m = 1$, $C = [[1, 8, 1]]$, $d = 3, 4$, $N = 3, 4$, $p = [[[1, 0], [0, 3, 0], [0, 3, 3, 0]]]$.

from the top level to $d - 1$. It also allows direct specification of the probability distribution for the number of literals in a clause at each modal depth. Thus, the probability distribution for the number of propositional atoms depends on both the modal depth and the number of literals in the clause.

As an example we present a set of tests with $m = 1$, $C = [[1, 8, 1]]$, $N = 3, 4$, $d = 3, 4$, and $p = [[[1, 0], [0, 3, 0], [0, 3, 3, 0]]]$; C represents the probability distribution for the number of literals in a clause, meaning "1/10, 8/10 and 1/10 clauses are of length 1, 2, 3 respectively"; p represents the probability distribution for the number of propositional atoms in a clause, meaning "1/1 and 0/1 1-literal clauses have 0 and 1 propositional literal respectively; 0/3, 3/3 and 0/3 2-literal clauses have 0, 1 and 2 propositional literals respectively; 0/6, 3/6, 3/6 and 0/6 3-literal clauses have 0, 1, 2 and 3 propositional literals respectively". In this example, both distributions do not vary with the modal depth. This set of tests introduces a small fraction of single-literal clauses that contain a modal literal (except at the greatest modal depth, where they contain, of course, a single propositional literal). The results of tests are given in Figure 2 (right). Notice that we can generate test sets of interesting difficulty even with modal depth 4.

Our generator also takes extreme care to not disturb the probablility distributions of the generated cluses as the result of rejecting clauses that have repeated or contradictory disjuncts. It first selects a "shape" for the formula and only when this is determined does it select the rest of the formula. This care is necessary because otherwise larger disjuncts would be preferentially selected because they have less chance of a repetition or contradiction.

Our generator is available for download at

> http://www-db.research.bell-labs.com/user/pfps/dlp.

It can output formulae in various syntaxes and can be used to capture statistics from various modal decision procedures.

References

1. E. Giunchiglia, F. Giunchiglia, R. Sebastiani, and A. Tacchella. SAT vs. Translation based decision procedures for modal logics: a comparative evaluation. *Journal of Applied Non-Classical Logics*, 10(2):145–172, 2000.
2. E. Giunchiglia, F. Giunchiglia, and A. Tacchella. SAT Based Decision Procedures for Classical Modal Logics. To appear in *Journal of Automated Reasoning.*, 2001.
3. F. Giunchiglia and R. Sebastiani. Building decision procedures for modal logics from propositional decision procedures - the case study of modal K. In *Proc. CADE13*, number 1104 in Lecture Notes in Artificial Intelligence, pages 583–597, Berlin, August 1996. Springer-Verlag.
4. I. Horrocks. Using an expressive description logic: FaCT or fiction? In *Sixth International Conference on Principles of Knowledge Representation and Reasoning (KR'98)*, pages 636–647, 1998.
5. Ian Horrocks, Peter F. Patel-Schneider, and Roberto Sebastiani. An analysis of empirical testing for modal decision procedures. *Logic Journal of the IGPL*, 8(3):293–323, 2000.
6. Ullrich Hustadt and Renate A. Schmidt. An empirical analysis of modal theorem provers. *Journal of Applied Non-Classical Logics*, 9(4):479–522, 1999.
7. F. Massacci. Design and Results of Tableaux-99 Non-Classical (Modal) System Competition. In *Automated Reasoning with Analytic Tableaux and Related Methods: International Conference (Tableaux'99)*, number 1617 in Lecture Notes in Artificial Intelligence, pages 14–18, Berlin, 1999. Springer-Verlag.
8. Peter F. Patel-Schneider. DLP system description. In E. Franconi, G. De Giacomo, R. M. MacGregor, W. Nutt, C. A. Welty, and F. Sebastiani, editors, *Collected Papers from the International Description Logics Workshop (DL'98)*, pages 87–89, 1998. Available as CEUR-WS/Vol-11 from http://SunSITE.Informatik.RWTH-Aachen.DE/Publications/CEUR-WS.

Decidable Classes of Inductive Theorems*

Jürgen Giesl[1] and Deepak Kapur[2]

[1] LuFG Informatik II, RWTH Aachen, Ahornstr. 55, 52074 Aachen, Germany,
`giesl@informatik.rwth-aachen.de`
[2] Computer Science Dept., University of New Mexico, Albuquerque, NM 87131, USA
`kapur@cs.unm.edu`

Abstract. Kapur and Subramaniam [8] defined syntactical classes of
equations where inductive validity is decidable. Thus, their validity can
be checked without any user interaction and hence, this allows an integration of (a restricted form of) induction in fully automated reasoning tools
such as model checkers. However, the results of [8] were only restricted
to equations. This paper extends the classes of conjectures considered in
[8] to a larger class of arbitrary quantifier-free formulas (e.g., conjectures
also containing negation, conjunction, disjunction, etc.).

1 Introduction

Inductive theorem provers usually require massive manual intervention and they
may waste huge amounts of time on proof attempts which fail due to the incompleteness of the prover. Therefore, induction has not yet been integrated in
fully automated reasoning systems (i.e., model checkers) used for hardware and
protocol verification, static and type analyses, byte-code verification, and proof-
carrying codes. Most such push-button systems use a combination of decision
procedures for theories such as Presburger arithmetic, propositional satisfiability,
and data structures including bit vectors, arrays, and lists. However, extending
these tools by the capability to perform induction proofs would be very desirable,
since induction is frequently needed to reason about structured and parameterized circuits (e.g., n-bit adders or multipliers), the timing behavior of circuits
with feedback loops, and code using loops and/or recursion.

For that reason, Kapur and Subramaniam proposed an approach for integrating induction schemes suggested by terminating function definitions with
decision procedures, and gave a syntactical characterization of a class of equations where inductive validity is decidable using decision procedures and the
cover set method for mechanizing induction [8,11]. For those equations, induction proofs can be accomplished without any user interaction and they only fail if
the conjecture is not valid. In Section 2, we give a simple characterization which
extends the class of decidable equations in [8]. Subsequently, we further extend
the approach to arbitrary quantifier-free formulas, i.e., we define classes of such

---

* Supported by the Deutsche Forschungsgemeinschaft Grant GI 274/4-1 and the National Science Foundation Grants nos. CCR-9996150 and CDA-9503064.

R. Goré, A. Leitsch, and T. Nipkow (Eds.): IJCAR 2001, LNAI 2083, pp. 469–484, 2001.

formulas where inductive validity is decidable. The crucial concept for this characterization are so-called *correctness predicates*. For a quantifier-free conjecture φ, c_φ is a correctness predicate iff for any tuple of (constructor) ground terms q^*, the truth of $c_\varphi(q^*)$ implies the truth of $\varphi[x^*/q^*]$ (cf. [6,9]). We present a technique for automatically generating correctness predicates in Section 3.

The truth of a correctness predicate is only sufficient, but not necessary for the truth of the corresponding conjecture. In Section 4 we examine for which equations φ the correctness predicate is *exact* (i.e., the truth of $c_\varphi(q^*)$ is both sufficient and necessary for the truth of $\varphi[x^*/q^*]$). We develop a characterization to recognize (a subclass of) these equations automatically. In Section 5 we show that the use of exact correctness predicates allows us to extend the decidable classes of inductive theorems from equations to arbitrary quantifier-free formulas.

Our results are also useful for conventional inductive theorem provers since exact correctness predicates can be used to simplify the proof of conjectures like $\mathsf{double}(y) = y \Rightarrow y = 0$ where inductive provers would fail otherwise.

Even though the paper focuses on constructor systems and the decidable theory of quantifier-free formulas on free constructors, we believe the approach extends to other decidable theories $\mathcal{T}$ as well (e.g., Presburger arithmetic).

2   Equations Where Inductive Validity Is Decidable

We use term rewrite systems $\mathcal{R}$ (TRSs) as our programming language [1]. In a TRS, all root symbols of left-hand sides are called *defined* and all other function symbols of $\mathcal{R}$ are *constructors*. We only consider constructor systems (CSs), i.e., TRSs where the left-hand sides contain no defined symbols below the root position, even though most of the results in this paper generalize to more general theory-based systems, called $\mathcal{T}$-based systems in [8], with a decidable theory $\mathcal{T}$, in which arguments to defined symbols are terms from $\mathcal{T}$. Moreover, we restrict ourselves to (ground-)convergent and sufficiently complete CSs $\mathcal{R}$, i.e., for every ground term t there exists a unique constructor ground term q such that $t \to_{\mathcal{R}}^* q$. (A term containing only variables and constructors is called a constructor term; a constructor term without variables is a constructor ground term.)

For induction proofs, we use the concept of *cover sets* [7,11]. A cover set is a finite set of pairs $\mathcal{C} = \{\langle s_1^*, \{t_{1,1}^*, \ldots, t_{1,n_1}^*\}\rangle, \ldots, \langle s_m^*, \{t_{m,1}^*, \ldots, t_{m,n_m}^*\}\rangle\}$, where s_i^* and $t_{i,j}^*$ are n-tuples of terms (for some $n \geq 0$). $\mathcal{C}$ is *complete* if for every n-tuple q^* of constructor ground terms, there is an s_i^* and a substitution σ such that $s_i^* \sigma = q^*$. Every cover set $\mathcal{C}$ induces a relation $<_{\mathcal{C}}$ on tuples of constructor ground terms: $p^* <_{\mathcal{C}} q^*$ iff there exists a pair $\langle s_i^*, \{t_{i,1}^*, \ldots, t_{i,n_i}^*\}\rangle \in \mathcal{C}$ such that $s_i^* \sigma = q^*$ and $t_{i,j}^* \sigma \to_{\mathcal{R}}^* p^*$. $\mathcal{C}$ is called *well-founded* iff $<_{\mathcal{C}}$ is well founded.[1]

A quantifier-free formula φ is *inductively valid* (or "valid" for short), denoted "$\mathcal{R} \models_{\mathrm{ind}} \varphi$", iff $\forall y^* \; \varphi$ holds in the initial model of the equations of $\mathcal{R}$ (where y^*

---

[1] $<_{\mathcal{C}}$ is well founded if there exists no infinite sequence $\ldots t_3 <_{\mathcal{C}} t_2 <_{\mathcal{C}} t_1 <_{\mathcal{C}} t_0$.

are the variables in φ).[2] For example, consider the following CS:

$$\mathsf{half}(0) \to 0, \quad \mathsf{half}(\mathsf{s}(0)) \to 0, \quad \mathsf{half}(\mathsf{s}(\mathsf{s}(x))) \to \mathsf{s}(\mathsf{half}(x)).$$

This function definition suggests the cover set $\mathcal{C}_{\mathsf{half}} = \{\langle 0, \varnothing \rangle, \langle \mathsf{s}(0), \varnothing \rangle,$ $\langle \mathsf{s}(\mathsf{s}(x)), \{x\}\rangle\}$. To prove φ by induction w.r.t. $\mathcal{C}_{\mathsf{half}}$ (using the induction variable y), one obtains the base formulas $\varphi[y/0]$ and $\varphi[y/\mathsf{s}(0)]$ and the step formula $\varphi[y/x] \Rightarrow \varphi[y/\mathsf{s}(\mathsf{s}(x))]$. Here, $\varphi[y/x]$ is the *induction hypothesis* and $\varphi[y/\mathsf{s}(\mathsf{s}(x))]$ is the *induction conclusion*. When proving a conjecture φ containing a term $f(y_1, \ldots, y_n)$, a successful heuristic for the choice of an induction relation is to perform induction w.r.t. $\mathcal{C}_f$ using the induction variables $y_1, \ldots, y_n$, cf. [2,11].

Kapur and Subramaniam [8] characterized classes of equations where inductive validity is decidable (the decision procedure consists of an induction proof attempt w.r.t. a particular cover set). The observation is that if each induction formula built according to some cover set $\mathcal{C}$ only contains terms from an underlying decidable theory, then validity of the original conjecture can be decided.

Def. 1 and Thm. 2 apply to general $\mathcal{T}$-based systems, but due to lack of space, we focus on the decidable quantifier-free theory of free constructors in this paper. Here, $r[s^*]$ abbreviates $r[y^*/s^*]$ where y^* contains all variables in r.

Definition 1 ($\mathcal{C}$-provability). *Let $\mathcal{R}$ be a convergent sufficiently complete CS and let $\mathcal{C}$ be a complete well-founded cover set. An equation $r_1 = r_2$ is $\mathcal{C}$-provable w.r.t. $\mathcal{R}$ iff r_2 is a constructor term, for every $\langle s_i^*, \{t_{i,1}^*, \ldots, t_{i,n}^*\}\rangle \in \mathcal{C}$, s_i^* and all $t_{i,j}^*$ are tuples of constructor terms, and there exists a constructor term context C_i such that $r_1[s_i^*] \to_{\mathcal{R}}^* C_i[r_1[t_{i,1}^*], \ldots, r_1[t_{i,n}^*]]$.*

As an example, let us extend the CS for half by the rules $\mathsf{double}(0) \to 0$ and $\mathsf{double}(\mathsf{s}(x)) \to \mathsf{s}(\mathsf{s}(\mathsf{double}(x)))$. Then the equation $\mathsf{double}(\mathsf{half}(y)) = y$ is $\mathcal{C}_{\mathsf{half}}$-provable. As required, the term y is a constructor term. Moreover, we obtain

$$r_1[s_1] = \mathsf{double}(\mathsf{half}(0)) \qquad \to_{\mathcal{R}}^* 0 \qquad\qquad \text{and thus, } C_1 = 0,$$
$$r_1[s_2] = \mathsf{double}(\mathsf{half}(\mathsf{s}(0))) \qquad \to_{\mathcal{R}}^* 0 \qquad\qquad \text{and thus, } C_2 = 0,$$
$$r_1[s_3] = \mathsf{double}(\mathsf{half}(\mathsf{s}(\mathsf{s}(x)))) \to_{\mathcal{R}}^* \mathsf{s}(\mathsf{s}(\mathsf{double}(\mathsf{half}(x)))) \text{ and thus, } C_3 = \mathsf{s}(\mathsf{s}(\square)).$$

Since $\mathcal{C}$-provability is decidable, Def. 1 characterizes a decidable class of conjectures. Instead of checking $\mathcal{C}$-provability directly, several sufficient conditions for $\mathcal{C}$-provability were given in [8]. We obtain the following theorem.

Theorem 2 (Decidability of inductive validity for equations). *Let $\mathcal{R}$ be a convergent sufficiently complete CS, let $\mathcal{C}$ be a complete well-founded cover set, and let $r_1 = r_2$ be a $\mathcal{C}$-provable equation. Then inductive validity of $r_1 = r_2$ is decidable (by attempting an induction proof w.r.t. $\mathcal{C}$).*

Proof. The decision procedure works by constructing the formulas

$$C_i[r_2[t_{i,1}^*], \ldots, r_2[t_{i,n}^*]] = r_2[s_i^*] \tag{1}$$

---

[2] $\mathcal{R} \models_{\mathrm{ind}} \varphi$ means that for all constructor ground terms q^*, $\varphi[y^*/q^*]$ follows from $\mathcal{R}$'s equations and axioms stating that different constructor ground terms are not equal.

for all $\langle s_i^*, \{t_{i,1}^*, \ldots, t_{i,n}^*\}\rangle \in \mathcal{C}$. As these equations only contain constructor terms, their validity is decidable.

It turns out that $r_1 = r_2$ is valid iff all these equations are valid. For the "if"-direction, notice that (1) implies the induction formula

$$r_1[t_{i,1}^*] = r_2[t_{i,1}^*] \wedge \ldots \wedge r_1[t_{i,n}^*] = r_2[t_{i,n}^*] \Rightarrow r_1[s_i^*] = r_2[s_i^*].$$

Thus, the validity of $r_1 = r_2$ follows by Noetherian induction. For the "only if"-direction, note that the validity of $r_1 = r_2$ implies the validity of (1).   □

Since $\mathsf{double}(\mathsf{half}(y)) = y$ is $\mathcal{C}_{\mathsf{half}}$-provable, the above decision procedure can determine its validity. It has to check the validity of the equations

$$C_1[r_2[t_1]] = r_2[s_1], \text{ i.e., } 0 = 0, \tag{2}$$
$$C_2[r_2[t_2]] = r_2[s_2], \text{ i.e., } 0 = \mathsf{s}(0), \tag{3}$$
$$C_3[r_2[t_3]] = r_2[s_3], \text{ i.e., } \mathsf{s}(\mathsf{s}(x)) = \mathsf{s}(\mathsf{s}(x)). \tag{4}$$

Since these equations only contain constructor terms, their validity is decidable. (Obviously, such an equation is valid iff both terms in the equation are syntactically identical.) While (2) and (4) are valid, the second equation (3) is not valid and thus, the conjecture $\mathsf{double}(\mathsf{half}(y)) = y$ is not valid either.

Our aim is to extend the result of Thm. 2 to more general formulas (i.e., not just equations), provided that all equations in these formulas are $\mathcal{C}$-provable. For example, we would like to consider formulas like $\mathsf{double}(\mathsf{half}(y)) = y \Rightarrow \mathsf{even}(y) = \mathsf{true}$ or $\mathsf{double}(y) = y \Rightarrow y = 0$. Equations appearing in these formulas are neither valid nor unsatisfiable; consequently, there is a need to characterize the subset of instantiations for the variables for which these equations are true. For this extension, we need the notion of correctness predicates.

3   Correctness Predicates

We present a technique which automatically generates algorithms for so-called *correctness predicates* c_φ for equations φ. For any tuple of constructor ground terms q^*, the truth of $c_\varphi(q^*)$ implies that $\varphi[y^*/q^*]$ is valid. Our definition of correctness predicates is similar to the definitions of [6,9], but its form is quite restricted since we are interested in ensuring that validity of correctness predicates is decidable and that exact correctness predicates can be generated which completely characterize the domain of values on which the conjecture holds.

We have seen that the proof of the conjecture $\mathsf{double}(\mathsf{half}(y)) = y$ can be attempted by induction w.r.t. the cover set $\mathcal{C}_{\mathsf{half}}$. If $y = 0$, the conjecture can be reduced to the equation (2) which is always true. In the case $y = \mathsf{s}(0)$ we obtain the equation (3) which is always false. Finally, in the step case where $y = \mathsf{s}(\mathsf{s}(x))$, we have to prove that the induction hypothesis $\mathsf{double}(\mathsf{half}(x)) = x$ implies the induction conclusion $\mathsf{double}(\mathsf{half}(\mathsf{s}(\mathsf{s}(x)))) = \mathsf{s}(\mathsf{s}(x))$. As shown in Section 2, $\mathsf{double}(\mathsf{half}(\mathsf{s}(\mathsf{s}(x))))$ evaluates to $\mathsf{s}(\mathsf{s}(\mathsf{double}(\mathsf{half}(x))))$. Due to the induction hypothesis, we can replace the subterm $\mathsf{double}(\mathsf{half}(x))$ by x. Thus, we obtain the equation (4) (which is always true). Hence, provided that the induction hypothesis is valid, the induction conclusion would also be valid. This gives rise to the following rules for the correctness predicate $c_{\text{double(half}(y))=y}$:

$$c_{\text{double(half}(y))=y}(0) \rightarrow \text{true}, \tag{5}$$
$$c_{\text{double(half}(y))=y}(\text{s}(0)) \rightarrow \text{false}, \tag{6}$$
$$c_{\text{double(half}(y))=y}(\text{s}(\text{s}(x))) \rightarrow c_{\text{double(half}(y))=y}(x). \tag{7}$$

Thus, we have synthesized the **even** algorithm. Note that the rule (7) is stronger than the following rule one would have gotten from the above analysis:

$$c_{\text{double(half}(y))=y}(\text{s}(\text{s}(x))) \rightarrow \text{true} \ \text{if} \ c_{\text{double(half}(y))=y}(x).$$

Since we want to generate unconditional rewrite rules for the definition of correctness predicates and to synthesize a complete definition, we use the form (7). As a result, the correctness predicate so generated may not be exact, and hence, provides only a sufficient condition for the conjecture to be valid.

In general, to prove a C-provable equation $r_1 = r_2$ w.r.t. a cover set C, for each pair $\langle s_i^*, \{t_{i,1}^*, \ldots, t_{i,n_i}^*\} \rangle \in C$ we must check whether the equation $C_i[r_2[t_{i,1}^*], \ldots, r_2[t_{i,n}^*]] = r_2[s_i^*]$ is valid, cf. Equation (1) in the proof of Thm. 2. In order to obtain correctness predicates as simple as the ones above, we have to demand that these equations are either valid for *all* instantiations or for *none*. This ensures that the right-hand sides of the rules for correctness predicates only have the form true, false, or recursive calls of correctness predicates.

Definition 3 (Radical equations). *Let $\mathcal{R}$ be a convergent sufficiently complete CS and let $C = \{\langle s_1^*, \{t_{1,1}^*, \ldots, t_{1,n_1}^*\} \rangle, \ldots, \langle s_m^*, \{t_{m,1}^*, \ldots, t_{m,n_m}^*\} \rangle\}$ be a complete well-founded cover set. An equation $r_1 = r_2$ is radical under C iff $r_1 = r_2$ is a C-provable equation where $r_1[s_i^*] \rightarrow_{\mathcal{R}}^* C_i[r_1[t_{i,1}^*], \ldots, r_1[t_{i,n_i}^*]]$ for a constructor term context C_i and for all $1 \le i \le m$ we have*

$$\mathcal{R} \models_{\text{ind}} \ C_i[r_2[t_{i,1}^*], \ldots, r_2[t_{i,n_i}^*]]] = r_2[s_i^*] \quad or$$
$$\mathcal{R} \models_{\text{ind}} \neg C_i[r_2[t_{i,1}^*], \ldots, r_2[t_{i,n_i}^*]]] = r_2[s_i^*].$$

Note that since all C_i, s_i^*, and t_i^* are constructor terms, it is decidable whether a C-provable equation is radical. The reason is that one only has to check whether an equation between two constructor terms is valid or unsatisfiable. Obviously, such an equation is unsatisfiable iff the two terms are not unifiable. For instance, the equation $\text{double(half}(y)) = y$ is radical under C_{half} since the terms in the equations (2) - (4) are either identical or not unifiable.

To ease the presentation, we will now restrict ourselves to cover sets where there is at most one induction hypothesis for every induction step case.[3] Thus,

---

[3] The definition of correctness predicates can be easily generalized to the case of multiple induction hypotheses. In fact, correctness predicates can be defined for arbitrary equations, i.e., they do not have to be C-provable or radical as required in this paper. However, these requirements are necessary in order to generate exact correctness predicates c_φ for arbitrary conjectures φ, such that validity of c_φ is decidable.

we only consider cover sets with pairs $\langle s_i^*, \{t_{i,1}^*, \ldots, t_{i,n_i}^*\}\rangle$ where $0 \le n_i \le 1$. Then we obtain the following definition of correctness predicates.

Definition 4 (Correctness Predicate). *Let* $\mathcal{R}$, $\mathcal{C}$, $r_1 = r_2$ *be as in Def. 3 where* $0 \le n_i \le 1$ *for all* $1 \le i \le m$ *and let* $r_1 = r_2$ *be radical under* $\mathcal{C}$. *Then the correctness predicate* $c_{r_1=r_2}$ *under* $\mathcal{C}$ *is defined by the following rules:*

$$c_{r_1=r_2}(s_i^*) \to \begin{cases} \text{true, } \textit{if } \mathcal{R} \models_{\text{ind}} C_i = r_2[s_i^*] \textit{ and } n_i = 0, & (8) \\ \text{false, } \textit{if } \mathcal{R} \models_{\text{ind}} \neg C_i = r_2[s_i^*] \textit{ and } n_i = 0, & (9) \end{cases}$$

$$c_{r_1=r_2}(s_i^*) \to \begin{cases} c_{r_1=r_2}(t_{i,1}^*), \textit{if } \mathcal{R} \models_{\text{ind}} C_i[r_2[t_{i,1}^*]] = r_2[s_i^*] \textit{ and } n_i = 1, & (10) \\ \text{false,} \quad \textit{if } \mathcal{R} \models_{\text{ind}} \neg C_i[r_2[t_{i,1}^*]] = r_2[s_i^*] \textit{ and } n_i = 1. & (11) \end{cases}$$

Thm. 5 proves that a correctness predicate indeed represents a sufficient, but not a necessary condition for the soundness of the corresponding equation.

Theorem 5 (Correctness predicates are sufficient, but not necessary). *Let* $\mathcal{R}$, $\mathcal{C}$, $r_1 = r_2$ *be as in Def. 4. Let* $c_{r_1=r_2}$ *be a correctness predicate for* $r_1 = r_2$ *under* $\mathcal{C}$ *and let* $\mathcal{R}$ *also contain the rules defining* $c_{r_1=r_2}$. *Then we have*

(a) $\mathcal{R} \models_{\text{ind}} c_{r_1=r_2}(y^*) = \text{true} \Rightarrow r_1 = r_2$.
(b) In general, we have $\mathcal{R} \not\models_{\text{ind}} r_1 = r_2 \Rightarrow c_{r_1=r_2}(y^*) = \text{true}$.

Proof.

(a) Let q^* be a tuple of constructor ground terms such that $\mathcal{R} \models_{\text{ind}} c_{r_1=r_2}(q^*) = \text{true}$. We prove $\mathcal{R} \models_{\text{ind}} r_1[q^*] = r_2[q^*]$ by induction w.r.t. $<_{\mathcal{C}}$. Due to the completeness of the cover set, there exists some $\langle s^*, \{t_1^*, \ldots, t_n^*\}\rangle \in \mathcal{C}$ and some substitution σ such that $q^* = s^*\sigma$ and since $r_1 = r_2$ is $\mathcal{C}$-provable (due to its radicality), we have $\mathcal{R} \models_{\text{ind}} r_1[s^*] = C[r_1[t_1^*], \ldots, r_1[t_n^*]]$.
If $n = 0$, then we also have $\mathcal{R} \models_{\text{ind}} C = r_2[s^*]$ and thus $\mathcal{R} \models_{\text{ind}} r_1[s^*] = r_2[s^*]$. If $n = 1$, we have $\mathcal{R} \models_{\text{ind}} C[r_2[t_1^*]] = r_2[s^*]$ and $\mathcal{R} \models_{\text{ind}} c_{r_1=r_2}(t_1^*\sigma) = \text{true}$. The induction hypothesis yields $\mathcal{R} \models_{\text{ind}} r_1[t_1^*\sigma] = r_2[t_1^*\sigma]$. From the validity of $r_1[s^*] = C[r_1[t_1^*]]$ and $C[r_2[t_1^*]] = r_2[s^*]$, $\mathcal{R} \models_{\text{ind}} r_1[s^*\sigma] = r_2[s^*\sigma]$.

(b) Consider the equation $\text{half}(y) = \text{s}(0)$ and induction w.r.t. the cover set $\mathcal{C}_{\text{half}}$. In the base cases $y = 0$ and $y = \text{s}(0)$ the resulting conjecture $0 = \text{s}(0)$ is unsatisfiable and in the step case, the induction conclusion $\text{half}(\text{s}(\text{s}(x))) = \text{s}(0)$ can be evaluated to $\text{s}(\text{half}(x)) = \text{s}(0)$. Applying the induction hypothesis $\text{half}(x) = \text{s}(0)$ yields $\text{s}(\text{s}(0)) = \text{s}(0)$ which is unsatisfiable. So the equation $\text{half}(y) = \text{s}(0)$ is radical under $\mathcal{C}_{\text{half}}$ and we obtain the rules $c_{\text{half}(y)=\text{s}(0)}(0) \to \text{false}$, $c_{\text{half}(y)=\text{s}(0)}(\text{s}(0)) \to \text{false}$, and $c_{\text{half}(y)=\text{s}(0)}(\text{s}(\text{s}(x))) \to \text{false}$. So $c_{\text{half}(y)=\text{s}(0)}$ is always false, but $\text{half}(y) = \text{s}(0)$ holds for $\text{s}^2(0)$ and $\text{s}^3(0)$. $\square$

In fact, a correctness predicate $c_\varphi(q^*)$ yields true iff the equation φ holds for both q^* *and* for all arguments p^* which are smaller than q^* w.r.t. the induction relation induced by the cover set. For that reason, the correctness predicate $c_{\text{half}(y)=\text{s}(0)}$ returns false for the arguments $\text{s}^2(0)$ and $\text{s}^3(0)$ although the conjecture is true, since it is false for the smaller arguments 0 and $\text{s}(0)$.

4 Conjectures with Exact Correctness Predicate

In this section we characterize equations $r_1 = r_2$ where the correctness predicate $c_{r_1=r_2}$ is *exact*, i.e., for all q^*, $c_{r_1=r_2}(q^*)$ is true iff $\mathcal{R} \models_{\text{ind}} r_1[q^*] = r_2[q^*]$. Exactness is ensured if in Def. 4, whenever Rule (10) is used, the induction conclusion $r_1[s_i^*] = r_2[s_i^*]$ is equivalent to $r_1[t_{i,1}^*] = r_2[t_{i,1}^*]$. As we have seen in Sect. 3, $c_{r_1=r_2}(q^*)$ only returns true if $r_1 = r_2$ is true for q^* and for all p^* smaller than q^* w.r.t. the induction relation induced by the cover set. Thus, $c_{r_1=r_2}$ is only exact if $r_1[q^*] = r_2[q^*]$ implies the validity of $r_1[p^*] = r_2[p^*]$ for all arguments $p^* <_\mathcal{C} q^*$. So $c_{r_1=r_2}$ only describes the exact set of instantiations where $r_1 = r_2$ is valid, if each induction conclusion implies all its induction hypotheses.

Consider again the proof of $\mathsf{double}(\mathsf{half}(y)) = y$ by induction w.r.t. $\mathcal{C}_{\mathsf{half}}$. We obtain the induction conclusion $\mathsf{double}(\mathsf{half}(\mathsf{s}(\mathsf{s}(x)))) = \mathsf{s}(\mathsf{s}(x))$ and the induction hypothesis $\mathsf{double}(\mathsf{half}(x)) = x$. Indeed, this conjecture has the desired property

$$\mathcal{R} \models_{\text{ind}} \mathsf{double}(\mathsf{half}(\mathsf{s}(\mathsf{s}(x)))) = \mathsf{s}(\mathsf{s}(x)) \;\Rightarrow\; \mathsf{double}(\mathsf{half}(x)) = x. \qquad (12)$$

To see this, note that in the first base case where $y = 0$, the left-hand side $\mathsf{double}(\mathsf{half}(0))$ evaluates to 0, which is smaller than or equal to the right-hand side 0 (if terms are compared by the subterm relation, for example). Similarly, in the second base case where $y = \mathsf{s}(0)$, the left-hand side evaluates to 0, which is again smaller than or equal to the right-hand side $\mathsf{s}(0)$. In the step case, the left hand side of the induction conclusion can be evaluated to

$$\mathsf{s}(\mathsf{s}(\underline{\mathsf{double}(\mathsf{half}(x))})) = \mathsf{s}(\mathsf{s}(\underline{x})).$$

This evaluated induction conclusion *contains* the induction hypothesis, since the underlined terms are the terms on both sides of the induction hypothesis. (This observation also forms the basis of the rippling technique [3].) Thus, when going from the induction hypothesis to the induction conclusion, both sides of the equation grow by the context $\mathsf{s}(\mathsf{s}(\square))$. In other words, in the induction base cases the left-hand side is at most as great as the right-hand side and afterwards, the left-hand side always grows at most as much as the right-hand side. Thus, if one ever reaches an instantiation t where $\mathsf{double}(\mathsf{half}(t)) = t$ is no longer true, then the reason is that $\mathsf{double}(\mathsf{half}(t))$ is *smaller* then t. But since $\mathsf{double}(\mathsf{half}(y))$ grows at most as fast as y, afterwards there can never be a number $s >_{\mathcal{C}_{\mathsf{half}}}$ t where $\mathsf{double}(\mathsf{half}(s)) = s$ is true again. Hence, if the induction hypothesis $\mathsf{double}(\mathsf{half}(x)) = x$ is false, then the induction conclusion $\mathsf{double}(\mathsf{half}(\mathsf{s}(\mathsf{s}(x)))) = \mathsf{s}(\mathsf{s}(x))$ is false as well (or, formulated as a contraposition, we have Property (12)).

The observation above leads to a general criterion. For many $\mathcal{C}$-provable equations $r_1 = r_2$, one does not only have $r_1[s_i^*] \to_\mathcal{R}^* C_i[r_1[t_{i,1}^*], \ldots, r_1[t_{i,n_i}^*]]$ for all $\langle s_i^*, \{t_{i,1}^*, \ldots, t_{i,n_i}^*\}\rangle \in \mathcal{C}$, but also $r_2[s_i^*] = D_i[r_2[t_{i,1}^*], \ldots, r_2[t_{i,n_i}^*]]$ for some constructor ground contexts C_i and D_i.

In our example, r_1 is $\mathsf{double}(\mathsf{half}(y))$ and r_2 is the term y. For the first pair of the cover set $\mathcal{C}_{\mathsf{half}}$, we have $C_1 = 0$ and $D_1 = 0$ and for the second pair we have $C_2 = 0$ and $D_2 = \mathsf{s}(0)$. For the third pair, we have $r_1[s_3^*] = \mathsf{double}(\mathsf{half}(\mathsf{s}(\mathsf{s}(x))))$, which can be evaluated to $s(s(\mathsf{double}(\mathsf{half}(x))))$ and as $t_{3,1}^* = x$, we obtain $C_3 = s(s(\Box))$. Since $r_2[s_3^*] = s(s(x))$, we also have $D_3 = s(s(\Box))$.

So r_1 grows by the context C_i and r_2 grows by the context D_i when going from the induction hypothesis $r_1[t_{i,1}^*] = r_2[t_{i,1}^*]$ to the induction conclusion $r_1[s_i^*] = r_2[s_i^*]$. Our aim is to ensure that whenever r_1 and r_2 are no longer $\mathcal{R}$-equal for some instantiation, then they will never become equal again for arguments which are greater w.r.t. the induction relation induced by the cover set. A sufficient requirement for this is that the contexts C_i added around r_1 are always at most as big as the contexts D_i added around r_2. To compare these contexts one can use an arbitrary ordering $\prec$ on constructor terms, i.e., any relation which is transitive and irreflexive. Moreover, we require $\prec$ to be monotonic (i.e., $s \prec t$ implies $f(\ldots s \ldots) \prec f(\ldots t \ldots)$ for all constructors f) and stable under substitutions (i.e., $s \prec t$ implies $s\sigma \prec t\sigma$). Then we only have to demand

$$C_i[x^*] \preceq D_i[x^*] \text{ for all } 1 \leq i \leq m.$$

As usual, "$\preceq$" denotes the union of "$\prec$" and "$=$" where "$=$" is syntactic equality.

Note that one may use any well-established technique for the generation of well-founded orderings such as the subterm ordering or the *recursive path ordering* $<_{\mathrm{rpo}}$ (cf. e.g. [5,10]) to synthesize a suitable ordering $\prec$ satisfying the above constraints. Moreover, since $\prec$ only has to be irreflexive, but not necessarily well founded, one can also use any ordering $>$ which results from the reversal of such a well-founded ordering $<$ (e.g., the superterm ordering or $>_{\mathrm{rpo}}$).

In our example we need a well-founded monotonic stable ordering $\prec$ where

$$C_1 = 0 \preceq 0 = D_1,$$
$$C_2 = 0 \preceq s(0) = D_2,$$
$$C_3[x] = s(s(x)) \preceq s(s(x)) = D_3[x].$$

Such an ordering can easily found by standard techniques for automated termination proofs. For example, the constraints are satisfied by the subterm ordering. Thus, one can automatically determine that $\mathsf{double}(\mathsf{half}(y)) = y$ is a conjecture whose correctness predicate is exact. As $c_{\mathsf{double}(\mathsf{half}(y))=y}$ is only true for even numbers, we have shown that indeed this conjecture is false for all odd ones.

In general, if $r_1 = r_2$ is an equation and $\mathcal{C}$ is a cover set such that the above conditions are satisfied by some ordering $\prec$, then we say that $r_1 = r_2$ *maintains* $\prec$ under the cover set $\mathcal{C}$ w.r.t. the underlying CS $\mathcal{R}$. The reason is that the relation $\prec$ between r_1 and r_2 is indeed maintained when going from an induction hypothesis to an induction conclusion. By using established (and decidable classes of) well-founded orderings $\prec$ from the area of term rewrite systems one immediately obtains a syntactical sufficient condition for maintenance of orderings, which can easily be checked automatically.

Definition 6 (Maintenance of orderings). *Let $\mathcal{R}$ be a convergent sufficiently complete CS and let $\mathcal{C} = \{\langle s_1^*, \{t_{1,1}^*, \ldots, t_{1,n_1}^*\}\rangle, \ldots, \langle s_m^*, \{t_{m,1}^*, \ldots, t_{m,n_m}^*\}\rangle\}$ be a complete well-founded cover set (where $0 \leq n_i \leq 1$ for all $1 \leq i \leq m$). Let*

$r_1 = r_2$ be C-provable and let C_i and D_i be constructor ground contexts where

$$r_1[s_i^*] \to_{\mathcal{R}}^* C_i[r_1[t_{i,1}^*], \ldots, r_1[t_{i,n_i}^*]] \quad and$$
$$r_2[s_i^*] = D_i[r_2[t_{i,1}^*], \ldots, r_2[t_{i,n_i}^*]].$$

Let $\prec$ be a monotonic ordering on constructor terms which is stable under substitutions. We say $r_1 = r_2$ maintains $\prec$ under the cover set C w.r.t. $\mathcal{R}$ iff $C_i[x^*] \preceq D_i[x^*]$ for all $1 \leq i \leq m$.

The following lemma proves that for equations which maintain an ordering, each induction conclusion indeed implies its induction hypothesis.

Lemma 7 (Equations where the reverse induction formulas hold). *Let $\mathcal{R}, C, \prec$ be as in Def. 6 and let $r_1 = r_2$ maintain $\prec$ under C w.r.t. $\mathcal{R}$. Then for all $1 \leq i \leq m$ with $n_i = 1$, $\mathcal{R} \models_{ind} r_1[s_i^*] = r_2[s_i^*] \Rightarrow r_1[t_{i,1}^*] = r_2[t_{i,1}^*]$.*

Proof. We first show that for all constructor ground terms q^*, we have

$$r_1[q^*]\downarrow_{\mathcal{R}} \preceq r_2[q^*]. \tag{13}$$

The proof of (13) is done by induction w.r.t. $<_C$. Due to the completeness of C, there must be a pair $\langle s_i^*, \{t_{i,1}^*, \ldots, t_{i,n_i}^*\}\rangle \in C$ such that $s_i^*\sigma = q^*$. If $n_i = 0$, then we have $r_1[q^*]\downarrow_{\mathcal{R}} = r_1[s_i^*\sigma]\downarrow_{\mathcal{R}} = C_i \preceq D_i = r_2[s_i^*\sigma] = r_2[q^*]$.

Otherwise, if $n_i = 1$, we have $r_1[q^*]\downarrow_{\mathcal{R}} = r_1[s_i^*\sigma]\downarrow_{\mathcal{R}} = C_i[r_1[t_{i,1}^*\sigma]\downarrow_{\mathcal{R}}] \preceq C_i[r_2[t_{i,1}^*\sigma]]$ by the induction hypothesis and monotonicity and stability of $\prec$. Furthermore, $C_i[r_2[t_{i,1}^*\sigma]] \preceq D_i[r_2[t_{i,1}^*\sigma]] = r_2[s_i^*\sigma] = r_2[q^*]$. So (13) is proved.

Now we can prove Lemma 7. Let σ substitute all variables of s_i^* by constructor ground terms such that $\mathcal{R} \models_{ind} r_1[s_i^*\sigma] = r_2[s_i^*\sigma]$. We assume that $\mathcal{R} \not\models_{ind} r_1[t_{i,1}^*\sigma] = r_2[t_{i,1}^*\sigma]$. By (13) we must have $r_1[t_{i,1}^*\sigma]\downarrow_{\mathcal{R}} \preceq r_2[t_{i,1}^*\sigma]$ and since the $\mathcal{R}$-normal forms of $r_1[t_{i,1}^*\sigma]$ and $r_2[t_{i,1}^*\sigma]$ are different by assumption this in fact implies $r_1[t_{i,1}^*\sigma]\downarrow_{\mathcal{R}} \prec r_2[t_{i,1}^*\sigma]$. Since $\prec$ is monotonic and stable we have

$$r_1[s_i^*\sigma]\downarrow_{\mathcal{R}} = C_i[r_1[t_{i,1}^*\sigma]\downarrow_{\mathcal{R}}] \prec C_i[r_2[t_{i,1}^*\sigma]] \preceq D_i[r_2[t_{i,1}^*\sigma]] = r_2[s_i^*\sigma].$$

But this contradicts $\mathcal{R} \models_{ind} r_1[s_i^*\sigma] = r_2[s_i^*\sigma]$ by the irreflexivity of $\prec$. □

Now we prove that if $r_1 = r_2$ maintains an ordering, then $c_{r_1=r_2}$ is indeed exact.

Theorem 8 (Equations where the correctness predicate is exact). *Let $\mathcal{R}, C, \prec$ be as in Def. 6 and let $r_1 = r_2$ be an equation which is radical and maintains some ordering $\prec$ under C w.r.t. $\mathcal{R}$. Moreover, let $c_{r_1=r_2}$ be a correctness predicate for $r_1 = r_2$ under C and let $\mathcal{R}$ also contain the rules defining $c_{r_1=r_2}$. Then $\mathcal{R} \models_{ind} r_1 = r_2 \Leftrightarrow c_{r_1=r_2}(y^*) = \mathsf{true}.$[4]*

---

[4] A more general version of this theorem can be proved in which a conjecture does not have to be radical, and further, it is not necessary for the induction scheme of a cover set to have at most one induction hypothesis in every subgoal.

Proof. Due to Thm. 5 (a) we only have to prove $\mathcal{R} \models_{\text{ind}} r_1[q^*] = r_2[q^*] \Rightarrow c_{r_1=r_2}(q^*) = \text{true}$ for all constructor ground term tuples q^*. Again, we use induction on $<_{\mathcal{C}}$. Let $\mathcal{R} \models_{\text{ind}} r_1[q^*] = r_2[q^*]$.

By the completeness of $\mathcal{C}$, there exists some $\langle s^*, \{t_1^*, \ldots, t_n^*\} \rangle \in \mathcal{C}$ and some substitution σ such that $q^* = s^*\sigma$. If $n = 0$, then we have the rule $c_{r_1=r_2}(s^*) \to \text{true}$ since the rule $c_{r_1=r_2}(s^*) \to \text{false}$ would only be generated if $\mathcal{R} \models_{\text{ind}} \neg r_1[s^*] = r_2[s^*]$. This implies $\mathcal{R} \models_{\text{ind}} c_{r_1=r_2}(q^*) = \text{true}$.

Otherwise, if $n = 1$, by Lemma 7 the truth of $r_1[s_i^*\sigma] = r_2[s_i^*\sigma]$ implies $\mathcal{R} \models_{\text{ind}} r_1[t_{i,1}^*\sigma] = r_2[t_{i,1}^*\sigma]$. So $\mathcal{R} \models_{\text{ind}} c_{r_1=r_2}(t_{i,1}^*\sigma)$ by the induction hypothesis. By the rule $c_{r_1=r_2}(s^*) \to c_{r_1=r_2}(t_1^*)$, we obtain $\mathcal{R} \models_{\text{ind}} c_{r_1=r_2}(s_i^*\sigma) = \text{true}$.   □

Let us consider the counterexample of Thm. 5 (b) again. When trying to prove $\text{half}(y) = \text{s}(0)$, we obtain $C_1 = 0$, $D_1 = \text{s}(0)$ and $C_2 = 0$, $D_2 = \text{s}(0)$. In the step case, the left-hand side $\text{half}(\text{s}(\text{s}(x)))$ evaluates to $\text{s}(\text{half}(x))$, i.e., we have $C_3 = \text{s}(\Box)$, whereas $D_3 = \Box$. There does not exist an ordering $\prec$ such that $C_i[x^*] \preceq D_i[x^*]$ for all i, since $C_1 \preceq D_1$ would imply $0 \prec \text{s}(0)$ and $C_3[0] \preceq D_3[0]$ would imply $\text{s}(0) \prec 0$ which contradicts the transitivity and irreflexivity of $\prec$. Thus, $\text{half}(y) = \text{s}(0)$ does not maintain any ordering under $\mathcal{C}_{\text{half}}$ and indeed, its correctness predicate is not exact as shown in Thm. 5 (b).

The above analysis of exactness of correctness predicates can be useful for fixing faulty conjectures, an objective for which correctness predicates were introduced by Protzen [9]. Since an exact correctness predicate precisely characterizes all instantiations on which the faulty conjecture is true, it can be used to fix the faulty conjecture into the "strongest theorem" possible.

5   Conjectures Where Inductive Validity Is Decidable

Now we extend Thm. 2 from equations to arbitrary quantifier-free formulas φ. We require that all equations $r_1 = r_2$ occurring in φ are radical and maintain some ordering under the same cover set $\mathcal{C}$.[5] Then by Thm. 8 their correctness predicates $c_{r_1=r_2}$ are sound and exact. For example, $\text{half}(y) = 0$ is radical and maintains the superterm ordering under $\mathcal{C}_{\text{half}}$. We obtain the correctness predicate

$$c_{\text{half}(y)=0}(0) \to \text{true}, \quad c_{\text{half}(y)=0}(\text{s}(0)) \to \text{true}, \quad c_{\text{half}(y)=0}(\text{s}(\text{s}(x))) \to \text{false}.$$

The last rule is due to the fact that the instantiated left-hand side $\text{half}(\text{s}(\text{s}(x)))$ evaluates to $\text{s}(\text{half}(x))$ and the replacement of the subterm $\text{half}(x)$ according to the induction hypothesis yields the equation $\text{s}(0) = 0$ which is unsatisfiable.

---

[5] Different equations in a conjecture may have to be proved using different cover sets; these cover sets can often be combined into a single cover set to generate a single induction scheme using merging and instantiation (cf. [2,7]). Further, it is not necessary for different equations to maintain the same monotonic ordering. For instance, in the running example of this section two different orderings are used in a conjecture.

Given a correctness predicate c_φ, we can generate $c_{\neg\varphi}$ by replacing the result true by false and the result false by true whereas right-hand sides of the form $c_\varphi(t^*)$ are replaced by $c_{\neg\varphi}(t^*)$. In the above example this yields

$$c_{\neg\mathsf{half}(y)=0}(0) \to \mathsf{false}, \quad c_{\neg\mathsf{half}(y)=0}(\mathsf{s}(0)) \to \mathsf{false}, \quad c_{\neg\mathsf{half}(y)=0}(\mathsf{s}(\mathsf{s}(x))) \to \mathsf{true}.$$

This correctness predicate is sound and exact for the conjecture $\neg\mathsf{half}(y) = 0$.

As stated before, exact correctness predicates can also be generated for non-radical equations, as well as for equations whose validity is decided using induction schemes with multiple induction hypotheses. Thus, inductive validity of a much larger class of literals (equations and negated equations) can be decided using arbitrary well-founded complete cover sets without the requirement of radicality. The restrictions to radical equations and to induction schemes involving at most one induction step in every subgoal are needed only for the decidability of conjunctions and disjunctions of conjectures as discussed below.

Given c_{φ_1} and c_{φ_2}, a straightforward idea to obtain rules for $c_{\varphi_1 \wedge \varphi_2}$ is as follows: If we have the rule $c_{\varphi_i}(s^*) \to \mathsf{false}$ for some $i \in \{1, 2\}$, then we also obtain the rule $c_{\varphi_1 \wedge \varphi_2}(s^*) \to \mathsf{false}$. If we have the rules $c_{\varphi_i}(s^*) \to \mathsf{true}$ for both $i \in \{1, 2\}$, then we obtain $c_{\varphi_1 \wedge \varphi_2}(s^*) \to \mathsf{true}$. Finally, if we have the rule $c_{\varphi_i}(s^*) \to c_{\varphi_i}(t^*)$ and either $c_{\varphi_j}(s^*) \to c_{\varphi_j}(t^*)$ or $c_{\varphi_j}(s^*) \to \mathsf{true}$ (for $i, j \in \{1, 2\}$, $i \neq j$), then we also obtain the rule $c_{\varphi_1 \wedge \varphi_2}(s^*) \to c_{\varphi_1 \wedge \varphi_2}(t^*)$. But as the following example illustrates, such a simplistic construction does not work.

Recall the rules (5) - (7) for $c_{\mathsf{double}(\mathsf{half}(y))=y}$. We would obtain the following correctness predicate for the formula $\varphi : \mathsf{double}(\mathsf{half}(y)) = y \wedge \neg\mathsf{half}(y) = 0$.

$$c_\varphi(0) \to \mathsf{false}, \quad c_\varphi(\mathsf{s}(0)) \to \mathsf{false}, \quad c_\varphi(\mathsf{s}(\mathsf{s}(x))) \to c_\varphi(x).$$

However, this correctness predicate is not exact, since it is always false, whereas φ is true for all even numbers greater than 0. Even worse, the resulting correctness predicate for the negated conjecture $\neg\varphi$ would not even be sound (since it would always be true whereas $\neg\varphi$ is false for 0 and all odd numbers).

The problem with the above construction of $c_{\varphi_1 \wedge \varphi_2}$ is the case where one rule $c_{\varphi_1}(s^*) \to c_{\varphi_1}(t^*)$ leads to a recursive call, but the other has the form $c_{\varphi_2}(s^*) \to \mathsf{true}$. If we use the rule $c_{\varphi_1 \wedge \varphi_2}(s^*) \to c_{\varphi_1 \wedge \varphi_2}(t^*)$, then we may lose the exactness of the correctness predicate, since it could be that $c_{\varphi_2}(t^*) \to^* \mathsf{false}$.

To avoid this problem, we will now construct so-called *basic* correctness predicates (denoted $b_{r_1=r_2}$) where for recursive pairs $\langle s^*, \{t^*\}\rangle \in \mathcal{C}$ we always have recursive rules $b_{r_1=r_2}(s^*) \to b_{r_1=r_2}(t^*)$, but never a rule with the result false.

Fortunately, if $r_1 = r_2$ is radical and maintains an ordering under $\mathcal{C}$, one can easily obtain a basic correctness predicate by simply extending the cover set $\mathcal{C}$ in an appropriate way. For that purpose we have to restrict ourselves to cover sets where for any two recursive pairs $\langle s_i^*, \{t_i^*\}\rangle, \langle s_j^*, \{t_j^*\}\rangle \in \mathcal{C}$ with $i \neq j$, the terms t_i^* and s_j^* do not unify (after renaming their variables). In other words, the arguments t_i^* in an induction hypothesis must not unify with the arguments s_j^* in any *other* induction conclusion. The cover set $\mathcal{C}_{\mathsf{half}} = \{\langle 0, \varnothing\rangle, \langle \mathsf{s}(0), \varnothing\rangle, \langle \mathsf{s}(\mathsf{s}(x)), \{x\}\rangle\}$ trivially satisfies this condition, since there is only one recursive pair. The motivation for this restriction is that for all chains $q_0^* <_\mathcal{C} q_1^* <_\mathcal{C} \ldots <_\mathcal{C} q_n^*$, it ensures

$c_\varphi(q_n^*) = \ldots = c_\varphi(q_1^*)$. So a change in the value of c_φ can only occur in the last value q_0^*, which corresponds to a base case (i.e., we might have $c_\varphi(q_1^*) \neq c_\varphi(q_0^*)$). Our aim is to extend $\mathcal{C}$ to a cover set $\mathcal{C}'$ where q_1^* is already a base case. Then for all chains $q_1^* <_{\mathcal{C}'} \ldots <_{\mathcal{C}'} q_n^*$ we have $c_\varphi(q_n^*) = \ldots = c_\varphi(q_1^*)$ and thus, we can indeed use the rule $c_\varphi(s^{*\prime}) \to c_\varphi(t^{*\prime})$ for all recursive pairs $\langle s^{*\prime}, \{t^{*\prime}\}\rangle$ of $\mathcal{C}'$.

The idea for the extension of cover sets is simply to unify the terms t_i^* of the induction hypotheses with the (variable-renamed) terms s_j^* in the left components of all pairs from $\mathcal{C}$. Let $\mu_{i,j}$ be the respective mgu's. Then every pair $\langle s_i^*, \{t_i^*\}\rangle$ is replaced by the new non-recursive pairs $\langle s_i^* \mu_{i,j}, \varnothing\rangle$ for $j \neq i$ and the instantiated recursive pair $\langle s_i^* \mu_{i,i}, \{t_i^* \mu_{i,i}\}\rangle$. For $\mathcal{C}_{\mathsf{half}}$ we obtain

$$\mathcal{C}'_{\mathsf{half}} = \{\langle 0, \varnothing\rangle, \langle \mathsf{s}(0), \varnothing\rangle, \langle \mathsf{s}(\mathsf{s}(0)), \varnothing\rangle, \langle \mathsf{s}(\mathsf{s}(\mathsf{s}(0))), \varnothing\rangle, \langle \mathsf{s}(\mathsf{s}(\mathsf{s}(\mathsf{s}(x)))), \{\mathsf{s}(\mathsf{s}(x))\}\rangle\}.$$

Definition 9 (Extending cover sets). *Let* $\mathcal{C} = \{\langle s_1^*, \{t_{1,1}^*, \ldots, t_{1,n_1}^*\}\rangle, \ldots, \langle s_m^*, \{t_{m,1}^*, \ldots, t_{m,n_m}^*\}\rangle\}$ *be a cover set with* $0 \leq n_i \leq 1$, *such that if* $n_i = n_j = 1$ *and* $i \neq j$ *then there do not exist substitutions* $\mu_{i,j}$ *with* $t_{i,1}^* \mu_{i,j} = s_j^* \nu \mu_{i,j}$ *for a variable renaming* ν. *Then the* extended *cover set* $\mathcal{C}'$ *is defined as follows:*

$$\mathcal{C}' = \{\langle s_i^*, \varnothing\rangle \mid n_i = 0\}$$
$$\cup \{\langle s_i^* \mu_{i,j}, \varnothing\rangle \mid n_i = 1, n_j = 0, \mu_{i,j} = mgu(t_{i,1}^*, s_j^* \nu) \text{ for a variable renaming } \nu\}$$
$$\cup \{\langle s_i^* \mu_{i,i}, \{t_{i,1}^* \mu_{i,i}\}\rangle \mid n_i = 1, \mu_{i,i} = mgu(t_{i,1}^*, s_i^* \nu) \text{ for a variable renaming } \nu\}.$$

Obviously, if $\mathcal{C}$ is complete and well founded, then the extension $\mathcal{C}'$ is complete and well founded, too. Moreover, if an equation $r_1 = r_2$ is radical and maintains an ordering under $\mathcal{C}$, then it is also radical and maintains the same ordering under the extension $\mathcal{C}'$. In this case we can construct the basic correctness predicate by taking the extension $\mathcal{C}'$ and by using the results true and false in its non-recursive cases and by using the rule $b_{r_1=r_2}(s^*) \to b_{r_1=r_2}(t^*)$ for all recursive pairs $\langle s^*, \{t^*\}\rangle$. Note that only *one* such extension step for cover sets $\mathcal{C}$ is already enough: If a correctness predicate b has a non-recursive rule $b(s^*) \to$ true or $b(s^*) \to$ false for a recursive pair $\langle s^*, \{t^*\}\rangle \in \mathcal{C}$, then a single extension step of $\mathcal{C}$ suffices to get recursive rules $b(s^{*\prime}) \to b(t^{*\prime})$ for all recursive pairs $\langle s^{*\prime}, \{t^{*\prime}\}\rangle$ of the extended cover set $\mathcal{C}'$. In our example we obtain

$$\begin{aligned}
b_{\mathsf{half}(y)=0}(0) &\to \mathsf{true}, & b_{\mathsf{double}(\mathsf{half}(y))=y}(0) &\to \mathsf{true},\\
b_{\mathsf{half}(y)=0}(\mathsf{s}(0)) &\to \mathsf{true}, & b_{\mathsf{double}(\mathsf{half}(y))=y}(\mathsf{s}(0)) &\to \mathsf{false},\\
b_{\mathsf{half}(y)=0}(\mathsf{s}^2(0)) &\to \mathsf{false}, & b_{\mathsf{double}(\mathsf{half}(y))=y}(\mathsf{s}^2(0)) &\to \mathsf{true},\\
b_{\mathsf{half}(y)=0}(\mathsf{s}^3(0)) &\to \mathsf{false}, & b_{\mathsf{double}(\mathsf{half}(y))=y}(\mathsf{s}^3(0)) &\to \mathsf{false},\\
b_{\mathsf{half}(y)=0}(\mathsf{s}^4(x)) &\to b_{\mathsf{half}(y)=0}(\mathsf{s}^2(x)). & b_{\mathsf{double}(\mathsf{half}(y))=y}(\mathsf{s}^4(x)) &\to b_{\mathsf{double}(\mathsf{half}(y))=y}(\mathsf{s}^2(x)).
\end{aligned}$$

Now indeed basic correctness predicates for conjunctions are constructed by using the result false if one of the conjuncts yields false and true if both conjuncts yield true. If one (and therefore, both) conjuncts have a recursive call, then the basic correctness predicate for the conjunction has a recursive call, too. So if φ

is again the formula $\mathsf{double}(\mathsf{half}(y)) = y \wedge \neg\mathsf{half}(y) = 0$, then we have

$$b_{\neg\mathsf{half}(y)=0}(0) \rightarrow \mathsf{false},$$
$$b_{\neg\mathsf{half}(y)=0}(\mathsf{s}(0)) \rightarrow \mathsf{false},$$
$$b_{\neg\mathsf{half}(y)=0}(\mathsf{s}^2(0)) \rightarrow \mathsf{true},$$
$$b_{\neg\mathsf{half}(y)=0}(\mathsf{s}^3(0)) \rightarrow \mathsf{true},$$
$$b_{\neg\mathsf{half}(y)=0}(\mathsf{s}^4(x)) \rightarrow b_{\neg\mathsf{half}(y)=0}(\mathsf{s}^2(x)).$$

$$b_\varphi(0) \rightarrow \mathsf{false},$$
$$b_\varphi(\mathsf{s}(0)) \rightarrow \mathsf{false},$$
$$b_\varphi(\mathsf{s}^2(0)) \rightarrow \mathsf{true},$$
$$b_\varphi(\mathsf{s}^3(0)) \rightarrow \mathsf{false},$$
$$b_\varphi(\mathsf{s}^4(x)) \rightarrow b_\varphi(\mathsf{s}^2(x)).$$

Definition 10 (Basic Correctness Predicates). *Let $\mathcal{R}$ be a convergent sufficiently complete CS and let $\mathcal{C}$ be a complete well-founded cover set such that for all $\langle s^*, \{t_1^*, \ldots, t_n^*\}\rangle \in \mathcal{C}$, we have $0 \leq n \leq 1$, and for two different pairs $\langle s^*, \{t^*\}\rangle, \langle s^{*\prime}, \{t^{*\prime}\}\rangle \in \mathcal{C}$, there does not exist a substitution μ with $t^*\mu = s^{*\prime}\nu\mu$ for a variable renaming ν. Let φ be a quantifier-free formula such that all equations in φ are radical and maintain some ordering under $\mathcal{C}$ w.r.t. $\mathcal{R}$.*

Let $\mathcal{C}' = \{\langle s_1^*, \{t_{1,1}^*, \ldots, t_{1,n_1}^*\}\rangle, \ldots, \langle s_m^*, \{t_{m,1}^*, \ldots, t_{m,n_m}^*\}\rangle\}$ be the extension of $\mathcal{C}$ and let $r_1[s_i^*] \rightarrow_{\mathcal{R}}^* C_i[r_1[t_{i,1}^*], \ldots, r_1[t_{i,n_i}^*]]$ for a constructor ground context C_i. Then the basic correctness predicate b_φ under $\mathcal{C}$ is defined by the following rules (analogous rules are used for formulas containing $\vee, \Rightarrow, \Leftrightarrow$):

$$b_{r_1=r_2}(s_i^*) \rightarrow \begin{cases} \mathsf{true}, & \text{if } \mathcal{R} \models_{\mathrm{ind}} C_i = r_2[s_i^*] \text{ and } n_i = 0, \\ \mathsf{false}, & \text{if } \mathcal{R} \models_{\mathrm{ind}} \neg C_i = r_2[s_i^*] \text{ and } n_i = 0, \\ b_{r_1=r_2}(t_{i,1}^*), & \text{if } n_i = 1, \end{cases}$$

$$b_{\neg\varphi'}(s_i^*) \rightarrow \begin{cases} \mathsf{true}, & \text{if we have the rule } b_{\varphi'}(s_i^*) \rightarrow \mathsf{false}, \\ \mathsf{false}, & \text{if we have the rule } b_{\varphi'}(s_i^*) \rightarrow \mathsf{true}, \\ b_{\neg\varphi'}(t_{i,1}^*), & \text{if we have the rule } b_{\varphi'}(s_i^*) \rightarrow b_{\varphi'}(t_{i,1}^*), \end{cases}$$

$$b_{\varphi_1 \wedge \varphi_2}(s_i^*) \rightarrow \begin{cases} \mathsf{true}, & \text{if } b_{\varphi_1}(s_i^*) \rightarrow \mathsf{true} \text{ and } b_{\varphi_2}(s_i^*) \rightarrow \mathsf{true}, \\ \mathsf{false}, & \text{if } b_{\varphi_1}(s_i^*) \rightarrow \mathsf{false} \text{ or } b_{\varphi_2}(s_i^*) \rightarrow \mathsf{false}, \\ b_{\varphi_1 \wedge \varphi_2}(t_{i,1}^*), & \text{if } b_{\varphi_1}(s_i^*) \rightarrow b_{\varphi_1}(t_{i,1}^*) \text{ and } b_{\varphi_2}(s_i^*) \rightarrow b_{\varphi_2}(t_{i,1}^*). \end{cases}$$

Now we can present the main theorem which shows that the inductive validity of arbitrary quantifier-free conjectures is decidable, if all their equations are radical and maintain an ordering under $\mathcal{C}$. The decision procedure works by constructing the basic correctness predicate and by checking whether it always yields true. The reason for the soundness of this approach is that basic correctness predicates are indeed sound and exact.

Theorem 11 (Decidability of inductive validity for arbitrary conjectures). *Let $\mathcal{R}, \mathcal{C}, \varphi$ be as in Def. 10. Then inductive validity of φ is decidable (by checking whether all non-recursive rules of b_φ have the right-hand side true, where b_φ is the basic correctness predicate for φ under $\mathcal{C}$).*

Proof. We have to show that b_φ is sound and exact, i.e., $\mathcal{R} \models_{\mathrm{ind}} \varphi \Leftrightarrow b_\varphi(y^*) = \mathsf{true}$ if $\mathcal{R}$ also contains the rules defining b_φ. We use an induction w.r.t. the structure of φ. First let φ be an equation $r_1 = r_2$.

Let q^* be a tuple of constructor ground terms. We prove $\mathcal{R} \models_{\mathrm{ind}} r_1[q^*] = r_2[q^*] \Leftrightarrow b_{r_1=r_2}(q^*) = \mathsf{true}$ by induction w.r.t. $<_{\mathcal{C}'}$. Since $\mathcal{C}$ is complete and well founded, obviously its extension C' is complete and well founded, too. Due to the completeness of C', there exists some $\langle s^*, \{t_1^*, \ldots, t_n^*\}\rangle \in C'$ and some substitution σ such that $q^* = s^*\sigma$. If $n = 0$, then the claim follows from radicality of $r_1 = r_2$ under C and thus, under C' as well.

If $n = 1$ and $\mathcal{R} \models_{\text{ind}} r_1[s^*\sigma] = r_2[s^*\sigma]$ then by Lemma 7 we also have $\mathcal{R} \models_{\text{ind}} r_1[t_1^*\sigma] = r_2[t_1^*\sigma]$ since $r_1 = r_2$ maintains an ordering under C and thus, under C' as well. The induction hypothesis yields $\mathcal{R} \models_{\text{ind}} b_{r_1=r_2}(t_1^*\sigma) = \text{true}$ and thus, $\mathcal{R} \models_{\text{ind}} b_{r_1=r_2}(s^*\sigma) = \text{true}$ as well.

Finally, let $n = 1$ and $\mathcal{R} \models_{\text{ind}} \neg r_1[s^*\sigma] = r_2[s^*\sigma]$. We have to show that this implies $\mathcal{R} \models_{\text{ind}} \neg r_1[t_1^*\sigma] = r_2[t_1^*\sigma]$. Then the induction hypothesis would yield $\mathcal{R} \models_{\text{ind}} b_{r_1=r_2}(t_1^*\sigma) = \text{false}$ and thus, $\mathcal{R} \models_{\text{ind}} b_{r_1=r_2}(s^*\sigma) = \text{false}$ as well.

Note that $s^* = s^{*\prime}\mu$ and $t_1^* = t_1^{*\prime}\mu$ for some $\langle s^{*\prime}, \{t_1^{*\prime}\}\rangle \in C$ by the definition of extensions. Moreover, by the requirement that arguments $t_1^{*\prime}$ of induction hypotheses may not unify with arguments of other induction conclusions we also have that $t_1^* = t_1^{*\prime}\mu = s^{*\prime}\nu\mu$ by the definition of extensions. Since $r_1 = r_2$ maintains an ordering under C we have $r_1[s^{*\prime}] \rightarrow^*_{\mathcal{R}} C_i'[r_1[t_1^{*\prime}]]$ for a constructor *ground* context C_i'. As $r_1[s^*] \rightarrow^*_{\mathcal{R}} C_i[r_1[t_1^*]]$, this means that $C_i' = C_i$ or, in other words, $r_1[s^{*\prime}] \rightarrow^*_{\mathcal{R}} C_i[r_1[t_1^{*\prime}]]$. Radicality of $r_1 = r_2$ under C implies that $\mathcal{R} \models_{\text{ind}} C_i[r_2[t_1^{*\prime}]] = r_2[s^{*\prime}]$ or $\mathcal{R} \models_{\text{ind}} \neg C_i[r_2[t_1^{*\prime}]] = r_2[s^{*\prime}]$.

First assume $\mathcal{R} \models_{\text{ind}} C_i[r_2[t_1^{*\prime}]] = r_2[s^{*\prime}]$. This implies $\mathcal{R} \models_{\text{ind}} (C_i[r_2[t_1^{*\prime}]] = r_2[s^{*\prime}])\mu$, i.e., $\mathcal{R} \models_{\text{ind}} C_i[r_2[t_1^*]] = r_2[s^*]$. If we had $\mathcal{R} \not\models_{\text{ind}} \neg r_1[t_1^*\sigma] = r_2[t_1^*\sigma]$ (i.e., $\mathcal{R} \models_{\text{ind}} (r_1[t_1^*] = r_2[t_1^*])\sigma\tau$ for some τ), then we would also have $\mathcal{R} \models_{\text{ind}} (C_i[r_1[t_1^*]] = r_2[s^*])\sigma\tau$. Since $r_1[s^*] \rightarrow^*_{\mathcal{R}} C_i[r_1[t_1^*]]$, this implies $\mathcal{R} \models_{\text{ind}} (r_1[s^*] = r_2[s^*])\sigma\tau$ in contradiction to the prerequisite $\mathcal{R} \models_{\text{ind}} \neg r_1[s^*\sigma] = r_2[s^*\sigma]$.

Thus, $\mathcal{R} \models_{\text{ind}} \neg C_i[r_2[t_1^*]] = r_2[s^{*\prime}]$. Again assume $\mathcal{R} \models_{\text{ind}} (r_1[t_1^*] = r_2[t_1^*])\sigma\tau$ for some τ. Since $t_1^*\sigma\tau = s^{*\prime}\nu\mu\sigma\tau$, we have $\mathcal{R} \models_{\text{ind}} (r_1[s^{*\prime}] = r_2[s^{*\prime}])\nu\mu\sigma\tau$ and since $r_1 = r_2$ maintains an ordering under C, this implies $\mathcal{R} \models_{\text{ind}} (r_1[t_1^{*\prime}] = r_2[t_1^{*\prime}])\nu\mu\sigma\tau$ by Lemma 7. By the prerequisite, this yields $\mathcal{R} \models_{\text{ind}} (\neg C_i[r_1[t_1^{*\prime}]] = r_2[s^{*\prime}])\nu\mu\sigma\tau$. However since $r_1[s^{*\prime}] \rightarrow^*_{\mathcal{R}} C_i[r_1[t_1^{*\prime}]]$, this is equivalent to $\mathcal{R} \models_{\text{ind}} (\neg r_1[s^{*\prime}] = r_2[s^{*\prime}])\nu\mu\sigma\tau$, which contradicts the assumption (as $t_1^*\sigma\tau = s^{*\prime}\nu\mu\sigma\tau$).

For formulas which are no equations, the claim immediately follows from the (outer) induction hypothesis. □

Note that the conditions in Thm. 11 (i.e., radicality and maintenance of orderings) can be checked automatically (by using orderings from the area of term rewrite systems which are amenable to automation). The set of all conjectures φ satisfying these conditions forms a class where inductive validity is decidable. To decide inductive validity of φ one simply constructs the rules for the basic correctness predicate b_φ (which can be done automatically) and one checks whether there is no rule of the form $b_\varphi(\ldots) \rightarrow \text{false}$.

So for a formula like $\text{double}(y) = y \Rightarrow y = 0$, one first checks whether this formula belongs to the class where inductive validity is decidable. For that purpose, one examines whether the conjecture contains a subterm $f(y^*)$ for pairwise disjoint variables y^* and an algorithm f and then one checks whether all equations in the conjecture are radical and maintain an ordering under C_f (using the induction variables y^*).

In our example, the equations $\mathsf{double}(y) = y$ and $y = 0$ indeed are both radical and they maintain the superterm ordering under C_{double}. So inductive validity of this conjecture is decidable. The decision procedure constructs the basic correctness predicate

$$b_{\mathsf{double}(y)=y\Rightarrow y=0}(0) \to \mathsf{true},$$
$$b_{\mathsf{double}(y)=y\Rightarrow y=0}(\mathsf{s}(0)) \to \mathsf{true},$$
$$b_{\mathsf{double}(y)=y\Rightarrow y=0}(\mathsf{s}(\mathsf{s}(x))) \to b_{\mathsf{double}(y)=y\Rightarrow y=0}(\mathsf{s}(x)),$$

and checks whether all non-recursive rules of $b_{\mathsf{double}(y)=y\Rightarrow y=0}$ have true on their right-hand side, which is obviously the case. Thus, the formula is valid.

Note that in this way we can *decide* the inductive validity of conjectures which were up to now hard problems for inductive theorem provers. In fact, virtually all existing inductive provers fail in verifying $\mathsf{double}(y) = y \Rightarrow y = 0$.[6] The reason is that the induction conclusion $\mathsf{double}(\mathsf{s}(x)) = \mathsf{s}(x) \Rightarrow \mathsf{s}(x) = 0$ can be evaluated to $\neg\mathsf{s}(\mathsf{double}(x)) = x$, but there is no way to apply the induction hypothesis $\mathsf{double}(x) = x \Rightarrow x = 0$ and thus, the proof of the induction step case does not succeed. On the other hand, by our decision procedure, validity of such conjectures can be shown without using any inductive theorem prover at all.

6   Conclusion

We presented a class of conjectures where inductive validity is decidable (by a very simple decision procedure). This allows an integration of inductive reasoning within fully automated tools like model checkers or compilers. First, we extended the results of [8] to a larger class of equations and subsequently, we extended the approach further to arbitrary quantifier-free conjectures. The main idea is to build correctness predicates for all equations occurring in a conjecture and we gave a criterion for checking whether these correctness predicates really describe the exact set of objects where the equation is valid. We showed how to construct (basic) correctness predicates for non-atomic formulas and by checking their defining rules, the inductive validity of such formulas can easily be decided.

We have used correctness predicates $c_{r_1=r_2}$ to describe the instances where an equation $r_1 = r_2$ is valid. However, in order to *combine* the correctness predicates $c_{r_1=r_2}$ and $c_{r_1'=r_2'}$ of two different equations (e.g., when building their conjunction), we have to restrict ourselves to *basic* correctness predicates and moreover, $c_{r_1=r_2}$ and $c_{r_1'=r_2'}$ must have been built w.r.t. "compatible" cover sets. In order to avoid these difficulties, an interesting alternative approach is to represent the set of instances where equations are valid by *tree automata* [4] instead of correctness predicates. As long as these sets of instances are *regular*, this indeed results in a very elegant method for deciding inductive validity (since regular languages are effectively closed under complement and intersection and since their emptiness is decidable). However, in general there are many equations where the set of instances which makes them valid is not regular. For example,

---

[6] This problem was pointed out to us by U. Kühler.

the equation $\mathsf{plus}(\mathsf{minus}(x,y),\mathsf{minus}(y,x)) = 0$ is valid iff x and y are equal. A correctness predicate describing this set can easily be constructed automatically, whereas this set is not regular and therefore cannot be described by (ordinary) tree automata. This indicates that the use of tree automata may be too restrictive compared to the use of (basic) correctness predicates. However, we intend to study the possibilities of using automata for deciding inductive validity further in future work.

In this paper, we focused on integrating induction schemes with a decision procedure for the quantifier-free theory of free constructors to obtain an extension of the decision procedure to quantifier-free formulas whose proofs (or disproofs) may require the use of induction. Kapur and Subramaniam [8] discussed an approach for integrating induction schemes into decidable quantifier-free theories including Presburger arithmetic, and they gave a decision procedure for inductive validity of a large class of equations involving $\mathcal{T}$-based function symbols, where $\mathcal{T}$ is a decidable quantifier-free theory. In future work, we intend to generalize the techniques developed in this paper from constructor systems to $\mathcal{T}$-based systems (including Presburger arithmetic) as well.

References

1. F. Baader & T. Nipkow, *Term Rewriting and All That*, Cambridge Univ. Pr., 1998.
2. R. S. Boyer and J Moore, *A Computational Logic*, Academic Press, 1979.
3. A. Bundy, A. Stevens, F. van Harmelen, A. Ireland, & A. Smaill, Rippling: A Heuristic for Guiding Inductive Proofs, *Artificial Intelligence*, 62:185-253, 1993.
4. H. Comon, M. Dauchet, R. Gilleron, F. Jacquemard, D. Lugiez, S. Tison, & M. Tommasi. Tree Automata and Applications. Draft, available from http://www.grappa.univ-lille3.fr/tata/, 1999.
5. N. Dershowitz, Termination of Rewriting, *J. Symb. Comp.*, 3:69–116, 1987.
6. M. Franova & Y. Kodratoff, Predicate Synthesis from Formal Specifications, in *Proc. ECAI 92*, 1992.
7. D. Kapur & M. Subramaniam, New Uses of Linear Arithmetic in Automated Theorem Proving by Induction, *Journal of Automated Reasoning*, 16:39–78, 1996.
8. D. Kapur & M. Subramaniam, Extending Decision Procedures with Induction Schemes, in *Proc. CADE-17*, LNAI 1831, pages 324-345, 2000.
9. M. Protzen, Patching Faulty Conjectures, *Proc. CADE-13*, LNAI 1104, 1996.
10. J. Steinbach, Simplification orderings: History of results, *Fundamenta Informaticae*, 24:47–87, 1995.
11. H. Zhang, D. Kapur, & M. S. Krishnamoorthy, A Mechanizable Induction Principle for Equational Specifications, in *Proc. CADE-9*, LNCS 310, 1988.

Automated Incremental Termination Proofs for Hierarchically Defined Term Rewriting Systems

Xavier Urbain

Laboratoire de Recherche en Informatique (LRI)
CNRS UMR 8623
Bât. 490, Université Paris-Sud, Centre d'Orsay
91405 Orsay Cedex, France
urbain@lri.fr

Abstract. We propose the notion of *rewriting modules* in order to provide a structural and hierarchical approach of TRS. We define then *relative dependency pairs* built upon these modules which allow us to perform termination proofs incrementally. Important results can be expressed in that new framework (regarding $C_{\mathcal{E}}$-termination for instance), and with help of π *extendable orderings*, we give effective new incremental methods for proving termination particularly suited for automation.

1 Introduction

Rewriting is used for specification, programming or in automated proofs. Yet, if structuring is a paradigm of 'clean' programming, a TRS is still considered in practice as a single set of rules and termination proofs are run on the whole system without taking any benefit of its possible modular structure.

Programs are (should be) developed in an incremental way: one defines some basic types or functions and then builds other functions which use the basic ones and so on. In other words, some functions are somehow created 'upon' previously defined ones. A hierarchical structure over functions naturally emerges from that incremental procedure. Since termination is a difficult issue—especially when it comes to automation of proofs—it would be useful to perform automated termination proofs incrementally, that is to use normalization information about 'basic' rules to show that adding 'new' ones will preserve termination.

Unfortunately, the termination property does not behave as well as we could expect when dealing with unions of TRS. As shown by Toyama [16] if two TRS are strongly normalizing, their union does not necessarily terminate, even if the two systems do not share any symbol. The significant work of Gramlich showed that projections were to blame and gave sufficient conditions for ensuring termination of unions of TRS possibly sharing constructors [8]. But with regard to hierarchical unions, that is unions of TRS sharing function symbols, fewer results are known. Dershowitz proposed [6] conditions over a certain kind of hierarchical unions—so called constructor based unions—but those conditions are too restrictive in practice and not very suited for automation.

R. Goré, A. Leitsch, and T. Nipkow (Eds.): IJCAR 2001, LNAI 2083, pp. 485–498, 2001.

Until 1997, the usual way of automatically proving termination of a TRS was to show that the relation described by that system was included in a reduction ordering, that is showing that each rule was decreasing w.r.t. the strict part of the ordering. Since that ordering was well-founded, termination of the system followed. However, constraints hence induced required search for orderings that computers had difficulties to perform (if any suitable reduction ordering ever existed) and proofs where more likely run 'by hand' with *ad hoc* techniques.

In 1997 Arts & Giesl introduced the notion of *dependency pairs* [3, 1],weakening constraints, and thus replacing the previously needed reduction ordering by a *weak* reduction ordering, more suited to automated search. Moreover, dependency pairs allow to prove termination of non-simply terminating TRS, that is of systems whose rules may be non-structural recursive like for instance $f(x, \dots) \to f(g(x), \dots)$ where g is not a constructor. The dependency pair approach consists of a structural analysis of rules, requiring for a suitable ordering that rules belong to its weak part and that only the so-called dependency pairs strictly decrease. A consistent approach in respect of our aim of modularity would then be to refine that structural analysis of rules depending on where the relevant rules lie within the hierarchy.

Thus, in order to give a general framework bringing modular structure of TRS to the fore, and so as to provide automated methods to prove termination incrementally, we will firstly recall some generalities, then in Sec. 3 we shall introduce *rewrite modules*, express unions of TRS in terms of modules and explain how to show out TRS hierarchical structure. Sec. 4 will deal with general termination difficulties in unions and will provide orderings with good properties with respect to projection (the so-called π *extendable orderings*). The framework of modules settled, *relative dependency pairs* will be defined in Sec. 5 and that powerful tool will lead us to Sec. 6 and thus to new criteria for modular termination (Thm 1 and Thm 2). Those results provide an incremental proof method illustrated with the complete example of Sec. 7. We shall then discuss in Sec. 8 how this work compares to others, especially to results of Arts & Giesl [2] and Dershowitz [6]. We will eventually conclude and give ideas on how we intend to apply and extend our framework.

2   Preliminaries

We recall usual notions about rewriting [7], and give our notations. A *signature* $\mathcal{F}$ is a finite set of *symbols* with arities. Let X be a countable set of *variables*; $T(\mathcal{F}, X)$ denotes the set of finite *terms* on $\mathcal{F}$ and X. Terms can be seen as trees: root position is then denoted by Λ, symbol at root position in a term t by $\Lambda(t)$, $t|_p$ denotes the subterm of t at position p. A *substitution* is a mapping σ from variables to terms s.t. $\{x \in X \mid \sigma(x) \neq x\}$ is finite. We use postfix notation for substitution applications. A substitution can easily be extended to endomorphisms of $T(\mathcal{F}, X)$: $f(t_1, \dots, t_n)\sigma = f(t_1\sigma, \dots, t_n\sigma)$. $t\sigma$ is then called an *instance* of t.

A *rewrite relation* is a binary relation $\to$ on terms which is monotonic and closed under substitution; $\to^*$ will denote its reflexive-transitive closure. A *term rewriting system* (TRS for short) over a signature[1] $\mathcal{F}$ and a set of variables X is a set $R(\mathcal{F})$ of *rewrite rules* $l \to r$. A TRS R defines a rewrite relation $\to_R$ the following way: $s \to_R t$ if there is a position p s.t. $s|p = l\sigma$ and $t = s[r\sigma]_p$ for a rule $l \to r \in R$ and a substitution σ. We then say that $s|_p$ is a *redex* and that s *reduces to t at position p with $l \to r$* or, if the rule of R is not relevant, *with R*, respectively denoted $s \xrightarrow[l\to r]{p} t$ and $s \xrightarrow[R]{p} t$.

We restrict ourselves to the study of *finitely branching* TRS, i.e. TRS s.t. the set of rules that can be applied to a term is always finite. A term is *strongly normalizable* (SN) if it cannot reduce indefinitely. A rewrite relation is *strongly normalizing* or *terminates* if any term is SN. Termination is usually proven with the help of *reduction orderings* [5] or quasi-orderings with dependency pairs. We briefly recall what we need. A *term ordering*, also known as *ordering pair* [10] is a pair $(\succeq, \succ)$ of relations over $T(\mathcal{F}, X)$ such that: $\succeq$ is a quasi-ordering, i.e. reflexive and transitive; $\succ \subseteq \succeq - \preceq$ is a strict ordering, i.e. irreflexive and transitive and s.t. $\succ \subseteq \succeq$, $\succ \cdot \succeq \subseteq \succ$ and $\succeq \cdot \succ \subseteq \succ$. A term ordering is said to be *well-founded* if there is no infinite strictly decreasing sequence $t_1 \succ t_2 \succ \cdots$; *stable* if both $\succ$ and $\succeq$ are stable under substitutions; *weakly* (resp. *strictly*) *monotonic* if for all terms t_1 and t_2, for all $f \in \mathcal{F}$, if $t_1 \succeq$ (resp. $\succ$) t_2 then $f(\ldots, t_1, \ldots) \succeq$ (resp. $\succ$) $f(\ldots, t_2, \ldots)$; A term ordering $(\succeq, \succ)$ is called a *weak (resp. strict) reduction ordering* if it is well-founded, stable and weakly (resp. strictly) monotonic.

3   Rewrite Modules

We define *hierarchical extensions* of TRS and introduce the notion of *rewrite modules*, thus providing a rather structural view of complex TRS that leads to an incremental approach.

3.1   Extensions & Modules

From an operational point of view, a module consists of new symbols together with rules that define them.

Definition 1. *A module extending a system $R_1(\mathcal{F}_1)$ is a couple $[\mathcal{F}_2 \mid R_2]$ s.t.:*

1. $\mathcal{F}_1 \cap \mathcal{F}_2 = \emptyset$;
2. R_2 *is a TRS over $\mathcal{F}_1 \cup \mathcal{F}_2$;*
3. *For each $l \to r \in R_2$, $\Lambda(l) \in \mathcal{F}_2$.*

System $R_1 \cup R_2$ over $\mathcal{F}_1 \cup \mathcal{F}_2$ is then a hierarchical extension *of system $R_1(\mathcal{F}_1)$ by module $[\mathcal{F}_2 \mid R_2]$; we denote $R_1(\mathcal{F}_1) \leftarrow [\mathcal{F}_2 \mid R_2]$ that extension.*

---

[1] We shall omit the signature if there is no ambiguity.

The usual notions of unions can be expressed in terms of modules in a straightforward way. We will say that $[\mathcal{F}_1 \mid R_1]$ extends $[\mathcal{F}_0 \mid R_0]$ regardless of $[\mathcal{F}_2 \mid R_2]$ if $\mathcal{F}_1 \cap \mathcal{F}_2 = \emptyset$, $[\mathcal{F}_0 \mid R_0] \leftarrow [\mathcal{F}_1 \mid R_1]$ and $[\mathcal{F}_0 \mid R_0] \leftarrow [\mathcal{F}_2 \mid R_2]$; such extension is indeed a union of composable TRS *[12,9,15]. We will talk about the* disjoint union $R_1 \cup R_2$ *if* $[\mathcal{F}_1 \mid R_1]$ *and* $[\mathcal{F}_2 \mid R_2]$ *extend* $[\emptyset \mid \emptyset]$. *The union will be* constructor sharing *if* $[\mathcal{F}_1 \mid R_1]$ *and* $[\mathcal{F}_2 \mid R_2]$ *extend* $[\mathcal{F}_0 \mid \emptyset]$.

A property P is said modular *for a specific kind of union if* R_1 *and* R_2 *having property* P *implies that the relevant union of* R_1 *and* R_2 *verifies* P.

3.2 Modular Splitting up of TRS

One may in practice provide a TRS as a hierarchy of modules and use it as it is. Every system $R(\mathcal{F})$ can anyway be automatically split up as an extension by *minimal* modules: the ones that cannot be seen as extensions of non-empty systems (see Example 1 below). This is done in two steps:

1. Build a graph $\mathcal{G}$ over symbols of $\mathcal{F}$ s.t. there is an arc from x to y if and only if there is a rule $l \to r \in R$ such that $x = \Lambda(l)$ and y occurs in l or in r.
2. Pack together symbols occuring in strongly connected parts of $\mathcal{G}$, that is symbols f and g such that $f \xrightarrow[\mathcal{G}]{}^* g$ and $g \xrightarrow[\mathcal{G}]{}^* f$.

Now, let us consider 'packs' of $\mathcal{G}$ as signatures of our modules, we just have to add rules the left-hand side root symbols of which are in those and we are done. Note that there is no cycle in the resulting hierarchy since 'mutually recursively' defined symbols were packed together.

Remark 1. For the sake of a better readability we may afterward gather constructors symbols that can be reached from the same packs.

Example 1. Let us consider addition and multiplication *à la* Peàno.

$$\mathcal{F} = \{s, 0, +, \times\}$$

$$R : \left\{ \underbrace{\begin{array}{l} x + 0 \to x, \\ x + s(y) \to s(x + y), \end{array}}_{R_+} \quad \underbrace{\begin{array}{l} x \times 0 \to 0, \\ x \times s(y) \to x + (x \times y). \end{array}}_{R_\times} \right.$$

A natural (for programmers) and modular way of seeing it is to consider addition built upon constructors, and multiplication upon addition and constructors. Describing this TRS in module framework consists in introducing $+$ with addition rules and *then* $\times$ with multiplication rules. The extension scheme is then:

$$[\{s, 0\} \mid \emptyset] \leftarrow [\{+\} \mid R_+] \leftarrow [\{\times\} \mid R_\times]$$

Note that those modules are minimal (in respect of Remark 1).

Further note that every system $R(\mathcal{F})$ can be seen that way as an extension of the empty system over *constructors* of $\mathcal{F}$ by the module consisting of *defined symbols*[2] (as signature part) together with *all rules* of R (as rules part).

---

[2] Constructors and defined symbols are here Arts & Giesl's notions.

4   Termination and Unions of TRS

Unfortunately, termination is not a modular property of TRS, even for disjoint unions, as shown by Toyama's famous counter-example [16].

Example 2 (Toyama). These two TRS (over disjoint signatures) are terminating:

$$R_1 : \{\, f(0,1,x) \to f(x,x,x) \qquad R_2 : \begin{cases} g(x,y) \to x \\ g(x,y) \to y. \end{cases}$$

Their union nevertheless allows infinite reductions, for instance:
$$f(g(0,1), g(0,1), g(0,1)) \xrightarrow[R_2]{} f(0, g(0,1), g(0,1)) \xrightarrow[R_2]{} f(0, 1, g(0,1))$$
$$\xrightarrow[R_1]{} f(g(0,1), g(0,1), g(0,1)) \xrightarrow[R_2]{} \cdots$$

One may observe that R_2 is not confluent, but similar counter-examples exist for confluent TRS [16]. However, problems come indeed with R_2 and its 'projective' behaviour as studied by Gramlich [8]. Propositions to contain projections lead to the definition of $C_{\mathcal{E}}$-termination[3] [15,8].

Definition 2. *Let G be a new (regarding all involved signatures) symbol; we denote π the projective TRS $\{G(x,y) \to x, G(x,y) \to y\}$.*

A system R is said $C_{\mathcal{E}}$-terminating if $R \cup \pi$ is terminating.

That notion gives precious results. In particular: $C_{\mathcal{E}}$-termination is a modular property for disjoint union of TRS [15,8], as well as for union of composable TRS as shown by Kurihara & Ohushi [9].

Note that before the introduction of dependency pairs termination proofs were usually performed using *simplification orderings* (reduction ordering with subterm property), thus restricting automated proofs to *simplifying* TRS.

Definition 3. *A TRS R over $\mathcal{F}$ is said* simply terminating *(or* simplifying*) if $R \cup \{f(x_1, \ldots, x_i, \ldots, x_n) \to x_i \mid 1 \le i \le n$ and $f \in \mathcal{F}\}$ is terminating.*

Proposition 1 (Gramlich [8]). *A simplifying TRS $C_{\mathcal{E}}$-terminates.*

Thanks to Proposition 1: lying between termination and *simple termination*, $C_{\mathcal{E}}$-termination is far from being too restrictive a property in practice. The example of Sec. 7 and, for instance, all examples by Arts & Giesl [1,2,3] (even those that do not simply terminate) $C_{\mathcal{E}}$-terminate. Actually, all known automated techniques prove indeed $C_{\mathcal{E}}$-termination.

In order to specify a class of orderings making the most of $C_{\mathcal{E}}$-termination, we define π extendable orderings.

Definition 4. *A term ordering $(\succeq, \succ)$ on $T(\mathcal{F}, X)$ is said to be π extendable if there is a reduction ordering $(\succeq', \succ')$ over $T(\mathcal{F} \cup \{G\}, X)$ s.t.: 1) The restriction of $(\succeq', \succ')$ on $T(\mathcal{F}, X)$ is exactly $(\succeq, \succ)$ and 2) $G(s,t) \succeq' s$ and $G(s,t) \succeq' t$ for all s and t in $T(\mathcal{F} \cup \{G\}, X)$.*

A π extendable ordering is a strong π extendable ordering if both $(\succeq, \succ)$ and a suitable $(\succeq', \succ')$ are strictly monotonic. Otherwise it is weak π extendable.

---

[3] We use here the terminology introduced by Olhebusch [14].

Consequently, if $(\succeq, \succ)$ is a strictly monotonic π extendable ordering and if $l \succ r$ for all rules $l \to r$ of R, then R $\mathcal{C}_\mathcal{E}$-terminates.

5   Relative Dependency Pairs

Now that we settled our modular approach, we would like to use it for proving termination. Most methods dealing with union of TRS use a notion of *rank*, that is rely on how signatures are interlaced within a term [8]. We define here *relative dependency pairs*, dependency pairs of modules with two aims in mind: Firstly to 'capture' signature swaps by selecting relevant subterms, secondly to 'forget' other defined symbols since the whole termination proof is supposed to be performed incrementally.

Definition 5. *Let $R_1(\mathcal{F}_1)$ and $[\mathcal{F}_2 \mid R_2]$ be such that $R_1(\mathcal{F}_1) \leftarrow [\mathcal{F}_2 \mid R_2]$.*
For each rule $l \to r \in R_2$, a pair $\langle l, r' \rangle$ where r' is a subterm of r such that $\Lambda(r') \in \mathcal{F}_2$ is called a dependency pair of module $[\mathcal{F}_2 \mid R_2]$.
The set of dependency pairs of all rules of a module M is denoted $DP(M)$.

Example 3 (Example 1 revisited). Dependency pairs of $[\mathcal{F}_\times \mid R_\times]$ consist in:

$$DP([\mathcal{F}_\times \mid R_\times]) : \{ \langle x \times s(y), x \times y \rangle \}$$

Notice that $\langle x \times s(y), (x \times y) + x \rangle$ is *not* a dependency pair, unlike in Arts & Giesl non-modular approach[4].

Remark 2. For a TRS seen as an extension of constructors by rules and defined symbols, relative dependency pairs and Arts & Giesl ones coincide.

Definition 6. *A dependency chain of a module $[\mathcal{F} \mid R]$ over R' (with $R \subseteq R'$) is a sequence of pairs $\ldots \langle s_j, t_j \rangle \ldots$ of $DP([\mathcal{F} \mid R])$ together with a substitution σ such that for any two successive pairs $\langle s_i, t_i \rangle$, $\langle s_{i+1}, t_{i+1} \rangle$:*

$$t_i \sigma \xrightarrow[R']{\neq \Lambda}{}^* s_{i+1}\sigma.$$

We shall consider *minimal* dependency chains, in the sense that each proper subterm of any left-hand side σ-instantiated is strongly normalizable by R'.

Proposition 2. *Let R be a TRS over $\mathcal{F}$, then R is $\mathcal{C}_\mathcal{E}$-strongly normalizing if and only if there is no infinite (dependency) chain of $[\mathcal{F} \mid R]$ over $R \cup \pi$.*

Proof. Straigthforward from Remark 2 and since $DP(R \cup \pi) = DP(R)$.

Corollary 1. *Let $(\succeq, \succ)$ be a (weak) π extendable ordering.*
If $l \succeq r$ for all $l \to r \in R$ and $s \succ t$ for all $\langle s, t \rangle \in DP(R)$, then R is $\mathcal{C}_\mathcal{E}$-strongly normalizing.

---

[4] We do not consider here the restricted case of innermost termination [2] where *usable rules* modify the set of dependency pairs.

6   Incremental Proofs

The modules framework is particularly efficient when it comes to modular and incremental proofs. We give our contributions for extensions of a system by, firstly one module, and secondly two disjoint modules. Thus, those apply to any hierarchical scheme. We obtain also effective reformulations of previous results.

6.1   One System and One Module

The basic case, typically a hierarchical extension of a system. Relative dependency pairs give us lighter conditions over systems to ensure termination.

Theorem 1. *Let $R_1(\mathcal{F}_1)$ and $[\mathcal{F}_2 \mid R_2]$ be such that $[\mathcal{F}_1 \mid R_1] \leftarrow [\mathcal{F}_2 \mid R_2]$.*

1. *If R_1 is $\mathcal{C}_\mathcal{E}$-strongly normalizing and*
2. *If there is no infinite dependency chain of $[\mathcal{F}_2 \mid R_2]$ over $R_1 \cup R_2$,*

Then $R_1 \cup R_2$ is strongly normalizing.

By contradiction: We will assume that there is an infinite dependency chain of $R_1 \cup R_2$, then we will conclude either on an infinite dependency chain of $[\mathcal{F}_2 \mid R_2]$ over $R_1 \cup R_2$, or on non $\mathcal{C}_\mathcal{E}$-termination of R_1 which contradicts premises.

Let us suppose $R_1 \cup R_2$ non terminating. Thus, there is an infinite dependency chain of $[\mathcal{F}_1 \cup \mathcal{F}_2 \mid R_1 \cup R_2]$ over $R_1 \cup R_2$. Among pairs of $DP([\mathcal{F}_1 \cup \mathcal{F}_2 \mid R_1 \cup R_2])$ are: Pairs of $[\mathcal{F}_1 \mid R_1]$; Pairs of $[\mathcal{F}_2 \mid R_2]$; Eventually pairs $\langle l, r' \rangle$ s.t. $l \rightarrow r \in R_2$ and r' is a subterm of r with $\Lambda(r')$ in $\mathcal{F}_1$. On the first two cases we have knowledge. In order to get rid of the third kind, we need the following lemma.

> **Lemma 1.** *Let $R_1(\mathcal{F}_1)$ and $[\mathcal{F}_2 \mid R_2]$ be such that $[\mathcal{F}_1 \mid R_1] \leftarrow [\mathcal{F}_2 \mid R_2]$. Then for each $\langle u_1, v_1 \rangle \in DP(R_1)$ and for each pair $\langle u_2, v_2 \rangle$ of $[\mathcal{F}_2 \mid R_2]$ there is no substitution σ such that: $v_1\sigma \xrightarrow{\neq\Lambda}^* u_2\sigma$.*

> *Proof.* Actually $\Lambda(u_2) \neq \Lambda(v_1)$ since $\Lambda(u_2\sigma) = \Lambda(u_2) \in \mathcal{F}_2$ and $\Lambda(v_1\sigma) = \Lambda(v_1) \in \mathcal{F}_1$.

According to Lemma 1, pairs of second and of third kinds can only follow pairs of second kind, pairs of first kind can only follow pairs of first and of third kind. Thus, we face three cases. The dependency chain either contains: Only pairs of second kind, that is pairs of $[\mathcal{F}_2 \mid R_2]$ or only pairs of first kind, that is pairs of $[\mathcal{F}_1 \mid R_1]$ or a finite number of pairs (maybe none) of second kind then *one* pair of third kind and then an infinity of pairs of first kind.

- First case: Infinite dependency chain of $[\mathcal{F}_2 \mid R_2]$ over $R_1 \cup R_2$ contradicts second premise of Theorem 1.
- Cases 2 & 3: In both cases we eventually get an infinite dependency chain of $[\mathcal{F}_1 \mid R_1]$ over $R_1 \cup R_2$. We are now going to show how that chain can be transformed into a chain of $[\mathcal{F}_1 \mid R_1]$ over $R_1 \cup \pi$. Doing so, we will

end up with an infinite chain of $[\mathcal{F}_1 \mid R_1]$ over $R_1 \cup \pi$, that is an infinite chain of $[\mathcal{F}_1 \cup \{G\} \mid R_1 \cup \pi]$ contradicting first premise stating that R_1 is $\mathcal{C}_\mathcal{E}$-terminating.

We actually use (here and also in proof of Theorem 2) a more general result (Lemma 2) whose proof provides a way to build such a suitable chain.

Lemma 2. *Let S_1 and S_2 two TRS over $\mathcal{F}_1$. Let $S_3(\mathcal{F}_1 \cup \mathcal{F}_2)$ be such that:*
- *$\mathcal{F}_1 \cap \mathcal{F}_2 = \emptyset$;*
- *For each $l \to r \in S_3$, $\Lambda(l) \in \mathcal{F}_2$.*

Then from any infinite minimal chain of $[\mathcal{F}_1 \mid S_2]$ over $S_1 \cup S_2 \cup S_3$, it is possible to build an infinite chain of $[\mathcal{F}_1 \mid S_2]$ over $S_1 \cup S_2 \cup \pi$ with the same sequence of pairs but new instantiation and rewriting steps.

The remaining part of Theorem 1 proof consists in applying Lemma 2 with $R_1 = S_1 = S_2$ and $R_2 = S_3$. Hence for any infinite dependency chains of $[\mathcal{F}_1 \mid R_1]$ over $R_1 \cup R_2$ we can build a corresponding chain of $[\mathcal{F}_1 \mid R_1]$ over $R_1 \cup \pi$, that is a dependency chain of $R_1 \cup \pi$. Since R_1 is supposed to be $\mathcal{C}_\mathcal{E}$-terminating, we raise a contradiction. Q.e.d.

Now we have got to prove Lemma 2. The proof is rather technical and use some interpretation of terms akin to Gramlich's [8]. We denote $T_\infty(\mathcal{F}, X)$ the set of infinite terms over signature $\mathcal{F}$ and a set X of variables.

Definition 7. *Let S be $S_1 \cup S_2 \cup S_3$. Let $>$ be an arbitrary yet total ordering over $T(\mathcal{F}_1 \cup \mathcal{F}_2, X)$.*

Interpretation $I_S(x) : T(\mathcal{F}_1 \cup \mathcal{F}_2, X) \to T_\infty(\mathcal{F}_1 \cup \{G : 2\} \cup \{\bot : 0\}, X)$ can be defined as:

$$I_S(x) = x \text{ if } x \in X$$
$$I_S(f(t_1 \ldots t_n)) = \begin{cases} f(I_S(t_1) \ldots I_S(t_n)) & \text{if } f \in \mathcal{F}_1, \\ Comb(Red(f(t_1 \ldots t_n))) & \text{if } f \in \mathcal{F}_2. \end{cases}$$
$$where \qquad Red(t) = \{I_S(t')/ \ t \xrightarrow[S_1 \cup S_2 \cup S_3]{} t'\},$$
$$Comb(\emptyset) = \bot,$$
$$Comb(\{a\} \cup set\,) = G(a, Comb(set)) \text{ where for all } x \subset set, a < x).$$

$Comb(E)$ is built from an unordered set E, ordering $>$ is then needed to remove any ambiguity.

Remark 3. Please note that interpreted terms appear in a 'comb-like' shape. Each 'tooth' being itself the interpretation of a one-step-reduced term, any of these can be reached by an appropriate $\xrightarrow[\pi_2]{*} \xrightarrow[\pi_1]{}$ reduction.

We give yet another few technical lemmas showing I_S good behaviour.

Lemma 3. *For each $t \in T(\mathcal{F}_1, X)$ and each substitution σ, $I(t\sigma) = tI(\sigma)$.*

Proof. Structural induction on t.

Lemma 4. *For all $t_1, \ldots, t_n$ in $T(\mathcal{F}_1 \cup \mathcal{F}_2, X)$ and for any context C over $\mathcal{F}_1$ with n holes, $I_S(C[t_1, \ldots, t_n]) = C[I_S(t_1), \ldots, I_S(t_n)]$.*

Proof. Structural induction on C.

Lemma 5. *For any term t strongly normalizable by $S_1 \cup S_2 \cup S_3$, $I(t)$ is finite.*

Proof. Trivial since we are interested in finitely branching TRS only.

Lemma 6. *For any s and t in $T(\mathcal{F}_1 \cup \mathcal{F}_2, X)$, and $l \to r \in S_1 \cup S_2$*

$$\text{if } s \xrightarrow[l \to r]{p} t \text{ then } I(s) \xrightarrow[S_1 \cup S_2 \cup \pi]{}{}^+ I(t).$$

Moreover, if $\Lambda(s) \in \mathcal{F}_1$ then $I(s) \xrightarrow[S_1 \cup S_2 \cup \pi]{\neq \Lambda}{}^+ I(t)$.

Proof. Two cases depending on symbols occurring along path from Λ to p.
• If there are only symbols of $\mathcal{F}_1$ then $s = C[s_1, \ldots, l\sigma, \ldots, s_n]$, $s|_p = l\sigma$ and C is a context with n holes over $\mathcal{F}_1$.

$$
\begin{aligned}
I(s) &= I(C[s_1, \ldots, l\sigma, \ldots, s_n]) \\
&= C[I(s_1), \ldots, I(l\sigma), \ldots, I(s_n)] \text{ (Lemma 4)} \\
&= C[I(s_1), \ldots, lI(\sigma), \ldots, I(s_n)] \text{ (Lemma 3)} \\
&\xrightarrow[S \cup S_2]{p} C[I(s_1), \ldots, rI(\sigma), \ldots, I(s_n)] \text{ (hypothesis)} \\
&= C[I(s_1), \ldots, I(r\sigma), \ldots, I(s_n)] \text{ (Lemma 3)} \\
&= I(C[s_1, \ldots, r\sigma, \ldots, s_n]) = I(t).
\end{aligned}
$$

• If symbols of $\mathcal{F}_2$ occur, then there is a smallest $p' < p$ (w.r.t. prefix ordering) s.t. $\Lambda(s|_{p'}) \in \mathcal{F}_2$. We may again assume w.l.o.g. that $s = C[s_1, \ldots, s', \ldots, s_n]$ where C is a context with n holes (possibly empty) over $\mathcal{F}_1$, $p = p'q$ and $s|_{p'} = s'$ with $s' = C'[l\sigma] \xrightarrow[S \cup S_2]{} C'[r\sigma] = t'$. Then

$$I(s) = I(C[s_1, \ldots, s', \ldots, s_n]) = C[I(s_1), \ldots, I(s'), \ldots, I(s_n)] \text{ (Lemma 4)}.$$

From Def. 7, $I(s') = \text{Comb}(\text{Red}(s'))$. But $s'|_q = l\sigma \xrightarrow[S_1 \cup S_2]{} r\sigma$, thus, by definition of Red, $r\sigma \in \text{Red}(l\sigma)$. Hence, $I(t')$ is a subterm of $I(s')$ and $I(s') \xrightarrow[\pi]{}{}^+ I(t')$. We get

$$C[I(s_1), \ldots, I(s'), \ldots, I(s_n)] \xrightarrow[\pi]{}{}^+ C[I(s_1), \ldots I(t'), \ldots, I(s_n)]$$

$$= I(C[s_1, \ldots, t', \ldots, s_n]) \text{ (Lemma 4)} = I(t).$$

Lemma 7. *For all s and t in $T(\mathcal{F}_1 \cup \mathcal{F}_2, X)$, if $s \xrightarrow[S_3]{p} t$ then $I(s) \xrightarrow[\pi]{}{}^+ I(t)$.*

Moreover, if $\Lambda(s) \in \mathcal{F}_1$ then $I(s) \xrightarrow[\pi]{\neq \Lambda}{}^+ I(t)$.

Proof. Similar to the proof of Lemma 6, case 2.

Proof of Lemma 2 Let $\langle u_1, v_1 \rangle, \langle u_2, v_2 \rangle, \ldots$ be a dependency chain of $[\mathcal{F}_1 \mid S_2]$ over $S_1 \cup S_2 \cup S_3$ with a substitution σ. Let σ' be the substitution such that for all x, $x\sigma' = I(x\sigma)$.

Since the considered chain is minimal, σ is strongly normalizable so, from Lemma 5, σ' is indeed a substitution to finite terms.

We show that $\langle u_1, v_1 \rangle, \langle u_2, v_2 \rangle, \ldots$ with σ' is a chain of $[\mathcal{F}_1 \mid S_2]$ over $S_1 \cup S_2 \cup \pi$. For that purpose we have to show that for each i, $v_i \sigma' \xrightarrow[S_1 \cup S_2 \cup \pi]{\neq \Lambda}{}^* u_{i+1}\sigma'$.

We know that $v_i \sigma \xrightarrow[S_1 \cup S_2 \cup S_3]{\neq \Lambda}{}^* u_{i+1}\sigma$. Let us consider any step $s \xrightarrow[S_1 \cup S_2 \cup S_3]{p} t$ from that reduction. Since we have $\Lambda(s) = \Lambda(t) = \Lambda(v_i) = \Lambda(u_{i+1}) \in \mathcal{F}_1$, either from Lemma 7 or Lemma 6 we get that $I(s) \xrightarrow[S_1 \cup S_2 \cup \pi]{\neq \Lambda}{}^* I(t)$. Thus, putting steps together we obtain $I(v_i \sigma) \xrightarrow[S_1 \cup S_2 \cup \pi]{\neq \Lambda}{}^* I(u_{i+1}\sigma)$. Since $I(v_i \sigma) = v_i \sigma'$ and $I(u_{i+1}\sigma) = u_{i+1}\sigma'$ from Lemma 3 we are done.

The following corollary enables us to compose our termination results, allowing this way incremental proofs.

Corollary 2. *Let $R_1(\mathcal{F}_1)$ and $[\mathcal{F}_2 \mid R_2]$ be such that $[\mathcal{F}_1 \mid R_1] \leftarrow [\mathcal{F}_2 \mid R_2]$.*

If R_1 is $\mathcal{C}_\mathcal{E}$-terminating and if there is no infinite dependency chain of $[\mathcal{F}_2 \mid R_2]$ over $R_1 \cup R_2 \cup \pi$, then $R_1 \cup R_2$ is $\mathcal{C}_\mathcal{E}$-terminating.

Proof. We apply Theorem 1 with $[\mathcal{F}_2 \cup \{G\} \mid R_2 \cup \pi]$ extending system R_1.

Those results can be turned into an effective method by Corollary 3.

Corollary 3. *Let $R_1(\mathcal{F}_1)$ and $[\mathcal{F}_2 \mid R_2]$ be such that $[\mathcal{F}_1 \mid R_1] \leftarrow [\mathcal{F}_2 \mid R_2]$.*

If R_1 is $\mathcal{C}_\mathcal{E}$-terminating and if there is a weak reduction ordering (resp. π extendable ordering) $(\succeq, \succ)$ such that: $R_1 \cup R_2 \subseteq \succeq$ and $DP([\mathcal{F}_2 \mid R_2]) \subseteq \succ$, then $R_1 \cup R_2$ is terminating (resp. $\mathcal{C}_\mathcal{E}$-terminating).

Proof. All chains of $[\mathcal{F}_2 \mid R_2]$ over $R_1 \cup R_2$ are actually finite. Since steps between pairs decrease w.r.t. $\succeq$ and DP steps decrease w.r.t. $\succ$, an infinite reduction would contradict the premise stating that $(\succeq, \succ)$ is well-founded.

6.2 One System and Two Modules

The other basic case, allowing us to apply our results to any hierarchical scheme.

Theorem 2. *Let $R_1(\mathcal{F}_1)$, $[\mathcal{F}_2 \mid R_2]$ and $[\mathcal{F}_3 \mid R_3]$ be such that: $[\mathcal{F}_1 \mid R_1] \leftarrow [\mathcal{F}_2 \mid R_2]$ and $[\mathcal{F}_1 \mid R_1] \leftarrow [\mathcal{F}_3 \mid R_3]$ with $\mathcal{F}_2 \cap \mathcal{F}_3 = \emptyset$.*

1. *If $R_1 \cup R_2$ is $\mathcal{C}_\mathcal{E}$-strongly normalizing and*
2. *If there is no infinite dependency chain of $[\mathcal{F}_3 \mid R_3]$ over $R_1 \cup R_3 \cup \pi$,*

Then $R_1 \cup R_2 \cup R_3$ is $\mathcal{C}_\mathcal{E}$-strongly normalizing.

Remark 4. Note that R_2 **does not interfere** with the premise over $[\mathcal{F}_3 \mid R_3]$.

Further note that the $\mathcal{C}_\mathcal{E}$-premise cannot be omitted: Toyama's counter-example (see Example 2) would then be a special case of the relevant extension.

As a corollary and using Prop. 2 we obtain a previous result by Kurihara & Ohuchi [9]: $\mathcal{C}_\mathcal{E}$-termination is modular for composable unions of TRS.

Theorem 2 applies to automated proof when used as the following corollary:

Corollary 4. *Let $R_1(\mathcal{F}_1)$, $[\mathcal{F}_2 \mid R_2]$ and $[\mathcal{F}_3 \mid R_3]$ be such that:*
$[\mathcal{F}_1 \mid R_1] \leftarrow [\mathcal{F}_2 \mid R_2]$ *and* $[\mathcal{F}_1 \mid R_1] \leftarrow [\mathcal{F}_3 \mid R_3]$ *with* $\mathcal{F}_2 \cap \mathcal{F}_3 = \emptyset$.

If $R_1 \cup R_2$ is $\mathcal{C}_\mathcal{E}$-terminating and if there is a π extendable weak reduction ordering $(\succeq, \succ)$ *s.t.* $R_1 \cup R_3 \subseteq \succeq$ *and* $DP([\mathcal{F}_3 \mid R_3]) \subseteq \succ$, *then* $R_1 \cup R_2 \cup R_3$ *is $\mathcal{C}_\mathcal{E}$-terminating.*

7   A Complete Example

Consider system $R_\#$ over signature $\mathcal{F}_\# = \{\# : constant;\ 0, 1 : postfix\ unary\}$ describing integers in binary notation: $R_\# = \{\#0 \to \#$. We want some arithmetic over them and define addition with $[\mathcal{F}_+ \mid R_+]$:

$$\mathcal{F}_+ \ \{+ : binary\},$$
$$R_+ \begin{cases} x + \# \to x, & x0 + y0 \to (x+y)0, & x0 + y1 \to (x+y)1, \\ \# + x \to x, & x1 + y0 \to (x+y)1, & x1 + y1 \to ((x+y) + \#1)0, \\ x + (y+z) \to (x+y) + z. \end{cases}$$

R_+ has critical pairs: its innermost termination does not implies termination and we cannot apply modularity criteria from Arts & Giesl [2]. We use relative dependency pairs and a polynomial interpretation.

$$DP([\mathcal{F}_+ \mid R_+]) : \begin{cases} \langle x0+y0, x+y \rangle, \ \langle x0+y1, x+y \rangle, \ \langle x1+y0, x+y \rangle, \\ \langle x1+y1, x+y \rangle, \ \langle x1+y1, (x+y) + \#1) \rangle, \\ \langle x+(y+z), x+y \rangle, \ \langle x+(y+z), (x+y) + z \rangle. \end{cases}$$

$[\![\#]\!] = 0$, $[\![0]\!](x) = [\![x]\!] + 1$, $[\![1]\!](x) = [\![x]\!] + 3$, $[\![+]\!](x, y) = [\![x]\!] + 2[\![y]\!] + 1$. Since $DP([\mathcal{F}_+ \mid R_+])$ strictly decreases and $R_\# \cup R_+ \cup \pi$ weakly decreases, by Cor. 3, $R_\# \cup R_+$ $\mathcal{C}_\mathcal{E}$-terminates. We might then add subtraction:

$$\mathcal{F}_- \ \{- : binary\},$$
$$R_- \begin{cases} x - \# \to x, & x1 - y1 \to (x-y)0, & x0 - y0 \to (x-y)0, \\ \# - x \to \#, & x1 - y0 \to (x-y)1, & x0 - y1 \to ((x-y) - \#1)1. \end{cases}$$

Relative dependency pairs and polynomial interpretation again suffice to prove that $R_\# \cup R_-$ $\mathcal{C}_\mathcal{E}$-terminates. We use: $[\![\#]\!] = 0$, $[\![0]\!](x) = [\![x]\!] + 1$, $[\![1]\!](x) = [\![x]\!] + 1$, $[\![-]\!](x, y) = [\![x]\!]$. Dependency pairs of $[\mathcal{F}_- \mid R_-]$ strictly decrease while $R_\# \cup R_-$ weakly decreases w.r.t. to that interpretation. Applying Cor. 4 we conclude that $R_\# \cup R_- \cup R_+$ $\mathcal{C}_\mathcal{E}$-terminates. In order to use comparison we need Booleans and provide $[\mathcal{F}_{\mathsf{Bool}} \mid R_{\mathsf{Bool}}]$.

$\mathcal{F}_{\mathsf{Bool}}$ $\{true, false : constants;\ \neg : unary;\ \wedge : infix\ binary;\ if : ternary\}$,

$R_{\mathsf{Bool}} \begin{cases} \neg(true) \rightarrow false, & \neg(false) \rightarrow true, & x \wedge true \rightarrow x, \\ x \wedge false \rightarrow false, & if(true, x, y) \rightarrow x, & if(false, x, y) \rightarrow y. \end{cases}$

That system has no dependency pair, hence $\mathcal{C}_\mathcal{E}$-terminates.

We now define comparisons in $[\mathcal{F}_{ge} \mid R_{ge}]$ extending $R_\#$ and R_{Bool}.

$\mathcal{F}_{ge}$ $\{ge : binary\}$,

$R_{ge} \begin{cases} ge(x0, y0) \rightarrow ge(x, y), & ge(\#, x1) \rightarrow false, & ge(x1, y0) \rightarrow ge(x, y), \\ ge(x1, y1) \rightarrow ge(x, y), & ge(x, \#) \rightarrow true, & ge(\#, x0) \rightarrow ge(\#, x), \\ ge(x0, y1) \rightarrow \neg ge(y, x). \end{cases}$

Termination of $R_\# \cup R_{\mathsf{Bool}} \cup R_{ge}$ is proven using RPO with precedence $\{ge > \neg > (true, false)\}$. RPO is a simplification ordering thus π extendable: that union $\mathcal{C}_\mathcal{E}$-terminates by Prop. 1. We may then apply Thm 2 to conclude that $R_\# \cup R_{\mathsf{Bool}} \cup R_{ge} \cup R_+ \cup R_-$ $\mathcal{C}_\mathcal{E}$-terminates.

Let us provide logarithm Log over intergers. It is technically easier firstly to define $Log'(x) = Log(x) + 1$ with convention $Log'(0) = 0$.

$\mathcal{F}_{Log'}$ $\{Log' : unary\}$, $R_{Log'} \begin{cases} Log'(\#) \rightarrow \#, & Log'(x1) \rightarrow Log'(x) + \#1, \\ Log'(x0) \rightarrow if(ge(x, \#1), Log'(x) + \#1, \#). \end{cases}$

We use again relative dependency pairs and a polynomial interpretation. $\mathrm{DP}([\mathcal{F}_{Log'} \mid R_{Log'}])$: $\{\ \langle Log'(x1), Log'(x)\rangle, \langle Log'(x0), Log'(x)\rangle\ \}$. For $[\![\#]\!] = 0$, $[\![0]\!](x) = [\![x]\!] + 1$, $[\![1]\!](x) = [\![x]\!] + 1$, $[\![+]\!](x, y) = [\![x]\!] + [\![y]\!]$ $[\![false]\!] = 0$, $[\![true]\!] = 0$, $[\![\neg]\!](x) = 0$, $[\![ge]\!](x) = 0$, $[\![if]\!](x, y, z) = [\![y]\!] + [\![z]\!]$, $[\![\wedge]\!](x, y) = [\![x]\!]$, $[\![Log']\!](x) = [\![x]\!]$, relative dependency pairs strictly decrease while rules of $R_\# \cup R_+ \cup R_{\mathsf{Bool}} \cup R_{ge} \cup R_{Log'}$ weakly decrease. By Cor. 4 we prove $\mathcal{C}_\mathcal{E}$-termination of $R_\# \cup R_+ \cup R_{\mathsf{Bool}} \cup R_{ge} \cup R_{Log'} \cup R_-$.

We may now compute logarithm:

$\mathcal{F}_{Log}$ $\{Log : unary\}$, R_{Log} $\{Log(x) \rightarrow Log'(x) - \#1$.

Since $[\mathcal{F}_{Log} \mid R_{Log}]$ has no dependency pair, applying Thm 2 we conclude on $\mathcal{C}_\mathcal{E}$-termination of $R_\# \cup R_+ \cup R_- \cup R_{\mathsf{Bool}} \cup R_{ge} \cup R_{Log'} \cup R_{Log}$.

8   Related Work

Our work has to compare firstly with Dershowitz's results [6]. Since we do not restrict ourselves to constructor based systems, and use a slightly more general definition of *hierarchical extensions*, we thus obtain more general conditions. Moreover our criteria (fully syntactical and applicable to most TRS met in practice) seem more suited for automation—because they were designed for it—than the finely tuned conditions of Dershowitz. Secondly, Arts & Giesl exploit the modular structure of dependency graphs [2]. Still, their criterion puts conditions over the whole system (whatever the extension might be). That is a drawback we wanted to get rid of, because it fundamentally acts as a break upon real incremental proving. As noticed in Rem. 4, criteria based on Thm 2 do not require anything from irrelevant sets of rules. Our framework furthermore provides for the general case the results they got for the special case of *innermost* rewriting.

9 Conclusion & Future Work

We proposed the notion of rewrite modules in order to express structural and hierarchical information about TRS. Different kinds of extensions can be seen as special cases of modules hierarchy schemes. Modules are a fertile ground for automated incremental termination proofs: we introduced relative dependency pairs and criteria that allow proofs to be split up an run in an incremental fashion. We eventually obtain as a corollary a former result from Kurihara & Ohuchi [9] that is directly and easily expressed in that framework. Relative dependency pairs approach with the help of π extendable orderings provide criteria particularly suited for automation. The resulting tests are finite and purely syntactical, thus implementable. Moreover that method weakens constraints over orderings to be found in two ways. Firstly, the (really) incremental scheme itself filters only relevant rules; secondly, the relative dependency pairs widen the class of suitable orderings over these relevant rules.

Marked pairs, dependency graphs and thus graphs refinements can be defined in the module framework; that extension is straightforward. Another extension, quite important in practice, is the application of modules to TRS with associative and commutative symbols. This could lead to relative pairs somewhat similar to AC-extended pairs [11].

Finally, modules and relative dependency pairs are now full parts of the C*i*ME2 [4] termination tool developped in our research team.

Acknowledgments. The author would like to thank Claude Marché for his most valuable comments and is grateful to Bernhard Gramlich for his judicious remarks.

References

1. T. Arts. *Automatically proving termination and innermost normalisation of term rewriting systems.* PhD thesis, Universiteit Utrecht, 1997.
2. T. Arts and J. Giesl. Modularity of termination using dependency pairs. In Nipkow [13], pages 226–240.
3. T. Arts and J. Giesl. Termination of term rewriting using dependency pairs. *Theoretical Comput. Sci.*, 236:133–178, 2000.
4. E. Contejean, C. Marché, B. Monate, and X. Urbain. Cime version 2, 2000. Pre-release available at http://www.lri.fr/~demons/cime.html.

5. N. Dershowitz. Termination of rewriting. *Journal of Symbolic Computation*, 3(1):69–115, Feb. 1987.
6. N. Dershowitz. Hierarchical termination. In N. Dershowitz and N. Lindenstrauss, editors, *Proceedings of the Fourth International Workshop on Conditional and Typed Rewriting Systems (Jerusalem, Israel, July 1994)*, volume 968, pages 89–105, Berlin, 1995. Springer-Verlag.
7. N. Dershowitz and J.-P. Jouannaud. Notations for rewriting. *EATCS Bulletin*, 43:162–172, 1990.
8. B. Gramlich. Generalized sufficient conditions for modular termination of rewriting. *Applicable Algebra in Engineering, Communication and Computing*, 5:131–158, 1994.
9. M. Kurihara and A. Ohuchi. Decomposable termination of composable term rewriting systems. *IEICE*, E78–D(4):314–320, Apr. 1995.
10. K. Kusakari, M. Nakamura, and Y. Toyama. Argument filtering transformation. In G. Nadathur, editor, *Principles and Practice of Declarative Programming, International Conference PPDP'99*, volume 1702 of *Lecture Notes in Computer Science*, pages 47–61, Paris, 1999. Springer-Verlag.
11. C. Marché and X. Urbain. Termination of associative-commutative rewriting by dependency pairs. In Nipkow [13], pages 241–255.
12. A. Middeldorp and Y. Toyama. Completeness of combinations of constructor systems. In *Proc. 4th Rewriting Techniques and Applications, LNCS 488*, Como, Italy, 1991.
13. T. Nipkow, editor. *9th International Conference on Rewriting Techniques and Applications*, volume 1379 of *Lecture Notes in Computer Science*, Tsukuba, Japan, Apr. 1998. Springer-Verlag.
14. E. Ohlebusch. On the modularity of termination of term rewriting systems. *Theoretical Comput. Sci.*, 136:333–360, 1994.
15. E. Ohlebusch. Modular properties of composable term rewriting systems. *Journal of Symbolic Computation*, 20:1–41, 1995.
16. Y. Toyama. Counterexamples to termination for the direct sum of term rewriting systems. *Inf. Process. Lett.*, 25:141–143, Apr. 1987.

Decidability and Complexity of Finitely Closable Linear Equational Theories

Christopher Lynch and Barbara Morawska

Department of Mathematics and Computer Science Box 5815, Clarkson University, Potsdam, NY 13699-5815, USA, {clynch,morawskb}@clarkson.edu**

Abstract. We define a subclass of the class of linear equational theories, called *finitely closable* linear theories. We consider unification problems with no repeated variables. We show the decidability of this subclass, and give an algorithm in PSPACE. If all function symbols are monadic, then the running time is in NP, and quadratic for unitary monadic finitely closable linear theories.

1 Introduction

The problem of E-unification[1] is an important problem for automated deduction, as well as other areas of computer science, such as formal verification and type inference. Given an equational theory E, an E-unifier of terms s and t is a substitution θ such that $s\theta$ and $t\theta$ are equivalent modulo E. In many applications it is necessary to find a *complete set of E-unifiers* of terms s and t, that is, to find a set of E-unifiers of s and t from which all other E-unifiers can be generated.

Unfortunately, E-unification is undecidable in general. In addition, for some equational theories there is no finite complete set of unifiers. Therefore, if it necessary to determine which classes of equational theories have a decidable algorithm and on which E-unification problems. Furthermore, the complexity of those algorithms should be analyzed.

There has been much work in finding particular equational theories with decidable E-unification problems and analyzing their complexity. There has been less work in identifying classes of equational theories with decidable E-unification problems. However, there has been some recent work in that area, but not all of it analyzes complexity. See [9] for some references.

Recently, we have developed a simple new method of E-unification and proved its soundness and completeness[6] for all equational theories. In [7] we have refined it for linear theories. The method is a generalization of the General Mutation inference rules for Syntactic Theories[2,3,4,5]. It is an inference procedure that does not always halt. However, the goal of developing this new method was to use it to find decidable classes of equational theories and analyze their complexity, which is what we do in this paper.

** This work was supported by NSF grant number CCR-9712388.

R. Goré, A. Leitsch, and T. Nipkow (Eds.): IJCAR 2001, LNAI 2083, pp. 499–513, 2001.

We consider *linear theories*, i.e., theories where in each equation no terms have repeated variables, although terms on opposite sides of an equation may share variables. This class of equational theories includes all theories with monadic functions symbols. We only consider E-unification problems whose set of goal equations contains no repeated variables. This is a restricted E-unification problem, but it contains the word problem, which is undecidable for equations on strings, and it also includes some existential problems.

The particular class we prove decidability of is what we call *finitely closable* theories. To use our algorithm, we must assume we know a finite set of terms, such that we can find a complete set of unifiers for each pair of those terms. If the terms that appear in each complete set of unifiers are already in the set, then we call the set finitely closed. When such a set exists, we have an algorithm to solve the E-unification problems mentioned in the previous paragraph. We show the algorithm is in PSPACE. However, for the case of monadic function symbols it is in NP, and furthermore it is quadratic if each complete set of unifiers mentioned above is unitary.

Of course, we have not mentioned, so far, how to find this finite set. We also show some ways in which such a finite set can be found.

The format of the paper is to give some preliminary definitions, then to present the algorithm which gives our decidability results and prove the complexity results. Finally we give a method for finding the finite set in some cases.

2   Preliminaries

We assume we are given a set of variables and a set of uninterpreted function symbols of various arities. An arity is a non-negative integer. *Terms* are defined recursively in the following way: each variable is a term, and if $t_1, \cdots, t_n$ are terms, and f is of arity $n \geq 0$, then $f(t_1, \cdots, t_n)$ is a term, and f is the symbol at the *root* of $f(t_1, \cdots, t_n)$. A term (or any object) without variables is called *ground*. If t is any object, then $Var(t)$ is the set of all variables in t.

We consider equations of the form $s \approx t$, where s and t are terms. Let E be a set of equations, and $u \approx v$ be an equation, then we write $E \models u \approx v$ (or $u =_E v$) if $u \approx v$ is true in any model of E. If G is a set of equations, then $E \models G$ means that $E \models e$ for all e in G. If all the function symbols in E are of arity no greater than one, then E is *monadic*.

A *substitution* is a mapping from the set of variables to the set of terms, such that it is almost everywhere the identity. We identify a substitution with its homomorphic extension. If θ is a substitution then $Dom(\theta) = \{x \mid x\theta \neq x\}$. The *range* of θ, $Ran(\theta)$ is $\{x\theta \mid x \in Dom(\theta)\}$. A substitution σ is *idempotent* if $\sigma\sigma = \sigma$. In this paper, all substitutions will be considered to be idempotent. A substitution θ is an E-*unifier* of an equation $u \approx v$ if $E \models u\theta \approx v\theta$. θ is an E-*unifier* of a set of equations G if θ is an E-unifier of all equations in G. Whenever an equation or a set of equations has an E-unifier, it also has an idempotent E-unifier. If θ is an E-unifier of $u \approx v$, we say that θ is *linear* if no variable appears more than twice in $Ran(\theta)$, and if a variable z appears twice in $Ran(\theta)$ then there is an x in u and a y in v such that z appears in $x\theta$ and z appears in $y\theta$. This implies that there are not two different variables x and w in u such that z appears in $x\theta$ and $w\theta$.

If σ and θ are substitutions, then we write $\sigma \leq_E \theta[Var(G)]$ if there is a substitution ρ such that $E \models x\sigma\rho \approx x\theta$ for all x appearing in G. If G is a set of equations, then a substitution θ is a *most general unifier of G*, written $\theta = mgu(G)$ if θ is an E unifier of G, and for all E unifiers σ of G, $\theta \leq_E \sigma[Var(G)]$. A complete set of E-unifiers of G, is a set of E-unifiers Θ of G such that for all E-unifiers σ of G, there is a θ in Θ such that $\theta \leq_E \sigma[Var(G)]$.

Given a unification problem we can either *solve* the unification problem or *decide* the unification problem. Given a goal G and a set of equations E, to *solve* the unification problem means to find a complete set of E-unifiers of G. To *decide* the unification problem simply means to answer true or false as to whether G has an E-unifier.

We say that a term t (or an equation or a set of equations) has *varity n* if each variable in t appears at most n times. An equation $s \approx t$ is linear if s and t are both of varity 1. Note that the equation $s \approx t$ is then of varity 2, but it might not be of varity 1. A set of equations is *linear* if each equation in the set is linear. For example, the axioms of group theory ($\{f(x, f(y, z)) \approx f(f(x, y), z), f(w, e) \approx w, f(u, i(u)) \approx e.$ are of varity 2.

3   Algorithm

We will be considering linear equational theories E. The goals G we are trying to solve are sets of equations with no repeated variables (varity 1). In this section we will give an E-unification algorithm, and in the next section we will prove the algorithm halts for E-closed sets T, defined below, and give the complexity of the algorithm.

Definition 1. *A set of terms T is called E-closed if it satisfies the following conditions:*

1. *every term in T is of varity 1;*
2. *no member of T is a variable;*
3. *if f is a symbol of arity $n \geq 0$ appearing in E, then $f(x_1 \cdots, x_n) \in T$;*
4. *T contains two new constants c and d, which are not symbols of E.*
5. *if s and t are renamings of terms in T, and $\theta \in CSU_E(s, t)$, then θ is linear, and for all x in $Var(s \approx t)$, whenever $x_i\theta$ is not a variable, there is a renaming ρ such that $x_i\theta\rho \in T$;*
6. *if t' is a nonvariable subterm of t, then there is a renaming ρ such that $t'\rho \in T$.*

In the definition of T we assume that we are able to calculate a complete set of E-unifiers for all pairs of terms in T. Each such T could have an associated table listing the complete set of unifiers for each pair of terms in T. If such a T exists, we will show that the E-unification problem for all goals G of varity 1 is solvable. But first we will show that if G contains symbols that are not in T, then T and its associated table of complete sets of unifiers can easily be extended to handle such goals. First T is extended so that whenever $u[c]$ is a member of T for some term u, then $u[f(x_1, \cdots, x_n)]$ is added to T for every new symbol f, of arity $n \geq 0$, appearing in G. Then the table of complete sets of E-unifiers is extended as follows.

Let $f(x_1, \cdots, x_n)$ and $g(y_1, \cdots, y_m)$ be terms in the extended T, such that f and g are different symbols, and at least one of f and g did not exist in E. If f is not a symbol in E, then let $u = c$, else let $u = f(x_1, \cdots, x_n)$. If g is not a symbol in E, then let $v = d$, else let $v = g(y_1, \cdots, y_m)$. Find the complete set of E-unifiers $\{\sigma_1, \cdots, \sigma_k\}$ of u and v. Let $\{\theta_1, \cdots, \theta_k\}$ be the set of substitutions such that each θ_i is created from σ_i by replacing each occurrence of c in the range of σ_i by $f(x_1, \cdots, x_n)$, and replacing every occurrence of d in the range of σ_i by $g(y_1, \cdots, y_m)$. Then $\{\theta_1, \cdots, \theta_k\}$ is a complete set of E-unifiers for $f(x_1, \cdots, x_n) \approx g(y_1, \cdots, y_m)$. Furthermore, all terms in the range of each θ_i have already been added to T.

Again, let f be a symbol in G that is not in E. Then, a complete set of E-unifiers for $f(x_1, \cdots, x_n) \approx f(y_1, \cdots, y_n)$ is $\{[x_1 \mapsto z_1, \cdots, x_n \mapsto z_n, y_1 \mapsto z_1, \cdots, y_n \mapsto z_n]\}$. All terms in the range of this substitution are variables.

Now we have an extended T which is E-closed over the symbols of $E \cup G$, and we have an extended table of complete sets of E-unifiers. For the rest of this paper, we will assume we are working with this extended set.

We give several examples of E-closed sets.

Example 1. Let E be the theory of associativity and commutativity, $\{f(f(x, y), z) \approx f(x, f(y, z)), f(x, y) \approx f(y, x)\}$. Let $T = \{f(x, y), c, d\}$. Then any pair of terms where one of them is c or d has no E-unifiers. So, to prove that T is E-closed we only need to check $CSU_E(f(x_1, x_2), f(y_1, y_2))$. In fact, $CSU_E(f(x_1, x_2), f(y_1, y_2)) = \{\sigma_1, \sigma_2, \sigma_3, \sigma_4, \sigma_5, \sigma_6, \sigma_7\}$, where

- $\sigma_1 = [x_1 \mapsto f(z_1, z_2), x_2 \mapsto f(z_3, z_4), y_1 \mapsto f(z_1, z_3), y_2 \mapsto f(z_2, z_4)]$
- $\sigma_2 = [x_1 \mapsto z_2, x_2 \mapsto f(z_3, z_4), y_1 \mapsto z_3, y_2 \mapsto f(z_2, z_4)]$
- $\sigma_3 = [x_1 \mapsto z_1, x_2 \mapsto f(z_3, z_4), y_1 \mapsto f(z_1, z_3), y_2 \mapsto z_4]$
- $\sigma_4 = [x_1 \mapsto f(z_1, z_2), x_2 \mapsto z_4, y_1 \mapsto z_1, y_2 \mapsto f(z_2, z_4)]$
- $\sigma_5 = [x_1 \mapsto f(z_1, z_2), x_2 \mapsto z_3, y_1 \mapsto f(z_1, z_3), y_2 \mapsto z_2]$
- $\sigma_6 = [x_1 \mapsto z_2, x_2 \mapsto z_3, y_1 \mapsto z_3, y_2 \mapsto z_2]$
- $\sigma_7 = [x_1 \mapsto z_1, x_2 \mapsto z_4, y_1 \mapsto z_1, y_2 \mapsto z_4]$

Notice that whenever a nonvariable term appears in the range of some σ_i, then a renaming of that term appears in T. Therefore, T is E-closed.

Example 2. Let E be the monadic theory $\{fgfx \approx gfgx\}$. Let $T = \{fx, gy, fgz, gfw, c, d\}$. Then again, any pair where one term is c or d is not unifiable. The complete set of unifiers of any term with a renaming of itself, such as fx_1 and fx_2, has as most general E-unifier, $[x_1 \mapsto z, y_1 \mapsto z]$. There are twelve more pairs that must be checked. For example $CSU_E(fx, gy) =$

$\{[x \mapsto gfz, y \mapsto fgz]\}$. Also $CSU_E(fx, gfy) = \{[x \mapsto gfz, y \mapsto gz]\}$. Also $CSU_E(fgx, gfy) = \{[x \mapsto fz, y \mapsto gz]\}$. We leave it to the interested reader to check the others. Notice that any term that appears in the range of a unifier is a renaming of something in T. So T is E-closed.

Example 3. Let E be the monadic theory $\{fggx \approx gffx\}$. Let $T = \{fx, gy, ffz, ggw, c, d\}$. Once again, any pair involving c or d is not E-unifiable. A pair of two renamings of the same term is as in the previous example. The pair of terms fx and gy has a most general E-unifier $[x \mapsto ggz, y \mapsto ffz]$. No other pair of terms is E-unifiable. Therefore, to show that T is E-closed, we only need to verify that a renaming of ggz and ffz are in T.

Example 4. Let $E = \{fx \approx x\}$. Let $T = \{fx, c, d\}$. Then $fx \approx fy$ has a most general E-unifier $[x \mapsto z, y \mapsto z]$. Also, $c \approx fx$ has a most general E-unifier $[x \mapsto c]$. The other complete sets of E-unifiers are easy. T is E-closed, because the only nonvariable terms which appear in the range of a unifier in a complete set of E-unifiers are c and d. Now, suppose we want to consider a goal containing a new monadic function symbol g. First, we add gy to T. Then we note that $c \approx fx$ has a most general E-unifier $[x \mapsto c]$. Therefore, $[x \mapsto gy]$ must be a most general E-unifier of $gy \approx fx$. So the extended set is also E-closed.

We define a function called H to calculate the height of a term in terms of the set T. The height is defined so that a term from T is considered as if it was a single symbol.

Definition 2. *Let T be an E-closed set of terms. $H(t)$ is defined recursively in terms of T.*

1. $H(x) = 0$, *if x is a variable;*
2. $K(s, \rho) = 1 + max\{H(x\rho) \mid x \in Var(s)\}$, *if ρ is a substitution;*
3. $H(t) = min\{K(s, \rho) \mid t = s\rho \text{ and } s \in T\}$ *if there exists an $s \in T$ and a ρ such that $s\rho = t$;*

Note that item 3 applies to a term t if the root symbol of f is in E, since we have said that $f(x_1, \cdots, x_n) \in T$ for all symbols f in E. If T is extended to include symbols in t as explained above, then item 3 always applies. If E is empty, then this definition gives the standard definition of the height of a term, which we denote $SH(t)$. The height of a term is the minimum number of applications of terms in T it takes to construct the term. If $H(t) = n$, we say that the T-*height of t* is n. If $SH(t) = m$, we say the *standard height of t* is m.

Example 5. For example, consider the set T to be $\{fx, gy, fgz, gfw, c, d\}$. Then the T-height $H(x) = 0$ and $H(fx) = H(gx) = H(fgx) = H(gfx) = 1$. The following set of terms are all of T-height 2:
$\{ffx, ggx, ffgx, gfgx, fgfx, gfgx, fgfgx, gffgx, fggfx, gfgfx\}$.

Let $h = max\{SH(t) \mid t \in T\}$. We can see from the definition that $H(t) \leq SH(t)$ and $SH(t) \leq h \times H(t)$.

As for height, we define the standard size of a term and the T-size of a term.

Definition 3. *Let T be an E-closed set. The T-size of a term t, $|t|$, is defined recursively as:*

1. *$|x| = 0$, for any variable x;*
2. *$|s|_\rho = 1 + \Sigma\{|x\rho| \mid x \in Var(s)\}$, if ρ is a substitution;*
3. *$|t| = min\{|s|_\rho \mid t = s\rho$ and $s \in T\}$ if there exists an $s \in T$ and a ρ such that $s\rho = t$;*

If $E = \emptyset$, then $|t|$ is the standard size of t. The T-size is related to the standard size in the same way as the T-height is related to the standard height.

If an E-closed set T is finite and G has no repeated variables, then we will prove that we can solve the E-unification problem for G. For the rest of this section, we will assume that T is closed and finite. Since G has no repeated variables, each equation in G can be solved separately without affecting the other results, so for simplicity we will assume that G is a single equation.

An equation $x \approx t$, where x is a variable, is called a *solved* equation.

Our algorithm is based on the following inference rule:

Suppose the goal is $u \approx v$. Let s and t be terms in T, and let ρ be a substitution such that $s\rho = u$ and $t\rho = v$, and such that $H(s, \rho) = H(u)$ and $H(t, \rho) = H(v)$. [1] We don't-know non-deterministically find a unifier $\sigma \in CSU_E(s, t)$. If $Var(s \approx t) = \{x_1, \cdots, x_n\}$ then the rule is the following:

Mutate

$$\frac{u \approx v}{\bigcup_{1 \leq i \leq n} x_i\rho \approx x_i\sigma}$$

Here is an example.

Example 6. Let $E = \{fgfx \approx gfgx\}$ and let T be the E-closed set $\{fx, gy, fgz, gfx, c, d\}$. Suppose that the goal is $fa \approx gb$. Then $CSU_E(fx, gy) = \{\sigma\}$, where $\sigma = [x \mapsto gfz, y \mapsto fgz]$. We also find a matcher $\rho = [x \mapsto a, y \mapsto b]$ such that $fa = fx\rho$ and $gb = gy\rho$. The Mutate inference rule applies:

$$\frac{fa \approx gb}{a \approx gfz, fgz \approx b}$$

This is because of the fact that $x\rho = a$, $x\sigma = gfz$, $y\sigma = gfz$, and $y\rho = b$.

It is obvious from this example that our inference rule is a generalization of the Mutate Rule from [7].

Consider a related example.

---

[1] This means that we use the same s, t and ρ as in the definition of T-height.

Example 7. Let E and T be as in the above example. Suppose that the goal is $fga \approx gfb$. Then $CSU_E(fgx, gfy) = \{\sigma\}$, where $\sigma = [x \mapsto fz, y \mapsto gz]$. We also find a matcher $\rho = [x \mapsto a, y \mapsto b]$ such that $fga = fgx\rho$ and $gfb = gfy\rho$. The Mutate inference rule applies:

$$\frac{fga \approx gfb}{a \approx fz, gz \approx b}$$

This is because of the fact that $x\rho = a$, $x\sigma = fz$, $y\sigma = gz$, and $y\rho = b$. In this example, if we chose $s = fx$, $t = gy$, and $\rho = [x \mapsto ga, y \mapsto fb]$, then it would have still been true that $s\rho = fga$ and $t\rho = gfb$. However, this would not have minimized the T-height, so it is not valid.

Mutate always applies to a goal $u \approx v$, because of the definition of T, as long as T is extended to cover all the symbols that appear in $u \approx v$ but do not appear in E, as explained above.

We also have an inference rule:

Clash

$$\frac{u \approx v \cup G}{\bot}$$

if there is an s and t with $s\rho = u$, $t\rho = v$, and s and t are not E-unifiable. If the symbol $\bot$ appears in a goal, then that goal will never yield an E-unifier. An example is:

Example 8. Let $E = \{fggx \approx gffx\}$. Let T be the E-closed set $\{fx, gy, ffz, ggw, c, d\}$. Suppose that the goal is $ffa \approx ga$. If $s = ffz$ and $t = gy$. Then $\rho = [z \mapsto a, y \mapsto a]$ is a matcher. But ffz and gy are not unifiable. The Clash rule applies:

$$\frac{ffa \approx ga}{\bot}$$

So ffa and ga are not E-unifiable. Interestingly, we could have chosen fx and gy from T. Those terms are E-unifiable. Therefore Mutate would have applied. If we kept applying the inference rules in that fashion, then we would not halt. That is why it is necessary to choose s and t to minimize the T-height, and why it is necessary that T is closed in order for this algorithm to halt.

We now prove the soundness of our inference rule.

Theorem 1. *Let s, t, u and v be terms, and let ρ, σ and θ be substitutions such that $s\rho = u$, $t\rho = v$, and $\sigma \in CSU_E(s, t)$. Suppose that for all $x \in Var(s \approx t)$, $x\rho\theta =_E x\sigma\theta$. Then $u\theta =_E v\theta$,*

Proof. Since $x\rho\theta =_E x\sigma\theta$ for all variables in s and t, then by the properties of substitutions: $s\rho\theta =_E s\sigma\theta$ and $t\rho\theta =_E t\sigma\theta$. Hence $u\theta = s\rho\theta =_E s\sigma\theta =_E t\sigma\theta =_E t\rho\theta = v\theta$. (Here the third equality holds because $\sigma \in CSU_E(s, t)$). $\qquad\square$

Now we prove the completeness of the rule.

Theorem 2. *Suppose there exists θ such that, $u\theta =_E v\theta$, and there is a matcher ρ, such that, $s\rho = u$ and $t\rho = v$, for some $s, t \in T$. Then there must be a substitution $\sigma \in CSU_E(s,t)$, such that $x\rho\theta =_E x\sigma\theta$ for all variables in $Var(s,t)$.*

Proof. Since $u\theta =_E v\theta$, and ρ is the matcher, $s\rho\theta = t\rho\theta$. Hence there must be a $\sigma \in CSU_E(s,t)$, such that, $\sigma\tau =_E \rho\theta$, Then $x\rho\theta =_E x\sigma\tau = x\sigma\sigma\tau$, since we assume every substitution is idempotent. Furthermore, $x\sigma\sigma\tau =_E x\sigma\rho\theta = x\sigma\theta$, because ρ does not apply to any variables in $Ran(\sigma)$.   □

Our algorithm is defined in terms of the Mutate inference rule:

$$\frac{u \approx v}{\bigcup_{1 \le i \le n} x_i\rho \approx x_i\sigma}$$

Recall that $u = s\rho$ and $v = t\rho$. Since s and t are from T, and we are assuming that $u \approx v$ has no repeated variables, we can divide the variables $\{x_1, \cdots, x_n\}$ into disjoint sets Y and Z such that Y contains all the variables in s, and Z contains all the variables in t.

Then the algorithm we will describe in this section is as follows. Suppose we want to solve the E-unification problem for a single equation $u \approx v$. If u is a variable, then we return the substitution $[u \mapsto v]$. If v is a variable we return $[v \mapsto u]$. Otherwise, find an s and t as required in the inference rule. Then for every $\sigma \in CSU_E(s,t)$ we will recursively solve $z\sigma \approx z\rho$ for all $z \in Z$. Assume these recursive calls to solve $z\sigma = z\rho$ all return an E-unifier. Then let θ' be the union of all the unifiers. Since $u \approx v$ will be assumed to have no repeated variable, and since each substitution in the complete set of unifiers of two terms in T will be linear, the union is well-defined.[2] Then we apply θ' to each equation $y_j\rho \approx y_j\sigma$. The result of the application of θ' will be $y_j\rho \approx y_j\sigma\theta'$, since θ' does not apply to any of the variables in the range of ρ. Let θ'' be the union of all of these unifiers obtained from recursive calls on $y_j\rho \approx y_j\sigma\theta'$. Then the unifier of $u \approx v$ is $\theta'\theta''$. If any of the recursive calls returns $\bot$, then *solve* will also return $\bot$. See the algorithm in Figure 1. We must prove that the algorithm will halt. We will prove it halts by giving a bound on the number of recursive calls. In order to do so, we also give a bound on the T-heights of the terms in the ranges of the E-unifiers which are generated.

We make the algorithm nondeterministic by using a choose function.[3] This makes it easier to define. We must take this into account when we analyze the complexity. The function *choose* will select one E-unifier out of a set of E-unifiers. The end of the algorithm results in one E-unifier. Each possible choice in this algorithm would supply a complete set of E-unifiers. This set of E-unifiers may contain some occurrences of $\bot$, since some choices may not give an E-unifier. Then just remove $\bot$ from the set.

In Figure 2, we give an example of performing the algorithm on the goal $fffu_1 \approx ggggu_2$, with the equational theory $E = \{fgfx \approx gfgx\}$ and $T =$

---

[2] The union of anything with $\bot$ is $\bot$.

[3] In a deterministic algorithm, choose would be replaced by a loop.

```
function solve(u ≈ v)

    if u is a variable
        return [u ↦ v]
    if v is a variable
        return [v ↦ u]
    find s and t, σ and ρ as in definition of inference rule
    if s and t are not unifiable
        return ⊥
    choose θ′ in CSU_E(s ≈ t)
    for i = 1 to q
        θ_i = solve(z_iσ ≈ z_iρ)
    θ′ = θ_1 ∪ ··· ∪ θ_q
    for j = 1 to r
        θ_j = solve(y_jρ ≈ y_jσθ′)
    θ″ = θ_1 ∪ ··· ∪ θ_r
    return θ′θ″
```

Fig. 1. Algorithm

$\{fx, gy, fgz, gfw\}$. In this example, after the inference rule, the right branch is always calculated first. That determines a unifier, which is applied to the left branch. Therefore, each left child is shown with the calculated unifier already applied.

4   Decidability and Complexity

We will prove that the size of the proof for $u \approx v$ is bounded. The proof is defined as a tree of equations, with $u \approx v$ at the root and for each node e, the children of e are obtained by our inference rule. As we explained in the algorithm, Mutate is applied as long as possible in a depth-first fashion, until we reach leaves of the form $x \approx t$ or $t \approx x$, where x is a variable and t is any term. This defines the mgu θ_i which is applied to the rest of the equations in the goal. The leaves are then counted as solved. Then another equation is selected and the process is repeated. The size of a proof is defined to be the number of non-leaf equations in the proof tree. We will show that if all non-constant function symbols are monadic, then the size of a proof tree of $u \approx v$ is less than or equal to $|u| \times |v|$.

Theorem 3. *Assume that E is a linear equational theory, containing only monadic function symbols, and that T is a finite E-closed set. The size of the proof-tree of a goal of varity 1, $u \approx v$, is less than or equal to $|u| \times |v|$. If x and*

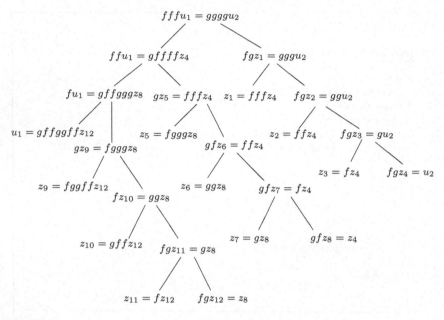

Fig. 2. Proof Tree

y are variables in u and v respectively, and θ is a unifier of u and v obtained in the proof, then $|x\theta| \leq |v|$ and $|y\theta| \leq |u|$.

Proof. The proof will be by induction on the sum of sizes of the terms in the equation $u \approx v$, i.e., $|u| + |v|$. The base case is when $|u| = 0$ or $|v| = 0$. In that case, $u \approx v$ is in normal form.[4] Therefore, the proof is of size 0, since we ignore leaf nodes in the tree.

Now assume that $|u| > 0$ and $|v| > 0$. Assume that the theorem is true for each equation with sum of term sizes smaller than $|u| + |v|$. First we must prove that induction is applicable, i.e. that the size is decreased with the application of the inference rule. An application of the rule with monadic terms will have the following form:

$$\frac{s\rho[x_u] \approx t\rho[y_v]}{x_s\rho[x_u] \approx x_s\sigma[z_1] \qquad y_t\sigma[z_1] \approx y_t\rho[y_v]}$$

$$\vdots$$

$$\theta_1$$

$$x_s\rho[x_u] \approx x_s\sigma[z_1]\theta_1 \quad : \text{new goal-equation}$$

$$\vdots$$

$$\theta_2$$

---

[4] Since $u \approx v$ has no repeated variables, it cannot be of the form $x \approx w[x]$ for some term w.

where $u \approx v$ is our goal, $s, t \in T$, $s\rho = u$, $t\rho = v$, x_u is the only variable in u, y_v is the only variable in v, x_s is the only variable in s, y_t is the only variable in t, z_1 is a variable possibly introduced by the unifier σ of $s[x_s]$ and $t[y_t]$.[5]

In order to apply induction, we need to establish that the size of an equation gets smaller with the application of the rule.

Claim. $|y_t\sigma| + |y_t\rho| < |s\rho| + |t\rho|$.

Proof of Claim. $|y_t\sigma| \leq 1$, because $y_t\sigma \in T$ or y_t is a variable. $|y_t\rho| = |t\rho| - 1$, because by definition: $|t\rho| = 1 + |y_t\rho|$. Hence $|y_t\sigma| + |y_t\rho| \leq 1 + |t\rho| - 1 = |t\rho| <$ $|t\rho| \leq |s\rho| + |t\rho|$, because $s\rho \geq 1$, since s is not a variable. □

Having proved this lemma, we can state, by the induction assumption, that the size of the proof-tree for $y_t\sigma \approx y_t\rho$ is less than or equal to $|y_t\sigma| \times |y_t\rho| \leq$ $1 \times (|t\rho| - 1) = |t\rho| - 1$. Also, $|z_1\theta_1| \leq |y_t\rho| = |t\rho| - 1$ and $|y_v\theta_1| \leq |y_t\sigma| \leq 1$, where θ_1 is the unifier obtained in the proof.

Claim. $|x_s\rho| + |x_s\sigma\theta_1| < |s\rho| + |t\rho|$.

Proof of Claim. By the definition of the size of term: $|x_s\rho| = |s\rho| - 1$. (This is because: $|s\rho| = 1 + |x_s\rho|$, where s is in T.) The size of the term: $|x_s\sigma[z_1]\theta_1| = 1 + |z_1\theta_1|$, because $x_s\sigma \in T$ or is a variable. We have shown that $|z_1\theta_1| \leq |t\rho| - 1$. Hence, $|x_s\sigma\theta_1| \leq 1 + |t\rho| - 1 = |t\rho|$. Taking together the sizes of these two terms, we get: $|x_s\rho| + |x_s\sigma\theta_1| \leq |s\rho| + |t\rho| - 1 < |s\rho| + |t\rho|$. □

If follows from this claim that the size of the proof tree for $x_s\rho \approx x_s\sigma\theta_1$ is less than or equal to $|x_s\rho| \times |x_s\sigma\theta_1| = (|s\rho| - 1) \times |t\rho|$. Also, $|x_u\theta_2| \leq |x_s\sigma\theta_1| \leq |t\rho|$ and $|z_2\theta_1| \leq |x_s\rho| = |s\rho| - 1$, where z_2 is a variable possibly introduced by the substitution θ_1.

Taking together these two statements, we can assess the size of the proof-tree for $s\rho \approx t\rho$. It is less than or equal to $1 + |t\rho| - 1 + ((|s\rho| - 1) \times |t\rho|) = |s\rho| \times |t\rho|$. Also, $|x_u\theta_1\theta_2| = |x_u\theta_2| \leq |t\rho|$ and $|y_v\theta_1\theta_2| = |y_v\theta_1[z_2]\theta_2| \leq |y_v\theta_1| + |z_2\theta_2| \leq 1 + |s\rho| - 1 = |s\rho|$. □

The theorem gives us the first major complexity result of the paper.

Theorem 4. *Let $u \approx v$ be a goal with no repeated variables. Let E be a linear equational theory, containing only monadic function symbols. Let n be the size of $u \approx v$, defined in the standard way. Then*

- *The nondeterministic algorithm in Figure 1 finds a set of E-unifiers for $u \approx v$ in nondeterministic time $O(n^2)$.*
- *Any E-unifier that is constructed is of size $O(n)$.*
- *If every pair of terms in T has a most general E-unifier, then the algorithm is deterministic, and runs in deterministic time $O(n^2)$.*

---

[5] Technically, we need to show that the new equations generated are of varity 1. We show this in the full paper[8].

In order to deal with the more general case of non-monadic terms, we will be considering height of a term and height of a proof-tree, in order to get an idea about the complexity of the procedure. The height of a term was defined earlier. The height of a proof tree is,the length of the longest branch in the proof-tree, excluding its leaf. We write the height of the proof-tree of $u \approx v$ as $H(u \approx v)$.

The general case of the application of our rule is as in the following diagram:

$$\frac{s\rho \approx t\rho}{\bigcup_{i=1}^{q} x_i^s \rho \approx x_i^s \sigma \qquad \bigcup_{i=1}^{r} y_i^t \sigma \approx y_i^t \rho}$$

$$\vdots$$

$$\theta_1$$

$$\bigcup_{i=1}^{q} x_i^s \rho \approx x_i^s \sigma \theta_1 \quad : \text{new goal-equations}$$

$$\vdots$$

$$\theta_2$$

where $u \approx v$ is our goal, $s, t \in T$, $s\rho = u$, $t\rho = v$, $x_1^u, \cdots, x_m^u$ are the variables in u, $y_1^v, \cdots, y_n^v$ are the variables in v, $x_1^s, \cdots, x_q^s$ are the variables in s, $y_1^t, \cdots, y_r^t$ are the variables in t, $z_1, \cdots z_p$ are variables possibly introduced by the unifier σ of s and t.

Theorem 5. *Assume T is a finite E-closed set. E is linear, and the goal $u \approx v$ is of varity 1, where u and v are not both variables. The height of a proof-tree of $u \approx v$ is less than or equal to $H(u) + H(v) - 1$. If $x_1^u, \cdots x_m^u$ and $y_1^v, \cdots, y_n^v$ are variables in u and v respectively, and θ is a unifier of u and v obtained in the proof, then $H(x_i^u \theta) \leq H(v)$ and $H(y_j^v \theta) \leq H(u)$.*

Proof. The proof will be by induction on $H(u) + H(v)$. The base case is when $H(u) = 0$ or $H(v) = 0$. In that case $u \approx v$ is in normal form. Therefore the proof is of height 0, since we ignore leaf nodes when calculating height.

Now assume that $H(u) > 0$ and $H(v) > 0$. Assume that the theorem is true for each equation with sum of heights smaller then $H(u) + H(v)$. First let us consider the right equation: $y_i^t \sigma \approx y_i^t \rho$.

Claim. $H(y_i^t \sigma) + H(y_i^t \rho) < H(s\rho) + H(t\rho)$

Proof of Claim. $H(y_i^t \sigma) \leq 1$, because $y_i^t \sigma$ is in T or is a variable. $H(y_i^t \rho) \leq H(t\rho) - 1$, because, according to the definition of height, $H(t\rho) = 1 + max\{H(y_i^t \rho)\}$. Hence $H(y_i^t \sigma) + H(y_i^t \rho) \leq 1 + H(t\rho) - 1 = H(t\rho) < H(s\rho) + H(t\rho)$. $\square$

By the induction assumption, if $H(y_i^t \sigma \approx y_i^t \rho) \neq 0$, then $H(y_i^t \sigma \approx y_i^t \rho) \leq H(y_i^t \sigma) + H(y_i^t \rho) - 1$, for every $i \in \{1, \cdots, r\}$. We know $H(y_i^t \sigma) \leq 1$, and we know $H(y_i^t \rho) \leq H(t\rho) - 1$. Hence, we know that the height of this proof-tree is: $H(y_i^t \sigma \approx y_i^t \rho) \leq 1 + H(t\rho) - 1 - 1 = H(t\rho) - 1$. If $H(y_i^t \sigma \approx y_i^t \rho) = 0$, then $H(y_i^t \sigma \approx y_i^t \rho) = 0 \leq H(t\rho) - 1$, since $H(t\rho) \geq 1$.

By induction we also know that:

- $H(z_j\theta_1) \le H(y_i^t\rho) \le H(t\rho) - 1$ for each z_j in $y_i^t\sigma$,

- $H(y_j^v\theta_1) \le H(y^t\sigma) \le 1$ for each y_j^v in $y_i^t\rho$.

Now, consider the left part of the proof-tree.

Claim. $H(x_i^s\rho) + H(x_i^s\sigma\theta_1) < H(s\rho) + H(t\rho)$

Proof of Claim. $H(x_i^s\rho) \le H(s\rho) - 1$, from the definition of height. $H(x_i^s\sigma\theta_1) \le H(x_i^s\sigma) + max\{H(z_i\theta_1)\}$, where $\{z_1, \cdots, z_k\}$ are the variables in $s\sigma$. $max\{H(z_i\theta_1)\} \le H(t\rho) - 1$, from the analysis of the right equation. Hence $H(x_i^s\sigma\theta_1) \le 1 + H(t\rho) - 1 = H(t\rho)$. Therefore, $H(x_i^s\rho) + H(x_i^s\sigma\theta_1) \le H(s\rho) - 1 + H(t\rho) < H(s\rho) + H(t\rho)$. □

Hence, by the induction assumption, we know that, if $H(x_i^s\rho \approx x_i^s\sigma\theta_1) \ne 0$, then $H(x_i^s\rho \approx x_i^s\sigma\theta_1) \le H(x_i^s\rho) + H(x_i^s\sigma\theta_1) - 1$. Now, $H(x_i^s\rho) \le H(s\rho) - 1$, because according to the definition of height of a term, $H(s\rho) = 1 + max\{x_i^s\rho\}$. Also, $H(x_i^s\sigma[z_1, \cdots, z_p]\theta_1) \le 1 + max\{H(z_i\theta_1)\} = H(t\rho)$, because $H(z_i\theta_1) \le H(t\rho) - 1$, by the previous lemma. Hence, the height of this proof-tree will be:

- $H(x_i^s\rho \approx x_i^s\sigma\theta_1) \le H(s\rho) - 1 + H(t\rho) - 1 = H(s\rho) + H(t\rho) - 2$.

If $H(x_i^s\rho \approx x_i^s\sigma\theta_1) = 0$ then $H(x_i^s\rho \approx x_i^s\sigma\theta_1) \le H(s\rho) + H(t\rho) - 2$, because $H(s\rho) \ge 1$ and $H(t\rho \ge 1)$.

The induction assumption also states that

- $H(x_j^u\theta_2) \le H(x_i^s\sigma\theta_1) \le 1 + max\{z_i\theta_1\} \le 1 + H(t\rho) - 1 = H(t\rho)$, for each x_j^u in $x^s\rho$, and
- $H(z_j'\theta_2) \le H(x_i^s\rho) \le H(s\rho) - 1$, for all z_j' in $x_i^s\sigma\theta_1$.

We can now prove the main claim:

The height of the proof-tree for $u \approx v$, i.e. for $s\rho \approx t\rho$, is then:
$H(s\rho \approx t\rho) \le 1 + max\{H(x_i^s\rho \approx x_i^s\sigma\theta_1), H(y_i^t\sigma \approx y_i^t\rho)\} \le 1 + max\{(H(s\rho) + H(t\rho) - 2), (H(t\rho) - 1)\} = 1 + H(s\rho) + H(t\rho) - 2 = H(s\rho) + H(t\rho) - 1$. This is because $H(s\rho) + H(t\rho) - 2 \ge H(t\rho) - 1$, because we assumed $H(s\rho) > 0$.

We only need to prove the claims about the heights of terms:
$H(x_j^u\theta_1\theta_2) = H(x_j^u\theta_2)$, because x_j^u cannot be in the domain of θ_1. By the assumption, $H(x_j^u\theta_2) \le H(t\rho)$.
$H(y_j^v\theta_1\theta_2) \le H(y_j^v\theta_1[z_1', \cdots, z_k']) + max\{H(z_j'\theta_2)\} \le 1 + H(s\rho) - 1 = H(s\rho)$.
□

This gives us the following complexity result.

Theorem 6. *Let $u \approx v$ be a goal with no repeated variables. Let E be a linear equational theory. Let n be the size of $u \approx v$, defined in the standard way. Then*

- *The nondeterministic algorithm in Figure 1 finds a set of E-unifiers for $u \approx v$ in PSPACE.*
- *The terms in the range of the E-unifier that is constructed are of height $O(n)$.*

5   Finding a Closed Set

We have shown that once you have an E-closed set, then unification problems of varity 1 are solvable, and we have given the complexity of the decision problem in several cases. That all assumes that we know of an E-closed set. That could be the case for some equational theories. But if we don't know whether there is an E-closed set, then in this section we give a method to produce one which will work for some equational theories.

First we show how to construct an E-closed set in an incremental way:

Let T_0 contain all terms of the form $f(x_1, \cdots, x_n)$, where f is a function symbol of arity $n \geq 0$ appearing in E, and $x_1, \cdots, x_n$ are fresh variables. Also, T_0 will contain two fresh constants c and d.

For $i \geq 0$, T_{i+1} is defined as the set of terms such that $t \in T_{i+1}$ if and only if t is a nonvariable such that there exists some u and v in T_i, a variable x appearing in u and a $\sigma \in CSU_E(u \approx v)$ such that t is a renaming of a subterm of $x\sigma$.

Let $T = \bigcup_{i \geq 0} T_i$. Then T is an E-closed set if the complete sets of unifiers for pairs of terms in T are linear. Of course, T might not be finite. But if T is finite, then this gives us a decision procedure for solving the E-unification problem when the goal has no repeated variables.

We still have not said how to find a complete set of E-unifiers for a pair of terms. This problem is undecidable in general, but in some cases it is possible to use a complete algorithm to generate the E-unifiers. One possibility is to use the complete procedure for linear equational theories presented in [7]. The inference system in that paper is a generalization of the General Mutate inference rules of [2,3,4,5], but it is complete for all linear equational theories. It uses a form of eager variable elimination which makes it more efficient.

The problem with using a complete inference system is that it may not halt when two terms are not E-unifiable. However, we also need to check cases of non-unifiability for our algorithm. But, inference rules, such as the ones in [7] can be extended to detect non-unifiability in some cases where the procedure would normally not halt. The inference rules are goal directed, in the sense, that it begins with the equation which must be E-unified. As in the algorithm in this paper, an inference rule will be applied to the goal yielding one or more subgoals. Also, as in this paper, one or more rules may apply at each point. So the algorithm amounts to the simultaneous construction of one or more proof-trees. In some cases, it happens that every proof tree contains an equation $u \approx v$ that is a descendant of a renaming of an equation $s \approx t$, such that $s\rho = u$ and

$t\rho = v$ for some ρ. In such cases, the algorithm will never halt, and therefore the initial equation is not E-unifiable.

6 Conclusion

Historically, much of the field of automated deduction has focused on inference procedures that search for a proof of a theorem, and not as much effort has been applied to finding methods of proving something is false. However, if these methods can be applied to verification problems and other applications, we believe it is necessary to identify classes of problems where automated theorem provers will halt, and to understand the complexity of these classes. This is a goal of our research.

The problems we considered in this paper are E-unification problems, since equational logic is useful for many applications. The procedure we give in this paper is an adaptation of a more general procedure for E-unification. However, on the class of problems we consider in this paper, we were able to show a measure on certain E-unification problems, such that the inference rules always reduce the measure; therefore it will halt and we can analyze how quickly it will halt, in order to examine the complexity.

Specifically, we introduce a subclass of linear equational theories, called *finitely closable*. We consider goals with no repeated variables. We show that this class is solvable in PSPACE in general. For monadic theories, it is in NP. For unitary monadic theories, it is solvable in $O(n^2)$.

We think this class is interesting. We also think this research raises many questions to be explored further. Which equational theories are in this class? What is a good procedure for finding a finite (or recursive) E-closed set? Can our complexity results be made better? How can this class be expanded?

References

1. F. Baader and T. Nipkow. *Term Rewriting and All That*. Cambridge, 1998.
2. C. Kirchner. Computing unification algorithms. In *Proceedings of the First Symposium on Logic in Computer Science*, Boston, 200-216, 1990.
3. C. Kirchner and H. Kirchner. *Rewriting, Solving, Proving*. http://www.loria.fr/~ckirchne/, 2000.
4. C. Kirchner and F. Klay. Syntactic Theories and Unification. In *LICS 5*, 270-277, 1990.
5. F. Klay. Undecidable Properties in Syntactic Theories. In *RTA 4*,ed. R. V. Book, LNCS vol. 488, 136-149, 1991.
6. C. Lynch and B. Morawska. Goal-directed E-unification. To appear in 12th International Conference on Rewriting Techniques and Applications.
7. C. Lynch and B. Morawska. Approximating E-unification. Submitted.
8. C. Lynch and B. Morawska. http://www.clarkson.edu/~clynch/papers/linear_full.ps/, 2001.
9. R. Nieuwenhuis and A. Rubio. Paramodulation-based Theorem Proving. To appear in *Handbook of Automated Reasoning*, 2001.

A New Meta-complexity Theorem for Bottom-Up Logic Programs

Harald Ganzinger[1] and David McAllester[2]

[1] MPI Informatik, D-66123 Saarbrücken, Germany, hg@mpi-sb.mpg.de
[2] AT&T Labs-Research, Florham Park NJ 07932, USA, dmac@research.att.com

Abstract. Nontrivial meta-complexity theorems, proved once for a programming language as a whole, facilitate the presentation and analysis of particular algorithms. This paper gives a new meta-complexity theorem for bottom-up logic programs that is both more general and more accurate than previous such theorems. The new theorem applies to algorithms not handled by previous meta-complexity theorems, greatly facilitating their analysis.

1 Introduction

McAllester has recently shown that the running time of a bottom-up logic program can be bounded by the number of "prefix firings" of its inference rules [10]. A prefix firing of a rule is a derivable instantiation of a prefix of the antecedents of that rule. This single nontrivial meta-complexity theorem simplifies the presentation and complexity analysis of a variety of parsing and static analysis algorithms. Many other algorithms, however, seem to fall outside of the range of this theorem. In particular, algorithms based on union-find or congruence closure can not be analyzed. A second meta-complexity theorem for the analysis of union-find based algorithms is also given in [10]. While this second theorem applies to a broader class of algorithms, it yields running time bounds that are often worse by logarithmic factors than algorithm-specific bounds — bounds proved without the use of a meta-complexity theorem. Here we prove a more accurate and more general meta-complexity theorem. The increased generality is achieved by proving the theorem for logic programs with priorities and deletions. Priorities and deletions allow the simulation of arbitrary classical control structures. So the new meta-complexity theorem has, in some sense, universal coverage. The new theorem yields improvements in meta-complexity-derived bounds for a variety of algorithms including union-find and congruence closure. As an example, section 5 presents an algorithm for determining the satisfiability of a ground set of Horn clauses with equality. The new meta-complexity theorem allows the simple derivation of a very tight running time bound for this algorithm. Proving the same bound for this problem without the the use of our new theorem appears to be significantly more difficult.

R. Goré, A. Leitsch, and T. Nipkow (Eds.): IJCAR 2001, LNAI 2083, pp. 514–528, 2001.

2 Inference Rules with Priorities and Deletions

We use the term *inference rule* to mean a first-order Horn clause, i.e., a formula of the form $A_1 \wedge \ldots \wedge A_n \to C$ where C and each A_i is a first-order atom. We will use *assertion* to mean a ground atom and use the term *data base* to mean a set of assertions. If R is a set of rules and D is a data base, then we let $R(D)$ denote the set of ground atoms derivable from the ground set D using the rules in R.

Here we are interested in expressing algorithms with prioritized inference rules with deletion. An inference rule with deletion is an expression of the form $A_1 \wedge \ldots \wedge A_n \to C$ where C is an atom and each A_i is either an atom or an expression of the form $[A]$ where A is an atom. Intuitively, the marking $[\ldots]$ means that the premise is to be deleted as soon as the rule is run. Deletion makes the behavior of the algorithm nondeterministic. For example, consider the following rules with deletion.

$$P \Rightarrow Q \qquad [Q] \Rightarrow S \qquad [Q] \Rightarrow W$$

Suppose the initial data base contains only P. The first rule fires adding the assertion Q. Now either the second or third rule can fire. Since each of these rules deletes Q, once one of them fires the other is blocked. Hence the final data base is nondeterministically either $\{P, S\}$ or $\{P, W\}$. When viewing rules with deletions as algorithms this nondeterminism is viewed as "don't care" nondeterminism — the choices are made arbitrarily and not backtracked. (In many cases this kind of don't-care nondeterminism can be justified by a suitable notion of redundancy, cf. Section 8.) Suppose now that we have additional rules through which W entails a large number of additional facts whereas the absence of W does not. Then, in order to obtain a more efficient run of the rules, we should prefer to fire the second rule rather than the third rule, which we could achieve by giving higher priority to the second rule. In summary, allowing for deletion makes deduction nondeterministic, and hence priorities are needed for indicating which choices are to be made in order to increase efficiency, or to avoid unwanted results.

The proof of the meta-complexity theorem requires that deletion be permanent — once an assertion is deleted further attempts to reassert it have no effect. (If deletion is based on a notion of redundancy such as the one proposed in [2], once an assertion has become redundant it remains so for the remainder of the computation.) To see the problem with deletion consider the simple pair of rules $[P] \to Q$ and $[Q] \to P$. If deletion is can be revoked by subsequent assertion, the rules can oscillate between a database containing P and a database containing Q and fail to terminate. To formalize this notion of permanent deletion we take a *state* of the computation to be a set S of literals (atoms and negations of atoms). The presence of a negated atom $\neg A$ in a state indicates that A should be considered deleted. Hence, we say that an atom A is *visible* in a state S if $A \in S$ and $\neg A \notin S$, while a negative literal $\neg A$ is called *visible* in S whenever $\neg A \in S$. If σ is a ground substitution and $[A]$ is a deleted antecedent then we define $\sigma([A])$ to be the ground atom $\sigma(A)$. Now let S be a state, and let r be an inference rule with deletions. We write $S \xrightarrow{r} S'$ if $S \neq S'$ and r is a rule $A_1 \wedge \ldots \wedge A_n \to C$ such that there exists a ground substitution σ defined on all the variables in r such that $\sigma(A_i)$ is visible in S, and S' is $S \cup \{\sigma(C), \neg\sigma(A_{i_1}), \ldots, \neg\sigma(A_{i_k})\}$ where $A_{i_1}, \ldots, A_{i_k}$ are the deleted antecedents of the rule. We say that a rule r is *applicable* at the state S if there exists a state S' (which must be different from S) such that $S \xrightarrow{r} S'$.

Now let R be a set of rules with deletions where each rule in R is associated with a positive rational number called its priority. We call R a *rule set with priorities and deletions*. For technical simplicity we may assume that priorities are unique in that no two rules have the same priority. We say that a state S is *visible* to a rule $r \in R$ if no higher priority rule in R is applicable at S. We write $S \xrightarrow{R} S'$ if there exists a rule $r \in R$ such that S is visible to r and $S \xrightarrow{r} S'$. We will say that a state S is *saturated* under R if it is a normal form, i.e., there is no S' such that $S \xrightarrow{R} S'$. An R-computation from a database D is a sequence S_0, $S_1, \ldots, S_T$ such that $S_0 = D$, $S_t \xrightarrow{R} S_{t+1}$. An R-computation is called *complete* if the final state S_T is saturated. If there is a complete R-computation from D ending in S_T then we say that S_T is an R-saturation of D. A rule set R is said to terminate on input database D if there is no infinite R-computation from D.

A *prefix firing* in an R-computation $\mathcal{C}$ is a triple $\langle r, \sigma, i \rangle$, where $r \in R$ is a rule $A_1 \wedge \ldots \wedge A_n \Rightarrow C$ such that the computation $\mathcal{C}$ contains a state S visible to r and σ is a ground substitution defined on the variables in the antecedent prefix $A_1, \ldots, A_i$ such that the $\sigma(A_j)$, for $1 \leq j \leq i$, are visible in S. Note that the set of prefix firings of a given rule is determined by the set of states visible to that rule. For any R-computation $\mathcal{C}$ we let $p(\mathcal{C})$ be the number of prefix firings in $\mathcal{C}$. We will call a rule range-restricted if every variable in the conclusion appears in some antecedent. Bottom-up logic programs are generally range-restricted and for simplicity we only consider range-restricted rules. In the following, by $|D|$ we denote the *size* of a database which is the number of nodes in its fully shared graphical representation by a dag.

Theorem 1. *For any given set R of range-restricted rules with priorities and deletions there exists an algorithm mapping an input database D to an R-saturation $R(D)$ of D whose running time is $O(|D|+\max_\mathcal{C} p(\mathcal{C}))$ where the maximization is over all R-computations $\mathcal{C}$ from D.*[1]

The theorem extends the one in [10] to inference rules with priorities and deletion showing essentially that no penalty has to be paid for these extensions. The complexity can again be linearly bounded by the number of prefix firings.

Before giving a proof of this theorem, in the next sections we will present a variety of applications. Before discussing those, the following example is given in order to clarify one of the more subtle issues behind our definitions. Consider

---

[1] Note that if there is no bound on the length of computations then the algorithm need not terminate.

the rules $r1$ and $r2$, where

$$r1: \ r(x,y), \ [p(x)] \Rightarrow s(x) \qquad r2: \ p(x), \ q(x,y) \Rightarrow r(x,y)$$

with priorities from left to right, on a database D consisting of facts $p(i)$, for $1 \le i \le n$, and $q(i,j)$, for $1 \le i,j \le n$. In any computation from D, whenever rule r2 produces an r-fact $r(i,j)$, in the next step r1 takes priority over r2, and the $p(i)$ is deleted so that no other fact $r(i,j')$ can by produced thereafter. Hence any computation takes at most $2n$ steps. However, the number of prefix firings of rule r2 is $n+n^2$, and that is the upper bound on the time complexity provided by the meta-complexity theorem above. A more refined meta-complexity theorem, based on refined notions of prefix firings, could be stated. However in this paper we deliberately confine ourselves to the simpler version. The additional technical complexity does not appear to be required for the examples that we are interested in at present.

3 A Union-Find Algorithm

This section presents an $O(n \log n)$ union-find algorithm given as a rule set with priorities and deletions. This union-find algorithm both gives an example of a use of theorem 1 and serves as a foundation for other algorithms given in later sections of this paper. The union-find algorithm in itself is perhaps not significantly simpler than classical presentations using pointers and recursive procedures. However its direct relation to the usual inference rules for Knuth-Bendix completion makes correctness arguments more straightforward.

The union-find algorithm is used to represent equivalence relations. In the inference rule union-find algorithm U in Figure 1 we assume a binary predicate union such that the assertion union(x, y) means that x and y are to be made equivalent — the procedure is to compute the least equivalence relation such that if the data base contains union(x, y) then x and y are equivalent. The find function is defined in terms of a more basic rewrite relation which we represent here as a set of assertions of the form $x \xrightarrow{f} y$. We define the "find" of x to be the normal form of x under the rewrite relation $\xrightarrow{f}$. Storing this relation explicitly as assertions in the data base defines in a more logical manner what is usually implemented with pointer structures.

The union-find inference system implements Knuth-Bendix completion for the simple case of equations between constants. The equations are represented by the union facts. The rules (F1) and (F2) compute the normal forms of terms. (U2)–(U4) orient equations into rewrite rules using an ordering that is dynamically determined by the weight computation in rules (U3) and (U4). If $\xrightarrow{f}*$ is the reflexive-transitive closure of $\xrightarrow{f}$, the weight of y is the number of nodes x such that $x \xrightarrow{f}* y$. An assertion $x \xrightarrow{f} y$ is to be added only for irreducible y for which $y \xrightarrow{f}! y$. The rule (F1)–(F2) are to run at a higher priority than any other rules mentioning the predicates find, $\xrightarrow{f}$ or $\xrightarrow{f}!$. This ensures that, at any state visible to other rules mentioning these relations, the relation $\xrightarrow{f}!$ is the fixed "normal form relation" determined by the $\xrightarrow{f}$ relation.

$$\text{(F1)} \frac{\text{find}(x)}{x \xrightarrow{f}! x, \quad \text{weight}(x,1)} \qquad\qquad \text{(F2)} \frac{[x \xrightarrow{f}! y] \quad y \xrightarrow{f} z}{x \xrightarrow{f}! z}$$

$$\text{(U1)} \frac{\text{union}(x,y)}{\text{find}(x), \ \text{find}(y)} \quad \text{(U2)} \frac{[\text{union}(x,y)] \quad x \xrightarrow{f}! z \quad y \xrightarrow{f}! z}{T} \quad \text{(U3)} \frac{[\text{union}(x,y)] \quad x \xrightarrow{f}! z_1 \quad y \xrightarrow{f}! z_2 \quad \text{weight}(z_1,w_1) \quad [\text{weight}(z_2,w_2)] \quad w_1 < w_2}{z_1 \xrightarrow{f} z_2, \ \text{weight}(z_2, w_1+w_2)} \quad \text{(U4)} \frac{[\text{union}(x,y)] \quad x \xrightarrow{f}! z_1 \quad y \xrightarrow{f}! z_2 \quad [\text{weight}(z_1,w_1)] \quad \text{weight}(z_2,w_2) \quad w_1 \ge w_2}{z_2 \xrightarrow{f} z_1, \ \text{weight}(z_1, w_1+w_2)}$$

Fig. 1. Module U for union-find. We write rules vertically as the antecedents followed by a horizontal line followed by the conclusions. The rules are listed in decreasing priority. Multiple conclusions $A_1,\ldots,A_k$ should be viewed as a single atom $a(A_1\ldots,A_k)$ with auxiliary rules (of highest priority) generating the individual conjuncts A_j.

All of the rules represented by (U1) run at higher priority than (U2), (U3), or (U4). This implies that in any state visible to (U2), (U3), or (U4) we have that find(x) and find(y) have been asserted and hence the normal forms of x and y have been computed and the weights have been initialized. Rule (U2) is given higher priority than either (U3) or (U4). This implies that at any state visible to (U3) or (U4) we have that x and y have distinct normal forms (otherwise the state would be visible to rule (U2) which would then delete the link assertion and assert the trivial "true" assertion T).

The use of addition in the rules (U3) and (U4) is outside of the formal language defined in section 2. However, the use of addition in the conclusion can be replaced by an additional final antecedent of the form $w_3 = w_1 + w_2$. Theorem 1 can be generalized to handle constraint antecedents provided that the set of assignments of values to the unassigned variables (those not appearing in earlier antecedents) can be computed in time proportional to the number of such assignments.

One should think of the rules (F1)–(U4) as a module U that takes as *input* assertions of the form find(x) and union(x,y) and produces as *output* assertions of the form $x \xrightarrow{f}! z$ such that z is the normal form of x in a canonical rewrite system generated from the equations union(x,y). The input can be extended dynamically by new assertions of the form find(x) and union(x,y) generated from additional rules that are *compatible* with U.

Definition 1. *A module consists of a rule set with priorities and deletions plus specified input and output predicate symbols. All other predicates of the module are called local. A rule set R will be called* compatible *with a module M provided that it does not mention local predicates of M, and*

- *no output predicate of M appears in any conclusion or deleted antecedent of a rule in R,*
- *no input predicate of M appears in any deleted antecedent of R,*
- *and every rule in R containing an output predicate of M in an antecedent has priority lower priority than all rules in M.*

An initial database D will be called compatible with a module M if it does not mention any predicates of M other than input predicates.

Theorem 2. *The union-find module U has the property that for any rule set R and initial database D, where R and D are both compatible with U, and any $(R \cup U)$-computation C from D, the total number of prefix firings in C of the rules in U is $O(m + n \log n)$ where m is the number of union assertions in C or produced by R, and n is the number of distinct terms appearing in union or find assertions.*

Proof. Note that each non-redundant union operation generates a single new assertion of the form $x \xrightarrow{J} y$ where the weight of y prior to the addition of this assertion is at least as large as the weight of x. This implies that weight at least doubles as one moves across any assertion of the form $x \xrightarrow{J} y$. So for a given x the set of y such that $x \xrightarrow{J}* y$ can have at most $\log n$ elements. (At most $n - 1$ rewrite rules $x \xrightarrow{J} y$ can be generated until all terms become equal.)

Now we show that each of the rules in U has an appropriate number of prefix firings. The rule (F1) has at most n firings. It follows from the above comments that there are at most $n \log n$ assertions ever generated of the form $x \xrightarrow{J}! y$. This, and the fact that out-degree of $\xrightarrow{J}$ is at most one, imply that rule (F2) has at most $n \log n$ prefix firings. Rule (U1) has at most m firings. All states containing the assertion union(x, y) and visible to any of the rules (U2), (U3), or (U4) must assign the same unique normal forms and weights to x and y. This implies that the rules (U2), (U3), and (U4) also have at most m prefix firings.

It is possible to give a more complex inference rule implementation of union-find that runs in $O(n\alpha(n))$ time where α is the inverse of Ackermann's function. However, many algorithms based on union-find run in $O(n \log n)$ time even when using an $O(n\alpha(n))$ implementation of union-find. For such applications an $O(n \log n)$ implementation of union-find suffices.

4   Congruence Closure

The congruence closure problem is to determine whether an equation $s = t$ between ground terms is provable from a given set of equations between ground

$$(\text{C1})\frac{\text{find}(\langle x,\ y\rangle)}{\text{find}(x),\ \text{find}(y),\ \text{init}_\Rightarrow(\langle x,\ y\rangle)} \qquad (\text{C2})\frac{[\text{init}_\Rightarrow(z)]\quad z\Rightarrow w}{T} \qquad (\text{C3})\frac{[\text{init}_\Rightarrow(z)]}{z\Rightarrow z}$$

$$(\text{C4})\frac{[\langle x,\ y\rangle\Rightarrow z]\quad x\xrightarrow{f}!\ x'\quad \langle x',\ y\rangle\Rightarrow z'}{\text{union}(z,\ z')} \qquad (\text{C5})\frac{[\langle x,\ y\rangle\Rightarrow z]\quad x\xrightarrow{f}!\ x'}{\langle x',\ y\rangle\Rightarrow z} \qquad (\text{C6})\frac{[\langle x,\ y\rangle\Rightarrow z]\quad y\xrightarrow{f}!\ y'\quad \langle x,\ y'\rangle\Rightarrow z'}{\text{union}(z,\ z')} \qquad (\text{C7})\frac{[\langle x,\ y\rangle\Rightarrow z]\quad y\xrightarrow{f}!\ y'}{\langle x,\ y'\rangle\Rightarrow z}$$

Fig. 2. Rules for congruence closure listed in order of decreasing priority

terms using the reflexivity, symmetry, transitivity and congruence rules for equality. Here we assume that expressions are represented using constants and a single pairing function. The congruence property of the pairing function states that if $u_1 = w_1$ and $u_2 = w_2$ then $\langle u_1,\ u_2\rangle = \langle w_1,\ w_2\rangle$.

The inference rules in Figure 2 are compatible with the union-find module U (assuming that rules in U have priority lower than the rules (C4)–(C7) that use the output predicate $\xrightarrow{f}!$ of U). Combined with U, they give an $O(n\log n)$ algorithm for congruence closure. The rules are related to rules given in [3] which view congruence closure as a form of ground completion. The ternary atoms $\langle \_,\_\rangle \Rightarrow \_$ represent the signatures in [5], or the definitions in [3]. Note that, by contrast to [3], we do not introduce new constants to denote the subterms of the input equations. The terms on the right side in $\Rightarrow$ play the role of these constants. This explains rule (C3) where $z \Rightarrow z$ gives us z as the handle to all terms that are semantically equal to z. These rules have lower priority than all rules in the union-find module and the priority between rules corresponding to the order in which the rules are given with (C1) having highest priority and (C7) having lowest. The precedence of (C2) over (C3), (C4) over (C6) and (C5) over (C7) ensure the invariant that for any pair $\langle x,\ y\rangle$ there is at most one assertion of the form $\langle x,\ y\rangle \Rightarrow z$. Furthermore, one can check that in any state visible to these rules, and hence where no union-find rules are still to be run, we have that if the state contains $\langle x,\ y\rangle \Rightarrow z$ then $\langle x,\ y\rangle$ and z have the same find value. Furthermore, the rules maintain the invariant that in states visible to (C4) through (C7), if the state contains $\text{find}(\langle x,\ y\rangle)$ then there exists an x' and y' such that $x \xrightarrow{f}! x'$ and $y \xrightarrow{f}! y'$ and the state contains $\langle x',\ y'\rangle \Rightarrow z$ where z is equivalent to (has the same find as) $\langle x,\ y\rangle$. In any final (saturated) state we must have that x' and y' are normal forms under the equivalence generated by the union assertions. This implies that in the final state, if we have $\text{find}(\langle x_1,\ y_1\rangle)$ and $\text{find}(\langle x_2,\ y_2\rangle)$ where x_1 and x_2 are equivalent and y_1 and y_2 are equivalent we must also have $\langle f_1, f_2 \rangle \Rightarrow z$ where f_1 is the common find of x_1 and x_2, f_2 is the common find of y_1 and y_2, and where z is equivalent to both input pairs, and hence the input pairs are equivalent to each other.

The union-find module satisfies the condition that for any given x there are at most $\log n$ terms y such that $x \xrightarrow{f}! y$. This implies that an initial assertion $\langle x, y \rangle \Rightarrow z$ can generate most $2 \log n$ "descendents" of the form $\langle x', y' \rangle \Rightarrow z$. This implies that at most $2n \log n$ assertions of this form are ever generated and this implies that each of the rules (C4), (C5), (C6), and (C7) have at most $2n \log n$ prefix firings. The other rules have at most $O(n)$ prefix firings. Hence we have the following theorem.

Let C be the module with input predicates union and find and output predicate $\xrightarrow{f}!$, resulting from combining the union-find module with the congruence closure rules.

Theorem 3. *If R is a rule set and D an initial database where both R and D are compatible with C then, in any computation from D of $R \cup C$, the total number of prefix firings of rules in C is $O(m + n \log n)$ where m is the number of input union assertions, that is, union assertions in D or generated by R, and n is the number of terms x appearing in* find *assertions (either in input* find *assertions or* find *assertions generated by (C1)).*

Note that in this case, n is proportional to the number of the different subterms in input find assertions. The complexity bound given by the theorem is the same as the one given in [5]. The latter paper, however, ignores the work needed for processing the input equations. Our inference rules do include this preprocessing and, therefore, come with an additional additive $O(m)$ term in the complexity bound.

5   Satisfiability of Ground Horn Clauses with Equality

We now extend congruence closure (in a compatible manner) to handle ground (object-level) Horn clauses represented as assertions in the input database D. (The meta-level Horn clauses are called inference rules.) More specifically we want to construct an algorithm for computing the deductive closure of a set of ground assertions of the form input$(\Phi \rightarrow A)$ where the possible expressions for Φ and A are defined by the following grammar where c ranges over constants and p ranges over binary predicate symbols including the special symbol $\doteq$ for denoting formal equality.

$$\Phi ::= A \mid \Phi_1 \wedge \Phi_2 \qquad A ::= T \mid p(t_1, t_2) \qquad t ::= c \mid \langle t_1, t_2 \rangle$$

The algorithm takes as input a set D of ground assertions of the form $\Phi \rightarrow A$. The algorithm uses the congruence closure module and all inference rules in this section run at priority higher than those in the congruence closure module.

We start with the following linear time module for ground Horn clauses without equality. The module may be viewed as a high-level implementation of the algorithm in [4]. The main idea in this set of rules is that atoms appearing in the antecedent of clauses are first detached from their clauses. This has the effect that even if an atom has many occurrences in antecedents of clauses it is nevertheless only derived once.

$$\text{(I1)} \frac{\text{input}(\Phi \rightarrow A)}{\substack{\text{antecedent}(\Phi), \text{ conclusion}(A), \\ \text{true}(\Phi \rightarrow A)}} \qquad \text{(I2)} \frac{\text{antecedent}(\Phi_1 \wedge \Phi_2)}{\text{antecedent}(\Phi_1), \text{ antecedent}(\Phi_2)}$$

$$\text{(I3)} \frac{}{\text{true}(T)} \qquad \text{(I4)} \frac{\substack{\text{true}(\Phi \rightarrow \Psi) \\ \text{true}(\Phi)}}{\text{true}(\Psi)} \qquad \text{(I5)} \frac{\substack{\text{antecedent}(\Phi_1 \wedge \Phi_2) \\ \text{true}(\Phi_1), \text{ true}(\Phi_2)}}{\text{true}(\Phi_1 \wedge \Phi_2)}$$

A natural way to extend these rules to handle equality would be to treat atoms themselves as terms and apply congruence closure. This would give a simple $O(m \log m)$ algorithm for conditional equations, where m is the size (in dag representation) of the set of clauses. If m is quadratic in the number n of different terms appearing in the set, that would give the bound $O(n^2 \log n)$. However, for the particular application given in section 6 a more refined bound, and more refined algorithm, is needed. We handle equality with the following rules.

$$\text{(I6)} \frac{\text{true}(\doteq(s, t))}{\text{union}(s, t)} \qquad \text{(I7)} \frac{\text{antecedent}(p(s, t))}{\substack{\text{find}(s), \text{ find}(t), \\ \text{push}(p(s, t))}} \qquad \text{(I8)} \frac{\text{conclusion}(p(s, t))}{\substack{\text{find}(s), \text{ find}(t), \\ \text{push}(p(s, t))}}$$

$$\text{(I9)} \frac{\substack{[\text{push}(p(s, t))] \\ s \xrightarrow{f}! s'}}{\substack{\text{true}(p(s', t) \rightarrow p(s, t)), \\ \text{true}(p(s, t) \rightarrow p(s', t)), \\ \text{push}(p(s', t))}} \qquad \text{(I10)} \frac{\substack{[\text{push}(p(s, t))] \\ t \xrightarrow{f}! t'}}{\substack{\text{true}(p(s, t') \rightarrow p(s, t)), \\ \text{true}(p(s, t) \rightarrow p(s, t')), \\ \text{push}(p(s, t'))}} \qquad \text{(I11)} \frac{[\text{push}(\doteq(s, s))]}{\text{true}(\doteq(s, s))}$$

The rules have priority in the order given with (I1) having highest priority and (I11) having lowest but with all rules at lower priority than any rules in the congruence closure module. We leave it to the reader to verify that any saturation of these rules contains a given conclusion if and only that conclusion follows from the input under the standard interpretation of equality. Here we focus on the run time (number of prefix firings) of these rules.

Let m be the number of antecedents as derived by rules (I1) and (I2), plus the number of clauses in the input, and let n and a, respectively, be the number of different terms and atoms appearing there. Clearly, a is in $O(m)$, and also in $O(n^2)$, with $m, n \leq |D|$. The number of prefix firings of rules (I1), (I2), (I3), (I5), (I7), and (I8) are all proportional to m. The number of prefix firings of (I4) is proportional to m plus the number of firings of (I9) and (I10). The number of prefix firings of (I6) is proportional to m plus the number of firings of rule (I11). The number of prefix firings of (I11) is bounded by to the number of prefix firings of (I7) and (I8) (which is m) plus the number of prefix firings of (I9) and (I10). So the total number of prefix firings is proportional to m plus the number of firings of the two rules (I9) and (I10). Since $\xrightarrow{l}!$ has out-degree at most one we immediately get that these two rules have at most n^2 firings where n is the number of subterms appearing in the input. By the properties of union-find we also get that those rules have at most $a \log n$ firings. The number of union operations generated by rule (I6) is at most $a \log n$, hence in $O(m \log n)$. The number of prefix firings inside the congruence-closure module is therefore $O((m + n) \log n)$. So the total number of prefix firings is $O(m + n \log n + \min(m \log n, n^2))$.

Theorem 4. *Satisfiability of ground Horn clauses with equality can be decided in time $O(|D| + n \log n + \min(m \log n, n^2))$ where m is the number of antecedents and input clauses and n is the number of terms.*

The above bound is better than $O(m \log m)$ in any family of problems where m is $\Omega(n^2)$. In that case, also $|D|$ is in $\Omega(n^2)$ and the algorithm becomes linear in the size of the input. In cases where the length of antecedents is bounded by a constant, m is proportional to the number of input clauses in D.

6   Henglein's Quadratic Typability Algorithm

Following the exposition by [10], the typability problem in a variant of the Abadi-Cardelli object calculus [1] considered by [8] can be taken to consist of a given set of assertions of the form $\sigma \leq \tau$ and $accepts(\sigma, l)$ and $notaccepts(\sigma, l)$, where σ and τ are type names and l is a message name. The instance is acceptable (solvable) provided that the following rules do not derive *fail*. We also assume that $type(\sigma)$ is derivable for those type terms σ that appear in $[not]accepts$- or $\leq$-facts in the input, or which are of the form $\tau.l$ with $accepts(\tau, l)$ appearing in the input. Moreover, we assume the standard reflexivity, symmetry, transitivity, and congruence properties of equality.

$$
\frac{type(\sigma)}{\sigma \sqsubseteq \sigma} \qquad
\frac{\begin{array}{c} \sigma \leq \tau \\ type(\rho) \\ \tau \sqsubseteq \rho \end{array}}{\sigma \sqsubseteq \rho} \qquad
\frac{\begin{array}{c} accepts(\sigma, l) \\ accepts(\tau, l) \\ \sigma \sqsubseteq \tau \end{array}}{\sigma.l \doteq \tau.l} \qquad
\frac{\begin{array}{c} type(\sigma) \\ accepts(\sigma, l) \\ notaccepts(\sigma, l) \end{array}}{fail}
$$

To determine solvability of a problem instance we can now simply build all ground Horn clauses that result from resolving the bodies of the first three rules with the [not]accepts- and the $\leq$-facts in the input, and also with the the type-facts derived from the input database. For the last rule we generate the ground instances by resolving with the type-facts and instantiating with all label terms in the input. If the size of the problem instance is m, the input contains $O(m)$ accepts- and input-facts and terms. Hence we obtain $O(m^2)$ resolvents which are ground clauses in which $O(m)$ terms appear. We now have that the input is solvable if and only if one cannot derive fail from these ground clauses together with the facts in the input.[2] Applying theorem 4 we get a novel and simple proof of Henglein's result that solvability is decidable in $O(m^2)$ time.

7   Proof of the Meta-complexity Theorem

In this section we prove Theorem 1. It turns out to be convenient to prove the theorem for a slightly more general language. We define a *literal-based rule* to be a rule of the form $A_1 \wedge \cdots \wedge A_n \to C$ where each A_i and C are literals, i.e., either atoms or negations of atoms. We write $S \xrightarrow{r} S'$ if there exists a ground substitution σ defined on all the variables in r such that for each antecedent A of r we have that $\sigma(A)$ is visible in S and S' is $S \cup \{\sigma(C)\}$ where C is the conclusion of r. We then define the notion of a rule is applicable to a state, a state being visible to a rule, an R-computation, and an R-saturated state, and a prefix firing as in the case for rules with deletions (in both cases we allow rules to have priorities). We now show that any rule set with priorities and deletions can be translated to a literal-based rule set in a way that allow computations to also be translated in a way that preserves the number of prefix firings up to a constant factor. In particular, we translate a rule with deletions of the form $A_1 \wedge \cdots \wedge A_n \to C$ to the following set of literal-based rules where p is a fresh predicate symbol, $x_1, \ldots, x_n$ are all variables in the rule, and $A_{i_1}, \ldots, A_{i_k}$ are all the deleted antecedents.

$$A_1 \wedge \cdots \wedge A_n \to p(x_1, \ldots, x_n)$$
$$p(x_1, \ldots, x_n) \to \neg A_{i_1}$$
$$\cdots$$
$$p(x_1, \ldots, x_n) \to \neg A_{i_k}$$
$$p(x_1, \ldots, x_n) \to C$$

The first rule above has the same priority as the translated rule. The other rules are called "transient" rules. Note that an "atomic" invocation of one of the original rules gets translated into a sequence of intermediate states where some,

---

[2]   One needs to show that further ground clauses are redundant. Suppose, for instance, we have $accepts(s, l)$ and $accepts(t, l)$ in the input. Then by the process just described we generate the ground clause $s \sqsubseteq t \to s.l \doteq t.l$. If we later derive $s \doteq s'$, the clause $s' \sqsubseteq t \to s'.l \doteq t.l$ is a consequence of the clause we already have.

but not all, of the deletions and insertions have been made. We must ensure that these "transient" states are not visible to other rules in the system. This can be done by assigning all transient rules higher priority than all rules in the set being translated. It now suffices to prove theorem 1 for rules over literals rather than rules with deletions.

We now perform source to source transformations on rules over literals to put the rules in a simplified form without increasing the number of prefix firings by more than a constant factor. First we convert rules such that they have at most three literals. If r is a rule over literals $A_1 \wedge A_2 \wedge \ldots \wedge A_n \rightarrow C$ with $n > 2$ then we replace r by the following set of rules where P_1, P_2, $\ldots$ P_n, are fresh predicate symbols and x_1, $\ldots$, x_{k_i} are the variables occurring in the first i antecedents. The predicate P_i represents the relation defined by the first i antecedents, and $\neg P_i$ represents the negation (retraction) of P_i.

$$A_1 \rightarrow P_1(x_1, \ldots, x_{k_1})$$
$$\neg A_1 \rightarrow \neg P_1(x_1, \ldots, x_{k_1})$$
$$P_1(x_1, \ldots, x_{k_1}) \wedge A_2 \rightarrow P_2(x_1, \ldots, x_{k_2})$$
$$P_2(x_1, \ldots, x_{k_2}) \wedge \neg P_1(x_1, \ldots, x_{k_1}) \rightarrow \neg P_2(x_1, \ldots, x_{k_2})$$
$$P_2(x_1, \ldots, x_{k_2}) \wedge \neg A_2 \rightarrow \neg P_2(x_1, \ldots, x_{k_2})$$
$$\vdots$$
$$P_{n-1}(x_1, \ldots, x_{k_1}) \wedge A_n \rightarrow P_n(x_1, \ldots, x_{k_2})$$
$$P_n(x_1, \ldots, x_{k_2}) \wedge \neg P_{n-1}(x_1, \ldots, x_{k_1}) \rightarrow \neg P_n(x_1, \ldots, x_{k_2})$$
$$P_n(x_1, \ldots, x_{k_2}) \wedge \neg A_n \rightarrow \neg P_n(x_1, \ldots, x_{k_2})$$
$$P_n(x_1, \ldots, x_{k_n}) \rightarrow C$$

Since we are now proving a version of theorem 1 for rules over literals we must consider the case where the rule being translated has negative antecedents. In that case the above rules might include rules with doubly negated antecedents. Such rules are simply dropped from the translation since negative literals can not be deleted (or overruled). The last rule above is given the same priority as the rule being translated. All other rules are given higher priority (but lower than the priority of any original rule with priority higher than the rule that is translated) where the priority is in the order given, i.e., the first rule has highest priority and so on. This is possible since the original rules have all different priorities.

To prove the version of theorem 1 for rules over literals it suffices to show that any computation of the translated rule set can be mapped back to a computation of the original rule set with no more than a constant factor reduction in the number of prefix firings. In particular, if the new rule set derives $P_i(x_1, \ldots, x_{k_i})$ then there must exist a single state in the computation of the original rule set where $A_1, \ldots, A_i$ all hold under the corresponding variable substitution. This follows from the observation that the priority assignment guarantees that in states visible to the rule deriving $P_i(x_1, \ldots, x_{k_i})$ the predicate P_{i-1} is guaranteed to have the appropriate meaning as a function of the predicates used in the original antecedents.

We have now shown that we can assume without loss of generality that each rule contains at most two antecedents. We now put the rules in an even more restricted form. For any rule r with two antecedents $A_1 \wedge A_2 \to C$ we replace r by the following set of rules where $x_1, \ldots, x_n$ are all variables occurring in A_1 but not A_2, $y_1, \ldots, y_m$ are all variables that occur in both A_1 and A_2, and $z_1, \ldots, z_k$ are all variables that occur in A_2 but not A_1. The predicates P, and Q, and the function symbols f, g, and h are all fresh.

$$A_1 \to P(f(x_1, \ldots, x_n), g(y_1, \ldots, y_m))$$
$$\neg A_1 \to \neg P(f(x_1, \ldots, x_n), g(y_1, \ldots, y_m))$$
$$A_2 \to Q(g(y_1, \ldots, y_m), h(z_1, \ldots, z_k))$$
$$\neg A_2 \to \neg Q(g(y_1, \ldots, y_m), h(z_1, \ldots, z_k))$$

$$P(f(x_1, \ldots, x_n), g(y_1, \ldots, y_m)) \wedge Q(g(y_1, \ldots, y_m), h(z_1, \ldots, z_k)) \to C$$

The rules are given priority in the order given with the last rule having the same priority as the rule being translated. Again the validity of the translation relies on the observation that in any state visible to a rule using one of the newly introduced predicates P and Q in antecedents, these predicates must have the intended meaning as a function of the underlying original predicates and hence there is a corresponding firing of the original rule.

Without loss of generality we now need only prove the theorem for prioritized inference rules over literals where each rule either has only a single antecedent or is of the form $P(t_1, t_2) \wedge Q(t_2, t_3) \to C$ where t_1, t_2, and t_3 do not share variables and where the rules maintain the invariant that for all derivable ground assertions of the form $P(s_1, s_2)$ we have that s_1 is a substitution instance of t_1 and s_2 is a substitution instance of s_2, and for all derivable ground assertions of the form $Q(s_2, s_3)$ we have that s_2 is a substitution instance of t_2 and s_3 is a substitution instance of t_3. For such rule sets we can use the algorithm shown below to compute an R-saturation of a given initial database D.

Algorithm to Compute $R(D)$:

Assume that D is in fully shared dag representation in which term equality can be checked in constant time. We maintain queues Q_p and R_p for each priority p in R. Initialize S to be D and place every element of D on every queue Q_p. Initialize all queues R_p to be empty.

While some queue is nonempty do the following:

> Let p be the highest priority such that either Q_p or R_p is nonempty. (The current state is visible to rules of priority p.) If Q_p is nonempty then remove a literal Φ from Q_p and if Φ is visible in S then notice Φ at priority p using the procedure given below. If Q_p is empty then remove a pair $\langle r, \sigma \rangle$ from R_p. (Here r is a rule of priority p and σ is a substitution assigning ground values to all variables of r such that for each antecedent of A of r we have $\sigma(A) \in S$.) If $\sigma(A)$ is visible in S for each antecedent A of R then let Ψ be the assertion $\sigma(C)$ where C is the conclusion of R, add Ψ to S, and place Ψ on all queues of the form $Q_{p'}$.

Algorithm to Notice Φ at priority p:

(The current state is visible to all rules of priority p and Φ is visible in S.)

1. For each single-antecedent rule of priority p of the form $A \rightarrow C$ determine whether there is a substitution σ such that $\sigma(A) = \Phi$ and, if so, add the pair $\langle r, \sigma \rangle$ to R_p.
2. For each two-antecedent rule of priority p of the form $P(t_1, t_2) \wedge Q(t_2, t_3) \rightarrow C$ do the following:
   (a) If Φ has the form $P(s_1, s_2)$ then for each s_3 such that $Q(s_2, s_3)$ is visible in S add the pair $\langle r, \sigma \rangle$ to R_p where σ is the substitution mapping t_1 to s_1, t_2 to s_2, and t_3 to s_3. (We are guaranteed that t_1, t_2, and t_3 do not share variables and that t_1 matches s_1, t_2 matches s_2, and t_3 matches s_3.)
   (b) If Φ has the form $Q(s_2, s_3)$ then for each s_1 such that $P(s_1, s_2)$ is visible in S add the pair $\langle r, \sigma \rangle$ to R_p where σ is defined as in (a). (Analogous guarantees also exist in this case.)

We leave it to the reader to verify the correctness and running time of this algorithm. The main feature of the algorithm is that the processing of a given rule r is restricted to states visible to r. By incrementally maintaining appropriate indices it is possible to run steps (2a) and (2b) in time proportional to the number of values of s_3 and s_1 defined in those steps respectively. For instance the set of substitutions s_3 such that $Q(s_2, s_3)$ is visible in S (cf. step 2a) has to be indexed using term s_2 as key, so that upon adding a new Q-atom to S the index can be updated in constant time. Note that since we have assumed dag representations for expressions under which equality testing is a unit time operation, all matching operations for patterns in R take unit time.

8   Future Work

We have presented a more refined concept of logic programming where the emphasis is on guaranteed execution time bounds linear in the number of prefix firings of its rules. We are optimistic that this logic will continue to prove itself useful in the design of algorithms. We have demonstrated some of the potential of the method by giving a novel algorithm for testing satisfiability of ground Horn clauses with equality and shown that in many cases its complexity is not worse that (unconditional) congruence closure. On top of this we have implemented an abstract version of Henglein's type analysis, confirming the quadratic upper bound that Henglein obtained before. In fact we believe that program analysis is a particularly fruitful area for applying our method. This point was illustrated in detail in [10]. With the methods in the present paper we are able to also deal with congruences that appear in such analyses in a logical way.

There are many directions into which this work should be extended. Theorem 4 should be generalized to cases of input clauses with variables. However, already in the given form it is useful for local (equational) theories in the sense of [7,9,6].

The relation between deletions and redundancy elimination, with redundancy in the sense of [2] as entailment from smaller atoms, should be explored. For instance, the rule (I9) deletes $\text{push}(s, t)$ if s can be reduced to s' after the reduced atom $\text{push}(s', t)$ has been generated. The deleted atom is "entailed" by the "smaller" assertions $\text{push}(s', t)$ and $s \xrightarrow{I}! s'$. The elimination of such redundancies is stable under enrichments of a state or deletions of other redundant atoms. Therefore, if only redundant premises are deleted, priorities are irrelevant for the correctness of the algorithm and only affect its complexity.

The concept of priorities for rules should be refined in an instance-based manner, allowing different instances of a rule to have different priorities. That would give one direct means of formalizing algorithms that would normally have to be defined via priority queues. For instance, minimal spanning trees can be computed by a two-rule program on top of union-find if rules referring to edges in graphs could be processed in an order related to their associated costs.

References

1. M. Abadi and L. Cardelli. *A Theory of Objects*. Springer, New York, Berlin, Heidelberg, 1996.
2. L. Bachmair and H. Ganzinger. Rewrite-based equational theorem proving with selection and simplification. *J. Logic and Computation*, 4(3):217–247, 1994. Revised version of Research Report MPI-I-91-208, 1991.
3. Leo Bachmair and Ashish Tiwari. Abstract congruence closure and specializations. In David McAllester, editor, *Automated Deduction – CADE-17, 17th International Conference on Automated Deduction*, LNAI 1831, pages 64–78, Pittsburgh, PA, USA, June 17–20, 2000. Springer-Verlag.
4. William F. Dowling and Jean H. Gallier. Linear-time algorithms for testing the satisfiability of propositional Horn formulae. *J. Logic Programming*, 3:267–284, 1984.
5. P. J. Downey, R. Sethi, and R. E. Tarjan. Variations on the common subexpressions problem. *J. Association for Computing Machinery*, 27(4):771–785, 1980.
6. H. Ganzinger. Relating semantic and proof-theoretic concepts for polynomial time decidability of uniform word problems. In *Proc. 16th IEEE Symposium on Logic in Computer Science*, IEEE Computer Society Press, 2001. To appear.
7. R. Givan and D. McAllester. New results on local inference relations. In *Principles of Knowledge Representation and reasoning: Proceedings of the Third International Conference (KR'92)*, pages 403–412. Morgan Kaufmann Press, 1992.
8. F. Henglein. Breaking through the n^3 barrier: Faster object type inference. *Theory and Practice of Object Systems*, 5(1):57–72, 1999. A preliminary version appeared in FOOL4.
9. D. McAllester. Automatic recognition of tractability in inference relations. *J. Association for Computing Machinery*, 40(2):284–303, 1993.
10. David McAllester. On the complexity analysis of static analyses. In A. Cortesi and R. Filé, editors, *Static Analysis — 6th International Symposium, SAS'99*, LNCS 1694, pages 312–329, Venice, Italy, September 1999. Springer-Verlag.

Canonical Propositional Gentzen-Type Systems

Arnon Avron and Iddo Lev

School of Computer Science
Tel-Aviv University
Ramat Aviv 69978, Israel
{aa,iddolev}@post.tau.ac.il

Abstract. Canonical propositional Gentzen-type systems are systems which in addition to the standard axioms and structural rules have only pure logical rules which have the subformula property, introduce exactly one occurrence of a connective in their conclusion, and no other occurrence of any connective is mentioned anywhere else in their formulation. We provide a constructive coherence criterion for the non-triviality of such systems, and show that a system of this kind admits cut elimination iff it is coherent. We show also that the semantics of such systems is provided by non-deterministic two-valued matrices (2-Nmatrices). 2-Nmatrices form a natural generalization of the classical two-valued matrix, and every coherent canonical system is sound and complete for one of them. Conversely, with any 2-Nmatrix it is possible to associate a coherent canonical Gentzen-type system which has for each connective at most one introduction rule for each side, and is sound and complete for that 2-Nmatrix. We show also that every coherent canonical Gentzen-type system either defines a fragment of the classical two-valued logic, or a logic which has no finite characteristic matrix.

1 Introduction

There is a long tradition starting from [Gen69] according to which the meaning of a connective is determined by the introduction and elimination rules which are associated with it.[1] The supporters of this thesis usually have in mind Natural Deduction systems of an ideal type. In this type of "canonical" systems each connective has its own introduction and elimination rules, which should meet the following conditions: in a rule for a connective $\diamond$, this connective should be mentioned exactly once, and no other connective should be involved. The rules should also be pure (in the sense of [Avr91]). Unfortunately, already the handling of classical negation requires rules which are not canonical in this sense. This problem was solved by Gentzen himself by moving to what is now known as Gentzen-type systems or sequential calculi. These calculi employ in their classical version multiple-conclusion two-sided sequents, and instead of introduction and elimination rules they use left introduction rules and right introduction rules. The intuitive notions of "canonical form of a rule" and "canonical system" can

---

[1] See e.g. [Hod86] and [Sun86] for discussions and references.

R. Goré, A. Leitsch, and T. Nipkow (Eds.): IJCAR 2001, LNAI 2083, pp. 529–544, 2001.

be adapted to such systems in a straightforward way, and it is well known that the usual classical connectives can indeed be fully characterized by canonical Gentzen-type rules. Moreover: although this can be done in several ways, in all of them the cut-elimination theorem obtains.

In this paper we shall considerably generalize these known facts. We shall define "canonical" Gentzen-type rules and systems in precise terms, and provide a constructive *coherence* criterion for their non-triviality. We then show that a canonical system admits cut-elimination iff it is coherent. Moreover: we show that any coherent set of canonical introduction rules for a connective ◇ completely determines the meaning of ◇. For this we shall need however to generalize the usual semantics of classical logic.

The structure of the rest of this paper is as follows: In section 2 we review some basic concepts related to logics. In section **??** we define canonical Gentzen-type systems, formulate the coherence criterion for their non-triviality, and investigate some special important types of them. In section 4 we introduce non-deterministic two-valued matrices (2-Nmatrices). These are a generalization of the classical two-valued (deterministic) matrices, and they provide the semantics of coherent canonical Gentzen-type systems. In the same section and in section 5 we show how to associate a 2-Nmatrix with every coherent canonical system **G**, so that **G** is sound and complete for that 2-Nmatrix. This allows us to prove that every system of this type admits cut elimination, and it defines either a fragment of the classical two-valued logic, or a logic which has no finite characteristic matrix. In section 6 we show that the connection works also in the other direction: with any 2-Nmatrix it is possible to associate a coherent canonical Gentzen-type system **G** which is sound and complete for that 2-Nmatrix. Moreover: for this we can confine ourselves to systems which have for any connective at most one left introduction rule and at most one right introduction rule, and these rules can be given a particularly concise normal form. We conclude the paper with some remarks and directions for further research.

2   Preliminaries

In what follows $\mathcal{L}$ is a propositional language with a finite set of connectives, $\mathcal{W}$ is its set of wffs, ψ, ϕ, τ denote arbitrary formulas (of $\mathcal{L}$), and Γ, Δ denote sets of formulas. We assume also that the atomic formulas of $\mathcal{L}$ are $p_1, p_2, \ldots$

Definition 1.

1. [Sco74] *A (Scott) consequence relation (scr for short) for $\mathcal{L}$ is a binary relation $\vdash$ between sets of formulas of $\mathcal{L}$ that satisfies the following conditions:*

   s-R   strong reflexivity: *if $\Gamma \cap \Delta \neq \emptyset$ then $\Gamma \vdash \Delta$.*
   M   monotonicity: *if $\Gamma \vdash \Delta$ and $\Gamma \subseteq \Gamma'$, $\Delta \subseteq \Delta'$ then $\Gamma' \vdash \Delta'$.*
   C   cut: *if $\Gamma \vdash \psi, \Delta$ and $\Gamma', \psi \vdash \Delta'$ then $\Gamma, \Gamma' \vdash \Delta, \Delta'$.*

2. *$\vdash$ is finitary if the following condition holds for all $\Gamma, \Delta \subseteq \mathcal{W}$: if $\Gamma \vdash \Delta$ then $\Gamma' \vdash \Delta'$ for some finite $\Gamma' \subseteq \Gamma$ and $\Delta' \subseteq \Delta$. $\vdash$ is uniform if for every*

uniform substitution σ and every Γ and Δ, if $\Gamma \vdash \Delta$ then $\sigma(\Gamma) \vdash \sigma(\Delta)$. $\vdash$ is consistent *(or non-trivial) if there exist non-empty Γ and Δ s.t. $\Gamma \not\vdash \Delta$.*

3. *A* propositional *logic is a pair $\langle \mathcal{L}, \vdash \rangle$, where $\mathcal{L}$ is a propositional language and $\vdash$ is a uniform consistent scr for $\mathcal{L}$.*

Note: There are exactly four inconsistent finitary scrs in any given language: the one in which $\Gamma \vdash \Delta$ iff Γ and Δ are non-empty; the one in which $\Gamma \vdash \Delta$ iff Γ is non-empty; the one in which $\Gamma \vdash \Delta$ iff Δ is non-empty; and the one in which $\Gamma \vdash \Delta$ for all Γ and Δ. All of them should be considered trivial, and are excluded from our definition of a *logic*.

3 Canonical Gentzen-Type Systems

Definition 2.

1. *A Gentzen-type system **G** is* standard *if its set of axioms includes the standard axioms $\Gamma, \psi \Rightarrow \psi, \Delta$ and it has all the standard structural rules (including cut).* [2]
2. *Let **G** be a standard Gentzen-type system. The scr $\vdash_{\mathbf{G}}$ which is induced by **G** is defined by: $\Gamma \vdash_{\mathbf{G}} \Delta$ iff the sequent $\Gamma \Rightarrow \Delta$ is provable in **G**.*
3. *A standard Gentzen-type system **G** is* consistent *if $\vdash_{\mathbf{G}}$ is consistent.*

From now on by a "calculus" we shall mean a standard Gentzen-type calculus, and Γ and Δ will denote *finite* sets of formulas.

In an ideal Gentzen-type system (of which the usual systems for classical logic provide the principal examples) every logical rule should be an introduction rule for one connective, it should introduce exactly one occurrence of that connective in its conclusion, and no other occurrence of that connective or any other connective should be mentioned anywhere else in its formulation. Moreover: the rule should be *pure* (i.e., there should be no side conditions limiting its application), and its side formulas should be immediate subformulas of the principal formula. The next definition formulates this idea in exact terms, and provides a method for describing such rules.

Definition 3.

1. *A* canonical rule *of arity n is an expression of the form $\{\Pi_i \Rightarrow \Sigma_i\}_{1 \leq i \leq m}/C$, where $m \geq 0$, C is either $\diamond(p_1, p_2, \ldots, p_n) \Rightarrow$ or $\Rightarrow \diamond(p_1, p_2, \ldots, p_n)$ for some connective $\diamond$ (of arity n), and for all $1 \leq i \leq m$, $\Pi_i \Rightarrow \Sigma_i$ is a clause such that $\Pi_i, \Sigma_i \subseteq \{p_1, p_2, \ldots, p_n\}$.* [3]

---

[2] This means that we can take Γ, Δ in a sequent $\Gamma \Rightarrow \Delta$ to be finite *sets* of formulas.

[3] By a clause we mean a sequent which consists of atomic formulas only. When propositional clauses are written in this way, resolution and cut amount to the same thing. $\{p_1, p_2, \ldots, p_n\}$ are, recall, the first n atomic formulas.

2. *An application of a canonical rule* $\{\Pi_i \Rightarrow \Sigma_i\}_{1\leq i\leq m} \;/\; \diamond(p_1,\ldots,p_n) \Rightarrow$ *is any inference step of the form:*

$$\frac{\{\Gamma_i, \Pi_i^* \Rightarrow \Delta_i, \Sigma_i^*\}_{1\leq i\leq m}}{\Gamma, \diamond(\psi_1,\ldots,\psi_n) \Rightarrow \Delta}$$

where Π_i^* *and* Σ_i^* *are obtained from* Π_i *and* Σ_i *(respectively) by substituting* ψ_j *for* p_j *(for all* $1 \leq j \leq n$*),* Γ_i, Δ_i *are any sets of formulas,* $\Gamma = \bigcup_{i=1}^m \Gamma_i$*, and* $\Delta = \bigcup_{i=1}^m \Delta_i$*. An application of a canonical rule with a conclusion of the form* $\Rightarrow \diamond(p_1,\ldots,p_n)$ *is defined similarly.*

Note: While sequents are written in a metalanguage for $\mathcal{L}$ (which includes the extra symbol $\Rightarrow$), a canonical rule is formulated in a meta-meta language of $\mathcal{L}$ (which includes one further extra symbol: $/$).

Example 1. The two usual introduction rules for the classical conjunction can be formulated as the following canonical rules: $\{p_1, p_2 \Rightarrow \;\} \;/\; p_1 \wedge p_2 \Rightarrow$ and $\{\Rightarrow p_1 \;,\; \Rightarrow p_2\} \;/\; \Rightarrow p_1 \wedge p_2$. Applications of these rules have the form:

$$\frac{\Gamma, \psi, \phi \Rightarrow \Delta}{\Gamma, \psi \wedge \phi \Rightarrow \Delta} \qquad \frac{\Gamma \Rightarrow \Delta, \psi \quad \Gamma' \Rightarrow \Delta', \phi}{\Gamma, \Gamma' \Rightarrow \Delta, \Delta', \psi \wedge \phi}$$

Definition 4. *A standard calculus is called* canonical *if in addition to the standard axioms and the standard structural rules it has only canonical logical rules.*

A given canonical calculus may be simplified in various ways. In later sections we will go deeper into questions concerning simplifications and normalizations of rules and calculi. For our immediate purposes we shall need only very obvious simplifications:

Definition 5. *A canonical rule is called* superfluous *if its set of premises is classically inconsistent (which is the case iff it is possible to obtain the empty clause from it using resolutions (= cuts)). A logical rule in a canonical calculus* **G** *is called* redundant *in* **G** *if its set of premises is a superset of the set of premises of another rule of* **G** *which has the same conclusion.*

Example 2. A rule with the set of premises $\{p_1, p_2 \Rightarrow \;,\; p_1 \Rightarrow p_2 \;,\; \Rightarrow p_1\}$ is superfluous. If a calculus **G** has the two rules $\{\Rightarrow p_1\}/\diamond(p_1, p_2) \Rightarrow$ and $\{\Rightarrow p_1 \;,\; \Rightarrow p_2\}/\diamond(p_1, p_2) \Rightarrow$ then the latter rule is redundant in **G**.

Proposition 1. *Let* **G** *be a canonical calculus, and let* **G**′ *be the calculus that is obtained from* **G** *by deleting superfluous and redundant rules. Then* **G**′ *is equivalent to* **G***. Moreover, every sequent that has a cut-free proof in* **G**′ *also has such a proof in* **G***.*

Proof. An application of a superfluous rule in **G** can be simulated in **G**′ by using cuts on its premises followed by a single weakening. The rest of the proposition is trivial.

For every canonical Gentzen-type system $\mathbf{G}$, the relation $\vdash_{\mathbf{G}}$ (see Definition 2) defined by $\mathbf{G}$ is obviously a uniform (and finitary) scr. However, in order to ensure that $\langle \mathcal{L}, \vdash_{\mathbf{G}} \rangle$ is a *logic*, we need to impose some constraints on the set of rules of $\mathbf{G}$. The following definition provides a constructive equivalent of the consistency condition:

Definition 6. *A canonical calculus* $\mathbf{G}$ *is called* coherent, *if for every two rules* $S_1 / \diamond (p_1, p_2, \ldots, p_n) \Rightarrow$ *and* $S_2 / \Rightarrow \diamond(p_1, p_2, \ldots, p_n)$ *of* $\mathbf{G}$, *the set of clauses* $S_1 \cup S_2$ *is classically inconsistent (and so the empty clause can be derived from it using cuts).*

Example 3. The two classical rules for conjunction described in Example 1 form a coherent set of rules. Here $S_1 = \{p_1, p_2 \Rightarrow \}$, $S_2 = \{ \Rightarrow p_1 , \Rightarrow p_2\}$ and so $S_1 \cup S_2$ is the classically inconsistent set $\{p_1, p_2 \Rightarrow , \Rightarrow p_1 , \Rightarrow p_2\}$.

Example 4. Let T be the famous "Tonk" connective of Prior ([Pri60]). It is defined in our framework by the following pair of rules: $\{p_1 \Rightarrow \} / p_1 T p_2 \Rightarrow$ and $\{ \Rightarrow p_2\} / \Rightarrow p_1 T p_2$. This pair is not coherent, since $\{p_1 \Rightarrow , \Rightarrow p_2\}$ is a classically consistent set of clauses. The resulting calculus is of course inconsistent.

Proposition 2. *Every consistent canonical calculus is coherent.*

Proof. Suppose there are two rules $S_1 / \diamond(p_1, \ldots, p_n) \Rightarrow$ and $S_2 / \Rightarrow \diamond(p_1, \ldots, p_n)$ such that $S_1 \cup S_2$ is classically consistent. Then there is a classical valuation v that satisfies $S_1 \cup S_2$. Let $\Pi' = \{p_i \mid 1 \leq i \leq n, \ v(p_i) = t\}$ and $\Sigma' = \{p_i \mid 1 \leq i \leq n, \ v(p_i) = f\}$. Let $S'_j = \{\Pi, \Pi' \Rightarrow \Sigma, \Sigma' \mid \Pi \Rightarrow \Sigma \in S_j\}$ for $j = 1, 2$. S'_1 and S'_2 are sets of standard axioms (because v satisfies $\Pi \Rightarrow \Sigma$, there is some $p_i \in \Sigma$ such that $v(p_i) = t$ or some $p_i \in \Pi$ such that $v(p_i) = f$. In the former case, $p_i \in \Pi'$, and in the latter case, $p_i \in \Sigma'$). By applying the first rule on S'_1 we obtain $\Pi', \diamond(p_1, \ldots, p_n) \Rightarrow \Sigma'$ and by applying the second rule on S'_2 we obtain $\Pi' \Rightarrow \Sigma', \diamond(p_1, \ldots, p_n)$. By cut, $\Pi' \Rightarrow \Sigma'$ is provable. Since $\Pi' \Rightarrow \Sigma'$ is a clause, $\Pi' \cap \Sigma' = \emptyset$, and the calculus is uniform, $p \Rightarrow q$ is provable for all $p \neq q$. The uniformity of the calculus and the closure under weakening entail that $\Gamma \Rightarrow \Delta$ is provable for every non-empty Γ and Δ. Hence the system is not consistent.

The converse of this proposition will be shown in Corollary 2 below. Note that coherence is also a necessary condition for cut elimination:

Proposition 3. *A canonical calculus which admits cut elimination is coherent.*

Proof. In canonical systems, *clauses* which are not axioms can be proved only by using cuts on non-atomic formulas. Thus, if a canonical calculus admits cut elimination it must be consistent and hence coherent.

Definition 7.

1. *A canonical rule of arity n is called* separated *if each of its premises is a unit clause, i.e. it has the form $p_i \Rightarrow$ or $\Rightarrow p_i$ for some $1 \leq i \leq n$.*
2. *A separated rule $\{\Pi_i \Rightarrow \Sigma_i\}_{1 \leq i \leq m}/C$ of arity n is called* full *if $m = n$ and for every $1 \leq i \leq n$, $\Pi_i \cup \Sigma_i = \{p_i\}$*

Example 5. $\{p_1, p_2 \Rightarrow\} / p_1 \wedge p_2 \Rightarrow$ is not separated. $\{p_1 \Rightarrow\} / p_1 \wedge p_2 \Rightarrow$ is separated, but not full. $\{p_1 \Rightarrow, \Rightarrow p_2\} / p_1 \wedge p_2 \Rightarrow$ is full.

Definition 8. *A canonical calculus is called* separated *(full) if all its logical rules are separated (full), and none of them is superfluous or redundant.*

Note that a full canonical calculus **G** is coherent simply iff no two rules of **G** for the same connective have the same set of premises but different conclusions.

As a first step in proving our general cut-elimination theorem we shall show that it suffices to prove it for full canonical calculi:

Proposition 4. *Every canonical calculus* **G** *has an equivalent full canonical calculus* **G'**. *Moreover: if* **G** *is coherent then so is* **G'**, *and every sequent that has a cut-free proof in* **G'** *has also a cut-free proof in* **G**.

Proof. We shall first explain the process of transforming a canonical calculus to an equivalent full canonical calculus by using an example. The transition is made in two stages: first, every canonical rule is split into separated rules, and then each of these rules is split into full rules.

Example 6. Take the classical introduction rules for conjunction:

$$[\wedge \Rightarrow] \quad \{p_1, p_2 \Rightarrow\} / p_1 \wedge p_2 \Rightarrow \qquad \{\Rightarrow p_1, \Rightarrow p_2\} / \Rightarrow p_1 \wedge p_2 \quad [\Rightarrow \wedge]$$

The second rule is already full. The first is neither full nor separated. We can replace it by the following pair of separated rules:

$$\{p_1 \Rightarrow\} / p_1 \wedge p_2 \Rightarrow \qquad \{p_2 \Rightarrow\} / p_1 \wedge p_2 \Rightarrow$$

This pair is equivalent to the original rule. Indeed, given the two rules, by applying the first to $\Gamma, \psi_1, \psi_2 \Rightarrow \Delta$ we obtain $\Gamma, \psi_1 \wedge \psi_2, \psi_2 \Rightarrow \Delta$. By applying the second to this sequent we obtain $\Gamma, \psi_1 \wedge \psi_2 \Rightarrow \Delta$. The other direction of the equivalence is obvious in view of weakening. Moreover: a given cut-free proof that uses the new rules can trivially be transformed into a cut-free proof that uses the original rule.

Next, the first of the two new rules can again be replaced by the following pair of full rules:

$$\{p_1 \Rightarrow, p_2 \Rightarrow\} / p_1 \wedge p_2 \Rightarrow \qquad \{p_1 \Rightarrow, \Rightarrow p_2\} / p_1 \wedge p_2 \Rightarrow$$

This pair is equivalent to the original (separated) rule. Indeed, using the two new rules, the original one can be simulated as follows. Given $\Gamma, \psi_1 \Rightarrow \Delta$, an application of the first of the two new rules to this sequent and to the standard axiom $\psi_2 \Rightarrow \psi_2$ yields $\Gamma, \psi_1 \wedge \psi_2 \Rightarrow \Delta, \psi_2$. By applying the second of the two new rules to this sequent and to $\Gamma, \psi_1 \Rightarrow \Delta$ we obtain $\Gamma, \psi_1 \wedge \psi_2 \Rightarrow \Delta$. The other direction of the equivalence is obvious, and again does not use the cut rule.

The second separated rule above is similarly split into the following full rules:

$$\{p_2 \Rightarrow , \ p_1 \Rightarrow \} \,/\, p_1 \wedge p_2 \Rightarrow \qquad\qquad \{p_2 \Rightarrow , \ \Rightarrow p_1\} \,/\, p_1 \wedge p_2 \Rightarrow$$

To conclude, from the first original classical two rules for conjunction, we obtain the following equivalent set of four full rules:

$$\{p_1 \Rightarrow , \ p_2 \Rightarrow \} \,/\, p_1 \wedge p_2 \Rightarrow \qquad\qquad \{p_1 \Rightarrow , \ \Rightarrow p_2\} \,/\, p_1 \wedge p_2 \Rightarrow$$

$$\{ \Rightarrow p_1, \ p_2 \Rightarrow \} \,/\, p_1 \wedge p_2 \Rightarrow \qquad\qquad \{ \Rightarrow p_1 , \ \Rightarrow p_2\} \,/\, \Rightarrow p_1 \wedge p_2$$

The general procedure for replacing a non-separated rule R by an equivalent set of separated ones is to put first in this set every separated rule which can be obtained by selecting exactly one formula from each premise of R (preserving its side). We then use Proposition 1 to remove any superfluous or redundant rule. The general procedure for splitting a given separated rule R of arity n into an equivalent set of full rules is to put in this set every full rule which has the same conclusion as R, and whose set of premises is an extension of that of R (thus if R has $m < n$ premises, it will be split into 2^{n-m} full rules). Using the methods employed in the last example, it is easy to see that these transformations preserve coherence of a system as well as its set of provable sequents, and that any cut-free proof in the resulting system can be simulated by a cut free proof in the original one (it is not difficult to directly show that the converse is also true, but this will be proved later using a semantic argument).

Example 7. Suppose we have a canonical rule for a ternary connective:

$$\{p_1, p_2 \Rightarrow , \ p_1 \Rightarrow p_2 , \ p_3 \Rightarrow p_2\} \,/\, \diamond (p_1, p_2, p_3) \Rightarrow$$

The first stage of the above process produces the following rules:

$$
\begin{aligned}
(1) \qquad & \{p_1 \Rightarrow , \ p_3 \Rightarrow \} \,/\, \diamond (p_1, p_2, p_3) \Rightarrow \\
(2),(4) \qquad & \{p_1 \Rightarrow , \ \Rightarrow p_2\} \,/\, \diamond (p_1, p_2, p_3) \Rightarrow \\
(3) \ & \{p_1 \Rightarrow , \ \Rightarrow p_2 , \ p_3 \Rightarrow \} \,/\, \diamond (p_1, p_2, p_3) \Rightarrow \\
(5) \ & \{p_2 \Rightarrow , \ p_1 \Rightarrow , \ p_3 \Rightarrow \} \,/\, \diamond (p_1, p_2, p_3) \Rightarrow \\
(6) \ & \{p_2 \Rightarrow , \ p_1 \Rightarrow , \ \Rightarrow p_2\} \,/\, \diamond (p_1, p_2, p_3) \Rightarrow \\
(7) \ & \{p_2 \Rightarrow , \ \Rightarrow p_2 , \ p_3 \Rightarrow \} \,/\, \diamond (p_1, p_2, p_3) \Rightarrow \\
(8) \qquad & \{p_2 \Rightarrow , \ \Rightarrow p_2\} \,/\, \diamond (p_1, p_2, p_3) \Rightarrow
\end{aligned}
$$

(6),(7),(8) are superfluous and we discard them. (3),(5) are redundant because of rules (1),(2), and we discard them as well. The next stage is to extend (1),(2) into full rules:

$$\{p_1 \Rightarrow , \ p_2 \Rightarrow , \ p_3 \Rightarrow \} \ / \ \diamond (p_1, p_2, p_3) \Rightarrow$$
$$\{p_1 \Rightarrow , \ \Rightarrow p_2 , \ p_3 \Rightarrow \} \ / \ \diamond (p_1, p_2, p_3) \Rightarrow$$
$$\{p_1 \Rightarrow , \ \Rightarrow p_2 , \ \Rightarrow p_3 \} \ / \ \diamond (p_1, p_2, p_3) \Rightarrow$$

Notation 1 Let **G** be a canonical calculus. $\mathbf{G}^F$ will denote the equivalent full calculus obtained by the process described in the proof of Proposition 4.

We shall later show that if **G** is coherent then $\mathbf{G}^F$ is the unique full canonical calculus that is equivalent to **G**.

Corollary 1. *If cut elimination obtains for every coherent full canonical calculus then it obtains for every coherent canonical calculus.*

Proof. This follows from Proposition 4.

It remains therefore to show that cut elimination obtains for every coherent full canonical calculus. For that, we shall need to use some semantic arguments. The corresponding semantics will be described next.

4   Semantics: Two-Valued Non-deterministic Matrices

For the semantics of coherent canonical calculi we need some structures which generalize the ordinary concept of a multi-valued matrix. The idea behind this generalization is to allow non-deterministic computations of truth-values. Thus the value that a valuation assigns in these structures to a complex formula is not always uniquely determined by the values that it assigns to its subformulas, but can be chosen non-deterministically from a certain non-empty set of options. The precise definition is as follows:

Definition 9. [AL00] *A non-deterministic matrix (Nmatrix for short) for $\mathcal{L}$ is a tuple $\mathcal{M} = \langle \mathcal{T}, \mathcal{D}, \mathcal{O} \rangle$, where $\mathcal{T}$ is a non-empty set of truth values, $\mathcal{D}$ is a non-empty proper subset of $\mathcal{T}$ (its designated values), and for every n-ary connective $\diamond$ of $\mathcal{L}$, $\mathcal{O}$ includes a corresponding n-ary function $\widetilde{\diamond}$ from $\mathcal{T}^n$ to $2^{\mathcal{T}} - \{\emptyset\}$. A valuation in $\mathcal{M}$ is a function $v : \mathcal{W} \to \mathcal{T}$ that satisfies the condition: if $\diamond$ is an n-ary connective, and $\psi_1, \ldots, \psi_n \in \mathcal{W}$, then $v(\diamond(\psi_1, \ldots, \psi_n)) \in \widetilde{\diamond}(v(\psi_1), \ldots, v(\psi_n))$. v satisfies a formula ψ in $\mathcal{M}$ ($v \models^{\mathcal{M}} \psi$) if $v(\psi) \in \mathcal{D}$. v is a model of Γ in $\mathcal{M}$ ($v \models^{\mathcal{M}} \Gamma$) if it satisfies every formula in Γ. Δ follows from Γ in $\mathcal{M}$ ($\Gamma \vdash_{\mathcal{M}} \Delta$) if for every model v of Γ in $\mathcal{M}$, $v \models^{\mathcal{M}} \phi$ for some $\phi \in \Delta$. If $\vdash$ is an scr then $\mathcal{M}$ is called a* characteristic Nmatrix *for $\vdash$ if $\vdash = \vdash_{\mathcal{M}}$.*

Notes:

1. Every (deterministic) matrix[4] can be identified with an Nmatrix whose functions in $\mathcal{O}$ always return singletons.

---
[4] See e.g. [Urq86].

2. It is easy to verify that if $\mathcal{M}$ is an Nmatrix for $\mathcal{L}$ then $\langle \mathcal{L}, \vdash_{\mathcal{M}} \rangle$ is a logic. In [AL00] it is proved that if $\mathcal{M}$ is finite then this logic is necessarily finitary (i.e., the compactness theorem obtains for it).

In this paper we shall use a special type of Nmatrices: those with exactly two truth values (which may be identified with the classical truth values). We shall show in fact that there is a strong connection between such Nmatrices and coherent canonical Gentzen-type calculi. [5]

Notation 2 A two-valued Nmatrix in which $\mathcal{T} = \{t, f\}$ and $\mathcal{D} = \{t\}$ will be called a *2-Nmatrix*.

Notation 3 For $x \in \{t, f\}$, denote: $-x = f$ if $x = t$, and $-x = t$ if $x = f$.

Notation 4 The expression Φ, a^x will denote $\Phi \cup \{a\}$ if $x = t$ and Φ if $x = f$ (note that Φ might be empty here).

Notation 5 The full canonical rule $\{p_i^{-x_i} \Rightarrow p_i^{x_i}\}_{1 \leq i \leq n} / \diamond(p_1, \ldots, p_n) \Rightarrow$ where $x_1, \ldots, x_n \in \{t, f\}$ will be denoted below by either $[\diamond(x_1, \ldots, x_n) : f]$ or by $[\diamond(x_1, \ldots, x_n) \Rightarrow]$. The rule with the same premises but with the complementary conclusion will be denoted by $[\diamond(x_1, \ldots, x_n) : t]$ or by $[\Rightarrow \diamond(x_1, \ldots, x_n)]$.

Definition 10. *Let* **G** *be a coherent canonical calculus. The 2-Nmatrix that is defined by* **G** *is the following: For each n-ary connective $\diamond$ and every $x_1, \ldots, x_n \in \{t, f\}$ we define*

$$\tilde{\diamond}(x_1, \ldots, x_n) = \begin{cases} \{y\} & \text{if } [\diamond(x_1, \ldots, x_n) : y] \text{ is a rule of } \mathbf{G}^F \text{ (for } y \in \{t, f\}) \\ \{t, f\} & \text{otherwise} \end{cases}$$

(This is well-defined since we assume that **G** (and hence $\mathbf{G}^F$) is coherent.)

Example 8. Suppose **G** has only one rule for the ternary connective $\diamond$: the one given at the beginning of Example 7. Then the three final rules obtained in that example determine the following interpretation of $\diamond$: $\tilde{\diamond}(f, f, f) = \tilde{\diamond}(f, t, f) = \tilde{\diamond}(f, t, t) = \{f\}$ (and $\tilde{\diamond}(x_1, x_2, x_3) = \{t, f\}$ for all other x_1, x_2, x_3).

Proposition 5. *Every coherent canonical calculus* **G** *is sound for the 2-Nmatrix that it defines.*

---

[5] Valuations in two-valued Nmatrices form a special type of what are called *bivaluations* in [Béz99]. Another related idea is Meyer's *metavaluations* (see e.g. [Dun86]).

Proof. By Proposition 4 and Definition 10, we may assume w.l.o.g. that $\mathbf{G}$ is full. Let $\mathcal{M}$ be the 2-Nmatrix that is defined by $\mathbf{G}$. We say that a valuation v in $\mathcal{M}$ *satisfies* a sequent $\Gamma \Rightarrow \Delta$ if $v(\psi) = t$ for some $\psi \in \Delta$, or $v(\psi) = f$ for some $\psi \in \Gamma$. It is easy to verify that:

(1) A valuation v satisfies a sequent $\Gamma, \psi^{-x} \Rightarrow \Delta, \psi^x$ iff either v satisfies $\Gamma \Rightarrow \Delta$, or $v(\psi) = x$.

(2) $\Gamma \vdash_{\mathcal{M}} \Delta$ iff every valuation v in $\mathcal{M}$ satisfies $\Gamma \Rightarrow \Delta$.

Consider now an application of the rule $[\diamond(x_1, \ldots, x_n) : y]$, and assume that v is a valuation which satisfies all the premises $\{\Gamma_i, \psi_i^{-x_i} \Rightarrow \Delta_i, \psi_i^{x_i}\}_{1 \leq i \leq n}$ of this application. Then v satisfies also its conclusion, $\Gamma, \psi^{-y} \Rightarrow \Delta, \psi^y$. Indeed, either v satisfies $\Gamma_i \Rightarrow \Delta_i$ for some i, and hence also $\Gamma \Rightarrow \Delta$, or else $v(\psi_i) = x_i$ for all $1 \leq i \leq n$, and since $\widetilde{\diamond}(x_1, \ldots, x_n) = \{y\}$, necessarily $v(\psi) = y$. In both cases (1) entails that v satisfies $\Gamma, \psi^{-y} \Rightarrow \Delta, \psi^y$. It follows by (2) that $[\diamond(x_1, \ldots, x_n) : y]$ is sound for $\mathcal{M}$.

Corollary 2. *A canonical calculus is consistent iff it is coherent.*

Proof. The "only if" part is Proposition 2. The converse follows from Proposition 5, and the fact that every Nmatrix induces a consistent logic.

Corollary 3. *The consistency of a canonical calculus is decidable.*

5   Completeness and Cut-Elimination

Notation 6 For $x \in \{t, f\}$, denote: $ite(x, A, B) = $ if x then A else B. Note: $ite(x, A, B) = ite(-x, B, A)$.

Theorem 7. *Every coherent full canonical calculus admits cut elimination, and it is complete for the 2-Nmatrix that it defines.*

Proof. By Proposition 5, we can prove completeness and cut elimination together by showing that if $\Gamma \Rightarrow \Delta$ does not have a cut-free proof in a coherent full canonical calculus $\mathbf{G}$, then $\Gamma \not\vdash_{\mathcal{M}} \Delta$, where $\mathcal{M}$ is the 2-Nmatrix defined by $\mathbf{G}$. For this extend first $\Gamma \Rightarrow \Delta$ to a sequent $\Gamma^* \Rightarrow \Delta^*$ with the following properties:

1. $\Gamma \subseteq \Gamma^*$ and $\Delta \subseteq \Delta^*$.
2. $\Gamma^* \Rightarrow \Delta^*$ does not have a cut-free proof in $\mathbf{G}$.
3. For every rule $[\diamond(x_1, \ldots, x_n) : y]$ in $\mathbf{G}$, if $\diamond(\psi_1, \ldots, \psi_n) \in ite(y, \Delta^*, \Gamma^*)$ then for some $1 \leq i \leq n$, $\psi_i \in ite(x_i, \Delta^*, \Gamma^*)$.

This extension is possible, because if a sequent $\Gamma' \Rightarrow \Delta'$ does not have a cut-free proof and $\diamond(\psi_1, \ldots, \psi_n) \in ite(y, \Delta', \Gamma')$ then for some $1 \leq i \leq n$, $\Gamma', \psi_i^{-x_i} \Rightarrow \Delta', \psi_i^{x_i}$ does not have a cut-free proof (because otherwise by adding an application of $[\diamond(x_1, \ldots, x_n) : y]$ to the proofs of these sequents we obtain a cut-free proof for $\Gamma', \psi^{-y} \Rightarrow \Delta', \psi^y$, which is exactly $\Gamma' \Rightarrow \Delta'$).

The refuting valuation is now defined as follows:
For atomic q, $v(q) = t$ iff $q \in \Gamma^*$.

$$v(\diamond(\psi_1, \ldots, \psi_n)) = \begin{cases} t & \text{if } \widetilde{\diamond}(v(\psi_1), \ldots, v(\psi_n)) = \{t\} \text{ or} \\ & [\widetilde{\diamond}(v(\psi_1), \ldots, v(\psi_n)) = \{t, f\} \text{ and } \diamond(\psi_1, \ldots, \psi_n) \in \Gamma^*] \\ f & \text{otherwise} \end{cases}$$

v is obviously a legal $\mathcal{M}$-valuation. We now show by induction on the complexity of a formula $\psi \in \Gamma^* \cup \Delta^*$ that if $\psi \in \Gamma^*$ then $v(\psi) = t$, and if $\psi \in \Delta^*$ then $v(\psi) = f$.

- Assume ψ is atomic. If $\psi \in \Gamma^*$ then $v(\psi) = t$ by definition. If $\psi \in \Delta^*$ then $\psi \notin \Gamma^*$ by property 2 of $\Gamma^* \Rightarrow \Delta^*$. Hence $v(\psi) = f$.
- Let $\psi = \diamond(\psi_1, \ldots, \psi_n)$ and let $x_i = v(\psi_i)$ for $1 \leq i \leq n$.
  Assume $\psi \in \Gamma^*$, but $v(\psi) = f$. According to the definition of v, this can happen only if $\widetilde{\diamond}(x_1, \ldots, x_n) = \{f\}$. It follows that $\psi \in ite(v(\psi), \Delta^*, \Gamma^*)$ in this case. Hence $\psi_i \in ite(x_i, \Delta^*, \Gamma^*)$ for some $1 \leq i \leq n$ by property 3 of $\Gamma^* \Rightarrow \Delta^*$ and the fact that by Definition 10, $\widetilde{\diamond}(x_1, \ldots, x_n) = \{f\}$ in $\mathcal{M}$ only if $[\diamond(x_1, \ldots, x_n) : f]$ is a rule of $\mathbf{G}$. On the other hand $\psi_i \in ite(x_i, \Gamma^*, \Delta^*)$ for all $1 \leq i \leq n$ by the induction hypothesis. This contradicts property 2 of $\Gamma^* \Rightarrow \Delta^*$.
  Now assume $\psi \in \Delta^*$, but $v(\psi) = t$. According to the definition of v, there are two possibilities here:
  1. $\widetilde{\diamond}(x_1, \ldots, x_n) = \{t\} = \{v(\psi)\}$. We get from this a contradiction like in the previous case.
  2. $\widetilde{\diamond}(x_1, \ldots, x_n) = \{t, f\}$ and $\psi \in \Gamma^*$. Since $\psi \in \Delta^*$ as well, this contradicts property 2 of $\Gamma^* \Rightarrow \Delta^*$.

By property 1 of $\Gamma^* \Rightarrow \Delta^*$ and what we have just proved, v is a model of Γ in $\mathcal{M}$ which does not satisfy any element of Δ. Hence $\Gamma \not\vdash_{\mathcal{M}} \Delta$.

Theorem 8. *A canonical calculus admits cut elimination iff it is coherent.*

Proof. The "only if" part is just Proposition 3. For the "if" part, suppose $\Gamma \Rightarrow \Delta$ has a proof in a coherent canonical calculus $\mathbf{G}$. Let $\mathcal{M}$ be the 2-Nmatrix that is defined by $\mathbf{G}$ (as well as $\mathbf{G}^F$). Then $\Gamma \vdash_{\mathcal{M}} \Delta$ by Proposition 5. Theorem 7 implies therefore that $\Gamma \Rightarrow \Delta$ has a cut-free proof in $\mathbf{G}^F$. Hence, by Proposition 4, it also has a cut-free proof in $\mathbf{G}$.

Theorem 9. *Every coherent canonical calculus is sound and complete for the 2-Nmatrix that it defines.*

Proof. This immediately follows from Theorem 7 and Proposition 5.

Theorem 10. *Let $\mathbf{G}$ be a consistent canonical calculus. Then either $\mathbf{G}$ defines a logic which is a fragment of classical logic, or it has no finite characteristic matrix.*

Proof. Assume that **G** is consistent. Then it defines a logic $L(\mathbf{G})$. By Corollary 2 and Theorem 9, $L(\mathbf{G})$ is induced by some 2-Nmatrix $\mathcal{S}$. If $\mathcal{S}$ includes only deterministic connectives (i.e.: connectives which return singletons for every combination of truth values), then **G** has a characteristic two-valued *matrix*, and so it is a fragment of classical logic. Otherwise $\mathcal{S}$ has at least one proper non-deterministic operation, and hence $L(\mathbf{G})$ has no finite characteristic matrix by Theorem 6 of [AL00].[6]

We turn now to some more corollaries of the completeness theorem. First, a result that was promised at the end of section **??**:

Corollary 4. *If* **G** *is a coherent canonical calculus then it has a unique equivalent full canonical calculus.*

Proof. Let **G** be a coherent canonical calculus, and let $\mathbf{G}'$ be a full canonical calculus that is equivalent to **G**. We need to show that $\mathbf{G}' = \mathbf{G}^F$. Suppose this is not the case. Let $\mathcal{M}^F$ and $\mathcal{M}'$ be the 2-Nmatrices that are defined by $\mathbf{G}^F$ and $\mathbf{G}'$ respectively. Since the two calculi are different then by Definition 10 there is some n-ary connective $\diamond$ and some $x_1, \ldots, x_n \in \{t, f\}$ such that the interpretation of $\diamond$ on $x_1, \ldots, x_n$ is different in $\mathcal{M}^F$ and in $\mathcal{M}'$. Suppose w.l.o.g. that in $\mathcal{M}^F$, $\widetilde{\diamond}(x_1, \ldots, x_n)$ is $\{f\}$, whereas in $\mathcal{M}'$ it is either $\{t\}$ or $\{t, f\}$. It is easy to see that $\{p_i \mid x_i = t\} \cup \{\diamond(p_1, \ldots, p_n)\} \vdash_{\mathcal{M}^F} \{p_i \mid x_i = f\}$, while this is not the case in $\vdash_{\mathcal{M}'}$. Hence, by Theorem 7, $\mathbf{G}^F$ and $\mathbf{G}'$ are not equivalent. This contradicts the fact that they are both equivalent to **G**.

Our next results compare strength of rules and introduce a normal form.

Definition 11. *Let R_1 and R_2 be two canonical rules (in the same language). We say that R_1 is at least as strong as R_2 if any application of R_2 can be simulated using R_1 together with the standard axioms and structural rules (including cut). We say that R_1 and R_2 are equivalent if each of them is at least as strong as the other.*

The characterization below of the strength of a rule can be summarized as follows: a rule is stronger when its set of premises is weaker!

Proposition 6. *A canonical rule S_1/C is at least as strong as the canonical rule S_2/C iff every clause in S_1 classically follows from S_2 (this is equivalent to saying that every clause in S_1 is subsumed by some clause that can be derived from the clauses of S_2 using resolutions).*

Proof. The "if" part can easily be proved directly. The converse can be shown by using 2-Nmatrices. We omit the details.

Corollary 5. *Two canonical rules S_1/C and S_2/C are equivalent if S_1 and S_2 are classically equivalent (as sets of clauses).*

---

[6] This theorem states that if $\mathcal{S}$ is a two-valued N-matrix which has at least one proper nondeterministic operation, then $\vdash_{\mathcal{S}}$ has no finite characteristic matrix.

This corollary naturally leads to the following economical normal form for canonical rules:

Definition 12. *A canonical rule is in* Resolution Normal Form *(RNF) if its set of premises S does not include a standard axiom, and any resolvent of two elements of S is subsumed by some other element of S.*

Corollary 6. *Every canonical rule has an equivalent canonical rule in RNF.*

An example of transforming a rule to *RNF* will be given in the next section.

6   Calculi for 2-Nmatrices

In the previous sections we associated with every coherent canonical calculus a 2-Nmatrix, for which it is sound and complete. In this section we go in the other direction, and associate with a given 2-Nmatrix coherent canonical calculi which are sound and complete for it. One way of doing so is rather obvious:

Definition 13. *Let $\mathcal{M}$ be a 2-Nmatrix. The full calculus that is defined by $\mathcal{M}$ is the canonical calculus $\mathbf{G}$ that has the rule $[\diamond(x_1, \ldots, x_n) : y]$ for each n-ary connective $\diamond$ and for every $x_1, \ldots, x_n, y \in \{t, f\}$ such that $\widetilde{\diamond}(x_1, \ldots, x_n) = \{y\}$.*

Note: The full calculus that is defined by a 2-Nmatrix is obviously coherent.

Proposition 7. *The full calculus that is defined by a 2-Nmatrix is sound and complete for it.*

Proof. The proof is similar to that of Proposition 5 and Theorem 7.

We introduce now for any given 2-Nmatrix a calculus of a more regular form.

Theorem 11. *Every 2-Nmatrix $\mathcal{M}$ has a sound and complete coherent canonical calculus which for every connective has at most one introduction rule on the left, and at most one introduction rule on the right.*

Proof. Let $\mathbf{G}(\mathcal{M})$ be the canonical calculus which for any n-ary connective $\diamond$ has the following rules (where $\widetilde{\diamond}$ is the interpretation of $\diamond$ in $\mathcal{M}$):

$$[\diamond \Rightarrow] \quad \{\{p_i \mid x_i = t\} \Rightarrow \{p_i \mid x_i = f\}\}_{t \in \widetilde{\diamond}(x_1, \ldots, x_n)} \ / \ \diamond(p_1, \ldots, p_n) \Rightarrow$$

$$[\Rightarrow \diamond] \quad \{\{p_i \mid x_i = t\} \Rightarrow \{p_i \mid x_i = f\}\}_{f \in \widetilde{\diamond}(x_1, \ldots, x_n)} \ / \ \Rightarrow \diamond(p_1, \ldots, p_n)$$

Note that if $t \in \widetilde{\diamond}(x_1, \ldots, x_n)$ for all $x_1, \ldots, x_n$, then the first rule is superfluous and can be discarded, while if $\widetilde{\diamond}(x_1, \ldots, x_n) = \{f\}$ for all $x_1, \ldots, x_n$ then that rule does not have any premises, i.e. it is a non-standard axiom (this type of axioms is permitted in canonical systems!). Similarly, if $f \in \widetilde{\diamond}(x_1, \ldots, x_n)$ for all $x_1, \ldots, x_n$ then the second rule can be discarded, while if $\widetilde{\diamond}(x_1, \ldots, x_n) = \{t\}$ for all $x_1, \ldots, x_n$ then that rule does not have any premises.

The soundness of $\mathbf{G}(\mathcal{M})$ is easy to verify. Take for example $[\diamond \Rightarrow]$. To show its soundness, assume that $\Gamma, \psi_1^{x_1}, \ldots, \psi_n^{x_n} \vdash_{\mathcal{M}} \Delta, \psi_1^{-x_1}, \ldots, \psi_n^{-x_n}$ for all $x_1, \ldots, x_n$ such that $t \in \tilde{\diamond}(x_1, \ldots, x_n)$, and let v be a model of $\Gamma \cup \{\diamond(\psi_1, \ldots, \psi_n)\}$ in $\mathcal{M}$. Then there are $y_1, \ldots, y_n \in \{t, f\}$ such that $t \in \tilde{\diamond}(y_1, \ldots, y_n)$ and $v(\psi_i) = y_i$ for all i. Since $\Gamma, \psi_1^{y_1}, \ldots, \psi_n^{y_n} \vdash_{\mathcal{M}} \Delta, \psi_1^{-y_1}, \ldots, \psi_n^{-y_n}$ by assumption, it follows that v is a model of one of the elements of Δ. The soundness of $[\Rightarrow \diamond]$ is proved similarly, while the proof of completeness is similar to that of Theorem 7.

Corollary 7. *Every coherent canonical calculus has an equivalent canonical calculus in which every connective has at most one introduction rule for each side.*

Example 9. Suppose we have the following interpretation for a binary connective $\diamond$, which makes it a very close relative of the classical conjunction:

$$\tilde{\diamond}(t, t) = \{t\}, \quad \tilde{\diamond}(t, f) = \{t, f\}, \quad \tilde{\diamond}(f, t) = \tilde{\diamond}(f, f) = \{f\}$$

The corresponding two rules as given in the proof of the last theorem are:

$$[\diamond \Rightarrow] \quad \{p_1, p_2 \Rightarrow \ , \ p_1 \Rightarrow p_2\} \ / \ \diamond(p_1, p_2) \Rightarrow$$

$$[\Rightarrow \diamond] \quad \{p_1 \Rightarrow p_2 \ , \ p_2 \Rightarrow p_1 \ , \ \Rightarrow p_1, p_2\} \ / \ \Rightarrow \diamond(p_1, p_2)$$

We next transform these two rules into rules in *RNF* as follows. Consider the set of premises of the second rule. Its closure under cut is:

$$\{p_1 \Rightarrow p_2 \ , \ p_2 \Rightarrow p_1 \ , \ \Rightarrow p_1, p_2 \ , \ \Rightarrow p_1 \ , \ \Rightarrow p_2 \ , \ p_1 \Rightarrow p_1 \ , \ p_2 \Rightarrow p_2\}$$

We now discard the last two standard axioms, and remove also the original three clauses since they are subsumed by $\Rightarrow p_1$ and $\Rightarrow p_2$. A similar process can be applied to the first rule. We are left with the simpler rules:

$$[\diamond \Rightarrow]' \quad \{p_1 \Rightarrow \ \} \ / \ \diamond(p_1, p_2) \Rightarrow$$

$$[\Rightarrow \diamond]' \quad \{\Rightarrow p_1 \ , \ \Rightarrow p_2\} \ / \ \Rightarrow \diamond(p_1, p_2)$$

Note that both rules are frequently used in the literature as introduction rules for classical conjunction.

Note: Given a 2-Nmatrix $\mathcal{M}$, the system $\mathbf{G}(\mathcal{M})$ which is constructed in the proof of Theorem 11 is a natural basis for a tableaux proof system for validity in $\mathcal{M}$. In fact, an application of a rule in $\mathbf{G}(\mathcal{M})$ backwards is like a step in a tableaux. For example: $[\diamond \Rightarrow]$ says that if $v(\diamond(\psi_1, \ldots, \psi_n)) = t$ then there are some $x_1, \ldots, x_n \in \{t, f\}$ such that $t \in \tilde{\diamond}(x_1, \ldots, x_n)$.

7   Conclusion

We defined canonical calculi which are the most natural type of multiple conclusion Gentzen-type systems, and showed that such calculi are non-trivial iff
they satisfy a certain constructive coherence condition. We introduced the semantics of two-valued non-deterministic matrices (2-Nmatrices) for such calculi,
and proved that the following are equivalent for any given logic L:

1. L is defined by some coherent, canonical Gentzen-type system.
2. L is defined by some cut-free, canonical Gentzen-type system.
3. L is the logic of some 2-Nmatrix.

One of the by-products of our work is a strong evidence for the thesis according to which the meaning of a connective is given by its introduction (and
"elimination") rules (in some appropriate deduction system). We have shown
that at least in the framework of multiple-conclusion consequence relations, any
reasonable set of canonical introduction rules completely determine the semantics of a connective. For this it is not even necessary that the left introduction
rules and the right introduction rules for a given connective precisely "match"
(in the sense of [Bel62] and [Sun86]). It suffices that there would be no conflict
between them (where this condition is defined in precise terms).

Obvious directions for further research are the following:

1. To extend the ideas and results to first order languages.
2. To develop an analogous framework and theory for single-conclusion consequence relations and Natural Deduction systems.
3. To generalize the framework to arbitrary finite n-valued Nmatrices, possibly
using sequents with n components like e.g. in [BFZ94][7] (see also the survey
papers [BFS00] and [Häh99] for more references and further details).

References

[AL00]   Arnon Avron and Iddo Lev, "Non-deterministic matrices," 2000. Submitted.
[Avr91]   Arnon Avron, "Simple consequence relations," *Information and Computa-
          tion*, vol. 92, no. 1, pp. 105–139, 1991.
[Bel62]   Nuel. D. Belnap, "Tonk, plonk and plink," *Analysis*, vol. 22, pp. 130–134,
          1962.
[Béz99]   Jean-Yves Béziau, "Classical negation can be expressed by one of its halves,"
          Logic Journal of the IGPL, vol. 7, pp. 145–151, 1999.

---

[7] This paper introduces a special type of canonical systems for sequents with n components (those that are induced according to a certain procedure by n-valued *deterministic* matrices), and proves a general cut elimination theorem for this type of
systems. In the case $n = 2$ this amounts to a cut elimination theorem for a certain
class of systems that correspond to fragments of classical logic.

[BFS00] Matthias Baaz, Christian G. Fermüller, and Gernot Salzer, "Automated deduction for many-valued logics," in *Handbook of Automated Reasoning* (A. Robinson and A. Voronkov, eds.), Elsevier Science Publishers, 2000.

[BFZ94] Matthias Baaz, Christian G. Fermüller, and Richard Zach, "Elimination of cuts in first-oder finite-valued logics," *Information Processing Cybernetics*, vol. 29, no. 6, pp. 333–355, 1994.

[Dun86] J. Michael Dunn, "Relevance logic and entailment," in [GG86], vol. III, ch. 3, pp. 117–224, 1986.

[Gen69] Gerhard Gentzen, "Investigations into logical deduction," in *The Collected Works of Gerhard Gentzen* (M. E. Szabo, ed.), pp. 68–131, North Holland, Amsterdam, 1969.

[GG86] Dov M. Gabbay and Franz Guenthner, *Handbook of Philosophical Logic*. D. Reidel Publishing company, 1986.

[Häh99] Reiner Hähnle, "Tableaux for multiple-valued logics," in *Handbook of Tableau Methods* (Marcello D'Agostino, Dov M. Gabbay, Reiner Hähnle, and Joachim Posegga, eds.), pp. 529–580, Kluwer Publishing Company, 1999.

[Hod86] Wilfrid Hodges, "Elementary predicate logic," in [GG86], vol. I, ch. 1, pp. 1–131, 1986.

[Pri60] A. N. Prior, "The runabout inference ticket," *Analysis*, vol. 21, pp. 38–9, 1960.

[Sco74] Dana S. Scott, "Completeness and axiomatization in many-valued logics," in *Proc. of the Tarski symposium*, vol. XXV of *Proc. of Symposia in Pure Mathematics*, (Rhode Island), pp. 411–435, American Mathematical Society, 1974.

[Sun86] Göran Sundholm, "Proof theory and meaning," in [GG86], vol. III, ch. 8, pp. 471–506, 1986.

[Urq86] Alasdair Urquhart, "Many-valued logic," in [GG86], vol. III, ch. 2, pp. 71–116, 1986.

Incremental Closure of Free Variable Tableaux

Martin Giese

Institut für Logik, Komplexität und Deduktionssysteme,
Universität Karlsruhe, Germany
giese@ira.uka.de

Abstract. This paper presents a technique for automated theorem proving with free variable tableaux that does not require backtracking. Most existing automated proof procedures using free variable tableaux require iterative deepening and backtracking over applied instantiations to guarantee completeness. If the correct instantiation is hard to find, this can lead to a significant amount of duplicated work. Incremental Closure is a way of organizing the search for closing instantiations that avoids this inefficiency.

1 Introduction

Since the 1980's, the technique of using free variables to postpone the choice of instantiations in the γ-expansions of tableau calculi for first-order logic is used in practically all implementations. These free variables have to be instantiated at some point in the proof search by unifying complementary literals on branches, and one faces the problem that doing this in a naïve way can lead to non-termination for some unsatisfiable sets of formulae, and thus to *incompleteness of the procedure*.

The most used way of regaining completeness employs backtracking and iterative deepening: A complexity limit for the proof is fixed, and a proof that does not exceed this complexity is sought for, using backtracking to explore the search space. If no proof is found, the limit is increased. Unfortunately, backtracking can lead to a large amount of duplicated work, because the prover forgets information which it might need again. On the other hand, the analytic free variable tableau calculus is *proof confluent*, meaning that any open tableaux for an unsatisfiable set of formulae may be closed by further expansion. This means that *the calculus* does not require backtracking, contrary to connection tableaux, for instance.

This is probably the main incentive to consider proof procedures that can do without backtracking. Another reason is that they are more suited for use in an integrated automated and interactive system: The user has more possibility of seeing what went wrong in a failed proof attempt, if all information about what has been tried so far is kept.

Lately, a number of tableau-based procedures has been proposed that work without backtracking. Most of these concentrate on overcoming the mentioned naïveté of simply closing a branch as soon as possible. In [Bil96] for instance,

R. Goré, A. Leitsch, and T. Nipkow (Eds.): IJCAR 2001, LNAI 2083, pp. 545–560, 2001.

instead of instantiating free variables in the tableau, the set of input clauses is extended by instantiated variants, leading to a kind of saturation process. A similar approach based on the connection calculus is presented in [BEF99]. In [Bec00], an ordering restriction on the sequence of generated tableaux is imposed that forbids cycles.

This paper describes an approach in which the free variables are never instantiated in the tableau, but the various possibilities are effectively considered in parallel. We use an *incremental* approach to compute an instantiation of the free variables that closes all branches *simultaneously*, hence the name *Incremental Closure*.[1]

After defining a few basic notions, we shall present the basic idea of the approach in Sect. 3. We describe a number of possible refinements in Sect. 4, and some experimental results are quoted in Sect. 5.

2   Preliminaries

We assume a fixed first-order signature throughout this paper. Let terms and first-order formulae (without equality) over that signature be defined in the usual way. A ground term is a term that does not contain variables.

A formula is in *negation normal form* (NNF), iff negation signs appear only in front of atomic formulae $p(t_1, \ldots, t_n)$. By the application of de Morgan's rules, any formula can be transformed into an equivalent NNF formula. A formula is in *skolemized negation normal form* (SNNF), iff it is in NNF and does not contain existential quantifiers. Any formula F can be transformed by skolemization into a formula F' in SNNF that is satisfiable iff F is satisfiable. A formula is *closed* if all variable occurrences in it are bound by a quantifier.

Definition 1. *An* instantiation *is a mapping from the set of all variables to ground terms. Let* Sub^0 *denote the set of all instantiations.*

This differs from the usual concept of a *ground substitution*, in that we require *all*, i.e. infinitely many variables to be mapped.

Definition 2. *A* goal *is a finite set of formulae. A* tableau *is a finite tree where every node has zero, one, or two children, and each node is labeled with a goal. A* leaf *is a node with no children. The* leaf goals *of a tableau are the goals that label its leaves.*

A tableau for a finite set of SNNF formulae S is defined inductively as follows:

1. *The tableau consisting of the root node labeled with the goal S is a tableau for S, called the* initial tableau.
2. *If there is a tableau for S that has a leaf n with goal $\{\alpha_1 \wedge \alpha_2\} \cup G$, then the tableau obtained by adding a new child n' with goal $\{\alpha_1, \alpha_2\} \cup G$ to n is also a tableau for S. (α-expansion)*

---

[1] A predecessor to this approach was sketched in the Postition Paper[Gie00b] under the name of 'Instance Streams'.

3. *If there is a tableau for S that has a leaf n with goal $\{\beta_1 \vee \beta_2\} \cup G$, then the tableau obtained by adding two new children n', resp. n'' with goals $\{\beta_1\} \cup G$, resp. $\{\beta_2\} \cup G$ to n is also a tableau for S. (β-expansion)*

4. *If there is a tableau for S that has a leaf n with goal $\{\forall x.\gamma_1\} \cup G$, then the tableau obtained by adding a new child n' with goal $\{[x/X]\gamma_1, \forall x.\gamma_1\} \cup G$ to n, where X did not previously occur in the tableau, is also a tableau for S. (γ-expansion)*

A complementary pair *is a pair ϕ, $\neg\psi$, where ϕ and ψ are unifiable atomic formulae. A goal G is* closed under *an instantiation σ, iff there is a complementary pair $\{\phi, \neg\psi\} \subseteq G$ with $\sigma(\phi) = \sigma(\psi)$. A tableau T is* closed under *an instantiation σ, iff each leaf goal of T is closed under σ. A tableau is* closable *iff it is closed under some instantiation.*

We use this somewhat unusual formulation of tableaux labeled with sets of formulae (Smullyan [Smu68] calls them *block tableaux*) because it helps in describing the procedure. Note that in an implementation, it is sufficient to keep the leaf goals in memory; they correspond to the branches in the usual definition.

Another deviation from the usual formulations of free variable tableau calculi in that there is no rule that instantiates the free variables introduced by a γ-rule. Instead, an instantiation that closes all branches simultaneously has to be found, to decide that a tableau is closable. This is an important aspect of the incremental closure technique. It is obvious, that the usual correctness and completeness proofs for free variable tableaux are also applicable to this formulation.

Proposition 1. *Let S be a set of closed formulae in SNNF. S is unsatisfiable iff there is a closable tableau for S.*

3   Incremental Closure

From Prop. 1, it is easy to derive a complete proof procedure:

```
T := initial tableau for S
while ( not closable(T) ) do
   if expandable(T) then
      select possible expansion of T
      expand T
   else
      answer 'satisfiable'
   end
end
answer 'unsatisfiable'
```

This is a complete proof procedure, provided the selection of tableau expansions is *fair*. Being fair means that if the procedure does not terminate, any extension step possible on a goal will at some point be applied on that goal or one of its descendants. In particular, in a non-terminating run, infinitely many instances of each γ-formula will ultimately be produced on each branch.

The main problem with this proof procedure is the test closable(T): In general, the right combination of complementary literals has to be found in the leaf goals to compute a simultaneous unifier. This is NP-complete in the size of the leaf goals.[2]

The problem of finding the right complementary pairs has to be solved in any free variable tableau proof procedure, backtracking or not. But although the worst-case complexity cannot be reduced, a speedup can be achieved by tuning the procedure to perform well in practical cases.

The approach presented here makes the procedure more efficient by computing closable(T) in an incremental fashion, based on the following observations:

- If a pair of complementary literals is unifiable, it will stay unifiable after any extension. This should make an incremental algorithm worthwhile.
- An instantiation has to be found for the free variables introduced by the γ-rule. These only occur in the proof tree below the corresponding γ formula. To take advantage of this locality, the algorithm should exploit the structure of the proof tree.

3.1   An Abstract Description

In this section we shall abstract away from concrete representations of instantiations, and assume that we can perform calculations on (potentially infinite) sets of instantiations. How to represent these in an actual implementation is discussed in Sect. 3.2.

Let

$$\text{unif}(\phi, \psi) := \{\sigma \in \text{Sub}^0 \mid \sigma(\phi) = \sigma(\psi)\}$$

be the set of instantiations that unify two atomic formulae. We define

$$\text{cl}(G) := \bigcup_{\phi, \neg\psi \in G} \text{unif}(\phi, \psi)$$

to be the set of instantiations under which a goal G is closed. For a node n of a tableau, let $lg(n)$ be the set of leaf goals associated with the leaves that are descendants of n. Use this to define

$$\text{cl}(n) := \bigcap_{G \in lg(n)} \text{cl}(G)$$

---

[2] Unifiability can be decided in linear time [PW78], so with indeterministic selection of complementary pairs, closable(T) is in NP. On the other hand, SAT for propositional clauses can be reduced to this problem by translating each clause to one leaf goal, mapping propositional symbols to free variables, such that a goal is closable under an instantiation to $\{0, 1\}$ iff the clause is satisfied by the corresponding interpretation. E.g., translate $A \vee \neg B$ to $\{p_A(0), \neg p_A(1), p_B(0), \neg p_B(1), p_A(A), \neg p_B(B)\}$.

to be the set of instantiations under which all leaves below n are closed. Obviously, $\mathrm{cl}(root)$, where $root$ is the root node, is the set of instantiations that close the whole tableau.

We can take advantage of the tableau structure by expressing $\mathrm{cl}(n)$ recursively: If a node n has only one child n', $\mathrm{cl}(n) = \mathrm{cl}(n')$, for two children n', n'', $\mathrm{cl}(n) = \mathrm{cl}(n') \cap \mathrm{cl}(n'')$. For a leaf n labeled with goal G, $\mathrm{cl}(n) = \mathrm{cl}(G)$.

We shall clarify these notions using the following tableau:

$$n_1 : \; \forall x.(qx \lor \neg px), \forall y.qy, \neg qb, pa$$
$$\mid$$
$$n_2 : \; \underline{qX \lor \neg pX}, \forall y.qy, \neg qb, pa, \forall x.(\ldots)$$

$$n_3 : \; \underline{qX}, \forall y.qy, \neg qb, pa, \forall x.(\ldots) \qquad\qquad n_4 : \; \underline{\neg pX}, \forall y.qy, \neg qb, pa, \forall x.(\ldots)$$

n_2 was constructed by applying a γ-expansion at n_1, and n_3, n_4 were introduced by a β-expansion at n_2. The newly introduced formulae are underlined in each goal. The goal at n_3 contains one complementary pair $qX, \neg qb$. So $\mathrm{cl}(n_3) = \mathrm{unif}(qX, qb) = \{\sigma \in \mathrm{Sub}^0 | \sigma(X) = b\}$, the set of instantiations that map X to b. Similarly, $\mathrm{cl}(n_4) = \{\sigma \in \mathrm{Sub}^0 | \sigma(X) = a\}$, because of the complementary pair $\neg pX, pa$. For $\mathrm{cl}(n_2)$ we have to find instantiations that close both leaf goals, $\mathrm{cl}(n_2) = \mathrm{cl}(n_3) \cap \mathrm{cl}(n_4)$. There are obviously no such instantiations, $\mathrm{cl}(n_2) = \emptyset$. The same holds for the root n_1, of course.

To get an incremental algorithm, we shall examine the values of cl change when a tableau expansion produces new complementary pairs. In general, one expansion step might lead to several new complementary pairs in one goal, or there might be two new goals, each of which can contain new complementary pairs. We shall examine the changes to cl induced by *one* new complementary pair $\phi, \neg\psi$ at *one* leaf l, called the *selected leaf*. If there are several new complementary pairs, these changes may be applied consecutively for each of them.

Let cl_0 denote the value of cl before taking into account $\phi, \neg\psi$, while cl is the updated value. Possible closing instantiations are never destroyed by an expansion step, so the sets cl can only grow when the tableau is expanded, i.e. $\mathrm{cl}(n) \supseteq \mathrm{cl}_0(n)$ for all nodes of the tableau. Define $\delta(n) := \mathrm{cl}(n) \setminus \mathrm{cl}_0(n)$ to be the set of *new* closing instantiations. Obviously, $\mathrm{cl}(n) = \mathrm{cl}_0(n)$, so $\delta(n) = \emptyset$, if the selected leaf l is not a descendant of n. In other words, $\delta(n)$ is non-empty only for nodes n on the path between l and the root of the tableau. For the selected leaf l, δ is given by

$$\delta(l) = \mathrm{unif}(\phi, \psi) \setminus \mathrm{cl}_0(n) \quad.$$

Using the recursive expression for $\mathrm{cl}(n)$, we can 'propagate' this change up the branch towards the root. We obtain $\delta(n) = \delta(n')$ for all nodes n with one child n'. For a node n with two children n' and n'', we assume that l lies below n'. This implies that $\mathrm{cl}(n'') = \mathrm{cl}_0(n'')$, so we have

$$\begin{aligned}
\delta(n) &= \mathrm{cl}(n) \setminus \mathrm{cl}_0(n) \\
&= (\mathrm{cl}(n') \cap \mathrm{cl}(n'')) \setminus (\mathrm{cl}_0(n') \cap \mathrm{cl}_0(n'')) \\
&= (\mathrm{cl}(n') \cap \mathrm{cl}_0(n'')) \setminus (\mathrm{cl}_0(n') \cap \mathrm{cl}_0(n'')) \\
&= (\mathrm{cl}(n') \setminus \mathrm{cl}_0(n')) \cap \mathrm{cl}_0(n'') \\
&= \delta(n') \cap \mathrm{cl}_0(n'')
\end{aligned}$$

The case where l lies below n'' is of course symmetrical.

The central idea of the incremental closure procedure is to keep track of the sets $\mathrm{cl}(n)$ and update them by propagating the additional closures $\delta(n)$ up the branch using this equation. As soon as $\delta(root) \neq \emptyset$, the tableau must be closable.

We shall continue the example from above to demonstrate the propagation of δ values.

$$n_1 : \ \forall x.(qx \vee \neg px), \forall y.qy, \neg qb, pa$$
$$n_2 : \ \underline{qX \vee \neg pX}, \forall y.qy, \neg qb, pa, \forall x.(\ldots)$$
$$n_3 : \ \underline{qX}, \forall y.qy, \neg qb, pa, \forall x.(\ldots) \qquad n_4 : \ \underline{\neg pX}, \forall y.qy, \neg qb, pa, \forall x.(\ldots)$$
$$n_5 : \ \underline{qY}, qX, \neg qb, pa, \forall x.(\ldots), \forall y.qy$$

There is a new node n_5 stemming from a γ-expansion at n_3. This leads to the new complementary pair $qY, \neg qb$. Not taking this into account leads to: $\mathrm{cl}_0(n_3) = \mathrm{cl}_0(n_5) = \{\sigma \in \mathrm{Sub}^0 | \sigma(X) = b\}$, $\mathrm{cl}_0(n_4) = \{\sigma \in \mathrm{Sub}^0 | \sigma(X) = a\}$, and $\mathrm{cl}_0(n_1) = \mathrm{cl}_0(n_2) = \emptyset$. These are the values we derived for cl on the previous page. Now, including $qY, \neg qb$, we get

$$\delta(n_5) = \mathrm{unif}(qY, qb) \setminus \mathrm{cl}_0(n_5) = \{\sigma \in \mathrm{Sub}^0 | \sigma(Y) = b \text{ and } \sigma(X) \neq b\}$$

This allows us to calculate δ for all nodes between n_5 and the root: After $\delta(n_3) = \delta(n_5)$, we have

$$\delta(n_2) = \delta(n_3) \cap \mathrm{cl}_0(n_4) = \{\sigma \in \mathrm{Sub}^0 | \sigma(Y) = b \text{ and } \sigma(X) = a\}$$

This in turn leads to $\delta(n_1) = \delta(n_2) \neq \emptyset$, so the proof is closable, namely by any instantiation that maps X to a and Y to b.

Still assuming we could calculate with infinite sets of instantiations, we shall now show how the computation and propagation of the δ values is organized. The procedure shall be described in a state-based way, but it turns out that operations on the state will typically be local. For that reason, we shall take an object oriented view.

Every leaf goal has an associated Sink object. A sink is an object capable of receiving a set of instantiations and performing some computation on it. This is realized by giving a put method to the Sink objects that takes a set of instantiations as parameter. The proof procedure will call this method after every expansion step with any set $\delta(n)$ of new closing instantiations coming from a new complementary pair i.e.

$$\text{goal.sink.put}(\text{unif}(\phi, \psi) \setminus \text{cl}_0(n))$$

We shall see further down how $\text{cl}_0(n)$ is extracted from the data structures.

There are two kinds of objects that act as sinks. One is the RootSink, which will receive $\delta(root)$. This contains a flag closable that records whether a non-empty set of instantiations has yet been received:

```
RootSink::put(S) is
  if S nonempty then
    closable := true
  end
end
```

The other kind of sink is provided by Merger objects which correspond to the splits in the tableau and are responsible for calculating the intersections $\delta(n) = \delta(n') \cap \text{cl}_0(n'')$.

The structure of a Merger is shown in the following diagram:

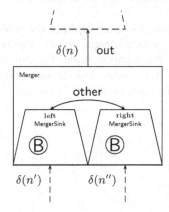

It consists of two MergerSink objects, one to receive $\delta(n')$ and one for $\delta(n'')$. The current set $\text{cl}(n')$, resp. $\text{cl}(n'')$ is stored in a buffer B in the corresponding input sink. Furthermore there is a reference out to an output sink, to which $\delta(n)$ will be passed on. The two sinks are mutually connected by an association other, so they can access each others buffers via other.B. Accordingly, the put method of the MergerSink object works as follows:

```
MergerSink::put(S) is
  J := S ∩ other.B   // δ(n) = δ(n') ∩ cl₀(n'')
  B := B ∪           // cl(n) = cl₀(n) ∩ δ(n)
  out.put(J)
end
```

It only remains to see how $\text{cl}_0(n)$ can be computed to determine $\delta(n) = \text{unif}(\phi, \psi) \setminus \text{cl}_0(n)$ for a new closure. There are two cases: If the goal is associated with the RootSink, $\text{cl}_0(n)$ must be empty, because the proof would otherwise be closed already. If it is associated with a MergerSink, this sink contains the current value of $\text{cl}_0(n)$ in its buffer B.

The proof procedure is now changed as follows:

```
T := initial tableau for S
associate RootSink r with goal of T
while ( not r.closable ) do
  if expandable(T) then
    select possible expansion of T
    expand T
    possibly generate new Merger
    handle new complementary pair
  else
    answer 'satisfiable'
  end
end
answer 'unsatisfiable'
```

At the initialization, a RootSink object is associated with the single goal of the tableau.

In the case of a β-expansion, i.e. a new split in the tableau, the step 'possibly generate new Merger' creates a new Merger object. The buffers B are initialized with the current value of cl_0 of the parent node. The output of the merger object is sent to the sink s of the parent node, and the new child nodes are associated with the input sinks of the merger, as shown in the following diagram:

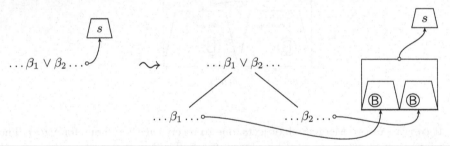

After the sinks have been updated, the procedure checks for new complementary pairs introduced by the expansion, calculates $\delta(n) = \mathrm{unif}(\phi, \psi) \setminus cl_0(n)$ for each of them, and sends $\delta(n)$ into the associated sink of the goal.

After all new closing instantiations have been passed to the sinks, the tableau is closable, if the closable flag of the root sink has been set.

3.2   Representation of Instantiation Sets

We have so far assumed that we can compute with infinite sets of instantiations. To get closer to a concrete implementation, we have to show how these may be represented with finite data structures. We shall briefly describe the representations used in the prototypical prover PrInS. (Prover with Instance Streams—referring to the streams of closing instantiations passed between the Sink objects.)

We use *syntactic equality constraints* to denote sets of instantiations: These are first-order formulae with equality as only predicate symbol, which are interpreted over the free term algebra. A constraint represents the set of instantiations that satisfy it. E.g. $\mathrm{unif}(p(X, b), p(a, Y))$ yields a constraint $X \equiv a \,\&\, Y \equiv b$, that is satisfied by all instantiations that map X to a and Y to b. As usual for unification, these constraints are kept in a 'solved form', that makes it easy to determine their satisfiability. The intersection of sets of instantiations required in the Mergers corresponds to the conjunction of constraints. In the updates of the MergerSinks' buffers B, set union is required which could be represented as disjunction of constraints. There is however no need to handle arbitrary disjunctive constraints: The buffers B can be implemented as lists of conjunctive constraints. The put method then looks as follows:

```
MergerSink::put(C) is
   foreach D in other.B do
      J := C & D
      if J satisfiable then
         out.put(J)
      end
   end
   add C to B
end
```

The constraints passed into the put methods are then purely conjunctive.

Finally, the set difference operation can be modeled either by taking negation into the constraint language, or by introducing subsumption checks at various places. We will not elaborate this here. See e.g. [Com91] for a survey on syntactic constraint solving methods.

4   Refinements

The Incremental Closure approach has the desirable property that it can easily be refined in a number of ways. We stress this point, because incremental closure is surely not the answer to all problems in automated theorem proving. It is therefore important to see how this new technique can be combined with successful existing approaches.

This section presents a number of possible refinements. While some of them are particular to the incremental closure technique, many are adaptations of refinements known from backtracking procedures.

4.1   Restriction of Instantiation Domains

On the abstract level, instead of passing instantiations for all free variables through the sink structure, it is possible to define the method to work with instantiations of only the free variables actually present at a certain tableau node.

For instance, in the following tableau:[3]

$$\vdots$$

$$\forall u, v. p(X, u, v) \lor q(u, v, Y)$$
$$p(X, U, V) \lor q(U, V, Y)$$

$$p(X, U, V) \qquad\qquad q(U, V, Y)$$

assume that the left branch may be closed for instantiations satisfying $X \equiv U$, and the right branch for $U \equiv Y$. The Merger corresponding to the split will find that both branches are closable with $X \equiv U \& U \equiv Y$. But U does not exist in the tableau above the γ-expansion, so this sub-tableau can be considered closable for all instantiations satisfying $X \equiv Y$.

In terms of constraints, the restriction to a subset of occurring variables corresponds to existential quantification: $X \equiv Y$ is equivalent to $\exists U, V. X \equiv U \& U \equiv Y$. This variation may be implemented by introducing a new kind of Sink object at every γ-expansion that computes the domain restriction.

This modification has several advantages:

- As in the example above, the existentially quantified constraints can often be simplified. Thus, they consume less space in the buffers B.
- Assume that a new combination of complementary literals leads to $X \equiv V \& V \equiv Y$ in the example. This would have to be handled separately in the original version, but domain restriction leads to $X \equiv Y$ as above. A subsumption check can be used to avoid further redundant computations.

4.2   Delete Propositionally Closable Branches

It occasionally happens that a sub-tableau is closable under *any* instantiation. This is the case in proofs of propositional formulae, where no free variables are required at all, but it can also happen with first-order formulae if there is a complementary pair that is unifiable without any further instantiation. It is useless to expand that part of the tableau any further, because no more closing instantiations can be found. The sub-tableau is called *propositionally closable*.

In the implementation using constraints, this corresponds to a constraint (equivalent to) *true* being passed through the sink structure. If this is detected, the corresponding goals and the Sink structure may be deleted to reduce memory consumption.

4.3   Using Buffers for Goal Selection

So far, the Sink structure built during a proof has only been used to check whether the tableau is closable. It turns out that it can also be useful for goal selection, i.e. deciding on which goal the next expansion step should take place.

---

[3] We shall adopt a more familiar and compact notation for tableaux here and in the sequel, by writing only the newly introduced formulae of each goal.

The buffers B contain representations of the closing instantiation for sub-tableaux. If, for instance, one subtree of a node has no closing instantiation yet, while the other does, expansions should take place in the first subtree, until at least one closing instantiation has also been found there. It is also possible to use the size of the buffers or of the constraints they contain for heuristics that tend to expand branches that seem harder to close.

4.4 Pruning

Pruning (see e.g. [BH98,BFN96]) is an important technique known from back-tracking procedures that can reduce the search space dramatically: The prover keeps track of the *ancestry* of formulae, i.e. the set of formulae in the tableau which were used to derive it.[4] If a branch is closed by unifying a particular complementary pair $\phi, \neg\psi$, the prover examines the β-expansions that occurred earlier on the branch. If for a particular expansion, neither the ancestry of ϕ, nor that of $\neg\psi$ contains the sub-formula $\beta_{1/2}$ introduced by the expansion, then the closure would have been possible without that expansion. Consequently, the expansion can be removed *a posteriori*, saving the work of closing the other branch introduced by it. Of course, the decision for that particular complementary pair might be revised in a backtracking step, and then the expansion has to be reintroduced.

We will now see how the pruning technique can be adapted to the incremental closure approach. We record for each formula an ancestry of β-expansions on which it depends. We can use references to the Merger objects for this, as there is exactly one of these for each β-expansion. In the abstract view of the procedure, we now compute with sets of pairs (σ, h) of instantiations with ancestries, instead of just sets of instantiations. We have to redefine the operations unif, $\cap$, $\cup$ and $\setminus$ to work with sets of such pairs. In particular, the $\cap$ operation in the Merger has to *combine* the histories of instantiations.

The 'pruning' takes place, when a Merger m receives an instantiation that does not have m in its ancestry: it can pass such an instantiation to the output sink independently of the contents of the buffer of the other branch:

```
MergerSink::put(S) is
    P := {(σ, h) ∈ S | this Merger ∉ h}
    out.put(P)
    S := S \ P;
    J := S ∩ other.B
    B := B ∪ S
    out.put(J)
end
```

As the complementary pair that led to this closing instantiation might not be the one that is ultimately needed to close the proof, the other branch may not, in general, be deleted. But the gain of passing a closing instantiation further up the Sink structure turns out to be very important in practice. Of course, in the case

---

[4] Actually, it suffices to record only the β-subformulae introduced by β-expansions.

of a propositional closure, it is even possible to delete the whole sub-tableau, giving an even greater advantage.

4.5   Constraints

A wide range of refinements and variations of the incremental closure method becomes possible if the prover is modified to work with *constrained formulae*. A constrained formula is a pair $\phi \ll C$ of a formula ϕ and a constraint C. The meaning of this is that ϕ may be used to close a branch only if the instantiation of the free variables of the tableau satisfies the constraint. Constrained formulae may be used to port tableau rules that normally require an instantiation to the incremental closure method, and also for restrictions of the search procedure that limit the permissible instantiations in some way.

Rules introducing constraints. In [Gie00a], a simplification rule using constraints was presented. Consider for instance a tableau branch containing the formulae

$$(1) : \forall y.(q(y) \wedge p(X))$$
$$(2) : p(a)$$

In a backtracking framework, (1) could be simplified with (2), by instantiating X with a, then replacing the occurrence of $p(X)$ by *true*, and finally rewriting the formula to $\forall y.q(y)$. The original, unsimplified formula (1) could be discarded. It would however be necessary to backtrack over the instantiation of X. Using constraints, one would not perform the instantiation explicitly; instead one would derive the constrained formula

$$(3) : \forall y.q(y) \ll X \equiv a$$

and replace the original formula by

$$(4) : \forall y.(q(y) \wedge p(X)) \ll X \not\equiv a \quad ,$$

using the constraint to keep track of the fact, that for instantiations with $X \equiv a$, formula (3) should be used instead of (4).

This approach blends perfectly with the incremental closure method. Only one change is needed: When a new complementary pair $\phi \ll C$, $\neg\psi \ll D$ is found, the constraints of the formulae have to be added to the unification constraint. One defines:

$$\mathrm{unif}(\phi \ll C, \psi \ll D) := \mathrm{unif}(\phi, \psi) \;\&\; C \;\&\; D \quad .$$

The same approach can be used to build an incremental closure version of hyper tableaux ([Bau98]) with (rigid) free variables: For a hyper tableau rule

$$p(a) \rightarrow q(y) \vee r(y)$$

and a literal $p(X)$, one can produce an expanded tableau

$$\vdots$$

$$\underset{q(Y) \ll X \equiv a \qquad r(Y) \ll X \equiv a}{\overset{p(X)}{\diagup \quad \diagdown}}$$

where Y is a new free variable.

Another use for constrained formulae is equality handling: In [NR01], Sect. 5, Nieuwenhuis and Rubio point out the importance of using ordering constraints to reduce the search space in automated equality reasoning, and [Bec94] presents a constraint based method for equality handling in tableaux that can be neatly integrated with the incremental closure approach.

To use ordering constraints, one simply has to extend the constraint language to contain an ordering predicate $\prec$ in addition to the syntactic equality $\equiv$. The semantics of constraints is given by fixing the interpretation of this predicate to some suitable reduction ordering.

Restrictions introducing constraints. Constraints can also be introduced to adapt certain proof search restrictions from backtracking procedures, in a similar way to what is described in [LS01], Sect. 8.

One example is *regularity*. In its simplest form, the regularity condition requires that no rule is applied that introduces a formula on a branch that already occurs on it. While this is easy to enforce for the (purely academic) case of ground tableaux, it requires a certain effort when free variables are used, because two formulae might become equal through an instantiation.

In the incremental closure framework, constraints can be used to ensure regularity. A goal containing $p(a)$ and $p(X) \vee q(X)$ might be expanded thus:

$$\begin{array}{c} p(a) \\ | \\ \underset{p(X) \ll X \not\equiv a \qquad\qquad q(X) \ll X \not\equiv a}{\overset{p(X) \vee q(X)}{\diagup \qquad\qquad\qquad \diagdown}} \end{array}$$

Then, if the instantiation $X \equiv a$ ultimately *does* lead to a proof of, say, the left branch, the ancestry of that instantiation cannot contain this β-split, so the pruning mechanism will take care that the redundant splitting expansion does no harm.

5   Experimental Results

In this section we shall quote some results obtained with an experimental implementation of the iterative closure procedure.

To cleanly separate the effects of incremental closure from those of various refinements, the technique was tested with a very simple implementation: No refinements like pruning or simplification were employed. Only the goal selection strategy from Sect. 4.3 was used. Formulae in goals are kept in a list and the first formula in this list is used for expansion. On the other hand, an equally simple backtracking prover was implemented using the same data-structures. Iterative deepening was applied on the number of γ-expansions per branch. The proof search of this backtracking prover is practically identical to that of leanT^AP[BP94].

Comparing the two provers on some simple probelms (harder problems require refinements in both cases) shows that the incremental closure prover is nearly always faster. The difference is particularly noticeable in cases that require heavy backtracking, i.e. where many complementary pairs are found that do *not* lead to a proof. This is the case, e.g. in SYN054+1 from the TPTP library, where the backtracking procedure requires 14317 rule applications and 37817 unifications versus 151 rule applications and 680 unifications with incremental closure. The problem family $p(c) \wedge \neg p(f^n(c)) \wedge \forall x.(p(x) \rightarrow p(f(x)))$ also shows this phenomenon very clearly, because there are many possible closures, and only few of them are correct for a low depth limit. For $n = 15$, the backtracking prover performs over 300,000 rule applications and over 1,000,000 unifications, while the incremental closure procedure requires 32 rule applications and 133 unifications.

The current implementation incorporates most of the refinements given in Sect. 4. It proves 161 of the 237 full-first-order theorems without equality in TPTP v.2.3.0, one of which, SYN067+1, is rated 0.67.

One would expect the described procedure to give rise to memory problems. But with resource limits of 300 s CPU and 160 MB heap, only about 30% of the failures were due to lack of memory. We hope to further reduce this amount by implementing more powerful rules and a better goal selection strategy, leading to shorter proofs. The idea is that without backtracking, one can afford to spend more time on individual rule applications, as these do not need to be repeated.

6   Conclusion, Related Work, and Future Research

We have presented an approach to eliminate backtracking from a proof procedure for free variable tableaux. It is built around the idea of incrementally computing instantiations that close sub-tableaux until one global instantiation is found that closes the whole tableau. We have demonstrated that this technique allows various refinements to be incorporated in the calculus and the procedure. Finally we have given some experimental results obtained by comparing the approach with a backtracking solution.

A similar approach has independently been described by B. Konev and T. Jebelean in [KJ00]. However, they hardly consider possibilies for refinements.

Further work includes experiments with integrated equality handling, search for specialized data structures, e.g. for the buffers B, better goal selection strategies and adaptations of further refinements from backtracking procedures. It might also be interesting to experiment with a parallelized implementation.

Acknowledgments. I thank Wolfgang Ahrendt, Elmar Habermalz, Reiner Hähnle and the anonymous referees for their numerous and helpful comments on drafts of this paper.

References

[Bau98]   Peter Baumgartner. Hyper Tableaux — The Next Generation. In Harrie de Swart, editor, *Proc. International Conference on Automated Reasoning with Analytic Tableaux and Related Methods, Oosterwijk, The Netherlands*, number 1397 in LNCS, pages 60–76. Springer-Verlag, 1998.

[Bec94]   Bernhard Beckert. A completion-based method for mixed universal and rigid *E*-unification. In Alan Bundy, editor, *Proc. 12th Conf. on Automated Deduction CADE, Nancy/France*, LNAI 814, pages 678–692. Springer, 1994.

[Bec00]   Bernhard Beckert. Depth-first proof search without backtracking for free variable clausal tableaux. In P. Baumgartner and H. Zhang, editors, *3rd Int. Workshop on First-Order Theorem Proving (FTP), St. Andrews, Scotland, TR 5/2000 Univ. of Koblenz*, pages 44–55, 2000.

[BEF99]   Peter Baumgartner, Norbert Eisinger, and Ulrich Furbach. A confluent connection calculus. In Harald Ganzinger, editor, *Proc. 16th International Conference on Automated Deduction, CADE-16, Trento, Italy*, volume 1632 of *LNCS*, pages 329–343. Springer-Verlag, 1999.

[BFN96]   Peter Baumgartner, Ulrich Furbach, and Ilkka Niemelä. Hyper tableaux. In José Júlio Alferes, Luís Moniz Pereira, and Ewa Orłowska, editors, *Proc. European Workshop: Logics in Artificial Intelligence, JELIA*, volume 1126 of *LNCS*, pages 1–17. Springer-Verlag, 1996.

[BH98]   Bernhard Beckert and Reiner Hähnle. Analytic tableaux. In W. Bibel and P. Schmitt, editors, *Automated Deduction: A Basis for Applications*, volume I, chapter 1, pages 11–41. Kluwer, 1998.

[Bil96]   Jean-Paul Billon. The disconnection method: a confluent integration of unification in the analytic framework. In P. Miglioli et al., editor, *Theorem Proving with Tableaux and Related Methods, TABLEAUX'96, Terrasini, Italy*, volume 1071 of *LNCS*, pages 110–126. Springer-Verlag, 1996.

[BP94]   Bernhard Beckert and Joachim Posegga. leanT^{A}P: Lean tableau-based theorem proving. extended abstract. In Alan Bundy, editor, *Proceedings, 12th International Conference on Automated Deduction (CADE), Nancy, France*, volume 814 of *LNCS 814*, pages 793–797. Springer-Verlag, 1994.

[Com91]   Hubert Comon. Disunification: a survey. In Jean-Louis Lassez and Gordon Plotkin, editors, *Computational Logic: Essays in Honor of Alan Robinson*, chapter 9, pages 322–359. MIT Press, Cambridge, MA, USA, 1991.

[Gie00a] Martin Giese. A first-order simplification rule with constraints. In Peter Baumgartner and Hantao Zhang, editors, *3rd Int. Workshop on First-Order Theorem Proving (FTP), St. Andrews, Scotland, TR 5/2000 Univ. of Koblenz*, pages 113–121, 2000.

[Gie00b] Martin Giese. Proof search without backtracking using instance streams, position paper. In Peter Baumgartner and Hantao Zhang, editors, *3rd Int. Workshop on First-Order Theorem Proving (FTP), St. Andrews, Scotland, TR 5/2000 Univ. of Koblenz*, pages 227–228, 2000.

[KJ00]   Boris Konev and Tudor Jebelean. Using meta-variables for natural deduction in theorema. In M. Kohlhase and M. Kerber, editors, *Proceedings of Calculemus-2000 Conference*. Electronic Notes in Computer Science, 2000.

[LS01]   Reinhold Letz and Gernot Stenz. Model elimination and connection tableau procedures. In Alan Robinson and Andrei Voronkov, editors, *Handbook of Automated Reasoning*. Elsevier Science, 2001. to appear.

[NR01]   Robert Nieuwenhuis and Albert Rubio. Paramodulation-based theorem proving. In Alan Robinson and Andrei Voronkov, editors, *Handbook of Automated Reasoning*. Elsevier Science, 2001. to appear.

[PW78]   M. S. Paterson and M. N. Wegman. Linear unification. *Journal of Computer and System Sciences*, 16(2):158–167, April 1978.

[Smu68]   Raymond M. Smullyan. *First-Order Logic*, volume 43 of *Ergebnisse der Mathematik und ihrer Grenzgebiete*. Springer-Verlag, New York, 1968.

Deriving Modular Programs from Short Proofs

Uwe Egly[1] and Stephan Schmitt[2]

[1] Institut für Informationssysteme 184/3
TU Wien
Favoritenstr. 9–11, A–1040 Wien, Austria
uwe@kr.tuwien.ac.at
[2] Department of Sciences and Engineering
Saint Louis University, Madrid Campus
Avd. del Valle 34, 28003 Madrid, Spain
schmitts@spmail.slu.edu

Abstract. We present a polynomial translation of dag sequent proofs into tree sequent proofs for first-order classical and intuitionistic logic. The basic idea is to interpret a reference in a dag proof as a lemma application, which is then simulated using an application of the cut rule. The result of this translation is a tree proof with cuts, which are only applied in order to "factorize" identical subproofs. We illustrate a central application of the presented cut-based translation, that is automated extraction of modular programs from first-order intuitionistic proofs.

1 Introduction

Sequent calculi are the common proof systems in interactive proof assistants, and are widely used to present machine found proofs in a way comprehensible by humans. Integrating an efficient automated deduction system for proving suitable subproblems, e.g., first-order intuitionistic formulae, into these proof assistants is a non-trivial task, because their underlying calculi are not suited for automated proof search. Automated deduction systems are typically based on redundancy-reduced calculi, like various kinds of connection and tableau calculi, special forms of (prefixed) sequent systems, or resolution. Consequently, automatically generated proofs have to be translated back into the underlying calculi of the proof assistants.

Automated proof search in sequent systems, however, can be performed in two directions: either from the end sequent towards the axioms (backward search), or from the axioms towards the end sequent (forward search). Although a naive implementation of forward proof search is hardly efficient, the picture changes if we respect the *subformula property*: Only such inferences are applicable which introduce a subformula of the formula to be proven into the sequent. Forward search in sequent calculi is often called *inverse* because it works in the inverse direction compared to the more usual backward search procedure. Many "inverse calculi" for different logics have been developed in the past by Maslov, Mints, Voronkov, Tammet and others (see [17] or [6] for further details

R. Goré, A. Leitsch, and T. Nipkow (Eds.): IJCAR 2001, LNAI 2083, pp. 561–577, 2001.
© Springer-Verlag Berlin Heidelberg 2001

and references). These calculi are well-suited for proof search because derived sequents can be used more than once as premises and thus, allow compact proof representations.

The proofs resulting from inverse proof calculi are essentially sequent proofs in *dag* form (dag = directed acyclic graph). In a dag sequent proof, every sequent that occurs identically multiple times is only proven once. The remaining non-proven occurrences of the sequent receive a *reference* to this proof in a way, such that the proof tree combined with all the references induces a directed acyclic graph. From this viewpoint, dag sequent proofs are closely related to the proof calculi underlying the interactive proof systems. However, these calculi are often based on a tree format for sequent proofs, which prevents a direct interpretation of the dag proofs by these systems. In order to integrate forward-directed proof search procedures into interactive proof assistants, one has to transform the resulting dag sequent proofs into tree sequent proofs.

In this paper, we present a *polynomial-time computable* translation of dag sequent proofs (even with *cuts*) into tree sequent proofs for first-order classical and intuitionistic logic. For this, we develop a general procedure which works uniformly for proofs in Gentzen-like sequent calculi $\mathcal{LK}$ and $\mathcal{LJ}$ [12] as well as in a multiple-conclusioned intuitionistic sequent calculus $\mathcal{LJ}_{mc}$ according to Fitting [11]. Our transformation is based on a simulation of dag proofs within the corresponding (tree) sequent calculi plus the *cut* rule. Each reference in a dag proof is interpreted as a lemma application, where the proof of the lemma occurs at the identical sequent the reference points to. The simulation can be described as follows: The sequent at the reference is encoded into a first-order formula, the formula image of the sequent, and introduced with an application of the cut rule. The lemma proof from the dag proof is used to prove the formula image. At the original references in the dag proof, we unfold the formula image of the identical sequent and thus, prove each reference with a short tree proof. The repeated process of cut introduction, unfolding formula images, and adding short proofs for references in the dag proof eventually leads to a tree proof. No additional search is involved, since the references in the dag proof indicate *which* sequents are subject to cut introduction. Moreover, the size of the resulting tree proof increases at most polynomially with respect to the size of the original dag proof.

We illustrate a main application of the presented cut-based translation, i.e. automated *extraction* of modular programs from intuitionistic sequent proofs. For this, we use the fact that the program extracted from an application of the cut rule can be interpreted as a procedure call. In our transformation, a cut is introduced to "factorize" subproofs of identical sequents in the original dag proof. If we regard the program extracted from this subproof as a procedure, its factorization using the cut rule realizes a (main) program which calls the procedure. When supporting interactive proof assistants with automated theorem provers for first-order intuitionistic logic (e.g., [19]), the design of the resulting programs strongly depends on the way the machine-found proofs are translated into the proof calculi of the systems. Standard cut-based transformations [15,4,18] on the one hand result in "unintuitive" programs with respect to the original specification, i.e. the formula to be proved valid. This is due to the fact that the cut rule is applied locally in order to overcome certain divergences between proof calculi. Permutation-based translations on the other hand may cause extensive duplications of subproofs [9,10], which means that the extracted programs contain identical code fragments multiple times. In contrast to that, our transformation produces "intuitive" programs, i.e. modular programs without duplications of identical code fragments, since the cut-rule is used as a procedural programming concept.

Section 2 introduces basic concepts and proof calculi we are using. In Section 3, we present and prove the polynomial transformation from dag sequent proofs to tree proofs with cuts. We show how our translation realizes the construction of procedural programs from proofs (Section 4). The conclusion contains a discussion of related work and some remarks on extensions of the translation method to other non-classical logics.

2   Definition and Notations

Throughout this paper, we use a first-order language consisting of *variables, constants, function symbols, predicate symbols, logical connectives, quantifiers* and *punctuation symbols*. *Terms* and *formulae* are defined according to the usual formation rules. For a first-order formula F, let $\mathcal{FV}(F)$ denote the set of all free variables in F.

Sequent calculi. A *sequent* S is an ordered pair of the form $\Gamma \vdash \Delta$, where Γ, Δ are finite *multisets* of first-order formulae. Γ is the *antecedent* of S and Δ is the *succedent* of S. The semantical meaning of a sequent $T \equiv A_1, \ldots, A_n \vdash B_1, \ldots, B_m$ is the same as the semantical meaning of the formula $F_T \equiv (\bigwedge_{i=1}^{n} A_i) \to (\bigvee_{i=1}^{m} B_i)$. The set of free variables for a sequent T is defined as $\mathcal{FV}(T) = \mathcal{FV}(F_T)$.

The *formula image* $\iota(T)$ of a sequent T is defined as $\iota(T) \equiv \forall c_1 \ldots \forall c_k. F_T$, where $\mathcal{FV}(T) = \{c_1, \ldots, c_k\}$. Sometimes, we use the schema $\forall_k(\Gamma_n^{\wedge} \to \Delta_m^{\vee})$ to abbreviate $\iota(T)$, where $\Gamma_n^{\wedge} \equiv \bigwedge_{i=1}^{n} A_i$ and $\Delta_m^{\vee} \equiv \bigvee_{i=1}^{m} B_i$.

As proof systems, we consider the cut-free classical sequent calculus $\mathcal{LK}$ as well as the two intuitionistic sequent calculi $\mathcal{LJ}_{mc}$ and $\mathcal{LJ}$. The calculus $\mathcal{LK}$ is depicted on the left hand side in Fig. 1. We consider *negation* as defined, e.g., $\neg A$ is $A \to \bot$, and use an additional $\bot$-*axiom*.

Structural rules such as contraction and weakening are not necessary for completeness of the above calculi, i.e. we use implicit contraction in the rules $\forall l, \exists r$, and $\to l$. We have added the weakening rules, since they are convenient for proof translations and allow shorter proofs (even in the propositional case) if dag proofs are considered [8].

The main difference between the intuitionistic sequent calculi $\mathcal{LJ}_{mc}$ and $\mathcal{LJ}$ is given by the fact that $\mathcal{LJ}$ sequents are restricted to at most one succedent formula whereas $\mathcal{LJ}_{mc}$ sequents are not. For $\mathcal{LJ}_{mc}$, we replace the rules $\to r, \forall r,$

$\mathcal{LK}$:

$$\overline{\Gamma,A \vdash A,\Delta}\ ax. \qquad \overline{\Gamma,\bot \vdash \Delta}\ \bot ax.$$

$$\frac{\Gamma \vdash \Delta}{\Gamma \vdash A,\Delta}\ wr \qquad \frac{\Gamma \vdash \Delta}{\Gamma,A \vdash \Delta}\ wl$$

$$\frac{\Gamma \vdash A,B,\Delta}{\Gamma \vdash A \vee B,\Delta}\ \vee r \qquad \frac{\Gamma,A \vdash \Delta \quad \Gamma,B \vdash \Delta}{\Gamma,A \vee B \vdash \Delta}\ \vee l$$

$$\frac{\Gamma \vdash A,\Delta \quad \Gamma \vdash B,\Delta}{\Gamma \vdash A \wedge B,\Delta}\ \wedge r \qquad \frac{\Gamma,A,B \vdash \Delta}{\Gamma,A \wedge B \vdash \Delta}\ \wedge l$$

$$\frac{\Gamma,A \vdash B,\Delta}{\Gamma \vdash A \rightarrow B,\Delta}\ \rightarrow r \qquad \frac{\Gamma \vdash A,\Delta \quad \Gamma,B \vdash \Delta}{\Gamma,A \rightarrow B \vdash \Delta}\ \rightarrow l$$

$$\frac{\Gamma \vdash A[x\backslash a],\Delta}{\Gamma \vdash \forall x.A,\Delta}\ \forall r\,a^* \qquad \frac{\Gamma,\forall x.A,A[x\backslash t] \vdash \Delta}{\Gamma,\forall x.A \vdash \Delta}\ \forall l\,t$$

$$\frac{\Gamma \vdash A[x\backslash t],\exists x.A,\Delta}{\Gamma \vdash \exists x.A,\Delta}\ \exists r\,t \qquad \frac{\Gamma,A[x\backslash a] \vdash \Delta}{\Gamma,\exists x.A \vdash \Delta}\ \exists l\,a^*$$

$\mathcal{LJ}_{mc}$:

$$\frac{\Gamma,A \vdash B}{\Gamma \vdash A \rightarrow B,\Delta}\ \rightarrow r$$

$$\frac{\Gamma \vdash A[x\backslash a]}{\Gamma \vdash \forall x.A,\Delta}\ \forall r\,a^*$$

$$\frac{\Gamma,A \rightarrow B \vdash A,\Delta \quad \Gamma,B \vdash \Delta}{\Gamma,A \rightarrow B \vdash \Delta}\ \rightarrow l$$

$\mathcal{LJ}$:

$$\frac{\Gamma \vdash A}{\Gamma \vdash A \vee B}\ \vee r\,1 \qquad \frac{\Gamma \vdash B}{\Gamma \vdash A \vee B}\ \vee r\,2$$

$$\frac{\Gamma \vdash A[x\backslash t]}{\Gamma \vdash \exists x.A}\ \exists r\,t$$

$$\frac{\Gamma,A \rightarrow B \vdash A \quad \Gamma,B \vdash \Delta}{\Gamma,A \rightarrow B \vdash \Delta}\ \rightarrow l$$

* a must not occur free in the conclusion of $\forall r$, $\exists l$ (eigenvariable condition).

Fig. 1. The sequent calculi $\mathcal{LK}$, $\mathcal{LJ}_{mc}$, and $\mathcal{LJ}$.

and $\rightarrow l$ with the rules shown on top of the right hand side in Fig. 1. For $\mathcal{LJ}$, we replace the rules $\vee r$, $\exists r$, and $\rightarrow l$ with the rules shown at the bottom of the right hand side in Fig. 1. Furthermore, $|\Delta| = 0$ must hold for the $ax.$ rule as well as for the remaining right (r) rules. For the rule $\bot ax.$ and all left (l) rules, it must hold $|\Delta| \leq 1$.

We will also consider $\mathcal{LK}$, $\mathcal{LJ}_{mc}$, and $\mathcal{LJ}$ extended by the following *cut* rule

$$\frac{\Gamma \vdash A,\Delta' \quad \Gamma,A \vdash \Delta}{\Gamma \vdash \Delta}\ cut$$

where $\Delta = \Delta'$ for $\mathcal{LK}$ and $\mathcal{LJ}_{mc}$. For $\mathcal{LJ}$, we have $|\Delta| \leq 1$ and $\Delta' = \emptyset$. The resulting calculi are denoted by $\mathcal{LK}+cut$, $\mathcal{LJ}_{mc}+cut$, and $\mathcal{LJ}+cut$, respectively. We use $\mathcal{LX}$ and $\mathcal{LX}+cut$ if we speak about properties that hold for all three sequent calculi above.

Sequent proofs. We distinguish between sequent proofs with *trees* as the underlying structure and proofs with *rooted directed acyclic graphs (dags)*. The former ones are more common whereas the latter ones allow more compact proofs.

A *tree proof* of a sequent E in each calculus $\mathcal{LX}$ is a finite (rooted and directed) tree with its nodes labeled with sequents such that the leaf nodes are labeled with axioms and the label of each non-leaf node is obtained from the label(s) of its successor node(s) by an application of a rule from $\mathcal{LX}$. The sequent E labeled at the root is called the *end sequent*. We say that an end sequent E is proven *from* a set $\mathcal{A}$ of sequents if $\mathcal{A}$ does not contain any axioms and there is a tree proof of E such that the sequents associated with the leaves of the tree are either axioms or elements from $\mathcal{A}$.

Dag proofs are considered as a special form of tree proofs.

Definition 1. *Let α be a tree proof of an end sequent E from $\mathcal{A} = \{A_1, \ldots, A_l\}$ in a calculus $\mathcal{LX}$, such that each A_i occurs at least twice in α. Then α is a dag proof of E, if the following conditions are satisfied:*

1. *There is a permutation π of $1, \ldots, l$ and a sequence of tree proofs $\alpha_{\pi(1)}, \ldots, \alpha_{\pi(l)}$ of end sequents $A_{\pi(1)}, \ldots, A_{\pi(l)}$ such that*

   a) *$\alpha_{\pi(1)}$ is a tree proof of $A_{\pi(1)}$ from $\emptyset$;*

   b) *$\alpha_{\pi(k)}$ $(1 < k \leq l)$ is a tree proof of $A_{\pi(k)}$ from $\{A_{\pi(1)}, \ldots, A_{\pi(k-1)}\}$;*

2. *each tree proof $\alpha_{\pi(i)}$ $(i \in \{1, \ldots, l\})$ occurs exactly once in α.*

Let $\mathcal{A}(\alpha)$ denote the set $\mathcal{A}$ associated with α. We call α a proper dag proof if $\mathcal{A}(\alpha) \neq \emptyset$.

It is immediately apparent that the graph induced by the construction in Definition 1 is acyclic. For each A_i (with possibly more than two occurrence in the dag proof), exactly one proof is present. The single occurrence of A_i with a proof is called a *reference sequent*; its proof is called *reference proof*. All occurrences of A_i without a proof are called *references* (to the reference sequent).

Example 1. Consider the formula $F_2 \equiv A_2 \wedge O_0 \wedge O_1 \wedge O_2 \rightarrow C$, where

$$O_i \equiv \begin{cases} A_0 \rightarrow C & \text{if } i = 0, \\ A_i \rightarrow (A_{i-1} \vee C) \vee A_{i-1} & \text{if } i = 1, 2, \end{cases}$$

and A_0, A_1, A_2, C are atoms. A dag proof α^d for F_2 in the calculus $\mathcal{LJ}$ is depicted in Fig. 2. The set $\mathcal{A}$ consists of the sequents $S_1 \equiv O_0, O_1, A_1 \vdash C$ and $S_2 \equiv O_0, A_0 \vdash C$. The proofs for the two reference sequents S_1 and S_2 are abbreviated by P_{S_1} and P_{S_2}. The two framed sequent occurrences in α^d refer to S_1 and S_2 (gray boxes), respectively, and thus, avoid the duplication of the proofs P_{S_1} and P_{S_2} in α^d.

Finally, we define two complexity measures for tree and dag proofs. Let α be a sequent proof in $\mathcal{LX}$. The *length* of α is defined by the number of sequent occurrences in α, written as $seq(\alpha)$. The *size* of α is defined by the number of character occurrences in α, written as $size(\alpha)$.

The following definition of a polynomial simulation is adapted from [7]: A calculus P_1 can *polynomially simulate* a calculus P_2 if there is a polynomial p such that the following holds: For every proof of a formula (or sequent) F in P_2 of size n, there is a proof of (the translation of) F in P_1, whose size is not greater than $p(n)$.

P_{S_1}

$$\cfrac{\cfrac{O_0, O_1, A_1 \vdash C \qquad \overline{O_0, O_1, C \vdash C}\;{}^{ax.}}{O_0, O_1, A_1 \vee C \vdash C}{}^{\vee l} \qquad \boxed{O_0, O_1, A_1 \vdash C}}{O_0, O_1, (A_1 \vee C) \vee A_1 \vdash C}{}^{\vee l}$$

$$\cfrac{\overline{A_2, O_0, O_1, O_2 \vdash A_2}\;{}^{ax.} \qquad \cfrac{\cfrac{O_0, O_1, (A_1 \vee C) \vee A_1 \vdash C}{A_2, O_0, O_1, (A_1 \vee C) \vee A_1 \vdash C}{}^{wl}}{}{}}{\cfrac{A_2, O_0, O_1, A_2 \to (A_1 \vee C) \vee A_1 \vdash C}{\vdash A_2 \wedge O_0 \wedge O_1 \wedge O_2 \to C}{}^{\to r, \wedge l, \wedge l, \wedge l}}{}^{\to l}$$

$\boxed{P_{S_1}}$ (repeating its end sequent $O_0, O_1, A_1 \vdash C$ from above):

P_{S_2}

$$\cfrac{\cfrac{O_0, A_0 \vdash C \qquad \overline{O_0, C \vdash C}\;{}^{ax.}}{O_0, A_0 \vee C \vdash C}{}^{\vee l} \qquad \boxed{O_0, A_0 \vdash C}}{O_0, (A_0 \vee C) \vee A_0 \vdash C}{}^{\vee l}$$

$$\cfrac{\overline{O_0, O_1, A_1 \vdash A_1}\;{}^{ax.} \qquad \cfrac{O_0, (A_0 \vee C) \vee A_0 \vdash C}{O_0, (A_0 \vee C) \vee A_0, A_1 \vdash C}{}^{wl}}{\underbrace{O_0, A_1 \to (A_0 \vee C) \vee A_0, A_1 \vdash C}_{O_1}}{}^{\to l}$$

$\boxed{P_{S_2}}$ (repeating its end sequent $O_0, A_0 \vdash C$ from above):

$$\underbrace{\cfrac{\overline{O_0, A_0 \vdash A_0}\;{}^{ax.} \qquad \overline{C, A_0 \vdash C}\;{}^{ax.}}{A_0 \to C, A_0 \vdash C}{}^{\to l}}_{O_0}$$

Fig. 2. Dag proof α^d for F_2 from Example 1.

3   Simulating Dag Proofs by Tree Proofs and Cut

We present a general method for translating first-order dag proofs in a calculus $\mathcal{LX}$ into tree proofs with cuts. Furthermore, we show that the size of the resulting tree proof is polynomial with respect to the size of the given dag proof.

The naive way for such a transformation would be to copy subproofs within a dag proof until all references have been completely eliminated, yielding the desired proof in tree format. Obviously, this approach would not be practicable since it could cause an exponential increase of proof size if the reference structure in the dag proof occurs in a nested manner. In order to avoid the exponential size increase in the worst case, we use applications of the cut rule, which take the formula images $\iota(D)$ of the reference sequents D as cut formulae. The following lemma follows immediately:

Lemma 1. Let $S \equiv \Gamma \vdash \Delta$ where $\Gamma \equiv G_1, \ldots, G_n$, $\Delta \equiv D_1, \ldots, D_m$, and $\mathcal{FV}(S)$ is $\{x_1, \ldots, x_k\}$. Then:

(i) The sequent $\vdash \iota(S)$ is provable from S in $\mathcal{LX}$ with $n + m + k - 1$ sequents.

(ii) $\Gamma, \iota(S) \vdash \Delta$ is provable in $\mathcal{LX}$ with $2 \cdot (m + n) + k - 1$ sequents.

In the next lemma, we construct a tree proof with cuts from a given dag proof. Since we provide a translation for arbitrary first-order proofs, the number of sequents in the resulting proof does not only depend on the number of sequents in the source proof, but also on the term structure and especially on the number of free variables, which occur in formulae of the source proof. Therefore, it is not surprising that the number of free variables influences the number of sequents occurring in the tree proof with cut.

Let α^d be a proper dag proof in $\mathcal{LX}$, i.e. $\mathcal{A}(\alpha^d) \neq \emptyset$, and let

$$m_v(\alpha^d) = \max\{|\mathcal{FV}(D)| \mid D \in \mathcal{A}(\alpha^d) \text{ is a reference sequent in } \alpha^d\}.$$

Let f_D be the number of formulae in the sequent D and let

$$m_f(\alpha^d) = \max\{f_D \mid D \text{ is a sequent in } \alpha^d\}.$$

For a tree proof α (with $\mathcal{A}(\alpha) = \emptyset$), we set $m_v(\alpha) = m_f(\alpha) = 0$. Finally, let $m_p(\alpha^d)$ denote the total number of occurrences of sequents from $\mathcal{A}(\alpha^d)$ in α^d. Observe that $m_p(\alpha) = 0$ for a tree proof α, and $m_p(\alpha^d) \geq 2$ for a proper dag proof α^d.

We prove that dag proofs *with cuts* can be translated to tree proofs with cuts with a polynomial increase of proof size. The use of source proofs with cuts is more general than it is required for our application.

Lemma 2. *Let α^d be a dag proof in $\mathcal{LX}$+cut. Then there exists a tree proof α of the same end sequent in $\mathcal{LX}$+cut with size $(\alpha) \leq \frac{5}{2} size \, (\alpha^d)^5$.*

Proof. Let α^d be a dag proof of the end sequent $S \equiv \Gamma \vdash \Delta$ in $\mathcal{LX}$+cut with an associated set $\mathcal{A}(\alpha^d)$. We proceed in two steps.

I. We show by induction on the number $k = |\mathcal{A}(\alpha^d)|$ that

$$seq \, (\alpha) \leq seq \, (\alpha^d) + 2 \cdot k \cdot m_p(\alpha^d) \cdot (m_f(\alpha^d) + m_v(\alpha^d) + \tfrac{k-1}{2}).$$

II. We estimate *size* (α) from *seq* (α) and the size of α^d.

I. The induction proof.
Base: $k = 0$. Then α^d is already in tree form, $\alpha = \alpha^d$, $\mathcal{A}(\alpha^d) = \emptyset$, and the relation on the number of sequents holds trivially.
Induction hypothesis. Assume $k > 0$ and, for each dag proof α^d in $\mathcal{LX}$+cut with $|\mathcal{A}(\alpha^d)| = k - 1$, there is a tree proof α in $\mathcal{LX}$+cut such that

$$seq \, (\alpha) \leq seq \, (\alpha^d) + 2 \cdot (k - 1) \cdot m_p(\alpha^d) \cdot (m_f(\alpha^d) + m_v(\alpha^d) + \tfrac{k-2}{2}).$$

Step. Consider a dag proof α_k^d with $k = |\mathcal{A}(\alpha_k^d)|$. Select arbitrarily from $\mathcal{A}(\alpha_k^d)$ a sequent $D_i \equiv \Gamma_i \vdash \Delta_i$ $(1 \leq i \leq k)$, whose proof is β_i^d, such that this proof is not contained in the proof of any other reference sequent (i.e. select a lower-most sequent). Let v_{D_i} denote $|\mathcal{FV}(D_i)|$ and let $p_{D_i} \geq 2$ be the number of occurrences of D_i in α_k^d. Observe that there are no references to D_i *within* β_i^d since a dag proof does not contain cycles. Cut out the proof β_i^d of D_i from α_k^d such that only D_i remains. Let γ_k^d denote this intermediate incomplete proof. Extend the antecedent of each sequent in γ_k^d by the formula image $\iota(D_i)$ of D_i. Since $\iota(D_i)$ is a formula without free variables, no eigenvariable conditions are violated.

Let $\tilde{\beta}_i$ be the tree proof of $\tilde{D}_i \equiv \Gamma_i, \iota(D_i) \vdash \Delta_i$ according to Lemma 1 (ii). Add a copy of $\tilde{\beta}_i$ over each occurrence of $\tilde{D}_i$ in γ_k^d. For the length of $\tilde{\beta}_i$, we obtain

$$seq\,(\tilde{\beta}_i) = 2 \cdot f_{D_i} + v_{D_i} - 1 < 2 \cdot m_f(\alpha_k^d) + m_v(\alpha_k^d).$$

Next, apply Lemma 1 (i) in order to extend the proof β_i^d of D_i to a proof of $\vdash \iota(D_i)$. Then we introduce all elements of Γ, Δ' from the end sequent S by applications of weakening. Observe that Δ' is empty in case of $\mathcal{LJ}+cut$, and $\Delta' = \Delta$ for $\mathcal{LK}+cut$ and $\mathcal{LJ}_{mc}+cut$. This results in the sequent $\Gamma \vdash \iota(D_i), \Delta'$. For the additional number of sequents $a(\beta_i^d)$, the following relation holds:

$$a(\beta_i^d) \leq f_{D_i} + v_{D_i} - 1 + f_S \leq 2 \cdot m_f(\alpha_k^d) + m_v(\alpha_k^d) - 1.$$

Finally, one additional sequent is introduced by the cut rule. Combining the modified subproofs yields the following proof which we call α_{k-1}^d:

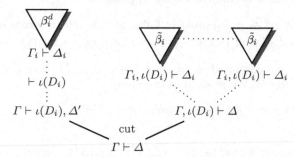

Observe that no new references have been introduced, but all references for D_i have been eliminated and therefore, $|\mathcal{A}(\alpha_{k-1}^d)| = k - 1$. Consequently, $p_{D_i} \leq m_p(\alpha_k^d)$. Moreover, $m_p(\alpha_{k-1}^d) < m_p(\alpha_k^d)$ and $m_v(\alpha_{k-1}^d) = m_v(\alpha_k^d)$ since no new free variables have been introduced by our manipulations. For $m_f(\alpha_{k-1}^d)$, we obtain $m_f(\alpha_{k-1}^d) \leq m_f(\alpha_k^d) + 1$ because $\iota(D_i)$ has been added to sequents in α_{k-1}^d.

Let us estimate the increase of length by the above manipulations resulting in α_{k-1}^d from α_k^d. From $seq\,(\alpha_{k-1}^d) = seq\,(\alpha_k^d) + p_{D_i} \cdot seq\,(\tilde{\beta}_i) + a(\beta_i^d) + 1$ and some calculations, we obtain the estimation

$$seq\,(\alpha_{k-1}^d) \leq seq\,(\alpha_k^d) + 2 \cdot p_{D_i} \cdot (m_f(\alpha_k^d) + m_v(\alpha_k^d)). \tag{1}$$

Since $|\mathcal{A}(\alpha_{k-1}^d)| = k - 1$, the induction hypothesis applies. We obtain:

$$seq\,(\alpha) \leq seq\,(\alpha_{k-1}^d) + 2 \cdot (k-1) \cdot m_p(\alpha_{k-1}^d) \cdot (m_f(\alpha_{k-1}^d) + m_v(\alpha_{k-1}^d) + \tfrac{k-2}{2}).$$

Using (1), $p_{D_i} \leq m_p(\alpha_k^d)$, $m_p(\alpha_{k-1}^d) < m_p(\alpha_k^d)$, $m_f(\alpha_{k-1}^d) \leq m_f(\alpha_k^d) + 1$, and $m_v(\alpha_{k-1}^d) = m_v(\alpha_k^d)$, the induction proof concludes as follows:

$$\begin{aligned} seq\,(\alpha) \leq\ & seq\,(\alpha_k^d) + 2 \cdot m_p(\alpha_k^d) \cdot (m_f(\alpha_k^d) + m_v(\alpha_k^d)) + \\ & 2 \cdot (k-1) \cdot m_p(\alpha_k^d) \cdot (m_f(\alpha_k^d) + 1 + m_v(\alpha_k^d) + \tfrac{k-2}{2}) \\ =\ & seq\,(\alpha_k^d) + 2 \cdot k \cdot m_p(\alpha_k^d) \cdot (m_f(\alpha_k^d) + m_v(\alpha_k^d) + \tfrac{k-1}{2}). \end{aligned}$$

II. The estimation of $size\,(\alpha)$.
Let us simplify some parameters from the above inequality. We can replace $2 \cdot k$ by $seq\,(\alpha_k^d)$ because, for each reference sequent, at least one additional reference must occur in α_k^d. With $m_p(\alpha_k^d) \leq seq\,(\alpha_k^d)$, we calculate

$$seq\,(\alpha) \leq seq\,(\alpha_k^d)^2 \cdot (m_f(\alpha_k^d) + m_v(\alpha_k^d) + \tfrac{k+1}{2}).$$

The size of any sequent in α is bounded by $(k+1) \cdot size\,(\alpha_k^d)$ because k additional formula images occur in any sequent, each of these images has size $\leq size\,(\alpha_k^d)$ and the original part of the sequent has size $\leq size\,(\alpha_k^d)$. Therefore, $size\,(\alpha) \leq (k+1) \cdot size\,(\alpha_k^d) \cdot seq\,(\alpha)$. With $seq\,(\alpha_k^d), (k+1), m_f(\alpha_k^d), m_v(\alpha_k^d) \leq size\,(\alpha_k^d)$, we obtain $size\,(\alpha) \leq \tfrac{5}{2} \cdot size\,(\alpha_k^d)^5$. □

From the constructive proof of Lemma 2, we can derive a recursive polynomial-time procedure which translates a given dag proof in $\mathcal{LX}$ into a tree proof in $\mathcal{LX}+cut$. In particular, no additional search is involved since the reference sequents in the dag proof completely guide the process of cut introduction. The estimation of the increase of proof size is rather generous; in most practical cases, the increase is much less than the worst-case bound. The resulting tree proof can then be interpreted directly by an interactive proof assistant. We conclude with

Theorem 1. *Tree $\mathcal{LX}+cut$ polynomially simulates dag $\mathcal{LX}+cut$.*

Refinements of the translation. So far, we have presented a general method to translate arbitrary first-order dag proofs in $\mathcal{LX}$ into tree proofs with cuts. It is also possible to introduce the cuts locally and not necessarily immediately above the end sequent of the proof. The cuts are introduced where they are needed, avoiding the extension of each sequent in the whole proof with all formula images. This results in a tree proof with an improved structure.

The local introduction of cuts works as follows: From a given dag proof of $\Phi \vdash \Psi$ in $\mathcal{LX}$, we first select an arbitrary reference sequent $D_i \equiv \Gamma_i \vdash \Delta_i$. Then, we search the top-most sequent $T \equiv \Gamma \vdash \Delta$ such that all occurrences of

D_i are within the proof of T. The cut rule is now introduced directly above T such that T becomes the conclusion of the cut. The sequents $\Gamma \vdash \iota(D_i), \Delta'$ and $\Gamma_i, \iota(D_i) \vdash \Delta_i$ are proven in the same way as in the proof of Lemma 2. Then, the following proof corresponds to α_{k-1}^d in the proof of Lemma 2, where $\Delta' = \emptyset$ in $\mathcal{LJ}+cut$ and $\Delta' = \Delta$ in $\mathcal{LJ}_{mc}+cut$ and $\mathcal{LK}+cut$:

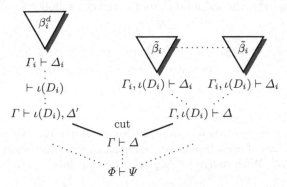

The following example illustrates this local cut introduction in the calculus $\mathcal{LJ}+cut$.

Example 2. We apply the translation to the $\mathcal{LJ}$ dag proof α^d (Fig. 2) for the formula F_2 from Example 1. For the two reference sequents S_1, S_2, we obtain as cut formulae the formula images $\iota(S_1) \equiv O_0 \wedge O_1 \wedge A_1 \rightarrow C$ and $\iota(S_2) \equiv O_0 \wedge A_0 \rightarrow C$. The resulting replacements leading to a tree proof α in $\mathcal{LJ}+cut$ are shown in Fig. 3. The subproof starts with the sequent $T_1 \equiv \Gamma_{T_1} \vdash C$ below the first cut introduction, where $\Gamma_{T_1} \equiv O_0, O_1, (A_1 \vee C) \vee A_1$. The second cut introduction takes place above the sequent $T_2 \equiv \Gamma_{T_2} \vdash C$ in *cut proof 1*, where $\Gamma_{T_2} \equiv O_0, (A_0 \vee C) \vee A_0$ (gray boxes in Fig. 3). The cut proof α can be completed by copying the first 7 inferences (starting from the end sequent) from the dag proof α^d in Fig. 2.

4   Cuts and the Structure of Programs

A central application for intuitionistic logic is the extraction of programs from proofs. Constructive program synthesis relies on the parallel process of program *construction* and program *verification*. Formalizing a logical specification within a constructive logic, e.g., Intuitionistic Type Theory (ITT) [16], this specification "formula" will first be proven valid using a sequent calculus for ITT. More precisely, one finds a *constructive* proof for the existence of a function f which maps input elements to output elements of the specified program. Then, in a second step, f will be *extracted* from the computational content of the proof according to the "proofs-as-programs" paradigm [2]. Hence, f forms a correctly verified *program term* with respect to the given specification. The whole process is performed interactively within an constructive program development systems, for example, the NuPRL system [3,1].

$$\boxed{\text{cut appl. 1}} \quad \cfrac{\cfrac{\cfrac{O_0, O_1, C, \iota(S_1) \vdash C}{O_0, O_1, A_1 \vee C, \iota(S_1) \vdash C} \text{ ax.}}{O_0, O_1, (A_1 \vee C) \vee A_1, \iota(S_1) \vdash C} \ \lor l}{} \quad \boxed{\text{cut appl. 1}}$$

$$\boxed{\text{cut proof 1}} \quad \Gamma_{T_1} \vdash \iota(S_1)$$

$$\cfrac{\Gamma_{T_1} \vdash \iota(S_1) \qquad O_0, O_1, (A_1 \vee C) \vee A_1, \iota(S_1) \vdash C}{\underbrace{O_0, O_1, (A_1 \vee C) \vee A_1 \vdash C}_{\Gamma_{T_1}}} \ \text{cut } \iota(S_1)$$

$\boxed{\text{cut proof 1}}$ (repeating its end sequent $\Gamma_{T_1} \vdash \iota(S_1)$ from above):

$$\boxed{\text{cut appl. 2}} \quad \cfrac{\cfrac{O_0, C, \iota(S_2) \vdash C}{O_0, A_0 \vee C, \iota(S_2) \vdash C} \text{ ax.}}{O_0, (A_0 \vee C) \vee A_0, \iota(S_2) \vdash C} \ \lor l \quad \boxed{\text{cut appl. 2}}$$

$$\boxed{\text{cut proof 2}} \quad \Gamma_{T_2} \vdash \iota(S_2)$$

$$\cfrac{O_0, (A_0 \vee C) \vee A_0 \vdash C}{} \ \text{cut } \iota(S_2)$$

$$\cfrac{\cfrac{\cfrac{O_0, O_1, A_1 \vdash A_1}{} \text{ ax.} \qquad \cfrac{\cfrac{O_0, (A_0 \vee C) \vee A_0 \vdash C}{O_0, (A_0 \vee C) \vee A_0, A_1 \vdash C} \ wl}{O_0, A_1 \to (A_0 \vee C) \vee A_0, A_1 \vdash C} \to l}{\vdash O_0 \wedge O_1 \wedge A_1 \to C} \ \to r, \wedge l, \wedge l}{\underbrace{\Gamma_{T_1} \vdash O_0 \wedge O_1 \wedge A_1 \to C}_{\iota(S_1)}} \ 3 \times wl$$

$\boxed{\text{cut proof 2}}$ (repeating its end sequent $\Gamma_{T_2} \vdash \iota(S_2)$ from above):

$$\cfrac{\cfrac{\cfrac{O_0, A_0 \vdash A_0 \ \text{ax.} \qquad C, A_0 \vdash C \ \text{ax.}}{A_0 \to C, A_0 \vdash C} \to l}{\vdash O_0 \wedge A_0 \to C} \ \to r, \wedge l}{\underbrace{\Gamma_{T_2} \vdash O_0 \wedge A_0 \to C}_{\iota(S_2)}} \ 2 \times wl$$

$\boxed{\text{cut appl. 1}}$ (two applications of wl have been introduced due to space restrictions):

$$\cfrac{\cfrac{\cfrac{O_0, O_1, A_1 \vdash O_0 \ \text{ax.} \qquad \cfrac{\cfrac{O_1, A_1 \vdash O_1 \ \text{ax.} \qquad O_1, A_1 \vdash A_1 \ \text{ax.}}{O_1, A_1 \vdash O_1 \wedge A_1} \land r}{O_0, O_1, A_1 \vdash O_1 \wedge A_1} \ wl}{\cfrac{O_0, O_1, A_1 \vdash O_0 \wedge O_1 \wedge A_1}{O_0, O_1, A_1, \iota(S_1) \vdash O_0 \wedge O_1 \wedge A_1} \ wl} \land r}{\underbrace{O_0, O_1, A_1, O_0 \wedge O_1 \wedge A_1 \to C \vdash C}_{\iota(S_1)}} \qquad O_0, O_1, A_1, C \vdash C \ \text{ax.}}{} \to l$$

$\boxed{\text{cut appl. 2}}$

$$\cfrac{\cfrac{\cfrac{O_0, A_0, \iota(S_2) \vdash O_0 \ \text{ax.} \qquad O_0, A_0, \iota(S_2) \vdash A_0 \ \text{ax.}}{O_0, A_0, \iota(S_2) \vdash O_0 \wedge A_0} \land r \qquad O_0, A_0, C \vdash C \ \text{ax.}}{\underbrace{O_0, A_0, O_0 \wedge A_0 \to C \vdash C}_{\iota(S_2)}} \to l}{}$$

Fig. 3. The resulting tree proof α with cuts from Example 2.

$\mathcal{LJ}$:

$$\frac{}{\Gamma, x:A \vdash A \ \lfloor\mathbf{ext}\ x\rfloor} \ ax. \qquad\qquad \frac{}{\Gamma, x:\bot \vdash A \ \lfloor\mathbf{ext}\ \mathrm{any}(x)\rfloor} \ \bot ax.$$

$$\frac{\Gamma \vdash A \ \lfloor\mathbf{ext}\ a\rfloor \quad \Gamma \vdash B \ \lfloor\mathbf{ext}\ b\rfloor}{\Gamma \vdash A \wedge B \ \lfloor\mathbf{ext}\ \langle a,b\rangle\rfloor} \ \wedge r \qquad\qquad \frac{\Gamma \vdash C \ \lfloor\mathbf{ext}\ t\rfloor}{\Gamma, x:T \vdash C \ \lfloor\mathbf{ext}\ t\rfloor} \ wl$$

$$\frac{\Gamma \vdash A \ \lfloor\mathbf{ext}\ a\rfloor}{\Gamma \vdash A \vee B \ \lfloor\mathbf{ext}\ \mathrm{inl}(a)\rfloor} \ \vee r\,l \qquad\qquad \frac{\Gamma, a:A, b:B \vdash C \ \lfloor\mathbf{ext}\ u\rfloor}{\Gamma, z:A \wedge B \vdash C \ \lfloor\mathbf{ext}\ \mathrm{let}\ \langle a,b\rangle = z\ \mathrm{in}\ u\rfloor} \ \wedge l$$

$$\frac{\Gamma \vdash B \ \lfloor\mathbf{ext}\ b\rfloor}{\Gamma \vdash A \vee B \ \lfloor\mathbf{ext}\ \mathrm{inr}(b)\rfloor} \ \vee r\,2 \qquad\qquad \frac{\Gamma, a:A \vdash C \ \lfloor\mathbf{ext}\ u\rfloor \quad \Gamma, b:B \vdash C \ \lfloor\mathbf{ext}\ v\rfloor}{\Gamma, z:A \vee B \vdash C \ \lfloor\mathbf{ext}\ \mathrm{case}\ z\ \mathrm{of}\ \mathrm{inl}(a) \mapsto u \mid \mathrm{inr}(b) \mapsto v\rfloor} \ \vee l$$

$$\frac{\Gamma, x:A \vdash B \ \lfloor\mathbf{ext}\ b\rfloor}{\Gamma \vdash A \to B \ \lfloor\mathbf{ext}\ \lambda x.b\rfloor} \ \to r \qquad\qquad \frac{\Gamma, pf:A \to B \vdash A \ \lfloor\mathbf{ext}\ a\rfloor \quad \Gamma, y:B \vdash C \ \lfloor\mathbf{ext}\ b\rfloor}{\Gamma, pf:A \to B \vdash C \ \lfloor\mathbf{ext}\ b[y\backslash pf\ a]\rfloor} \ \to l$$

$$\frac{\Gamma, x':T \vdash A[x\backslash x'] \ \lfloor\mathbf{ext}\ a\rfloor}{\Gamma \vdash \forall x:T.A \ \lfloor\mathbf{ext}\ \lambda x'.a\rfloor} \ \forall r\,x'\,{}^{*} \qquad\qquad \frac{\Gamma, pf:\forall x:T.A, y:A[x\backslash t] \vdash C \ \lfloor\mathbf{ext}\ u\rfloor}{\Gamma, pf:\forall x:T.A \vdash C \ \lfloor\mathbf{ext}\ u[y\backslash pf\ t]\rfloor} \ \forall l\,t\,{}^{**}$$

$$\frac{\Gamma \vdash A[x\backslash t] \ \lfloor\mathbf{ext}\ a\rfloor}{\Gamma \vdash \exists x:T.A \ \lfloor\mathbf{ext}\ \langle t,a\rangle\rfloor} \ \exists r\,t\,{}^{**} \qquad\qquad \frac{\Gamma, x':T, a:A[x\backslash x'] \vdash C \ \lfloor\mathbf{ext}\ u\rfloor}{\Gamma, z:\exists x:T.A \vdash C \ \lfloor\mathbf{ext}\ \mathrm{let}\ \langle x',a\rangle = z\ \mathrm{in}\ u\rfloor} \ \exists l\,x'\,{}^{*}$$

$\mathcal{LJ}+cut$:

$$\frac{\Gamma \vdash T \ \lfloor\mathbf{ext}\ s\rfloor \quad \Gamma, x:T \vdash C \ \lfloor\mathbf{ext}\ t\rfloor}{\Gamma \vdash C \ \lfloor\mathbf{ext}\ (\lambda x.t)\ s\rfloor} \ cut\,x\,T$$

* a must not occur free in the conclusion of $\forall r$, $\exists l$ (eigenvariable condition).

** t must satisfy the declaration condition for $\forall l$ and $\exists r$: All free variables in t must be declared in Γ.

Fig. 4. The calculus $\mathcal{LJ}$ with extract term annotations.

Logic and type theory. Whereas ITT is designed as a higher-order logic in order to express all kinds of reasoning about programs and mathematics, its first-order fragment corresponds to first-order intuitionistic logic. According to the "formulas-as-types paradigm" [13], every (intuitionistic) formula corresponds to a type. Elements of these types can be constructed by proving the corresponding formulae in $\mathcal{LJ}$ using the "Curry-Howard-isomorphism" [5,20].

In order to model this process *within* the calculus $\mathcal{LJ}$, we need to provide two kinds of modifications. First, every calculus rule will be annotated with some term pattern for *program (term) extraction*. For instance, the $\wedge r$ rule will look like

$$\frac{\Gamma \vdash A \ \lfloor\mathbf{ext}\ a\rfloor \quad \Gamma \vdash B \ \lfloor\mathbf{ext}\ b\rfloor}{\Gamma \vdash A \wedge B \ \lfloor\mathbf{ext}\ \langle a,b\rangle\rfloor} \ \wedge r$$

and is interpreted as follows: If term a can be extracted from the proof for $\Gamma \vdash A$, that is, a has been constructed as an element of data type A, and term b can be extracted from the proof for $\Gamma \vdash B$, then the pair $\langle a,b\rangle$ will be extracted from $\Gamma \vdash A \wedge B$.

For the second modification to integrate program extraction into $\mathcal{LJ}$, we have to represent type elements as part of sequents. Every free (first-order) variable in a sequent $S \equiv \Gamma \vdash C$ has to be *declared* in the *hypotheses* Γ. That means, the variable denotes an (object) parameter which has already been constructed during the proof. For simplicity, we assume a general object type T for all first-order variables and parameters, i.e. we do not consider multi-typed first-order logic. Then, a *parameter declaration* is an expression of the form $c : T$, where c is a first-order parameter and T is our general object type. Moreover, every formula F in Γ interpreted as a type is part of a declaration and thus, declares an (object) variable z of type F. Hence, a *formula declaration* is an expression of the form $z : F$, where z is a variable and F is a (first-order) formula.

We redefine a sequent as an ordered pair $\Gamma \vdash C$, where C is a formula and Γ is a *set* of parameter and formula declarations, where all objects are pairwisely distinct. Let $S \equiv \Gamma \vdash C$ be a sequent with the free variables $\mathcal{FV}(S) = \{c_1, \ldots, c_m\}$, and let $x_1 : S_1, x_2 : S_2, \ldots, x_n : S_n$ be the formula declarations in Γ. Then, the *closed sequent* S_c of S is defined as

$$S_c \equiv c_1 : T, \ldots, c_m : T, x_1 : S_1, x_2 : S_2, \ldots, x_n : S_n \vdash C.$$

The set $\{c_1, \ldots, c_m\}$ is also called the set of *declared parameters* in S_c. In the following, we assume exclusively closed sequents in sequent proofs. For this, we will use "sequent" instead of "closed sequent" when it is clear from the context.

The modifications in $\mathcal{LJ}$ take place at the quantifier rules. First-order formulae of the form $\mathcal{Q}x. F$ will be extended by the object type to $\mathcal{Q}x : T. F$, for $\mathcal{Q} \in \{\forall, \exists\}$. The rules $\exists l\, x'$ and $\forall r\, x'$ add a new declaration $x' : T$ to Γ for the introduced eigenvariable x'. The rules $\exists r\, t$ and $\forall l\, t$ have to respect the *declaration condition*: All free variables in t must already have been declared in Γ. That is, all (first-order) objects must have been constructed in the proof before using them for quantifier instantiations.

The complete sequent calculi $\mathcal{LJ}$ and $\mathcal{LJ}+cut$ with declarations and extract term annotations are shown in Fig. 4. We use the λ-calculus as "programming" language for the extract terms. The syntax used in Fig. 4 is for improving readability only. It is derived from the programming language ML and corresponds to the standard display form in the NuPRL system [1]. An actual program construction from a closed sequent $\Gamma \vdash C$ works as follows: First, we prove the validity of $\Gamma \vdash C$. Then, we extract the program term by stepwise instantiation of the annotated term patterns at each applied inference rule in the proof (starting from the axiom rules).

Extracting procedural programs from $\mathcal{LJ}+cut$ proofs. The cut-based transformation from Section 3 converts a dag proof α^d in $\mathcal{LJ}$ into a tree proof α in $\mathcal{LJ}+cut$. The proof α can be transformed easily into a proof for the extended calculus $\mathcal{LJ}$ with closed sequents, containing formula and parameter declarations. Then, program extraction can be performed using the appropriate term annotations at the inference rules. [1] In particular, a modular structured program

---

[1] As pointed out by a referee, formulations of sequent calculi are computationally sensitive. The use of multisets and implicit contraction in $\mathcal{LJ}$ from Section 2 suppresses computational anomalies when switching to the annotated calculus in Fig. 4 (see [21,23] for details).

Table 1. Formula declarations for subformulae and cut formulae of Example 3.

Subformulae of $F_2 \equiv A_2 \wedge O_o \wedge O_1 \wedge O_2 \to C$:

$c : C$ $x_1 : A_2 \wedge O_0$

for $i = 0, 1, 2$ $a_i : A_i$ and $pf_i : O_i$ $x_2 : A_2 \wedge O_0 \wedge O_1$

for $i = 1, 2$ $z_i : A_{i-1} \vee C$ and $y_i : (A_{i-1} \vee C) \vee A_{i-1}$ $x_3 : A_2 \wedge O_0 \wedge O_1 \wedge O_2$

Cut formula $\iota(S_1) \equiv O_0 \wedge O_1 \wedge A_1 \to C$ Cut formula $\iota(S_2) \equiv O_0 \wedge A_0 \to C$

$S_1' : O_1 \wedge A_1$ $S_2 : O_0 \wedge A_0$

$S_1 : O_0 \wedge O_1 \wedge A_1$ $pf_{S_2} : O_0 \wedge A_0 \to C$

$pf_{S_1} : O_0 \wedge O_1 \wedge A_1 \to C$

can be automatically synthesized from the sequent proof by taking advantage of the introduced cut rules.

In our transformation, a cut rule will only be applied in order to factorize subproofs of identical sequents. That is, a *cut introduction* provides a proof for (the formula image of) a certain sequent S, whereas the *cut applications* "apply" this proof whenever an identical sequent S occurs in the remaining proof. As this process is known as lemma application in proof theory, it can be transferred to *procedural programming concepts* on the side of program extraction. Consider the cut rule with its extract terms:

$$\frac{\Gamma \vdash T \;\; \lfloor \textbf{ext } s \rfloor \quad \Gamma, x : T \vdash C \;\; \lfloor \textbf{ext } t \rfloor}{\Gamma \vdash C \;\; \lfloor \textbf{ext } (\lambda x . t) s \rfloor} \; cut \; x \; T$$

In the resulting program $(\lambda x . t) s$, the term s is extracted from the proof of the cut formula (lemma) T and thus, s represents a *procedure*. Instead of repeating the code for s within the (main) program t at every occurrence of x, the abstraction $\lambda x . t$ avoids duplication. Program evaluation, for instance in the NuPRL system, is based on *lazy evaluation*. In our case, β-reduction is first applied to $(\lambda x . t) s$. Then, lazy evaluation proceeds on the result $t[x \backslash s]$. Thus, every occurrence of x in t simulates a call-by-name of the procedure s (see [3] for details). From this viewpoint, our cut-based proof transformation allows the automatic generation of procedural and hence, short programs with the effect that the procedures will not be unfolded before runtime.

Example 3. Reconsider the cut proof for F_2 from Example 2, as shown in Fig. 3 together with the first 7 inferences (starting from the end sequent) of Fig. 2. We now develop the extract term for this proof by using the annotated sequent calculus from Fig. 4. The formula declarations for the subformulae of F_2 as well as for the two cut formulae $\iota(S_1)$ and $\iota(S_2)$ are shown in Table 1. The resulting program terms extracted from the proof fragments of Fig. 3 and Fig. 2 are depicted in Table 2. The procedures in the final program term are the two cut proofs cp_1 and cp_2, where cp_2 occurs within cp_1 due to the local introduction of cuts in our refined translation. The two cut introductions are represented

Table 2. Procedural program terms extracted from the cut proof of Example 3.

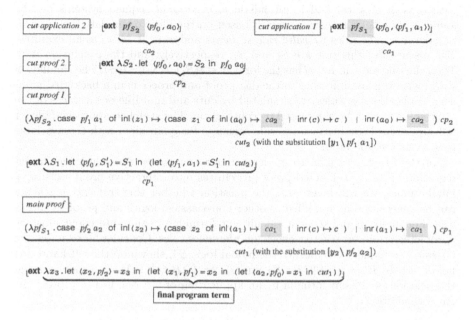

by the subterms cut_1 and cut_2, respectively. The λ-abstraction in cut_1 avoids duplication of the code for cp_1 (two gray boxes), i.e. replacing the function variable pf_{S_1} in the subterm ca_1 with cp_1 (see gray box in *cut application 1*). Similarly, the λ-abstraction in cut_2 avoids duplication of the code for cp_2. Since cut_2 occurs within the procedure cp_1, the two cut introductions prevent the code for cp_2 to be copied four times, two times in each copy of cp_1, and thus, yield a procedural and short program.

5   Conclusion, Related and Future Work

We have established a polynomial simulation of dag proofs by tree proofs with cuts for classical and intuitionistic sequent calculi. This simulation leads to a polyomial-time computable translation procedure which is used to integrate forward-directed automated theorem provers into interactive proof development systems based on the usual classical and intuitionistic sequent calculi (in tree form). The outputs of these theorem provers correspond directly to sequent proofs in dag form and thus, guide the translation without involving any additional search. We have presented a central application of our translation, namely the generation of modular programs from first-order intuitionistic proofs. Here, the introduction of cuts is interpreted as a procedural programming concept with a call-by-name of the procedures when evaluating the resulting programs.

A completely different approach for translating dag proofs into tree proofs with cuts in classical and intuitionistic *propositional* sequent systems can be found in [14]. Letz' transformation is based on the idea to simulate the applications of inference rules by *valid rule sequents* and cuts. Take $\vee l$ as an example. The premises of this rule are S_1 and S_2, respectively, and the conclusion is S. Then the rule sequent for $\vee l$ has the form $R \equiv \iota(S_1), \iota(S_2) \vdash \iota(S)$. The simulation starts with the last inference in the dag proof and proceeds in a backward manner. Applications of rules are simulated by cuts and multiple occurrences of the same subproof can be eliminated by contraction. The sketched approach fails for full first-order logic because the rule sequents for quantifiers with eigenvariable conditions are not valid.

In the future, we plan to implement our translation procedure for the integration of forward-directed proof procedures into interactive proof assistants. Furthermore, we will investigate the question whether this translation schema can be generalized and applied to other non-classical logics and proof calculi as well. For many modal sequent calculi, cut formulae in the antecedent must be "protected" from destructive effects of inferences rules. Consider for example the $\Box r$ rule in a sequent calculus for the modal logic $S4$, shown on the left hand side below, where $\Gamma^* = \{\Box B | \Box B \in \Gamma\}$ and $\Delta^* = \{\Diamond B | \Diamond B \in \Delta\}$. In order to avoid the deletion of the cut formula by an application of $\Box r$, one has to protect the cut formula by $\Box$.

$$\frac{\Gamma^* \vdash F, \Delta^*}{\Gamma \vdash \Box F, \Delta} \ \Box r \qquad\qquad \frac{\Gamma, \iota(S), \forall_k(\Gamma_n^\wedge \to \Delta_m^\vee) \vdash \Delta}{\Gamma, \iota(S) \vdash \Delta} \ \Box l$$

The formula image $\iota(S)$ for a sequent $S \equiv \Gamma \vdash \Delta$ is then of the form $\Box\forall_k(\Gamma_n^\wedge \to \Delta_m^\vee)$. The proof of the sequent $\Gamma, \iota(S) \vdash \Delta$ ends with an application of the $\Box l$ rule (see right hand side above). For sequent calculi for S4, a modified translation schema exists. We will investigate the applicability of the translation schema to other kinds of non-classical logics with classical as well as intuitionistic logic as basis.

References

1. S. F. ALLEN, R. L. CONSTABLE ET AL. The NuPRL Open Logical Environment. In *CADE-17*, LNAI 1831, pp. 170–176, Springer, 2000.

2. J. L. BATES AND R. L. CONSTABLE. Proofs as Programs. *ACM Transactions on Programming Languages and Systems*, 7(1):113–136, January 1985.

3. R. L. CONSTABLE, S. F. ALLEN ET AL. *Implementing Mathematics with the NuPRL Proof Development System*. Prentice Hall, 1986.

4. H. B. CURRY. *Foundations of Mathematical Logic*. Dover, Dover edition, 1977.

5. H. B. CURRY, R. FEYS, AND W. CRAIG. *Combinatory Logic*, volume 1. North–Holland, Amsterdam, 1958.

6. A. DEGTYAREV AND A. VORONKOV. The Inverse Method. In J. A. Robinson and A. Voronkov, editors, *Handbook of Automated Reasoning*, pp. 177–268, Elsevier Science Publishers, 2001.

7. E. EDER. *Relative Complexities of First Order Calculi*. Vieweg, 1992.

8. U. EGLY. On Different Intuitionistic Calculi and Embeddings from INT to S4. *Studia Logica*, 2001. To appear

9. U. EGLY AND S. SCHMITT. Intuitionistic Proof Transformations and their Application to Constructive Program Synthesis. In AISC'98, LNAI 1476, pp. 132–144, 1998.

10. U. EGLY AND S. SCHMITT. On Intuitionistic Proof Transformations, Their Complexity and Application to Constructive Program Synthesis. *Fundamenta Informaticae*, 39:59–83, 1999.

11. M. C. FITTING. *Intuitionistic Logic, Model Theory and Forcing*. Studies in logic and the foundations of mathematics. North–Holland, 1969.

12. G. GENTZEN. Untersuchungen über das logische Schließen. *Mathematische Zeitschrift*, 39:176–210, 405–431, 1935.

13. W. HOWARD. The Formulas-as-Types Notion of Construction. In *To H. B. Curry: Essays in Combinatory Logic, Lambda Calculus and Formalism.*, pp. 479–490, Academic Press 1980.

14. R. LETZ. Polynomial Simulation of Sequent Systems by Tree Sequent Systems. Technical report, Institut für Informatik, TU München, 1992.

15. S. MAEHARA. Eine Darstellung der intuitionistischen Logik in der klassischen. *Nagoya Mathematical Journal*, 7:45–64, 1954.

16. P. MARTIN-LÖF. *Intuitionistic Type Theory*, volume 1 of *Studies in Proof Theory Lecture Notes*. Bibliopolis, Napoli, 1984.

17. G. MINTS. Resolution Strategies for the Intuitionistic Logic. In B. Mayoh, E. Tyugu, and J. Penyam, editors, *Constraint Programming*, Nato ASI Series, pp. 289–311. Springer Verlag, 1994.

18. S. SCHMITT AND C. KREITZ. On Transforming Intuitionistic Matrix Proofs into Standard-sequent Proofs. In 4th *TABLEAUX Workshop*, LNAI 918, pp. 106–121, 1995.

19. S. SCHMITT, L. LORIGO, C. KREITZ, AND A. NOGIN. JProver: Integrating Connection-based Theorem Proving into Interactive Proof Assistants. In *IJCAR*, this volume, 2001.

20. W. W. TAIT. Intensional Interpretation of Functionals of Finite Type. *Journal of Symbolic Logic*, 32(2):187–199, 1967.

21. A. S. TROELSTRA Marginalia on Sequent Calculi. *Studia Logica*, 62:291–303,1999.

22. A. S. TROELSTRA AND H. SCHWICHTENBERG. *Basic Proof Theory*. Second Edition. Cambridge Univ. Press, 2000.

23. R. VESTERGAARD Revisting Kreisel: A Computational Anomaly in the Troelstra-Schwichtenberg G3i System. Technical report, 1999. `http://iml.univ-mrs.fr:80/~vester/Writings/index.html`

A General Method for Using Schematizations in Automated Deduction

Nicolas Peltier

Centre National de la Recherche Scientifique
Laboratoire LEIBNIZ-IMAG
46, Avenue Félix Viallet 38031 Grenoble Cedex - FRANCE
E-mail: Nicolas.Peltier@imag.fr
Phone: (33) 4 76 57 48 05

Abstract. We propose a new method for using recurrent schematizations in Theorem Proving. We provide techniques for detecting cycles in proofs (via proof generalization), and we show how to take advantage of the expressive power of schematizations in order to avoid generating such cycles explicitly. This may shorten proofs and avoid divergence in some cases. These techniques are more general than existing ones, and unlike them, they can be used with any kind of proof procedure (using tableaux-based approaches as well as resolution-based ones).

1 Motivations

The concept of term schematization has been originally introduced by Chen, Hsiang and Kong (see for example [1,2]). The intended goal was to denote, using recurrent expressions, infinite sets of structurally similar terms, obtained by iterating a given context along some particular paths. Schematizations have been developed in order to avoid non-termination and divergence, in existing symbolic computation procedures, for example in Rewriting (e.g. to handle the case in which the Knuth-Bendix procedure diverges), in Logic Programming (to express finitely the set of solutions of a non-terminating query) or in Automated Deduction (for giving a finite description of an infinite derivation). Since then, improvements have been proposed to the initial language, and several different formalisms are now available, with various complexities and expressive powers [1,3,11,10,5]. However, only few papers have been dealing with the actual *use* of these formalisms in Automated Deduction. Since the unification problem is decidable for most of them, it is of course possible to integrate them into existing proof procedures. This enables the user to specify his/her problem into a more expressive language. However, in order to take full advantage of the expressive power of schematizations, it is necessary to go further, and to define rules allowing to integrate *automatically* such schematizations into the clause set at hand. This should indeed avoid repeated applications of the same sequence of rules, thus reducing the search space and avoiding divergence. For example, a theorem prover should be able to *deduce* automatically the formula F', given

R. Goré, A. Leitsch, and T. Nipkow (Eds.): IJCAR 2001, LNAI 2083, pp. 578–592, 2001.

the formula F above, instead of blindly enumerating the set of ground facts $\{even(0), odd(succ(0)), even(succ(succ(0))), \ldots\}$.

For this purpose, some techniques have been proposed, in order to detect and avoid loops during the proof search.

In [11], a rule is defined in order to avoid repeated applications of the resolution rule. Its principle is the following. Given a binary self-resolving clause of the form $P\lambda \leftarrow P$ and a clause $H \leftarrow B_1, \ldots, B_k$ such that P and H have a m.g.u. μ, it is possible to deduce a clause of the form $P\lambda^n\mu \leftarrow B_1\mu, \ldots, B_k\mu$, provided that λ satisfies some conditions allowing to express the atom $P\lambda^n\mu$ finitely, using R-terms. For example, given the clause $\neg P(x) \vee P(f(x))$ and $P(a)$, it is possible to compute the clause $P(f^n(a))$ (where n denotes a new integer variable). In [12], the more expressive formalism of primal grammar is used in order to control self-application of binary clauses. The proposed method deals with clauses of the form $A\lambda_1 \rightarrow A\lambda_2$ (where λ_1, λ_2 satisfy some additional conditions) and computes the set of resolvents $A\lambda_1^n \rightarrow A\lambda_2^n$ that can be deduced from such clauses. In [8], a similar approach is introduced for extending existing model building procedures. An inductive inference rule is defined in order to simulate repeated applications of the resolution rule on self-resolving clauses, and to give a finite description of Herbrand models of clause sets for which the original method did not terminate.

Still, all these techniques are very restricted. First, they do not take into account the use of the equality predicate and paramodulation rule (as well as similar macro-inference rules coming from specific axioms). Consider for example the clause $f(x, g(y)) = f(h(x), y)$. One should be able to derive the clause: $f(x, g^n(y)) = f(h^n(x), y)$, which is *not* possible using the techniques above. Second, it does not allow crossed recursion. Consider for example the following set of clauses: $S = \{\neg P(x) \vee \neg Q(x) \vee P(f(x)), \neg P(x) \vee \neg Q(x) \vee Q(f(x)), P(a), Q(a)\}$. It is clear that for all non negative integers n, the clauses $P(f^n(a))$ and $Q(f^n(a))$ are logical consequences of S. Therefore, it should be possible to deduce the clauses $(\forall n)P(f^n(a))$ and $(\forall n)Q(f^n(a))$ explicitly. However, the reader can easily check that none of the techniques mentioned above can deduce such clauses. Intuitively speaking, this is due to the fact that crossed recursion is needed to deduce $P(f^n(a))$ and $Q(f^n(a))$ (i.e. $P(f^{n-1}(a))$ is needed to prove $Q(f^n(a))$ and conversely).

Third, the above techniques may only be used if a self-resolving clause **explicitly occurs** in the clause set. This is a very restrictive condition if the method is to be integrated into a calculus that do not share the lemma building property of the resolution method, or even if restriction strategies (such as semantic resolution) are used to prune the search space (which is very often the case in practice, for obvious efficiency reasons).

Consider for example, the clause set: $S = \{c_1 : P(a), c_2 : (\forall x)\neg P(x) \vee \neg Q(x), c_3 : (\forall x)Q(x) \vee P(f(x))\}$

Here, the clause $c_4 : (\forall x)\neg P(x) \vee P(f(x))$ does *not* occur in S, but can be *deduced* from S by applying the resolution rule between the clauses c_2 and c_3. Then, the inductive rule can be applied, yielding the clause c_5 :

$(\forall x, n) \neg P(x) \lor P(f^n(x))$. By applying the resolution rule between c_5 and c_1 we generate the clause $c_6 : P(f^n(a))$ and then (with c_2) the clause $c_7 : \neg Q(f^n(a))$, which leads to termination (if an appropriate redundancy checking mechanism such as subsumption is used).

However, if the clause c_4 is not generated, then the generation of the clauses $(\forall x, n) \neg P(x) \lor P(f^n(x))$, $P(f^n(a))$ and $\neg Q(f^n(a))$ would be **impossible**, which leads to non-termination of the resolution process.

Unfortunately, most of the proof procedures implemented in running systems would not generate such clause as c_4. Of course, provers based on the tableaux or connection method, or on model elimination, will never generate *explicitly* such a formula (instead, they would for example compute the terms $P(f(a)), P(f(f(a))), P(f(f(f(a))))$, etc. from $P(a)$). However, even if we restrict ourselves to resolution-based provers, it is well known that restriction conditions are often added to the resolution rule in order to prune the search space and make the system more efficient. Such restrictions will frequently avoid the generation of such clauses as c_4. For example, most (trivial) semantic restrictions (such as positive or negative hyper-resolution) would prevent the application of the resolution rule between c_2 and c_3. Again, only the ground clauses $P(f(a)), P(f(f(a))), \ldots$ (or $\neg Q(a), \neg Q(f(a)), \ldots$) will be generated.

This leads to a very paradoxal and unnatural situation: restricting the application of the inference rules can actually make the whole proof search process *non-terminating*.

The calculus presented in [6] in the context of semantic tableau *cannot* deal with these examples because it still relies on the explicit generation of self-resolvent clauses[1] (called cycle unification clauses in [6]). The calculus described in [13] in the context of HyperTableaux is closer to our approach. It is applicable on tableaux-based procedures, and can handle examples as the ones presented above. Moreover it also deals with more powerful schematizations techniques where the integer exponents may be non-linear expressions (which entails that unification problems may become undecidable). However it does not use proof generalisation techniques. Moreover only some very specific cycles of order 1 (see below) are considered.

Therefore, more general and sophisticated techniques are needed in order to overcome these limitations and to control cycles that are much more complex than previously investigated ones. This is particularly important if we want to integrate the use of term schematizations into tableau-based procedures. The present paper presents a solution to this problem. The proposed techniques allow to integrate the use of term schematizations into all existing proof procedures and in particular into those that does not generate explicitly self-resolving clauses. Due to space restrictions, proofs are not included in the paper. Some of the proofs and other examples can be found in [9].

---

[1] Since this is a tableau calculus, this entails that the rule can be applied only on input clauses.

2   Basic Notions and Definitions

2.1   Terms Schematizations

As mentioned before, there exist several languages for denoting sets of structurally similar terms. Among all the existing formalisms, we have chosen the language of *terms with integer exponents* (I-terms for short), which have been introduced by Comon [3] and extended to terms with several "holes" in [8]. We consider this language to be a good compromise between "expressive power" and "simplicity".

In this section, we recall the definition of I-terms (syntax and semantics). Let Σ be a set of *function symbols* (including constants), ϑ be a set of *ordinary variables* and ϑ_N be a set of integer variables. Let *arity* be a function mapping each symbol f in Σ to a natural number (the *arity* of f).

The set of *arithmetic expressions* $\mathcal{N}$ defined as usual. Ground arithmetic expressions of the form $s^n(0)$ (resp. $s^n(x)$) will be simply denoted by n (resp. $x + n$).

The set of *ordinary terms* $\tau(\Sigma, \vartheta)$, the set of *I-terms* $\tau_I(\Sigma, \vartheta)$ and the set of *terms with several holes* $\tau_\diamond(\Sigma, \vartheta)$ are the smallest sets satisfying the following conditions:

- $\vartheta \subseteq \tau_I(\Sigma, \vartheta)$.
- $\vartheta \subseteq \tau(\Sigma, \vartheta)$.
- $\diamond \in \tau_\diamond(\Sigma, \vartheta)$.
- If $f \in \Sigma$, $arity(f) = n$ and $(t_1, \ldots, t_n) \in \tau_I(\Sigma, \vartheta)^n$, then $f(t_1, \ldots, t_n) \in \tau_I(\Sigma, \vartheta)$ (if $n = 0$, i.e. if f is a constant symbol, $f(t_1, \ldots, t_n)$ is to be read as the term f).
- If $f \in \Sigma$, $arity(f) = n$ and $(t_1, \ldots, t_n) \in \tau(\Sigma, \vartheta)^n$, then $f(t_1, \ldots, t_n) \in \tau(\Sigma, \vartheta)$.
- If $f \in \Sigma$, $arity(f) = n$, $(t_1, \ldots, t_n) \in (\tau_I(\Sigma, \vartheta) \cup \tau_\diamond(\Sigma, \vartheta))^n$ and if there exists $i \in [1..n]$ such that $t_i \in \tau_\diamond(\Sigma, \vartheta)$ then $f(t_1, \ldots, t_n) \in \tau_\diamond(\Sigma, \vartheta)$.
- If $t \in \tau_\diamond(\Sigma, \vartheta)$, $n \in \mathcal{N}$, $s \in \tau_I(\Sigma, \vartheta)$ then $t^n.s \in \tau_I(\Sigma, \vartheta)$.

Let $t \in \tau_\diamond(\Sigma, \vartheta)$ and let $s \in \tau_I(\Sigma, \vartheta)$. We denote by $t(s)$ the term obtained by replacing each occurrence of $\diamond$ in t by s. We denote by $\mathcal{R}$ the following system of rewriting rules: $\{t^0.s \to s, t^{s(n)}.s \to t(t^n.s)\}$.

It is immediate to see that $\mathcal{R}$ is terminating and confluent. We denote by $t \downarrow_\mathcal{R}$ the normal form of t w.r.t. $\mathcal{R}$.

2.2   Formulae

Let ϑ_B be a set of *boolean variables*[2], and let Ω be a set of *predicate symbols*.

Definition 1. *The set of first-order formulae $\psi_I(\Omega, \Sigma, \vartheta)$ is the least set such that:*

- $\vartheta_B \subseteq \psi_I(\Omega, \Sigma, \vartheta)$.

---

[2] The use of boolean variables in the context of the present paper will become clear when we will be defining the proof generalization techniques (see Section 4.1).

- If $p \in \Omega$, $arity(p) = n$, $(t_1, \ldots, t_n) \in \tau_I(\Sigma, \vartheta)^n$, then $p(t_1, \ldots, t_n) \in \psi_I(\Omega, \Sigma, \vartheta)$.
- If $F \in \psi_I(\Omega, \Sigma, \vartheta)$, then $\neg F$ is in $\psi_I(\Omega, \Sigma, \vartheta)$.
- If $F_1, F_2 \in \psi_I(\Omega, \Sigma, \vartheta)$, then $F_1 \wedge F_2$ and $F_1 \vee F_2$ are in $\psi_I(\Omega, \Sigma, \vartheta)$.
- If $F \in \psi_I(\Omega, \Sigma, \vartheta)$ and $x \in (\vartheta \cup \vartheta_N)$, then $(\forall x)F$ and $(\exists x)F$ are in $\psi_I(\Omega, \Sigma, \vartheta)$.

Formulae are interpreted as usual. Any term of the form $t^n.s$ is interpreted as $t^n.s \downarrow_{\mathcal{R}}$.

We now introduce the notion of *mixed unification problem*, that will be used later to encode conditions about formulae: A *mixed unification problem* (m.u.p. for short) is either *false* or a (possibly empty) finite conjunction of equations of the form: $F \doteq G$ where F, G are first-order formulae or $t \doteq s$ where t, s are terms (note that empty conjunction are interpreted as *true*).

M.u.p. are said to be "mixed" because they mix two kinds of equations: equations between terms (as in standard unification problems) and equations between first-order formulae. Both kinds of equations are interpreted syntactically, i.e. $\doteq$ means syntactical identity between expressions (up to a renaming of bounded variables).

A substitution σ is said to be a *solution* of $\mathcal{P} \equiv \bigwedge_{i=1}^n t_i \doteq s_i$ iff for all $i \in [1..n]$, $t_i \sigma$ is syntactically equivalent to $s_i \sigma$ (up to a renaming of quantified variables). Since semantic properties of first-order formulae are not taken into account in the definition, a m.u.p. can be solved using standard unification rules *plus* specific rules to handle the case of quantified formulae (see [9] for details).

3   The Calculus

Before presenting our technique, we need to introduce a calculus, allowing to construct and denote proofs. The chosen calculus is a very basic version of tableaux (without unification). It should be emphasized that it is not intended to be used for actually *generating* proofs. Instead, the proofs should be obtained using more efficient *existing* calculi (such as an efficient tableaux calculus, resolution, model elimination etc.) and then *translated* into our formalism (this can be done in an efficient way). This technique avoids the necessity to define several versions of the algorithm, one for each proof procedure. Naturally, from a practical point of view, explicit translation of the proof should probably be avoided in an implementation.

The following definition introduces the standard notion of *tree* labelled by formulae.

A *position* is a (possibly infinite and possibly empty) sequence of natural integers. "$p.q$" denote the concatenation of the positions p, q and ϵ denotes the empty position. $\prec$ denotes the prefix ordering between positions, i.e. $p \prec q$ iff there exists a non empty position p' such that $p.p' = q$.

A *formula tree* is a function mapping each finite position p to a multiset of formulae such that, for any pair of finite positions (p, q), if $p \prec q$ and $\mathcal{T}(p) = \emptyset$

then $T(q) = \emptyset$. The notion of branch, closed branch, etc. is defined as usual. For any formula tree T and for any finite position p in T, we denote by $T_{|p}$ the formula tree defined as follows: for any position q, we have, by definition:

$$T_{|p}(q) = T(p.q)$$

If T and T' are two formula trees and p is a finite position, we denote by $T[p \leftarrow T']$ the formula tree defined as follows:

$$T[p \leftarrow T'](q) =_{def} T(p) \cup T'(p') \text{ if } q = p.p'$$

$$T[p \leftarrow T'](q) =_{def} T(q) \text{ otherwise}$$

Roughly speaking, we call "proof tree" a formula tree in which each node is generated from an existing one using the usual inference rules of tableaux (γ-rule, $\vee$-rule, $\wedge$-rule, etc.) including the "cut" rule. The following is the formal definition of this notion.

Definition 2. *A* proof tree *is a formula tree T such that for any finite position p such that $T(p) \neq \emptyset$, one of the following condition holds.*

- *Either for all integer i, $\mathcal{I}(p.i) = \emptyset$. In this case p is said to be a* leaf node *for T.*
- *Or $T(p) = S \cup \{F_1 \vee F_2\}$, $T(p.i) = S \cup \{F_i\}$ if $i = 1, 2$ and $T(p.i) = \emptyset$ else (p is said to be a $\vee$-node on $F_1 \vee F_2$).*
- *Or $T(p) = S \cup \{F_1 \wedge F_2\}$, $T(p.1) = S \cup \{F_1, F_2\}$ and $T(p.i) = \emptyset$ if $i \neq 1$ ($\wedge$-node on $F_1 \wedge F_2$).*
- *Or $T(p) = S \cup \{(\forall x)F\}$, $x \in \vartheta$, t is a ground I-term, $T(p.1) = S \cup \{(\forall x)F, F\{x \rightarrow t\}\}$ and $T(p.i) = \emptyset$ if $i \neq 1$ ($\forall$-node on $(\forall x)F$ and t).*
- *Or $T(p) = S \cup \{(\forall x)F\}$, $x \in \vartheta_N$, t is a ground arithmetic expression, $T(p.1) = S \cup \{(\forall x)F, F\{x \rightarrow t\}\}$ and $T(p.i) = \emptyset$ if $i \neq 1$ ($\forall$-node on $(\forall x)F$ and t).*
- *Or $T(p) = S \cup \{(\exists x)F\}$, t is a constant symbol not occurring in $T(p)$, $T(p.1) = S \cup \{F\{x \rightarrow t\}\}$ and $T(p.i) = \emptyset$ if $i \neq 1$ ($\exists$-node on $(\exists x)F$).*
- *Or $T(p) = S \cup \{\neg(F_1 \wedge F_2)\}$, $T(p.1) = S \cup \{(\neg F_1) \vee (\neg F_2)\}$ and $T(p.i) = \emptyset$ if $i \neq 1$ ($\neg$-$\wedge$-node on $\neg(F_1 \wedge F_2)$).*
- *Or $T(p) = S \cup \{\neg(F_1 \vee F_2)\}$, $T(p.1) = S \cup \{(\neg F_1) \wedge (\neg F_2)\}$ and $T(p.i) = \emptyset$ if $i \neq 1$ ($\neg$-$\vee$-node on $\neg(F_1 \vee F_2)$).*
- *Or $T(p) = S \cup \{\neg\neg F\}$, $T(p.1) = S \cup \{F\}$ and $T(p.i) = \emptyset$ if $i \neq 1$ ($\neg$-$\neg$-node on $\neg\neg F$).*
- *Or $T(p) = S \cup \{\neg(\forall x)F\}$, $T(p.1) = S \cup \{(\exists x)F\}$ and $T(p.i) = \emptyset$ if $i \neq 1$ ($\neg$-$\forall$-node on $\neg(\forall x)F$).*
- *Or $T(p) = S \cup \{\neg(\exists x)F\}$, $T(p.1) = S \cup \{(\forall x)F\}$ and $T(p.i) = \emptyset$ if $i \neq 1$ ($\neg$-$\exists$-node on $\neg(\exists x)F$).*
- *Or $T(p) = S \cup \{F, \neg F'\}$, $T(p.1) = \{\bot\}$ and $T(p.i) = \emptyset$ if $i \neq 1$, if $F \downarrow_{\mathcal{R}} = F' \downarrow_{\mathcal{R}}$ (closing node on F and $\neg F$).*
- *Or $T(p) = S$, $T(p.1) = S \cup \{F\}$, $T(p.2) = S \cup \{\neg F\}$ and $T(p.i) = \emptyset$ if $i \neq 1, 2$ and F is any formula (cut node on F).*

A proof tree *of a set of first-order formula S is a proof tree T such that $T(\epsilon) = S$.*

A proof tree is said to be complete *iff it contains no non-closed leaf.*

A proof tree $\mathcal{T}$ is said to be fair *iff no rule is infinitely delayed (i.e. if a rule is applicable at a given finite position p in $\mathcal{T}$ on a formula F, then it must be applied on the same formula in all non-closed branches of the form $p.q$ in $\mathcal{T}$).*

It is well-known that this calculus is sound and refutationnally complete. Of course, allowing integer variables in the formulae makes the calculus incomplete. Indeed, the satisfiability problem is actually undecidable for formulae containing integer variables (since they may encode Peano arithmetic).

It is well-known that the proofs constructed using any existing proof procedure such as resolution, tableaux (with unification), sequent calculus, the connection method etc., can be translated into a closed proof tree. Due to the use of the cut rule, this can be done in polynomial time, w.r.t. the size of the original proof which makes this approach effective. This is also true if the paramodulation rule is added to the calculus, since it can be simulated by repeated application of the substitution axioms.

4   The Inductive Rule

We now have all what we need to define our new inductive rule. Intuitively speaking, its principle can be summarized as follows:

1. Try - during the search for a proof - to detect a potential *cycle* in the proof tree, i.e. a sequence of inference rules that can be applied repeatedly, hence that can lead to non termination. Potential cycles are detected using *proof generalization*, by constructing the *most general form* of the corresponding sequence of inference rules.
2. Construct (using terms with integer exponents) the *general form* of the formulae deduced during the repeated applications of the sequence of inference rules.

This technique permits to express finitely infinite branches. Thus, divergence is avoided in some case.

4.1   Proof Generalization

We first describe our proof generalization algorithm. It takes as input a proof tree $\mathcal{T}$ and computes a new formula tree $\mathcal{T}^g$ such that $\mathcal{T}$ is an instance of $\mathcal{T}^g$, and a m.u.p. $\mathcal{P}^g$ expressing constraints on the variables occurring in $\mathcal{T}^g$ so that $\mathcal{T}^g$ is a proof tree.

For any formula F of the form $(Qx)G$ (where $Q = \exists, \forall$), we denote by $\Delta(F)$ the formula obtained from G by replacing all terms (resp. formulae) not containing the variable x by new, pairwise distinct variables in ϑ (resp. ϑ_B) (this is a kind of variable abstraction (as originally introduced by Baader and Schulz), performed on all terms not containing x).

Let $\mathcal{T}$ be a finite proof tree. Let Φ be a function mapping all formulae F to distinct variables in ϑ_B. For any finite position p, we inductively define the pair $(\mathcal{T}^g(p), \mathcal{P}_p)$ as follows (starting from the leaf to the root).

- If $T(p)$ is empty then $T^g(p) =_{def} \emptyset$ and $\mathcal{P} =_{def}$ *true*.
- Assume that for all integers i, $(T^g_{p.i}, \mathcal{P}_{p.i})$ is constructed. Then we define $(T^g(p), \mathcal{P}_p)$ as follows.
  $T^g(p) =_{def} \{\Phi(F) \mid F \in T(p)\}$. Moreover:
  - If p is a leaf, then $\mathcal{P}_p =_{def} \top$.
  - If p is a $\neg\text{-}\neg$-node on $\neg\neg F$, then $\mathcal{P}_p =_{def} \mathcal{P}_{p.1} \wedge \Phi(\neg\neg F) \doteq \neg\neg\Phi(F)$.
  - If p is a $\neg\text{-}\vee$-node on $\neg(F_1 \vee F_2)$, then $\mathcal{P}_p =_{def} \mathcal{P}_{p.1} \wedge \Phi(\neg(F_1 \vee F_2)) \doteq \neg(\Phi(F_1) \vee \Phi(F_2)) \wedge \Phi(\neg F_1 \wedge \neg F_2) \doteq \neg\Phi(F_1) \wedge \neg\Phi(F_2))$.
  - If p is a $\neg\text{-}\wedge$-node on $\neg(F_1 \wedge F_2)$, then $\mathcal{P}_p =_{def} \mathcal{P}_{p.1} \wedge \Phi(\neg(F_1 \wedge F_2)) \doteq \neg(\Phi(F_1) \wedge \Phi(F_2)) \wedge \Phi(\neg F_1 \vee \neg F_2) \doteq \neg\Phi(F_1) \vee \neg\Phi(F_2)$.
  - If p is a $\neg\text{-}\forall$-node on $\neg(\forall x)F$, then $\mathcal{P}_p =_{def} \mathcal{P}_{p.1} \wedge \Phi(\neg(\forall x)F) \doteq \neg(\forall x)F' \wedge F' \doteq \Delta(F) \wedge \Phi((\exists x)F) \doteq (\exists x)F'$.
  - If p is a $\neg\text{-}\exists$-node on $\neg(\exists x)F$, then $\mathcal{P}_p =_{def} \mathcal{P}_{p.1} \wedge \Phi(\neg(\exists x)F) \doteq \neg(\exists x)F' \wedge F' \doteq \Delta(F) \wedge \Phi((\exists x)F) \doteq (\exists x)F'$.
  - If p is a $\vee$-node on $F_1 \vee F_2$, then $\mathcal{P}_p =_{def} \mathcal{P}_{p.1} \wedge \mathcal{P}_{p.2} \wedge \Phi(F) \doteq \Phi(F_1) \vee \Phi(F_2)$.
  - If p is a $\wedge$-node on $F_1 \wedge F_2$, then $\mathcal{P}_p =_{def} \mathcal{P}_{p.1} \wedge \Phi(F) \doteq \Phi(F_1) \wedge \Phi(F_2)$.
  - If p is a closing node on $F, \neg F$, then $\mathcal{P}_p =_{def} \neg\Phi(F) \doteq \Phi(\neg F)$.
  - If p is a cut node on F, then $\mathcal{P}_p =_{def} \Phi(\neg F) \doteq \neg\Phi(F) \wedge \mathcal{P}_{p.1} \wedge \mathcal{P}_{p.2}$.
  - If p is a $\forall$-node on $(\forall x)F$, then $\mathcal{P}_p =_{def} \Phi((\forall x)F) \doteq (\forall x)\Delta(F) \wedge \Phi(F\{x \to t\}) \doteq \Delta(F)\{x \to y\} \wedge \mathcal{P}_{p.1}$ where y is a new variable.
  - If p is a $\exists$-node on $(\exists x)F$, then $\mathcal{P}_p =_{def} \Phi((\exists x)F) \doteq (\exists x)\Delta(F) \wedge \Phi(F\{x \to t\}) \doteq \Delta(F)\{x \to f(x_1, \ldots, x_n)\} \wedge \mathcal{P}_{p.1}\{y \to f(x_1, \ldots, x_n)\}$ where f is a new function symbol and $x_1, \ldots, x_n$ are the free variables in $T^g(p)$.

$(T^g, \mathcal{P}_\epsilon)$ is called a *generalization* of T.

Lemma 1. *Let T be a proof tree. Let S be a set of quantifier-free formulae. Let $(T^g, \mathcal{P})$ be a generalization of T.*

1. *For any solution σ of $\mathcal{P}$, $T^g\sigma$ is a proof tree.*
2. *There exists a solution θ of $\mathcal{P}$ such that $T^g\theta = T$.*

Example 1. Let us consider the formula $(\forall x, y)P(g(a), x, y) \to P(g(a), f(x), g(y)) \wedge P(g(a), a, b)$. We build the following (obviously not complete) proof tree[3].

```
p(g(a),a,b) and all(x,all(y,not p(g(a),x,y) or p(g(a),f(x),g(y))))
  p(g(a),a,b)
  all(x,all(y,not p(g(a),x,y) or p(g(a),f(x),g(y))))
```

---

[3] This proof tree (as all other examples in this paper) has been constructed with the interactive theorem prover μ-IPS$_{ATINF}$ (a micro **I**nteractive **P**rover with **S**chematization) described in Section 5.

```
all(y,not p(g(a),a,y) or p(g(a),f(a),g(y)))
    not p(g(a),a,b) or p(g(a),f(a),g(b))
        not p(g(a),a,b)
            false
        p(g(a),f(a),g(b))
```

The corresponding generalization of this proof tree is (after solution of the constraints and simplification):

```
p(_X,_Y,_Z) and all(x,all(y,not p(_X,x,y)or p(_X,f(x),g(y))))
    p(_X,_Y,_Z)
    all(x,all(y,not p(_X,x,y)or p(_X,f(x),g(y))))
        all(y,not p(_X,_Y,y) or p(_X,f(_Y),g(y)))
            not p(_X,_Y,_Z) or p(_X,f(_Y),g(_Z))
                not p(_X,_Y,_Z)
                    false
                p(_X,f(_Y),g(_Z))
```

Following standard Prolog conventions, expressions starting with an underscore denotes variables.

4.2   Detecting Cycles and Computing Schematizations

In this section, we identify criteria sufficient to detect potential cycles and to ensure that the re-construction of the infinite branch using term schematizations is possible.

A substitution θ is said to be *simply recursive* iff for all variables $x \in \mathcal{D}(\theta)$:

1. x occurs in $x\theta$;
2. x does not occur in a subterm of the form $t^n.s$ in $x\theta$;
3. and for any $y \in \mathcal{D}(\theta) \setminus \{x\}$, y does not occur in $x\theta$.

In contrast to the similar notions of directly recursive [11] or mono-cyclic substitution [12], we do not require the substitution θ to be variable-preserving. However, we will *not* rename the variables introduced by the substitution θ at each step, which will insure that the obtained set of terms can be denoted finitely using I-terms. Similarly, we also assume that all cycles are of length 1 (i.e. we do not allow substitutions of the form $x \to f(y), y \to g(x)$). This does not entail any loss of generality, because, as shown in [12], cycles of length greater than 1 may be reduced to cycles of length 1 (using unfolding).

Lemma 2. *If θ is simply recursive, then for any natural number n and for all variables $x \in \mathcal{D}(\theta)$, we have $x\theta^n \equiv t^n.x$ where $t = x\theta\{x \to \diamond\}$.*

Two Kinds of Proof Cycles

In this section, we give the definition of proof cycles and we identify syntactical criteria allowing to detect them. Informally speaking, a proof cycle is a sequence of inference rules (or a subtree) that can be repeatedly applied, leading to a proof tree of arbitrary size (hence to divergence). Our goal is to " eliminate" these cycles, by expressing these arbitrary proof trees in a *symbolic* way (using integer variables and term schematizations). To this purpose, an important restriction is that we explicitly require that these proof trees should contains only a *fixed* number of (distinct) open branches (the conditions in the Definition of proof cycles below will insure that this property is always satisfied).

Indeed, having to consider proof trees containing N open branches, when N is not a fixed integer, but a term, whose value depends on the value of the variables in the proof tree would require a specific schematization language for denoting those trees and would therefore lead to a much more difficult treatment of trees. This additional source of complexity could be possibly superior to the advantage of using term schematizations (see also Section 6).

Proof Cycles of Order 1 Proof cycles of order 1 are the most simple possible cycles. They occur when a given branch contains a sequence of applications of inference rules that can be generalized in such a way that these rules can be applied again on the *conclusion* of the sequence. This is well illustrated by Figure 1: since the actual value of the term $f(a)$ is *not relevant* for the application of the inference rules in the considered sequence, we may apply also these rules on $f(f(a))$, which leads to a loop.

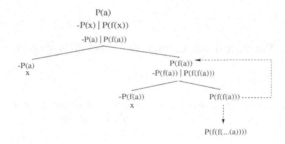

Fig. 1. A simple example of a proof cycle of order 1

Since we want to deal with proof trees having a fixed number of branches, we explicitly require that all branches parallel to the considered one in the generalized cycle must not contain any variable. As we shall see, this ensures that all these branches can be merged into a single one.

Definition 3. *Let T be a proof tree. A cycle of order 1 in T w.r.t. S_{init}, θ, η is a pair (p, q) such that $p \prec q$ and satisfying the following conditions:*

1. *$(T^g, \mathcal{P}^g)$ is a generalization of T.*
2. *σ is the most general solution of $\mathcal{P}^g$.*
3. *$T^g(p)\sigma$ is of the form $S \cup S_{init}$.*
4. *$T^g(q)\sigma$ is of the form $S \cup S_{init} \cup S_{init}\theta \cup S'$.*

5. θ is simply recursive.
6. $S\sigma$ does not contain any variable in $\mathcal{D}(\theta)$.
7. For any position q' such that $p \prec q'$ and $q \parallel q'$, $\mathcal{T}^g(q')\sigma$ contains no variable in $\mathcal{D}(\theta)$.
8. η is the substitution such that $\mathcal{T}^g\sigma\eta = \mathcal{T}$ (η must exist by Lemma 1, point 2).
9. q is a leaf in $\mathcal{T}$.

Proof cycles of order 2 Proof cycles of order 2 are slightly more complicated. They occur when a sequence of inference rules leads to a conclusion C that is "similar" (in a sense to be formally defined: they are two instances of the same scheme) to the negation of a formula occurring in an open branch in the corresponding subtree. Then, this sequence of inference rules may be repeatedly applied from the leaf containing C. This will leads to a finite number of open branches, since the branch containing $\neg C$ will be closed.

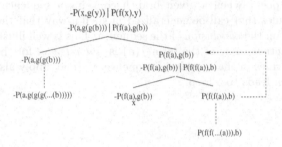

Fig. 2. A simple example of a proof cycle of order 2

Definition 4. *Let $\mathcal{T}$ be a proof tree. A cycle of order 2 in $\mathcal{T}$ w.r.t. S_{init}, θ, η is a tuple (p, q, q') such that $p \prec q$, $p \prec q'$, $q \parallel q'$ and satisfying the following conditions:*

1. $(\mathcal{T}^g, \mathcal{P}^g)$ is the generalization of $\mathcal{T}$.
2. σ is the most general solution of $\mathcal{P}^g$.
3. $\mathcal{T}^g(p)\sigma = S$.
4. $\mathcal{T}^g(q')\sigma = S \cup S'' \cup \{\neg F \mid F \in S_{init}\theta'\}$.
5. $\mathcal{T}^g(q)\sigma$ is of the form $S \cup S_{init} \cup S_{init}\theta \cup S'$.
6. θ, θ' are simply recursive and of disjoint domains.
7. $S\sigma$ and $S''\sigma$ do not contain any variable in $\mathcal{D}(\theta)$.
8. For any position q'' such that $p \prec q'$ and $q \parallel q''$, $q'' \parallel q'$, $\mathcal{T}^g(q')\sigma$ contains no variable in $\mathcal{D}(\theta)$.
9. η is the substitution such that $\mathcal{T}^g\sigma\eta = \mathcal{T}$ (η must exist by Lemma 1, point 2).
10. q and q' are two leaves in $\mathcal{T}$.

We can now add a **Cycle Detection** rule into our calculus. This is done by the following:

Definition 5. *A proof tree $\mathcal{T}$ is said to be an extended proof tree iff for all positions p in $\mathcal{T}$ such that $\mathcal{T}(p) \neq \emptyset$, either one of the conditions of Definition 2 holds or one of the following conditions holds:*

- (q,p) *is a cycle of order 1 w.r.t.* S_{init}, θ, η, *and*
  - $\mathcal{T}(p.1) = \mathcal{T}(p) \cup (S_{init}\theta^n)\eta$,
  - *and* $\mathcal{T}(p.i) = \emptyset$ *if* $i \neq 1$;
- *or* (q, p, q') *is a cycle of order 2 w.r.t.* S_{init}, θ, η *and:*
  - $\mathcal{T}(p.1) = \mathcal{T}(p) \cup (S_{init}\theta'^n)\eta$,
  - $\mathcal{T}(p.2) = \mathcal{T}(p) \cup (S_{init}\theta^n)\eta$,
  - $\mathcal{T}(p.i) = \emptyset$ *if* $i \neq 1$;

By Lemma 2, $S_{init}\theta^n$ can be denoted by a set of first-order formulae on I-terms. Next theorem can be considered as the "main result" in this paper:

Theorem 1. *The Cycle Detection rule is correct, i.e. if a formula F has an extended closed proof tree then it must be unsatisfiable.*

Example 2. (continued) Let us apply the cycle detection rule on the proof tree obtained in Example 1. We get the following enriched proof tree:

```
p(g(a),a,b) and all(x,all(y,not p(g(a),x,y) or p(g(a),f(x),g(y))))
   p(g(a),a,b)
   all(x,all(y,not p(g(a),x,y)or p(g(a),f(x),g(y))))
      all(y,not p(g(a),a,y) or p(g(a),f(a),g(y)))
         not p(g(a),a,b) or p(g(a),f(a),g(b))
            not p(g(a),a,b)
               false
            p(g(a),f(a),g(b))
               p(g(a),f(<>)^n.a,g(<>)^n.b)
% <> denotes the "hole"
```

Keeping the notations of Definition 3, θ is the substitution (which is clearly simply recursive): $\{ \_Y \rightarrow f(\_Y), \_Z \rightarrow g(\_Z) \}$ and η is: $\{ \_Y \rightarrow a, \_Z \rightarrow b\}$

5 Implementation and Examples of Application

The techniques and algorithms described in this paper has been implemented into a small theorem prover, called μ-IPS$_{ATINF}$. Through μ-IPS$_{ATINF}$ does provide some features for automatic theorem proving (using unification, iterative depth-first search, or breath-first search, with very simple selection strategies allowing to restrict or to delay the application of the most costly rules), its intend is mainly to allow a user to construct proof trees interactively and to experiment with the techniques proposed in the present paper, namely proof generalization, cycle detection and computation of term schematizations. The integration of these techniques into an efficient, "real-world" prover and the corresponding experimentations remains to be done (to the best of our knowledge, no existing running prover integrates similar techniques). μ-IPS$_{ATINF}$ is implemented in B-Prolog (see for example http://www.probp.com/).

We now give a slightly more complicated example, coming from a resolution proof, and including crossed recursion and equational reasoning. We show how the corresponding proof can be transformed into our calculus and how the corresponding cycle can be detected and eliminated.

$$1\ f(0) = succ(0) \quad 3\ g(f(x)) \neq g(succ(x)) \vee \neg p(u,x) \vee p(u,succ(x))$$
$$2\ p(y,0) \qquad\qquad 4\ \neg p(0,succ(x)) \vee f(succ(x)) = succ(f(x))$$

Assume that the following derivation δ has been constructed:

5 $g(succ(0)) \neq g(succ(0)) \vee \neg p(u,0) \vee p(u,succ(0))$ (paramodulation, 1, 3)
6 $\neg p(u,0) \vee p(u,succ(0))$ (resolution, 5, reflexivity axiom)
7 $p(u,succ(0))$ (resolution, 2, 6)
8 $f(succ(0)) = succ(f(0))$ (resolution, 4, 7)
9 $f(succ(0)) = succ(succ(0))$ (paramodulation, 1, 8)

First, δ is automatically translated into our calculus. Application of the resolution rule on ground unit clauses can be handled by the standard rules (closing and $\vee$-rules). Any application of the resolution rule on non ground clauses or non unit clause may be simulated by an application of the "cut" rule[4]. The application of the paramodulation rule is replaced by several steps of resolution with the appropriate substitutivity axioms. Note that this technique can be applied on any similar macro-inference rule.

We obtain (after generalization and simplification), the following proof tree (in this particular case, the generalized proof tree is almost identical to the original one, excepted that the term 0 is replaced by a variable _X). For the sake of clarity and conciseness, we dropped the irrelevant equality axioms and merged some of the resolution steps into one single "cut" rules.

```
f(_X) = succ(_X)
all(y,p(y,_X))
all(x,all(u,not(g(f(x)) = g(succ(x))) or not(p(u,x)) or p(u,succ(x)))))
all(x,not(p(_X,succ(x)) or f(succ(x)) = succ(f(x))))
  not(all(u,p(u,succ(_X)))) % an application of the cut rule for cl. 7
    exists(u,not(p(u,succ(_X))))
      not(p(c,succ(_X)))
        all(u,not(g(f(_X)) = g(succ(_X))) or not(p(u,_X)) or p(u,succ(_X)))
          not(g(f(_X)) = g(succ(_X))) or not(p(c,_X)) or p(c,succ(_X))
            not(g(f(_X)) = g(succ(_X)))
              all(x,all(y,g(x) = g(y) or not(x = y))) % an equality axiom
                g(f(_X)) = g(succ(_X)) or not(f(_X) = succ(_X))
                  g(f(_X)) = g(succ(_X))
                    false
                  not(f(_X) = succ(_X))
                    false
            not(p(c,_X))
              p(c_X,_X)
                false
```

---

[4] More precisely, a resolution step between $P(t) \vee R$ and $\neg P(t') \vee R'$ is replaced by an application of the cut rule on the clause $(\forall \boldsymbol{x})(R \vee R')\theta$ where θ is the m.g.u. of t and t' and $\boldsymbol{x}$ is the vector of variables in $(R \vee R')\theta$.

```
        p(c,succ(_X))
          false
all(u,p(u,succ(_X)))      % thus clause 7 must be true
   not(p(_X,succ(_X)) or f(succ(_X)) = succ(f(_X)))
       not(p(_X,succ(_X)))
         p(_X,succ(_X))
           false
     f(succ(_X)) = succ(f(_X)))
        f(succ(_X)) = succ(succ(_X)) % using equality axioms as previously
```

Here, we obtain a proof cycle of order 1 between $f(succ(\_X)) =$ $succ(succ(\_X))$ and $f(\_X) = succ(\_X)$ (on one hand) and $all(y,p(y,\_X))$ and $all(u,p(u,succ(\_X)))$ on the other hand. The reader can check that we can apply the cycle detection rule (of order 1), in order to derive the clauses: $(\forall n)f(succ(\diamond)^n.0) = f(succ(succ(\diamond)^n.0))$ and $(\forall u, n)p(u, succ(\diamond)^n.0)$.

6   Conclusion

We have presented some techniques for using term schematizations in Automated Deduction. They are particularly useful for integrating term schematizations into proof calculi that does not have the lemma building property of the resolution method (such as the tableaux or connection calculus) or into calculi that combines features from both approaches, such as model elimination procedures, or model generation theorem provers (see for example [7]). This could help to reduce the search space, and possibly to improve the termination behavior of these methods. The corresponding algorithms have been implemented in the context of an interactive theorem prover μ-IPS$_{ATINF}$. We believe that the Cycle Detection rule could be added into a theorem prover with a "reasonable" computation cost. Indeed, the construction of the generalized proof tree can be done in linear time w.r.t. the size of the original proof (it can actually be done dynamically during the construction of the proof) and only require a small amount of additional memory. Moreover, the Cycle Detection rule is also polynomial w.r.t. the size of the considered proof tree. Heuristics could be used to decide whether the system should try to apply the Cycle Detection rule (for example if it detects a "potential cycle" such as $p(a), p(f(a)), \ldots$). The prize to be payed, of course, is that the theorem prover should be able to deal with the more expressive formalism. Hence, unification algorithms for term schemes (using for example the procedures described in [1,3,12,8,4]) have to be added into the prover. These algorithms are in general significantly more costly than the standard unification procedure. However, this could be overcome by the additional expressive power of term schematizations (otherwise, it would be necessary to further restrict the considered class of term schematization in order to ensure that unification problems may be solved effectively). We consider that these questions could deserve to be investigated more deeply. The possibility of dealing with cycles containing an *indefinite* number of distinct open branches could be investigated. This would require to extend the existing term schematization techniques in order to denote infinite sets of *formulae* and infinite sets of *tableaux* instead of terms. Some completeness or incompleteness results could also be investigated for the

Cycle Detection rule, i.e. one should try to identify classes of formula F for which *any* logical consequence C of F (where C possibly contains integer variables) can be deduced using the usual inference rules enriched by the Cycle Detection rule. Another way of future research is to investigate the effect of the adding of the Cycle Detection rule on the termination behavior of the calculus. Does there exists classes of formulae for which the calculus enriched by the Cycle Detection rule (and by appropriate strategies and redundancy checking mechanisms) can be proven to be a decision procedure?

References

1. H. Chen and J. Hsiang. Logic programming with recurrence domains. In *Automata, Languages and Programming (ICALP'91)*, pages 20–34. Springer, LNCS 510, 1991.
2. H. Chen, J. Hsiang, and H. Kong. On finite representations of infinite sequences of terms. In *Conditional and Typed Rewriting Systems, 2nd International Workshop*, pages 100–114. Springer, LNCS 516, June 1990.
3. H. Comon. On unification of terms with integer exponents. *Mathematical System Theory*, 28:67–88, 1995.
4. M. Hermann. Divergence des systèmes de réécriture et schématisation des ensembles infinis de termes. Habilitation, Université de Nancy I, and CRIN-CNRS Inria-Lorraine, Nancy, France, March 1994.
5. M. Hermann and R. Galbavý. Unification of Infinite Sets of Terms schematized by Primal Grammars. *Theorical Computer Science*, 176(1–2):111–158, 1997.
6. S. Klingenbeck. *Counter Examples in Semantic Tableaux*. PhD thesis, University of Karlsruhe, 1996.
7. R. Manthey and F. Bry. SATCHMO: A theorem prover implemented in Prolog. In *Proc. of CADE-9*, pages 415–434. Springer, LNCS 310, 1988.
8. N. Peltier. Increasing the capabilities of model building by constraint solving with terms with integer exponents. *Journal of Symbolic Computation*, 24:59–101, 1997.
9. N. Peltier. Using Term Schematizations in Automated Deduction. Research Report. Available on `http://www-leibniz.imag.fr/ATINF/Nicolas.Peltier/`, 2001.
10. G. Salzer. *Unification of Meta-Terms*. PhD thesis, Technische Universität Wien., 1991.
11. G. Salzer. The unification of infinite sets of terms and its applications. In *Logic Programming and Automated Reasoning (LPAR'92)*, pages 409–429. Springer, LNAI 624, July 1992.
12. G. Salzer. Primal grammar and unification modulo a binary clause. In *Proc. of CADE-12*, pages 72–86. Springer, 1994. LNAI 814.
13. F. Stolzenburg. Loop-detection in hyper-tableaux by powerful model generation. *Journal of Universal Computer Science*, 5(3):135–155, 1999.

Approximating Dependency Graphs Using Tree Automata Techniques

Aart Middeldorp

Institute of Information Sciences and Electronics
University of Tsukuba, Tsukuba 305-8573, Japan
ami@is.tsukuba.ac.jp

Abstract. The dependency pair method of Arts and Giesl is the most powerful technique for proving termination of term rewrite systems automatically. We show that the method can be improved by using tree automata techniques to obtain better approximations of the dependency graph. This graph determines the ordering constraints that need to be solved in order to conclude termination. We further show that by using our approximations the dependency pair method provides a decision procedure for termination of right-ground rewrite systems.

1 Introduction

In the area of term rewriting termination has been studied for several decades and many powerful techniques have been developed. Three general directions can be distinguished:

1. Syntactic methods that compare terms by constructing an explicit well-founded order. These methods are fully automatable but have limited power. Well-known examples are the recursive path order of Dershowitz [10] and the Knuth-Bendix order [18].
2. Semantic methods that compare terms by interpreting them in some well-founded domain. These methods can be very powerful in theory but their implementations rely on heuristics that greatly reduce this power. Well-known examples are the polynomial interpretations of Lankford [22] and the semantic path order of Kamin and Lévy [17].
3. Transformation methods which do not attempt to prove termination directly but rather transform the given rewrite system into another rewrite system such that termination of the latter system is easier to prove and implies termination of the former system. Examples include the transformation order of Bellegarde and Lescanne [6], and Zantema's distribution elimination [27] and semantic labelling [28]. Transformations differ in their degree of automation.

Since termination is an undecidable property of rewrite systems, even for systems that consist of a single rewrite rule, no method will work in all cases. Recently a new *automatable* technique emerged: the dependency pair method of Arts and Giesl. In this method a rewrite system is transformed into a set of ordering

R. Goré, A. Leitsch, and T. Nipkow (Eds.): IJCAR 2001, LNAI 2083, pp. 593–610, 2001.

constraints such that termination of the rewrite system is equivalent to the solvability of the constraints. The generated constraints are typically solved by standard techniques (polynomial interpretations, path orders), even when these techniques are not applicable to the original rewrite system. The power of the dependency pair method has been amply illustrated in a sequence of papers by Arts and Giesl [2,3,4].

The ordering constraints in the dependency pair method are generated by analyzing the cycles in the *dependency graph*. This graph summarizes the relationships between the dependency pairs of the rewrite system. More precisely, there is an arrow from dependency pair $s \rightarrow t$ to dependency pair $u \rightarrow v$ in the dependency graph if some instance of t rewrites to some instance of u. Since this is undecidable in general, the dependency graph has to be estimated by a decidable approximation. Arts and Giesl proposed the following simple algorithm for this purpose: there is an arrow in the so-called estimated dependency graph from $s \rightarrow t$ to $u \rightarrow v$ if the term obtained from t by replacing all outermost defined symbols by variables and a subsequent linearization unifies with u.

The approximation of Arts and Giesl often results in an unnecessarily large graph and hence a large number of constraints. Sometimes, as examples in this paper will demonstrate, this causes the failure of the termination proof. The aim of this paper is to show that by using tree automata techniques we obtain a much better estimation of the dependency graph. Our approach is based on the following two ingredients:

1. The powerful framework of Durand and Middeldorp for the study of decidable call-by-need computations in orthogonal term rewriting. This framework is parameterized by so-called approximation mappings. An approximation mapping abstracts from certain parts of the terms in the rewrite rules such that the set of terms that rewrite to a term in an arbitrary regular tree language is again regular.

2. The folklore result that it is decidable whether the set of ground instances of an arbitrary term intersects with a regular tree language. This result is well-known for linear terms but it also holds for non-linear terms.

We show that by adopting the so-called *nv* approximation we always obtain an estimation of the dependency graph which is at least as good as the one of Arts and Giesl but often better. Interestingly, we can automatically prove termination of rewrite systems outside the class of so-called DP quasi-simply terminating systems. This class, proposed by Giesl and Ohlebusch [14], consists of all rewrite systems "where an automated termination proof using dependency pairs is potentially feasible".

The remainder of the paper is organized as follows. In the next section we briefly recall the basics of the dependency pair technique. Section 3 contains background material on tree automata. In Section 4 we define new approximations of the dependency graph. We compare our approximations with the one of Arts in Giesl in Section 5. We also include a comparison with the approximation of Kusakari and Toyama [19,21]. In Section 6 we show that by using our approximations the dependency pair method provides a decision procedure for termination of right-ground rewrite systems.

2   Dependency Pairs

We assume familiarity with the basics of term rewriting ([5]). A term rewrite system (TRS for short) consists of rewrite rules $l \to r$ that satisfy $l \notin \mathcal{V}$ and $\mathsf{Var}(r) \subseteq \mathsf{Var}(l)$. If these conditions are not imposed we find it useful to speak of extended TRSs (eTRSs). Such systems arise naturally when we approximate TRSs or orient the rewrite rules from right to left, as explained in Section 4. Note that eTRSs which are not TRSs can never be terminating, but in this paper we will make clear that such eTRSs are very useful for automatically proving termination of TRSs.

Below we recall the basic notions and results of the dependency pair technique of Arts and Giesl. We refer to [2,4,12] for motivations and further refinements. We adopt the notation of [13,20]. Let $\mathcal{R}$ be a TRS over a signature $\mathcal{F}$. As usual, root symbols of left-hand sides of rewrite rules are called defined. Let $\mathcal{F}^{\sharp}$ denote the union of $\mathcal{F}$ and $\{f^{\sharp} \mid f$ is a defined symbol of $\mathcal{R}\}$ where $f^{\sharp}$ has the same arity as f. Given a term $t = f(t_1, \ldots, t_n) \in \mathcal{T}(\mathcal{F}, \mathcal{V})$ with f defined, we write $t^{\sharp}$ for the term $f^{\sharp}(t_1, \ldots, t_n)$. If $l \to r \in \mathcal{R}$ and t is a subterm of r with defined root symbol then the rewrite rule $l^{\sharp} \to t^{\sharp}$ is called a dependency pair of $\mathcal{R}$. The set of all dependency pairs of $\mathcal{R}$ is denoted by $\mathsf{DP}(\mathcal{R})$. In examples we often write F for $f^{\sharp}$.

An argument filtering for a signature $\mathcal{F}$ is a mapping π that associates with every n-ary function symbol an argument position $i \in \{1, \ldots, n\}$ or a (possibly empty) list $[i_1, \ldots, i_m]$ of argument positions with $1 \leqslant i_1 < \cdots < i_m \leqslant n$. The signature $\mathcal{F}_{\pi}$ consists of all function symbols f such that $\pi(f)$ is some list $[i_1, \ldots, i_m]$, where in $\mathcal{F}_{\pi}$ the arity of f is m. Every argument filtering π induces a mapping from $\mathcal{T}(\mathcal{F}, \mathcal{V})$ to $\mathcal{T}(\mathcal{F}_{\pi}, \mathcal{V})$, also denoted by π:

$$\pi(t) = \begin{cases} t & \text{if } t \text{ is a variable,} \\ \pi(t_i) & \text{if } t = f(t_1, \ldots, t_n) \text{ and } \pi(f) = i, \\ f(\pi(t_{i_1}), \ldots, \pi(t_{i_m})) & \text{if } t = f(t_1, \ldots, t_n) \text{ and } \pi(f) = [i_1, \ldots, i_m]. \end{cases}$$

Thus, an argument filtering is used to replace function symbols by one of their arguments or to eliminate certain arguments of function symbols.

A preorder is a transitive and reflexive relation. A rewrite preorder is a preorder $\succsim$ on terms that is closed under contexts and substitutions. A reduction pair consists of a rewrite preorder $\succsim$ and a compatible well-founded order $>$ which is closed under substitutions. Here compatibility means that the inclusion $\succsim \cdot > \, \subseteq \, >$ or the inclusion $> \cdot \succsim \, \subseteq \, >$ holds. The following theorem presents the basic dependency pair approach.

Theorem 1 (Arts and Giesl [4]). *A TRS $\mathcal{R}$ over a signature $\mathcal{F}$ is terminating if and only if there exists an argument filtering π for $\mathcal{F}^{\sharp}$ and a reduction pair $(\succsim, >)$ such that $\pi(\mathcal{R}) \subseteq \succsim$ and $\pi(\mathsf{DP}(\mathcal{R})) \subseteq \, >$.* □

Because rewrite rules are just pairs of terms, $\pi(\mathcal{R}) \subseteq \succsim$ is a shorthand for $\pi(l) \succsim \pi(r)$ for every rewrite rule $l \to r \in \mathcal{R}$. From now on we assume that all (e)TRSs are finite.

Rather than considering all dependency pairs at the same time, like in the above theorem, it is advantageous to treat groups of dependency pairs separately. These groups correspond to *cycles* in the *dependency graph* $\mathsf{DG}(\mathcal{R})$ of $\mathcal{R}$. The nodes of $\mathsf{DG}(\mathcal{R})$ are the dependency pairs of $\mathcal{R}$ and there is an arrow from $s \to t$ to $u \to v$ if and only if there exist substitutions σ and τ such that $t\sigma \to_{\mathcal{R}}^{*} u\tau$. (By renaming variables in different occurrences of dependency pairs we may assume that $\sigma = \tau$.) A *cycle* is a non-empty subset $\mathcal{C}$ of dependency pairs of $\mathsf{DP}(\mathcal{R})$ if for every two (not necessarily distinct) pairs $s \to t$ and $u \to v$ in $\mathcal{C}$ there exists a non-empty path in $\mathcal{C}$ from $s \to t$ to $u \to v$.

Theorem 2 (Arts and Giesl [2]). *A TRS $\mathcal{R}$ is terminating if and only if for every cycle $\mathcal{C}$ in $\mathsf{DG}(\mathcal{R})$ there exists an argument filtering π and a reduction pair $(\succsim, >)$ such that $\pi(\mathcal{R}) \subseteq \succsim$, $\pi(\mathcal{C}) \subseteq \succsim \cup >$, and $\pi(\mathcal{C}) \cap > \neq \varnothing$.* □

Note that $\pi(\mathcal{C}) \cap > \neq \varnothing$ denotes the situation that $\pi(s) > \pi(t)$ for at least one dependency pair $s \to t \in \mathcal{C}$.

Since it is undecidable whether there exists substitutions σ, τ such that $t\sigma \to_{\mathcal{R}}^{*} u\tau$, the dependency graph cannot be computed in general. Hence, in order to mechanize the termination criterion of Theorem 2 one has to approximate the dependency graph. To this end, Arts and Giesl proposed a simple algorithm.

Definition 3. *Let $\mathcal{R}$ be a TRS. The nodes of the estimated dependency graph $\mathsf{EDG}(\mathcal{R})$ are the dependency pairs of $\mathcal{R}$ and there is an arrow from $s \to t$ to $u \to v$ if and only if $\mathsf{REN}(\mathsf{CAP}(t))$ and u are unifiable. Here CAP replaces all outermost subterms with a defined root symbol by distinct fresh variables and REN replaces all occurrences of variables by distinct fresh variables.*

Lemma 4 (Arts and Giesl [4]). *Let $\mathcal{R}$ be a TRS.*

1. $\mathsf{EDG}(\mathcal{R})$ *is computable.*
2. $\mathsf{DG}(\mathcal{R}) \subseteq \mathsf{EDG}(\mathcal{R})$. □

3   Tree Automata

We briefly recall some basic definitions and results concerning tree automata. Much more information can be found in [8].

A (finite bottom-up) tree automaton is a quadruple $\mathcal{A} = (\mathcal{F}, Q, Q_f, \Delta)$ consisting of a finite signature $\mathcal{F}$, a finite set Q of states, disjoint from $\mathcal{F}$, a subset $Q_f \subseteq Q$ of final states, and a set of transition rules Δ. Every transition rule has the form $f(q_1, \ldots, q_n) \to q$ with $f \in \mathcal{F}$ and $q_1, \ldots, q_n, q \in Q$. So a tree automaton $\mathcal{A} = (\mathcal{F}, Q, Q_f, \Delta)$ is simply a finite ground TRS $(\mathcal{F} \cup Q, \Delta)$ whose rewrite rules have a special shape, together with a subset Q_f of Q. The induced rewrite relation on $\mathcal{T}(\mathcal{F} \cup Q)$ is denoted by $\to_{\mathcal{A}}$. A ground term $t \in \mathcal{T}(\mathcal{F})$ is accepted by $\mathcal{A}$ if $t \rightarrow_{\mathcal{A}}^* q$ for some $q \in Q_f$. The set of all such terms is denoted by $L(\mathcal{A})$. A subset $L \subseteq \mathcal{T}(\mathcal{F})$ is called regular if there exists a tree automaton $\mathcal{A} = (\mathcal{F}, Q, Q_f, \Delta)$ such that $L = L(\mathcal{A})$. It is well-known that every regular language is accepted by a deterministic tree automaton without inaccessible states. A deterministic automaton has no two different rules with the same left-hand side. A state is inaccessible if no ground term reduces to it. In this paper we make use of the additional properties mentioned below.

Lemma 5. *The set of ground instances of a linear term is regular.* □

We write $\Sigma(t)$ for the set of ground instances of the term t. The next result states that it is decidable whether a ground instance of an arbitrary term is accepted by a given tree automaton. For a linear term t this is obvious since (1) $\Sigma(t)$ is regular by Lemma 5, (2) regular languages are effectively closed under intersection, and (3) emptiness is decidable for regular languages. The point is that the problem remains decidable for non-linear terms. This extension will turn out to be very important for automatically proving termination of TRSs that rely on non-linearity (i.e., by linearizing the rewrite rules the TRS becomes non-terminating).

Theorem 6 (Tison [26]). *The following problem is decidable:*

> instance: tree automaton $\mathcal{A}$, term t
> question: $\Sigma(t) \cap L(\mathcal{A}) = \varnothing$?

Proof. First we transform $\mathcal{A}$ into an equivalent deterministic tree automaton $\mathcal{B} = (\mathcal{F}, Q, Q_f, \Delta)$ without inaccessible states. We claim that $\Sigma(t) \cap L(\mathcal{A}) \neq \varnothing$ if and only if there exists a mapping $\sigma \colon \mathcal{V}ar(t) \rightarrow Q$ such that $t\sigma \in L(\mathcal{B})$.

$\Rightarrow$ Suppose $\Sigma(t) \cap L(\mathcal{A}) \neq \varnothing$. So there exists a substitution $\tau \colon \mathcal{V}ar(t) \rightarrow \mathcal{T}(\mathcal{F})$ such that $t\tau \in L(\mathcal{A}) = L(\mathcal{B})$. Hence $t\tau \rightarrow_{\Delta}^* q$ for some $q \in Q_f$. In this sequence every subterm $\tau(x)$ of $t\tau$ is reduced to some state. Because $\mathcal{B}$ is deterministic, different occurrences of $\tau(x)$ in $t\tau$ reduce to the same state, say $q_x \in Q$. Define the mapping $\sigma \colon \mathcal{V}ar(t) \rightarrow Q$ by $\sigma(x) = q_x$ for every $x \in \mathcal{V}ar(t)$. Clearly $t\tau \rightarrow_{\Delta}^* t\sigma \rightarrow_{\Delta}^* q$ and hence $t\sigma \in L(\mathcal{B})$.

$\Leftarrow$ Suppose $t\sigma \in L(\mathcal{B})$ for some mapping $\sigma \colon \mathcal{V}ar(t) \rightarrow Q$. So $t\sigma \rightarrow_{\Delta}^* q$ for some $q \in Q_f$. Since all states of $\mathcal{B}$ are accessible, there exists a substitution $\tau \colon \mathcal{V}ar(t) \rightarrow \mathcal{T}(\mathcal{F})$ such that $\tau(x) \rightarrow_{\Delta}^* \sigma(x)$ for all $x \in \mathcal{V}ar(t)$. Hence $t\tau \rightarrow_{\Delta}^* t\sigma \rightarrow_{\Delta}^* q$ and thus $t\tau \in L(\mathcal{B}) = L(\mathcal{A})$. In other words, $\Sigma(t) \cap L(\mathcal{A}) \neq \varnothing$.

Since there are only finitely many mappings from $\mathcal{V}ar(t)$ to Q, this yields a decision procedure. □

We stress that for a linear term t there is no need to perform the expensive determinization of $\mathcal{A}$.

4   Approximations

In this section we define new approximations of the dependency graph. Our approximations are based on the framework of Durand and Middeldorp [11] for the study of decidable call-by-need computations in orthogonal term rewriting.

If $\mathcal{R}$ is an eTRS over a signature $\mathcal{F}$ and $L \subseteq \mathcal{T}(\mathcal{F})$ then $(\rightarrow^*_{\mathcal{R}})[L]$ denotes the set of all terms $s \in \mathcal{T}(\mathcal{F})$ such that $s \rightarrow^*_{\mathcal{R}} t$ for some term $t \in L$.

Definition 7. *An* approximation mapping *is a mapping α from eTRSs to eTRSs with the property that $\rightarrow_{\mathcal{R}} \subseteq \rightarrow^*_{\alpha(\mathcal{R})}$ for every eTRS $\mathcal{R}$. In the following we write $\mathcal{R}_\alpha$ instead of $\alpha(\mathcal{R})$. We say that α is* regularity preserving *if $(\rightarrow^*_{\mathcal{R}_\alpha})[L]$ is regular for all eTRSs $\mathcal{R}$ and regular L.*

In [11] an approximation mapping α is also required to satisfy the condition that the ground normal forms of $\mathcal{R}$ and $\mathcal{R}_\alpha$ coincide, but we do not need that condition here. Next we define three approximation mappings that are known to be regularity preserving. Our definitions are slightly different from the ones found in the literature because we have to deal with possibly non-left-linear TRSs.

Definition 8. *Let $\mathcal{R}$ be an eTRS. The* strong approximation *$\mathcal{R}_s$ is obtained from $\mathcal{R}$ by replacing the right-hand side and all occurrences of variables in the left-hand side of every rewrite rule by distinct fresh variables, i.e., $\mathcal{R}_s = \{\mathsf{REN}(l) \rightarrow x \mid l \rightarrow r \in \mathcal{R}$ and x is a fresh variable$\}$. The* nv approximation *$\mathcal{R}_{nv}$ is obtained from $\mathcal{R}$ by replacing all occurrences of variables in the rewrite rules by distinct fresh variables: $\mathcal{R}_{nv} = \{\mathsf{REN}(l) \rightarrow \mathsf{REN}(r) \mid l \rightarrow r \in \mathcal{R}\}$. An eTRS is called* growing *if for every rewrite rule $l \rightarrow r$ the variables in $\mathcal{V}ar(l) \cap \mathcal{V}ar(r)$ occur at depth 1 in l. The* growing approximation *$\mathcal{R}_g$ is defined as any left-linear growing eTRS that is obtained from $\mathcal{R}$ by linearizing the left-hand sides and renaming the variables in the right-hand sides that occur at a depth greater than 1 in the corresponding left-hand sides.*

For instance, if $\mathcal{R}$ contains the rewrite rule $\mathsf{f}(x, \mathsf{g}(x), y) \rightarrow \mathsf{f}(x, x, \mathsf{g}(y))$ then $\mathcal{R}_s$ contains $\mathsf{f}(x, \mathsf{g}(x'), y) \rightarrow z$, $\mathcal{R}_{nv}$ contains $\mathsf{f}(x, \mathsf{g}(x'), y) \rightarrow \mathsf{f}(x'', x''', \mathsf{g}(y'))$, and $\mathcal{R}_g$ contains $\mathsf{f}(x, \mathsf{g}(x'), y) \rightarrow \mathsf{f}(x, x, \mathsf{g}(y))$ or $\mathsf{f}(x', \mathsf{g}(x), y) \rightarrow \mathsf{f}(x'', x'', \mathsf{g}(y))$. (The former is preferred as it is closer to the original rule. The ambiguity in the definition of $\mathcal{R}_g$ causes no problems in the sequel.)

Theorem 9. *The approximation mappings* s, nv, *and* g *are regularity preserving.*   □

Nagaya and Toyama [24] proved the above result for the growing approximation; the tree automaton that recognizes $(\rightarrow^*_{\mathcal{R}_g})[L]$ is defined as the limit of a finite saturation process. This saturation process is similar to the ones defined in Comon [7] and Jacquemard [16], but by working exclusively with deterministic tree automata, non-right-linear rewrite rules can be handled. For the strong and nv approximation simpler constructions using ground tree transducers are possible (see e.g. Durand and Middeldorp [11]).

Recently, Takai *et al.* [25] introduced the class of left-linear inverse finite path overlapping rewrite systems and showed that the preceding theorem is true for the corresponding approximation mapping. Growing rewrite systems constitute a proper subclass of the class of inverse finite path overlapping rewrite systems. Since the definition of this class is rather difficult and the construction in the proof of regularity preservingness very complicated, we do not consider the inverse finite path overlapping approximation here. We note however that our results easily extend.

Definition 10. *Let $\mathcal{R}$ be a TRS and α an approximation mapping. The nodes of the α-approximated dependency graph $\mathsf{DG}_\alpha(\mathcal{R})$ are the dependency pairs of $\mathcal{R}$ and there is an arrow from $s \to t$ to $u \to v$ if and only if both $\Sigma(t) \cap (\to^*_{\mathcal{R}_\alpha})[\Sigma(\mathsf{REN}(u))] \neq \varnothing$ and $\Sigma(u) \cap (\to^*_{(\mathcal{R}^{-1})_\alpha})[\Sigma(\mathsf{REN}(t))] \neq \varnothing$.*

So we draw arrow from $s \to t$ to $u \to v$ if a ground instance of t rewrites in $\mathcal{R}_\alpha$ to a ground instance of $\mathsf{REN}(u)$ *and* a ground instance of u rewrites in $(\mathcal{R}^{-1})_\alpha$ to a ground instance of $\mathsf{REN}(t)$. The reason for having both conditions is that (1) for decidability t or u should be made linear and (2) depending on α and $\mathcal{R}$, $\mathcal{R}_\alpha$ may better approximate $\mathcal{R}$ than $(\mathcal{R}^{-1})_\alpha$ approximates $\mathcal{R}^{-1}$, or vice-versa. Also, the more conditions one imposes, the closer one gets to the real dependency graph.

Lemma 11. *Let $\mathcal{R}$ be a TRS and α an approximation mapping.*

1. *If α is regularity preserving then $\mathsf{DG}_\alpha(\mathcal{R})$ is computable.*
2. *$\mathsf{DG}(\mathcal{R}) \subseteq \mathsf{DG}_\alpha(\mathcal{R})$.*

Proof.

1. Let $s \to t$ and $u \to v$ be dependency pairs of $\mathcal{R}$. Because $\mathsf{REN}(u)$ is a linear term, $\Sigma(\mathsf{REN}(u))$ is regular (Lemma 5). Since α is regularity preserving, $(\to^*_{\mathcal{R}_\alpha})[\Sigma(\mathsf{REN}(u))]$ is regular. Hence, according to Theorem 6, it is decidable whether $\Sigma(t)$ intersects with $(\to^*_{\mathcal{R}_\alpha})[\Sigma(\mathsf{REN}(u))]$. By the same reasoning it follows that it is decidable whether $\Sigma(u)$ and $(\to^*_{(\mathcal{R}^{-1})_\alpha})[\Sigma(\mathsf{REN}(t))]$ intersect. Hence it is decidable whether there exists an arrow from $s \to t$ to $u \to v$ in $\mathsf{DG}_\alpha(\mathcal{R})$.

2. Suppose there is an arrow from dependency pair $s \to t$ to dependency pair $u \to v$ in $\mathsf{DG}(\mathcal{R})$. So $t\sigma \to^*_\mathcal{R} u\tau$ for some substitutions σ and τ. We may assume without loss of generality that $t\sigma$ and $u\tau$ are ground terms. Hence $t\sigma \in \Sigma(t) \subseteq \Sigma(\mathsf{REN}(t))$ and $u\tau \in \Sigma(u) \subseteq \Sigma(\mathsf{REN}(u))$. Consequently, $\Sigma(t) \cap (\to^*_\mathcal{R})[\Sigma(\mathsf{REN}(u))] \neq \varnothing$ and $\Sigma(u) \cap (\to^*_{\mathcal{R}^{-1}})[\Sigma(\mathsf{REN}(t))] \neq \varnothing$. Because α is an approximation mapping, $\to^*_\mathcal{R} \subseteq \to^*_{\mathcal{R}_\alpha}$ and $\to^*_{\mathcal{R}^{-1}} \subseteq \to^*_{(\mathcal{R}^{-1})_\alpha}$. Therefore $\Sigma(t) \cap (\to^*_{\mathcal{R}_\alpha})[\Sigma(\mathsf{REN}(u))] \neq \varnothing$ and $\Sigma(u) \cap (\to^*_{(\mathcal{R}^{-1})_\alpha})[\Sigma(\mathsf{REN}(t))] \neq \varnothing$. In other words, there exists an arrow from $s \to t$ to $u \to v$ in $\mathsf{DG}_\alpha(\mathcal{R})$. $\qquad\square$

It should be clear that a better approximation mapping results in a better approximation of the dependency graph. Hence we have the following result.

Lemma 12. $DG_g(\mathcal{R}) \subseteq DG_{nv}(\mathcal{R}) \subseteq DG_s(\mathcal{R})$ *for every TRS* $\mathcal{R}$.   □

The reason for considering the strong and nv approximations in this paper is that DG_s and DG_{nv} are easier to compute than DG_g, cf. the paragraph following Theorem 9.

5 Comparison

In this section we compare our α-approximated dependency graph with the estimated dependency graph of Arts and Giesl and the approximation of the dependency graph defined by Kusakari and Toyama [19,21].

The first two examples show that the s-approximated dependency graph and the estimated dependency graph are incomparable in general.

Example 13. Consider the TRS $\mathcal{R}$ consisting of the two rewrite rules

$$f(g(a)) \rightarrow f(a)$$
$$a \rightarrow b$$

There are two dependency pairs:

$$F(g(a)) \rightarrow F(a) \quad (1)$$
$$F(g(a)) \rightarrow A \quad (2)$$

Because $REN(CAP(F(a))) = F(x)$ unifies with $F(g(a))$, $EDG(\mathcal{R})$ contains two arrows:

$$(1) \longrightarrow (2)$$

We have $(\mathcal{R}^{-1})_s = \{f(a) \rightarrow x, b \rightarrow x\}$. Hence $(\rightarrow^*_{(\mathcal{R}^{-1})_s})[\{F(a)\}]$ consists of all terms of the form $f^n(a)$, $f^n(b)$, $F(f^n(a))$, $F(f^n(b))$ with $n \geqslant 0$. The term $F(g(a))$ clearly does not belong to this set and hence there are no arrows in $DG_s(\mathcal{R})$.

Example 14. Consider the TRS $\mathcal{R}$ consisting of the single rewrite rule

$$f(x, x) \rightarrow f(a, b)$$

There is one dependency pair:

$$F(x, x) \rightarrow F(a, b)$$

Because $REN(CAP(F(a, b))) = F(a, b)$ and $F(x, x)$ are not unifiable, $EDG(\mathcal{R})$ contains no arrows. However, both $\Sigma(F(a, b)) \cap (\rightarrow^*_{\mathcal{R}_s})[\Sigma(REN(F(x, x)))]$ and $\Sigma(F(x, x)) \cap (\rightarrow^*_{(\mathcal{R}^{-1})_s})[\Sigma(REN(F(a, b)))]$ are non-empty, as witnessed by the terms $F(a, b)$ and $F(f(a, b), f(a, b))$.

The non-left-linearity in the preceding example is essential. This is shown in Lemma 16 below. In the proof we make use of the following lemma. Here $\leftarrow^*_{\mathcal{R}_s}$ is the inverse of the relation $\rightarrow^*_{\mathcal{R}_s}$ (which is different from $\rightarrow^*_{(\mathcal{R}^{-1})_s}$).

Lemma 15. $(\leftarrow^*_{\mathcal{R}_s})[\Sigma(\mathsf{REN}(t))] \subseteq \Sigma(\mathsf{REN}(\mathsf{CAP}(t)))$ *for every TRS $\mathcal{R}$ and term t.*

Proof. Let $\mathcal{F}$ be the signature of $\mathcal{R}$. We use induction on the structure of t. If t is a variable or if the root symbol of t is a defined symbol then $\mathsf{CAP}(t)$ is a variable and hence $\Sigma(\mathsf{REN}(\mathsf{CAP}(t))) = \mathcal{T}(\mathcal{F})$ and thus trivially $(\leftarrow^*_{\mathcal{R}_s})[\Sigma(\mathsf{REN}(t))] \subseteq \Sigma(\mathsf{REN}(\mathsf{CAP}(t)))$. Suppose $t = f(t_1, \ldots, t_n)$ with f a constructor. Because the left-hand side of every rule in $\mathcal{R}_s$ starts with a defined symbol and the arguments of $\mathsf{REN}(t)$ do not share variables, $(\leftarrow^*_{\mathcal{R}_s})[\Sigma(\mathsf{REN}(t))] = \{f(s_1, \ldots, s_n) \mid s_i \in \Sigma(\mathsf{REN}(t_i))\}$. Also $\Sigma(\mathsf{REN}(\mathsf{CAP}(t))) = \{f(s_1, \ldots, s_n) \mid s_i \in \Sigma(\mathsf{REN}(\mathsf{CAP}(t_i)))\}$. Hence the desired inclusion follows from the induction hypothesis. □

The previous lemma does not hold for eTRSs. For instance, consider the eTRS $\mathcal{R} = \{x \to \mathsf{a}\}$ over the signature consisting of the constants a and b. If $t = \mathsf{b}$ then $(\leftarrow^*_{\mathcal{R}_s})[\Sigma(\mathsf{REN}(t))] = \{\mathsf{a}, \mathsf{b}\}$ and $\Sigma(\mathsf{REN}(\mathsf{CAP}(t))) = \{\mathsf{b}\}$.

Lemma 16. *If $\mathcal{R}$ is a left-linear TRS then $\mathsf{DG_s}(\mathcal{R}) \subseteq \mathsf{EDG}(\mathcal{R})$.*

Proof. Suppose there is an arrow from dependency pair $s \to t$ to dependency pair $u \to v$ in $\mathsf{DG_s}(\mathcal{R})$. By definition, $\Sigma(t) \cap (\to^*_{\mathcal{R}_s})[\Sigma(\mathsf{REN}(u))] \neq \varnothing$. Since $\mathcal{R}$ is left-linear, u is a linear term and thus $\Sigma(\mathsf{REN}(u)) = \Sigma(u)$. Hence there exist ground substitutions σ and τ such that $t\sigma \to^*_{\mathcal{R}_s} u\tau$. Clearly $t\sigma \in \Sigma(\mathsf{REN}(t))$. According to the preceding lemma $u\tau \in \Sigma(\mathsf{REN}(\mathsf{CAP}(t)))$. Since $\mathsf{REN}(\mathsf{CAP}(t))$ and u do not share variables, they are unifiable and thus there exists an arrow from $s \to t$ to $u \to v$ in $\mathsf{EDG}(\mathcal{R})$. □

Actually, with the strong approximation we can never benefit from non-linearity. This is formally expressed in the following lemma.

Lemma 17. *Let $\mathcal{R}$ be a nonempty eTRS, t a term, and L a set of ground terms. The following statements are equivalent:*

1. $\Sigma(t) \cap (\to^*_{\mathcal{R}_s})[L] \neq \varnothing$,
2. $\Sigma(\mathsf{REN}(t)) \cap (\to^*_{\mathcal{R}_s})[L] \neq \varnothing$.

Proof.

$\Rightarrow$ Obvious since $\Sigma(t) \subseteq \Sigma(\mathsf{REN}(t))$.

$\Leftarrow$ Let Δ be an arbitrary ground redex and define the substitution $\sigma = \{x \mapsto \Delta \mid x \in \mathcal{V}ar(t)\}$. Because in $\mathcal{R}_s$ a redex can be rewritten to any term, $t\sigma \to^*_{\mathcal{R}_s} t'$ for every $t' \in \Sigma(\mathsf{REN}(t))$. Hence, if $t' \in \Sigma(\mathsf{REN}(t)) \cap (\to^*_{\mathcal{R}_s})[L]$ then $t\sigma \in \Sigma(t) \cap (\to^*_{\mathcal{R}_s})[L]$. □

As a consequence, the strong approximation is not all that useful for approximating dependency graphs. For the nv approximation matters are quite different. Our next result states that the nv-approximated dependency graph is always a subgraph of the estimated dependency graph. In order to prove this, we need the following preliminary result.

Lemma 18. $(\mathcal{R}_{\mathrm{nv}})^{-1} = (\mathcal{R}^{-1})_{\mathrm{nv}}$ *for every eTRS $\mathcal{R}$.*

Proof. Since $\mathcal{R}_{\mathrm{nv}} = \{\mathsf{REN}(l) \to \mathsf{REN}(r) \mid l \to r \in \mathcal{R}\}$, the result is obvious. $\qquad\square$

We stress that the above lemma is not true for the strong and growing approximations. For the strong approximation the TRSs of Examples 13 and 14 serve as counterexample.

Theorem 19. $\mathsf{DG}_{\mathrm{nv}}(\mathcal{R}) \subseteq \mathsf{EDG}(\mathcal{R})$ *for every TRS* $\mathcal{R}$.

Proof. Suppose there is an arrow from dependency pair $s \to t$ to dependency pair $u \to v$ in $\mathsf{DG}_{\mathrm{nv}}(\mathcal{R})$. By definition, $\Sigma(u) \cap (\to^{*}_{(\mathcal{R}^{-1})_{\mathrm{nv}}})[\Sigma(\mathsf{REN}(t))] \neq \varnothing$. According to Lemmata 18 and 15, and using the observation that $\to^{*}_{\mathcal{R}_{\mathrm{nv}}}$ is a subrelation of $\to^{*}_{\mathcal{R}_{\mathrm{s}}}$, $(\to^{*}_{(\mathcal{R}^{-1})_{\mathrm{nv}}})[\Sigma(\mathsf{REN}(t))] = (\leftarrow^{*}_{\mathcal{R}_{\mathrm{nv}}})[\Sigma(\mathsf{REN}(t))] \subseteq (\leftarrow^{*}_{\mathcal{R}_{\mathrm{s}}})[\Sigma(\mathsf{REN}(t))] \subseteq \Sigma(\mathsf{REN}(\mathsf{CAP}(t)))$. Hence $\Sigma(u) \cap \Sigma(\mathsf{REN}(\mathsf{CAP}(t))) \neq \varnothing$ and hence u and $\mathsf{REN}(\mathsf{CAP}(t))$ are unifiable. Therefore the arrow from $s \to t$ to $u \to v$ also exists in $\mathsf{EDG}(\mathcal{R})$. $\qquad\square$

The next example shows that the nv-approximated dependency graph is in general a proper subgraph of the estimated dependency graph.

Example 20. Consider the TRS $\mathcal{R}$ consisting of the two rewrite rules

$$\mathsf{f}(\mathsf{a}, \mathsf{b}, x) \to \mathsf{f}(x, x, x)$$
$$\mathsf{a} \to \mathsf{c}$$

There is one dependency pair:

$$\mathsf{F}(\mathsf{a}, \mathsf{b}, x) \to \mathsf{F}(x, x, x)$$

Since $\mathsf{REN}(\mathsf{CAP}(\mathsf{F}(x, x, x))) = \mathsf{F}(x_1, x_2, x_3)$ unifies with $\mathsf{F}(\mathsf{a}, \mathsf{b}, x)$, $\mathsf{EDG}(\mathcal{R})$ contains a cycle. We have $\Sigma(\mathsf{REN}(\mathsf{F}(\mathsf{a}, \mathsf{b}, x))) = \{\mathsf{F}(\mathsf{a}, \mathsf{b}, t) \mid t \in \mathcal{T}(\mathcal{F})\}$ and $\mathcal{R}_{\mathrm{nv}} = \{\mathsf{f}(\mathsf{a}, \mathsf{b}, x) \to \mathsf{f}(x_1, x_2, x_3), \mathsf{a} \to \mathsf{c}\}$. Consequently $(\to^{*}_{\mathcal{R}_{\mathrm{nv}}})[\Sigma(\mathsf{REN}(\mathsf{F}(\mathsf{a}, \mathsf{b}, x)))] = \Sigma(\mathsf{REN}(\mathsf{F}(\mathsf{a}, \mathsf{b}, x)))$ and since no instance of $\mathsf{F}(x, x, x)$ belongs to this set, $\mathsf{DG}_{\mathrm{nv}}(\mathcal{R})$ contains no arrow. Therefore $\mathcal{R}$ is trivially terminating.

The TRS in the above example is not *DP quasi-simply terminating*. The class of DP quasi-simply terminating TRSs was introduced by Giesl and Ohlebusch [14] and supposed to "capture all TRSs where an automated termination proof using dependency pairs is potentially feasible". We note that the various refinements of the dependency pair method (narrowing, rewriting, instantiation; see Giesl and Arts [12]) are not applicable and moreover that proving innermost termination (which is easy with the standard dependency pair technique) is insufficient for termination as the TRS does not belong to a known class for which termination and innermost termination coincide.

The next example shows a TRS that cannot be proved terminating with the nv approximation but whose (automatic) termination proof becomes easy with the growing approximation.

Example 21. Consider the TRS $\mathcal{R}$ consisting of the three rewrite rules

$$\begin{aligned}
f(x, a) &\rightarrow f(x, g(x, b)) \\
g(h(x), y) &\rightarrow g(x, h(y)) \\
g(a, y) &\rightarrow y
\end{aligned}$$

There are three dependency pairs:

$$\begin{aligned}
F(x, a) &\rightarrow F(x, g(x, b)) & (1) \\
F(x, a) &\rightarrow G(x, b) & (2) \\
G(h(x), y) &\rightarrow G(x, h(y)) & (3)
\end{aligned}$$

One easily verifies that $\mathsf{DG_{nv}}(\mathcal{R})$ contains two cycles:

$$\circlearrowright (1) \longrightarrow (2) \longrightarrow (3) \circlearrowright$$

In particular, $F(a, g(a, b)) \rightarrow_{\mathcal{R}_{nv}} F(a, a)$ which explains the arrows from (1) to (1) and (2). The problematic cycle $\{(1)\}$ does not exist in $\mathsf{DG_g}(\mathcal{R})$ because no ground instance of $F(x, g(x, b))$ rewrites in $\mathcal{R}_g$ to a ground instance of $F(x, a)$:

$$(1) \qquad (2) \longrightarrow (3) \circlearrowright$$

As a consequence, the resulting ordering constraints (obtained from Theorem 2) are easily satisfied (e.g. by taking $\pi(f) = 1$ in combination with the lexicographic path order with precedence $G > h$ and $g > h$).[1]

In the final part of this section we compare our α-approximated dependency graph with the approximation of the dependency graph defined by Kusakari and Toyama [19,21]. Their approximation relies on the concepts of ω-reduction and Ω-reduction. The first concept stems from Huet and Lévy [15].

Let $\mathcal{R}$ be a TRS over a signature $\mathcal{F}$. Let Ω be a fresh constant. The set of ground terms over the extended signature $\mathcal{F} \cup \{\Omega\}$ is denoted by $\mathcal{T}_\Omega(\mathcal{F})$. Given a term $t \in \mathcal{T}(\mathcal{F}, \mathcal{V})$, the term in $\mathcal{T}_\Omega(\mathcal{F})$ obtained from t by replacing all variables by Ω is denoted by t_Ω. The prefix order $\geqslant$ on $\mathcal{T}_\Omega(\mathcal{F})$ is defined by the following two clauses:

- $t \geqslant \Omega$ for every $t \in \mathcal{T}_\Omega(\mathcal{F})$,
- $f(s_1, \ldots, s_n) \geqslant f(t_1, \ldots, t_n)$ if $s_i \geqslant t_i$ for every $1 \leqslant i \leqslant n$.

Two terms $s, t \in \mathcal{T}_\Omega(\mathcal{F})$ are compatible, denoted by $s \uparrow t$, if there exists a term $u \in \mathcal{T}_\Omega(\mathcal{F})$ such that both $u \geqslant s$ and $u \geqslant t$. Finally, ω-reduction is the relation $\rightarrow_\omega$ on $\mathcal{T}_\Omega(\mathcal{F})$ defined as follows: $s \rightarrow_\omega t$ if and only if $s = C[s']$ and $t = C[\Omega]$ such that $\Omega \neq s' \uparrow l_\Omega$ for some $l \rightarrow r \in \mathcal{R}$. It is easy to prove that

---

[1] Again, the TRS is not DP quasi-simply terminating. Unlike the previous example, proving innermost termination is sufficient for termination, but the estimated innermost dependency graph coincides with $\mathsf{EDG}(\mathcal{R}) = \mathsf{DG_{nv}}(\mathcal{R})$ and the narrowing refinement for innermost termination fails to make the requirements for an automatic proof easier.

ω-reduction is terminating and confluent. Hence every term $t \in \mathcal{T}_\Omega(\mathcal{F})$ has a unique normal form, which is denoted by $\omega(t)$. It is well-known that ω-reduction is closely related to the strong approximation. Below we make use of the following well-known facts (for all terms $s, t \in \mathcal{T}_\Omega(\mathcal{F})$):

- $\omega(t) \leqslant t$,
- if $s \leqslant t$ then $\omega(s) \leqslant \omega(t)$.

The concept of Ω-reduction corresponds to the nv approximation and is defined as follows: $s \to_\Omega t$ if and only if $s = C[s']$ and $t = C[r_\Omega]$ for some $l \to r \in \mathcal{R}$ such that $\Omega \neq s' \uparrow l_\Omega$. Unlike ω-reduction, Ω-reduction is in general neither confluent nor terminating.

Lemma 22. *Let $\mathcal{R}$ be a TRS. If $s \to_{\mathcal{R}_{nv}}^* t$ and $s' \leqslant s$ then $s' \to_\Omega^* t'$ for some $t' \leqslant t$.*

Proof. Induction on the length of $s \to_{\mathcal{R}_{nv}}^* t$, using the easy to prove fact that if $s \to_{\mathcal{R}_{nv}} t$ and $s' \leqslant s$ then $s' \to_\Omega^{\bar{=}} t'$ for some $t' \leqslant t$. □

We now have all ingredients to define Kusakari and Toyama's approximation of the dependency graph. Actually, their definition applies to AC rewriting, an extension that we do not consider in this paper. The definition below is the specialization to ordinary term rewriting.

Definition 23. *Let $\mathcal{R}$ be a TRS. For every $n \geqslant 0$ we define the graph $\mathrm{DG}_\Omega^n(\mathcal{R})$ as follows. Its nodes are the dependency pairs of $\mathcal{R}$ and there is an arrow from $s \to t$ to $u \to v$ if and only if there exists a term $t' \in \mathcal{T}_\Omega(\mathcal{F})$ such that $t' \uparrow u_\Omega$ and either $t_\Omega \to_\Omega^m t'$ with $m < n$ or $t_\Omega \to_\Omega^n \cdot \to_\omega^! t'$. (Note that the latter condition is equivalent to $t_\Omega \to_\Omega^n t''$ and $t' = \omega(t'')$ for some term $t'' \in \mathcal{T}_\Omega(\mathcal{F})$.)*

Lemma 24 (Kusakari and Toyama [19,21]). *Let $\mathcal{R}$ be a TRS and $n \geqslant 0$.*

1. *$\mathrm{DG}_\Omega^n(\mathcal{R})$ is computable.*
2. *$\mathrm{DG}_\Omega^n(\mathcal{R}) \subseteq \mathrm{DG}(\mathcal{R})$.*   □

It is not difficult to show that $\mathrm{EDG}(\mathcal{R})$ and $\mathrm{DG}_\Omega^n(\mathcal{R})$ are incomparable in general, for all $n \geqslant 0$ (contradicting the remark in Kusakari and Toyama [21] that their algorithm for approximating the dependency graph is more powerful than the one of Arts and Giesl). For instance, for the TRS $\mathcal{R}$ of Example 14 $\mathrm{DG}_\Omega^n(\mathcal{R})$ contains a cycle for every $n \geqslant 0$ whereas $\mathrm{EDG}(\mathcal{R})$ is empty. The same holds for $\mathrm{DG}_s(\mathcal{R})$. However, it is easy to prove that $\mathrm{DG}_s(\mathcal{R})$ is always a subgraph of $\mathrm{DG}_\Omega^0(\mathcal{R})$ and sometimes a proper subgraph, like in Example 13 where $\mathrm{DG}_\Omega^0(\mathcal{R})$ coincides with $\mathrm{EDG}(\mathcal{R})$. Below we compare Kusakari and Toyama's approximation with our nv-approximated dependency graph.

Lemma 25. *Let $\mathcal{R}$ be a TRS. If $n > 0$ then $\mathrm{DG}_\Omega^n(\mathcal{R}) \subseteq \mathrm{DG}_\Omega^{n-1}(\mathcal{R})$.*

Proof. Suppose there is an arrow from dependency pair $s \to t$ to dependency pair $u \to v$ in $\mathsf{DG}^n_{\mathrm{nv}}(\mathcal{R})$. So there exists a term $t' \in \mathcal{T}_\Omega(\mathcal{F})$ such that $t' \uparrow u_\Omega$ and either $t_\Omega \to^m_\Omega t'$ with $m < n$ or $t_\Omega \to^n_\Omega \cdot \to^!_\omega t'$. First suppose that $t_\Omega \to^m_\Omega t'$ with $m < n$. If $m < n - 1$ then the arrow from $s \to t$ to $u \to v$ also exists in $\mathsf{DG}^{n-1}(\mathcal{R})$. If $m = n - 1$ then we reason as follows. Since $\omega(t') \leqslant t'$, $\omega(t') \uparrow u_\Omega$. Clearly $t_\Omega \to^{n-1}_\Omega t' \to^!_\omega \omega(t')$. Hence the arrow from $s \to t$ to $u \to v$ exists in $\mathsf{DG}^{n-1}(\mathcal{R})$. Finally consider the case that $t_\Omega \to^n_\Omega \cdot \to^!_\omega t'$. So there exists terms t_1 and t_2 such that $t_\Omega \to^{n-1}_\Omega t_1 \to_\Omega t_2 \to^!_\omega t'$. Clearly $t_1 \to_\omega t'_2$ with $t'_2 \leqslant t_2$. We have $\omega(t_1) = \omega(t'_2) \leqslant \omega(t_2) = t'$. Hence $\omega(t_1)$ and u_Ω are compatible. Since $t_\Omega \to^{n-1}_\Omega t_1 \to^!_\omega \omega(t_1)$, the arrow from $s \to t$ to $u \to v$ exists in $\mathsf{DG}^{n-1}_\Omega(\mathcal{R})$. $\square$

The following result is the key to showing that our nv-approximated dependency graph is a subgraph of $\mathsf{DG}^n_\Omega(\mathcal{R})$, for all $n \geqslant 0$.

Lemma 26. *Let $s \to t$ and $u \to v$ be dependency pairs of $\mathcal{R}$. If $\Sigma(\mathsf{REN}(t)) \cap (\to^*_{\mathcal{R}_{\mathrm{nv}}})[\Sigma(\mathsf{REN}(u))] \neq \varnothing$ then there is an arrow from $s \to t$ to $u \to v$ in $\mathsf{DG}^n_\Omega(\mathcal{R})$ for all $n \geqslant 0$.*

Proof. Suppose $\Sigma(\mathsf{REN}(t)) \cap (\to^*_{\mathcal{R}_{\mathrm{nv}}})[\Sigma(\mathsf{REN}(u))] \neq \varnothing$. So $\mathsf{REN}(t)\sigma \to^*_{\mathcal{R}_{\mathrm{nv}}} \mathsf{REN}(u)\tau$ for some ground substitutions σ and τ. Since $t_\Omega \leqslant \mathsf{REN}(t)\sigma$, an application of Lemma 22 yields $t_\Omega \to^*_\Omega u'$ for some term $u' \leqslant \mathsf{REN}(u)\tau$. Because also $u_\Omega \leqslant \mathsf{REN}(u)\tau$, $u' \uparrow u_\Omega$. Let m be the length of the Ω-reduction sequence from t_Ω to u'. It follows that there is an arrow from $s \to t$ to $u \to v$ in $\mathsf{DG}^n_\Omega(\mathcal{R})$ for all $n > m$. According to Lemma 25, the arrow exists also in $\mathsf{DG}^n_\Omega(\mathcal{R})$ for $n \leqslant m$. $\square$

Theorem 27. $\mathsf{DG}_{\mathrm{nv}}(\mathcal{R}) \subseteq \mathsf{DG}^n_\Omega(\mathcal{R})$ *for every TRS $\mathcal{R}$ and $n \geqslant 0$.*

Proof. Suppose there is an arrow from dependency pair $s \to t$ to dependency pair $u \to v$ in $\mathsf{DG}_{\mathrm{nv}}(\mathcal{R})$. By definition, $\Sigma(t) \cap (\to^*_{\mathcal{R}_{\mathrm{nv}}})[\Sigma(\mathsf{REN}(u))] \neq \varnothing$. Since $\Sigma(t) \subseteq \Sigma(\mathsf{REN}(t))$, also $\Sigma(\mathsf{REN}(t)) \cap (\to^*_{\mathcal{R}_{\mathrm{nv}}})[\Sigma(\mathsf{REN}(u))] \neq \varnothing$. According to Lemma 26 the arrow from $s \to t$ to $u \to v$ exists in $\mathsf{DG}^n_\Omega(\mathcal{R})$ for all $n \geqslant 0$. $\square$

The reverse inclusion does not hold. Consider for instance the TRS $\mathcal{R}$ of Example 20. Since $\mathsf{F}(\Omega, \Omega, \Omega)$ is compatible with $\mathsf{F}(\mathsf{a}, \mathsf{b}, \Omega)$, $\mathsf{DG}^n_\Omega(\mathcal{R})$ contains a cycle for all $n \geqslant 0$.

In retrospect, Kusakari and Toyama's approximation suffers from the following two problems: (1) since all variables are replaced by Ω, TRSs that are terminating because of non-linearity cannot be handled appropriately, and (2) there is no need to bound the number of Ω-reduction steps, rather, by avoiding such a bound we can make effective use of tree automata techniques.

6 Decidable Classes

Termination is known to be decidable for several subclasses of TRSs. In this section we investigate whether these decidability results can be obtained with the dependency pair technique. The best known class of TRSs with a decidable termination problem is the class of right-ground TRSs (Dershowitz [9]). The following easy result states that in principle the dependency pair technique is very suitable for deciding termination of right-ground TRSs.

Theorem 28. *A right-ground TRS $\mathcal{R}$ is terminating if and only if $\mathsf{DG}(\mathcal{R})$ contains no cycles.*

Proof.

$\Rightarrow$ Suppose $\mathsf{DG}(\mathcal{R})$ contains a cycle $\mathcal{C} = \{s_i \to t_i \mid 1 \leqslant i \leqslant n\}$. We show that $\mathcal{R}$ is non-terminating. Without loss of generality we assume that $\mathcal{C}$ is minimal. Since $\mathcal{R}$ is right-ground, there exist substitutions σ_i such that $t_i \to_{\mathcal{R}}^* s_{i+1}\sigma_i$ for all $1 \leqslant i \leqslant n$ and with $s_{n+1} = s_1$. By definition of dependency pairs, for every $1 \leqslant i \leqslant n$ there exists a rewrite rule $l_i \to r_i \in \mathcal{R}$ and a subterm u_i of r_i such that $s_i = l_i^{\sharp}$ and $t_i = u_i^{\sharp}$. Let C_i be the context such that $r_i = C_i[u_i]$. Since all steps in $u_i^{\sharp} \to_{\mathcal{R}}^* l_{i+1}^{\sharp}\sigma_i$ take place below the root position, we also have $u_i \to_{\mathcal{R}}^* l_{i+1}\sigma_i$ and thus $r_i = C_i[u_i] \to_{\mathcal{R}}^* C_i[l_{i+1}\sigma_i]$ (with $l_{n+1} = l_1$). Therefore

$$l_1 \to_{\mathcal{R}} r_1 \to_{\mathcal{R}}^* C_1[l_2\sigma_1] \to_{\mathcal{R}} C_1[r_2] \to_{\mathcal{R}}^* C_1[C_2[l_3\sigma_2]]$$
$$\to_{\mathcal{R}} \cdots \to_{\mathcal{R}}^* C_1[C_2[\cdots[C_n[l_1\sigma_n]]\cdots]]$$

which gives rise to an infinite rewrite sequence.

$\Leftarrow$ If there are no cycles in $\mathsf{DG}(\mathcal{R})$ then the conditions of Theorem 2 are trivially satisfied and thus $\mathcal{R}$ is terminating. $\qquad\square$

So as far as termination of right-ground TRSs is concerned, the only thing that matters is a good approximation of the dependency graph. Next we consider how the various approximations of the dependency graph deal with right-ground TRSs.

Theorem 29. *For every left-linear right-ground TRS $\mathcal{R}$, $\mathsf{DG}(\mathcal{R}) = \mathsf{DG}_{\mathrm{nv}}(\mathcal{R})$.*

Proof. According to Lemma 11 it suffices to show that $\mathsf{DG}_{\mathrm{nv}}(\mathcal{R}) \subseteq \mathsf{DG}(\mathcal{R})$. So suppose there is an arrow from dependency pair $s \to t$ to dependency pair $u \to v$ in $\mathsf{DG}_{\mathrm{nv}}(\mathcal{R})$. Hence $\Sigma(t) \cap (\to_{\mathcal{R}_{\mathrm{nv}}}^*)[\Sigma(\mathsf{REN}(u))] \neq \varnothing$. Because $\mathcal{R}$ is left-linear and right-ground, $\mathcal{R}_{\mathrm{nv}} = \mathcal{R}$, t is a ground term, and u is linear. Hence $t \in (\to_{\mathcal{R}}^*)[\Sigma(u)]$ and thus $t \to_{\mathcal{R}}^* u\sigma$ for some ground substitution σ. Therefore the arrow from $s \to t$ to $u \to v$ also exists in $\mathsf{DG}(\mathcal{R})$. $\qquad\square$

The following example shows that without the left-linearity condition $\mathsf{DG}(\mathcal{R})$ and $\mathsf{DG}_{\mathrm{nv}}(\mathcal{R})$ may differ.

Example 30. Consider the right-ground TRS $\mathcal{R}$ consisting of the three rewrite rules

$$\begin{aligned}
\mathsf{f}(\mathsf{a}) &\to \mathsf{g}(\mathsf{h}(\mathsf{a}, \mathsf{b})) \\
\mathsf{g}(\mathsf{g}(\mathsf{a})) &\to \mathsf{f}(\mathsf{b}) \\
\mathsf{h}(x, x) &\to \mathsf{g}(\mathsf{a})
\end{aligned}$$

There are four dependency pairs:

$$\mathsf{F}(\mathsf{a}) \;\rightarrow\; \mathsf{G}(\mathsf{h}(\mathsf{a},\mathsf{b})) \qquad (1) \qquad\qquad \mathsf{G}(\mathsf{g}(\mathsf{a})) \;\rightarrow\; \mathsf{F}(\mathsf{b}) \qquad (3)$$
$$\mathsf{F}(\mathsf{a}) \;\rightarrow\; \mathsf{H}(\mathsf{a},\mathsf{b}) \qquad\;\; (2) \qquad\qquad \mathsf{H}(x,x) \;\rightarrow\; \mathsf{G}(\mathsf{a}) \qquad (4)$$

Because $\mathcal{R}_{\mathrm{nv}}$ contains the rewrite rule $\mathsf{h}(x,y) \rightarrow \mathsf{g}(\mathsf{a})$ and $(\mathcal{R}_{\mathrm{nv}})^{-1} = (\mathcal{R}^{-1})_{\mathrm{nv}}$, $\mathsf{DG}_{\mathrm{nv}}(\mathcal{R})$ contains an arrow from (1) to (3). However, since $\mathsf{G}(\mathsf{h}(\mathsf{a},\mathsf{b}))$ does not rewrite to $\mathsf{G}(\mathsf{g}(\mathsf{a}))$ in $\mathcal{R}$, this arrow does not exist in $\mathsf{DG}(\mathcal{R})$.

Note that in the previous example $\mathcal{R}_{\mathrm{g}}$ also contains the rule $\mathsf{h}(x,y) \rightarrow \mathsf{g}(\mathsf{a})$ but the corresponding rule in $(\mathcal{R}^{-1})_{\mathrm{g}}$ is $\mathsf{g}(\mathsf{a}) \rightarrow \mathsf{h}(x,x)$ and therefore $\mathsf{G}(\mathsf{g}(\mathsf{a}))$ does not belong to $(\rightarrow^{*}_{(\mathcal{R}^{-1})_{\mathrm{g}}})[\{\mathsf{G}(\mathsf{h}(\mathsf{a},\mathsf{b}))\}]$. Hence there is no arrow from (1) to (3) in $\mathsf{DG}_{\mathrm{g}}(\mathcal{R})$. This holds in general.

Theorem 31. *For every right-ground TRS $\mathcal{R}$, $\mathsf{DG}(\mathcal{R}) = \mathsf{DG}_{\mathrm{g}}(\mathcal{R})$.*

Proof. According to Lemma 11 it suffices to show that $\mathsf{DG}_{\mathrm{g}}(\mathcal{R}) \subseteq \mathsf{DG}(\mathcal{R})$. So suppose there is an arrow from dependency pair $s \rightarrow t$ to dependency pair $u \rightarrow v$ in $\mathsf{DG}_{\mathrm{g}}(\mathcal{R})$. Hence $\Sigma(u) \cap (\rightarrow^{*}_{(\mathcal{R}^{-1})_{\mathrm{g}}})[\Sigma(\mathsf{REN}(t))] \neq \varnothing$. Because $\mathcal{R}$ is right-ground, $(\mathcal{R}^{-1})_{\mathrm{g}} = \mathcal{R}^{-1}$ and t is a ground term. Hence $\Sigma(u) \cap (\leftarrow^{*}_{\mathcal{R}})[\{t\}] \neq \varnothing$ and thus $t \rightarrow^{*}_{\mathcal{R}} u\sigma$ for some ground substitution σ. Therefore the arrow from $s \rightarrow t$ to $u \rightarrow v$ also exists in $\mathsf{DG}(\mathcal{R})$. $\qquad\qquad\square$

The above results provide an easy decision procedure for termination of right-ground TRSs $\mathcal{R}$: Compute the dependency graph of $\mathcal{R}$ using the growing (nv, if $\mathcal{R}$ is left-linear) approximation and determine whether there are any cycles. We stress that the above results are not true for the estimated dependency graph.

Recently, Nagaya and Toyama [24] obtained the following decidability result.

Theorem 32. *Termination is decidable for almost orthogonal growing TRSs.*
$$\qquad\qquad\qquad\qquad\qquad\qquad\qquad\qquad\qquad\qquad\qquad\qquad\qquad\qquad\qquad\square$$

It should be noted that this result does not cover the preceding results due to the almost orthogonality requirement. (A TRS is called almost orthogonal if it is left-linear and all critical pairs are trivial overlays.) On the other hand, although it is very easy to prove that $\mathsf{DG}(\mathcal{R}) = \mathsf{DG}_{\mathrm{g}}(\mathcal{R})$ for every left-linear growing TRS $\mathcal{R}$, the dependency pair approach does not seem to give an easy decision procedure since the dependency graph may contain cycles, as shown in the following example.

Example 33. Consider the (almost) orthogonal growing TRS $\mathcal{R}$ consisting of the two rewrite rules

$$f(x) \;\rightarrow\; g(x)$$
$$g(a) \;\rightarrow\; f(b)$$

There are two dependency pairs:

$$\mathsf{F}(x) \;\rightarrow\; \mathsf{G}(x) \qquad (1)$$
$$\mathsf{G}(\mathsf{a}) \;\rightarrow\; \mathsf{F}(\mathsf{b}) \qquad (2)$$

One easily verifies that $\mathsf{DG}(\mathcal{R})$ contains a cycle:

$$(1) \underset{\longleftarrow}{\overset{\longrightarrow}{}} (2)$$

However, $\mathcal{R}$ is clearly terminating (and it is very easy to solve the constraints stemming from the dependency pair technique in Theorem 2).

7   Conclusion

In this paper we have shown that simple tree automata techniques are useful to obtain better approximations of the dependency graph and hence we can automatically prove termination of a larger class of TRSs. More sophisticated tree automata techniques have been developed for dealing with non-linearity, see [8, Chapter 4], but we are not aware of any preservation results for the corresponding language classes and hence it is unclear whether these techniques could further improve automatic termination techniques.

Obviously, our α-approximated dependency graphs are harder to compute than the estimated dependency graph of Arts and Giesl. Consequently, we do not propose to eliminate the estimated dependency graph. Rather, our approximations should be tried only if tools based on the estimated dependency graph (like [1]) fail to prove termination or maybe in parallel to the search for suitable argument filterings and orderings to satisfy the resulting constraints. Clearly experimentation is needed to determine when to invoke our approximations. Currently we are working on an implementation of our algorithms.

It is worthwhile to investigate whether our approach can be extended to AC termination ([21,23]) and to innermost termination ([4]). For AC termination we do not expect any problems, but innermost termination seems more difficult. The reason is that the existence of an arrow from $s \to t$ to $u \to v$ in the *innermost* dependency graph does not only depend on whether a ground instance $t\sigma$ of t innermost rewrites to a ground instance $u\tau$ of u, but $s\sigma$ and $u\tau$ are additionally required to be normal forms. The latter condition is easily verified by tree automata techniques but it is unclear how to deal with the synchronization between the two conditions.

Since there are numerous examples of terminating TRSs whose dependency graphs do contain cycles, it goes without saying that the work reported in this paper is not the final answer to the problem of proving termination of rewrite systems automatically.

Acknowledgements. I thank Seitaro Yuuki for useful discussions. The paper benefitted from detailed comments of Jürgen Giesl and the anonymous referees.

References

1. T. Arts. System description: The dependency pair method. In *Proc. 11th RTA*, volume 1833 of *LNCS*, pages 261–264, 2000.

2. T. Arts and J. Giesl. Modularity of termination using dependency pairs. In *Proc. 9th RTA*, volume 1379 of *LNCS*, pages 226–240, 1998.
3. T. Arts and J. Giesl. Applying rewriting techniques to the verification of Erlang processes. In *Proc. 13th CSL*, volume 1862 of *LNCS*, pages 457–471, 2000.
4. T. Arts and J. Giesl. Termination of term rewriting using dependency pairs. *Theoretical Computer Science*, 236:133–178, 2000.
5. F. Baader and T. Nipkow. *Term Rewriting and All That*. Cambridge University Press, 1998.
6. F. Bellegarde and P. Lescanne. Termination by completion. *Applicable Algebra in Engineering, Communication and Computing*, 1:79–96, 1990.
7. H. Comon. Sequentiality, monadic second-order logic and tree automata. *Information and Computation*, 157:25–51, 2000.
8. H. Comon, M. Dauchet, R. Gilleron, F. Jacquemard, D. Lugiez, S. Tison, and M. Tommasi. Tree automata techniques and applications, 1999. Draft, available from http://www.grappa.univ-lille3.fr/tata/.
9. N. Dershowitz. Termination of linear rewriting systems (preliminary version). In *Proc. 8th ICALP*, volume 115 of *LNCS*, pages 448–458, 1981.
10. N. Dershowitz. Orderings for term-rewriting systems. *Theoretical Computer Science*, 17:279–301, 1982.
11. I. Durand and A. Middeldorp. Decidable call by need computations in term rewriting. In *Proc. 14th CADE*, volume 1249 of *LNAI*, pages 4–18, 1997.
12. J. Giesl and T. Arts. Verification of Erlang processes by dependency pairs. *Applicable Algebra in Engineering, Communication and Computing*, 2001. To appear.
13. J. Giesl and A. Middeldorp. Eliminating dummy elimination. In *Proc. 17th CADE*, volume 1831 of *LNAI*, pages 309–323, 2000.
14. J. Giesl and E. Ohlebusch. Pushing the frontiers of combining rewrite systems farther outwards. In *Proc. FroCoS'98*, volume 7 of *Studies in Logic and Computation*, pages 141–160. Wiley, 2000.
15. G. Huet and J.-J. Lévy. Computations in orthogonal rewriting systems, I and II. In *Computational Logic, Essays in Honor of Alan Robinson*, pages 396–443. The MIT Press, 1991. Original version: Report 359, Inria, 1979.
16. F. Jacquemard. Decidable approximations of term rewriting systems. In *Proc. 7th RTA*, volume 1103 of *LNCS*, pages 362–376, 1996.
17. S. Kamin and J.J. Lévy. Two generalizations of the recursive path ordering. Unpublished manuscript, University of Illinois, 1980.
18. D.E. Knuth and P. Bendix. Simple word problems in universal algebras. In *Computational Problems in Abstract Algebra*, pages 263–297. Pergamon Press, 1970.
19. K. Kusakari. *Termination, AC-Termination and Dependency Pairs of Term Rewriting Systems*. PhD thesis, JAIST, 2000.
20. K. Kusakari, M. Nakamura, and Y. Toyama. Argument filtering transformation. In *Proc. 1st PPDP*, volume 1702 of *LNCS*, pages 48–62, 1999.
21. K. Kusakari and Y. Toyama. On proving AC-termination by AC-dependency pairs. Research Report IS-RR-98-0026F, School of Information Science, JAIST, 1998.
22. D. Lankford. On proving term rewriting systems are noetherian. Report MTP-3, Louisiana Technical University, 1979.
23. C. Marché and X. Urbain. Termination of associative-commutative rewriting by dependency pairs. In *Proc. 9th RTA*, volume 1379 of *LNCS*, pages 241–255, 1998.
24. T. Nagaya and Y. Toyama. Decidability for left-linear growing term rewriting systems. In *Proc. 10th RTA*, volume 1631 of *LNCS*, pages 256–270, 1999.

25. T. Takai, Y. Kaji, and H. Seki. Right-linear finite path overlapping term rewriting systems effectively preserve recognizability. In *Proc. 11th RTA*, volume 1833 of *LNCS*, pages 246–260, 2000.
26. S. Tison. Tree automata and term rewrite systems, July 2000. Invited tutorial at the 11th RTA.
27. H. Zantema. Termination of term rewriting: Interpretation and type elimination. *Journal of Symbolic Computation*, 17:23–50, 1994.
28. H. Zantema. Termination of term rewriting by semantic labelling. *Fundamenta Informaticae*, 24:89–105, 1995.

On the Use of Weak Automata for Deciding Linear Arithmetic with Integer and Real Variables*

Bernard Boigelot, Sébastien Jodogne, and Pierre Wolper

Université de Liège,
Institut Montefiore, B28,
4000 Liège, Belgium
{boigelot,jodogne,pw}@montefiore.ulg.ac.be,
http://www.montefiore.ulg.ac.be/~{boigelot,jodogne,pw}/

Abstract. This paper considers finite-automata based algorithms for handling linear arithmetic with both real and integer variables. Previous work has shown that this theory can be dealt with by using finite automata on infinite words, but this involves some difficult and delicate to implement algorithms. The contribution of this paper is to show, using topological arguments, that only a restricted class of automata on infinite words are necessary for handling real and integer linear arithmetic. This allows the use of substantially simpler algorithms and opens the path to the implementation of a usable system for handling this combined theory.

1 Introduction

Among the techniques used to develop algorithms for deciding or checking logical formulas, finite automata have played an important role in a variety of cases. Classical examples are the use of infinite-word finite automata by Büchi [Büc62] for obtaining decision procedures for the first and second-order monadic theories of one successor as well as the use of tree automata by Rabin [Rab69] for deciding the second-order monadic theory of n successors. More recent examples are the use of automata for obtaining decision and model-checking procedures for temporal and modal logics [VW86a,VW86b,VW94,KVW00]. In this last setting, automata-based procedures have the advantage of moving the combinatorial aspects of the procedures to the context of automata, which are simple graph-like structures well adapted to algorithmic development. This separation of concerns between the logical and the algorithmic has been quite fruitful for instance in the implementation of model checkers for linear-time temporal logic [CVWY90, Hol97].

As already noticed by Büchi [Büc60,Büc62], automata-based approaches are not limited to sequential and modal logics, but can also be used for Presburger arithmetic. To achieve this, one adopts the usual encoding of integers in a base

---

* This work was partially funded by a grant of the "Communauté française de Belgique - Direction de la recherche scientifique - Actions de recherche concertées".

R. Goré, A. Leitsch, and T. Nipkow (Eds.): IJCAR 2001, LNAI 2083, pp. 611–625, 2001.

$r \geq 2$, thus representing an integer as a word over the alphabet $\{0, \ldots, r-1\}$. By extension, n-component integer vectors are represented by words over the alphabet $\{0, \ldots, r-1\}^n$ and a finite automaton operating over this alphabet represents a set of integer vectors. Given that addition and order are easily represented by finite automata and that these automata are closed under Boolean operations as well as projection, one easily obtains a decision procedure for Presburger arithmetic. This idea was first explored at the theoretical level, yielding for instance the very nice result that base-independent finite-automaton representable sets are exactly the Presburger sets [Cob69,Sem77,BHMV94]. Later, it has been proposed as a practical means of deciding and manipulating Presburger formulas [BC96,Boi98,SKR98,WB00]. The intuition behind this applied use of automata for Presburger arithmetic is that finite automata play with respect to Presburger arithmetic a role similar to the one of Binary Decision Diagrams (BDDs) with respect to Boolean logic. These ideas have been implemented in the LASH tool [LASH], which has been used successfully in the context of verifying systems with unbounded integer variables.

It almost immediately comes to mind that if a finite word over the alphabet $\{0, \ldots, r-1\}$ can represent an integer, an infinite word over the same alphabet extended with a fractional part separator (the usual dot) can represent a real number. Finite automata on infinite words can thus represent sets of real vectors, and serve as a means of obtaining a decision procedure for real additive arithmetic. Furthermore, since numbers with empty fractional parts can easily be recognized by automata, the same technique can be used to obtain a decision procedure for a theory combining the integers and the reals. This is not presently handled by any tool, but can be of practical use, for instance in the verification of timed systems using integer variables [BBR97]. However, turning this into an effective implemented system is not as easy as it might first seem. Indeed, projecting and complementing finite automata on infinite words is significantly more difficult than for automata on finite words. Projection yields nondeterministic automata and complementing or determinizing infinite-word automata is a notoriously difficult problem. A number of algorithms have been proposed for this [Büc62,SVW87,Saf88,KV97], but even though their theoretical complexity remains simply exponential as in the finite-word case, it moves up from $2^{O(n)}$ to $2^{O(n \log n)}$ and none of the proposed algorithms are as easy to implement and fine-tune as the simple Rabin-Scott subset construction used in the finite-word case.

However, it is intuitively surprising that handling reals is so much more difficult than handling integers, especially in light of the fact that the usual polyhedra-based approach to handling arithmetic is both of lower complexity and easier to implement for the reals than for the integers [FR79]. One would expect that handling reals with automata should be no more difficult than handling integers[1]. The conclusion that comes out of these observations is that

---

[1] Note that one cannot expect reals to be easier to handle with automata than integers since, by nature, this representation includes explicit information about the existence of integer values satisfying the represented formula.

infinite-word automata constructed from linear arithmetic formulas must have a special structure that makes them easier to manipulate than general automata on infinite words. That this special structure exists and that it can exploited to obtain simpler algorithms is precisely the subject of this paper.

As a starting point, let us look at the topological characterization of the sets definable by linear arithmetic formulas. Let us first consider a formula involving solely real variables. If the formula is quantifier free, it is a Boolean combination of linear constraints and thus defines a set which is a finite Boolean combination of open and closed sets. Now, since real linear arithmetic admits quantifier elimination, the same property also holds for quantified formulas. Then, looking at classes of automata on infinite words, one notices that the most restricted one that can accept Boolean combinations of open and closed sets is the class of deterministic weak automata [SW74,Sta83]. These accept all ω-regular sets in the Borel class $F_\sigma \cap G_\delta$ and hence also finite Boolean combinations of open and closed sets. So, with some care about moving from the topology on vectors to the topology on their encoding as words, one can conclude that the sets representable by arithmetic formulas involving only real variables can always be accepted by deterministic weak automata on infinite words. If integers are also involved in the formula, there is no established quantifier elimination result for the combined theory and one cannot readily conclude the same. A first result in this paper closes this loophole. It establishes that sets definable by quantified linear arithmetic formulas involving both real and integer variables are within $F_\sigma \cap G_\delta$ and thus are representable by deterministic weak automata. Rather than using a quantifier elimination type argument to establish this, our proof relies on separating the integer and fractional parts of variables and on topological properties of $F_\sigma \cap G_\delta$.

The problematic part of the operations on automata needed to decide a first-order theory is the sequence of projections and complementations needed to eliminate a string of quantifiers alternating between existential and universal ones. The second result of this paper shows that for sets defined in linear arithmetic this can be done with constructions that are simple adaptations of the ones used for automata on finite words. Indeed, deterministic weak automata can be viewed as either Büchi or co-Büchi automata. The interesting fact is that co-Büchi automata can be determinized by the "breakpoint" construction [MH84, KV97], which basically amounts to a product of subset constructions. Thus, one has a simple construction to project and determinize a weak automaton, yielding a deterministic co-Büchi automaton, which is easily complemented into a deterministic Büchi automaton. In the general case, another round of projection will lead to a nondeterministic Büchi automaton, for which a general determinization procedure has to be used. However, we have the result that for automata obtained from linear arithmetic formulas, the represented sets stay within those accepted by deterministic weak automata. We prove that this implies that the automata obtained after determinization will always be weak.

Note that this cannot be directly concluded from the fact that the represented sets stay within those representable by deterministic weak automata.

Indeed, even though the represented sets can be accepted by deterministic weak automata, the automata that are obtained by the determinization procedure might not have this form. Fortunately, we can prove that this is impossible. For this, we go back to the link between automata and the topology of the sets of infinite words they accept. The argument is that ω-regular sets in $F_\sigma \cap G_\delta$ have a topological property that forces the automata accepting them to be inherently weak, i.e. not to have strongly connected components containing both accepting and non accepting cycles.

As a consequence of our results, we obtain a decision procedure for the theory combining integer and real linear arithmetic that is suitable for implementation. The fact that this theory is decidable was known [BBR97], but the results of this paper move us much closer to an implemented tool that can handle it effectively.

2   Automata-Theoretic and Topological Background

In this section we recall some automata-theoretic and topological concepts that are used in the paper.

2.1   Automata on Infinite Words

An infinite word (or ω-word) w over an alphabet Σ is a mapping $w : \mathbb{N} \to \Sigma$ from the natural numbers to Σ. A Büchi automaton on infinite words is a five-tuple $A = (Q, \Sigma, \delta, q_0, F)$, where

- Q is a finite set of states;
- Σ is the input alphabet;
- δ is the transition function and is of the form $\delta : Q \times \Sigma \to 2^Q$ if the automaton is nondeterministic and of the form $\delta : Q \times \Sigma \to Q$ if the automaton is deterministic;
- q_0 is the initial state;
- F is a set of accepting states.

A run π of a Büchi automaton $A = (Q, \Sigma, \delta, q_0, F)$ on an ω-word w is a mapping $\pi : \mathbb{N} \to Q$ that satisfies the following conditions:

- $\pi(0) = q_0$, i.e. the run starts in the initial state;
- For all $i \geq 0$, $\pi(i + 1) \in \delta(\pi(i), w(i))$ (nondeterministic automata) or $\pi(i + 1) = \delta(\pi(i), w(i))$ (deterministic automata), i.e. the run respects the transition function.

Let $inf(\pi)$ be the set of states that occur infinitely often in a run π. A run π is said to be accepting if $inf(\pi) \cap F \neq \emptyset$. An ω-word w is accepted by a Büchi automaton if that automaton has some accepting run on w. The language $L_\omega(A)$ of infinite words defined by a Büchi automaton A is the set of ω-words it accepts.

A co-Büchi automaton is defined exactly as a Büchi automaton except that its accepting runs are those for which $inf(\pi) \cap F = \emptyset$.

We will also use the notion of *weak* automata [MSS86]. For a Büchi automaton $A = (Q, \Sigma, \delta, q_0, F)$ to be *weak*, there has to be a partition of its state set Q into disjoint subsets $Q_1, \ldots, Q_m$ such that

- for each of the Q_i either $Q_i \subseteq F$ or $Q_i \cap F = \emptyset$, and
- there is a partial order $\leq$ on the sets $Q_1, \ldots, Q_m$ such that for every $q \in Q_i$ and $q' \in Q_j$ for which, for some $a \in \Sigma$, $q' \in \delta(q, a)$ ($q' = \delta(q, a)$ in the deterministic case), $Q_j \leq Q_i$.

For more details, a survey of automata on infinite words can be found in [Tho90].

2.2 Topology

Given a set S, a distance $d(x, y)$ defined on this set induces a topology on subsets of S. A neighborhood $N_\varepsilon(x)$ of a point $x \in S$ is the set $N_\varepsilon(x) = \{y \mid d(x, y) < \varepsilon\}$. A set $C \subseteq S$ is said to be open if for all $x \in C$, there exists $\varepsilon > 0$ such that the neighborhood $N_\varepsilon(x)$ is contained in C. A closed set is a set whose complement with respect to S is open. We will be referring to the first few levels of the Borel hierarchy which are shown in Figure 1. The notations used are the following:

- F are the closed sets,
- G are the open sets,
- F_σ is the class of countable unions of closed sets,
- G_δ is the class of countable intersections of open sets,
- $F_{\sigma\delta}$ is the class of countable intersections of F_σ sets,
- $G_{\delta\sigma}$ is the class of countable unions of G_δ sets,
- $\mathcal{B}(X)$ represents the finite Boolean combinations of sets in X.

An arrow between classes indicates proper inclusion.

3 Topological Characterization of Arithmetic Sets

We consider the theory $\langle \mathbb{R}, \mathbb{Z}, +, \leq \rangle$, where $+$ represents the predicate $x + y = z$. Since any linear equality or order constraint can be encoded into this theory, we refer to it as additive or linear arithmetic over the reals and integers. It is the extension of Presburger arithmetic that includes both real and integer variables. In this section, we prove that the sets representable in this theory belong to the topological class $F_\sigma \cap G_\delta$ defined relatively to the Euclidean distance between vectors. This result is formalized by the following theorem.

Theorem 1. *Let $S \subseteq \mathbb{R}^n$, with $n > 0$, be a set defined in the theory $\langle \mathbb{R}, \mathbb{Z}, +, \leq \rangle$. This set belongs to the topological class $F_\sigma \cap G_\delta$ induced by the distance*

$$d(\boldsymbol{x}, \boldsymbol{y}) = \left(\sum_{i=1}^{n} (x_i - y_i)^2 \right)^{1/2}.$$

Proof. Since $\langle \mathbb{R}, \mathbb{Z}, +, \leq \rangle$ is closed under negation, it is actually sufficient to show that each formula of this theory defines a set that belongs to F_σ, i.e., a set that can be expressed as a countable union of closed sets.

⋮

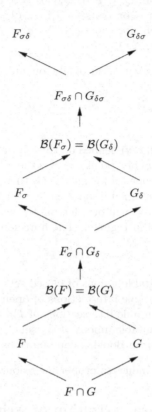

Fig. 1. The first few levels of the Borel hierarchy.

Let φ be a formula of $\langle \mathbb{R}, \mathbb{Z}, +, \leq \rangle$. To simplify our argument, we will assume that all free variables of φ are reals. This can be done without loss of generality since quantified variables can range over both $\mathbb{R}$ and $\mathbb{Z}$. We introduce $u < v$ as a shorthand for $u \leq v \wedge \neg(u = v)$.

The first step of our proof consists of modifying φ in the following way. We replace each variable x that appears in φ by two variables x_I and x_F representing respectively the integer and the fractional part of x. Formally, this operation replaces each occurrence in φ of a free variable x by the sum $x_I + x_F$ while adding to φ the constraints $0 \leq x_F$ and $x_F < 1$, and transforms the quantified variables of φ according to the following rules:

$$(\exists x \in \mathbb{R})\phi \longrightarrow (\exists x_I \in \mathbb{Z})(\exists x_F \in \mathbb{R})(0 \leq x_F \wedge x_F < 1 \wedge \phi[x/x_I + x_F])$$
$$(\forall x \in \mathbb{R})\phi \longrightarrow (\forall x_I \in \mathbb{Z})(\forall x_F \in \mathbb{R})(x_F < 0 \vee 1 \leq x_F \vee \phi[x/x_I + x_F])$$
$$(Qx \in \mathbb{Z})\phi \longrightarrow (Qx_I \in \mathbb{Z})\phi[x/x_I],$$

where $Q \in \{\exists, \forall\}$, ϕ is a subformula, and $\phi[x/y]$ denotes the result of replacing by y each occurrence of x in ϕ. The transformation has no influence on the set represented by φ, except that the integer and fractional parts of each value are now represented by two distinct variables.

Now, the atomic formulas of φ are of the form $p = q + r$, $p = q$ or $p \leq q$, where p, q and r are either integer variables, sums of an integer and of a fractional variable, or integer constants. The second step consists of expanding these atomic formulas so as to send into distinct atoms the occurrences of the integer and of the fractional variables. This is easily done with the help of simple arithmetic rules, for the truth value of the atomic formulas that involve both types of variables has only to be preserved for values of the fractional variables that belong to the interval $[0, 1[$. The less trivial expansion rules[2] are given below:

$$(x_I + x_F) = (y_I + y_F) \longrightarrow x_I = y_I \wedge x_F = y_F$$
$$(x_I + x_F) \leq (y_I + y_F) \longrightarrow x_I < y_I \vee (x_I = y_I \wedge x_F \leq y_F)$$
$$(x_I + x_F) = (y_I + y_F) + (z_I + z_F) \longrightarrow (x_I = y_I + z_I \wedge x_F = y_F + z_F)$$
$$\vee (x_I = y_I + z_I + 1 \wedge x_F = y_F + z_F - 1)$$
$$(x_I + x_F) = (y_I + y_F) + z_I \longrightarrow x_I = y_I + z_I \wedge x_F = y_F$$
$$x_I = (y_I + y_F) + (z_I + z_F) \longrightarrow (x_I = y_I + z_I \wedge y_F + z_F = 0)$$
$$\vee (x_I = y_I + z_I + 1 \wedge y_F + z_F = 1)$$

After the transformation, each atomic formula of φ is either a formula ϕ_I involving only integer variables or a formula ϕ_F over fractional variables. We now distribute existential (resp. universal) quantifiers over disjunctions (resp. conjunctions), after rewriting their argument into disjunctive (resp. conjunctive) normal form, and then apply the simplification rules

$$(Qx_I \in \mathbb{Z})(\phi_I \, \alpha \, \phi_F) \longrightarrow (Qx_I \in \mathbb{Z})(\phi_I) \, \alpha \, \phi_F$$
$$(Qx_F \in \mathbb{R})(\phi_I \, \alpha \, \phi_F) \longrightarrow \phi_I \, \alpha \, (Qx_F \in \mathbb{R})(\phi_F),$$

where $Q \in \{\exists, \forall\}$ and $\alpha \in \{\vee, \wedge\}$.

Repeating this operation, we eventually get a formula φ that takes the form of a finite Boolean combination $\mathcal{B}(\phi_I^{(1)}, \phi_I^{(2)}, \ldots, \phi_I^{(m)}, \phi_F^{(1)}, \phi_F^{(2)}, \ldots, \phi_F^{(m')})$ of subformulas $\phi_I^{(i)}$ and $\phi_F^{(i)}$ that involve respectively only integer and fractional variables.

Let $x_I^{(1)}, x_I^{(2)}, \ldots, x_I^{(k)}$ be the free integer variables of φ. For each assignment of values to these variables, the subformulas $\phi_I^{(i)}$ are each identically true or false, hence we have

$$\varphi \equiv \bigvee_{(a_1, \ldots, a_k) \in \mathbb{Z}^k} \left((x_I^{(1)}, \ldots, x_I^{(k)}) = (a_1, \ldots, a_k) \wedge \mathcal{B}_{(a_1, \ldots, a_k)}(\phi_F^{(1)}, \ldots, \phi_F^{(m')}) \right).$$

Each subformula $\phi_F^{(i)}$ belongs to the theory $\langle \mathbb{R}, +, \leq, 1 \rangle$, which admits the elimination of quantifiers [FR79]. The sets of reals vectors satisfying these formulas are thus finite Boolean combinations of linear constraints with open or

---

[2] In these rules, the expression $p = q + r + s$ is introduced as a shorthand for $(\exists u)(u = q + r \wedge p = u + s)$, where the quantifier is defined over the appropriate domain.

closed boundaries. It follows that, for each $(a_1, \ldots, a_k) \in \mathbb{Z}^k$, the set described by $\mathcal{B}_{(a_1,\ldots,a_k)}(\phi_F^{(1)}, \ldots, \phi_F^{(m')})$ is a finite Boolean combination of open and closed sets and, since any open set is a countable union of closed sets, is within F_σ. Therefore, the set described by φ is a countable union of F_σ sets and is also within F_σ.

4   Representing Sets of Integers and Reals with Finite Automata

In this section, we recall the finite-state representation of sets of real vectors as introduced in [BBR97].

In order to make a finite automaton recognize numbers, one needs to establish a mapping between these and words. Our encoding scheme corresponds to the usual notation for reals and relies on an arbitrary integer base $r > 1$. We encode a number x in base r, most significant digit first, by words of the form $w_I \star w_F$, where w_I encodes the integer part x_I of x as a finite word over $\{0, \ldots, r-1\}$, the special symbol "$\star$" is a separator, and w_F encodes the fractional part x_F of x as an infinite word over $\{0, \ldots, r-1\}$. Negative numbers are represented by their r's complement. The length p of $|w_I|$, which we refer to as the *integer-part length* of w, is not fixed but must be large enough for $-r^{p-1} \le x_I < r^{p-1}$ to hold.

According to this scheme, each number has an infinite number of encodings, since their integer-part length can be increased unboundedly. In addition, the rational numbers whose denominator has only prime factors that are also factors of r have two distinct encodings with the same integer-part length. For example, in base 10, the number $11/2$ has the encodings $005 \star 5(0)^\omega$ and $005 \star 4(9)^\omega$, "ω" denoting infinite repetition.

To encode a vector of real numbers, we represent each of its components by words of identical integer-part length. This length can be chosen arbitrarily, provided that it is sufficient for encoding the vector component with the highest magnitude. An encoding of a vector $x \in \mathbb{R}^n$ can indifferently be viewed either as a n-tuple of words of identical integer-part length over the alphabet $\{0, \ldots, r-1, \star\}$, or as a single word w over the alphabet $\{0, \ldots, r-1\}^n \cup \{\star\}$.

Since a real vector has an infinite number of possible encodings, we have to choose which of these the automata will recognize. A natural choice is to accept all encodings. This leads to the following definition.

Definition 1. *Let $n > 0$ and $r > 1$ be integers. A* Real Vector Automaton *(RVA) A in base r for vectors in $\mathbb{R}^n$ is a Büchi automaton over the alphabet $\{0, \ldots, r-1\}^n \cup \{\star\}$, such that*

- *Every word accepted by A is an encoding in base r of a vector in $\mathbb{R}^n$, and*
- *For every vector $x \in \mathbb{R}^n$, A accepts either all the encodings of x in base r, or none of them.*

An RVA is said to *represent* the set of vectors encoded by the words that belong to its accepted language. Efficient algorithms have been developed for constructing RVA representing the sets of solutions of systems of linear equations and inequations [BRW98]. Since it is immediate to constrain a number to be an integer with an RVA and since, using existing algorithms for infinite-word automata, one can apply Boolean operations as well as projection to RVA, it follows that one can construct an RVA for any formula of the arithmetic theory we are considering.

5   Weak Automata and Their Properties

If one examines the constructions given in [BRW98] to build RVA for linear equations and inequations, one notices that they have the property that all states within the same strongly connected component are either accepting or nonaccepting. This implies that these automata are *weak* in the sense of [MSS86] (see Section 2).

Weak automata have a number of interesting properties. A first one is that they can be represented both as Büchi and co-Büchi. Indeed, a weak Büchi automaton $A = (Q, \Sigma, \delta, q_0, F)$ is equivalent to the co-Büchi automaton $A = (Q, \Sigma, \delta, q_0, Q \setminus F)$, since a run eventually remains within a single component Q_i in which all states have the same status with respect to being accepting. A consequence of this is that weak automata can be determinized by the fairly simple "breakpoint" construction [MH84,KV97] that can be used for co-Büchi automata. This construction is the following.

Let $A = (Q, \Sigma, \delta, q_0, F)$ be a nondeterministic co-Büchi automaton. The deterministic co-Büchi automaton $A' = (Q', \Sigma, \delta', q'_0, F')$ defined as follows accepts the same ω-language.

- $Q' = 2^Q \times 2^Q$, the states of A' are pairs of sets of states of A.
- $q'_0 = (\{q_0\}, \emptyset)$.
- For $(S, R) \in Q'$ and $a \in \Sigma$, the transition function is defined by
  - if $R = \emptyset$, then $\delta((S, R), a) = (T, T \setminus F)$ where $T = \{q \mid \exists p \in S \text{ and } q \in \delta(p, a)\}$, T is obtained from S as in the classical subset construction, and the second component of the pair of sets of states is obtained from T by eliminating states in F;
  - if $R \neq \emptyset$, then $\delta((S, R), a) = (T, U \setminus F)$ where $T = \{q \mid \exists p \in S \text{ and } q \in \delta(p, a)\}$, and $U = \{q \mid \exists p \in R \text{ and } q \in \delta(p, a)\}$, the subset construction set is now applied to both S and R and states in F are removed from U.
- $F' = 2^Q \times \{\emptyset\}$.

When the automaton A' is in a state (S, R), R represents the states of A that can be reached by a run that has not gone through a state in F since that last "breakpoint", i.e. state of the form $(S, \emptyset)$. So, for a given word, A has a run that does not go infinitely often through a state in F iff A' has a run that does not go infinitely often through a state in F'. Notice that the difficulty that exists for determinizing Büchi automata, which is to make sure that the *same* run repeatedly reaches an accepting state disappears since, for co-Büchi automata, we are just looking for a run that eventually avoids accepting states.

It is interesting to notice that the construction implies that all reachable states (S, R) of A' satisfy $R \subseteq S$. The breakpoint construction can thus be implemented as a subset construction in which the states in R are simply tagged. One can thus expect it to behave in practice very similarly to the traditional subset construction for finite-word automata.

Another property of weak automata that will be of particular interest to us is the topological characterization of the sets of words that they can accept. Consider the topology on the set of ω-words induced by the distance

$$d(w, w') = \begin{cases} \frac{1}{|common(w,w')|+1} & \text{if } w \neq w' \\ 0 & \text{if } w = w', \end{cases}$$

where $|common(w, w')|$ denotes the length of the longest common prefix of w and w'. In this topology, weak deterministic automata accept exactly the ω-regular languages that are in $F_\sigma \cap G_\delta$. This follows from the results on the Staiger-Wagner class of automata [SW74,Sta83], which coincides with the class of deterministic weak automata, as can be inferred from [SW74] and is shown explicitly in [MS97]. Given the result proved in Section 3, it is tempting to conclude that the encodings of sets definable in the theory $\langle \mathbb{R}, \mathbb{Z}, +, \leq \rangle$ can always be accepted by weak deterministic automata. This conclusion is correct, but requires shifting the result from the topology on numbers to the topology on words, which we will do in the next section. In the meantime, we need one more result in order to be able to benefit algorithmically from the fact that we are dealing with $F_\sigma \cap G_\delta$ sets, i.e. that any deterministic automaton accepting a $F_\sigma \cap G_\delta$ set is essentially a weak automaton.

Consider the following definition.

Definition 2. *A Büchi automaton is inherently weak if none of the reachable strongly connected components of its transition graph contains both accepting (including at least one accepting state) and non accepting (not including any accepting state) cycles.*

Clearly, if an automaton is inherently weak, it can directly be transformed into a weak automaton. The partition of the state set is its partition into strongly connected components and all the states of a component are made accepting or not, depending on whether the cycles in that component are accepting or not.

We will now prove the following.

Theorem 2. *Any deterministic Büchi automaton that accepts a language in $F_\sigma \cap G_\delta$ is inherently weak.*

To prove this, we use the fact that the language accepted by an automaton that is not inherently weak must have the following *dense oscillating sequence* property.

Definition 3. *A language $L \subseteq \Sigma^\omega$ has the* dense oscillating sequence *property if, $w_1, w_2, w_3, \ldots$ being words and $\varepsilon_1, \varepsilon_2, \varepsilon_3, \ldots$ being distances, one has that*

$\exists w_1 \forall \varepsilon_1 \exists w_2 \forall \varepsilon_2 \ldots$ such that $d(w_i, w_{i+1}) \leq \varepsilon_i$ for all $i \geq 1$, $w_i \in L$ for all odd i, and $w_i \notin L$ for all even i.

The fact that the language accepted by an automaton that is not inherently weak has the dense oscillating sequence property is an immediate consequence of the fact that such an automaton has a reachable strongly connected component containing both accepting and non accepting cycles. Given this, it is sufficient to prove the following lemma in order to establish Theorem 2.

Lemma 1. *An ω-regular language that has the dense oscillating sequence property cannot be accepted by a weak deterministic automaton and hence is not in $F_\sigma \cap G_\delta$.*

Proof. We proceed by contradiction. Assume that a language L having the dense oscillating sequence property is accepted by a weak deterministic automaton A. Consider the first word w_1 in a dense oscillating sequence for L. This word eventually reaches an accepting component Q_{i_1} of the partition of the state set of A and will stay within this component. Since ε_1 can be chosen freely, it can be taken small enough for the run of A on w_2 to also reach the component Q_{i_1} before it starts to differ from w_1. Since w_2 is not in L, the run of A on w_2 has to eventually leave the component Q_{i_1} and will eventually reach and stay within a non accepting component $Q_{i_2} < Q_{i_1}$. Repeating a similar argument, one can conclude that the run of A on w_3 eventually reaches and stays within an accepting component $Q_{i_3} < Q_{i_2}$. Carrying on with this line of reasoning, one concludes that the state set of A must contain an infinite decreasing sequence of distinct components, which is impossible given that it is finite.

6 Deciding Linear Arithmetic with Real and Integer Variables

We first show that the result of Section 3 also applies to the sets of words encoding sets defined in $\langle \mathbb{R}, \mathbb{Z}, +, \leq \rangle$. In order to do so, we need to establish that the topological class $F_\sigma \cap G_\delta$ defined over sets of reals is mapped to its ω-word counterpart by the encoding relation described in Section 4.

Theorem 3. *Let $n > 0$ and $r > 1$ be integers, and let $L(S) \subseteq (\{0, \ldots, r-1\}^n \cup \{\star\})^\omega$ be the set of all the encodings in base r of the vectors belonging to the set $S \subseteq \mathbb{R}^n$. If the set S belongs to $F_\sigma \cap G_\delta$ (with respect to Euclidean distance), then the language $L(S)$ belongs to $F_\sigma \cap G_\delta$ (with respect to ω-word distance).*

Proof. Not all words over the alphabet $\{0, \ldots, r-1\}^n \cup \{\star\}$ encode a real vector. Let V be the set of all the valid encodings of vectors in base r. Its complement $\overline{V}$ can be partitioned into a set $\overline{V}_0$ containing only words in which the separator "$\star$" does not appear, and a set $\overline{V}_+$ containing words in which "$\star$" occurs at least once.

The set $\overline{V}_0 \cup V$ is closed. Indeed, each element of its complement is a word that does not encode validly a vector and that contains at least one separator.

Such a word admits a neighborhood entirely composed of words satisfying the same property, which entails that the complement of $\overline{V}_0 \cup V$ is open. In the same way, one obtains that the set $\overline{V}_+ \cup V$ is open.

Let now consider an open set $S \subseteq \mathbb{R}^n$. The language $L' = L(S) \cup \overline{V}_+$ is open. Indeed, each word $w \in L(S)$ has a neighborhood entirely composed of words in $L(S)$ (formed by the encodings of vectors that belong to a neighborhood of the vector encoded by w), and of words that do not encode vectors but contain at least one separator. Moreover, each word $w \in \overline{V}_+$ admits a neighborhood fully composed of words in $\overline{V}_+$. Since $L(S) = L' \cap (\overline{V}_0 \cup V)$, we have that $L(S)$ is the intersection of an open and of a closed set.

The same line of reasoning can be followed with a closed set $S \subseteq \mathbb{R}^n$. The language $L'' = L(S) \cup \overline{V}_0$ is easily shown to be closed, which, since $L(S) = L'' \cap (\overline{V}_+ \cup V)$, implies that $L(S)$ is the intersection of a closed and of an open set.

We are now ready to address the case of a set $S \subseteq \mathbb{R}^n$ that belongs to $F_\sigma \cap G_\delta$. Since S is in F_σ, it can be expressed as a countable union of closed sets $S_1, S_2, \ldots$. The languages $L(S_1), L(S_2), \ldots$ are Boolean combinations of open and of closed sets, and thus belong to the topological class F_σ. Therefore, $L(S) = L(S_1) \cup L(S_2) \cup \cdots$ is a countable union of sets in F_σ, and thus belongs itself to F_σ. Now, since S is in G_δ, it can also be expressed as a countable intersection of open sets $S_1', S_2', \ldots$. The languages $L(S_1'), L(S_2'), \ldots$ belong to the topological class G_δ. Hence, $L(S) = L(S_1') \cap L(S_2') \cap \cdots$ is a countable intersection of sets in G_δ, and thus belongs itself to G_δ. This concludes our proof of the theorem.

Knowing that the encodings of sets definable in the theory $\langle \mathbb{R}, \mathbb{Z}, +, \leq \rangle$ are in $F_\sigma \cap G_\delta$, we use the results of Section 5 to conclude the following.

Theorem 4. *Every deterministic RVA representing a set definable in $\langle \mathbb{R}, \mathbb{Z}, +, \leq \rangle$ is inherently weak.*

This property has the important consequence that the construction and the manipulation of RVA obtained from arithmetic formulas can be performed effectively by algorithms operating on weak automata. Precisely, to obtain an RVA for an arithmetic formula one can proceed as follows.

For equations and inequations, one uses the constructions given in [BRW98] to build weak RVA. Computing the intersection, union, and Cartesian product of sets represented by RVA simply reduces to performing similar operations with the languages accepted by the underlying automata, which can be done by simple product constructions. These operations preserve the weak nature of the automata. To complement a weak RVA, one determinizes it using the breakpoint construction, which is guaranteed to yield an inherently weak automaton (Theorem 4) that is easily converted to a weak one. This deterministic weak RVA is then complemented by inverting the accepting or non-accepting status of each of its components, and then removing from its accepted language the words that do not encode validly a vector (which is done by means of an intersection operation).

Applying an existential quantifier to a weak RVA is first done by removing from each transition label the symbol corresponding to the vector component that is projected out. This produces a non-deterministic weak automaton that may only accept some encodings of each vector in the quantified set, but generally not all of them. The second step thus consists of modifying the automaton so as to make it accept every encoding of each vector that it recognizes. Since different encodings of a same vector differ only in the number of times that their leading symbol is repeated, this operation can be carried out by the same procedure as the one used with finite-word number automata [Boi98]. This operation does not affect the weak nature of the automaton, which can then be determinized by the breakpoint construction, which has to produce an inherently weak RVA easily converted to a weak automaton.

Thus, in order to decide whether a formula of $\langle \mathbb{R}, \mathbb{Z}, +, \leq \rangle$ is satisfiable, one simply builds an RVA representing its set of solutions, and then check whether this automaton accepts a nonempty language. This also makes it possible to check the inclusion or the equivalence of sets represented by RVA. The main result of this paper is that, at every point, the constructed automaton remains weak and thus only the simple breakpoint construction is needed as a determinization procedure.

7   Conclusions

A probably unusual aspect of this paper is that it does not introduce new algorithms, but rather shows that existing algorithms can be used in a situation where *a priori* they could not be expected to operate correctly. To put it in other words, the contribution is not the algorithm but the proof of its correctness.

The critical reader might be wondering if all this is really necessary. After all, algorithms for complementing Büchi automata exist, either through determinization [Saf88] or directly [Büc62,SVW87,Kla91,KV97] and the more recent of these are even fairly simple and potentially implementable. There are no perfectly objective grounds on which to evaluate "simplicity" and "ease of implementation", but it is not difficult to convince oneself that the breakpoint construction for determinizing weak automata is simpler than anything proposed for determinizing or complementing Büchi automata. Indeed, it is but one step of the probably simplest complementation procedure proposed so far, that of [KV97]. Furthermore, there is a complexity improvement from $2^{O(n \log n)}$ to $2^{O(n)}$; experience with the subset construction as used for instance in the LASH tool [LASH] indicates that the breakpoint construction is likely to operate very well in practice; and being able to work with deterministic automata allows minimization [Löd01], which leads to a normal form.

An implementation and some experiments would of course substantiate the claims to simplicity and ease of implementation. It is planned in the context of the LASH tool and will be made available [LASH]. However, this paper is not about an implementation, but about the fact that, with the help of what might appear to be pure theory, one can obtain very interesting conclusions about algorithms for handling the theory $\langle \mathbb{R}, \mathbb{Z}, +, \leq \rangle$.

References

[BBR97]   B. Boigelot, L. Bronne, and S. Rassart. An improved reachability analysis method for strongly linear hybrid systems. In *Proc. 9th Int. Conf.on Computer Aided Verification*, volume 1254 of *Lecture Notes in Computer Science*, pages 167–178, Haifa, June 1997. Springer-Verlag.

[BC96]   A. Boudet and H. Comon. Diophantine equations, Presburger arithmetic and finite automata. In *Proceedings of CAAP'96*, number 1059 in Lecture Notes in Computer Science, pages 30–43. Springer-Verlag, 1996.

[BHMV94]   V. Bruyère, G. Hansel, C. Michaux, and R. Villemaire. Logic and p-recognizable sets of integers. *Bulletin of the Belgian Mathematical Society*, 1(2):191–238, March 1994.

[Boi98]   B. Boigelot. *Symbolic Methods for Exploring Infinite State Spaces*. PhD thesis, Université de Liège, 1998.

[BRW98]   Bernard Boigelot, Stéphane Rassart, and Pierre Wolper. On the expressiveness of real and integer arithmetic automata. In *Proc. 25th Colloq. on Automata, Programming, and Languages (ICALP)*, volume 1443 of *Lecture Notes in Computer Science*, pages 152–163. Springer-Verlag, July 1998.

[Büc60]   J. R. Büchi. Weak second-order arithmetic and finite automata. *Zeitschrift Math. Logik und Grundlagen der Mathematik*, 6:66–92, 1960.

[Büc62]   J.R. Büchi. On a decision method in restricted second order arithmetic. In *Proc. Internat. Congr. Logic, Method and Philos. Sci. 1960*, pages 1–12, Stanford, 1962. Stanford University Press.

[Cob69]   A. Cobham. On the base-dependence of sets of numbers recognizable by finite automata. *Mathematical Systems Theory*, 3:186–192, 1969.

[CVWY90]   Constantin Courcoubetis, Moshe Y. Vardi, Pierre Wolper, and Mihalis Yannakakis. Memory efficient algorithms for the verification of temporal properties. In *Proc. 2nd Workshop on Computer Aided Verification*, volume 531 of *Lecture Notes in Computer Science*, pages 233–242, Rutgers, June 1990. Springer-Verlag.

[FR79]   J. Ferrante and C. W. Rackoff. *The Computational Complexity of Logical Theories*, volume 718 of *Lecture Notes in Mathematics*. Springer-Verlag, Berlin-Heidelberg-New York, 1979.

[Hol97]   Gerard J. Holzmann. The model checker SPIN. *IEEE Transactions on Software Engineering*, 23(5):279–295, May 1997. Special Issue: Formal Methods in Software Practice.

[Kla91]   N. Klarlund. Progress measures for complementation of ω-automata with applications to temporal logic. In *Proceedings of the 32nd IEEE Symposium on Foundations of Computer Science*, San Juan, October 1991.

[KV97]   O. Kupferman and M. Vardi. Weak alternating automata are not that weak. In *Proc. 5th Israeli Symposium on Theory of Computing and Systems*, pages 147–158. IEEE Computer Society Press, 1997.

[KVW00]   Orna Kupferman, Moshe Y. Vardi, and Pierre Wolper. An automata-theoretic approach to branching-time model checking. *Journal of the ACM*, 47(2):312–360, March 2000.

[LASH]   The Liège Automata-based Symbolic Handler (LASH). Available at
            http://www.montefiore.ulg.ac.be/~boigelot/research/lash/.
[Löd01]   C. Löding. Efficient minimization of deterministic weak ω-automata, 2001.
            Submitted for publication.
[MH84]   S. Miyano and T. Hayashi. Alternating finite automata on ω-words. *The-
            oretical Computer Science*, 32:321–330, 1984.
[MS97]   O. Maler and L. Staiger. On syntactic congruences for ω-languages. *The-
            oretical Computer Science*, 183(1):93–112, 1997.
[MSS86]   D.E. Muller, A. Saoudi, and P.E. Schupp. Alternating automata, the
            weak monadic theory of the tree and its complexity. In *Proc. 13th Int.
            Colloquium on Automata, Languages and Programming*. Springer-Verlag,
            1986.
[Rab69]   M.O. Rabin. Decidability of second order theories and automata on infi-
            nite trees. *Transaction of the AMS*, 141:1–35, 1969.
[Saf88]   S. Safra. On the complexity of omega-automata. In *Proceedings of the 29th
            IEEE Symposium on Foundations of Computer Science*, White Plains,
            October 1988.
[Sem77]   A. L. Semenov. Presburgerness of predicates regular in two number sys-
            tems. *Siberian Mathematical Journal*, 18:289–299, 1977.
[SKR98]   T. R. Shiple, J. H. Kukula, and R. K. Ranjan. A comparison of Presburger
            engines for EFSM reachability. In *Proceedings of the 10th Intl. Conf. on
            Computer-Aided Verification*, volume 1427 of *Lecture Notes in Computer
            Science*, pages 280–292, Vancouver, June/July 1998. Springer-Verlag.
[Sta83]   L. Staiger. Finite-state ω-languages. *Journal of Computer and System
            Sciences*, 27(3):434–448, 1983.
[SVW87]   A. Prasad Sistla, Moshe Y. Vardi, and Pierre Wolper. The complemen-
            tation problem for Büchi automata with applications to temporal logic.
            Theoretical Computer Science, 49:217–237, 1987.
[SW74]   L. Staiger and K. Wagner. Automatentheoretische und automatenfreie
            Charakterisierungen topologischer Klassen regulärer Folgenmengen. *Elek-
            tron. Informationsverarbeitung und Kybernetik EIK*, 10:379–392, 1974.
[Tho90]   Wolfgang Thomas. Automata on infinite objects. In J. Van Leeuwen,
            editor, *Handbook of Theoretical Computer Science – Volume B: Formal
            Models and Semantics*, chapter 4, pages 133–191. Elsevier, Amsterdam,
            1990.
[VW86a]   Moshe Y. Vardi and Pierre Wolper. An automata-theoretic approach to
            automatic program verification. In *Proceedings of the First Symposium
            on Logic in Computer Science*, pages 322–331, Cambridge, June 1986.
[VW86b]   Moshe Y. Vardi and Pierre Wolper. Automata-theoretic techniques for
            modal logics of programs. *Journal of Computer and System Science*,
            32(2):183–221, April 1986.
[VW94]   Moshe Y. Vardi and Pierre Wolper. Reasoning about infinite computa-
            tions. *Information and Computation*, 115(1):1–37, November 1994.
[WB00]   Pierre Wolper and Bernard Boigelot. On the construction of automata
            from linear arithmetic constraints. In *Proc. 6th International Conference
            on Tools and Algorithms for the Construction and Analysis of Systems*,
            volume 1785 of *Lecture Notes in Computer Science*, pages 1–19, Berlin,
            March 2000. Springer-Verlag.

A Sequent Calculus for First-Order Dynamic Logic with Trace Modalities

Bernhard Beckert and Steffen Schlager

University of Karlsruhe
Institute for Logic, Complexity and Deduction Systems
D-76128 Karlsruhe, Germany
beckert@ira.uka.de, schlager@ira.uka.de

Abstract. The modalities of Dynamic Logic refer to the final state of a program execution and allow to specify programs with pre- and post-conditions. In this paper, we extend Dynamic Logic with additional trace modalities "throughout" and "at least once", which refer to *all* the states a program reaches. They allow one to specify and verify invariants and safety constraints that have to be valid throughout the execution of a program. We give a sound and (relatively) complete sequent calculus for this extended Dynamic Logic.

1 Introduction

We present a sequent calculus for an extended version of Dynamic Logic (DL) that has additional modalities "throughout" and "at least once" referring to the intermediate states of program execution.

Dynamic Logic [10,5,9,6] can be seen as an extension of Hoare logic [2]. It is a first-order modal logic with modalities $[\alpha]$ and $\langle\alpha\rangle$ for every program α. These modalities refer to the worlds (called states in the DL framework) in which the program α terminates when started in the current world. The formula $[\alpha]\phi$ expresses that ϕ holds in *all* final states of α, and $\langle\alpha\rangle\phi$ expresses that ϕ holds in *some* final state of α. In versions of DL with a non-deterministic programming language there can be several such final states (worlds). Here we consider a Deterministic Dynamic Logic (DDL) with a deterministic *while* programming language [4,7]. For deterministic programs there is exactly one final world (if α terminates) or there is no final world (if α does not terminate). The formula $\phi \to \langle\alpha\rangle\psi$ is valid if, for every state s satisfying pre-condition ϕ, a run of the program α starting in s terminates, and in the terminating state the post-condition ψ holds. The formula $\phi \to [\alpha]\psi$ expresses the same, except that termination of α is not required, i.e., ψ only has to hold *if* α terminates.

Thus, $\phi \to [\alpha]\psi$ is similar to the Hoare triple $\{\phi\}\alpha\{\psi\}$. But in contrast to Hoare logic, the set of formulas of DL is closed under the usual logical operators. In Hoare logic, the formulas ϕ and ψ are pure first-order formulas, whereas in DL they can contain programs. That is, DL allows one to involve programs in the formalisation of pre- and post-conditions. The advantage of using programs

R. Goré, A. Leitsch, and T. Nipkow (Eds.): IJCAR 2001, LNAI 2083, pp. 626–641, 2001.
© Springer-Verlag Berlin Heidelberg 2001

is that one can easily specify, for example, that some data structure is not cyclic, which is impossible in pure first-order logic.

In some regard, however, standard DL (and DDL) still lacks expressivity: The semantics of a program is a relation between states; formulas can only describe the input/output behaviour of programs. Standard DL cannot be used to reason about program behaviour not manifested in the input/output relation. It is inadequate for reasoning about non-terminating programs and for verifying invariants or constraints that must be valid throughout program execution.

We overcome this deficiency and increase the expressivity of DDL by adding two new modalities $[\![\alpha]\!]$ ("throughout") and $\langle\!\langle\alpha\rangle\!\rangle$ ("at least once"). In the extended logic, which we call (Deterministic) Dynamic Logic with Trace Modalities (DLT), the semantics of a program is the sequence of all states its execution passes through when started in the current state (its *trace*). It is possible in DLT to specify properties of the intermediate states of terminating and non-terminating programs. And such properties (typically safety constraints) can be verified using the calculus presented in Section 4. This is of great importance as safety constraints occur in many application domains of program verification.

Previous work in this area includes Pratt's Process Logic [10,11], which is an extension of *propositional* DL with trace modalities (DLT can be seen as a first-order Process Logic). Also, Temporal Logics have modalities that allow one to talk about intermediate states. There, however, the program is fixed and considered to be part of the structure over which the formulas are interpreted. Temporal Logics, thus, do not have the compositionality of Dynamic Logics.

The calculus for DDL described in [7] (which is based on the one given in [4]) has been implemented in the software verification systems KIV [12] and VSE [8]. It has successfully been used to verify software systems of considerable size.

The work reported here has been carried out as part of the KeY project [1].[1] The goal of KeY is to enhance a commercial CASE tool with functionality for formal specification and deductive verification and, thus, to integrate formal methods into real-world software development processes. In the KeY project, a version of DL for the JAVA CARD programming language [3] is used for verification. Deduction in DL (and DLT) is based on symbolic program execution and simple program transformations and is, thus, close to a programmer's understanding of a program's semantics. Our motivation for considering trace modalities was that in typical real-world specifications as they are done with the help of CASE tools, there are often program parts for which invariants and safety constraints are given, but for which the user did not bother to give a full specification with pre- and post-conditions.

We define the syntax of DLT in Section 2 and its semantics in Section 3. In Section 4, we describe our sequent calculus for DLT. Theorems stating soundness and (relative) completeness are presented in Section 5 (due to space restrictions, the proofs are only sketched, they can be found in [13]). In Section 6, we give an example for verifying that a non-terminating program preserves a certain invariant. Finally, in Section 7, we discuss future work.

---

[1] More information on KeY can be found at i12www.ira.uka.de/~key.

2   Syntax of DL with Trace Modalities

In first-order DL, states are not abstract points (as in propositional DL) but valuations of variables. Atomic programs are assignments of the form $x := t$. Executing $x := t$ changes the program state by assigning the value of the term t to the variable x. The value of a term t depends on the current state s (namely the value that s gives to the variables occurring in t). The function symbols are interpreted using a fixed first-order structure. This *domain of computation*, over which quantification is allowed, can be considered to define the data structures used in the programs. The logic DLT as well as the calculus presented in Section 4 are basically independent of the domain actually used. The only restriction is that the domain must be sufficiently expressive. In the following, for the sake of simplicity, we use arithmetic as the single domain. In practice, there will be additional function and predicate symbols and different types of variables ranging over different sorts of a many-sorted domain (different data structures).

The arithmetic *signature* $\Sigma_\mathbb{N}$ contains (a) the constant 0 (zero) and the unary function symbol s (successor) as constructors (in the following we abbreviate terms of the form $s(\cdots s(0) \cdots)$ with their decimal representation, e.g. "2" abbreviates "$s(s(0))$"), (b) the binary function symbols $+$ (addition) and $*$ (multiplication), and (c) the binary predicate symbols $\leq$ (less or equal than) and $\doteq$ (equality). In addition, there is an infinite set *Var* of *object variables*, which are also used as program variables. The set $Term_\mathbb{N}$ of *terms* over $\Sigma_\mathbb{N}$ is built as usual in first-order predicate logic from the variables in *Var* and the function symbols in $\Sigma_\mathbb{N}$. The formulas of first-order predicate logic without modal operators (FOL-formulas) over $\Sigma_\mathbb{N}$ are constructed as usual from the terms in $Term_\mathbb{N}$ and the predicate symbols in $\Sigma_\mathbb{N}$, using the classical connectives $\wedge$ (conjunction), $\vee$ (disjunction), $\rightarrow$ (implication), and $\neg$ (negation), and the quantifiers $\forall$ and $\exists$.

We proceed to define what the programs of the deterministic programming language of DDL and DLT are. The programming constructs for forming complex programs from the atomic assignments are the concatenation of programs, if-then-else conditionals, and while loops (the two latter program constructs use quantifier-free FOL-formulas as conditions).

Definition 1. *The programs of DLT are recursively defined by: (i) If $x \in Var$ and $t \in Term_\mathbb{N}$, then $x := t$ is a program (assignment). (ii) If α and β are programs, then $\alpha;\beta$ is a program (concatenation). (iii) If α and β are programs and ϵ is a quantifier-free FOL-formula, then if ϵ then α else β is a program (conditional). (iv) If α is a program and ϵ is a quantifier-free FOL-formula, then while ϵ do α is a program (loop).*

The programs of DLT form a computationally complete programming language. For every partial recursive function $f : \mathbb{N} \rightarrow \mathbb{N}$ there is a program $\alpha_f(x)$ that computes f, i.e., if $\alpha_f(x)$ is started in an arbitrary state in which the value of x is some $n \in \mathbb{N}$, then it terminates in a state in which the value of x is $f(n)$.

Now, we define the formulas of DLT. Note, that the first four conditions in Definition 2 are the same as in the definition of FOL-formulas. Only the last condition is new, which adds the modalities (and programs) to the formulas.

Definition 2. *The set of DLT-formulas is recursively defined by: (i) true and false are DLT-formulas. (ii) If $t_1, t_2 \in Term_{\mathbb{N}}$, then $t_1 \leq t_2$ and $t_1 \doteq t_2$ are DLT-formulas. (iii) If ϕ, ψ are DLT-formulas, then so are $\neg\phi$, $\phi \vee \psi$, $\phi \wedge \psi$, and $\phi \rightarrow \psi$. (iv) If ϕ is a DLT-formula and $x \in Var$, then $\exists x\,\phi$, $\forall x\,\phi$ are DLT-formulas. (v) If ϕ is a DLT-formula and α is a program (Def. 1), then $[\alpha]\phi$, $\langle\alpha\rangle\phi$, $[\![\alpha]\!]\phi$, and $\langle\!\langle\alpha\rangle\!\rangle\phi$ are DLT-formulas.*

Definition 3. *A sequent is of the form $\phi_1, \ldots, \phi_m \vdash \psi_1, \ldots, \psi_n$ $(m, n \geq 0)$, where the ϕ_i and ψ_j are DLT-formulas. The order of the ϕ_i resp. the ψ_j is irrelevant, i.e., $\phi_1, \ldots, \phi_m$ and $\psi_1, \ldots, \psi_n$ are treated as multi-sets.*

Definition 4. *A variable $x \in Var$ is bound in a DLT-formula ϕ if it occurs inside the scope of (i) a quantification $\forall x$ resp. $\exists x$, or (ii) a modality $[\alpha]$, $\langle\alpha\rangle$, $[\![\alpha]\!]$, or $\langle\!\langle\alpha\rangle\!\rangle$ containing an assignment $x := t$. The variable x is free in ϕ if there is an occurrence of x in ϕ that is neither bound by a quantifier nor a modality.*

Definition 5. *A substitution assigns to each object variable in Var a term in $Term_{\mathbb{N}}$. A substitution σ is applied to a DLT-formula ϕ by replacing all free occurrences of variables x in ϕ by $\sigma(x)$.*

If a substitution $\{x/t\}$ instantiates only a single variable x, its application to a formula ϕ or a formula multi-set Γ is denoted by ϕ_x^t resp. Γ_x^t.

A substitution σ is admissible w.r.t. a DLT-formula ϕ if there are no variables x and y such that x is free in ϕ, y occurs in $\sigma(x)$, and, after replacing $\sigma(x)$ for some free occurrence of x in ϕ, the occurrence of y in $\sigma(x)$ is bound in $\sigma(\phi)$.

3 Semantics of DL with Trace Modalities

Since we use arithmetic as the only domain of computation, the semantics of DLT is defined using a single fixed model, namely $\langle \mathbb{N}, I_{\mathbb{N}} \rangle$. It consists of the universe $\mathbb{N}$ of natural numbers and the canonical interpretation function $I_{\mathbb{N}}$ assigning the function and predicate symbols of $\Sigma_{\mathbb{N}}$ their natural meaning in arithmetic.

The states (worlds) of the model (only) differ in the value assigned to the object variables. Therefore, the states can be defined to *be* variable assignments.

Definition 6. *A state s assigns to each variable $x \in Var$ a number $s(x) \in \mathbb{N}$.*

Let $x \in Var$ and $n \in \mathbb{N}$; then $s' = s\{x \leftarrow n\}$ is the state that is identical to s except that x is assigned n, i.e., $s'(x) = n$ and $s'(y) = s(y)$ for all $x \neq y$.

The truth value of DLT-formulas in a state s is given by a valuation function val_s that assigns to each term $t \in Term_{\mathbb{N}}$ a natural number $val_s(t) \in \mathbb{N}$ and to each formula one of the truth values $\underline{t}$ and $\underline{f}$. This function is defined step by step. For variables $x \in Var$, it is defined by $val_s(x) = s(x)$. It is extended to terms and FOL-formulas as usual in first-order predicate logic (note, that the way in which function symbols are interpreted depends on the interpretation function of the domain of computation, which in our case is $I_{\mathbb{N}}$). Below, we describe how val_s is defined for programs (Def. 7) and, finally, is extended to DLT-formulas (Def. 8).

In DDL, where the modalities only refer to the final state of a program execution, the semantics of a program α is a reachability relation on states: A state s' is α-reachable from s if α terminates in s' when started in s. In DLT the situation is different. The additional modalities refer to the intermediate states as well. Since the programs are deterministic, their intermediate states form a sequence. Thus, the semantics of a program α w.r.t. a state s is the—finite or infinite—sequence of all states that α reaches when started in s, called the *trace* of α. It includes the initial state s (and the final state in case α terminates).

Definition 7. *A* trace *is a non-empty, finite or infinite sequence of states.*

The last element of a finite trace T is denoted with $last(T)$.

The concatenation *of traces T_1 and T_2 is defined by: $T_1 \circ T_2 = T_1$ if T_1 is infinite, and $T_1 \circ T_2 = (s_1^1, \ldots, s_k^1, s_2^2, s_3^2, \ldots)$ if $T_1 = (s_1^1, \ldots, s_k^1)$ is finite and $T_2 = (s_1^2, s_2^2, s_3^2, \ldots)$ (the first state of T_2 is omitted in the concatenation).*

Given a state s, the valuation function val_s *assigns a trace to each program as follows:*

- $val_s(x := t) = (s, s\{x \leftarrow val_s(t)\})$.
- $val_s(\alpha;\beta) = val_s(\alpha) \circ val_{last(val_s(\alpha))}(\beta)$.
- $val_s(\text{if } \epsilon \text{ then } \alpha \text{ else } \beta)$ *is defined to be equal to $val_s(\alpha)$ if $val_s(\epsilon) = \underline{t}$ and to be equal to $val_s(\beta)$ if $val_s(\epsilon) = \underline{f}$.*
- $val_s(\text{while } \epsilon \text{ do } \alpha)$ *is defined as follows (there are three cases). Let s_n be the initial state of the n-th iteration of the loop body α, i.e., $s_1 = s$ and, for $n \geq 1$, $s_{n+1} = last(val_{s_n}(\alpha))$ if s_n is defined and $val_{s_n}(\alpha)$ is finite (otherwise s_{n+1} remains undefined).*

Case 1 (the loop terminates): If for some $n \in \mathbb{N}$, (i) $val_{s_i}(\alpha)$ is finite for all $i \leq n$, (ii) $val_{s_i}(\epsilon) = \underline{t}$ for all $i \leq n$, and (iii) $val_{s_{n+1}}(\epsilon) = \underline{f}$, then we define $val_s(\text{while } \epsilon \text{ do } \alpha)$ to be the finite sequence $val_{s_1}(\alpha) \circ \cdots \circ val_{s_n}(\alpha)$.

Case 2 (each iteration terminates but the condition ϵ remains true such that the loop does not terminate): If for all $n \geq 1$, (i) $val_{s_n}(\alpha)$ is finite and (ii) $val_{s_n}(\epsilon) = \underline{t}$, then we define $val_s(\text{while } \epsilon \text{ do } \alpha)$ to be the infinite sequence $val_{s_1}(\alpha) \circ val_{s_2}(\alpha) \circ \cdots$.

Case 3 (some iteration does not terminate): If for some $n \in \mathbb{N}$, (i) $val_{s_i}(\alpha)$ is finite for $i < n$, (ii) $val_{s_n}(\alpha)$ is infinite, and (iii) $val_{s_i}(\epsilon) = \underline{t}$ for all $i \leq n$, then $val_s(\text{while } \epsilon \text{ do } \alpha)$ is the infinite sequence $val_{s_1}(\alpha) \circ \cdots \circ val_{s_n}(\alpha)$.

Definition 8. *Given a state s, the* valuation function val_s *assigns to a DLT-formula ϕ one of the truth values $\underline{t}$ and $\underline{f}$ as follows: (i) If ϕ is true, false, or an atomic formula, or its principal logical operator is one of the classical operators $\wedge$, $\vee$, $\rightarrow$, $\neg$, or one of the quantifiers $\forall$, $\exists$, then $val_s(\phi)$ is recursively defined as usual in first-order predicate logic. (ii) $val_s([\alpha]\phi) = \underline{t}$ iff $val_s(\alpha)$ is infinite or $val_{s'}(\phi) = \underline{t}$ where $s' = last(val_s(\alpha))$. (iii) $val_s(\langle\alpha\rangle\phi) = \underline{t}$ iff $val_s(\alpha)$ is finite and $val_{s'}(\phi) = \underline{t}$ where $s' = last(val_s(\alpha))$. (iv) $val_s([\![\alpha]\!]\phi) = \underline{t}$ iff $val_{s'}(\phi) = \underline{t}$ for all $s' \in val_s(\alpha)$. (v) $val_s(\langle\!\langle\alpha\rangle\!\rangle\phi) = \underline{t}$ iff $val_{s'}(\phi) = \underline{t}$ for at least one $s' \in val_s(\alpha)$.*

Table 1. The elementary rules of the calculus.

Axioms

$$\frac{}{\Gamma, \phi \vdash \phi, \Delta} \ \text{(R1)} \qquad \frac{}{\Gamma \vdash true, \Delta} \ \text{(R2)} \qquad \frac{}{\Gamma, false \vdash \Delta} \ \text{(R3)}$$

Rules for classical logical operators and quantifiers

$$\frac{\Gamma, \phi \vdash \Delta}{\Gamma \vdash \neg\phi, \Delta} \ \text{(R4)} \qquad\qquad \frac{\Gamma \vdash \phi, \Delta}{\Gamma, \neg\phi \vdash \Delta} \ \text{(R5)} \qquad\qquad \frac{\Gamma \vdash \phi, \Delta \quad \Gamma \vdash \psi, \Delta}{\Gamma \vdash \phi \wedge \psi, \Delta} \ \text{(R6)}$$

$$\frac{\Gamma, \phi, \psi \vdash \Delta}{\Gamma, \phi \wedge \psi \vdash \Delta} \ \text{(R7)} \qquad\qquad \frac{\Gamma \vdash \phi, \psi, \Delta}{\Gamma \vdash \phi \vee \psi, \Delta} \ \text{(R8)} \qquad\qquad \frac{\Gamma, \phi \vdash \Delta \quad \Gamma, \psi \vdash \Delta}{\Gamma, \phi \vee \psi \vdash \Delta} \ \text{(R9)}$$

$$\frac{\Gamma, \phi \vdash \psi, \Delta}{\Gamma \vdash \phi \rightarrow \psi, \Delta} \ \text{(R10)} \qquad \frac{\Gamma \vdash \phi, \Delta \quad \Gamma, \psi \vdash \Delta}{\Gamma, \phi \rightarrow \psi \vdash \Delta} \ \text{(R11)} \qquad \frac{\Gamma \vdash \phi_x^{x'}, \Delta}{\Gamma \vdash \forall x\, \phi, \Delta} \ \text{(R12)}$$

$$x' \text{ is new w.r.t. } \phi, \Gamma, \Delta$$

$$\frac{\Gamma, \forall x\, \phi, \phi_x^t \vdash \Delta}{\Gamma, \forall x\, \phi \vdash \Delta} \ \text{(R13)} \qquad \frac{\Gamma, \phi_x^{x'} \vdash \Delta}{\Gamma, \exists x\, \phi \vdash \Delta} \ \text{(R14)} \qquad \frac{\Gamma \vdash \phi_x^t, \exists x\, \phi, \Delta}{\Gamma \vdash \exists x\, \phi, \Delta} \ \text{(R15)}$$

where $\{x/t\}$ is x' is new   where $\{x/t\}$ is
admissible w.r.t. ϕ w.r.t. ϕ, Γ, Δ admissible w.r.t. ϕ

Weakening and Cut

$$\frac{\Gamma \vdash \Delta}{\Gamma \vdash \phi, \Delta} \ \text{(R16)} \qquad\qquad \frac{\Gamma \vdash \Delta}{\Gamma, \phi \vdash \Delta} \ \text{(R17)} \qquad\qquad \frac{\Gamma, \phi \vdash \Delta \quad \Gamma \vdash \phi, \Delta}{\Gamma \vdash \Delta} \ \text{(R18)}$$

Definition 9. *If* $val_s(\phi) = \underline{t}$, *then* ϕ *is said to be* true *in the state* s*; otherwise it is* false *in* s. *A formula is* valid *if it is true in all states.*

A sequent $\Gamma \vdash \Delta$ *is* valid *iff the DLT-formula* $\bigwedge \Gamma \rightarrow \bigvee \Delta$ *is valid.*

4   A Sequent Calculus for DL with Trace Modalities

In this section, we present a sequent calculus for DLT, which we call $\mathcal{C}_{\text{DLT}}$. It is sound and relatively complete, i.e., complete up to the handling of arithmetic (see Section 5). The set of those $\mathcal{C}_{\text{DLT}}$-rules in which the additional modalities $[\![\cdot]\!]$ and $\langle\!\langle\cdot\rangle\!\rangle$ do not occur forms a sound and (relatively) complete calculus for DDL. This restriction of $\mathcal{C}_{\text{DLT}}$ is similar to the DDL-calculus described in [7].

Most rules of the calculus are analytic and therefore could be applied automatically. The rules that require user interaction are: (a) the rules for handling while loops (where a loop invariant has to be provided), (b) the induction rule (where a useful induction hypothesis has to be found), (c) the cut rule (where the right case distinction has to be used), and (d) the quantifier rules (where the right instantiation has to be found).

In the rule schemata, Γ, Δ denote arbitrary, possibly empty multi-sets of formulas, and ϕ, ψ denote arbitrary formulas. As usual, the sequents above the horizontal line in a schema are its premises and the single sequent below the horizontal line is its conclusion. Note, however, that in practice the rules are applied from bottom to top. Proof construction starts with the original proof

Table 2. The rules for handling arithmetic.

Oracle rules	$\dfrac{}{\Gamma \vdash \Delta}$ (R19)	$\dfrac{\Gamma_1',\ \Gamma_2 \vdash \Delta}{\Gamma_1,\ \Gamma_2 \vdash \Delta}$ (R20)
	where $\bigwedge \Gamma \to \bigvee \Delta$ is a valid arithmetical FOL-formula	where $\bigwedge \Gamma_1 \to \bigwedge \Gamma_1'$ is a valid arithmetical FOL-formula

Induction
$$\frac{\Gamma \vdash \phi(0),\ \Delta \quad \Gamma,\ \phi(n) \vdash \phi(s(n)),\ \Delta}{\Gamma \vdash \forall n\, \phi(n),\ \Delta}\ \text{(R21)}$$
where n does not occur in Γ, Δ

obligation at the bottom. Therefore, if a constraint is attached to a rule that requires a variable to be "new", it has to be new w.r.t. the *conclusion*.

Definition 10. *The calculus* $\mathcal{C}_{\mathrm{DLT}}$ *consists of the rules* (R1) *to* (R51) *shown in Tables 1–4. A sequent is* derivable *(with* $\mathcal{C}_{\mathrm{DLT}}$ *) if it is an instance of the conclusion of a rule schema and all corresponding instances of the premisses of that rule schema are derivable sequents. In particular, all sequents are derivable that are instances of the conclusion of a rule that has no premisses* (R1, R2, R3, R19).

The Elementary Rules. The elementary rules of $\mathcal{C}_{\mathrm{DLT}}$ are shown in Table 1. The table contains rules for axioms (which have no premisses and make it possible to close a branch in the proof tree), rules for the propositional operators and the quantifiers, weakening rules, and the cut rule. These rules form a sound and complete calculus for first-order predicate logic.

Rules for Handling Arithmetic. Our calculus is basically independent of the domain of computation resp. data structures that are used. We therefore abstract from the problem of handling the data structure(s) and just assume that an oracle is available that can decide the validity of FOL-formulas in the domain of computation (note that the oracle only decides pure FOL-formulas). In the case of arithmetic, the oracle is represented by rule (R19) in Table 2. Rule (R20) is an alternative formalisation of the oracle that is often more useful.

Of course, the FOL-formulas that are valid in arithmetic are not even enumerable. Therefore, in practice, the oracle can only be approximated, and rules (R19) and (R20) must be replaced by a rule (or set of rules) for computing resp. enumerating a *subset* of all valid FOL-formulas (in particular, these rules must include equality handling). This is not harmful to "practical completeness". Rule sets for arithmetic are available, which—as experience shows—allow to derive all valid FOL-formulas that occur during the verification of actual programs.

Typically, an approximation of the computation domain oracle contains a rule for structural induction. In the case of arithmetic, that is rule (R21). This rule, however, is not only used to approximate the arithmetic oracle but is indispensable for completeness. It not only applies to FOL-formulas but also to

Table 3. Rules for the modal operators.

Assignment

$$\frac{\Gamma_x^{x'},\ x \doteq t_x^{x'} \vdash \phi,\ \Delta_x^{x'}}{\Gamma \vdash [x := t]\phi,\ \Delta}\ (R22)$$

where x' is new w.r.t. t, ϕ, Γ, Δ

$$\frac{\Gamma_x^{x'},\ x \doteq t_x^{x'} \vdash \phi,\ \Delta_x^{x'}}{\Gamma \vdash \langle x := t\rangle\phi,\ \Delta}\ (R23)$$

where x' is new w.r.t. t, ϕ, Γ, Δ

$$\frac{\Gamma \vdash \phi,\ \Delta \quad \Gamma_x^{x'},\ x \doteq t_x^{x'} \vdash \phi,\ \Delta_x^{x'}}{\Gamma \vdash [\![x := t]\!]\phi,\ \Delta}\ (R24)$$

where x' is new w.r.t. t, ϕ, Γ, Δ

$$\frac{\Gamma_x^{x'},\ x \doteq t_x^{x'} \vdash \phi_x^{x'},\ \Delta_x^{x'}}{\Gamma \vdash \langle\!\langle x := t\rangle\!\rangle\phi,\ \Delta}\ (R25)$$

where x' is new w.r.t. t, ϕ, Γ, Δ

Concatenation

$$\frac{\Gamma \vdash [\alpha][\beta]\phi,\ \Delta}{\Gamma \vdash [\alpha;\beta]\phi,\ \Delta}\ (R26)$$

$$\frac{\Gamma \vdash \langle\alpha\rangle\langle\beta\rangle\phi,\ \Delta}{\Gamma \vdash \langle\alpha;\beta\rangle\phi,\ \Delta}\ (R27)$$

$$\frac{\Gamma \vdash [\![\alpha]\!]\phi,\ \Delta \quad \Gamma \vdash [\alpha][\![\beta]\!]\phi,\ \Delta}{\Gamma \vdash [\![\alpha;\beta]\!]\phi,\ \Delta}\ (R28)$$

$$\frac{\Gamma \vdash \langle\!\langle\alpha\rangle\!\rangle\phi,\ \langle\alpha\rangle\langle\!\langle\beta\rangle\!\rangle\phi,\ \Delta}{\Gamma \vdash \langle\!\langle\alpha;\beta\rangle\!\rangle\phi,\ \Delta}\ (R29)$$

If-then-else

$$\frac{\Gamma,\ \epsilon \vdash [\alpha]\phi,\ \Delta \quad \Gamma,\ \neg\epsilon \vdash [\beta]\phi,\ \Delta}{\Gamma \vdash [\text{if } \epsilon \text{ then } \alpha \text{ else } \beta]\phi,\ \Delta}\ (R30)$$

$$\frac{\Gamma,\ \epsilon \vdash \langle\alpha\rangle\phi,\ \Delta \quad \Gamma,\ \neg\epsilon \vdash \langle\beta\rangle\phi,\ \Delta}{\Gamma \vdash \langle\text{if } \epsilon \text{ then } \alpha \text{ else } \beta\rangle\phi,\ \Delta}\ (R31)$$

$$\frac{\Gamma,\ \epsilon \vdash [\![\alpha]\!]\phi,\ \Delta \quad \Gamma,\ \neg\epsilon \vdash [\![\beta]\!]\phi,\ \Delta}{\Gamma \vdash [\![\text{if } \epsilon \text{ then } \alpha \text{ else } \beta]\!]\phi,\ \Delta}\ (R32)$$

$$\frac{\Gamma,\ \epsilon \vdash \langle\!\langle\alpha\rangle\!\rangle\phi,\ \Delta \quad \Gamma,\ \neg\epsilon \vdash \langle\!\langle\beta\rangle\!\rangle\phi,\ \Delta}{\Gamma \vdash \langle\!\langle\text{if } \epsilon \text{ then } \alpha \text{ else } \beta\rangle\!\rangle\phi,\ \Delta}\ (R33)$$

While

$$\frac{\Gamma \vdash Inv,\ \Delta \quad Inv,\ \epsilon \vdash [\alpha]Inv \quad Inv,\ \neg\epsilon \vdash \phi}{\Gamma \vdash [\text{while } \epsilon \text{ do } \alpha]\phi,\ \Delta}\ (R34)$$

where Inv is an arbitrary DLT-formula

$$\frac{\Gamma \vdash \epsilon,\ \Delta \quad \Gamma \vdash \langle\alpha\rangle\langle\text{while } \epsilon \text{ do } \alpha\rangle\phi,\ \Delta}{\Gamma \vdash \langle\text{while } \epsilon \text{ do } \alpha\rangle\phi,\ \Delta}\ (R35)$$

$$\frac{\Gamma \vdash \neg\epsilon,\ \Delta \quad \Gamma \vdash \phi,\ \Delta}{\Gamma \vdash \langle\text{while } \epsilon \text{ do } \alpha\rangle\phi,\ \Delta}\ (R36)$$

$$\frac{\Gamma \vdash Inv,\ \Delta \quad Inv,\ \epsilon \vdash [\alpha]Inv \quad Inv,\ \epsilon \vdash [\![\alpha]\!]\phi \quad Inv,\ \neg\epsilon \vdash \phi}{\Gamma \vdash [\![\text{while } \epsilon \text{ do } \alpha]\!]\phi,\ \Delta}\ (R37)$$

where Inv is an arbitrary DLT-formula

$$\frac{\Gamma \vdash \epsilon,\ \Delta \quad \Gamma \vdash \langle\alpha\rangle\langle\!\langle\text{while } \epsilon \text{ do } \alpha\rangle\!\rangle\phi,\ \Delta}{\Gamma \vdash \langle\!\langle\text{while } \epsilon \text{ do } \alpha\rangle\!\rangle\phi,\ \Delta}\ (R38)$$

$$\frac{\Gamma,\ \neg\epsilon \vdash \phi,\ \Delta \quad \Gamma,\ \epsilon \vdash \langle\!\langle\alpha\rangle\!\rangle\phi,\ \Delta}{\Gamma \vdash \langle\!\langle\text{while } \epsilon \text{ do } \alpha\rangle\!\rangle\phi,\ \Delta}\ (R39)$$

DLT-formulas containing programs; and it is needed for handling the modalities $\langle\cdot\rangle$ and $\langle\!\langle\cdot\rangle\!\rangle$ when they contain while loops (see Section 4).

Rules for Modalities and Programs. The rules for the modal operators and the programs they contain are shown in Table 3. As is easy to see, they basically perform a symbolic program execution.

There is a rule for each combination of program construct (assignment, concatenation, if-then-else, while loop) and modality ($[\cdot]$, $\langle\cdot\rangle$, $[\![\cdot]\!]$, $\langle\!\langle\cdot\rangle\!\rangle$). To keep the description of our calculus compact we only give rules for the case where the modal formula is on the right side of a sequent. That is sufficient for completeness because using the cut rule (R18) and the rules for negated modalities (R48) to (R51) (see Table 4), every modal formula on the left side of a sequent can be turned into an equivalent formula on the right side of the sequent. For example, from the proof obligation $[\alpha]\phi \vdash$ we get the proof obligation $\vdash \neg[\alpha]\phi$ with the cut rule, which then can be turned into $\vdash \langle\!\langle\alpha\rangle\!\rangle\neg\phi$ applying rule (R50).

Rules for Assignments. The rules for the modalities $[\cdot]$ (R22) and $\langle\cdot\rangle$ (R23) are the traditional assignment rules of calculi for first-order DL. They introduce a new variable x' representing the old value of x before the assignment $x := t$ is executed. In the premisses of the assignment rules, both x and x' occur because the premisses express the relation between the old and the new value of x without using an explicit assignment. Since assignments always terminate, there is no difference between the two rules. Note, that the premiss and the conclusion of these rules are not necessarily equivalent (as a new symbol is introduced). But if one is valid then the other is valid as well.

Example 1. Consider the valid sequent $x \doteq 5 \vdash \langle x := x+1\rangle x \doteq 6$. Applying rule (R23) yields the new sequent $x' \doteq 5,\ x \doteq x'+1 \vdash x \doteq 6$. It can be read as: "If the old value of x is 5 and its new value is its old value plus 1, then the new value of x is 6." This exactly captures the meaning of the original sequent.

Assignments $x := t$ are atomic programs. By definition, their semantics is a trace consisting of the initial state s and the final state $s' = s\{x \leftarrow val_s(t)\}$. Therefore, the meaning of $[x := t]\phi$ is that ϕ is true in both s and s', which is what the two premisses of rule (R24) express. The formula $\langle\!\langle x := t\rangle\!\rangle\phi$, on the other hand, is true (in s) if ϕ is true in at least one of the two states. Note, that the two formulas $\phi_x^{x'}$ and ϕ in the premiss of rule (R25), which express that ϕ is true in s resp. s', are implicitly disjunctively connected.

Example 2. We use rule (R24) to show that $x \doteq 5 \vdash [x := x+1]x \leq 6$ is a valid sequent. This results in the two new proof obligations $x \doteq 5 \vdash x \leq 6$ and $x' \leq 5,\ x \doteq x'+1 \vdash x \leq 6$. They state that $x \leq 6$ is true in both the initial and the final state of the assignment.

Let $even(x)$ be an abbreviation for the FOL-formula $\exists y\,(x \doteq 2*y)$. To prove the validity of $\vdash \langle\!\langle x := x+1\rangle\!\rangle even(x)$, we apply rule (R25) and get the new proof obligation $x \doteq x'+1 \vdash even(x),\ even(x')$, which is obviously valid.

Rules for Concatenation. Again, the rules for the modalities $[\cdot]$ (R26) and $\langle\cdot\rangle$ (R27) are the traditional rules for first-order DL. They are based on the equivalences $[\alpha;\beta]\phi \leftrightarrow [\alpha][\beta]\phi$ resp. $\langle\alpha;\beta\rangle\phi \leftrightarrow \langle\alpha\rangle\langle\beta\rangle\phi$.

In the case of the $[\cdot]$ modality, the concatenation rule (R28) branches. To show that a formula ϕ is true throughout the execution of $\alpha;\beta$, one has to prove (a) that ϕ is true throughout the execution of α, i.e. $[\alpha]\phi$, and (b) provided α terminates, that ϕ is true throughout the execution of β that is started in the final state of α, i.e. $[\alpha][\beta]\phi$.

The concatenation rule for $\langle\!\langle\cdot\rangle\!\rangle$ (R29) does not branch. A formula ϕ is true at least once during the execution of $\alpha;\beta$ if (a) it is true at least once during the execution of α, *or* (b) α terminates and ϕ is true at least once during the execution of β that is started in the final state of α.[2]

Rules for If-then-else. The rules for if-then-else conditionals have the same form for all four modalities, and for the modalities $[\cdot]$ and $\langle\cdot\rangle$ they are the same as in calculi for standard DDL.

Rules for While Loops. The rules for while loops in the modalities $[\cdot]$ and $[\![\cdot]\!]$, (R34) resp. (R37), use a loop *invariant*, i.e., a DLT-formula that must be true before and after each execution of the loop body. Three premisses of (R37) are the same as the premisses of (R34). The first one expresses that the invariant *Inv* holds in the current state, i.e., before the loop is started. The second premiss expresses that *Inv* is indeed an invariant, i.e., if it holds before executing the loop body α, then it holds again if and when α terminates. And the third premiss expresses that ϕ—the formula that supposedly holds after resp. throughout executing the loop—is a logical consequence of the invariant and the negation of the loop condition ϵ, i.e., is true when the loop terminates. For the $[\cdot]$ modality, this last premiss is only needed for the case that ϵ is false from the beginning and the loop body α is never executed. The rule for $[\![\cdot]\!]$ (R37) has an additional premiss, which requires to show that ϕ remains true throughout the execution of α if the invariant is true at the beginning (this latter condition follows from the other premisses).

Example 3. Let α be the loop `while` *true* `do` $x := 0$. Then, because α does *not* terminate, the sequent $x \doteq 0 \vdash [\alpha; x := 1]x \doteq 0$ is valid. To prove that, we apply rule (R28), which results in the two new proof obligations $x \doteq 0 \vdash [\alpha]x \doteq 0$ and $x \doteq 0 \vdash [\alpha][\![x := 1]\!]x \doteq 0$. Both are easy to derive with the rules for while loops, namely the former one with rule (R37) and the invariant $x \doteq 0$ and the latter one with rule (R34) and the invariant *true*.

The modalities $\langle\cdot\rangle$ and $\langle\!\langle\cdot\rangle\!\rangle$ are handled in a different way. Two rules are provided for each of them. One rule, (R35) resp. (R38), allows us to "unwind" the loop, i.e., to symbolically execute it once, provided that the loop condition ϵ is true in the current state. The other rule, (R36) resp. (R39), is used if "unwinding" the loop is not useful. For the $\langle\cdot\rangle$ modality that is the case if ϵ is false and the loop terminates immediately. Rule (R39) for the $\langle\!\langle\cdot\rangle\!\rangle$ modality applies in case the formula ϕ—which supposedly is true at least once during the execution of the loop—becomes true before or during the first execution of the loop body. The rules for $\langle\cdot\rangle$ and $\langle\!\langle\cdot\rangle\!\rangle$ only work in combination with the induction rule, as the following example demonstrates.

---

[2] For non-deterministic versions of DL, rule (R29) is only sound provided that the following semantics is chosen for the $\langle\!\langle\cdot\rangle\!\rangle$ modality: $\langle\!\langle\alpha\rangle\!\rangle\phi$ is true iff ϕ is true at least once in *some* of the (several) traces of α. If, however, a non-deterministic semantics is chosen where ϕ must be true at least once in *every* trace of α (as Pratt did for the propositional case [11]), then rule (R29) is *not* correct, and indeed we failed to find a sound rule for that kind of semantics.

Table 4. Miscellaneous rules.

Generalisation

$$\frac{\phi \vdash \psi}{[\alpha]\phi \vdash [\alpha]\psi} \ (R40) \qquad \frac{\phi \vdash \psi}{\langle\alpha\rangle\phi \vdash \langle\alpha\rangle\psi} \ (R41) \qquad \frac{\phi \vdash \psi}{[\![\alpha]\!]\phi \vdash [\![\alpha]\!]\psi} \ (R42) \qquad \frac{\phi \vdash \psi}{\langle\!\langle\alpha\rangle\!\rangle\phi \vdash \langle\!\langle\alpha\rangle\!\rangle\psi} \ (R43)$$

Quantifier/modality rules

$$\frac{\Gamma, \ \forall x_1 \ldots \forall x_k \ \phi, \ [\alpha]\phi \ \vdash \ \Delta}{\Gamma, \ \forall x_1 \ldots \forall x_k \ \phi \ \vdash \ \Delta} \ (R44)$$
where $Var(\alpha) \subseteq \{x_1, \ldots, x_k\}$

$$\frac{\Gamma \ \vdash \ \langle\alpha\rangle\phi, \ \exists x_1 \ldots \exists x_k \ \phi, \ \Delta}{\Gamma \ \vdash \ \exists x_1 \ldots \exists x_k \ \phi, \ \Delta} \ (R45)$$
where $Var(\alpha) \subseteq \{x_1, \ldots, x_k\}$

$$\frac{\Gamma, \ \forall x_1 \ldots \forall x_k \ \phi, \ [\![\alpha]\!]\phi \ \vdash \ \Delta}{\Gamma, \ \forall x_1 \ldots \forall x_k \ \phi \ \vdash \ \Delta} \ (R46)$$
where $Var(\alpha) \subseteq \{x_1, \ldots, x_k\}$

$$\frac{\Gamma \ \vdash \ \langle\!\langle\alpha\rangle\!\rangle\phi, \ \exists x_1 \ldots \exists x_k \ \phi, \ \Delta}{\Gamma \ \vdash \ \exists x_1 \ldots \exists x_k \ \phi, \ \Delta} \ (R47)$$
where $Var(\alpha) \subseteq \{x_1, \ldots, x_k\}$

Rules for negated modalities

$$\frac{\Gamma \vdash \langle\alpha\rangle\neg\phi, \ \Delta}{\Gamma \vdash \neg[\alpha]\phi, \ \Delta} \ (R48) \quad \frac{\Gamma \vdash [\alpha]\neg\phi, \ \Delta}{\Gamma \vdash \neg\langle\alpha\rangle\phi, \ \Delta} \ (R49) \quad \frac{\Gamma \vdash \langle\!\langle\alpha\rangle\!\rangle\neg\phi, \ \Delta}{\Gamma \vdash \neg[\![\alpha]\!]\phi, \ \Delta} \ (R50) \quad \frac{\Gamma \vdash [\![\alpha]\!]\neg\phi, \ \Delta}{\Gamma \vdash \neg\langle\!\langle\alpha\rangle\!\rangle\phi, \ \Delta} \ (R51)$$

Example 4. Consider the sequent $x \doteq 0 \vdash \langle\!\langle \text{while } true \text{ do } x := x + 1 \rangle\!\rangle x \doteq k$. It states that, if the value of x is 0 initially, then during the execution of the non-terminating loop, x will at least once have the value k. To show that this sequent is valid, we first use the induction rule to prove that $\vdash \forall n \ \phi(n)$ is valid, where $\phi(n) = (x \leq k \wedge n + x \doteq k) \rightarrow \langle\!\langle \text{while } true \text{ do } x := x + 1 \rangle\!\rangle x \doteq k$, from which then the original proof obligation can be derived instantiating n with k. The first premiss of the induction rule, $\vdash \phi(0)$, can easily be derived with rule (R39) as $x \doteq k$ is immediately true in case $n = 0$. The second premiss, $\phi(n) \vdash \phi(n + 1)$, can be derived by first applying the cut rule to distinguish the cases $x < k$ and $x \doteq k$. In the first case, the unwind rule (R38) can be used successfully; and the second case is again easily covered with rule (R39).

Miscellaneous Other Rules. There are three types of miscellaneous other rules (see Table 4). (a) The generalisation rules (R40) to (R43) permit to derive $Op \ \phi \vdash Op \ \psi$ from $\phi \vdash \psi$ where Op is any of the four modal operators. (b) Rules (R44) to (R47) allow to replace (universal) quantifications by modalities. They are similar to the quantifier instantiation rules (R13) and (R15) and are based on the fact that, for example, $[\alpha(x)]\phi$ is true in a state s if $\forall x \ \phi$ is true in s and x is the only variable in $\alpha(x)$. (c) Rules (R48) to (R51) implement the equivalences $\neg[\alpha]\phi \leftrightarrow \langle\alpha\rangle\neg\phi$ and $\neg[\![\alpha]\!]\phi \leftrightarrow \langle\!\langle\alpha\rangle\!\rangle\neg\phi$.

5   Soundness and Relative Completeness

Soundness of the calculus $\mathcal{C}_{\text{DLT}}$ (Corollary 1) is based on the following theorem, which states that all rules preserve validity of the derived sequents.

Theorem 1. *For all rule schemata of the calculus $\mathcal{C}_{\mathrm{DLT}}$, (R1) to (R51), the following holds: If all premisses of a rule schema instance are valid sequents, then its conclusion is a valid sequent.*

Corollary 1. *If a sequent $\Gamma \vdash \Delta$ is derivable with the calculus $\mathcal{C}_{\mathrm{DLT}}$, then it is valid, i.e., $\bigwedge \Gamma \rightarrow \bigvee \Delta$ is a valid formula.*

Proving Theorem 1 is not difficult. The proof is, however, quite large as soundness has to be shown separately for each rule. For the assignment rules, the proof is based on a substitution lemma and is technically involved.

The calculus $\mathcal{C}_{\mathrm{DLT}}$ is *relatively* complete; that is, it is complete up to the handling of the domain of computation (the data structures). It is complete if an oracle rule for the domain is available—in our case one of the oracle rules for arithmetic, (R19) and (R20). If the domain is extended with other data types, $\mathcal{C}_{\mathrm{DLT}}$ remains relatively complete; and it is still complete if rules for handling the extended domain of computation are added.

Theorem 2. *If a sequent is valid, then it is derivable with $\mathcal{C}_{\mathrm{DLT}}$.*

Corollary 2. *If ϕ is a valid DLT-formula, then the sequent $\vdash \phi$ is derivable.*

Due to space restrictions, the proof of Theorem 2, which is quite complex, cannot be given here (it can be found in [13]). The proof technique is the same as that used by Harel [4] to prove relative completeness of his sequent calculus for first-order DL. The following lemmata are central to the completeness proof.

Lemma 1. *For every DLT-formula ϕ_{DLT} there is an (arithmetical) FOL-formula ϕ_{FOL} that is equivalent to ϕ_{DLT}, i.e., $val_s(\phi_{\mathrm{DLT}}) = val_s(\phi_{\mathrm{FOL}})$ for all states s.*

The above lemma states that DLT is not more expressive than first-order arithmetic. This holds as arithmetic—our domain of computation—is expressive enough to encode the behaviour of programs. In particular, using Gödelisation, arithmetic allows one to encode program states (i.e., the values of all the variables occurring in a program) and finite traces into a single number. Note that the lemma states a property of the logic DLT that is independent of the calculus.

Lemma 1 implies that a DLT-formula ϕ_{DLT} could be decided by constructing an equivalent FOL-formula ϕ_{FOL} and then invoking the computation domain oracle—if such an oracle were actually available. But even with a good approximation of an arithmetic oracle, that is not practical (the formula ϕ_{FOL} would be too complex to prove automatically or interactively). And, indeed, the calculus $\mathcal{C}_{\mathrm{DLT}}$ does not work that way.

It may be surprising that the (relative) completeness of $\mathcal{C}_{\mathrm{DLT}}$ requires an expressive computation domain and is lost if a simpler domain and less expressive data structures are used. The reason is that in a simpler domain it may not be possible to express the required invariants resp. induction hypotheses to handle while loops.

Lemma 2. *Let ϕ and ψ be FOL-formulas, let α be a program, and let M_α be any of the modalities $[\alpha], \langle\alpha\rangle, [\![\alpha]\!], \langle\!\langle\alpha\rangle\!\rangle$.*

If the sequent $\phi \vdash M_\alpha \psi$ is valid, then it is derivable with $\mathcal{C}_{DLT}$.

This lemma is at the core of the completeness of $\mathcal{C}_{DLT}$. It is proven by induction on the complexity of the program α, and the proof would not go through if the calculus would lack important rules (not all rules are indispensable; some can be derived from other rules, they are included for convenience.).

Besides Lemmata 1 and 2, the completeness proof makes use of the fact that the calculus has the necessary rules (a) for the operators of classical logic (in particular all propositional tautologies can be derived), and (b) for generalisation, (R40) to (R43).

6 Extended Example

Consider the program "while *true* do if $y \doteq 1$ then α else β" where α abbreviates the sub-program "$x := x + 1$; if $x \doteq 2$ then $y := 0$ else $y := 1$" and β stands for "$x := 0$; $y := 1$". The program consists of a non-terminating while loop. The loop body changes the value of x between 0 and 2 and the value of y between 0 and 1. We want to prove that $0 \leq x \leq 2$ is true in all states reached by this program, if it is started in a state where $val_\partial(x) = 0$ and $val_s(y) = 1$ (we use $0 \leq x \leq 2$ as an abbreviation for $0 \leq x \wedge x \leq 2$). The proof is shown in Figure 1. Its initial proof obligation is the sequent (1). First, the while loop is eliminated applying rule (R37) with the invariant

$$Inv \quad := \quad 0 \leq y \leq 1 \wedge (y \doteq 0 \rightarrow x \doteq 1 \vee x \doteq 2) \wedge (y \doteq 1 \rightarrow x \doteq 0) .$$

The formula $0 \leq x \leq 2$, which is a logical consequence of Inv, does not describe the behaviour of the loop in sufficient detail and, therefore, is not a suitable invariant itself. The result of applying rule (R37) to (1) are the four new proof obligations (2)–(5). Proof obligation (2) can immediately be derived with rule (R19). And, applying rule (R5) to (5) yields a sequent (5') with *true* on the right, which can be derived with rule (R2).

In the sequel, we concentrate on the proof of (4). Proof obligation (3) can be derived in a similar way; its derivation is omitted due to lack of space.

The next step is the application of rule (R32) to (4) to symbolically execute the if-then-else statement. The result are the two proof obligations (6) and (7). Eliminating the concatenations in (6) and (7) with applications of rule (R28) yields (8) and (9) resp. (10) and (11). Next, we simplify (and weaken) the left sides of (8)–(11) with the arithmetic rule (R20) (this is not really necessary but the sequents get shorter and easier to understand). The result are the sequents (12)–(15), respectively. The derivations of proof obligations (12), (14), and (15) need no further explanation and are shown in Figure 1. To derive (13), we apply (R22) and get (16). The if-then-else statement is symbolically executed with rule (R32), which results in (17) and (18). Proof obligation (17) is derived by applying rule (R24), which yields (19) and (20). It is easy to check that (19)

$$
\dfrac{\dfrac{*}{x \doteq 0 \vdash\ 0 \le x \le 2}\ \text{(R19)}\quad \dfrac{*}{x' \doteq 0,\, x \doteq x'+1\ \vdash\ 0 \le x \le 2}\ \substack{\text{(R19)}\\[2pt]\text{(R24)}}}{x \doteq 0 \vdash\ [x := x+1]0 \le x \le 2\quad(12)}
$$

$$
\dfrac{\dfrac{\dfrac{*}{x \doteq 0 \vdash\ 0 \le x \le 2}\ \text{(R19)}\quad \dfrac{*}{x \doteq 0,\, y \doteq 1 \vdash\ 0 \le x \le 2}\ \substack{\text{(R19)}\\[2pt]\text{(R24)}}}{x \doteq 0 \vdash\ [y := 1]0 \le x \le 2}}{\vdash\ [x := 0][y := 1]0 \le x \le 2\quad(15)}\ \text{(R22)}
$$

$$
\dfrac{\dfrac{*}{x \doteq 1 \vee x \doteq 2 \vdash\ 0 \le x \le 2}\ \text{(R19)}\quad \dfrac{*}{x' \doteq 1 \vee x' \doteq 2,\, x \doteq 0 \vdash\ 0 \le x \le 2}\ \substack{\text{(R19)}\\[2pt]\text{(R24)}}}{x \doteq 1 \vee x \doteq 2 \vdash\ [x := 0]0 \le x \le 2\quad(14)}
$$

$$
\dfrac{\dfrac{\dfrac{*}{(19)}\text{(R19)}\quad\dfrac{*}{(20)}\substack{\text{(R19)}\\[2pt]\text{(R24)}}}{(17)}\quad\dfrac{\dfrac{*}{(19')}\text{(R19)}\quad\dfrac{*}{(20')}\substack{\text{(R19)}\\[2pt]\text{(R24)}}}{(18)}}{\dfrac{(16)}{\ }\text{(R32)}}\ \text{(R22)}
$$

$$
\dfrac{*}{(2)}\text{(R19)}\quad \vdots\ (3)\quad
\dfrac{\dfrac{(12)}{(8)}\text{(R20)}}{\ }\quad
\dfrac{\dfrac{(13)}{(9)}\text{(R20)}}{(6)}\text{(R28)}\quad
\dfrac{\dfrac{(14)}{(10)}\text{(R20)}}{\ }\quad
\dfrac{\dfrac{(15)}{(11)}\text{(R20)}}{(7)}\text{(R28)}\ (R32)\quad
\dfrac{\dfrac{*}{(5')}\text{(R2)}}{(5)}\substack{\text{(R5)}}
$$

$$
\dfrac{(4)}{(1)}\ \text{(R37)}
$$

$$
x \doteq 0,\ y \doteq 1\ \vdash\ [\textbf{while } true \textbf{ do if } y \doteq 1 \textbf{ then } \alpha \textbf{ else } \beta]0 \le x \le 2 \tag{1}
$$

$$
x \doteq 0,\ y \doteq 1\ \vdash\ Inv \tag{2}
$$

$$
Inv,\ true\ \vdash\ [\textbf{if } y \doteq 1 \textbf{ then } \alpha \textbf{ else } \beta]Inv \tag{3}
$$

$$
Inv,\ true\ \vdash\ [\textbf{if } y \doteq 1 \textbf{ then } \alpha \textbf{ else } \beta]0 \le x \le 2 \tag{4}
$$

$$
Inv,\ \neg true\ \vdash\ 0 \le x \le 2. \tag{5}
$$

$$
Inv,\ true,\ y \doteq 1\ \vdash\ [x := x+1;\ \textbf{if } x \doteq 2 \textbf{ then } y := 0 \textbf{ else } y := 1]0 \le x \le 2 \tag{6}
$$

$$
Inv,\ true,\ \neg y \doteq 1\ \vdash\ [x := 0;\ y := 1]0 \le x \le 2 \tag{7}
$$

$$
Inv,\ true,\ y \doteq 1\ \vdash\ [x := x+1]0 \le x \le 2 \tag{8}
$$

$$
Inv,\ true,\ y \doteq 1\ \vdash\ [x := x+1][\textbf{if } x \doteq 2 \textbf{ then } y := 0 \textbf{ else } y := 1]0 \le x \le 2 \tag{9}
$$

$$
Inv,\ true,\ \neg y \doteq 1\ \vdash\ [x := 0]0 \le x \le 2 \tag{10}
$$

$$
Inv,\ true,\ \neg y \doteq 1\ \vdash\ [x := 0][y := 1]0 \le x \le 2. \tag{11}
$$

$$
x \doteq 0\ \vdash\ [x := x+1]0 \le x \le 2 \tag{12}
$$

$$
x \doteq 0\ \vdash\ [x := x+1][\textbf{if } x \doteq 2 \textbf{ then } y := 0 \textbf{ else } y := 1]0 \le x \le 2 \tag{13}
$$

$$
x \doteq 1 \vee x \doteq 2\ \vdash\ [x := 0]0 \le x \le 2 \tag{14}
$$

$$
\vdash\ [x := 0][y := 1]0 \le x \le 2 \tag{15}
$$

$$
x' \doteq 0,\ x \doteq x'+1\ \vdash\ [\textbf{if } x \doteq 2 \textbf{ then } y := 0 \textbf{ else } y := 1]0 \le x \le 2 \tag{16}
$$

$$
x' \doteq 0,\ x \doteq x'+1,\ x \doteq 2\ \vdash\ [y := 0]0 \le x \le 2 \tag{17}
$$

$$
x' \doteq 0,\ x \doteq x'+1,\ \neg x \doteq 2\ \vdash\ [y := 1]0 \le x \le 2 \tag{18}
$$

$$
x' \doteq 0,\ x \doteq x'+1,\ x \doteq 2\ \vdash\ 0 \le x \le 2 \tag{19}
$$

$$
x' \doteq 0,\ x \doteq x'+1,\ x \doteq 2,\ y \doteq 0\ \vdash\ 0 \le x \le 2 \tag{20}
$$

Fig. 1. The derivation described in Section 6.

and (20) are valid FOL-sequents and can therefore be derived with the oracle rule for arithmetic (R19).

Applying rule (R24) to (18) yields similar FOL-sequents (19′) and (20′), which differ from (19) and (20) in that they contain $\neg x \doteq 2$ instead of $x \doteq 2$ and $y \doteq 1$ instead of $y \doteq 0$. They, too, can be derived with the oracle (R19).

7 Future Work

Future work includes an implementation of our calculus $\mathcal{C}_{\mathrm{DLT}}$, which would allow us to carry out case studies going beyond the simple examples shown in this paper and to test the usefulness of DLT in practice.

A useful extension of $\mathcal{C}_{\mathrm{DLT}}$ for practical applications may be special rules for formulas of the form $[\alpha]\phi \wedge [\![\alpha]\!]\psi$, such that splitting the two conjuncts is avoided and they do not have to be handled in separate—but similar—sub-proofs.

Also, it may be useful to consider (a) a non-deterministic version of DLT, and (b) extensions of DLT with further modalities such as "α preserves ϕ", which expresses that, once ϕ becomes true in the trace of α, it remains true throughout the rest of the trace. It seems, however, to be difficult to give a (relatively) complete calculus for this modality.

Acknowledgement. We thank W. Ahrendt, E. Habermalz, W. Menzel, and P. H. Schmitt for fruitful discussions and comments.

References

1. W. Ahrendt, T. Baar, B. Beckert, M. Giese, E. Habermalz, R. Hähnle, W. Menzel, and P. H. Schmitt. The KeY approach: Integrating object oriented design and formal verification. In M. Ojeda-Aciego, I. P. de Guzman, G. Brewka, and L. M. Pereira, editors, *Proceedings, Logics in Artificial Intelligence (JELIA), Malaga, Spain*, LNCS 1919. Springer, 2000.
2. K. R. Apt. Ten years of Hoare logic: A survey – part I. *ACM Transactions on Programming Languages and Systems*, 1981.
3. B. Beckert. A Dynamic Logic for the formal verification of Java Card programs. In *Proceedings, Java Card Workshop (JCW), Cannes, France*, LNCS 2014. Springer, 2001.
4. D. Harel. *First-order Dynamic Logic*. LNCS 68. Springer, 1979.
5. D. Harel. Dynamic Logic. In D. Gabbay and F. Guenthner, editors, *Handbook of Philosophical Logic, Volume II: Extensions of Classical Logic*. Reidel, 1984.
6. D. Harel, D. Kozen, and J. Tiuryn. *Dynamic Logic*. MIT Press, 2000.
7. M. Heisel, W. Reif, and W. Stephan. A Dynamic Logic for program verification. In A. Meyer and M. Taitslin, editors, *Proceedings, Logic at Botic, Pereslavl-Zalessky, Russia*, LNCS 363. Springer, 1989.
8. D. Hutter, B. Langenstein, C. Sengler, J. H. Siekmann, and W. Stephan. Deduction in the Verification Support Environment (VSE). In M.-C. Gaudel and J. Woodcock, editors, *Proceedings, International Symposium of Formal Methods Europe (FME), Oxford, UK*, LNCS 1051. Springer, 1996.

9. D. Kozen and J. Tiuryn. Logic of programs. In J. van Leeuwen, editor, *Handbook of Theoretical Computer Science*, chapter 14, pages 89–133. Elsevier, 1990.
10. V. R. Pratt. Semantical considerations on Floyd-Hoare logic. In *Proceedings, 18th IEEE Symposium on Foundation of Computer Science*, pages 109–121, 1977.
11. V. R. Pratt. Process logic: Preliminary report. In *Proceedings, ACM Symposium on Principles of Programming Languages (POPL), San Antonio/TX, USA*, 1979.
12. W. Reif. The KIV-approach to software verification. In M. Broy and S. Jähnichen, editors, *KORSO: Methods, Languages, and Tools for the Construction of Correct Software – Final Report*, LNCS 1009. Springer, 1995.
13. S. Schlager. Erweiterung der Dynamischen Logik um temporallogische Operatoren. Studienarbeit, Fakultät für Informatik, Universität Karlsruhe, 2000. In German. Available at: `ftp://i12ftp.ira.uka.de/pub/beckert/schlager.ps.gz`.

Flaw Detection in Formal Specifications

Wolfgang Reif, G. Schellhorn, and Andreas Thums

Lehrstuhl für Softwaretechnik und Programmiersprachen,
Universität Augsburg, D-86135 Augsburg
{reif, schellhorn, thums}@informatik.uni-augsburg.de

Abstract. In verification of finite domain models (model checking) counterexamples help the user to identify, why a proof attempt has failed. In this paper we present an approach to construct counterexamples for first-order goals over infinite data types, which are defined by algebraic specifications. The approach avoids the implementation of a new calculus, by integrating counterexample search with the interactive theorem proving strategy. The paper demonstrates, that this integrations requires only a few modifications to the theorem proving strategy.

1 Introduction

It is common knowledge, that most of time in theorem proving is not spent on successful, but on unsuccessful proof attempts. Usually the reason is simply that the theorem under consideration is wrong. For data types with a finite domain, the counterexamples generated by model checkers help the proof engineer to detect the errors. Also tools which generate finite models are available. Unfortunately, there is no such mechanism for data types with an infinite domain. Additionally, proofs for infinite data types usually have to be found interactively and decisions made during the proof attempt may be incorrect.

In this paper we will demonstrate a method which generates counterexamples for infinite data types in algebraic specifications. The method is implemented in the interactive theorem prover KIV [7,2] which supports algebraic specifications in the style of CASL [5] as well as state based specifications (using imperative programs or ASMs [3]). It avoids the need to implement a new calculus by adaption of the existing proof calculus to the construction of counterexamples.

Our paper is organized as follows: Section 2 will first give an informal overview of the method, which is based on reduction of an unprovable goal to *false*. The capabilities of the algorithm will be demonstrated with an example and we will discuss the limitations of the approach.

After introducing some necessary notation (Sect. 3), Sect. 4 describes how conjectures may be falsified by reduction, thereby proving the existence of a counterexample. The backtracing algorithm, that computes the actual variable assignment and the model condition is given in Sect. 5. A number of important implementation issues are discussed in section 6. Some examples and a comparison to related work is given in Sect. 7. The paper concludes with Sect. 8.

R. Goré, A. Leitsch, and T. Nipkow (Eds.): IJCAR 2001, LNAI 2083, pp. 642–657, 2001.

2   Overview

To demonstrate the capabilities of the algorithm, let us start with an example:

Example 1

We assume a data type **list** *over a parameter type* **elem**. *Lists are generated by the constructors* nil *for the empty list, and* $e + l$, *which adds an element* e *to a list* l. *The elements are assumed to be totally ordered with* $\leq$. head *and* tail *select* e *resp.* l *from a list* $e + l$, length(l) *computes the length of a list,* append(l_1, l_2) *appends two lists,* member(e, l) *tests, if the element* e *is in* l, *and* sort(l) *states that* l *is (nonstrictly) sorted with respect to the ordering on elements. Then the conjecture*

$$\varphi \equiv \quad sort(l_1) \land l_2 = \text{append}(l_1, head(l_3) + nil) \land length(l_3) \geq 2 * length(l_1)$$
$$\land \; l_3 \neq nil \land member(head(l_1), tail(l_3)) \to sort(l_2)$$

does not hold in all models of the specification. A counterexample consists of the variable assignment $l_1 := e_1 + nil, l_2 := e_1 + e_2 + nil, l_3 = e_2 + append(l_4, e_1 + l_5)$ *(*e_1, e_2 *are variables of type* **elem**) *and the model condition* $\neg\, e_1 \leq e_2$.

The example has explicitly been chosen to be somewhat contrived for two reasons: First, it should demonstrate several aspects of the counterexample search. Second, we want to stress, that formulas encountered during interactive proof attempts are even *more complex* than the example given above and counterexamples are totally non-obvious. Our algorithm finds the counterexample within 5 seconds, which is faster than we (and we hope the reader) can guess one by inspecting the goal.

In contrast to a proof, where it is tried to reduce the conjecture to true, the counterexample search tries to reduce the conjecture to false. If this is successful, the conjecture is not provable and a counterexample exists. Because a conjecture is usually not false in general, we try to find a variable assignment under which the conjecture can be reduced to false.

The variable assignment is usually built up by incrementally instantiating variables of generated sorts with constructors, resulting in constructor terms. A counterexample for Example 1, which uses constructor terms only, would set l_3 to $e_2 + e_1 + nil$. Often this incremental construction can be shortcut by combining it with the application of proof rules. By imposing some restrictions (see Sect. 4) on the ones that may be used, the more abstract counterexample shown above can be derived.

Incrementally instantiating a variable in a conjecture gives several subgoals, one for each constructor. Since each subgoal could lead to a counterexample, we have implemented a heuristic search strategy to choose a case, which (hopefully) quickly leads to a counterexample.

Construction of a variable assignment terminates (at the latest) when all variables of generated sorts have been instantiated by constructor terms. The remaining formula may be *false*, but we may also end up with a *model condition*. Model conditions occur for three reasons: First, as in the example above, we may have parameter sorts, and the model condition may impose a restriction on the models of the parameter: in the example above $\neg\, e_1 \leq e_2$ ensures that there exist two different elements in **elem**: in models, where **elem** contains just one element, the conjecture would be true. A second reason for model conditions is underspecification: E.g., if $pred(0)$, the predecessor of zero is not specified (over natural numbers), then the counterexample search may finish with a model condition like $pred(0) > 3$. Finally, the interactive proof strategy may be too weak, to deduce automatically, that a fully instantiated goal is equivalent to false. This case usually arises, when not enough simplifier rules have been provided for the data structure under consideration. Since these would be needed to automate (successful) proofs anyway, we simply restart the counterexample search after adding the required rules.

We have found that the success of counterexample generation depends mainly on the efficiency of the heuristic search strategy that computes variable assignments. If the strategy is successful, we usually end up with trivial model conditions, like the ones shown above. Therefore we have not invested in the automation of the check, whether the model condition is consistent with the specification. Instead we simply ask the user to acknowledge satisfiability.

After the original goal has been reduced to *false* (or a model condition) the variable assignment for the original goal has to be deduced. This is done by a backtracing algorithm, that analyzes the reduction backwards from *false* to the beginning of the counterexample search (see Sect. 5). Because the attempt to compute a counterexample will usually not be made on an initial conjecture, the counterexample will also be traced back further in the original proof attempt.

3   Formal Basics

The basis for our approach are first-order algebraic specifications, consisting of a finite, many-sorted signature with function symbols F, a finite set of axioms (first-order formulas) and a set of constructors $C \subseteq F$ (typically, constructors are given by clauses like *nat* **generated by** *0,suc*). A sort is called a *target sort*, if it is generated by some constructors, otherwise it is called a *parameter sort*. A *constructor term* is a term that uses only constructor functions and variables of parameter sorts. If syntactically different constructor terms always represent different elements, the data type is called a *free* or *freely generated* data type. **nat** is a free data type while *int* **generated by** *0, suc, pred* is not free, because $i = pred(suc(i))$.

We assume loose semantics, i.e. an algebra $\mathcal{A}$ is in the class $Mod(SP)$ of models of a specification SP, iff the axioms hold in $\mathcal{A}$ and if every element contained in the carrier set of some target sort can be represented as the semantic value of a constructor term, given a suitable valuation for the parameter variables.

In the following, we will use x, y as (metavariables for) variables, f, g as (constructor) function symbols τ, ρ as terms, and φ, ψ, χ as formulas. Free variables $free(\varphi)$ are defined as usual. $gen(\varphi)$ are the free variables of target sorts (also called *target variables*), $param(\varphi)$ are all other free variables. φ_x^t denotes substitution in formulas.

Sequents $\Gamma \vdash \Delta$ consists of two lists $\Gamma = \varphi_1, \ldots, \varphi_n$ and $\Delta = \psi_1, \ldots, \psi_m$ of formulas. $\Gamma \vdash \Delta$ abbreviates the formula $\varphi_1 \wedge \ldots \wedge \varphi_n \rightarrow \psi_1 \vee \ldots \vee \psi_m$.

We will use a sequent calculus for first-order logic (like the one defined in [13]) to derive proofs for conjectures. Proofs in this calculus are trees, in which every node contains a sequent. Starting with a proof tree of one node, which contains the initial goal c, a fixed, predefined set of proof rules of the form $\frac{p_1 \ldots p_n}{c}$ is applied to reduce the goal to simpler subgoals. c is called the *conclusion*, $p_1 \ldots p_n$ are called the *premises* of the rule. By recursive application of further rules to the resulting premises, a proof tree is built up. The conclusion c is proven if all premises in the tree are reduced to the trivially true premise *true*.

Given these preliminaries on specifications and proofs, we can now define the central notion of a *counterexample*:

Definition 1 (counterexample)
Let φ be a formula over a specification SP with $\underline{x} = \text{gen}(\varphi)$. Then a pair $\mathcal{C} = (\chi, \underline{y} = \underline{t})$ is called a counterexample for the formula φ iff $\underline{y}$ is a vector of target variables, $\underline{t}$ is a list of terms such that $y_i \notin t_j$ for $i \leq j$, $\text{gen}(\chi) = \emptyset$ and there is a model $\mathcal{A}$ such that

$$\mathcal{A} \in \text{Mod}(SP \cup Cl_\exists \, \chi) \tag{1}$$

and

$$SP \models (\chi \wedge \underline{y} = \underline{t}) \rightarrow \neg \, \varphi \tag{2}$$

A counterexample for φ is called a strong counterexample, if $\underline{y} \subseteq \underline{x}$.

The definition requests that there is a model $\mathcal{A}$ of the specification SP, in which χ is satisfiable (1) and that for every model of the specification either χ does not hold or $\neg \, \varphi$ is true under the variable assignment $\underline{y} = \underline{t}$ (2). The restriction $y_i \notin t_j$ for $i \leq j$ ensures that no cyclic dependencies between variables y_i and terms t_j exists.

We can now prove that with this definition every non-provable formula has a counterexample:

Theorem 1
A formula $\neg \, \varphi$ is satisfiable over SP iff there exists a (strong) counterexample for φ.

Proof 1
If a formula $\neg \, \varphi$ is satisfiable, there exists a model $\mathcal{A} \in \text{Mod}(SP)$ and a valuation v where $\mathcal{A}, v \models \neg \, \varphi$. Now, let $\text{gen}(\varphi) = \underline{x}$. Since the model is generated we have constructor terms $\underline{t}$ and valuations v_i, such that $v_i(x_i) = v_i(t_i)$. By suitably renaming variables in $\underline{t}$ we can find a common valuation v', such that $\mathcal{A}, v' \models \neg \, \varphi$ and $v'(x_i) = v'(t_i)$ for every i. Setting $\chi := \neg \, \varphi_{\underline{x}}^t$ and $\mathcal{C} := (\chi, \underline{x} = \underline{t})$, we have found a counterexample: the substitution theorem of first-order logic implies $\mathcal{A}, v' \models \chi$, and this implies condition (1). Obviously, we also have $SP \models (\neg \, \varphi_{\underline{x}}^t \wedge \underline{x} = \underline{t}) \rightarrow \neg \, \varphi$, i.e. condition (2) is satisfied too.

Conversely, assume that a formula φ has a counterexample. Choose a model $\mathcal{A} \in Mod(SP)$ and a valuation v such that $\mathcal{A}, v \models \chi$. This is possible because of (1). Then the constraint $y_i \notin t_j$ for $i \leq j$ allows to sequentially modify $v(y_1), \ldots, v(y_n)$, such that for the resulting, modified valuation v' $\mathcal{A}, v' \models \underline{y} = \underline{t}$ holds. Since χ does not contain target variables, we still have $\mathcal{A}, v' \models \chi$. Finally, property (2) implies $\mathcal{A}, v' \models \neg \varphi$, i.e. $\neg \varphi$ is satisfiable.

The proof shows, that we could restrict the terms assigned to the variables to be constructor terms. We have not done so, to allow more abstract counterexamples as in Example 1.

Finally we would like to note, that for efficiency reasons our algorithm will not construct *strong* counterexamples. Instead it will sometimes assign values to variables, that are only present in intermediate goals. This is not problematic, since it is easy to prove that any counterexample can be transformed to a strong counterexample by iterative application of the following post processing step:

Lemma 1
Let $(\chi, \underline{y} = \underline{t})$ be a counterexample for φ with $x = t \in \underline{y} = \underline{t}$ and $x \notin \text{free}(\varphi)$. Let $\underline{y}' = \underline{t}'$ be $\underline{y} = \underline{t}$ with $x = t$ removed. Then $(\chi, (\underline{y}' = \underline{t}')^t_x)$ is a counterexample for φ too.

4   Falsifying Conjectures

We found that with a simple restriction sequent calculus is already suitable to do satisfiability reasoning instead of proving theorems. This restriction, called invertibility, will be discussed in the first subsection. A second advantage of using sequent calculus is that we do not have to use a brute-force search for the correct instantiation which is needed for the counterexample. Instead we use an incremental instantiation strategy called structural expansion as discussed in the second subsection. Finally, the last subsection exemplifies, how the combined proof strategy falsifies a conjecture.

4.1   Existence of a Counterexample

To find counterexamples we cannot use every rule of the sequent calculus due to the fact that some of them weaken the conjecture. This means that some rules reduce a provable conjecture (with no counterexample) to one which is no longer provable (and therefore has a counterexample). We call rules which do not weaken conjectures *invertible rules*.

Definition 2 (invertible rule)

A rule $\frac{p_1 \cdots p_n}{c}$ is called invertible, iff for every model $\mathcal{A}$ of the specification:

$$\text{from } \mathcal{A} \models c \text{ follows } \mathcal{A} \models p_i \text{ for } 1 \leq i \leq n \qquad (3)$$

This means, that if the conclusion of a rule is valid in every model $\mathcal{A}$ of the specification then every premise p_i is valid in that model. This condition is sufficient to show the existence of a counterexample.

Lemma 2 (existence of a counterexample)
Let T be a proof tree consisting of invertible rules, c the conclusion and $p_1 \ldots p_n$ the premises of T.

If there exists an i, such that $\neg\, p_i$ is satisfiable then there exists a counterexample for the conclusion c.

Proof 2
This lemma is proven by contradiction to the validity of the conclusion c. Assuming c is valid, definition 2 and transitivity imply that all p_i are valid. This contradicts the existence of an i, such that $\neg\, p_i$ is satisfiable.
$\square$

Fortunately, almost every rule of the sequent calculus is invertible, especially all rules that eliminate propositional connectives. The main exception is of course the rule *weakening* which drops formulas from the sequent. This rule must therefore be avoided when searching for a counterexample. The rules *all right* (shown below, x_0 is a new variable) and *exists left* to eliminate quantifiers are also invertible. Two other critical rules are the quantifier instantiation rules *all left* and *exist right*. These rules are invertible only, if we use a version that does not drop the quantifier as shown below for *all left* .

$$\frac{\Gamma \vdash \varphi_x^{x_0}, \Delta}{\Gamma \vdash \forall\, x.\varphi, \Delta} \text{ all right} \qquad \frac{\forall\, x.\varphi, \varphi_x^t, \Gamma \vdash \Delta}{\forall\, x.\varphi, \Gamma \vdash \Delta} \text{ all left}$$

4.2   Searching Counterexamples

To compute the variable assignment for a counterexample, we could in principle use a generate and test algorithm, which would enumerate all constructor terms. E.g. for a formula involving two variables m, n over natural numbers, we could try to instantiate the variables with all pairs $(suc^i(0),\ suc^j(0))$ of constructor terms. But this would be very inefficient. Consider e.g. a goal of the form φ (m,n) $\wedge$ m $\leq$ n. Then for n = 0 it is totally redundant to consider any instance of m except 0, since the formula obviously has no counterexample for other cases of m.

Therefore a more elaborate search for the existence of a counterexample is needed, which exploits constraints (like m $\leq$ n) present in the conjecture to avoid the enumeration of as many cases as possible. Therefore the variables are instantiated stepwise with respect to the constructors and thereafter proof rules simplify the resulting formula. If, for example, variable n is instantiated with 0 (and $suc(n_1)$), then the conjecture can immediately be simplified to $\varphi(0,\ 0)$ (by the fact m $\leq$ 0 $\leftrightarrow$ m = 0), thereby avoiding any enumeration of instances of m. This method of stepwise instantiation is called *structural expansion*.

Definition 3 (structural expansion)

Let φ be a formula with a free variable x and $sort(x) = s$ generated by $f_1 \ldots f_n$. Furthermore let $\underline{x}_1 \ldots \underline{x}_n$ be proper argument vectors for $f_1 \ldots f_n$ that are not free in φ. Then the step from $\varphi(x)$ to

$$(\varphi(f_1(\underline{x}_1)) \vee \ldots \vee \varphi(f_n(\underline{x}_n)))$$

is called structural expansion.

Structural expansion spans a search tree and partly instantiates the free variables of the conjecture. Because the structural expansion stepwise enumerates every constructor term – for a variable every possible prefix is generated – the complete search space is generated. This is true because the data types are generated by the constructors, i.e. every value of a variable can be represented by a constructor term. The structural expansion may be included in the sequent calculus through an additional rule called *constructor cut*.

Definition 4 (constructor cut)

The constructor cut rule has the form

$$\frac{(\Gamma \vdash \Delta)_x^{f_1(\underline{x}_1)} \quad \ldots \quad (\Gamma \vdash \Delta)_x^{f_n(\underline{x}_n)}}{\Gamma \vdash \Delta} \quad \text{constructor cut}$$

where $x \in gen(\varphi)$ and $sort(x) = s$ generated by $f_1, \ldots f_n$.

To use this rule in the counterexample search it has to be correct and invertible.

Lemma 3 (correctness and invertibility of *constructor cut*)

The rule constructor cut *(see Def. 4) is correct and invertible.*

Because the *constructor cut* rule is itself a weak form of the induction principle, the proof is easy using structural induction.

Adding the *constructor cut* rule to the proof search and avoiding non invertible rules are the only changes needed to adapt the usual sequent calculus to counterexample search.

To have a complete proof strategy in the sense, that every vector of constructor terms $\underline{t}$ is considered, we have to use a *fair* search strategy, that assures, that every branch is considered ultimately as well as every variable is ultimately expanded with the *constructor cut* rule. Then every constructor term instance of every variable will ultimately be tried as a potential counterexample. A simple fair search strategy would be breadth-first search on the branches of the proof tree, combined with expanding one of the variables that has been expanded the fewest times before. A more elaborate strategy is discussed in Sect. 6.2.

4.3   Generating a Counterexample Proof

We will now exemplify the counterexample search.

Example 2

The following conjecture

$$\varphi \equiv \text{sort}(l_1) \wedge \text{sort}(l_2) \rightarrow \text{sort}(\text{append}(l_1, l_2))$$

does not hold in all models of the specification. A counterexample consists of the variable assignment $l_1 := e_1 + nil, l_2 := e_2 + nil$ *and the* model condition $\neg$ $e_1 \leq e_2$.

The following proof tree gives an impression how the counterexample generation works.

Example 3

$$
\cfrac{
\cfrac{
\cfrac{
\cfrac{
\cfrac{
\cfrac{e_1 \leq e_2}{\text{sort}(e_2 + nil) \rightarrow \text{sort}(\text{append}(e_1 + nil, e_2 + nil))} \quad \cdots}
{\cdots \quad \text{sort}(e_2 + l_2') \rightarrow \text{sort}(\text{append}(e_1 + nil, e_2 + l_2'))} \text{ con. cut } l_2'}
{\text{sort}(l_2) \rightarrow \text{sort}(\text{append}(e_1 + nil, l_2))} \text{ con. cut } l_2}
{\text{sort}(e_1 + nil) \wedge \text{sort}(l_2) \rightarrow \text{sort}(\text{append}(e_1 + nil, l_2))} \quad \cdots}
{\cdots \quad \text{sort}(e_1 + l_1') \wedge \text{sort}(l_2) \rightarrow \text{sort}(\text{append}(e_1 + l_1', l_2))} \text{ con. cut } l_1'}
{\text{sort}(l_1) \wedge \text{sort}(l_2) \rightarrow \text{sort}(\text{append}(l_1, l_2))} \text{ con. cut } l_1
$$

with labels: simplify, con. cut l_2', con. cut l_2, simplify, con. cut l_1', con. cut l_1

Starting with the conjecture (conclusion of the proof tree) the rule *constructor cut* stepwise introduces the constructor terms which build up the counterexample. Dots indicate branches of the proof tree, which are not relevant to the outcome. Applications of *constructor cut* are alternated with applications of sequent calculus rules. In our case only the *simplifier* rule is needed, which simplifies goals using rewrite rules: in the example the recursive definitions of *append* and *sort* are used as rewrite rules to simplify e.g. $sort(e_1 + nil)$ to true. The search stops at the formula $e_1 \leq e_2$, because there are no more target variables to instantiate and the predicate may not be automatically decided. Under the assumption that $\neg\, e_1 \leq e_2$ is satisfiable – which the user acknowledges to be true – there exists a counterexample for the conclusion.

5   The Counterexample

In section 4 we discussed how to prove the existence of a counterexample. In this section we will discuss how to reconstruct the variable assignment from a failed proof attempt, by tracing an initial variable assignments back through the proof attempt. The first subsection considers the proof generated by the counterexample search which uses invertible rules only whereas the second subsection extends the approach to arbitrary sequent calculus proofs.

5.1   Computing the Counterexample

We will compute the counterexample backwards beginning at a premise ψ. According to Sect. 4, this premise is either *false*, or a satisfiable formula without free target variables. In either case, $\mathcal{C} = (\neg \psi, \emptyset)$ is a counter example. The task is now to compute a counterexample for every goal contained in the branch of the considered premise by *backtracing* through it. To adapt the counterexample from φ' to the previous formula φ in the backtrace path the rule which deduces φ' from φ has to be considered. Some rules of the sequent calculus have the property that the variable assignment has to be adjusted. We define rules where the assignment needs no adjustment *strong invertible rules*.

Definition 5 (strong invertible rule)

A rule $\dfrac{p_1 \cdots p_n}{c}$ is called *strong invertible*, iff $\mathcal{A} \models c \rightarrow p_1 \wedge \ldots \wedge p_n$ holds for every model $\mathcal{A}$ of the specification.

It is easy to prove that this definition implies:

Lemma 4 (backtrace through strong invertible rules)

Let $\dfrac{p_1 \cdots p_n}{c}$ be a strong invertible rule and $\mathcal{C} = (\chi, \underline{x} = \underline{t})$ a counterexample for a premise p_i. Then $\mathcal{C}$ is also a counterexample for the conclusion c.

In the sequent calculus most rules that are invertible, are also strong invertible. Especially all propositional rules, and the quantifier rules are strong invertible as well as the rule to apply a lemma. The only invertible rules, which are not strong invertible are *insert equation*, *structural induction*, and *constructor cut*.

Example 4 (insert equation)
The insert equation *rule*

$$\frac{\Gamma_x^\tau \vdash \Delta_x^\tau}{x = \tau, \Gamma \vdash \Delta} \quad \text{insert equation}$$

with $x \notin vars(\tau)$ is an invertible rule. Let

$$\frac{\vdash \mathrm{prime}(4)}{x = 4 \vdash \mathrm{prime}(x)} \quad \text{insert equation}$$

be an instantiation of the rule where the predicate $\mathrm{prime}(x)$ is true iff x is a prime number. Because 4 is not prime there exists a counterexample $\mathcal{C} = (true, \emptyset)$ with no conditions for this premise. But for a model $\mathcal{A}$ and the variable assignment $v(x) = 3$ the antecedent of the conclusion may be refuted, therefore in general $\mathcal{C}$ is not a counterexample for the conclusion.

The example shows, that the counterexample for the premise is not a counterexample for the conclusion. We have to adapt the counterexample by adding the dropped equation: $C' = (true, x = 4)$. In general the adaption for the three invertible (but not strong invertible) rules are as follows:

Theorem 2 (adaption for *insert equation*)
Let $C = (\chi, \underline{x} = \underline{t})$ be a counterexample for the premise of the insert equation rule. Then $C' = (\chi, (\underline{x} = \underline{t})^{x_0}_x \land x = \tau)$ (where x_0 is a new variable) is a counterexample for the conclusion.

The *structural induction* is similar to the *constructor cut* rule (see Def. 4), but adds an induction hypothesis of the form $\forall\, y.(\Gamma \rightarrow \Delta)^{x_i}_x$ for every x_i in $\underline{x}$ (y are all free variables of the conclusion $\Gamma \vdash \Delta$ except the induction variable x).

Theorem 3 (adaption for *structural induction* and *constructor cut*)
Let $C = (\chi, \underline{x} = \underline{t})$ be a counterexample for the premise with $f_i(\underline{x_i})$ of the structural induction or the constructor cut rule. Then $C' = (\chi, (\underline{x} = \underline{t})^{x_0}_x) \land x = f_i(\underline{x_i}))$, where x_0 is a new variable is a counterexample for the conclusion.

The correctness proofs for this adaption is not too difficult, details are given in [10]. In the following, we will shows the backtrace algorithm applied on Ex. 3.

Example 5
The backtrace begins at the premise. Because the negation of $e_1 \leq e_2$ is satisfiable the counterexample for the premise is $C = (\neg\, e_1 \leq e_2, \emptyset)$. At the simplifier rule no adaption is necessary, but at the constructor cut rule $(l'_2 = nil)$ the counterexample is adapted to $C = (\neg\, e_1 \leq e_2, l'_2 = nil)$ and stepwise further to $C = (\neg\, e_1 \leq e_2, (l'_2 = nil, l_2 = e_2 + l_2))$ which also holds for the premise of the simplification rule, $C = (\neg\, e_1 \leq e_2, (l'_2 = nil, l_2 = e_2 + l_2, l'_1 = nil))$, and $C = (\neg\, e_1 \leq e_2, (l'_2 = nil, l_2 = e_2 + l_2, l'_1 = nil, l_1 = e_1 + l'_1))$. Finally, neither l'_1 nor l'_2 appear in the conclusion, the counterexample may be simplified to the strong counterexample $C = (\neg\, e_1 \leq e_2, (l_2 = e_2 + nil, l_1 = e_1 + nil))$ using lemma 1.

Because the search for a counterexample applies only invertible rules and the counterexample can be backtraced over these rules, a variable assignment for the free variables of disproved conjectures may be derived.

5.2   Earliest Point of Failure

In interactive proof systems the user does not only look at the conjecture he wants to prove, but also at formulas which will be derived during a proof. Therefore the counterexample search will usually not be started at the conjecture under consideration – why should one try to prove a conjecture assumed to be wrong? – but at some 'fishy' goal within the proof tree. If a counterexample can be found for that goal the users interest is whether some proof decisions were incorrect or whether nevertheless the conjecture is faulty. Therefore the counterexample has to be backtraced in the original proof attempt tree as well.

Because the counterexample search already uses the (restricted) proof rules, the backtrace algorithm of Sect. 5.1 may also be used for backtracing in the original proof, as long as only invertible rules have been used. For non-invertible rules (like weakening formulas) the best that may be done are two proof attempts: the first tries to show that the counterexample can be propagated through the rule (by proving $(\chi \wedge \underline{x} = \underline{t}) \rightarrow \neg \varphi$), the second tries to show that it surely cannot be propagated (by proving $(\chi \wedge \underline{x} = \underline{t}) \rightarrow \varphi$). If the first proof attempt succeeds, backtracing can proceed. If the second attempt succeeds, the earliest point of failure has been found, i.e. application of the rule is at least one of the reasons why the proof attempt failed. In KIV two incomplete proof attempts are done just using the built-in simplifier (these do not cost much time). If both fail, more elaborate proof attempts are optional, but rarely used. Since wrong conjectures are much more common than wrong proof decisions, one usually skips non-invertible rules and tries to compute a counterexample for the conclusion first. Only if this computed counterexample is found not to be an actual counterexample for the conclusion, because it was incorrectly propagated through a non-invertible rule, more elaborate checks are necessary.

6   Implementation

The previous sections have shown the approach for the computation of counterexamples on infinite data types. This approach is tightly integrated in the proof engine of the KIV system. When the user proves a conjecture and gets stuck at some subgoal he might start the counterexample search. This strategy takes the subgoal and tries to reduce it to false with a special set of counterexample heuristics. These heuristics expand a free target variable structurally and simplify the resulting premises. Thereafter the premises are weighted (see Sect. 6.2) and a 'good' premise, i.e. one that hopefully quickly leads to a counterexample, is chosen and structurally expanded again. This process continues as long as one premise has no more free target variables. If this premise is *false* or the user agrees, that the negation is satisfiable, the backtrace algorithm evaluates the variable assignment for the free target variables of the subgoal. Thereafter the counterexample may also be backtraced in the original proof attempt to the earliest point of failure (see Sect. 5.2).

In this section we will first discuss implementation details that are specific to the the sequent calculus and the heuristics to automate proofs used in KIV. Second we will give some heuristics to get an efficient search strategy for the counterexample proofs and finally shortly discuss extensions to Dynamic Logic and to higher-order logic.

6.1   KIV Specific Rules and Heuristics

To simplify formulas with axioms and theorems from the underlying specification KIV uses a *simplifier*. The simplifier does propositional reasoning and applies theorems from the specification as forward, elimination or conditional rewrite rules. Details on the various types of rules can be found in [9]. Additional heuristics do quantifier instantiations and unfolding of function definitions.

To optimize the proof search for non-free data types, a variant of the structural expansion rule is implemented. The modified rule adds preconditions to each premise. If variable x is expanded to $f(\underline{x})$ in a premise, inequations $f(\underline{x}) \neq x_i$ for each argument $x_i \in \underline{x}$ that has the same sort as x are added. This ensures, that the constructor does not behave like identity on some argument. E.g. when expanding a variable s for a set to $s' \cup \{a\}$, the inequality $s' \cup \{a\} \neq s'$ is added, which ensures that the element a was not already present in s'. This halves the search tree because the case of the identity is omitted.

6.2   Search Strategies for the Counterexample

Instantiating variables using structural expansion has to be done in a fair way: Every premise and every variable of each premise has to be eventually considered. KIV uses an A^* search strategy to implement a fair search. The strategy computes a weight for each open goal and always continues the counterexample search at the branch with the smallest weight. To be fair, it has to be made sure, that there is a $\delta > 0$, such that the weight of goals increases at least by δ, when another rule is applied [11]. This is easily achieved by adding some weight proportional to the branch length. Further criteria rely on the assumption, that less complicated goals lead more quickly to a counterexample.

Therefore the complexity is measured by considering the number of formulas and different variables of the goal, the number of times, the structural expansion was applied, and the complexity of the symbols corresponding to the *specification hierarchy*. KIV supports structured algebraic specifications [9] with the usual operations (union, enrichment, renaming, parameterization and actualization) which form this hierarchy. E.g. the specification of **list** (see Ex. 1) may be an enrichment of a list specification with the definition of the *sort* predicate, which may in turn be a generic specification with **elem** as parameter. Usually axioms, that define operations in specifications higher up in the hierarchy are based on symbols below in the hierarchy (like the definition of *sort* is based on elementary list constructors). Therefore symbols higher up in the hierarchy carry more weight than those lower in the hierarchy.

The expansion of a variable should reduce the complexity of the current formula. Therefore variables get a weight corresponding to the specification hierarchy of the sort, the number of occurrences (since expansion expands all occurrences), and, to get a fair strategy, a negative weight proportional to the number of times the variables have been expanded. Then the variable with the highest weight is chosen.

The exact factors used in the weights were chosen by evaluating a number of experiments. They do a good job in practice.

6.3   Extending the Method

Because the presented method for counterexample search uses a proof calculus, the approach may easily be extended. The KIV system supports program verification of imperative programs in Dynamic Logic [8]. To extend the counterexample search to Dynamic Logic, we only had to check that the program rules (mainly rules for the symbolic execution of programs) are (strong) invertible. We found that almost all are strongly invertible, with only one exception: In Hoare's invariant rule, the introduction of an incorrect program invariant leads to a not invertible rule. Therefore this rule is not used in the counterexample search for programs and backtracing through the proof stops at the invariant rule. The counterexample search for program verification is also implemented and works with good results. The basic strategy is to expand the input variables and then to execute the programs until the program formulas are reduced to pure predicate logic formulas. Then the usual counterexample search can proceed.

Another extension of the proof calculus would be to apply it to higher-order goals (KIV already uses a higher-order logic). The new proof rules (beta reduction and rules for the axioms of choice and extensionality) are not problematic, since they are strongly invertible, but higher-order variables present a problem, which we could not solve yet: since functions are not generated data types, we do not have a simple mechanism available, that can find instances of function variables.

7   Related Work and Examples

A large number of tools has been developed which try to construct finite models [12,15,14], even for first-order specifications [4]. Since our specifications usually do not have finite models at all, these tools are not directly applicable. Nevertheless an interesting alternative to our approach would be to use them by defining suitable abstractions (e.g. by collapsing all data structures with more than a fixed number of constructors into a single element). Model generator tools could also be used to check model conditions over parameter specifications, since these often can be satisfied with finite models.

The idea to incrementally search for constructor terms to generate the search space for counterexamples also appears in [6] and in [1]. The first approach is restricted to the special case of initial specifications over free data types with complete recursive definitions. In this case the satisfiability of $\neg \varphi_{\underline{x}}^{t}$ can be decided by unfolding definitions and the inequality between syntactically different constructor terms. The special case allows to prove a completeness result for the counterexample search.

The second approach in [1] is restricted to freely generated data types, and tries to integrate counterexample search with the calculus of a model generation prover.

Both approaches differ from ours in that they try to develop a fully automatic calculus for the counterexample search, while we have developed an approach that is tightly integrated with the paradigm of interactive theorem proving. Nevertheless our approach is able to solve the problems given in [6] within a few seconds. No residual formula remain, the solutions are found automatically.

Summarizing, we have developed a practically applicable approach to counterexample generation, which can deal with arbitrary data types. Below we will give a small number of examples for graphs. Graphs g are generated from the empty graph $\emptyset$ by inserting nodes a and edges $a \mapsto b$ with $g +_v a$ and $g +_e (a \mapsto b)$. Adding an edge $(a \mapsto b)$ implicitly adds both nodes a and b. The data type is not free, since $g +_v a +_v a = g +_v a$. A path in a graph will be written $[a_1, \ldots, a_n]$. Here are some conjectures over graphs and the corresponding counterexamples:

1. every graph g is acyclic:
   $C = (true, (g = \emptyset +_e (a \mapsto a)))$
2. every path x in a graph g is a shortest path (i.e. no path y is shorter):
   $C = (true, (g = \emptyset +_e (a \mapsto b), x = [a, b], y = [a]))$
3. if g has no edge of the form $(a \mapsto a)$, then it is acyclic:
   $C = (a \neq b, (g = \emptyset +_e (a \mapsto b) +_e (b \mapsto a)))$
4. two paths x and y in an acyclic graph g with same source and same destination are equal:
   $C = (a_0 \neq a_1 \wedge a_1 \neq a_2 \wedge a_0 \neq a_2, (x = [a_0, a_1], y = [a_0, a_2, a_1],$
   $\quad g = \emptyset +_e (a_0 \mapsto a_2) +_e (a_2 \mapsto a_1) +_e (a_0 \mapsto a_1)))$

Our approach solves the first two examples automatically, the other two require that the user acknowledges the model condition. The examples are typical for KIV case studies, which often use abstract data types like the one above. They are atypical, since the goals are small and only one data type with only a few axioms is involved. More examples over larger specifications can be found in [10].

8   Conclusion

We have presented an approach for the generation of counterexamples in algebraic specifications. The counterexample search is based on a sequent calculus for first-order logic (and on Dynamic Logic for program verification). The approach extends other approaches limited to free data types, and has been implemented into the specification and verification tool KIV.

The use of the existing proof calculus lead to a relative short implementation time, and allowed to reuse the already existing, well-tested and efficient strategies for theorem proving in the search for a counterexample.

The counterexample mechanism already has been successfully applied in several applications as a supporting feature to aid the proof engineer in detecting flaws in conjectures.

The applications have shown that the satisfiability of model conditions usually is decided easily by the user. Nevertheless an interesting topic for further research would be to further automate this test. Some cases are easy to automate, e.g. terms like *pred(0)* or *head(nil)* could easily be checked to be unspecified by inspecting the specification of natural numbers or lists. Model conditions for parameters like $\neg\, e_1 \leq e_2$ could be solved either by a decision procedure for total orders or by a model generation tool (finite models often are sufficient for this case). Care has to be taken, that the automation does not waste a lot of unnecessary time, since usually the variable assignment computed is already sufficient to deduce why a conjecture is wrong.

Another topic for further research are restrictions on the search space for non-free data types. Currently only the restriction mentioned at the end of Sect. 6.1 has been implemented, but commutativity constraints like $g +_v a +_v b = g +_v b +_v a$ (for graphs) or cancellation laws like $n + 1 - 1 = n$ (for integers) can also be exploited to reduce the search space.

References

1. W. Ahrendt. A basis for model computation in free data types. In *CADE-17 Workshop on Model Computation - Principles, Algorithms, Applications*, 2000.
2. M. Balser, W. Reif, G. Schellhorn, K. Stenzel, and A. Thums. Formal system development with KIV. In T. Maibaum, editor, *Fundamental Approaches to Software Engineering*, number 1783 in LNCS. Springer, 2000.
3. M. Gurevich. Evolving algebras 1993: Lipari guide. In E. Börger, editor, *Specification and Validation Methods*. Oxford University Press, 1995.
4. D. Jackson. Automating first-order relational logic. In *Proceedings of the ACM SIGSOFT 8th International Symposium on the Foundations of Software Engineering (FSE-00)*, volume 25, 6 of *ACM Software Engineering Notes*, pages pp. 130 – 139. ACM press, 2000.
5. P. D. Mosses. CoFI : The common framework initiative for algebraic specification. In H. Ehrig, F. v. Henke, J. Meseguer, and M. Wirsing, editors, *Specification and Semantics*. Dagstuhl-Seminar-Report 151, 1996. http://www.brics.dk/Projects/CoFI.
6. M. Protzen. Disproving conjectures. In D. Kapar, editor, *Automated Deduction – CADE 11*, volume 607 of *Lecture Notes in Computer Science*. Springer-Verlag, 1992.
7. W. Reif. The KIV-approach to Software Verification. In M. Broy and S. Jähnichen, editors, *KORSO: Methods, Languages, and Tools for the Construction of Correct Software – Final Report*, LNCS 1009. Springer, Berlin, 1995.
8. W. Reif, G. Schellhorn, and K. Stenzel. Interactive Correctness Proofs for Software Modules Using KIV. In *COMPASS'95 – Tenth Annual Conference on Computer Assurance*, Gaithersburg (MD), USA, 1995. IEEE press.
9. W. Reif, G. Schellhorn, K. Stenzel, and M. Balser. Structured specifications and interactive proofs with KIV. In W. Bibel and P. Schmitt, editors, *Automated Deduction—A Basis for Applications*. Kluwer, Dordrecht, 1998.
10. W. Reif, G. Schellhorn, and A. Thums. Fehlersuche in formalen Spezifikationen. Technischer Bericht 2000-06, Fakultät für Informatik, Universität Ulm, Germany, 2000. (in German).
11. S. Russell and P. Norwig. *Artificial Intelligence: A Modern Approach*. Prentice Hall, 1995.
12. J. Slaney. FINDER: Finite domain enumerator. *Lecture Notes in Computer Science*, 814:798ff, 1994.

13. V. Sperschneider and G. Antoniou. *Logic: A Foundation for Computer Science.* Addison Wesley, 1991.
14. H. Zhang. SATO: An efficient propositional prover. In *CADE-97*, 1997.
15. J. Zhang and H. Zhang. SEM: A system for enumerating models. In *IJCAI-95*, 1995.

CCE: Testing Ground Joinability

Jürgen Avenhaus and Bernd Löchner

FB Informatik, Universität Kaiserslautern, Kaiserslautern, Germany
{avenhaus,loechner}@informatik.uni-kl.de

Abstract. Redundancy elimination is a key feature for the efficiency of
saturation based theorem provers. Ground joinability is a good candidate
for a redundancy criterion but is rarely used in practice since the available
algorithms are not believed to have a good cost-benefit ratio. In order to
have a framework for the evaluation and the design of new methods for
testing ground joinability we developed the system CCE.

1 Introduction

Redundancy elimination is a key feature for the efficiency of saturation based theorem provers [BG94]. In equational theorem proving one therefore needs methods
for detecting that an equation $s = t$ is redundant w. r. t. $(R, E, \succ)$. Here $\succ$ denotes a reduction ordering, R a rewrite system with $R \subseteq \succ$, and E a set of
equations which are not orientable by $\succ$. The stronger the chosen notion of redundancy is the harder it becomes to decide, it may even become undecidable.
Ground joinability of $s = t$ w. r. t. $(R, E, \succ)$ is a good candidate for such a criterion since it can be approximated to different degrees. One reason that it is
rarely used at the moment might be that existing algorithms are not believed
to have a good balance between computational cost and detection strength in
practice. We therefore have developed the system CCE to have a framework
for the evaluation of existing and the design of new redundancy tests based on
ground joinability.

In this paper we deal with the following problem: Given $s = t$ and $(R, E, \succ)$,
is $s = t$ ground joinable (w. r. t. to the extended signature semantics)? If not,
we are interested in a set $\{e_1 \,|\, c_1, \ldots, e_n \,|\, c_n\}$ of constrained equations that gives
precise information for the reason why: Each ground instance $\sigma(s) = \sigma(t)$ that
is not provably joinable in $(R, E, \succ)$ is *covered* by some $e_i \,|\, c_i$. Our system CCE
("Covering Constrained Equations") computes such coverings for an equation
$s = t$. If $\varnothing$ is a covering of $s = t$ then $s = t$ is ground joinable. A non-empty
covering may also be helpful for improving the efficiency of theorem proving.
Note that testing ground confluence is in general computationally much easier
than deciding whether $\varnothing$ is a covering for $s = t$ in case that $(R, E, \succ)$ is not
ground confluent (which is normally the case in theorem proving).

A simple sufficient approach to this problem goes back to [MN90]. In [AHL00]
we experimented with that test for $\succ$ being a lexicographic path ordering (LPO)
or a Knuth-Bendix ordering (KBO). We could show that this kind of redundancy

R. Goré, A. Leitsch, and T. Nipkow (Eds.): IJCAR 2001, LNAI 2083, pp. 658–662, 2001.

elimination can result in considerable speed-ups of equational theorem proving, especially if AC-operators are involved in the problem.

The problem whether $s = t$ is ground joinable in $(R, E, \succ)$ is decidable if $\succ$ is an LPO [CNNR98]. The decision procedure is more involved than the method of [MN90] since it relies on a solver for LPO constraints. One may conjecture that the decision algorithm of [CNNR98] is too expensive to be useful as a redundancy criterion in theorem proving in practice. But it is natural to ask the following questions:

1. What is the potential of this way of redundancy elimination, regardless of its cost?
2. Can one weaken the decision procedure to get cheap sufficient tests for ground joinability?
3. What is the trade-off between cost and sharpness of such weakenings?
4. How do these weakenings compare with the method of [MN90]?

Motivated by these questions we have implemented our system CCE. At the moment it contains various methods based on [CNNR98] and [MN90]. Furthermore, it provides some variants of a solver for LPO constraints based on the recent [NR99]. For KBO-constraints see [KV00]. The system is written in ANSI-C and shares most of its code with our theorem prover WALDMEISTER [HJL99].

2   Notations

We use standard notations whenever possible. We start with a fixed alphabet $\mathcal{F}$ with operators $f \in \mathcal{F}$ of fixed arity and a fixed set $\mathcal{V}$ of variables. Then $\mathrm{Term}(\mathcal{F}, \mathcal{V})$ is the set of terms and $\mathrm{Term}(\mathcal{F})$ is the set of ground terms. Let $\mathcal{F}^e = \mathcal{F} \cup \{0, \mathsf{succ}\}$ be the extended signature where $0, \mathsf{succ} \notin \mathcal{F}$ and $0, \mathsf{succ}$ of arity 0 and 1, respectively. Then σ is an (extended) ground substitution if $\sigma(x) \in \mathrm{Term}(\mathcal{F}^e)$ for all $x \in \mathrm{dom}(\sigma)$. An equation e is $s = t$ with $s, t \in \mathrm{Term}(\mathcal{F}, \mathcal{V})$. Let $\succ$ be a reduction ordering on $\mathrm{Term}(\mathcal{F}^e, \mathcal{V})$ which is total on $\mathrm{Term}(\mathcal{F}^e)$. For an LPO we require $f >_{\mathcal{F}} \mathsf{succ} >_{\mathcal{F}} 0$ in the precedence for each $f \in \mathcal{F}$. We write $(R, E, \succ)$ to denote sets of equations R and E with $R \subseteq \succ$. Then $\rightarrow = \rightarrow_{R,E,\succ}$ is the rewrite relation induced by R and by instances of equations in E that are orientable by $\succ$. We write $s \downarrow t$ to denote joinability in $(R, E, \succ)$ and we write $s \Downarrow t$ if $\sigma(s) \downarrow \sigma(t)$ for each ground substitution σ. In this case $s = t$ is called (extended) ground joinable. Then $s = t$ is ground joinable w. r. t. each signature extension $\mathcal{F}_0$ of $\mathcal{F}$. A constraint c is a Boolean expression (using $\wedge, \vee, \neg$) of atoms $s = t$ and $s > t$ where $s, t \in \mathrm{Term}(\mathcal{F}, \mathcal{V})$ or c is $\top$ (denoting $x = x$) or $\bot$ (denoting $\neg x = x$). We write $\sigma \models c$ if the ground substitution σ satisfies c. A constrained equation has the form $e \,|\, c$. We call $s = t \,|\, c$ trivial if either $s \equiv t$ or c is unsatisfiable.

3   Coverings of Constrained Equations

We first make precise what we mean by a covering.

Definition 1. *Let $c \mid c$ be a constrained equation where e is $s = t$. A covering of $e \mid c$ is a set CE of constrained equations such that for each ground substitution σ such that $\sigma \models c$ either $\sigma(s) \downarrow \sigma(t)$ or $\sigma(e) \overset{*}{\to} \sigma(e')$ and $\sigma \models c'$ for some $e' \mid c'$ in CE. CE' is a* refinement *of CE if CE' is a covering for each $e \mid c$ in CE.*

3.1   Computing Coverings with Constraints on Variables

The idea of [MN90] is to consider all constraints of the form $x_{\pi(i)} \; \varrho_i \; x_{\pi(i+1)}$ for all $i = 1, \ldots, n$. Here $\mathrm{Var}(s,t) = \{x_1, \ldots, x_n\}$ is the set of variables of $s = t$, π is a permutation and $\varrho_i \in \{\succ, =\}$. If $s = t$ is joinable under all these constraints, then $s = t$ is ground joinable. The appealing aspect of this method is that it is quite easy to implement both for LPO and KBO. The implementations of the orderings have to be slightly extended to admit the constraints on the variables. No solver for ordering constraints is needed. Unfortunately, the number of cases grows exponentially with the number of variables. We have therefore refined the method such that a small subset $V_0 \subseteq \mathrm{Var}(s,t)$ is initially chosen and is successively extended if ground joinability could not be shown. This was essential to make the method fast in practice. A further drawback is that the chosen constraints may be too weak to make a variable comparable to another subterm.

3.2   Computing Coverings with Arbitrary Constraints

Employing full ordering constraints enables a more powerful test. The set of coverings can be refined incrementally by analyzing potential rewrite steps. In [CNNR98] a confluence tree is used to organize the refinements:

$CE \cup \{e \mid c\} \vdash CE \cup CE_0$ iff CE_0 results from $e \mid c$ by

(1) Constrained Rewriting or (2) Decomposition or (3) Instantiation.

On a more concrete level of our implementation, we basically have three macros reflecting practical considerations. The first macro is Normalization, i. e., $e \mid c$ is simplified to $e' \mid c$ by $e \overset{*}{\to}_c e'$ where $\to_c$ means rewriting in $(R, E, \succ_c)$ where $\succ_c$ is $\succ$ enhanced by c as described in [NR99]. The second macro CRew is just Constrained Rewriting as in [CNNR98] which introduces new constraints. The third macro is Splitting $e \mid c$ into $e_i \mid c_i$ for $i = 1, \ldots, n$, where c_i is in solved form. It combines Decomposition and Instantiation.

We have implemented two strategies: In strategy CRew-Split macro CRew has priority over macro Splitting. In strategy Split-CRew it is vice versa. In both strategies the macro Normalization has highest priority.

3.3   Constraint Solving

The macro Splitting as well as the test for satisfiability is realized by a constraint solver. It is well known that checking LPO-constraints for satisfiability is NP-hard [N93]. So it is interesting to see how costly it is in practice. From the algorithms that are known from the literature we chose the method recently devised in [NR99]. It is there shown to be far superior in comparison to [N93].

The algorithm works roughly as follows: The given constraint c is *decomposed* in a way which is analogous to the definition of the LPO. By a restricted form of transitivity new *consequences* are added. To ensure the termination of this process a special notion of *redundancy* is used which can also be used to detect early unsatisfiability. The resulting normal forms are called *solved forms*. The constraint c is satisfiable iff there is a solved form which contains no *variable cycle* such as $(x > g(y)) \wedge (y > h(x))$.

As is proposed in [NR99] we use a backtracking approach to implement this technique. This allows a rather quick test for satisfiability; continuing the backtrack search enables collecting all cycle-free solved forms. This is important for our application. There are some variations in organizing the search, such as adding new consequences as early or as late as possible, applying the checks for redundancy/unsatisfiability or variable cycles in different frequencies and so on. A specialized memoization technique enables us to share such redundancy information between different branches of the search. Our implementation in ANSI-C is about ten times faster than the PROLOG-program of [NR99].

4   Experimental Evaluation

For evaluating the different algorithms we used test-sets generated by WALD-MEISTER during proving 268 examples taken from the TPTP library [SS98]. The recorded equations had passed different redundancy tests and the prover would directly profit from showing them ground joinable. In Table 1 we give the data for six representative examples and for the whole test set. Besides the number of tests a problem contains we show the number of equations which could be shown to be ground joinable by any of the three methods. The figures given in per cent tell how many of them were found by the respective method.

As we can see, the potential of ground joinability as a redundancy test lies between 2 and 30 per cent depending on the domain of the proof task. The method Split-CRew is in general stronger than CRew-Split which takes much

Table 1. Comparison of three different methods for testing ground joinability: Variable constraints [MN90] vs. two variants of confluence trees [CNNR98], cf. Sect. 3.2 (running times for the tests in seconds on a SPARC UltraII/333 MHz).

Problem	number of all tests	found ground joinable	var. constraints strength	time	Split-CRew strength	time	CRew-Split strength	time
BOO007-2	2 097	337	82 %	1.5	96 %	4.7	90 %	36.4
GRP187-1	1 599	103	63 %	1.2	73 %	9.8	61 %	41.0
LAT023-1	566	119	78 %	0.7	92 %	2.0	82 %	15.9
RNG009-7	978	22	18 %	0.1	95 %	0.7	86 %	1.5
RNG027-5	455	61	43 %	13.7	69 %	9.3	89 %	414.0
ROB006-1	1 704	512	26 %	0.5	96 %	5.6	83 %	27.3
268 Ex.	109 632	12 011	61 %	254.0	83 %	999.0	76 %	11 264.0

more time. The exception is the domain of non-associative rings: Here CRew-Split is stronger at the price of a really large running time. By limiting the branching width or the depth of the confluence tree we made weakenings of both methods. This influenced especially the method CRew-Split: With a limit of 8 for the height of the tree the running time drops down to a quarter while the strength is still at 60 % (considering all 268 problems). With a limit of 6 the strength is at 50 % and the running time descends from over 11 000 to about 1 500 seconds For Split-CRew the influence is not so pronounced, but here too, with weakening the test the running time drops faster than the strength of the test. For example, limiting the branching width to 7 the strength is still at 77 % with the running time going down from about 1 000 to 700 seconds. On the over all, the method of [MN90] has the best cost-benefit ratio, but in some domains it is rather weak.

The constraint solver is pretty fast in practice, at least it can cope well with the problems generated by the modules using the solver. We made detailed measurements for the six examples of Table 1. More than 90 % of the calls are finished within one millisecond and only a tiny fraction needs more than 10 milliseconds. Analyzing the longer runs, we noted that there is still room for improvements. Especially, when the solver is used in the collecting mode and the investigated constraint is highly nonlinear, identical solved forms may be generated dozens of times.

References

[AHL00]   J. Avenhaus, T. Hillenbrand, B. Löchner (2000). On Using Ground Join-able Equations in Equational Theorem Proving. In *Proc. FTP 2000*, Fachberichte Informatik 5/2000, Universität Koblenz-Landau. Extended version submitted for publication.

[BG94]   L. Bachmair and H. Ganzinger (1994). Rewrite-based Equational Theorem Proving with Selection and Simplification. J. Logic and Computation, **4**, pages 217–247.

[CNNR98]   H. Comon, P. Narendran, R. Nieuwenhuis, and M. Rusinowitch (1998). Decision Problems in Ordered Rewriting. In *Proc. 13th LICS*, IEEE Computer Society, pages 276–286.

[HJL99]   T. Hillenbrand, A. Jaeger, and B. Löchner (1999). System description: WALDMEISTER – Improvements in Performance and Ease of Use. In *Proc. CADE-16*, LNAI 1632, pages 232–236.

[KV00]   K. Korovin and A. Voronkov (2000). A Decision Procedure for the Existential Theory of Term Algebras with the Knuth-Bendix Ordering. In *Proc. LICS 2000*, IEEE Computer Society, pages 291–302.

[MN90]   U. Martin and T. Nipkow (1990). Ordered rewriting and confluence. In *Proc. CADE-10*, LNCS 449, pages 366–380.

[N93]   Robert Nieuwenhuis (1993). Simple LPO constraint solving methods. Information Processing Letters, **47**, pages 65–69.

[NR99]   R. Nieuwenhuis and J. M. Rivero (1999). Solved Forms for Path Ordering Constraints. In *Proc. 10th RTA*, LNCS 1631, pages 1–15.

[SS98]   G. Sutcliffe and G. Suttner (1998). The TPTP Problem Library. CNF Release v.1.2.1. J. Automated Reasoning, **21**, pages 177–203.

System Description: RDL
Rewrite and Decision Procedure Laboratory

Alessandro Armando[1], Luca Compagna[1], and Silvio Ranise[1,2]

[1] DIST–Università degli Studi di Genova, via all'Opera Pia 13, 16145, Genova, Italy,
{armando, compa, silvio}@dist.unige.it
[2] LORIA-INRIA-Lorraine, 615, rue du Jardin Botanique, BP 101, 54602 Villers les
Nancy Cedex, France

1 Introduction

RDL[1] simplifies clauses in a quantifier-free first-order logic with equality using a
tight integration between rewriting and decision procedures. On the one hand,
this kind of integration is considered the key ingredient for the success of state-
of-the-art verification systems, such as ACL2 [10], STEP [8], Tecton [9], and
Simplify [7]. On the other hand, obtaining a principled and effective integration
poses some difficult problems. Firstly, there are no formal accounts of the incorporation of decision procedures in rewriting. This makes it difficult to reason
about basic properties such as soundness and termination of the implementation of the proposed schema. Secondly, most integration schemas are targeted
to a given decision procedure and they do not allow to easily plug new decision
procedures in the rewriting activity. Thirdly, only a tiny portion of the proof
obligations arising in many practical verification efforts falls exactly into the
theory decided by the available decision procedure. **RDL** solves the problems
above as follows:

1. **RDL** is based on CCR (Constraint Contextual Rewriting) [1,2], a formally
   specified integration schema between (ordered) conditional rewriting and a
   satisfiability decision procedure [11]. **RDL** inherits the properties of soundness [1] and termination [2] of CCR. It is also fully automatic.
2. **RDL** is an open system which can be modularly extended with new decision
   procedures provided these offer certain interface functionalities (see [2] for
   details).
   In its current version, **RDL** offers *'plug-and-play' decision procedures* for the
   theories of Universal Presburger Arithmetic over Integers (UPAI), Universal
   Theory of Equality (UTE), and UPAI extended with uninterpreted function
   symbols [13].
3. **RDL** implements instances of a *generic extension schema* for decision procedures [3]. The key ingredient of such a schema is a *lemma speculation
   mechanism* which 'reduces' the validity problem of a given theory to the

---

[1] The system is available via the *Constraint Contextual Rewriting Project Home Page*
at http://www.mrg.dist.unige.it/ccr.

R. Goré, A. Leitsch, and T. Nipkow (Eds.): IJCAR 2001, LNAI 2083, pp. 663–669, 2001.
© Springer-Verlag Berlin Heidelberg 2001

validity problem of one of its sub-theories for which a decision procedure is available. The proposed mechanism is capable of generating lemmas which are entailed by the union of the theory decided by the available decision procedure and the facts stored in the current context. Three instances of the extension schema lifting a decision procedure for UPAI are available. First, *augmentation* copes with user-defined functions whose properties can be expressed by conditional lemmas. Second, *affinization* is a mechanism for the 'on-the-fly' generation of lemmas to handle a significant class of formulae in the theory of Universal Arithmetic over Integers (UAI). Third, a *combination* of augmentation and affinization puts together the flexibility of the former with the automation of the latter. Finally, **RDL** can be extended with new lemma speculation mechanisms provided these meet certain requirements (see [3] for details).

Since extensions of quantifier-free first-order logic with equality are useful in practically all verification efforts, **RDL** can be seen as an open reasoning module which can be integrated in larger verification systems. In fact, most state-of-the-art verification systems feature similar components, e.g. ACL2's simplifier, STEP validity checker, Tecton's integration of contextual rewriting and a decision procedure for UPAI, and *Simplify* developed within the Extended Static Checking project.

2 A Motivating Example

Consider the problem of showing the termination of a function to normalize conditional expressions in propositional logic as described in Chap. IV of [5]. The argument in the proof of termination is based on exhibiting a measure function that decreases (according to a given ordering) at each function's recursive call. ms (reported in [12]) is one such function: $ms(a)=1$ and $ms(\mathsf{If}(x,y,z))=ms(x) + ms(x)ms(y)+ ms(x)ms(z)$, where If is the ternary constructor for non-atomic conditional expressions, a is an atomic conditional expression, x, y, and z are conditional expressions, and juxtaposition denotes multiplication. One of the proof obligation formalizing the 'decreaseness' argument above is

$$ms(\mathsf{If}(u, \mathsf{If}(v,y,z), \mathsf{If}(w,y,z))) < ms(\mathsf{If}(\mathsf{If}(u,v,w),y,z)), \tag{1}$$

where $<$ is the 'less-than' relation over integers. In order to prove the validity of (1), we check the unsatisfiability of its negation. By rewriting the l.h.s. and the r.h.s. (of the negation of (1)) with the definition of ms, we obtain:

$$\begin{aligned} u + uv + uvy + uvz + uw + uwy + uwz \geq \\ u + uv + uw + uy + uvy + uwy + uz + uvz + uwz \end{aligned} \tag{2}$$

where u, v, y, w and z abbreviates $ms(u)$, $ms(v)$, $ms(y)$, and $ms(z)$, respectively. Then, we perform all the possible cancellations in (2) and we obtain:

$$ms(u)ms(y) + ms(u)ms(z) \leq 0. \tag{3}$$

We assume the availability of the following two facts:

$$ms(E) > 0, \qquad\qquad (4)$$
$$(X > 0 \wedge Y > 0) \Rightarrow XY > 0. \qquad\qquad (5)$$

for each conditional expression E and for each pair of numbers X and Y. Then, consider the following two instances of (5), obtained by matching the conclusion of (5) with the first and second summand in the l.h.s. of (3):

$$(ms(u) > 0 \wedge ms(y) > 0) \Rightarrow ms(u)ms(y) > 0 \qquad\qquad (6)$$
$$(ms(u) > 0 \wedge ms(z) > 0) \Rightarrow ms(u)ms(z) > 0. \qquad\qquad (7)$$

In order to relieve the hypotheses of (6) and (7), it is sufficient to instantiate (4) three times, namely $ms(u) > 0$, $ms(y) > 0$, and $ms(z) > 0$. Finally, it is trivial to detect the unsatisfiability of (3) and the conclusions of (6) and (7).

Three (cooperating) reasoning capabilities are required to automate the above reasoning: (*i*) *rewriting*, (*ii*) *satisfiability checking* and *normalization in a given theory*, and (*iii*) *ground lemma speculation* (in a sense that will be made clear later). The first is used to simplify formulae, e.g. unfolding the definition of ms in (1). The second presents two aspects: the simplification of a literal, e.g. canceling out common terms in (2), and the check for the unsatisfiability of (conjunctions of) literals, e.g. the conclusions of lemmas (6) and (7) with (3). The third is the capability of supplying instances of valid facts to (partially) interpret user-defined function symbols occurring in the current formula, e.g. two instances of (4) are used to relieve hypotheses of (6).

3   Architecture

RDL features a tight integration (based on CCR) of three modules implementing the reasoning capabilities mentioned above: a module for *ordered conditional rewriting*, a *satisfiability decision procedure*, and a module for *lemma speculation*.

In the following, let cl be the clause to be simplified and p be a literal in cl which is going to be rewritten. The *context* C associated to p is the conjunction of the negation of the literals occurring in cl except p. Let T be the theory decided by the decision procedure.

The decision procedure. For efficiency reasons, this module is state-based, incremental, and resettable [11]. The context C is stored by a specialized data structure in the state of the decision procedure. There are three functionalities. First, `cs-unsat` characterizes a set of inconsistent (in T) contexts whose inconsistency can be checked by means of computationally inexpensive checks. Second, given a literal l and the current context C, `cs-simp` computes the new context C' resulting from the addition of l to C in such a way that C' is entailed by the conjunction of l and C in T. Third, `cs-norm` computes a normal representation p' of p w.r.t. T and the information stored in C. This functionality must be compatible with rewriting, i.e. it is required that $p' \prec p$ where $\prec$ denotes a total term ordering on ground literals. As an example, a decision procedure for UPAI can implement cs-norm by collecting like terms in literals whose top-most predicate symbol is $<$ and rewriting the resulting literal by using the equalities entailed by the current context.

Constraint Contextual Rewriting. The rewriter provides the functionality ccr. It handles conditional rules of the form $h_1 \wedge \cdots \wedge h_n \Rightarrow (l = r)$, where l and r are terms, and $h_1, ..., h_n$ are literals. Assume $r\sigma \prec l\sigma$ for a ground substitution σ (otherwise, if $l\sigma$ is different from $r\sigma$, swap l with r in the following). Given $p[l\sigma]$, ccr returns $p[r\sigma]$ if $h_1\sigma, ..., h_n\sigma$, and $p[r\sigma]$ are smaller (w.r.t. $\prec$) than $p[l\sigma]$, and for $i = 1, ..., n$ either $h_i\sigma$ is (recursively) rewritten to *true* by invoking ccr[2] or by checking whether $h_i\sigma$ is entailed by C (this is done by invoking cs-unsat so to check that the negation of $h_i\sigma$ is inconsistent with C). There are two other means of rewriting. Firstly, p is rewritten to *false* (*true*) if cs-unsat checks that p (the negation of p, resp.) is inconsistent with C. Secondly, p is rewritten to p' if p' has been obtained by invoking cs-norm.

Lemma speculation. Three instances of the lemma speculation mechanism described in [3] are implemented in **RDL**. All the instances share the goal of feeding the decision procedure with new facts about function symbols which are otherwise uninterpreted in T. More precisely, they inspect the context C and return a set of ground facts entailed by C using T as the background theory. Furthermore, these facts must enjoy some properties to ensure termination (see [3,2] for details).

The simplest form of lemma speculation is augment [6,1,2,3], which consists of selecting and instantiating lemmas from a set of available valid formulae in order to obtain ground facts whose conclusions can be readily used by the decision procedure. As an example, consider a decision procedure for UPAI implemented by means of the Fourier-Motzkin method. Here the basic operation is to eliminate one variable from a set of inequalities by means of a linear combination of two inequalities. Then, augment finds instances of the conclusions among the conditional lemmas which can promote further variable eliminations. There are two crucial problems. Firstly, we must relieve hypotheses of lemmas in order to be able to send their conclusions to the decision procedure. We solve this problem by rewriting each hypothesis to *true* (if possible). This is done by invoking ccr and it implies that the rewriter and the decision procedure are mutually recursive. The other problem is the presence of extra variables in the hypotheses (w.r.t. the conclusion) of lemmas. **RDL** avoids this problem by requiring that the conclusion contains all the variables occurring in the lemma and that all the variables get instantiated by matching the conclusion of the lemma against the largest (according to $\prec$) literal in C. As an example of how augment works, recall that (6) and (7) are generated by matching the conclusion of lemma (5) against (3) twice.

If a suitable set of lemmas is defined, augment increases dramatically the effectiveness of the decision procedure. Unfortunately, devising such a suitable

---

[2] **RDL** performs no case splitting while releaving hypotheses of conditional lemmas.

set is a time consuming activity. This problem can be solved in some important special cases. In the actual version of **RDL**, `affinize` implements the 'on-the-fly' generation of lemmas about multiplication over integers. To understand how `affinize` works, consider the non-linear inequality $XY \leq -1$ (where X and Y range over integers). By resorting to its geometrical interpretation, it is easy to verify that $XY \leq -1$ is equivalent to $(X \geq 1 \wedge Y \leq -1) \vee (X \leq -1 \wedge Y \geq 1)$. To avoid case splitting, we observe that the semi-planes represented by $X \geq 1$ and $X \leq -1$ as well as those represented by $Y \leq -1$ and $Y \geq 1$ are non-intersecting. This allows to derive the following four lemmas: $X \geq 1 \Rightarrow Y \leq -1$, $X \leq -1 \Rightarrow Y \geq 1$, $Y \geq 1 \Rightarrow X \leq -1$, and $Y \leq -1 \Rightarrow X \geq 1$. This process can be generalized to non-linear inequalities which can be put in the form $XY \leq K$ (where K is an integer) by factorization. The generated (conditional) lemmas are used as for `augment`.

On the one hand `affinize` can be seen as a significant improvement over `augment` since it does not require any user intervention. On the other hand it fails to apply when inequalities cannot be transformed into a form suitable for affinization. **RDL** combines augmentation and affinization by considering the function symbols occurring in the context C, i.e. the top-most function symbol of the largest (according to $\prec$) literal in C triggers the invocation of either affinization or augmentation.

4 Experiments

RDL must be judged w.r.t. its effectiveness in simplifying (and possibly checking the validity of) proof obligations arising in practical verification efforts where decision procedures play a crucial role. Hence, standard benchmarks for theorem provers (e.g. TPTP) are not in the scope of **RDL**. We are currently building a corpus of proof obligations extracted from the literature as well as examples available for similar components integrated in verification systems. The problems selected for the corpus are *representative* of disparate verification scenarios and are considered *difficult* for current state-of-the-art verification systems.

Table 1 reports the results of our computer experiments. PROBLEM lists the available lemmas[3] (if any) and the formula to be decided. $\vdash$ is the binary relation characterizing the deductive capability of **RDL** (we have that $\vdash$ is contained in $\models_T$, where T is the theory decided by the available decision procedure extended with the available facts). The last column record the successful attempt (time is expressed in msec) to solve a problem by **RDL**.[4]

RDL solves problems 1 and 2 with a decision procedure for UTE. In the former, the decision procedure is used to derive equalities entailed by the context which are used as rewrite rules and enable the use of the available lemma. The ordered rewriting engine implemented by **RDL** is a key feature to successfully solve problem 2 since this form of rewriting allows to handle usually non-orientable

---

[3] Capitalized letters denote implicitly universally quantified variables.
[4] Benchmarks run on a 600 MHz Pentium III running Linux. **RDL** is implemented in Prolog and it was compiled using Sicstus Prolog, version 3.8.

Table 1. Experimental Results

#	PROBLEM	RDL
1	$f(A) = f(B) \Rightarrow (r(g(A,B),A) = A) \vdash$ $r(g(y,z),x) = x \vee \neg(g(x,y) = g(y,z)) \vee \neg(y = x)$	26
2	$A * B = B * A, (\neg(C = 0)) \Rightarrow (rem(C * D, C) = 0) \vdash$ $rem(y * z, x) = 0 \vee \neg(x * y = z * y) \vee x = 0$	109
3	$(A > 0) \Rightarrow (rem(A * B, A) = 0) \vdash rem(x * y, x) = 0 \vee x \leq 0$	12
4	$min(A) \leq max(A) \vdash$ $\neg(k \geq 0) \vee \neg(l \geq 0) \vee \neg(l \leq min(b)) \vee \neg(0 < k) \vee l < max(b) + k$	12
5	$(memb(A,B)) \Rightarrow (len(del(A,B)) < len(B)) \vdash$ $\neg(w \geq 0) \vee \neg(k \geq 0) \vee \neg(z \geq 0) \vee \neg(v \geq 0) \vee \neg(memb(z,b))$ $\vee \neg(w + len(b) \leq k) \vee w + len(del(z,b)) < k + v$	17
6	$(0 < A) \Rightarrow (B \leq A * B), 0 < ms(C) \vdash$ $ms(c) + ms(d)^2 + ms(b)^2 < ms(c) + ms(b)^2 + 2ms(d)^2 * ms(b) + ms(d)^4$	72
7	$A \geq 4 \Rightarrow (A^2 \leq 2^A) \vdash \neg(c \geq 4) \vee \neg(b \leq c^2) \vee \neg(2^c < b)$	14
8	$(max(A,B) = A) \Rightarrow (min(A,B) = B), (p(C)) \Rightarrow (f(C) \leq g(C)) \vdash$ $\neg(p(x)) \vee \neg(z \leq f(max(x,y))) \vee \neg(0 < min(x,y)) \vee \neg(x \leq max(x,y)) \vee$ $\neg(max(x,y) \leq x) \vee z < g(x) + y$	114
9	$0 < ms(C) \vdash$ $ms(c) + ms(d)^2 + ms(b)^2 < ms(c) + ms(b)^2 + 2ms(d)^2 * ms(b) + ms(d)^4$	63
10	$\vdash x \geq 0 \Rightarrow x^2 - x + 1 \neq 0$	40

rewrite rules such as $A * B = B * A$. **RDL** solves problem 3 with a decision procedure for UPAI. In fact, the available lemma is applied once its instantiated condition, namely $x > 0$, is relieved by the decision procedure (it is straightforward to check the inconsistency of $x > 0$ and the literal $x \leq 0$ in the context). **RDL** solves problems 4, 5, 6, and 7 with a decision procedure for UPAI and augment. In particular, the formula of problem 6 is a non-linear formula whose validity is successfully established by **RDL** in a similar way of the example in Section 2. **RDL** solves problem 8 with the combination of a decision procedure for UPAI and for UTE. **RDL** solves problems 9 and 10 with the combination of a decision procedure for UPAI, augment and affinize. The lemma about multiplication (i.e. $0 < I \Rightarrow J \leq I * J$) is supplied in problem 6 but it is not in problem 9. Only the combination of augment and affinize can solve problem 9. Finally, problem 10 shows the importance of the context in which proof obligations are proved (since **RDL** does not case-split). In fact, without $x \geq 0$ augment and affinize would not be able to solve problem 10. This shows the importance of the context in which proof obligations are proved (since **RDL** does not case-split).

As a matter of fact, the online version of STeP fails to solve all of the problems reported in Table 1. However, most of the problems are successfully solved by the improved version of STeP described in [4]. *Simplify* successfully solves problems 1 to 8 thanks to a Nelson-Oppen combination of decision procedure and an incomplete matching algorithm which is capable of instantiating (valid) universally quantified clauses. However, it does not solve problems 9 and 10 since it is unable to handle non-linear facts without user-supplied lemmas (such as, e.g., $0 < I \Rightarrow J \leq I * J$ in problem 6). Finally, SVC fails to solve all the problems involving augmentation and affinization since it does not provide a mechanism to take into account facts which partially interpret user-defined function symbols.

References

1. A. Armando and S. Ranise. Constraint Contextual Rewriting. In *Proc. of the 2nd Intl. Workshop on First Order Theorem Proving (FTP'98), Vienna (Austria)*, pages 65–75, 1998.
2. A. Armando and S. Ranise. Termination of Constraint Contextual Rewriting. In *Proc. of the 3rd Intl. Workshop on Frontiers of Combining Systems (FroCos'2000), LNCS 1794*, pages 47–61, March 2000.
3. A. Armando and S. Ranise. A Practical Extension Mechanism for Decision Procedures: the Case Study of Universal Presburger Arithmetic. *J. of Universal Computer Science (Special Issue: Formal Methods and Tools)*, 7(2):124–140, 2001.
4. N. S. Bjørner. *Integrating Decision Procedures for Temporal Verification*. PhD thesis, Computer Science Department, Stanford University, 1998.
5. R.S. Boyer and J S. Moore. *A Computational Logic*. Academic Press, 1979.
6. R.S. Boyer and J S. Moore. Integrating Decision Procedures into Heuristic Theorem Provers: A Case Study of Linear Arithmetic. *Machine Intelligence*, 11:83–124, 1988.
7. D. L. Detlefs, G. Nelson, and J. Saxe. Simplify: the ESC Theorem Prover. Technical report, DEC, 1996.
8. Z. Manna *et. al.* STEP: The Stanford Temporal Prover. Technical Report CS-TR-94-1518, Stanford University, June 1994.
9. D. Kapur, D.R. Musser, and X. Nie. An Overview of the Tecton Proof System. *Theoretical Computer Science*, Vol. 133, October 1994.
10. M. Kaufmann and J S. Moore. Industrial Strength Theorem Prover for a Logic Based on Common Lisp. *IEEE Trans. on Software Engineering*, 23(4):203–213, April 1997.
11. G. Nelson and D. Oppen. Fast Decision Procedures Based on Congruence Closure. *J. of the ACM*, 27(2):356–364, April 1980.
12. L. C Paulson. Proving termination of normalization functions for conditional expressions. *J. of Automated Reasoning*, pages 63–74, 1986.
13. R.E. Shostak. Deciding Combination of Theories. *Journal of the ACM*, 31(1):1–12, 1984.

lolliCoP – A Linear Logic Implementation of a Lean Connection-Method Theorem Prover for First-Order Classical Logic

Joshua S. Hodas and Naoyuki Tamura*

Department of Computer Science
Harvey Mudd College
Claremont, CA 91711, USA
hodas@cs.hmc.edu,tamura@kobe-u.ac.jp

Abstract. When Prolog programs that manipulate lists to manage a collection of resources are rewritten to take advantage of the linear logic resource management provided by the logic programming language Lolli, they can obtain dramatic speedup. Thus far this has been demonstrated only for "toy" applications, such as n-queens. In this paper we present such a reimplementation of the lean connection-calculus prover leanCoP and obtain a theorem prover for first-order classical logic which rivals or outperforms state-of-the-art provers on a significant body of problems.

1 Introduction

The development of logic programming languages based on intuitionistic [11] and linear logic [6] has been predicated on two principal assumptions. The first, and the one most argued in public, has been that, given the increased expressivity, programs written in these languages are more perspicuous, more natural, and easier to reason about formally. The second assumption, which the designers have largely kept to themselves, is that by moving the handling of various program features into the logic, and hence from the term level to the formula level, we would expose them to the compiler, and, thus, to optimization. In the end, we believed, this would yield programs that executed more efficiently than the equivalent program written in more traditional logic programming languages. Until now, this view has been downplayed as most of these new languages have thus far been implemented only in relatively inefficient, interpreted systems.

With the recent development of compilers for languages such as λ-Prolog [13] and Lolli [7], however, we are beginning to see this belief justified. In the case of Lolli, we are focused on logic programs which have used a term-level list as a sort of bag from which items are selected according to some rules. In earlier work we showed that when such code is rewritten in Lolli, allowing the elements in the list to instead be stored in the proof context –with the underlying rules

---

* This paper reports work done while the second author was on a sabbatical-leave from Kobe University.

R. Goré, A. Leitsch, and T. Nipkow (Eds.): IJCAR 2001, LNAI 2083, pp. 670–684, 2001.

of linear logic managing their consumption– substantial speedups can occur. To date, however, that speedup has been demonstrated only on the execution of simple, "toy" applications, such as an n-queens problem solver [7].

Now we have turned our attention to a more sophisticated application: theorem proving. We have reimplemented the leanCoP connection-calculus theorem prover of Otten and Bibel [14] in Lolli. This "lean" theorem prover has been shown to have remarkably good performance relative to state-of-the-art systems, particularly considering that it is implemented in just a half-page of Prolog code. The reimplemented prover, which we call lolliCoP, is of comparable size, and, when compiled under LLP (the reference Lolli compiler [7]), provides a speedup of 40% over leanCoP. On many of the hardest problems that both can solve, it is roughly the same speed as the OTTER theorem prover [8]. (Both lean-CoP and lolliCoP solve a number of problems that OTTER cannot. Conversely, OTTER solves many problems that they cannot. On simpler problems that both solve, Otter is generally much faster than leanCoPand lolliCoP.)

While this is a substantial improvement, it is not the full story. LLP is a relatively naive, first-generation compiler and run-time system. Whereas, it is being compared to a program compiled in a far more mature and optimized Prolog compiler (SICStus Prolog 3.7.1). When we adjust for this difference, we find that lolliCoP is more than twice as fast as leanCoP, and solves (within a limited time allowance) more problems from the test library. Also, when the program is rewritten in Lolli, two simple improvements become obvious. When these changes are made to the program, performance improves by a further factor of three, and the number of problems solved expands even further.

1.1   Organization

The remainder of this paper is organized as follows: Section 2 gives a brief introduction to the connection calculus for first-order classical logic; Section 3 describes the leanCoP theorem prover; Section 4 gives a brief introduction to linear logic, Lolli, and the LLP compiler; Section 5 introduces lolliCoP; Section 6 presents the results and analysis of various performance tests and comparisons; and, Section 7 presents the two optimizations mentioned above.

2   Connection-Calculus Theorem Proving

The connection calculus [2] is a matrix proof procedure for clausal first-order classical logic. (Variations have been proposed for other logics, but this is its primary application.) The calculus, which uses a positive representation, proving matrices of clauses in disjunctive normal form, has been utilized in a number of theorem proving systems, including KOMET [3], SETHEO and E-SETHEO [9, 12]. It features two principal rules, *extension* and *reduction*. The *extension* step, which corresponds roughly to backchaining, consists of matching the complement of a literal in the active goal clause with the head of some clause in the matrix. The body of that clause is then proved, as is the remainder of the original clause.

$$\frac{\Gamma \mid \emptyset \vdash A_1, \ldots, A_n}{\Gamma} \; (start)$$
(provided $C \in \Gamma$, $C[t/x] = \{A_1, \ldots, A_n\}$ for some t, $n \geq 0$) $\dfrac{}{\Gamma \mid \Pi \vdash} \; (extension_0)$

$$\frac{\Gamma \mid L_i, \Pi \vdash L_{11}, \ldots, L_{1m} \quad C, \Gamma \mid \Pi \vdash L_1, \ldots, L_{i-1}, L_{i+1}, \ldots, L_n}{C, \Gamma \mid \Pi \vdash L_1, \ldots, L_n} \; (extension_1)$$
(provided C is ground, $C = \{\overline{L_i}, L_{11}, \ldots, L_{1m}\}$, $1 \leq i \leq n$, and $m \geq 0$)

$$\frac{C, \Gamma \mid L_i, \Pi \vdash L_{11}, \ldots, L_{1m} \quad C, \Gamma \mid \Pi \vdash L_1, \ldots, L_{i-1}, L_{i+1}, \ldots, L_n}{C, \Gamma \mid \Pi \vdash L_1, \ldots, L_n} \; (extension_2)$$
(provided C is not ground, $C[t/x] = \{\overline{L_i}, L_{11}, \ldots, L_{1m}\}$ for some t, $1 \leq i \leq n$, and $m \geq 0$)

$$\frac{\Gamma \mid \overline{L_i}, \Pi \vdash L_1, \ldots, L_{i-1}, L_{i+1}, \ldots, L_n}{\Gamma \mid \overline{L_i}, \Pi \vdash L_1, \ldots, L_n} \; (reduction, 1 \leq i \leq n)$$

Fig. 1. A deduction system for the derivation relation of the Connection Calculus

For the duration of the proof of the body of the matching clause, however, the literal that matched is added to a secondary data structure called the *path*. If at a later point the complement of a literal being matched occurs in the path, that literal need not be proved. This short-circuiting of the proof constitutes the *reduction* step. Search terminates when the goal clause is empty. Finally, note that in the *extension* step, if the clause matched is ground, it is removed from the matrix during the subproof.

Figure 1 shows a deduction system for the derivation relation. Two versions of the *extension* rule are given, depending on whether the matched clause is ground or not. A third version handles the termination case. In the core rules of this system, the left-hand side of the derivation has two parts: the matrix, Γ, is a multiset of clauses; the path, Π, is a multiset of literals. The goal clause on the right-hand side is a sequence of literals. Note that the calculus is more general than necessary. We can, without loss of completeness, restrict the selection of a literal from the goal clause to the leftmost literal (i.e., restrict $i = 1$).

A derivation is a deduction tree rooted at an application of the *start* rule, for some positive clause C, with instances of *extension$_0$* and premiseless instances of *reduction* at the leaves. In an implementation, the choice of terms t in the *start* and *extension$_2$* rules would be delayed via unification in the usual manner. Otten and Bibel provide an alternate, isomorphic, formulation of the calculus by way of an operational semantics in which substitutions are made explicit [14].

3   The leanCoP Theorem Prover

The leanCoP theorem prover of Otten and Bibel [14] is a Prolog program, shown in Figure 2, providing a direct encoding of the calculus shown in Figure 1. In this implementation clauses, paths, and matrices are represented as Prolog lists. Atomic formulas are represented with Prolog terms. A negated atom is represented by applying the unary - operator to the corresponding term. Prolog

```
prove(Mat) :- prove(Mat,1).

prove(Mat,PathLim) :-
    append(MatA,[Cla|MatB],Mat), \+member(-_,Cla),
    append(MatA,MatB,Mat1), prove([!],[[-!|Cla]|Mat1],[],PathLim).
prove(Mat,PathLim) :-
    \+ground(Mat), PathLim1 is PathLim+1, prove(Mat,PathLim1).

prove([],_,_,_).
prove([Lit|Cla],Mat,Path,PathLim) :-
    (-NegLit=Lit; -Lit=NegLit) ->
    ( member_oc(NegLit,Path) ;
      append(MatA,[Cla1|MatB],Mat), copy_term(Cla1,Cla2),
      append_oc(ClaA,[NegLit|ClaB],Cla2), append(ClaA,ClaB,Cla3),
      ( Cla1==Cla2 -> append(MatB,MatA,Mat1)
                    ; length(Path,K), K<PathLim,
                      append(MatB,[Cla1|MatA],Mat1)
      ), prove(Cla3,Mat1,[Lit|Path],PathLim)
    ), prove(Cla,Mat,Path,PathLim).
```

Fig. 2. The leanCoP theorem prover of Otten and Bibel

variables are used to represent object variables. This last fact causes some complications, discussed below.

The first evident difference between the calculus and its implementation is that an extra value, an integer path-depth limit, is added to each of the Prolog predicates. It is used to implement iterative deepening based on the maximum allowed path length, which is necessary to insure completeness in the first-order case, due to Prolog's depth-first search strategy. When **prove/1** is called, it sets the initial path limit to 1 and calls **prove/2**, which in turn selects (without loss of generality) a purely positive start clause.

The selection of the clause, **Cla**, is done using a trick of Prolog: since the predicate **append(A,B,C)** holds if the list C results from appending list B to list A, **append(A,[D|B],C)** (in which [D|B] is a list that has D as it's first item, followed by the list B) will hold if D is an element of C and if, further, A is the list of items preceding it and B is the list of items following it. Thus Prolog can, in one predicate, select an element from an arbitrary position in a list and identify all the remaining elements in the list, which result from appending A and B.

This technique is used to select literals from clauses and clauses from matrices throughout leanCoP. While it is an interesting trick, it relies on significant manipulation and construction of list structures on the heap. It is precisely certain uses of this trick which will be replaced by linear logic resource management at the formula level in lolliCoP.

To insure that the selected clause is purely positive, the code checks that the clause contains no negated terms (terms of the form -_, where the underscore is a wildcard). This is done using Prolog's negation-as-failure operator: \+. Once this is confirmed, the proof is started using a dummy (unit) goal clause, !, which will cause the selected clause to become the goal clause in the next step. This is done to avoid duplicating some bookkeeping code already present in the general case in `prove/4`, which implements the core of the prover. Note that the similarity of appearance to the Prolog cut operator is coincidental.

Should the call to `prove/4` at the end of the first clause of `prove/2` fail, then, provided this is not a purely propositional problem (That is, if it is not true that the entire matrix is ground.) the second clause of `prove/2` will cause the entire process to repeat, but with a path-depth limit one larger.

The first clause of `prove/4` implements the termination case, $extension_0$, and is straightforward. The second implements the remaining rules. This clause begins by selecting, without loss of completeness, the first literal, `Lit`, from the goal clause. If the complement of this literal as computed by the first line of the body of the clause matches a literal in the `Path`, then the system attempts to apply an instance of the *reduction* rule, jumping to the last line of the clause, where it recursively proves the remainder of the goal using the same matrix and path, under the substitution resulting from the matching process. (That is, free variables in literals in the goal and the path may have become instantiated.)

If a match to the complement of the literal is not found on the path, that is, if all attempts to apply instances of *reduction* have failed, then this is treated as either $extension_1$ or $extension_2$, depending on whether or not the clause selected next is ground. A clause is selected by the technique described above. Then a literal matching the complement of the goal literal is selected from the clause. (If this fails then the program backtracks and selects another clause.) The test `Cla1==Cla2` is used, as explained below, to determine if the selected clause is ground, and the matrix for the subproof is constructed accordingly, either with or without the chosen clause. If the path limit has not been reached, the prover recursively proves the body of the selected clause under the new path assumption and substitution, and, if it succeeds, goes on to prove the remainder of the current goal clause. As the depth-first prover is complete for propositional logic, the path limit check is not done if the selected clause is ground.

Note, `P -> Q ; R` is an extra-logical control structure corresponding to an `if-then-else` statement, The difference between this and `((P,Q) ; (\+P,R))` is that the latter allows for backtracking and retrying the test under another substitution, whereas the former allows the test to be computed only once and an absolute choice is made at that point. It can also be written without `R`, as is done in some cases here. Such use is, in essence, a hidden use of the Prolog cut operator, which is used for pruning search.

As mentioned above, the use of Prolog terms to represent atomic formulas introduces complications. This is because the free variables of a term, intended to represent the implicitly quantified variables of the atoms, can become bound if the term is compared (unified) with another term. In order to avoid the variables in clauses in the matrix from being so bound, when a clause is selected from the matrix, a copy with a fresh set of variables is produced using `copy_term`, and that copy is the clause that is used. Thus, the comparison `Cla1==Cla2`, which checks for syntactic identity, succeeds only if there were no variables in the

$$\frac{}{\Gamma; B \longrightarrow B}\ identity \qquad \frac{}{\Gamma; \Delta \longrightarrow \top}\ \top_R \qquad \frac{}{\Gamma; \emptyset \longrightarrow 1}\ 1_R$$

$$\frac{\Gamma; \Delta, B \longrightarrow C}{\Gamma; \Delta \longrightarrow B \multimap C}\ \multimap_R \qquad \frac{\Gamma, B; \Delta \longrightarrow C}{\Gamma; \Delta \longrightarrow B \Rightarrow C}\ \Rightarrow_R$$

$$\frac{\Gamma; \Delta \longrightarrow B \quad \Gamma; \Delta \longrightarrow C}{\Gamma; \Delta \longrightarrow B\,\&\,C}\ \&_R \qquad \frac{\Gamma; \Delta_1 \longrightarrow B_1 \quad \Gamma; \Delta_2 \longrightarrow B_2}{\Gamma; \Delta_1, \Delta_2 \longrightarrow B_1 \otimes B_2}\ \otimes_R$$

$$\frac{\Gamma; \Delta \longrightarrow B[x \mapsto t]}{\Gamma; \Delta \longrightarrow \exists x.B}\ \exists_R \qquad \frac{\Gamma; \Delta \longrightarrow B_i}{\Gamma; \Delta \longrightarrow B_1 \oplus B_2}\ \oplus_{R_i}$$

$$\frac{\Gamma, B; \Delta, B \longrightarrow C}{\Gamma, B; \Delta \longrightarrow C}\ absorb \qquad \frac{\Gamma; \Delta, B[x \mapsto t] \longrightarrow C}{\Gamma; \Delta, \forall x.B \longrightarrow C}\ \forall_L$$

$$\frac{\Gamma; \emptyset \longrightarrow B \quad \Gamma; \Delta, C \longrightarrow E}{\Gamma; \Delta, B \Rightarrow C \longrightarrow E}\ \Rightarrow_L \qquad \frac{\Gamma; \Delta_1 \longrightarrow B \quad \Gamma; \Delta_2, C \longrightarrow E}{\Gamma; \Delta_1, \Delta_2, B \multimap C \longrightarrow E}\ \multimap_L$$

Fig. 3. A proof system for a fragment of linear logic

original term Cla1 (since they would have been modified by copy_term), and, hence, if that term was ground.

Because Prolog unification is unsound, as it lacks the "occurs check" for barring the construction of cyclic unifiers, if the prover is to be sound we must force sound unification when comparing literals. In Eclipse Prolog, used in the original leanCoP paper, this is done with a global switch, affecting all unification in the system. In SICStus Prolog, used for the tests in this paper, it is done with the predicate unify_with_occurs_check. This predicate is used within the member_oc and append_oc predicates, whose definitions have been elided in the code above.

Many of these complications could have been avoided by using λ-Prolog, which supports the use of λ-terms as data for representing name-binding structures, and whose unification algorithm is sound [11].

4 A Brief Introduction to Linear Logic Programming

Linear logic was first proposed by Girard in 1987 [4]. Figure 3 gives a Gentzen sequent calculus for part of the fragment of intuitionistic linear logic which forms the foundation of the logic programming language Lolli, named for the linear logic implication operator, ⊸, known as lollipop. The calculus is not the standard one, but for this fragment is equivalent to it, and is easier to explain in the context of logic programming. In these sequents, the left-hand side has two parts: the context Γ holds assumptions that can be freely reused and discarded, as in traditional logics, while the assumptions in Δ, in contrast, must be used exactly once in a given branch of a tree. The two implication operators, ⇒, and ⊸, are used to add assumptions to the unrestricted and linear contexts, respectively. In Lolli they are written => and -o.

In the absence of contraction and weakening (that is, the ability to freely reuse or discard assumptions, respectively), all of the other logical operators split into two variants as well. For example, the conjunction operator splits into *tensor*, ⊗, and *with*, &. In proving a conjunction formed with ⊗, the current set of restricted assumptions, Δ, is split between the two conjuncts: those not used in proving the first conjunct must be used while proving the second. To prove a & conjunction, the set of assumptions is copied to both sides: each conjunct's proof must use all of the assumptions. In Lolli, the ⊗ conjunction is represented by the familiar ",". This is a natural mapping, as we expect the effect of a succession of goals to be cumulative: each has available to it the resources not yet used by its predecessors. The & conjunction, which is less used, is written "&".

Thus, a query showing that two dollars are needed to buy pizza and soda when each costs a dollar can be written in Lolli as:

```
?- (dollar -o pizza) => (dollar -o soda) =>
        (dollar -o dollar -o (pizza,soda))
```

which would succeed. In contrast, a single, ordinary dollar would be insufficient, as in the failing query:

```
?- (dollar -o pizza) => (dollar -o soda) =>  (dollar -o (pizza,soda))
```

If we wished to allow ourselves a single, infinitely reusable dollar, we would write:

```
?- (dollar -o pizza) => (dollar -o soda) =>  (dollar => (pizza,soda))
```

which would also succeed. Finally, the puzzling query:

```
?- (dollar -o pizza) => (dollar -o soda) =>  (dollar -o (pizza & soda))
```

would also succeed. It says that with a dollar it is possible to buy soda and possible to buy pizza, but not both at the same time.

It is important to note that while the implication operators add clauses to a program while it is running, they are not the same as the Prolog **assert** mechanism. First, the addition is scoped over the subgoal on the right of the implication, whereas a clause **asserted** in Prolog remains until it is **retracted**. So, for example, the following query will fail:

```
?- (dollar => dollar), dollar.
```

Assumed clauses also go out of scope if search backtracks out of the subordinate goal. Second, whereas **assert** automatically universalizes any free variables in an added clause, in Lolli clauses added with implication can contain free logic variables, which may get bound when the clause is used to prove some goal. Therefore, whereas the Prolog query:

```
?- assert(p(X)), p(a), p(b).
```

will succeed, because X is universalized, the seemingly similar Lolli query:

```
?- p(X) => (p(a), p(b)).
```

will fail, because the attempt to prove p(a) causes the variable X to become instantiated to a. If we desire the other behavior, we must quantify explicitly:

```
?- (forall X\p(X)) => (p(a), p(b)).
```

What's more, any action that causes the variable X to become instantiated will affect instances of that variable in added assumptions. For example, the query:

```
?- p(X) => r(a) => (r(X), p(b)).
```

will fail, since proving r(X) causes the variable X to be instantiated to a, both in that position, and in the assumption p(X). Our implementation of lolliCoP will rely crucially on all these behaviors.

Though there are two forms of disjunction in linear logic, only one, "⊕" is used in Lolli. It corresponds to the traditional one and is therefore written with a semicolon in Lolli as in Prolog.

There are also two forms of truth, ⊤, and 1. The latter, which Lolli calls "true", can only be proved if all the linear assumptions have already been used. In contrast, ⊤ is provable even if some resources are, as yet, unused. Thus if a ⊤ occurs as one of the conjuncts in a ⊗ conjunction, then the conjunction may succeed even if the other conjuncts do not use all the linear resources. The ⊤ is seen to consume the leftovers. Therefore, Lolli calls this operator "erase".

It is beyond the scope of this paper to demonstrate the applications of all these operators. Many good examples can be found in the literature, particularly in the papers on Lygon and Lolli [5,6]. The proof theory of this fragment has also been developed extensively [6]. Of crucial importance is that there is a straightforward goal-directed proof procedure (conceptually similar to the one used for Prolog) that is sound and complete for this fragment of linear logic.

5   The lolliCoP Theorem Prover

Figure 4 gives the code for lolliCoP, a reimplementation of leanCoP in Lolli/LLP.[1] The basic premise of its design is that, rather than being passed around as a list, the matrix will be loaded as assumptions into the proof context and accessed directly. In addition, ground clauses will be added as linear resources, since the calculus dictates that in any given branch of the proof, a ground clause should be removed from the matrix once it is used. Non-ground clauses are added to the intuitionistic (unbounded) context. In either case (ground or non-ground) these assumptions are stored as clauses for the special predicate cl/1. Literals in the path are also stored as assumptions added to the program. They are unbounded assumptions added as clauses of the special predicate path. While Lolli supports the λ-terms of λ-Prolog, LLP does not. Therefore, clauses are still represented as lists of literals, which are represented as terms as before.

The proof procedure begins with a call to prove/1 with a matrix to be proved. This predicate first reverses the order of the clauses, so that when they are added recursively the resultant context will be searched in their original order. It then calls pr/1 to load the matrix into the unrestricted and linear proof contexts, as appropriate. First, however, it checks whether the entire matrix is ground

---

[1] Because the LLP parser is written in Prolog, LLP uses -<> for –∘, rather than –o.

```
prove(Mat) :- reverse(Mat,Mat1),
              (ground(Mat) -> propositional => pr(Mat1)
                            ; pr(Mat1)
              ).

pr([])        :- p(1).
pr([Cla|Mat]) :- (ground(Cla) -> (cl(Cla) -<> pr(Mat))
                              ; (cl(Cla)  => pr(Mat))
                 ).

p(PathLim) :- cl(Cla), \+member(-_,Cla),
              copy_term(Cla,Cla1), prove(Cla1,PathLim).

p(PathLim) :- \+propositional,
              PathLim1 is PathLim+1, p(PathLim1).

prove([],_) :- erase.
prove([Lit|Cla],PathLim) :-
  (-NegLit=Lit; -Lit=NegLit) ->
    ( path(NegLit), erase ;
      cl(Cla1), copy_term(Cla1,Cla2), append(ClaA,[NegLit|ClaB],Cla2),
      append(ClaA,ClaB,Cla3), (Cla1==Cla2 -> true ; PathLim>0),
      PathLim1 is PathLim-1, path(Lit) => prove(Cla3,PathLim1)
    ) & prove(Cla,PathLim).
```

Fig. 4. The lolliCoP theorem prover

or not. If it is, a flag predicate is assumed (using =>) to indicate that this is a propositional problem, and that iterative deepening is not necessary.

The predicate pr/1 takes the first clause out of the given matrix, adds it to the current context as either a linear or unlimited assumption, as appropriate, and then calls itself recursively as the goal nested under the implication. Thus, each call to this predicate will be executed in a context which contains the assumptions added by all the previous calls. When the end of the given matrix is reached, the first clause of pr/1 calls p/1 with an initial path-length limit of 1, so that a start clause can be selected, and the proof search begun.

The clauses for p/1 take the place of the clauses for prove/2 in leanCoP. They are responsible for managing the iterative deepening, and for selecting the start clause for the search. A clause is selected just by attempting to prove the predicate cl/1 which will succeed by matching one of the clauses from the matrix which has been added to the program. This is significantly simpler than the process in leanCoP. Once the program finds a purely positive start clause, it is copied and its proof is attempted at the current path-length limit. Should that process fail for all possible choices of start clause, the second clause of p/1 is invoked. It checks to see that this is not a purely propositional problem, and if it is not, makes a recursive call with the path-length limit one higher.

The predicate `prove/2` takes the role of `prove/4` in leanCoP; because the matrix and path are stored in the proof context, they no longer need to be passed around as arguments. The first clause, corresponding to $extension_0$, here has a body consisting of the `erase` ($\top$) operator. Its purpose is to discard any linear assumptions (i.e. ground clauses in the matrix) that were not used in this branch of the proof. This is necessary since we are building a prover for classical logic, in which assumptions can be discarded.

The second clause of this predicate is, as before, the core of the prover, covering the remaining three rules. It begins by selecting a literal from the goal clause and forming its complement. If a literal matching the complement occurs as an argument to one of the assumed `path/1` clauses, then this is an instance of the *reduction* rule and this branch is terminated. As with the $extension_0$ rule, `erase` is used to discard unused assumptions.

Otherwise, the predicate `cl/1` extracts a clause from the matrix, which is then copied and checked to see if it contains a match for the complement of the goal literal. If the clause is ground or if the path-length limit has not been reached, the current literal is added to the path and `prove/2` is called recursively as a subordinate goal (within the scope of the assumption added to the path) to prove the body of the selected clause.

If this was an instance of the *reduction* rule, or if it was an instance of $extension_1$ or $extension_2$ and the proof of the body of the matching clause succeeded, the call to `prove/2` finishes with a recursive call to prove the rest of the current goal clause. Because this must be done using the same matrix and path that were used in the other branch of the proof, the two branches are joined with a & conjunction. Thus the context is copied independently to the two branches.

It is important to notice that, other than checking whether the path-length limit has been reached, there is no difference between the cases when the selected clause is ground or not. If it was ground, it was added to the context using linear implication, and, since it has been used (to prove the `cl/1` predicate), it has automatically been removed from the program, and, hence, the matrix. Also, lolliCoP uses a different method for checking path length against the limit: the limit is simply decremented each time a literal is added to the path. This is done because there is no way to access the whole path to check its length, but has the advantage of being significantly more efficient as well.

It is also important to note that, as mentioned before, we rely on the fact that free variables in assumptions retain their identity as logic variables and may become instantiated subsequently. In particular, the literals added to the path may contain instances of free variables from the goal clause from which they derive. Anything which causes these variables to become instantiated will similarly affect those occurrences in these assumptions. Thus, this technique could not be implemented using Prolog's `assert` mechanism. In any case, `assert`ed clauses are generally not as fast as compiled ones.

6   Performance Analysis

We have tested lolliCoP on the 2200 clausal form problems in the TPTP library version 2.3.0 [15,8]. These consist of 2193 problems known to be unsatisfiable (or valid using positive representation) and 7 propositional problems known to be satisfiable (or invalid). Each problem is rated from 0.00 to 1.00 relative to its difficulty. A rating of "?" means the difficulty is unknown. No reordering of clauses or literals has been done.

The tests were performed on a Linux system with a 550MHz Pentium III processor and 128M bytes of memory. The programs were compiled with version 0.50 of LLP which generated abstract machine code executed by an emulator written in C. The time limit for all proof attempts was 300 seconds.

Table 1. Overall performance of OTTER, leanCoP, and lolliCoP

	Total	OTTER	leanCoP	lolliCoP	lolliCoP$_2$
Solved	2200	1602 (73%)	810 (37%)	822 (37%)	880 (40%)
0 to < 1 second	1209	541	554	614	
1 to < 10 seconds	142	135	124	117	
10 to <100 seconds	209	93	91	94	
100 to <200 seconds	31	18	25	34	
200 to <300 seconds	11	23	28	21	
Problems rated 0.00	1308	1230 (94%)	713 (55%)	716 (55%)	737 (56%)
Problems rated >0.00	733	249 (34%)	76 (10%)	83 (11%)	118 (16%)
Problems rated ?	159	123 (77%)	21 (13%)	23 (14%)	25 (16%)

The overall performance of OTTER 3.1 (with MACE 1.4) [8,10], leanCoP [14], and lolliCoP, in terms of the number of problems solved, are shown in Table 1. The table also includes data for an improved version of lolliCoP, called lolliCoP$_2$, discussed in the next section. The results for leanCoP were obtained in the same environment as those for lolliCoP, using SICStus Prolog 3.7.1, and are better than those reported by the authors [14]. The results for OTTER 3.1 (with MACE 1.4), which is not publicly available, are taken from a report by its developers [8]. These results were produced on a 400MHz Pentium II, which is somewhat slower than the machine we used.

It is interesting to note that lolliCoP solved 57 problems, and lolliCoP$_2$ 77, which OTTER can not solve. Most of these (47 for lolliCoP and 63 for lolliCoP$_2$) are rated higher than 0.00. It should also be noted that leanCoP solved ten problems that neither lolliCoP nor lolliCoP$_2$ solved. Since nine of these were rated 0.0, and given the structural similarities of the systems, we believe this to be due to serendipitous advantages with respect to clause ordering, since leanCoP orders clauses slightly differently. Fig. 5 depicts the overlap of problems solved by each system.

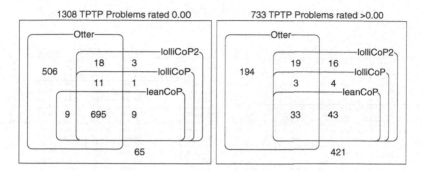

Fig. 5. Performance of OTTER, leanCoP and lolliCoP classified by problem rating

6.1 Performance Comparison

In order to produce a more detailed comparison, we tested all the systems on the 118 problems rated greater than 0.0 which lolliCoP$_2$ can solve. Because OTTER 3.1 is not yet available, we used OTTER 3.0.6 instead. All tests were made on the same 550MHz Pentium III. Table 2 gives the results of this comparison. (OTTER results labeled "error" refer to an empty set-of-support.)

As mentioned in the introduction, although the table shows lolliCoP as almost consistently outpacing leanCoP these results do not tell the entire story. Because LLP is a first-generation implementation, the code generator is not nearly as sophisticated as SICStus', nor is its runtime system. To adjust for this factor we also executed a version of leanCoP using the LLP compiler and runtime system (since Lolli is a superset of Prolog). In this test, looking only at the problems that it succeeded in solving, leanCoP took 2.3 times as long as lolliCoP, providing a more accurate measure of the benefits accrued from the logical treatment.

Table 3a compares the performance of all four systems on the 33 problems that they can all solve. Total CPU time is shown, along with a speedup ratio relative to leanCoP (under SICStus). On just these problems, lolliCoP has almost the same performance as OTTER. However, comparing the result of 36 problems solved by both OTTER and lolliCoP, OTTER is 71% faster as shown in Table 3b. Finally, Table 3c shows a similar analysis for the 76 problems that lolliCoP and leanCoP can both solve.

7 Improvements to the lolliCoP Prover

In the design of leanCoP, Otten and Bibel seem to have been focused primarily on keeping the code as short as possible. In the process of reimplementing the system in Lolli, a simple but significant performance improvement became apparent, which we discuss here.

The most obvious inefficiency in the system as described thus far is that `copy_term` is called in order to create a new set of logic variables in a selected

Table 2. Problems solved by lolliCoP$_2$ and rated higher than 0.00

Problem	Rating	OTTER	leanCoP	lolliCoP	lolliCoP$_2$	Problem	Rating	OTTER	leanCoP	lolliCoP	lolliCoP$_2$
BOO012-1	(0.17)	3.44	8.13	7.33	1.28	NUM009-1	(0.12)	3.40	75.63	49.37	4.44
BOO012-3	(0.33)	17.39	237.81	63.00	9.43	NUM283-1.005	(0.20)	0.43	0.28	0.20	0.17
CAT002-4	(0.17)	2.67	>300	>300	231.07	NUM284-1.014	(0.20)	0.89	180.54	147.55	129.91
CAT003-2	(0.50)	>300	15.83	12.05	4.26	PLA004-1	(0.40)	>300	4.00	3.06	2.46
CAT003-3	(0.11)	>300	2.43	1.70	0.34	PLA004-2	(0.40)	>300	5.97	5.09	3.90
CAT012-4	(0.17)	0.26	19.95	14.88	4.57	PLA005-1	(0.40)	>300	0.44	0.36	0.24
COL002-3	(0.33)	>300	0.01	0.03	0.01	PLA005-2	(0.40)	>300	0.10	0.06	0.03
COL075-1	(0.50)	>300	>300	275.77	60.29	PLA007-1	(0.40)	>300	0.14	0.13	0.08
FLD002-3	(0.67)	1.20	201.23	162.01	43.73	PLA008-1	(0.40)	>300	251.50	204.21	142.95
FLD003-1	(0.67)	>300	>300	264.10	70.76	PLA009-1	(0.40)	>300	0.06	0.05	0.03
FLD004-1	(0.67)	>300	>300	>300	201.74	PLA009-2	(0.40)	>300	2.10	1.73	1.20
FLD009-3	(0.33)	>300	>300	272.94	72.59	PLA010-1	(0.40)	>300	250.53	203.04	142.21
FLD013-1	(0.67)	>300	0.46	0.48	0.15	PLA011-1	(0.40)	>300	0.14	0.09	0.05
FLD013-2	(0.67)	>300	>300	>300	106.83	PLA011-2	(0.40)	>300	0.45	0.36	0.24
FLD013-3	(0.33)	>300	>300	>300	153.85	PLA012-1	(0.40)	>300	66.06	52.21	36.79
FLD013-4	(0.33)	3.19	>300	>300	270.74	PLA013-1	(0.40)	>300	0.24	0.18	0.11
FLD016-3	(0.33)	11.90	>300	>300	155.71	PLA014-1	(0.40)	>300	2.06	1.61	1.21
FLD018-1	(0.33)	>300	>300	>300	101.86	PLA014-2	(0.40)	>300	2.13	1.75	1.33
FLD019-1	(0.33)	>300	>300	>300	196.80	PLA016-1	(0.40)	>300	0.07	0.07	0.04
FLD022-3	(0.33)	12.45	>300	>300	155.83	PLA019-1	(0.40)	>300	0.06	0.05	0.03
FLD023-1	(0.33)	>300	0.62	0.48	0.13	PLA021-1	(0.40)	>300	0.18	0.13	0.07
FLD025-1	(0.67)	>300	0.45	0.49	0.15	PLA022-1	(0.40)	>300	0.40	0.32	0.24
FLD025-3	(0.33)	>300	>300	>300	130.73	PLA022-2	(0.40)	>300	0.03	0.02	0.01
FLD028-3	(0.33)	13.45	>300	>300	187.66	PLA023-1	(0.40)	>300	72.74	57.73	40.71
FLD030-1	(0.33)	0.41	0.03	0.02	0.01	PUZ034-1.004	(0.67)	error	15.87	12.42	9.83
FLD030-2	(0.33)	>300	0.44	0.35	0.11	RNG006-2	(0.20)	4.69	0.26	0.35	0.06
FLD031-1	(0.33)	>300	>300	>300	268.48	RNG040-1	(0.11)	0.05	0.01	0.01	0.01
FLD032-1	(0.33)	>300	>300	>300	247.57	RNG040-2	(0.22)	0.10	0.21	0.19	0.04
FLD035-3	(0.33)	14.11	>300	>300	257.05	RNG041-1	(0.22)	0.16	43.86	36.25	6.37
FLD036-3	(0.33)	13.73	>300	>300	135.22	SET014-2	(0.33)	176.24	174.31	134.35	27.97
FLD037-1	(0.33)	>300	1.64	1.25	0.32	SET016-7	(0.12)	>300	10.99	8.29	1.05
FLD060-1	(0.67)	>300	0.59	0.51	0.15	SET018-7	(0.12)	>300	11.13	8.37	1.06
FLD060-2	(0.67)	>300	>300	>300	127.11	SET041-3	(0.44)	>300	59.88	45.36	4.88
FLD061-1	(0.67)	>300	0.66	0.58	0.17	SET060-6	(0.12)	0.19	0.04	0.03	0.00
FLD061-2	(0.67)	>300	>300	>300	155.31	SET060-7	(0.12)	0.33	0.05	0.03	0.00
FLD064-1	(0.67)	>300	>300	>300	114.86	SET083-7	(0.12)	24.39	40.34	34.70	5.41
FLD067-1	(0.33)	>300	1.47	1.20	0.31	SET085-6	(0.12)	12.72	>300	>300	65.58
FLD067-3	(0.33)	20.08	186.68	150.37	40.97	SET085-7	(0.25)	65.79	46.01	33.55	5.22
FLD069-1	(0.33)	>300	>300	>300	125.96	SET119-7	(0.25)	177.97	60.35	48.50	6.71
FLD070-1	(0.33)	>300	2.52	0.68	0.18	SET120-7	(0.25)	181.62	60.23	48.46	6.71
FLD071-3	(0.33)	2.52	0.36	0.34	0.08	SET121-7	(0.25)	178.42	72.81	55.77	7.63
GEO026-3	(0.11)	2.15	20.34	19.16	2.35	SET122-7	(0.25)	180.13	72.83	55.82	7.64
GEO030-3	(0.44)	8.04	>300	271.72	30.90	SET152-6	(0.12)	0.45	3.50	2.60	0.38
GEO032-3	(0.25)	1.16	>300	292.07	32.02	SET153-6	(0.12)	>300	0.70	0.56	0.10
GEO033-3	(0.38)	4.81	>300	>300	39.41	SET187-6	(0.38)	>300	18.01	13.53	2.27
GEO041-3	(0.22)	0.21	42.28	32.90	3.60	SET196-6	(0.12)	10.59	>300	>300	196.13
GEO051-3	(0.25)	7.26	>300	>300	56.82	SET197-6	(0.12)	10.63	>300	>300	196.06
GEO064-3	(0.12)	0.33	>300	>300	55.07	SET199-6	(0.25)	>300	>300	>300	203.63
GEO065-3	(0.12)	0.34	>300	>300	55.11	SET231-6	(0.12)	>300	12.86	9.74	1.63
GEO066-3	(0.12)	0.32	>300	>300	55.14	SET234-6	(0.25)	>300	>300	>300	251.18
GRP008-1	(0.22)	0.69	1.00	0.72	0.12	SET252-6	(0.25)	61.53	>300	>300	202.60
HEN007-6	(0.17)	0.12	>300	>300	211.72	SET253-6	(0.25)	>300	>300	>300	203.24
LCL045-1	(0.20)	98.03	1.31	0.90	0.50	SET451-6	(0.12)	>300	>300	>300	281.67
LCL097-1	(0.20)	0.26	0.67	0.20	0.12	SET553-6	(0.25)	36.81	>300	>300	204.46
LCL111-1	(0.20)	0.13	0.20	0.14	0.07	SYN048-1	(0.20)	0.00	0.00	0.00	0.00
LCL130-1	(0.20)	0.01	0.03	0.01	0.02	SYN074-1	(0.11)	0.87	>300	>300	74.43
LCL195-1	(0.20)	error	18.76	15.23	6.93	SYN075-1	(0.11)	0.17	>300	266.47	49.18
LCL230-1	(0.40)	error	209.09	133.76	61.13	SYN102-1.007:007	(0.33)	1.00	39.38	39.70	22.95
LCL231-1	(0.40)	error	>300	189.14	85.31	SYN311-1	(0.20)	error	123.36	99.86	45.68

clause, even when the clause is ground, since that test is not made till later on. Given the size of some of the clauses in the problems in the TPTP library, this can be quite inefficient. While the obvious solution would be to move the use of `copy_term` into the body of the if-then-else along with the path-limit check, Lolli affords a more creative solution.

In lolliCoP we already check whether each clause is ground or not at the time the clauses are added into the proof context in `pr/1`. We can further take advantage of that check by not only adding the clauses differently, but by adding different sorts of clauses. In lolliCoP a clause c (ground or not) is represented by the Lolli clause `cl($c$)`. We can continue to represent ground clauses in the

Table 3. Comparison of OTTER, leanCoP, and lolliCoP

(a) 33 problems solved by OTTER, leanCoP, and lolliCoP

	OTTER	leanCoP	lolliCoP	lolliCoP$_2$
Total CPU time	1143.03	1590.66	1139.41	338.47
Average CPU time	34.64	48.20	34.53	10.26
Speedup Ratio	1.39	1.00	1.40	4.70

(b) 36 problems solved by OTTER and lolliCoP   (c) 76 problems solved by leanCoP and lolliCoP

	OTTER	lolliCoP	lolliCoP$_2$
Total CPU time	1152.40	1969.67	450.57
Average CPU time	32.01	54.71	12.52
Speedup Ratio	1.71	1.00	4.37

	leanCoP	lolliCoP	lolliCoP$_2$
Total CPU time	2757.83	2038.58	853.24
Average CPU time	36.29	26.82	11.23
Speedup Ratio	1.00	1.35	3.23

same way, but when c is non-ground, instead represent it by the Lolli clause: `cl(C1) :- copy_term(`c`,C1)`. When this clause is used, it will return not the original clause, c, but a copy of it. To be precise, we replace the second clause of `pr/1` with a clause of the form:

```
pr([C|Mat]) :-
     (ground(C) -> (cl(C) -<> pr(Mat)
       ; (forall C1\ cl(C1) :- copy_term(C,C1)) =>  pr(Mat)).
```

Note the use of explicit quantification over the variable C1.

In lolliCoP$_2$ the loaded clauses are further modified to take a second parameter, the path-depth limit. The Lolli clauses for ground clauses simply ignore this parameter. The ones for non-ground clauses check it first and proceed only if the limit has not yet been reached. In this version of the prover there is no check whatsoever for the ground status of a clause in the core (`prove/2`). This removes the potentially significant computational cost of checking the ground status each time a clause is selected: an operation linear in the size of the selected clause. Space constraints keep us from including the full program.

Taken together these small improvements actually triple the performance of the system. While the first optimization can be added, awkwardly, to leanCoP, it is not possible to do away entirely with the groundness check in that setting.

8   Conclusion

Lean theorem proving began with leanTAP [1], which provided an existence proof that it was possible to implement interesting theorem proving techniques using clear short Prolog programs. It was not expected, however, to provide particularly powerful systems. Recently, leanCoP showed that these programs can be at once perspicuous and powerful.

However, to the extent that these programs rely on the use of term-level Prolog data structures to maintain their proof contexts, they require the use of list manipulation predicates that are neither particularly fast nor clear. In this paper we have shown that by representing the proof context within the proof context of the meta-language, we can obtain a program that is at once clearer, simpler, and faster.

Source code for the examples in this paper, as well as the LLP compiler can be found at http://www.cs.hmc.edu/~hodas/research/lollicop.

References

1. B. Beckert and J. Posegga. leanTAP: lean tableau-based theorem proving. In *12th CADE*, pages 793–797. Springer-Verlag LNAI 814, 1994.
2. W. Bibel. *Deduction: Automated Logic*. Academic Press, 1993.
3. W. Bibel, S. Brüning, U. Egly, and T. Rath. KoMET. In *12th CADE*, pages 783–787. Springer-Verlag LNAI 814, 1994.
4. J.-Y. Girard. Linear logic. *Theoretical Computer Science*, 50:1–102, 1987.
5. James Harland, David Pym, and Michael Winikoff. Programming in Lygon: An overview. In M. Wirsing and M. Nivat, editors, *Algebraic Methodology and Software Technology*, pages 391–405, Munich, Germany, 1996. Springer-Verlag LNCS 1101.
6. J. S. Hodas and D. Miller. Logic programming in a fragment of intuitionistic linear logic. *Information and Computation*, 110(2):327–365, 1994. Extended abstraction in the Proceedings of the Sixth Annual Symposium on Logic in Computer Science, Amsterdam, July 15–18, 1991.
7. J. S. Hodas, K. Watkins, N. Tamura, and K.-S. Kang. Efficient implementation of a linear logic programming language. In *Proceedings of the 1998 Joint International Conference and Symposium on Logic Programming*, pages 145–159, June 1998.
8. Argonne National Laboratory. Otter and MACE on TPTP v2.3.0. Web page at http://www-unix.msc.anl.gov/AR/otter/tptp230.html, May 2000.
9. R. Letz, J. Schumann, S. Bayerl, and W. Bibel. SETHEO: a high-performance theorem prover. *Journal of Automated Reasoning*, 8(2):183–212, 1992.
10. W. MacCune. OTTER 3.0 reference manual and guide. Technical Report ANL-94/6, Argonne National Laboratory, 1994.
11. D. Miller, G. Nadathur, F. Pfenning, and A. Scedrov. Uniform proofs as a foundation for logic programming. *Annals of Pure and Applied Logic*, 51:125–157, 1991.
12. M. Moser, O. Ibens, R. Letz, J. Steinbach, C. Goller, J. Schumann, and K. Mayr. SETHEO and E-SETHEO—the CADE-13 systems. *Journal of Automated Reasoning*, 18:237–246, 1997.
13. G. Nadathur and D. J. Mitchell. Teyjus—a compiler and abstract machine based implementation of lambda Prolog. In *6th CADE*, pages 287–291. Springer-Verlag LNCS 1632, 1999.
14. J. Otten and W. Bibel. leanCoP: lean connection-based theorem proving. In *Proceedings of the Third International Workshop on First-Order Theorem Proving*, pages 152–157. University of Koblenz, 2000. Electronically available, along with submitted journal-length version, at http://www.intellektik.informatik.tu-darmstadt.de/~jeotten/leanCoP/.
15. G. Sutcliffe and C. Suttner. The TPTP problem library—CNF release v1.2.1. *Journal of Automated Reasoning*, 21:177–203, 1998.

MUSCADET 2.3: A Knowledge-Based Theorem Prover Based on Natural Deduction

Dominique Pastre

Crip5 - Université René Descartes - Paris
pastre@math-info.univ-paris5.fr

1 Introduction

The MUSCADET theorem prover is a knowledge-based system. It is based on natural deduction, following the terminology of Bledsoe ([1], [2]), and uses methods which resemble those used by humans. It is composed of an inference engine, which interprets and executes rules, and of one or several bases of facts, which are the internal representations of "theorems to be proved".

Rules are either universal and put into the system, or built by the system itself by metarules from data (definitions and lemmas) given by the user. They are in the form *if <list of conditions>, then <list of actions>*. Actions may be "super-actions" which are defined by packs of rules.

The representation of a "theorem to be proved" (or a sub-theorem) is a description of its state during the proof. It is composed of objects that were created, of hypotheses, of a conclusion to be proved, of rules called active rules, possibly of sub-theorems, etc. At the beginning, it is only composed of a conclusion, which is the initial statement of the theorem to be proved, and of a list of "active" rules, relevant for this theorem, and which were built automatically.

Rules may add new hypotheses, modify the conclusion, create objects, create sub-theorems or build new rules which are local for a (sub-)theorem. A theorem is proved, for example, if the conclusion to be proved was added as a new hypothesis or if there is an existential conclusion $\exists X p(X)$ and a hypothesis $p(a)$.

2 Example: Transitivity of Inclusion

Prove the transitivity of inclusion, that is the theorem
$$\forall A \forall B \forall C (A \subset B \land B \subset C \Rightarrow A \subset C)$$
with the definition of inclusion $A \subset B \Leftrightarrow \forall X (X \in A \Rightarrow X \in B)$
To prove this theorem MUSCADET creates objects a, b and c by applying three times the rule
    Rule $\forall$: if the conclusion is $\forall X p(X)$,
        then create a new object x and the new conclusion is $p(x)$
and the new conclusion is $a \subset b \land b \subset c \Rightarrow a \subset c$. Then the rule
    Rule $\Rightarrow$: if the conclusion is $H \Rightarrow C$,
        then add the hypothesis H and the new conclusion is C
replaces the conclusion by $a \subset c$ and adds the two hypotheses $a \subset b$ and $b \subset c$.

R. Goré, A. Leitsch, and T. Nipkow (Eds.): IJCAR 2001, LNAI 2083, pp. 685–689, 2001.

In effect, hypotheses H are analyzed before being added: a super-action $addhyp(H)$ contains, among others, the rule

> To addhyp(H): if H is a conjunction,
>> then successively add all the elements of the conjunction

The conclusion is then replaced by its definition $\forall X (X \in a \Rightarrow X \in c)$ by applying the rule

> Rule defconcl: if the conclusion is C
>> and there exists a definition of the form $C \Leftrightarrow D$,
> then   the new conclusion is D

By the preceding rules $\forall$ and $\Rightarrow$, there is then a new object x, a new hypothesis $x \in a$, and the conclusion is now $x \in c$. The following rule

> Rule $\subset$: if there are hypotheses $A \subset B$ and $X \in A$,
>> then add the hypothesis $X \in B$

is a rule that was automatically built by MUSCADET from the definition of inclusion. Here it is applied twice, adds the hypotheses $x \in b$ then $x \in c$, which is the conclusion to be proved. The proof ends by applying the rule

> Rule stop1:   if the conclusion C is also a hypothesis,
>> then set the conclusion to true

MUSCADET is also able to work in second order predicate calculus, and the preceding example may be written $transitive(\subset)$ with the definition of the transitivity $transitive(R) \Leftrightarrow \forall A \forall B \forall C (R(A, B) \land R(B, C) \Rightarrow R(A, C))$

After the conclusion $transitive(\subset)$ has been replaced by its definition, the proof is the same as above.

3   From MUSCADET1 to MUSCADET2

The first version of MUSCADET, which is now called MUSCADET1, was described and analyzed in [4], [5], [6]. The inference engine of MUSCADET1 was written in PASCAL, and knowledge (rules, metarules and super-actions) was written in a language that was considered simple and declarative. MUSCADET1 produced good results; it was evaluated for several years but its use was limited. In particular, the language was not adapted to the expression of procedural strategies.

The current version, called MUSCADET2, is completely written in PROLOG. The reason for this is that it is possible to use the same language to express declarative knowledge such as rules, definitions, hypotheses, etc., more procedural knowledge such as proof strategies, and the inference engine itself. The inference engine contains only few predicates since it is completed by the PROLOG interpreter. This leads to more flexibility, more facilities for writing, and even more efficiency. Moreover it was possible to carry out many improvements and to write new strategies, which were not possible in the first version. It was also possible to use, without having to implement them, all the facilities of expression of PROLOG.

MUSCADET2   was   able   to   work   on   problems   of   the   TPTP Problem   Library   (Thousands   of   Problems   for   Theorem   Provers, http://www.cs.jcu.edu.au/~tptp). It participated in competitions CASC-16 and CASC-17. It could of course only compete in the "first order" divisions, that is FOF (FEQ and NEQ) and SEM, since it does not work with clauses. The results (http://www.cs.jcu.edu.au/~tptp/CASC) show the complementarity of MUSCADET2 with regard to provers based on the resolution principle.

4   Machine Representations

PROLOG is not only used as implementation language of MUSCADET2, but also as representation language to represent mathematical statements, facts and rules. Rules express declarative knowledge. Elementary actions and some strategies define procedural actions. Super-actions group packs of rules for a given goal. The inference engine is composed of the PROLOG interpreter and of some clauses which process the application of rules.

Expression of mathematical statements. The logical connectives and, or, not, =>, <=> are defined as infix operators with precedences in the order as the connectives are written down in mathematics. They are right associative. The quantifiers are used as binary prefix operators, that is for_all(X, <...X>) and exists(X, < ...X>). The example theorem introduced in section 2 is written, with the infix inc operator,

   for_all(A,for_all(B,for_all(C,A inc B and B inc C => A inc C).

The proof of the theorem T will be requested by the PROLOG call prove(T). The definition of inclusion and intersection are, with the infix elt operator for the member relation,

      A inc B <=> for_all(X,(X elt A => X elt B))

and   A inter B = [X, X elt A and X elt B]

Expression of facts. The fact that the property C is the *conclusion* of the *(sub-)theorem to be proved* with number N is represented by the unit PROLOG clause concl(N,C). (concl was declared dynamic). Some other properties are handled in the same manner, such as to be a *hypothesis* (hyp(N,H)), an *object* (obj(N,O)), a *sub-theorem* (sousth(N,N1) or any other property that seems useful.

Expression of rules and super-actions. Here are the machine expressions of the rule ⇒ and partly of the super-action *addhyp* of section 2.

```
  rule(N,=>) :- concl(N, A => B), addhyp(N, A), newconcl(N,B)
  addhyp(N, H) :- ( H = A and B -> addhyp(N, A), addhyp(N,B)
                  ; hyp(N, H) ->  true
                  ; H=for_all(_,_) -> create_nam_ru(rulehyp,Name),
                                      buildrules(H,_,N,Name,[])
                  ...
                  ; assert(hyp(N, H)),   % default action
                  ) .
```

The parameter N helps to apply a rule to the (sub-)theorem of number N.

5   How to Use MUSCADET2

MUSCADET2 is available at the address
http://www.math-info.univ-paris5.fr/ pastre/muscadet/muscadet.html
The PROLOG used is SWI-Prolog, version 3.2.9, which is freeware downloaded
at the following address
http://www.swi.psy.uva.nl/projects/SWI-Prolog/download.html.

Direct proof. The predicate prove may be directly called with the statement
of the theorem to be proved as a parameter. The definitions of mathematical
concepts have to be given before if necessary. For instance, for the first example,
introduce the definition of the inclusion and ask for building new rules by
```
:- op(200,xfy,inc).
assert(definition(A inc B <=> forall(X,(X elt A => X elt B)))).
buildrules.
```
Then call
```
prove(for_all(A,for_all(B,for_all(C,
                        A inc B and B inc C => A inc C)).
```
MUSCADET2 then proves the theorem and displays the trace of the proof. It ends
by writing that the theorem is proved and gives the length of time for the proof.

Work with files and libraries. You may also work with files containing a list
of definitions and a list of theorems to be proved or work with the TPTP problem
library. MUSCADET2 accepts the TPTP syntax but translates statements into
the syntax that was previously described and which is used for the trace.
As MUSCADET must know if a statement is a definition or a lemma, it analyses
TPTP axioms and hypotheses and classifies them either as definitions or as
lemmas.

6   Elimination of Functional Symbols

Strategies of MUSCADET are designed to work with mathematical or logical
predicates rather than with functional symbols. Nevertheless, MUSCADET accepts statements written with functions, but it "eliminates" them by giving a
name to functional expressions which will replace this expression in the predicative formula. So, $p(f(a))$ will be replaced by $f(a) : b$ and $p(b)$. The symbol
":" is used to express that b is the object $f(a)$, and the formula $f(a) : b$ will be
handled as if it was a predicative formula $p_f(a, b)$.

For formulas with variables it is a little more complicated. A statement of
the form $p(f(X))$ where f is a functional symbol means *for the only Y equal to
f(X) p(Y) is true*. It is equivalent to the two following statements $\forall Y(f(X) :
Y \Rightarrow p(Y))$ and $\exists Y(f(X) : Y \wedge p(Y))$. Depending on the context, one or the
other of these two statements is preferable. The reasons for this are developed
in [4]. So, $p(f(X))$ is replaced by only(f(X):Y, p(Y)) which will be handled
by specific rules.

7 Metarules

Metarules automatically build rules from the definitions and lemmas. These rules are more operational than the definitions and lemmas themselves.

Other metarules build the list of active rules, which is the list of rules that are pertinent to the theorem to be proved. Rules will be tried in the listed order. If this order is important, it will have to be stated by metarules.

8 Some Strategies

The strategies are rather classic ones. They come from natural deduction. Sometimes it is necessary to avoid carrying out some treatments too early in order to avoid possible infinite branches or too much splitting.

There are no universal hypotheses since the super-action *addhyp*, instead of adding them, considers them as lemmas and creates new rules which are local for the current (sub-)theorem.

Existential hypotheses lead to the creation of new mathematical objects, but this is done very carefully in order to avoid generating infinitely many objects in only one direction.

Disjunctive hypotheses lead to splitting but, as for existential hypotheses, this is done not too early and one by one, in order to avoid multiplying splitting needlessly.

9 Perspectives

The strategies of MUSCADET2 will continue to be improved and refined. The building of rules will be developed in order to perform new actions while avoiding infinite chains of created objects.

For domains that require mathematical heuristics as data, the work already done in [4] will be taken up again, but this time expressing all this knowledge by first order statements and writing new metarules capable of deducing effective rules.

References

1. Bledsoe, W.W. Splitting and reduction heuristics in automatic theorem proving, Journal of Artificial Intelligence 2 (1971), 55-77
2. Bledsoe, W.W., Non-resolution theorem proving, Journal of Artificial Intelligence 9 (1977), 1-35
3. Pastre D., Automatic theorem Proving in Set theory, Journal of Artificial Intelligence 10 (1978), 1-27
4. Pastre D., MUSCADET: An Automatic theorem Proving System using Knowledge and Metaknowledge in Mathematics - Journal of Artificial Intelligence 38.3 (1989)
5. Pastre D., Automated theorem Proving in Mathematics, Annals on Artificial Intelligence and Mathematics 8.3-4 (1993), 425-447
6. Pastre D., Entre le déclaratif et le procédural : l'expression des connaissances dans le système MUSCADET, Revue d'Intelligence Artificielle 8.4 (1995), 361-381

Hilberticus - A Tool Deciding an Elementary Sublanguage of Set Theory

Jörg Lücke

Univ. Dortmund, Fachbereich Informatik, LS 5, 44227 Dortmund, Germany
luecke@ls5.cs.uni-dortmund.de, http://www.hilberticus.de

Abstract. We present a tool deciding a fragment of set theory. It is designed to be easily accessible via the internet and intuitively usable by anyone who is working with sets to describe and solve problems. The tool supplies features which are well-suited for teaching purposes as well. It offers a self explaining user interface, a parser reflecting the common operator bindings, parse tree visualization, and the possibility to generate Venn diagrams as examples or counterexamples for a given formula. The implemented decision procedure which is based on the semantics of class theory is particularly suitable for this.

Keywords: Set Theory, Decision Procedures, First-Order Logic.

1 Motivation

The language of set theory is one of the most common formal languages. It is used in research fields ranging from mathematics, computer science over natural science and engineering up to economy. This success is probably due to the fact that set theory can be used informally in most of its applications and that the well-known Venn diagrams can illustrate relations and combinations of sets. In contrast to this, in logics the precise definition of the syntax and semantics of languages of set theory is crucial and it provides methods which allow the automatic decision whether formulas of set theory are true, satisfiable, or inconsistent. However, the number of actual users of such methods is far smaller than the number of users of set theory. This is probably the case because an average user of the language of set theory is often not aware of such methods and because existing implementations are not easy to find, to install, and often difficult to learn even if the user should be familiar with the notation of logic. The Hilberticus tool is designed to overcome these problems and it supplies the user with a graphical feedback in the form of Venn diagrams which eases its use and illustrates examples and counterexamples of a satisfiable formula in a convenient way. An alpha version of the tool is accessible via www.hilberticus.de. The name *Hilberticus* is a synthesis of the names *Hilbert* and *abacus*.

2 Syntax and Semantics of the Language SL

Syntax of **SL**. The language **SL** (**S**et **L**anguage) contains the parentheses symbols ')' and '(', the logical connectives ¬, ∧, ∨, ⇒, and ⇔, the predicate symbols

R. Goré, A. Leitsch, and T. Nipkow (Eds.): IJCAR 2001, LNAI 2083, pp. 690–695, 2001.
© Springer-Verlag Berlin Heidelberg 2001

=, ⊆, and ⊂, the function symbols ∩, ∪, and \, the constant symbols ∅ and $\mathcal{D}$, and a countable set of class variables A, B etc. The formulas of **SL** are defined in the usual way. A string t of the above symbols is called a *term* if t is a constant symbol, if t is a variable, or if t is of the form $(t_1 \bullet t_2)$ where t_1 and t_2 are terms and where $\bullet$ denotes a function symbol. A string Φ is called an *atomic formula* if Φ is of the form $(t_1 = t_2)$, $(t_1 \subseteq t_2)$, or $(t_1 \subset t_2)$ (where t_1 and t_2 are terms). Finally, we call Φ a *formula* if it is a propositional combination of atomic formulas. To write and read **SL**-formulas more conveniently we allow the suppression of parentheses using binding priorities for function symbols and logical connectives. The binding of the function symbols is defined according to the list ∩, ∪, \ which is ordered descendingly by binding priority. Similarly, the binding of the logical connectives is defined according to the list ¬, ∧, ∨, ⇒, ⇔, where ⇒ and ⇔ have the same priority and associate to the right, in contrast to all other binary symbols.

Semantics of **SL**. The semantics of **SL** is taken from [1,2]. There a constructive definition of set theory is developed which is based on the Zermelo-Fraenkel axioms. We denote by $\sigma : V \rightarrow \mathcal{D}$ an *assignment* from the set of variables V into the collection of all classes $\mathcal{D}$. The interpretation I of the predicate and function symbols =, ⊆, ⊂ and ∩, ∪, \ is standard. The constant symbols ∅ and $\mathcal{D}$ are interpreted as empty and universal class respectively. The domain $\mathcal{D}$ of all classes and the interpretation I make up the *model* of interest which we will denote by $\mathcal{M}$. A formula Φ is *satisfiable* in $\mathcal{M}$ if it is true for some assignment σ and Φ is *true* in $\mathcal{M}$ if it is true for all assignments σ.

3   The Decision Procedure and Its Implementation

The decision procedure consists of two steps. Firstly, a formula of **SL** is transformed to a formula of the more fundamental language **SBL** (Set Basis Language) using the calculus **L∈** of [1,2]. Secondly, the formula in **SBL** is decided via the transformation to a propositional formula. The language **SBL** is a first-order language without function symbols and with '∈' as the only (uninterpreted) predicate symbol. The calculus **SBL** is the usual predicate calculus for this signature and we call a formula *valid* if it is true in all models of **SBL**.

The calculus **L∈**. Descriptions of set theory are typically based on a first-order language with '∈' and '=' as the only predicate symbols. As soon as the formulas of this first-order language are becoming large and unreadable abbreviations are introduced and this not only for formulas but also for terms denoting sets or set-like objects. The *abstraction terms* '$\{v \mid \Phi\}$' are used for this purpose. Generally, the use of abstraction terms is only informal which can lead to problems (as discussed in [1]). In **L∈** such problems are avoided by the introduction of abstraction terms as a part of the language together with axioms and inference rules for their manipulation. A number of theorems show the equivalence of set theories based on **L∈** and set theories based on a usual predicate logic. In the following we focus on some elements of **L∈** and formulate relevant results in a way appropriate to the scope of this paper. For a thorough investigation of **L∈** (including the notation of frames) we refer to [1,2]. For our purposes the following derivations are of interest [2]:

$$
\vdash_{L\in} \{v|\Phi\} \in t \Leftrightarrow \exists w\,(w=\{v|\Phi\} \land w \in t),
$$

$$
\vdash_{L\in} t \in \{v|\Phi(v)\} \Leftrightarrow \Phi(t), \qquad \vdash_{L\in} t_1=t_2 \Leftrightarrow \forall v\,(v \in t_1 \Leftrightarrow v \in t_2) \tag{1}
$$

By successively applying these formulas it is always possible to eliminate all abstraction terms of an **LE**-formula in favor of an equivalent **SBL**-formula. The fact that the additional axioms and rules for the abstraction terms do only serve as abbreviations in our case is reflected by the following theorem.

Theorem 1. *If Φ is a formula in* **LE** *and $\tilde{\Phi}$ the formula in* **SBL** *derived by the application of the formulas of (1) to Φ, then Φ is derivable in* **LE** *iff $\tilde{\Phi}$ is derivable in* **SBL**,

$$
\vdash_{L\in} \Phi \quad \textit{iff} \quad \vdash_{SBL} \tilde{\Phi}. \tag{2}
$$

We can now use the calculus **LE** to introduce the following abbreviations:

$$
A \subseteq B \Leftrightarrow_{def} \forall x\,(x \in A \Rightarrow x \in B), \qquad A \cup B =_{def} \{y|\ y \in A \lor y \in B\},
$$

$$
A \subset B \Leftrightarrow_{def} A \subseteq B \land \neg(A=B), \qquad A \cap B =_{def} \{y|\ y \in A \land y \in B\}, \tag{3}
$$

$$
\emptyset =_{def} \{x|\neg(x=x)\}, \quad \mathcal{D} =_{def} \{x|\ x=x\}, \quad A \setminus B =_{def} \{y|\ y \in A \land \neg(y \in B)\}
$$

To translate a formula in **SL** to a formula in **SBL** we replace the predicates and functions of **SL** by the corresponding definitions and subsequently apply the formulas of (1) as shown below. The semantics of **SL** (Sect.2) is the same as the semantics of **LE** defined in [1]. Due to the soundness and completeness of the calculus **LE** [1] a formula Φ of **SL** is thus true in the underlying model $\mathcal{M}$ if it is derivable in **LE**. Together with (1) and the soundness and completeness of the predicate calculus we obtain the result:

Theorem 2. *If Φ is a formula in* **SL** *and $\tilde{\Phi}$ the formula derived after the expansion of Φ by the use of the abbreviations of (3) and the successive elimination of abstraction terms by the use of the formulas of (1) then Φ is true in $\mathcal{M}$ iff $\tilde{\Phi}$ is valid in* **SBL**.

Theorem 2 reduces the decidability problem of **SL** to the decidability problem of the calculus **SBL**.

Transformation to a propositional formula. As mentioned earlier the calculus **SBL** is a predicate calculus without function symbols and with '$\in$' as the only predicate symbol. Such a calculus is undecidable in general (see for instance [3]). The translation procedure described above produces, however, formulas lying in a certain subset of **SBL** which we call **SBL²**. The variables of a formula of this subset can always be divided into one set of variables occurring only on the left-hand side of the '$\in$' predicate (element variables) and one set of variables occurring only on the right-hand side of the '$\in$' predicate (class variables). Furthermore only quantifiers for element variables are introduced in the course of the translation from **SL** to **SBL** because due to the syntax of **SL** the first formula of (1) is not applied. A formula in $\mathbf{SBL}^2$ can be transformed to a propositional formula $\mathcal{F}(\rho_1, \ldots, \rho_m)$ where $\rho_1, \ldots, \rho_m$ are $\mathbf{SBL}^2$-formulas of the form:

$$\underbrace{\forall x(x \in A_1 \vee \ldots \vee x \in A_n)}_{\rho_1}, \underbrace{\forall x(x \in A_1 \vee \ldots \vee \neg(x \in A_n))}_{\rho_2}, \ldots, \underbrace{\forall x(\neg(x \in A_1) \vee \ldots \vee \neg(x \in A_n))}_{\rho_m}$$

This is possible because due to the restricted quantification a universal quantifier $\forall x$ can be moved (after elimination of $\Rightarrow$ and $\Leftrightarrow$) to the right until a disjunction of the form $(x \in A_1 \vee \ldots \vee x \in A_k)$ is encountered (where $k \leq n$, and atomic formulas are possibly negated). This can be achieved by the use of the formulas below where $\Phi_{(x)}, \Psi_{(x)}$ denote formulas containing x as a free variable and Φ, Ψ are formulas *not* containing x as a free variable.

$$\underset{\text{SBL}}{\vdash} \quad \forall x(\Phi_{(x)} \wedge \Psi_{(x)}) \Leftrightarrow ((\forall x\ \Phi_{(x)}) \wedge (\forall x\ \Psi_{(x)})),$$

$$\underset{\text{SBL}}{\vdash} \quad \forall x(\Phi \vee \Psi_{(x)}) \Leftrightarrow (\Phi \vee (\forall x\ \Psi_{(x)})), \quad \underset{\text{SBL}}{\vdash} \quad \Phi \Leftrightarrow (\Phi \vee \Psi) \wedge (\Phi \vee \neg \Psi) \tag{4}$$

In Example 1 below we demonstrate the complete transformation of an **SL**-formula to a propositional formula. We use the abbreviations of (3), the second formula of (1), the formulas of (4), and the notation:

$$P_x^A := x \in A, \quad \overline{P_x^A} := \neg(x \in A), \quad [P_x^A, \ldots, P_x^C] := P_x^A \vee \ldots \vee P_x^C.$$

Example 1

$A \cap B \subseteq C \Rightarrow (A \subseteq C \vee B \subseteq C)$

$\{v | v \in A \wedge v \in B\} \subseteq C \Rightarrow (A \subseteq C \vee B \subseteq C)$

$(\forall x(x \in \{v | v \in A \wedge v \in B\} \Rightarrow x \in C)) \Rightarrow ((\forall y(y \in A \Rightarrow y \in C)) \vee (\forall z(z \in B \Rightarrow z \in C)))$

$(\forall x((x \in A \wedge x \in B) \Rightarrow x \in C)) \Rightarrow ((\forall y(y \in A \Rightarrow y \in C)) \vee (\forall z(z \in B \Rightarrow z \in C)))$

$(\forall x((P_x^A \wedge P_x^B) \Rightarrow P_x^C)) \Rightarrow ((\forall y(P_y^A \Rightarrow P_y^C)) \vee (\forall z(P_z^B \Rightarrow P_z^C)))$

$\neg(\forall x(\neg(P_x^A \wedge P_x^B) \vee P_x^C)) \vee ((\forall y(\overline{P_y^A} \vee P_y^C)) \vee (\forall z(\overline{P_z^B} \vee P_z^C)))$

$\neg(\forall x[\overline{P_x^A}, \overline{P_x^B}, P_x^C]) \vee (\forall y[\overline{P_y^A}, P_y^C]) \vee (\forall z[\overline{P_z^B}, P_z^C])$

$\neg(\forall x[\overline{P_x^A}, \overline{P_x^B}, P_x^C]) \vee (\forall y([\overline{P_y^A}, P_y^C, P_y^B] \wedge [\overline{P_y^A}, P_y^C, \overline{P_y^B}])) \vee (\forall z([\overline{P_z^B}, P_z^C, P_z^A] \wedge [\overline{P_z^B}, P_z^C, \overline{P_z^A}]))$

$\neg(\forall x[\overline{P_x^A}, \overline{P_x^B}, P_x^C]) \vee (\forall y[\overline{P_y^A}, P_y^C, P_y^B] \wedge \forall y[\overline{P_y^A}, P_y^C, \overline{P_y^B}]) \vee (\forall z[\overline{P_z^B}, P_z^C, P_z^A] \wedge \forall z[\overline{P_z^B}, P_z^C, \overline{P_z^A}])$

$\underbrace{\neg(\forall x[\overline{P_x^A}, \overline{P_x^B}, P_x^C])}_{\rho_7} \vee (\underbrace{\forall x[\overline{P_x^A}, P_x^B, P_x^C]}_{\rho_5} \wedge \underbrace{\forall x[\overline{P_x^A}, \overline{P_x^B}, P_x^C]}_{\rho_7}) \vee (\underbrace{\forall x[P_x^A, \overline{P_x^B}, P_x^C]}_{\rho_3} \wedge \underbrace{\forall x[\overline{P_x^A}, \overline{P_x^B}, P_x^C]}_{\rho_7})$

The resulting propositional formula is $\mathcal{F} = \neg \rho_7 \vee (\rho_5 \wedge \rho_7) \vee (\rho_3 \wedge \rho_7)$. In general the formula can have $m = 2^n$ ρ-arguments if n is the number of different variables occurring in the **SL**-formula. The ρ-arguments are independent of one another except that all of them cannot be true simultaneously. The validity of a formula in $\mathbf{SBL}^2$ can be decided by a Boolean valuation of the propositional formula $\mathcal{F}$ excluding the case that all ρ-arguments are simultaneously true. Note at this point that a formula $\mathcal{F}$ in which all possible ρ-arguments occur represents the worst case. In the great majority of cases only a small subset of the theoretically possible ρ-arguments actually occur in $\mathcal{F}$ making the evaluation of $\mathcal{F}$ much more efficient (see Example 1). In the case of a satisfiable **SL**-formula it is possible to construct a collection of finite sets from a Boolean valuation $b : \{\rho_1, \ldots, \rho_m\} \rightarrow \{false, true\}^m$ of the corresponding formula $\mathcal{F}$. These sets can subsequently serve to generate Venn diagrams as counterexamples (or examples) of the **SL**-formula. A description of such a generation can be found on the homepage of the tool.

The implementation. The Hilberticus tool decides whether an **SL**-formula is true, satisfiable, or inconsistent in the model $\mathcal{M}$ described in Section 2. Given an **SL**-formula the tool generates a strictly typed abstract syntax tree (AST) using the syntax and the priorities of functions and logical connectives described in Section 2. Then the tree is transformed to an AST representing the corresponding **SBL**-formula which is subsequently decided according to the described procedure. The tool is written in Java© to be easily accessible. It is made up of different modules which are tested and bound together within the Electronic Tool Integration platform (ETI) [4], the experimental platform of the *Int. J. STTT*. The integrated parser was generated using SableCC, a suitable Java© compiler compiler [5]. The ASTs of **SL**- and **SBL**-formulas can be visualized using the PLGraph class library which is supplied by the ETI platform.

4   Related and Future Work

The language **SL** is a sublanguage of a language first described in [6] and later named Multi-level Syllogistic (**MLS**), see for instance [7]. Later on, various variants of (**MLS**) have been shown to be decidable. The most recent decision procedures use semantic tableaux [7]. The semantics of the languages is based on Zermelo-Fraenkel set theory or parts of it. The sublanguage of **MLS** which is the most similar to **SL** is called **2LS** and is described together with a decision procedure in [8]. As mentioned earlier the Hilberticus tool is the first implementation of the decision procedure described in Section 3. It was chosen because it supplies a convenient way to obtain finite sets for Venn diagram generation and because of the calculus **L𝕮** offering a natural possibility to extend the language. The use of abstraction terms makes it an easy task to introduce new function symbols. With a test implementation containing a generalized decision procedure we were able to find the incorrect formula $\bigcup(M{\cap}N) = \bigcup M{\cap}\bigcup N$ in [9, p.545], a book which is used as reference for all kinds of mathematical formulas. We are currently working on a version of the tool which uses tableaux based procedures such as mentioned above and the described translation to **SBL** in combination with decision (or verification) procedures for predicate and monadic logic. In this context the ETI platform supplies the ideal environment for the integration, comparison, and testing of these translation and decision procedures.

References

1. Glubrecht, J.-M., Oberschelp, A., Todt, G.: Klassenlogik. BI Wissenschaftsv. (1983)
2. Oberschelp, A.: Allgemeine Mengenlehre. BI Wissenschaftsverlag (1994)
3. Enderton, H.B.: A mathematical introduction to logic. Acad. Press (1972)
4. Steffen, B., Margaria, T., Braun, V.: The Electronic Tool Integration platform: concepts and design. Int. J. STTT (1997) 1:9-30, eti.cs.uni-dortmund.de
5. Gagnon, E. M., Hendren, L. J.: SableCC. In Proc. TOOLS'98, IEEE
6. Ferro, A., Omodeo, E., Schwartz, J.T.: Decision Procedures for Elementary Sublanguages of Set Theory I. Comm. Pure Appl. Math. Vol. XXXIII (1980) 599–608
7. Cantone, D., Ferro, A.: Techniques of Computable Set Theory with Applications to Proof Verification. Comm. Pure Appl. Math. Vol. XLVIII (1995) 901–945
8. Cantone, D., Omodeo, E., Policriti, A.: Set Theory for Computing. Springer (2001)
9. Bronstein, I. N., Semendjajew, K. A.: Taschenbuch der Mathematik. (25. Aufl., 1991)

STRIP: Structural Sharing for Efficient Proof-Search

D. Larchey-Wendling, D. Méry, and Didier Galmiche

LORIA UMR 7503 - Université Henri Poincaré
Campus Scientifique, BP 239 Vandœuvre-lès-Nancy, France

Abstract. The STRIP system is a theorem prover for intuitionistic
propositional logic with two main characteristics: it deals with the duplication of formulae during proof-search from a fine and explicit management of formulae (as resources) based on a structural sharing and it
builds, for a given formula, either a proof or a countermodel.

1 Introduction

In recent years there was a renewed interest in proof-search for constructive logics like intuitionistic logic (IL), mainly because of research in intuitionistic type
theories and their relationships with programming through proof-search. Dif-
ferents methods (based on resolution, connections, translation in classical logic,
constraints calculus) and implementations have been already designed for IL but
our aim in this work is to focus on two main problems: to avoid the duplication
of formulae during proof-search and to efficiently build countermodels in case of
non-provability. Firstly, we consider the propositional fragment of IL (IPL) but
our main goal is to define structural solutions general enough to be applicable
to other substructural or intermediate logics [1]. A good and efficient explicit
management of formulae (as resources), both in the logical system and in the
implementation, is important to have reliable and efficient implementation techniques of logical calculi (and connected proof-search methods), for instance in
imperative programming languages like C or Java. We have already studied this
point in [3] for the contraction-free sequent calculus LJT [2] for which there are
refinements in order to solve the duplication problem [1,2,5]. The STRIP system,
available at `http://www.loria.fr/~larchey/STRIP` decides the provability of
a given IPL sequent and then builds a proof or a countermodel (as a Kripke
tree). It is based on a new logical system, named SLJ [4] and on a structural
solution of the duplication problem (without the introduction of new formulae
and variables like in [2,5]). In order to illustrate and emphasize the interest and
the results of structural sharing and its implementation we have compared, from
various IPL formulae, the STRIP system with Porgi[1], a similar prover for IPL
written in SML, and then with the **ft**[2] system that is not based on LJT but is
written in C like our system.

---

[1] available at `http://www.cis.ksu.edu/~allen/porgi.html`
[2] available at `http://www.sics.se/isl/ft.html`

R. Goré, A. Leitsch, and T. Nipkow (Eds.): IJCAR 2001, LNAI 2083, pp. 696–700, 2001.

$$\frac{\Gamma, A, \boxed{B} \to C \vdash \boxed{B} \qquad \cdots}{\Gamma, (A \to B) \to C \vdash G} \; [(\to)\to] \qquad\qquad \frac{\Gamma, A \to \boxed{C}, B \to \boxed{C} \vdash G}{\Gamma, (A \lor B) \to C \vdash G} \; [(\lor)\to]$$

Fig. 1. Rules and duplication

2 Formulae Duplication and Sharing

In LJT [2], two kinds of formulae duplication appear even though the system is contraction-free: these are illustrated in the rules of figure 1. Let us give a brief overview of the results and techniques presented in [4]. The duplication on the lhs is treated as in [5], introducing a mark and a so-called *boxed sequent*: $\Gamma, A, B^\star \to C \vdash \blacksquare$ with the intended meaning of $\Gamma, A, B \to C \vdash B$.

The duplication on the rhs is addressed on a *structural* way. Whereas the lhs part of a sequent is usually considered to be a *flat* list a formulae, we use a list of trees, i.e. a *formulae-indexed forest*. Thus, the sequents are represented by specific trees in which for-mulae are paths from roots to leaves and log-

$$(A \lor B) \to C \mid A \to C, B \to C$$

Fig. 2. Logical rule

ical operations are operations on the tree leaves. The problem of duplication is then a problem of structural sharing in such trees. Similar ideas can be applied to a refutability system in order to generate countermodels in case of non-provability [6]. By such an approach, there is no formulae duplication anymore: each subformula is used at most once in a proof-search branch. During proof-search the structure of the forest changes but not its size. The STRIP system, that provides proofs or countermodels for IPL formulae, is based on structural sharing techniques with the following results: no dynamic memory allocation, a finer control on the resources and a $\mathcal{O}(n \log n)$-space algorithm for provability [4].

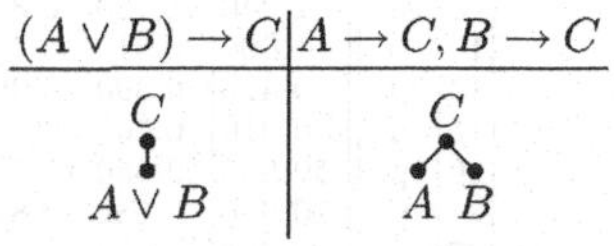

3 Structures and Strategies

The formulae-indexed forest data structure has to be implemented in such a way that the *administrative*, *logical* operations, and those related to *strategies* take the less time possible. The leaves are chained into a list to provide fast access to active formulae (which are those indexing leaves). The STRIP system includes two different implementations of this structure depending on the way to deal with the operations of cutting or pasting subtrees. In the first one, called *lrmost*, one memorizes for each node the indexes of the leftmost and rightmost leaves under this node. In the second one, called *index-scope*, one computes for each node its scope that is the greatest index among the indexes that could potentially be under the current node. Moreover the system proposes two proof-search methods (or *strategies* for the choice of the leaf to develop at each step). The strategy, called *first-leaf*, chooses the first active leaf from a left-to-right search in the set

4 $((A \to (B \vee (B \to C))) \to C) \to C$
6 $A \to B \to ((A \to B \to C) \to C) \to (A \to B \to C)$
7 $(((A \wedge B) \vee C) \to (C \vee (C \wedge D))) \to (\neg A \vee ((A \vee B) \to C))$
8 $((A \leftrightarrow B \vee A \leftrightarrow C \vee B \leftrightarrow C) \to (A \wedge B \wedge C)) \to (A \wedge B \wedge C)$
13 $\neg\neg((\neg A \to B) \to (\neg A \to \neg B) \to A)$
14 $(\neg(A \to (B \vee C))) \to (B \vee C) \to A$
15 $\neg\neg A \vee (A \to \neg\neg B \vee (B \to \neg\neg C \vee (C \to (\neg\neg D \to D) \vee \neg D \vee \neg\neg D)))$
20 $(((G \to A) \to J) \to D \to E) \to (((H \to B) \to I) \to C \to J)$
    $\to (A \to H) \to F \to G \to (((C \to B) \to I) \to D) \to (A \to C)$
    $\to (((F \to A) \to B) \to I) \to E$
21 $\neg\neg(((A \leftrightarrow B) \leftrightarrow C) \leftrightarrow (A \leftrightarrow (B \leftrightarrow C)))$
22 $((\neg\neg(\neg A \vee \neg B) \to (\neg A \vee \neg B)) \to (\neg\neg(\neg A \vee \neg B) \vee \neg(\neg A \vee \neg B)))$
    $\to (\neg\neg(\neg A \vee \neg B) \vee \neg(\neg A \vee \neg B))$
24 Pigeonhole 2-3
25 Pigeonhole 3-4

		Time in 10^{-4} seconds				Number of operations					
	p/u	T_{porgi}	T_{fl}^{lr}	T_{fl}^{is}	T_{rp}^{lr}	T_{rp}^{is}	SS_{fl}	SS_{rp}	SC_{fl}^{lr}	SC_{fl}^{is}	SC_{rp}^{lr}	SC_{rp}^{is}
4	p	0.220	0.280	0.170	0.290	0.210	9	9	71	41	76	46
6	u	0.480	0.390	0.270	0.290	0.230	14	9	104	76	87	71
7	u	0.540	0.280	0.170	0.310	0.240	11	11	82	48	93	59
8	u	1.240	1.180	0.840	0.780	0.570	42	24	253	155	191	131
13	p	0.600	0.570	0.460	0.470	0.340	20	15	133	81	124	75
14	p	0.420	0.390	0.300	0.260	0.200	13	9	99	55	69	44
15	u	0.910	0.560	0.440	0.630	0.510	26	26	158	117	190	149
20	p	10.500	16.880	13.700	2.720	2.220	573	69	4253	3129	900	757
21	p	10.910	3.680	2.820	3.500	2.650	126	97	723	463	918	680
22	u	3.500	15.980	12.660	8.930	7.130	479	220	4104	2935	2677	2157
24	p	21.570	1.390	1.380	1.520	1.450	100	99	452	424	586	560
25	p	225300	26.310	25.550	32.860	31.850	2248	2238	10432	10180	15488	15256

Fig. 3. Comparison : STRIP vs Porgi

of leaves. The strategy, called *rule-prec*, also considers such a left-to-right search but the set of leaves is split in different groups having different priorities and then it selects the first active leaf in the group with the highest priority. With the *rule-prec* strategy one always builds a countermodel in case of non-provability because the invertible rules are applied before the non-invertible rules.

4   Results and Comparisons

For a given sequent, STRIP can decide its provability and then build a proof or a countermodel (as a Kripke tree). The user can select the forest implementation (`lrmost` or `index-scope`) and the strategy (`first-leaf` or `rule-prec`). The system provides various statistics about the search like the number of rules applications (in order to evaluate the efficiency of strategies) or the number of performed operations (in order to determine if the forest implementation induces a huge overhead). We have compared the STRIP system with two provers, namely Porgi [8] and **ft** [7].

Porgi is a proof-or-refutation generator, written in SML, that is based on LJT with a strategy closed to the `rule-prec` strategy. It has a very simple lexical analyzer and thus we have only been able to test it with small formulae.

Some comparisons between Porgi and STRIP are given in figure 3 with formulae that are provable (p) or unprovable (u). The left part of the table presents exe-

Dom.	Size	STRIP	ft	ft/STRIP
\multicolumn — $\neg\neg\forall x\,(p(x) \vee \neg p(x))$				
7	257	1.5 ms	0.41 s	270
8	291	3.1 ms	2.64 s	850
9	325	6.4 ms	26.50 s	4100
10	359	12.8 ms	6 m 28 s	30000
21	733	31 s		
22	767	62 s		
23	801	1 m 50 s		
$\neg\neg(A \wedge B \to C)$ where $\begin{cases} A \equiv \exists x\,(p(x) \wedge \forall y(q(y) \to r(x,y))) \\ B \equiv \neg\exists x\,(q(x) \wedge \forall y(p(y) \to r(x,y))) \\ C \equiv \exists x\,(p(x) \wedge \neg q(x)) \end{cases}$				
4	1254	8.4 ms	130 ms	15
5	1870	90.0 ms	4.62 s	50
6	2610	1.53 s	4 m 19 s	170
7	3474	29.5 s	+ 1 h20 m	
8	4462	11 m 26 s		
$\neg\neg\left[\neg\exists x \forall y \exists z\, p(x,y,z) \leftrightarrow \forall x \exists y \forall z\, \neg p(x,y,z)\right]$				
2	495	0.42 ms	10 ms	24
3	1597	13.6 ms	+ 9 h	$+\,10^6$
4	3743	1.07 s		
Pigeonhole x-y				
5-6	1788	2.23 s	0.33 s	0.14
6-7	2918	64 s	16 s	0.25
7-8	4446	32 m 44 s	19 m 27 s	0.60

Fig. 4. Comparison : STRIP vs **ft**

cution times for both systems (T_y^x in seconds for 10^4 loops CPU time only). The exponent 'lr' (resp. 'is') corresponds to the `lrmost` (resp. `index-scope`) implementation of the forest. The suffix 'fl' (resp 'rp') corresponds to the `first-leaf` (resp. `rule-prec`) strategy. The right part presents measures of the space size and search costs for STRIP proof-search. The expressions SS_y and SC_y^x respectively represent the number of logical rules applications and the total number of operations during the proof-search.

The STRIP system is always more efficient with the `rule-prec` strategy, which is almost the same as the one implemented in Porgi. Moreover, the difference is significant for the pigeon-hole formulae. We can also compare the strategies. In examples 20 and 22, we see that `first-leaf` is much less efficient than `rule-prec` because of the size of the proof-search space and thus in this case STRIP is slower than Porgi. The pigeon-hole examples illustrate the actual impact of sharing techniques and that some tasks, like forest management, are implemented much more efficiently in our system. From the SC statistics, i.e. measures of the search cost, we observe that the `index-scope` implementation of the forest is always better than the `lrmost` one but that the difference does not grow over factor 2, whichever strategy is chosen. However it is clear that efficient management of formulae becomes crucial on larger scales.

The **ft** system is suited for first-order logic but has a propositional subsystem for IPL. This system is not based on LJT and does not build countermodels but it is written in C like STRIP. Some comparisons between STRIP and **ft** are given in figure 4. In order to analyze the impact of sharing techniques on proof-search, we have considered instances on finite domains of provable first-order formulae and also pigeonhole examples. The Dom. column represents the size of the domain or the number of pigeons and the Size column represents the size of the generated formulae[3]. We observe that, in general, STRIP is much faster than **ft** for the first kind of examples. But for the pigeonhole examples both systems are on par with a slight but decreasing advantage to **ft**. In fact they include very few implications ($\rightarrow$) and the structural sharing is such that left implication rules may cut down the problems by large amounts. In this case, the choice of a light strategy is important. We observe that with the `rule-prec` strategy, STRIP spends 85 % of its computation time looking for the active formulae. With some cyclic variants of `first-leaf`, we can cut-down this time to 65 % which is not optimum but nevertheless better, thus being close to **ft** and even better on the larger case (7-8). Anyway, we see that the greater the pigeon problem is, the better STRIP behaves, compared to **ft**.

Further work will be devoted to other tests and comparisons but regarding our positive results, our main goal is to apply or extend these implementation techniques (forest representation, structural sharing, forest implementation) to other substructural or intermediate logics [1] and thus to provide efficient provers that build proofs or countermodels for such logics.

References

1. A. Avellone, M. Ferrari, and P. Miglioli. Duplication-free tableau calculi and related cut-free sequent calculi for the interpolable propositional intermediate logics. *Logic Journal of the IGPL*, 7(4):447–480, 1999.
2. R. Dyckhoff. Contraction-free sequent calculi for intuitionistic logic. *Journal of Symbolic Logic*, 57:795–807, 1992.
3. D. Galmiche and D. Larchey-Wendling. Formulae-as-resources management for an intuitionistic theorem prover. In *5th Workshop on Logic, Language, Information and Computation, WoLLIC'98*, Sao Paulo, Brazil, July 1998.
4. D. Galmiche and D. Larchey-Wendling. Structural sharing and efficient proof-search in propositional intuitionistic logic. In *Asian Computing Science Conference, ASIAN'99, LNCS 1742*, pages 101–112, Phuket, Thailand, December 1999.
5. J. Hudelmaier. An O(n log n)-space decision procedure for intuitionistic propositional logic. *Journal of Logic and Computation*, 3(1):63–75, 1993.
6. L. Pinto and R. Dyckhoff. Loop-free construction of counter-models for intuitionistic propositional logic. In Behara and al., editors, *Symposia Gaussiana*, pages 225–232, 1995.
7. D. Sahlin, T. Franzén, and S. Haridi. An intuitionistic predicate logic theorem prover. *Journal of Logic and Computation*, 2(5):619–656, 1993.
8. A. Stoughton. Porgi: a proof-or-refutation generator for intuitionistic propositional logic. In *CADE Workshop on Proof-search in Type-theoretic Languages*, pages 109–116, Rutgers University, New Brunswick, USA, 1996.

---

[3] The Size grows in n^p where p is the number of variables and n the size of the domain. For the pigeonhole, the size grows in n^3.

RACER System Description

Volker Haarslev and Ralf Möller

University of Hamburg, Computer Science Department
Vogt-Kölln-Str. 30, 22527 Hamburg, Germany
http://kogs-www.informatik.uni-hamburg.de/~haarslev|moeller/

Abstract. RACER implements a TBox and ABox reasoner for the logic $\mathcal{SHIQ}$. RACER was the first full-fledged ABox description logic system for a very expressive logic and is based on optimized sound and complete algorithms. RACER also implements a decision procedure for modal logic satisfiability problems (possibly with global axioms).

1 Introduction

The description logic (DL) $\mathcal{SHIQ}$ [18] extends the logic $\mathcal{ALCNH}_{R^+}$ [9] by additionally providing qualified number restrictions and inverse roles. $\mathcal{ALCNH}_{R^+}$ was the logic supported by RACE (Reasoner for ABoxes and Concept Expressions), the precursor of RACER (Renamed ABox and Concept Expression Reasoner). Using the $\mathcal{ALCNH}_{R^+}$ naming scheme, $\mathcal{SHIQ}$ could be called $\mathcal{ALCQHI}_{R^+}$ (pronunciation: ALC-choir).

$\mathcal{ALCQHI}_{R^+}$ is briefly introduced as follows. We assume a set of concept names C, a set of role names R, and a set of individual names O. The mutually disjoint subsets P and T of R denote non-transitive and transitive roles, respectively ($R = P \cup T$). $\mathcal{ALCQHI}_{R^+}$ is introduced in Figure 1 using a standard Tarski-style semantics. The term $\top$ ($\bot$) is used as an abbreviation for $C \sqcup \neg C$ ($C \sqcap \neg C$).

If $R, S \in R$ are role names, then $R \sqsubseteq S$ is called a *role inclusion axiom*. A *role hierarchy* $\mathcal{R}$ is a finite set of role inclusion axioms. Then, we define $\sqsubseteq^*$ as the reflexive transitive closure of $\sqsubseteq$ over such a role hierarchy $\mathcal{R}$. Given $\sqsubseteq^*$, the set of roles $R^{\downarrow} = \{S \in R \mid S \sqsubseteq^* R\}$ defines the *sub-roles* of a role R. We also define the set $S := \{R \in P \mid R^{\downarrow} \cap T = \emptyset\}$ of *simple* roles that are neither transitive nor have a transitive role as sub-role.

The concept language of $\mathcal{ALCQHI}_{R^+}$ syntactically restricts the combination of number restrictions and transitive roles. Number restrictions are only allowed for simple roles. This restriction is motivated by a known undecidability result in case of an unrestricted syntax [17]. In concepts, instead of a role name R (or S), the inverse role R^{-1} (or S^{-1}) may be used.

If C and D are concepts, then $C \sqsubseteq D$ is a terminological axiom (*generalized concept inclusion* or *GCI*). A finite set of terminological axioms $\mathcal{T}_\mathcal{R}$ is called a *terminology* or *TBox* w.r.t. to a given role hierarchy $\mathcal{R}$.[1] An *ABox* $\mathcal{A}$ is a finite set of assertional axioms as defined in Figure 1.

---

[1] The reference to $\mathcal{R}$ is omitted in the following if we use $\mathcal{T}$.

R. Goré, A. Leitsch, and T. Nipkow (Eds.): IJCAR 2001, LNAI 2083, pp. 701–705, 2001.
© Springer-Verlag Berlin Heidelberg 2001

Syntax	Semantics
Concepts	
A	$A^{\mathcal{I}} \subseteq \Delta^{\mathcal{I}}$
$\neg C$	$\Delta^{\mathcal{I}} \setminus C^{\mathcal{I}}$
$C \sqcap D$	$C^{\mathcal{I}} \cap D^{\mathcal{I}}$
$C \sqcup D$	$C^{\mathcal{I}} \cup D^{\mathcal{I}}$
$\exists R . C$	$\{a \in \Delta^{\mathcal{I}} \mid \exists b \in \Delta^{\mathcal{I}} : (a, b) \in R^{\mathcal{I}}, b \in C^{\mathcal{I}}\}$
$\forall R . C$	$\{a \in \Delta^{\mathcal{I}} \mid \forall b \in \Delta^{\mathcal{I}} : (a, b) \in R^{\mathcal{I}} \Rightarrow b \in C^{\mathcal{I}}\}$
$\exists_{\geq n} S . C$	$\{a \in \Delta^{\mathcal{I}} \mid \,\|\{y \mid (x, y) \in S^{\mathcal{I}}, y \in C^{\mathcal{I}}\}\| \geq n\}$
$\exists_{\leq n} S . C$	$\{a \in \Delta^{\mathcal{I}} \mid \,\|\{y \mid (x, y) \in S^{\mathcal{I}}, y \in C^{\mathcal{I}}\}\| \leq n\}$
Roles	
R	$R^{\mathcal{I}} \subseteq \Delta^{\mathcal{I}} \times \Delta^{\mathcal{I}}$

A is a concept name and $\|\cdot\|$ denotes the cardinality of a set. Furthermore, we assume that $R \in R$ and $S \in S$.

Axioms	
Syntax	Satisfied if
$R \in T$	$R^{\mathcal{I}} = (R^{\mathcal{I}})^{+}$
$R \sqsubseteq S$	$R^{\mathcal{I}} \subseteq S^{\mathcal{I}}$
$C \sqsubseteq D$	$C^{\mathcal{I}} \subseteq D^{\mathcal{I}}$

Assertions	
Syntax	Satisfied if
$a : C$	$a^{\mathcal{I}} \in C^{\mathcal{I}}$
$(a, b) : R$	$(a^{\mathcal{I}}, b^{\mathcal{I}}) \in R^{\mathcal{I}}$

Fig. 1. Syntax and Semantics of $\mathcal{ALCQHI}_{R^+}$.

An interpretation $\mathcal{I}$ is a *model* of a concept C (or *satisfies* a concept C) iff $C^{\mathcal{I}} \neq \emptyset$ and for all $R \in R$ it holds that iff $(x, y) \in R^{\mathcal{I}}$ then $(y, x) \in (R^{-1})^{\mathcal{I}}$ An interpretation is a model of a TBox T iff it satisfies all axioms in T. See Figure 1 for the satisfiability conditions. An interpretation is a model of an ABox $\mathcal{A}$ w.r.t. a TBox iff it is a model of T and satisfies all assertions in $\mathcal{A}$. Different individuals are mapped to different domain objects (unique name assumption).

2   Inference Services

In the following we define several inference services offered by RACER.

A *concept* is called *consistent* (w.r.t. a TBox T) iff there exists a model of C (that is also a model of T and $\mathcal{R}$). An *ABox* $\mathcal{A}$ is *consistent* (w.r.t. a TBox T) iff $\mathcal{A}$ has model $\mathcal{I}$ (which is also a model of T). A *knowledge base* $(T, \mathcal{A})$ is called *consistent* iff there exists a model for $\mathcal{A}$ which is also a model for T. A concept, ABox, or knowledge base that is not consistent is called *inconsistent*.

A concept D *subsumes* a concept C (w.r.t. a TBox T) iff $C^{\mathcal{I}} \subseteq D^{\mathcal{I}}$ for all interpretations $\mathcal{I}$ (that are models of T). If D subsumes C, then C is said to be *subsumed by* D.

Besides these basic problems, some additional inference services are provided by description logic systems. A basic reasoning service is to compute the subsumption relationship between concept names (i.e. elements from C). This inference is needed to build a hierarchy of concept names w.r.t. specificity. The problem of computing the most-specific concept names mentioned in T that subsume a certain concept is known as computing the *parents* of a concept. The *children* are the most-general concept names mentioned in T that are subsumed by a certain concept. We use the name *concept ancestors* (*concept descendants*) for the transitive closure of the parents (children) relation. The computation of the parents and children of every concept name is also called *classification* of the TBox. Another important inference service for practical knowledge representation is to check whether a certain concept name is inconsistent. Usually, inconsistent concept names are the consequence of modeling errors. Checking the consistency of all concept names mentioned in a TBox without computing the parents and children is called a TBox *coherence check*.

If the description logic supports full negation, consistency and subsumption can be mutually reduced to each other since D subsumes C (w.r.t. a TBox $\mathcal{T}$) iff C ⊓ ¬D is inconsistent (w.r.t. $\mathcal{T}$) and C is inconsistent (w.r.t. $\mathcal{T}$) iff C is subsumed by ⊥ (w.r.t. $\mathcal{T}$). Consistency of concepts can be reduced to ABox consistency as follows: A concept C is consistent (w.r.t. a TBox $\mathcal{T}$) iff the ABox {a:C} is consistent (w.r.t. $\mathcal{T}$). An individual i is an *instance* of a concept C (w.r.t. a TBox $\mathcal{T}$ and an ABox $\mathcal{A}$) iff $i^{\mathcal{I}} \in C^{\mathcal{I}}$ for all models $\mathcal{I}$ (of $\mathcal{T}$ and $\mathcal{A}$). Again, for description logics that support full negation for concepts, the instance problem can be reduced to the problem of deciding if the ABox $\mathcal{A} \cup \{a: \neg C\}$ is inconsistent (w.r.t. $\mathcal{T}$). This test is also called *instance checking*. The most-specific concept names mentioned in a TBox $\mathcal{T}$ that an individual is an instance of are called the *direct types* of the individual w.r.t. a knowledge base $(\mathcal{T}, \mathcal{A})$. The direct types inference problems can be reduced to subsequent instance problems. The *retrieval* inference problem is to find all individuals mentioned in an ABox that are an instance of a certain concept C. The set of *fillers* of a role R for an individual i w.r.t. a knowledge base $(\mathcal{T}, \mathcal{A})$ is defined as $\{x \mid (\mathcal{T}, \mathcal{A}) \models (i,x):R\}$ where $(\mathcal{T}, \mathcal{A}) \models ax$ means that all models of $\mathcal{T}$ and $\mathcal{A}$ are also models of ax. The set of *roles* between two individuals i and j w.r.t. a knowledge base $(\mathcal{T}, \mathcal{A})$ is defined as $\{R \mid (\mathcal{T}, \mathcal{A}) \models (i,j):R\}$.

As in other systems, there are some auxiliary queries supported: retrieval of the concept names or individuals mentioned in a knowledge base, retrieval of the set of roles, retrieval of the role parents and children (defined analogously to the concept parents and children, see above), retrieval of the set of individuals in the domain and in the range of a role, etc. As a distinguishing feature to other systems, which is important for many applications, we would like to emphasize that RACER supports multiple TBoxes and ABoxes. Assertions can be added to ABoxes after queries have been answered. In addition, RACER also provides support for retraction of assertions in particular ABoxes. The inference services supported by RACER for TBoxes and ABoxes are described in detail in [11].

3   The RACER Architecture

The ABox consistency algorithm implemented in the RACER system is based on the tableaux calculus of its precursor RACE [9]. For dealing with qualified number restrictions and inverse roles, the techniques introduced in the tableaux calculus for $\mathcal{SHIQ}$ [18] are employed.

However, optimized search techniques are required in order to guarantee good average-case performance. The RACER architecture incorporates the following standard optimization techniques: dependency-directed backtracking [22] and

V. Haarslev and R. Möller

DPLL-style semantic branching (see [6] for an overview of the literature). Among a set of new optimization techniques, the integration of these techniques into DL reasoners for concept consistency has been described in [15]. The implementation of these techniques in the ABox reasoner RACER differs from the implementation of other DL systems, which provide only concept consistency (and TBox) reasoning. The latter systems have to consider only so-called "labels" (sets of concepts) whereas an ABox prover such as RACER has to explicitly deal with individuals (nominals). ABox optimizations are also explained in [8].

The techniques for TBox reasoning described in [3] (marking and propagation as well as lazy unfolding) are also supported by RACER. As indicated in [7], the architecture of RACER is inspired by recent results on optimization techniques for TBox reasoning [16], namely transformations of axioms (GCIs) [19], model caching [8] and model merging [15] (including so-called deep model merging and model merging for ABoxes [13]). RACER also provides additional support for very large TBoxes (see [10]).

RACER is implemented in Common Lisp and is available for research purposes as a server program which can be installed under Linux and Windows (http://kogs-www.informatik.uni-hamburg.de/~race). Specific licenses are not required. Client programs can connect to the RACER DL server via a very fast TCP/IP interface based on sockets. Client-side interfaces for Java and Common Lisp are available. A C/C++ interface is available soon.

4 Applications

An application of RACER for ontology engineering is described in [10]. The theory behind another application of RACER in the domain of telecommunication systems is explained in [2]. RACER has also be used for solving modal logic satisfiability problems [8] and for database integration tasks. The Java interface has been developed in order to support a TBox learning application (see [1]).

5 Outlook

The integration of techniques for representing "concrete domains" (e.g. linear inequalities between real numbers) on the role fillers of an individual has been investigated in [14]. In addition, optimization techniques for dealing with qualified number restrictions [12] will be integrated into RACER in the next release.

References

1. J. Alvarez. Tbox acquisition and information theory. In Baader and Sattler [4], pages 11–20.
2. C. Areces, W. Bouma, and M. de Rijke. Description logics and feature interaction. In Lambrix et al. [20], pages 33–36.
3. F. Baader, E. Franconi, B. Hollunder, B. Nebel, and H.J. Profitlich. An empirical analysis of optimization techniques for terminological representation systems. *Applied Intelligence*, 2(4):109–138, 1994.
4. F. Baader and U. Sattler, editors. *Proceedings of the International Workshop on Description Logics (DL'2000), Aachen, Germany*, August 2000.

5. A.G. Cohn, F. Giunchiglia, and B. Selman, editors. *International Conference on Principles of Knowledge Representation and Reasoning (KR'2000)*, April 2000.
6. J.W. Freeman. *Improvements to propositional satisfiability search algorithms.* PhD thesis, University of Pennsylvania, Computer and Information Science, 1995.
7. V. Haarslev and R. Möller. An empirical evaluation of optimization strategies for ABox reasoning in expressive description logics. In Lambrix et al. [20], pages 115–119.
8. V. Haarslev and R. Möller. Consistency testing: The RACE experience. In R. Dyckhoff, editor, *Proceedings, Automated Reasoning with Analytic Tableaux and Related Methods*, number 1847 in Lecture Notes in Artificial Intelligence, pages 57–61. Springer-Verlag, April 2000.
9. V. Haarslev and R. Möller. Expressive ABox reasoning with number restrictions, role hierarchies, and transitively closed roles. In Cohn et al. [5], pages 273–284.
10. V. Haarslev and R. Möller. High performance reasoning with very large knowledge bases: A practical case study. In B. Nebel H. Levesque, editor, *International Joint Conference on Artificial Intelligence (IJCAI'2001), August 4th - 10th, 2001, Seattle, Washington, USA*. Morgan-Kaufmann, August 2001.
11. V. Haarslev and R. Möller. RACER user's guide and reference manual version 1.5. Technical report, University of Hamburg, Computer Science Department, 2001.
12. V. Haarslev and R. Möller. Signature calculus: Optimizing reasoning with number restrictions. Technical report, University of Hamburg, Computer Science Department, June 2001.
13. V. Haarslev, R. Möller, and A.-Y. Turhan. Exploiting pseudo models for tbox and abox reasoning in expressive description logics. In Massacci [21]. In this volume.
14. V. Haarslev, R. Möller, and M. Wessel. The description logic $\mathcal{ALCNH}_{R+}$ extended with concrete domains. In Massacci [21]. In this volume.
15. I. Horrocks. *Optimising Tableaux Decision Procedures for Description Logics.* PhD thesis, University of Manchester, 1997.
16. I. Horrocks and P. Patel-Schneider. Optimising description logic subsumption. *Journal of Logic and Computation*, 9(3):267–293, June 1999.
17. I. Horrocks, U. Sattler, and S. Tobies. Practical reasoning for expressive description logics. In Harald Ganzinger, David McAllester, and Andrei Voronkov, editors, *Proceedings of the 6th International Conference on Logic for Programming and Automated Reasoning (LPAR'99)*, number 1705 in Lecture Notes in Artificial Intelligence, pages 161–180. Springer-Verlag, September 1999.
18. I. Horrocks, U. Sattler, and S. Tobies. Reasoning with individuals for the description logic SHIQ. In David MacAllester, editor, *Proceedings of the 17th International Conference on Automated Deduction (CADE-17)*, number 1831 in Lecture Notes in Computer Science, Germany, 2000. Springer-Verlag.
19. I. Horrocks and S. Tobies. Reasoning with axioms: Theory and practice. In Cohn et al. [5], pages 285–296.
20. P. Lambrix et al., editor. *Proceedings of the International Workshop on Description Logics (DL'99), July 30 - August 1, 1999, Linköping, Sweden*, June 1999.
21. F. Massacci, editor. *International Joint Conference on Automated Reasoning (IJCAR'2001), June 18-23, 2001, Siena, Italy.*, Lecture Notes in Artificial Intelligence. Springer-Verlag, June 2001.
22. R.M. Stallman and G.J. Sussman. Forward reasoning and dependency-directed backtracking in a system for computer-aided circuit analysis. *Artificial Intelligence*, 9(2):135–196, 1977.

Author Index

Anger, Christian 325
Armando, Alessandro 663
Armelín, Pablo A. 289
Audemard, Gilles 427
Avenhaus, Jürgen 658
Avron, Arnon 529

Baader, Franz 92
Beckert, Bernhard 626
Beeson, Michael 318
Benedetti, Marco 331
Boigelot, Bernard 611
Bonacina, Maria Paola 448

Cerrito, Serenella 137
Cialdea Mayer, Marta 137
Compagna, Luca 663

Doutre, Sylvie 272

Egly, Uwe 561

Fariñas del Cerro, Luis 453
Fauthoux, David 453
Fiedler, Armin 416
Formisano, Andrea 152

Galmiche, Didier 696
Ganzinger, Harald 242, 514
Gasquet, Olivier 453
Giese, Martin 545
Giesl, Jürgen 469
Giunchiglia, Enrico 347, 364

Haarslev, Volker 29, 61, 701
Hähnle, Reiner 182
Happe, Jens 459
Henocque, Laurent 427
Herzig, Andreas 453
Hillenbrand, Thomas 257
Hodas, Joshua S. 670
Hodgson, Kahlil 443

Jodogne, Sébastien 611
Jones, Neil D. 1

Kapur, Deepak 469
Konczak, Kathrin 325
Kreitz, Christoph 421

Larchey-Wendling, D. 696
Letz, Reinhold 381
Lev, Iddo 529
Linke, Thomas 325
Löchner, Bernd 658
Longin, Dominique 453
Lorigo, Lori 421
Lücke, Jörg 690
Luther, Marko 386
Lutz, Carsten 45, 121
Lynch, Christopher 499

Maratea, Massimo 347
Massacci, Fabio 453
McAllester, David 514
Mengin, Jérôme 272
Méry, D. 696
Middeldorp, Aart 593
Möller, Ralf 29, 61, 701
Morawska, Barbara 499
Murray, Neil V. 182

Narizzano, Massimo 364
Nieuwenhuis, Robert 242, 257
Nivela, Pilar 242
de Nivelle, Hans 211
Nogin, Aleksey 421

Omodeo, Eugenio G. 152

Pastre, Dominique 685
Patel-Schneider, Peter F. 464
Paulson, Lawrence C. 5
Peltier, Nicolas 578
Pientka, Brigitte 401
Pliuškevičius, Regimantas 107
Pratt-Hartmann, Ian 211
Pym, David J. 289

Ranise, Silvio 663
Reif, Wolfgang 642
Riazanov, Alexandre 257, 376
Rosenthal, Erik 182

Sattler, Ulrike   76
Schellhorn, G.   642
Schlager, Steffen   626
Schmitt, Stephan   421, 561
Schulz, Stephan   370
Sebastiani, Roberto   464
Slaney, John   443
Stenz, Gernot   381
Stuber, Jürgen   195
Sturm, Holger   121
Szeider, Stefan   168

Tacchella, Armando   347, 364
Tamura, Naoyuki   670
Temperini, Marco   152
Thums, Andreas   642

Tobies, Stephan   92
Turhan, Anni-Yasmin   61

Urbain, Xavier   485

Vardi, Moshe Y.   76
Voronkov, Andrei   13, 257, 376

Waldmann, Uwe   226
Wang, Kewen   305
Wessel, Michael   29
Wolper, Pierre   611
Wolter, Frank   121

Zakharyaschev, Michael   121
Zambonin, Davide   347

Lecture Notes in Artificial Intelligence (LNAI)

Vol. 1866: J. Cussens, A. Frisch (Eds.), Inductive Logic Programming. Proceedings, 2000. X, 265 pages. 2000.

Vol. 1867: B. Ganter, G.W. Mineau (Eds.), Conceptual Structures: Logical, Linguistic, and Computational Issues. Proceedings, 2000. XI, 569 pages. 2000.

Vol. 1881: C. Zhang, V.-W. Soo (Eds.), Design and Applications of Intelligent Agents. Proceedings, 2000. X, 183 pages. 2000.

Vol. 1886: R. Mizoguchi, J. Slaney (Eds.), PRICAI 2000: Topics in Artificial Intelligence. Proceedings, 2000. XX, 835 pages. 2000.

Vol. 1898: E. Blanzieri, L. Portinale (Eds.), Advances in Case-Based Reasoning. Proceedings, 2000. XII, 530 pages. 2000.

Vol. 1889: M. Anderson, P. Cheng, V. Haarslev (Eds.), Theory and Application of Diagrams. Proceedings, 2000. XII, 504 pages. 2000.

Vol. 1891: A.L. Oliveira (Ed.), Grammatical Inference: Algorithms and Applications. Proceedings, 2000. VIII, 313 pages. 2000.

Vol. 1902: P. Sojka, I. Kopeček, K. Pala (Eds.), Text, Speech and Dialogue. Proceedings, 2000. XIII, 463 pages. 2000.

Vol. 1904: S.A. Cerri, D. Dochev (Eds.), Artificial Intelligence: Methodology, Systems, and Applications. Proceedings, 2000. XII, 366 pages. 2000.

Vol. 1910: D.A. Zighed, J. Komorowski, J. Zytkow (Eds.), Principles of Data Mining and Knowledge Discovery. Proceedings, 2000. XV, 701 pages. 2000.

Vol. 1916: F. Dignum, M. Greaves (Eds.), Issues in Agent Communication. X, 351 pages. 2000.

Vol. 1919: M. Ojeda-Aciego, I.P. de Guzman, G. Brewka, L. Moniz Pereira (Eds.), Logics in Artificial Intelligence. Proceedings, 2000. XI, 407 pages. 2000.

Vol. 1925: J. Cussens, S. Džeroski (Eds.), Learning Language in Logic. X, 301 pages 2000.

Vol. 1930: J.A. Campbell, E. Roanes-Lozano (Eds.), Artificial Intelligence and Symbolic Computation. Proceedings, 2000. X, 253 pages. 2001.

Vol. 1932: Z.W. Raś, S. Ohsuga (Eds.), Foundations of Intelligent Systems. Proceedings, 2000. XII, 646 pages.

Vol. 1934: J.S. White (Ed.), Envisioning Machuine Translation in the Information Future. Proceedings, 2000. XV, 254 pages. 2000.

Vol. 1937: R. Dieng, O. Corby (Eds.), Knowledge Engineering and Knowledge Management. Proceedings, 2000. XIII, 457 pages. 2000.

Vol. 1952: M.C. Monard, J. Simão Sichman (Eds.), Advances in Artificial Intelligence. Proceedings, 2000. XV, 498 pages. 2000.

Vol. 1955: M. Parigot, A. Voronkov (Eds.), Logic for Programming and Automated Reasoning. Proceedings, 2000. XIII, 487 pages. 2000.

Vol. 1967: S. Arikawa, S. Morishita (Eds.), Discovery Science. Proceedings, 2000. XII, 332 pages. 2000.

Vol. 1968: H. Arimura, S. Jain, A. Sharma (Eds.), Algorithmic Learning Theory. Proceedings, 2000. XI, 335 pages. 2000.

Vol. 1972: A. Omicini, R. Tolksdorf, F. Zambonelli (Eds.), Engineering Societies in the Agents World. Proceedings, 2000. IX, 143 pages. 2000.

Vol. 1979: S. Moss, P. Davidsson (Eds.), Multi-Agent-Based Simulation. Proceedings, 2000. VIII, 267 pages. 2001.

Vol. 1991: F. Dignum, C. Sierra (Eds.), Agent Mediated Electronic Commerce. VIII, 241 pages. 2001.

Vol. 1994: J. Lind, Iterative Software Engineering for Multiagent Systems. XVII, 286 pages. 2001.

Vol. 2003: F. Dignum, U. Cortés (Eds.), Agent-Mediated Electronic Commerce III. XII, 193 pages. 2001.

Vol. 2007: J.F. Roddick, K. Hornsby (Eds.), Temporal, Spatial, and Spatio-Temporal Data Mining. Proceedings, 2000. VII, 165 pages. 2001.

Vol. 2014: M. Moortgat (Ed.), Logical Aspects of Computational Linguistics. Proceedings, 1998. X, 287 pages. 2001.

Vol. 2019: P. Stone, T. Balch, G. Kraetzschmar (Eds.), RoboCup 2000: Robot Soccer World Cup IV. XVII, 658 pages. 2001.

Vol. 2033: J. Liu, Y. Ye (Eds.), E-Commerce Agents. VI, 347 pages. 2001.

Vol. 2035: D. Cheung, G.J. Williams, Q. Li (Eds.), Advances in Knowledge Discovery and Data Mining – PAKDD 2001. Proceedings, 2001. XVIII, 596 pages. 2001.

Vol. 2039: M. Schumacher, Objective Coordination in Multi-Agent System Engineering. XIV, 149 pages. 2001.

Vol. 2056: E. Stroulia, S. Matwin (Eds.), Advances in Artificial Intelligence. Proceedings, 2001. XII, 366 pages. 2001.

Vol. 2062: A. Nareyek, Constraint-Based Agents. XIV, 178 pages. 2001.

Vol. 2070: L. Monostori, J. Váncza, M. Ali (Eds.), Engineering of Intelligent Systems. Proceedings, 2001. XVIII, 951 pages. 2001.

Vol. 2083: R. Goré, A. Leitsch, T. Nipkow (Eds.), Automated Reasoning. Proceedings, 2001. XV, 708 pages. 2001.